CLIMATES OF THE STATES

VOL. I — EASTERN STATES

CLIMATES of the STATES

in two volumes

VOLUME I — EASTERN STATES
plus PUERTO RICO and the U.S. VIRGIN ISLANDS

– *A Practical Reference Containing Basic Climatological Data of the United States*

*by Officials of the National Oceanic and Atmospheric Administration,
U.S. Department of Commerce*

WATER INFORMATION CENTER, INC.

PERIODICALS

Water Newsletter
Research and Development News
Ground Water Newsletter

BOOKS

Geraghty and Miller *Water Atlas of the United States*
Todd *The Water Encyclopedia*
van der Leeden *Ground Water — A Selected Bibliography*
Giefer and Todd *Water Publications of State Agencies*
Soil Conservation Service *Drainage of Agricultural Land*
Gray *Handbook on the Principles of Hydrology*
National Water Commission *Water Policies for the Future*
Officials of NOAA *Climates of the States*

Published 1974 in two volumes by Water Information Center, Inc., Water Research Building, Manhasset Isle, Port Washington, N.Y. 11050.

Library of Congress Catalog Card Number: 73-93482

ISBN: 0-912394-09-9

Printed in the United States of America.

FOREWORD

Climates of the States combines into two convenient volumes the fifty-one individual state climatological reports issued during the last decade by the National Oceanic and Atmospheric Administration of the U.S. Department of Commerce. Gathered together for handy reference are descriptions, maps and tabular data covering the climate of each of the fifty states plus Puerto Rico and the U.S. Virgin Islands. Each state section contains a general summary of climatic conditions followed by detailed tables of freeze data; normals of temperature and precipitation by climatic divisions and stations; normals, means, and extremes of selected individual stations; and maps showing temperature, precipitation and locations of stations. Also included are miscellaneous data on snowfall, sunshine and occurrence of tropical storms. Altogether the two volumes contain 395 tables and 310 maps.

It is believed that this publication will prove to be a valuable reference for professionals and students in climatology, hydrology, the environmental and agricultural sciences, and other persons interested in finding out what climatic conditions prevail in specific sections of the United States.

Acknowledgement is given to the cooperation of officials of the National Climatic Center of the National Oceanic and Atmospheric Administration in preparing this publication. Mr. William H. Haggard, Director of the National Climatic Center, kindly granted permission to reproduce the material. Gilbert E. Stegall, Acting Chief of the Information Services Division of the National Climatic Center assisted with securing of out-of-print material.

Frits van der Leeden
Associate
Geraghty & Miller, Inc.
Port Washington, N.Y.

Fred L. Troise
Director of Publications
Water Information Center, Inc.

PUBLISHER'S NOTE

Figures and letters following a station name, such as 12 SSW, indicate distance in miles and direction from the Post Office.

The following units are generally used: Temperature in °F., precipitation and snowfall in inches, wind movement in miles per hour, and relative humidity in percent. Degree day totals are the sums of the negative departures of average daily temperature from 65° F. Below zero temperatures are preceded by a minus sign (—).

CONTENTS

Appendix

THE CLIMATE OF
ALABAMA

by
Arthur R. Long

February 1959

The surface of Alabama rises as a rolling plain from the Gulf of Mexico in the southwest to foothills in the central part of the State. Thence there is a rise to the Appalachian Mountains which extend into the northeastern counties. Ridges from the Appalachians extend southward through the eastern counties, with elevations along these ridges as much as 600 to 800 feet above sea level in the southeast. The general elevation of the high northeastern area is about 800 feet above sea level, but some mountain summits rise to over 2,000 feet, the highest (Mount Cheaha in southwestern Cleburne County) being 2,407 feet.

The climate is temperate, becoming largely subtropical near the coast. The summers are long, hot, and humid, with little day-to-day temperature change. In the northeastern counties higher altitudes help make the summer nights more comfortable. From late June through middle August, approximately a third of the evenings are made comfortable by local afternoon thundershowers which bring cool breezes over areas at and near where they occur.

In the coldest months of December, January, and February, there are frequent shifts between mild air, which has been moistened and warmed by the Gulf, and dry, cool continental air. Severely cold weather seldom occurs. Even in the northern third of the State, temperatures of zero or lower are rare and occur only when there is snow on the ground. In the Tennessee Valley in the extreme north, they occur in cold spells about an average of once in 6 or 7 years. Zero temperatures have been recorded only once in the south (February 13, 1899). In the Tennessee Valley, temperatures of 10° or lower

occur on an average of about two times per winter. Across the middle of the State, such temperatures occur about once a year, while they are recorded about once in 10 years in the coastal counties.

Since cold air on clear nights collects in low places, there is considerable irregularity in the distribution of last spring or first fall freezes in all sections.

Precipitation is nearly all in the form of rain. Snow falls in the northern counties on an average of about twice each winter. The average fall in that area is only about 3 inches per year, and since this includes unusually heavy snows in a few individual winters, some winters have little or none. To most dwellers of southern Alabama, snow is a curiosity. From late June through the first half of August, nearly all precipitation is from local thundershowers which occur mostly in the afternoons. During late August and in September, summer conditions of atmospheric temperature and moisture persist, but thundershowers become less frequent. However, late night and early morning thundershowers, characteristic of late summer on the coast, continue in the coastal counties until mid-September. Rains during October are nearly always from showers or thundershowers occurring ahead of temperature drops. Such changes become more frequent and more pronounced as winter approaches. Dry, sunny weather prevails most of the time in September and October, but from August through early October, heavy general rain may occur with a tropical disturbance or hurricane moving inland from the Gulf of Mexico. All types and intensities of rain, except the heat thundershowers of summer may occur at any time from December through March or early April. From late April through early June, rain is mostly in the form of thunder-showers occurring in advance of approaching cool waves.

Rainfall, when received regularly and in average amounts, is adequate for most agricultural needs. More rainfall is needed to maintain adequate soil moisture in Alabama than in more northern states where temperatures are lower. Since summer rain is heavier near the coast than elsewhere and winter rain is heavier in the north, the middle areas of the State get somewhat less precipitation for the year as a whole than the other areas.

Droughts may occur any time during the growing season from late April through October. Relatively long periods with little or no rain are more likely to occur in late summer and autumn than at any other time, while a secondary maximum of such periods occurs in May and June. In most sections periods of 3 weeks or longer without as much as 0.10 inch of rain occur on an average of about once in 12 years in May and once in 9 years in June. The frequency for September is once in 6 years. For October, usually the driest month, such dry spells occur about once in 3 years over most of the State, except about once in 2 years in the extreme southeast and once in 5 years in the extreme northeast. The normally light rain of October is desirable in harvesting cotton, corn, and other crops, but rain during that month is important (since it is still warm) to the winter gardens and cover crops that are being started. Pasture grasses also need at that time moisture to maintain growth.

Only once in about 20 years on an average does a period of 3 weeks or longer with less than 0.10 inch of rain occurs in July. However, the abundant sunshine and the high temperatures of July make it nearly as droughty as the other summer months from the standpoint of lack of soil moisture.

Over a 10-year period, from 1948 to 1957, the daily level of soil moisture at Montgomery was estimated, using the daily values of rainfall, sunshine, wind speed, relative humidity, and mean temperature. It was found that soil moisture depletion, below the usual early spring saturation level, was an inch or more on an average of 17 days in August, 16 in May and June, 14 in September, 13 in July, 11 in October, and 6 in April. Likewise, a deficiency of 2 inches or more of soil moisture prevailed on an average of 13 days in June, August, and September, 12 in May, 8 in July, 5 in October, and 1 in April. To keep the soil moisture from dropping to more than an inch below saturation at Montgomery, for instance, would require an average of 2.66 inches of irrigation water in August, 2.65 in June, 2.18 in May, 1.99 in July, 1.80 in September, 0.74 in October, and 0.62 in April, making an average of 12.64 inches for the season. The most unfavorable year during that 10-year period was 1954, when 18.02 inches of irrigation water would have been required to keep the soil within 1 inch of saturation during the growing season. The most favorable year in the period was 1949 when only 8.27 inches of irrigation water could have kept soil moisture up to that level.

The effect of a theoretical soil moisture deficit varies according to the type of soil, since some soils can hold more moisture than others. The effect also varies according to the type of crops which have different moisture requirements.

Using estimates based on approximate monthly averages of wind, sunshine, humidity, and temperature, the frequencies of drought days were determined for 4 stations in the Tennessee Valley drainage area during the 30-year period from 1927 to 1956. The results gave frequencies not much different from the Montgomery data. However, irrigation requirements to maintain moisture within an inch of saturation appear somewhat less than for Montgomery, averaging about 11 inches per season.

Similar drought data are not yet available for the counties nearer the coast where summer showers are more frequent, where they occur over a longer season than elsewhere over the State, and where temperature and humidity conditions are not as desiccating. Spring and autumn droughts are more frequent in the extreme southeast than in the north and central portions of Alabama.

Severe local droughts occur nearly every year, but severe statewide droughts are practically unknown.

Rivers in Alabama overflow about once a year on an average. Most floods occur from rains in late winter and early spring, with March the month of greatest flood frequency. The lower Tombigbee overflows most often, and in some stretches may stay over the banks most of the time in wet winter and spring seasons.

Nearly all tornadoes occur during the season from November through early May. There are occasional exceptions, such as occurred on June 28, 1957, when a tropical storm was moving northeastward across northern Mississippi, 15 tornadoes occurred in southern Alabama. The greatest frequency is in March and April. In a 35-year period from 1916 through 1950, 188 reported tornadoes caused 713 deaths. Since 1950, more complete reports have been available from over the State, resulting in an apparent increase in these storms. In the 7 years since 1950, there have been 114 reported tornadoes causing 62 deaths. The area covered by the average tornado is small, however, and the chance of a particular point in Alabama being hit, would be only once in about 15 to 20 thousand years.

Destructive tropical hurricanes visit the coastal area on an average of about once in 7 years between July and November. Windstorm damage, such as uprooted trees and damage to signs and poorly constructed buildings, may occur in local thundersqualls any time of the year. The fastest wind ever recorded at Birmingham was 65 miles per hour, at Montgomery 60 miles per hour, and at Mobile 98 miles per hour, the latter during a hurricane in 1916.

Thunderstorms in the north and central sections occur on an average of 1 day each month in winter, about 13 days in July, and about 60 days during the year. At Mobile (representative of the coastal area), December, January, and February average 2 days of thunderstorms each; July 19 days, and the year, 87 days.

Hail occurs almost entirely from February through May, although in the northern counties there are rare occurrences of damaging hail in June. Because of the immaturity of plants prior to June, hail seldom causes much crop damage. Damage to buildings and automobiles usually occurs in small areas with local thunderstorms. The worst and most extensive hailstorm in the past 30 years in Alabama was in April 1946, when hail, wind, and water did roof and other damage totalling $2,000,000 along a path 3 miles wide and 100 miles long in the northeastern portion of the State.

Heavy fog occurs mostly in Winter. It occurs 5 days per year on an average in Birmingham, 8 days per year at Montgomery, and 31 days per year at Mobile near the coast.

In winter winds from a northerly direction are most frequent. In summer the wind is quite variable, but most often from southerly directions.

Cotton, corn, sugar cane, sorghums, peanuts, and early season truck crops are particular adaptations to the climate of the region.

The large amount of stream flow from winter and early spring rainfall is an important com-plement of the coal deposits in Alabama, and many industries use water power in the east-central portion. The full capacity is used in the wettest months, while the year round capac-ity averages about one-half that amount. In addition, a large amount of power is produced by dams on the Tennessee River from water com-ing mostly from North Carolina, Tennessee, and Virginia. The new Woodruff Dam on the Chat-tahoochee River between Georgia and northwest-ern Florida is an important hydro-electric pro-ject using water from Alabama and Georgia.

All of the State's large artificial lakes are important areas for water sports during the long summers.

BIBLIOGRAPHY

(1) Climatic Summary of the United States (Bulletin W) 1930 edition, Sections 100 and 101. U. S. Weather Bureau

(2) Climatic Summary of the United States, Alabama - Supplement for 1931 through 1952 (Bulletin W Supplement). U. S. Weather Bureau

(3) Climatological Data - Alabama. U. S. Weather Bureau

(4) Climatological Data National Summary. U. S. Weather Bureau

(5) Hourly Precipitation Data - Alabama. U. S. Weather Bureau

(6) Local Climatological Data, U. S. Weather Bureau, for Birmingham, Mobile and Montgomery, Alabama.

FREEZE DATA

STATION	Freeze threshold temperature	Mean date of last Spring occurrence	Mean date of first Fall occurrence	Mean No. of days between dates	Years of record Spring	No. of occurrences in Spring	Years of record Fall	No. of occurrences in Fall	STATION	Freeze threshold temperature	Mean date of last Spring occurrence	Mean date of first Fall occurrence	Mean No. of days between dates	Years of record Spring	No. of occurrences in Spring	Years of record Fall	No. of occurrences in Fall
ANDALUSIA	32	03-04	12-03	274	13	13	12	11	GENEVA	32	03-21	11-11	235	13	13	14	14
	28	02-07	12-13	309	13	12	12	9		28	02-16	11-28	285	13	13	14	14
	24	01-19	12-20	335	12	7	12	5		24	02-01	12-15	318	13	11	14	8
	20	01-15	12-26	345	12	5	12	3		20	01-15	12-18	337	13	6	14	7
	16	01-08	⊕	⊕	12	3	12	1		16	01-03	12-26	357	13	2	14	3
ANNISTON CAA AP	32	03-30	11-06	221	29	29	28	28	GOODWATER	32	03-30	11-09	224	30	30	30	30
	28	03-11	11-20	254	29	29	28	28		28	03-15	11-21	251	30	30	30	30
	24	02-22	12-06	287	29	28	28	22		24	02-19	12-06	290	30	28	29	23
	20	02-11	12-18	310	29	26	28	14		20	02-08	12-17	312	30	26	28	16
	16	01-22	12-25	337	29	16	28	9		16	01-27	12-21	329	30	20	28	11
ATMORE STATE FARM	32	03-01	11-22	267	10	10	10	10	GREENSBORO	32	03-15	11-16	246	30	30	28	27
	28	02-08	12-08	303	10	9	10	9		28	02-26	12-03	280	30	29	28	25
	24	01-18	12-18	335	10	6	10	5		24	02-08	12-19	313	30	27	28	16
	20	01-11	12-19	342	9	3	10	5		20	01-23	12-25	335	30	19	28	10
	16	⊕	⊕	⊕	9	1	10	0		16	01-12	12-29	351	30	14	28	5
AUBURN	32	03-21	11-16	239	30	30	30	29	HALEYVILLE	32	04-01	11-08	221	14	14	15	15
	28	03-01	12-02	277	30	30	30	27		28	03-22	11-19	242	14	14	15	15
	24	02-11	12-14	307	29	27	30	19		24	03-01	12-03	277	14	14	15	14
	20	02-03	12-23	323	29	23	30	10		20	02-11	12-14	305	14	13	15	10
	16	01-13	12-27	348	29	12	30	6		16	01-26	12-19	327	14	9	15	6
BIRMINGHAM	32	03-19	11-14	241	30	30	30	29	HIGHLAND HOME	32	03-09	11-24	260	30	30	29	27
	28	03-04	12-03	274	30	30	30	28		28	02-23	12-09	290	30	28	28	21
	24	02-16	12-13	300	30	28	30	21		24	02-06	12-19	316	30	26	28	15
	20	01-28	12-22	328	30	19	30	12		20	01-19	12-24	339	30	15	28	8
	16	01-16	12-26	344	30	14	30	8		16	01-08	⊕	⊕	30	11	28	3
BREWTON 3 SSE	32	03-23	11-07	229	24	24	23	23	LIVINGSTON	32	03-28	11-03	220	13	13	14	14
	28	02-27	11-19	265	24	24	22	21		28	03-03	11-16	257	13	13	14	14
	24	02-06	12-08	305	24	22	22	16		24	02-10	12-04	297	13	12	14	13
	20	01-21	12-20	334	23	15	22	8		20	01-22	12-14	326	13	9	14	9
	16	01-07	12-26	353	23	7	22	4		16	01-14	12-20	341	13	6	14	5
CENTREVILLE	32	03-30	11-05	220	30	30	30	30	MADISON	32	04-04	11-04	215	30	30	30	30
	28	03-14	11-21	252	30	30	30	29		28	03-20	11-14	240	30	30	30	29
	24	02-17	12-04	290	29	27	30	23		24	03-02	11-26	269	30	29	30	29
	20	02-04	12-16	315	29	24	30	17		20	02-19	12-12	296	29	27	30	23
	16	01-16	12-22	340	29	15	30	12		16	02-09	12-19	314	30	26	30	16
CITRONELLE	32	02-28	11-29	275	30	29	28	27	MOBILE WB CITY	32	02-17	12-12	298	30	28	30	22
	28	02-19	12-10	295	30	28	28	20		28	01-25	12-19	328	29	19	30	15
	24	01-31	12-19	322	30	20	28	15		24	01-14	12-24	344	30	14	30	11
	20	01-16	12-26	344	30	14	28	7		20	01-06	⊕	⊕	30	9	30	2
	16	01-05	⊕	⊕	30	8	28	2		16	⊕	⊕	⊕	30	6	30	0
CLANTON	32	03-25	11-08	229	30	30	30	30	MONTGOMERY WB CITY	32	02-27	12-03	279	30	28	30	28
	28	03-02	11-20	263	30	30	30	29		28	02-11	12-16	308	30	26	30	18
	24	02-11	12-12	305	30	27	30	20		24	01-25	12-22	332	30	18	30	11
	20	01-29	12-21	326	30	21	30	13		20	01-14	12-26	346	30	14	30	8
	16	01-16	12-27	345	30	15	30	6		16	01-05	⊕	⊕	30	8	30	3
CORDOVA	32	04-04	11-01	210	13	13	12	12	MUSCLE SHOALS CAA AP	32	03-24	11-09	230	10	10	10	10
	28	03-29	11-11	227	11	11	12	12		28	03-06	11-13	252	10	10	10	10
	24	03-11	11-23	257	11	11	12	12		24	02-13	12-10	300	10	10	10	7
	20	02-14	12-04	294	11	10	12	11		20	02-11	12-20	312	10	9	10	5
	16	01-31	12-20	323	11	8	12	6		16	01-24	12-24	334	10	7	10	3
DECATUR	32	03-24	11-11	232	30	30	30	30	ONEONTA	32	03-29	11-03	219	10	10	9	9
	28	03-10	11-22	257	30	30	30	29		28	03-14	11-13	243	10	10	9	9
	24	02-24	12-04	284	30	29	30	27		24	02-26	12-02	279	10	10	9	8
	20	02-10	12-19	312	30	26	30	18		20	02-08	12-17	312	9	8	9	7
	16	01-20	12-24	339	30	15	30	11		16	01-18	12-23	339	9	4	9	3
EUFAULA	32	03-12	11-14	247	30	30	30	30	OZARK	32	03-03	11-26	268	30	28	30	28
	28	02-21	11-28	280	30	29	30	27		28	02-14	12-08	297	30	26	30	22
	24	01-31	12-19	323	30	22	29	13		24	01-28	12-21	327	30	21	30	12
	20	01-18	12-24	340	30	16	29	10		20	01-18	12-24	340	30	15	30	10
	16	01-06	⊕	⊕	30	8	29	2		16	01-03	⊕	⊕	30	5	29	3
FAYETTE	32	04-01	11-02	216	14	14	15	15	PRATTVILLE	32	03-18	11-15	241	30	30	30	30
	28	03-20	11-15	241	14	14	15	15		28	02-27	11-24	270	30	30	30	29
	24	02-25	11-23	271	14	14	15	14		24	02-09	12-13	307	30	26	30	20
	20	02-09	12-09	303	14	13	15	12		20	01-20	12-22	336	30	17	30	11
	16	01-29	12-24	329	14	12	15	5		16	01-11	12-28	351	30	11	30	5
GADSDEN	32	04-01	11-04	217	30	30	29	29	PUSHMATAHA	32	03-21	11-10	234	30	30	28	28
	28	03-15	11-17	248	29	29	29	28		28	03-04	11-24	265	29	28	26	25
	24	02-26	12-04	281	29	28	29	23		24	02-14	12-10	299	29	27	26	20
	20	02-05	12-13	311	29	25	28	19		20	01-26	12-19	327	28	19	26	13
	16	01-21	12-23	336	29	17	28	11		16	01-13	12-25	346	28	13	26	8

FREEZE DATA

STATION	Freeze threshold temperature	Mean date of last Spring occurrence	Mean date of first Fall occurrence	Mean No. of days between dates	Years of record Spring	No. of occurrences in Spring	Years of record Fall	No. of occurrences in Fall	STATION	Freeze threshold temperature	Mean date of last Spring occurrence	Mean date of first Fall occurrence	Mean No. of days between dates	Years of record Spring	No. of occurrences in Spring	Years of record Fall	No. of occurrences in Fall
ROBERTSDALE 7 E	32	03-19	11-21	246	18	18	18	18	THOMASVILLE	32	03-17	11-14	242	29	29	29	28
	28	02-18	12-02	286	18	16	18	17		28	02-25	12-07	285	28	27	27	22
	24	01-31	12-13	316	17	12	18	11		24	02-04	12-18	317	28	23	27	15
	20	01-18	12-18	334	17	9	18	10		20	01-22	12-24	336	27	16	27	8
	16	01-07	⊕	⊕	17	4	18	2		16	01-10	12-28	352	27	11	27	5
SAINT BERNARD	32	04-07	10-28	204	30	30	29	29	TUSCALOOSA CAA AP	32	03-16	11-10	239	22	22	22	22
	28	03-24	11-10	231	30	29	29	29		28	03-04	11-26	266	22	22	22	20
	24	03-07	11-22	261	30	30	29	27		24	02-16	12-11	297	22	21	22	16
	20	02-17	12-09	295	30	28	29	22		20	01-26	12-23	331	22	15	22	8
	16	02-08	12-18	313	29	25	29	15		16	01-11	12-27	350	22	9	22	5
SCOTTSBORO	32	04-04	11-01	210	28	28	26	26	UNION SPRINGS	32	03-10	11-22	257	29	29	28	27
	28	03-19	11-11	237	28	28	26	26		28	02-23	12-09	289	29	28	28	21
	24	03-06	11-26	265	28	28	25	22		24	02-09	12-18	313	29	25	27	11
	20	02-15	12-10	298	28	26	24	19		20	01-24	12-22	333	27	17	27	10
	16	01-27	12-20	327	26	19	24	11		16	01-11	12-27	350	27	12	27	5
SELMA	32	03-08	11-25	262	30	30	30	29	VALLEY HEAD	32	04-11	10-28	200	30	30	29	29
	28	02-13	12-10	299	30	28	30	22		28	03-25	11-10	230	30	30	29	29
	24	01-28	12-20	326	30	21	30	13		24	03-14	11-20	251	30	30	29	28
	20	01-12	12-24	345	30	14	30	10		20	02-22	12-02	284	29	28	28	23
	16	01-04	⊕	⊕	30	7	30	4		16	02-09	12-12	306	27	23	28	20
TALLADEGA	32	03-30	11-06	221	29	29	30	30	WETUMPKA	32	03-15	11-18	248	30	30	30	29
	28	03-14	11-19	251	29	29	29	28		28	02-21	12-02	285	30	28	30	24
	24	02-24	12-03	282	29	28	29	25		24	02-05	12-15	314	30	26	30	18
	20	02-07	12-12	308	29	26	29	22		20	01-17	12-23	340	30	16	30	12
	16	01-18	12-22	338	29	15	28	10		16	01-06	12-28	356	30	9	30	6

Data in the above table are based on the period 1921-1950, or that portion of this period for which data are available.

⊕ When the frequency of occurrence in either spring or fall is one year in ten, or less, mean dates are not given.

Means have been adjusted to take into account years of non-occurrence.

A freeze is a numerical substitute for the former term "killing frost" and is the occurrence of a minimum temperature at or below the threshold temperature of 32°, 28, etc.

Freeze data tabulations in greater detail are available and can be reproduced at cost.

*MEAN TEMPERATURE AND PRECIPITATION
ALABAMA

STATION	JANUARY		FEBRUARY		MARCH		APRIL		MAY		JUNE		JULY		AUGUST		SEPTEMBER		OCTOBER		NOVEMBER		DECEMBER		ANNUAL	
	Temperature	Precipitation	Temperature	Precipitation	Temperature	Precipitation	Temperature	Precipitation	Temperature	Precipitation	Temperature	Precipitation	Temperature	Precipitation	Temperature	Precipitation	Temperature	Precipitation	Temperature	Precipitation	Temperature	Precipitation	Temperature	Precipitation	Temperature	Precipitation
NORTHERN VALLEY																										
DECATUR	44.9	5.96	46.7	5.51	53.8	5.77	62.5	4.33	71.0	3.12	79.1	3.15	81.2	4.45	80.7	3.56	75.1	2.86	64.5	2.34	51.9	3.74	44.9	5.13	63.0	49.92
MADISON	43.4	5.70	44.9	5.45	52.0	6.11	60.9	4.41	69.8	3.46	77.9	3.39	80.1	4.93	79.7	3.70	74.0	3.15	63.3	2.80	50.6	3.83	43.6	5.14	61.7	52.07
DIVISION	43.8	5.68	45.5	5.60	52.6	5.91	61.4	4.27	69.9	3.38	78.0	3.48	80.4	4.50	79.9	3.49	74.0	3.09	63.0	2.68	50.7	3.73	44.0	5.08	61.9	50.89
APPALACHIAN MNTN																										
ALBERTVILLE 2 SE	43.5	5.84	44.8	5.61	51.7	6.42	60.1	4.41	68.8	3.62	76.5	4.01	78.6	4.91	77.7	4.16	72.6	2.65	62.2	2.86	49.9	3.90	43.2	5.41	60.8	53.80
BESSEMER 5 SSW		5.15		5.46		6.22		4.66		3.38		4.05		4.44		4.76		6.95		2.44		3.33		5.48		56.32
BIRMINGHAM WB AP	45.2	4.98	47.7	5.17	53.8	6.31	61.9	4.65	69.5	3.74	77.4	4.21	79.6	5.11	79.1	4.55	74.8	2.71	63.7	2.86	52.0	3.98	45.7	5.25	62.5	53.52
BRIDGEPORT 2 W		5.89		5.51		6.03		4.24		3.66		3.54		5.34		3.28		3.02		2.69		3.66		5.25		52.11
LEEDS		5.74		5.44		6.18		4.99		3.20		3.94		5.95		5.07		2.95		2.84		3.72		5.89		55.91
SAINT BERNARD	42.0	6.06	45.3	5.86	52.1	6.21	60.6	4.34	68.8	3.50	76.7	3.64	78.9	4.92	78.3	4.17	72.9	2.64	62.3	3.18	50.0	4.05	43.3	5.44	60.9	54.01
SCOTTSBORO	44.3	6.11	45.8	5.98	52.6	6.40	60.9	4.72	69.2	3.55	77.0	4.12	79.4	5.75	78.6	3.84	73.2	2.83	62.5	3.12	50.5	3.84	43.8	5.60	61.5	55.86
VALLEY HEAD 3 S	42.9	5.70	44.5	5.39	50.9	5.97	59.2	4.67	66.7	3.70	75.6	3.87	77.9	4.55	77.2	3.99	72.1	2.66	61.7	2.93	49.5	3.64	43.3	5.00	60.1	52.07
DIVISION	44.6	5.79	46.1	5.67	52.7	6.23	61.0	4.65	69.4	3.51	77.1	3.83	79.2	4.98	78.5	4.32	73.4	2.77	63.1	2.81	50.9	3.72	44.4	5.44	61.7	53.72
UPPER PLAINS																										
CENTREVILLE	48.1	5.32	49.7	5.63	56.3	6.77	63.8	5.38	72.0	4.24	79.2	4.41	81.4	5.54	81.1	4.57	75.8	2.99	65.4	2.04	53.7	3.36	47.4	5.77	64.5	56.02
CLANTON	47.8	5.26	49.1	5.33	55.6	6.54	63.3	5.93	71.9	4.03	79.4	4.33	80.9	5.93	80.2	5.18	75.3	2.80	64.8	1.97	53.1	3.48	47.2	5.40	64.1	56.18
DANCY		4.83		4.68		5.40		4.33		3.73		3.30		5.36		2.82		2.81		2.07		3.51		5.04		47.88
PRATTVILLE	49.6	4.67	51.2	4.93	57.4	6.07	64.1	5.52	71.8	3.63	78.8	4.08	80.7	5.68	80.2	5.22	75.8	3.08	65.8	2.04	54.4	3.38	49.0	5.15	64.9	53.45
TUSCALOOSA LOCK AND DAM	47.6	5.41	49.6	5.40	56.2	6.09	63.9	4.36	72.1	3.99	79.9	3.67	81.9	4.93	81.5	4.29	76.3	2.86	65.4	2.58	55.6	3.63	49.3	5.56	64.9	52.77
DIVISION	47.6	5.51	49.1	5.44	55.7	6.20	63.3	4.95	71.3	3.76	78.9	4.00	80.8	5.31	80.4	4.46	75.3	2.86	65.0	2.34	53.1	3.53	47.1	5.44	64.0	53.80
EASTERN VALLEY																										
CALERA 2 SW		5.21		5.44		6.61		5.43		3.61		3.93		5.84		4.95		2.95		2.31		3.48		5.42		55.18
GADSDEN		6.13		5.85		6.19		5.00		3.48		3.74		5.09		4.60		2.93		2.77		3.56		5.61		54.95
TALLADEGA	47.6	4.81	49.0	5.45	55.5	6.10	62.8	4.85	70.9	3.44	78.3	4.39	80.3	4.84	79.8	4.13	75.0	2.61	64.5	2.46	53.0	3.21	46.8	5.19	63.6	51.48
DIVISION	46.2	5.48	47.6	5.49	54.3	6.24	62.1	4.93	70.3	3.37	77.8	3.98	80.0	5.17	79.4	4.46	74.4	2.82	63.6	2.49	51.9	3.36	45.6	5.48	62.8	53.27
PIEDMONT PLATEAU																										
AUBURN 3 SW	49.4	4.56	50.6	4.91	56.5	6.36	63.8	5.11	72.1	3.65	79.0	3.71	80.4	5.28	79.9	4.79	75.9	3.22	66.8	1.91	55.1	3.54	48.9	5.22	64.9	52.26
DADEVILLE		4.96		5.10		6.32		5.25		3.59		3.80		5.66		4.55		3.05		2.00		3.40		5.42		53.10
MARTIN DAM		4.73		4.85		5.83		5.02		3.36		3.72		5.35		4.16		2.95		1.79		3.26		5.15		50.17
WETUMPKA		4.70		4.91		5.87		5.19		3.45		4.49		5.09		4.85		2.93		1.97		3.31		5.06		51.82
YATES HYDRO PLANT		4.70		4.76		6.26		4.99		3.65		4.06		5.49		4.10		3.07		1.78		3.32		5.04		51.22
DIVISION	48.8	4.92	50.1	5.02	56.3	6.12	63.6	5.31	71.6	3.57	78.6	3.95	80.4	5.56	79.8	4.64	75.5	3.00	65.7	2.06	54.2	3.42	48.1	5.15	64.4	52.72
PRAIRIE																										
GREENSBORO	49.9	4.92	51.5	5.29	57.7	6.57	65.2	5.20	73.2	4.40	80.4	3.12	82.0	5.14	81.9	4.15	77.5	3.17	67.9	1.98	56.0	3.30	49.8	5.45	66.1	52.69
MARION 1 N		4.55		5.27		6.81		5.05		3.80		3.68		5.33		4.56		3.01		1.96		3.17		5.25		52.44
MILSTEAD		4.60		4.70		6.02		5.17		3.67		3.66		5.52		4.92		3.07		1.83		3.52		5.32		52.00
MONTGOMERY WB AP	49.2	4.60	51.6	4.73	57.1	6.50	64.7	4.81	72.5	3.46	79.6	4.69	81.2	5.76	80.9	4.75	77.1	3.51	66.4	2.36	55.2	3.95	49.4	4.54	65.4	53.66
PRIMROSE FARM		4.17		4.22		5.24		4.82		3.42		4.31		5.63		4.57		3.20		1.79		3.14		4.76		49.27
SELMA	51.6	4.62	53.2	5.03	59.2	6.01	65.8	5.75	73.5	4.00	80.3	3.49	81.7	5.00	81.5	4.66	77.2	2.32	67.6	1.88	56.6	3.14	50.9	5.49	66.6	51.39
UNION SPRINGS 5 S	51.2	4.39	52.5	4.66	58.1	6.08	65.1	5.29	72.9	3.43	79.5	3.81	80.9	6.47	80.7	4.84	76.6	2.59	67.3	1.78	56.3	2.89	50.5	4.68	66.0	50.91
DIVISION	50.4	4.48	52.0	4.84	58.1	6.17	65.1	5.13	73.1	3.68	80.0	3.66	81.6	5.49	81.3	4.14	76.9	2.86	67.2	1.85	55.7	3.23	49.9	5.19	65.9	50.72
COASTAL PLAIN																										
BRANTLEY		4.97		4.75		6.28		5.21		4.15		4.20		6.91		5.61		3.55		1.79		3.24		4.87		55.53
BREWTON 3 SSE	52.2	4.80	53.7	4.55	58.9	6.31	65.2	5.86	72.6	4.62	78.9	4.88	80.5	8.03	80.5	6.25	76.2	4.55	66.7	2.08	56.1	3.41	51.9	4.85	66.1	60.19
BRUNDIDGE		4.48		4.36		6.13		5.14		3.61		3.97		6.21		5.62		3.49		1.69		3.00		4.46		52.16
CLAYTON		4.93		4.83		6.18		5.54		4.15		4.02		6.44		5.30		4.04		1.85		3.11		4.92		55.31
COFFEE SPRINGS 2 NW		4.16		4.57		5.75		5.52		4.16		4.44		7.07		5.76		4.12		1.60		3.01		4.68		54.84
DOTHAN CAA AIRPORT		3.95		4.52		5.58		4.72		3.37		4.07		6.70		5.63		4.22		1.71		2.83		4.41		51.71
ELBA		4.70		4.51		5.83		5.34		4.24		4.53		6.87		5.43		3.77		1.87		3.63		5.09		55.81
EUFAULA	52.0	4.45	53.3	4.35	59.2	5.78	66.2	5.51	73.9	4.20	80.6	4.01	81.8	6.68	81.3	5.09	77.2	3.27	67.4	1.61	56.6	2.75	51.3	5.20	66.7	52.90
FRISCO CITY		4.65		4.62		6.83		5.86		4.24		4.79		6.80		5.44		4.28		1.97		4.17		5.88		59.53
GENEVA		3.92		4.24		5.87		5.43		4.07		4.23		6.72		6.25		5.10		1.43		3.22		4.47		54.95
GREENVILLE	51.7	4.86	53.4	4.95	59.2	6.77	66.2	5.54	73.9	4.39	80.4	4.27	81.5	7.12	81.0	5.54	76.8	3.64	67.6	2.07	56.6	3.93	51.4	5.58	66.6	58.66
HIGHLAND HOME	51.7	5.09	53.1	5.11	59.0	6.43	65.8	5.55	73.4	3.94	79.6	4.87	81.2	6.92	80.8	5.32	76.7	2.81	67.6	1.63	56.6	3.48	51.4	5.36	66.4	56.51
JACKSON LOCK 1		4.61		4.55		6.51		5.37		4.30		3.84		6.81		3.94		3.29		2.03		3.91		5.37		54.53
MILLERS FERRY		4.45		5.01		6.34		5.65		3.92		3.83		5.74		4.49		3.07		1.73		3.34		6.07		53.64
NEWTON		4.02		4.37		5.26		4.51		3.81		3.84		5.49		4.94		4.19		1.08		2.67		4.50		48.68
OZARK	52.9	4.22	54.3	4.59	59.9	5.56	66.9	5.16	74.2	4.25	80.1	4.42	80.9	6.54	80.6	6.25	77.0	4.21	68.8	1.43	58.0	3.06	52.8	4.87	67.2	54.56
PENNINGTON LOCK 2		4.96		5.27		6.38		5.23		4.47		3.49		6.49		3.70		2.90		1.91		3.54		6.18		54.52
PUSHMATAHA	49.9	4.37	51.4	4.66	57.4	5.67	64.2	4.84	71.8	2.99	78.8	3.70	80.5	5.53	80.4	3.04	75.6	2.99	65.9	1.70	54.8	2.97	49.2	4.89	65.0	47.35
RIVER FALLS		5.42		4.94		6.61		5.87		4.34		3.89		7.62		5.96		4.20		1.96		3.92		5.23		59.96
THOMASVILLE	50.6	4.90	52.3	5.34	58.2	6.55	65.1	5.78	72.8	4.42	79.6	3.48	80.8	6.43	80.8	4.64	76.3	3.13	67.2	2.07	55.9	3.61	50.5	5.70	65.8	56.05
TROY	51.1	4.69	52.6	4.37	58.4	6.70	65.5	5.54	73.3	4.19	79.6	3.24	80.7	6.40	80.3	5.75	76.3	3.71	67.4	1.79	56.1	3.37	50.8	4.84	66.0	54.59
WHITFIELD LOCK 3		4.66		4.65		6.19		4.96		3.60		3.31		5.47		3.26		2.76		1.85		3.36		5.65		49.72
DIVISION	51.9	4.61	53.4	4.65	59.1	6.33	65.9	5.62	73.5	4.15	79.9	4.28	81.0	6.81	80.8	5.47	76.7	3.91	67.6	1.82	56.6	3.51	51.5	5.26	66.5	56.42
GULF																										
CITRONELLE	53.5	4.87	55.2	4.38	60.5	6.91	67.1	5.99	74.2	5.21	80.0	4.97	81.0	8.33	81.4	5.92	77.6	4.82	69.3	2.33	58.4	4.49	53.4	5.99	67.6	64.21
FAIRHOPE	54.0	4.23	55.3	3.89	60.2	6.30	66.8	6.01	74.2	5.01	80.0	6.19	81.2	9.46	81.4	6.27	77.7	6.33	69.3	2.58	58.5	3.30	54.0	5.21	67.7	64.78
MOBILE WB AIRPORT	52.7	5.02	55.0	4.81	59.5	7.62	66.7	5.10	73.4	4.64	79.6	6.09	80.7	8.97	80.8	6.31	77.7	5.78	69.1	3.75	58.7	4.07	53.5	5.41	67.3	67.57
DIVISION	53.9	4.45	55.4	4.16	60.4	6.73	67.0	6.03	74.3	4.98	80.2	5.72	81.3	9.33	81.5	6.35	77.8	5.68	69.4	2.60	58.7	3.57	54.0	5.40	67.8	65.00

*** Averages for period 1931-1955, except for stations marked WB which are "normals" based on period 1921-1950. Divisional means may not be the arithmetical average of individual stations published, since additional data from shorter period stations are used to obtain better areal representation.**

CONFIDENCE LIMITS

In the absence of trend or record changes, the chances are 9 out of 10 that the true **mean will lie in the interval** formed by adding and subtracting the values in the following table from the means for any station in the State:

1.8	1.16	1.5	.91	1.5	.97	.8	1.16	.7	.85	.5	.77	.5	1.16	.5	.65	1.0	.61	.9	.67	1.1	1.08	1.3	.79	.3	3.04

COMPARATIVE DATA

Data in the following table are the mean temperature and average precipitation for Greensboro, Alabama for the period 1906-1930 and are included in this publication for comparative purposes:

47.9	4.95	49.6	4.67	56.7	5.57	64.3	4.83	71.5	4.59	79.2	3.62	80.4	5.63	80.2	3.88	76.8	3.23	65.7	2.52	55.3	3.41	47.8	5.55	64.6	52.45

NORMALS, MEANS, AND EXTREMES

BIRMINGHAM, ALABAMA — MUNICIPAL AIRPORT

LATITUDE 33° 34' N
LONGITUDE 86° 45' W
ELEVATION (ground) 610 Feet

Temperature

Month	Normal Daily max	Normal Daily min	Normal Monthly	Record highest	Year	Record lowest	Year
J	55.5	34.9	45.2	81	1949	1	1940
F	58.3	37.1	47.7	82	1918	-10	1899
M	65.3	42.3	53.8	90	1907	-12	1899
A	73.9	49.8	61.9	91	1943	27	1950+
M	81.5	57.5	69.5	99	1898	35	1944
J	88.7	66.1	77.4	106	1931	46	1956
J	90.2	69.0	79.6	107	1930	52	1947
A	89.9	68.2	79.1	104	1935	51	1930
S	86.4	63.1	74.8	106	1925	39	1949
O	76.8	50.6	63.7	94	1954+	27	1925
N	64.0	40.0	52.0	84	1935	5	1950
D	56.2	35.1	45.7	80	1951	5	1901
Year	73.9	51.1	62.5	107 July 1930		-10 Feb 1899	

Normal degree days (by month): 623, 491, 378, 128, 30, 0, 0, 0, 13, 123, 396, 598 — Year 2780

Precipitation (inches)

Month	Normal total	Max monthly	Year	Min monthly	Year	Max in 24 hrs	Year
J	4.98	13.37	1937	1.16	1949	5.81	1949
F	5.17	15.86	1903	1.15	1898	5.60	1903
M	6.31	13.14	1929	0.32	1951	5.89	1951
A	4.65	10.54	1900	0.03	1938	5.66	1938
M	3.74	12.09	1900	0.11	1920	4.29	1920
J	4.21	20.12	1916	0.47	1931	4.11	1900
J	5.11	13.83	1901	0.61	1952	8.84	1916
A	4.55	13.19	1906	T	1924	4.15	1941
S	2.71	10.95	1934	0.01	1955	7.59	1906
O	2.86	15.25	1932	0.01	1924	5.02	1929
N	3.98	13.85	1928	0.91	1924	7.76	1942
D	5.25						
Year	53.52	20.12 June 1916		0.00 Oct. 1924		8.84 July 1916	

Snow, Sleet (inches)

Month	Mean total	Max monthly	Year	Max in 24 hrs	Year
J	0.9	11.8	1936	11.0	1936
F	0.3	5.0	1901	5.0	1901
M	0.2	6.5	1924	6.5	1924
A	T	T	1920+	T	1920+
M–S	0.0	0.0		0.0	
N	T	1.4	1955+	1.4	1955+
D	0.3	5.5	1929	5.5	1929
Year	1.7	11.8 Jan. 1936		11.0 Jan. 1936	

Relative humidity (%)

Month	12:00 Mid. CST	6:00 a.m. CST	12:00 N. CST	6:00 p.m. CST
J	80	81	61	67
F	80	79	58	62
M	78	77	52	56
A	77	77	50	54
M	77	80	53	62
J	84	84	56	68
J	86	84	56	66
A	86	84	52	66
S	83	84	49	64
O	79	80	52	63
N	79	82	61	69
D	81	82		
Year	81	81	54	63

Wind

Month	Mean hourly speed	Prevailing direction	Fastest mile Speed	Direction	Year
J	8.3	S	42	SE	1946
F	8.5	S	51	SW	1951
M	8.9	S	65	SW	1955
A	8.1	S	56	SW	1956
M	6.7	S	56	SW	1957
J	5.8	NE			
J	5.4	NE	54	NW	1945
A	5.3	NE	50	NW	1956
S	6.1	ENE	50	SE	1951
O	6.4	N	43	W	1955
N	7.2	S	52	N	1944
D	7.7	S	41	SE	1954
Year	7.0	S	65	SW	Mar. 1955+

Sunshine / Sky cover / Mean number of days

Month	Pct. possible sunshine	Mean sky cover	Clear	Partly cloudy	Cloudy	Precip .01"+	Snow 1.0"+	Thunderstorms	Heavy fog	90°+	Max 32° below	Min 32° below	Min 0° below
J	45	6.2	9	7	15	11	*	1	1	0	1	11	17
F	50	5.9	10	6	12	11	*	2	1	*	*	8	*
M	56	5.5	11	8	12	11	*	4	2	*	*	4	0
A	63	5.2	11	9	10	10		5	4	0	0	*	0
M	65	5.2	9	14	9	10		7	5	3	0	0	0
J	68	5.2	7	15	7	9		11	1	13	0	0	0
J	62	5.7	7	15	8	10		14	1	17	0	0	0
A	65	5.3	13	8	7			14	1	17	0	0	0
S	67	4.8	16	7	7			6	1	9	0	*	0
O	64	4.7	14	7	9			2	1	1	0	*	0
N	59	6.1	10	7	14	11	*	1	1	0	*	5	0
D	44									0		10	0
Year	60	5.3	130	116	119	118	*	65	5	60	1	38	*

MOBILE, ALABAMA — BATES FIELD

LATITUDE 30° 41' N
LONGITUDE 80° 15' W
ELEVATION (ground) 211 feet

Temperature

Month	Normal Daily max	Normal Daily min	Normal Monthly	Record highest	Year	Record lowest	Year
J	62.2	43.2	52.7	84	1949	14	1948
F	64.7	45.2	55.0	84	1944	11	1951
M	69.1	49.9	59.5	90	1946	21	1943
A	76.3	56.9	66.6	91	1943	36	1950+
M	82.8	64.0	73.4	100	1953+	45	1952
J	88.5	70.6	79.6	102	1952	56	1956
J	89.4	72.0	80.7	104	1952	60	1949
A	89.7	71.7	80.8	102	1951+	59	1955
S	86.6	68.7	77.7	98	1954	42	1949
O	79.4	58.8	69.1	93	1954	28	1957
N	69.0	48.3	58.7	85	1946	18	1950
D	63.0	43.9	53.5	80	1943	11	1945
Year	76.7	57.8	67.3	104 July 1952		11 Feb. 1951	

Normal degree days (by month): 416, 304, 222, 47, 0, 0, 0, 0, 0, 28, 219, 376 — Year 1612

Precipitation (inches)

Month	Normal total	Max monthly	Year	Min monthly	Year	Max in 24 hrs	Year
J	5.02	8.52	1947	1.18	1954	3.44	1952
F	4.81	8.12	1951	1.31	1948	6.52	1951
M	7.62	15.58	1946	1.47	1955	6.52	1951
A	5.10	17.69	1955	0.48	1955	13.36	1955
M	4.84	11.17	1946	1.39	1951	4.33	1957
J	6.09	12.70	1942	1.94	1954	3.42	1950
J	8.97	19.29	1949	2.83	1947	3.92	1951
A	6.31	11.46	1944	2.78	1943	4.45	1955
S	5.78	13.61	1957	1.73	1953	5.79	1956
O	3.75	4.19	1956	0.05	1946	1.90	1951
N	4.07	13.65	1948	0.44	1949	3.11	1951
D	5.41	11.38	1953	2.25	1946	5.28	1950
Year	67.57	19.29 July 1949		0.05 Oct. 1946		13.36 Apr. 1955	

Snow, Sleet (inches)

Month	Mean total	Max monthly	Year	Max in 24 hrs	Year
J	0.2	3.5	1955	3.5	1955
F	0.1	T	1951+	T	1951+
M	0.0	1.6	1954	1.6	1954
A–N	0.0	0.0		0.0	
D	T	T	1952+	T	1952+
Year	0.3	3.5 Jan. 1955		3.5 Jan. 1955	

Relative humidity (%)

Month	12:00 Mid. CST	6:00 a.m. CST	12:00 N. CST	6:00 p.m. CST
J	85	87	63	78
F	84	87	60	74
M	84	86	57	71
A	86	87	56	70
M	86	87	56	69
J	89	87	58	71
J	91	90	63	77
A	90	91	60	77
S	88	90	61	77
O	83	87	57	75
N	82	84	55	75
D	84	86	62	77
Year	86	87	59	74

Wind

Month	Mean hourly speed	Prevailing direction
J	11.5	N
F	11.5	N
M	11.7	S
A	10.9	S
M	9.6	S
J	8.6	S
J	7.8	NE
A	7.9	NE
S	9.3	N
O	10.5	N
N	10.8	N
Year	9.9	S

Sunshine / Sky cover / Mean number of days

Month	Mean sky cover	Clear	Partly cloudy	Cloudy	Precip .01"+	Snow 1.0"+	Thunderstorms	Heavy fog	90°+	Max 32° below	Min 32° below	Min 0° below
J	6.6	9	8	16	10	0	2	5	0	0	5	0
F	6.5	7	8	14	10	*	3	5	0	*	3	0
M	6.0	6	8	14	10	0	6	3	0	0	1	0
A	5.6	9	11	11	8	0	6	2	*	0	0	0
M	5.4	10	12	9	8	0	8	*	5	0	0	0
J	5.6	8	14	8	12	0	13	1	17	0	0	0
J	6.6	3	15	13	17	0	19	1	19	0	0	0
A	5.2	9	16	11	13	0	15	1	21	0	0	0
S	5.9	9	11	9	10	0	8	3	12	0	0	0
O	4.8	17	7	9	8	0	3	3	1	0	0	0
N	6.1	13	8	15	12	0	2	5	0	0	2	0
D		8	8			0			0	0	4	0
Year	5.7	110	122	133	123	*	87	31	72	*	15	0

Means and extremes in the above table are from the existing or comparable location(s). Annual extremes have been exceeded at prior locations as follows:
Lowest temperature -1 in February 1899; maximum monthly snowfall 6.0 in February 1895; maximum snowfall in 24 hours 6.0 in February 1895.

LATITUDE 32° 18' N
LONGITUDE 86° 24' W
ELEVATION (ground) 198 Feet

NORMALS, MEANS, AND EXTREMES

Month	Temp Normal Daily max	Temp Normal Daily min	Temp Normal Monthly	Temp Extreme Record highest	Year	Temp Extreme Record lowest	Year	Normal degree days	Precip Normal total	Precip Max monthly	Year	Precip Min monthly	Year	Precip Max 24 hrs	Year	Snow Mean total	Snow Max monthly	Year	Snow Max 24 hrs	Year
(Yrs of record)	(b)	(b)	(b)	85	85	85	85	(b)	(b)	85	85	85	85	85	85	85	85	85	85	85
J	59.7	38.7	49.2	83	1949	5	1886	517	4.60	17.78	1892	0.49	1927	9.98	1892	0.2	4.1	1881	3.5	1881
F	62.6	40.6	51.6	84	1918	-5	1899	388	4.73	11.76	1903	1.32	1938	5.21	1919	0.2	4.1	1914	4.1	1914
M	68.6	45.5	57.1	90	1929+	20	1932	288	6.50	16.51	1909	0.72	1887	6.06	1919	T	T	1924	1.4	1924
A	76.6	52.7	64.7	92	1942+	30	1881	90	4.81	15.94	1912	0.38	1915	5.97	1876	T	T	1910	T	1910
M	84.2	60.7	72.5	99	1916	43	1889	0	3.46	10.25	1873	0.50	1898	4.24	1929	0.0	0.0		0.0	
J	90.5	68.7	79.6	106	1881	48		0	4.69	15.59	1928	0.45	1908	7.81	1928	0.0	0.0		0.0	
J	91.3	71.0	81.2	107	1881	61	1947+	0	5.76	13.42	1899	0.43	1902	5.73	1916	0.0	0.0		0.0	
A	91.0	70.7	80.9	104	1954	58	1891	0	4.75	15.58	1939	0.44	1925	7.22	1925	0.0	0.0		0.0	
S	88.0	66.2	77.1	106	1925	45	1916+	0	3.51	12.00	1917	0.12	1923	8.81	1953	0.0	0.0		0.0	
O	78.9	53.9	66.4	100	1954	26	1952	69	2.36	10.29	1879	T	1904	4.74	1934	0.0	0.0		0.0	
N	67.3	43.1	55.2	86	1935	13	1950	304	3.95	20.10	1948	0.19	1924	6.82	1948	T	T	1950+	T	
D	60.1	38.6	49.4	83	1951	8	1880	491	4.54	10.32	1941	0.49	1889	5.09	1901	0.2	11.0	1886	11.0	1886
Year	76.6	54.2	65.4	107 July 1881		-5 Feb. 1899		2137	53.66	20.10 Nov. 1948		T Oct. 1904		9.98 Jan. 1892		0.6	11.0 Dec. 1886		11.0 Dec. 1886	

Month	RH 12:00 Mid. CST	RH 6:00 a.m. CST	RH 12:00 M. CST	RH 6:00 p.m. CST	Wind Mean hourly speed	Wind Prevailing direction	Wind Fastest mile Speed	Direction	Year	Mean sky cover sunrise to sunset	Pct. of possible sunshine	Days Clear	Days Partly cloudy	Days Cloudy	Precip .01 inch or more	Snow/Sleet 1.0 inch or more	Thunderstorms	Heavy fog	Max 90° and above	Max 32° and below	Min 32° and below	Min 0° and below
(Yrs of record)	14	70	40	70	14	13	46	46	46	85	46	85	85	85	85	85	85	85	85	85	85	85
J	83	84	61	67	7.7	NW	38	S	1951	5.9	50	9	9	13	11	1	2	2	0	*	7	0
F	80	82	58	63	8.2	S	47	W	1946+	5.8	54	9	7	12	10	1	2	1	0	*	4	*
M	78	81	53	58	8.3	S	60	SW	1952	5.5	61	11	9	11	10	*	4	*	0	0	1	0
A	81	80	52	56	7.4	S	54	SE	1952	4.8	69	12	9	9	8	*	5	*	*	0	*	0
M	83	79	51	57	6.0	S	50	W	1957	4.7	72	12	13	8	9	0	6	*	5	0	0	0
J	84	81	53	61	5.9	SW	51	NW	1953	5.1	72	9	9	8	11	0	9	*	17	0	0	0
J	87	85	58	68	5.8	SW	51	E	1954	5.5	65	8	14	9	12	0	11	*	20	0	0	0
A	86	87	57	69	5.3	E	43	SW	1956	5.3	69	9	14	8	12	0	9	*	18	0	0	0
S	85	85	54	65	6.0	NE	40	NE	1956	4.5	68	13	9	8	8	0	4	*	10	0	0	0
O	85	84	50	61	5.5	NE	40	W	1955	3.9	70	16	8	7	6	0	1	1	1	0	0	0
N	83	84	53	61	6.5	NW	46	S	1957	4.5	63	14	7	9	7	*	1	1	0	*	2	0
D	83	84	61	67	6.8	NW	46	SW	1956	5.7	48	10	7	14	10	1	1	2	0	*	6	0
Year	83	83	55	63	6.6	S	60	SW	Mar. 1952	5.1	65	132	117	116	113	*	54	8	71	1	20	*

REFERENCE NOTES APPLYING TO ALL "NORMALS, MEANS, AND EXTREMES" TABLES.

(a) Length of record, years.
(b) Normal values are based on the period 1921-1950, and are means adjusted to represent observations taken at the present standard location.

- No record.
† Airport data.
‡ City Office data.
+ Also on earlier dates, months, or years.
T Trace, an amount too small to measure.
* Less than one-half.

Sky cover is expressed in a range of 0 for no clouds or obscuring phenomena to 10 for complete sky cover. The number of clear days is based on average cloudiness 0-3 tenths; partly cloudy days on 4-7 tenths; and cloudy days on 8-10 tenths. Monthly degree day totals are the sum of the negative departures of average daily temperatures from 65°F. Sleet was included in snowfall totals beginning with July 1948. Heavy fog also includes data referred to at various times in the past as "Dense" or "Thick". The upper visibility limit for heavy fog is 1/4 mile. Data in these tables are based on records through 1957.

Mean Maximum Temperature (°F.), January

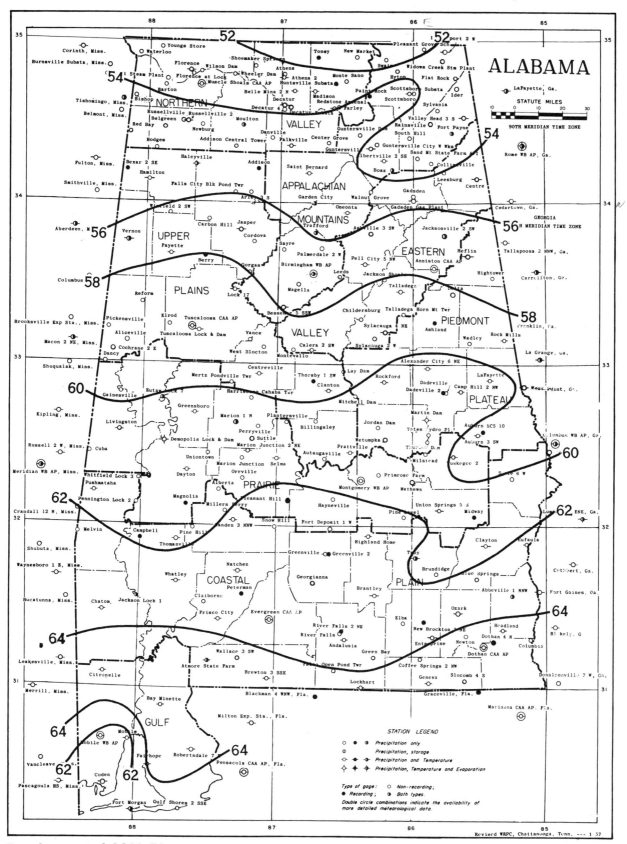

ALABAMA

STATUTE MILES

90TH MERIDIAN TIME ZONE

STATION LEGEND

○ ● ◐ Precipitation only

Ⓟ Precipitation, storage

◇ ◆ Precipitation and Temperature

✧ ✦ Precipitation, Temperature and Evaporation

Type of gage: ○ Non-recording;
● Recording; ◐ Both types.
Double circle combinations indicate the availability of
more detailed meteorological data.

Revised WRPC, Chattanooga, Tenn. --- 1 57

Based on period 1931-52

Isolines are drawn through points of approximately equal value. Caution should be used
in interpolating on these maps, particularly in mountainous areas.

Mean Minimum Temperature (°F.), January

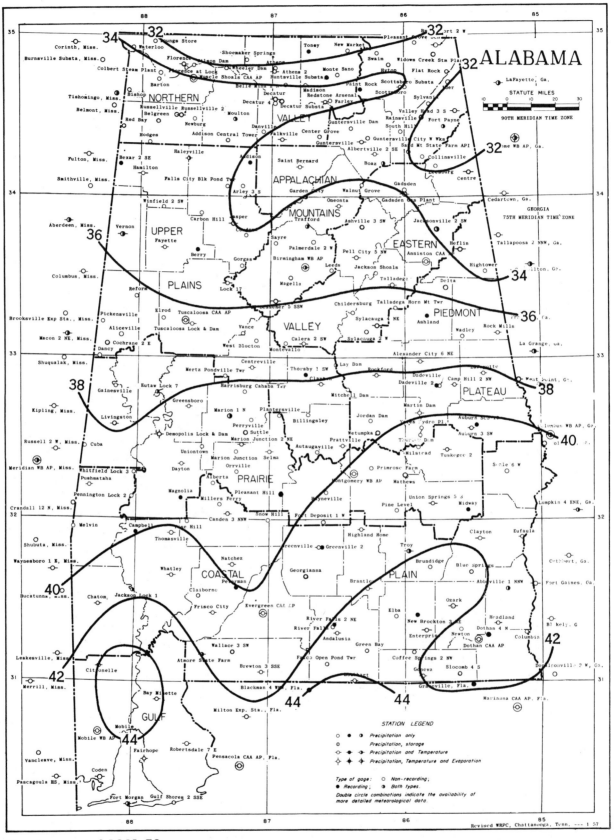

Based on period 1931-52

Isolines are drawn through points of approximately equal value. Caution should be used in interpolating on these maps, particularly in mountainous areas.

Mean Maximum Temperature (°F.), July

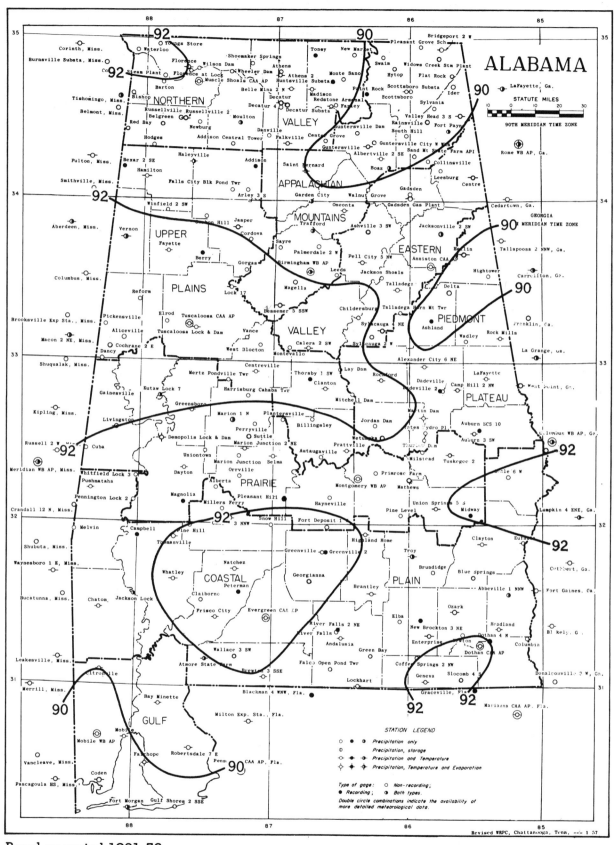

Based on period 1931-52

Isolines are drawn through points of approximately equal value. Caution should be used in interpolating on these maps, particularly in mountainous areas.

Mean Minimum Temperature (°F.), July

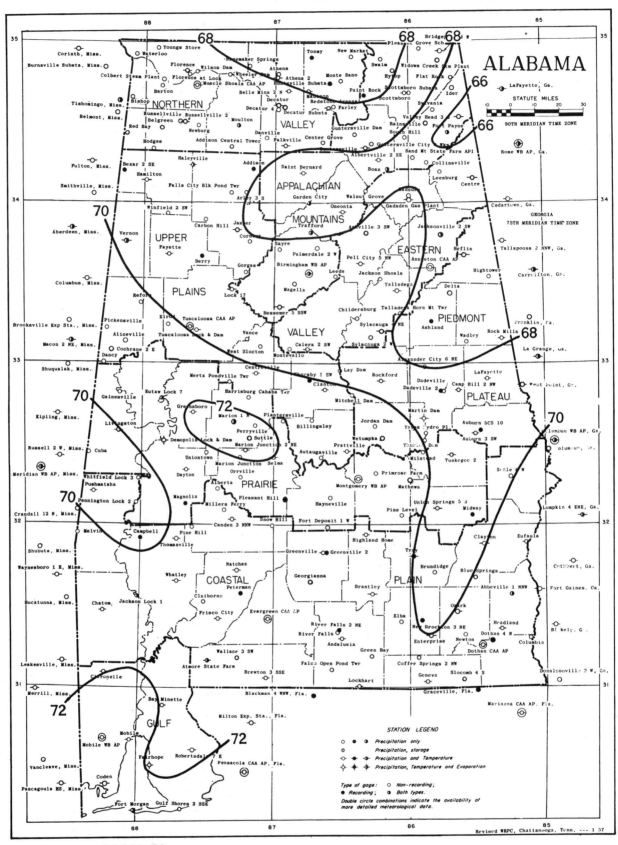

Based on period 1931-52

Isolines are drawn through points of approximately equal value. Caution should be used in interpolating on these maps, particularly in mountainous areas.

Mean Annual Precipitation, Inches

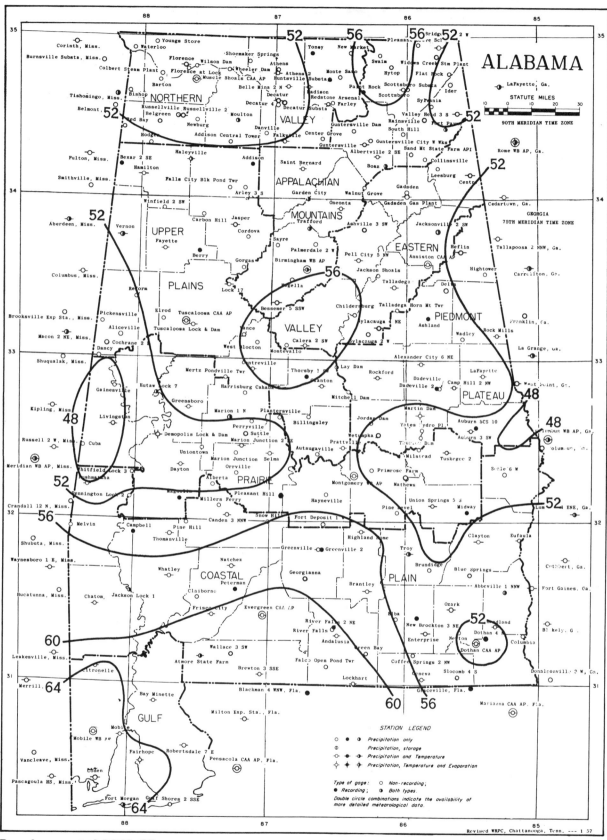

Based on period 1931-55

Isolines are drawn through points of approximately equal value. Caution should be used in interpolating on these maps, particularly in mountainous areas.

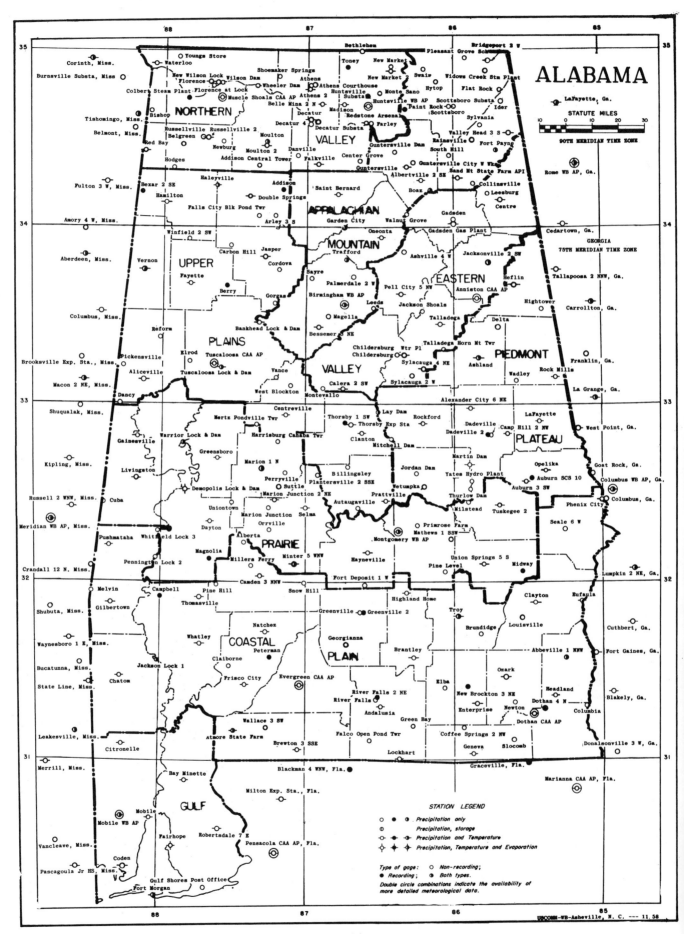

ALABAMA

STATUTE MILES

90TH MERIDIAN TIME ZONE

STATION LEGEND

- ○ ● Precipitation only
- Precipitation, storage
- Precipitation and Temperature
- Precipitation, Temperature and Evaporation

Type of gage: ○ Non-recording;
● Recording; ● Both types.
Double circle combinations indicate the availability of
more detailed meteorological data.

USCOMM-WB-Asheville, N. C. --- 11.58

THE CLIMATE OF
CONNECTICUT

by
A. Boyd Pack

November 1959

Connecticut occupies the southwestern portion of the region known as New England. The State extends for 90 miles in an east-west direction and 75 miles from north to south. The total area of 5,009 square miles makes Connecticut the third smallest state in the Nation.

The topography of Connecticut is predominantly hilly. The highest terrain is found in the northwest portion of the State, with elevations of 1,000 to 2,000 feet. The southwestern quarter and most of the eastern half have elevations of 300 to 1,000 feet. The State of Connecticut is bisected by the Connecticut River which rises in Canada. Smaller river basins in the State with their headwaters in the southern half of Massachusetts include the Housatonic in the west and the Shetucket, Quinebaug, and Thames in the east. The narrow river valleys and steep hillsides in much of the western highlands make for destructive flash floods during periods of unusually heavy or intense rainfall.

The entire southern border of Connecticut is washed by the waters of Long Island Sound. The coastline of approximately 100 miles is indented by small coves and the mouths of numerous rivers and streams. Beaches are found along the greater length.

The chief characteristics of Connecticut's climate may be summarized as follows: (1) equable distribution of precipitation among the four seasons, (2) large ranges of temperature both daily and annual, (3) great differences in the same season or month of different years, and (4) considerable diversity of the weather over short periods of time.

Connecticut lies in the "prevailing westerlies", the belt of generally eastward air movement which encircles the globe in middle latitudes. Embedded in this circulation are extensive masses of air originating in higher and lower latitudes and interacting to produce low-pressure storm systems. A large number of storm centers and air-mass fronts pass near or over Connecticut during a year.

Three types of air affect this State: (1) cold, dry air pouring down from subartic North America, (2) warm, moist air streaming up on a long overland journey from the Gulf of Mexico and subtropical waters of the Atlantic, and (3) cool, damp air moving in from the North Atlantic. Because the flow of air is usually from continental areas, Connecticut is more influenced by the first two types than it is by the third. In other words, the adjacent ocean constitutes an important modifying factor, particularly on the immediate coast, but does not dominate the climate.

The procession of contrasting air masses and the relatively frequent passage of storms bring about a roughly twice weekly alternation from fair to cloudy or storm conditions, usually attended by abrupt changes in temperature, moisture,

– 15 –

sunshine, and wind direction and speed. There is no regular or persistent rhythm to this sequence; it is sometimes interrupted by periods during which the weather pattern continues much the same for several days, and infrequently for a few weeks.

Connecticut's weather is better known for day to day variety rather than monotony. Changeability is also one of its features on a longer time scale. That is, the same month or season will exhibit varying characteristics over the years. A "normal" month, season or year is the exception rather than the rule.

The basic climate, as just outlined, obviously does not result from the predominance of any single controlling weather regime. It is the integrated effect of a large variety of weather patterns. Hence, weather averages in Connecticut are not sufficient for important planning purposes and should be supplemented by more detailed climatological analysis.

Despite the small size of Connecticut there is a difference of about 6°F. in mean annual temperature from north to south. In the higher elevations of the northwestern corner the mean annual value is near 45°F., while along the immediate coast it is about 51°F. The greater area of the State ranges from 47° to 49°F. in the eastern and western highlands to near 50°F. in the central valley. The extreme temperatures of record in Connecticut are 105°F. at Waterbury on July 22, 1926, and -32°F. at Falls Village on February 16, 1943.

The greater contrast of temperature over the State occurs during the winter season. The average minimum temperature in January and February is 13° to 14°F. in northwestern Litchfield County, as compared with an average of about 21°F. in coastal sections and 18°F. in the central valley. The number of days with minimum temperatures of zero or below average about 10 per year at the higher elevations, about 5 in the lower uplands and central valley, and 2 or less along the shore of Long Island Sound. On the average about 70 days with maximum temperature 32°F. or lower can be expected in the colder regions of Connecticut, as compared to 25 to 30 days in the central valley and coastal sections.

Summer temperatures are comparatively uniform over the State. The maximum temperature in July averages from 82° to 84°F., except where altitude or nearness to the ocean reduces the average to near 80°F. The central valley experiences the greatest number of hot days. Temperatures of 90°F. or higher occur on an average on about 10 days per year with a variation of from 5 or less during cool summers to as many as 25 in exceptionally warm summers. At the higher elevations and near the coast the average number is approximately 3 days per year, with a range of from none to about 15 in cold and hot summers, respectively. In much of the western and eastern highlands the occurrence of 90°F. temperatures is a little less frequent than in the central valley.

Temperatures of 100°F. or higher are rare. While most of the long-term weather observing stations in Connecticut have recorded extremes this high, it is only in an occasional summer that 100°F. or higher occurs generally over the State.

During the warmest month of the summer the average minimum temperature ranges from about 56°F. in the cool northwestern corner of the State to about 63°F. in the warmer coastal sections. Over most of the State the average July minimum temperature is within a degree or two of 60°F.

The period free from temperatures of 32°F. or lower has an average length of 155 to 170 days over the greater portion of Connecticut. In the northwest as well as in local areas of the western and eastern highlands, the freeze-free season lasts about 125 to 135 days. Along the immediate coast approximately 190 days will elapse between the last spring and first fall freeze.

In the major crop areas the average growing season begins about mid-April for grasses and hardy crops and about mid-May for the frost sensitive crops. It comes to an end for most crops in early October in the interior and by late October along the immediate coast. Due to elevation, special exposure to nocturnal cooling, and other factors, these dates vary considerably in local areas. There is also a good deal of variation among different years in the length of the freeze-free period.

Precipitation tends to become evenly distributed throughout the year in all parts of Connecticut. Low-pressure centers and their accompanying air mass fronts are the principal year-round producers of precipitation. Storms moving up the Atlantic coast generally yield the heaviest amounts of rain and snow. In the summer bands and patches of thunderstorms and convective showers add considerable precipitation and make up the difference resulting from decreased activity of low-pressure storm centers. Thunderstorms are of brief duration and often scattered in comparison with the general storms, but they yield the heaviest local rainfall.

Variations in precipitation from month to month are sometimes extreme. A month yielding 5 inches or more may be preceded or followed by one with less than 2 inches of precipitation, in any season. Months with less than 1 inch are known to occur, as well as those with precipitation in excess of 10 inches. Such large fluctuations, however, are not characteristic of the precipitation supply in Connecticut. Consequently, prolonged droughts and widespread floods are infrequent.

Annual precipitation averages 44 to 48 inches over most of the State. The amount varies from 40 to 42 inches in the north-central portion (north-northeast of Hartford) to near 50 inches in small areas of the northwest. There is a tendency for the annual precipitation total to decrease from the northwest to the southeast.

Considerable variation in annual precipitation occurs over short distances in northwestern Connecticut. This reflects topographic differences in the area where valleys around 500 feet elevation mingle with hills of 1,500 to 2,000 feet elevation. The annual precipitation increases from about 42 inches in the Housatonic River Valley to 50 inches at high elevations less than 10 miles to the east.

While there are no pronounced wet and dry months as in other climates, February and October are relatively dry. The average total precipitation for each of these months is 3 inches or slightly less in comparison with 3.5 to 4 inches in the other 10 months. Measurable precipitation falls on an average of 1 day in 3, with the yearly total approximating 120 days. Periods of 5 days or more of successive daily precipitation occur a few times during most years. On the other hand, extended periods of little or no precipitation are observed nearly every summer or fall, usually lasting from 10 to 20 days.

Rare meteorological occasions have produced rainfall totals of 4.5 inches or more in all parts of Connecticut within a 24-hour period. Such heavy rainfalls have occurred most often during the summer or fall months. The average annual snowfall increases from the coast to the northwestern corner of the State, with the greater area of the State receiving from 35 to 45 inches. Along the coast the annual total ranges from about 25 inches in the southeastern sector to near 35 inches in west and central sections. In mild winters coastal weather stations record a total of 15 inches or less, while in snowy winters totals of 60 inches or more have been observed. In the highlands of northwestern Connecticut the amount increases with elevation to about 80 inches. Total snowfalls of 150 inches or more have been recorded at the higher elevations in particularly snowy winters.

Most of the snow falls in January and February, but in the majority of winters substantial amounts fall in December or March storms as well. Except for the northwestern highlands, snowfalls of more than 1 inch are quite rare before mid-November and after April 15. In about 1 year out of 4, the April snowfall will total 10 inches or more in the northwest.

The average number of days per year with snow on the ground similarly shows an increase from the shore to the northwest. During an average winter a measurable snow cover is present most of the time from late December through the early half of March in the greater portion of the State. In the immediate coastal areas a snow cover does not last more than a few days unless a heavy snowstorm is followed by prolonged cold temperatures.

During the colder months the prevailing wind is northwest to north over Connecticut, while from April through September southwest or south winds predominate. The mean hourly speed ranges from about 7 miles per hour in the summer and early fall to about 10 miles per hour in the winter and spring seasons.

An important feature of the climate is the sea breeze along the coast. During the summer and late spring this onshore wind blows from cool ocean during the afternoon and penetrates inland from 5 to 10 miles. It occurs often enough to give lower mean summer maximum temperatures in a narrow coastal belt than prevail over interior lowlands.

Thunderstorms occur on an average of 20 to 30 days per year, with the greatest frequency during the summer months and in the afternoon or evening hours of the day. Often these storms are accompanied by destructive hail and/or wind with considerable damage suffered by crops and property over wide areas of the State. Nearly every winter a damaging storm of glaze or freezing rain occurs. Power and telephone lines are extensively disrupted, trees broken, and highway traffic badly crippled.

Aside from infrequent tornadoes and hurricanes, coastal storms or "northeasters" are the most serious weather hazard in Connecticut. They generate very strong winds and heavy rain and produce the greatest snowstorms in the winter. If these storms occur at the time of high tide, heavy water damage results along the shore.

In occasional years a tornado or storm with tornadic characteristics strikes some part of the State. According to historical and Weather Bureau records, 20 tornadoes have hit Connecticut in the past century, but less than 10 were especially damaging. The central valley appears to be the most likely part of the State to be struck, and the summer months the most likely season.

Storms of tropical origin occasionally affect Connecticut during the summer or fall months, as they move on a path well out over the ocean. However, hurricanes have been known to strike the State full force with recent occurrences in 1938, 1944, 1954 and 1955, resulting in enormous property damage and maximal loss of life.

The Connecticut River shows an annual rise in early spring as the result of the melting of high elevation snow in northern and central New England. Melting of the snow cover combined with heavy rainfall is one of the principal causes of flooding in the Connecticut River and tributaries resulting in a maximum annual flood frequency in March, April, and May. Flooding also occurs occasionally in the lower Connecticut River as well as the other small river basins in the State during the winter and early spring from heavy rains combined with melting snow in Connecticut and southern Massachusetts or falling on frozen ground. A secondary period of flooding (occasionally of major proportions) is caused by heavy rains which may be associated with hurricanes or storms of tropical origin in late summer or fall, normally the low water season.

The percentage of possible sunshine averages 55 to 60 percent, ranging from 45 percent in the interior during the months of November through January to near 65 percent along the coast in the summer. The average number of clear days per year is between 100 and 125, with the greatest number per month usually occurring in September and October. An average of about 140 cloudy days occur per year. One or more prolonged periods of sunless skies are commonly observed during the winter and early spring seasons.

Heavy or dense fog is observed on an average of about 25 days per year in both coastal and inland sections. In the former section heavy fog is most common during the late winter and spring seasons, while inland the late summer and fall is the period of maximum occurrence.

The humidity tends to be lowest in the spring and highest in the late summer and early fall. While an occasional summer day is uncomfortable from the combined effects of high temperature and high humidity, the frequency of such days is much less in Connecticut than in the Southern or Midwestern States.

Connecticut is primarily an industrial state, although agriculture is of considerable local and regional importance. The agriculture is of an intensive type with about 20 percent of the land area devoted to farming. High yields of crops are obtained, and the State ranks high in annual receipts per acre from farm marketings.

The climate plays a significant role in the State's agriculture. The patterns of temperature and precipitation are favorable to a wide variety of crops as well as to dairying and poultry raising, which are the two most important agricultural enterprises in Connecticut. A summer mean temperature of 70°F. or lower in the highlands either side of the central valley favors the growth of pastures and hay crops. The production of a good hay crop is important because livestock must be barn-fed for almost 6 months out of the year.

Tobaccos for cigar manufacture and potatoes are very valuable crops grown in the northern portion of the State's Connecticut River Valley. While the former is a crop favored by warm weather and the latter by cool weather, moderate summer temperatures and adequate precipitation promote high yields of both. Humidity conditions from July through September are advantageous to the air-curing process of the tobacco. There are times, however, when persistent high humidity requires preventative measures against fungus spoilage of the tobacco leaves.

Apples, peaches, strawberries, and commercial truck crops are produced principally in the central valley and the coastal uplands. Altogether they represent about 7 percent of the State's agricultural income. The production of nursery and greenhouse plants has become increasingly important in recent years. The climate is satisfactory for a great variety of ornamental flowers, shrubs and trees. Field corn is grown mainly in dairying areas of the highlands, and the production of small grains is of minor importance. These two crops are used for livestock feed.

Forests cover about 65 percent of Connecticut and represent the dominant vegetation. The dominant species comprise two large zones running east and west, covering the southern and north-central portions of the State. Two additional but small zones are recognized in the northwestern corner of Connecticut, reflecting the effects of elevation and the variation in climatic elements described earlier. The forests are valuable to the State's economy, not only for the wood products and support of wood processing industries, but as a scenic attraction in the autumn and as a preventative factor in the control of erosion and floods.

The climate has been an important element in the growth and development of industry in Connecticut. Water is an indispensable element for the successful functioning of an industrial society.

The ample rainfall, dependable runoff, and ground water supplies have made Connecticut desirable for the location of a great variety of industries. Comfortable summer temperatures and winters that are vigorous, but not unduly severe, have also made the State tolerable for the many aspects of an industrial economy.

The tourist and vacation trade represents a considerable part of the economy. The climate is generally agreeable for many recreational activities. Pleasant temperatures and frequent sunny days prevail during the summer and early fall months at both seaside and inland resorts. While an occasional winter will permit some winter sports activities, the commercial importance of these is much less than in the northern New England States.

In summary, the climate contributes greatly to Connecticut's prominence as an industrial, agricultural, and recreational area. It is a rich natural asset, invigorating to its citizens and favorable for further economic development of the State.

GENERAL REFERENCES

National Planning Association: The Economic State of New England. 1954.

U. S. Dept. of Agriculture: Atlas of American Agriculture. 1936.

------: Climate and Man. Yearbook of Agriculture for 1941.

U. S. Dept. of Commerce, Weather Bureau: Climatic Summary of the United States, Bulle. W. 1930. Section 86 (Massachusetts, Connecticut and Rhode Island). Climatic Summary of the United States-Supplement for 1931 through 1952, No. 11-23, New England.

------: Climatological Data, New England. (issued monthly and annually, Jan. 1888 -- present; published under various other titles previous to Jan. 1921).

------: Local Climatological Data with Comparative Data. Issued monthly and annually for Bridgeport, New Haven, and Hartford, Conn.

Upton, W.: Characteristics of the New England Climate. Annals Harvard Astron. Observatory. 1890.

SPECIALIZED REFERENCES

Brooks, C. F. : New England Snowfall. Monthly Weather Review. Vol. 45, 1917.

------: The Rainfall of New England. General Statement. Journal New Eng. Water Works Assoc., Vol 44, 1930.

Church, P. E. : A Geographical Study of New England Temperatures. Geographical Review, Vol 26, 1936.

Eustis, R. S. : Winds over New England in Relation to Topography. Bulle. Amer. Met. Soc., Vol. 23, 1942.

Goodnough, X. H.: Rainfall in New England. Jour. New Eng. Water Works Assoc., Vols. 29, 1915; 35, 1921; 40, 1926.

Kirk, J. M. : The Weather and Climate of Connecticut. Conn. Geological and Natural History Survey, Bulle. 61, 1939.

Perley, S.: Historic Storms of New England. 1891.

------: Special Lists

Stone, R. G. : Distribution of snow depths over New York and New England. Trans. Amer. Geophy. Union. 1940.

------: The average length of the season with snow cover of various depths in New England. Trans. Amer. Geophy. Union. 1944.

Westveld, Marinus, et al: Natural Forest Vegetation Zones in New England. Jour. of Forestry, Vol. 54, 332-338. 1956.

Weber, J. H. : The Rainfall of New England. Historical Statement. Annual Rainfall. Seasonal Rainfall. Mean Monthly Rainfall of Southern New England. Maximum and Minimum Rainfall of Southern New England. Jour. New Eng. Water Works Assoc., Vol. 44, 1930.

White, G. V. : Rainfall in New England. Jour. New Eng. Water Works Assoc. Vols. 56, 1942; 57, 1943.

Brown, R. A. : Twisters in New England. Unpublished compilation of historical records of tornadoes. 1957.

(1) Climatic Summary of the United States (Bulletin W) 1930 edition, Section 86. U. S. Weather Bureau

(2) Climat Summary of the United States, Connecticut-Supplement for 1931 through 1952 (Bulletin W Supplement). U. S. Weather Bureau

(3) Climatological Data - Connecticut. U. S. Weather Bureau

(4) Climatological Data National Summary. U. S. Weather Bureau

(5) Hourly Precipitation Data - Connecticut. U. S. Weather Bureau

(6) Local Climatological Data, U. S. Weather Bureau, for Bridgeport, Hartford, Middletown and New Haven, Connecticut.

FREEZE DATA

STATION	Freeze threshold temperature	Mean date of last Spring occurrence	Mean date of first Fall occurrence	Mean No. of days between dates	Years of record Spring	No. of occurrences in Spring	Years of record Fall	No. of occurrences in Fall	STATION	Freeze threshold temperature	Mean date of last Spring occurrence	Mean date of first Fall occurrence	Mean No. of days between dates	Years of record Spring	No. of occurrences in Spring	Years of record Fall	No. of occurrences in Fall
		CONNECTICUT									RHODE ISLAND						
HARTFORD	32	04-22	10-19	186	30	30	30	30	BLOCK ISLAND WB CITY	32	04-09	11-16	221	30	30	30	30
	28	04-06	11-02	210	30	30	30	30		28	03-23	11-27	249	30	30	30	30
	24	03-24	11-17	239	30	30	30	30		24	03-14	12-03	264	30	30	30	30
	20	03-13	11-29	261	30	30	30	30		20	03-08	12-09	276	30	30	30	27
	16	03-08	12-07	273	30	30	30	29		16	02-28	12-15	290	30	30	30	24
NEW HAVEN WB AP	32	04-15	10-27	195	30	30	30	30	KINGSTON	32	05-08	10-05	150	30	30	30	30
	28	03-28	11-10	226	30	30	30	30		28	04-24	10-16	176	30	30	30	30
	24	03-18	11-23	249	30	30	30	30		24	04-08	10-28	203	30	30	29	29
	20	03-10	12-03	268	30	30	30	29		20	03-24	11-16	237	30	30	29	29
	16	03-05	12-10	280	30	30	30	27		16	03-12	11-26	259	30	30	29	29
Putnam	32	05-15	09-24	133	15	15	17	17	PROVIDENCE WB CITY	32	04-13	10-27	197	30	30	30	30
	28	04-29	10-08	162	15	15	17	17		28	04-01	11-11	224	29	29	30	30
	24	04-13	10-19	189	15	15	17	17		24	03-18	11-24	251	29	29	30	30
	20	03-30	11-06	221	15	15	16	16		20	03-12	12-02	265	29	29	30	30
	16	03-19	11-18	244	15	15	16	16		16	03-05	12-07	276	29	29	30	29

Data in the above table are based on the period 1921-1950, or that portion of this period for which data are available.

Means have been adjusted to take into account years of non-occurrence.

A freeze is a numerical substitute for the former term "killing frost" and is the occurrence of a minimum temperature at or below the threshold temperature of 32°, 28°, etc.

Freeze data tabulations in greater detail are available and can be reproduced at cost.

*MEAN TEMPERATURE AND PRECIPITATION

STATION	JANUARY Temperature	Precipitation	FEBRUARY Temperature	Precipitation	MARCH Temperature	Precipitation	APRIL Temperature	Precipitation	MAY Temperature	Precipitation	JUNE Temperature	Precipitation	JULY Temperature	Precipitation	AUGUST Temperature	Precipitation	SEPTEMBER Temperature	Precipitation	OCTOBER Temperature	Precipitation	NOVEMBER Temperature	Precipitation	DECEMBER Temperature	Precipitation	ANNUAL Temperature	Precipitation
CONNECTICUT																										
NORTHWEST																										
CREAM HILL	25.0	3.55	25.4	2.99	33.7	3.88	45.4	3.76	56.9	4.23	65.4	4.70	70.4	4.59	68.5	4.24	61.2	4.24	51.6	3.34	39.3	4.14	27.5	3.60	47.5	47.26
FALLS VILLAGE	25.5	3.09	26.0	2.38	34.6	3.39	45.7	3.55	57.2	3.97	65.7	4.58	70.3	4.01	68.4	3.92	60.9	4.14	50.5	3.03	39.8	3.74	27.7	2.93	47.7	42.73
DIVISION	24.7	3.56	25.1	2.91	33.6	3.91	45.1	3.78	56.6	4.31	64.9	4.78	69.8	4.20	67.8	4.33	60.4	4.21	50.5	3.37	38.7	4.26	27.3	3.57	47.0	47.19
CENTRAL																										
COLLINSVILLE 1 S		4.05		3.25		4.66		4.14		4.30		4.30		3.87		4.91		3.96		3.37		4.55		3.97		49.33
HARTFORD BRAINARD FLD	27.0	3.74	27.5	3.03	36.9	3.53	47.4	3.55	58.9	3.77	67.7	3.76	72.7	3.93	70.4	3.67	63.1	3.41	52.6	2.70	41.7	3.85	30.1	3.49	49.7	42.43
HARTFORD WB AIRPORT	27.0	3.15	28.1	2.57	37.2	3.81	48.0	3.56	59.7	3.66	68.9	3.62	73.8	3.56	71.4	3.54	63.8	3.44	52.9	2.80	41.3	3.48	29.6	3.29	50.1	40.48
MIDDLETOWN WB	26.8	3.91	27.1	3.01	36.0	3.92	46.3	3.96	57.8	4.14	66.7	3.80	71.6	3.23	69.7	3.54	63.0	3.49	52.3	2.81	41.3	4.40	29.8	3.93	49.0	44.14
MIDDLETOWN 4 W	28.3	4.19	28.7	3.21	36.5	4.71	47.4	4.38	58.7	4.48	67.3	4.18	72.7	3.72	70.6	4.51	63.2	4.40	53.8	3.48	42.4	5.06	31.0	4.20	50.1	50.52
STORRS	26.5	3.62	26.7	2.82	34.5	4.34	45.2	3.84	56.4	4.04	65.1	3.65	70.2	3.56	68.4	5.13	61.2	4.07	51.7	3.35	40.6	4.23	29.1	3.56	48.0	46.21
DIVISION	28.0	3.80	28.4	2.92	36.5	4.33	47.1	3.92	58.1	4.03	66.8	3.82	72.1	3.69	70.0	4.62	62.7	4.04	53.0	3.29	41.7	4.47	30.4	3.79	49.6	46.72
COASTAL																										
BRIDGEPORT WB AP	29.2	3.43	29.0	2.97	36.9	3.60	46.3	3.49	57.2	3.60	66.9	3.47	72.8	3.97	71.7	4.43	65.2	3.55	54.4	2.83	43.5	3.59	32.3	3.08	50.5	42.01
LAKE KONOMOC		4.36	0	3.44		4.99		4.18		4.19		3.55		3.73		4.83		4.49		3.86		4.97		4.29		50.88
NEW HAVEN WB AIRPORT	29.1	3.89	29.1	3.30	37.1	4.12	46.1	3.89	56.7	3.87	65.8	3.81	71.2	3.66	69.8	4.11	63.6	3.46	53.3	3.00	42.9	3.94	31.9	3.94	49.7	44.99
DIVISION	30.1	3.98	30.4	3.09	37.8	4.71	47.9	3.91	58.4	4.00	67.3	3.61	72.8	3.59	71.1	4.95	64.3	4.08	54.4	3.42	43.4	4.30	32.4	3.85	50.9	47.49
RHODE ISLAND																										
RHODE ISLAND																										
BLOCK ISLAND WB AP	31.9	3.67	30.9	3.25	37.1	3.54	44.9	3.37	54.2	2.96	63.0	2.86	69.1	2.55	69.0	3.46	63.5	2.98	54.5	3.10	45.3	3.53	35.1	3.36	49.9	38.63
KINGSTON	29.5	4.12	29.7	3.27	36.7	4.26	45.6	3.83	55.7	3.49	64.2	3.06	71.0	3.06	68.9	4.66	62.7	3.21	52.5	3.21	42.2	4.74	31.7	3.75	49.1	44.70
PROVIDENCE WB AIRPORT	28.7	3.75	28.6	2.84	36.8	3.58	46.0	3.37	56.8	3.02	65.6	3.17	71.5	2.55	70.3	4.33	63.6	3.50	52.7	2.83	42.6	4.40	31.6	3.45	50.4	42.99
DIVISION	30.7	3.99	30.5	3.15	37.4	4.16	46.6	3.78	56.7	3.31	65.4	3.02	71.5	2.55	70.3	4.33	63.6	3.50	54.3	3.15	43.9	4.44	33.3	3.61	50.4	42.99

⨯ Averages for period 1931-1955, except for stations marked WB which are "normals" based on period 1921-1950. Divisional means may not be the arithmetical average of individual stations published, since additional data from shorter period stations are used to obtain better areal representation.

CONFIDENCE LIMITS

In the absence of trend or record changes, the chances are 9 out of 10 that the true mean will lie in the interval formed by adding and subtracting the values in the following table from the means for any station in the State. Because of the wider variation in mean precipitation, the corresponding monthly means and annual mean must be substituted for "p" in the precipitation table below to obtain mean precipitation confidence limits.

1.7	.54 p	1.4	.52 p	1.6	.68 p	1.0	.64 p	.9	.55 p	.7	.69 p	.5	.66 p	.8	.70 p	.8	.70 p	1.0	.66 p	1.0	.74 p	1.2	.50 p	.4	2.67 p

COMPARATIVE DATA

Data in the following table are the mean temperature and average precipitation for Cream Hill, Connecticut for the period 1906-1930 and are included in this publication for comparative purposes:

23.7	3.54	23.4	3.38	32.7	3.44	42.4	3.60	54.9	4.01	63.8	3.85	69.1	4.08	66.9	4.36	61.0	4.04	50.4	3.69	37.9	3.63	26.5	3.43	46.1	45.05

NORMALS, MEANS, AND EXTREMES

BRIDGEPORT, CONNECTICUT — MUNICIPAL AIRPORT

LATITUDE 41° 10' N
LONGITUDE 73° 08' W
ELEVATION (ground) 7 feet

Temperature and Degree Days

Month	Normal Daily Max	Normal Daily Min	Normal Monthly	Extreme Record Highest	Year	Extreme Record Lowest	Year	Normal degree days
(Length of record, yrs.)	(b)	(b)	(b)	57	57	57	57	(b)
J	36.8	21.8	29.2	68	1932	-14	1904	1110
F	36.7	21.2	29.0	70	1930	-20	1934	1008
M	45.0	28.8	36.9	83	1921+	3	1934+	871
A	55.2	37.4	46.3	93	1915	11	1923	561
M	66.5	47.7	57.2	95	1930+	28	1947	249
J	75.9	57.9	66.9	96	1945	34	1945	38
J	81.6	63.9	72.8	103	1957	44	1947	0
A	80.4	62.9	71.7	101	1900	43	1923	0
S	74.2	56.1	65.2	99	1928	32	1947	66
O	64.0	44.7	54.4	90	1924	20	1936	334
N	51.6	35.2	43.5	77	1909	9	1924	645
D	39.6	26.0	33.3	67	1908	-12	1917	1014
Year	58.9	41.9	50.5	103	July 1957	-20	Feb. 1934	5896

Precipitation and Snow, Sleet

Month	Precip Normal total	Max monthly	Year	Min monthly	Year	Max in 24 hrs	Year	Snow Mean total	Snow Max monthly	Year	Snow Max 24 hrs	Year
J	3.43	7.88	1936	0.51	1955	1.56	1949	9.0	30.3	1923	9.7	1952
F	2.97	6.32	1896	0.85	1901	1.80	1951	10.8	47.0	1934	13.8	1949
M	3.60	9.54	1899	0.29	1915	2.72	1953	5.9	19.4	1941	11.7	1956
A	3.49	9.41	1901	0.69	1896	2.10	1953	1.1	8.1	1924	3.7	1950
M	3.60	10.18	1898	0.49	1903	2.12	1949	0.0	0.0		0.0	
J	3.47	8.48	1903	0.40	1957	1.82	1952	0.0	0.0		0.0	
J	3.97	18.77	1897	0.45	1944	3.45	1953	0.0	0.0		0.0	
A	4.43	12.02	1927	0.20	1899	3.97	1955	0.0	0.0		0.0	
S	3.55	14.15	1938	0.09	1941	5.84	1954	0.0	0.0		0.0	
O	2.83	10.72	1955	0.30	1924	5.84	1955	T	T		T	
N	3.59	7.60	1898	0.81	1933	4.07	1954	1.6	14.1	1938	5.4	1925+
D	3.08	9.85	1901	0.82	1935	2.05	1954	6.8	25.8	1945	5.6	1957
Year	43.01	18.77	July 1897	0.09	Sept. 1941	5.84	Oct. 1955	35.2	47.0	Feb. 1934	13.8	Feb. 1949

Relative humidity, Wind, Sky cover

Month	RH 1:00 a.m.	RH 7:00 a.m.	RH 1:00 p.m.	RH 7:00 p.m.	Wind Mean hourly	Prevailing direction	Fastest mile Speed	Direction	Year	Mean sky cover sunrise to sunset
J	77	75	64	70	9.9	NE	40	NW	1950	6.0
F	74	74	57	68	9.4	NW	47	WNW	1951	5.5
M	74	72	57	65	11.4	NW	45	SSW	1950	5.6
A	74	71	56	67	10.5	NW	45	SSW	1949	5.9
M	82	79	57	70	10.8	SW	32	W	1952	5.8
J	85	79	57	71	8.3	SW	40	WSW	1950	5.1
J	86	79		73	7.4	SW	40	NW	1949	5.1
A	86	81		73	7.5	SW	32	W	1952	5.0
S	86	85		75	8.2	NE	30	SSE	1952	5.0
O	81	79		72	9.6	NE	38	SE	1951	5.0
N	79	79		69	10.6	NE	38	E	1953	5.9
D	75	74		69	10.2	WNW	35	N	1951+	6.1
Year	80	76	58	70	9.6	NE	62	E	Nov. 1950	5.5

Mean number of days

Month	Clear	Partly cloudy	Cloudy	Precip .01 in. or more	Snow, Sleet 1.0 in. or more	Thunderstorms	Heavy fog	Max 90° and above	Max 32° and below	Min 32° and below	Min 0° and below
J	7	8	16	12	2	0	4	0	9	22	0
F	8	8	13	10	1	0	4	0	6	20	0
M	8	9	14	13	1	2	3	0	3	16	0
A	7	9	14	12	0	4	3	0	1	3	0
M	6	11	14	13	0	4	3	0	0	0	0
J	8	11	11	7	0	4	2	2	0	0	0
J	9	12	10	9	0	7	2	5	0	0	0
A	9	11	11	9	0	6	2	2	0	0	0
S	12	9	9	7	0	3	2	2	0	0	0
O	10	8	11	9	0	0	2	0	0	0	0
N	8	8	11	9	0	0	2	0	1	9	0
D	8	7	13	10	1	0	2	0	1	22	0
Year	103	110	152	119	5	24	26	9	12	92	0

HARTFORD, CONNECTICUT — BRADLEY FIELD

LATITUDE 41° 56' N
LONGITUDE 72° 41' W
ELEVATION (ground) 169 feet

Temperature and Degree Days

Month	Normal Daily Max	Normal Daily Min	Normal Monthly	Extreme Record Highest	Year	Extreme Record Lowest	Year	Normal degree days
(Length of record, yrs.)	(b)	(b)	(b)	53	53	53	53	(b)
J	36.1	17.9	27.0	70	1932	-17	1957	1178
F	37.9	18.2	28.1	72	1954	-24	1943	1033
M	47.4	27.0	37.2	86	1945	-4	1948	862
A	60.0	36.0	48.0	91	1938	11	1923	510
M	72.3	47.0	59.7	94	1914	28	1956+	183
J	80.9	56.8	68.9	101	1952	38	1957	21
J	85.6	62.0	73.8	100	1936	48	1957+	0
A	82.9	59.9	71.4	101	1955	38	1940	8
S	75.5	52.1	63.8	101	1953	30	1957+	90
O	64.4	41.2	52.9	91	1927	6	1924	375
N	51.4	32.1	41.8	83	1950	-8	1938	711
D	39.0	20.2	29.6	67	1946	-18	1917	1097
Year	61.1	39.1	50.1	101	Aug. 1955+	-24	Feb. 1943	6068

Precipitation and Snow, Sleet

Month	Precip Normal total	Max monthly	Year	Min monthly	Year	Max in 24 hrs	Year	Snow Mean total	Snow Max monthly	Year	Snow Max 24 hrs	Year
J	3.15	7.77	1953	0.91	1955	3.32	1940	10.7	40.1	1923	13.6	1923+
F	2.57	5.72	1957	1.54	1957	3.04	1916	11.9	32.7	1926	14.0	1949
M	3.81	9.21	1953	0.29	1915	2.92	1953	6.9	43.3	1956	19.0	1956
A	3.56	7.66	1942	0.65	1942	3.73	1940	1.4	9.7	1924	9.7	1924
M	3.66	7.04	1955	0.78	1965	3.68	1909	0.0	0.0		0.0	
J	3.62	8.08	1912	0.66	1912	2.52	1920	0.0	0.0		0.0	
J	3.56	11.24	1938	0.54	1924	2.87	1928	0.0	0.0		0.0	
A	3.54	21.87	1955	0.93	1955	12.12	1955	0.0	0.0		0.0	
S	3.44	11.81	1938	0.20	1914	6.72	1938	0.0	0.0		0.0	
O	2.80	11.36	1955	0.18	1924	5.19	1937	T	0.1	1925	0.1	1925
N	3.48	7.36	1927	0.87	1937	3.55	1933	1.9	17.1	1938	9.9	1945
D	3.29	6.88	1936	0.78	1955	2.35	1942	7.5	45.3	1945	18.2	1945
Year	40.48	21.87	Aug. 1955	0.18	Oct. 1924	12.12	Aug. 1955	40.3	45.3	Dec. 1945	19.0	Feb. 1949

Relative humidity, Wind, Sunshine, Sky cover

Month	RH 1:00 a.m.	RH 7:00 a.m.	RH 1:00 p.m.	RH 7:00 p.m.	Wind Mean hourly	Prevailing direction	Fastest mile Speed	Direction	Year	Pct of possible sunshine	Mean sky cover sunrise to sunset
J	73	61	67	67	8.9	N	51	SW	1913	46	6.3
F	72	59	65	65	9.2	N	49	S	1912	46	5.8
M	74	55	63	63	9.6	S	55	SE	1953	56	5.8
A	72	52	61	61	9.7	S	57	NE	1956	56	6.1
M	79	56	56	67	8.1	S	41	NW	1957	54	5.8
J	83	56			7.5	S	52	N	1946	60	6.3
J	86				7.6	S	42	SW	1941	62	5.8
A	89				7.3	NE	56	NE	1954	60	5.6
S	89				7.9	N	62	N	1944	57	5.5
O	84				8.6	N	57	N	1950	55	5.3
N	81	60		69	8.7	NW	70	NW	1915	46	6.2
D	76	60		68	8.4	N	54	W	1915	45	6.3
Year	81	56	67	67	8.4	S	70	E	Nov. 1950	55	5.9

Mean number of days

Month	Clear	Partly cloudy	Cloudy	Precip .01 in. or more	Snow, Sleet 1.0 in. or more	Thunderstorms	Heavy fog	Max 90° and above	Max 32° and below	Min 32° and below	Min 0° and below
J	8	8	15	13	3	*	3	0	11	29	1
F	8	8	13	11	3	*	2	0	8	25	1
M	10	9	13	12	2	1	2	*	2	20	*
A	9	10	13	12	1	2	2	0	*	6	0
M	8	11	11	10	0	3	2	*	0	*	0
J	9	12	11	10	0	5	2	4	0	0	0
J	11	11	11	10	0	7	2	3	0	0	0
A	9	11	11	9	0	5	5	1	0	0	0
S	12	9	10	8	0	2	4	*	0	0	0
O	12	9	11	8	0	1	4	0	0	3	0
N	9	8	13	10	1	*	3	0	1	14	0
D	8	8	15	11	2	*	3	0	8	25	1
Year	111	106	148	128	10	28	43	10	31	120	3

NORMALS, MEANS, AND EXTREMES

LATITUDE 41° 16'
LONGITUDE 72° 53'
ELEVATION (ground) 6 feet

Month	Temp Normal Daily Max (b)	Temp Normal Daily Min (b)	Temp Normal Monthly (b)	Ext Record Highest	Year	Record Lowest	Year	Normal Degree Days (b)	Precip Normal Total (b)	Precip Max Monthly	Year	Precip Min Monthly	Year	Precip Max in 24 hrs	Year	Snow Mean Total	Snow Max Monthly	Year	Snow Max in 24 hrs	Year	RH 7:00 a.m. EST	RH 1:00 p.m. EST	RH 7:00 p.m. EST	Wind Mean Hourly	Prevailing Dir	Fastest Mile Speed	Fastest Mile Dir	Year	Mean Sky Cover Sunrise–Sunset	Pct Possible Sunshine	Days Clear	Partly Cloudy	Cloudy	Precip .01 in or more	Snow/Sleet 1.0 in or more	Thunderstorms	Heavy Fog	Max 90° and above	Max 32° and below	Min 32° and below	Min 0° and below
(a) length of record	85	85	85	85		85		(b)	(b)	85		85		85		85	85		85		69	40	63	85	85	85			85	58	85	85	85	85	85	85	24	85	85	85	85
J	37.0	21.2	29.1	67	1932	-14	1873	1113	3.89	8.59	1955	0.63	1940	2.74	1940	9.6	31.2	1923	13.7	1873	75	65	69	9.2		50	NW	1890	5.7	52	10	9	12	13	3	*	3	0	10	26	1
F	37.1	21.0	29.1	69	1930	-15	1934	1005	3.30	6.98	1901	0.54	1878	4.13	1878	10.4	46.3	1934	23.2	1888	74	63	68	9.5		47	E	1897	5.3	61	10	9	10	12	3	1	2	0	8	24	1
M	45.6	28.5	37.1	84	1921	0	1885	865	4.12	10.78	1953	0.25	1915	4.78	1876	6.9	44.9	1888	28.0	1888	73	63	67	9.8		47	N	1924	5.2	60	10	9	12	12	2	1	3	0	3	19	*
A	55.4	36.8	46.1	91	1915	13	1923	567	3.89	9.03	1901	1.19	1876	5.90	1876	1.4	10.0	1874	8.0	1915	71	60	67	9.5		47	NE	1929	5.3	58	9	11	12	12	1	1	2	*	*	4	0
M	66.4	46.9	56.7	95	1930	30	1882	261	3.87	8.03	1898	0.18	1887	3.32	1883	T	1.2	1876	1.2	1876	72	60	70	8.2		40	N	1898	5.3	61	10	11	11	12	*	3	4	1	0	*	0
J	75.1	56.5	65.8	99	1934	40	1946	52	3.81	13.96	1949	0.12	1936	7.50	1936	0.0	0.0	–	0.0	–	75	63	73	7.4		40	NW	1920	5.1	61	10	11	10	11	0	5	3	2	0	0	0
J	80.1	62.3	71.2	101	1926	47	1945	0	3.66	17.08	1889	0.52	1876	7.00	1876	0.0	0.0	–	0.0	–	77	63	74	7.1		41	S	1897	5.0	66	11	11	8	11	0	6	2	1	0	0	0
A	78.8	60.7	69.8	100	1948	45	1908	18	4.11	12.99	1874	0.26	1882	8.73	1874	0.0	0.0	–	0.0	–	79	65	75	7.0		47	S	1893	4.9	64	11	11	9	11	0	5	1	1	0	0	0
S	73.1	54.0	63.6	100	1881	32	1888	93	3.46	14.52	1938	0.17	1878	6.69	1878	0.0	0.0	–	0.0	–	80	62	76	7.6		62	S	1903	4.7	63	12	9	10	9	0	2	3	*	0	0	0
O	63.3	43.3	53.3	89	1908	24	1879	363	3.00	10.84	1913	0.23	1924	5.19	1906	T	2.0	1876	2.0	1876	78	62	70	8.3		54	SE	1904	4.7	63	12	9	10	9	*	1	2	0	0	*	0
N	51.6	34.2	42.9	75	1928	2	1875	663	3.94	8.14	1944	0.67	1890	3.69	1947	1.7	23.3	1898	16.0	1898	77	63	63	8.9		57	E	1950	5.5	54	10	9	11	11	1	*	2	0	1	11	*
D	39.9	23.8	31.9	68	1889	-12	1917	1026	3.94	8.34	1936	0.96	1935	4.25	1948	7.3	27.3	1904	15.0	1948	75	64	69	9.1		54	NW	1915	5.5	54	10	9	12	12	2	*	2	0	6	23	*
Year	56.6	40.8	49.7	101	July 1926	-15	Feb. 1934	6026	44.99	17.08	July 1889	0.12	June 1949	8.73	Aug. 1874	37.3	46.3	Feb. 1934	28.0	Mar. 1888	76	62	71	8.5		62	S	Sept. 1903	5.2	60	125	117	123	131	11	25	29	4	27	108	2

REFERENCE NOTES APPLYING TO ALL "NORMALS, MEANS, AND EXTREMES" TABLES.

(a) Length of record, years.
(b) Normal values are based on the period 1921-1950, and are means adjusted to represent observations taken at the present standard location.
* Less than one-half.

– No record.
+ Airport data.
++ City Office data.
+++ Also on earlier dates, months, or years.
T Trace, an amount too small to measure.

Sky cover is expressed in a range of 0 for no clouds or obscuring phenomena to 10 for complete sky cover. The number of clear days is based on average cloudiness 0-3 tenths; partly cloudy days on 4-7 tenths; and cloudy days on 8-10 tenths. Monthly degree day totals are the sum of the negative departures of average daily temperatures from 65°F. Sleet was included in snowfall totals beginning with July 1948. Heavy fog also includes data referred to at various times in the past as "Dense" or "Thick". The upper visibility limit for heavy fog is 1/4 mile. Data in these tables are based on records through 1957.

Mean Annual Precipitation, Inches

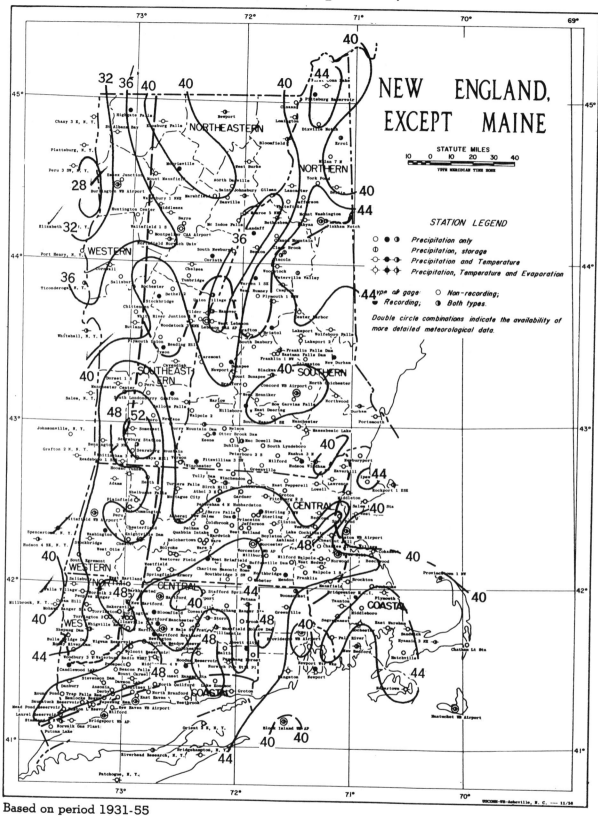

Based on period 1931-55

Isolines are drawn through points of approximately equal value. Caution should be used in interpolating on these maps, particularly in mountainous areas.

Mean Maximum Temperature (°F.), January

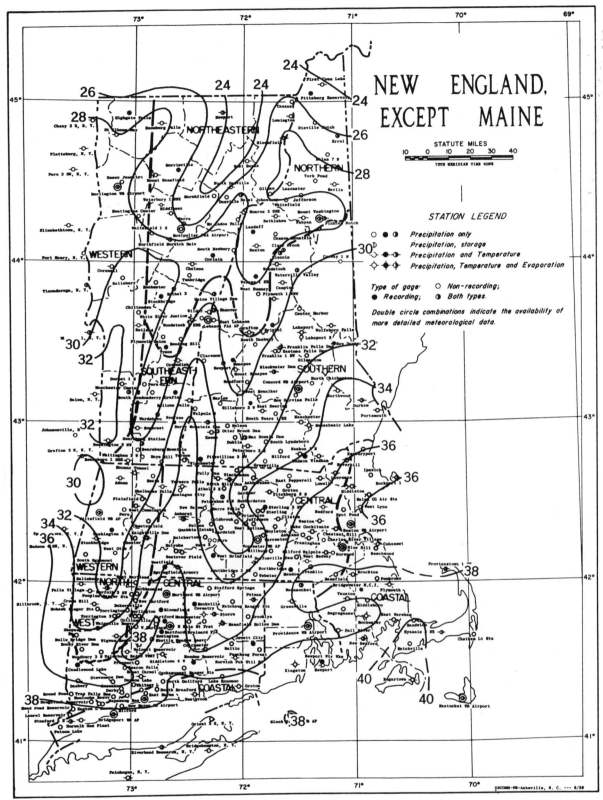

Based on period 1931-52

Isolines are drawn through points of approximately equal value. Caution should be used in interpolating on these maps, particularly in mountainous areas.

Mean Minimum Temperature (°F.), January

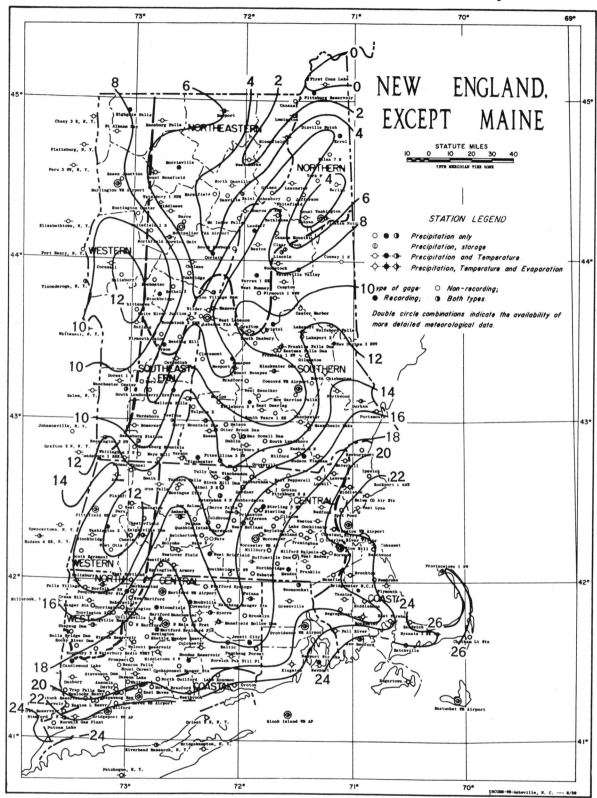

NEW ENGLAND, EXCEPT MAINE

STATUTE MILES

STATION LEGEND

Precipitation only
Precipitation, storage
Precipitation and Temperature
Precipitation, Temperature and Evaporation

Type of gage: O Non-recording;
● Recording; ◑ Both types.

Double circle combinations indicate the availability of
more detailed meteorological data.

Based on period 1931-52

Isolines are drawn through points of approximately equal value. Caution should be used
in interpolating on these maps, particularly in mountainous areas.

Mean Maximum Temperature (°F.), July

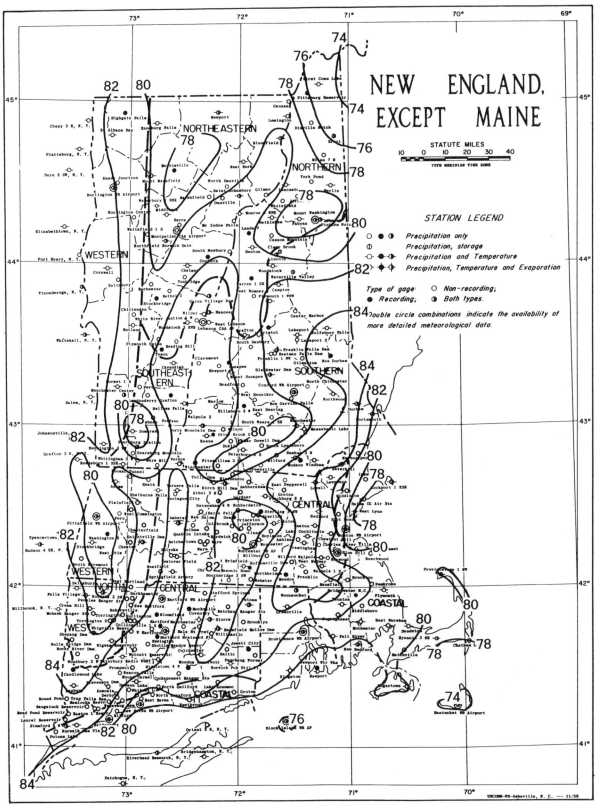

NEW ENGLAND, EXCEPT MAINE

STATUTE MILES
10 0 10 20 30 40
75TH MERIDIAN TIME ZONE

STATION LEGEND

Precipitation only
Precipitation, storage
Precipitation and Temperature
Precipitation, Temperature and Evaporation

Type of gage: O Non-recording;
               ● Recording;   ◑ Both types.

Double circle combinations indicate the availability of more detailed meteorological data.

Based on period 1931-52

Isolines are drawn through points of approximately equal value. Caution should be used in interpolating on these maps, particularly in mountainous areas.

USCOMM-WB-Asheville, N. C. --- 11/56

— 26 —

Mean Minimum Temperature (°F.), July

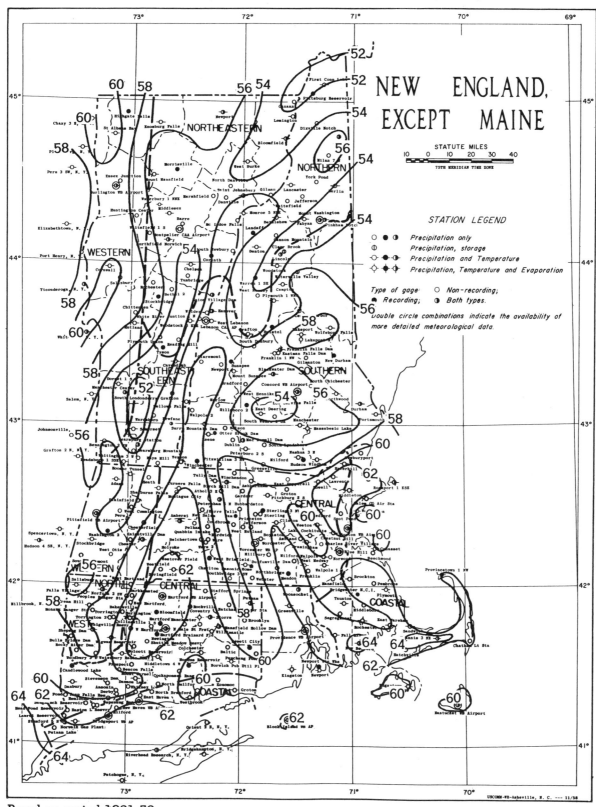

NEW ENGLAND, EXCEPT MAINE

STATUTE MILES
10 0 10 20 30 40
75TH MERIDIAN TIME ZONE

STATION LEGEND

○ ● ◑ Precipitation only
◍ Precipitation, storage
◌ ◒ ◓ Precipitation and Temperature
◈ ◆ ◈ Precipitation, Temperature and Evaporation

Type of gage: ○ Non-recording; ● Recording; ◑ Both types.
Double circle combinations indicate the availability of more detailed meteorological data.

USCOMM-WB-Asheville, N. C. --- 11/58

Based on period 1931-52

Isolines are drawn through points of approximately equal value. Caution should be used in interpolating on these maps, particularly in mountainous areas.

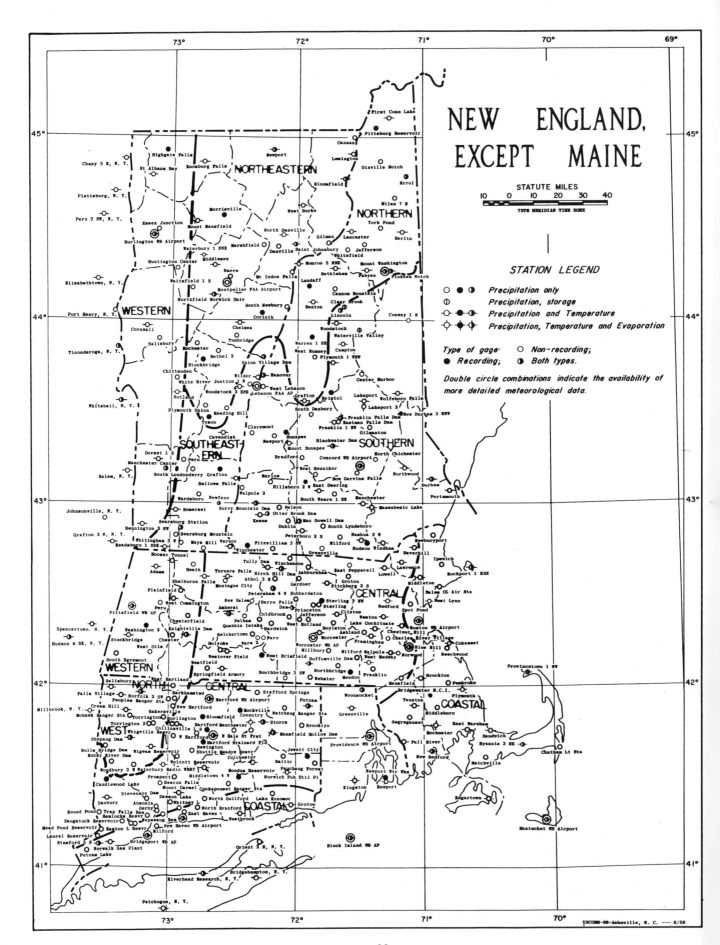

NEW ENGLAND, EXCEPT MAINE

STATUTE MILES

10 0 10 20 30 40

75TH MERIDIAN TIME ZONE

STATION LEGEND

○ ● ◐	Precipitation only
◑	Precipitation, storage
○- ●- ◐-	Precipitation and Temperature
○- ●- ◐-	Precipitation, Temperature and Evaporation

Type of gage: ○ Non-recording;

● Recording;   ◐ Both types.

Double circle combinations indicate the availability of more detailed meteorological data.

USCOMM-WB-Asheville, N. C. --- 6/59

THE CLIMATE OF
DELAWARE

by
Howard H. Engelbrecht

December 1959

The State of Delaware is located on the east coast of the United States midway between the North and the South. It is bounded on the east by the Atlantic Ocean, and by the Delaware Bay and Delaware River, which separate it from New Jersey for a distance of about 85 miles. The northern arc-shaped portion of the State is bounded by Pennsylvania for a distance of 25 miles. On the west and south it is bounded by Maryland for distances of 85 and 35 miles, respectively.

Delaware lies in a north-south position, spanning a distance of 96 miles. The width increases from 9 miles in the northern portion to 35 miles in the extreme southern portion. The State occupies the eastern and northern portion of the Delmarva Peninsula which is bounded by the Chesapeake Bay on the west and the Delaware Bay and Atlantic Ocean on the east.

The total area of Delaware is 2,057 square miles. It is the second smallest state of the United States with respect to size, exceeding only Rhode Island.

Although Delaware ranks as one of the smallest states and does not encompass a wide range of physiographic features, its north-south orientation with highest elevations in the extreme northern portion most distant from the Bays and Ocean, contribute to significant climatic differences between northern and southern portions of the State. It lies in two rather well-defined physiographic belts which parallel the Atlantic coast--the Coastal Plain and the Piedmont Province.

The land rises more or less gradually from the Atlantic Ocean across the Coastal Plain, which makes up virtually the entire State except for about 120 square miles in the extreme northern portion. Elevations in the southernmost county, Sussex County, are generally below 50 feet above mean sea level, but rise to 70 feet at a point near the midpoint of the southern boundary. Large portions of the State along the Atlantic coast and the Delaware Bay and Delaware River are low and marshy. Small streams and tidal estuaries comprise the drainage of the State. In the southwestern portion the flow is southwestward into Chesapeake Bay, while in the northern and eastern portions it is eastward into Delaware Bay and the Atlantic Ocean.

The Piedmont Province includes only about 120 square miles or approximately 5 percent of the land area of the State. It is characterized by undulating, hilly terrain. From its southern boundary, which is known as the Fall Line, low undulating hills gradually increase in elevation toward the Pennsylvania boundary. The highest elevation reached in extreme northern portions is 438 feet above mean sea level.

There are a number of factors which control the climates of Delaware. The most important factors include (1) the distribution of land and water masses, (2) mountain barriers, (3) topographic features, (4) semipermanent pressure centers, (5) prevailing winds at the surface and at upper levels, (6) storm tracks, including tropical and extratropical cyclones, (7) latitude, (8) altitude, and (9) ocean currents.

Since the flow of the atmosphere in temperate latitudes is from west to east the distribution of land and water masses, i.e., the expansive North American Continent situated immediately to the west, predisposes the Delaware area to a continental type of climate. This type of climate in middle latitudes is marked by well-defined seasons. Winter is the dormant season for plant growth and is one of low temperature rather than drought. In spring and fall the changeableness of the weather is a striking characteristic. It is occasioned by a rapid succession of warm and cold periods associated with storms, which generally move from a westerly direction over the eastern portion of the United States. Summers are warm to hot. The higher atmospheric humidity along the sea coast causes the summer heat to be more oppressive or sultry and the winter cold more raw and penetrating than in drier climates of the interior.

The topography of the eastern United States is characterized by the Appalachian Mountains, which extend along a northeast-southwest axis about 150 miles to the northwest of Delaware. To the west and northwest of Delaware, these mountains range in height from 2,000 to 3,000 feet above mean sea level and contribute to some slight tempering of the cold air masses which move rapidly out of the interior of the continent over the Delaware region in the winter.

Over 95 percent of the land area of the State is more or less flat and without topographic features; however, the extreme northern portion, about 120 square miles, which lies on the Piedmont is undulating and hilly with elevations rising to 438 feet above mean sea level. This increase in elevation no doubt contributes to a slight decrease in local temperatures under certain circumstances.

A semipermanent high pressure area with a clockwise circulation virtually overspreads the entire Atlantic Ocean at middle latitudes and exerts a pronounced effect on the weather regimes of the east coast as well as contiguous regions. During the winter season the Atlantic High (or Azores High as it is sometimes called) maintains an average position between latitudes 30°N. and 33°N. and longitudes 25° W. and 35° W. and overspreads the eastern portion of the south Atlantic Ocean. As the summer season approaches, the Atlantic High moves westward and slightly northward to a mean position between latitudes 32°N. and 35°N. and longitudes 40°W. and 45°W. During this period it becomes more intense and widespread as the semipermanent low of the north Atlantic Ocean becomes smaller and weaker. In the summer location the Atlantic High dominates the flow of air over the eastern United States much of the time. A persistence of the Atlantic High in a westerly position in the vicinity of Bermuda results in a prolonged flow of moist, warm tropical air over the entire eastern United States. Weather in this type of air mass consists of scattered thunderstorms, considerable daytime cloudiness, and hot, sultry conditions. In the westerly position the High exerts blocking action on Lows which are forced to travel across more northerly latitudes. Persistence of this High over the eastern United States frequently results in drought conditions over the Delaware region, as the dry, subsiding air of the High prevents the formation of precipitation.

Prevailing surface winds in northern Delaware blow from the northwesterly quadrant in all months except June when southerly winds prevail. However, during the periods of May and July through September winds come from the southwesterly quadrant a high proportion of the time. In southern Delaware surface winds prevail from the southwesterly quadrant from May through September and from the northwesterly quadrant from October through April.

Average wind speeds are highest during the period January through April, largely due to the rapid succession of well-developed storm systems which migrate from a westerly to easterly direction. During this period average winds speeds of about 10 miles per hour prevail. From July through October winds are somewhat lighter, averaging from 7 to 9 miles per hour.

During the fall, winter, and spring seasons, it is not unusual to experience brief windstorms associated with violent, fast-moving cold fronts with gusts from 50 to 60 miles per hour. In the summer, rare occurrences of violent windstorms are associated with severe thunderstorms. During the hurricane season from June through October, it is estimated that wind speeds of more than 75 miles per hour could occur anywhere in Delaware during the rare event of a hurricane traversing or passing very near the State.

With respect to the upper level winds, Delaware lies well within the belt of the westerlies which form a continuous, wave-like band around the northern hemisphere. More specifically, Delaware lies slightly to the south of the axis of the mean position of the westerlies from October through May, but as the summer season approaches the belt of the prevailing westerlies moves well to the north of Delaware.

Delaware lies in the mean zone of the westerlies in the winter and slightly south of the tracks followed by most of the migrating cyclones in their movement from some point in the United States to the region of semipermanent low pressure in the Iceland or North Atlantic area. Cyclones which enter the north Pacific coast or develop north of the Colorado area or in the central United States tend to follow paths to the north of Delaware in their easterly migration; however, cyclones which have their origin in the south Pacific coastal region, Texas, or the Gulf or South Atlantic States have a greater tendency to follow a track through the Delaware region, except some east Gulf or south Atlantic types which may pass to the south of Delaware. Storms of the south Pacific coast, Texas, east Gulf, and sometimes of the south Atlantic bring the heaviest widespread rains to the Delaware area, except for hurricanes or extremely local severe thunderstorms.

There is no record of a hurricane entering the State from the Atlantic coast, but a few have crossed over the State from the south or southwest. Some hurricanes have passed northward a short distance off the Delaware coast.

The difference in latitude of northern Delaware and southern Delaware contributes in some part to the difference in mean temperature between these two regions of the State. The mean temperature difference of 3° to 4° between northern and southern portions in winter and 1° to 2° in summer is largely but not entirely, due to the variation in solar radiational heating. The greater yearly range in mean temperature in northern Delaware, 42°, as compared to 40° in southern Delaware is due, in part, to the increase in the yearly variation of solar radiation at higher latitudes.

In the extreme northern portion where elevations range from 300 to 400 feet on the higher hills, altitude is a controlling factor, although a small one, and reduces temperatures by approximately 1° on the average as compared to the nearby lower terrain.

In order for ocean currents to have a direct temperature control, the winds must be prevailing onshore. The relatively frequent occurrence of easterly winds associated with cyclonic storms to the southeast bring about advection of air off the mild waters and consequently tend to raise the normal winter temperatures and lower the summer temperatures. Therefore, mean winter temperatures of Delaware are roughly 5° higher than for regions of the continental interior at the same latitude.

The effect of the Gulf Stream and the Atlantic Ocean is to lower mean temperatures along the Atlantic coastal area by about 1° to 2° in summer as compared to mean temperatures at the same latitude along the western boundary of the State. To the north, as the width of the State diminishes and distance from the Atlantic Ocean increases, this summer depression of the mean temperature along the coast virtually vanishes. More specifically, average daily maximum temperatures are 3° to 4° lower along the Atlantic coast and average daily minimum temperatures 1° to 2° higher than at points in western Delaware at the same latitude. In northern Delaware the average daily maximum temperatures are from 1° to 2° lower along the coast and average daily minimum temperatures from 1° to 2° higher than for locations near the western boundary of the State. The greater maximum temperature depression in southern coastal areas arises largely from a greater amount of insolation and a more pronounced sea breeze. On some hot summer days the maximum temperatures on the coast may be as much as 10° to 15° lower than at points inland. It is this climatic condition that makes the Delaware coast a popular resort and vacation area in summer.

In winter there is virtually no difference between mean temperatures at coastal and inland points at the same latitude. However, the extreme low temperatures virtually always occur at inland points. Extreme high temperatures for the various winter months, on the other hand, occur more or less at equal frequencies between coastal and inland points.

The climate of Delaware is "humid, temperate" with hot summers and mild winters. The mean annual temperature ranges from 55°F. in New Castle County to 56°F. over Kent and Sussex Counties.

The winter climate is intermediate between the cold of the Northeast and the mild weather of the South. Extremely cold air masses from the interior of the continent are moderated somewhat by passage over the Appalachian Mountains. The average frost penetration ranges from about 5 inches in southern Delaware to about 10 inches in northern Delaware.

Summer weather is characterized by considerable warm weather, including at least several hot, humid periods. However, nights are usually quite comfortable. The average length of the growing (frost-free) season, based on minimum temperatures higher than 32°F., ranges from about 175 to 195 days.

The lowest temperature of record for Delaware, -17°F., at Millsboro on January 17, 1893, came in a month which is quite likely the coldest ever known in Delaware and surrounding areas.

Noteworthy is the fact that Millsboro also has the distinction of having the highest temperature, 110°F., which occurred on July 21, 1930. An extended heat wave during the period July 18-31, 1930, brought maximum temperatures of 100°F. or higher on 3 to 4 consecutive days at many places.

Temperatures of 90°F. or higher occur on 15 to 20 days per year on the average in coastal areas and on 25 to 30 days at inland points. The first day with a 90° temperature or higher occurs in May or early June. An average of one day with a max-imum temperature of 90° or higher occurs in May. The last day of each summer with a maximum temperature of 90° or higher usually falls in September and infrequently in early October. September has an average of 1 to 2 days per year with a maximum temperature of 90° or higher.

The average number of days per year with a minimum temperature of 32°F. or lower ranges from 90 to 100 days for coastal locations to 100 to 110 days at inland points. Temperatures of 0°F. or lower occur in the months of December, January, or February, on the average of 1 out of 4 to 6 years at inland locations and 1 out of 6 to 8 years in coastal locations.

The average annual precipitation ranges from 44 inches in northern Delaware to 47 inches in southern Delaware. The monthly distribution is fairly uniform throughout the year, with July and August the months with heaviest amounts. The greater precipitation in July and August is attributed to the heavy intensities or larger quantities of precipitation associated with the infrequent to rare tropical-type storms or hurricanes.

Precipitation in the summer season is less dependable and more variable than in winter. Although rainfall amounts are generally sufficient to grow good crops, the unequal distribution of summer showers and occasional dry periods at critical stages of crop development make irrigation necessary for maximum crop yields in some years. The seasonal increase in evapotranspiration during the summer results in a rapid loss of soil moisture and contributes to the development of drought conditions.

The greatest daily precipitation of record for Delaware, 7.42 inches, occurred at Dover on August 13, 1919. Rain was quite general over Maryland and Delaware on August 13, 1919, but was particularly heavy on the Delmar Peninsula where amounts ranged generally from 2 to 9 inches. The heavy rain was accompanied by strong winds and lightning and resulted in losses estimated at $2 to $3 million. Rain fell in torrents, washing fields and roads badly, and in cases, undermined and washed out bridges. Lightning disrupted operation of some industrial plants, and strong winds downed telephone and utility poles. Crops were flattened in the fields by the wind and rain. Livestock and poultry were drowned or in some instances died from exposure. Several persons were injured or killed. The year 1919 was probably the wettest year of record in Delaware based on records since 1895 with an average of 55 inches of precipitation over the State.

Flooding occurs infrequently, usually causes only minor damage, and results largely from tides pushed by strong easterly winds. The passage through the area of storms of tropical origin, usually during the late summer or fall, with their high winds and intense rains constitute the most serious flood threat.

The mean snowfall is 18 inches in northern Delaware and 14 inches in southern Delaware. The snow season runs from December through March, with a few light flurries in some years as early as November or late October and as late as early April. Heaviest snowfalls in Delaware generally occur in February and March. Among the outstanding snowstorms is the occurrence of February 14, 1899, when 30 to 36 inches of snow was reported on the level at Odessa. Snow fell continuously without interruption. Roads and streets were impassable. Strong winds accompanied the snow with occasioanl drifts reported up to 16 feet deep. The annual average number of days with snow cover of 1 inch or more varies from about 5 days in southern portions to about 15 days in northern portions.

Thunderstorms occur at a given station on the average of 30 to 33 days per year. The Atlantic coastal region has fewer thunderstorms than interior portions on the average. They have been observed in every month of the year; however, occurrences in the four winter months from November through February are extremely rare. From 1 to 2 thunderstorm days are noted each year at a given station on the average in March and also in October. The 4-month period, May through August, makes up the thunderstorm season with an average of about 6 thunderstorm days in each month at a given station. As many as 15 days per month with thunderstorms at a given station have occurred. In extreme years thunderstorms have occurred on 50 or more days at a given station. As few as 20 thunderstorm days per year have been recorded. July is the month with the greatest frequency of thunderstorms on the average; however, there is a pronounced variability in the number of thunderstorms in July from year to year.

Hail is uncommon in Delaware. It occurs in some portion of the State on 1 or 2 days per year on the average and usually results in little or no damage to crops. The most heavily insured crops in Delaware are peaches and apples. In rare instances damaging hailstorms occur over small areas and cause heavy damage to crops, particularly fruits. In Delaware the ratio of hail to thunderstorm days is comparatively low for the United States and ranges from 2 to 3 percent, with the higher ratios in northern Delaware. Hail generally occurs in the spring and early summer months, with April and May the favored months.

During the period from 1880 to 1958, inclusive, a total of 13 tornadoes were recorded in Delaware. Most likely a few more occurred, but did not come to the attention of the Weather Bureau. Of the 13 on record, about 6 were particularly destructive, while the others caused little or no damage. One life was lost in Delaware during a tornado in the Wilmington area on August 21, 1888. Total property damage from all known tornadoes in Delaware since 1880 is probably well under $1 million. The frequency of occurrence of tornadoes in Delaware is estimated at about 1 in 2 to 3 years on the average.

Average relative humidity in Delaware is lowest in winter and early spring and highest in the late summer and early fall. More specifically, February and March appear to have the lowest average relative humidities, or about 60 to 65 percent, and August, September, and October, the higher average relative humidities, or about 75 to 80 percent. Relative humidities of 80 percent or higher occur about 25 to 30 percent of the time in February and March and 45 to 50 percent of the time in August, September, and October.

Relative humidities of 40 percent or lower occur about 10 to 15 percent of the time in February, March, April, and May, but only about 3 to 5 percent of the time during the period from August through January.

REFERENCES

(1) U. S. Weather Bureau, Normal Weather Maps (Northern Hemisphere Sea Level Pressure), Technical Paper No. 21, 74 pages, U. S. Weather Bureau, Washington, October 1952.

(2) Winston, Jay S., The Annual Course of Zonal Wind Speed at 700 mb, Bulletin of the American Meteorological Society, Volume 35, No. 10, December 1954, pp. 468-471.

(3) Bowie, Edward H., and R. Hanson Weightman, Types of Storms of the United States and Their Average Movements, Monthly Weather Review, Supplement No. 1, Washington, D. C., 1914.

(4) U. S. Geological Survey and U. S. Corps of Engineers, Topographic Charts for Maryland and Delaware.

(5) Weather Bureau Technical Paper No. 15 – Maximum Station Precipitation for 1, 2, 3, 6, 12, and 24 Hours.

(6) Weather Bureau Technical Paper No. 16 – Maximum 24-Hour Precipitation in the United States. Washington, D. C. 1952.

(7) Weather Bureau Technical Paper No. 25 – Rainfall Intensity-Duration-Frequency Curves. For selected stations in the United States, Alaska, Hawaiian Islands, and Puerto Rico.

(8) Weather Bureau Technical Paper No. 29 – Rainfall Intensity-Frequency Regime. Washington, D. C.

BIBLIOGRAPHY

(A) Climatic Summary of the United States (Bulletin W) 1930 edition, Section 92. U. S. Weather Bureau

(B) Climatic Summary of the United States, Maryland and Delaware - Supplement for 1931 through 1952 (Bulletin W Supplement). U. S. Weather Bureau

(C) Climatological Data - Delaware. U. S. Weather Bureau

(D) Climatological Data National Summary. U. S. Weather Bureau

(E) Hourly Precipitation Data - Delaware. U. S. Weather Bureau

(F) Local Climatological Data, U. S. Weather Bureau, for Wilmington, Delaware.

FREEZE DATA

STATION	Freeze threshold temperature	Mean date of last Spring occurrence	Mean date of first Fall occurrence	Mean No. of days between dates	Years of record Spring	No. of occurrences in Spring	Years of record Fall	No. of occurrences in Fall
BRIDGEVILLE	32	04-23	10-21	181	26	26	27	27
	28	04-06	11-01	209	26	26	27	27
	24	03-21	11-16	240	26	26	27	27
	20	03-10	12-01	266	26	26	27	27
	16	02-24	12-09	288	26	25	27	26
DELAWARE CITY REEDY	32	04-14	10-26	195	29	29	30	30
	28	03-29	11-13	229	29	29	30	30
	24	03-16	11-27	256	29	29	30	30
	20	03-07	12-07	276	29	29	30	28
	16	02-26	12-15	292	29	28	30	27
DOVER	32	04-15	10-26	195	30	30	30	30
	28	03-28	11-14	230	30	30	30	30
	24	03-16	11-27	256	30	30	30	30
	20	03-03	12-09	281	30	30	30	28
	16	02-23	12-14	294	30	29	30	26

MARYLAND

STATION	Freeze threshold temperature	Mean date of last Spring occurrence	Mean date of first Fall occurrence	Mean No. of days between dates	Years of record Spring	No. of occurrences in Spring	Years of record Fall	No. of occurrences in Fall
ABERDEEN PHILLIPS FL	32	04-15	10-25	192	30	30	30	30
	28	03-28	11-09	227	30	30	30	30
	24	03-14	12-01	262	30	27	30	21
	20	02-24	12-17	296	30	22	30	11
	16	02-08	12-25	320	30	16	30	5
ANNAPOLIS USN ACAD	32	04-04	11-15	225	30	30	30	27
	28	03-16	12-03	262	30	27	30	20
	24	02-29	12-18	301	30	20	30	10
	20	02-05	12-24	322	30	15	30	6
	16	01-19	@	@	30	8	30	2
BALTIMORE WB CITY	32	03-28	11-17	234	30	30	30	30
	28	03-17	11-28	257	30	30	30	30
	24	03-10	12-09	274	30	30	30	29
	20	02-27	12-15	292	30	30	30	24
	16	02-13	12-23	313	30	26	30	17
BENTLEY SPRINGS	32	04-26	10-16	173	27	27	29	29
	28	04-14	10-31	200	27	27	29	29
	24	03-30	11-13	228	26	26	29	29
	20	03-18	11-27	254	26	26	29	29
	16	03-08	12-07	274	26	26	29	28
CAMBRIDGE	32	04-16	10-29	196	28	28	27	27
	28	03-23	11-12	233	28	28	27	26
	24	02-27	12-07	283	28	22	27	15
	20	02-07	12-18	314	28	15	26	9
	16	01-19	12-23	338	28	8	26	6
CHARLOTTE HALL	32	04-15	10-28	195	12	12	12	12
	28	03-31	11-11	225	11	11	12	12
	24	03-22	11-27	250	11	11	12	12
	20	03-08	12-07	273	11	11	11	11
	16	02-21	12-18	301	11	10	11	8
CHELTENHAM	32	04-21	10-19	181	30	30	30	30
	28	04-08	11-03	209	30	30	30	30
	24	03-18	11-16	243	30	27	30	29
	20	02-22	12-08	289	30	22	30	16
	16	02-09	12-20	314	30	17	30	9
CHESTERTOWN	32	04-13	11-01	202	14	14	15	15
	28	03-28	11-17	234	14	14	15	14
	24	03-09	12-04	269	14	11	15	11
	20	02-14	@	@	14	9	15	2
	16	01-30	@	@	14	6	15	2

STATION	Freeze threshold temperature	Mean date of last Spring occurrence	Mean date of first Fall occurrence	Mean No. of days between dates	Years of record Spring	No. of occurrences in Spring	Years of record Fall	No. of occurrences in Fall
MILFORD	32	04-20	10-24	187	26	26	26	26
	28	04-04	11-07	217	26	26	26	26
	24	03-20	11-19	244	26	26	26	26
	20	03-08	12-05	272	26	26	26	24
	16	02-25	12-13	291	26	26	25	22
MILLSBORO	32	04-25	10-22	179	30	30	29	29
	28	04-10	11-05	209	30	30	29	29
	24	03-23	11-19	241	30	30	29	29
	20	03-09	12-01	267	30	30	29	28
	16	02-26	12-10	287	30	30	29	28
NEWARK PUMP STA	32	04-25	10-16	175	10	10	10	10
	28	04-12	10-25	196	10	10	10	10
	24	03-27	11-16	234	10	10	10	10
	20	03-13	11-26	258	10	10	10	10
	16	03-02	12-13	286	10	10	10	10
WILMINGTON P R	32	04-18	10-26	191	30	30	30	30
	28	04-03	11-11	221	30	30	30	30
	24	03-21	11-25	249	30	30	30	30
	20	03-14	12-04	266	30	30	30	30
	16	03-03	12-13	286	30	30	30	27

MARYLAND

STATION	Freeze threshold temperature	Mean date of last Spring occurrence	Mean date of first Fall occurrence	Mean No. of days between dates	Years of record Spring	No. of occurrences in Spring	Years of record Fall	No. of occurrences in Fall
CHEWSVILLE BRDGEPORT	32	05-03	10-11	161	30	30	30	30
	28	04-20	10-21	184	30	30	30	30
	24	04-05	11-05	213	30	30	30	29
	20	03-11	11-29	263	30	26	30	23
	16	02-16	12-13	300	30	19	30	13
CLEAR SPRING	32	04-24	10-17	176	22	22	13	13
	28	04-08	10-30	205	22	22	13	13
	24	03-20	11-10	236	22	21	13	12
	20	02-28	12-08	283	22	17	13	7
	16	02-06	12-19	316	22	11	13	4
COLEMAN	32	04-06	11-07	215	29	29	30	30
	28	03-23	11-28	250	29	29	30	24
	24	03-04	12-13	284	29	24	30	14
	20	02-12	12-23	314	29	17	30	7
	16	01-28	@	@	28	11	30	4
COLLEGE PARK	32	04-29	10-15	168	30	30	30	30
	28	04-15	10-26	194	30	30	30	30
	24	03-24	11-13	234	30	29	30	28
	20	03-06	11-30	269	30	26	30	21
	16	02-14	12-16	305	30	19	30	12
CONOWINGO DAM	32	04-19	10-18	183	15	15	15	15
	28	04-06	11-08	217	15	15	15	15
	24	03-23	11-18	240	15	15	15	15
	20	03-11	11-29	263	15	15	15	15
	16	03-05	12-12	282	15	15	15	14
CRISFIELD	32	03-31	11-17	231	26	26	30	28
	28	03-06	12-04	273	25	21	30	20
	24	02-12	12-20	311	25	15	30	8
	20	01-19	@	@	25	7	30	3
	16	@	@	@	25	2	30	1
CUMBERLAND	32	04-24	10-14	172	29	29	29	29
	28	04-13	10-29	199	29	29	29	29
	24	03-18	11-13	240	29	27	29	28
	20	02-27	12-06	282	29	22	29	18
	16	02-16	12-21	309	29	19	29	8
EASTON POL BARCKS	32	04-17	10-27	193	30	30	30	30
	28	04-01	11-08	221	30	30	30	30
	24	03-07	12-02	270	30	25	30	21
	20	02-14	12-17	306	30	19	30	11
	16	01-21	12-24	337	30	9	30	6

STATION	Freeze threshold temperature	Mean date of last Spring occurrence	Mean date of first Fall occurrence	Mean No. of days between dates	Years of record Spring	No. of occurrences in Spring	Years of record Fall	No. of occurrences in Fall
ELKTON	32	04-22	10-19	180	23	23	24	24
	28	04-11	10-27	198	23	23	24	24
	24	03-25	11-13	233	23	23	24	24
	20	03-13	11-25	257	23	23	24	24
	16	03-03	12-09	281	23	23	24	22
EMMITSBURG	32	04-18	10-28	193	30	30	30	30
	28	04-04	11-14	224	30	30	30	28
	24	03-15	12-02	263	30	27	30	21
	20	03-04	12-20	291	30	25	30	9
	16	02-08	12-23	318	30	16	30	7
FALLSTON	32	04-20	10-25	188	30	30	30	30
	28	04-11	11-08	212	30	30	30	29
	24	03-17	11-21	249	30	28	30	27
	20	03-04	12-16	287	30	25	30	12
	16	02-07	12-23	318	30	16	30	7
FREDERICK WB AP	32	04-24	10-17	176	30	30	30	30
	28	04-12	10-30	201	30	30	30	29
	24	03-25	11-20	240	30	30	30	26
	20	03-04	12-06	277	30	25	30	19
	16	02-15	12-16	304	30	20	30	16
FRIENDSVILLE	32	05-19	09-30	134	26	26	26	26
	28	04-30	10-15	168	26	26	26	26
	24	04-17	10-24	190	26	26	26	26
	20	03-30	11-06	221	26	26	26	26
	16	03-22	11-21	244	26	26	26	26
FROSTBURG	32	05-05	10-05	153	30	30	30	30
	28	04-21	10-22	184	30	30	30	30
	24	04-11	11-06	209	30	30	30	30
	20	03-29	11-20	237	30	30	30	30
	16	03-14	11-28	259	30	30	30	30
GERMTN BOYDS PLSVLE	32	04-28	10-17	172	28	28	28	28
	28	04-12	10-29	200	27	27	28	28
	24	03-31	11-11	225	27	27	26	26
	20	03-12	11-24	258	27	27	25	25
	16	03-04	12-06	277	27	27	26	25
GLEN DALE BELL STA	32	05-04	10-11	160	30	30	29	29
	28	04-21	10-21	183	30	30	30	30
	24	04-05	11-02	212	30	30	30	30
	20	03-23	11-16	238	30	30	30	30
	16	03-09	11-28	264	30	30	30	30
HUNTINGTOWN	32	04-19	10-26	191	14	14	14	14
	28	04-04	11-05	215	14	14	14	14
	24	03-21	11-26	250	14	14	14	14
	20	03-09	12-07	273	14	14	14	14
	16	02-23	12-13	293	14	13	14	11
KEEDYSVILLE	32	04-30	10-12	165	30	30	30	30
	28	04-21	10-25	188	30	30	30	30
	24	04-01	11-06	220	30	30	30	30
	20	03-17	11-23	251	30	30	30	30
	16	03-06	12-07	276	30	30	30	30
LA PLATA	32	04-19	10-22	186	29	29	29	29
	28	04-06	11-05	213	29	29	29	29
	24	03-17	11-18	245	29	29	29	29
	20	03-08	12-01	268	29	29	29	29
	16	02-25	12-12	290	29	28	29	26
LAUREL 3 W	32	04-20	10-21	184	25	25	26	26
	28	04-08	11-02	208	24	24	26	26
	24	03-20	11-21	246	24	24	26	26
	20	03-15	12-01	261	24	24	26	26
	16	03-03	12-11	283	24	24	25	22
MILLINGTON	32	04-19	10-19	183	30	30	30	30
	28	04-06	11-02	211	30	30	30	30
	24	03-24	11-17	238	30	30	30	30
	20	03-09	12-01	268	30	30	30	30
	16	02-27	12-11	287	30	30	30	28
OAKLAND	32	05-25	09-21	119	30	30	30	30
	28	05-12	10-06	147	30	30	30	30
	24	04-27	10-16	172	30	30	30	30
	20	04-17	10-24	190	30	30	30	30
	16	03-25	11-10	230	30	30	30	30
OWINGS FERRY LANDING	32	04-15	10-26	194	30	30	30	30
	28	03-30	11-09	224	30	30	30	30
	24	03-17	11-25	253	30	30	30	30
	20	03-05	12-05	275	30	30	30	29
	16	02-25	12-13	292	30	28	30	26
OXFORD	32	03-26	11-19	238	13	13	13	13
	28	03-15	12-03	263	13	13	13	13
	24	03-06	12-09	278	13	13	13	13
	20	02-16	12-21	308	13	12	13	9
	16	01-25	12-29	338	13	8	13	4
PICARDY	32	04-29	10-10	164	24	24	24	24
	28	04-18	10-23	188	24	24	24	24
	24	03-27	11-07	225	23	23	24	24
	20	03-17	11-23	251	23	23	24	24
	16	03-08	12-01	269	23	23	24	23
PRINCESS ANNE	32	04-20	10-22	186	26	26	26	26
	28	04-03	11-04	215	26	26	26	26
	24	03-17	11-18	246	26	26	26	26
	20	03-04	12-03	274	26	26	26	25
	16	02-15	12-14	301	26	26	26	22
RIDGELY DENTON	32	04-16	10-24	190	28	28	30	30
	28	03-30	11-09	221	28	28	30	30
	24	03-16	11-23	252	28	28	30	30
	20	03-04	12-08	279	27	27	29	27
	16	02-23	12-15	294	27	27	29	25
SALISBURY	32	04-20	10-26	189	28	28	29	29
	28	04-01	11-09	222	28	28	29	29
	24	03-15	11-23	252	28	28	29	28
	20	03-03	12-06	278	28	28	29	27
	16	02-18	12-15	300	27	26	28	25
SNOW HILL	32	04-27	10-22	178	18	18	18	18
	28	04-12	11-05	207	18	18	18	18
	24	03-19	11-18	244	17	17	18	18
	20	03-08	11-30	267	17	17	18	18
	16	02-20	12-11	294	17	17	17	15
SOLOMONS	32	03-31	11-21	235	29	29	26	26
	28	03-17	11-30	258	28	28	27	27
	24	03-06	12-13	283	28	28	27	23
	20	02-23	12-20	300	28	27	27	17
	16	02-01	12-25	328	28	21	27	13
TAKOMA PARK MISS AVE	32	04-16	10-31	198	30	30	30	30
	28	03-28	11-14	231	30	30	30	30
	24	03-14	11-25	256	30	30	29	29
	20	03-06	12-08	277	30	30	29	28
	16	02-23	12-14	294	30	29	29	24
WESTERNPORT	32	04-30	10-12	165	30	30	30	30
	28	04-17	10-26	193	30	30	30	30
	24	03-27	11-12	230	30	30	30	30
	20	03-15	11-26	257	30	30	30	30
	16	03-06	12-08	277	30	30	30	28
WESTMINSTER	32	04-23	10-19	179	30	30	30	30
	28	04-13	10-30	200	30	30	30	30
	24	03-22	11-15	238	29	29	30	30
	20	03-15	11-29	259	29	29	29	29
	16	03-06	12-08	277	29	29	29	29
WOODSTOCK	32	05-01	10-10	162	30	30	30	30
	28	04-18	10-27	193	30	30	30	30
	24	04-01	11-05	218	30	30	30	30
	20	03-19	11-20	247	30	30	30	30
	16	03-05	12-03	273	30	30	30	30

Data in the above table are based on the period 1921-1950, or that portion of this period for which data are available.

Means have been adjusted to take into account years of non-occurrence.

A freeze is a numerical substitute for the former term "killing frost" and is the occurrence of a minimum temperature at or below the threshold temperature of 32°, 28°, etc.

Freeze data tabulations in greater detail are available and can be reproduced at cost.

*MEAN TEMPERATURE AND PRECIPITATION

STATION	Jan Temp	Jan Precip	Feb Temp	Feb Precip	Mar Temp	Mar Precip	Apr Temp	Apr Precip	May Temp	May Precip	Jun Temp	Jun Precip	Jul Temp	Jul Precip	Aug Temp	Aug Precip	Sep Temp	Sep Precip	Oct Temp	Oct Precip	Nov Temp	Nov Precip	Dec Temp	Dec Precip	Ann Temp	Ann Precip
DELAWARE																										
NORTHERN																										
WILMGTON NCASTLE WB AP	33.3	3.56	33.7	2.98	42.5	3.61	51.8	3.64	62.8	3.81	71.8	4.02	75.9	4.49	73.8	5.28	68.0	3.80	56.2	2.99	45.5	3.33	35.1	2.99	54.2	44.50
WILMINGTON CITY HALL		3.66		2.74		4.16		3.55		4.30		3.95		3.84		5.39		3.18		2.91		3.60		3.08		44.36
WILMINGTON PORTER RESVR		3.63		2.65		4.09		3.73		4.27		4.02		4.29		5.52		3.28		2.96		3.71		3.09		45.26
DIVISION	34.3	3.65	34.2	2.72	42.2	4.02	52.0	3.57	63.0	4.22	71.5	3.90	76.1	4.17	74.2	5.34	68.0	3.27	57.3	3.01	45.9	3.51	35.7	3.06	54.5	44.44
SOUTHERN																										
BRIDGEVILLE 1 NW	37.1	3.73	37.0	2.79	44.2	4.00	53.2	3.54	63.6	4.00	72.1	3.42	76.2	4.69	74.4	5.65	68.3	4.05	58.0	3.09	47.3	3.19	37.8	2.93	55.8	45.08
DOVER	36.8	3.97	36.7	2.98	44.3	4.18	53.7	3.56	64.4	4.39	72.9	3.61	77.2	4.50	75.4	5.75	69.3	3.70	58.7	3.15	47.8	3.49	37.9	3.12	56.3	46.40
DIVISION	37.4	4.00	37.1	2.99	44.3	4.20	53.3	3.61	63.8	4.10	72.3	3.59	76.6	4.49	74.9	5.74	68.9	4.13	58.5	3.25	47.8	3.54	38.1	3.11	56.1	46.75
MARYLAND																										
SOUTHERN EASTERN S																										
CRISFIELD	40.0	3.64	40.0	3.03	47.2	3.98	56.8	3.64	67.0	3.70	75.4	3.47	79.4	4.99	78.0	5.07	72.7	3.91	62.4	3.03	51.2	3.23	41.4	2.73	59.3	44.42
SALISBURY		3.90		3.06		3.79		3.33		3.73		3.63		4.22		6.29		4.62		3.20		3.23		2.94		45.94
SNOW HILL	38.8	4.07	38.6	3.29	45.2	4.48	53.8	3.60	64.1	3.61	72.6	3.51	76.8	4.74	75.2	5.79	69.1	4.67	58.7	3.42	48.2	3.58	39.3	3.31	56.7	48.07
DIVISION	38.9	3.77	38.9	3.04	45.6	4.11	54.5	3.43	64.7	3.54	73.2	3.50	77.4	4.59	75.9	5.55	70.1	4.31	59.7	3.13	49.0	3.33	39.7	2.96	57.3	45.27
CENTRAL EASTERN SH																										
CAMBRIDGE 4 W		3.53		2.90		3.83		3.52		3.81		3.68		4.87		5.49		4.05		3.05		3.62		3.02		45.37
EASTON POLICE BRKS	36.9	3.67	37.0	2.89	44.7	3.93	54.0	3.59	64.5	4.03	73.0	3.48	77.2	4.33	75.4	4.97	69.1	4.03	58.2	3.04	47.7	3.24	37.6	3.11	56.3	44.31
DIVISION	37.4	3.70	37.3	2.83	44.9	3.87	54.3	3.44	64.8	4.09	73.1	3.62	77.2	4.57	75.6	5.19	69.4	4.05	58.9	3.01	48.1	3.34	38.3	2.99	56.6	44.70
LOWER SOUTHERN																										
LA PLATA 1 W		4.02		2.81		3.73		3.65		4.27		3.98		5.10		5.18		4.22		3.62		3.13		3.19		46.90
OWINGS FERRY LANDING	37.5	3.69	37.6	2.56	45.2	3.47	54.8	3.57	64.8	7.29	73.0	3.35	77.0	4.02	75.1	5.45	69.0	4.17	58.8	3.09	48.2	3.22	38.4	3.00	56.6	43.88
SOLOMONS	39.1	3.71	39.0	2.68	45.8	3.62	55.2	3.49	66.0	3.89	74.7	3.50	79.0	5.23	77.9	5.11	72.2	3.65	61.3	2.96	50.3	3.13	40.4	2.89	58.4	44.85
DIVISION	37.8	3.75	38.0	2.69	45.4	3.66	54.8	3.55	65.0	4.07	73.2	3.52	77.2	4.87	75.6	5.36	69.5	3.98	59.1	3.22	48.3	3.18	38.6	3.00	56.9	44.85
UPPER SOUTHERN																										
ANNAPOLIS U S N ACADEMY	36.5	3.24	36.5	2.46	43.6	3.58	53.5	3.41	64.2	3.99	73.2	3.58	77.6	3.90	76.0	4.59	69.9	3.46	59.3	2.59	48.1	2.76	38.2	2.84	56.4	40.40
BALTIMORE WB AIRPORT	34.2	3.66	35.0	2.99	43.2	3.63	52.4	3.72	62.9	4.01	72.2	3.52	76.3	3.94	74.4	4.38	67.9	3.46	56.5	3.37	45.6	2.96	35.7	2.95	54.7	42.59
COLLEGE PARK	36.5	3.40	36.7	2.63	44.6	3.74	54.1	3.36	64.6	4.54	72.7	3.95	76.8	3.88	75.0	5.32	68.5	3.96	57.6	3.26	46.6	3.21	37.0	3.00	55.9	44.16
GLENN DALE BELL STA	35.6	3.13	35.8	2.74	43.8	3.86	53.3	3.50	63.8	4.47	72.1	4.05	76.0	4.30	74.2	5.10	67.8	4.03	56.8	3.40	45.8	3.31	36.0	2.90	55.1	44.79
WASHINGTON WB CITY DC	36.5	3.41	37.5	2.64	45.7	3.36	54.7	3.20	65.2	3.91	73.9	3.42	77.8	4.11	75.8	4.49	69.9	4.36	58.4	2.97	48.0	2.82	38.2	2.75	56.8	41.44
DIVISION	36.7	3.41	36.8	2.64	44.4	3.68	54.2	3.42	64.9	4.22	73.4	3.83	77.7	3.98	75.9	5.00	69.4	3.87	58.4	2.97	47.6	3.17	38.0	3.04	56.5	43.23
NORTHERN EASTERN S																										
COLEMAN 3 WNN	35.6	3.76	36.0	2.79	43.8	3.68	53.9	3.43	64.5	4.52	73.0	3.67	77.5	3.85	75.9	4.94	69.8	3.37	59.3	3.03	47.4	4.31	37.2	3.02	56.2	43.57
MILLINGTON	36.0	3.87	36.0	2.85	44.1	3.89	53.7	3.36	64.1	4.03	72.3	3.36	76.7	3.92	74.9	4.94	68.7	3.36	58.1	2.94	47.1	3.37	37.1	3.05	53.2	42.94
ROCK HALL 3 N	35.9	3.47	36.0	2.74	43.7	3.77	53.6	3.52	64.4	4.72	72.9	3.45	77.3	4.35	75.2	4.86	69.0	3.48	58.3	3.01	47.3	3.16	37.1	3.03	55.9	43.56
DIVISION	35.9	3.79	36.0	2.84	43.8	3.85	53.6	3.46	64.3	4.07	72.7	3.51	77.1	4.25	75.2	4.92	69.0	3.47	58.3	3.03	47.2	3.29	37.1	3.05	55.9	43.53
NORTHERN CENTRAL																										
ABERDEEN PHILLIPS FLD	33.9	3.39	33.9	2.62	41.9	3.70	52.0	3.20	63.1	4.01	71.6	3.33	76.4	3.67	74.5	3.67	67.8	3.17	59.3	2.72	45.4	3.09	35.3	2.88	54.6	40.42
BALTIMORE WB CITY	36.6	3.66	37.3	2.99	45.3	3.63	54.3	3.72	65.3	4.01	74.3	3.52	78.5	3.94	76.4	4.38	70.4	3.46	59.3	3.37	48.7	2.96	38.8	2.95	57.1	42.59
ELKTON	34.1	3.61	34.4	2.63	44.6	4.21	52.4	3.56	63.6	4.54	71.8	3.83	76.4	4.16	74.5	5.13	67.8	3.28	56.6	3.17	45.3	3.60	35.2	3.11	54.5	45.07
FREDERICK WB AIRPORT	32.7	2.88	34.1	2.58	43.3	3.31	52.9	3.50	63.9	3.60	72.6	3.96	76.7	3.65	74.6	4.03	68.5	3.56	56.8	3.29	45.4	2.95	35.0	2.92	54.7	40.23
TOWSON		3.69		2.87		4.27		3.86		4.31		3.74		3.75		4.65		3.57		3.63		3.46		3.38		45.18
WESTMINSTER	33.7	3.43	33.9	2.67	41.9	4.16	51.9	3.71	62.6	3.97	70.8	4.24	75.2	4.05	73.2	5.08	66.6	3.35	55.9	3.55	44.6	3.39	34.5	3.33	53.7	44.93
WOODSTOCK	34.2	3.35	34.4	2.64	42.4	3.67	52.4	3.52	63.1	4.22	71.1	3.79	75.2	3.78	73.5	4.37	66.4	3.76	55.6	3.35	44.3	3.22	34.6	2.97	53.9	42.64
DIVISION	34.2	3.41	34.5	2.66	42.4	3.93	52.4	3.57	63.2	4.31	71.4	3.86	75.7	4.01	73.8	4.75	67.2	3.56	56.4	3.35	45.1	3.30	35.1	3.15	54.3	43.86
APPALACHIAN MOUNTA																										
CHEWSVILLE BRIDGEPORT	32.7	2.46	33.2	1.89	41.0	3.25	51.0	3.04	61.7	4.05	70.1	3.61	74.4	3.42	72.3	4.21	65.4	3.05	54.8	3.07	43.4	2.73	33.6	2.69	52.8	37.47
FROSTBURG	31.4	3.23	31.5	2.26	38.7	4.06	49.4	3.50	60.0	4.23	68.0	4.54	71.6	3.56	70.0	4.15	63.4	3.12	53.5	3.19	41.6	2.72	31.9	2.81	50.9	41.37
HANCOCK FRUIT LAB		2.77		1.95		3.57		3.07		3.88		3.69		3.48		3.40		2.98		3.41		2.73		2.57		37.50
KEEDYSVILLE	34.1	2.59	34.8	1.98	42.8	3.20	52.8	3.12	63.7	3.90	72.0	3.45	76.2	3.46	74.4	4.13	67.3	3.46	56.7	3.26	44.9	2.90	35.0	2.91	54.6	38.78
PICARDY		2.89		2.01		3.47		2.95		3.71		3.66		4.03		2.82		3.09		2.65		2.53				37.18
TONOLOWAY	33.1	2.72	33.7	1.97	41.5	3.63	51.4	3.12	61.7	3.80	70.0	3.90	74.1	3.36	71.8	3.50	65.3	2.96	54.5	3.28	43.0	2.63	33.6	2.64	52.8	37.51
WESTERN PORT	34.4	3.16	35.0	2.30	42.5	3.88	52.7	3.36	62.7	4.23	70.4	4.40	74.2	3.58	72.7	4.24	66.4	3.25	56.2	2.83	44.5	2.42	34.9	2.59	53.9	40.24
DIVISION	33.3	2.83	33.8	2.07	41.5	3.62	51.8	3.23	62.3	3.95	70.4	3.88	74.4	3.57	72.5	3.93	65.7	3.09	55.2	3.09	43.6	2.69	33.9	2.67	53.2	38.62
ALLEGHENY PLATEAU																										
OAKLAND 1 SE	29.9	4.33	29.9	3.22	36.8	4.73	47.0	4.16	56.9	4.66	65.1	4.85	68.2	4.66	66.0	4.53	60.8	3.32	50.9	3.08	39.1	3.25	30.3	3.62	48.5	48.41
SINES DEEP CREEK	28.7	4.13	28.5	3.40	35.5	4.78	45.9	4.20	56.2	4.64	64.3	5.08	67.7	4.74	66.0	4.47	59.4	3.35	49.3	2.99	37.8	3.16	29.2	3.66	47.4	48.60
DIVISION	28.9	3.93	28.7	2.91	35.7	4.44	46.2	3.94	56.6	4.48	64.7	4.71	68.0	4.32	66.7	4.45	60.1	3.22	50.1	3.03	38.5	2.94	29.4	3.32	47.8	45.69

* Averages for period 1931-1955, except for stations marked WB which are "normals" based on period 1921-1950. Divisional means may not be the arithmetical average of individual stations published, since additional data from shorter period stations are used to obtain better areal representation.

CONFIDENCE LIMITS

In the absence of trend or record changes, the chances are 9 out of 10 that the true mean will lie in the interval formed by adding and subtracting the values in the following table from the means for any station in the State. Because of the wider variation in mean precipitation, the corresponding monthly means and annual mean must be substituted for "p" in the precipitation table below to obtain mean precipitation confidence limits.

1.6	.61 p	1.4	.39 p	1.4	.63 p	.9	.55 p	.8	.78 p	.7	.68 p	.5	1.00 p	.7	1.14 p	.9	1.19 p	.9	.72 p	.8	.76 p	1.1	.48 p	.4	2.54 p

COMPARATIVE DATA

Data in the following table are the mean temperature and average precipitation for Dover, Delaware for the period 1906-1930 and are included in this publication for comparative purposes:

34.9	3.31	35.7	3.15	44.1	3.59	53.2	3.65	62.9	3.47	71.9	3.73	76.5	4.55	74.1	4.97	69.3	3.41	57.8	2.54	46.9	2.81	37.1	3.48	55.4	42.66

NORMALS, MEANS, AND EXTREMES

LATITUDE 39° 40' N
LONGITUDE 75° 36' W
ELEVATION (ground) 78 Feet

WILMINGTON, DELAWARE
NEW CASTLE COUNTY AIRPORT

Temperature · Precipitation · Snow

Month	(a)	Normal Daily maximum (b)	Normal Daily minimum (b)	Normal Monthly (b)	Record highest	Year	Record lowest	Year	Normal degree days (b)	Precip Normal total (b)	Precip Max monthly	Year	Precip Min monthly	Year	Precip Max in 24 hrs	Year	Snow/Sleet Mean total	Snow Max monthly	Year	Snow Max in 24 hrs	Year
					10		10				10		10		10		10	10		10	
J		41.8	24.7	33.3	75	1950	-4	1957	983	3.56	5.55	1949	0.59	1955	1.56	1956	6.0	13.0	1948	7.3	1956
F		42.8	24.6	33.7	74	1954+	-2	1948	876	2.98	4.21	1954	1.84	1954	1.90	1956	3.9	14.8	1956	9.7	1955
M		53.0	32.0	42.5	86	1948	10	1950	698	3.81	5.72	1953	2.84	1957	1.65	1957	3.0	12.3	1956	7.7	1956
A		63.2	40.4	51.8	88	1957	23	1950	398	3.64	5.57	1952	1.81	1950	1.67	1957	T	0.0	1957+	T	1957+
M		74.6	51.0	62.8	93	1956+	34	1957	110	3.61	7.35	1948	1.18	1950	2.01	1952	0.0	0.0		0.0	
J		83.1	60.4	71.8	99	1952	45	1957+	6	4.02	6.34	1955	0.44	1954	2.67	1955	0.0	0.0		0.0	
J		86.7	65.0	75.9	102	1957+	50	1952	0	4.49	7.40	1952	0.16	1955	6.24	1952	0.0	0.0		0.0	
A		84.6	62.9	73.8	101	1955+	48	1952+	0	5.28	12.09	1955	1.89	1953	4.00	1955	0.0	0.0		0.0	
S		78.8	57.1	68.0	100	1953	37	1957+	47	3.80	7.06	1950	1.16	1951	2.29	1950	0.0	0.0		0.0	
O		67.2	45.2	56.2	91	1951	28	1952	282	2.99	3.87	1953	0.46	1952	2.94	1953	T	T	1957	T	1957
N		55.3	35.7	45.5	85	1950	14	1955	585	3.33	7.32	1951	1.15	1949	3.83	1956	1.9	11.9	1953	11.9	1957
D		43.9	26.2	35.1	71	1951	6	1951+	927	2.99	5.79	1951	0.19	1955	1.99	1955	3.0	11.2	1957	10.3	1957
Year		64.6	43.8	54.2	102	July 1957+	-4	Jan. 1957	4910	44.50	12.09	Aug. 1955	0.16	July 1955	6.24	July 1952	17.8	14.8	Feb. 1948	11.9	Nov. 1953

Relative humidity · Wind · Sky cover · Mean number of days

Month	RH 1:00 a.m. EST	RH 7:00 a.m. EST	RH 1:00 p.m. EST	RH 7:00 p.m. EST	Wind Mean hourly speed	Prevailing direction	Mean sky cover sunrise to sunset	Clear	Partly cloudy	Cloudy	Precip .01 in or more	Snow/Sleet 1.0 in or more	Thunderstorms	Heavy fog	Max 90° and above	Max 32° and below	Min 32° and below	Min 0° and below
(a)	10	10	10	10	9	10	10	10	10	10	10	10	10	10	10	10	10	10
J	75	77	62	71	9.0	NW	7.0	6	7	18	13	2	*	5	0	5	25	*
F	76	77	59	69	9.4	NW	6.4	7	7	14	10	1	*	5	0	2	21	0
M	73	75	54	65	10.8	NW	6.6	9	8	14	13	1	1	4	0	1	16	0
A	74	74	52	65	10.0	NW	6.5	7	8	15	12	0	2	3	0	0	3	0
M	77	77	52	68	8.3	S	6.7	5	12	13	12	0	3	3	1	0	0	0
J	82	78	54	69	7.5	NW	5.8	8	11	11	10	0	3	3	6	0	0	0
J	86	79	51	68	7.3	NW	5.9	7	13	11	9	0	6	2	11	0	0	0
A	88	84	55	72	7.5	NW	5.5	9	11	11	9	0	6	4	6	0	0	0
S	87	86	57	74	7.5	NNW	5.4	10	12	8	9	0	3	5	1	0	0	0
O	85	85	54	74	7.6	NW	5.4	12	7	12	8	0	1	5	*	0	1	0
N	80	82	57	72	8.5	NW	6.1	9	7	14	10	*	1	4	0	*	12	0
D	77	78	60	72	8.5	NW	6.5	7	9	15	11	1	0	4	0	4	22	*
Year	81	79	56	70	8.6	NW	6.2	96	108	161	127	5	32	47	24	11	100	*

Means and extremes in the above table are from the existing or comparable location(s). Annual extremes have been exceeded at prior locations as follows: Highest temperature 107 in August 1918; lowest temperature -15 in February 1934; maximum monthly precipitation 14.91 in August 1911; minimum monthly precipitation 0.06 in October 1924; maximum precipitation in 24 hours 6.53 in August 1945; maximum monthly snowfall 27.0 in January 1935; maximum snowfall in 24 hours 22.0 in December 1909.

REFERENCE NOTES APPLYING TO ALL "NORMALS, MEANS, AND EXTREMES" TABLES.

(a) Length of record, years.
(b) Normal values are based on the period 1921-1950, and are means adjusted to represent observations taken at the present standard location.
* Less than one-half.

- No record.
† Airport data.
‡ City Office data.
+ Also on earlier dates, months, or years.
T Trace, an amount too small to measure.

Sky cover is expressed in a range of 0 for no clouds or obscuring phenomena to 10 for complete sky cover. The number of clear days is based on average cloudiness 0-3 tenths; partly cloudy days on 4-7 tenths; and cloudy days on 8-10 tenths. Monthly degree day totals are the sum of the negative departures of average daily temperatures from 65°F. Sleet was included in snowfall totals beginning with July 1948. Heavy fog also includes data referred to at various times in the past as "Dense" or "Thick". The upper visibility limit for heavy fog is 1/4 mile. Data in these tables are based on records through 1957.

NORMALS, MEANS, AND EXTREMES

LATITUDE 39° 11' N
LONGITUDE 76° 40' W
ELEVATION (ground) 146 feet

BALTIMORE, MARYLAND
FRIENDSHIP INTERNATIONAL AIRPORT

Month	Temperature Normal Daily max	Daily min	Monthly	Extremes Record highest	Year	Record lowest	Year	Normal degree days	Precip. Normal total	Max monthly	Year	Min monthly	Year	Max in 24 hrs	Year	Snow/Sleet Mean total	Max monthly	Year	Max in 24 hrs	Year	Rel. hum. 1:00 a.m.	7:00 a.m.	1:00 p.m.	7:00 p.m.	Wind mean hourly	Prev. dir.	Fastest speed	Dir.	Year	Pct sun	Mean sky cover	Clear	Partly cloudy	Cloudy	Precip .01+	Snow 1.0+	Tstorms	Heavy fog	Max 90+	Max 32−	Min 32−	Min 0−
J	42.8	25.6	34.2	70	1954	−4	1957	955	3.66	4.56	1953	0.29	1955	1.67	1953	6.0			6.0	1956	74	77	60	68	10.5	NW	60	W	1954	46	6.7	8	5	16	12	2	*	4	0	4	24	*
F	43.9	26.1	35.0	76	1954	6	1955	840	2.99	3.38	1956	1.21	1954	1.75	1956	3.2	13.7	1954	2.5	1955	74	77	54	64	11.0	NW	68	W	1956	54	6.3	8	10	12	12	1	0	4	0	2	20	0
M	53.3	33.1	43.2	79	1955	14	1956	676	3.63	6.80	1953	2.27	1957	1.63	1953	1.2	6.2	1955	7.8	1952	74	78	54	63	11.5	NW	80	SE	1952	51	6.3	8	8	15	12	1	1	4	0	1	14	0
A	63.2	41.5	52.4	90	1957	25	1956	378	3.72	8.15	1952	2.18	1957	2.80	1952	0.1	0.7	1957	6.4	1956	74	74	51	60	12.0	NW	60	W	1954	51	6.1	8	8	14	11	0	2	2	*	0	3	0
M	73.2	52.5	62.9	95	1956	34	1956	115	4.01	8.80	1956	0.55	1956	2.44	1957	0.0	0.0		0.0		74	74	51	60	10.5	NW	65	SW	1955	57	6.2	8	11	12	11	0	5	2	1	0	0	0
J	82.6	61.7	72.2	99	1954	48	1954	5	3.52	8.80	1954	0.15	1954	2.38	1955	0.0	0.0		0.0		77	77	53	64	9.6	SW	65	W	1952+	57	6.2	8	11	10	10	0	6	1	8	0	0	0
J	86.6	66.0	76.3	102	1957+	52	1952	0	3.94	7.06	1955	0.30	1955	5.86	1952	0.0	0.0		0.0		81	76	48	62	8.7	SW	46	NW	1952	69	5.5	10	14	7	11	0	6	1	12	0	0	0
A	84.7	64.0	74.4	100	1955	50	1955	0	4.38	45.88	1951	0.77	1951	5.82	1955	0.0	0.0		0.0		85	82	54	69	8.1	S	54	NE	1955	62	5.5	10	14	11	10	0	4	1	6	0	0	0
S	78.1	57.7	67.9	99	1954+	37	1952	50	3.46	45.68	1952	0.56	1952	3.96	1952	0.0	0.0		0.0		86	84	53	71	9.0	S	56	S	1952	62	5.0	12	10	8	6	0	3	2	3	0	0	0
O	67.1	45.9	56.5	92	1950	28	1952	278	3.37	7.68	1955	0.51	1955	3.43	1952	T	T	1954+	T	1954+	81	84	53	71	10.0	WNW	73	SE	1954	59	5.0	13	7	10	10	0	1	4	*	0	1	0
N	55.4	35.7	45.6	c77	1955	13	1955	582	2.96	4.49	1953	1.32	1953	3.43	1953	1.5	5.9	1953	5.5	1953	77	80	55	68	10.0	WNW	58	E	1952	48	6.4	9	8	14	9	1	*	3	0	1	13	0
D	44.1	27.2	35.7	74	1951	−4	1951	908	2.95	4.49	1955	0.20	1955	e1.60	1957	2.7	9.5	1957	8.4	1957	76	78	60	70	10.3	NW	57	W	1953	48	6.4	8	8	15	9	1	*	4	0	3	20	*
Year	64.6	44.8	54.7	102	July 1957+	−4	Jan. 1957	4787	42.59	18.35	Aug. 1955	0.15	June 1954	7.82	Aug. 1955	15.1	13.7	Jan. 1954	8.4	Dec. 1957	78	78	54	65	10.3	NW	80	N	June 1952+	57	5.9	106	114	145	115	6	28	31	30	10	95	*

LATITUDE 39° 17' N
LONGITUDE 76° 37' W
ELEVATION (ground) 14 feet

BALTIMORE, MARYLAND
CUSTOM HOUSE

Month	Temperature Normal Daily max	Daily min	Monthly	Extremes Record highest	Year	Record lowest	Year	Normal degree days	Precip. Normal total	Max monthly	Year	Min monthly	Year	Max in 24 hrs	Year	Snow/Sleet Mean total	Max monthly	Year	Max in 24 hrs	Year	Rel. hum. 7:00 a.m.	12:00 N	7:00 p.m.	Wind mean hourly	Prev. dir.	Fastest speed	Dir.	Year	Pct sun	Mean sky cover	Clear	Partly cloudy	Cloudy	Precip .01+	Snow 1.0+	Tstorms	Heavy fog	Max 90+	Max 32−	Min 32−	Min 0−
J	43.6	29.6	36.6	79	1950+	−7	1881	880	3.66	6.81	1915	0.33	1901	3.73	1944	6.1	31.3	1922	24.5	1922	72	58	66	10.3	SW	56	NW	1949	48	5.9	9	9	13	12	2	*	2	0	5	20	*
F	44.6	29.9	37.3	83	1930	−5	1934+	776	2.99	7.07	1896	0.65	1896	3.48	1896	6.6	33.9	1899	15.5	1899	70	58	63	10.8	SW	66	NW	1949	56	5.6	9	9	10	11	1	*	2	0	4	18	*
M	53.7	36.8	45.3	94	1945	15	1873	611	3.63	7.94	1891	0.46	1910	3.51	1881	4.8	25.6	1892	22.0	1942	69	53	61	11.5	NW	66	W	1942	56	5.6	10	10	11	12	1	1	1	*	1	11	0
A	63.3	45.3	54.3	94	1941+	34	1876	326	3.72	8.70	1889	0.85	1899	3.99	1937	0.6	9.4	1924	T	1924	65	50	61	11.3	SW	65	W	1948	59	5.4	10	11	10	12	*	2	*	*	0	1	0
M	74.4	56.2	65.3	98	1941+	46	1913	73	4.01	9.36	1948	0.76	1955	3.89	1952	T	T	1949+	T	1949+	67	51	61	10.0	SW	57	NE	1937	59	5.4	10	11	10	11	0	5	*	1	0	0	0
J	83.1	65.5	74.3	105	1934	54	1933	0	3.52	9.36	1948	0.37	1948	4.47	1954	0.0	0.0		0.0		70	53	64	9.5	SW	61	NE	1948	64	5.3	11	13	8	11	0	6	*	6	0	0	0
J	86.9	70.0	78.5	107	1936	54	1933	0	3.94	11.03	1889	0.64	1930	4.13	1952	0.0	0.0		0.0		71	52	65	8.8	SW	61	SW	1931	65	5.1	10	12	9	12	0	8	*	10	0	0	0
A	84.6	68.2	76.4	105	1918	51	1890	0	4.38	17.69	1955	0.64	1877	7.82	1955	0.0	0.0		0.0		74	55	69	8.9	SW	60	SW	1937	63	5.1	11	11	9	11	0	6	*	6	0	0	0
S	78.6	62.2	70.4	101	1881	39	1888	29	3.37	12.41	1934	1.09	1884	6.07	1912	0.0	0.0		0.0		75	55	69	9.8	SW	54	NW	1948	63	4.4	13	10	9	9	0	3	1	2	0	0	0
O	67.8	50.7	59.3	97	1936+	30	1936+	207	3.37	7.75	1942+	1.05	1924	5.30	1922	0.0	0.0		2.5	1925	74	53	67	10.0	SW	49	NW	1950	63	4.7	13	10	9	9	0	1	2	*	0	*	0
N	56.2	41.1	48.7	87	1950	12	1929	489	2.96	9.12	1922	0.44	1922	4.19	1937	0.6	2.5	1938	8.5	1938	72	58	65	9.9	SW	57	SW	1931	55	5.7	9	10	10	9	1	*	*	0	*	5	0
D	45.4	32.1	38.8	75	1951	−3	1880	812	2.95	7.10	1936	0.15	1955	3.18	1948	3.9	17.1	1904	11.5	1904	71	58	65	9.9	SW	56	S	1932	49	5.7	9	10	12	11	1	*	3	0	3	17	0
Year	65.2	49.0	57.1	107	July 1936	−7	Feb. 1934+	4203	42.59	17.69	Aug. 1955	0.05	Oct. 1924	7.82	Aug. 1955	22.7	33.9	Feb. 1899	24.5	Jan. 1922	71	54	64	9.9	SW	66	W	Mar. 1949	59	5.4	121	123	121	133	6	32	15	25	13	72	*

− 37 −

NORMALS, MEANS, AND EXTREMES

FREDERICK, MARYLAND MUNICIPAL AIRPORT
LATITUDE 39° 25' N LONGITUDE 77° 23' W ELEVATION (ground) 294 Feet

Month	Temp Normal Daily max	Daily min	Monthly	Extremes Record highest	Year	Record lowest	Year	Normal degree days	Precip Normal total	Max monthly	Year	Min monthly	Year	Max in 24 hrs	Year	Snow/Sleet Mean total	Max monthly	Year	Max in 24 hrs	Year	R.H. 1:00 p.m. EST	Precip .01 in or more	Snow 1.0 in or more	Max 90° and above	Max 32° and below	Min 32° and below	Min 0° and below
(a)	(b)	(b)	(b)	16		16		(b)	(b)					18		16	16		16		13	16	16	16	16	16	16
J	41.4	24.0	32.7	76	1950	-10	1957	1001	2.88	5.37	1955	0.35	1949	2.27	1948	6.4	14.3	1945	9.5	1945	64	11	2	0	4	25	1
F	43.2	25.0	34.1	80	1949	-5	1948	865	2.58	5.88	1954	1.41	1951	1.56	1942	4.5	12.1	1947	9.8	1947	60	11	1	0	2	23	*
M	54.1	32.5	43.3	88	1945	8	1943	673	3.31	5.89	1949	1.00	1954	1.73	1950	4.2	17.5	1942	17.5	1942	56	12	1	0	1	18	0
A	65.0	40.7	52.9	92	1942	20	1948	368	3.50	10.53	1952	0.86	1953	2.72	1952	T	T	1953+	T	1953+	55	12	0	*	0	7	0
M	76.1	51.6	63.9	95	1944+	28	1956+	106	3.60	8.16	1953	1.52	1953	2.08	1942	0.0	0.0		0.0		58	13	0	1	0	1	0
J	84.1	61.0	72.6	100	1954	41	1944	0	3.98	11.26	1951	0.85	1951	3.41	1946	0.0	0.0		0.0		57	10	0	8	0	0	0
J	87.9	65.4	76.7	102	1954	42	1943	0	3.65	8.82	1943	0.84	1943	3.28	1956	0.0	0.0		0.0		56	11	0	12	0	0	0
A	85.9	63.3	74.6	100	1953+	43	1942	0	4.03	9.59	1955	1.53	1955	5.56	1955	0.0	0.0		0.0		56	10	0	9	0	0	0
S	79.9	57.1	68.5	101	1953	28	1947	47	3.56	7.16	1950	0.99	1950	2.11	1942	0.0	0.0		0.0		58	9	0	3	0	1	0
O	68.3	45.2	56.8	92	1951	22	1952	278	3.29	8.31	1942	0.82	1945	2.87	1945	0.0	0.0		0.0		57	9	0	*	0	4	0
N	55.1	35.6	45.4	83	1950	11	1950	588	2.95	5.69	1952	1.02	1946	2.85	1952	1.0	5.0	1953	4.6	1955+	60	11	*	0	*	15	0
D	43.4	26.6	35.0	72	1956+	-13	1942	930	2.92	6.12	1955	0.15	1955	1.97	1950	3.7	13.8	1951	7.0	1948	63	8	1	0	4	25	1
Year	65.4	44.0	54.7	102 July 1954		-13 Dec 1942		4854	40.23	11.26 June 1951		0.15 Dec 1955		5.58 Aug 1955		19.8	17.5 Mar 1942		17.5 Mar 1942		59	126	5	33	12	119	1

WASHINGTON, D. C. 24TH AND M STREETS, N. W.
LATITUDE 38° 54' N LONGITUDE 77° 03' W ELEVATION (ground) 72 feet

Month	Temp Normal Daily max	Daily min	Monthly	Extremes Record highest	Year	Record lowest	Year	Normal degree days	Precip Normal total	Max monthly	Year	Min monthly	Year	Max in 24 hrs	Year	Snow/Sleet Mean total	Max monthly	Year	Max in 24 hrs	Year	Wind Mean hourly speed	Prevailing direction	Fastest mile Speed	Direction	Year	Pct possible sunshine	Mean sky cover	Clear	Partly cloudy	Cloudy	Precip .01 in or more	Snow 1.0 in or more	Thunderstorms	Heavy fog	Max 90° and above	Max 32° and below	Min 32° and below	Min 0° and below
(a)	(b)	(b)	(b)	86		86		(b)	(b)	87		87		87		70	70		70		26	78	86			64	75	87	87	87	87	87	70	85	87	87	87	87
J	44.0	28.9	36.5	80	1950	-14	1881	884	3.41	7.83	1937	0.30	1955	2.98	1915	6.0	31.5	1922	25.0	1922	8.0	NW	47	NW	1936	46	6.1	9	9	14	10	2	*	1	0	5	21	*
F	45.6	29.3	37.5	84	1930	-15	1930	770	2.64	6.84	1884	0.62	1896	2.29	1896	5.4	35.2	1899	14.1	1936	8.5	NW	47	NW	1934+	54	5.6	9	10	11	10	2	*	1	0	4	19	*
M	55.0	36.4	45.7	93	1907	4	1873	606	3.36	8.84	1891	0.57	1910	2.60	1909	3.6	19.3	1914	11.5	1942	8.4	NW	56	NW	1914	56	5.8	10	10	10	11	1	1	1	0	1	12	0
A	66.5	46.5	56.5	95	1915	15	1923	314	3.20	9.13	1889	0.54	1910	3.21	1924	T	T	1923+	T	1923+	8.4	NW	50	SW	1924	57	5.5	10	11	10	11	*	2	*	*	0	3	0
M	75.5	54.9	65.2	97	1941	33	1906	80	3.91	10.69	1889	0.86	1924	3.50	1889	0.0	0.0		0.0		6.4	S	47	NW	1913	61	5.5	11	12	8	11	0	5	1	1	0	0	0
J	83.7	64.1	73.9	102	1874	43	1897	0	3.42	10.94	1900	0.62	1886	4.16	1886	0.0	0.0		0.0		5.8	S	54	NW	1913	63	5.2	11	12	8	11	0	6	1	7	0	0	0
J	87.1	68.4	77.8	106	1930	52	1907	0	4.11	10.63	1886	0.82	1872	5.80	1878	0.0	0.0		0.0		5.7	S	49	NE	1916	64	5.1	11	12	9	11	0	6	2	17	0	0	0
A	85.0	66.6	75.8	106	1918	49	1934	0	4.49	14.41	1928	0.14	1930	7.31	1928	0.0	0.0		0.0		5.4	S	54	SE	1896	62	4.8	13	10	9	10	0	6	2	10	0	0	0
S	79.1	60.7	69.9	104	1881	36	1894	32	4.36	17.45	1934	0.28	1884	6.66	1874	0.0	0.0		0.0		5.8	S	54	SE	1952	61	4.7	14	9	9	8	0	3	2	3	0	1	0
O	68.2	48.6	58.4	96	1941	26	1893	231	2.97	8.81	1877	0.53	1930	5.66	1884	0.1	2.0	1925	2.2	1938	6.3	NW	47	NW	1927	61	4.5	12	8	11	8	*	1	2	*	0	1	0
N	56.5	39.4	48.0	87	1950	11	1929	510	2.82	7.18	1877	0.19	1917	3.98	1943	0.8	7.0	1938	2.2	1938	7.3	NW	47	NW	1916+	53	5.4	10	9	11	9	1	*	2	0	*	8	0
D	45.6	30.6	38.2	74	1951+	-13	1880	831	2.75	7.56	1901	0.15	1889	2.70	1880	3.2	14.5	1932	12.0	1932	7.3	NW	40	NW	1919+	46	5.9	10	9	13	10	1	*	2	0	3	19	*
Year	65.9	47.7	56.8	106 July 1930+		-15 Feb 1899		4258	41.44	17.45 Sept 1934		0.14 Sept 1884		7.31 Aug 1928		19.5	35.2 Feb 1899		25.0 Jan 1922		7.1	NW	62	SE	Sept 1896	58	5.4	124	118	123	124	6	33	12	28	13	82	*

(a) Length of record, years.
(b) Normal values are based on the period 1921-1950, and are means adjusted to represent observations taken at the present standard location.
* Less than one-half.
+ No record.
† Airport data.
‡ City Office data.
‡‡ Also on earlier dates, months, or years.
T Trace, an amount too small to measure.

REFERENCE NOTES APPLYING TO ALL "NORMALS, MEANS, AND EXTREMES" TABLES.

Sky cover is expressed in a range of 0 for no clouds or obscuring phenomena to 10 for complete sky cover. The number of clear days is based on average cloudiness 0-3 tenths; partly cloudy days on 4-7 tenths; and cloudy days on 8-10 tenths. Monthly degree day totals are the sum of the negative departures of average daily temperatures from 65°F. Sleet was included in snowfall totals beginning with July 1948. Heavy fog also includes data referred to at various times in the past as "Dense" or "Thick". The upper visibility limit for heavy fog is 1/4 mile. Data in these tables are based on records through 1957.

Mean Annual Precipitation, Inches

MARYLAND AND DELAWARE

Based on period 1931-55

Isolines are drawn through points of approximately equal value. Caution should be used in interpolating on these maps, particularly in mountainous areas.

Mean Maximum Temperature (°F.), January

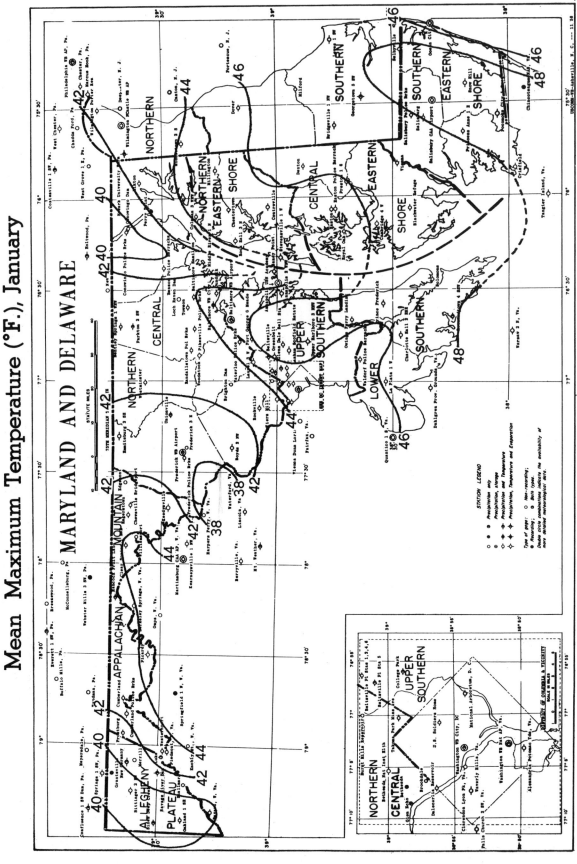

MARYLAND AND DELAWARE

Based on period 1931-52

Isolines are drawn through points of approximately equal value. Caution should be used in interpolating on these maps, particularly in mountainous areas.

Mean Minimum Temperature (°F.), January

MARYLAND AND DELAWARE

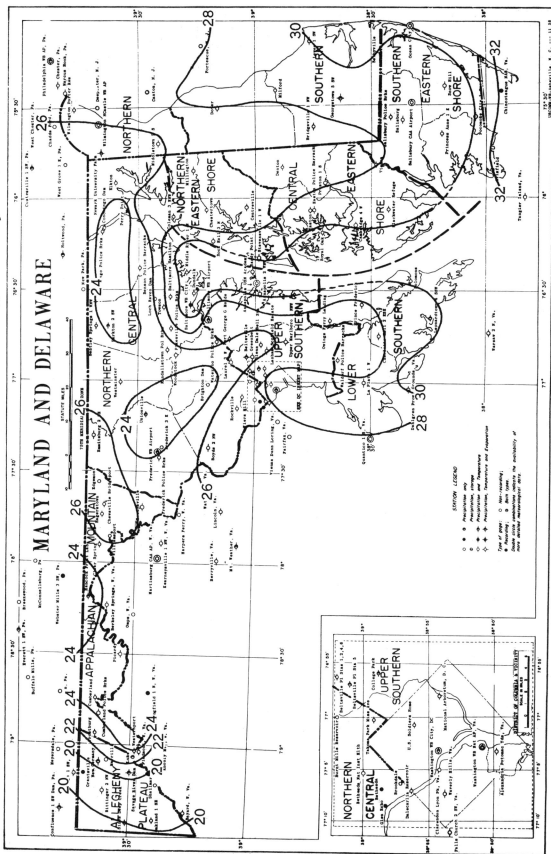

Based on period 1931-52

Isolines are drawn through points of approximately equal value. Caution should be used in interpolating on these maps, particularly in mountainous areas.

Mean Maximum Temperature (°F.), July

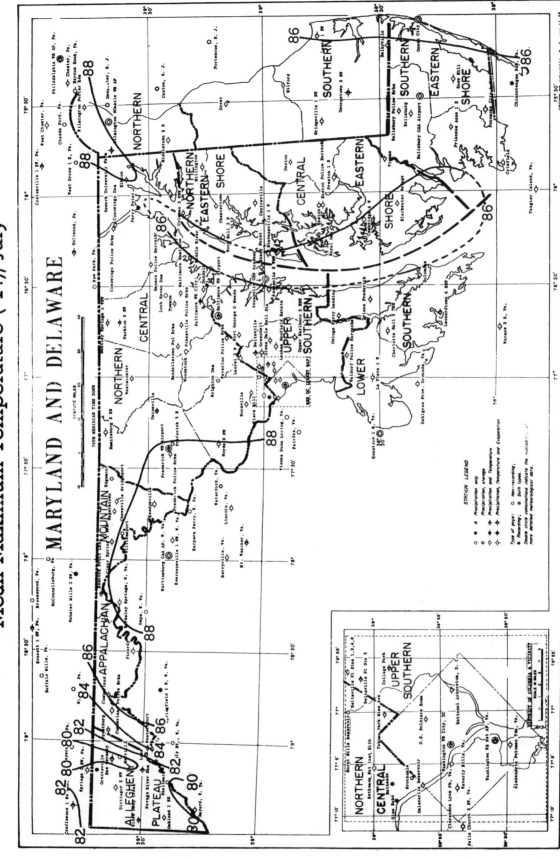

MARYLAND AND DELAWARE

Based on period 1931-52

Isolines are drawn through points of approximately equal value. Caution should be used in interpolating on these maps, particularly in mountainous areas.

Mean Minimum Temperature (°F.), July

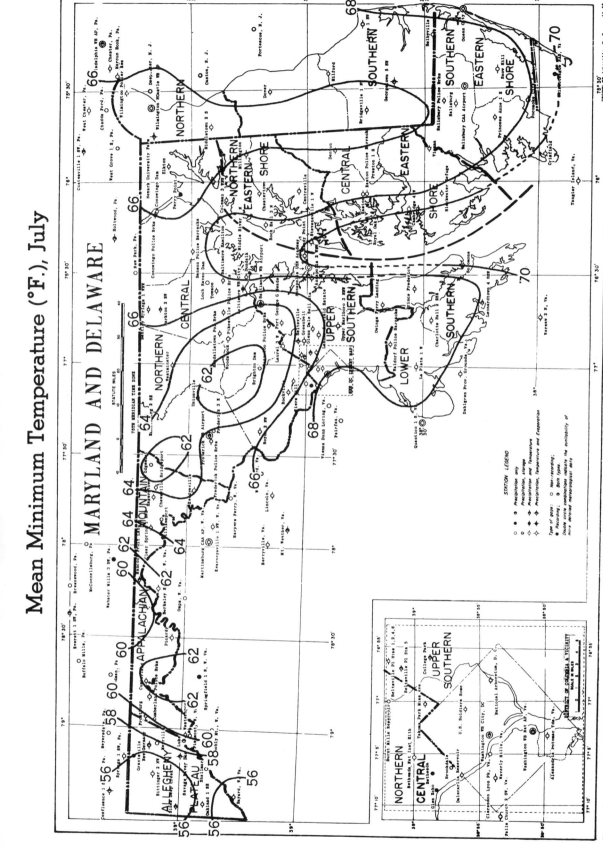

MARYLAND AND DELAWARE

Based on period 1931-52

Isolines are drawn through points of approximately equal value. Caution should be used in interpolating on these maps, particularly in mountainous areas.

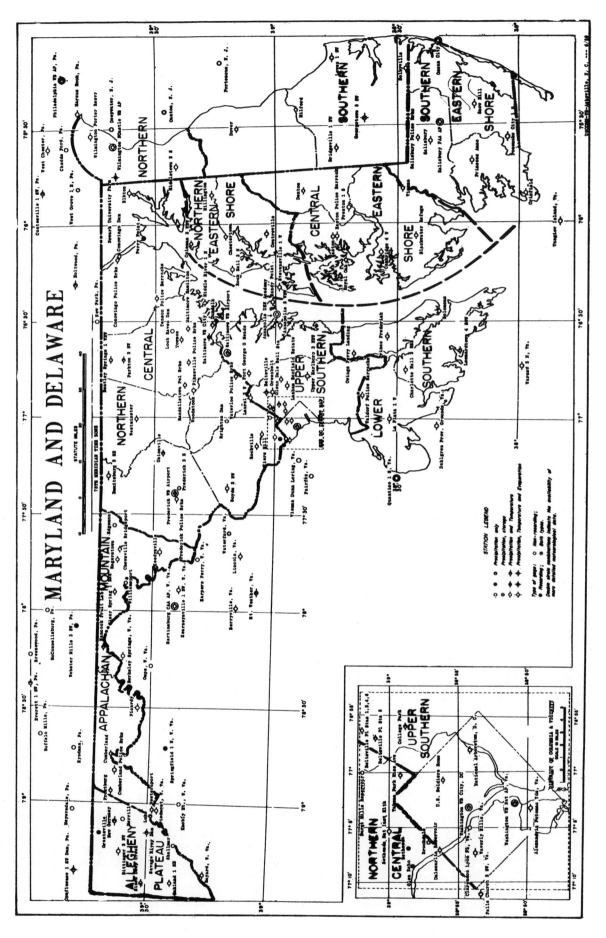

MARYLAND AND DELAWARE

THE CLIMATE OF

FLORIDA

by
James T. Bradley

June 1972

PHYSICAL DESCRIPTION. Florida, situated between latitudes 24° 30' and 31° N. and longitudes 80° and 87° 30' W., is largely a lowland peninsula comprising about 54,100 square miles of land area and is surrounded on three sides by the waters of the Atlantic Ocean and the Gulf of Mexico. Countless shallow lakes, which exist particularly on the peninsula and range in size from small cypress ponds to that of Lake Okeechobee, account for approximately 4,400 square miles of additional water area.

No point in the State is more than 70 miles from salt water, and the highest natural land in the Northwest Division is only 345 feet above sea level. Coastal areas are low and flat and are indented by many small bays or inlets. Many small islands dot the shorelines. The elevation of most of the interior ranges from 50 to 100 feet above sea level, though gentle hills in the interior of the peninsula and across the northern and western portions of the State rise above 200 feet.

A large portion of the southern one-third of the peninsula is the swampland known as the Everglades. An ill-defined divide of low, rolling hills, extending north-to-south near the middle of the peninsula and terminating north of Lake Okeechobee, gives rise to most peninsula streams, chains of lakes, and many springs. Stream gradients are slight and often insufficient to handle the runoff following heavy rainfall. Consequently, there are sizable areas of swamp and marshland near these

streams.

Soils are generally sandy and low in natural fertility, the main exception being a large area of peat and muck soils in the Everglades. About one-third of Florida's soils can be classified as uplands or ridge soils that are generally well- to excessively well-drained. Soils in the remaining two-thirds of the State, including the muck soils, generally have imperfect to very poor natural drainage. Large areas of Florida are underlain by compact subsoils that intensify the effects of both wet and dry weather.

GENERAL CLIMATIC FEATURES. Climate is probably Florida's greatest natural resource. General climatic conditions range from a zone of transition between temperate and subtropical conditions in the extreme northern interior portion of the State to the tropical conditions found on the Florida Keys. The chief factors of climatic control are: latitude, proximity to the Atlantic Ocean and Gulf of Mexico, and numerous inland lakes.

Summers throughout the State are long, warm, and relatively humid; winters, although punctuated with periodic invasions of cool to occasionally cold air from the north, are mild because of the southern latitude and relatively warm adjacent ocean waters. The Gulf Stream, which flows around the western tip of Cuba, through the Straits of Florida, and northward along the lower east coast, exerts a warming influence to the southern east coast largely because the predominate wind direction is from the east. Coastal stations throughout the State average slightly warmer in winter and cooler in summer than do inland stations at the same latitude.

Florida enjoys abundant rainfall. Except for the northwestern portion of the State, the average year can be divided into two seasons--the so-called "rainy season" and the long, relatively dry season. On the peninsula, generally more than one-half of the precipitation for an average year can be expected to fall during the 4-month period, June through September. In northwest Florida, there is a secondary rainfall maximum in late winter and in early spring.

Some notable climatic extremes are: highest recorded temperature, 109° at Monticello on June 29, 1931; lowest recorded temperature, -2° at Tallahassee on February 13, 1899; and the greatest 24-hour rainfall, 38.7 inches at Yankeetown on September 5-6, 1950.

TEMPERATURE. Mean annual temperatures range from the upper 60s in the northern portions of the State to the middle 70s on the southern mainland, but reach to nearly 78° at Key West. Summertime mean temperatures are about the same throughout the State, 81° to 82°; during the coolest months, temperatures average about 13° lower in northern than in southern Florida. July and August temperature averages are the warmest in all areas, and December and January temperature averages are the coolest in the northern and central portions of the State. January and February, on the average, are the coolest months in the extreme south and on the Keys.

Maximum temperatures during the warmest months average near 90° along the coast and slightly above 90° in the interior; minima average in the low 70s, but are slightly higher along the immediate coast and on the Keys than inland. During June, July, and August, maximum temperatures exceed 90° about 2 days in 3 in all interior areas; in May and September, 90° or higher can be expected about 1 day in 3 in the northern interior and about 1 day in 2 in the southern interior. Extreme heat waves, characteristic of continental locations, are felt occasionally--but in a modified form--over the northern interior portions of Florida. Temperatures of 100° or higher are infrequent in northern, rare in central, and practically unknown in southern Florida.

The summer heat is tempered by sea breezes along the coast and by frequent afternoon or early evening thunderstorms in all areas. During the warm season, sea breezes are felt almost daily within several miles of the coast and occasionally 20 to 30 miles inland. Thundershowers, which on the average occur about one-half of the days in summer, frequently are accompanied by as much as a rapid 10°- to 20°-drop in temperature, resulting in comfortable weather for the remainder of the day. Gentle breezes occur almost daily in all areas and serve to mitigate further the oppressiveness that otherwise would accompany the prevailing summer temperature and humidity conditions. Because most of the large-scale wind patterns affecting Florida have passed over water surfaces, hot drying winds seldom occur.

Although average minimum temperatures during the coolest months range from the middle 40s in the north to the middle 50s in the south, no place on the mainland is entirely safe from frost or freezing. An occasional cold wave of the more severe type brings minima, ranging from 15° to 20° over the northern portions to freezing or below over the southern portions of the peninsula. These cold waves, except in rare instances, seldom last more than 2 or 3 days at a time. It is extremely rare for temperatures to remain below freezing throughout the day at any place in the State.

On the first night of a cold wave, there usually is considerable wind which, because of the continual mixing of the air, prevents marked temperature differences between high and low ground. By the second night, winds usually have subsided and radiational cooling under clear skies accelerates the temperature fall after sundown. On such occasions, marked differences in temperature are noticeable at places not far apart, depending upon such factors as topography and proximity to bodies of water. These facts are of primary concern in selecting sites for growing plants intolerant of cold.

Some winters--occasionally several in succession--pass without widespread freezing to the southern portions of the peninsula; others may bring several severe cold waves. Winters with more than one severe cold wave, interspersed with periods of relative warmth, are especially distressing to the agricultural industry because the later freezes almost always find vegetation in a tender stage of growth and highly susceptible to additional cold damage.

Limited weather records available for the 19th century indicate severe freezes occurred in February 1835, January 1857, December 1870, December 1880, January 1886, December 1894, February 1895, January 1898, and February 1899. Probably the 1835 freeze was the most severe. Freezes of lesser severity are also known to have occurred, but temperature records are sparse up to 1886. Noteworthy cold spells during the 20th century are: January 1905, December 1906, December 1909, February 1917, January 1928, December 1934, January 1940, February 1947, and the freezes of the 1957-58 winter season. The freeze of December 1962 was as severe as those in late 1890s, while the historic early freeze of November 1970 and the long-duration freeze of January 1971 combined to be almost as severe.

Since 1937, the Federal-State Frost Warning

Service in Lakeland, a cooperative venture of the National Weather Service and the State of Florida, has been securing temperature data at a great many peninsula stations during the winter seasons. Additional detailed information, including temperature-duration data, is published in the annual and periodic summary reports of the Service.

PRECIPITATION. Rainfall in Florida is quite varied both in annual amount and in seasonal distribution. Individual station annual averages, based upon the 30-year period, 1931-1960, range from about 50 to 65 inches. On the Keys, annual averages are only about 40 inches. High annual rainfall is measured at stations in the extreme northwestern counties and in the southern end of the peninsula. Rainfall varies greatly from year to year, with wet years sometimes doubling the amounts received during a dry year. Many localities have received more than 80 inches in a calendar year, and a few stations have measured more than 100 inches. In contrast, almost all localities have received less than 40 inches in a calendar year.

The distribution of rainfall within the year is quite uneven. In the summer rainy season, there is close to a 50-50 chance that some rain will fall on any given day. During the remainder of the year, the chances of rainfall are much less, with some rain being likely to fall on 1 or 2 days in a week.

The seasonal distribution changes somewhat from north to south. In the northwestern portions of the State, there are two high points--late winter or early spring and again during summer--and one pronounced low point--October; a secondary low point occurs in April and May. On the peninsula, the most striking features of the seasonal distribution are the dominance of summer rainfall (generally more than one-half of the average annual amount falls in the 4-month period, June through September) and the rather abrupt start and end of the summer rainy season (June average rainfall tends to be nearly double the amount of May, and in the fall, the average for the last month of the wet season tends to be about double the amount of the following month). October averages as the driest month in northwest Florida, but, in general, is among the wettest months on the southeast coast and Keys.

The start and end of the rainy season varies considerably from year to year. According to climatic records, the season has begun as soon as early May and has been as delayed as late June. Late September or early October usually marks the end of the wet season, except along a narrow strip of the entire east coast where relatively large October rainfalls are frequently noted. The tendency for relatively large October rainfall diminishes quite rapidly westward.

Most of the summer rainfall is derived from "local" showers or thundershowers. Many sta-

tions average more than 80 thundershowers per year, and some average more than 100. Showers are often heavy, usually lasting only 1 or 2 hours, and generally occur near the hottest part of the day. The more severe thundershowers are occasionally attended by hail or locally strong winds which may inflict serious local damage to crops and property. Day-long summer rains are usually associated with tropical disturbances and are infrequent. Even in the wet season, the rainfall duration is generally less than 10 percent of the time.

Because most summer rains are local in character, large differences in monthly and annual amounts at nearby stations are common, but these differences disappear when a comparison is made on the basis of long-period averages. However, large differences in the long-period averages do exist within short distances. For example, the normal annual rainfalls for Miami Beach and for the Miami Airport are 46.26 and 59.76 inches, respectively, yet it is less than 10 airline miles distance from the Beach to the Airport. Similar conditions undoubtedly exist elsewhere along the immediate coast.

Most localities have, at one time or another, experienced 2-hour rainfalls in excess of 3 inches and 24-hour amounts of near or greater than 10 inches. Nearly all localities have had within a single month from one-third to one-half as much rain as will fall during an entire average year. Occasionally, tropical storms produce copious rainfall over relatively large areas. A detailed survey of the September 1950 hurricane, conducted by the U. S. Corps of Engineers Florida District, headquartered at Jacksonville, indicated an amount that was near 34 inches fell in a 24-hour period within the vicinity of Cedar Key. The 38.70 inches of rainfall that fell during that 24-hour period at Yankeetown on September 5-6, 1950, during this hurricane is the record 24-hour rainfall for the Nation. Because of water disposal problems, heavy rains can be just as serious as droughts.

DROUGHTS. Florida is not immune from drought, even though annual rainfall amounts are relatively large. Prolonged periods of deficient rainfall are occasionally experienced even during the time of the expected rainy season. Several such dry periods, in the course of 1 or 2 years, can lead to significantly lowered water tables and lake levels which, in turn, may cause serious water shortages for those communities that depend upon lakes and shallow wells for their water supply. The worst drought in over 40 years along the Lower East Coast Division occurred in 1971. In that Division, the lowest 12-month rainfall of record, 34.59 inches, was set during the period from July 1970 to June 1971. The level of Lake Okeechobee dropped to 10.3 feet, only 0.16 of a foot above the record minimum of 10.14 feet.

Because a large part of the State's agricultural produce is planted, grown, and marketed during fall, winter, and spring (normally the driest part of the year), growers of high-per-acre-value crops have long concluded that it is almost mandatory to provide supplemental irrigation for crop success. The flat topography of the area where many of these crops are grown is well suited to subsurface irrigation by water table control. However, heavy rains can occur during these growing seasons, and growers have found it necessary to provide water-removal facilities in addition to those for irrigation.

Statewide droughts during summer are rare, but it is not unusual during a drought in one portion of the State for other portions to receive generous rainfall. In a few instances, individual stations have experienced periods of a month or more without rainfall.

SNOW. Snowfall in Florida is unusual, although measurable amounts have fallen in the northern portions at irregular intervals, and a trace of snow has been recorded as far south as Fort Myers. The greatest recorded snowfalls in Florida occurred on February 13, both in 1899 and 1958. In 1899, 4 inches were measured at Lake Butler in Union County and one-half inch at Bartow in Polk County. On the night of February 12-13, 1958, most of northwest Florida west of the Suwannee River received 2 to 3 inches of snow; areas east of the Suwannee River and north of about latitude 30° received only 1 to 2 inches of snow on that same night. The 3 inches measured at Tallahassee on this same date is the greatest ever measured at that station since records began in 1886; the 1.5 inches measured at Jacksonville approximates the only other recorded measurable amount, 1.9 inches, on February 13, 1899.

WIND. Prevailing winds over the southern peninsula are southeast and east. Over the remainder of the State, wind directions are influenced locally by convectional forces inland and by the land-and-sea-breeze-effect near the coast. Consequently, prevailing directions are somewhat erratic, but, in general, follow a pattern from the north in winter and from the south in summer. The windiest months are March and April. High local winds of short duration occur occasionally in connection with thunderstorms in summer and with cold fronts moving across the State in other seasons.

Tornadoes, funnel clouds, and waterspouts also occur, averaging 10 to 15 in a year. Tornadoes have occurred in all seasons, but are most frequent in spring; they also occur in connection with tropical storms. Generally, tornado paths in Florida are short, and damage has not been extensive. Occasionally, waterspouts come inland, but they usually dissipate soon after reaching land and affect only very small areas. The frequency of occurrence of severe local storms is given in a 1969 ESSA Technical Memorandum publication edited by Pautz (see Bibliography).

TROPICAL STORMS. Storms that produce high winds and are often destructive are usually tropical in origin. Florida, jutting out into the ocean between the subtropical Atlantic and the Gulf of Mexico, is the most exposed of all States to these storms. In particular, hurricanes can approach from the Atlantic Ocean to the east, from the Caribbean Sea to the south, and from the Gulf of Mexico to the west.

From 1885 to 1971 inclusive, 160 tropical storms of all intensities have entered or affected Florida significantly. About 55 percent of these storms are known to have been of hurricane intensity. Table 1 shows the distribution of these storms throughout the past 87 years. The average number has been 1.7 storms in a year, but individual years may vary from none to as many as five. The State has never gone more than 2 years without a tropical storm. The longest period since 1885 without a major hurricane was 9 years, from 1951 to 1959, inclusive.

Hurricane Donna, which crossed the Keys and then moved northeastward across the State from near Fort Myers to near Daytona Beach on September 9-10, 1960, is thought to be the most destructive storm ever experienced in Florida. This storm caused an estimated $300 million in damage in Florida. Only 13 fatalities resulted from this very intense storm, indicating the great value of the modern storm warning service now available in hurricane-threatened areas. Three hurricanes--Cleo, Dora, and Isabel--in 1964 caused the greatest damage for any 1 year, $348,875,000.

The chances of hurricane-force winds reaching a particular locality in any given year are shown in table 2. The probability varies from 1 in 100 at Jacksonville to 1 in 7 at Key West and Miami. Only 10 or 11 storms of hurricane intensity in 87 years have passed inland on the west coast in the area from Cedar Key to Fort Myers. On the extreme northeast coast from Saint Augustine to Jacksonville, no tropical storm of full hurricane force was recorded until hurricane Dora on September 9, 1964. Until that time, the city of Jacksonville had the unique distinction of being the only large city in Florida, and indeed the only one on the Atlantic coast from Boston southward, that had never sustained winds of hurricane force in a tropical storm in modern times.

The vulnerability of the State to tropical storms varies with the progress of the hurricane season. In August and early September, tropical storms normally approach the State from the east or southeast, but as the season progresses into late September and October, the region of maximum hurricane activity (insofar as Florida is concerned) shifts to the western Caribbean. Most of those storms that move into Florida approach the State from the south or southwest, entering the Keys, the Miami area, or along the west coast.

The most intense hurricane of modern times to affect the State of Florida occurred on Labor Day in 1935. The lowest sea-level pressure ever measured in the Western Hemisphere, 26.35 inches, was recorded at that time. Engineers have calculated that winds of between 200 to 250 m.p.h. would have been required to account for some of the damage that was sustained during this severe hurricane. During the hurricane of August 1949, winds at West Palm Beach reached 110, with gusts to 125 m.p.h., before the anemometer was blown away. The highest sustained speed was estimated at 120, with gusts to 130 m.p.h. A privately owned anemometer, the accuracy of which is unknown, recorded gusts of 155 m.p.h.

The highest winds of a hurricane are seldom measured because these usually occur at isolated places where no anemometers are installed. It seems likely that winds of 150 m.p.h., which occasionally accompany major hurricanes, occur in Florida perhaps once in 7 years.

Some of the world's heaviest rainfalls have occurred within tropical cyclones. Rainfall over 20 inches in 24 hours is not uncommon! The intensity of the rainfall, however, does not seem to bear any relation to the intensity of the wind circulation. For example, a storm that entered the west coast of Florida in October 1941 was never of hurricane intensity; nevertheless, over a 3-day period, about 35 inches of rain fell at Trenton. The 24-hour amount for this same storm was about 30 inches. Another Florida hurricane of 1947 caused a rainfall of about 6 inches in 1 hour at Hialeah. Such extremes, however, are relatively rare; the average hurricane rainfall in Florida usually does not exceed 6 to 8 inches in a 24-hour period.

OTHER CLIMATIC ELEMENTS. The climate of Florida is humid. Inland areas with greater temperature extremes enjoy slightly lower relative humidity, especially during times of hot weather. On the average, variations in relative humidity from one place to another are small; humidities range from about 50 to 65 percent during the afternoon hours to about 85 to 95 percent during the night and early morning hours.

Heavy fogs are usually confined to the night and early morning hours in the late fall, winter, and early spring months. On the average, they occur about 35 to 40 days in a year over the extreme northern portion; about 25 to 30 days in a year over the central portion; and less than 10 days in a year over the extreme southern portion of the State. These fogs usually dissipate or thin soon after sunrise; heavy daytime fog is seldom observed in Florida.

Florida has been nicknamed the Sunshine State. Sunshine measurements made at widely separated stations in the State indicate the sun shines about two-thirds of the possible sunlight hours during the year, ranging from slightly more than 60 percent of possible in December and January to more than 70 percent of possible in April and May. In general, southern Florida enjoys a higher percentage of possible sunshine hours than does northern Florida. The length of day operates to Florida's advantage. In winter, when sunshine is highly valued, the sun can shine longer in Florida than in the more northern latitudes. In summer, the situation reverses itself with longer days returning to the north.

Meteorological conditions that aggravate air pollution do not often occur at any one place in the State and are probably the least frequent in southeastern Florida. The air over the State is usually sufficiently unstable--a condition conducive to the development of cumulus clouds and thunderstorms--to disperse pollutants to higher levels. This fact, plus the relative constancy of the easterly trade winds in southeastern Florida, greatly reduces the general pollution problem in the State. However, communitywide pollution, in which the cumulative effects of all sources of pollution become significant, is becoming more prevalent as population increases.

CLIMATE AND THE ECONOMY. Florida's economy rests largely upon agriculture, industry, and the tourist trade, each of which has been or is being developed in relation to the general climatic conditions found in the State. The principal agricultural products are: citrus and other tropical fruits; all varieties of truck crops; general farm crops such as tobacco, corn, peanuts, cotton, and small grains; nursery products; flowers (especially gladioli and chrysanthemums); sugarcane; sweet and Irish potatoes; tung nuts; pecans; and honey and wax. Large areas in the central and southern portions of the State are utilized by the cattle industry because the long growing season allows forage production throughout most of the year.

The State contains three distinct agricultural districts. Production in the southern Florida district consists of many kinds of winter vegetables and those tropical fruits that will not tolerate the winter conditions of the northern and western districts; some citrus is also grown in southern Florida. In the northern and western districts, general farm crops predominate, although many spring and fall truck crops are also grown. Citrus and early spring and late fall truck crops predominate on the central peninsula, although large areas in that portion of the State, where soils or microclimatic conditions are not suitable for such crops, are used for cattle grazing. Practically all truck crop production in the northern and central portions of the State is conducted on a calculated "climatic risk" basis where the dangers of adverse weather, such as freezing, are weighed against the advantages of an early market. There are some commercial crops grown in Florida in each month of the year.

Increasing industrial activity in the State also has its climatic implications. From the standpoint of raw material production, the State's largest industry in terms of payroll—foods processing—stems from a climate favorable to the growing of fruits and vegetables. Similarly, the paper and pulp industry has found that the long growing season enables the pine forests to renew themselves more rapidly than in more northern latitudes. Other industries, such as chemical and aircraft, have found that the Florida climate permits plant construction and operations to meet both indoor and outdoor operational requirements.

NOTES FOR VISITORS. Statistics alone cannot always convey the difference between climatic conditions in two locations. Some differences observed by visitors to the State are the bright sunshine, sharp shadows, and violent rainshowers. First-time Florida visitors who take pictures tend to overexpose most of their photographs because of the bright sunlight. The lack of middle and upper level cloudiness gives bright contrasts to shadows, while at night, the few low-level fair-weather cumulus clouds produce scenes around the bright moon that befit the ideal tropical scene.

Rainfall is abundant, with most places—except on the Keys—receiving over 50 inches in a year. However, the rainfall is generally in the form of short-lived showers. The typical duration of such showers is only about one-quarter that of similar ones in more temperate climates. As a result, one should seldom postpone plans because it is raining at the present time. Because of the abundance of rainfall, lush tropical rainforest growth may be found almost to the shoulders of highways.

One of the most popular outdoor activities is water recreation. Along the east coast, water temperature at Miami Beach ranges from 72° in January to 89° in July. In general, along the entire Atlantic seaboard of Florida, the sea-surface temperature averages 78°, ranging from 74° in February to 83° in August. The sea-surface temperature on the west coast of Florida averages 77°, ranging from 70° in February to 84° in August.

Wintertime minimum temperatures are deceptive. While stations in northern Florida record 10 to 20 days in a year with minimum temperatures of 32° or below, there have been only 5 days in the past 72 years at Jacksonville where the maximum temperature for the day has failed to climb above freezing. This means that the coat you might wear at 7 a.m. will become heavy by 10 a.m., and, if you are driving, that you may well be in your shirt sleeves by noon.

While sunshine hours in Miami are 66 percent of possible in December as compared to 51 percent in New York City, greater difference is reflected in the amount of solar radiation that leads to temperature contrasts. New York City receives only 116 langleys (a unit of solar radiation) on a horizonal surface each day during December. In contrast, Miami receives 317 langleys, almost three times as much solar radiation.

Wintertime is generally a time of relatively low rainfall. There are only rare occurrences of the overcast, drizzly days that are typical of winters in more northern latitudes.

Summers are hot and humid in Florida. In fact, the humidity throughout the year is so high each night that one can expect to find his car covered with dew every morning unless it is placed under shelter. The Temperature-Humidity-Index (THI) climbs rapidly to 79 throughout the State in early June and stays between 79 and 81 during most of the afternoon hours until late September. At a THI of 79, nearly all people will feel uncomfortable; as the Index passes 80, discomfort becomes more pronounced. However, the extensive use of air conditioning and the informal dress of Floridians have greatly alleviated the effects of the hot, humid summertime weather. In addition, thundershowers, which on the average occur on about one-half of the summer days, are frequently accompanied by a rapid 10° to 20° drop in temperature, resulting in comfortable weather for the remainder of the day. Thunderstorms in Florida are nature's air conditioners.

Visitors can experience a tropical climate in Florida. A tropical climate is defined as one in which the average temperature of the coldest month is 64.4° or above. The climate found in Florida along the east coast from Vero Beach southward and along the west coast from Punta Gorda southward fits that definition.

TABLE 1. --FREQUENCY OF TROPICAL STORMS BY YEARS IN FLORIDA

Year	Of Known Hurricane Intensity	Not or Of Doubtful Hurricane Intensity	Total	Year	Of Known Hurricane Intensity	Not or Of Doubtful Hurricane Intensity	Total
1885	3	1	4	1930	0	1	1
1886	3	1	4	1931	0	0	0
1887	1	1	2	1932	1	1	2
1888	2	1	3	1933	2	2	4
1889	1	2	3	1934	0	0	0
1890	0	0	0	1935	3	0	3
1891	1	1	2	1936	1	2	3
1892	0	2	2	1937	0	3	3
1893	3	2	5	1938	0	1	1
1894	2	1	3	1939	1	1	2
1895	1	3	4	1940	0	0	0
1896	3	0	3	1941	1	1	2
1897	0	1	1	1942	0	0	0
1898	2	0	2	1943	0	0	0
1899	1	2	3	1944	1	0	1
1900	0	1	1	1945	2	1	3
1901	0	2	2	1946	1	1	2
1902	0	1	1	1947	2	1	3
1903	1	0	1	1948	2	0	2
1904	1	0	1	1949	1	0	1
1905	0	0	0	1950	2	2	4
1906	3	1	4	1951	0	1	1
1907	0	1	1	1952	0	1	1
1908	0	0	0	1953	1	2	3
1909	1	1	2	1954	0	0	0
1910	1	0	1	1955	0	0	0
1911	1	1	2	1956	1	0	1
1912	1	0	1	1957	0	3	3
1913	0	0	0	1958	0	1	1
1914	0	0	0	1959	0	2	2
1915	1	1	2	1960	1	2	3
1916	3	0	3	1961	0	0	0
1917	1	0	1	1962	0	0	0
1918	0	0	0	1963	0	1	1
1919	1	1	2	1964	3	2	5
1920	0	1	1	1965	1	1	2
1921	1	0	1	1966	2	0	2
1922	0	0	0	1967	0	0	0
1923	0	1	1	1968	4	0	4
1924	2	1	3	1969	1	1	2
1925	1	0	1	1970	1	1	2
1926	3	0	3	1971	0	0	0
1927	0	0	0				
1928	3	0	3				
1929	1	0	1	Total	84	66	150

TABLE 2. --CHANCES OF HURRICANE FORCE WINDS IN ANY GIVEN YEAR

City	Chances	City	Chances
Jacksonville	1 in 100	Key West	1 in 8
Daytona Beach	1 in 50	Fort Myers	1 in 11
Melbourne-Vero Beach	1 in 20	Tampa-St. Petersburg	1 in 25
Palm Beach	1 in 7	Apalachicola-St. Marks	1 in 17
Miami	1 in 6	Pensacola	1 in 8

ACKNOWLEDGMENTS. That portion of this publication relating to the climatic features of hurricanes was prepared by the staff of the National Hurricane Center at Miami.

BIBLIOGRAPHY

Dunn, Gordon E., and Staff, "Florida Hurricanes," ESSA Technical Memorandum WBTM-SR-38, Fort Worth, Tex., Nov. 1967, 26 pp.

Farrer, L. A., "Design Rainfall: Central and Southern Florida Project," Proceedings American Society of Civil Engineers, Vol. 80, Separate No. 519, Oct. 1954, pp. 1-14.

Hefny, M. B., "A Climatic Study of Florida," Bulletin of the Faculty of Arts, Fouad I University, Cairo, Egypt, Vol. XIV, Part II, Dec. 1952, pp. 63-78.

Johnson, Warren O., "Minimum Temperatures in the Agricultural Areas of Peninsular Florida; Summary of 30 Winter Seasons 1937-67," Institute of Food Agricultural Sciences, IFAS Publication No. 9, University of Florida, Gainesville, Fla., Sept. 1970, 154 pp.

Morris, Alan, The Florida Business Handbook, The Capital, Tallahassee, Fla., 1958, 117 pp.

NOAA, Environmental Data Service, "Apalachicola, Daytona Beach, Fort Myers, Jacksonville, Key West, Lakeland, Miami, Orlando, Pensacola, Tallahassee, Tampa, and West Palm Beach, Fla.," Local Climatological Data, Asheville, N.C., monthly plus annual summary.

_____, Climatological Data--Florida, Asheville, N.C., monthly plus annual summary.

_____, Hourly Precipitation Data--Florida, Asheville, N.C., monthly plus annual summary.

_____, "Monthly Averages for State Climatic Divisions, 1931-1960, Florida," Climatography of the United States No. 85-6, Washington, D.C., 1963, 3 pp.

Reuss, L. A., "Florida's Land Resources and Land Use," University of Florida Bulletin 555, Nov. 1954, 52 pp.

Simpson, R. H., and Lawrence, Miles B., "Atlantic Hurricane Frequencies Along the U.S. Coastline," NOAA Technical Memorandum NWS SR-58, Fort Worth, Tex., June 1971, 14 pp.

Staff, SELS, Pautz, Maurice E., ed., "Severe Local Storm Occurrences 1955-1967," ESSA Technical Memorandum WBTM-FCST-12, Silver Spring, Md., Sept. 1969, 77 pp.

U.S. Weather Bureau, Climatic Summary of the United States (Bulletin W), Sections 104 and 105, Washington, D.C., 1930, 49 pp.

_____, "Climatic Summary of the United States--Supplement for 1931 Through 1952, Florida" (Bulletin W Supplement), Climatography of the United States No. 11-6, Washington, D.C., 1953, 36 pp.

_____, "Climatic Summary of the United States--Supplement for 1951 Through 1960, Florida" (Bulletin W Supplement), Climatography of the United States No. 86-6, Washington, D.C., 1964, 61 pp.

_____, "Evaporation Maps for the United States," Technical Paper No. 37, Washington, D.C., 1959, 13 pp.

_____, "Maximum Station Precipitation for 1, 2, 3, 6, 12, and 24 Hours, Part III: Florida," Technical Paper No. 15, Washington, D.C., 1952, 92 pp.

_____, "Rainfall Frequency Atlas of the United States for Durations from 30 Minutes to 24 Hours and Return Periods from 1 to 100 Years," Technical Paper No. 40, Washington, D.C., May 1961, 115 pp.

_____, "Rainfall Intensity-Duration-Frequency Curves for Selected Stations in the United States, Alaska, Hawaiian Islands, and Puerto Rico," Technical Paper No. 25, Washington, D.C., Dec. 1955, 53 pp.

_____, "Rainfall Intensity-Frequency Regime, Part 2--Southeastern United States," Technical Paper No. 29, Washington, D.C., Mar. 1958, 51 pp.

_____, "Two- to Ten-Day Precipitation for Return Periods of 2 to 100 Years in the Contiguous United States," Technical Paper No. 49, Washington, D.C., 1964, 29 pp.

FREEZE DATA

STATION	Freeze threshold temperature	Mean date of last Spring occurrence	Mean date of first Fall occurrence	Mean No. of days between dates	Years of record Spring	No. of occurrences in Spring	Years of record Fall	No. of occurrences in Fall
APALACHICOLA WSO CITY	32	02-02	12-21	322	30	25	30	13
	28	01-16	12-27	345	30	13	30	7
	24	01-04	@	@	30	7	30	4
	20	@	@	@	30	2	30	0
	16	@	@	@	30	0	30	0
ARCADIA	32	02-01	12-18	321	29	24	30	17
	28	01-11	12-26	350	29	9	30	9
	24	@	@	@	29	3	30	2
	20	@	@	@	29	1	30	0
	16	@	@	@	29	0	30	0
AVON PARK	32	01-16	12-26	344	30	12	30	8
	28	@	@	@	30	4	30	1
	24	@	@	@	30	0	30	1
	20	@	@	@	30	0	30	0
BARTOW	32	01-26	12-18	326	30	21	30	16
	28	01-04	12-27	357	30	5	30	6
	24	@	@	@	30	3	30	1
	20	@	@	@	30	0	30	0
BELLE GLADE EXP STA	32	01-30	12-25	329	25	16	26	10
	28	01-09	12-28	353	25	7	26	5
	24	@	@	@	25	1	26	0
	20	@	@	@	25	0	26	0
BLOUNTSTOWN	32	02-24	11-22	271	30	29	29	27
	28	02-06	12-07	304	30	25	29	20
	24	01-19	12-20	335	28	14	29	12
	20	01-04	12-28	358	28	6	29	6
	16	@	@	@	28	3	29	0
BRADENTON	32	01-15	12-21	340	30	14	30	13
	28	@	@	@	30	4	30	2
	24	@	@	@	30	1	30	0
	20	@	@	@	30	0	30	0
BROOKSVILLE CHIN HILL	32	02-06	12-14	311	30	23	30	20
	28	01-17	12-24	341	30	15	30	8
	24	01-05	@	@	30	6	30	1
	20	@	@	@	30	0	30	1
	16	@	@	@	30	0	30	0
CARRABELLE 1 NNW	32	02-21	12-05	287	30	29	29	23
	28	02-06	12-14	311	29	24	29	17
	24	01-17	12-23	340	27	12	29	11
	20	01-08	@	@	27	7	29	4
	16	@	@	@	27	3	29	0
CEDAR KEY 1 WSW	32	01-31	12-19	322	30	21	28	16
	28	01-10	12-24	347	30	9	28	9
	24	01-04	@	@	30	7	27	2
	20	@	@	@	30	3	27	0
	16	@	@	@	30	0	27	0
CLERMONT 6 SSW	32	01-17	12-25	342	29	13	29	11
	28	01-06	@	@	29	7	29	4
	24	@	@	@	29	2	29	0
	20	@	@	@	29	0	29	0
CRESCENT CITY	32	02-07	12-15	311	30	25	29	19
	28	01-15	12-24	343	30	13	29	8
	24	@	@	@	30	3	29	0
	20	@	@	@	30	1	29	0
	16	@	@	@	30	0	29	0
DAYTONA BEACH	32	01-28	12-20	326	28	16	28	13
	28	01-07	12-26	353	28	8	28	6
	24	@	@	@	28	3	28	2
	20	@	@	@	28	1	28	0
	16	@	@	@	28	0	28	0

STATION	Freeze threshold temperature	Mean date of last Spring occurrence	Mean date of first Fall occurrence	Mean No. of days between dates	Years of record Spring	No. of occurrences in Spring	Years of record Fall	No. of occurrences in Fall
DE FUNIAK SPRINGS	32	02-26	11-25	272	30	29	29	27
	28	02-16	12-10	297	30	29	29	21
	24	01-26	12-19	327	30	19	29	13
	20	01-10	12-27	351	30	12	29	8
	16	01-04	@	@	30	6	29	1
DELAND 1 SSE	32	02-15	12-14	302	30	28	30	21
	28	01-24	12-21	331	30	17	30	14
	24	01-04	@	@	30	5	29	3
	20	@	@	@	30	2	29	1
	16	@	@	@	30	0	29	0
EVERGLADES	32	01-03	@	@	22	4	23	1
	28	@	@	@	22	1	23	1
	24	@	@	@	22	1	23	0
	20	@	@	@	22	0	23	0
FEDERAL POINT	32	02-02	12-19	321	29	22	29	15
	28	01-09	12-26	351	29	8	29	6
	24	@	@	@	29	2	29	1
	20	@	@	@	29	0	29	0
FELLSMERE 7 SSW	32	01-23	12-24	335	30	16	29	12
	28	01-06	@	@	30	7	28	1
	24	@	@	@	30	1	28	1
	20	@	@	@	30	0	28	0
FERNANDINA BEACH	32	02-09	12-16	311	29	24	30	17
	28	01-23	12-25	336	29	18	30	8
	24	01-06	12-28	357	29	6	30	5
	20	@	@	@	29	2	30	0
	16	@	@	@	29	0	30	0
FORT LAUDERDALE	32	01-03	@	@	29	5	29	2
	28	@	@	@	29	0	29	0
FORT MYERS	32	@	@	@	30	4	30	1
	28	@	@	@	30	0	30	0
FORT PIERCE	32	01-06	@	@	30	6	30	3
	28	@	@	@	30	0	30	0
GAINESVILLE UNIV	32	02-14	12-06	295	30	27	30	21
	28	01-30	12-16	320	30	21	30	18
	24	01-08	12-26	352	30	9	30	8
	20	@	@	@	30	4	30	2
	16	@	@	@	30	1	30	1
GLEN ST MARY 1 W	32	03-08	11-22	260	30	30	30	29
	28	02-11	12-05	298	30	28	29	21
	24	01-26	12-22	330	30	20	29	13
	20	01-12	12-28	350	30	12	29	5
	16	@	@	@	30	2	29	0
HOMESTEAD EXP STA	32	01-17	12-26	344	29	12	25	6
	28	@	@	@	29	3	26	1
	24	@	@	@	29	0	26	0

FREEZE DATA

STATION	Freeze threshold temperature	Mean date of last Spring occurrence	Mean date of first Fall occurrence	Mean No. of days between dates	Years of record Spring	No. of occurrences in Spring	Years of record Fall	No. of occurrences in Fall
INVERNESS	32	02-12	12-06	298	28	26	29	22
	28	01-21	12-15	328	28	16	29	14
	24	*	12-23	*	28	4	27	8
	20	*	*	*	28	0	27	1
	16	*	*	*	28	0	27	0
JACKSONVILLE	32	02-06	12-16	313	30	24	30	15
	28	01-22	12-21	333	30	17	30	11
	24	01-06	12-28	356	30	8	30	5
	20	*	*	*	30	2	30	0
	16	*	*	*	30	0	30	0
JASPER 9 ESE	32	03-06	11-22	262	7	7	10	9
	28	02-05	12-06	304	6	5	9	7
	24	01-23	12-21	332	6	3	9	5
	20	01-15	12-25	344	6	2	9	4
	16	*	*	*	6	0	9	0
KEY WEST WSO AP	32	*	*	*	30	0	30	0
KISSIMMEE	32	01-25	12-24	334	27	18	26	11
	28	01-06	*	*	26	5	26	3
	24	*	*	*	26	1	26	0
	20	*	*	*	26	0	26	0
LAKE CITY 2 E	32	02-22	12-01	282	29	28	30	25
	28	02-05	12-17	315	29	24	30	17
	24	01-17	12-25	343	29	14	30	7
	20	01-06	*	*	29	7	30	4
	16	*	*	*	29	1	30	0
LAKELAND WSO CITY	32	01-10	12-25	349	30	9	29	8
	28	*	*	*	30	4	29	3
	24	*	*	*	30	2	29	0
	20	*	*	*	30	0	29	0
MADISON	32	02-19	12-02	286	30	29	30	27
	28	01-31	12-17	320	30	24	30	14
	24	01-16	12-24	341	30	14	30	10
	20	01-08	12-28	354	30	10	30	5
	16	*	*	*	30	4	30	0
MARIANNA SCH FOR BOYS	32	03-03	11-24	266	29	29	29	29
	28	02-05	12-11	310	29	25	29	19
	24	01-21	12-22	336	28	19	29	13
	20	01-08	12-27	354	28	9	29	5
	16	*	*	*	28	3	29	1
MIAMI BEACH	32	*	*	*	19	0	20	0
MONTICELLO 3 W	32	02-26	11-26	274	29	28	29	26
	28	02-05	12-13	311	29	24	29	16
	24	01-19	12-24	339	29	16	29	9
	20	01-05	12-27	356	29	8	29	5
	16	01-02	*	*	29	5	29	0
MOORE HAVEN LOCK 1	32	01-23	12-25	336	30	19	30	10
	28	*	*	*	30	3	30	3
	24	*	*	*	30	1	30	1
	20	*	*	*	30	0	30	0
MT PLEASANT	32	02-26	11-22	269	30	29	29	27
	28	02-17	12-05	291	30	27	29	22
	24	01-27	12-19	326	30	21	28	15
	20	01-13	12-25	347	30	13	28	9
	16	01-04	*	*	30	5	28	2
NEW SMYRNA BEACH 3 NW	32	02-04	12-20	319	26	20	27	14
	28	01-08	12-28	354	26	7	27	4
	24	*	*	*	26	3	27	1
	20	*	*	*	26	0	27	0
NICEVILLE	32	03-03	11-18	261	12	11	10	10
	28	02-19	11-30	284	11	10	10	9
	24	01-25	12-16	325	8	4	10	5
	20	01-15	12-25	344	9	4	10	3
	16	*	*	*	8	1	10	0
OCALA	32	02-13	12-06	296	30	27	30	21
	28	01-24	12-14	324	30	18	30	18
	24	01-08	12-29	355	30	9	30	5
	20	*	*	*	30	3	30	1
	16	*	*	*	30	0	30	0
OKEECHOBEE HRCN GATE	32	01-16	12-25	343	27	10	24	7
	28	01-06	*	*	27	5	24	2
	24	*	*	*	27	0	24	1
	20	*	*	*	27	0	24	0
ORLANDO WSO AP	32	01-31	12-17	319	30	23	30	19
	28	01-11	12-28	351	30	10	30	5
	24	*	*	*	30	1	30	2
	20	*	*	*	30	1	30	0
	16	*	*	*	30	0	30	0
PANAMA CITY	32	02-17	12-02	288	23	22	23	21
	28	01-31	12-17	320	23	17	23	11
	24	01-16	12-21	339	23	12	23	8
	20	*	*	*	23	3	23	3
	16	*	*	*	23	1	23	0
PENSACOLA	32	02-18	12-15	300	30	28	30	19
	28	01-21	12-21	334	30	18	30	13
	24	01-15	12-27	346	30	14	30	6
	20	01-06	*	*	30	9	30	3
	16	*	*	*	30	2	30	0
PLANT CITY	32	02-18	12-09	294	27	26	26	20
	28	01-24	12-21	331	26	16	26	13
	24	01-06	*	*	26	6	26	3
	20	*	*	*	26	2	26	0
	16	*	*	*	26	0	26	0
PUNTA GORDA	32	01-04	*	*	28	6	30	2
	28	*	*	*	28	1	30	1
	24	*	*	*	28	0	30	0
QUINCY 3 SSW	32	03-07	11-21	259	30	30	30	29
	28	02-16	12-07	294	30	26	30	23
	24	01-26	12-20	328	30	19	30	14
	20	01-10	12-26	350	30	11	30	8
	16	*	*	*	30	4	30	2
RAIFORD ST PRISON	32	02-22	12-01	282	24	24	24	21
	28	01-29	12-14	319	23	15	23	15
	24	01-11	12-24	347	21	7	23	8
	20	01-05	*	*	21	4	23	1
	16	*	*	*	21	0	23	0
ST AUGUSTINE	32	02-05	12-18	316	30	23	29	15
	28	01-22	12-25	337	30	17	29	8
	24	01-04	*	*	30	5	29	3
	20	*	*	*	30	2	29	0
	16	*	*	*	30	0	29	0

FREEZE DATA

STATION	Freeze threshold temperature	Mean date of last Spring occurrence	Mean date of first Fall occurrence	Mean No. of days between dates	Years of record Spring	No. of occurrences in Spring	Years of record Fall	No. of occurrences in Fall	STATION	Freeze threshold temperature	Mean date of last Spring occurrence	Mean date of first Fall occurrence	Mean No. of days between dates	Years of record Spring	No. of occurrences in Spring	Years of record Fall	No. of occurrences in Fall
ST LEO	32	01-24	12-20	330	30	18	30	15	TARPON SPGS SWG PLT	32	01-20	12-18	332	29	17	28	16
	28	01-09	12-27	352	30	9	30	6		28	01-08	⊕	⊕	29	9	28	4
	24	⊕	⊕	⊕	30	3	30	0		24	⊕	⊕	⊕	29	3	28	0
	20	⊕	⊕	⊕	30	1	30	0		20	⊕	⊕	⊕	29	0	28	0
	16	⊕	⊕	⊕	30	0	30	0		16	⊕	⊕	⊕	29	0	28	0
ST PETERSBURG	32	01-06	⊕	⊕	30	5	30	2	TAVERNIER	32	⊕	⊕	⊕	14	0	13	0
	28	⊕	⊕	⊕	30	1	30	0									
	24	⊕	⊕	⊕	30	0	30	0									
SANFORD EXP STA	32	01-21	12-24	337	29	16	30	11	TITUSVILLE 3 NW	32	01-28	12-23	329	27	18	28	10
	28	01-05	⊕	⊕	29	5	30	4		28	01-06	12-27	355	27	7	28	5
	24	⊕	⊕	⊕	29	1	30	0		24	⊕	⊕	⊕	27	2	28	1
	20	⊕	⊕	⊕	29	1	30	0		20	⊕	⊕	⊕	27	0	28	0
	16	⊕	⊕	⊕	29	0	30	0									
TALLAHASSEE WSO AP	32	02-26	12-03	280	30	29	30	23									
	28	02-04	12-16	316	30	25	30	16									
	24	01-18	12-22	338	30	15	30	11									
	20	01-06	12-27	356	30	7	30	7	WEST PALM BEACH	32	⊕	⊕	⊕	21	1	22	1
	16	01-02	⊕	⊕	30	5	30	0		28	⊕	⊕	⊕	21	0	22	0
TAMPA WSO AP	32	01-10	12-26	349	30	10	30	6									
	28	⊕	⊕	⊕	30	4	30	2									
	24	⊕	⊕	⊕	30	0	30	0									

Data in the above table are based on the period 1921-1950, or that portion of this period for which data are available.

⊕ When the frequency of occurrence in either spring or fall is one year in ten, or less, mean dates are not given.

Means have been adjusted to take into account years of non-occurrence.

A freeze is a numerical substitute for the former term "killing frost" and is the occurrence of a minimum temperature at or below the threshold temperature of 32°, 28°, etc.

Freeze data tabulations in greater detail are available and can be reproduced at cost.

*NORMALS BY CLIMATOLOGICAL DIVISIONS

Taken from "Climatography of the United States No. 81-4, Decennial Census of U. S. Climate"

TEMPERATURE (°F)   PRECIPITATION (In.)

STATIONS (By Divisions)	JAN	FEB	MAR	APR	MAY	JUNE	JULY	AUG	SEPT	OCT	NOV	DEC	ANN	JAN	FEB	MAR	APR	MAY	JUNE	JULY	AUG	SEPT	OCT	NOV	DEC	ANN
NORTHWEST																										
APALACHICOLA WSO	55.1	56.8	61.0	67.5	74.8	80.2	81.5	81.5	78.9	71.2	61.3	55.8	68.8	3.14	3.91	4.52	4.30	2.88	5.30	7.93	7.74	8.53	2.44	2.58	2.96	56.23
BLOUNTSTOWN	53.1	55.5	60.4	67.7	75.2	80.6	81.7	81.5	78.1	69.2	58.6	53.3	67.9	3.85	4.06	5.28	5.08	5.12	5.74	7.91	7.78	4.86	2.52	2.94	3.64	58.78
CARYVILLE														4.19	4.06	5.91	5.19	4.41	4.73	8.08	6.71	5.16	2.08	3.01	4.29	57.82
DE FUNIAK SPRINGS	54.6	56.5	61.4	68.6	75.8	81.0	81.8	81.9	78.3	70.1	59.5	54.5	68.7	4.45	4.63	6.45	5.69	4.93	5.82	9.39	8.06	6.49	2.45	3.34	4.61	66.31
MARIANNA SCH FOR BOYS	54.4	56.3	61.1	67.7	74.7	80.0	81.0	80.8	77.6	69.0	59.2	54.4	68.0	3.78	4.34	5.70	5.02	4.30	4.82	7.67	6.47	4.81	2.08	3.27	4.07	56.33
MONTICELLO 3 W	53.9	55.4	60.4	67.2	74.6	79.7	80.8	80.6	77.4	69.4	59.3	54.1	67.7	3.44	4.01	5.62	4.37	3.70	5.84	7.09	6.45	5.88	2.56	2.28	3.15	54.39
NICEVILLE														4.00	4.07	5.39	5.31	4.00	5.51	8.81	8.06	6.85	2.93	3.31	4.29	62.53
PANAMA CITY 2	55.2	56.9	61.3	67.9	75.1	80.7	82.0	82.0	79.1	71.0	60.8	55.6	69.0	3.56	4.08	5.32	4.65	3.02	4.46	8.21	7.90	6.67	2.70	3.30	4.14	58.01
PENSACOLA WSO	53.5	56.1	61.0	67.9	75.5	81.1	81.7	81.5	78.2	70.4	59.5	54.3	68.4	4.22	4.25	6.04	5.25	4.56	5.43	8.02	6.97	7.69	2.98	3.24	4.22	62.87
QUINCY 3 SSW	54.0	55.6	60.3	66.8	74.1	79.4	80.4	80.3	77.3	69.3	59.5	54.2	67.6	3.51	4.10	5.23	4.89	4.28	5.22	6.90	5.76	5.03	2.52	2.47	3.72	53.63
TALLAHASSEE WSO	53.9	55.6	60.6	67.5	74.9	80.2	81.3	81.1	78.1	69.6	59.2	54.1	68.0	3.42	4.18	5.18	4.64	4.10	6.54	8.05	6.93	5.51	2.43	2.44	3.44	56.86
DIVISION	53.9	55.8	60.5	67.2	74.5	79.9	81.1	81.1	77.9	69.6	59.4	54.3	67.9	3.79	4.12	5.54	4.92	4.16	5.45	7.96	7.09	6.15	2.60	2.92	3.85	58.55
NORTH																										
CEDAR KEY 1 WSW	58.3	60.0	64.0	70.5	77.4	81.9	82.5	82.7	81.1	74.5	65.6	59.8	71.5	2.47	2.81	3.62	2.95	2.02	4.19	8.08	7.40	6.38	3.07	1.38	2.19	46.56
CRESCENT CITY														2.00	3.10	3.93	3.65	3.74	6.71	6.99	7.58	6.95	4.55	1.99	2.54	53.73
FEDERAL POINT														2.27	2.98	3.77	3.28	5.78	7.54	7.39	7.33	5.49	1.82	2.42	53.20	
FERNANDINA BEACH	56.6	58.0	62.1	68.6	75.3	80.4	82.0	82.3	79.6	72.0	63.2	57.1	69.8	2.26	2.80	3.66	3.08	3.16	5.51	6.17	5.86	8.40	5.00	1.74	2.34	49.98
GAINESVILLE 2 WSW	58.0	59.6	63.6	69.5	75.8	80.1	81.1	81.2	79.2	72.1	63.7	58.5	70.2	2.60	3.27	4.11	3.72	3.48	6.64	7.49	7.69	4.95	4.23	1.75	2.52	52.45
GLEN ST MARY 1 W	55.5	57.4	61.6	67.7	74.1	79.2	80.8	80.8	78.0	69.9	60.9	55.4	68.4	2.83	3.51	4.22	3.18	3.63	6.96	8.65	6.59	6.32	4.20	1.99	2.89	54.97
JACKSONVILLE WSO	55.9	57.5	62.2	68.7	75.8	80.8	82.6	82.3	79.4	71.0	61.7	56.1	69.5	2.45	2.91	3.49	3.55	3.47	6.33	7.68	6.85	7.56	5.16	1.69	2.22	53.36
LAKE CITY 2 E	56.1	58.3	62.5	68.6	75.4	80.1	81.1	81.2	78.4	70.5	61.4	56.3	69.2	2.77	3.50	4.00	3.46	3.39	5.91	7.57	6.71	5.45	3.60	2.05	2.60	51.01
MADISON	55.4	57.0	61.7	68.4	75.7	80.5	81.4	81.2	78.2	70.2	60.6	55.4	68.8	2.96	3.85	4.85	3.96	2.85	5.51	7.32	6.65	5.58	2.89	2.20	2.94	51.57
PALATKA	59.2	61.0	65.0	70.6	76.8	81.2	82.5	82.5	80.3	73.2	64.8	59.4	71.4	2.17	3.27	3.95	3.52	3.34	6.54	7.54	7.27	7.13	4.92	1.77	2.44	53.86
SAINT AUGUSTINE	57.8	59.4	63.1	69.0	75.1	79.7	81.0	81.1	79.4	72.8	64.2	58.6	70.1	2.35	3.06	4.05	3.25	2.85	5.35	6.21	5.88	7.77	6.56	2.48	2.57	52.38
DIVISION	57.0	58.7	62.8	68.8	75.3	80.1	81.4	81.5	79.1	71.6	62.8	57.4	69.7	2.53	3.24	4.02	3.44	3.35	6.13	7.62	6.90	6.55	4.40	1.95	2.52	52.65
NORTH CENTRAL																										
BROOKSVILLE CHIN HILL	61.2	62.8	66.3	71.3	76.7	80.4	81.0	81.1	79.7	73.9	66.4	62.0	71.9	2.33	3.07	4.44	3.55	3.53	7.82	9.49	8.97	7.43	3.33	1.84	2.25	58.05
CLERMONT 6 SSW	61.6	63.1	66.8	71.9	77.5	81.1	82.1	82.5	80.8	74.8	67.1	62.1	72.6	2.00	2.63	3.87	3.74	3.38	7.10	8.79	6.55	6.49	3.11	1.53	2.04	51.23
DAYTONA BEACH WSO	59.2	60.6	63.8	68.9	74.2	78.4	80.1	80.7	79.2	73.4	65.3	60.3	70.3	1.96	2.75	3.56	2.97	2.85	5.81	6.74	6.37	7.00	5.61	2.33	1.95	49.90
DELAND 1 SSE	59.8	61.5	65.5	70.2	75.9	80.2	81.7	81.8	79.9	73.2	65.3	60.6	71.3	2.03	2.93	3.75	3.38	3.43	7.85	8.38	7.45	6.88	4.93	1.86	2.07	54.70
INVERNESS	59.8	61.8	65.3	70.6	76.4	80.5	81.6	81.9	80.1	73.3	65.0	60.2	71.4	2.22	3.10	4.44	3.16	3.67	7.49	9.75	8.86	6.65	3.46	1.48	2.07	56.35
ISLEWORTH														2.03	2.62	3.63	3.13	4.32	7.91	7.45	7.06	7.44	3.64	1.53	1.92	52.68
ORLANDO WSO	60.4	61.9	65.9	71.2	77.6	81.3	82.5	82.8	80.0	74.0	65.9	60.9	72.0	2.00	2.42	3.41	3.42	3.57	6.96	8.00	6.94	7.23	3.96	1.57	1.89	51.37
SAINT LEO	61.2	62.8	66.4	71.5	76.9	80.6	81.5	81.8	80.1	74.1	66.4	62.0	72.1	2.16	3.05	4.54	3.65	3.66	8.65	9.17	8.12	7.40	3.17	1.74	2.08	57.39
SANFORD EXP STATION	60.9	62.3	66.0	70.9	76.5	80.5	81.9	82.1	80.2	74.4	66.8	62.1	72.1	2.04	2.40	3.69	3.39	3.24	7.02	8.89	7.13	7.33	4.48	1.73	2.01	53.35
TITUSVILLE 3 NW	61.7	63.0	66.3	71.0	76.4	80.4	81.7	81.9	80.3	74.8	67.7	62.6	72.3	1.87	2.36	3.66	3.42	3.18	7.73	7.67	6.95	8.75	6.18	2.18	2.01	55.96
DIVISION	60.5	62.1	65.7	70.8	76.4	80.3	81.6	81.7	80.0	74.0	66.2	61.3	71.7	2.10	2.72	3.92	3.39	3.40	7.07	8.16	7.12	7.21	4.27	1.85	2.03	53.24
SOUTH CENTRAL																										
ARCADIA	62.9	64.4	67.8	72.1	76.9	80.2	81.5	82.0	80.5	75.0	67.9	63.8	72.9	1.97	2.50	2.97	3.09	3.67	8.50	8.91	7.18	8.37	3.74	1.49	1.60	53.99
AVON PARK	63.6	65.0	68.3	72.8	77.6	81.1	82.3	82.6	81.2	75.9	69.0	64.5	73.7	2.14	2.57	3.16	3.74	3.98	8.51	8.49	7.33	7.90	4.17	1.68	1.52	55.19
BARTOW	61.7	63.6	67.1	72.0	77.2	80.7	82.0	82.5	80.9	74.8	67.1	62.6	72.7	2.24	2.77	3.84	3.81	4.88	8.41	8.81	7.51	7.31	3.21	1.81	1.82	54.42
ELLSMERE 7 SSW	63.2	64.5	67.4	71.9	76.3	80.3	81.8	82.0	80.5	75.6	68.8	64.4	73.1	2.12	2.44	3.63	3.55	4.18	7.53	7.29	7.53	9.74	7.52	2.25	1.87	59.65
FORT PIERCE	64.8	65.7	68.4	72.6	76.7	80.0	81.6	81.9	81.0	76.7	70.4	66.3	73.8	1.90	2.44	3.49	4.32	4.19	6.07	5.23	6.01	8.46	8.27	2.75	2.14	55.27
KISSIMMEE NO. 2	60.9	62.6	66.2	71.2	76.4	80.2	81.4	81.8	80.0	74.2	66.7	61.9	72.0	1.91	2.44	4.03	3.34	3.61	7.75	8.03	6.83	7.25	3.97	1.74	1.90	52.80
LAKE ALFRED EXP STA	61.3	62.9	66.6	71.4	76.4	80.6	81.8	82.0	80.2	74.4	66.7	61.9	72.2	2.01	2.64	3.77	3.53	4.20	7.96	7.71	6.97	7.25	3.19	1.44	2.11	52.78
LAKELAND WSO	61.7	63.1	66.5	71.5	76.9	80.6	81.6	81.9	80.3	74.4	67.2	62.8	72.3	2.05	2.51	4.25	3.51	3.54	7.20	8.30	7.08	6.55	2.93	1.59	1.86	51.37
MOUNTAIN LAKE	61.9	63.3	66.9	71.6	76.5	80.2	81.4	81.9	80.3	74.4	67.2	62.8	72.4	2.20	2.55	3.76	3.32	4.00	7.78	8.55	7.75	6.93	3.66	1.59	1.72	53.81
PLANT CITY	61.5	62.9	66.1	70.8	75.7	79.7	81.0	81.4	80.1	74.0	66.6	62.4	71.9	2.20	2.87	4.26	3.61	3.44	8.00	8.93	8.58	7.51	3.22	1.65	2.04	56.31
ST PETERSBURG	63.3	64.8	67.9	73.1	78.4	81.8	82.7	83.0	81.7	76.3	68.9	64.3	73.9	2.46	3.01	3.67	3.21	2.58	6.32	9.22	8.96	8.37	3.85	1.66	2.10	55.41
TAMPA WSO	61.2	62.7	66.0	71.4	76.8	80.6	81.6	82.0	80.5	74.7	66.8	62.3	72.2	2.13	2.84	3.75	2.84	2.85	7.28	8.62	8.24	6.89	2.78	1.46	1.89	51.57
TARPON SPGS SEWAGE PL	61.1	62.8	65.9	71.1	76.6	80.8	82.0	82.3	81.0	74.9	66.9	62.0	72.3	2.46	2.94	4.25	3.36	2.38	6.00	9.20	9.22	7.68	3.27	1.66	2.14	54.56
DIVISION	62.4	63.8	67.0	71.8	76.8	80.4	81.6	82.0	80.6	75.1	67.9	63.4	72.7	2.09	2.64	3.59	3.38	3.54	7.27	8.14	7.63	7.79	4.14	1.77	1.86	53.84
EVERGLADES + SW COAST																										
BELLE GLADE EXP STA	63.2	64.4	67.0	70.8	74.8	78.5	80.1	80.5	79.5	75.2	68.8	64.6	72.3	1.81	1.71	3.17	3.29	4.50	9.02	8.17	8.46	9.19	5.12	2.48	1.75	58.67
EVERGLADES	66.6	67.6	70.1	74.0	77.6	81.1	82.3	83.0	82.1	78.3	72.2	67.8	75.2	1.70	1.48	2.22	2.59	5.18	8.86	8.48	7.34	9.50	4.50	1.46	1.38	54.69
FORT MYERS WSO	63.5	65.2	68.2	72.8	77.4	80.8	82.2	82.7	81.3	76.1	69.2	65.0	73.7	1.52	2.21	2.62	2.63	5.13	9.08	7.38	8.50	4.09	1.20	1.29	53.34	
LABELLE														1.56	2.07	2.99	3.25	4.01	8.28	8.23	7.45	7.64	4.06	1.23	1.40	52.17
MOORE HAVEN LOCK 1	63.4	64.7	67.8	72.0	76.4	79.8	81.1	81.7	80.6	76.1	69.1	64.8	73.1	1.62	1.80	2.78	3.31	4.46	6.86	7.78	6.92	7.72	4.38	1.33	1.38	50.34
DIVISION	64.7	65.9	68.8	72.9	77.0	80.3	81.7	82.2	81.2	76.8	70.3	66.0	74.0	1.63	1.91	2.65	2.98	4.32	8.29	8.18	7.47	8.47	4.39	1.53	1.42	53.24

*NORMALS BY CLIMATOLOGICAL DIVISIONS

Taken from "Climatography of the United States No. 81-4, Decennial Census of U. S. Climate"

TEMPERATURE (°F)   PRECIPITATION (In.)

STATIONS (By Divisions)	JAN	FEB	MAR	APR	MAY	JUNE	JULY	AUG	SEPT	OCT	NOV	DEC	ANN	JAN	FEB	MAR	APR	MAY	JUNE	JULY	AUG	SEPT	OCT	NOV	DEC	ANN
LOWER EAST COAST																										
FORT LAUDERDALE	67.8	68.4	70.7	74.3	77.5	80.4	81.8	82.6	81.5	77.9	72.6	69.0	75.4	2.20	2.06	2.84	4.19	5.29	7.42	5.96	6.88	8.98	8.39	3.18	2.90	60.29
HOMESTEAD EXP STA	65.6	66.5	69.2	72.8	75.9	79.2	80.2	80.7	80.0	76.3	70.7	66.9	73.7	1.75	1.71	2.38	3.69	6.46	8.77	8.81	8.29	10.61	8.72	2.28	1.22	64.69
MIAMI BEACH	69.1	69.6	71.6	74.9	78.2	81.1	82.3	82.9	81.7	78.4	73.8	70.3	76.2	1.68	1.65	1.95	2.92	4.54	5.63	4.45	5.06	7.36	6.71	2.53	1.78	46.26
MIAMI WSO	66.9	67.9	70.5	74.2	77.6	80.8	81.8	82.3	81.3	77.8	72.4	68.1	75.1	2.03	1.87	2.27	3.88	6.44	7.37	6.75	6.97	9.47	8.21	2.83	1.67	59.76
MIAMI 12 SSW	66.5	67.5	70.4	74.2	77.5	80.4	81.6	82.0	81.0	77.2	71.6	67.7	74.8	2.05	1.80	2.44	3.75	6.13	7.00	6.58	6.25	9.03	8.23	2.59	1.63	57.48
WEST PALM BEACH WSO	66.9	67.6	69.9	73.9	77.6	81.0	82.6	83.0	82.1	78.2	72.5	68.2	75.3	2.48	2.35	3.44	4.34	5.11	7.53	6.66	6.74	9.66	7.96	2.86	2.57	61.70
DIVISION	66.9	67.6	70.0	73.6	77.0	80.2	81.5	82.0	81.0	77.4	72.0	68.2	74.8	2.16	2.02	2.82	3.90	5.49	7.44	6.65	6.82	9.47	8.15	2.84	2.17	60.00
KEYS																										
KEY WEST WSO	69.6	70.4	72.5	75.8	79.0	81.8	83.3	83.6	82.3	79.0	74.1	70.6	76.8	1.53	1.98	1.77	2.48	2.73	3.97	4.16	4.33	6.73	5.82	2.80	1.69	39.99
KEY WEST	70.4	71.3	73.6	77.1	80.2	82.8	84.0	84.3	83.0	79.6	74.9	71.4	77.7	1.49	2.00	1.73	2.51	2.77	4.01	4.16	4.25	6.53	5.87	2.81	1.71	39.84
DIVISION	70.2	71.0	73.3	76.6	79.8	82.4	83.5	83.8	82.5	79.2	74.5	71.2	77.3	1.71	1.88	1.92	2.29	3.10	4.33	4.54	4.68	7.05	6.73	2.43	1.88	42.55

* Normals for the period 1931-1960. Divisional normals may not be the arithmetical average of individual stations published, since additional data for shorter period stations are used to obtain better areal representation.

CONFIDENCE - LIMITS

In absence of trend or record changes, the chances are 9 out of 10 that the true mean will lie in the interval formed by adding and subtracting the values in the following table from the means for any station in the State. Because of the wider variation in mean precipitation, the corresponding monthly means and annual mean must be substituted for "p" in the precipitation table below to obtain mean precipitation confidence limits.

| 1.3 | 1.3 | 1.1 | .5 | .5 | .4 | .4 | .3 | .4 | .5 | .8 | 1.2 | .4 | $39\sqrt{p}$ | $37\sqrt{p}$ | $44\sqrt{p}$ | $48\sqrt{p}$ | $47\sqrt{p}$ | $48\sqrt{p}$ | $38\sqrt{p}$ | $43\sqrt{p}$ | $53\sqrt{p}$ | $59\sqrt{p}$ | $44\sqrt{p}$ | $35\sqrt{p}$ | $44\sqrt{p}$ |

COMPARATIVE DATA

Data in the following table are the mean temperature and average precipitation for St. Leo's Abbey, Florida, for the period 1901-1930 and are included in this publication for comparative purposes.

| 60.3 | 61.7 | 66.3 | 70.6 | 75.8 | 79.2 | 80.5 | 80.7 | 79.2 | 73.2 | 65.4 | 60.3 | 71.1 | 2.87 | 2.54 | 2.90 | 2.20 | 4.44 | 8.19 | 8.22 | 8.48 | 6.91 | 3.70 | 2.20 | 2.51 | 55.16 |

NORMALS, MEANS, AND EXTREMES

APALACHICOLA, FLORIDA — POST OFFICE BLDG.

Station: APALACHICOLA, FLORIDA   Standard time used: EASTERN   Latitude: 29° 44' N   Longitude: 84° 59' W   Elevation (ground): 13 feet

Means and extremes above are from existing and comparable exposures. Annual extremes have been exceeded at other sites in the locality as follows:
Maximum monthly precipitation 27.73 in September 1924.

\# To R compass points only.
$ Through 1964.
% Through 1958.

DAYTONA BEACH, FLORIDA — REGIONAL AIRPORT

Station: DAYTONA BEACH, FLORIDA   Standard time used: EASTERN   Latitude: 29° 11' N   Longitude: 81° 03' W   Elevation (ground): 31 feet

Means and extremes above are from existing and comparable exposures. Annual extremes have been exceeded at other sites in the locality as follows:
Lowest temperature 18 in January 1940; maximum monthly precipitation 24.82 in October 1924; maximum precipitation 12.85 in October 1924.

NORMALS, MEANS, AND EXTREMES

Station: FORT MYERS, FLORIDA — PAGE FIELD

Standard time used: EASTERN Latitude: 26° 35' N Longitude: 81° 52' W Elevation (ground): 15 feet

Month	Daily max	Daily min	Monthly	Rec. highest	Year	Rec. lowest	Year	Heating degree days (Base 65°)	Precip. Normal total
J	74.8	52.2	63.5	88	1966	28	1966	146	1.52
F	76.3	54.8	65.2	92	1962	32	1967	101	2.21
M	79.6	56.0	67.8	94	1967	36	1943	62	2.62
A	84.1	61.1	72.8	95	1968	44	1968	0	2.64
M	88.6	66.1	77.4	98	1971	54	1971	0	3.85
J	90.5	71.0	80.8	97	1964	64	1971+	0	8.96
J	91.0	73.3	82.2	97	1968+	70	1968	0	9.38
A	91.6	73.8	82.4	96	1958	70	1950+	0	8.50
S	89.7	72.9	81.3	95	1963	70	1967	0	8.58
O	85.2	67.0	76.1	93	1964	52	1964	0	4.09
N	79.7	58.7	69.2	89	1951	34	1970	24	1.20
D	76.0	54.0	65.0	87	1967	26	1962	109	1.29
YR	83.9	63.4	73.7	98 MAY 1962		26 DEC. 1962		442	53.34

Means and extremes above are from existing and comparable exposures. Annual extremes have been exceeded at other sites in the locality as follows: Highest temperature 101 in July 1942; lowest temperature 24 in December 1894; maximum monthly precipitation 26.91 in June 1912; maximum precipitation in 24 hours 11.70 in June 1901. Maximum monthly snowfall T in February 1899; maximum snowfall in 24 hours T in February 1899.

Station: JACKSONVILLE, FLORIDA — INTERNATIONAL AIRPORT

Standard time used: EASTERN Latitude: 30° 30' N Longitude: 81° 42' W Elevation (ground): 26 feet

Month	Daily max	Daily min	Monthly	Rec. highest	Year	Rec. lowest	Year	Heating degree days (Base 65°)	Precip. Normal total
J	66.8	45.0	55.9	85	1947+	19	1971+	332	2.45
F	68.5	46.5	57.5	88	1962	19	1943	246	2.91
M	51.1...	51.1	62.2	91	1968	25	1943	174	3.55
A	79.4	57.8	68.7	95	1967	35	1944	21	3.47
M	85.6	65.1	75.8	100	1954+	45	1971	0	3.73
J	90.5	71.1	80.8	103	1954+	56	1966	0	6.33
J	92.0	73.2	82.6	105	1942	65	1965+	0	7.68
A	91.4	73.4	82.4	102	1954	64	1950+	0	6.85
S	87.0	71.4	79.4	100	1944	50	1967	0	7.55
O	80.2	62.1	71.0	96	1951	38	1957	12	5.16
N	72.2	51.2	61.7	89	1961	21	1970	144	1.69
D	66.7	45.5	56.1	84	1954	12	1962	310	2.22
YR	79.6	59.4	69.5	105 JUL. 1942		12 DEC. 1962		1239	53.36

Means and extremes above are from existing and comparable exposures. Annual extremes have been exceeded at other sites in the locality as follows: Lowest temperature 10 in February 1899; maximum monthly precipitation 23.32 in June 1932; maximum snowfall in 24 hours 1.9 in February 1899.

To 8 compass points only.

NORMALS, MEANS, AND EXTREMES

KEY WEST, FLORIDA — INTERNATIONAL AIRPORT

Standard time used: EASTERN Latitude: 24° 33' N Longitude: 81° 45' W Elevation (ground): 4 feet

Month	Normal Daily maximum	Normal Daily minimum	Normal Monthly	Extremes Record highest	Year	Extremes Record lowest	Year	Normal heating degree days (Base 65°)	Precipitation Normal total	Precipitation Max. monthly	Year	Min. monthly	Year	Max. 24 hrs.	Year	Snow Mean total	Max. monthly	Year	Max. 24 hrs.	Year
(a)	(b)	(b)	(b)	19	19	19	19	(b)	(b)	23		23		23		23	23		23	
J	73.7	65.4	69.6	85	1960+	46	1971	40	1.53	9.27	1953	0.03	1960	4.43	1958	0.0	0.0		0.0	
F	74.6	66.4	70.4	85	1952	47	1958	31	1.98	4.46	1966	0.02	1948	2.54	1965	0.0	0.0		0.0	
M	76.5	68.4	72.5	87	1949	53	1968	9	1.77	4.41	1971	T	1971	3.10	1953	0.0	0.0		0.0	
A	79.0	72.0	75.8	89	1965	55	1965	0	2.48	12.83	1959	0.00	1959	3.15	1950	0.0	0.0		0.0	
M	82.7	75.3	79.0	91	1953+	66	1960+	0	2.73	12.90	1960	0.12	1945	4.72	1960	0.0	0.0		0.0	
J	85.6	78.0	81.8	94	1952+	69	1952+	0	3.97	13.67	1966	0.90	1961	4.00	1966	0.0	0.0		0.0	
J	87.4	79.1	83.3	95	1951	69	1970	0	4.16	11.69	1970	0.54	1961	3.05	1970	0.0	0.0		0.0	
A	88.0	79.1	83.6	95	1957	68	1952	0	4.33	11.34	1945	2.25	1969	7.23	1952	0.0	0.0		0.0	
S	86.5	78.1	82.3	93	1962	70	1951	0	6.73	18.45	1963	1.70	1951	6.65	1963	0.0	0.0		0.0	
O	82.5	75.4	79.0	91	1959+	60	1957	0	5.82	21.57	1969	1.82	1963	8.47	1969	0.0	0.0		0.0	
N	78.1	70.7	74.5	88	1964+	46	1957	28	2.49	9.01	1959	0.13	1961	7.33	1959	0.0	0.0		0.0	
D	74.5	66.7	70.6	85	1961	46	1962+	57	1.69	4.84	1968	0.18	1968	4.60	1959	0.0	0.0		0.0	
YR	80.8	72.9	76.8	95	AUG.1957+	46	JAN.1971+	108	39.99	21.57	OCT.1969	0.00	APR.1959	8.89	MAY 1960	0.0	0.0		0.0	

BASED ON RECORDS FROM AIRPORT LOCATIONS FOR 1-1-45 TO 6-30-53 AND 7-1-57 TO DATE.

Means and extremes above are from existing and comparable exposures. Annual extremes have been exceeded at other sites in the locality as follows: Highest temperature 97 in August 1956; lowest temperature 41 in January 1886; maximum monthly precipitation 23.56 in October 1933; maximum precipitation in 24 hours 19.88 in November 1954.

\# To 8 compass points only.

LAKELAND, FLORIDA — CITY HALL

Standard time used: EASTERN Latitude: 28° 02' N Longitude: 81° 57' W Elevation (ground): 214 feet

Month	Normal Daily maximum	Normal Daily minimum	Normal Monthly	Extremes Record highest	Year	Extremes Record lowest	Year	Normal heating degree days (Base 65°)	Precipitation Normal total	Precipitation Max. monthly	Year	Min. monthly	Year	Max. 24 hrs.	Year	Snow Mean total	Max. monthly	Year	Max. 24 hrs.	Year
(a)	(b)	(b)	(b)	31	31	31	31	(b)	(b)	31		31		31		31	31		31	
J	71.2	52.2	61.7	85	1947+	25	1970	195	2.05	8.74	1948	0.12	1950	3.88	1945	T	T	1958	T	1958
F	72.8	53.3	63.1	88	1962+	28	1970+	146	2.51	6.59	1963	0.17	1945	3.92	1971	T	T	1951	T	1951
M	76.7	56.3	66.5	91	1943	30	1943	99	4.25	12.40	1960	0.78	1956	4.96	1960	0.0	0.0		0.0	
A	81.3	61.7	71.5	95	1968	40	1950	0	3.51	8.48	1967	0.00	1967	4.41	1941	0.0	0.0		0.0	
M	86.8	66.8	76.8	99	1945	54	1971+	0	3.54	14.86	1968	0.13	1953	4.36	1968	0.0	0.0		0.0	
J	89.9	71.2	80.6	100	1945	63	1949	0	7.20	15.67	1968	1.87	1948	10.12	1945	0.0	0.0		0.0	
J	90.5	72.6	81.6	101	1942	66	1947+	0	8.30	15.67	1960	3.09	1961	6.66	1960	0.0	0.0		0.0	
A	90.6	72.8	81.8	98	1961	63	1957	0	7.08	15.57	1948	2.04	1942	6.20	1942	0.0	0.0		0.0	
S	88.1	72.2	80.2	97	1944	61	1947	0	6.55	11.68	1947	1.26	1944	6.34	1964	0.0	0.0		0.0	
O	82.9	66.1	74.5	94	1959+	43	1957	0	2.93	6.72	1952	T	1942	4.99	1960	0.0	0.0		0.0	
N	75.9	57.8	66.9	89	1946+	28	1970	57	1.59	5.94	1941	0.25	1957	3.97	1963	0.0	0.0		0.0	
D	71.7	53.1	62.4	85	1971	20	1962	164	1.86	5.33	1941	0.09	1944	4.88	1944	T	T	1958+	T	1958+
YR	81.5	63.1	72.3	101	JUL.1942	20	DEC.1962	661	51.37	15.67	JUL.1960	0.00	APR.1967	10.12	JUN.1945	T	T	JAN.1958+	T	JAN.1958+

% Through 1964. The station did not operate 24 hours daily. Fog and thunderstorm data may be incomplete.

$ Through 1964.

\# To 8 compass points only.

NORMALS, MEANS, AND EXTREMES

Station: MIAMI, FLORIDA — INTERNATIONAL AIRPORT

Standard time used: EASTERN Latitude: 25° 48' N Longitude: 80° 16' W Elevation (ground): 7 feet

Temperature (°F)

Month	Daily maximum	Daily minimum	Monthly	Record highest	Year	Record lowest	Year	Normal heating degree days (Base 65°)
J	75.8	57.9	66.9	86	1971+	35	1971+	74
F	77.0	58.8	67.9	87	1967	36	1967	56
M	79.8	61.1	70.5	90	1971	37	1968	19
A	82.6	65.8	74.2	92	1971	46	1971+	0
M	85.4	69.7	77.6	93	1967	61	1967	0
J	88.0	73.5	80.8	94	1967	67	1967	0
J	88.8	74.7	81.8	96	1969	70	1965	0
A	89.7	74.9	82.3	96	1970	70	1967	0
S	88.0	74.6	81.3	96	1968+	70	1960+	0
O	84.7	70.9	77.8	90	1969+	56	1968	0
N	80.2	64.6	72.4	87	1969	40	1968	0
D	77.1	59.1	68.1	85	1968+	34	1968	65
YR	83.1	67.1	75.1	96	APR. 1971+	34	DEC. 1968	214

Precipitation (inches)

Month	Normal total	Maximum monthly	Year	Minimum monthly	Year	Maximum in 24 hrs.	Year
J	2.03	6.66	1951	0.04	1969	2.50	1958
F	1.87	6.55	1958	0.01	1966	5.73	1966
M	2.27	7.22	1949	0.02	1949	7.07	1949
A	3.88	10.21	1956	0.07	1960	5.18	1956
M	6.44	18.54	1968	0.44	1968	8.42	1968
J	6.75	22.36	1967	1.81	1945	7.63	1966
J	6.97	13.51	1947	1.77	1963	4.55	1952
A	9.47	10.88	1943	1.65	1954	4.92	1964
S	7.58	24.40	1960	2.63	1951	7.58	1960
O	9.21	21.08	1952	1.50	1962	9.95	1948
N	2.83	13.15	1959	0.13	1970	7.93	1959
D	1.67	6.39	1968+			4.38	1964
YR	59.76	24.40	SEP. 1960	0.01	FEB. 1944	9.95	OCT. 1948

Means and extremes above are from existing and comparable exposures. Annual extremes have been exceeded at other sites in the locality as follows:
Highest temperature 100 in July 1942; lowest temperature 28 in January 1940; maximum precipitation in 24 hours 12.58 in April 1942. Old City Office:
Lowest temperature 27 in February 1917; maximum precipitation in 24 hours 15.10 in November 1925.

Station: ORLANDO, FLORIDA — HERNDON AIRPORT

Standard time used: EASTERN Latitude: 28° 33' N Longitude: 81° 20' W Elevation (ground): 108 feet

Temperature (°F)

Month	Daily maximum	Daily minimum	Monthly	Record highest	Year	Record lowest	Year	Normal heating degree days (Base 65°)
J	70.7	50.0	60.4	86	1966	24	1966	220
F	72.0	50.7	61.9	88	1970+	28	1970+	165
M	76.7	55.0	65.9	92	1968	31	1968	105
A	81.7	60.8	71.2	96	1971	44	1971	0
M	87.1	67.2	77.4	97	1970+	54	1967	0
J	91.1	71.4	81.3	97	1970+	63	1965	0
J	92.0	73.0	82.5	98	1969	68	1965	0
A	92.0	73.5	82.8	98	1969	65	1964	0
S	88.6	73.5	80.0	96	1970	64	1964	0
O	74.0	65.3	74.0	94	1971	47	1968	72
N	75.6	56.2	62.6	88	1970	30	1970	198
D	70.6	51.2	60.9	86	1971	29	1971	766
YR	81.8	62.2	72.0	98	JUL. 1969	24	JAN. 1966	766

Precipitation (inches)

Month	Normal total	Maximum monthly	Year	Minimum monthly	Year	Maximum in 24 hrs.	Year
J	2.00	6.44	1948	0.15	1950	3.35	1964
F	2.50	6.77	1970	0.10	1944	4.88	1970
M	3.41	10.54	1960	0.16	1956	5.03	1960
A	3.42	6.18	1953	0.28	1967+	2.79	1953
M	3.57	8.58	1957	0.43	1961	3.14	1957
J	6.96	18.28	1948	1.97	1948	6.50	1945
J	8.00	19.57	1960	3.83	1960	8.19	1960
A	6.94	15.19	1953	3.20	1953	5.29	1953
S	7.23	15.17	1958	1.65	1958	7.74	1958
O	3.35	14.51	1950	0.35	1951	4.03	1950
N	1.77	6.39	1957	0.03	1967	3.52	1957
D	1.89	4.66	1969	T	1969	3.61	1969
YR	51.37	19.57	JUL. 1960	T	DEC. 1944	9.67	SEP. 1945

Ø For period December 1963 through the current year.
Means and extremes above are from existing and comparable exposures. Annual extremes have been exceeded at other sites in the locality as follows:
Highest temperature 103 in September 1921; lowest temperature 20 in December 1962.

NORMALS, MEANS, AND EXTREMES

Station: **PENSACOLA, FLORIDA**   **HAGLER FIELD**   Standard time used: **CENTRAL**   Latitude: **30° 28' N**   Longitude: **87° 12' W**   Elevation (ground): **112** feet

	Temperature							Normal heating degree days (Base 65°)	Precipitation Ø							Snow, Ice pellets						Relative humidity				Wind &					Pct. of possible sunshine	Mean sky cover sunrise to sunset	Mean number of days											Average daily solar radiation-langleys
	Normal			Extremes Ø																		Hour 00	Hour 06	Hour 12	Hour 18	Mean speed	Prevailing direction	Fastest mile					Sunrise to sunset			Precipitation		Thunderstorms	Heavy fog	Temperatures				
Month	Daily maximum	Daily minimum	Monthly	Record highest	Year	Record lowest	Year		Normal total	Maximum monthly	Year	Minimum monthly	Year	Maximum in 24 hrs.	Year	Mean total	Maximum monthly	Year	Maximum in 24 hrs.	Year		(Local time)						Speed	Direction #	Year			Clear	Partly cloudy	Cloudy	.01 inch or more	Snow, Ice pellets 1.0 inch or more			Max. 90° and above	Max. 32° and below	Min. 32° and below	Min. 0° and below	
(a)	(b)	(b)	(b)	8		8		(b)	(b)	8		8		8		8	8		8		6	6	6	6	8		6	6		5	5	5	5	5	8	8	2	8	8	8	8	8
J	62.3	44.6	53.5	78	1971	11	1966	400	4.22	11.83	1964	1.22	1968	3.69	1964	T	T	1964	T	1964	79	81	64	73	8.8		48	S	1965	48	6.9	6	6	19	11	0	1	6	0	*	7	0	
F	65.2	46.9	56.1	77	1967	19	1970	277	4.25	11.66	1966	2.78	1968	4.19	1967	T	T	1969+	T	1969+	75	79	57	66	9.4		41	E	1969	53	5.7	9	7	12	11	0	5	5	0	0	4	0	
M	70.2	51.8	61.0	83	1967	27	1968	183	6.04	9.26	1969	0.87	1967	4.33	1969	0.0	0.0		0.0		79	81	58	68	9.3		45	SE	1969	61	5.9	10	7	14	10	0	5	7	0	0	1	0	
A	76.5	59.3	67.9	87	1967+	37	1971	36	5.25	15.52	1964	0.67	1971	7.51	1964	0.0	0.0		0.0		84	86	58	69	9.1		57	SE	1966	63	6.1	8	8	14	7	0	3	4	0	0	0	0	
M	84.0	66.9	75.5	96	1964	51	1971	0	4.56	8.35	1969	0.30	1965	4.74	1970	0.0	0.0		0.0		83	85	56	65	8.3		39	E	1966	67	5.7	8	11	11	7	0	5	3	2	0	0	0	
J	89.3	72.8	81.1	96	1970*	57	1966	0	5.43	10.00	1970	2.00	1966	6.77	1970	0.0	0.0		0.0		83	86	60	69	7.1		51	W	1968	67	5.3	8	15	7	10	0	14	0	13	0	0	0	
J	89.2	74.1	81.7	98	1970	61	1967	0	8.02	13.98	1969	1.69	1970	3.50	1969	0.0	0.0		0.0		86	88	65	71	6.6		56	E	1966	57	6.9	3	14	14	15	0	15	1	13	0	0	0	
A	89.2	73.8	81.5	96	1964	63	1967	.0	6.97	13.09	1967	6.25	1966	3.05	1966	0.0	0.0		0.0		87	90	67	75	6.6		59	SW	1967	58	6.8	4	12	15	15	0	16	1	10	0	0	0	
S	85.9	70.5	78.2	97	1964	43	1967	0	7.69	10.28	1967	2.38	1970	10.02	1967	0.0	0.0		0.0		83	86	63	72	7.7		47	E	1965	60	5.7	8	10	12	8	0	8	1	4	0	0	0	
O	79.6	61.1	70.4	90	1971+	38	1968+	19	2.98	12.01	1970	0.00	1963	4.98	1967	0.0	0.0		0.0		79	83	55	68	8.0		54	E	1967	71	4.9	12	9	10	5	0	4	2	*	0	0	0	
N	69.3	49.7	59.5	84	1971	26	1970	195	3.24	5.34	1963	0.55	1965	2.96	1963	0.0	0.0		0.0		79	82	55	72	8.0		43	NW	1963	64	4.3	15	8	7	7	0	1	2	0	0	1	0	
D	63.2	45.3	54.3	80	1971+	24	1968+	353	4.22	6.53	1967	3.46	1965	4.52	1964	0.0	T	1963	T	1963	82	83	67	78	8.8		40	N	1968+	49	6.3	8	6	17	11	0	1	9	0	0	3	0	
YR	77.0	59.7	68.4	98	JUL. 1970	11	JAN. 1966	1463	62.87	15.52	APR. 1964	0.00	OCT. 1963	10.02	SEP. 1967	T	T	FEB. 1969+	T	FEB. 1969+	81	84	60	71	8.1		59	SW	1967	60	5.9	100	113	152	118	0	74	38	45	*	16	0	

Ø Extremes for period October 1963 through the current year.
Means and extremes above are from existing and comparable exposures. Annual extremes have been exceeded at other sites in the locality as follows: Highest temperature 103 in August 1947; lowest temperature 7 in February 1899; maximum monthly precipitation 21.43 in August 1935; maximum precipitation in 24 hours 17.07 in October 1934; maximum monthly snowfall 3.0 in February 1895; maximum snowfall in 24 hours 3.0 in February 1895; fastest mile of wind 114 from East in September 1926.

To 8 compass points only. Data through April 1969.

Station: **TALLAHASSEE, FLORIDA**   **MUNICIPAL AIRPORT**   Standard time used: **EASTERN**   Latitude: **30° 23' N**   Longitude: **84° 22' W**   Elevation (ground): **55** feet   Year: **1971**

	Temperature							Normal heating degree days (Base 65°)	Precipitation							Snow, Ice pellets						Relative humidity				Wind &					Pct. of possible sunshine	Mean sky cover sunrise to sunset	Mean number of days											Average daily solar radiation-langleys
	Normal			Extremes																		Hour 01	Hour 07	Hour 13	Hour 19	Mean speed	Prevailing direction	Fastest mile					Sunrise to sunset			Precipitation		Thunderstorms	Heavy fog	Temperatures				
Month	Daily maximum	Daily minimum	Monthly	Record highest	Year	Record lowest	Year		Normal total	Maximum monthly	Year	Minimum monthly	Year	Maximum in 24 hrs.	Year	Mean total	Maximum monthly	Year	Maximum in 24 hrs.	Year		(Local time)						Speed	Direction	Year			Clear	Partly cloudy	Cloudy	.01 inch or more	Snow, Ice pellets 1.0 inch or more			Max. 90° and above	Max. 32° and below	Min. 32° and below	Min. 0° and below	
(a)	(b)	(b)	(b)	11		11		(b)	(b)	11		11		11		10	11		11		10	10	10	10	10	2	12	12		10	10	10	10	10	10	10	10	10	10	10	10	4
J	65.1	42.7	53.9	80	1971+	11	1971	375	3.42	9.27	1964	0.40	1969	2.81	1963	T	T	1968+	T	1968+	86	87	59	72	8.1	N	46	23	1963		6.4	8	7	16	10	0	1	7	0	*	10	0	229
F	67.0	44.2	55.6	85	1962	14	1971	286	4.18	11.50	1964	2.43	1962	5.60	1964	T	T	1968+	T	1968+	84	87	54	65	8.6	N	40	09	1969+		5.9	8	7	13	10	0	2	4	0	0	8	0	305
M	72.2	49.0	60.6	90	1967	23	1971	202	5.18	11.49	1970	1.29	1967	7.16	1962	0.0	0.0		0.0		85	88	51	59	9.0	S	40	27	1964		5.8	9	7	15	9	0	4	5	0	0	4	0	380
A	79.0	55.9	67.5	95	1968	29	1971	36	4.64	7.14	1965	1.06	1968	4.73	1964	0.0	0.0		0.0		83	92	48	59	9.0	S	35	27	1961		5.4	9	12	9	7	0	4	2	0	0	0	0	463
M	86.4	63.3	74.9	98	1962	34	1971	0	4.10	8.23	1966	T	1965	3.27	1966	0.0	0.0		0.0		87	92	49	58	7.1	E	40	29	1961		5.5	4	12	9	7	0	7	5	10	0	0	0	495
J	90.5	69.9	80.2	100	1969+	54	1966	0	6.54	12.62	1965	2.96	1968	6.75	1966	0.0	0.0		0.0		88	90	54	65	6.8	S	44	03	1966		5.9	1	15	14	12	0	14	3	20	0	0	0	524
J	90.5	72.0	81.3	100	1966	57	1967	0	8.05	20.12	1964	4.87	1962	8.94	1964	0.0	0.0		0.0		91	92	61	75	5.9	SW	38	23	1963		6.8	2	17	12	19	0	21	2	21	0	0	0	418
A	90.3	71.8	81.1	99	1968	61	1969	0	6.93	10.75	1969	4.88	1969	3.39	1961	0.0	0.0		0.0		91	93	61	75	5.8	E	58	33	1962		6.3	4	16	11	15	0	18	2	21	0	0	0	407
S	87.2	68.9	78.1	99	1962	40	1967	0	5.51	15.92	1969	1.57	1971	9.47	1969	0.0	0.0		0.0		89	92	57	72	7.0	ENE	44	36	1961		5.9	4	16	10	10	0	7	2	15	0	0	0	389
O	80.6	58.6	69.6	93	1971	32	1968+	28	2.43	10.48	1964	T	1963	5.95	1964	0.0	0.0		0.0		88	91	51	72	7.4	N	30	34	1961		4.6	14	8	9	5	0	2	2	0	0	*	0	334
N	71.1	47.3	59.2	88	1961	13	1970	198	2.44	7.42	1962	0.88	1971	2.92	1962	0.0	0.0		0.0		87	89	51	75	7.0	N	37	29	1963		4.5	13	9	8	7	0	1	6	0	0	5	0	283
D	65.4	42.8	54.1	84	1971	10	1962	360	3.44	12.65	1964	2.44	1962	9.26	1964	T	T	1962	T	1962	87	88	57	76	7.2	N	32	15	1969		5.8	9	8	14	9	0	2	7	0	0	9	0	242
YR	78.8	57.2	68.0	100	JUN. 1969+	10	DEC. 1962	1485	56.86	20.12	JUL. 1965+	T	MAY 1965+	9.47	SEP. 1969	T	T	FEB. 1968+	T	FEB. 1968+	87	90	54	69	7.3	N	58	02	1962		5.7	98	136	131	119	0	83	52	90	*	37	0	372

Means and extremes above are from existing and comparable exposures. Annual extremes have been exceeded at other sites in the locality as follows: Highest temperature 104 in June 1933 and earlier; lowest temperature -2 in February 1899; maximum monthly precipitation 23.85 in September 1924; minimum monthly precipitation 0.00 in December 1889; maximum monthly snowfall 2.8 in February 1958; maximum snowfall in 24 hours 2.8 in February 1958.

NORMALS, MEANS, AND EXTREMES

Station: TAMPA, FLORIDA — INTERNATIONAL AIRPORT **Standard time used:** EASTERN **Latitude:** 27° 58' N **Longitude:** 82° 32' W **Elevation (ground):** 19 feet

Month	Normal Daily maximum	Normal Daily minimum	Normal Monthly	Record highest	Year	Record lowest	Year	Normal heating degree days (Base 65°)	Precipitation Normal total
J	71.3	51.0	61.2	84	1971	23	1971	202	2.13
F	72.8	52.6	62.7	88	1964	27	1967+	148	2.84
M	76.0	56.0	66.0	87	1971	35	1971	102	3.75
A	81.4	61.3	71.4	92	1965+	40	1971	0	2.84
M	87.0	66.6	76.8	96	1967	49	1971	0	2.85
J	89.7	71.7	80.7	97	1964	61	1971	0	7.28
J	89.7	73.4	81.6	97	1964	63	1970	0	8.62
A	90.3	73.7	82.0	97	1970	68	1969	0	8.24
S	88.7	72.3	80.5	95	1966	64	1967	0	6.89
O	83.8	65.6	74.7	90	1967+	40	1964	0	2.78
N	76.8	56.8	66.8	87	1971+	27	1970	60	1.46
D	72.5	52.1	62.3	83	1971	23	1971+	171	1.89
YR	81.6	62.8	72.2	97	AUG. 1970+	23	JAN. 1971+	683	51.57

Station: WEST PALM BEACH, FLORIDA — PALM BEACH INTERNATIONAL AP **Standard time used:** EASTERN **Latitude:** 26° 41' N **Longitude:** 80° 06' W **Elevation (ground):** 15 feet

Month	Normal Daily maximum	Normal Daily minimum	Normal Monthly	Record highest	Year	Record lowest	Year	Normal heating degree days (Base 65°)	Precipitation Normal total
J	75.5	58.3	66.9	86	1965	29	1970	87	2.48
F	76.5	58.3	67.6	87	1964	35	1967	64	2.35
M	78.0	61.1	69.9	92	1965	31	1968	31	3.44
A	81.1	65.4	73.9	99	1971	45	1971	0	4.34
M	84.5	69.2	77.6	95	1971	56	1971	0	5.11
J	87.6	72.7	81.0	96	1969	62	1965	0	6.66
J	90.8	74.3	82.6	96	1970	68	1965	0	6.74
A	91.4	74.8	83.0	92	1971+	68	1969	0	6.84
S	89.7	74.5	82.1	93	1971	68	1967	0	9.06
O	85.2	71.1	78.2	93	1971	46	1970	0	8.71
N	80.2	64.8	72.5	89	1971+	37	1970	6	2.86
D	76.9	59.4	68.2	85	1971	33	1968	65	2.57
YR	83.5	67.0	75.3	99	APR. 1971	29	JAN. 1970	253	61.70

Ø For period November 1963 through the current year.
Means and extremes above are from existing and comparable exposures. Annual extremes have been exceeded at other sites in the locality as follows: Highest temperature 98 in June 1952+; lowest temperature 18 in December 1962; maximum monthly snowfall 0.1 in February 1889; fastest mile wind 84 in September 1935.

Ø For period February 1964 through the current year.
Means and extremes above are from existing and comparable exposures. Annual extremes have been exceeded at other sites in the locality as follows: Highest temperature 101° in July 1942.

REFERENCE NOTES APPLYING TO ALL "NORMALS, MEANS, AND EXTREMES" TABLES

(a) Length of record, years, based on January data. Other months may be for more or fewer years if there have been breaks in the record. Climatological standard normals (1931-1960).

(b) Less than one half.

* Also on earlier dates, months, or years.

+ Trace, an amount too small to measure.

T Below zero temperatures are preceded by a minus sign.
The prevailing direction for wind in the Normals, Means, and Extremes table is from records through 1963.

Unless otherwise indicated, dimensional units used in this bulletin are: temperature in degrees F.; precipitation, including snowfall, in inches; wind movement in miles per hour; and relative humidity in percent. Heating degree day totals are the sums of negative departures of average daily temperatures from 65° F. Cooling degree day totals are the sums of positive departures of average daily temperatures from 65° F. Sleet was included in snowfall totals beginning with July 1948. The term "Ice pellets" includes solid grains of ice (sleet) and particles consisting of snow pellets encased in a thin layer of ice. Heavy fog reduces visibility to 1/4 mile or less.

Sky cover is expressed in a range of 0 for no clouds or obscuring phenomena to 10 for complete sky cover. The number of clear days is based on average cloudiness 0-3, partly cloudy days 4-7, and cloudy days 8-10 tenths.

Figures instead of letters in a direction column indicate direction in tens of degrees from true North; i.e., 09 - East, 18 - South, 27 - West, 36 - North. Resultant wind is the vector sum of wind directions and speeds divided by the number of observations. If figures appear in the direction column under "Fastest mile" the corresponding speeds are fastest observed 1-minute values.

‡ 70' at Alaskan station. The langley

MEAN ANNUAL PRECIPITATION, INCHES

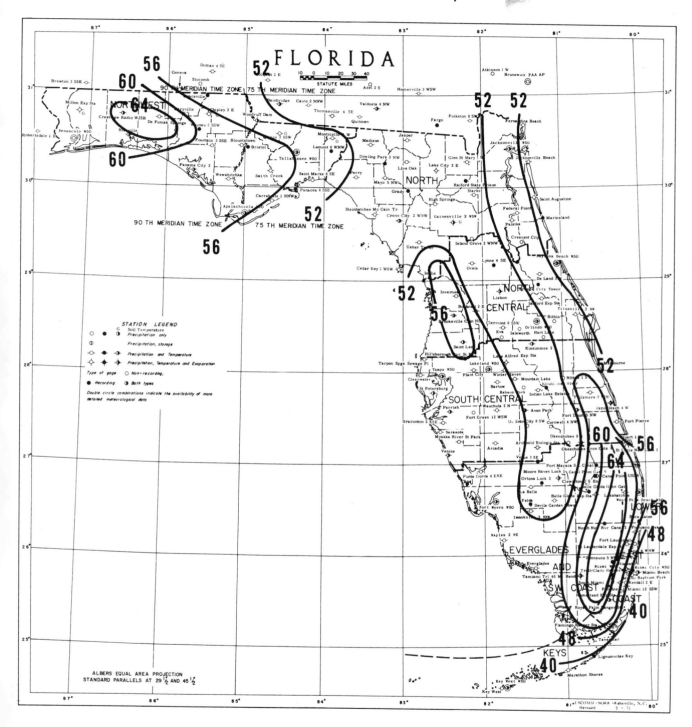

Data are based on the period 1931-55. Isolines are drawn through points of approximately equal value. Caution should be used in interpolating on these maps.

MEAN MAXIMUM TEMPERATURE (°F.), JANUARY

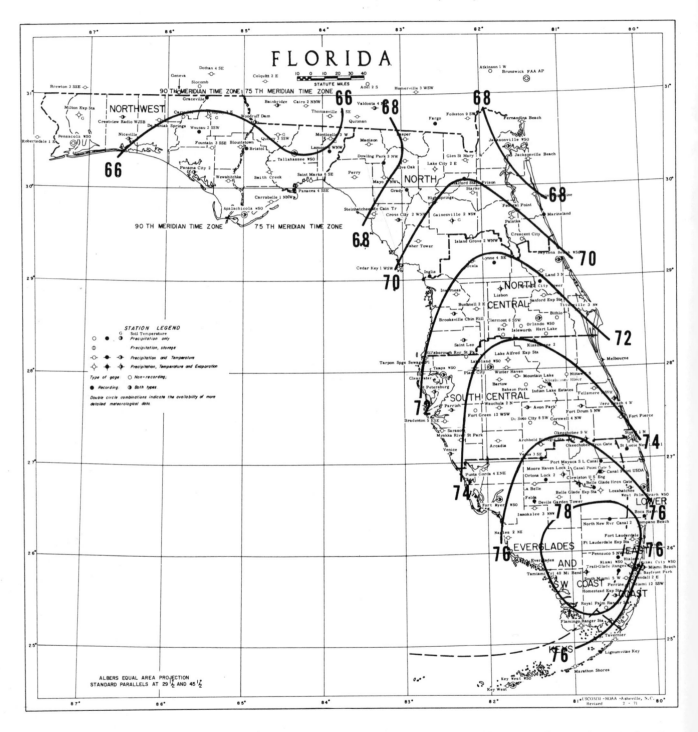

Data are based on the period 1931-52. Isolines are drawn through points of approximately equal value. Caution should be used in interpolating on these maps.

MEAN MINIMUM TEMPERATURE (°F.), JANUARY

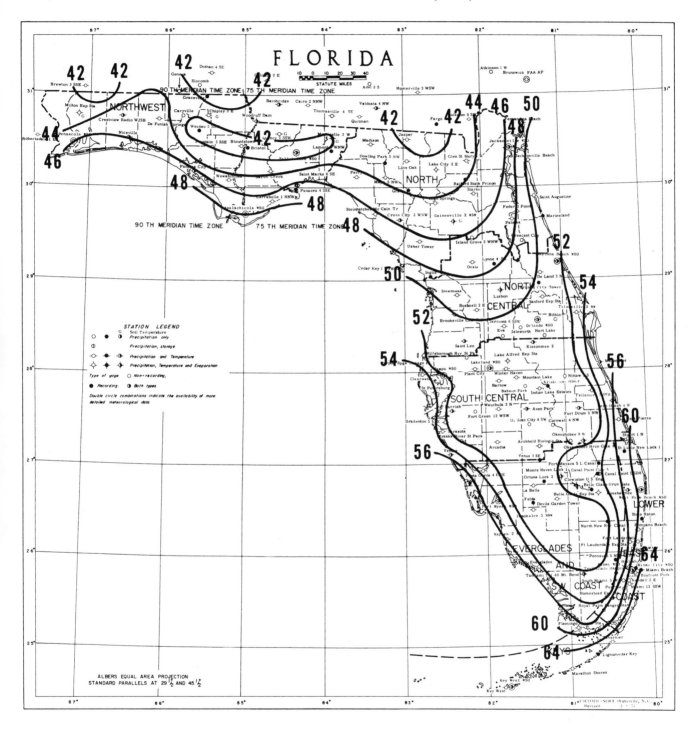

Data are based on the period 1931-52. Isolines are drawn through points of approximately equal value. Caution should be used in interpolating on these maps.

MEAN MAXIMUM TEMPERATURE (°F.), JULY

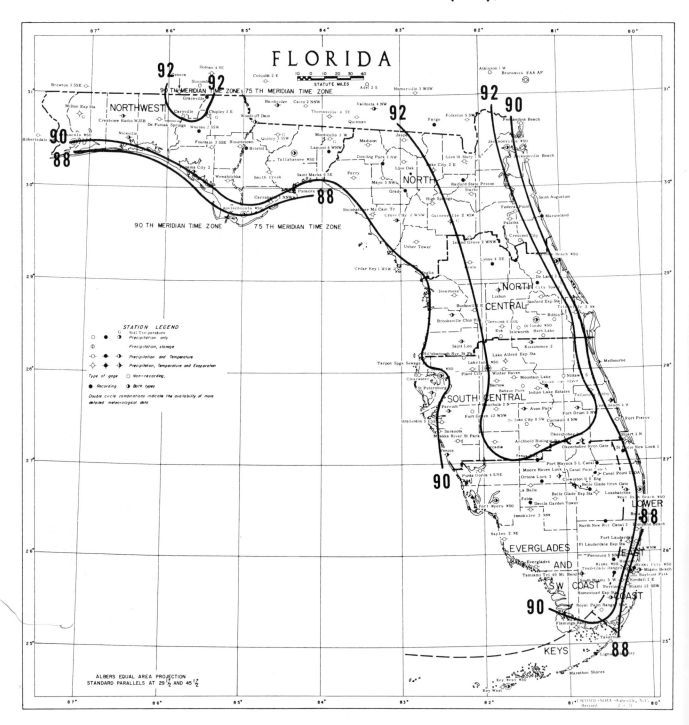

Data are based on the period 1931-52. Isolines are drawn through points of approximately equal value. Caution should be used in interpolating on these maps.

MEAN MINIMUM TEMPERATURE (°F.), JULY

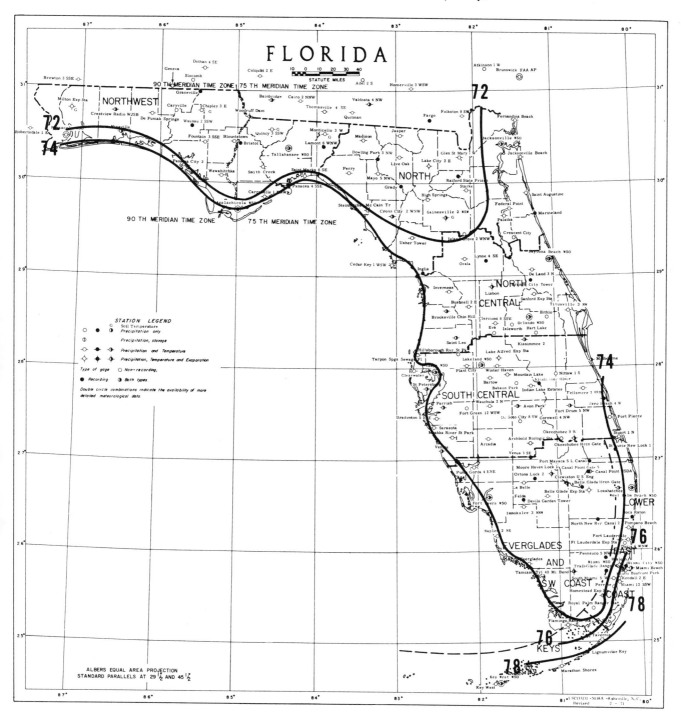

FLORIDA

Data are based on the period 1931-52. Isolines are drawn through points of approximately equal value. Caution should be used in interpolating on these maps.

FLORIDA

STATUTE MILES
10 0 10 20 30 40

NORTHWEST

90 TH MERIDIAN TIME ZONE | 75 TH MERIDIAN TIME ZONE

90 TH MERIDIAN TIME ZONE   75 TH MERIDIAN TIME ZONE

NORTH

NORTH CENTRAL

STATION LEGEND

○ ● ◑ Soil Temperature
          Precipitation only
◑   Precipitation, storage
◐ ◓ ◑ Precipitation and Temperature
◇ ◆ ◈ Precipitation, Temperature and Evaporation

Type of gage ○ Non-recording,

● Recording ◑ Both types

Double circle combinations indicate the availability of more
detailed meteorological data.

SOUTH CENTRAL

EVERGLADES

AND

SW COAST

LOWER

EAST

COAST

KEYS

ALBERS EQUAL AREA PROJECTION
STANDARD PARALLELS AT 29 ½ AND 45 ½

NOAA - NWS - Asheville, N.C.
Revised   2 - 71

THE CLIMATE OF
GEORGIA

by
Horace S. Carter

June 1969

Georgia is located roughly between latitudes 30° and 35° N. and longitudes 81° and 86° W. From north to south its length is 320 miles, and its maximum width is about 250 miles. With an area of almost 59,000 square miles, it is the largest State east of the Mississippi River. Its elevation ranges from near sea level along the southeast coast to almost 5,000 feet at its highest point in the northeast.

Georgia's land area is made up of four principal physiographic provinces: The Blue Ridge or Mountain Province, the Valley and Ridge Province, the Piedmont Province, and the Coastal Plain Province.

The Blue Ridge or Mountain Province is located in the northeastern part of the State. The terrain in this area is characterized by forest-covered mountains and narrow valleys with rapidly flowing streams. The average elevation of the area is less than 2,000 feet, but the higher mountains reach altitudes between 4,000 and 5,000 feet above sea level. There are no large cities in this section, but the many water power and reservoir sites make the area one of the most popular for recreation and water sports.

The Valley and Ridge Province, located in northwest Georgia, is composed of wide, flat, cultivated valleys separated by narrow, steep, wooded ridges that run more or less northeast-southwest. The elevation of the valleys ranges mostly between 500 and 800 feet above sea level, with the ridges rising to heights of 600 to 2,000 feet.

The Piedmont Plateau Province is a wide area extending from the foothills of the Appalachian Mountains to the Coastal Plain and comprising nearly one-third of the area of the State. The terrain is mostly hilly in the north to rolling in the south, where it merges with the Coastal Plain. Elevations range from near 1,200 feet in the north to less than 500 feet in the south. The soils of the Piedmont are predominantly sandy loams to clay loams and are well suited for the production of cotton, corn, small grains, and many other crops. Peaches are grown commercially in many sections of the Piedmont.

The boundary between the Piedmont Province and the Coastal Plain is called the Fall Line, because of the steep fall of rivers as they cross this boundary. The Fall Line extends across the State from west-southwest to east-northeast, following roughly a line from Columbus to Macon to Augusta. The Fall Line marks the head of navigation on the large rivers and is the site of waterpower dams at several places across the State.

The Coastal Plain Province includes all of Georgia south of the Fall Line and comprises about three-fifths of the total area of the State. The terrain is slightly rolling to level and ranges in altitude from near sea level along the coast to a maximum of 600 feet. The low-lying coastal sections are rather marshy and the large slow-moving streams are bordered by wide, swampy, densely wooded areas. Most of the soils of the Coastal Plain are sandy and are well adapted to a wide variety of agricultural products. Most of the State's tobacco, peanuts, and truck crops are grown in this area, as well as much cotton, corn, and small grains. The State's biggest pecan production and a sizable volume of peaches also come from the upper and middle Coastal Plain.

Georgia streams are divided into two main groups -- those flowing southeastward into the Atlantic and those flowing southward directly into the Gulf of Mexico, or indirectly into the Gulf through the Alabama-Mobile and Tennessee River systems. The Chattahoochee Ridge marks the dividing line between the parts of the State that are drained into the Atlantic and into the Gulf. It enters the State in the north at the boundary between Rabun and Towns Counties and runs southwestward through the City of Atlanta, the drainage of which is partly into the Gulf and partly into the Atlantic. From Atlanta, the divide runs south-southeastward through Fort Valley and then southeastward through the Okefenokee Swamp between the Suwanee and St. Mary River basins into Florida. The main streams in the Atlantic drainage system are the Savannah and Altamaha Rivers. The Savannah and its headwater stream form the boundary between Georgia and South Carolina throughout its entire length. The Altamaha drains a large area of central Georgia. The Chattahoochee and Flint River systems constitute the major streams of west Georgia, which drain directly into the Gulf of Mexico.

Water resources appear to be adequate for present uses. Among the larger users of water are the numerous steam and hydro-electric power plants and paper mills. Agricultural irrigation is becoming increasingly important as a consumptive use of water. The State's many streams and reservoirs are widely used for boating and fishing.

The greatest number and most damaging floods occur during winter and spring. The flood-producing rains are usually associated with slow-moving low pressure centers that pass through or near the State during these seasons. About half of the major river rises have occurred in March and April. Flooding has been relatively infrequent during the warm season, but occasionally major floods occur during this period as a result of tropical storms or prolonged heavy thunderstorm activity.

Georgia's climate is determined primarily by its latitude, the proximity of the Gulf of Mexico and the Atlantic Ocean, and by the altitude.

Average annual rainfall in Georgia ranges from more than 75 inches in the extreme northeast corner to about 40 inches in a small area of the East Central Division. From this driest part of the State, rainfall increases toward the south and southwest to an average of about 53 inches along the lower east coast and to about 54 inches in extreme southwest Georgia. Total rainfall varies greatly from year to year in all parts of the State, and most stations with several years of record show more than twice as much rain in their wettest year as in their driest. It is not at all unusual for these extreme variations to occur in successive years. The largest total rainfall recorded at a station in Georgia in one calendar year was 122.16 inches at Flat Top, in Fannin County, in 1959. On the other extreme, Swainsboro, in east-central Georgia measured only 17.14 inches in 1954. This is less than the 18.00 inches that fell at St. George in a 17-hour period on August 28, 1911, and considerably less than the record monthly total of 30.23 inches that was measured at Blakely during July 1916. The distribution of rainfall throughout the year is also highly variable in all parts of the State, but the extremes occur at different seasons in different areas. Most of the State shows two maxima and two minima in the annual rainfall curve. One maximum occurs in winter and early spring and the other in midsummer. The driest season for all the State is autumn, with most areas showing a secondary minimum about May. In the northern third of the State, the cool season rainfall maximum predominates, with either January or March normally the wettest month. This is due to the greater influence in that area of the cyclonic storms that move across the country with regularity during winter and early spring. The mountains of north Georgia add enough lift to the moist air that is drawn into the forward side of these storms from the Gulf to add materially to the total annual rainfall of the area. Most sections of central and south Georgia have their greatest rainfall in midsummer, with a secondary maximum about March. The lower east coastal area has its highest normal rainfall in September, due to the occasional extremely heavy rains that occur with late summer and autumn tropical storms. October is normally the driest month in most of the State, except in the southeast where November is usually drier. The mild, sunny weather of autumn is usually ideal for harvesting the many agricultural crops that are grown in the State.

In spite of the apparent abundance of rainfall in Georgia, irregular distribution results in the occurrence of damaging dry spells in some parts of the State almost every year. The use of irrigation to supplement natural rainfall for agricultural purposes has increased rapidly in recent years. Water supplies for this purpose come from streams and farm ponds throughout the State, and from artesian wells in the Coastal Plain area.

Snowfall is light in Georgia and of no significance at all in most of the State. Only in the extreme northern mountains is the average annual fall as much as 5 inches.

Due to its latitude and proximity to the warm waters of the Gulf of Mexico and Atlantic Ocean, most of Georgia has warm, humid summers and short, mild winters. However, in the northern part of the State, altitude becomes the more predominant influence with resulting cool summers and colder, but not severe, winters. All four seasons are apparent, but spring is usually short and blustery with rather frequent periods of storminess of varying intensity. In autumn long periods of mild, sunny weather are the rule for all of Georgia.

Average summer temperatures range from about 73° F. in the extreme north to nearly 82° F. in parts of south Georgia. There is little difference in summer averages over the southern two-thirds of the State, where they range between 80° and 82° F. Summer days are characteristically warm and humid in this area, with high temperatures exceeding 90° F. on most days and reaching 100° F. during most years. Temperatures usually drop to the middle or low 70's, or even below 70° F. by early morning, giving some relief from the daytime warmth. The flow of moist air from the Gulf over the warm land surface results in frequent afternoon thundershowers in south and central Georgia during summer. These showers not only provide most of the summer rainfall, but oftentime bring welcome relief from the afternoon heat. The highest temperature ever officially recorded in Georgia was 112° F. at Louisville on July 24, 1952. All parts of the State have experienced 100° F. weather at one time or another during the period of official records, but such occurrences are highly unusual in the mountain section of the north.

Winter temperatures show more variation from north to south than do those of summer. There is also a much greater variation in winter from day to day in all sections of the State. The average temperature for the three winter months ranges from 41° F. in the north to about 56° F. on the lower east coast, with the increase being almost uniform from north to south. All of Georgia experiences freezing temperatures almost every year, but the frequency of such occurrences varies greatly from the mountains to the coast. The average annual number of days with a temperature of 32° F. or less ranges from 110 in the north to about 10 in the lower coastal region. The lowest official temperature of record in Georgia is 17° F. below zero. This occurred at a CCC Camp in Floyd County on January 27, 1940.

Subzero weather has occurred as far south as Blakely in the west, Fitzgerald in central Georgia, and Waynesboro in the east. All sections of the State have experienced temperatures as low as 15° F. Georgia winters are characterized by frequent and sometimes large fluctuations in temperature. The cold snaps, which usually occur with regularity from mid-November to mid-March, alternate with longer periods of mild weather. Daytime temperatures almost always rise to above freezing in the southern three-fourths of the State, even during the coldest weather. There is approximately 4 months difference in the average length of the freeze-free growing season from north to south, ranging from about 170 days in the northernmost areas to near 300 days on the lower coast.

Relative humidity averages are moderately high in most of Georgia, as would be expected from its location in relation to the Gulf of Mexico and the Atlantic Ocean and from the high frequency of wind flow from the direction of these warm waters. Year-round averages at about 7:00 a.m. are approximately 85 percent, or slightly higher in the south. By 1:00 p.m. the average drops to about 55 percent, again being a little higher in some areas and a little lower in others. Monthly averages for both morning and afternoon are higher in summer than in other seasons in all sections of the State. The range from highest monthly average to the lowest monthly average is usually about 10 percent for both morning and afternoon readings.

Several tornadoes may be expected in Georgia each year, with resulting property damage in the thousands and sometimes millions of dollars. These storms have occurred during every month of the year, but have their highest frequency in spring. Approximately 50 percent of Georgia's tornadoes have occurred in March and April. During the 15-year period, 1953-1967, Georgia had an average of 18 reported tornadoes per year. About one-sixth of Georgia's tornadoes has resulted in property losses of $100,000 or more, and approximately 1 out of 4 has caused the death of one or more persons. Local windstorms, other than tornadoes, occur frequently in spring and early summer. These storms usually occur in connection with thunderstorms, the more severe of which may also produce hail. The southeast Georgia coast has been battered by hurricane winds on a few occasions; but, since most of these storms do not reach the State or move into the State after having traveled over land areas, they usually produce only moderate winds and heavy to copious rains. Most of the record rainfalls of south Georgia, including the State's 24-hour record total of

18 inches at St. George, occurred in connection with hurricanes. Tropical storm rainfall contributes materially to the precipitation normals for the late summer and fall months in southeast Georgia and to a lesser extent in other areas of the State.

Georgia has a great variety of recreational facilities. Its geographical changes from mountains to sea-coast enable it to have both winter and summer resorts. Good hunting and fishing abound throughout the State. Mountain trout, river perch, lake bass, and deepsea tarpon are all native to Georgia. Quail, wild turkey, deer, bear, and fox provide great hunting sport. Along the northern boundary is a 3,000 square mile region of forested mountains, deep lakes, and clear mountain streams. Accomodations are available here for hunting, fishing, or a cool summer vacation. Miles of beaches and island resorts are present along the Georgia sea coast, where mild winter weather prevails. Year-round use of these seaside attractions has produced extensive accomodations to supply recreation and entertainment at all price ranges. Georgia has more than 35 state parks, 8 wildlife refuges, and 6 national parks.

REFERENCES

(1) The Availability and Use of Water in Georgia-1956, by M. T. Thomson, S. M. Herrick, Eugene Brown and others, Department of Mines, Mining and Geology, Bulletin Number 65.

(2) Water in Georgia-1956. Prepared by the Georgia Water Use and Conservation Committee.

(3) Soils of Georgia - By R. L. Carter and Joel Giddens. Bulletin of the University of Georgia College of Agriculture Experiment Stations.

(4) Late Spring and Early Fall Freeze in Georgia-By Horace S. Carter, Bulletin N.S. 41 Georgia Agricultural Experiment Stations.

(5) Agricultural Drought in Georgia-1957 - By C. H. M. Van Bavel and John R. Carreker. Technical Bulletin N. S. 15, Georgia Agricultural Experiment Stations.

(6) Industrial Survey of Georgia -1958. Compiled by Georgia State Chamber of Commerce.

(7) Climate of the States - Georgia, 1941 Agricultural Yearbook Separate No. 1831.

(8) Tornadoes in Georgia - Mimeographed Release by Weather Bureau State Climatologist, Athens, Georgia.

(9) Floods in Georgia - By R. W. Carter, 1951, U. S. Geological Survey Circular 100.

(10) Precipitation in Georgia - 1963. By Horace S. Carter, Bulletin N. S. 102, Georgia Agricultural Experiment Stations.

(11) Relative Humidity in Georgia - 1966. By Horace S. Carter. Technical Memorandum No. 26, Weather Bureau Southern Region, ESSA.

(12) Georgia Temperatures - 1968. By Horace S. Carter, Unpublished Manuscript.

(13) Georgia Tropical Storms - 1968. By Horace S. Carter and Henry R. McQueen. Unpublished Manuscript.

BIBLIOGRAPHY

(A) Climatic Summaries of Resort Areas. (1) Warm Springs, Georgia and (2) The Golden Isles of Georgia.

(B) Climatic Summary of the United States (Bulletin W) 1930 edition, Sections 102 and 103. U. S. Weather Bureau.

(C) Climatic Summary of the United States, Georgia-Supplement for 1931 through 1952 (Bulletin W Supplement). U. S. Weather Bureau

(D) Climatic Summary of the United States, Georgia - Supplement for 1951 through 1960 (Bulletin W Supplement) Climatography of the United States, Series 86-9. U. S. Weather Bureau.

(E) Climatological Data - Georgia. U. S. Weather Bureau

(F) Climatological Data National Summary. U. S. Weather Bureau

(G) Climatological Summaries for Substations (41 locations).

(H) Hourly Precipitation Data - Georgia. U. S. Weather Bureau

(I) Local Climatological Data, U. S. Weather Bureau for Athens, Atlanta, Augusta, Columbus, Dawsonville, Macon, Rome, Savannah, and Thomasville, Georgia.

FREEZE DATA

STATION	Freeze threshold temperature	Mean date of last Spring occurrence	Mean date of first Fall occurrence	Mean No. of days between dates	Years of record Spring	No. of occurrences in Spring	Years of record Fall	No. of occurrences in Fall	STATION	Freeze threshold temperature	Mean date of last Spring occurrence	Mean date of first Fall occurrence	Mean No. of days between dates	Years of record Spring	No. of occurrences in Spring	Years of record Fall	No. of occurrences in Fall
ALAPAHA EXP STATION	32	03-13	11-11	243	30	30	30	29	BROOKLET 1 W	32	03-10	11-19	254	30	30	30	29
	28	02-23	11-27	277	30	29	30	25		28	02-12	12-02	293	30	28	30	28
	24	02-04	12-09	308	30	25	28	21		24	01-30	12-10	314	30	23	30	22
	20	01-17	12-21	338	29	15	30	13		20	01-17	12-25	342	30	16	30	10
	16	01-07	⊕	⊕	29	7	30	3		16	01-05	⊕	⊕	30	4	30	3
ALBANY 3 SE	32	02-26	11-25	272	30	29	30	29	BRUNSWICK	32	02-07	12-09	305	30	28	30	23
	28	02-11	12-07	299	30	24	30	23		28	01-25	12-18	327	30	18	30	18
	24	01-24	12-16	326	30	19	30	18		24	01-12	12-27	349	30	11	30	7
	20	01-13	12-25	346	30	13	30	8		20	⊕	⊕	⊕	30	3	30	1
	16	⊕	⊕	⊕	30	3	30	1		16	⊕	⊕	⊕	30	0	30	0
ALMA FAA AIRPORT	32	03-12	11-13	246	22	22	23	23	BRUNSWICK FAA AP	32	02-16	11-30	287	10	10	10	10
	28	02-24	11-27	276	22	22	23	23		28	01-28	12-08	314	10	8	10	9
	24	01-30	12-10	314	22	17	23	18		24	01-19	12-26	341	10	6	10	3
	20	01-17	12-19	336	22	12	23	11		20	⊕	⊕	⊕	10	1	10	1
	16	01-06	12-28	356	22	6	23	3		16	⊕	⊕	⊕	10	0	10	0
AMERICUS 4 ENE	32	03-15	11-15	245	30	30	30	29	CAIRO 2 NNW	32	03-06	11-24	263	10	10	10	10
	28	02-24	12-02	281	30	30	30	28		28	02-18	11-30	285	10	10	10	10
	24	02-03	12-11	311	30	24	30	22		24	01-28	12-09	315	9	7	10	8
	20	01-24	12-20	330	30	18	30	14		20	01-14	⊕	⊕	10	5	10	1
	16	01-09	12-27	352	30	10	30	6		16	⊕	⊕	⊕	10	0	10	0
ATHENS	32	03-29	11-10	226	30	30	30	30	CAMILLA	32	03-11	11-16	250	10	10	10	10
	28	03-11	11-23	257	30	30	29	28		28	02-19	11-30	284	9	9	10	10
	24	02-19	12-07	291	30	29	29	25		24	01-26	12-14	322	9	6	10	7
	20	02-05	12-14	312	30	23	30	18		20	01-16	12-23	341	9	5	10	3
	16	01-22	12-25	337	30	16	30	10		16	01-09	⊕	⊕	9	2	9	0
ATHENS WB AIRPORT	32	03-31	11-07	221	10	10	10	10	CAMP STEWART	32	03-15	11-16	246	19	19	18	18
	28	03-07	11-11	249	10	10	10	10		28	02-15	11-30	288	19	18	18	18
	24	02-12	11-26	287	10	10	10	10		24	01-24	12-15	325	19	14	18	13
	20	02-05	12-09	307	10	8	10	8		20	01-17	12-26	343	19	6	18	6
	16	01-22	12-26	338	10	5	10	4		16	01-03	⊕	⊕	19	2	19	0
ATLANTA WB AIRPORT	32	03-19	11-10	236	10	10	10	10	CARLTON BRIDGE	32	04-17	10-25	191	10	10	10	10
	28	03-01	11-25	269	10	10	10	10		28	04-06	10-31	208	10	10	10	10
	24	02-07	12-02	298	10	10	10	10		24	03-20	11-06	231	10	10	10	10
	20	01-31	12-21	324	10	7	10	5		20	03-05	11-16	256	10	10	10	10
	16	01-30	12-26	330	10	7	10	4		16	02-05	12-02	300	10	9	10	10
ATLANTA WB CITY	32	03-21	11-18	242	24	24	23	23	CARTERSVILLE	32	04-05	11-02	211	19	19	21	21
	28	03-05	11-29	269	24	24	23	22		28	03-21	11-12	236	19	19	21	21
	24	02-15	12-13	301	24	23	23	16		24	03-04	11-22	263	20	20	22	22
	20	02-02	12-19	320	24	19	23	13		20	02-09	12-08	302	19	17	22	18
	16	01-20	12-25	339	24	14	23	7		16	02-02	12-18	319	19	15	22	12
AUGUSTA WB AIRPORT	32	03-30	11-04	219	10	10	10	10	CLAYTON 1 W	32	04-15	10-26	194	28	28	29	29
	28	03-08	11-09	246	10	10	10	10		28	04-05	11-03	212	29	29	29	29
	24	02-22	11-25	276	10	10	10	10		24	03-19	11-17	243	29	29	28	28
	20	02-01	12-09	311	10	8	10	9		20	02-21	11-27	279	29	28	28	27
	16	01-19	12-22	337	10	6	10	5		16	02-13	12-12	302	29	26	28	19
AUGUSTA	32	03-13	11-17	249	29	29	30	30	COLUMBUS	32	03-10	11-20	255	25	25	24	23
	28	02-16	12-04	291	29	28	30	25		28	02-16	12-04	291	25	24	24	19
	24	02-04	12-14	313	29	24	30	19		24	01-28	12-16	322	25	16	24	14
	20	01-16	12-24	342	29	14	30	11		20	01-13	12-26	347	25	9	24	7
	16	01-04	⊕	⊕	29	3	30	2		16	01-03	12-28	359	25	3	24	3
BAINBRIDGE	32	03-10	11-14	249	30	30	30	30	COLUMBUS WB AIRPORT	32	03-27	11-06	224	10	10	10	10
	28	02-17	11-25	281	30	29	30	28		28	03-05	11-20	260	10	10	10	10
	24	01-29	12-12	317	30	21	30	20		24	02-10	11-30	293	10	10	10	10
	20	01-14	12-21	341	30	15	30	14		20	01-27	12-15	322	10	7	10	7
	16	01-04	⊕	⊕	30	5	30	3		16	01-14	12-27	347	10	5	10	3
BLAIRSVILLE EXP STA	32	04-26	10-15	172	28	28	29	29	CORDELE WATER WORKS	32	03-09	11-17	253	22	22	23	23
	28	04-16	10-29	196	28	28	29	29		28	02-18	12-02	287	22	21	23	22
	24	03-31	11-03	217	28	28	29	29		24	01-23	12-14	325	20	12	23	15
	20	03-18	11-14	241	28	28	30	29		20	01-14	12-25	345	20	8	22	5
	16	02-28	11-26	271	28	27	29	28		16	01-04	12-27	357	20	3	22	3
BLAKELY	32	03-08	11-21	258	30	30	30	29	CORNELIA	32	04-06	11-01	209	28	28	27	27
	28	02-14	12-06	295	30	28	30	25		28	03-21	11-08	232	28	28	27	27
	24	01-25	12-14	323	30	19	30	18		24	03-07	11-24	262	28	28	27	26
	20	01-19	12-24	339	30	15	30	10		20	02-21	12-03	285	28	27	27	25
	16	01-07	⊕	⊕	30	6	30	3		16	02-09	12-18	312	28	24	27	15

STATION	Freeze threshold temperature	Mean date of last Spring occurrence	Mean date of first Fall occurrence	Mean No. of days between dates	Years of record Spring	No. of occurrences in Spring	Years of record Fall	No. of occurrences in Fall
COVINGTON	32	03-30	11-05	220	28	28	29	29
	28	03-16	11-15	244	28	28	28	28
	24	02-23	11-26	276	28	27	29	28
	20	02-06	12-09	306	28	23	29	22
	16	01-21	12-21	334	28	15	28	11
CUTHBERT	32	03-09	11-23	259	14	14	15	15
	28	02-08	12-06	301	14	13	14	13
	24	01-27	12-08	315	14	8	14	11
	20	01-19	12-25	340	14	7	12	4
	16	01-08	12-26	352	14	3	12	3
DAHLONEGA	32	04-05	11-03	212	30	30	30	30
	28	03-22	11-16	239	30	30	30	29
	24	03-05	11-27	267	30	30	30	27
	20	02-15	12-05	293	30	27	30	24
	16	02-02	12-14	315	30	23	30	18
DALTON	32	04-05	11-02	211	24	24	26	26
	28	03-18	11-13	240	24	24	26	26
	24	02-25	11-25	273	24	24	26	25
	20	02-10	12-06	299	24	23	26	22
	16	01-30	12-16	320	24	17	26	15
DUBLIN 3 S	32	03-17	11-10	238	30	30	29	29
	28	02-21	11-23	275	30	30	30	30
	24	02-05	12-08	306	30	25	30	24
	20	01-22	12-20	332	30	18	29	14
	16	01-08	12-28	354	30	9	29	3
EASTMAN 1 W	32	03-09	11-20	256	30	30	27	26
	28	02-16	12-06	293	30	29	28	23
	24	02-02	12-11	312	30	25	28	20
	20	01-20	12-25	339	30	16	28	9
	16	01-08	⊕	⊕	30	8	28	2
EXPERIMENT	32	04-02	11-05	217	10	10	10	10
	28	03-13	11-15	247	10	10	10	10
	24	02-17	11-29	285	10	10	10	10
	20	02-05	12-08	306	10	8	10	9
	16	01-29	12-22	327	10	6	10	5
FAIRVIEW	32	03-24	11-13	234	22	22	22	22
	28	03-06	11-24	263	22	22	22	21
	24	02-13	12-13	303	21	19	22	15
	20	01-30	12-19	323	21	16	22	10
	16	01-15	12-26	345	21	10	22	6
FITZGERALD	32	03-06	11-20	259	30	30	27	26
	28	02-10	12-04	297	30	28	27	24
	24	01-28	12-13	319	30	21	27	18
	20	01-13	12-25	346	30	12	27	10
	16	01-06	⊕	⊕	30	5	27	1
FOLKSTON 9 SW	32	03-06	11-24	263	13	13	14	14
	28	02-08	12-06	301	13	12	15	12
	24	01-22	12-16	328	14	9	15	9
	20	01-14	⊕	⊕	14	7	14	1
	16	⊕	⊕	⊕	14	0	13	0
FORT GAINES	32	03-14	11-12	243	28	28	29	29
	28	02-18	11-27	282	28	27	29	27
	24	02-03	12-12	312	28	23	28	19
	20	01-22	12-20	332	28	16	28	13
	16	01-12	12-28	350	27	11	28	5
FORT VALLEY 2 E	32	03-17	11-13	241	30	30	30	30
	28	03-01	11-30	274	30	30	30	29
	24	02-06	12-12	309	30	25	30	20
	20	01-28	12-18	324	30	22	30	16
	16	01-14	12-27	347	30	13	30	7
GAINESVILLE	32	04-04	11-02	212	30	30	29	29
	28	03-20	11-12	237	30	30	29	29
	24	03-03	11-25	267	30	30	29	28
	20	02-13	12-08	298	30	28	29	24
	16	01-31	12-20	323	30	22	29	15
GLENNVILLE	32	03-07	11-21	259	30	30	30	29
	28	02-13	12-03	293	30	29	30	27
	24	01-23	12-13	324	30	19	30	21
	20	01-12	12-24	346	30	12	30	10
	16	⊕	⊕	⊕	30	3	30	2
GREENSBORO	32	03-27	11-08	226	29	29	29	29
	28	03-14	11-20	251	29	29	29	29
	24	02-26	11-28	275	29	29	29	26
	20	02-06	12-12	309	29	24	29	19
	16	01-25	12-23	332	29	18	29	10
SUNNYSIDE	32	03-26	11-13	232	30	30	30	30
	28	03-07	11-27	265	30	30	30	30
	24	02-16	12-08	295	30	30	30	25
	20	02-02	12-16	317	30	23	30	17
	16	01-25	12-25	334	30	18	30	10
HARTWELL	32	04-01	11-03	216	29	29	28	28
	28	03-18	11-13	240	29	29	28	28
	24	02-20	11-27	280	29	29	28	27
	20	02-11	12-12	304	29	27	28	20
	16	01-29	12-21	326	28	20	28	14
HAWKINSVILLE	32	03-14	11-12	243	30	30	30	30
	28	02-24	11-25	274	30	30	30	29
	24	02-07	12-08	304	30	26	30	23
	20	01-22	12-19	331	30	17	30	16
	16	01-09	12-27	352	30	10	30	5
HOGGARDS MILL	32	03-20	11-08	233	19	19	18	18
	28	03-01	11-23	267	19	19	18	17
	24	01-30	12-15	319	19	14	18	13
	20	01-16	12-19	337	19	10	18	8
	16	01-08	12-29	355	19	6	18	2
JASPER 1 NNW	32	04-07	11-02	209	20	20	20	20
	28	03-29	11-10	226	20	20	20	20
	24	03-03	11-27	269	20	20	20	19
	20	02-18	12-08	293	20	18	20	16
	16	02-02	12-18	319	20	16	21	12
LA FAYETTE	32	04-08	10-29	204	26	26	27	27
	28	03-28	11-06	223	25	25	27	27
	24	03-09	11-22	258	24	24	26	26
	20	02-19	12-01	285	24	22	25	23
	16	02-02	12-13	314	24	18	25	19
LA GRANGE	32	04-07	11-03	210	10	10	10	10
	28	03-15	11-17	247	10	10	10	10
	24	02-24	11-29	278	10	10	10	10
	20	02-09	12-05	299	10	8	10	9
	16	01-28	12-24	330	10	6	10	5
LOUISVILLE 3S	32	03-21	11-10	234	29	29	28	28
	28	02-27	11-19	265	29	29	29	29
	24	02-07	12-04	300	30	25	29	25
	20	01-23	12-17	328	30	19	29	17
	16	01-11	12-25	348	30	12	29	8
LUMBER CITY	32	03-17	11-09	237	11	11	15	15
	28	03-02	11-22	265	11	11	15	15
	24	02-10	11-30	293	10	10	15	13
	20	01-26	12-14	322	11	8	15	9
	16	01-15	12-23	342	10	5	15	4
MACON WB AIRPORT	32	03-12	11-07	240	10	10	10	10
	28	02-24	11-22	271	10	10	10	10
	24	02-03	12-02	302	10	7	10	6
	20	01-28	12-19	325	10	6	10	6
	16	01-12	⊕	⊕	10	4	10	1
MACON	32	03-14	11-18	249	22	22	22	22
	28	02-19	12-05	289	23	22	21	20
	24	02-03	12-18	318	23	19	22	11
	20	01-17	12-25	342	23	10	22	7
	16	01-10	⊕	⊕	23	7	22	2

FREEZE DATA

STATION	Freeze threshold temperature	Mean date of last Spring occurrence	Mean date of first Fall occurrence	Mean No. of days between dates	Years of record Spring	No. of occurrences in Spring	Years of record Fall	No. of occurrences in Fall
MILLEDGEVILLE	32	03-24	11-06	227	29	29	29	29
	28	03-09	11-18	254	29	29	29	29
	24	02-11	12-01	293	29	25	29	26
	20	01-30	12-14	318	29	22	28	17
	16	01-16	12-24	342	29	13	29	9
MILLEN 4 N	32	03-20	11-10	235	28	28	29	29
	28	03-02	11-19	262	28	28	30	30
	24	02-07	12-05	301	30	25	30	26
	20	01-23	12-18	329	30	18	29	16
	16	01-10	12-25	349	29	9	30	9
MONTICELLO	32	03-30	11-10	225	29	29	28	28
	28	03-19	11-18	244	28	28	28	27
	24	02-23	12-03	283	28	27	27	23
	20	02-11	12-10	302	26	23	26	18
	16	01-27	12-18	325	26	18	26	12
MOULTRIE 2 ESE	32	02-26	11-24	271	29	28	28	28
	28	02-06	12-07	304	29	27	28	21
	24	01-27	12-14	321	29	19	28	18
	20	01-11	12-28	351	28	11	28	5
	16	⊕	⊕	⊕	28	2	28	0
NEWNAN	32	03-24	11-11	232	30	30	29	29
	28	03-10	11-20	255	30	30	29	28
	24	02-17	12-05	291	30	29	29	25
	20	02-07	12-12	308	30	27	29	20
	16	01-24	12-22	332	30	18	29	13
QUITMAN	32	03-07	11-19	257	30	30	29	28
	28	02-13	12-05	295	30	26	28	22
	24	01-28	12-18	324	29	19	28	14
	20	01-13	12-24	345	29	12	28	9
	16	⊕	⊕	⊕	30	2	29	1
ROME	32	04-04	11-04	214	30	30	29	29
	28	03-15	11-15	245	30	30	29	29
	24	02-23	11-28	278	30	29	29	29
	20	02-07	12-10	306	30	26	29	22
	16	01-26	12-21	329	30	21	29	14
ROME WB AIRPORT	32	04-12	10-29	200	10	10	10	10
	28	03-27	11-03	221	10	10	10	10
	24	03-04	11-14	255	10	10	10	10
	20	02-17	11-23	279	10	10	10	10
	16	01-30	12-16	320	10	7	10	8
SAVANNAH USDA PL GRDN	32	03-29	11-04	220	10	10	10	10
	28	03-15	11-10	240	10	10	10	10
	24	02-21	11-27	279	10	10	10	10
	20	02-01	12-06	308	10	8	10	10
	16	01-14	12-24	344	10	5	10	3
SAVANNAH WB AP	32	02-27	11-29	275	30	30	30	27
	28	02-05	12-10	308	30	25	30	21
	24	01-22	12-17	329	30	18	30	16
	20	01-08	12-25	351	30	7	30	8
	16	⊕	⊕	⊕	30	1	30	1
TALBOTTON	32	03-23	11-11	233	29	29	25	25
	28	03-07	11-22	260	29	29	26	26
	24	02-16	12-07	294	29	27	28	21
	20	02-01	12-15	317	29	21	28	18
	16	01-18	12-26	342	29	15	28	8
TALLAPOOSA 2 N	32	04-17	10-25	191	30	30	30	30
	28	04-03	10-31	211	30	30	29	29
	24	03-10	11-14	249	30	30	29	29
	20	02-22	11-26	277	30	29	28	26
	16	02-08	12-09	304	30	26	28	22
THOMASVILLE 4 SE	32	03-03	11-25	267	30	30	30	27
	28	02-14	12-06	295	30	26	30	22
	24	01-23	12-14	325	30	19	30	18
	20	01-11	12-25	348	30	11	30	8
	16	⊕	⊕	⊕	30	2	30	1
TIFTON EXP STATION	32	03-16	11-16	245	10	10	10	10
	28	02-17	11-30	286	10	10	10	10
	24	02-03	12-04	304	10	8	10	10
	20	01-24	12-26	336	10	7	10	3
	16	01-07	⊕		10	2	10	0
TOCCOA	32	03-30	11-08	223	30	30	30	30
	28	03-15	11-22	252	30	30	29	29
	24	02-23	12-04	284	30	29	30	27
	20	02-09	12-09	303	30	26	30	23
	16	01-31	12-25	328	30	21	30	10
WARRENTON	32	03-23	11-12	234	30	30	30	30
	28	03-07	11-23	261	30	30	30	29
	24	02-13	12-06	296	29	27	30	26
	20	01-29	12-15	320	29	21	30	19
	16	01-15	12-26	345	29	12	30	8
WASHINGTON	32	03-27	11-10	228	28	28	28	28
	28	03-09	11-20	256	29	29	28	27
	24	02-16	12-04	291	29	27	28	24
	20	02-01	12-13	315	29	23	26	19
	16	01-18	12-23	339	29	14	26	10
WAYCROSS 4 NE	32	03-07	11-16	254	30	30	30	30
	28	02-19	11-27	281	30	29	30	28
	24	01-26	12-10	318	30	21	30	23
	20	01-11	12-23	346	30	11	30	11
	16	⊕	⊕	⊕	30	3	30	1
WEST POINT	32	03-28	11-06	223	30	30	30	30
	28	03-16	11-13	242	30	30	29	29
	24	02-20	11-28	281	30	30	29	28
	20	02-04	12-11	310	30	24	29	20
	16	01-19	12-22	337	30	16	29	12

Data in the above table are based on the period 1931-1960, or that portion of this period for which data are available.

⊕ When the frequency of occurrence in either spring or fall is one year in ten, or less, mean dates are not given.

Means have been adjusted to take into account years of non-occurrence.

A freeze is a numerical substitute for the former term "killing frost" and is the occurrence of a minimum temperature at or below the threshold temperature of 32°, 28°, etc.

Freeze data tabulations in greater detail are available and can be reproduced at cost.

*NORMALS BY CLIMATOLOGICAL DIVISIONS

Taken from "Climatography of the United States No. 81-4,   Decennial Census of U. S. Climate"

TEMPERATURE (°F)   PRECIPITATION (In.)

STATIONS (By Divisions)	JAN	FEB	MAR	APR	MAY	JUNE	JULY	AUG	SEPT	OCT	NOV	DEC	ANN	JAN	FEB	MAR	APR	MAY	JUNE	JULY	AUG	SEPT	OCT	NOV	DEC	ANN
NORTHWEST																										
RESACA	·	·	·	·	·	·	·	·	·	·	·	·	·	5.75	5.62	5.71	4.83	3.77	3.65	4.88	4.08	3.17	3.07	3.67	5.34	53.54
ROME	44.9	46.8	53.0	62.2	70.2	77.6	80.1	79.4	74.1	63.1	51.5	44.6	62.3	5.41	5.39	6.01	4.61	3.57	3.97	4.69	4.04	3.45	2.75	3.46	5.03	52.38
ROME WB AIRPORT	42.1	44.4	50.4	59.8	68.7	76.4	79.0	78.4	72.4	61.2	49.2	42.4	60.4	5.51	5.30	5.87	4.59	3.80	3.76	4.89	3.76	3.30	2.93	3.42	5.13	52.26
DIVISION	43.8	45.8	51.8	61.1	69.5	77.0	79.5	78.8	73.3	62.4	50.8	43.9	61.5	5.28	5.31	5.76	4.57	3.70	3.75	4.66	4.24	3.23	2.86	3.43	4.99	51.78
NORTH CENTRAL																										
ATHENS	46.0	47.5	53.1	62.1	70.4	77.4	79.3	78.6	73.5	63.2	52.7	45.7	62.5	4.83	4.64	5.20	4.44	3.55	3.74	4.88	3.55	2.69	2.97	2.84	4.44	47.77
ATHENS WB AIRPORT	44.6	46.1	51.6	61.1	70.1	77.6	79.7	79.0	73.5	63.0	51.5	44.6	61.9	4.89	4.70	5.10	4.39	3.61	3.88	4.99	3.63	3.02	2.86	2.93	4.53	48.53
ATLANTA WB AIRPORT	44.7	46.1	51.4	60.2	69.1	76.6	78.9	78.2	73.1	62.4	51.2	44.8	61.4	4.44	4.51	5.37	4.47	3.16	3.83	4.72	3.60	3.26	2.44	2.96	4.38	47.14
BLAIRSVILLE EXP STA	40.2	41.1	46.5	55.4	63.7	70.9	73.6	73.0	67.5	57.6	46.5	40.1	56.3	5.17	5.49	5.76	4.65	4.01	3.77	5.29	4.40	3.31	3.15	3.43	4.94	53.37
CANTON	·	·	·	·	·	·	·	·	·	·	·	·	·	5.16	5.22	5.99	4.61	3.50	3.59	5.15	4.09	3.28	2.78	3.41	5.18	51.96
DAHLONEGA	42.8	44.3	50.1	59.5	67.2	74.3	76.5	75.8	70.6	60.6	49.6	43.0	59.5	6.09	6.04	6.53	5.05	4.01	3.74	6.17	5.07	3.56	3.30	4.04	6.23	59.83
GAINESVILLE	43.6	45.1	51.0	60.2	68.2	75.6	77.9	77.3	71.8	61.4	50.4	43.5	60.5	5.11	5.07	5.71	4.55	3.72	3.78	5.33	3.97	3.38	3.14	3.45	5.11	52.32
DIVISION	43.9	45.3	51.2	60.3	68.5	75.8	77.9	77.3	72.0	61.8	50.7	43.9	60.7	5.07	5.18	5.57	4.65	3.65	3.75	5.10	4.00	3.23	2.99	3.26	4.96	51.31
NORTHEAST																										
CARLTON BRIDGE	44.0	45.5	51.0	60.5	69.1	77.0	79.4	78.7	73.1	61.8	50.5	43.5	61.2	4.66	4.47	5.00	4.09	3.41	3.76	4.33	3.62	2.87	2.87	2.85	4.48	46.41
CLAYTON 1 W	41.9	42.8	48.2	56.8	64.5	71.9	74.3	73.6	68.3	58.8	48.3	41.8	57.6	6.54	6.47	7.18	5.98	4.55	4.88	6.48	6.08	4.64	4.22	4.65	6.45	68.12
CORNELIA	42.9	44.4	50.0	59.3	67.0	74.3	76.7	76.0	70.7	60.8	50.0	43.0	59.6	5.67	5.56	5.91	5.03	3.92	4.32	5.88	4.99	3.73	3.45	3.97	5.75	58.18
HARTWELL	45.4	46.8	52.5	61.8	70.4	78.0	79.8	78.9	73.8	63.5	52.4	45.1	62.4	4.73	4.56	5.25	4.16	3.16	3.35	4.97	3.99	3.62	3.01	3.34	4.59	48.73
TOCCOA	45.2	46.6	52.3	61.4	69.3	76.7	78.6	77.8	72.3	62.4	52.1	45.1	61.7	5.41	5.55	6.29	4.53	3.90	4.04	5.69	5.46	4.09	3.49	3.81	5.55	57.81
WASHINGTON	46.7	48.6	54.2	63.0	71.2	78.4	80.3	79.9	75.0	64.8	53.6	46.5	63.5	4.33	4.29	4.98	4.11	3.26	3.78	5.17	3.86	3.11	2.58	2.71	3.86	46.04
DIVISION	44.3	45.8	51.4	60.5	68.6	76.1	78.2	77.5	72.2	62.0	51.1	44.2	61.0	5.22	5.15	5.76	4.63	3.71	4.02	5.42	4.67	3.68	3.27	3.57	5.11	54.21
WEST CENTRAL																										
COLUMBUS WB AIRPORT	47.8	49.7	55.3	63.7	72.1	79.4	81.1	80.7	76.0	65.4	54.1	47.7	64.4	4.06	4.63	6.01	4.59	3.50	3.68	5.87	4.64	2.77	1.77	2.59	4.56	48.67
CONCORD	·	·	·	·	·	·	·	·	·	·	·	·	·	4.26	4.53	6.12	4.73	3.37	3.87	5.55	3.73	3.16	2.15	2.78	4.42	48.67
GOAT ROCK	·	·	·	·	·	·	·	·	·	·	·	·	·	4.07	4.19	5.74	4.85	3.24	3.29	5.23	3.67	2.96	1.74	2.79	4.38	46.15
MONTEZUMA	·	·	·	·	·	·	·	·	·	·	·	·	·	3.78	4.29	5.19	4.38	3.02	3.54	5.80	4.42	3.35	2.35	2.44	4.19	46.75
NEWNAN	45.9	47.9	53.6	62.2	70.7	77.6	79.4	78.9	73.9	64.0	52.9	46.1	62.8	4.95	4.90	5.81	4.86	3.41	3.95	5.16	4.38	3.22	2.38	3.15	4.68	50.85
SUNNYSIDE	47.6	49.2	54.7	63.4	71.3	77.9	79.5	79.2	74.7	65.5	54.7	47.7	63.8	4.53	4.67	5.92	4.65	3.11	4.15	5.34	4.08	3.35	2.24	2.82	4.39	49.25
TALLAPOOSA 2 N	43.8	45.6	51.1	59.7	67.5	74.7	77.2	76.6	71.2	60.4	49.7	43.7	60.1	4.86	4.84	5.63	4.89	3.61	3.78	4.90	4.04	3.19	2.47	3.10	4.75	50.06
WEST POINT	47.7	49.3	55.0	63.2	71.4	78.4	80.2	79.6	74.6	64.4	53.6	47.3	63.7	4.43	4.87	6.19	4.98	3.64	3.74	5.74	4.63	3.30	2.28	3.08	4.78	51.66
WOODBURY	·	·	·	·	·	·	·	·	·	·	·	·	·	4.19	4.60	5.80	4.66	3.42	3.88	5.41	4.03	3.62	2.17	2.73	4.54	49.05
DIVISION	47.2	49.0	54.6	63.0	71.2	78.0	79.7	79.2	74.3	64.4	53.5	47.1	63.4	4.44	4.69	5.91	4.67	3.42	3.90	5.58	4.24	3.33	2.22	2.84	4.56	49.80
CENTRAL																										
COVINGTON	45.6	47.5	53.1	62.1	70.7	78.0	80.0	79.4	74.1	63.6	52.4	45.5	62.7	4.72	4.57	5.52	4.26	3.32	3.54	4.44	3.43	2.83	2.48	2.99	4.33	46.43
DUBLIN 3 S	50.4	52.2	57.6	65.6	73.5	80.0	81.6	81.0	76.5	66.5	56.2	49.9	65.9	3.58	4.14	4.64	3.75	3.59	3.86	5.18	4.47	3.58	2.67	2.08	3.80	45.34
EASTMAN 1 W	51.3	52.9	58.3	66.4	74.6	81.0	82.3	81.7	77.2	67.6	57.4	51.0	66.8	3.36	4.00	5.08	4.42	3.40	3.72	6.03	5.21	3.68	2.55	2.15	3.76	47.36
FORT VALLEY 2 E	48.4	50.2	55.9	64.4	72.8	79.4	80.6	80.3	75.5	65.6	54.6	48.1	64.7	3.70	4.57	5.36	4.26	3.44	4.32	5.57	3.94	3.39	2.58	2.69	4.21	48.03
HAWKINSVILLE	50.7	52.2	57.8	65.4	73.5	80.0	81.7	81.4	77.0	67.0	56.5	50.3	66.1	3.97	4.33	4.93	4.17	3.49	4.23	5.90	4.17	3.76	2.33	2.28	3.99	47.55
MACON WB AIRPORT	49.2	51.1	56.9	65.6	73.9	80.7	81.9	81.3	76.4	66.2	55.4	49.0	65.6	3.37	4.26	4.94	3.75	3.32	3.34	5.64	4.18	2.77	2.01	2.45	4.05	44.08
MILLEDGEVILLE	47.1	48.8	54.5	63.1	71.7	79.0	80.8	80.2	75.2	64.6	53.6	46.6	63.8	3.87	4.35	4.82	3.80	3.72	3.31	5.44	3.89	3.01	2.03	2.59	4.07	44.90
SILOAM	46.5	48.2	53.8	62.6	70.8	78.3	80.1	79.4	74.4	64.3	53.2	46.3	63.2	4.52	4.16	5.40	4.29	4.01	3.17	4.91	4.30	3.28	2.45	2.76	4.21	47.46
SPARTA 2 NNW	·	·	·	·	·	·	·	·	·	·	·	·	·	3.71	4.20	4.89	3.94	3.51	3.28	4.67	3.82	3.16	2.17	2.47	3.89	43.71
DIVISION	48.5	50.2	55.8	64.2	72.4	79.3	80.9	80.4	75.6	65.5	54.8	48.2	64.7	3.93	4.31	5.13	4.11	3.55	3.71	5.34	4.15	3.37	2.36	2.51	4.10	46.57
EAST CENTRAL																										
AUGUSTA WB AIRPORT	47.6	49.3	54.9	63.2	71.6	78.9	80.7	80.2	75.3	65.2	54.1	47.2	64.0	2.99	3.52	4.17	3.56	2.99	3.04	4.50	3.89	3.00	2.01	2.18	3.33	39.18
BROOKLET 1 W	51.7	53.5	58.9	66.8	74.0	79.8	81.3	80.9	76.8	67.7	58.2	51.3	66.7	2.55	3.49	3.81	3.51	3.38	4.44	5.57	5.42	4.56	2.28	2.03	2.78	43.82
DOVER	·	·	·	·	·	·	·	·	·	·	·	·	·	2.68	3.80	4.13	3.49	3.19	4.62	5.57	5.39	4.71	2.43	2.20	2.96	45.17
LOUISVILLE 3S	49.8	51.4	56.9	65.2	73.2	80.0	81.7	81.1	76.0	66.1	55.9	49.2	65.5	3.33	4.08	4.54	3.67	3.42	3.52	4.55	4.34	3.37	2.74	2.16	3.61	43.33
MIDVILLE	·	·	·	·	·	·	·	·	·	·	·	·	·	3.11	3.76	4.27	3.56	3.71	3.60	4.96	4.42	3.83	2.18	2.07	3.29	42.76
MILLEN 4 N	50.7	52.4	55.9	65.9	73.7	80.4	82.0	81.7	76.9	67.0	56.8	50.0	66.3	3.20	3.86	4.23	3.57	3.68	4.48	4.61	4.47	4.14	2.48	2.17	3.72	44.61
WARRENTON	47.8	49.4	55.0	63.6	71.7	78.7	80.4	79.6	74.7	64.6	54.2	47.7	64.0	4.06	4.38	4.77	4.13	3.19	3.29	4.41	4.02	3.55	2.67	2.58	3.99	45.04
DIVISION	49.9	51.5	57.0	65.3	73.2	79.8	81.5	80.9	76.2	66.3	56.1	49.4	65.6	3.17	3.87	4.34	3.65	3.36	3.76	4.88	4.48	3.80	2.46	2.20	3.46	43.43

* Normals for the period 1931-1960. Divisional normals may not be the arithmetical averages of individual stations published, since additional data for shorter period stations are used to obtain better areal representation.

*NORMALS BY CLIMATOLOGICAL DIVISIONS

Taken from "Climatography of the United States No. 81-4, Decennial Census of U. S. Climate"

TEMPERATURE (°F)   PRECIPITATION (In.)

STATIONS (By Divisions)	JAN	FEB	MAR	APR	MAY	JUNE	JULY	AUG	SEPT	OCT	NOV	DEC	ANN	JAN	FEB	MAR	APR	MAY	JUNE	JULY	AUG	SEPT	OCT	NOV	DEC	ANN
SOUTHWEST																										
ALBANY 3 SE	52.6	54.5	60.0	67.4	75.3	81.4	82.5	82.2	78.2	68.9	58.3	52.5	67.8	3.46	4.22	5.21	4.59	3.87	4.05	5.69	4.81	3.56	2.11	2.52	3.75	47.84
AMERICUS 4 ENE	50.5	52.2	57.4	65.4	73.5	79.9	81.0	81.0	76.8	67.3	56.6	50.4	66.0	4.03	4.56	5.38	4.37	3.39	4.40	5.85	4.34	3.49	2.15	2.55	4.31	48.82
BAINBRIDGE	53.2	55.1	60.2	67.2	74.8	80.6	81.6	81.3	77.8	68.7	58.1	53.3	67.7	3.81	4.04	5.27	4.83	3.77	4.60	6.75	5.50	5.00	2.01	2.37	3.84	51.79
BLAKELY	52.3	54.0	59.0	66.2	74.1	80.1	81.1	80.9	77.2	68.4	57.7	52.1	66.9	4.01	4.80	5.78	5.14	3.90	3.98	6.93	6.07	4.39	1.90	2.91	4.40	54.21
CAIRO 2 NNW	.	.	.	.	.	.	.	.	.	.	.	.	.	3.52	3.88	5.28	4.96	3.49	4.84	6.73	6.09	4.63	2.24	2.35	3.30	51.31
FORT GAINES	51.7	53.4	58.6	65.8	73.8	80.1	81.2	81.0	76.9	67.8	57.2	51.5	66.6	4.25	4.87	5.34	4.91	3.70	4.25	6.44	5.75	4.24	1.85	2.68	4.79	53.07
THOMASVILLE 4 SE	53.9	55.4	60.3	67.1	74.3	79.7	80.8	80.7	77.4	68.8	58.9	53.8	67.6	3.58	4.02	5.28	4.65	3.85	4.94	6.73	5.54	4.90	2.18	2.10	3.32	51.09
THOMASVILLE WB CITY	53.7	55.3	60.4	67.8	75.4	80.8	82.0	81.9	78.4	69.7	59.4	53.9	68.2	3.58	3.99	5.26	4.64	3.84	4.82	6.81	5.56	4.82	2.27	2.08	3.34	51.01
DIVISION	52.5	54.2	59.4	66.7	74.4	80.4	81.4	81.2	77.5	68.5	58.0	52.5	67.2	3.78	4.32	5.32	4.71	3.71	4.43	6.33	5.29	4.27	2.05	2.51	4.01	50.73
SOUTH CENTRAL																										
ABBEVILLE	.	.	.	.	.	.	.	.	.	.	.	.	.	2.96	3.85	4.52	3.69	3.40	3.75	5.71	4.62	3.70	2.08	2.11	3.27	43.64
ALAPAHA EXP STATION	52.0	53.7	58.5	65.9	73.6	79.4	80.9	80.5	76.7	67.7	57.7	51.9	66.5	3.20	3.88	4.73	4.25	3.52	4.25	5.95	5.43	3.86	2.19	1.68	3.11	46.05
FITZGERALD	52.9	54.4	59.6	67.0	74.9	80.7	82.0	81.5	77.4	68.2	58.4	52.5	67.5	3.13	3.69	4.63	4.08	3.34	3.50	5.51	5.17	3.84	2.30	2.10	3.38	44.67
LUMBER CITY	.	.	.	.	.	.	.	.	.	.	.	.	.	3.03	3.88	4.56	3.93	3.82	4.69	6.28	5.49	4.32	2.70	1.91	3.22	47.83
MOULTRIE 2 ESE	54.4	55.9	60.8	67.8	75.1	80.5	81.6	81.4	77.7	69.4	59.7	54.2	68.2	3.51	4.01	4.95	4.54	3.37	4.43	6.37	5.55	4.24	2.26	2.08	3.30	48.61
QUITMAN	54.3	56.0	60.9	67.6	74.7	80.2	81.3	81.1	77.9	69.5	59.9	54.4	68.2	3.35	4.09	4.93	4.09	3.70	5.04	7.28	6.28	4.82	2.15	2.29	2.97	50.99
TIFTON EXP STATION	51.7	53.2	58.3	66.0	73.9	79.7	81.0	80.8	77.1	68.1	57.9	51.8	66.6	3.39	3.83	4.66	4.26	3.39	4.11	6.30	5.02	3.74	2.04	1.80	3.23	45.77
DIVISION	52.9	54.4	59.4	66.7	74.3	80.1	81.4	81.1	77.3	68.3	58.4	52.6	67.2	3.19	3.83	4.67	4.09	3.54	4.48	6.28	5.39	4.14	2.30	1.97	3.18	47.06
SOUTHEAST																										
BRUNSWICK	55.4	56.7	61.3	68.3	75.5	80.8	82.4	82.3	78.9	70.9	61.8	55.6	69.2	2.30	2.95	3.78	3.22	3.62	5.49	7.26	7.04	9.06	4.51	1.53	2.56	53.32
GLENNVILLE	52.6	54.2	59.1	66.6	73.9	79.8	81.1	80.8	76.8	68.1	58.6	52.3	67.0	2.75	3.16	4.01	3.90	3.21	4.83	6.75	5.45	2.62	1.99	2.86	46.73
SAVANNAH USDA PL GRDN	51.3	52.8	57.7	64.6	72.0	78.4	80.2	80.0	75.9	66.4	56.6	50.9	65.6	2.60	3.13	3.82	3.09	4.49	5.80	7.55	6.28	7.02	2.93	2.09	2.76	51.56
SAVANNAH WB AP	51.7	53.1	58.3	65.7	73.4	79.7	81.3	81.0	76.7	67.2	57.3	51.4	66.4	2.78	3.68	3.97	3.70	3.77	5.09	6.61	6.62	5.25	2.58	2.05	2.81	48.91
WAYCROSS 4 NE	54.5	56.0	60.7	67.5	74.7	80.3	81.8	81.7	77.9	69.3	59.7	54.1	68.2	2.72	3.44	4.07	3.44	3.57	4.97	7.04	5.64	5.35	2.88	2.01	2.72	47.85
DIVISION	53.2	54.6	59.5	66.6	74.0	79.8	81.4	81.2	77.4	68.7	59.1	53.0	67.4	2.50	3.13	3.84	3.19	3.36	5.12	6.83	6.13	6.49	3.25	1.75	2.65	48.24

* Normals for the period 1931-1960. Divisional normals may not be the arithmetical averages of individual stations published, since additional data for shorter period stations are used to obtain better areal representation.

TEMPERATURE   PRECIPITATION

Jan.	Feb.	Mar.	Apr.	May	June	July	Aug.	Sept.	Oct.	Nov.	Dec.	Annual	Jan.	Feb.	Mar.	Apr.	May	June	July	Aug.	Sept.	Oct.	Nov.	Dec.	Annual

CONFIDENCE LIMITS

In the absence of trend or record changes, the chances are 9 out of 10 that the true mean will lie in the interval formed by adding and subtracting the values in the following table from the means for any station in the State. Because of the wider variation in mean precipitation, the corresponding monthly means and annual mean must be substituted for "p" in the precipitation table below to obtain mean precipitation confidence limits.

| 1.5 | 1.3 | 1.4 | .6 | .6 | .5 | .4 | .4 | .8 | .8 | .9 | 1.3 | .3 | $.40\sqrt{p}$ | $.35\sqrt{p}$ | $.41\sqrt{p}$ | $.39\sqrt{p}$ | $.39\sqrt{p}$ | $.36\sqrt{p}$ | $.29\sqrt{p}$ | $.39\sqrt{p}$ | $.47\sqrt{p}$ | $.52\sqrt{p}$ | $.43\sqrt{p}$ | $.38\sqrt{p}$ | $.40\sqrt{p}$ |

COMPARATIVE DATA

Data in the following table are the mean temperature and average precipitation for Milledgeville, Georgia, for the period 1906-1930 and are included in this publication for comparative purposes.

| 47.3 | 48.9 | 56.4 | 64.5 | 72.3 | 79.4 | 81.2 | 80.8 | 76.5 | 64.9 | 53.9 | 47.5 | 64.5 | 4.24 | 4.31 | 4.96 | 3.60 | 3.21 | 4.30 | 6.66 | 4.38 | 3.93 | 2.93 | 2.64 | 3.82 | 48.98 |

NORMALS, MEANS, AND EXTREMES

ATHENS, GEORGIA — CLARKE COUNTY AIRPORT

LATITUDE 33° 57' N
LONGITUDE 83° 19' W
ELEVATION (ground) 802 Feet

Month	Temperature — Normal Daily maximum	Normal Daily minimum	Normal Monthly	Extremes Record highest	Year	Extremes Record lowest	Year	Normal degree days	Precip. Normal total	Precip. Max monthly	Year	Precip. Min monthly	Year	Precip. Max in 24 hrs	Year	Snow,Sleet Mean total	Max monthly	Year	Max in 24 hrs	Year	RH 1AM	7AM	1PM	7PM	Wind Mean hourly speed	Prevailing direction	Fastest mile Speed	Direction	Year	Mean sky cover	Pct. sunshine	Clear	Partly cloudy	Cloudy	Precip .01+	Snow 1.0+	Thunderstorms	Heavy fog	Max 90+	Max 32 below	Min 32 below	Min 0 below
(a)	(b)	(b)	(b)	25	25	25	25	(b)	(b)	25	25	25	25	25	25	25	25	25	25	13	13	13	13	13	8	13	13			20		25	25	25	25	25	25	13	25	25	25	25
J	54.6	34.5	44.6	80	1949	1	1966	642	4.89	9.47	1960	1.83	1965	3.29	1966	0.7	5.0	1948	5.0	1948	77	82	58	66	9.0	NW	52	25	1959	6.2		9	7	15	11	*	1	4	0	*	14	0
F	57.0	36.1	46.1	79	1962	1	1958	529	4.70	9.24	1961	1.45	1968	2.79	1966	0.6	5.0	1952	5.0	1952	74	81	53	58	9.0	NW	52	20	1961	6.0		8	7	13	11	*	1	3	0	*	11	0
M	63.4	39.7	51.6	86	1963	5	1960	431	5.10	10.93	1964	2.27	1964	4.41	1964	0.3	8.4	1960	5.0	1960	73	81	51	57	9.5	NW	38	34	1956	5.9		10	8	14	10	*	3	3	0	*	6	0
A	73.3	48.9	61.1	91	1965	13	1950	141	4.39	9.54	1964	0.69	1950	4.61	1963	0.0	0.0		0.0		73	83	46	52	9.1	NW	47	23	1957	5.6		10	8	12	9	0	4	2	0	*	1	0
M	82.2	57.9	70.1	97	1962	27	1963+	22	3.61	11.34	1959	0.55	1959	5.54	1959	0.0	0.0		0.0		82	86	54	62	7.5	ENE	35	16	1968	5.6		8	9	10	9	2	6	1	0	*	0	0
J	89.2	65.9	77.6	104	1954	38	1956	0	3.88	13.21	1967	0.87	1967	5.93	1967	0.0	0.0		0.0		86	86	56	64	7.0	SW	40	25	1957	5.6		8	10	12	8	4	12	1	0	12	0	0
J	90.2	69.2	79.7	103	1952	55	1967	0	4.99	10.53	1947	0.09	1951	4.14	1964	0.0	0.0		0.0		89	91	60	70	6.5	SW	35	30	1965	6.1		6	13	12	7	15	12	2	0	15	0	0
A	89.6	68.4	79.0	105	1954	54	1968	0	3.63	7.43	1951	0.09	1958	5.34	1951	0.0	0.0		0.0		88	92	58	70	6.9	SW	37	16	1959	5.5		9	13	10	7	16	9	3	0	16	0	0
S	84.4	62.6	73.5	99	1954	36	1967	9	3.02	6.56	1951	0.52	1954	4.70	1964	0.0	0.0		0.0		86	89	54	69	6.9	NE	29	05	1952	5.5		15	10	7	7	3	7	2	0	*	0	0
O	75.1	50.8	63.0	98	1961	24	1952	115	2.05	6.73	1963	0.33	1950	4.05	1948	0.0	0.0		0.0		84	89	53	64	7.0	NE	29	07	1956	4.3		15	10	7	8	1	3	1	0	*	1	0
N	63.3	39.6	51.2	86	1961	7	1950	405	2.93	14.98	1945	0.33	1950	4.05	1948	0.1	2.2	1968	2.2	1968	79	85	57	67	7.8	NW	43	29	1956	5.3		11	7	12	9	0	2	2	0	0	6	0
D	54.8	34.4	44.6	79	1964+	1	1962	632	4.53	8.45	1945	1.03	1965	2.85	1961	0.1	2.0	1963	2.0	1963	77	82	57	67	8.2	ENE	43	29	1957	5.9		10	7	11	11	*	1	4	0	*	15	0
YR	73.1	50.6	61.9	105	AUG. 1954	-1	JAN. 1966	2929	48.53	14.98	NOV. 1948	T	OCT. 1963	9.93	JUN. 1967	1.8	8.4	MAR. 1960	5.0	MAR. 1960+	81	86	56	65	7.9	NW	52	20	FEB. 1961+	5.6		111	109	145	111	1	51	34	0	52	53	*

Means and extremes in the above table are from existing or comparable location(s). Annual extremes have been exceeded at other locations as follows:
Highest temperature 108 in July 1930 and earlier date(s); lowest temperature -3 in February 1899; maximum monthly precipitation 18.43 in August 1908; maximum monthly snowfall 10.0 in March 1942; maximum snowfall in 24 hours 10.0 in March 1942.
The prevailing direction for wind in the Normals, Means, and Extremes table is from records through 1963.

ATLANTA, GEORGIA — MUNICIPAL AIRPORT

LATITUDE 33° 39' N
LONGITUDE 84° 26' W
ELEVATION (ground) 1010 Feet

Month	Temperature — Normal Daily maximum	Normal Daily minimum	Normal Monthly	Extremes Record highest	Year	Extremes Record lowest	Year	Normal degree days	Precip. Normal total	Precip. Max monthly	Year	Precip. Min monthly	Year	Precip. Max in 24 hrs	Year	Snow,Sleet Mean total	Max monthly	Year	Max in 24 hrs	Year	RH 1AM	7AM	1PM	7PM	Wind Mean hourly speed	Prevailing direction	Fastest mile Speed	Direction	Year	Mean sky cover	Pct. sunshine	Clear	Partly cloudy	Cloudy	Precip .01+	Snow 1.0+	Thunderstorms	Heavy fog	Max 90+	Max 32 below	Min 32 below	Min 0 below	Avg daily solar radiation - langleys
(a)	(b)	(b)	(b)	8	8	8	8	(b)	(b)	34	34	34	34	34	34	34	34	34	34	8	8	8	8	30	14	26	26			34	34	34	34	34	34	34	34	34	8	8	8	8	18
J	52.0	37.3	44.7	72	1967	-3	1966+	639	4.44	10.82	1936	1.42	1941	3.67	1946	0.8	8.3	1940	8.3	1940	76	79	60	64	10.8	NW	54	NW	1945	6.4	48	8	7	16	11	*	1	4	5	0	2	18	*	218
F	53.8	38.4	46.1	78	1962	1	1963	529	4.51	12.77	1961	0.99	1943	5.27	1961	0.3	4.0	1952	3.9	1952	79	78	56	55	11.3	NW	54	NW	1946	6.1	57	7	7	14	12	*	2	3	3	0	*	14	0	288
M	60.3	42.5	51.4	83	1967	21	1968+	437	5.37	11.51	1942	2.73	1937	4.82	1942	0.3	3.8	1960	4.0	1960	79	78	53	53	11.1	NW	66	NW	1947	6.0	59	9	8	15	12	*	4	5	1	0	*	6	0	388
A	70.1	50.2	60.2	93	1962	31	1966+	168	4.37	7.83	1936	1.45	1936	5.13	1948	0.0	0.0		0.0		79	84	56	59	10.3	NW	50	SW	1956	5.5	65	10	10	11	10	0	5	6	1	0	0	0	0	481
M	78.9	59.2	69.1	93	1962	41	1963	22	3.16	7.52	1948	0.32	1950	3.41	1943	0.0	0.0		0.0		79	87	61	66	8.8	NW	70	NE	1953	5.5	67	7	13	11	10	0	9	2	4	0	0	0	0	538
J	85.7	66.5	76.6	98	1964	49	1966	0	3.83	7.52	1944	0.74	1944			0.0	0.0		0.0		84	87	61	66	8.0	NW		NE		5.8	67	6	13	11	10	0	11	1	4	0	0	0	0	550
J	87.0	70.7	78.9	96	1966	53	1967	0	4.72	11.26	1952	1.20	1952	5.54	1948	0.0	0.0		0.0		86	90	65	71	7.5	SW	56	SE	1954	6.3	66	5	14	12	12	0	11	2	5	0	0	0	0	526
A	86.6	69.8	78.2	98	1968	56	1968	0	3.60	8.69	1967	0.88	1963	5.05	1940	0.0	0.0		0.0		86	90	62	69	7.1	NW	62	W	1941	6.3	65	6	13	12	10	0	10	2	5	0	0	0	0	495
S	81.8	64.3	73.1	93	1962	36	1967	18	3.26	7.32	1953	0.26	1954	5.46	1956	0.0	0.0		0.0		85	89	60	63	7.3	NW	49	N	1952	5.4	68	10	10	10	8	0	8	1	1	0	0	0	0	413
O	72.4	51.2	62.4	84	1968+	29	1968+	127	2.44	7.53	1966	T	1963	3.27	1937	T	0.0		1.0	1968	77	83	50	55	8.1	NW	47	NW	1954+	4.3	68	11	7	6	6	*	2	2	0	0	0	*	0	353
N	60.9	41.5	51.2	84	1961	1	1964	414	2.96	15.72	1948	0.41	1939	4.11	1935	0.2	1.0	1968	1.0	1968	78	83	55	60	9.2	NW	46	NW	1947	5.3	59	12	7	12	6	0	2	2	0	0	*	6	0	260
D	54.2	37.1	44.8	72	1967	3	1962	626	4.38	9.92	1961	1.08	1955	3.85	1942	0.2	2.0	1963	2.2	1963	76	81	60	67	9.9	NW	63	W	1942	6.2	50	9	7	15	10	*	1	2	0	0	*	15	-15	209
YR	70.2	52.6	61.4	98	AUG. 1968+	-3	JAN. 1966+	2983	47.14	15.72	NOV. 1948	T	OCT. 1963	5.67	FEB. 1961	1.8	8.3	JAN. 1940	8.3	JAN. 1940	78	83	57	63	9.2	NW	70	NE	JUN. 1953	5.7	61	111	108	146	115	1	50	29	17	3	61	*	388

Means and extremes in the above table are from existing or comparable location(s). Annual extremes have been exceeded at other locations as follows:
Highest temperature 103 in July 1952; lowest temperature -9 in February 1899; maximum monthly precipitation 15.82 in January 1883; maximum precipitation in 24 hours 7.36 in March 1886; maximum monthly snowfall 11.6 in February 1895.
The prevailing direction for wind in the Normals, Means, and Extremes table is from records.

NORMALS, MEANS, AND EXTREMES

AUGUSTA, GEORGIA — BUSH FIELD

LONGITUDE 81° 58' W
ELEVATION (ground) 136 Feet

Ø For period February 1964 through 1968.

Means and extremes in the above table are from existing or comparable location(s). Annual extremes have been exceeded at other locations as follows: Highest temperature 106 in July 1952; lowest temperature 3 in February 1899; maximum monthly precipitation 14.00 in July 1906; maximum precipitation in 24 hours 9.82 in October 1929; maximum monthly snowfall 10.5 in February 1914; maximum snowfall 10.5 in February 1914.

The prevailing direction for wind in the Normals, Means, and Extremes table is from records through 1963.

COLUMBUS, GEORGIA — MUSCOGEE COUNTY AIRPORT

LATITUDE 32° 31' N
LONGITUDE 84° 56' W
ELEVATION (ground) 385 Feet

Means and extremes in the above table are from the existing or comparable location(s). Annual extremes have been exceeded at other locations as follows: Highest temperature 106 in September 1925; lowest temperature -3 in February 1899.
Highest temperature 106 in September 1925; lowest temperature -3 in February 1899.

– 81 –

NORMALS, MEANS, AND EXTREMES

MACON, GEORGIA — LEWIS B. WILSON AIRPORT

LATITUDE 32° 42' N
LONGITUDE 83° 39' W
ELEVATION (ground) 354 Feet

Month	Normal Daily maximum	Normal Daily minimum	Normal Monthly	Extremes Record highest	Year	Extremes Record lowest	Year	Normal degree days	Precip. Normal total	Max. monthly	Year	Min. monthly	Year	Max. in 24 hrs.	Year	Snow/Sleet Mean total	Max. monthly	Year	Max. in 24 hrs.	Year
J	60.2	38.2	49.2	78	1967	3	1966	505	3.37	8.30	1964	0.69	1954	4.44	1962	0.2	3.7	1955	3.7	1955
F	62.7	39.4	51.1	80	1965	14	1967	403	4.26	9.12	1949	1.39	1968	2.95	1952	0.2	T	1968	1.9	1968
M	68.5	44.8	56.9	88	1967	22	1968	295	4.94	9.69	1962	1.26	1958	3.45	1958	T	T	1968+	T	1968+
A	78.2	52.9	65.6	91	1968+	35	1966	63	3.75	8.42	1964	0.97	1966	3.05	1966	0.0	0.0		0.0	
M	86.5	61.2	73.9	99	1967	47	1966	0	3.32	11.77	1957	0.32	1957	4.41	1957	0.0	0.0		0.0	
J	92.6	68.7	80.7	98	1968	54	1967	0	3.34	9.06	1965	0.99	1954	4.97	1965	0.0	0.0		0.0	
J	92.8	71.0	81.9	101	1966	56	1967	0	5.64	9.04	1953	1.21	1952	2.96	1952	0.0	0.0		0.0	
A	92.2	70.3	81.3	104	1968	56	1966+	0	4.18	6.29	1959	1.39	1968	2.96	1959	0.0	0.0		0.0	
S	87.5	65.3	76.4	96	1966+	35	1967	71	2.77	8.82	1953	0.64	1958	4.60	1956	0.0	0.0		0.0	
O	78.0	53.8	66.2	89	1968+	29	1965	297	2.01	9.39	1959	0.00	1963+	4.63	1966	0.0	0.0		0.0	
N	68.0	42.7	55.4	85	1964	22	1967	297	2.45	5.89	1957	0.43	1956	1.63	1968	T	0.5	1950	0.5	1950
D	60.1	37.9	49.0	79	1967+	18	1967+	502	4.05	9.43	1953	0.58	1955	3.40	1956	T	0.5	1963	0.5	1963
YR	77.4	53.9	65.6	104	AUG. 1968	3	JAN. 1966	2136	44.08	11.77	MAY 1957	0.00	OCT. 1963	4.97	JUN. 1965	0.4	3.7	JAN. 1955	3.7	JAN. 1955

Relative humidity (Standard time used: EASTERN) — annual: 1 AM 81, 7 AM 85, 1 PM 53, 7 PM 62
Wind — Mean hourly speed 8.1; Prevailing direction WNW; Fastest mile 70 S (AUG. 1961)
Mean sky cover (sunrise to sunset) 5.7; Pct. of possible sunshine 63
Days sunrise to sunset — Clear 115; Partly cloudy 103; Cloudy 147
Mean number of days (annual) — Precipitation .01 inch or more 112; Snow/Sleet 1.0 inch or more *; Thunderstorms 56; Heavy fog 23; 90° and above 66; Max. 32° and below 1; Min. 32° and below 63; Min. 0° and below 0

Means and extremes in the above table are from the existing or comparable location(s). Annual extremes have been exceeded at other locations as follows:
Highest temperature 106 in June 1954; maximum monthly precipitation 20.52 in August 1928; maximum precipitation in 24 hours 8.36 in August 1928; maximum monthly snowfall 6.9 in February 1914; maximum snowfall in 24 hours 6.9 in February 1914.

ROME, GEORGIA — RUSSELL FIELD

LATITUDE 34° 21' N
LONGITUDE 85° 10' W
ELEVATION (ground) 637 Feet

Month	Normal Daily maximum	Normal Daily minimum	Normal Monthly	Extremes Record highest	Year	Extremes Record lowest	Year	Normal degree days	Precip. Normal total	Max. monthly	Year	Min. monthly	Year	Max. in 24 hrs.	Year	Snow/Sleet Mean total	Max. monthly	Year	Max. in 24 hrs.	Year
J	52.8	31.3	42.1	80	1950	-5	1963	710	5.51	12.65	1947	1.97	1956	4.15	1954	0.6	3.3	1962	2.4	1962
F	55.9	32.9	44.4	80	1962+	-1	1958	577	5.30	11.22	1961	1.22	1968	3.84	1955	0.6	8.3	1968	6.8	1968
M	64.2	38.1	50.4	86	1960	11	1960	468	5.87	11.48	1964	1.48	1964	7.53	1951	0.4	T	1960	T	1960
A	73.1	46.5	59.8	91	1955	23	1960	177	4.59	10.37	1964	1.68	1950	4.57	1963	T	0.0		0.0	
M	81.2	55.3	68.7	96	1962	33	1963	34	3.80	9.54	1959	0.61	1951	4.35	1964	0.0	0.0		0.0	
J	89.0	63.7	76.4	102	1954+	43	1966	0	3.76	6.76	1949	0.56	1968	2.39	1947	0.0	0.0		0.0	
J	90.6	67.4	79.0	106	1952	53	1967+	0	4.89	11.15	1953	0.54	1947	3.94	1953	0.0	0.0		0.0	
A	90.3	67.8	78.1	105	1947	49	1967	0	3.76	11.22	1961	0.52	1953	4.78	1967	0.0	0.0		0.0	
S	85.3	59.5	72.4	101	1954+	32	1967	24	3.30	9.97	1957	0.48	1963	6.35	1963	0.0	0.0		0.0	
O	75.1	47.3	61.2	96	1954	17	1952	161	2.93	6.39	1966	0.00	1964	5.16	1964	T	T	1954	T	1954
N	62.3	36.0	49.2	87	1961	3	1950	474	3.42	17.37	1948	0.81	1949	5.81	1948	T	0.8	1950	0.8	1950
D	53.4	31.3	42.4	78	1951	-5	1962	701	5.13	13.46	1961	1.07	1955	7.53	1961	0.4	8.3	1963	8.2	1963
YR	72.7	48.0	60.4	106	JUL. 1952	-5	JAN. 1963	3326	52.26	17.37	NOV. 1948	T	OCT. 1963	7.53	MAR. 1951	2.0	8.3	MAR. 1960	8.2	DEC. 1963

Relative humidity (Standard time used: EASTERN) — annual: 1 AM 80, 7 AM 85, 1 PM 52, 7 PM 68
Mean sky cover (sunrise to sunset) 5.7
Days sunrise to sunset — Clear 105; Partly cloudy 117; Cloudy 143
Mean number of days (annual) — Precipitation .01 inch or more 122; Snow/Sleet 1.0 inch or more 1; Thunderstorms 61; Heavy fog 26; 90° and above 61; Max. 32° and below 1; Min. 32° and below 82; Min. 0° and below *

Means and extremes in the above table are from the existing or comparable location(s). Annual extremes have been exceeded at other locations as follows:
Highest temperature 109 in July 1913; lowest temperature -7 in February 1899.
The prevailing direction for wind in the Normals, Means, and Extremes table is from records through 1963.

NORMALS, MEANS, AND EXTREMES

LATITUDE 32° 08' N
LONGITUDE 81° 12' W
ELEVATION (ground) 46 Feet

SAVANNAH, GEORGIA
TRAVIS FIELD

Month	Temperature Normal — Daily maximum	Daily minimum	Monthly	Extremes — Record highest	Year	Record lowest	Year	Normal degree days	Precipitation — Normal total	Maximum monthly	Year	Minimum monthly	Year	Maximum in 24 hrs.	Year	Snow,Sleet — Mean total	Maximum monthly	Year	Maximum in 24 hrs.	Year	Rel. hum. 1 AM	7 AM	1 PM	7 PM	Wind — Mean hourly speed	Prevailing direction	Fastest mile Speed	Direction	Year	Mean sky cover sunrise to sunset	Pct. of possible sunshine	Clear	Partly cloudy	Cloudy	Days Precip .01"+	Snow/Sleet 1.0"+	Thunderstorms	Heavy fog	Max 90+	Max 32−	Min 32−	Min 0−
(a)	(b)	(b)	(b)	4		4		(b)	(b)	18		18		18		18	18		18		4	4	4	4	18	13	18	18	18	18	18	18	18	18	18	18	18	18	4	4	4	4
J	62.5	40.9	51.7	77	1967+	9	1966	437	2.78	7.18	1957	0.51	1957	2.80	1967	T	T	1969+	T	1968+	83	86	55	70	8.9	WNW	46	NW	1955	5.9	56	10	7	14	9	T	1	4	0	0	14	0
F	64.4	41.7	53.1	84	1966	16	1967	353	3.68	7.92	1964	1.16	1968+	3.46	1964	0.3	3.6	1968	3.6	1968	78	82	52	64	9.7	NE	44	WNW	1960	6.0	57	8	7	13	9	*	1	3	0	0	10	0
M	69.6	46.9	58.3	90	1967	26	1966	254	3.97	9.57	1959	0.18	1955	4.65	1959	T	T	1968+	T	1968+	79	83	45	58	9.6	WNW	39	SW	1956	5.9	63	9	9	13	10	0	3	3	*	0	5	0
A	77.3	54.1	65.7	95	1967	33	1967	45	3.70	7.74	1961	1.38	1967	3.66	1961	0.0	0.0		0.0		81	85	49	61	9.4	SSE	40	N	1963	5.4	69	10	9	11	7	0	3	3	0	0	0	0
M	84.8	62.0	73.4	98	1967	46	1967	0	3.77	10.08	1957	0.51	1953	3.80	1957	0.0	0.0		0.0		85	87	53	66	8.2	SW	42	N	1965	5.6	69	10	10	11	9	0	6	3	2	0	0	0
J	90.2	69.1	79.7	95	1968+	53	1966	0	5.09	14.39	1954	0.84	1954	4.06	1963	0.0	0.0		0.0		89	90	58	73	7.8	SW	66	E	1953	6.1	65	7	11	12	11	0	11	2	7	0	0	0
J	91.2	71.4	81.3	99	1968	61	1966	0	6.61	20.10	1964	1.92	1952	6.36	1957	0.0	0.0		0.0		91	91	60	75	7.5	SW	51	SE	1951	6.4	65	5	15	11	15	0	16	1	9	0	0	0
A	90.8	71.2	81.0	100	1968	61	1966	0	6.62	12.80	1961	1.53	1954	5.78	1964	0.0	0.0		0.0		91	92	61	78	7.1	SW	58	NW	1954	5.9	66	6	15	10	12	0	13	2	17	0	0	0
S	85.9	67.4	76.7	94	1966	43	1967	3	5.25	13.47	1953	0.48	1968	5.87	1968	0.0	0.0		0.0		89	91	58	76	8.0	NE	56	NE	1959	6.3	59	6	11	10	8	0	6	4	18	0	0	0
O	78.1	56.3	67.2	90	1968	33	1965	47	2.58	8.54	1963	0.02	1963	3.57	1959	0.0	0.0		0.0		83	85	46	74	8.0	NNE	37	NE	1952	4.8	65	13	8	10	6	0	1	3	*	0	0	0
N	69.1	45.5	57.3	85	1964	25	1968	246	2.05	4.57	1957	0.15	1966	2.37	1951	0.0	0.0		0.0		81	85	52	70	8.0	NE	34	NW	1957	5.9	57	7	7	14	6	0	1	4	0	0	5	0
D	62.6	40.2	51.4	83	1967	19	1967	437	2.81	5.50	1953	0.40	1955	3.47	1964	T	T	1962+	T	1962+	81	85	52	70	8.3	SW	37	NW	1962	5.9	57	10	11	10	8	0	*	4	0	0	12	0
YR	77.2	55.6	66.4	100	AUG. 1968	9	JAN. 1966	1819	48.91	20.10	JUL. 1964	0.02	OCT. 1963	6.36	JUL. 1957	0.3	3.6	FEB. 1968	3.6	FEB. 1968	84	87	53	69	8.4	SW	66	E	JUN. 1953	5.8	63	106	116	143	110	*	65	37	57	*	46	0

Means and extremes in the above table are from existing or comparable location(s). Annual extremes have been exceeded at other locations as follows: Highest temperature 105 in July 1879; lowest temperature 8 in February 1899; maximum monthly precipitation 22.88 in September 1924; minimum monthly precipitation Trace in December 1889; maximum precipitation in 24 hours 11.44 in September 1928; fastest mile wind 90 in August 1940.

REFERENCE NOTES APPLYING TO ALL "NORMALS, MEANS, AND EXTREMES" TABLES

(a) Length of record, years.
(b) Climatological standard normals (1931-1960).
* Less than one half.
+ Also on earlier dates, months or years.
T Trace, an amount too small to measure.
Below-zero temperatures are preceded by a minus sign.
¢ Through 1963.
$ Through 1966.
# To 8 compass points only.

Unless otherwise indicated, dimensional units used in this bulletin are: temperature in degrees F.; precipitation, including snowfall, in inches; wind movement in miles per hour; and relative humidity in percent. Degree day totals are the sums of the negative departures of average daily temperatures from 65°F. Sleet was included in snowfall totals beginning with July 1948. Heavy fog reduces visibility to 1/4 mile or less.

Sky cover is expressed in a range of 0 for no clouds or obscuring phenomena to 10 for complete sky cover. The number of clear days is based on average cloudiness 0-3; partly cloudy days 4-7; and cloudy days 8-10 tenths.

Solar radiation data are the averages of direct and diffuse radiation on a horizontal surface. The langley denotes one gram calorie per square centimeter. Averages in the lower table for some months may be for more than the listed number of years.

& Figures instead of letters in a direction column indicate direction in tens of degrees from true North; i.e., 09 - East, 18 - South, 27 - West, 36 - North, and 00 - Calm. Resultant wind is the vector sum of wind directions and speeds divided by the number of observations. If figures appear in the direction column under "Fastest mile" the corresponding speeds are fastest observed 1-minute values.

% Through 1964. The station did not operate 24 hours daily. Fog and thunderstorm data therefore may be incomplete.

Mean Maximum Temperature (°F.), January

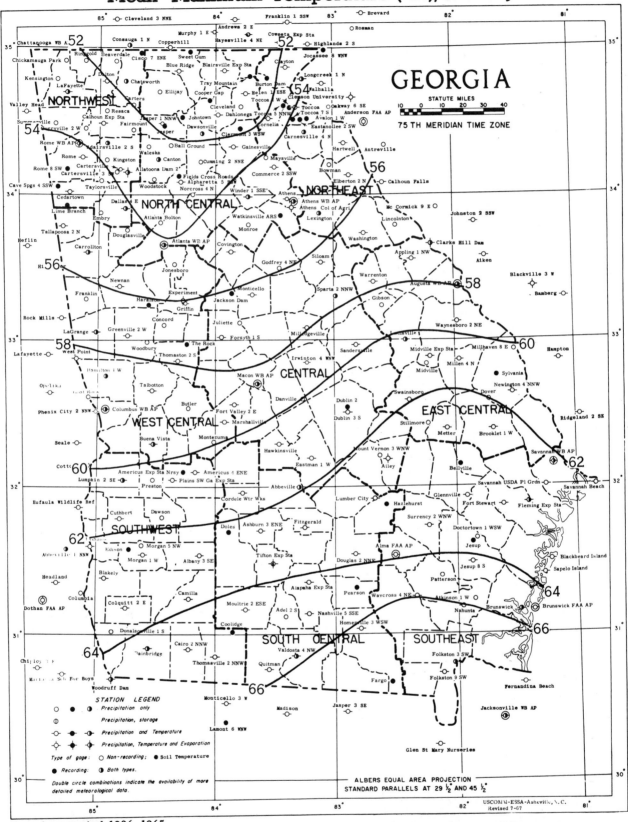

Based on period 1936-1965

Isolines are drawn through points of approximately equal values. Caution should be used in interpolating on these maps, particularly in mountainous areas.

Mean Minimum Temperature (°F.), January

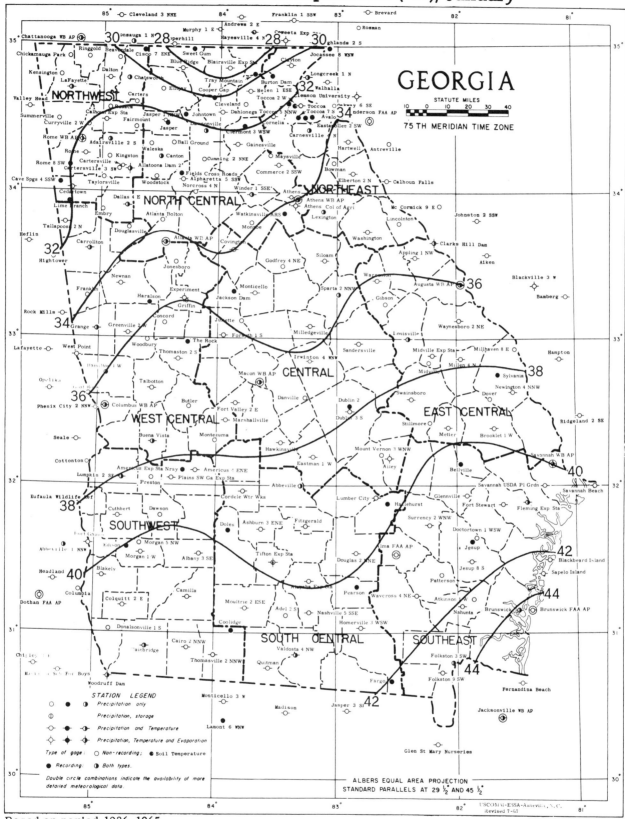

Based on period 1936-1965
Isolines are drawn through points of approximately equal values. Caution should be used in interpolating on these maps, particularly in mountainous areas.

Mean Maximum Temperature (°F.), July

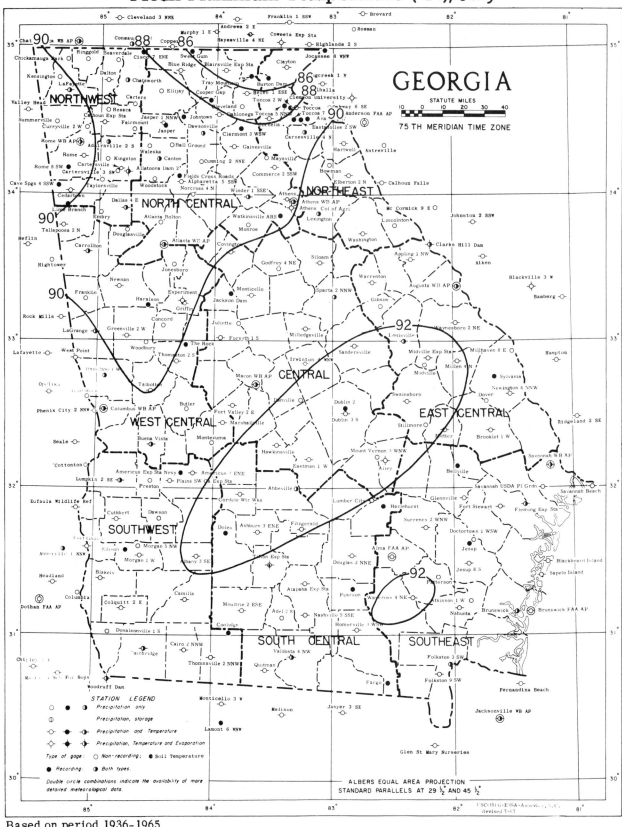

Based on period 1936-1965

    Isolines are drawn through points of approximately equal values. Caution should be used in interpolating on these maps, particularly in mountainous areas.

Mean Minimum Temperature (°F.), July

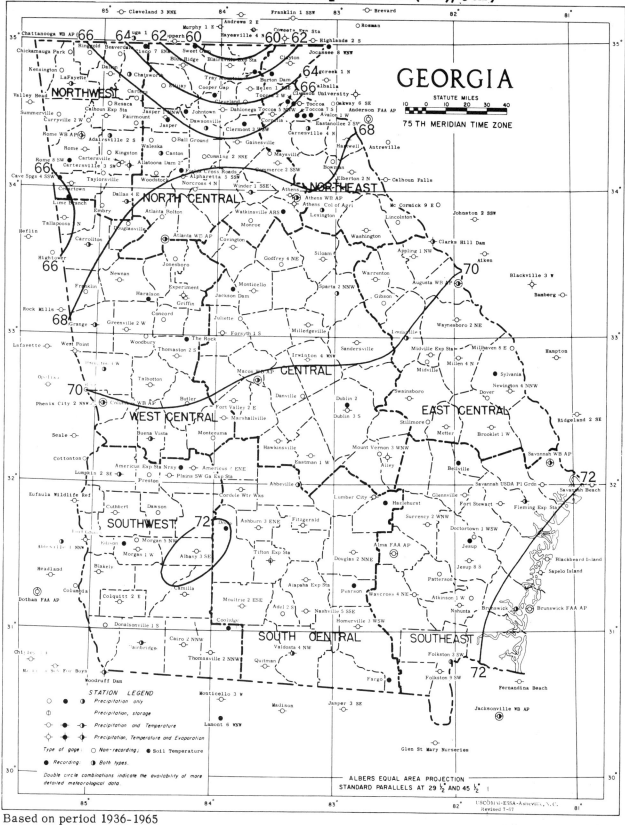

Based on period 1936-1965

Isolines are drawn through points of approximately equal values. Caution should be used in interpolating on these maps, particularly in mountainous areas.

Mean Annual Precipitation, Inches

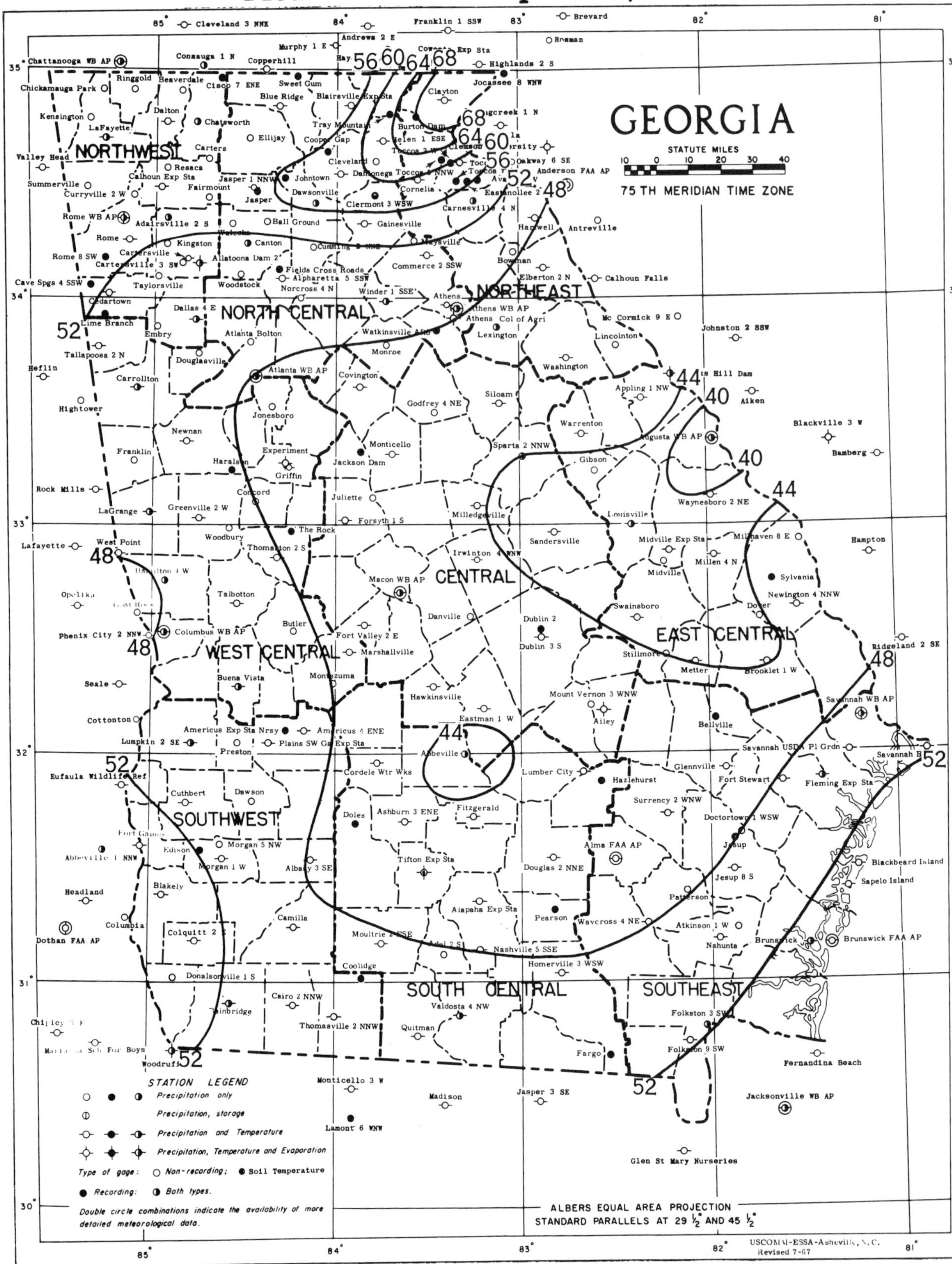

GEORGIA

STATUTE MILES

10 0 10 20 30 40

75 TH MERIDIAN TIME ZONE

Regions: NORTHWEST, NORTH CENTRAL, NORTHEAST, WEST CENTRAL, CENTRAL, EAST CENTRAL, SOUTHWEST, SOUTH CENTRAL, SOUTHEAST

Isoline values shown: 56, 60, 64, 68, 64, 60, 56, 52, 48, 44, 40, 52

STATION LEGEND

- Precipitation only
- Precipitation, storage
- Precipitation and Temperature
- Precipitation, Temperature and Evaporation

Type of gage: ○ Non-recording; ● Soil Temperature

● Recording; ◑ Both types.

Double circle combinations indicate the availability of more detailed meteorological data.

ALBERS EQUAL AREA PROJECTION
STANDARD PARALLELS AT 29 ½ AND 45 ½

USCOMM-ESSA-Asheville, N. C.
Revised 7-67

Based on period 1936-1965
Isolines are drawn through points of approximately equal values. Caution should be used in interpolating on these maps, particularly in mountainous areas.

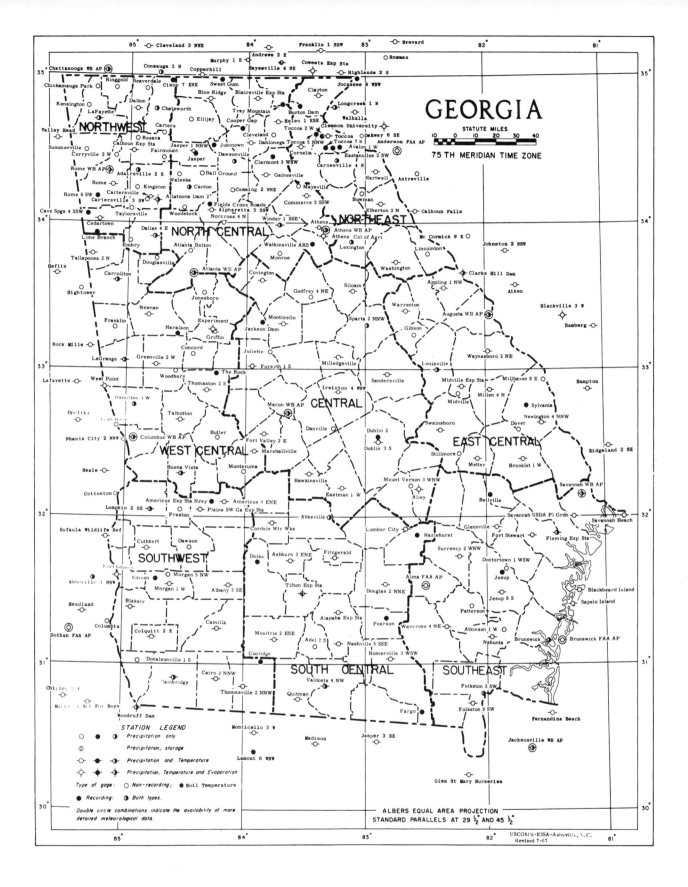

GEORGIA

STATUTE MILES

10 0   10 20 30 40

75 TH MERIDIAN TIME ZONE

NORTHWEST

NORTH CENTRAL

NORTHEAST

CENTRAL

WEST CENTRAL

EAST CENTRAL

SOUTHWEST

SOUTH CENTRAL

SOUTHEAST

STATION LEGEND

○ ● ◑ Precipitation only

◐ Precipitation, storage

◒ ◉ Precipitation and Temperature

◇ ◆ Precipitation, Temperature and Evaporation

Type of gage: ○ Non-recording; ● Soil Temperature

● Recording; ◑ Both types.

Double circle combinations indicate the availability of more
detailed meteorological data.

ALBERS EQUAL AREA PROJECTION
STANDARD PARALLELS AT 29 ½ AND 45 ½

USCOMM-ESSA-Asheville, N.C.
Revised 7-67

THE CLIMATE OF

ILLINOIS

by
William L. Denmark

June 1969

INTRODUCTION: - Illinois lies midway between the Continental Divide and the Atlantic Ocean and some 500 miles north of the Gulf of Mexico. Its climate is typically continental with cold winters, warm summers, and frequent short period fluctuations in temperature, humidity, cloudiness, and wind direction. The excellent soil and well distributed annual precipitation of 32 to 48 inches favor a very high standard of agricultural production.

PHYSIOGRAPHY:-The irregularly-shaped area of the State has a width of less than 200 miles at most points, but extends for 385 miles in the north-south direction. Except for a few low hills in the extreme south and a small unglaciated area in the extreme northwest, the terrain is flat. Differences in elevation have no significant influence on the climate. River drainage is mainly toward the Mississippi River, which forms the entire western boundary of the State. From north to south the principal rivers entering the Mississippi are the Rock, Illinois, Kaskaskia, and the Big Muddy. The Illinois River has been canalized from its upper reaches to Chicago, and is a vital link in the waterway connecting the Great Lakes and Gulf of Mexico. Approximately one-seventh of the State area drains southeastward into the Wabash and Ohio Rivers. Only a small area drains into Lake Michigan.

The deep, loess soils of the central and northern portions were originally in prairie. Forests in the extreme south, and scattered areas of forest in the north, have long since been cut down, but reforestation is active in parts of the south. The soils of the north and central portions are quite fertile, but farther south they are less fertile, have been seriously eroded, and are often underlain by hardpan. Artificial drainage of cropland is common in all portions of the State.

CLIMATIC PATTERNS: - Without the protection of natural barriers, such as mountain ranges, Illinois experiences the full sweep of the winds which are constantly bringing in the climates of other areas. Southeast and easterly winds bring mild and wet weather; southerly winds are warm and showery; westerly winds are dry with moderate temperatures, and winds from the northwest and north are cool and dry. Winds are controlled by the storm systems and weather fronts which move eastward and northeastward through this area.

Storm systems move through the State most frequently during the winter and spring months and cause a maximum of cloudiness during those seasons. Summer-season storm systems tend to be weaker and to stay farther north, leaving Illinois with much sunshine interspersed with thunderstorm situations of comparatively short duration. The retreat of the sun in autumn is associated with variable periods of pleasant dry weather of the Indian summer variety. This season ends rather abruptly with the returning storminess which usually begins in November.

TEMPERATURE: - Because Illinois extends so far in a north-south direction, the contrasts in winter temperature conditions are rather strong. The extreme north has frequent snows and temperatures drop to below zero several times each winter. The soil freezes to a depth of about 3 feet and occasionally remains snow-covered for weeks at a time. In the extreme south snow falls only occasionally and leaves after a few days, while temperatures drop to zero on an average of only about 1 day each winter. The soil freezes, but only to a depth of 8 to 12 inches, with great variation in the duration of soil-frost periods. The north-south range in winter mean temperatures is approximately 14° F.

During the summer season the sun heats the entire State quite strongly and uniformly. The north-south range of mean temperatures in July is only about 6° F. The annual average of days with temperatures of 90° F. or higher is near 20 in the north and near 50 in the south and west-central. Summer also brings periods of uncomfortably hot and humid weather, which are most persistent in the south. In the north the heat is usually broken after a few days by the arrival of cool air from Canada, but this cooling does not always penetrate to the southern portions of the State. The highest temperature of record is 117° F., observed July 14, 1954, at East St. Louis; the lowest -35° F., January 22, 1930, at Mount Carroll.

PRECIPITATION: - Latitude is the principal control for both temperature and precipitation, with the northern counties averaging cooler and drier than the south. Distance from the Gulf of Mexico and lower airmass temperatures both tend to reduce the amounts of precipitation in the northern portion. Annual precipitation is approximately one and one-half times as great in the extreme south as in the extreme north, but most of the excess in the southern portion falls during winter and early spring. Mean total precipitation for the 4-month period of December through March ranges from near 7 inches in the extreme northwest to more than 14 inches in the extreme southeast. The extra precipitation during the cold season in the south is of no benefit to agriculture, but causes harmful erosion and leaching of minerals from the soil. Precipitation during the warm season is more uniform. Totals for the 6-month period of April through September range from 21 to 24 inches throughout the State. The driest month is February. The wettest months are May and June.

Precipitation during fall, winter, and spring tends to fall uniformly over large areas. In contrast, summer rainfall occurs principally as brief showers affecting relatively small areas. The erratic occurrence of summer showers results in uneven distribution. The high rates of summer rainfall also cause runoff and soil erosion. Summer showers are usually thundery and may be accompanied by hail or destructive windstorms.

Floods occur nearly every year in at least some part of the State. The spring and early summer flood season results from a tendency for heavy general rainfall at that time of the year. The extreme north frequently has late winter or early spring flooding with the breakup of river ice, especially if there is an appreciable snow cover which is taken off by rain. River stages tend to decline during late summer, but local flash floods in minor streams, due to heavy thunderstorm rains, are common throughout the warm season. The interior rivers in the central and south have flat beds and sluggish currents so that they rise slowly and remain in flood for relatively long periods.

SNOWFALL: - The annual average of snowfall ranges from near 30 inches in the extreme north to only 10 inches in the extreme south. In the extreme north the most likely form of winter precipitation is snow. In contrast, more than 90 percent of Cairo's winter precipitation falls as rain. In a large number of winter storm situations, only a slight change in the temperature pattern would suffice to change rain to snow or vice versa. For this reason, Illinois snowfall records show great variability. The seasonal snowfall at Chicago has ranged from 10 to 68 inches. Cairo has had 48 inches of snow in a single winter; yet one-third of Cairo winters have less than 5 inches of snow. Snowfalls of 1 inch or more occur on an average of 10 to 12 days per year in the extreme north and decrease to 3 or 4 in the far south. The two northern divisions average about 50 days annually when the ground is covered with 1 inch or more of snow, and this average decreases to about 15 days in the two southern divisions.

SEVERE STORMS: - Heavy snows of 4-6 inches or more average one or two per year in the north and less frequently in the south. Strong winds will drift snow and make driving hazardous. Even major highways may become impassable for a time. Moderate to heavy ice storms average about once very four or five years and can be quite damaging to utility lines and trees, as well as being a serious traffic hazard. Thunderstorms average about 35 to 50 annually, but most are quite harmless. On occasion they provide the source for hail, damaging winds, and tornadoes. Hail falls on an average of two or three days annually in the same locality, but usually causes little damage.

Illinois is a part of the plains area of the United States which is particularly favorable for tornado formation. The State has a number of undesirable records in its severe storm history. More tornado deaths have occurred in Illinois than in any other state, more than 1,000 from 1915-68. Illinois was in the path of the most deadly tornado in U. S. history as it crossed the State from Missouri to Indiana on March 19, 1925. The death toll was 606 in Illinois. There were 101 Illinoisans killed on May 26, 1917, when the State provided the major portion of the longest tornado path on record, 293 miles.

The story has a much brighter side. Nearly 70 percent of the deaths in Illinois were the result of two tornadoes, both years before modern methods existed for detection, tracking, and warning. Despite the rapid increase in population density, only three Illinois' tornadoes of the last quarter century have claimed more than 20 lives each. Two of these occurred on the same day, April 21, 1967. The years of 1953-68 provide the best record of tornado acitivity. During this period the annual average had been 11 tornado days, 22 tornadoes, and 7 tornado deaths. The mathematical probability of a tornado striking a particular point in Illinois is very slight, about once in 500 years.

More than 65 percent of Illinois tornadoes occur during the months of March, April, May, and June. This "tornado season" is marked by a rapid increase in activity during March, a peak in April and May, and a decline during June. Tornadoes have occurred during each of the twelve months of the year.

WEATHER AND AGRICULTURE: - The major field crops in order of importance are corn, soybeans, wheat, hay, and oats. Illinois is the country's leading producer of soybeans and, in some years, corn. Peach and apple orchards are numerous in the south. Vegetables for the fresh market are widely scattered, but with some concentration near Chicago and St. Louis. Vegetables for processing are grown largely in the northern half of the State and include green peas, sweet corn, asparagus, tomatoes, snap beans, and pumpkins. Hogs, beef cattle, and milk are the principal livestock products.

The mean freeze-free period ranges from 160 to 190 days over the State. The mean date of the last spring freeze ranges from April 5 in the extreme south to May 5 in the extreme north. The mean date of the first fall freeze ranges from October 5 in the extreme northwest to October 25 in the extreme south. Serious freeze damage to major crops is not common, but careful evaluation of the freeze hazard is quite important in the commercial production of fruits and vegetables.

Normal July and August rainfall is insufficient to meet the demands of a vigorously growing field crop. Subsoil moisture must be stored during the previous fall through spring for best crop production during most seasons. Major droughts are infrequent. Rather prolonged dry periods during a portion of the growing season are not unusual. Such periods may result in reduction from potential crop yields. It is unusual for such conditions to exist over the entire State at the same time or season.

Growing field crops are subject to damage from hail during June, July, and August. Hail at any given locality will occur on an average of less than once a year. Not all hailstorms have stones of sufficient quantity or size to produce extensive crop damage.

In the extreme south the soils commonly have a lower water-holding capacity due to impermeable claypan subsoils, and there is an increased tendency toward midsummer heat and drought conditions.

WEATHER AND INDUSTRY: - Illinois industry is well developed and located mainly near Lake Michigan or along the principal rivers. Most industry depends on land transportation which is favored by the level terrain but is sometimes snarled by winter snow and ice. Few industries are restricted by climate, but there are some limitations on water supply. Adequate ground water for manufacturing operations is limited to a few locations. Large areas in the central and south depend on artificial reservoirs for industrial and municipal water supply. The design and use of such reservoirs are strongly affected by the climate-linked factors of precipitation, runoff, evaporation and evapotranspiration, erosion, and silting.

WEATHER AND PEOPLE: - The principal assets of the Illinois climate are its adequate, but seldom excessive, rainfall and the lack of severe extremes. A similar climate prevails throughout the heavily populated and productive section of the United States which extends from the upper Mississippi Valley eastward to the Middle Atlantic states. The daily and seasonal variability promotes health and vigor.

INFLUENCE OF LAKE MICHIGAN: - Because prevailing winds are westerly and storm systems move from the same direction, the influence of the lake on Illinois weather is not large. However, the area under the lake influence includes the Chicago area and other communities where approximately one-half of the State population lives. When the wind blows from the lake toward the shore, which it does for approximately one-fourth of the time during spring and summer and for about one-eighth of the time during fall and winter, the result is a moderation of tempera-ture. In addition to the general occurrence of onshore winds, there is the local "sea breeze" effect on summer afternoons which is usually observable in a narrow strip near the lake shore.

BIBLIOGRAPHY

(1) Climatic Summary of the United States (Bulletin W) 1930 edition, Sections 56, 57, and 58, U. S. Weather Bureau.

(2) Climatic Summary of the United States, Illinois, Climatography of the United States, Series 11-9, (Bulletin W Supplement), 1931-52. U. S. Weather Bureau.

(3) Climatic Summary of the United States, Illinois, Climatography of the United States, Series 86-9, (Bulletin W Supplement), 1951-60. U. S. Weather Bureau.

(4) Climatological Data - Illinois. (Monthly and annual series) U. S. Weather Bureau.

(5) Climatological Data National Summary (Monthly and annual series) U. S. Weather Bureau.

(6) Hourly Precipitation Data - Illinois (Monthly and annual series). U. S. Weather Bureau.

(7) Summary of Hourly Observations, Climatography of the United States, No. 30-11, for Chicago, Moline, Springfield, Illinois; No. 30-23 for St. Louis, Missouri; No. -12 for Evansville, Indiana.

(8) Summary of Hourly Observations, Climatography of the United States No. 82-11, for Chicago (O'Hare and Midway), Moline, Springfield, Illinois; No. 82-23 for St. Louis, Missouri; No. 82-12 for Evansville, Indiana.

(9) Local Climatological Data, U. S. Weather Bureau (Monthly and annual series), for Cairo, Chicago, Moline, Peoria, and Springfield, Illinois; Evansville, Indiana; Burlington and Dubuque, Iowa; St. Louis, Missouri.

(10) Climatic Guide for Chicago, Illinois, Area, Climatography of the United States No. 40-11, U. S. Weather Bureau.

(11) Hail Climatology of Illinois - F. A. Huff & S. A. Changnon, Jr. Report of Investigation #38, Illinois State Water Survey, Urbana, Illinois 1959.

(12) Weather and Climate of Chicago - Cox and Armington, The Geographic Society of Chicago, Bulletin No. 4.

(13) Climate of Illinois - John L. Page, University of Illinois, Agricultural Experiment Station, Bulletin No. 532, Urbana, Illinois. 1949

ADDITIONAL REFERENCES

(a) Weather Bureau Technical Paper No. 15 - Maximum Station Precipitations for 1, 2, 3, 6, 12, and 24 Hours.

(b) Weather Bureau Technical Paper No. 16 - Maximum 24-Hour Precipitation in the United States, Washington, D. C. 1952.

(c) Weather Bureau Technical Paper No. 25 - Rainfall Intensity - Duration - Frequency Curves. For selected stations in the United States, Alaska, Hawaiian Islands, and Puerto Rico.

(d) Weather Bureau Technical Paper No. 29 - Rainfall Intensity - Frequency Regime. Washington, D. C.

FREEZE DATA

STATION	Freeze threshold temperature	Mean date of last Spring occurrence	Mean date of first Fall occurrence	Mean No. of days between dates	Years of record Spring	No. of occurrences in Spring	Years of record Fall	No. of occurrences in Fall	STATION	Freeze threshold temperature	Mean date of last Spring occurrence	Mean date of first Fall occurrence	Mean No. of days between dates	Years of record Spring	No. of occurrences in Spring	Years of record Fall	No. of occurrences in Fall
ALEDO	32	04-25	10-16	174	30	30	30	30	DECATUR	32	04-22	10-20	181	30	30	30	30
	28	04-11	10-23	195	30	30	30	30		28	04-09	10-29	203	30	30	30	30
	24	03-31	11-01	215	30	30	30	30		24	03-27	11-08	226	30	30	30	30
	20	03-24	11-13	234	30	30	30	30		20	03-16	11-19	248	30	30	30	29
	16	03-13	11-23	255	30	30	30	29		16	03-06	11-29	268	30	30	30	28
ANNA 1 E	32	04-07	10-30	206	30	30	30	30	DIXON	32	04-29	10-07	161	30	30	30	30
	28	03-30	11-07	222	30	30	30	30		28	04-19	10-20	184	30	30	30	30
	24	03-16	11-17	246	30	30	30	29		24	04-06	10-31	208	30	30	30	30
	20	03-04	11-25	266	30	30	30	29		20	03-24	11-10	231	30	30	30	30
	16	02-22	12-08	289	30	30	30	27		16	03-15	11-21	251	30	30	30	30
AURORA COLLEGE	32	05-02	10-07	158	30	30	30	30	DU QUOIN 2 S	32	04-14	10-22	191	30	30	30	30
	28	04-22	10-23	184	30	30	30	30		28	03-31	10-31	214	29	29	30	30
	24	04-05	11-02	211	30	30	30	30		24	03-17	11-13	241	29	29	30	30
	20	03-28	11-08	225	30	30	30	30		20	03-05	11-24	264	29	29	30	28
	16	03-16	11-19	248	30	30	30	30		16	02-26	12-06	283	29	29	30	28
BELLEVILLE SO ILL UNIV	32	04-12	10-30	201	10	10	10	10	EAST ST LOUIS PARKS CO	32	04-05	10-28	206	16	16	17	17
	28	03-27	11-04	222	10	10	10	10		28	03-27	11-03	221	16	16	17	17
	24	03-16	11-14	243	10	10	10	10		24	03-15	11-18	248	16	16	18	18
	20	03-04	11-24	265	10	10	10	9		20	03-09	11-29	265	18	18	18	18
	16	02-22	12-08	289	10	10	10	7		16	02-24	12-07	286	16	16	18	16
BLOOMINGTON NORMAL	32	04-25	10-17	175	30	30	30	30	EFFINGHAM	32	04-22	10-16	177	30	30	29	29
	28	04-11	10-27	199	30	30	30	30		28	04-10	10-28	201	29	29	30	30
	24	03-31	11-03	217	30	30	30	30		24	03-23	11-08	230	28	28	30	30
	20	03-21	11-16	240	30	30	29	29		20	03-11	11-19	253	28	28	29	29
	16	03-10	11-28	263	30	30	30	29		16	03-04	11-30	271	28	28	30	28
CAIRO WB CITY	32	03-24	11-09	230	30	30	30	30	FAIRFIELD RADIO WFIW	32	04-14	10-24	193	30	30	29	29
	28	03-14	11-21	252	30	30	30	30		28	04-01	11-02	215	30	30	29	29
	24	02-28	12-03	278	30	30	30	29		24	03-19	11-12	238	30	30	29	28
	20	02-18	12-10	295	30	29	30	25		20	03-09	11-22	258	30	30	29	27
	16	02-06	12-18	315	30	27	30	18		16	02-25	12-05	283	30	30	29	26
CARBONDALE SEWAGE PL	32	04-14	10-17	186	30	30	30	30	FLORA 2 NW	32	04-15	10-22	190	28	28	28	28
	28	03-31	11-01	215	29	29	30	30		28	03-30	10-31	215	28	28	29	29
	24	03-18	11-10	237	29	29	30	30		24	03-20	11-10	235	29	29	29	29
	20	03-04	11-23	264	29	29	30	28		20	03-08	11-25	262	29	29	29	27
	16	02-19	12-05	289	29	29	30	27		16	03-01	12-05	279	29	28	30	28
CARLINVILLE	32	04-20	10-16	179	30	30	29	29	FREEPORT	32	05-05	10-04	152	30	30	30	30
	28	04-07	10-28	204	29	29	29	29		28	04-24	10-15	174	30	30	30	30
	24	03-27	11-06	224	29	29	29	29		24	04-14	10-27	196	30	30	30	30
	20	03-15	11-16	246	29	29	30	29		20	03-30	11-07	222	30	30	30	30
	16	03-06	11-28	267	29	29	29	28		16	03-19	11-18	244	30	30	30	30
CARLYLE	32	04-13	10-23	193	20	20	21	21	GALESBURG	32	04-25	10-17	175	10	10	10	10
	28	03-29	11-01	217	19	19	21	21		28	04-14	10-27	196	10	10	10	10
	24	03-17	11-11	239	19	19	21	21		24	03-31	11-03	217	10	10	10	10
	20	03-07	11-24	262	19	19	21	20		20	03-28	11-13	230	10	10	10	10
	16	03-01	12-03	277	19	19	21	20		16	03-12	11-25	258	10	10	10	10
CHICAGO UNIVERSITY	32	04-21	10-27	189	30	30	30	30	GALVA	32	04-29	10-12	166	28	28	30	30
	28	04-04	11-07	217	30	30	30	30		28	04-13	10-24	194	29	29	30	30
	24	03-25	11-15	235	30	30	30	30		24	04-03	11-03	214	30	30	30	30
	20	03-15	11-23	253	30	30	30	30		20	03-25	11-13	233	30	30	30	30
	16	03-07	11-30	268	30	30	30	28		16	03-13	11-22	254	30	30	30	30
CHICAGO WB AIRPORT	32	04-19	10-28	192	10	10	10	10	GOLCONDA DAM 51	32	04-11	10-30	202	10	10	10	10
	28	04-03	11-03	214	10	10	10	10		28	04-02	11-05	217	10	10	10	10
	24	03-23	11-18	240	10	10	10	10		24	03-14	11-09	240	10	10	10	10
	20	03-16	11-21	250	10	10	10	10		20	02-25	11-19	267	10	10	10	10
	16	03-09	11-29	265	10	10	10	9		16	02-17	12-01	287	10	10	10	9
CHICAGO WB CITY	32	04-08	11-09	215	10	10	10	10	GREENVILLE 1 E	32	04-15	10-25	193	29	29	28	28
	28	03-27	11-13	231	10	10	10	10		28	04-01	10-30	212	29	29	27	27
	24	03-22	11-22	245	10	10	10	10		24	03-21	11-12	236	29	29	27	27
	20	03-11	11-28	262	10	10	10	10		20	03-09	11-24	260	27	27	27	26
	16	03-05	12-04	274	10	10	10	8		16	03-03	12-01	273	28	28	27	25
DANVILLE	32	05-01	10-08	160	29	29	29	29	GRIGGSVILLE	32	04-16	10-22	189	30	30	30	30
	28	04-18	10-24	189	30	30	29	29		28	04-05	10-30	208	30	30	30	30
	24	03-31	11-02	216	30	30	29	29		24	03-25	11-08	228	30	30	30	30
	20	03-21	11-14	238	30	30	29	29		20	03-16	11-20	249	30	30	30	29
	16	03-10	11-24	259	30	30	29	29		16	03-07	11-28	266	30	30	30	29

FREEZE DATA

STATION	Freeze threshold temperature	Mean date of last Spring occurrence	Mean date of first Fall occurrence	Mean No. of days between dates	Years of record Spring	No. of occurrences in Spring	Years of record Fall	No. of occurrences in Fall
HARRISBURG	32	04-17	10-19	185	30	30	30	30
	28	04-01	10-30	212	30	30	30	30
	24	03-18	11-10	237	30	30	30	30
	20	03-03	11-22	264	30	30	30	29
	16	02-21	12-07	289	30	30	30	25
HAVANA	32	04-22	10-15	176	30	30	30	30
	28	04-08	10-29	204	30	30	30	30
	24	03-28	11-06	223	30	30	30	30
	20	03-17	11-16	244	30	30	30	30
	16	03-08	11-28	265	30	30	30	29
HILLSBORO	32	04-18	10-19	184	29	29	29	29
	28	04-05	10-30	208	30	30	30	30
	24	03-24	11-09	230	29	29	30	30
	20	03-11	11-21	255	30	30	29	29
	16	03-05	11-29	269	30	30	29	27
HOOPESTON	32	04-28	10-15	170	30	30	30	30
	28	04-16	10-27	194	30	30	30	30
	24	04-01	11-05	218	30	30	30	30
	20	03-18	11-17	244	30	30	30	30
	16	03-09	11-28	264	30	30	30	29
JACKSONVILLE	32	04-22	10-16	177	30	30	30	30
	28	04-09	10-27	201	30	30	30	30
	24	03-31	11-03	217	30	30	30	30
	20	03-16	11-15	244	30	30	30	30
	16	03-07	11-26	264	30	30	30	29
JERSEYVILLE 2 SW	32	04-20	10-19	182	10	10	10	10
	28	04-11	10-28	200	10	10	10	10
	24	03-30	11-03	218	9	9	10	10
	20	03-15	11-07	237	10	10	10	10
	16	03-08	11-23	260	10	10	10	10
KANKAKEE 3 SW	32	04-28	10-16	171	30	30	30	30
	28	04-13	10-29	199	30	30	30	30
	24	03-30	11-06	221	30	30	30	30
	20	03-21	11-16	240	30	30	29	29
	16	03-10	11-26	261	30	30	29	28
LACON 2 E	32	04-27	10-13	169	24	24	25	25
	28	04-12	10-26	197	24	24	24	24
	24	03-31	11-05	219	23	23	24	24
	20	03-20	11-15	240	23	23	24	24
	16	03-10	11-24	259	23	23	24	23
LA HARPE	32	04-24	10-14	173	30	30	30	30
	28	04-10	10-26	199	30	30	30	30
	24	04-01	11-03	216	30	30	30	30
	20	03-19	11-13	239	30	30	30	30
	16	03-10	11-24	259	29	29	30	29
LA SALLE PERU	32	04-27	10-13	169	30	30	30	30
	28	04-11	10-28	200	30	30	30	30
	24	04-01	11-05	218	30	30	30	30
	20	03-17	11-17	245	30	30	30	30
	16	03-11	11-28	262	30	30	30	29
LINCOLN	32	04-25	10-18	176	30	30	30	30
	28	04-11	10-27	199	30	30	30	30
	24	03-29	11-07	223	30	30	30	30
	20	03-16	11-15	244	30	30	29	29
	16	03-07	11-28	266	30	30	29	28
MARENGO	32	05-05	10-06	154	29	29	30	30
	28	04-25	10-17	175	29	29	29	29
	24	04-08	11-01	207	29	29	29	29
	20	03-28	11-14	231	29	29	29	29
	16	03-18	11-21	248	28	28	29	29
MATTOON	32	04-20	10-18	181	10	10	10	10
	28	04-05	10-30	208	10	10	10	10
	24	03-28	11-02	219	10	10	10	10
	20	03-13	11-12	244	10	10	10	10
	16	03-06	12-03	272	10	10	10	9
MCLEANSBORO	32	04-13	10-27	197	30	30	29	29
	28	04-01	11-02	217	30	30	29	29
	24	03-18	11-13	240	30	30	30	30
	20	03-06	11-25	264	30	30	30	29
	16	02-25	12-07	285	30	30	30	27
MINONK	32	04-29	10-13	167	30	30	30	30
	28	04-13	10-24	194	30	30	30	30
	24	04-03	11-01	212	30	30	30	30
	20	03-24	11-11	232	29	29	29	29
	16	03-12	11-23	256	27	27	29	28
MOLINE WB AIRPORT	32	04-23	10-14	174	10	10	10	10
	28	04-12	10-29	200	10	10	10	10
	24	03-31	11-02	216	10	10	10	10
	20	03-21	11-14	238	10	10	10	10
	16	03-16	11-23	252	10	10	10	10
MONMOUTH	32	04-24	10-15	174	30	30	30	30
	28	04-10	10-28	201	30	30	30	30
	24	03-29	11-05	221	30	30	30	30
	20	03-21	11-13	237	30	30	30	30
	16	03-12	11-25	258	30	30	30	29
MORRIS 5 N	32	05-02	10-09	160	30	30	30	30
	28	04-20	10-23	186	30	30	28	28
	24	04-07	11-03	210	30	30	30	30
	20	03-25	11-12	232	30	30	30	30
	16	03-14	11-23	254	30	30	30	29
MORRISON	32	04-26	10-09	166	30	30	30	30
	28	04-17	10-22	188	30	30	30	30
	24	04-02	11-01	213	30	30	30	30
	20	03-24	11-11	232	30	30	30	30
	16	03-14	11-22	253	30	30	30	30
MORRISONVILLE 4 SE	32	04-22	10-15	176	30	30	30	30
	28	04-07	10-29	205	30	30	30	30
	24	03-26	11-05	224	30	30	30	30
	20	03-13	11-17	249	30	30	30	29
	16	03-06	11-28	267	30	30	29	27
MT CARMEL	32	04-12	10-27	198	21	21	20	20
	28	03-30	11-04	219	21	21	20	20
	24	03-17	11-16	244	21	21	20	20
	20	03-05	11-29	269	21	21	20	19
	16	02-26	12-09	286	21	21	19	17
MOUNT CARROLL	32	05-03	10-04	154	30	30	30	30
	28	04-23	10-16	176	30	30	30	30
	24	04-10	10-29	202	30	30	30	30
	20	03-28	11-07	224	30	30	30	30
	16	03-19	11-19	245	30	30	30	30
MOUNT VERNON	32	04-15	10-23	191	30	30	30	30
	28	04-04	11-01	211	30	30	30	30
	24	03-22	11-10	233	30	30	30	30
	20	03-09	11-21	257	30	30	30	28
	16	03-01	12-03	277	30	30	30	27
CREAL SPRINGS	32	04-21	10-15	177	30	30	30	30
	28	04-05	10-28	206	30	30	30	30
	24	03-25	11-04	224	30	30	30	30
	20	03-13	11-18	250	30	30	30	30
	16	02-26	12-01	278	30	30	30	27
OLNEY 2 S	32	04-18	10-23	188	28	28	28	28
	28	04-04	10-29	208	28	28	27	27
	24	03-21	11-09	233	28	28	27	27
	20	03-08	11-18	255	28	28	28	28
	16	03-01	12-03	277	29	29	28	26
OTTAWA	32	04-21	10-22	184	10	10	10	10
	28	04-10	10-31	204	10	10	10	10
	24	03-28	11-10	227	10	10	10	10
	20	03-18	11-17	244	10	10	10	10
	16	03-15	11-25	255	10	10	10	9

FREEZE DATA

STATION	Freeze threshold temperature	Mean date of last Spring occurrence	Mean date of first Fall occurrence	Mean No. of days between dates	Years of record Spring	No. of occurrences in Spring	Years of record Fall	No. of occurrences in Fall	STATION	Freeze threshold temperature	Mean date of last Spring occurrence	Mean date of first Fall occurrence	Mean No. of days between dates	Years of record Spring	No. of occurrences in Spring	Years of record Fall	No. of occurrences in Fall
PALESTINE	32	04-18	10-16	181	30	30	29	29	SPRINGFIELD WB AIRPORT	32	04-20	10-23	186	10	10	10	10
	28	04-05	10-30	208	30	30	29	29		28	04-08	10-29	204	10	10	10	10
	24	03-24	11-07	228	30	30	29	29		24	03-29	11-05	221	10	10	10	10
	20	03-11	11-19	253	30	30	30	30		20	03-15	11-16	246	10	10	10	10
	16	03-04	12-01	272	30	30	30	29		16	03-08	11-28	265	10	10	10	10
PANA	32	04-20	10-23	186	30	30	30	30	SPRINGFIELD WB CITY	32	04-07	10-30	206	23	23	23	23
	28	04-08	10-30	205	30	30	30	30		28	03-28	11-11	228	24	24	23	23
	24	03-24	11-07	228	30	30	30	30		24	03-20	11-21	246	24	24	23	22
	20	03-13	11-21	253	30	30	30	29		20	03-08	11-29	266	24	24	23	22
	16	03-04	11-28	269	30	30	30	28		16	03-03	12-04	276	24	24	23	22
PARIS WATERWORKS	32	04-25	10-21	179	28	28	30	30	STOCKTON	32	05-05	10-05	153	10	10	10	10
	28	04-09	10-29	203	29	29	29	29		28	04-21	10-19	181	10	10	10	10
	24	03-25	11-07	227	30	30	30	30		24	04-13	11-03	204	10	10	9	9
	20	03-16	11-17	246	30	30	30	30		20	03-29	11-08	224	10	10	9	9
	16	03-06	11-30	269	30	30	30	29		16	03-20	11-16	241	10	10	10	9
PEORIA WB AIRPORT	32	04-22	10-20	181	30	30	30	30	SYCAMORE	32	05-05	10-06	154	30	30	29	29
	28	04-09	10-31	205	30	30	30	30		28	04-23	10-17	177	30	30	30	30
	24	03-28	11-08	225	30	30	30	30		24	04-06	10-30	207	30	30	30	30
	20	03-16	11-20	249	30	30	30	29		20	03-30	11-10	225	29	29	30	30
	16	03-08	11-26	263	30	30	30	29		16	03-19	11-20	246	29	29	30	30
PONTIAC	32	04-22	10-21	182	30	30	30	30	TUSCOLA	32	04-25	10-19	177	10	10	10	10
	28	04-08	11-01	207	30	30	30	30		28	04-12	10-28	199	10	10	10	10
	24	03-25	11-11	231	30	30	30	30		24	03-31	11-03	217	10	10	10	10
	20	03-17	11-21	249	30	30	30	29		20	03-13	11-12	244	10	10	10	10
	16	03-09	11-28	264	30	30	30	29		16	03-09	11-28	264	10	10	10	10
QUINCY	32	04-12	10-25	196	30	30	30	30	URBANA	32	04-21	10-23	185	30	30	30	30
	28	04-02	10-31	212	30	30	30	30		28	04-05	11-01	210	30	30	30	30
	24	03-26	11-11	230	30	30	30	30		24	03-24	11-11	232	30	30	30	30
	20	03-13	11-22	254	30	30	30	29		20	03-12	11-23	256	30	30	30	29
	16	03-05	11-28	268	30	30	30	29		16	03-05	12-01	271	30	30	30	29
QUINCY FAA AIRPORT	32	04-21	10-20	182	10	10	10	10	VANDALIA FAA AIRPORT	32	04-22	10-11	172	10	10	10	10
	28	04-09	10-30	204	10	10	10	10		28	04-07	10-24	200	10	10	10	10
	24	04-04	11-02	212	10	10	10	10		24	03-27	11-01	219	10	10	10	10
	20	03-21	11-10	234	10	10	10	10		20	03-13	11-13	245	9	9	10	10
	16	03-10	11-26	261	10	10	10	10		16	02-26	11-25	272	9	9	10	9
RANTOUL	32	04-24	10-23	182	10	10	10	10	WALNUT	32	04-28	10-13	168	30	30	30	30
	28	04-07	10-29	205	10	10	10	10		28	04-14	10-27	196	30	30	30	30
	24	03-27	11-06	224	10	10	10	10		24	04-03	11-03	214	30	30	29	29
	20	03-17	11-21	249	10	10	10	10		20	03-24	11-16	237	30	30	29	29
	16	03-08	11-28	265	10	10	10	9		16	03-15	11-26	256	30	30	29	28
ROBERTS 3 N	32	04-23	10-12	172	10	10	10	10	WATSEKA 2 NW	32	04-25	10-13	171	10	10	10	10
	28	04-13	10-28	198	10	10	10	10		28	04-10	10-31	204	10	10	10	10
	24	03-30	11-07	222	10	10	10	10		24	03-27	11-02	220	10	10	10	10
	20	03-22	11-14	237	10	10	10	10		20	03-18	11-14	241	10	10	10	10
	16	03-13	11-25	257	10	10	10	10		16	03-09	11-24	260	10	10	10	10
ROCKFORD	32	04-30	10-07	160	27	27	26	26	WAUKEGAN 2 WNW	32	05-01	10-14	166	30	30	30	30
	28	04-20	10-19	182	27	27	26	26		28	04-20	10-26	189	30	30	30	30
	24	04-06	11-02	210	27	27	26	26		24	04-03	11-07	218	30	30	30	30
	20	03-28	11-12	229	27	27	26	26		20	03-23	11-16	238	30	30	30	30
	16	03-17	11-21	249	27	27	26	26		16	03-16	11-23	252	30	30	30	30
ROCKFORD WB AIRPORT	32	04-22	10-11	172	10	10	10	10	WHEATON 3 SE	32	04-26	10-12	169	10	10	10	10
	28	04-14	10-23	192	10	10	10	10		28	04-16	10-29	196	10	10	10	10
	24	04-06	11-01	209	10	10	10	10		24	04-02	11-02	214	10	10	10	10
	20	03-27	11-11	229	10	10	10	10		20	03-23	11-13	235	10	10	10	10
	16	03-16	11-17	246	10	10	10	10		16	03-18	11-24	251	10	10	10	10
RUSHVILLE	32	04-20	10-22	185	30	30	30	30	WHITE HALL 1 E	32	04-19	10-20	184	30	30	29	29
	28	04-08	10-28	203	30	30	30	30		28	04-06	10-30	207	30	30	29	29
	24	03-28	11-05	222	30	30	30	30		24	03-25	11-07	227	30	30	30	30
	20	03-17	11-17	245	30	30	30	29		20	03-13	11-19	251	30	30	30	30
	16	03-09	11-27	263	30	30	30	29		16	03-06	11-28	267	30	30	30	29
SPARTA	32	04-10	10-25	198	30	30	30	30	WINDSOR	32	04-25	10-18	176	29	29	29	29
	28	03-28	11-06	223	30	30	30	30		28	04-11	10-29	201	29	29	29	29
	24	03-16	11-12	241	30	30	30	30		24	03-28	11-08	225	29	29	29	29
	20	03-08	11-28	265	30	30	30	29		20	03-16	11-19	248	29	29	29	28
	16	02-27	12-08	284	30	30	30	27		16	03-06	11-30	269	29	29	29	28

Data in the above table are based on the period 1931-1960, or that portion of this period for which data are available.
⊕ When the frequency of occurrence in either spring or fall is one year in ten, or less, mean dates are not given.
Means have been adjusted to take into account years of non-occurrence.
A freeze is a numerical substitute for the former term "killing frost" and is the occurrence of a minimum temperature at or below the threshold temperature of 32°, 28°, etc.
Freeze data tabulations in greater detail are available and can be reproduced at cost.

*NORMALS BY CLIMATOLOGICAL DIVISIONS

Taken from "Climatography of the United States No. 81-4, Decennial Census of U. S. Climate"

STATIONS (By Divisions)	TEMPERATURE (°F)													PRECIPITATION (In.)												
	JAN	FEB	MAR	APR	MAY	JUNE	JULY	AUG	SEPT	OCT	NOV	DEC	ANN	JAN	FEB	MAR	APR	MAY	JUNE	JULY	AUG	SEPT	OCT	NOV	DEC	ANN
NORTHWEST																										
ALEDO	24.7	27.7	37.5	51.3	62.2	71.9	76.2	74.2	66.3	55.4	39.7	28.6	51.3	1.61	1.38	2.65	3.50	4.14	4.85	3.80	2.93	3.23	2.68	1.98	1.78	34.53
DIXON	24.0	26.6	36.2	50.0	61.1	70.7	74.9	73.4	65.4	54.2	39.1	27.5	50.3	1.74	1.38	2.29	2.96	3.63	4.39	3.44	3.65	3.29	2.67	2.35	1.86	33.65
FREEPORT	21.3	24.0	33.8	48.2	59.4	69.2	73.3	71.4	63.2	52.1	37.0	25.3	48.2	1.74	1.17	2.11	2.85	3.76	4.55	4.02	3.95	3.38	2.25	2.41	1.71	33.90
GALENA	•	•	•	•	•	•	•	•	•	•	•	•	•	1.61	1.08	2.03	2.73	3.86	4.49	3.59	4.11	3.20	2.41	2.24	1.70	33.05
GALVA	24.9	27.6	37.2	51.0	61.8	71.6	75.8	73.9	66.2	55.4	39.7	28.6	51.1	1.65	1.50	2.55	3.39	4.12	4.52	3.39	3.21	3.01	2.85	2.08	1.66	33.93
GENESEO	•	•	•	•	•	•	•	•	•	•	•	•	•	1.62	1.39	2.41	3.15	3.96	4.52	3.21	4.01	3.18	2.62	2.06	1.57	33.70
KEITHSBURG 1 NW	•	•	•	•	•	•	•	•	•	•	•	•	•	1.41	1.24	2.20	2.91	3.50	4.06	3.24	2.64	2.97	2.13	1.67	1.35	29.32
MOLINE WB AIRPORT	22.6	25.7	35.4	50.0	61.2	71.4	75.6	73.6	65.3	54.8	39.2	26.9	50.1	1.61	1.35	2.39	3.17	3.80	4.37	3.26	3.53	3.25	2.46	1.95	1.65	32.79
MORRISON	23.7	26.8	36.3	50.3	61.1	71.0	75.4	73.6	65.5	54.5	39.0	27.6	50.4	1.79	1.44	2.35	3.18	4.04	4.46	3.44	3.99	3.46	2.54	2.23	1.80	34.72
MOUNT CARROLL	22.6	25.2	35.0	49.1	60.0	69.9	73.9	72.1	64.0	53.0	37.8	26.4	49.1	1.81	1.34	2.23	3.01	4.03	4.82	3.70	3.64	3.38	2.45	2.46	1.89	34.76
PAW PAW	•	•	•	•	•	•	•	•	•	•	•	•	•	1.83	1.52	2.48	3.28	3.99	4.43	3.40	3.60	3.17	3.00	2.23	1.85	34.78
ROCHELLE 1 W	•	•	•	•	•	•	•	•	•	•	•	•	•	1.61	1.25	2.20	3.04	3.54	4.23	3.36	3.63	3.06	2.53	2.04	1.66	32.15
ROCKFORD WB AIRPORT	22.0	24.4	34.0	47.8	59.1	69.7	74.2	72.5	63.8	52.5	37.1	25.6	49.6	1.98	1.44	2.46	3.05	3.83	4.30	4.14	4.14	3.51	2.70	2.37	1.70	35.62
TISKILWA	•	•	•	•	•	•	•	•	•	•	•	•	•	1.78	1.56	2.56	3.44	4.14	4.04	3.37	3.53	2.83	2.79	1.95	1.74	33.73
WALNUT	24.8	27.3	36.8	50.6	62.0	72.2	76.5	74.4	66.1	55.0	39.6	28.3	51.1	1.76	1.42	2.31	3.31	4.14	4.28	3.19	3.84	3.52	2.71	2.00	1.63	34.11
DIVISION	23.6	26.2	35.9	49.8	61.0	70.8	75.1	73.2	65.2	54.2	38.7	27.3	50.0	1.76	1.37	2.35	3.14	3.94	4.47	3.67	3.65	3.30	2.63	2.23	1.74	33.95
NORTHEAST																										
AURORA COLLEGE	24.3	26.3	35.5	48.6	59.2	69.2	73.7	72.4	64.3	53.3	38.4	27.3	49.4	1.96	1.54	2.50	2.89	3.74	4.28	3.15	3.47	3.26	2.87	2.23	1.91	33.80
CHICAGO O HARE AIRPORT	24.2	26.2	34.7	47.2	57.8	68.0	72.7	71.3	63.6	53.1	38.1	27.4	48.7	1.82	1.51	2.68	2.94	3.76	3.96	3.39	3.21	2.74	2.75	2.17	1.83	32.76
CHICAGO UNIVERSITY	27.1	28.6	36.3	48.1	58.4	68.8	74.2	73.4	66.2	55.7	40.8	30.2	50.7	1.80	1.58	2.69	3.08	3.87	3.79	3.09	3.19	2.77	2.71	2.08	1.80	32.45
CHICAGO WB AIRPORT	26.0	27.7	36.3	49.0	60.0	70.5	75.6	74.2	66.1	55.1	39.9	29.1	50.8	1.86	1.60	2.74	3.04	3.73	4.07	3.37	3.16	2.73	2.78	2.20	1.90	33.18
CHICAGO WB CITY	27.9	29.3	37.0	48.7	59.3	69.9	75.5	74.7	67.2	56.6	41.5	31.1	51.6	1.92	1.63	2.80	3.06	3.78	4.00	3.24	3.43	2.72	2.81	2.19	1.91	33.49
DE KALB	22.8	25.0	34.3	48.0	59.2	69.5	73.7	72.1	63.9	52.8	37.5	26.1	48.7	1.84	1.45	2.43	2.92	4.09	4.52	3.77	3.87	3.30	2.89	2.25	1.89	35.22
ELGIN	•	•	•	•	•	•	•	•	•	•	•	•	•	1.90	1.52	2.60	3.19	3.91	4.37	3.37	3.38	3.02	2.70	2.27	1.91	34.14
MORRIS 5 N	25.9	27.6	37.3	50.3	61.2	70.9	75.0	73.2	65.9	54.9	40.1	28.7	50.9	1.52	1.40	2.28	3.55	3.71	3.49	3.02	3.06	2.80	2.50	2.02	1.64	30.99
OTTAWA	26.4	28.6	37.9	51.2	61.9	71.9	76.1	74.5	66.5	55.5	40.6	29.5	51.7	1.80	1.57	2.58	3.73	3.74	3.96	3.57	3.53	3.01	2.42	1.94	1.84	33.69
PERU SEWAGE PLANT	26.5	28.9	38.2	51.6	62.6	72.8	77.0	75.3	67.3	55.9	40.8	29.7	52.2	1.81	1.57	2.60	3.56	3.83	3.69	3.39	3.23	3.00	2.50	1.88	1.73	32.79
WAUKEGAN 2 WNW	24.3	26.0	34.3	46.3	56.7	67.2	72.5	71.6	64.3	53.4	38.9	27.7	48.6	1.96	1.48	2.38	3.21	3.71	3.88	3.18	3.30	3.01	2.36	2.37	1.97	32.81
DIVISION	25.3	27.1	36.1	49.1	59.8	70.0	74.6	73.2	65.4	54.4	39.4	28.3	50.2	1.86	1.56	2.50	3.20	3.84	3.97	3.27	3.42	3.02	2.65	2.15	1.83	33.44
WEST																										
GALESBURG	•	•	•	•	•	•	•	•	•	•	•	•	•	1.90	1.63	2.85	3.46	4.24	4.75	3.51	3.20	3.05	2.91	2.14	1.82	35.46
GOLDEN 1 NW	•	•	•	•	•	•	•	•	•	•	•	•	•	1.80	1.52	2.76	3.47	4.13	4.44	3.84	3.34	3.31	2.86	2.09	1.81	35.37
LA HARPE	26.3	29.4	38.9	52.1	62.7	72.5	76.6	74.8	67.1	56.4	40.9	30.0	52.3	1.79	1.47	2.80	3.59	3.95	4.96	3.54	3.79	3.33	2.82	2.05	1.85	35.94
MACOMB	•	•	•	•	•	•	•	•	•	•	•	•	•	1.76	1.49	2.80	3.52	4.20	4.87	3.23	3.59	3.28	2.76	2.07	1.76	35.33
MONMOUTH	25.2	28.3	37.9	51.3	62.0	71.8	76.0	74.2	66.4	55.6	39.9	29.0	51.5	1.95	1.65	2.73	3.42	4.12	5.20	3.61	3.30	3.29	2.74	2.14	1.75	35.90
QUINCY	28.9	32.4	41.6	54.5	65.2	75.1	79.6	77.8	69.7	58.0	43.0	32.5	54.9	1.76	1.60	2.56	3.47	4.18	4.67	3.58	3.88	3.53	2.85	2.07	1.82	35.97
QUINCY MEMORIAL BRIDGE	•	•	•	•	•	•	•	•	•	•	•	•	•	1.77	1.58	2.54	3.50	4.20	4.59	3.75	3.85	3.52	2.81	2.08	1.78	35.97
RUSHVILLE	28.0	31.1	39.9	52.9	63.5	73.3	77.5	75.6	67.9	57.1	41.6	31.0	53.3	1.82	1.56	2.70	3.78	4.47	4.39	3.33	3.62	3.05	2.69	2.39	2.00	35.80
WARSAW	•	•	•	•	•	•	•	•	•	•	•	•	•	1.74	1.52	2.70	3.45	3.84	4.95	3.59	3.47	3.35	2.76	2.04	1.70	35.11
DIVISION	27.0	30.2	39.4	52.5	63.2	73.1	77.3	75.5	67.7	56.7	41.2	30.5	52.9	1.81	1.56	2.70	3.58	4.19	4.88	3.53	3.56	3.26	2.75	2.20	1.83	35.78
CENTRAL																										
BLOOMINGTON NORMAL	27.6	30.2	39.1	52.2	62.9	72.8	76.7	75.0	67.6	56.7	41.2	30.5	52.7	2.13	1.79	2.79	3.94	4.08	4.61	3.15	3.18	3.07	2.80	2.61	2.05	36.20
CLINTON 1 SSW	•	•	•	•	•	•	•	•	•	•	•	•	•	2.02	1.99	3.01	3.87	4.03	4.98	3.69	3.23	3.24	3.13	2.67	1.84	37.70
DECATUR	29.8	32.4	40.9	53.4	63.9	73.6	77.5	75.9	68.7	57.7	42.5	32.4	54.1	2.14	2.17	2.90	3.93	4.30	4.70	3.07	2.90	3.15	3.18	2.66	2.03	37.13
HAVANA	27.7	30.8	40.0	53.1	63.7	73.5	77.3	75.6	68.0	56.5	41.4	30.8	53.2	1.75	1.68	2.64	3.67	4.09	4.22	3.49	3.28	3.02	2.62	2.24	1.79	34.49
HAVANA POWER STATION	27.7	30.8	40.0	53.1	63.7	73.5	77.3	75.6	68.0	56.5	41.4	30.8	53.2	1.82	1.77	2.84	3.78	3.96	4.14	3.53	3.31	3.01	2.62	2.33	1.92	35.03
LINCOLN	28.5	31.1	40.0	52.8	63.5	73.2	77.2	75.4	67.7	56.5	41.6	31.2	53.2	2.02	1.95	2.91	3.80	3.96	4.51	3.25	3.12	2.98	2.78	2.39	2.14	35.81
MINONK	•	•	•	•	•	•	•	•	•	•	•	•	•	1.73	1.64	2.56	3.58	4.03	3.83	3.34	3.28	2.85	2.54	2.11	1.92	33.41
PEORIA WB AIRPORT	25.7	28.4	37.6	50.8	61.5	71.7	76.0	74.3	66.4	55.3	39.7	29.1	51.4	1.88	1.71	2.85	3.97	4.27	4.08	3.54	2.88	3.05	2.53	2.14	1.94	34.84
PRINCEVILLE 1 N	•	•	•	•	•	•	•	•	•	•	•	•	•	1.55	1.46	2.48	3.36	4.23	3.91	3.39	2.65	3.04	2.30	1.92	1.56	31.85
DIVISION	27.5	30.2	39.2	52.2	62.8	72.7	76.7	75.0	67.4	56.3	41.0	30.4	52.9	1.94	1.77	2.77	3.76	4.09	4.21	3.32	3.21	3.01	2.70	2.31	1.95	35.51
EAST																										
DANVILLE	29.2	31.6	39.8	52.0	62.3	72.2	75.9	74.2	66.9	55.9	41.6	31.5	52.8	2.21	2.11	2.77	3.28	4.33	4.57	3.67	3.12	3.24	3.06	2.93	2.15	37.44
HOOPESTON	28.1	30.4	38.9	51.4	62.2	72.2	76.0	74.2	67.0	55.9	40.9	30.4	52.3	2.10	2.08	3.01	3.60	4.04	4.56	3.82	2.79	3.11	3.21	2.74	1.98	37.04
KANKAKEE 3 SW	26.7	28.7	37.7	50.3	61.0	71.1	75.2	73.4	66.1	55.2	40.3	29.5	51.3	1.69	1.79	2.41	3.75	4.16	4.08	3.11	3.43	2.51	2.72	2.09	1.77	33.51
PIPER CITY 3 SE	•	•	•	•	•	•	•	•	•	•	•	•	•	1.76	1.85	2.75	3.90	3.99	3.93	3.30	3.11	2.72	2.75	2.09	1.74	33.89
PONTIAC	27.4	29.5	38.6	51.6	62.3	72.1	76.1	74.2	66.9	56.2	41.0	30.2	52.2	1.73	1.67	2.44	3.87	3.76	4.09	3.73	3.07	2.84	2.66	2.07	1.90	33.83
URBANA	28.7	31.0	39.5	51.9	62.5	72.4	76.2	74.4	67.3	56.6	41.3	31.1	52.7	2.16	2.09	3.17	3.54	4.22	4.54	3.49	3.04	3.04	3.01	2.62	2.08	37.00
WATSEKA 2 NW	•	•	•	•	•	•	•	•	•	•	•	•	•	1.99	1.89	2.93	3.92	4.54	3.91	4.00	3.06	2.73	2.77	2.33	1.94	36.01
DIVISION	27.8	29.9	38.6	51.2	61.9	71.9	75.8	74.0	66.7	55.8	40.8	30.3	52.0	1.96	1.95	2.76	3.69	4.11	4.30	3.57	3.08	2.87	2.88	2.44	1.96	35.41
WEST SOUTHWEST																										
BEARDSTOWN	•	•	•	•	•	•	•	•	•	•	•	•	•	1.81	1.68	2.70	3.77	4.09	4.35	3.59	2.95	3.02	2.71	2.46	1.85	34.98
CARLINVILLE	30.8	33.8	42.1	54.2	64.1	73.9	78.0	76.5	69.1	58.2	43.3	33.5	54.8	2.17	2.13	3.15	3.92	4.09	4.93	3.04	3.19	3.02	2.97	2.89	2.02	37.52
EDWARDSVILLE	•	•	•	•	•	•	•	•	•	•	•	•	•	2.38	2.35	3.62	4.19	4.87	4.69	3.84	3.51	3.32	3.16	3.13	2.16	41.22
GRAFTON	•	•	•	•	•	•	•	•	•	•	•	•	•	2.12	2.20	2.98	3.59	4.08	4.06	3.32	3.22	2.85	2.76	2.81	2.03	36.02
GRIGGSVILLE	29.1	32.5	41.3	54.1	64.6	74.2	78.5	76.7	69.1	58.2	42.9	32.5	54.5	•	•	•	•	•	•	•	•	•	•	•	•	•
HILLSBORO	31.6	34.5	42.8	54.7	64.3	73.8	77.7	76.3	69.0	58.3	43.6	32.7	54.9	2.27	2.30	3.17	3.83	4.22	4.63	3.40	3.46	2.87	3.09	2.84	2.02	38.10
JACKSONVILLE	29.4	32.6	41.3	53.9	64.0	73.8	77.4	75.8	68.3	57.3	42.7	32.4	54.1	1.78	1.70	2.75	3.52	4.26	4.27	3.09	3.03	3.03	3.00	2.52	1.86	34.81
MORRISONVILLE 4 SE	30.2	32.7	41.3	53.2	63.4	73.2	76.9	75.3	68.1	57.2	42.4	32.6	53.9	1.90	2.02	2.71	3.45	3.71	4.51	2.97	3.60	2.79	2.76	2.55	1.81	34.78
PANA	30.5	33.3	41.8	53.9	64.2	74.1	77.9	76.0	68.6	57.8	42.7	32.9	54.5	2.15	2.26	2.93	3.59	4.30	4.54	2.86	4.07	2.99	2.91	2.75	2.13	37.48
SPRINGFIELD WB AIRPORT	28.4	31.6	40.2	53.4	63.9	74.0	78.0	75.2	67.6	56.7	41.8	32.0	53.6	1.89	1.82	2.88	3.59	3.88	4.45	3.49	2.74	2.93	2.91	2.36	1.89	34.83
WHITE HALL 1 E	30.6	33.8	42.2	54.5	64.4	74.1	78.1	76.6	69.1	58.4	43.6	33.5	54.9	1.68	1.60	2.66	3.43	4.06	4.39	3.25	3.11	2.81	3.06	2.49	1.77	34.31
DIVISION	30.5	33.5	42.0	54.2	64.3	74.0	77.9	76.3	68.9	58.0	43.1	33.2	54.7	2.02	1.99	2.96	3.67	4.14	4.52	3.18	3.34	2.90	2.94	2.65	1.95	36.31

*NORMALS BY CLIMATOLOGICAL DIVISIONS

Taken from "Climatography of the United States No. 81-4, Decennial Census of U. S. Climate"

TEMPERATURE (°F)   PRECIPITATION (In.)

STATIONS (By Divisions)	JAN	FEB	MAR	APR	MAY	JUNE	JULY	AUG	SEPT	OCT	NOV	DEC	ANN	JAN	FEB	MAR	APR	MAY	JUNE	JULY	AUG	SEPT	OCT	NOV	DEC	ANN
EAST SOUTHEAST																										
CHARLESTON	•	•	•	•	•	•	•	•	•	•	•	•	•	2.40	2.13	3.34	3.51	4.45	4.61	3.39	3.34	3.17	2.90	3.14	2.19	38.57
EFFINGHAM	31.0	33.4	41.6	53.5	63.5	73.4	77.1	75.6	68.2	57.3	42.7	33.1	54.2	2.44	2.34	3.48	3.63	4.28	4.65	3.62	3.30	3.11	3.05	3.38	2.33	39.61
FLORA 2 NW	33.3	35.7	44.0	55.7	65.5	74.8	78.3	76.7	69.5	58.9	44.7	35.3	56.0	•	•	•	•	•	•	•	•	•	•	•	•	•
OLNEY 2 S	33.6	36.1	44.1	55.6	65.2	74.7	78.2	76.9	70.0	59.0	45.0	35.5	56.2	•	•	•	•	•	•	•	•	•	•	•	•	•
PALESTINE	32.6	35.1	43.1	55.0	64.9	74.7	78.2	76.6	69.2	58.0	43.9	34.5	55.5	3.16	2.42	3.53	3.86	4.44	4.37	3.40	3.04	3.42	2.93	3.47	2.93	40.97
PARIS WATERWORKS	30.3	32.7	40.8	53.1	63.6	73.3	77.1	75.5	68.3	57.3	42.3	32.4	53.9	2.74	2.22	3.24	3.41	4.43	4.97	3.42	2.88	3.18	2.88	3.08	2.49	38.94
SALEM	•	•	•	•	•	•	•	•	•	•	•	•	•	2.74	2.50	3.45	3.87	4.35	4.36	3.57	3.25	3.17	3.05	3.31	2.70	40.32
TUSCOLA	•	•	•	•	•	•	•	•	•	•	•	•	•	2.25	2.16	3.35	3.63	4.19	4.75	3.06	3.52	3.25	3.10	2.95	2.21	38.42
WINDSOR	30.2	32.8	40.9	53.1	63.5	73.3	76.9	75.2	68.1	57.1	42.3	32.5	53.8	2.56	2.44	3.37	3.84	4.44	5.04	3.49	3.46	3.22	3.17	3.01	2.44	40.48
DIVISION	31.5	34.0	42.2	54.1	64.2	73.9	77.6	76.0	68.7	57.8	43.3	33.3	54.8	2.78	2.38	3.46	3.79	4.42	4.48	3.48	3.23	3.16	2.94	3.27	2.55	39.65
SOUTHWEST																										
ANNA 1 E	35.5	38.4	46.1	57.3	66.3	75.2	78.6	77.5	70.7	60.2	46.4	37.7	57.5	4.10	3.55	4.79	4.77	5.20	4.39	3.22	4.08	3.68	3.27	3.97	3.54	48.56
CAIRO WB CITY	37.4	40.7	48.2	59.4	69.0	77.9	81.1	79.9	72.5	61.9	47.9	39.5	59.6	4.46	3.67	4.79	4.07	4.39	4.13	3.19	3.10	3.01	2.88	3.87	3.67	45.23
CARBONDALE SEWAGE PL	35.5	38.4	46.2	57.3	66.4	75.6	78.9	77.8	70.5	59.6	46.0	37.7	57.5	3.66	3.05	4.19	4.36	4.63	4.19	3.23	3.86	3.47	3.15	3.62	2.98	44.39
CHESTER	•	•	•	•	•	•	•	•	•	•	•	•	•	2.73	2.40	3.34	4.06	4.36	3.88	2.95	3.53	2.97	2.60	3.04	2.44	38.30
CREAL SPRINGS	35.4	38.2	46.0	57.1	65.9	74.8	78.1	77.3	70.3	59.7	46.1	37.4	57.2	4.38	3.50	4.68	4.67	4.79	4.19	3.35	3.61	3.45	3.05	3.84	4.31	46.82
DU QUOIN 2 S	34.8	37.7	45.6	57.0	66.5	75.6	79.0	77.8	70.6	59.7	46.0	37.1	57.3	3.15	2.71	3.79	4.01	4.15	3.89	3.22	3.64	3.34	3.15	3.49	2.86	41.40
MARION 4 NNE	35.4	38.2	46.0	57.1	65.9	74.8	78.1	77.3	70.3	59.7	46.1	37.4	57.2	4.38	3.50	4.68	4.67	4.79	4.19	3.35	3.61	3.45	3.05	3.84	4.31	46.82
NASHVILLE 4 NE	•	•	•	•	•	•	•	•	•	•	•	•	•	2.69	2.46	3.34	4.06	4.28	4.06	3.04	3.81	3.22	3.07	3.15	2.54	39.72
SPARTA	34.7	37.8	45.8	57.5	66.5	75.9	79.6	78.5	71.5	60.8	46.2	37.1	57.7	2.73	2.41	3.40	4.07	4.44	3.85	3.29	3.70	3.23	3.17	3.24	2.35	39.88
WATERLOO 1 WSW	•	•	•	•	•	•	•	•	•	•	•	•	•	2.33	2.24	3.13	3.75	4.39	4.03	3.44	3.35	3.32	3.08	3.28	2.32	38.66
DIVISION	34.8	37.8	45.7	57.1	66.4	75.6	79.1	77.9	70.7	59.9	45.9	37.1	57.4	3.37	2.89	3.94	4.21	4.47	4.30	3.34	3.80	3.42	3.12	3.52	2.88	43.45
SOUTHEAST																										
BENTON FOREST SERVICE	•	•	•	•	•	•	•	•	•	•	•	•	•	3.40	2.59	3.69	3.94	4.07	3.58	3.07	3.80	3.40	2.84	3.28	2.82	40.48
BROOKPORT DAM 52	•	•	•	•	•	•	•	•	•	•	•	•	•	4.93	3.89	4.84	4.23	4.24	3.99	3.65	3.30	3.49	2.71	3.93	3.49	46.69
CARMI 6 NW	•	•	•	•	•	•	•	•	•	•	•	•	•	3.84	3.13	4.16	3.90	4.29	3.52	3.28	3.47	3.30	2.72	3.36	3.05	42.02
FAIRFIELD RADIO WFIW	34.3	37.0	44.8	56.3	65.7	75.2	78.7	77.4	70.3	59.5	45.5	36.4	56.8	•	•	•	•	•	•	•	•	•	•	•	•	•
GOLCONDA DAM 51	•	•	•	•	•	•	•	•	•	•	•	•	•	4.40	3.57	4.77	4.02	3.97	3.67	3.65	3.37	3.16	2.58	3.75	3.39	44.30
HARRISBURG	36.4	39.0	46.8	57.8	67.1	76.0	79.5	78.2	71.0	60.2	46.9	39.2	58.1	3.94	3.04	4.33	4.13	4.25	3.69	3.29	3.54	2.93	2.85	3.32	2.97	42.28
MCLEANSBORO	35.1	37.8	45.7	57.2	66.4	75.6	79.1	78.0	71.0	60.1	46.1	37.1	57.4	3.68	2.82	3.87	4.04	4.32	3.71	3.36	3.62	3.49	2.91	3.51	3.00	42.33
MOUNT VERNON	33.4	36.3	44.3	56.1	65.8	75.3	79.0	77.8	70.5	59.3	44.8	35.7	56.5	3.27	2.63	3.77	4.46	4.36	3.97	3.26	3.83	3.46	3.09	3.55	2.84	42.49
SHAWNEETOWN NEW TOWN	•	•	•	•	•	•	•	•	•	•	•	•	•	4.39	3.20	4.79	4.36	4.19	3.85	3.76	3.82	2.98	2.72	3.57	3.04	44.67
DIVISION	34.9	37.6	45.5	57.0	66.3	75.6	79.1	77.9	70.8	59.8	45.9	37.0	57.3	3.79	2.94	4.05	4.21	4.42	3.92	3.52	3.50	3.31	2.91	3.62	3.08	43.15

* Normals for the period 1931-1960. Divisional normals may not be the arithmetical averages of individual stations published, since additional data for shorter period stations are used to obtain better areal representation.

TEMPERATURE   PRECIPITATION

Jan.	Feb.	Mar.	Apr.	May	June	July	Aug.	Sept.	Oct.	Nov.	Dec.	Annual	Jan.	Feb.	Mar.	Apr.	May	June	July	Aug.	Sept.	Oct.	Nov.	Dec.	Annual

CONFIDENCE LIMITS

In the absence of trend or record changes, the chances are 9 out of 10 that the true mean will lie in the interval formed by adding and subtracting the values in the following table from the means for any station in the State. Because of the wider variation in mean precipitation, the corresponding monthly means and annual mean must be substituted for "p" in the precipitation table below to obtain mean precipitation confidence limits.

Jan.	Feb.	Mar.	Apr.	May	June	July	Aug.	Sept.	Oct.	Nov.	Dec.	Annual	Jan.	Feb.	Mar.	Apr.	May	June	July	Aug.	Sept.	Oct.	Nov.	Dec.	Annual
1.7	1.6	1.6	1.0	1.1	1.0	.8	.8	1.0	1.2	1.2	1.5	.5	$.33\sqrt{p}$	$.24\sqrt{p}$	$.32\sqrt{p}$	$.28\sqrt{p}$	$.36\sqrt{p}$	$.36\sqrt{p}$	$.35\sqrt{p}$	$.36\sqrt{p}$	$.40\sqrt{p}$	$.34\sqrt{p}$	$.30\sqrt{p}$	$.23\sqrt{p}$	$.33\sqrt{p}$

COMPARATIVE DATA

Data in the following table are the mean temperature and average precipitation for Urbana, (University of Illinois) Illinois, for the period 1906-1930 and are included in this publication for comparative purposes.

Jan.	Feb.	Mar.	Apr.	May	June	July	Aug.	Sept.	Oct.	Nov.	Dec.	Annual	Jan.	Feb.	Mar.	Apr.	May	June	July	Aug.	Sept.	Oct.	Nov.	Dec.	Annual
26.2	29.6	40.0	50.5	60.9	70.1	74.9	72.7	66.4	54.3	41.6	29.7	51.4	2.28	1.73	3.23	3.91	3.97	3.45	2.96	3.72	3.61	2.67	2.36	2.23	36.12

POST OFFICE BUILDING

Month	(b)	Normal Daily maximum	Normal Daily minimum	Normal Monthly	Extremes Record highest	Year	Extremes Record lowest ∅	Year	Normal degree days	Precip. Normal total	Precip. Max monthly	Year	Precip. Min monthly	Year	Precip. Max in 24 hrs	Year	Snow Mean total	Snow Max monthly	Year	Snow Max in 24 hrs	Year	RH 6 AM	RH Noon	Wind Mean hourly speed	Prevailing dir.	Fastest mile Speed	Dir.	Year	Mean sky cover	Pct. sunshine	Clear	Partly cloudy	Cloudy	Precip .01+	Snow 1.0+	Thunderstorms	Heavy fog	Max 90+	Max 32−	Min 32−	Min 0−
(a)	(b)	(b)	(b)	(b)	26	26	26	26	(b)	(b)	26		26		26		26	26		26		22	22	22	22	26	26	26	26	26	26	26	26	26	26	22	22	26	26	26	26	26
J		44.6	30.2	37.4	75	1950	−5	1963	856	4.46	14.95	1950	0.33	1943	6.09	1949	11.3		1951	6.3	1951	80	67	9.8	SW	43	SW	1965	6.6	46	8	6	17	10		2	1	0	5	20	*
F		48.5	32.9	40.7	76	1962			680	4.67	8.88	1950	0.81	1944	4.79	1950	8.9		1960	5.8	1960	79	64	9.8	NE	56	SW	1956	6.3	54	8	6	14	9	1	2	1	0	3	15	*
M		57.0	39.4	48.2	82	1963+	6		539	4.79	12.67	1964	1.12	1966	6.58	1964	20.3		1960	8.0	1967	78	59	10.2	SW	56	NW	1960	6.2	58	9	7	15	12	1	4	*	0	1	8	0
A		68.9	49.9	59.4	89	1965	30		195	4.07	9.45	1966	0.95	1949	3.79	1944	0.2		1962	0.2	1962	77	54	10.6	SW	59	SW	1957	6.2	58	7	9	14	12	0	5	*	0	0	*	0
M		78.5	59.4	69.0	98	1953	39		47	4.39	11.15	1948	1.97	1948	6.16	1948	0.0			0.0		77	54	10.2	SW	59	SW	1953	6.1	62	8	11	11	11	0	6	*	1	0	0	0
J		87.5	68.2	77.9	104	1934	51		0	4.13	10.24	1951	0.76	1952	5.91	1961	0.0			0.0		83	57	7.4	SW	50	NE	1952	5.7	67	8	13	10	10	0	7	*	3	0	0	0
J		90.3	71.8	81.1	104	1954+	54	1947	0	3.19	8.19	1955	0.84	1945	2.79	1952	0.0			0.0		84	59	6.5	SW	45	SW	1952	5.7	76	11	8	9	9	0	6	*	4	0	0	0
A		89.2	70.6	79.9	98	1953	53	1952	0	3.10	8.13	1952	0.81	1955	7.56	1952	0.0			0.0		87	57	6.2	SW	44	NW	1952	5.4	75	12	7	9	8	0	4	1	3	0	0	0
S		82.4	62.5	72.5	104	1954+	42	1967+	36	3.01	9.39	1965	0.47	1953	2.96	1965	0.0			0.0		85	53	7.0	NE	35	SW	1968+	4.9	71	11	8	10	7	0	3	1	*	0	*	0
O		71.9	51.9	61.9	92	1963	29	1965+	164	2.88	13.05	1954	T	1964	4.33	1957	0.4		1957	5.2	1957	80	58	7.3	S	53	SW	1968+	4.5	64	9	7	13	9	*	2	1	0	0	0	0
N		56.3	39.5	47.9	82	1955	6	1964	513	3.87	6.55	1949	0.70	1954	4.33	1954	5.3		1950	6.6	1950	79	58	9.1	S	53	SW	1968+	5.8	55	9	8	13	9	*	2	1	0	0	6	*
D		46.4	32.6	39.5	73	1951	−10	1962	791	3.67	7.99	1951	0.66	1955	2.64	1957	7.5		1950	6.6	1950	79	66	9.3	S	43	S	1957+	6.5	48	8	7	16	11	1	1	1	0	3	17	*
YR	(b)	68.5	50.7	59.6	104	JUL 1954+	−5	JAN 1963+	3821	45.23	14.95	JAN 1950	T	OCT 1964	7.56	AUG 1952	9.0	20.3	MAR 1960	8.0	MAR 1967	82	59	8.5	SW	60	NW	MAR 1964	5.8	65	111	102	152	114	3	53	7	12	51	66	*

LATITUDE 41° 47' N
LONGITUDE 87° 45' W
ELEVATION (ground) 607 Feet

Means and extremes in the above table are from the existing or comparable location(s). Annual extremes have been exceeded at other locations as follows:
Highest temperature 106 in August 1930 and earlier; lowest temperature −16 in January 1918 and earlier; maximum monthly precipitation 15.70 in June 1928;
maximum monthly snowfall 24.2 in December 1917; maximum snowfall 24 hours 17.0 in December 1917; fastest mile of wind 62 from North in May 1937.

CHICAGO, ILLINOIS — MIDWAY AIRPORT

Month	(b)	Normal Daily maximum	Normal Daily minimum	Normal Monthly	Extremes Record highest	Year	Extremes Record lowest ∅	Year	Normal degree days	Precip. Normal total	Precip. Max monthly	Year	Precip. Min monthly	Year	Precip. Max in 24 hrs	Year	Snow Mean total	Snow Max monthly	Year	Snow Max in 24 hrs	Year	RH Mid.	RH 6 AM	RH Noon	RH 6 PM	Wind Mean hourly speed	Prevailing dir.	Fastest mile Speed	Dir.	Year	Mean sky cover	Pct. sunshine	Clear	Partly cloudy	Cloudy	Precip .01+	Snow 1.0+	Thunderstorms	Heavy fog	Max 90+	Max 32−	Min 32−	Min 0−
(a)	(b)	5	5	5	5	5	5	5	(b)	(b)	26		26		26		26	26		26		5	5	5	5	26	19	26	26	26	26	26	26	26	26	26	26	26	26	5	5	5	5
J		33.4	19.0	26.0	65	1967	−16	1966	1209	1.86	4.09	1965	0.26	1961	2.86	1960	9.4	28.9	1967	19.8	1967	72	71	65	67	11.4	W	50	SW	1950	7.0	44	6	7	18	10	3	*	2	0	5	27	5
F		34.7	20.6	26.7	59	1965	−5	1965	1044	1.60	3.35	1950	0.33	1958	1.54	1960	8.1	22.5	1960	8.9	1956	76	68	62	66	11.6	W	51	SW	1967	6.7	47	7	6	15	9	3	*	1	0	4	26	4
M		43.5	29.0	36.3	78	1968+	5	1960	890	2.74	5.00	1954	0.33	1958	2.50	1948	7.1	22.3	1965	11.8	1954	74	68	59	62	12.1	W	54	NW	1967	6.6	51	6	8	17	12	3	2	1	0	2	26	2
A		57.0	40.5	48.8	84	1967+	26	1964	480	3.04	8.33	1947	0.46	1961	4.08	1947	0.8	6.8	1957	5.4	1961	74	68	58	60	11.7	SW	54	NW	1951	6.3	52	6	9	16	13	2	4	*	0	0	3	0
M		69.1	50.9	60.0	93	1967+	40		211	3.73	7.59	1945	0.78	1958	3.63	1966	T	0.8	1966+	0.2	1966+	73	67	59	63	10.4	SSW	54	NE	1950	6.3	61	6	11	14	13	*	5	*	1	0	0	0
J		79.5	60.9	70.2	96	1967	50	1959	48	4.07	6.89	1950	0.30	1962	4.58	1959	0.0	0.0		0.0		73	67	53	54	9.2	SW	50	SW	1953	6.0	67	7	12	11	11	0	7	*	6	0	0	0
J		84.1	67.1	75.6	98	1966	50	1957	0	3.37	8.98	1957	1.33	1945	6.24	1957	0.0	0.0		0.0		72	76	56	56	8.2	SW	46	NW	1959	5.3	70	10	12	9	9	0	6	*	7	0	0	0
A		82.4	66.1	74.2	98	1964	40	1962	0	3.16	6.79	1965	0.85	1962	3.83	1968	0.0	0.0		0.0		76	74	56	56	8.0	SW	54	NW	1954	5.3	68	10	10	11	8	0	5	1	3	0	0	0
S		74.8	57.4	66.1	94	1963	27	1961	81	2.73	14.17	1961	0.20	1964	2.63	1961	0.0	0.0		0.0		74	79	53	58	8.9	S	48	NW	1949	5.1	64	11	10	9	8	0	4	1	1	0	0	0
O		63.4	46.7	55.1	91	1963	17	1964	326	2.78	12.06	1954	0.89	1964	2.93	1954	0.4	4.4	1967	4.4	1967	74	76	59	62	9.5	SW	45	SW	1952	5.1	64	12	11	8	7	*	2	1	*	0	2	0
N		47.1	33.2	40.2	78	1964	−10	1966	753	2.20	5.05	1966	0.20	1966	2.63	1965	2.6	14.3	1951	8.0	1951	71	76	62	68	11.4	SW	60	SW	1952	5.7	42	6	7	17	10	3	1	1	0	1	14	0
D		35.7	22.5	29.1	63	1966			1113	1.90	6.67	1949	0.30	1962	2.79	1960	9.8	33.3	1951	12.5	1951	73	73	54	60	11.2	SW	50	SW	1948	6.0	41	7	6	18	12	3	*	2	0	11	24	6
YR	(b)	58.7	42.8	50.8	98	JUL 1966+	−16	JAN 1966	6155	33.18	14.17	SEP 1961	0.20	OCT 1964	6.24	JUL 1957	38.3	33.3	DEC 1951	19.8	JAN 1967	70	75	58	60	10.3	W	60	SW	NOV 1952	6.2	57	94	105	166	120	12	36	14	19	45	117	6

∅ For period August 1963 through 1967.
Means and extremes in the above table are from the existing location. Annual extremes have been exceeded at other locations as follows:
Highest temperature 105 in July 1934; lowest temperature −23 in December 1872; minimum monthly precipitation 0.06 in February 1877;
maximum monthly snowfall 42.5 in January 1918; fastest mile of wind 87 from Northeast in February 1894.

NORMALS, MEANS, AND EXTREMES

LATITUDE 41° 27' N
LONGITUDE 90° 31' W
ELEVATION (ground) 582 Feet

Means and extremes in the above table are from existing locations. Annual extremes have been exceeded at other locations as follows:
Highest temperature 106 in August 1936.

LATITUDE 40° 40' N
LONGITUDE 89° 41' W
ELEVATION (ground) 652 Feet

Means and extremes in the above table are from the existing or comparable location(s). Annual extremes have been exceeded at other locations as follows:
Highest temperature 113 in July 1936; lowest temperature -27 in January 1884; maximum precipitation in 24 hours 5.52 in May 1927; maximum monthly snowfall 26.5 in February 1900.
Highest temperature 18.0 in February 1900.

GREATER ROCKFORD AIRPORT

ø For period September 1963 through 1967.
Means and extremes in the above table are from the existing location. Annual extremes have been exceeded at other locations as follows:
Highest temperature 112 in July 1936; lowest temperature –25 in January 1924, and February 1933; maximum monthly snowfall 36.1 in January 1918.

LATITUDE 39° 50' N
LONGITUDE 89° 40' W
ELEVATION (ground) 588 Feet

SPRINGFIELD, ILLINOIS
CAPITAL AIRPORT

Means and extremes are from the existing location. Annual extremes have been exceeded at other locations as follows:
Highest temperature 112 in July 1954; lowest temperature –24 in February 1905; maximum monthly precipitation 15.16 in September 1926;
minimum monthly precipitation 0.02 in November 1904; maximum precipitation 15.0 in June 1917;
maximum monthly snowfall 24.4 in February 1900; maximum snowfall in 24 hours 15.0 in February 1900.

REFERENCE NOTES APPLYING TO ALL "NORMALS, MEANS, AND EXTREMES" TABLES

(a) Length of record, years.
(b) Climatological standard normals (1931–1960).
* Less than one half.
+ Also on earlier dates, months or years.
T Trace, an amount too small to measure.
Below-zero temperatures are preceded by a minus sign.
The prevailing direction for wind in the Normals, Means, and Extremes table is from records through 1963.
To 8 compass points only.
¢ Through 1964.

Unless otherwise indicated, dimensional units used in this bulletin are: temperature in degrees F.; precipitation, including snowfall, in inches; wind movement in miles per hour; and relative humidity in percent. Degree day totals are the sums of the negative departures of average daily temperatures from 65° F. Sleet was included in snowfall totals beginning with July 1948. Heavy fog reduces visibility to 1/4 mile or less.

Sky cover is expressed in a range of 0 for no clouds or obscuring phenomena to 10 for complete sky cover. The number of clear days is based on average cloudiness 0–3; partly cloudy days 4–7; and cloudy days 8–10 tenths.

& Figures instead of letters in a direction column indicate direction in tens of degrees from true North; i.e., 09–East, 18–South, 27–West, 36–North, and 00–Calm. Resultant wind is the vector sum of wind directions and speeds divided by the number of observations. If figures appear in the direction column under "Fastest mile" the corresponding speeds are fastest observed 1-minute values.

g Through 1964. The station did not operate 24 hours daily. Fog and thunderstorm data therefore may be incomplete.

– 101 –

Mean Annual Precipitation, Inches

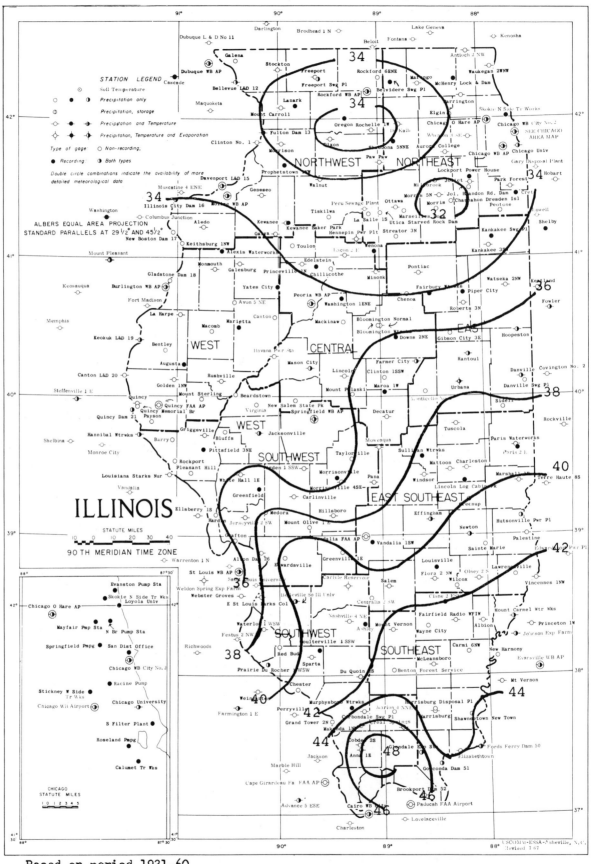

Based on period 1931-60
Isolines are drawn through points of approximately equal value.
Caution should be used in interpolating on these maps.

Mean Maximum Temperature (°F.), January

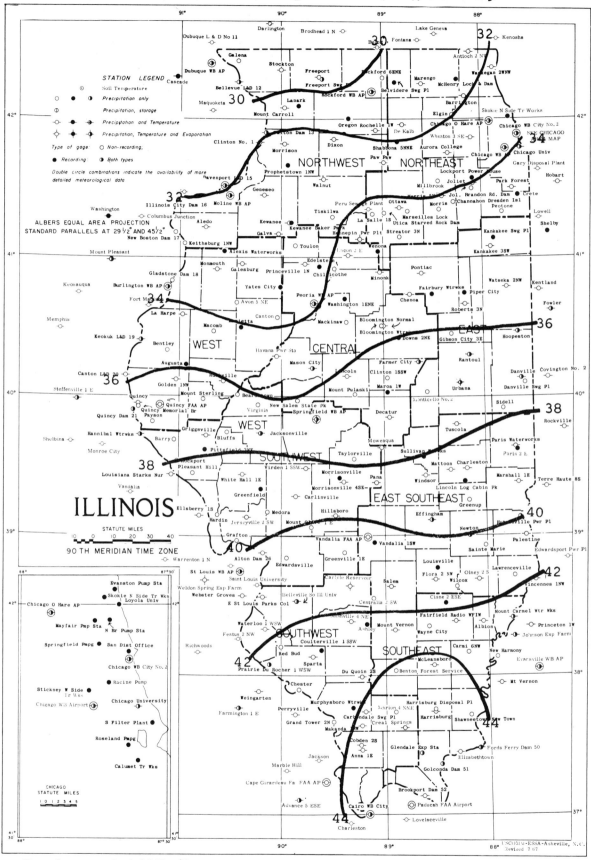

Based on period 1931-60
   Isolines are drawn through points of approximately equal value.
   Caution should be used in interpolating on these maps.

Mean Minimum Temperature (°F.), January

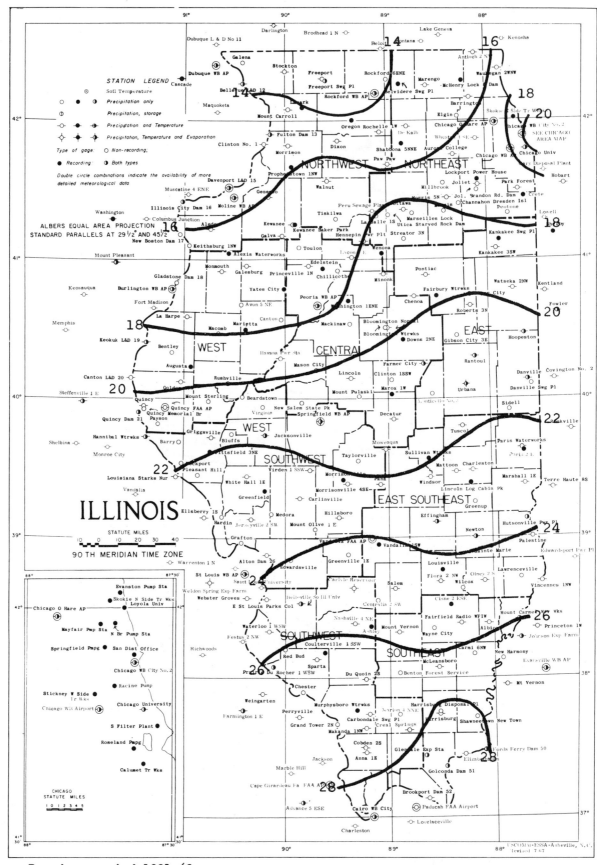

Based on period 1931-60
Isolines are drawn through points of approximately equal value.
Caution should be used in interpolating on these maps.

Mean Maximum Temperature (°F.), July

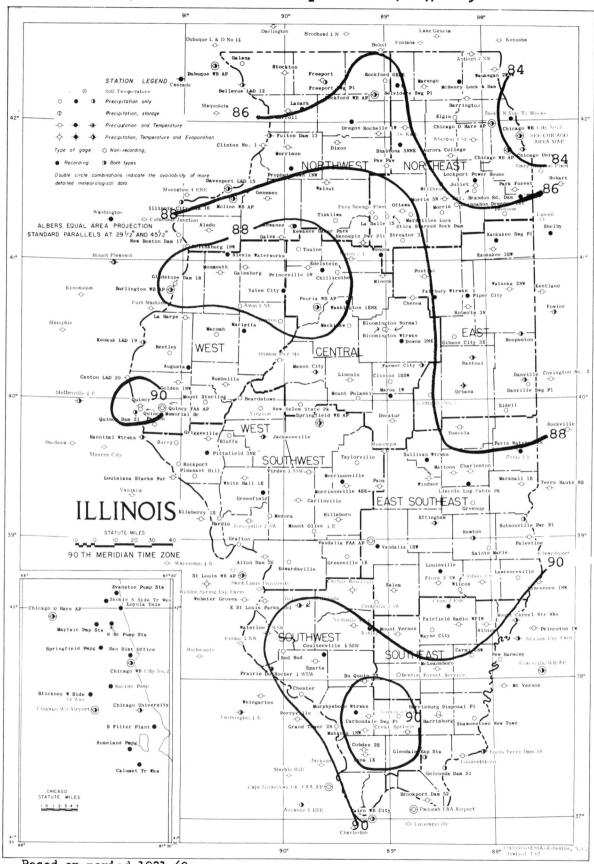

Based on period 1931-60
Isolines are drawn through points of approximately equal value.
Caution should be used in interpolating on these maps.

Mean Minimum Temperature (°F.), July

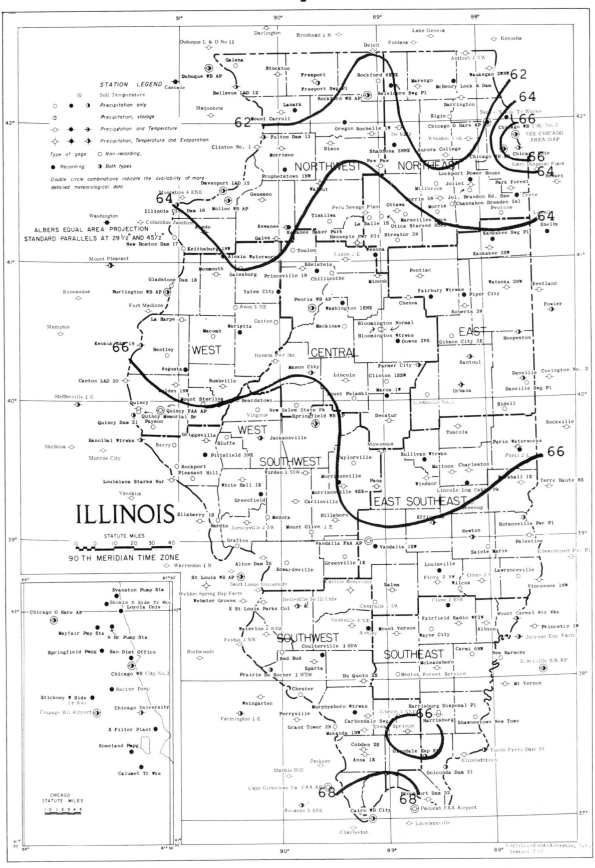

Based on period 1931-60
Isolines are drawn through points of approximately equal value.
Caution should be used in interpolating on these maps.

ILLINOIS

STATION LEGEND

Soil Temperature
Precipitation only
Precipitation, storage
Precipitation and Temperature
Precipitation, Temperature and Evaporation

Type of gage: ○ Non-recording;
● Recording; ◑ Both types

Double circle combinations indicate the availability of more
detailed meteorological data

ALBERS EQUAL AREA PROJECTION
STANDARD PARALLELS AT 29½° AND 45½°

STATUTE MILES
10 0   10   20   30   40

90TH MERIDIAN TIME ZONE

CHICAGO
STATUTE MILES
1 0 1 2 3 4 5

USC&GS—ESSA—Asheville, N.C.
Revised 12 - 68

THE CLIMATE OF
INDIANA

by
Lawrence A. Schaal

October 1959

Indiana has an invigorating climate of warm summers and cool winters, because of its location in the middle latitudes in the interior of a large continent. Imposed on the well known daily and seasonal changes of temperature are changes occurring every few days as surges of polar air move southeastward or air of tropical origin move northeastward. These outbreaks are more frequent and pronounced in the winter than in the summer. A winter may be unusually cold or a summer cool if the influence of polar air is rather continuous. Likewise, a summer may be unusually warm or a winter mild if air of tropical origin predominates. The action between these two air masses with a contrast in temperature and density fosters the development of low pressure centers which in moving generally eastward frequently pass through or near Indiana, resulting in normally abundant rain. The cyclones are least active and frequently pass north of Indiana in midsummer. Thunderstorms, often local in areal coverage, are important at such times when evaporation and loss of moisture from the soil and vegetation exceeds rainfall. Major climatological variations within the State are caused by differences of latitude, elevation, terrain, soil, and lakes.

The effect of the Great Lakes and more specifically, Lake Michigan, on the climate of northern Indiana is well defined in the climatological data. This is most pronounced just inland from the Lake Michigan shore and diminishes to insignificance in central Indiana. The result of cold air passing over the warmer lake water of Lake Michigan induces precipitation in the lee of Lake Michigan in fall and winter. Average daily minimum temperatures in the fall are higher and daily maximum temperatures in the spring are lower in northwestern Indiana than farther south. Winter precipitation, especially snowfall, is several times greater in the counties of Lake, Porter, and LaPorte as the result of this phenomena. Lake related snowfall and cloudiness often extends to central Indiana in the winter. Very local severe snowstorms have occurred just inland from Lake Michigan. The storm of February 15-18, 1958, is an example.

Another important variable in the composition of Indiana weather is the topography of the State. Elevations range from a little more than 300 feet at the mouth of the Wabash in the southwest corner of the State to a little over 1,200 feet in the east-central portion (Randolph County) and northeastern section (Steuben County). Differences of terrain effect the climate considerably. South-central Indiana is unglaciated and has the most rugged relief. The Kankakee Valley in the northwest has but little slope to the west and drains what was formerly marshlands, and now muck-land farming is practiced extensively. Many small lakes abound in northeastern Indiana among numerous glacial moraines and hills. Most of the north, central, and southwest is rolling country and excellent for farming.

Variations of temperature and precipitation occur in short distances where terrain is hilly. On calm, clear nights the valley bottoms have lower temperatures than the slopes and tops of the surrounding hills. The location of the climatic station is particularly important when estimating the overall climate of these areas. A temperature of 35°F. below zero occurred at Greensburg on February 2, 1951. This is the lowest temperature of record in the State, despite the fact that average temperatures are much lower in northern Indiana. Other factors resulting in the cold February night, were light winds, clear skies, and snow cover. On the same night Scottsburg and Salem with somewhat the same type of instrument exposure recorded 32°F. below zero.

The highest temperature on record is held by a northern station. The temperature on July 14, 1936, reached 116°F. at Collegeville. Mean maximum as well as mean minimum temperatures decrease from south to north with latitude and decrease from west to east with elevation. Near Lake Michigan temperatures average higher than expected for the latitude in the fall and winter and lower than expected for the latitude in the spring and summer.

The date of the last freezing temperature in the spring and the first in the fall varies greatly from year to year. Two-thirds of the time it occurs within a 20- to 24-day period centered at the mean date. The average date of the last freezing temperature in the spring ranges from the first week of April in the Ohio River Valley of the southwest to the second week of May in the extreme northeast. The usual trend of a later date toward the north is reversed in extreme northwestern Indiana, where the average date is about April 30 near Lake Michigan. In the fall the average date of the first temperature of 32°F. or colder is from October 7 in the extreme northeast to October 26 along the Ohio River in the southwest.

Spring freezes are later in valleys and hollows and fall freezes are earlier. Longer freeze-free periods occur on ridges and hills. Southern Indiana has much of this type of terrain. The average date of last spring freeze and first fall freeze, therefore, varies considerably from place to place. The gradual slope upward from southwestern Indiana to northeastern Indiana results in lower minimum temperatures and shorter growing seasons in the east compared to the west at the same latitude. In the Kankakee Valley, peat or muck lands experience late spring and early fall frosts because of the radiative characteristic of the soil.

Average annual rainfall ranges from 36 inches in northern Indiana to 43 inches in southern Indiana. July rainfall averages about the same in all areas. The greater precipitation in the south compared with the north comes in the winter months. Southern Indiana has the greatest rainfall in March and the least in October. The wettest month in northern and central Indiana is June and the driest is February. Climatic data indicates that 2 inches of rain in one hour occurs about once in 23 years at Ft. Wayne and about once in 10 years at Evansville.

Annual precipitation is adequate, but an uneven distribution within a year occasionally limits crops and ground water supplies. A drought occasionally occurs in the summer when evaporation is highest and dependence on rainfall is greatest for crops. Approximately one-third of the annual

rainfall flows to the Mississippi or Great Lakes, mainly during cool weather. The soil usually becomes saturated with water during the winter, and when combined with frozen ground most water runs off to streams and lakes. Ground water storage is generally abundant in the north, where glacial deposits cover ancient lake beds or streams. An underlying bed of limestone in much of southern Indiana with shallower soils limits ground water storage.

The soils of Indiana account for some variations in moisture needs for farming. Droughts develop more rapidly in some of the sands of northern Indiana. On the other hand, drainage is rapid in rainy periods and flooding is seldom a problem. Some of the clays in east-central Indiana are "tight", with poor drainage and frequent ponding, but distress of crops from lack of rain is slow in coming. Much of south-central Indiana has a poor retention capacity for water because of the underlying limestone. Such areas need frequent rains.

Most of the State is drained by the Wabash River system. The total drainage area of the Wabash is 33,000 square miles, of which 24,000 square miles is in Indiana. Other river basins are the Maumee in the extreme northeast, the St. Joseph (Lake Michigan) and Kankakee (Illinois River) in the north-central and northwest, and some Ohio River drainage in the extreme south and southeast.

Floods occur in some part of the State nearly every year and have occurred in every month of the year. The season of greatest flood frequency is during the winter and spring months. The primary cause of floods is prolonged periods of heavy rains, although occasionally the rains falling on a snow cover and the formation of ice jams are an added factor. The most common type of flood producing storm in the area are those having a quasi-stationary front oriented from west-southwest to east-northeast with a series of waves or perturbations moving to the east along the front.

The most devastating of the floods in the State occurred in March 1913. Since then, important flood years include 1916, 1930, 1937, 1943, 1950, 1958, and 1959. In earlier years, it is known that large floods occurred in 1828, January 1847, August 1875, and March-April 1904.

Average annual snowfall increases from about 10 inches in southern Indiana to 40 inches in the northern portion of the State and higher in the three county areas along Lake Michigan. From year to year snowfall varies greatly, depending both on temperatures and the frequency of winter storms. At a given latitude in central and southern Indiana snowfall is greatest toward the east because of higher elevations.

Cloudiness is least in the fall and greatest in the winter. The north is cloudier than the south, particularly in the winter when the Great Lakes have the greatest effect upon the weather.

Average relative humidity differs very little at night over Indiana. During the day relative humidity is usually lower in the south than in the north. This is true for all seasons. However, the simultaneous occurrence of high temperatures and high relative humidity is most frequent in the south.

Prevailing winds are from the southwest quadrant throughout most of the year. Winds from the northern quadrant occur in the winter and persist for a longer time in the north. Along the shore of Lake Michigan the sea-breeze effect is observed in the summer when winds in central United States are light or calm. Vertical currents from the heating of land during the day causes wind near the ground to flow from over water to land reducing the maximum temperature of the day. At night the breezes are in the opposite direction or from the land to water because of land cooling. These breezes are important in limiting extremely high temperatures of a summer day and account for rapid changes in short distances within a mile or so of the lake shore. Winds meet less friction passing over water so off-lake winds have a considerably higher speed than those off or over land.

Evaporation from a water surface in a small tank ranges from 6 to 7 inches at Valparaiso to 8 inches at Evansville in July. Evaporation is 3 or 4 inches in the north in April and October, a little higher in the south.

Severe storms which damage property and cause loss of life are most frequent in the spring. Indiana, with 277 tornadoes during the past 43 years, ranks 14th in the Nation in the number of these storms. About one-half of the tornadoes occur between 2 p.m. and 6 p.m. and nearly three-fourths between 10 a.m. and 10 p.m. Several deaths are caused by lightning each year, and property damage is greatest from thunderstorm winds. Hail occasionally causes loss of crops in very local areas.

The length of Indiana from a latitude of 38° to nearly 42°N. results in various activities suitable to the climate. Corn, fall wheat, spring oats, and soybeans are the most important crops in the State. Tomatoes, strawberries, and melons are grown commercially in the southwest where the growing season is long. Tobacco is grown in the extreme southeast. Fruit is grown near Lake Michigan and in the lower Wabash River Basin where chances of late spring freezes are a minimum. Potatoes and mint are suited to northern Indiana climate. Lespedeza is a legume which grows well in southern Indiana. Many industries of northern and central Indiana depend on abundant ground water supplies and replenishing rains. Research in future years will reveal other ways to take advantage of the State's climate.

REFERENCES

Climatological Summary - Richmond, Indiana. U. S. Weather Bureau and Richmond's Committee of 100 Inc.

Climatological Summary - Anderson, Lafayette, Madison, Marion, Muncie, Richmond, Rochester, Valparaiso, Indiana.

Maximum Station Precipitation for 1, 2, 3, 6, 12 and 24 hours, Technical Paper No. 15. Part XX: Indiana, U. S. Weather Bureau.

Summary of Hourly Observations - (Climatography of the United States No. 30-12) Evansville, Ft. Wayne, and Indianapolis.

Climate of the States - Indiana, Agricultural Yearbook.

Climate of Indiana by S. S. Visher.

Rainfall Intensity - Duration - Frequency Curves, Technical Paper No. 25, U. S. Weather Bureau.

BIBLIOGRAPHY

(1) Climatic Summary of the United States (Bulletin W) 1930 edition, Sections 66 and 67. U. S. Weather Bureau

(2) Climatic Summary of the United States, Indiana - Supplement for 1931 through 1952 (Bulletin W Supplement). U. S. Weather Bureau

(3) Climatological Data - Indiana. U. S. Weather Bureau

(4) Climatological Data National Summary. U. S. Weather Bureau

(5) Hourly Precipitation Data - Indiana. U. S. Weather Bureau

(6) Local Climatological Data, U. S. Weather Bureau, for Evansville, Fort Wayne, Indianapolis, South Bend, Indiana.

FREEZE DATA

STATION	Freeze threshold temperature	Mean date of last Spring occurrence	Mean date of first Fall occurrence	Mean No. of days between dates	Years of record Spring	No. of occurrences in Spring	Years of record Fall	No. of occurrences in Fall
ALBION	32	05-07	10-11	157	30	30	30	30
	28	04-20	10-27	190	30	30	30	30
	24	04-06	11-07	215	30	30	30	30
	20	03-24	11-18	239	30	30	30	30
	16	03-14	11-29	260	30	30	30	29
ANDERSON QUARTZ PLT	32	04-26	10-20	178	30	30	30	30
	28	04-09	11-02	208	30	30	30	30
	24	03-28	11-13	230	30	30	30	30
	20	03-16	11-25	254	30	30	30	29
	16	03-05	12-05	276	30	29	30	27
ANGOLA	32	05-06	10-16	163	30	30	30	30
	28	04-18	10-30	195	30	30	30	30
	24	04-03	11-11	222	30	30	30	30
	20	03-25	11-22	242	30	30	30	30
	16	03-13	11-29	261	30	30	30	29
BERNE	32	05-04	10-10	160	30	30	30	30
	28	04-19	10-26	190	30	30	30	30
	24	04-01	11-07	220	30	30	30	30
	20	03-18	11-21	248	30	30	30	30
	16	03-09	12-02	269	30	30	30	28
BLOOMINGTON IND UNIV	32	04-21	10-20	182	30	30	30	30
	28	04-05	11-02	211	30	30	30	30
	24	03-23	11-13	235	30	30	30	30
	20	03-12	11-26	259	30	30	30	29
	16	03-03	12-08	280	30	30	30	28
BLUFFTON WATER WORKS	32	05-05	10-09	157	29	29	30	30
	28	04-26	10-20	177	29	29	30	30
	24	04-03	11-04	215	30	30	30	30
	20	03-21	11-12	237	30	30	30	30
	16	03-12	11-27	260	30	30	29	28
BROOKVILLE 1 S	32	05-02	10-06	157	26	26	24	24
	28	04-22	10-19	180	25	25	24	24
	24	04-07	11-01	208	25	25	25	25
	20	03-19	11-08	233	25	25	25	25
	16	03-08	11-23	260	25	25	25	24
CAMBRIDGE CITY	32	05-06	10-07	153	30	30	30	30
	28	04-20	10-18	181	30	30	30	30
	24	04-06	10-28	206	30	30	30	30
	20	03-19	11-12	238	30	30	30	30
	16	03-08	11-27	264	30	30	30	28
COLLEGEVILLE ST JOS	32	05-03	10-09	158	30	30	30	30
	28	04-19	10-24	189	30	30	30	30
	24	04-05	11-05	214	30	30	30	30
	20	03-21	11-16	240	30	30	30	30
	16	03-12	11-26	260	30	30	30	29
COLUMBUS	32	04-27	10-10	167	30	30	30	30
	28	04-16	10-25	192	29	29	30	30
	24	03-29	11-04	220	29	29	30	30
	20	03-19	11-16	242	29	29	30	30
	16	03-04	11-28	269	29	29	30	29
CRAWFORDSVILLE PR PL	32	05-01	10-11	163	30	30	30	30
	28	04-17	10-25	191	30	30	30	30
	24	03-30	11-04	219	30	30	30	30
	20	03-19	11-18	244	30	30	30	29
	16	03-06	12-01	270	30	30	30	27
DELPHI	32	05-01	10-10	161	30	30	30	30
	28	04-18	10-25	191	30	30	30	30
	24	03-28	11-06	224	30	30	30	30
	20	03-16	11-20	248	30	30	30	30
	16	03-07	11-30	268	30	30	30	29
ELLISTON	32	04-25	10-16	174	30	30	30	30
	28	04-11	10-25	197	30	30	30	30
	24	03-26	11-08	226	30	30	30	30
	20	03-14	11-18	249	30	30	30	30
	16	03-02	12-05	277	30	30	30	27
EVANSVILLE WB AP	32	04-02	11-04	216	30	30	30	30
	28	03-22	11-13	235	30	30	30	30
	24	03-09	11-26	261	30	30	30	28
	20	02-28	12-07	282	30	30	30	28
	16	02-16	12-14	301	30	28	30	25
FARMERSBURG 3 SW	32	04-22	10-19	181	30	30	30	30
	28	04-04	10-30	209	30	30	30	30
	24	03-21	11-08	232	30	30	30	30
	20	03-08	11-22	259	30	30	30	28
	16	02-27	12-06	281	29	29	30	27
FT WAYNE WB BAER AP	32	04-24	10-20	179	30	30	30	30
	28	04-12	11-04	206	30	30	30	30
	24	03-29	11-17	233	30	30	30	30
	20	03-17	11-25	254	30	30	30	30
	16	03-08	12-05	272	30	30	30	28
FOWLER POWER SUBSTA	32	04-29	10-17	171	20	20	20	20
	28	04-14	10-30	199	20	20	20	20
	24	03-30	11-07	222	20	20	19	19
	20	03-17	11-20	247	20	20	18	18
	16	03-10	11-28	264	20	20	18	17
FRANKFORT DISP PLANT	32	05-02	10-09	160	29	29	30	30
	28	04-16	10-23	190	27	27	29	29
	24	03-29	11-01	216	27	27	28	28
	20	03-19	11-16	241	27	27	28	28
	16	03-09	12-01	267	27	27	28	26
GOSHEN COLLEGE	32	05-11	10-07	149	30	30	29	29
	28	04-27	10-20	177	30	30	30	30
	24	04-08	11-07	213	30	30	30	30
	20	03-24	11-16	237	30	30	30	30
	16	03-13	11-30	262	30	30	29	29
GREENCASTLE 1 E	32	04-27	10-19	175	27	27	30	30
	28	04-10	10-31	205	29	29	30	30
	24	03-25	11-09	229	28	28	30	30
	20	03-16	11-20	249	28	28	30	30
	16	03-06	12-04	273	28	28	30	27
GREENFIELD	32	04-27	10-16	171	28	28	30	30
	28	04-14	10-29	197	28	28	30	30
	24	03-30	11-06	221	28	28	30	30
	20	03-14	11-20	251	28	28	29	29
	16	03-05	12-01	271	28	27	29	26
GREENSBURG	32	04-25	10-23	181	30	30	30	30
	28	04-08	11-02	208	30	30	30	30
	24	03-24	11-13	233	30	30	30	30
	20	03-14	11-27	258	30	30	30	28
	16	03-02	12-06	279	30	30	30	27
HENRYVILLE STATE FOR	32	05-02	10-06	157	30	30	30	30
	28	04-19	10-20	184	30	30	30	30
	24	04-02	11-01	213	30	30	30	30
	20	03-23	11-13	235	30	30	30	30
	16	03-08	11-25	262	30	30	30	28
HOWE MILITARY SCHOOL	32	05-10	10-06	149	29	29	29	29
	28	04-28	10-20	175	28	28	29	29
	24	04-12	11-07	209	28	28	29	29
	20	03-27	11-17	235	28	28	29	29
	16	03-16	12-02	261	28	28	29	28
HUNTINGBURG AP	32	04-22	10-17	179	29	29	29	29
	28	04-04	10-27	206	28	28	29	29
	24	03-22	11-11	235	29	29	29	29
	20	03-08	11-24	260	28	28	29	27
	16	02-27	12-07	283	27	27	29	26
HUNTINGTON	32	05-06	10-08	155	29	29	29	29
	28	04-22	10-21	182	30	30	29	29
	24	04-07	11-04	210	29	29	28	28
	20	03-24	11-12	234	29	29	27	27
	16	03-12	11-28	261	29	29	27	26
INDIANAPOLIS WB CITY	32	04-17	10-27	193	30	30	30	30
	28	04-01	11-09	221	30	30	30	30
	24	03-21	11-20	244	30	30	30	30
	20	03-11	11-30	264	30	30	30	28
	16	02-28	12-07	282	30	30	30	27
JEFFERSONVILLE	32	04-13	10-27	198	28	28	29	29
	28	03-26	11-07	226	27	27	29	29
	24	03-14	11-21	252	27	27	28	27
	20	03-06	12-03	273	27	27	28	27
	16	02-18	12-13	298	27	26	28	24

FREEZE DATA

STATION	Freeze threshold temperature	Mean date of last Spring occurrence	Mean date of first Fall occurrence	Mean No. of days between dates	Years of record Spring	No. of occurrences in Spring	Years of record Fall	No. of occurrences in Fall
KOKOMO POST OFFICE	32	04-24	10-22	181	30	30	30	30
	28	04-07	11-04	211	30	30	30	30
	24	03-27	11-13	231	30	30	30	30
	20	03-16	11-23	253	30	30	30	29
	16	03-06	12-06	275	30	30	30	26
LA PORTE	32	05-07	10-08	155	30	30	30	30
	28	04-22	10-26	188	30	30	30	30
	24	04-06	11-06	214	30	30	30	30
	20	03-27	11-19	238	30	30	30	30
	16	03-16	11-26	255	30	30	30	29
MADISON	32	04-19	10-26	190	29	29	29	29
	28	04-04	11-05	215	29	29	29	29
	24	03-22	11-18	241	29	29	29	29
	20	03-08	11-29	266	29	29	27	24
	16	02-27	12-10	286	29	29	27	23
MARION COLLEGE	32	05-01	10-17	168	30	30	30	30
	28	04-15	10-27	195	30	30	30	30
	24	03-31	11-07	221	30	30	30	30
	20	03-17	11-18	246	30	30	30	30
	16	03-11	12-01	266	30	30	30	28
MARKLAND DAM 39	32	04-24	10-18	177	29	29	30	30
	28	04-07	10-29	205	29	29	30	30
	24	03-23	11-13	235	29	29	30	30
	20	03-12	11-27	260	29	29	30	29
	16	03-02	12-07	280	29	29	30	28
MOORES HILL	32	04-23	10-19	178	30	30	30	30
	28	04-10	10-28	201	30	30	30	30
	24	03-29	11-08	224	30	30	30	30
	20	03-15	11-23	252	30	30	30	30
	16	03-06	12-03	272	30	30	30	29
MOUNT VERNON WTR WKS	32	04-07	10-25	201	30	30	30	30
	28	03-24	11-10	231	30	30	30	30
	24	03-11	11-20	254	30	30	30	29
	20	03-04	12-07	279	30	30	30	26
	16	02-21	12-13	295	30	29	30	24
NASHVILLE STATE PARK	32	04-24	10-19	178	10	10	10	10
	28	04-08	10-28	204	10	10	10	10
	24	03-27	11-11	229	10	10	10	10
	20	03-18	11-19	247	9	9	10	10
	16	03-04	12-05	276	7	7	10	10
NEW HARMONY 66 BRGE	32	04-07	10-27	204	9	9	9	9
	28	03-29	11-02	218	9	9	9	9
	24	03-15	11-19	249	9	9	10	10
	20	03-06	12-06	275	9	9	10	10
	16	02-23	12-13	293	9	9	10	8
NOTRE DAME MOREAU SM	32	04-30	10-19	172	30	30	29	29
	28	04-16	11-05	203	30	30	29	29
	24	04-02	11-14	226	30	30	30	30
	20	03-20	11-26	252	30	30	30	30
	16	03-11	12-03	267	30	30	30	29
OOLITIC PURDUE EX FM	32	04-24	10-17	176	29	29	30	30
	28	04-08	11-02	201	29	29	30	30
	24	03-28	11-09	225	30	30	30	30
	20	03-14	11-19	249	30	30	30	29
	16	03-03	12-04	276	30	30	30	27
PAOLI	32	04-26	10-13	170	30	30	30	30
	28	04-11	10-25	197	30	30	30	30
	24	03-27	11-05	223	30	30	30	30
	20	03-17	11-18	246	30	30	30	28
	16	03-05	12-01	271	30	30	30	28
PLYMOUTH PWR SUB STA	32	05-07	10-08	154	28	28	28	28
	28	04-23	10-20	180	28	28	26	26
	24	04-05	11-06	215	28	28	26	26
	20	03-25	11-17	237	27	27	27	27
	16	03-13	11-27	259	27	27	27	27
PRINCETON	32	04-16	10-25	192	29	29	30	30
	28	03-29	11-04	220	29	29	30	30
	24	03-18	11-13	240	29	29	30	30
	20	03-05	11-28	268	29	29	30	28
	16	02-24	12-12	291	29	29	30	25
RICHMOND WATER WORKS	32	05-01	10-11	163	30	30	30	30
	28	04-16	10-25	192	30	30	30	30
	24	04-02	11-05	216	30	30	30	30
	20	03-17	11-16	244	30	30	30	30
	16	03-06	11-30	269	30	30	29	28
ROCKVILLE	32	04-23	10-19	180	30	30	30	30
	28	04-06	11-02	210	30	30	30	30
	24	03-24	11-09	229	30	30	30	30
	20	03-14	11-22	253	30	30	30	30
	16	03-02	12-04	277	30	30	29	28
Rushville Sewage Plant	32	05-05	10-09	157	30	30	30	30
	28	04-24	10-19	178	30	30	30	30
	24	04-06	11-01	209	30	30	30	30
	20	03-19	11-13	239	30	30	30	30
	16	03-06	11-28	267	30	30	30	30
SALAMONIA	32	05-08	10-07	152	30	30	29	29
	28	04-27	10-20	177	30	30	29	29
	24	04-07	11-03	210	30	30	29	29
	20	03-22	11-10	233	30	30	29	29
	16	03-11	11-25	260	30	30	29	28
SALEM 1 W	32	04-24	10-15	175	30	30	29	29
	28	04-07	10-29	205	30	30	29	29
	24	03-23	11-07	229	30	30	29	28
	20	03-11	11-23	257	30	30	29	28
	16	03-02	12-04	277	30	30	29	27
SCOTTSBURG	32	04-25	10-15	174	29	29	27	27
	28	04-12	10-25	196	29	29	27	27
	24	03-27	11-06	224	29	29	27	27
	20	03-10	11-20	255	29	29	27	26
	16	02-28	12-03	278	29	29	27	25
SEYMOUR 1 N	32	04-25	10-15	173	30	30	30	30
	28	04-10	10-26	199	30	30	30	30
	24	03-23	11-09	230	30	30	30	30
	20	03-10	11-20	255	30	30	30	30
	16	03-03	12-03	275	30	30	30	27
SHELBYVILLE POWER PL	32	04-26	10-18	175	30	30	29	29
	28	04-08	10-28	203	29	29	29	29
	24	03-26	11-10	228	29	29	29	29
	20	03-14	11-24	255	29	29	29	29
	16	03-03	12-04	277	29	29	28	26
SHOALS HIWAY 50 BRDG	32	04-27	10-13	169	30	30	30	30
	28	04-11	10-25	197	30	30	30	30
	24	03-26	11-04	223	30	30	30	30
	20	03-15	11-17	247	30	30	30	29
	16	03-03	11-30	272	30	30	30	28
SOUTH BEND WB AP	32	05-03	10-16	165	29	29	28	28
	28	04-22	10-31	192	29	29	28	28
	24	04-05	11-10	220	29	29	28	28
	20	03-25	11-21	241	29	29	28	28
	16	03-13	11-29	261	29	29	28	27
TELL CITY POWER PLT	32	04-09	10-26	200	12	12	12	12
	28	03-28	11-06	223	12	12	12	12
	24	03-13	11-25	257	12	12	12	12
	20	03-09	12-09	275	11	11	12	11
	16	02-27	12-14	290	11	11	12	10
TERRE HAUTE WB AP	32	04-11	10-28	201	30	30	30	30
	28	03-28	11-08	225	30	30	30	30
	24	03-17	11-19	247	30	30	30	29
	20	03-07	11-30	268	30	30	30	29
	16	02-27	12-09	285	30	30	30	27
VALPARAISO WATER WKS	32	04-30	10-16	169	30	30	30	30
	28	04-15	10-31	199	30	30	30	30
	24	03-31	11-07	221	30	30	30	30
	20	03-21	11-19	243	30	30	30	30
	16	03-12	11-29	261	30	30	30	30
VEEDERSBURG	32	05-02	10-11	163	30	30	30	30
	28	04-16	10-25	192	29	29	30	30
	24	03-28	11-05	222	29	29	30	30
	20	03-20	11-15	240	29	29	30	30
	16	03-07	11-29	267	29	29	30	29

FREEZE DATA

STATION	Freeze threshold temperature	Mean date of last Spring occurrence	Mean date of first Fall occurrence	Mean No. of days between dates	Years of record Spring	No. of occurrences in Spring	Years of record Fall	No. of occurrences in Fall
VINCENNES	32	04-14	10-23	192	30	30	30	30
	28	03-27	11-07	225	30	30	30	30
	24	03-19	11-15	242	30	30	30	30
	20	03-08	12-04	271	30	30	30	28
	16	02-27	12-11	287	30	30	30	26
WABASH	32	05-04	10-11	160	25	25	25	25
	28	04-22	10-23	184	25	25	25	25
	24	04-06	11-04	212	25	25	25	25
	20	03-20	11-13	238	24	24	24	24
	16	03-11	11-25	259	24	24	24	24
WARSAW WATER WORKS	32	05-02	10-12	163	28	28	29	29
	28	04-16	10-27	194	28	28	29	29
	24	04-02	11-08	221	27	27	29	29
	20	03-24	11-21	242	28	28	29	29
	16	03-14	11-30	261	29	29	29	28
WASHINGTON	32	04-13	10-23	194	30	30	30	30
	28	03-31	11-05	219	30	30	30	30
	24	03-21	11-13	237	30	30	30	30
	20	03-07	11-28	267	30	30	30	28
	16	02-27	12-08	284	30	30	29	25
W LAFAYETTE PURDUE U	32	04-21	10-23	184	30	30	30	30
	28	04-06	11-05	214	30	30	30	30
	24	03-22	11-13	236	30	30	30	30
	20	03-12	11-24	258	30	30	30	30
	16	03-04	12-05	278	30	30	30	27
WHEATFIELD	32	05-10	10-04	147	30	30	30	30
	28	04-27	10-19	175	30	30	30	30
	24	04-07	10-31	207	30	30	30	30
	20	03-26	11-10	230	30	30	30	30
	16	03-15	11-23	254	30	30	30	30
WHITESTOWN	32	05-02	10-10	161	30	30	30	30
	28	04-16	10-25	191	30	30	30	30
	24	03-31	11-04	218	30	30	30	30
	20	03-20	11-17	242	30	30	30	30
	16	03-10	11-28	263	30	30	30	28
WHITING	32	04-18	10-23	189	29	29	28	28
	28	04-05	11-07	216	29	29	28	28
	24	03-25	11-16	236	29	29	28	28
	20	03-13	11-23	255	29	29	27	27
	16	03-05	12-01	271	29	29	26	25
WINAMAC	32	05-04	10-12	161	30	30	30	30
	28	04-18	10-30	194	30	30	30	30
	24	04-03	11-05	216	30	30	30	30
	20	03-22	11-19	242	30	30	30	30
	16	03-12	11-29	262	29	29	30	29

Data in the above table are based on the period 1921-1950, or that portion of this period for which data are available.

Means have been adjusted to take into account years of non-occurrence.

A freeze is a numerical substitute for the former term "killing frost" and is the occurrence of a minimum temperature at or below the threshold temperature of 32°, 28°, etc.

Freeze data tabulations in greater detail are available and can be reproduced at cost.

NOTE: Addition to mean Temperature and Precipitation table on Page 7

*MEAN TEMPERATURE AND PRECIPITATION

STATION	JANUARY Temperature	JANUARY Precipitation	FEBRUARY Temperature	FEBRUARY Precipitation	MARCH Temperature	MARCH Precipitation	APRIL Temperature	APRIL Precipitation	MAY Temperature	MAY Precipitation	JUNE Temperature	JUNE Precipitation	JULY Temperature	JULY Precipitation	AUGUST Temperature	AUGUST Precipitation	SEPTEMBER Temperature	SEPTEMBER Precipitation	OCTOBER Temperature	OCTOBER Precipitation	NOVEMBER Temperature	NOVEMBER Precipitation	DECEMBER Temperature	DECEMBER Precipitation	ANNUAL Temperature	ANNUAL Precipitation
Central Division																										
Rushville Sewage Plant	30.3	3.47	31.5	2.56	39.8	3.82	50.6	3.74	61.2	3.95	71.3	3.89	74.5	3.48	72.8	2.90	66.4	4.09	55.2	2.72	41.3	3.04	31.4	2.56	52.2	40.22

CONFIDENCE LIMITS

In the absence of trend or record changes, the chances are 9 out of 10 that the true mean will lie in the interval formed by adding and subtracting the values in the following table from the means for any station in the State. Because of the wider variation in mean precipitation, the corresponding monthly means and annual mean must be substituted for "p" in the precipitation table below to obtain mean precipitation confidence limits.

1.8	$.42\sqrt{p}$	1.7	$.29\sqrt{p}$	1.8	$.33\sqrt{p}$	1.2	$.30\sqrt{p}$	1.2	$.39\sqrt{p}$	1.1	$.38\sqrt{p}$	1.0	$.34\sqrt{p}$	.9	$.33\sqrt{p}$	1.2	$.40\sqrt{p}$	1.2	$.35\sqrt{p}$	1.2	$.30\sqrt{p}$	1.7	$.28\sqrt{p}$	.5	$.35\sqrt{p}$

COMPARATIVE DATA

Data in the following table are the mean temperature and average precipitation for Whitestown, Indiana for the period 1906-1930 and are included in this publication for comparative purposes:

27.1	3.10	29.5	1.93	39.6	4.00	50.4	4.04	60.9	4.17	70.2	3.83	74.5	3.73	72.3	3.58	65.9	3.69	53.6	3.01	41.1	2.61	30.0	2.80	51.2	40.49

•MEAN TEMPERATURE AND PRECIPITATION

STATION	JANUARY		FEBRUARY		MARCH		APRIL		MAY		JUNE		JULY		AUGUST		SEPTEMBER		OCTOBER		NOVEMBER		DECEMBER		ANNUAL	
	Temperature	Precipitation	Temperature	Precipitation	Temperature	Precipitation	Temperature	Precipitation	Temperature	Precipitation	Temperature	Precipitation	Temperature	Precipitation	Temperature	Precipitation	Temperature	Precipitation	Temperature	Precipitation	Temperature	Precipitation	Temperature	Precipitation	Temperature	Precipitation
NORTHWEST DIVISION																										
COLLEGEVILLE ST JOS COL	27.8	2.05	29.2	1.92	38.2	2.99	50.0	3.92	60.9	4.26	71.6	3.94	75.7	3.33	73.2	3.37	65.8	3.34	55.0	3.20	40.2	2.59	29.5	1.98	51.4	36.89
FOWLER	28.2	2.14	30.2	2.00	38.9	2.97	50.6	3.43	61.4	4.09	72.1	4.20	76.3	3.59	73.9	2.70	66.9	3.33	55.9	3.05	40.6	2.66	29.6	1.90	52.1	36.06
HOBART	27.3	1.92	28.6	1.78	37.2	2.92	48.8	3.50	59.5	4.23	70.4	4.37	74.8	3.00	72.9	3.68	65.6	3.19	54.6	3.34	40.4	2.73	29.3	2.13	50.8	36.79
LA PORTE	25.8	3.08	27.4	2.83	36.0	3.91	48.0	4.79	59.0	5.74	69.8	5.59	74.5	4.14	72.4	4.13	64.7	4.39	53.6	4.66	39.2	3.51	28.2	3.26	49.9	50.03
MONTICELLO		2.20		1.84		2.83		3.70		4.11		3.96		2.91		2.97		2.99		3.28		2.62		1.81		35.22
VALPARAISO WATERWORKS	26.5	2.13	27.8	2.09	36.2	3.23	48.1	3.87	58.5	4.24	69.1	4.39	73.6	2.90	71.8	3.89	64.7	3.19	54.4	3.61	39.9	2.81	28.7	2.16	49.9	38.51
WHEATFIELD	26.5	2.05	27.9	1.95	36.9	2.95	48.7	3.81	59.5	3.94	70.1	4.03	74.3	2.92	72.0	3.55	64.6	3.17	53.6	3.05	38.9	2.65	28.0	2.17	50.1	36.24
WHITING	28.4	1.88	29.7	1.54	37.9	2.61	49.1	3.18	59.8	4.15	71.1	4.03	76.5	2.92	75.1	3.31	68.1	3.02	57.2	3.22	42.1	2.25	30.8	1.84	52.2	33.95
WINAMAC	27.8	2.09	29.3	1.93	37.8	3.01	49.7	3.46	60.3	4.05	70.6	4.44	74.8	2.99	72.4	3.52	65.2	3.45	54.5	3.18	40.2	2.62	29.5	1.98	51.0	36.72
DIVISION	27.1	2.11	28.9	1.99	37.5	3.06	49.3	3.73	59.9	4.37	70.6	4.32	75.1	3.24	73.1	3.48	65.8	3.23	55.1	3.39	40.2	2.64	29.3	2.13	51.0	37.69
NORTH CENTRAL DIVISION																										
DELPHI	30.0	2.34	31.7	1.94	40.1	2.95	51.6	3.60	62.2	4.02	72.4	3.80	76.4	3.17	74.0	2.99	66.7	3.30	56.1	3.16	42.0	2.58	31.5	1.94	52.9	35.79
GOSHEN COLLEGE	26.9	1.97	27.7	1.81	36.2	2.74	48.1	3.52	59.1	3.63	69.6	3.88	73.7	3.61	71.8	3.74	64.3	3.12	53.3	3.05	39.6	2.32	28.8	1.90	49.9	35.29
LOGANSPORT CICOTT ST BR		2.48		2.08		3.13		3.88		4.36		4.46		3.62		3.30		3.58		3.26		2.86		2.12		39.1[illegible]
PLYMOUTH POWER SUBSTA	27.5	2.12	28.9	1.85	37.5	2.97	49.4	3.39	60.2	4.13	70.8	4.28	74.8	3.26	72.4	3.14	64.8	3.14	54.1	3.59	38.3	2.53	29.3	2.12	50.7	36.52
ROCHESTER	28.1	2.07	29.3	1.82	37.8	2.67	49.7	3.43	60.8	4.05	71.3	4.11	75.4	3.66	73.1	3.38	65.7	3.43	54.9	3.27	40.5	2.58	29.8	1.74	51.4	36.21
SOUTH BEND WB AP	24.6	2.00	26.4	1.56	35.7	2.95	47.3	3.40	58.4	3.84	68.7	3.83	73.4	3.04	71.4	3.25	64.5	3.71	52.9	2.99	38.7	2.74	27.8	2.28	49.1	35.59
WABASH	28.4	2.61	29.3	2.08	37.9	3.01	49.5	3.49	60.4	4.05	71.1	4.04	74.8	3.61	72.6	3.73	65.1	2.94	54.4	3.22	40.3	2.86	29.9	1.96	51.1	37.60
WARSAW	27.7	2.12	28.5	1.92	37.2	3.04	49.2	3.56	60.5	3.85	71.1	4.21	75.5	3.46	73.3	3.32	65.9	3.28	54.8	3.41	40.4	2.53	29.5	1.94	51.1	36.64
DIVISION	28.0	2.17	29.1	1.88	37.6	2.86	49.4	3.49	60.3	3.93	70.8	4.04	75.0	3.45	72.7	3.38	65.3	3.17	54.5	3.21	40.2	2.56	29.7	1.98	51.1	36.12
NORTHEAST DIVISION																										
ALBION	25.9	2.18	26.6	1.55	35.2	2.32	47.7	2.58	58.8	3.25	69.2	3.78	73.2	3.44	71.2	3.51	64.0	2.90	52.9	2.91	38.5	2.45	27.9	1.81	49.3	32.68
ANGOLA	26.4	2.15	26.7	1.97	35.1	3.02	47.4	3.51	59.1	3.70	69.6	4.01	73.9	3.06	72.0	3.23	64.6	3.27	53.7	2.84	39.3	2.62	28.5	2.09	49.7	35.47
BERNE	29.2	2.58	30.1	1.93	38.4	3.42	49.6	3.34	60.7	3.81	71.0	4.00	74.9	3.58	72.6	2.84	65.7	3.17	54.6	2.66	40.9	2.49	30.6	2.02	51.5	35.84
BLUFFTON WATERWORKS	29.1	2.76	30.2	2.16	38.7	3.60	50.0	3.45	61.0	3.56	71.2	4.22	75.0	3.46	73.1	2.83	66.2	3.19	54.8	2.69	41.0	2.70	30.4	2.20	51.7	36.82
COLUMBIA CITY		2.39		2.10		2.98		3.21		3.69		3.93		3.61		3.26		2.77		3.17		2.63		1.91		35.65
FORT WAYNE WB AP	26.3	2.54	28.0	1.77	36.8	3.16	47.8	3.25	59.0	3.44	69.0	3.56	73.5	3.28	71.4	2.80	64.6	3.07	53.2	2.62	39.7	2.46	28.8	2.26	49.9	34.21
HUNTINGTON	28.4	2.87	29.6	2.11	37.9	3.53	49.5	3.48	60.5	4.07	71.0	3.88	74.9	3.22	72.9	3.17	65.6	2.92	54.5	3.20	40.5	2.65	29.9	2.20	51.3	37.30
DIVISION	27.6	2.43	28.7	1.97	37.0	3.09	48.8	3.31	60.0	3.73	70.5	3.94	74.4	3.39	72.4	3.20	65.2	3.02	54.2	2.96	40.0	2.60	29.4	2.04	50.7	35.68
WEST CENTRAL DIVISION																										
COVINGTON		2.36		2.08		3.09		3.42		4.48		4.36		3.45		3.07		2.98		3.11		2.95		2.00		37.35
CRAWFORDSVILLE PWR PL	30.5	2.72	32.0	2.11	40.6	3.30	51.9	3.76	62.6	4.57	72.9	4.41	76.8	3.28	74.8	3.28	67.4	3.73	56.3	2.83	41.6	2.89	31.6	2.23	53.3	39.11
GREENCASTLE 1 E	31.1	3.10	32.7	2.20	41.2	3.60	52.5	3.57	62.6	4.62	72.6	5.13	76.1	3.46	74.3	3.30	67.3	3.95	56.9	2.91	42.4	3.29	32.4	2.53	53.5	41.66
ROCKVILLE	30.6	3.02	32.5	2.28	40.9	3.85	52.5	3.61	63.2	5.06	73.2	4.93	77.0	3.14	74.9	3.27	67.6	3.30	56.7	3.07	42.1	2.98	32.2	2.40	53.6	40.91
DIVISION	30.6	2.99	32.5	2.26	40.9	3.43	52.3	3.60	62.7	4.50	72.8	4.66	76.7	3.44	74.6	3.25	67.3	3.41	56.7	3.05	41.9	3.00	31.8	2.32	53.4	39.91
CENTRAL DIVISION																										
ANDERSON SEWAGE PLANT	31.3	2.55	32.8	1.93	41.5	3.34	52.5	3.53	63.0	4.01	72.7	3.90	76.1	3.40	74.3	3.30	68.5	3.34	57.4	2.45	43.0	2.60	32.7	2.16	53.8	36.71
COLUMBUS	33.3	4.02	34.7	2.72	42.9	4.32	53.6	4.01	63.9	3.89	73.7	3.80	76.9	2.64	75.1	3.30	68.1	3.58	57.1	2.60	43.5	3.26	34.0	2.97	54.7	41.11
FRANKFORT DISPOSAL PL	28.9	2.57	30.3	2.10	39.1	3.21	50.4	3.72	61.1	4.48	71.2	4.04	75.0	3.30	72.7	2.98	65.6	3.31	54.7	2.88	40.5	2.65	30.5	1.97	51.7	37.21
GREENFIELD	31.0	3.01	32.7	2.29	41.1	3.81	52.0	3.68	62.7	4.08	72.6	4.39	76.4	3.40	74.5	2.69	67.6	3.60	56.5	2.50	42.2	2.89	32.4	2.31	53.5	38.65
GREENSBURG 3 SW	31.7	3.81	33.0	2.77	41.3	4.40	52.6	3.67	63.2	3.90	72.8	4.16	76.5	3.35	75.0	2.60	68.0	3.19	56.8	2.72	42.4	3.23	32.7	2.85	53.8	40.65
INDIANAPOLIS WB AP	28.8	3.15	31.5	2.08	40.1	3.89	50.8	3.85	61.4	3.85	71.4	4.21	76.0	3.03	74.0	3.52	67.2	3.67	55.9	2.52	41.5	3.13	31.1	2.79	52.5	39.69
KOKOMO SEWAGE PLANT	30.0	2.83	31.8	2.34	40.4	3.35	52.0	3.87	63.2	4.52	73.8	3.56	77.8	3.11	75.2	2.85	68.0	3.88	56.8	2.93	42.1	2.91	31.5	2.32	53.6	38.47
MARION 2 N	28.9	2.71	30.0	2.03	38.6	3.36	49.9	3.63	61.0	4.19	71.4	3.51	75.1	3.36	73.1	3.18	66.0	3.55	55.1	2.61	40.8	2.77	30.3	2.14	51.7	37.04
MARTINSVILLE		3.39		2.56		3.91		3.66		3.92		4.70		3.46		3.26		3.90		2.41		3.24		2.40		40.81
NOBLESVILLE		2.75		1.97		3.50		3.44		3.82		3.94		3.04		2.82		3.22		2.57		2.90		2.11		36.08
SHELBYVILLE NURSERY	32.5	3.35	34.1	2.42	42.4	3.73	53.2	3.67	63.6	3.90	73.4	4.21	76.8	3.45	74.8	3.35	68.0	3.82	57.1	2.44	43.2	3.20	33.4	2.60	54.4	40.14
WHITESTOWN	29.4	2.92	31.1	2.12	39.5	3.78	50.8	3.44	61.6	4.16	71.6	3.91	75.4	2.81	73.1	2.78	65.7	3.22	54.7	2.82	40.6	2.88	30.7	2.33	52.0	37.17
DIVISION	30.9	3.23	32.6	2.37	40.8	3.76	51.9	3.68	62.4	4.08	72.4	4.11	76.0	3.24	74.0	3.01	67.1	3.52	56.1	2.58	41.9	3.06	31.9	2.43	53.2	39.07
EAST CENTRAL DIVISION																										
CAMBRIDGE CITY	30.4	3.33	31.7	2.59	39.9	3.85	50.7	4.03	61.1	3.97	71.0	3.93	74.5	3.09	72.5	3.02	65.3	3.72	54.3	2.84	41.1	3.25	31.3	2.84	52.0	40.46
MUNCIE 4 SE		2.81		2.18		3.53		3.64		4.20		4.22		3.52		3.42		3.58		2.74		2.91		2.36		39.11
RICHMOND WATERWORKS	30.2	3.22	31.0	2.22	39.2	3.46	49.8	3.18	60.1	3.54	69.8	4.36	73.1	3.47	71.0	3.02	63.8	3.37	53.2	2.72	40.5	2.75	31.2	2.49	51.1	37.80
SALAMONIA 7 W	29.9	3.22	30.9	2.53	39.4	3.89	50.3	3.89	61.0	4.00	71.3	4.19	74.9	3.67	72.9	2.78	66.4	3.41	55.4	2.75	41.4	2.83	31.1	2.57	52.1	39.73
WINCHESTER AIRPORT	30.0		30.9		39.2		50.2		61.2		71.5		75.1		73.0		66.3		55.5		41.5		31.3		52.1	
DIVISION	30.1	3.12	31.3	2.38	39.6	3.66	50.4	3.75	61.0	3.94	71.0	4.26	74.5	3.53	72.4	3.04	65.5	3.51	54.8	2.75	41.1	2.94	31.2	2.51	51.9	39.39
SOUTHWEST DIVISION																										
CYPRESS DAM 48		4.47		3.34		4.84		4.22		3.91		3.19		3.01		3.12		3.27		2.68		3.16		2.93		42.14
EDWARDSPORT POWER PLANT	32.4	3.70	34.4	2.48	42.8	4.07	54.3	3.91	65.1	3.76	75.4	4.18	78.8	3.50	76.9	3.11	69.6	3.37	58.2	2.61	43.4	3.14	33.7	2.79	55.4	40.62
ELLISTON	31.8	3.64	33.8	2.50	42.3	4.06	53.6	4.01	64.1	4.37	74.2	4.49	77.9	3.14	76.1	3.40	68.7	3.53	57.3	2.62	42.6	3.38	33.0	2.71	54.6	41.85
EVANSVILLE WB AP	34.7	3.93	37.5	2.95	46.6	4.29	57.1	3.79	65.5	3.90	74.6	3.87	78.2	3.00	76.3	3.04	70.5	3.40	59.3	2.82	46.0	3.20	36.9	3.18	56.9	41.37
HUNTINGBURG AIRPORT	35.2	4.56	37.7	3.09	46.1	4.92	56.5	4.30	66.0	4.10	75.3	4.24	78.5	3.32	77.3	3.55	70.2	3.34	59.2	2.80	45.3	3.18	36.3	2.94	57.0	44.34
MOUNT VERNON	35.7	4.49	37.8	3.26	45.7	4.82	56.1	4.18	65.5	3.96	75.0	3.20	78.3	3.12	75.9	3.30	70.1	3.07	59.5	2.79	46.2	3.00	37.1	2.94	57.0	42.13
NEWBURGH DAM 47		4.25		3.29		4.90		4.22		3.78		3.35		2.93		3.49		3.16		2.50		3.20		2.86		41.93
PRINCETON 1 W	34.9	4.16	37.3	2.95	45.7	4.43	56.5	4.36	65.9	4.43	75.6	4.04	78.8	3.20	77.2	3.35	70.4	3.13	59.5	2.66	45.7	3.21	36.3	3.05	57.0	42.97
SHOALS HIWAY 50 BRIDGE	33.5	4.16	35.0	2.78	43.0	4.68	54.1	3.81	64.2	3.90	74.0	4.09	77.5	3.06	75.5	3.20	68.1	3.56	56.9	2.68	43.5	3.58	34.5	2.72	55.0	42.22
VINCENNES	34.2	3.75	36.2	2.54	44.6	4.16	55.8	4.32	65.9	4.26	75.9	4.16	79.4	3.04	77.5	3.12	70.3	3.48	59.1	2.80	44.8	3.44	35.2	2.97	56.6	42.04
WASHINGTON	34.5	4.08	36.7	2.66	44.8	4.27	55.6	4.30	65.1	3.98	74.7	4.75	78.3	3.01	76.8	3.03	70.1	3.35	59.3	2.71	45.0	3.28	35.5	2.94	56.4	42.36
DIVISION	34.1	4.06	36.3	2.85	44.6	4.36	55.5	4.17	65.4	4.13	75.0	4.15	78.4	3.21	76.8	3.24	69.7	3.32	58.8	2.68	44.6	3.27	35.2	2.81	56.2	42.25
SOUTH CENTRAL DIVISION																										
BLOOMINGTON INDIANA U	33.2	4.02	35.2	2.70	43.5	4.28	54.6	4.07	64.6	4.29	73.9	4.87	77.5	3.30	75.7	3.32	68.9	3.98	58.2	2.79	44.0	3.41	34.2	3.01	55.3	44.04
EVANS LANDING DAM 43		4.56		3.19		4.80		3.98		3.67		3.89		3.40		3.12		2.70		2.21		3.03		3.22		41.77
LEAVENWORTH DAM 44		4.58		3.23		5.13		4.26		3.94		4.09		4.06		3.19		2.93		2.55		3.44		3.11		44.51
OOLITIC PURDUE EXP FARM	33.7	4.51	35.7	3.09	44.0	4.93	54.6	4.23	64.4	4.33	73.8	4.66	77.5	3.32	75.8	3.17	68.5	3.69	57.7	2.71	44.1	3.55	34.7	3.31	55.4	45.50
PAOLI	34.2	4.60	35.9	3.19	43.8	4.78	54.4	4.17	64.0	3.95	73.2	4.65	76.8	3.65	75.4	3.10	68.3	3.18	57.4	2.78	44.2	3.28	35.0	3.09	55.2	44.42
SALEM	34.3	4.52	36.2	3.14	44.4	5.01	54.6	4.09	64.2	3.89	73.4	4.43	77.1	3.95	75.4	3.31	68.6	3.27	57.7	2.79	44.4	3.17	35.2	3.16	55.5	44.73
SEYMOUR 2 N	34.0	4.07	35.6	2.83	43.8	4.32	54.6	3.71	64.6	3.49	74.1	3.87	77.7	3.06	76.1	2.82	69.0	3.17	57.9	2.52	44.3	3.21	34.7	2.83	55.5	39.90
WILLIAMS POWER PLANT		4.02		2.84		4.50		3.82		4.10		4.14		3.13		3.15		3.33		2.66		3.39		2.98		42.06
DIVISION	34.1	4.33	35.9	3.03	43.9	4.69	54.6	4.12	64.5	4.00	73.8	4.38	77.2	3.43	75.8	3.18	68.7	3.36	57.9	2.69	44.3	3.34	35.1	3.09	55.5	43.64
SOUTHEAST DIVISION																										
BROOKVILLE 1 S	32.2	3.51	33.1	2.47	41.2	3.85	51.7	3.37	62.0	3.71	71.9	3.95	75.3	3.47	73.7	2.75	66.8	3.71	55.2	2.47	42.6	2.88	32.8	2.61	53.2	38.75
HENRYVILLE ST FOREST	33.5	4.31	35.0	3.37	43.2	4.81	53.5	3.93	63.3	3.56	72.9	4.07	76.6	3.36	75.3	3.23	68.1	2.79	56.6	2.54	43.4	3.03	34.3	3.12	54.6	42.12
JEFFERSONVILLE	37.0	4.55	38.6	3.39	46.6	5.16	56.8	4.05	66.5	3.64	75.8	3.73	79.1	3.31	77.5	3.33	70.6	2.53	59.3	2.37	46.6	3.08	37.7	3.37	57.7	42.51
MADISON SEWAGE PLANT	35.7	4.60	37.4	3.27	45.4	5.23	56.0	4.20	66.0	3.93	75.2	4.23	78.7	3.43	77.6	3.37	71.1	2.95	60.2	2.58	46.2	3.36	36.8	3.11	57.2	44.26
MARKLAND DAM 39	35.1	4.09	36.4	3.20	43.9	4.82	53.8	3.96	63.6	3.35	73.0	4.04	76.7	3.75	75.4	3.52	68.8	2.86	58.2	2.27	45.1	2.92	36.0	2.98	55.5	41.76
MOORES HILL	33.1	4.05	34.6	3.20	42.9	4.51	53.6	3.72	63.6	3.80	72.9	3.97	76.7	3.26	75.3	2.79	68.9	2.90	57.8	2.40	43.7	3.26	34.0	2.93	54.8	40.79
NORTH VERNON	33.9	4.55	35.7	3.24	43.8	4.90	54.2	4.04	63.9	4.17	73.1	4.69	76.7	3.17	75.3	3.31	68.5	3.04	57.5	2.75	44.0	3.45	34.7	3.15	55.1	44.46
SCOTTSBURG	34.3	4.28	35.7	3.12	44.2	4.77	54.4	4.13	64.2	3.76	73.6	4.52	77.2	3.86	75.9	3.57	69.0	2.99	57.8	2.78	44.3	3.06	35.1	2.98	55.5	43.82
DIVISION	34.4	4.24	35.8	3.16	43.9	4.76	54.2	3.93	64.1	3.75	73.5	4.14	77.1	3.47	75.7	3.22	69.0	2.98	57.8	2.53	44.5	3.14	35.2	3.02	55.4	42.34

Averages for period 1931-1955, except for stations marked WB which are "normals" based on period 1921-1950. Divisional means may not be the arithmetical average of individual stations published, since additional data from shorter period stations are used to obtain better areal representation.

NORMALS, MEANS, AND EXTREMES

EVANSVILLE, INDIANA
DRESS MEMORIAL AIRPORT

LATITUDE 38° 03' N
LONGITUDE 87° 32' W
ELEVATION (ground) 383 Feet

Evansville — Temperature, Degree Days, Precipitation, Snow

Month	Daily max	Daily min	Monthly	Rec. highest	Year	Rec. lowest	Year	Normal degree days	Precip. normal total	Max monthly precip	Year	Min monthly precip	Year	Max precip 24 hrs	Year
J	43.4	26.0	34.7	76	1942	-15	1943	939	3.93	13.50	1950	.59	1943	2.99	1950
F	46.5	28.5	37.5	73	1951	-23	1954	770	2.95	7.25	1956	.27	1947	3.13	1945
M	56.8	36.4	46.6	83	1943	1	1942	589	4.29	11.43	1945	.89	1941	5.16	1943
A	67.8	46.3	57.1	89	1942	24	1950#	251	3.79	7.99	1955	2.16	1946	3.95	1955
M	76.5	54.5	65.5	94	1953	33	1944	90	3.90	8.22	1957	1.74	1951	4.29	1951
J	85.4	63.8	74.6	104	1954	41	1956	6	3.87	9.30	1943	1.15	1953	3.15	1943
J	89.1	67.3	78.2	105	1954#	47	1947	0	3.00	7.18	1950	.99	1944	3.12	1950
A	87.0	65.5	76.3	102	1943	46	1946	0	3.04	5.55	1946	.13	1943	2.61	1945
S	81.8	59.1	70.5	100	1954	31	1942	59	3.40	9.89	1945	.56	1953	3.45	1941
O	71.0	47.6	59.3	94	1953	21	1952	215	2.82	8.33	1941	.27	1944	2.32	1944
N	55.6	36.3	46.0	82	1955#	-1	1950	570	3.18	8.49	1949	.97	1949	2.68	1957
D	45.6	28.2	36.9	75	1951	0	1945	871	3.18	7.95	1951	.72	1955	2.03	1951
Year	67.2	46.6	56.9	105 JULY 1954#		-23 FEB 1951		4360	41.37	13.50 JAN 1950		.13 AUG 1943		5.16 MAR 1943	

Month	Snow mean total	Snow max monthly	Year	Snow max 24 hrs	Year
J	3.6	12.3	1951	8.2	1948
F	2.2	11.3	1948	6.6	1948
M	1.6	6.7	1947	9.6	1947
A	.1	1.8	1953	1.8	1953
M	.0	.0		.0	
J	.0	.0		.0	
J	.0	.0		.0	
A	.0	.0		.0	
S	.0	T	1957#	.0	1957#
O	T	T	1950	T	1950
N	.4	2.4	1950	1.9	1950
D	2.2	6.5	1945#	4.3	1952
Year	10.1	12.3 JAN 1951		9.6 MAR 1947	

Evansville — Humidity, Wind, Sky, Mean number of days

Month	RH Midnight	RH 6AM	RH Noon	RH 6PM	Mean hourly wind	Prevailing dir.	Fastest mile speed	Dir.	Year	Pct. sunshine	Mean sky cover
J	81	84	69	75	9.9	SSW	45	SW	1948	37	7.4
F	79	82	64	71	10.0	SSW	59	—	1956	55	6.6
M	77	80	57	61	10.8	NNW	56	SW	1945	55	6.6
A	77	80	55	60	10.3	SW	55	SE	1955	58	6.3
M	83	83	55	60	8.4	SSW	57	NW	1957	66	5.9
J	84	85	55	62	7.6	SSW	53	NW	1943	73	5.5
J	85	88	54	64	6.6	SW	49	NW	1941	75	5.5
A	86	89	52	58	6.2	NNW	37	SW	1943	78	4.7
S	85	87	51	61	6.2	SSW	47	SW	1946	73	4.7
O	78	83	59	69	7.3	SSW	54	N	1940	67	6.1
N	81	83	66	69	9.3	SSW	—	—	1957	51	6.9
D	83	84	75	75	9.3	SSW	—	—		40	6.9
Year	82	84	58	65	8.6	SSW	59	W	FEB 1956	62	6.0

Means and extremes in the above table are from the existing or comparable location(s). Annual extremes have been exceeded at prior locations as follows: Highest temperature 108 in July 1936; maximum monthly precipitation 14.78 in January 1937; minimum monthly precipitation 0.01 in March 1910; maximum precipitation in 24 hours 6.94 in October 1910; maximum monthly snowfall 41.0 in January 1918; maximum snowfall in 24 hours 20.0 in January 1918.

FORT WAYNE, INDIANA
BAER FIELD

LATITUDE 41° 00' N
LONGITUDE 85° 12' W
ELEVATION (ground) 801 Feet

Fort Wayne — Temperature, Degree Days, Precipitation, Snow

Month	Daily max	Daily min	Monthly	Rec. highest	Year	Rec. lowest	Year	Normal degree days	Precip. normal total	Max monthly precip	Year	Min monthly precip	Year	Max precip 24 hrs	Year
J	33.4	19.1	26.3	69	1950	-9	1957	1200	2.54	9.72	1950	1.61	1956	2.64	1950
F	35.4	20.6	28.0	69	1954	-17	1951	1036	1.77	4.43	1950	.42	1947	1.86	1956
M	45.4	28.1	36.8	85	1953	-7	1955	874	3.16	5.29	1955	1.38	1957	2.15	1955
A	57.8	37.8	47.8	85	1948	14	1957	516	3.25	7.11	1957	1.61	1950	1.76	1948
M	69.6	48.4	59.0	100	1953	27	1947	226	3.44	6.85	1952	1.26	1953	2.53	1949
J	79.4	58.5	69.0	103	1954	38	1956	53	3.56	6.67	1957	—	—	3.29	1957
J	84.3	62.6	73.5	103	1954	45	1950	7	3.28	6.34	1951	2.63	1956	3.47	1955
A	82.6	60.8	71.4	103	1947	45	1956#	17	2.80	5.61	1950	1.45	1956	4.05	1950
S	75.1	54.1	64.6	100	1953	20	1951	107	3.07	5.23	1950	1.11	1956	4.60	1947
O	63.1	43.1	53.1	90	1951	-1	1952	377	2.62	9.26	1954#	.68	1956	2.96	1956#
N	47.1	32.3	39.7	75	1950	-8	1950	759	2.46	5.28	1950	.90	1949	1.79	1950
D	35.4	22.2	28.8	64	1952#	-17	1951	1122	2.26	4.63	1957	.54	1955	1.80	1952
Year	59.0	40.7	49.9	103 JULY 1954		-17 FEB 1951		6287	34.21	9.72 JAN 1950		.42 FEB 1947		4.60 SEPT 1950	

Month	Snow mean total	Snow max monthly	Year	Snow max 24 hrs	Year
J	7.5	11.7	1957	5.9	1951
F	6.3	10.2	1956	7.7	1952
M	4.7	8.3	1955	7.1	1955
A	1.9	10.6	1954#	6.4	1957
M	T	.0		T	1954#
J	.0	.0		.0	
J	.0				
A	.0				
S	.0				
O	.7	.7	1952	.7	1952
N	3.2	14.1	1950	5.0	1951
D	5.5	11.4	1951	7.7	1952
Year	29.4	14.1 NOV 1950		7.7 DEC 1952#	

Fort Wayne — Humidity, Wind, Sky

Month	RH Midnight	RH 6AM	RH Noon	RH 6PM	Mean hourly wind	Prevailing dir.	Fastest mile speed	Dir.	Year	Pct. sunshine	Mean sky cover
J	83	84	75	81	10.8	SW	59	SW	1949	35	8.0
F	82	84	70	77	10.7	SW	44	SW	1953	43	7.3
M	80	84	63	70	12.1	SW	65	W	1948	52	7.0
A	80	83	59	61	11.8	SW	55	S	1955	52	6.4
M	81	84	54	61	10.2	SW	57	SE	1957	63	6.3
J	82	84	56	58	8.9	SW	61	NW	1954#	71	5.6
J	85	88	53	61	7.9	SW	47	N	1957#	72	5.2
A	86	89	52	64	7.0	SW	40	S	1947	67	5.2
S	82	89	54	66	8.4	S	45	W	1953	63	5.4
O	82	85	59	76	8.7	SW	52	SW	1948	41	6.8
N	83	85	73	80	11.3	SW	—	—		27	7.5
D	81	85	—	—	11.0	SW	—	—		—	—
Year	82	85	60	65	9.9	SW	65	SE	JUNE 1948#	57	6.5

Means and extremes in the above table are from the existing or comparable location(s). Annual extremes have been exceeded at prior locations as follows: Highest temperature 106 in July 1936 and earlier date(s); lowest temperature -24 in January 1918; maximum monthly precipitation 10.09 in August 1926; minimum monthly precipitation 0.20 in February 1920; maximum precipitation in 24 hours 4.93 in August 1926; maximum monthly snowfall 25.4 in January 1918; maximum snowfall in 24 hours 12.4 in February 1912.

NORMALS, MEANS, AND EXTREMES

INDIANAPOLIS, INDIANA — WEIR COOK AIRPORT

LATITUDE 39° 44' N LONGITUDE 86° 16' W ELEVATION (ground) 793 Feet

Month	Temperature Normal — Daily maximum	Daily minimum	Monthly	Extremes Record highest	Year	Record lowest	Year	Normal degree days	Precipitation Normal total	Max monthly	Year	Min monthly	Year	Max in 24 hrs	Year	Snow/Sleet Mean total	Max monthly	Year	Max in 24 hrs	Year
(a)	(b)	(b)	(b)	20		20		(b)	(b)	18		18		15		26	26		15	
J	37.1	20.5	28.8	71	1950	-12	1940	1122	3.15	12.69	1950	.21	1944	3.47	1950	4.1	15.0	1939	5.6	1951
F	39.8	22.1	31.5	72	1954	-19	1951	938	2.08	5.32	1950	.37	1947	2.32	1950	3.7	10.3	1940	4.5	1944
M	49.8	30.4	40.1	81	1945	-6	1939	772	3.89	7.76	1945	1.03	1941	2.48	1957	3.3	8.9	1947	3.1	1948
A	61.8	40.1	50.8	88	1942	16	1940	432	3.85	7.90	1944	1.07	1946	2.34	1944	.6	4.0	1940	.0	1953
M	72.9	49.9	61.4	93	1942	29	1940	176	3.85	10.10	1943	1.93	1943	2.62	1946	T	T	1940	.0	
J	82.7	60.0	71.4	102	1954	39	1956	30	4.21	9.74	1942	1.15	1941	3.71	1957	.0	.0		.0	
J	87.9	64.1	76.0	104	1954	44	1942	0	3.03	6.19	1953	.99	1941	3.75	1953	.0	.0		.0	
A	85.5	62.5	74.0	100	1954	42	1946	0	3.52	6.64	1949	1.04	1948	2.72	1949	.0	.0		.0	
S	78.8	55.6	67.2	100	1954#	28	1942	79	3.67	6.01	1955	.96	1954	2.45	1955	.0	T		.0	
O	67.3	44.4	55.9	90	1954#	17	1952	306	2.52	8.36	1941	.61	1944	2.50	1951	T	T	1957#	T	1957#
N	50.5	32.5	41.5	81	1950	-2	1950	705	3.13	5.52	1950	.89	1940	3.02	1955	1.0	7.9	1932	3.6	1955
D	38.8	23.4	31.1	64	1956#	-15	1951	1051	2.79	6.70	1957	.93	1945	1.80	1957	4.3	15.0	1945	5.6	1957
Year	62.7	42.2	52.5	104	JULY 1954	-19	FEB 1951	5611	39.69	12.69	JAN 1950	.21	JAN 1944	3.75	JULY 1953	18.0	15.0	JAN 1939	5.6	JAN 1951#

Means and extremes in the above table are from the existing or comparable location(s). Annual extremes have been exceeded at prior locations as follows:
Highest temperature 107 in July 1934; lowest temperature -25 in January 1884.

SOUTH BEND, INDIANA — ST. JOSEPH COUNTY AIRPORT

LATITUDE 41° 42' N LONGITUDE 86° 19' W ELEVATION (ground) 768 Feet

Month	Temperature Normal — Daily maximum	Daily minimum	Monthly	Extremes Record highest	Year	Record lowest	Year	Normal degree days	Precipitation Normal total	Max monthly	Year	Min monthly	Year	Max in 24 hrs	Year	Snow/Sleet Mean total	Max monthly	Year	Max in 24 hrs	Year
(a)	(b)	(b)	(b)	18		18		(b)	(b)	18		18		18		18	18		18	
J	31.9	17.2	24.6	68	1950	-22	1943	1252	2.00	4.62	1950	.44	1945	1.70	1945	11.8	25.4	1957	8.3	1957
F	34.0	18.8	26.4	71	1954	-13	1951	1081	1.56	4.01	1955	.87	1947	2.64	1955	10.2	14.2	1942	7.4	1942
M	43.7	27.7	35.7	81	1945	-13	1943	908	2.95	5.56	1944	1.59	1956	2.26	1956	6.1	14.2	1954	5.8	1954
A	57.9	36.7	47.3	91	1942	13	1954#	531	3.40	9.20	1947	1.27	1949	3.14	1947	1.1	6.5	1957	.6	1957
M	69.5	47.2	58.4	95	1942	23	1954	248	3.84	8.28	1943	2.20	1955	2.11	1955	.1	.6	1940	.6	1940
J	79.6	57.7	68.7	100	1953	35	1956	62	3.83	6.28	1950	2.20	1951	2.86	1956	.0	.0		.0	
J	84.6	62.1	73.4	101	1941	45	1950#	5	3.04	6.20	1946	.02	1946	2.94	1946	.0	.0		.0	
A	82.4	60.3	71.4	100	1947	41	1946	13	3.25	6.45	1955	.32	1950	2.55	1950	.0	.0		.0	
S	75.7	53.9	64.5	99	1953	23	1942	101	3.71	7.02	1947	.53	1957	2.40	1957	T	1.2	1942	1.0	1942
O	62.7	43.1	52.9	88	1953	16	1952#	381	2.99	9.75	1954	.42	1953	3.18	1953	1.2	6.4	1954	4.2	1954
N	46.2	31.2	38.7	84	1950	-1	1950	783	2.74	4.27	1951	1.64	1951	1.91	1951	7.2	23.4	1951	10.3	1951
D	34.5	21.0	27.8	64	1949	-16	1951	1153	2.28	5.01	1949	.60	1943	2.29	1943	14.0	30.7	1951	12.5	1951
Year	58.5	39.7	49.1	101	JULY 1941	-22	JAN 1943	6524	35.59	9.75	OCT 1954	.02	JULY 1946	3.18	OCT 1954	51.4	30.7	DEC 1951	12.5	DEC 1951

Means and extremes in the above table are from the existing or comparable location(s). Annual extremes have been exceeded at prior locations as follows:
Highest temperature 109 in July 1934; maximum precipitation in 24 hours 3.50 in June 1903; maximum monthly snowfall 35.5 in February 1908.

REFERENCE NOTES APPLYING TO ALL "NORMALS, MEANS, AND EXTREMES" TABLES.

(a) Length of record, years.
(b) Normal values are based on the period 1921-1950, and are means adjusted to represent observations taken at the present standard location.
      Less than one-half.

- No record.
† Airport data.
+ City Office data.
# Also on earlier dates, months, or years.
T Trace, an amount too small to measure.

Sky cover is expressed in a range of 0 for no clouds or obscuring phenomena to 10 for complete sky cover. The number of clear days is based on average cloudiness 0-3 tenths; partly cloudy days on 4-7 tenths; and cloudy days on 8-10 tenths. Monthly degree day totals are the sum of the negative departures of average daily temperatures from 65°. Sleet was included in snowfall totals beginning with July 1948. Heavy fog also includes data referred to at various times in the past as "Dense" or "Thick". The upper visibility limit for heavy fog is 1/4 mile. Data in these tables are based on records through 1957.

Mean Maximum Temperature (°F.), January

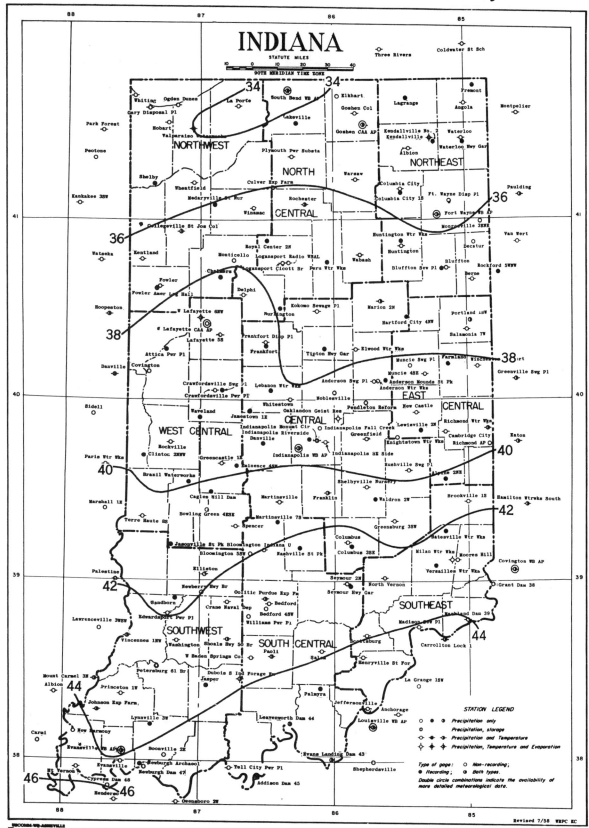

Based on period 1931-52

Isolines are drawn through points of approximately equal value.   Caution should be used in interpolating on these maps.

Mean Minimum Temperature (°F.), January

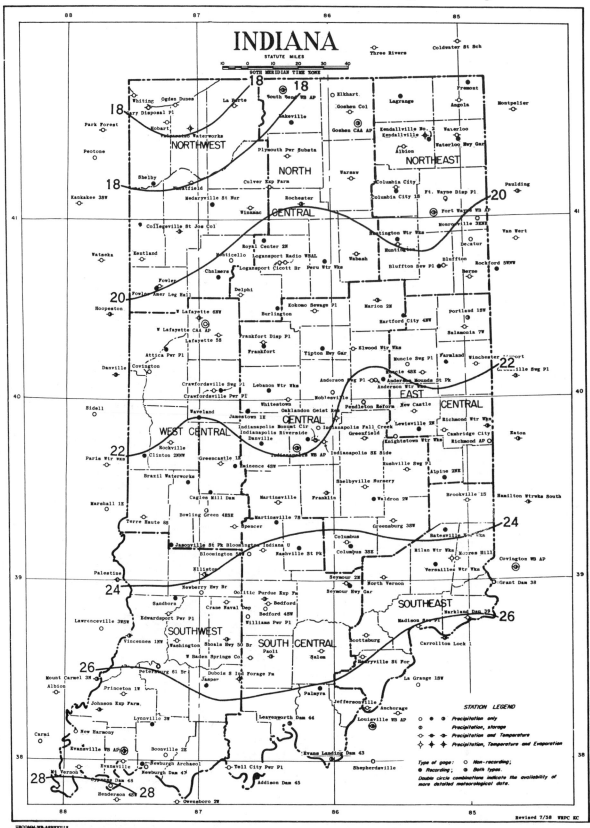

Based on period 1931-52

Isolines are drawn through points of approximately equal value. Caution should be used in interpolating on these maps.

Mean Maximum Temperature (°F.), July

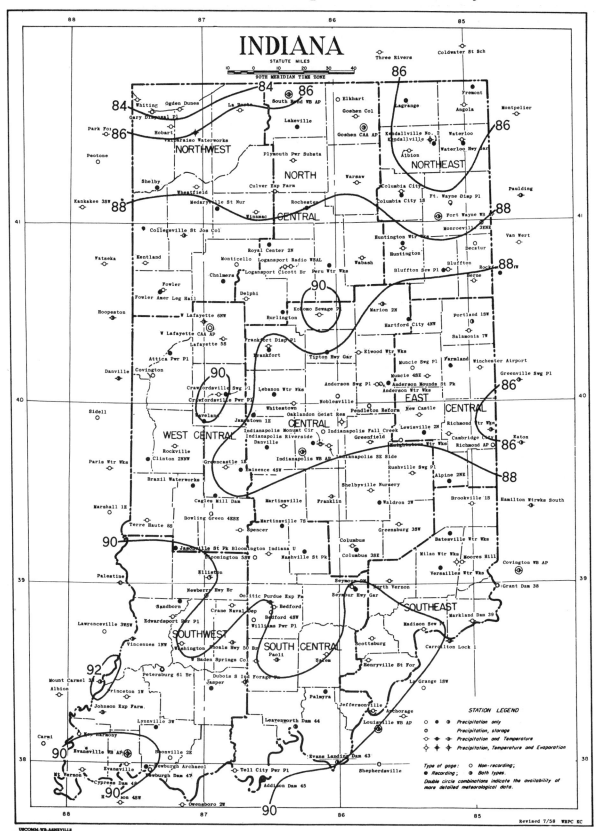

Based on period 1931-52

Isolines are drawn through points of approximately equal value. Caution should be used in interpolating on these maps.

Mean Minimum Temperature (°F.), July

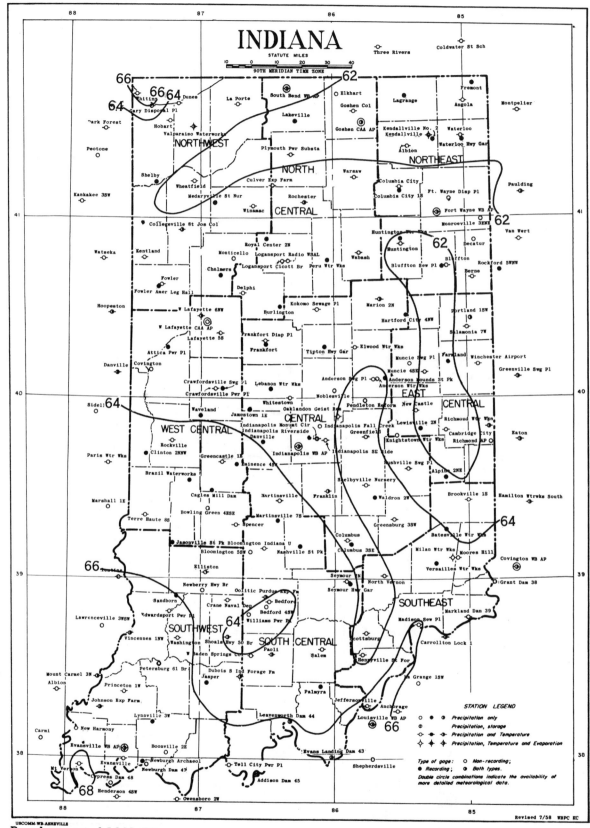

Based on period 1931-52

Isolines are drawn through points of approximately equal value. Caution should be used in interpolating on these maps,

Mean Annual Precipitation, Inches

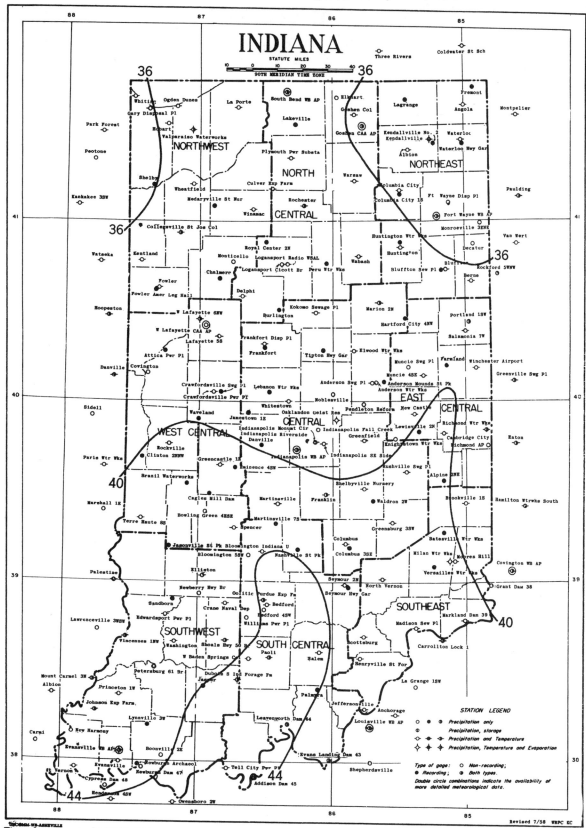

INDIANA

STATUTE MILES

90TH MERIDIAN TIME ZONE

STATION LEGEND

○ ◑ ● Precipitation only

⊕ Precipitation, storage

◐ ◑ ◑ Precipitation and Temperature

✦ ✦ ✦ Precipitation, Temperature and Evaporation

Type of gage: ○ Non-recording;
● Recording; ◑ Both types.
Double circle combinations indicate the availability of
more detailed meteorological data.

Revised 7/58 WRPC KC

Based on period 1931-55

Isolines are drawn through points of approximately equal value. Caution should be used
in interpolating on these maps,

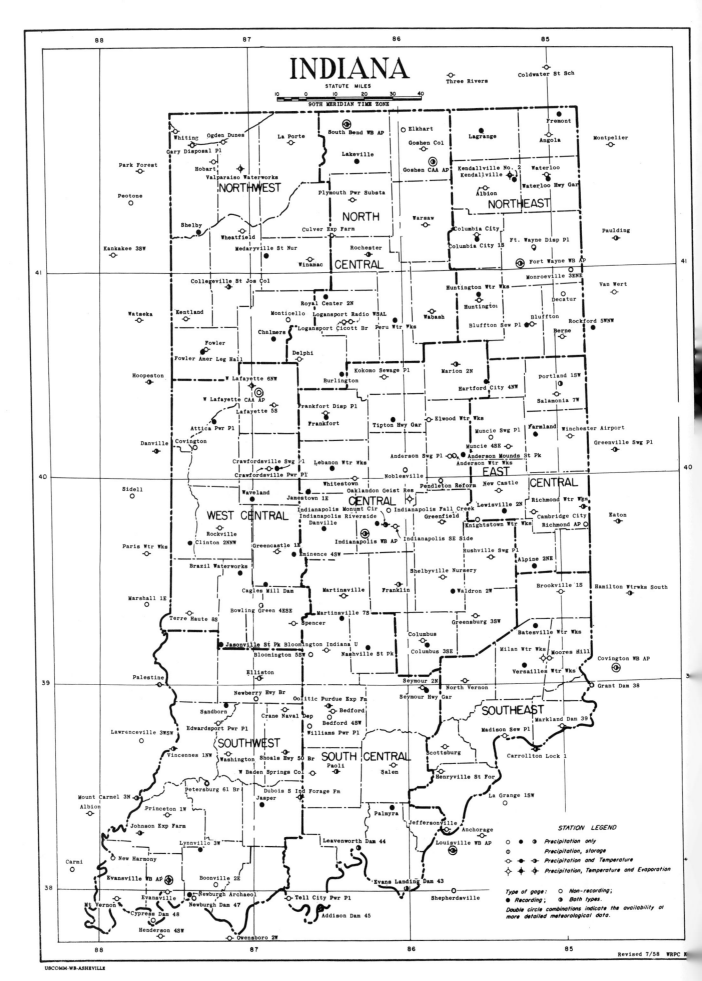

INDIANA

STATUTE MILES
90TH MERIDIAN TIME ZONE

NORTHWEST

NORTH

CENTRAL

NORTHEAST

WEST CENTRAL

CENTRAL

EAST CENTRAL

SOUTHWEST

SOUTH CENTRAL

SOUTHEAST

Three Rivers · Coldwater St Sch
Whiting · Ogden Dunes · South Bend WB AP · Elkhart · Lagrange · Fremont
Gary Disposal Pl · La Porte · Goshen Col · Angola · Montpelier
Park Forest · Hobart · Lakeville · Goshen CAA AP · Kendallville No. 2 · Waterloo
Peotone · Valparaiso Waterworks · Kendallville · Waterloo Hwy Gar
Plymouth Pwr Substa · Albion · Paulding
Shelby · Warsaw · NORTHEAST
Kankakee 3SW · Wheatfield · Culver Exp Farm · Columbia City · Ft. Wayne Disp Pl · Paulding
Medaryville St Nur · Rochester · Columbia City 1S · Fort Wayne WB AP
Winamac · Monroeville 3ENE · Van Wert
Collegeville St Jos Col · Huntington Wtr Wks · Decatur
Watseka · Kentland · Royal Center 2N · Wabash · Huntington · Bluffton
Monticello · Logansport Radio WSAL · Bluffton Sew Pl · Berne · Rockford 5WNW
Fowler · Chalmers · Logansport Cicott Br · Peru Wtr Wks
Fowler Amer Leg Hall · Delphi · Kokomo Sewage Pl · Marion 2N · Portland 1SW
Hoopeston · W Lafayette 6NW · Burlington · Hartford City 4NW · Salamonia 7W
W Lafayette CAA AP · Frankfort Disp Pl · Elwood Wtr Wks · Farmland · Winchester Airport
Lafayette 5S · Tipton Hwy Gar · Muncie Swg Pl · Greenville Swg Pl
Attica Pwr Pl · Frankfort · Muncie 4SE
Danville · Covington · Anderson Swg Pl · Anderson Mounds St Pk
Crawfordsville Swg Pl · Lebanon Wtr Wks · Anderson Wtr Wks
Sidell · Crawfordsville Pwr Pl · Noblesville · New Castle · Richmond Wtr Wks
Waveland · Whitestown · Oaklandon Geist Res · Pendleton Reform · Lewisville 2N · Eaton
Jamestown 1E · Indianapolis Monumt Cir · Indianapolis Fall Creek · Cambridge City
Rockville · Indianapolis Riverside · Greenfield · Richmond AP
Paris Wtr Wks · Clinton 2NNW · Greencastle 1E · Danville · Knightstown Wtr Wks
Brazil Waterworks · Eminence 4SW · Indianapolis WB AP · Indianapolis SE Side · Rushville Swg Pl
Marshall 1E · Cagles Mill Dam · Martinsville · Franklin · Alpine 2NE
Bowling Green 4ESE · Shelbyville Nursery · Brookville 1S · Hamilton Wtrwks South
Terre Haute 8S · Spencer · Waldron 2W · Batesville Wtr Wks
Martinsville 7S · Columbus · Greensburg 3SW · Milan Wtr Wks · Moores Hill
Jasonville St Pk · Bloomington Indiana U · Columbus 3SE · Covington WB AP
Bloomington 5SW · Nashville St Pk · Versailles Wtr Wks
Palestine · Elliston · Seymour 2N · North Vernon · Grant Dam 38
Newberry Hwy Br · Seymour Hwy Gar · Markland Dam 39
Sandborn · Oolitic Purdue Exp Fm · Bedford · Madison Sew Pl
Lawrenceville 3WSW · Edwardsport Pwr Pl · Crane Naval Dep · Bedford 4SW · Scottsburg · Carrollton Lock 1
Vincennes 1NW · Williams Pwr Pl · Henryville St For
Washington · Shoals Hwy 50 Br · Paoli · Salem
Mount Carmel 3N · W Baden Springs Col · La Grange 1SW
Albion · Petersburg 61 Br · Dubois S Ind Forage Fm · Palmyra
Princeton 1W · Jasper · Jeffersonville · Anchorage
Johnson Exp Farm · Louisville WB AP
Carmi · Lynnville 3W · Leavenworth Dam 44
New Harmony · Evans Landing Dam 43
Evansville WB AP · Boonville 2E · Tell City Pwr Pl · Shepherdsville
Mt Vernon · Evansville · Newburgh Archaeol · Addison Dam 45
Cypress Dam 48 · Newburgh Dam 47
Henderson 4SW · Owensboro 2W

STATION LEGEND

		Precipitation only
		Precipitation, storage
		Precipitation and Temperature
		Precipitation, Temperature and Evaporation

Type of gage: ○ Non-recording;
● Recording; ◉ Both types.
Double circle combinations indicate the availability of
more detailed meteorological data.

Revised 7/58 WRPC

THE CLIMATE OF
KENTUCKY

by
O.K Anderson

August 1959

Kentucky has a land surface of 40,109 square miles. It is essentially an eroded plateau that slopes downward gradually to the southwest, with elevations ranging from about 400 feet above sea level at the western edge to 1,000 feet in the central districts to above 4,000 feet near the southeastern border. There are seven major physiographic or natural regions.

The Bluegrass Region comprises about one-fifth of the State. The central area of this region is undulating to gently rolling. The outer area is more rolling and less uniform. Separating the two areas is a terrain that is hilly, with winding ridges and valleys and steep slopes.

The Knobs Region, named for its conical and flat-topped hills, comprises about one-tenth of the State. It forms a narrow crescent encircling the Bluegrass on the east, south, and west. Towards the Bluegrass the terrain is flat to rolling with scattered knobs and wide valleys, while the outer margin is rough.

The Eastern Mountains, also called the Cumberland Plateau, extends over the entire eastern fourth of the State. Ridges are high and sharp-crested; there is little level land, and valleys are narrow. In the southeast the Pine and Cumberland Mountains comprise the highest and most rugged part of Kentucky.

The Pennyroyal Region or the Mississippean Plateaus Region is one of the three largest regions. Much of the surface is quite uniform, but as a whole is rather diverse. Much of the terrain is undulating to rolling. In some places it is hilly or cavernous. Subsurface drainage has created limestone sinks and karst terrain in much of the area.

The Western Coal Field is a small region. This area has extensive bottom lands in the valleys of the Ohio, Green, and Tradewater Rivers and many of their tributaries. There is also some undulating to gently rolling uplands.

The Cumberland-Tennessee Rivers Area is the smallest region. The topography is hilly and rough, except for the wide bottoms along the two major streams.

The Jackson Purchase is the extreme western area of Kentucky. In both elevation and relief it is lower than the other regions of the State, but it also has a varied surface. It is largely an upland plain which is mostly undulating to gently rolling, but is also level in places and hilly in others.

The climate of Kentucky is essentially continental in character, with rather wide extremes of temperature and precipitation. The State lies within the path of storms, in the belt of the westerly winds. The temperature generally varies as the storms move across the State. Thus in winter and summer, there are occasional cold and hot spells of short duration. In the spring and fall, the systems have a smaller frequency, temperatures are more consistent, and fewer extremes are experienced. Precipitation occurs with the systems which generally move from the west to east, or from summer thunderstorms. However, the greater portion of precipitation is due to the moisture-bearing low-pressure formations which move from southwest to northeast from the western Gulf of Mexico and frequently cross Kentucky. With warm moist tropical air predominating during the summer months, relative humidity remains consistently high during that season.

The mean annual temperature ranges from 54° F. in the extreme north to 59° F. in the southwestern counties. July is usually the warmest month and January, the coldest. Extreme summer temperatures nearly always reach 100° or higher at most locations, but the frequencies of these high temperatures are low. Minimum temperatures of 0° or below can be expected during the months of December, January, and February at most locations, but the number of days with such temperatures is relatively small. Because of the State's geographic locations with reference to the center of the continent, the mid-winter cold waves from the Canadian Northwest usually have their intensity considerably modified by the time they reach Kentucky. In summer when the high pressure off the Florida coast is displaced westward from its normal position, extended periods of hot, sultry weather will occur. The spring and fall months are usually pleasant.

Precipitation is generally plentiful for all agricultural and industrial activities. The fall season is generally the driest and the spring season the wettest. Approximately half of the average annual total occurs during the warm months of April to September. The Average annual total in the State ranges from 36 inches in northern counties to 50 inches in the southern. Thunderstorms with high intensity rainfall are common during the spring and summer months, and rainfall during these storms in a 24-hour period frequently exceeds 2 to 3 inches, occasionally reaching 5 to 6 inches. Flash floods frequently result from the high intensity showers. Snowfall occurrence also varies from year to year but is common from November through March. Some snow has also been reported in the months of October and April. In some sections, the ground seldom remains covered with snow for more than a few days. The average annual snowfall for the State ranges from 6 to 10 inches in the southwest to 15 to 20 inches in the southeast.

Winds in the State have an average velocity of 7 to 12 miles per hour, and the prevailing direction is from south to southwest for the year. During the fall season some areas show a prevailing direction having a northerly component. The highest wind speeds usually range from 50 to 70 miles per hour, but in some storms (generally squalls attending thunderstorms), winds in gusts may occasionally exceed these speeds. A number of years may pass without a tornado, or several may visit the State in a single year. On the average, about one per year occurs somewhere in the State.

Thunderstorms may occur in any month, but they occur most frequently during the months March through September. The mean number of days with thunderstorms ranges approximately between 45 and 60. They are occasionally attended by damaging hail, but the area thus affected is nearly always small.

Heavy fogs are rather rare in the State. The average number of days with heavy fogs varies between 8 and 17 during the year with the majority occurring during the months of September through March inclusive.

The average date of the last spring freeze ranges from April 4 in the extreme west to May 5 in the mountain region in the extreme southeast; that of the first fall freeze, from October 11 in the Pennyroyal Region to October 30 near the lower Ohio River. The average length of the freeze-free period varies from 166 days on the southeastern plateau to 210 near the lower Ohio River.

The average number of days with clear and with partly cloudy skies is about the same and ranges between 115 and 120 days over the State. The number of days with cloudy skies averages about 130. The extreme northern section shows the greatest number of days with cloudy skies. The percentage of possible sunshine averages 35 to 50 for the winter months, 50 to 65 in the spring, 65 to 75 in the summer, and 55 to 65 in the fall. The largest percentage of possible sunshine is recorded in the extreme western section of the State.

The Ohio River forms the northern boundary and the Mississippi the western. All of the State is in the Ohio River Basin, except for a small section in the Jackson Purchase area that drains directly into the Mississippi. Kentucky lies in the path of rain producing lows moving from the west Gulf area northeastward. The flood season is in the winter and spring. Numerous flash floods occur from excessive rains and thunderstorms, particularly in the mountains of the eastern portion.

Kentucky is primarily an agricultural state. Three-fourths of the total land area is utilized for agriculture. About one-fourth of the farm acreage is used for harvested crops. Another fourth is in woodland, and the remainder in pasture, farmsteads, and other uses. A large proportion of cropland is in pasture. Kentucky's agriculture varies from subsistence and part-time farming in the eastern mountains; to specialized tobacco and livestock farming in the central part of the State; to "Corn Belt" conditions in the Lower Ohio River Valley, particularly in the interior of the Western Coal Field; and to tobacco, livestock, small grain, cotton, and part-time farming in the southwest and Jackson Purchase. Tobacco is the most important single source of income; however, income from livestock exceeds that from crops. The variation in agriculture is primarily a result of the variation in soils, which are many. Practically all of them have developed under forest cover and essentially the same climate, but there are distinctions due chiefly to differences in parent materials, the topography which these materials occupy, and the length of time the materials have been exposed to soil forming processes. The best soils occupy gentle to moderate slopes. On flat uplands, drainage is usually impeded, and these wetter soils are less productive than their better drained counterparts. During the growing season there is sufficient rain for staple crops, and occasionally there is too much (especially in the spring months), delaying planting and cultivation. Widespread drought conditions sometimes prevail, but, even in such periods, local rains prevent complete crop failures.

Forests in the State cover nearly 11-1/2 million acres or about 45 percent of the total land area. They occur throughout the State, but are concentrated in the counties of the eastern third and south-central portions. Kentucky's forests are growing more timber than is being removed each year. This resource is primarily a result of the temperature and precipitation regime of the State.

Kentucky, long known as a leader in the tobacco and distilling industries, is now the home of a large group of diversified industries. Kentucky's mineral resources, coal, fluorspar, oil, and limestone play a vital role in the State's economy. Natural resources exist which are specifically suited to provide fuel, power, and production materials for metal-working, machinery, and chemical industries. Cheap water transportation on the river systems of the State is a major factor in the industrial economy. The favorable and moderate climate of Kentucky has been a definite factor, directly or indirectly, in the development of the State's industry in varied fields.

Because of its geographic location, Kentucky, enjoys a preferred market location between the highly urbanized and industrialized North and the rapid industrializing South. A very substantial portion of the industrial and consumer market of the United States can be reached in a short time from most points in Kentucky. Cheap water transportation again is one of the big factors. Climate also plays a large role in the favorable position, in that extreme severe weather which can paralyze transportation is extremely rare. Local trading centers and markets are also in favorable position because of the temperate climate with the few days of temperature extremes.

Recreational facilities in Kentucky are many and varied. Fishing, hunting, and water activities are the major recreational activities. The numerous natural streams, lakes, and rivers and the man-made lakes (such as formed by Kentucky Dam and Wolf Creek Dam) are the areas for much of the activity. The forest areas and the caves at Mammoth Cave are other attractions. Large areas have been set aside as National and State Parks. Climate again is a favorable factor for recreational facilities in that most locations can be utilized for a large portion of the year.

REFERENCES

1. Kentucky, Land Use Suitability Map - Agricultural and Industrial Development Board of Kentucky
2. Historic Kentucky Highways - Kentucky Division of Publicity
3. Editor & Publisher, October 31, 1953
4. Kentucky Retail Market Areas and Trading Centers - F. G. Coolsen and W. S. Myers, University of Kentucky, College of Commerce in cooperation with AID Board of Kentucky
5. Kentucky - Kentucky Utilities Company Brochure
6. Desk Book of Kentucky Economic Statistics - Agricultural and Industrial Development Board of Kentucky
7. Stream Flow Data in Kentucky - U. S. Geological Survey
8. Kentucky Official Highway Map - Division of Publicity
9. Chemical Character of Surface Waters of Kentucky 1949-1951 - Agricultural and Industrial Development Board of Kentucky
10. Industrial Possibilities of the Forest Resource of Kentucky - Agricultural and Industrial Development Board of Kentucky
11. Weather Bureau Technical Paper No. 15 - Maximum Station Precipitation for 1, 2, 3, 6, 12, and 24 Hours
12. Weather Bureau Technical Paper No. 16 - Maximum 24-Hour Precipitation in the United States. Washington, D. C. 1952
13. Weather Bureau Technical Paper No. 25 - Rainfall Intensity-Duration-Frequency Curves. For selected stations in the United States, Alaska, Hawaiian Islands, and Puerto Rico
14. Weather Bureau Technical Paper No. 29 - Rainfall Intensity-Frequency Regime. Washington, D. C.

BIBLIOGRAPHY

A   Climatic Summary of the United States (Bulletin W) 1930 edition, Sections 74 and 75. U. S. Weather Bureau
B   Climatic Summary of the United States, Kentucky - Supplement for 1931 through 1952 (Bulletin W Supplement). U. S. Weather Bureau
C   Climatological Data - Kentucky. U. S. Weather Bureau
D   Climatological Data National Summary. U. S. Weather Bureau
E   Hourly Precipitation Data - Kentucky. U. S. Weather Bureau
F   Local Climatological Data, U. S. Weather Bureau for Lexington and Louisville, Kentucky

FREEZE DATA

STATION	Freeze threshold temperature	Mean date of last Spring occurrence	Mean date of first Fall occurrence	Mean No. of days between dates	Years of record Spring	No. of occurrences in Spring	Years of record Fall	No. of occurrences in Fall
ANCHORAGE	32	04-25	10-15	173	30	30	30	30
	28	04-13	10-27	198	30	30	30	30
	24	03-28	11-08	226	30	30	30	30
	20	03-15	11-15	245	30	30	30	30
	16	03-04	12-02	273	30	30	30	28
ASHLAND DAM 29	32	04-20	10-22	185	30	30	30	30
	28	04-07	11-05	212	30	30	30	30
	24	03-16	11-17	247	30	30	30	30
	20	03-06	12-02	271	30	30	30	28
	16	02-23	12-11	291	30	29	30	25
BEAVER DAM	32	04-20	10-14	177	25	25	25	25
	28	04-03	10-26	206	25	25	25	25
	24	03-16	11-08	237	25	25	24	24
	20	03-08	11-22	259	25	25	24	22
	16	02-27	12-06	282	25	25	24	21
BEREA COLLEGE	32	04-17	10-19	185	28	28	30	30
	28	04-04	11-01	211	29	29	30	30
	24	03-24	11-13	234	29	29	30	30
	20	03-10	11-24	259	29	29	30	29
	16	02-27	12-08	284	29	28	30	28
BOWLING GREEN WB AP	32	04-08	10-28	204	30	30	30	30
	28	03-25	11-09	229	30	30	30	30
	24	03-10	11-23	258	30	30	30	29
	20	03-03	12-02	274	30	30	30	27
	16	02-12	12-14	305	30	27	30	23
BURGIN DIX DAM	32	04-13	10-24	195	27	27	27	27
	28	03-30	11-06	222	26	26	27	27
	24	03-18	11-18	245	26	26	26	26
	20	03-07	11-30	268	26	26	26	25
	16	02-24	12-11	290	27	27	26	23
FRANKFORT LOCK 4	32	04-17	10-23	190	30	30	30	30
	28	04-03	11-04	215	30	30	30	30
	24	03-19	11-14	240	30	30	30	30
	20	03-07	11-28	266	30	30	30	29
	16	02-24	12-09	289	30	30	30	27
GREENSBURG	32	04-22	10-11	172	29	29	29	29
	28	04-09	10-24	198	30	30	29	29
	24	03-25	11-04	224	30	30	28	28
	20	03-10	11-19	254	30	30	28	27
	16	03-01	12-03	277	30	30	28	24
GREENVILLE 2 W	32	04-20	10-15	179	30	30	30	30
	28	04-03	10-27	207	30	30	29	29
	24	03-25	11-09	229	30	30	29	29
	20	03-12	11-22	255	30	30	29	28
	16	02-26	12-05	282	29	29	29	26
HEIDELBERG LOCK 14	32	04-23	10-18	178	17	17	18	18
	28	04-09	10-29	203	17	17	18	18
	24	03-25	11-08	228	17	17	18	18
	20	03-13	11-25	258	17	17	18	18
	16	03-05	12-02	272	17	17	18	18
HENDERSON 4 SW	32	04-12	10-28	198	19	19	20	20
	28	03-30	11-05	220	19	19	20	20
	24	03-17	11-19	246	19	19	20	20
	20	03-06	11-27	266	19	19	20	19
	16	02-24	12-11	290	19	19	20	17
HOPKINSVILLE 2 E	32	04-12	10-23	194	30	30	30	30
	28	03-29	11-06	222	30	30	30	30
	24	03-13	11-13	245	30	30	30	30
	20	03-04	11-28	269	30	30	30	28
	16	02-21	12-12	294	30	29	30	26
IRVINGTON	32	04-14	10-21	190	30	30	30	30
	28	03-28	11-08	226	30	30	30	30
	24	03-17	11-18	246	30	30	30	28
	20	03-05	12-02	272	30	30	30	28
	16	02-23	12-11	291	30	29	30	26
LEITCHFIELD	32	04-14	10-25	195	30	30	30	30
	28	03-30	11-06	222	30	30	30	30
	24	03-19	11-19	245	30	30	30	29
	20	03-05	12-01	271	30	30	30	28
	16	02-25	12-10	289	30	30	30	27
LEXINGTON WB AP	32	04-13	10-28	198	30	30	30	30
	28	03-30	11-07	222	30	30	30	30
	24	03-20	11-20	245	30	30	30	29
	20	03-09	12-01	267	30	30	30	28
	16	02-25	12-10	288	30	30	30	27
LOUISVILLE WB CITY	32	04-01	11-07	220	30	30	30	30
	28	03-23	11-17	239	29	29	30	30
	24	03-10	11-29	264	30	30	30	28
	20	03-02	12-09	283	30	30	30	26
	16	02-20	12-15	299	30	29	30	24
MAYFIELD SUB STA	32	04-09	10-27	200	29	29	29	29
	28	03-28	11-07	224	28	28	29	29
	24	03-13	11-19	251	28	28	29	29
	20	03-01	12-04	279	28	28	29	25
	16	02-15	12-14	302	28	26	29	24
MAYSVILLE DAM 33	32	04-20	10-19	182	30	30	30	30
	28	04-08	11-03	209	30	30	30	30
	24	03-22	11-18	241	30	30	30	30
	20	03-08	11-27	264	30	29	30	28
	16	02-27	12-07	283	30	28	30	27
MIDDLESBORO	32	04-22	10-17	178	28	28	26	26
	28	04-08	10-31	206	25	25	26	26
	24	03-22	11-12	234	26	26	25	25
	20	03-06	11-23	262	26	25	25	24
	16	02-27	12-05	280	25	24	25	22
MT STERLING	32	04-22	10-19	181	26	26	26	26
	28	04-11	10-28	200	26	26	25	25
	24	03-24	11-13	235	26	26	25	25
	20	03-10	11-24	259	27	27	25	25
	16	03-01	12-03	278	28	28	26	24
OWENSBORO 2 W	32	04-13	10-21	191	30	30	30	30
	28	03-27	11-03	222	30	30	30	30
	24	03-15	11-15	246	30	30	30	30
	20	03-04	11-29	270	29	29	30	27
	16	02-23	12-08	288	29	29	30	26
PADUCAH CAA AP	32	04-04	10-30	209	28	28	29	29
	28	03-20	11-09	234	26	26	29	29
	24	03-10	11-22	257	26	26	29	29
	20	02-25	12-04	282	26	26	29	28
	16	02-16	12-18	305	26	26	28	20
PIKEVILLE	32	04-14	10-31	200	14	14	15	15
	28	04-04	11-09	220	13	13	15	15
	24	03-16	11-23	252	14	14	15	15
	20	03-06	12-04	273	14	14	15	15
	16	02-20	12-14	297	13	12	15	11
PRINCETON	32	04-15	10-21	190	30	30	30	30
	28	04-01	11-03	216	30	30	30	30
	24	03-19	11-11	237	30	30	30	30
	20	03-05	12-01	272	30	30	30	28
	16	02-20	12-10	294	30	28	30	24
RICHMOND E ST COLL	32	04-11	10-28	201	27	27	28	28
	28	04-02	11-10	222	27	27	28	28
	24	03-17	11-20	248	27	27	28	28
	20	03-08	12-01	268	27	27	28	26
	16	02-23	12-11	291	27	26	27	23
RUSSELLVILLE	32	04-13	10-21	192	29	29	28	28
	28	04-01	11-01	214	29	29	28	28
	24	03-18	11-15	243	30	30	27	27
	20	03-04	11-30	271	29	29	27	25
	16	02-25	12-07	285	28	28	26	23

FREEZE DATA

STATION	Freeze threshold temperature	Mean date of last Spring occurrence	Mean date of first Fall occurrence	Mean No. of days between dates	Years of record Spring	No. of occurrences in Spring	Years of record Fall	No. of occurrences in Fall	STATION	Freeze threshold temperature	Mean date of last Spring occurrence	Mean date of first Fall occurrence	Mean No. of days between dates	Years of record Spring	No. of occurrences in Spring	Years of record Fall	No. of occurrences in Fall
ST JOHN BETHLEHEM AC	32	04-20	10-19	181	30	30	30	30	WILLIAMSBURG	32	04-19	10-21	185	30	30	30	30
	28	04-09	10-28	202	30	30	30	30		28	04-02	10-31	212	30	30	30	30
	24	03-23	11-12	234	29	29	30	30		24	03-19	11-12	238	30	30	30	30
	20	03-10	11-24	259	29	29	28	26		20	03-08	11-22	259	29	29	30	29
	16	02-28	12-06	281	29	29	27	25		16	02-26	12-05	281	29	28	29	27
SHELBYVILLE 2 W	32	04-25	10-15	173	30	30	30	30	WILLIAMSTOWN	32	04-20	10-24	188	30	30	29	29
	28	04-05	10-29	207	30	30	29	29		28	04-07	11-05	212	30	30	28	28
	24	03-25	11-11	231	30	30	29	28		24	03-27	11-15	234	30	30	28	28
	20	03-13	11-24	256	30	30	30	28		20	03-12	11-26	259	30	30	29	29
	16	03-04	12-05	277	30	30	30	28		16	03-05	12-06	275	29	29	29	27

Data in the above table are based on the period 1921-1950, or that portion
of this period for which data are available.

Means have been adjusted to take into account years of non-occurrence.

A freeze is a numerical substitute for the former term "killing frost" and
is the occurrence of a minimum temperature at or below the threshold temperature of 32°, 28°, etc.

Freeze data tabulations in greater detail are available and can be reproduced
at cost.

*MEAN TEMPERATURE AND PRECIPITATION

STATION	JANUARY		FEBRUARY		MARCH		APRIL		MAY		JUNE		JULY		AUGUST		SEPTEMBER		OCTOBER		NOVEMBER		DECEMBER		ANNUAL	
	Temperature	Precipitation	Temperature	Precipitation	Temperature	Precipitation	Temperature	Precipitation	Temperature	Precipitation	Temperature	Precipitation	Temperature	Precipitation	Temperature	Precipitation	Temperature	Precipitation	Temperature	Precipitation	Temperature	Precipitaion	Temperature	Precipitation	Temperature	Precipitation
WESTERN																										
FORDS FERRY DAM 50		4.69		3.36		5.26		4.13		3.47		3.72		2.76		3.23		3.18		2.51		3.33		3.14		42.78
GREENVILLE 2 W	37.5	5.10	39.6	3.69	47.5	5.31	57.1	4.30	65.4	3.77	74.6	4.09	77.9	4.17	76.7	3.55	70.1	3.10	59.1	2.50	46.8	3.35	38.5	3.92	57.6	46.85
HENDERSON 4 SW		4.91	38.8	3.50	46.8	5.55	56.9	4.50	66.1	3.87	75.7	3.67	78.9	3.21	77.3	3.27	70.7	3.20	60.1	2.90	46.6	3.32	37.6	3.63	57.7	45.53
HOPKINSVILLE	38.1	5.46	40.2	3.94	47.9	5.38	57.7	4.17	66.6	4.00	75.6	4.15	78.8	3.65	77.8	3.41	71.2	3.13	60.1	2.52	47.0	3.53	39.0	4.08	58.3	47.42
LOVELACEVILLE	38.1	5.04	40.6	4.23	48.7	5.53	58.4	4.47	67.1	3.97	76.4	3.62	79.3	3.16	78.4	3.44	71.4	3.13	60.8	3.15	47.4	3.24	39.1	3.99	58.8	46.97
MAYFIELD 2 S	38.1	4.68	40.7	3.89	48.4	5.34	58.3	4.13	67.1	4.26	76.1	3.90	79.1	4.51	78.3	3.13	71.9	3.30	61.4	3.01	48.0	3.91	40.0	4.39	59.0	48.45
MURRAY	38.9	5.27	41.3	4.07	49.2	5.74	58.7	4.21	67.4	4.08	76.2	3.32	79.3	3.64	78.6	3.39	71.4	3.30	60.9	2.87	47.3	3.63	40.0	4.21	59.1	47.73
OWENSBORO 2 W	36.6	4.82	38.5	3.26	46.5	5.32	56.5	4.12	65.8	3.83	75.3	3.90	78.5	3.35	76.8	3.42	70.0	3.21	59.3	2.50	45.9	3.35	37.3	3.21	57.2	44.29
OWENSBORO DAM 46		4.80		3.36		5.24		4.19		3.76		3.80		3.06		3.49		3.14		2.43		3.29		3.14		43.70
PADUCAH		5.00		3.90		5.28		4.38		4.02		3.70		3.03		3.30		3.36		2.75		3.72		3.55		45.99
PRINCETON	38.2	5.20	40.1	3.98	48.0	5.48	57.9	4.11	66.7	3.97	75.7	3.94	79.0	3.31	77.9	3.39	71.1	3.08	60.1	2.72	46.9	3.46	38.8	3.82	58.3	46.46
RUMSEY LOCK 2		4.90		3.55		5.33		4.07		3.98		3.78		4.05		3.36		2.84		2.29		3.50		3.42		45.07
RUSSELLVILLE	37.4	5.44	38.9	4.13	46.9	5.24	57.3	3.97	66.2	3.62	75.6	3.88	78.6	4.39	77.2	3.37	70.5	3.60	59.1	2.20	45.6	3.52	37.9	4.19	57.6	47.55
UNIONTOWN DAM 49		4.36		3.07		4.75		4.12		3.81		3.27		3.29		3.32		3.29		2.62		3.18		2.92		42.00
DIVISION	37.6	5.09	39.8	3.81	47.8	5.40	57.7	4.19	66.6	3.94	75.8	3.79	78.9	3.63	77.8	3.32	71.0	3.22	60.2	2.66	47.0	3.52	38.6	3.78	58.2	46.35
CENTRAL																										
ADDISON DAM 45		4.81		3.65		5.21		4.03		3.97		3.73		3.51		3.24		3.26		2.40		3.14		3.42		44.37
ANCHORAGE	34.8	4.50	36.5	2.90	44.3	4.87	54.3	4.09	63.6	4.02	72.6	4.03	76.1	3.67	74.7	3.54	68.4	3.03	57.7	2.54	44.4	3.13	35.6	3.31	55.3	43.63
BOWLING GREEN CAA AP	38.7	5.82	40.5	4.03	48.5	5.63	58.3	4.04	67.3	3.81	76.4	3.94	79.4	4.27	78.0	3.71	71.6	3.19	60.2	2.34	47.3	3.47	39.2	4.42	58.8	48.67
BROWNSVILLE LOCK 6		5.53		4.11		5.39		4.12		3.72		4.32		4.02		4.19		3.32		2.51		3.65		4.20		49.08
GREENSBURG	37.6	5.47	39.2	3.74	46.8	5.58	56.6	3.82	65.5	4.32	74.5	4.84	77.6	4.23	76.1	3.95	69.7	3.12	58.1	2.38	45.8	3.38	38.2	3.92	57.1	48.75
IRVINGTON	36.7	4.59	38.6	3.53	46.5	5.02	56.6	4.02	65.6	3.87	74.5	4.24	77.9	3.41	76.4	3.20	70.1	2.84	59.2	2.41	46.1	3.11	37.6	3.48	57.2	43.72
LEITCHFIELD	37.4	5.61	39.2	3.95	47.0	5.39	56.8	4.19	65.8	3.93	74.6	4.40	77.9	3.86	76.6	3.56	70.3	3.08	59.5	2.62	46.6	3.58	38.3	8	57.5	48.25
LOUISVILLE WB AP	34.9	4.10	37.2	2.99	45.6	4.67	56.0	4.01	65.3	3.93	74.2	4.06	77.9	3.08	76.1	3.06	70.2	2.70	58.6	2.45	45.7	3.12	36.9	3.30	56.5	41.47
ST JOHN BETHLEHEM ACAD	36.0	5.54	37.6	3.77	45.3	5.12	55.5	4.34	64.7	4.09	73.8	4.13	77.0	3.96	75.5	3.68	69.4	2.77	58.3	2.46	45.0	3.28	36.5	3.86	56.2	47.00
SALVISA LOCK 6		5.46		3.76		5.33		3.96		3.83		4.39		3.84		4.05		3.27		2.24		3.37		3.46		46.96
TYRONE LOCK 5		4.92		3.47		5.15		4.02		3.62		3.89		4.02		3.79		2.88		2.28		3.32		3.57		44.93
WOODBURY LOCK 4		5.39		3.83		5.24		4.05		3.71		3.98		3.77		3.62		3.38		2.44		3.52		3.87		46.80
DIVISION	37.0	5.31	38.8	3.71	46.6	5.31	56.5	4.06	65.6	3.93	74.5	4.15	77.7	4.04	76.3	3.64	70.0	3.07	58.9	2.43	45.9	3.37	37.9	3.87	57.1	46.89
BLUE GRASS																										
BEREA COLLEGE	38.2	5.08	39.3	3.64	47.0	4.96	56.8	3.81	65.5	3.83	73.9	4.80	77.0	4.99	75.8	4.17	69.8	2.76	59.1	2.21	46.4	3.28	38.6	3.56	57.3	47.09
BRENT DAM 36		3.80		2.83		4.33		3.53		3.44		4.13		3.51		3.32		2.88		2.20		2.83		2.70		39.50
CARROLLTON LOCK 1		4.03		2.83		4.56		3.75		3.29		3.84		3.43		3.60		2.89		2.32		2.96		2.67		40.17
COLLEGE HILL LOCK 11		4.61		3.41		4.67		3.49		3.65		4.26		4.85		4.03		2.72		1.97		3.17		3.35		44.18
COVINGTON WB AIRPORT	31.8	3.40	33.9	2.55	41.7	4.04	52.1	3.59	62.5	3.45	71.8	4.04	75.1	3.75	73.5	3.34	67.3	2.97	56.0	2.17	43.4	3.03	33.6	2.83	53.6	39.16
CYNTHIANA		4.33		3.08		4.89		3.88		3.33		3.68		3.61		3.30		2.68		1.94		3.12		2.99		40.83
DANVILLE		5.28		3.76		4.87		3.80		4.14		4.60		4.32		3.90		3.00		2.16		3.22		3.49		46.54
FALMOUTH		4.02		3.07		4.91		3.95		3.76		4.15		4.01		3.14		2.74		2.17		3.30		3.02		42.24
FARMERS	36.8	4.68	37.9	3.46	45.5	4.92	55.2	3.90	64.4	4.02	72.6	4.43	75.7	4.55	74.1	4.18	68.0	3.04	57.0	2.20	45.0	3.17	36.3	3.31	55.7	45.86
FLEMINGSBURG		4.70		3.20		5.19		3.98		3.94		4.36		4.51		3.97		3.26		2.08		3.43		3.39		46.01
FORD LOCK 10		4.60		3.37		4.54		3.62		3.77		4.44		5.00		3.74		2.76		2.04		3.12		3.30		44.30
FRANKFORT LOCK 4	36.6	4.50	37.9	3.14	45.4	4.83	55.0	4.09	64.5	3.88	74.2	3.81	77.7	4.38	76.2	3.60	69.8	2.72	58.5	2.18	45.8	3.25	37.4	3.17	56.6	43.55
GEST LOCK 3		4.22		2.93		4.84		3.88		3.53		3.77		4.32		3.35		2.86		2.21		3.09		2.97		41.97
GRANT DAM 38		3.85		2.71		4.14		3.43		3.06		3.54		3.48		2.69		2.71		2.22		2.82		2.54		37.19
HIGH BRIDGE LOCK 7		4.95		3.55		4.79		3.72		3.72		4.61		4.28		3.50		2.95		3.34		3.17		3.28		45.86
LANCASTER		5.41		3.91		5.47		3.65		4.15		4.67		5.06		3.76		2.92		2.29		3.54		3.73		48.56
LEXINGTON WB AIRPORT	32.5	4.50	34.5	3.50	42.4	4.46	52.9	3.76	62.6	3.56	72.2	4.21	75.7	4.25	74.4	3.37	68.9	2.83	57.3	2.44	43.8	3.15	34.9	3.68	54.4	43.71
LITTLE HICKMAN LOCK 8		5.06		3.70		5.36		3.58		3.93		4.79		4.71		3.83		2.77		2.02		3.16		3.42		46.33
LOCKPORT LOCK 2		4.67		3.26		5.06		4.40		3.93		4.15		3.86		3.26		3.03		2.38		3.36		3.32		44.68
MAYSVILLE DAM 33	35.7	4.34	36.5	3.24	44.1	4.81	54.1	3.84	63.4	3.62	72.8	4.12	76.3	4.32	74.7	4.24	68.3	3.12	57.4	2.32	44.9	3.02	36.2	3.20	55.4	44.19
MOUNT STERLING		4.84		3.41		5.07		4.10		4.07		4.49		5.13		3.73		2.97		2.30		3.00		3.54		46.65
ONEONTA DAM 35		3.72		2.70		4.20		3.60		3.45		3.93		3.68		2.83		2.85		2.04		2.76		2.63		38.39
SHELBYVILLE 2 W	35.9	5.07	37.2	3.50	44.9	5.59	54.7	4.33	64.2	4.06	73.4	4.49	77.2	3.75	75.7	3.67	69.2	2.90	58.7	2.42	45.1	3.51	36.3	3.63	56.0	46.92
VALLEY VIEW LOCK 9		4.98		3.65		4.74		3.48		3.71		4.44		5.05		4.04		3.05		1.93		3.29		3.42		45.78
WILLIAMSTOWN 5 WSW	35.4	4.43	36.4	3.25	44.3	5.02	55.1	4.02	65.2	3.87	74.2	4.02	77.6	3.79	76.1	3.30	69.8	3.02	58.9	2.30	45.1	3.14	35.8	3.01	56.1	43.17
DIVISION	36.4	4.68	37.7	3.35	45.4	4.95	55.3	3.89	64.8	3.78	73.7	4.24	77.0	4.36	75.6	3.70	69.5	2.89	58.5	2.22	45.5	3.18	37.1	3.29	56.4	44.53
EASTERN																										
ASHLAND DAM 29	37.7	3.93	38.6	2.88	46.0	4.38	56.2	3.42	65.3	3.88	73.9	4.00	77.1	4.05	75.8	3.49	69.9	2.62	58.9	1.86	46.6	2.73	38.3	2.82	57.0	40.06
GREENUP DAM 30		3.86		2.88		4.33		3.59		3.86		4.11		4.37		3.44		2.97		2.26		2.63		2.83		41.13
HEIDELBERG LOCK 14		4.76		3.81		5.00		3.48		3.78		4.32		5.17		4.08		2.62		2.02		3.30		3.22		45.56
PIKEVILLE		3.73		3.57		4.67		3.37		4.08		4.47		4.89		3.89		2.85		2.04		2.58		3.20		43.34
RAVENNA LOCK 12		4.50		3.30		4.73		3.63		3.56		4.08		5.09		4.47		2.95		1.93		3.26		2.96		44.46
VANCEBURG DAM 32		4.25		3.21		4.74		3.71		4.04		4.25		4.18		3.86		2.62		2.28		3.00		3.12		43.26
WILLIAMSBURG	40.5	4.70	41.1	4.19	48.6	4.97	57.5	3.67	66.1	4.03	74.5	3.75	77.3	4.81	76.2	3.80	70.1	2.54	59.1	2.15	46.8	3.22	39.9	3.65	58.1	45.48
WILLOW LOCK 13		5.09		4.09		5.24		3.51		3.73		4.33		4.91		4.32		2.76		2.07		3.46		3.57		47.08
DIVISION	38.7	4.56	39.7	3.73	46.9	4.92	56.5	3.66	65.1	3.99	73.4	4.19	76.4	4.77	75.2	4.00	69.4	2.74	58.3	2.16	46.1	3.18	38.6	3.44	57.0	45.34

* Averages for period 1931-1955, except for stations marked WB which are "normals" based on period 1921-1950. Divisional means may not be the arithmetical average of individual stations published, since additional data from shorter period stations are used to obtain better areal representation.

CONFIDENCE LIMITS

In the absence of trend or record changes, the chances are 9 out of 10 that the true mean will lie in the interval formed by adding and subtracting the values in the following table from the means for any station in the State. Because of the wider variation in mean precipitation, the corresponding monthly means and annual mean must be substituted for "p" in the precipitation table below to obtain mean precipitation confidence limits.

1.9	$.45\sqrt{p}$	1.7	$.33\sqrt{p}$	1.9	$.36\sqrt{p}$	1.0	$.29\sqrt{p}$	1.0	$.38\sqrt{p}$	1.0	$.38\sqrt{p}$	.9	$.39\sqrt{p}$	.9	$.34\sqrt{p}$	1.1	$.32\sqrt{p}$	1.1	$.38\sqrt{p}$	1.1	$.35\sqrt{p}$	1.4	$.33\sqrt{p}$	.4	$.36\sqrt{p}$

COMPARATIVE DATA

Data in the following table are the mean temperature and average precipitation for Greensburg, Kentucky, for the period 1906-1930 and are included in this publication for comparative purposes.

35.6	5.35	37.4	3.80	46.0	4.89	55.4	4.04	63.8	4.82	72.2	4.46	75.8	3.85	74.7	4.33	69.4	3.36	57.0	3.57	45.7	3.75	37.1	4.35	55.8	50.57

NORMALS, MEANS, AND EXTREMES

LEXINGTON, KENTUCKY — BLUE GRASS FIELD

LATITUDE 38° 02' N
LONGITUDE 84° 36' W
ELEVATION (ground) 979 Feet

Temperature

Month	Normal Daily max	Normal Daily min	Normal Monthly	Extremes Record highest	Year	Extremes Record lowest	Year
(a)	(b)	(b)	(b)	74		74	
J	42.7	25.1	33.9	80	1943	-15	1936
F	45.2	26.7	36.0	76	1945	-20	1899
M	54.5	33.8	44.3	86	1929	-1	1873
A	65.5	43.4	54.5	91	1942	15	1876
M	75.5	52.8	64.1	96	1941	30	1876
J	83.8	62.1	73.0	104	1936	40	1910
J	87.4	65.7	76.6	108	1936	47	1947
A	85.5	64.1	75.0	105	1954	45	1946
S	80.5	58.4	69.5	105	1954	32	1899
O	69.1	46.6	57.9	93	1941	21	1925
N	55.3	35.5	44.8	80	1950	-9	1929
D	44.5	27.3	35.9	74	1942	-20	1917
Year	65.7	45.1	55.4	108	July 1936	-20	Feb.1899

Precipitation and Snow

Month	Normal degree days	Precip Normal total	Max monthly	Year	Max 24 hr	Year	Min monthly	Year	Snow Mean total	Snow Max monthly	Year	Snow Max 24 hr	Year
(a)	(b)	(b)	74		74		74		70	74		74	
J	964	4.50	16.85	1950	3.50	1913	0.77	1931	6.1	21.7	1918	13.4	1943
F	812	3.46	11.06	1883	3.04	1909	0.52	1883	4.4	14.7	1894	8.0	1894
M	651	3.76	9.91	1890	3.85	1910	0.46	1910	1.8	16.3	1896	9.5	1896
A	325	3.56	7.86	1887	2.86	1948	0.60	1896	0.4	4.0	1901	3.0	1901
M	113	4.21	11.03	1882	2.87	1945	1.18	1936	0.1	0.4	1894	6.0	1894
J	8	4.50	10.82	1928	5.50	1928			0.0	0.0		0.0	
J	0	4.25	11.24	1875	4.02	1875	0.45	1930	0.0	0.0		0.0	
A	0	3.37	8.96	1932	8.06	1932	0.62	1932	0.0	0.0		0.0	
S	50	2.83	7.95	1919	2.48	1925	0.33	1895	0.0	0.0		0.0	
O	246	2.44	7.95	1919	2.35	1904	0.11	1924	0.1	2.8	1925	2.3	1925
N	606	3.15	8.50	1919	3.83	1925	0.53	1904	1.2	9.7	1950	9.5	1950
D	902	3.68	9.02	1879			0.80	1925	3.8	19.4	1917		1917
Year	4677	43.71	16.85	Jan.1950	8.06	Aug.1932	0.11	Oct.1924	19.1	21.7	Jan.1918	13.4	Jan.1943

Relative humidity, Wind, Sunshine, Sky cover

Month	RH 12:00 mid CST	RH 6:00 a.m. CST	RH 12:00 CST	RH 6:00 p.m. CST	Wind Mean hourly	Prevailing direction	Fastest mile Speed	Direction	Year	Pct of possible sunshine	Mean sky cover
(a)	11	58	28	24	53	11	45	45	45	34	59
J	81	82	73	74	13.5	S	52	W	1928	36	6.8
F	81	81	68	69	13.4	S	47	SW	1918+	40	6.5
M	75	79	63	63	13.9	SSW	53	SW	1909	47	6.6
A	75	75	58	61	13.0	S	50	SW	1909	53	5.8
M	84	79	56	65	10.7	S	50	SW	1924	59	5.5
J	84	79	56	65	9.4	SSW	47	SW	1917+	62	5.0
J	84	79	56	65	8.8	SSW	49	NW	1929	64	4.7
A	83	81	55	64	8.4	S	53	S	1918	63	4.7
S	83	81	55	64	9.3	S	50	S	1899	62	4.5
O	78	78	61	68	10.5	S	40	NW	1904	46	4.7
N	79	83	73	72	12.9	S	52	NW	1929	34	5.8
D	79	84	79	72	12.9	S	47	S	1922	34	6.7
Year	80	80	62	66	11.4	S	56	S	Aug.1918	52	5.6

Mean number of days

Month	Sunrise-sunset Clear	Partly cloudy	Cloudy	Precip .01 in or more	Snow/Sleet 1.0 in or more	Thunderstorms	Heavy fog	Max 90° and above	Max 32° and below	Min 32° and below	Min 0° and below
(a)	70	70	70	74	74	52	52	73	73	73	73
J	7	7	17	13	2	1	2	0	7	22	1
F	7	8	14	12	1	1	1	0	5	19	1
M	8	9	14	13	1	1	1	0	1	14	0
A	9	10	11	12	0	3	0	0	0	3	0
M	10	11	10	12	0	6	0	1	0	0	0
J	11	12	7	12	0	9	0	5	0	0	0
J	12	13	6	11	0	9	1	9	0	0	0
A	13	13	6	10	0	7	1	7	0	0	0
S	13	14	7	9	0	4	1	3	0	0	0
O	14	14	9	8	0	1	1	0	0	2	0
N	11	13	13	10	1	1	1	0	1	11	0
D	8	6	18	12	1	2	2	0	5	20	2
Year	106	127	132	133	6	47	12	25	20	91	2

LOUISVILLE, KENTUCKY — STANDIFORD FIELD

LATITUDE 38° 11' N
LONGITUDE 85° 44' W
ELEVATION (ground) 474 feet

Temperature

Month	Normal Daily max	Normal Daily min	Normal Monthly	Extremes Record highest	Year	Extremes Record lowest	Year
(a)	(b)	(b)	(b)	85		85	
J	43.6	26.2	34.9	79	1943	-20	1884
F	46.3	28.0	37.2	78	1932	-19	1951
M	56.0	35.2	45.6	88	1929+	-3	1899+
A	67.0	45.0	56.0	91	1925+	21	1875
M	76.6	54.0	65.3	97	1911	31	1910
J	85.2	63.2	74.2	103	1944	43	1910
J	89.1	66.6	77.9	107	1936+	49	1947
A	87.5	66.1	76.1	105	1918+	45	1946
S	81.9	58.4	70.2	104	1954	33	1949
O	70.5	46.7	58.6	92	1953	23	1952+
N	55.3	36.1	45.7	83	1950	-7	1950
D	45.5	28.2	36.9	74	1875	-20	1884
Year	67.0	46.0	56.5	107	July 1936+	-20	Jan.1884

Precipitation and Snow

Month	Normal degree days	Precip Normal total	Max monthly	Year	Max 24 hr	Year	Min monthly	Year	Snow Mean total	Snow Max monthly	Year	Snow Max 24 hr	Year
(a)	(b)	(b)	85		85		85		73	73		73	
J	933	4.10	19.17	1937	4.25	1937	0.82	1896	4.1	24.6	1918	10.4	1918
F	778	2.99	9.84	1884	5.00	1909	0.40	1947	3.5	17.4	1910	10.9	1910
M	611	4.67	10.23	1945	5.80	1943	0.12	1910	2.0	14.0	1896	10.0	1887
A	285	4.01	9.08	1872	4.06	1880	0.25	1896	T	3.2	1917	3.2	1917
M	94	3.93	9.58	1950	3.97	1950	0.63	1932	0.0	1.0	1898	0.0	1898
J	5	4.06	8.48	1942	3.02	1872	0.35	1936	0.0	0.0		0.0	
J	0	3.06	16.46	1875	5.50	1896	0.25	1930	0.0	0.0		0.0	
A	0	3.08	10.53	1888	3.78	1879	0.15	1881	0.0	0.0		0.0	
S	51	2.70	5.90	1884	4.22	1954	0.27	1953	0.0	0.0		0.0	
O	232	2.45	8.05	1888	5.06	1910	0.07	1908	0.1	3.5	1925	3.0	1925
N	579	3.12	9.12	1957	3.58	1948	0.25	1904	0.5	7.6	1936	7.4	1936
D	871	3.30	8.43	1915	4.41	1915	1.11	1955	3.0	24.6	1917	15.0	1917
Year	4439	41.47	19.17	Jan.1937	5.80	Mar.1943	0.07	Oct.1908	13.4	24.6	Jan.1918+	15.0	Dec.1917

Relative humidity, Wind, Sunshine, Sky cover

Month	RH 12:00 mid CST	RH 6:00 a.m. CST	RH 12:00 CST	RH 6:00 p.m. CST	Wind Mean hourly	Prevailing direction	Fastest mile Speed	Direction	Year	Pct of possible sunshine	Mean sky cover
(a)	17	70	40	70	85	9	45	45	45	57	53
J	78	78	69	77	9.9	S	53	S	1934	41	6.8
F	78	76	65	81	10.3	NE	62	W	1918	47	6.1
M	75	75	60	82	10.8	SW	62	SW	1896	53	5.9
A	72	74	56	81	10.1	SE	66	NW	1943	57	5.2
M	73	72	54	77	7.5	S	66	NW	1915	64	
J	76	83	54	78	7.7	SW	66	NW	1931	64	
J	77	83	57	84	7.0	N	57	SW	1927	73	4.8
A	81	81	52	69	6.6	S	63	NW	1930	69	4.6
S	80	82	53		7.1	NW	57	NW	1954	68	4.5
O	75	81	60	63	7.4	SE	60	SE	1952	51	5.7
N	77	77	59	69	9.4	S	60	SE	1953	39	
D	78	78	64		9.0	S	61	NW			
Year	77	77	57	61	8.7	S	68	NW	May 1915	59	5.6

Mean number of days

Month	Sunrise-sunset Clear	Partly cloudy	Cloudy	Precip .01 in or more	Snow/Sleet 1.0 in or more	Thunderstorms	Heavy fog	Max 90° and above	Max 32° and below	Min 32° and below	Min 0° and below
(a)	86	86	86	85	73	65	65	85	85	85	85
J	7	8	16	12	1	1	2	0	6	20	1
F	7	8	13	12	1	1	1	0	4	17	*
M	8	9	14	12	1	3	*	0	1	11	0
A	9	11	12	11	0	6	*	*	0	1	0
M	10	13	10	11	0	9	*	1	0	0	0
J	12	12	7	9	0	8	1	7	0	0	0
J	12	12	7	9	0	8	*	12	0	0	0
A	13	12	7	8	0	7	*	10	0	0	0
S	13	10	8	9	0	4	1	1	0	0	0
O	10	7	9	8	*	2	1	*	0	1	0
N	7	7	13	9	*	1	2	0	1	8	*
D	7	13	16	11	1	*	2	0	4	18	*
Year	115	132	132	122	4	46	9	34	16	76	1

REFERENCE NOTES APPLYING TO ALL "NORMALS, MEANS, AND EXTREMES" TABLES.

(a) Length of record, years.
(b) Normal values are based on the period 1921-1950, and are means adjusted to represent observations taken at the present standard location.
* Less than one-half.

† No record.
‡ Airport data.
◦ City Office data.
+ Also on earlier dates, months, or years.
T Trace, an amount too small to measure.

Sky cover is expressed in a range of 0 for no clouds or obscuring phenomena to 10 for complete sky cover. The number of clear days is based on average cloudiness 0-3 tenths; partly cloudy days on 4-7 tenths; and cloudy days on 8-10 tenths. Monthly degree day totals are the sum of the negative departures of average daily temperatures from 65°F. Sleet was included in snowfall totals beginning with July 1948. Heavy fog also includes data referred to at various times in the past as "Dense" or "Thick". The upper limit for heavy fog is 1/4 mile. Data in these tables are based on records through 1957.

Mean Maximum Temperature (°F.), January

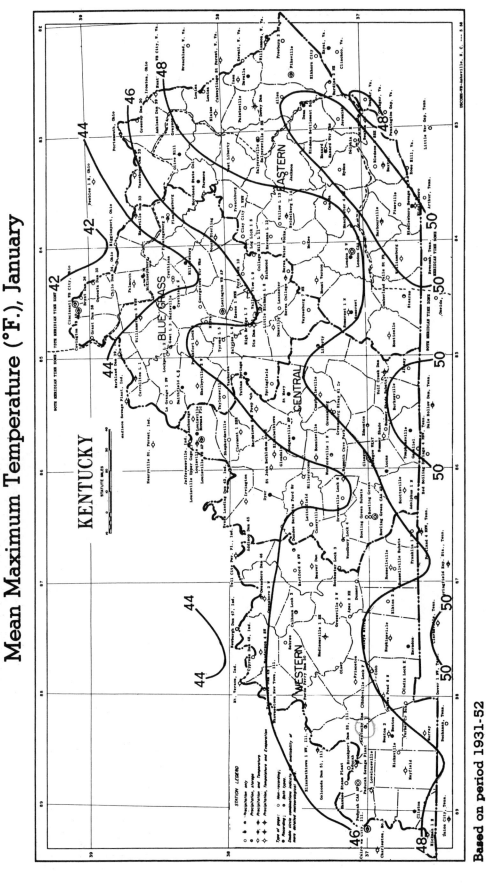

KENTUCKY

STATUTE MILES

Based on period 1931-52

Isolines are drawn through points of approximately equal value. Caution should be used in interpolating on these maps, particularly in mountainous areas.

— 130 —

Mean Minimum Temperature (°F.), January

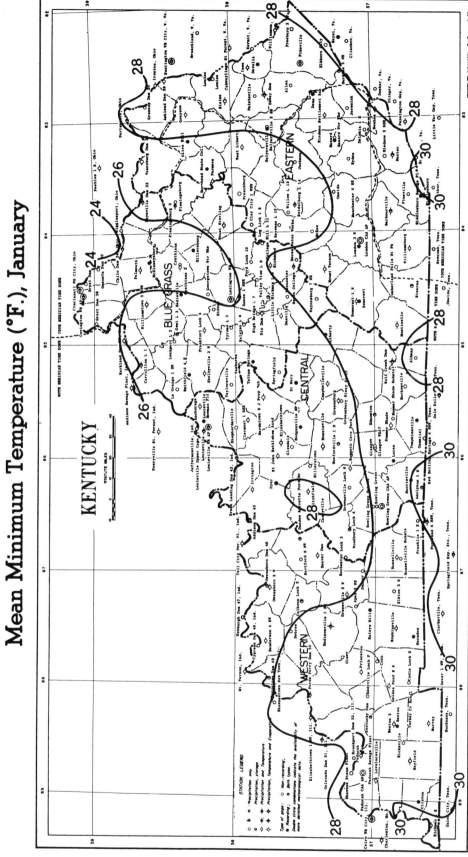

Based on period 1931-52

Isolines are drawn through points of approximately equal value. Caution should be used in interpolating on these maps, particularly in mountainous areas.

- 131 -

Mean Maximum Temperature (°F.), July

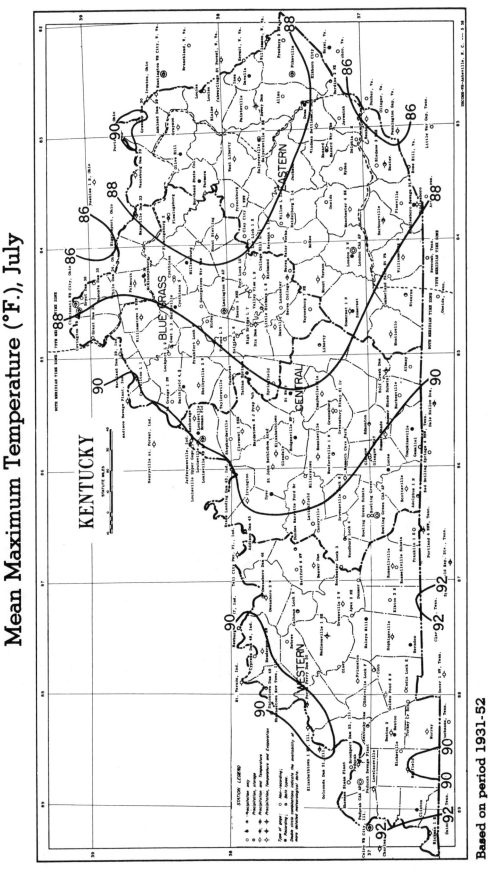

Based on period 1931-52

Isolines are drawn through points of approximately equal value. Caution should be used in interpolating on these maps, particularly in mountainous areas.

Mean Minimum Temperature (°F.), July

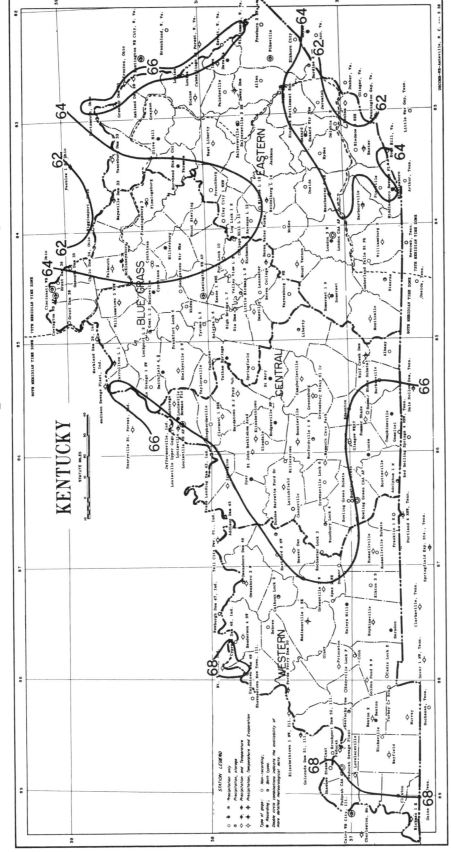

Based on period 1931-52

Isolines are drawn through points of approximately equal value. Caution should be used in interpolating on these maps, particularly in mountainous areas.

KENTUCKY

– 133 –

Mean Annual Precipitation, Inches

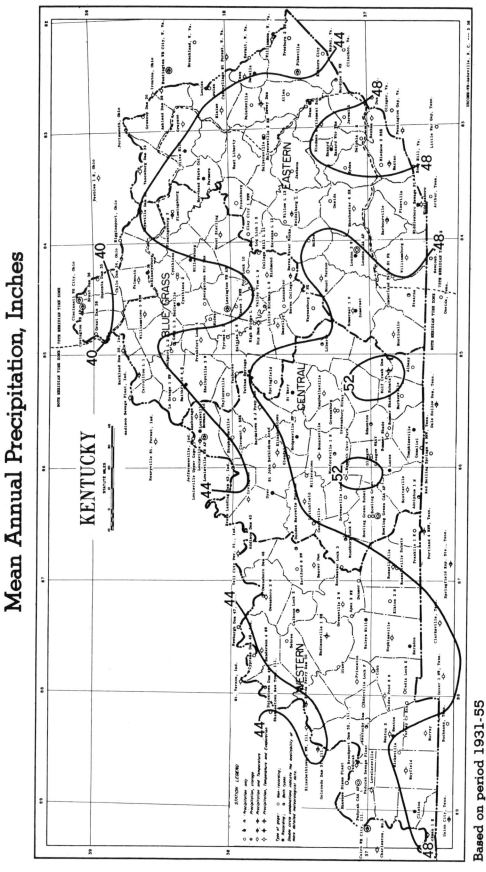

Based on period 1931-55

Isolines are drawn through points of approximately equal value. Caution should be used in interpolating on these maps, particularly in mountainous areas.

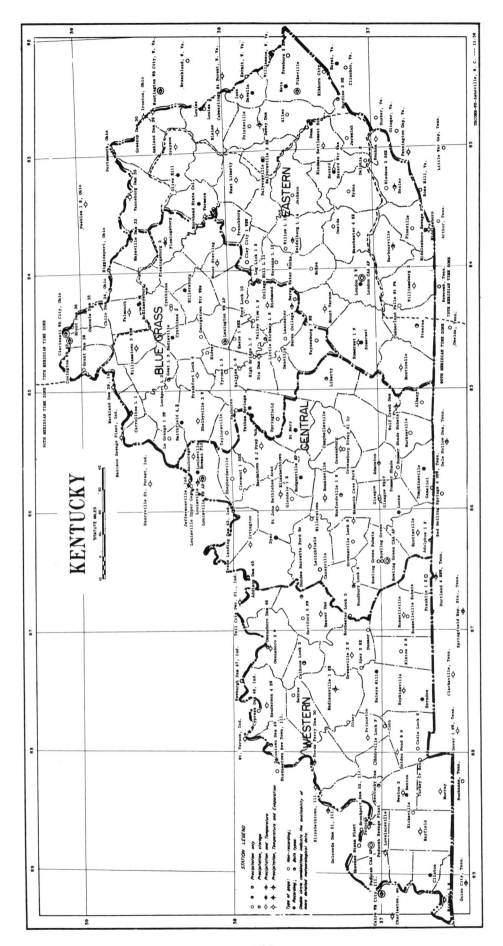

KENTUCKY

THE CLIMATE OF

MAINE

by
Robert E. Lautzenheiser

May 1972

PHYSICAL DESCRIPTION. Maine occupies 33,215 square miles, almost exactly one-half of New England's total area. From near the 43d parallel, the State extends northward over 300 miles, spanning a full 4-1/2° of latitude. Its width from the 67th meridian extends westward over 4° of longitude, a span of over 200 miles.

The terrain is hilly. Elevations are generally less than 500 feet above sea level over the southeastern one-half of the State. The northwestern one-half is a plateau ranging in elevation from 1,000 to 1,500 feet, but sloping downward to 500 feet in the northeast from 1,000 feet in the north (Aroostook County). A number of mountain peaks, extensions of the Appalachian chain, rise to heights from 3,000 to 5,000 feet mostly in the western and central portions of the State. Mt. Katahdin is the highest point. Its summit, at 5,268 feet, rises nearly 4,500 feet from a relatively low base elevation. Mt. Katahdin is thus as imposing a sight as many of the peaks in the Rockies which rise from a high base plateau.

The great glaciers of the ice age were all-important in the physical formation of the State. They left "horsebacks" or ridges of glacial deposits, some as long as 150 miles in length. These ridges furnish both natural highway routes and abundant material for roadbuilding. The glaciers formed or left over 1,600 beautiful lakes, spread abundantly over the entire State. The largest of these lakes is Moosehead. The total water area of the State exceeds 2,200 square miles. Some flatland is found near the coast, especially near the mouths of the Androscoggin and Kennebec Rivers. Other tracts of flatland, often marshy, lie near lakes. Large acreages that are not too rolling are suitable for agricultural development. Aroostook County, especially, has vast tracts of agricultural land. In total, Maine has about 2,000 square miles now used as farmland. Erosion can be a problem, especially in the Saint John Valley, but improved farming methods and reforestation help to control this problem.

The coastal portion of the State has many inlets, bays, channels, fine harbors, rocky islands, and promontories which provide a treasure of scenic beauty. The extreme irregularity of the coast stretches the total coastline to about 2,400 miles, more than 10 times the distance from Kittery to Eastport. The southwestern portion of the coast has many fine beaches. The midcoastal portion has many rugged hills and small mountains, some of which rise abruptly from the water, such as Mount Desert Island. The coastal portion adds much to Maine's scenic appeal.

More than five-sixths of the State is forest land. Much of this, especially in the north, remains without settlements or highways other than a few private logging roads. Game in the unspoiled forests and fish in the lakes and streams attract sportsmen from near and far.

GENERAL CLIMATIC FEATURES. Maine's chief climatic characteristics include: (1) Changeableness of the weather, (2) large ranges of temperature, both diurnal and annual, (3) great differences between the same seasons in different years, (4) equable distribution of precipitation, and (5) considerable diversity from place to place. The regional climatic influences are modified in Maine by varying distances from the ocean, by elevations, and by types of terrain. These modifying factors divide the State into three natural climatological divisions.

Maine lies in the "prevailing westerlies"--the belt of generally eastward air movement which encircles the globe in the middle latitudes. Embedded in this circulation are extensive masses of air originating in higher or lower latitudes and interacting to produce storm systems. Relative to most other sections of the country, a large number of such storms pass over or near Maine. The majority of air masses affecting this State belong to three types: (1) Cold, dry air pouring down from subarctic North America, (2) warm, moist air streaming up on a long overland journey from the Gulf of Mexico and from subtropical waters eastward, and (3) cool, damp air moving in from the North Atlantic. Because the atmospheric flow is usually offshore, Maine is influenced more by the first two types than it is by the third. In other words, the adjacent ocean constitutes an important modifying factor on the immediate coast, but does not dominate the climate statewide.

The procession of contrasting air masses and the relatively frequent passage of storms bring about a roughly twice-weekly alternation from fair to cloudy or stormy conditions, attended by often abrupt changes in temperature, moisture, sunshine, wind direction, and wind speed. There is no regular or persistent rhythm to this sequence. It is interrupted by periods of time during which the same weather patterns continue for several days, infrequently for several weeks. Maine weather, however, is distinguished for variety rather than for monotony. Changeability is also one of its features on a longer time-scale; that is, the same month or season will exhibit varying characteristics over the years--sometimes in close alternation, sometimes arranged in similar groups for successive years. A "normal" month, season, or year is indeed the exception rather than the rule.

The basic climate, as outlined above, obviously does not result from the predominance of any single controlling weather regime, but is rather the integrated effect of a variety of weather patterns. Hence, "weather averages" in Maine usually are not sufficient for important planning purposes without further climatological analysis.

CLIMATOLOGICAL DIVISION. The Northern Division contains slightly more than one-half of the State's area, with its southern boundary nearly parallel to the coast. It represents that area of the State least affected by ocean influences and

most affected by higher elevations. In contrast, the Coastal Division is a strip roughly 20 to 30 miles in width. It is most affected by maritime influences and has the lowest average elevation above sea level. The remainder, known conveniently as the Southern Interior Division, covers nearly one-third of the State's area.

TEMPERATURE. The average annual temperature ranges from near 40° in the Northern Division, to 44° in the Southern Interior Division, and to nearly 45° in the Coastal Division. Within the large Northern Division, there is a range from near 37° to about 43° from north to south. Temperature averages vary within the three Divisions from causes other than latitude. Elevation, slope, and other environmental aspects, including some urbanization, each has an effect. The highest temperature of record in the State is 105°, which was observed at North Bridgton on July 10, 1911, and at other locations and dates. The lowest of record is -48° at Van Buren on January 19, 1925.

Summer temperatures are delightfully cool and are reasonably uniform over the State. The July long-period average temperature reaches 70° only at a few stations, mostly in the Southern Interior Division. Hot days, with maxima of 90° or higher, are extremely rare along much of the immediate coast. From 2 to 7 days is the average number of hot days elsewhere, with the greatest number in the western portion of the Southern Interior Division. Frequency of 90° temperature-occurrence days varies from year to year from none in the coolest summers up to 25 days at warmest stations in the warmest summers. The average daily temperature range in the wintertime is approximately 20° over much of the State. In summertime, the average temperature range is greater, 30° in the central portion of the Northern Division; the range may exceed 40° during cool, dry weather, especially in valleys and marshes. Frost may be a threat even in the warmer months in these susceptible areas.

Average temperatures vary from place to place much more in winter than in summer. They range in January from less than 10° in the extreme north to over 20° along the Atlantic coast. Midwinter temperatures in Maine vary little from those at similar latitudes in the North Central States. An exception is the Coastal Division where temperature averages compare with those found 200 miles or more farther south in the central part of the Nation. Days with subzero readings average from 40 to 60 per year at the colder stations in the Northern Division to only 10 to 20 at stations in the Coastal Division. Along the immediate coast, subzero occurrences are even less frequent.

The average length of the growing season for vegetation subject to injury from freezing temperature is about 140 to 160 days in the Coastal Division, 120 to 140 days in the Southern Interior Division, and 100 to 120 days in most of the Northern Division. An average freeze-free growing season of slightly less than 100 days occurs in extreme northern and northwestern Maine. Marshes or other susceptible low areas may have a shorter season than indicated above. The average date of the last freezing temperature in spring is near the end of April at coastal stations and near the end of May in extreme northern Maine. The freeze-free season usually ends in September, except in the Coastal Division where it extends into October.

PRECIPITATION. Maine has precipitation rather evenly distributed throughout the year. The distribution is most regular in the Southern Interior Division. Along the Atlantic coast, summer thunderstorm activity is somewhat suppressed by the effects of the cool ocean, while winter precipitation is increased by coastal storms or "northeasters." These combined effects give this coastal area more precipitation in the winter than in the summer months. Monthly totals are about 4 inches during the winter as compared to 3 inches during the summer. In the Northern Division, these effects are reversed with increased thunderstorm activity in summer and with very little effect of coastal storms in winter. Precipitation totals in this Division are greater in summer, with the difference being about 1 inch. The averages for each winter month are about 2-1/2 to 3 inches compared to 3-1/2 to 4 inches for each summer month.

Storm systems are the principal year-round moisture producers. Such systems are less active in the summer, but bands or patches of thunderstorm or shower activity take over much of this function. Though brief and often of small areal extent, thunderstorms produce the heaviest local rainfall rates for short intervals. Many stations have received from 1 to 2 inches in an hour. Minor washouts of roads and soils may occur.

Variation in precipitation totals from month to month or for the same month in different years may be extreme. Monthly totals range from negligible amounts up to 10 inches or more. Totals may also vary within the State in the same month. In August 1958, for example, Rumford recorded only 1.52 inches, while Caribou had 8.45 inches, setting a new record for a monthly total for the latter station. Such large fluctuations are rare, however, as most monthly totals fall in the range of 50 to 200 percent of the average amounts. Prolonged droughts are infrequent; irrigation water is available for use during the fairly common short dry spells in summer.

Total precipitation averages near 40 inches yearly in the Northern Division, slightly higher in the Southern Interior Division, and over 44 inches in the Coastal Division. Local influences cause considerable variation in totals from station to station, even within a Division.

A valley station may receive several inches less than one situated on a slope. Moisture-bearing winds, when forced upward by the slope, drop increased rainfall.

Winter precipitation occurs mostly as snow, except in the Coastal Division where considerable rain or wet snow falls; stations in this Division, more than stations farther inland, are subject to occasional glazing, or "ice storm" conditions. Freezing rain coats streets, roads, and all exposed surfaces; on rare occasions, a heavy load of ice builds on trees and wires, causing extensive damage.

Measurable amounts of precipitation fall an average of 1 day in 3 over much of the State. This frequency increases to near 1 day in 2 over the extreme northern portion. For example, at Caribou, the average is over 4 days in 10. However, amounts as great as 6 inches of rain in 24 hours are rare occurrences in Maine. Many stations have never recorded that much in a single day. Brunswick, however, received 8.05 inches from hurricane Edna on September 11, 1954.

SNOWFALL. As a rule, average seasonal snowfall amounts increase northwestward from the coast. The Coastal Division snowfall totals range from 50 to 80 inches. The Southern Interior Division receives from 60 to 90 inches. The Northern Division totals range on the average from 90 to 110 inches. The largest long-period average for any station with official records is a 27-year average of 118 inches at Jackman in northwestern Maine, between Moosehead Lake and the Canadian border. Stations in extreme northern Maine tend to receive slightly less snow than those in the southwestern portion of the Northern Division. Local topography has a marked influence on snowfall, causing large variations within a short distance. In general, an increase of 1 inch of seasonal snowfall will be measured for each 25-foot increase in elevation.

The number of days with 1 inch or more of snowfall varies from about 20 in a season at stations near the coast to as high as 30 or more at some northern stations. Most winters will have several snowstorms of 5 inches or more. Storms of this magnitude temporarily disrupt transportation and communications. On December 29-30, 1962, a single storm dropped 46 inches at Ripogenus Dam (Piscataquis County). Heavy individual snowstorms, dropping up to 30 inches of snow, have occurred along the coast. Many stations through the State have experienced over 20 inches in a single day.

January is usually the snowiest month. Many stations average over 20 inches of snow in that month. The snowfall season usually begins in late October or in November and lasts into April and sometimes into May. Seasonal totals in the north do not vary as markedly as along the coast.

In the north, the seasonal totals for those years with the heaviest snowfalls are usually less than double the least. Greater fluctuations occur near the coast. For example, Eastport has recorded 30 inches or less in some seasons. However, in the 1906-07 season, 188 inches of snow was measured, over six times more than that of the lowest snowfall total recorded in a season.

Snow cover lasts throughout the season in the north. Along the coast, however, the snow cover may melt entirely in midwinter and then be replaced by a new cover. The average length of the longest continuous cover of 1 inch or more ranges from about 50 days near the coast to more than 4 months in the northwest where prevailing low temperatures and wooded terrain prevent rapid melting. The average date of the maximum snow depth varies from early February along the coast to late February or even to early March in the extreme northwest. These dates range widely from winter to winter along the coast where the greatest snow depth may come at any time during the season. Water stored in the snow cover provides the State's watersheds with an important part of their annual water supply. Melting is usually gradual enough to prevent serious flooding.

FLOODS. Rivers flow generally in a southward direction from the mountains in the interior of Maine, from the Canadian border, or from New Hampshire to the Atlantic coast. The principal rivers are the Saco, Androscoggin, Kennebec, and Penobscot Rivers. The Saint John River, rising in northern Maine, forms the border between Maine and the Canadian province of New Brunswick for a distance before the river enters Canada. The State is generally forested and contains extensive areas of lakes and ponds which tend to retard flood runoff.

Widespread major flooding is relatively infrequent. The greatest frequency of floods occurs in the early spring when substantial rains and melting snow combine to produce heavy runoff. Because the rivers flow from north to south and because the coastal areas have decidedly higher springtime temperatures than the elevated and forested headwater areas, the rivers in the lower reaches are usually free of ice and the snow is depleted before thawing starts in the upper reaches. Thus, in most years, the spring runoff occurs without serious flooding. The very destructive flood of March 1936 was caused by several days of excessive rainfall upon a heavy snow cover.

In addition to the 1936 flood, major springtime floods occurred in 1895, 1896, 1917, 1923, and 1953. Although not of such general magnitude, rather widespread floods occurred as the aftermath of heavy rains alone during the fall of 1907, 1909, 1927, and 1950. Occasional flash floods occur on small streams during the summertime from thunderstorms, but usually these floods have little effect on main stream flows.

OTHER CLIMATIC FEATURES. The amount of possible sunshine averages from 50 to 60 percent in most of the southern one-half of the State. This percentage varies along the coast from near 50 at Eastport to 60 percent at Portland. At higher elevations and over much of northern Maine, the average is near 45 percent. The average annual number of clear days ranges from 80 to 120 days in the southern one-half and from about 50 to 90 days in the northern one-half of the State.

Heavy fog is frequent and sometimes persistent along the coast, particularly in the eastern portion of the coast where it may occur on an average of 1 day out of 6. Fog frequency and duration diminish inland. But short-duration heavy ground fogs of early morning occur frequently at susceptible places inland. These, plus a few occurrences of other heavy fog, may produce a frequency at some inland locations approaching that along the coast. The number of days with heavy fog varies over the State, ranging from about 25 to 60 days in a year.

Prolonged dry spells, quite frequent in late summer or fall, create serious forest-fire hazards. Low humidities and lack of precipitation during some late summers cause the forest litter to become extremely inflammable. The hazard is particularly great in resort and recreation areas. In recent years, serious fires have occurred.

WINDS AND STORMS. On a yearly basis, the wind direction is mostly from the west. During winter, north to northwest winds tend to prevail. In the summer, they are more often from the southwest or south. Topography has a strong influence on the prevailing direction. Parts of a major river valley, for example, may have a prevailing wind paralleling the valley. Along the coast in spring and summer, the sea breeze is important. Onshore local winds, blowing from the cool ocean, may come as far inland as 10 miles. They tend to retard spring growth, but are pleasingly cooling in summer.

Coastal storms or "northeasters" sometimes seriously affect the Coastal Division. They generate very strong winds and heavy rain or snow, sometimes glaze or "ice storm conditions." They can produce abnormally high wind-driven tides, affecting beaches and coastal installations. In winter, these storms produce some of the heavier snowfalls along the coast. Occasionally, in summer or fall, a storm of tropical origin affects Maine. Usually the storm will be similar to the northeasters. But a few such storms may retain near or full hurricane force. For example, in 1954, two hurricanes affected Maine within a period of less than 2 weeks. The first, Carol, traveled northward along the Maine-New Hampshire border on August 31. Wind speeds were no longer at full hurricane force, but substantial property and crop damage resulted in western Maine. Then Edna entered the coastline near Eastport on September 11. The principal damage from Edna was the result of heavy flooding and washing rains. Ordinarily, maximum loss is concentrated along the shore. Fortunately, storms of tropical origin do not affect Maine at all in most years. Two or more storms in 1 year should be expected about 1 year in 20.

Tornadoes are a phenomena not common in Maine. Yet, they are not as rare in Maine as have been generally believed. It is likely that several occur on the average each year. Fortunately, most tornadoes are very small, affecting a very localized area. Because of the preponderance of forests or other unsettled areas, a large percentage of tornadoes in Maine are neither seen, recorded, nor do appreciable damage. As a reminder that these storms can and do visit the State, a huge tornado occurred in the Allagash region of northern Maine on August 15, 1958. The funnel winds devastated the forest in that region in a path 20 miles long and 300 to 400 yards wide. Had this storm occurred in urban areas, it could have rivaled the famous Worcester, Mass., tornado of 1953. About 80 percent of Maine's tornadoes occur between May 15 and September 15; about 90 percent strike between 1 and 7 p.m. The peak month is July, and the peak hour of occurrence is from 2 to 3 p.m. However, the chance of a tornado striking any given spot is extremely small.

Thunderstorms and hailstorms have a similar frequency maximum from midspring to early fall. Thunderstorms occur in a range from 10 to 20 days a year in the Coastal Division and from 15 to 30 days a year elsewhere. The most severe storms are attended by hail. Hailstorms can severely injure or even ruin field crops, break glass, dent automobiles, and damage other vulnerable exposed objects. Fortunately, the size of an area pelted by hail is usually small. Glaze and ice storms of winter can produce perilous conditions for transportation. These storms are usually of brief duration. A few widespread, prolonged ice storms have occurred. Besides affecting travel and transport, they break trees and limbs, utility lines, and utility poles. In designing structures such as steel towers, designers must consider the possible ice load. An ice load also magnifies the wind stress by increasing the area exposed to the wind.

CLIMATE AND ECONOMY. Activities in Maine are profoundly influenced by climate. Tree growth is especially favored. About 85 percent of the State is covered by forests which constitute a major scenic attraction and provide material for a one-half of a billion-dollar forest products industry. This includes lumbering, papermaking, wood products manufacturing, and related industries. Forest products industries employ nearly one-third of all male workers in the State. The ample supply of rainfall provides water for the growth of trees, for a system of waterways to transport the felled timber, and for the huge amounts required to meet the needs of wood products manufacturing. Other major industries include textiles, shoes, and shipbuilding. A great diversity of smaller industries also takes advantage of the ample water supplies.

Climate plays a significant role in the State's agriculture by favoring the production of high-value specialized crops. Maine ranks very high in the Nation in cash receipts per acre from farm marketing. The principal crop is potatoes, with Maine producing more annually than any other State. The long summer days, favorable precipitation, and temperature combine with large tracts of suitable soil in Aroostook County to make this the Nation's leading potato producing area. Other important farm products include peas for freezing and canning, corn, oats, and hay. Many truck gardens are found in the coastal plains. Blueberry production is on a large commercial scale, especially along the eastern portion of the Coastal Division. Apples are the most prolific of the tree fruits. The production of quality apples is an important commercial pursuit. Top-quality maple syrup is produced in commercial quantity. Poultry raising, especially broiler production, is a principal activity, exceeding in importance dairying.

Climate is particularly important to the tourist and vacation trade, a major industry. This trade amounts to more than one-third of a billion dollars annually. Much of this trade is concentrated in the summer when pleasant temperatures prevail at coastal and lake resorts. Abundant game and teeming lakes and streams also draw sportsmen. Skiing and other winter sports are developing into an important winter attraction, made possible by the abundant snowfall in the State.

In summary, the climate of Maine contributes greatly to its industrial, agricultural, and vacation activities. Its climate is a rich, natural asset that invigorates persons and is favorable to further economic development of the State.

BIBLIOGRAPHY

Brown, Rodger A., "Twisters in New England," unpublished, Antioch College, Yellow Springs, Ohio, Mar. 1957, 55 pp.

Brooks, Charles F., "New England Snowfall," Monthly Weather Review, Vol. 45, No. 6, June 1917, pp. 271-285.

Church, P. E., "A Geographical Study of New England Temperatures," The Geographical Review, Vol. 26, No. 2, Apr. 1936, pp. 283-292.

Cooper, G. R., and Lautzenheiser, R. E., "Freezes in Maine," Bulletin No. 679, Maine Agricultural Experiment Station, University of Maine, Orono, Maine, Sept. 1969, 34 pp.

Department of Commerce, NOAA, Environmental Data Service, and Department of Agriculture, Statistical Reporting Service, Weekly Weather and Crop Bulletin-National Summary, Washington, D.C., weekly.

Dunn, Gordon E., and Miller, Banner I., Atlantic Hurricanes, Louisiana State University Press, Baton Rouge, La., rev. ed., 1964, 377 pp.

Eustis, R. S., "Winds over New England in Relation to Topography," Bulletin of the American Meteorological Society, Vol. 23, No. 10, Dec. 1942, pp. 383-387.

Flora, Snowden Dwight, Hailstorms of the United States, University of Oklahoma Press, Norman, Okla., 1956, 201 pp.

_____, Tornadoes of the United States, University of Oklahoma Press, Norman, Okla., rev. ed., 1954, 221 pp.

Fobes, Charles Bartlett, "Climatic Divisions of Maine," Bulletin No. 40, Maine Technology Experiment Station, University of Maine, Orono, Maine, 1946, 44 pp.

_____, "Lightning Fires in the Forests of Northern Maine," Journal of Forestry, Vol. 42, No. 4, Apr. 1944, pp. 291-292.

Galway, Joseph Gerard, "A Statistical Study of New England Snowfall," unpublished, U.S. Weather Bureau, Apr. 1954, 140 pp.

Goodnough, X. H., "Rainfall in New England," Journal of the New England Water Works Association, Vol. 29, No. 3, Sept. 1915, pp. 237-438; Vol. 35, No. 3, Sept. 1921, pp. 228-293; and Vol. 40, No. 2, June 1926, pp. 178-247.

Lautzenheiser, R. E., "Snowfall, Snowfall Frequencies and Snow Cover Data for New England," ESSA Technical Memorandum EDSTM-12, Silver Spring, Md., Dec. 1969, 16 pp.

NOAA, Environmental Data Service, "Caribou and Portland, Maine," Local Climatological Data, Asheville, N.C., monthly plus annual summary.

_____, Climatological Data--National Summary, Asheville, N.C., monthly plus annual summary.

_____, Climatological Data--New England, Asheville, N.C., monthly plus annual summary.

_____, "Climatological Substation Summaries for Augusta, Bangor, Eastport, Farmington, Greenville, Houlton, Lewiston, Machias, Rockland, Winthrop, and Woodland, Maine," Climatography of the United States No. 20-17, Boston, Mass., irregular.

_____, Hourly Precipitation Data--New England, Asheville, N.C., monthly plus annual summary.

_____, Storm Data, Asheville, N.C., monthly.

NOAA, National Weather Service, "Tabulations of Frequencies of Various Climatic Elements for Various Stations," available at library of Climatologist for Maine, Boston, Mass., on microfilm.

Palmer, Robert S., "Agricultural Drought in New England," Technical Bulletin 97, Agricultural Experiment Station, University of New Hampshire, Durham, N.H., Aug. 1958, 51 pp.

Perley, Sidney, Historic Storms of New England, The Salem Press Publishing and Printing Co., Salem, Mass., 1891, 341 pp.

Stone, R. G., "The Distribution of the Average Depth of Snow on Ground in New York and New England. II: Curves of Average Depth and Variability," Transactions, American Geophysical Union, 21st Annual Meeting, Apr. 24-27, 1940, pp. 672-692.

_____, "The Average Length of the Season With Snow Cover of Various Depths in New England," Transactions, American Geophysical Union, 25th Annual Meeting, June 1-2, 1944, pp. 874-881.

U.S. Department of Agriculture, Atlas of American Agriculture, Washington, D.C., 1936, 215 pp.

_____, Climate and Man--1941 Yearbook of Agriculture, Washington, D.C., 1941, 1,248 pp.

_____, Soil--The Yearbook of Agriculture 1957, Washington, D.C., 1957, 784 pp.

U.S. Weather Bureau, Climatic Summary of the United States (Bulletin W), Section 85--Maine, Washington, D.C., 1934, 21 pp.

_____, "Climatic Summary of the United States--Supplement for 1931 through 1952, New England" (Bulletin W Supplement), Climatography of the United States No. 11-23, Washington, D.C., 1958, 92 pp.

_____, "Climatic Summary of the United States--Supplement for 1951 through 1960, New England" (Bulletin W Supplement), Climatography of the United States No. 86-23, Washington, D.C., 1964, 142 pp.

_____, "Evaporation Maps for the United States, " Technical Paper No. 37, Washington, D.C., 1959, 13 pp.

_____, "Heating Degree Day Normals, New England," Climatography of the United States No. 83-23, Washington, D.C., 1963, 1 pp.

_____, "Maximum Station Precipitation for 1, 2, 3, 6, 12, and 24 Hours, Part VI: New England," Technical Paper No. 15, Washington, D.C., 1954, 123 pp.

_____, "Maximum 24-Hour Precipitation in the United States," Technical Paper No. 16, Washington, D.C., Jan. 1952, 284 pp.

_____, "Monthly Averages for State Climatic Divisions, 1931-1960, New England," Climatography of the United States No. 85-23, Washington, D.C., 1963, 6 pp.

_____, "Monthly Normals of Temperature, Precipitation, and Heating Degree Days, New England," Climatography of the United States No. 81-23, Washington, D.C., 1962, 6 pp.

_____, "Rainfall Frequency Atlas of the United States for Durations from 30 Minutes to 24 Hours and Return Periods from 1 to 100 Years," Technical Paper No. 40, Washington, D.C., May 1961, 115 pp.

_____, "Rainfall Intensity-Duration-Frequency Curves for Selected Stations in the United States, Alaska, Hawaiian Islands, and Puerto Rico," Technical Paper No. 25, Washington, D.C. Dec. 1955, 53 pp.

_____, "Rainfall Intensity-Frequency Regime, Part 4--Northeastern United States," Technical Paper No. 29, Washington, D.C., May 1959, 35 pp.

_____, "Summary of Hourly Observations, Portland, Maine, 1951-1960," Climatography of the United States No. 82-17, Washington, D.C., 1963, 16 pp.

_____, "Tropical Cyclones of the North Atlantic Ocean, Tracks and Frequencies of Hurricanes and Tropical Storms, 1871-1963," Technical Paper No. 55, Washington, D.C., 1965, 148 pp.

BIBLIOGRAPHY - Continued

_____, "Two to Ten-Day Precipitation for Return Periods of 2 to 100 Years in the Contiguous United States," Technical Paper No. 49, Washington, D.C., 1964, 29 pp.

White, George V., "Rainfall in New England," Journal of New England Water Works Association, Vol. 56, No. 4, Dec. 1942, pp. 405-502; Vol. 57, No. 1, Mar. 1943, pp. 15-62.

FREEZE DATA

Station	Data Years	Temp.	SPRING DATES BY WHICH THE CHANCE OF OCCURRENCE OF INDICATED TEMPERATURE (OR LOWER) DECREASES TO:					AVG. DAYS FREEZE-FREE	FALL DATES BY WHICH THE CHANCE OF OCCURRENCE OF INDICATED TEMPERATURE (OR LOWER) INCREASES TO:				
			90% 9 in 10	75% 3 in 4	50% 1 in 2 Avg.Date	25% 1 in 4	10% 1 in 10		10% or 1 in 10	25% or 1 in 4	50% or 1 in 2 Avg.Date	75% or 3 in 4	90% or 9 in 10
Augusta	24	32°	Apr 18	Apr 25	May 2	May 9	May 16	159	Sept 24	Oct 1	Oct 8	Oct 15	Oct 22
	(1945–68)	28°	Apr 5	Apr 12	Apr 19	Apr 26	May 3	185	Oct 7	Oct 14	Oct 21	Oct 28	Nov 4
		24°	Mar 23	Mar 30	Apr 6	Apr 13	Apr 20	213	Oct 22	Oct 29	Nov 5	Nov 12	Nov 19
		20°	Mar 13	Mar 20	Mar 27	Apr 3	Apr 10	240	Nov 8	Nov 15	Nov 22	Nov 29	Dec 6
		16°	Mar 6	Mar 13	Mar 20	Mar 27	Apr 3	256	Nov 17	Nov 24	Dec 1	Dec 8	Dec 15
		0°	Feb 2	Feb 10	Feb 19	Feb 28	Mar 8	317	Dec 18	Dec 25	Jan 2	Jan 10	Jan 17
Bangor	16	32°	Apr 17	Apr 24	May 1	May 8	May 15	156	Sept 20	Sept 27	Oct 4	Oct 11	Oct 18
	(1953–68)	28°	Apr 6	Apr 13	Apr 20	Apr 27	May 4	182	Oct 5	Oct 12	Oct 19	Oct 26	Nov 2
		24°	Mar 26	Apr 2	Apr 9	Apr 16	Apr 23	209	Oct 21	Oct 28	Nov 4	Nov 11	Nov 18
		20°	Mar 15	Mar 22	Mar 29	Apr 5	Apr 12	231	Nov 1	Nov 8	Nov 15	Nov 22	Nov 29
		16°	Mar 9	Mar 16	Mar 23	Mar 30	Apr 6	251	Nov 15	Nov 22	Nov 29	Dec 6	Dec 13
		0°	Feb 8	Feb 16	Feb 25	Mar 6	Mar 14	305	Dec 12	Dec 19	Dec 27	Jan 4	Jan 11
Bar Harbor	38	32°	Apr 25	May 2	May 9	May 16	May 23	156	Sept 28	Oct 5	Oct 12	Oct 19	Oct 26
	(1931–68)	28°	Apr 12	Apr 19	Apr 26	May 3	May 10	182	Oct 11	Oct 18	Oct 25	Nov 1	Nov 8
		24°	Mar 26	Apr 2	Apr 9	Apr 16	Apr 23	212	Oct 24	Oct 31	Nov 7	Nov 14	Nov 21
		20°	Mar 13	Mar 20	Mar 27	Apr 3	Apr 10	238	Nov 6	Nov 13	Nov 20	Nov 27	Dec 4
		16°	Mar 7	Mar 14	Mar 21	Mar 28	Apr 4	254	Nov 16	Nov 23	Nov 30	Dec 7	Dec 14
		0°	Jan 28	Feb 7	Feb 18	Mar 1	Mar 11	313	Dec 9	Dec 18	Dec 28	Jan 7	Jan 16
Belfast	24	32°	Apr 26	May 3	May 10	May 17	May 24	143	Sept 16	Sept 23	Sept 30	Oct 7	Oct 14
	(1945–68)	28°	Apr 12	Apr 19	Apr 26	May 3	May 10	173	Oct 2	Oct 9	Oct 16	Oct 23	Oct 30
		24°	Mar 28	Apr 4	Apr 11	Apr 18	Apr 25	206	Oct 20	Oct 27	Nov 3	Nov 10	Nov 17
		20°	Mar 17	Mar 24	Mar 31	Apr 7	Apr 14	234	Nov 6	Nov 13	Nov 20	Nov 27	Dec 4
		16°	Mar 8	Mar 15	Mar 22	Mar 29	Apr 5	252	Nov 15	Nov 22	Nov 29	Dec 6	Dec 13
		0°	Feb 5	Feb 15	Feb 26	Mar 9	Mar 19	303	Dec 7	Dec 16	Dec 26	Jan 5	Jan 14
Bridgewater	11	32°	May 24	May 31	Jun 7	Jun 14	Jun 21	88	Aug 19	Aug 26	Sept 3	Sept 10	Sept 17
	(1958–68)	28°	May 4	May 11	May 18	May 25	Jun 1	126	Sept 7	Sept 14	Sept 21	Sept 28	Oct 5
		24°	Apr 24	May 1	May 8	May 15	May 22	151	Sept 22	Sept 29	Oct 6	Oct 13	Oct 20
		20°	Apr 9	Apr 16	Apr 23	Apr 30	May 7	182	Oct 8	Oct 15	Oct 22	Oct 29	Nov 5
		16°	Mar 30	Apr 6	Apr 13	Apr 20	Apr 27	211	Oct 27	Nov 3	Nov 10	Nov 17	Nov 24
		0°	Mar 6	Mar 13	Mar 20	Mar 27	Apr 3	260	Nov 22	Nov 28	Dec 5	Dec 12	Dec 18
Bridgton	14	32°	Apr 28	May 5	May 12	May 19	May 26	148	Sept 23	Sept 30	Oct 7	Oct 14	Oct 21
	(1955–68)	28°	Apr 11	Apr 18	Apr 25	May 2	May 9	176	Oct 4	Oct 11	Oct 18	Oct 25	Nov 1
		24°	Apr 1	Apr 8	Apr 15	Apr 22	Apr 29	203	Oct 21	Oct 28	Nov 4	Nov 11	Nov 18
		20°	Mar 20	Mar 27	Apr 3	Apr 10	Apr 17	226	Nov 1	Nov 8	Nov 15	Nov 22	Nov 29
		16°	Mar 10	Mar 17	Mar 24	Mar 31	Apr 7	246	Nov 11	Nov 18	Nov 25	Dec 2	Dec 9
		0°	Feb 10	Feb 18	Feb 27	Mar 8	Mar 16	298	Dec 7	Dec 14	Dec 22	Dec 30	Jan 6
Brunswick	17	32°	Apr 18	Apr 25	May 2	May 9	May 16	163	Sept 28	Oct 5	Oct 12	Oct 19	Oct 26
	(1952–68)	28°	Mar 31	Apr 7	Apr 14	Apr 21	Apr 28	192	Oct 9	Oct 16	Oct 23	Oct 30	Nov 6
		24°	Mar 20	Mar 27	Apr 3	Apr 10	Apr 17	221	Oct 27	Nov 3	Nov 10	Nov 17	Nov 24
		20°	Mar 11	Mar 18	Mar 25	Apr 1	Apr 8	238	Nov 4	Nov 11	Nov 18	Nov 25	Dec 2
		16°	Mar 6	Mar 13	Mar 20	Mar 27	Apr 3	256	Nov 17	Nov 24	Dec 1	Dec 8	Dec 15
		0°	Feb 1	Feb 11	Feb 22	Mar 5	Mar 15	308	Dec 8	Dec 17	Dec 27	Jan 6	Jan 15
Caribou	30	32°	May 5	May 12	May 19	May 26	Jun 2	125	Sept 7	Sept 14	Sept 21	Sept 28	Oct 5
	(1939–68)	28°	Apr 22	Apr 29	May 6	May 13	May 20	149	Sept 18	Sept 25	Oct 2	Oct 9	Oct 16
		24°	Apr 6	Apr 13	Apr 20	Apr 27	May 4	182	Oct 5	Oct 12	Oct 19	Oct 26	Nov 2
		20°	Mar 29	Apr 5	Apr 12	Apr 19	Apr 26	206	Oct 21	Oct 28	Nov 4	Nov 11	Nov 18
		16°	Mar 21	Mar 28	Apr 4	Apr 11	Apr 18	223	Oct 30	Nov 6	Nov 13	Nov 20	Nov 29
		0°	Feb 28	Mar 7	Mar 14	Mar 21	Mar 28	270	Nov 26	Dec 2	Dec 9	Dec 16	Dec 22
Corinna	22	32°	May 12	May 19	May 26	Jun 2	Jun 9	113	Sept 2	Sept 9	Sept 16	Sept 23	Sept 30
	(1947–68)	28°	Apr 24	May 1	May 8	May 15	May 22	142	Sept 13	Sept 20	Sept 27	Oct 4	Oct 11
		24°	Apr 12	Apr 19	Apr 26	May 3	May 10	175	Oct 4	Oct 11	Oct 18	Oct 25	Nov 1
		20°	Mar 24	Mar 31	Apr 7	Apr 14	Apr 21	207	Oct 17	Oct 24	Oct 31	Nov 7	Nov 14
		16°	Mar 17	Mar 24	Mar 31	Apr 7	Apr 14	226	Oct 29	Nov 5	Nov 12	Nov 19	Nov 26
		0°	Feb 21	Mar 1	Mar 10	Mar 19	Mar 27	277	Nov 27	Dec 4	Dec 12	Dec 20	Dec 27
Eastport	38	32°	Apr 17	Apr 24	May 1	May 8	May 15	174	Oct 8	Oct 15	Oct 22	Oct 29	Nov 5
	(1931–68)	28°	Apr 4	Apr 11	Apr 18	Apr 25	May 2	200	Oct 21	Oct 28	Nov 4	Nov 11	Nov 18
		24°	Mar 20	Mar 27	Apr 3	Apr 10	Apr 17	227	Nov 2	Nov 9	Nov 16	Nov 23	Nov 30
		20°	Mar 12	Mar 19	Mar 26	Apr 2	Apr 9	245	Nov 12	Nov 19	Nov 26	Dec 3	Dec 10
		16°	Mar 6	Mar 13	Mar 20	Mar 27	Apr 3	259	Nov 20	Nov 27	Dec 4	Dec 11	Dec 18
		0°	Jan 20	Jan 30	Feb 10	Feb 21	Mar 3	324	Dec 12	Dec 21	Dec 31	Jan 10	Jan 19
E. Sangerville	22	32°	May 10	May 17	May 24	May 31	Jun 7	116	Sept 3	Sept 10	Sept 17	Sept 24	Oct 1
	(1947–68)	28°	Apr 26	May 3	May 10	May 17	May 24	143	Sept 16	Sept 23	Sept 30	Oct 7	Oct 14
		24°	Apr 9	Apr 16	Apr 23	Apr 30	May 7	180	Oct 6	Oct 13	Oct 20	Oct 27	Nov 3
		20°	Mar 26	Apr 2	Apr 9	Apr 16	Apr 23	206	Oct 18	Oct 25	Nov 1	Nov 8	Nov 15
		16°	Mar 18	Mar 25	Apr 1	Apr 8	Apr 15	230	Nov 3	Nov 10	Nov 17	Nov 24	Dec 1
		0°	Feb 22	Mar 2	Mar 11	Mar 20	Mar 28	278	Nov 29	Dec 6	Dec 14	Dec 22	Dec 29

FREEZE DATA

			SPRING DATES BY WHICH THE CHANCE OF OCCURRENCE OF INDICATED TEMPERATURE (OR LOWER) DECREASES TO:						FALL DATES BY WHICH THE CHANCE OF OCCURRENCE OF INDICATED TEMPERATURE (OR LOWER) INCREASES TO:				
			90%	75%	50% 1 in 2	25%	10%	AVG. DAYS	10% or	25% or	50% or 1 in 2	75% or	90% or
Station	Years	Temp.	9 in 10	3 in 4	Avg.Date	1 in 4	1 in 10	FREEZE-FREE	1 in 10	1 in 4	Avg.Date	3 in 4	9 in 10
E. Winthrop (1942-61)	20	32°	Apr 26	May 3	May 10	May 17	May 24	140	Sept 13	Sept 20	Sept 27	Oct 4	Oct 11
		28°	Apr 15	Apr 22	Apr 29	May 6	May 13	168	Sept 30	Oct 7	Oct 14	Oct 21	Oct 28
		24°	Mar 30	Apr 6	Apr 13	Apr 20	Apr 27	197	Oct 13	Oct 20	Oct 27	Nov 3	Nov 10
		20°	Mar 19	Mar 26	Apr 2	Apr 9	Apr 16	225	Oct 30	Nov 6	Nov 13	Nov 20	Nov 27
		16°	Mar 12	Mar 19	Mar 26	Apr 2	Apr 9	247	Nov 14	Nov 21	Nov 28	Dec 5	Dec 12
		0°	Feb 16	Feb 24	Mar 5	Mar 14	Mar 22	291	Dec 6	Dec 13	Dec 21	Dec 29	Jan 5
Farmington (1931-68)	38	32°	May 6	May 13	May 20	May 27	Jun 3	123	Sept 6	Sept 13	Sept 20	Sept 27	Oct 4
		28°	Apr 22	Apr 29	May 6	May 13	May 20	148	Sept 17	Sept 24	Oct 1	Oct 8	Oct 15
		24°	Apr 6	Apr 13	Apr 20	Apr 27	May 4	182	Oct 5	Oct 12	Oct 19	Oct 26	Nov 2
		20°	Mar 23	Mar 30	Apr 6	Apr 13	Apr 20	206	Oct 15	Oct 22	Oct 29	Nov 5	Nov 12
		16°	Mar 15	Mar 22	Mar 29	Apr 5	Apr 12	231	Nov 2	Nov 8	Nov 15	Nov 22	Nov 29
		0°	Feb 21	Mar 1	Mar 10	Mar 19	Mar 27	281	Dec 1	Dec 8	Dec 16	Dec 24	Dec 31
Fort Kent (1945-68)	24	32°	May 16	May 23	May 30	Jun 6	Jun 13	107	Sept 2	Sept 9	Sept 16	Sept 23	Sept 30
		28°	Apr 28	May 5	May 12	May 19	May 26	139	Sept 14	Sept 21	Sept 28	Oct 5	Oct 12
		24°	Apr 19	Apr 26	May 3	May 10	May 17	163	Sept 29	Oct 6	Oct 13	Oct 20	Oct 27
		20°	Apr 6	Apr 13	Apr 20	Apr 27	May 4	184	Oct 7	Oct 14	Oct 21	Oct 28	Nov 4
		16°	Mar 27	Apr 3	Apr 10	Apr 17	Apr 24	211	Oct 24	Oct 31	Nov 7	Nov 14	Nov 21
		0°	Mar 8	Mar 15	Mar 22	Mar 29	Apr 5	258	Nov 22	Nov 28	Dec 5	Dec 12	Dec 18
Gardiner (1931-68)	38	32°	May 2	May 9	May 16	May 23	May 30	133	Sept 12	Sept 19	Sept 26	Oct 3	Oct 10
		28°	Apr 18	Apr 25	May 2	May 9	May 16	159	Sept 23	Sept 30	Oct 7	Oct 14	Oct 21
		24°	Apr 4	Apr 11	Apr 18	Apr 25	May 2	186	Oct 7	Oct 14	Oct 21	Oct 28	Nov 4
		20°	Mar 21	Mar 28	Apr 4	Apr 11	Apr 18	215	Oct 22	Oct 29	Nov 5	Nov 12	Nov 19
		16°	Mar 13	Mar 20	Mar 27	Apr 3	Apr 10	239	Nov 7	Nov 14	Nov 21	Nov 28	Dec 5
		0°	Feb 14	Feb 22	Mar 3	Mar 14	Mar 20	290	Dec 3	Dec 10	Dec 18	Dec 26	Jan 2
Greenville (1931-68)	38	32°	May 15	May 22	May 29	Jun 5	Jun 12	110	Sept 2	Sept 9	Sept 16	Sept 23	Sept 30
		28°	May 1	May 8	May 15	May 22	May 29	135	Sept 13	Sept 20	Sept 27	Oct 4	Oct 11
		24°	Apr 19	Apr 26	May 3	May 10	May 17	164	Sept 30	Oct 7	Oct 14	Oct 21	Oct 28
		20°	Apr 5	Apr 12	Apr 19	Apr 26	May 3	192	Oct 4	Oct 11	Oct 18	Oct 25	Nov 1
		16°	Mar 25	Apr 1	Apr 8	Apr 15	Apr 22	219	Oct 30	Nov 6	Nov 13	Nov 20	Nov 27
		0°	Mar 5	Mar 12	Mar 19	Mar 26	Apr 2	266	Nov 27	Dec 3	Dec 10	Dec 17	Dec 23
Hiram (1931-68)	38	32°	Apr 28	May 5	May 12	May 19	May 26	137	Sept 12	Sept 19	Sept 26	Oct 3	Oct 10
		28°	Apr 15	Apr 22	Apr 29	May 6	May 13	161	Sept 23	Sept 30	Oct 7	Oct 14	Oct 21
		24°	Mar 31	Apr 7	Apr 14	Apr 21	Apr 28	194	Oct 11	Oct 18	Oct 25	Nov 1	Nov 8
		20°	Mar 20	Mar 27	Apr 3	Apr 10	Apr 17	215	Oct 21	Oct 28	Nov 4	Nov 11	Nov 18
		16°	Mar 13	Mar 20	Mar 27	Apr 3	Apr 10	236	Nov 4	Nov 11	Nov 18	Nov 25	Dec 2
		0°	Feb 16	Feb 24	Mar 5	Mar 14	Mar 22	285	Nov 30	Dec 7	Dec 15	Dec 23	Dec 30
Houlton (1931-68)	38	32°	May 7	May 14	May 21	May 28	Jun 4	123	Sept 7	Sept 14	Sept 21	Sept 28	Oct 5
		28°	Apr 23	Apr 30	May 7	May 14	May 21	148	Sept 18	Sept 25	Oct 2	Oct 9	Oct 16
		24°	Apr 7	Apr 14	Apr 21	Apr 28	May 5	179	Oct 3	Oct 10	Oct 17	Oct 24	Oct 31
		20°	Mar 27	Apr 3	Apr 10	Apr 17	Apr 24	207	Oct 20	Oct 27	Nov 3	Nov 10	Nov 17
		16°	Mar 16	Mar 23	Mar 30	Apr 6	Apr 13	229	Oct 31	Nov 7	Nov 14	Nov 21	Nov 28
		0°	Feb 27	Mar 6	Mar 13	Mar 20	Mar 27	272	Nov 27	Dec 3	Dec 10	Dec 17	Dec 23
Houlton (FAA) (1946-68)	22	32°	May 12	May 19	May 26	Jun 2	Jun 9	113	Sept 2	Sept 9	Sept 16	Sept 23	Sept 30
		28°	Apr 27	May 4	May 11	May 18	May 25	141	Sept 15	Sept 22	Sept 29	Oct 6	Oct 13
		24°	Apr 13	Apr 20	Apr 27	May 4	May 11	170	Sept 30	Oct 7	Oct 14	Oct 21	Oct 28
		20°	Mar 31	Apr 7	Apr 14	Apr 21	Apr 28	198	Oct 15	Oct 22	Oct 29	Nov 5	Nov 12
		16°	Mar 19	Mar 26	Apr 2	Apr 9	Apr 16	224	Oct 29	Nov 5	Nov 12	Nov 19	Nov 26
		0°	Mar 2	Mar 9	Mar 16	Mar 23	Mar 30	268	Nov 26	Dec 2	Dec 9	Dec 16	Dec 22
Jackman (1951-68)	18	32°	May 15	May 22	May 29	Jun 5	Jun 12	109	Sept 1	Sept 8	Sept 15	Sept 22	Sept 29
		28°	Apr 28	May 5	May 12	May 19	May 26	139	Sept 14	Sept 21	Sept 28	Oct 5	Oct 12
		24°	Apr 12	Apr 19	Apr 26	May 3	May 10	176	Oct 5	Oct 12	Oct 19	Oct 26	Nov 2
		20°	Mar 31	Apr 7	Apr 14	Apr 21	Apr 28	200	Oct 17	Oct 24	Oct 31	Nov 7	Nov 14
		16°	Mar 27	Apr 3	Apr 10	Apr 17	Apr 24	214	Oct 27	Nov 3	Nov 10	Nov 17	Nov 24
		0°	Mar 9	Mar 16	Mar 23	Mar 30	Apr 6	260	Nov 25	Dec 1	Dec 8	Dec 15	Dec 21
Jonesboro (1949-68)	20	32°	May 2	May 9	May 16	May 23	May 30	133	Sept 12	Sept 19	Sept 26	Oct 3	Oct 10
		28°	Apr 20	Apr 27	May 4	May 11	May 18	157	Sept 24	Oct 1	Oct 8	Oct 15	Oct 22
		24°	Apr 1	Apr 8	Apr 15	Apr 22	Apr 29	191	Oct 9	Oct 16	Oct 23	Oct 30	Nov 6
		20°	Mar 18	Mar 25	Apr 1	Apr 8	Apr 15	222	Oct 26	Nov 2	Nov 9	Nov 16	Nov 23
		16°	Mar 11	Mar 18	Mar 25	Apr 1	Apr 8	239	Nov 5	Nov 12	Nov 19	Nov 26	Dec 3
		0°	Feb 6	Feb 16	Feb 27	Mar 10	Mar 20	29	Dec 3	Dec 12	Dec 22	Jan 1	Jan 1
Lewiston (1931-68)	38	32°	Apr 15	Apr 22	Apr 29	May 6	May 13	168	Sept 30	Oct 7	Oct 14	Oct 21	Oct 28
		28°	Mar 31	Apr 7	Apr 14	Apr 21	Apr 28	197	Oct 14	Oct 21	Oct 28	Nov 4	Nov 11
		24°	Mar 19	Mar 26	Apr 2	Apr 9	Apr 16	225	Oct 30	Nov 6	Nov 13	Nov 20	Nov 27
		20°	Mar 11	Mar 18	Mar 25	Apr 1	Apr 8	245	Nov 11	Nov 18	Nov 25	Dec 2	Dec 9
		16°	Mar 4	Mar 11	Mar 18	Mar 25	Apr 1	259	Nov 18	Nov 25	Dec 2	Dec 9	Dec 16
		0°	Feb 4	Feb 12	Feb 21	Mar 2	Mar 10	311	Dec 14	Dec 21	Dec 29	Jan 6	Jan 13

FREEZE DATA

			SPRING DATES BY WHICH THE CHANCE OF OCCURRENCE OF INDICATED TEMPERATURE (OR LOWER) DECREASES TO:						FALL DATES BY WHICH THE CHANCE OF OCCURRENCE OF INDICATED TEMPERATURE (OR LOWER) INCREASES TO:				
Station	Years	Temp.	90% 9 in 10	75% 3 in 4	50% 1 in 2 Avg. Date	25% 1 in 4	10% 1 in 10	AVG. DAYS FREEZE-FREE	10% or 1 in 10	25% or 1 in 4	50% or 1 in 2 Avg. Date	75% or 3 in 4	90% or 9 in 10
Long Falls Dam	16 (1953-68)	32°	May 15	May 22	May 29	Jun 5	Jun 12	113	Sept 5	Sept 12	Sept 19	Sept 26	Oct 3
		28°	Apr 27	May 4	May 11	May 18	May 25	144	Sept 18	Sept 25	Oct 2	Oct 9	Oct 16
		24°	Apr 16	Apr 23	Apr 30	May 7	May 14	175	Oct 8	Oct 15	Oct 22	Oct 29	Nov 5
		20°	Apr 2	Apr 9	Apr 16	Apr 23	Apr 30	197	Oct 16	Oct 23	Oct 30	Nov 6	Nov 13
		16°	Mar 23	Mar 30	Apr 6	Apr 13	Apr 20	223	Nov 1	Nov 8	Nov 15	Nov 22	Nov 29
		0°	Mar 8	Mar 15	Mar 22	Mar 29	Apr 5	263	Nov 26	Dec 3	Dec 10	Dec 17	Dec 23
Machias	15 (1954-68)	32°	May 3	May 10	May 17	May 24	May 31	132	Sept 12	Sept 19	Sept 26	Oct 3	Oct 10
		28°	Apr 20	Apr 27	May 4	May 11	May 18	160	Sept 27	Oct 4	Oct 11	Oct 18	Oct 25
		24°	Apr 7	Apr 14	Apr 21	Apr 28	May 5	187	Oct 11	Oct 18	Oct 25	Nov 1	Nov 8
		20°	Mar 24	Mar 31	Apr 7	Apr 14	Apr 21	217	Oct 27	Nov 3	Nov 10	Nov 17	Nov 24
		16°	Mar 15	Mar 22	Mar 29	Apr 5	Apr 12	235	Nov 5	Nov 12	Nov 19	Nov 26	Dec 3
		0°	Feb 12	Feb 22	Mar 5	Mar 16	Mar 26	294	Dec 5	Dec 14	Dec 24	Jan 3	Jan 12
Madison	38 (1931-68)	32°	May 6	May 13	May 20	May 27	Jun 3	127	Sept 10	Sept 17	Sept 24	Oct 1	Oct 8
		28°	Apr 23	Apr 30	May 7	May 14	May 21	153	Sept 23	Sept 30	Oct 7	Oct 14	Oct 21
		24°	Apr 9	Apr 16	Apr 23	Apr 30	May 7	180	Oct 6	Oct 13	Oct 20	Oct 27	Nov 3
		20°	Mar 25	Apr 1	Apr 8	Apr 15	Apr 22	206	Oct 17	Oct 24	Oct 31	Nov 7	Nov 14
		16°	Mar 15	Mar 22	Mar 29	Apr 5	Apr 12	232	Nov 2	Nov 9	Nov 16	Nov 23	Nov 30
		0°	Feb 19	Feb 27	Mar 8	Mar 17	Mar 25	279	Nov 27	Dec 4	Dec 12	Dec 20	Dec 27
Millinocket	38 (1931-68)	32°	May 3	May 10	May 17	May 24	May 31	130	Sept 10	Sept 17	Sept 24	Oct 1	Oct 8
		28°	Apr 23	Apr 30	May 7	May 14	May 21	152	Sept 22	Sept 29	Oct 6	Oct 13	Oct 20
		24°	Apr 9	Apr 16	Apr 23	Apr 30	May 7	181	Oct 7	Oct 14	Oct 21	Oct 28	Nov 4
		20°	Mar 26	Apr 2	Apr 9	Apr 16	Apr 23	208	Oct 20	Oct 27	Nov 3	Nov 10	Nov 16
		16°	Mar 14	Mar 21	Mar 28	Apr 4	Apr 11	235	Nov 4	Nov 11	Nov 18	Nov 24	Dec 1
		0°	Feb 25	Mar 4	Mar 11	Mar 18	Mar 25	277	Nov 30	Dec 6	Dec 13	Dec 20	Dec 26
Millinocket (FAA)	24 (1945-68)	32°	May 4	May 11	May 18	May 25	Jun 1	125	Sept 6	Sept 13	Sept 20	Sept 27	Oct 4
		28°	Apr 22	Apr 29	May 6	May 13	May 20	152	Sept 21	Sept 28	Oct 5	Oct 12	Oct 19
		24°	Apr 9	Apr 16	Apr 23	Apr 30	May 7	180	Oct 6	Oct 13	Oct 20	Oct 27	Nov 3
		20°	Mar 25	Apr 1	Apr 8	Apr 15	Apr 22	208	Oct 19	Oct 26	Nov 2	Nov 9	Nov 16
		16°	Mar 15	Mar 22	Mar 29	Apr 5	Apr 12	234	Nov 4	Nov 11	Nov 18	Nov 25	Dec 2
		0°	Feb 25	Mar 4	Mar 11	Mar 18	Mar 25	278	Dec 1	Dec 7	Dec 14	Dec 21	Dec 28
Millinocket Dam	10 (1947-56)	32°	May 20	May 27	Jun 3	Jun 10	Jun 17	104	Sept 1	Sept 8	Sept 15	Sept 22	Sept 29
		28°	May 5	May 12	May 19	May 26	Jun 2	128	Sept 10	Sept 17	Sept 24	Oct 1	Oct 8
		24°	Apr 23	Apr 30	May 7	May 14	May 21	163	Oct 3	Oct 10	Oct 17	Oct 24	Oct 31
		20°	Apr 4	Apr 11	Apr 18	Apr 25	May 2	198	Oct 19	Oct 26	Nov 2	Nov 9	Nov 16
		16°	Mar 22	Mar 29	Apr 5	Apr 12	Apr 19	219	Oct 27	Nov 3	Nov 10	Nov 17	Nov 24
		0°	Mar 6	Mar 13	Mar 20	Mar 27	Apr 3	265	Nov 27	Dec 3	Dec 10	Dec 17	Dec 23
N. Bridgton	31 (1921-68)	32°	May 3	May 10	May 17	May 24	May 31	135	Sept 15	Sept 22	Sept 29	Oct 6	Oct 13
		28°	Apr 20	Apr 27	May 4	May 11	May 18	155	Sept 22	Sept 29	Oct 6	Oct 13	Oct 20
		24°	Apr 5	Apr 12	Apr 19	Apr 26	May 3	187	Oct 9	Oct 16	Oct 23	Oct 30	Nov 6
		20°	Mar 23	Mar 30	Apr 6	Apr 13	Apr 20	214	Oct 23	Oct 30	Nov 6	Nov 13	Nov 20
		16°	Mar 13	Mar 20	Mar 27	Apr 3	Apr 10	236	Nov 4	Nov 11	Nov 18	Nov 25	Dec 2
		0°	Feb 19	Feb 27	Mar 8	Mar 17	Mar 25	281	Nov 29	Dec 6	Dec 14	Dec 22	Dec 29
Old Town	21 (1948-68)	32°	May 8	May 15	May 22	May 29	Jun 5	118	Sept 3	Sept 10	Sept 17	Sept 24	Oct 1
		28°	Apr 22	Apr 29	May 6	May 13	May 20	148	Sept 17	Sept 24	Oct 1	Oct 8	Oct 15
		24°	Apr 7	Apr 14	Apr 21	Apr 28	May 5	180	Oct 4	Oct 11	Oct 18	Oct 25	Nov 1
		20°	Mar 23	Mar 30	Apr 6	Apr 13	Apr 20	204	Oct 13	Oct 20	Oct 27	Nov 3	Nov 10
		16°	Mar 12	Mar 19	Mar 26	Apr 2	Apr 9	232	Oct 30	Nov 6	Nov 13	Nov 20	Nov 27
		0°	Feb 20	Feb 28	Mar 9	Mar 18	Mar 26	285	Dec 4	Dec 11	Dec 19	Dec 27	Jan 3
Oquossoc	15 (1954-68)	32°	May 20	May 27	Jun 3	Jun 10	Jun 17	99	Aug 27	Sept 3	Sept 10	Sept 17	Sept 24
		28°	May 9	May 16	May 23	May 30	Jun 6	123	Sept 9	Sept 16	Sept 23	Sept 30	Oct 7
		24°	Apr 26	May 3	May 10	May 17	May 24	152	Sept 25	Oct 2	Oct 9	Oct 16	Oct 23
		20°	Apr 9	Apr 16	Apr 23	Apr 30	May 7	185	Oct 11	Oct 18	Oct 25	Nov 1	Nov 8
		16°	Mar 31	Apr 7	Apr 14	Apr 21	Apr 28	209	Oct 26	Nov 2	Nov 9	Nov 16	Nov 23
		0°	Mar 12	Mar 19	Mar 26	Apr 2	Apr 9	253	Nov 20	Nov 27	Dec 4	Dec 11	Dec 17
Orono	38 (1931-68)	32°	Apr 29	May 6	May 13	May 20	May 27	135	Sept 11	Sept 18	Sept 25	Oct 2	Oct 9
		28°	Apr 14	Apr 21	Apr 28	May 5	May 12	164	Sept 25	Oct 2	Oct 9	Oct 16	Oct 23
		24°	Mar 30	Apr 6	Apr 13	Apr 20	Apr 27	196	Oct 12	Oct 19	Oct 26	Nov 2	Nov 9
		20°	Mar 17	Mar 24	Mar 31	Apr 7	Apr 14	227	Oct 30	Nov 6	Nov 13	Nov 20	Nov 27
		16°	Mar 9	Mar 16	Mar 23	Mar 30	Apr 6	246	Nov 5	Nov 12	Nov 20	Nov 28	Dec 8
		0°	Feb 11	Feb 19	Feb 28	Mar 9	Mar 17	295	Dec 5	Dec 12	Dec 20	Dec 28	Jan 4
Portland	28 (1941-68)	32°	Apr 29	May 6	May 13	May 20	May 27	136	Sept 12	Sept 19	Sept 26	Oct 3	Oct 10
		28°	Apr 15	Apr 22	Apr 29	May 6	May 13	161	Sept 23	Sept 30	Oct 7	Oct 14	Oct 21
		24°	Mar 30	Apr 6	Apr 13	Apr 20	Apr 27	194	Oct 10	Oct 17	Oct 24	Oct 31	Nov 7
		20°	Mar 15	Mar 22	Mar 29	Apr 5	Apr 12	225	Oct 26	Nov 2	Nov 9	Nov 16	Nov 23
		16°	Mar 9	Mar 16	Mar 23	Mar 30	Apr 6	246	Nov 10	Nov 17	Nov 24	Dec 1	Dec 8
		0°	Feb 6	Feb 16	Feb 27	Mar 10	Mar 20	300	Dec 5	Dec 14	Dec 24	Jan 3	Jan 12
Presque Isle	38 (1931-68)	32°	May 10	May 17	May 24	May 31	Jun 7	116	Sept 3	Sept 10	Sept 17	Sept 24	Oct 1
		28°	Apr 27	May 4	May 11	May 18	May 25	141	Sept 15	Sept 22	Sept 29	Oct 6	Oct 13
		24°	Apr 12	Apr 19	Apr 26	May 3	May 10	173	Oct 2	Oct 9	Oct 16	Oct 23	Oct 30
		20°	Apr 3	Apr 10	Apr 17	Apr 24	May 1	196	Oct 16	Oct 23	Oct 30	Nov 6	Nov 13
		16°	Mar 24	Mar 31	Apr 7	Apr 14	Apr 21	219	Oct 29	Nov 5	Nov 12	Nov 19	Nov 26
		0°	Feb 28	Mar 7	Mar 14	Mar 21	Mar 28	266	Nov 22	Nov 28	Dec 5	Dec 12	Dec 18

FREEZE DATA

			SPRING DATES BY WHICH THE CHANCE OF OCCURRENCE OF INDICATED TEMPERATURE (OR LOWER) DECREASES TO:					FALL DATES BY WHICH THE CHANCE OF OCCURRENCE OF INDICATED TEMPERATURE (OR LOWER) INCREASES TO:				
Station Years	Temp.	90% 9 in 10	75% 3 in 4	50% 1 in 2 Avg.Date	25% 1 in 4	10% 1 in 10	AVG. DAYS FREEZE-FREE	10% or 1 in 10	25% or 1 in 4	50% 1 in 2 Avg.Date	75% or 3 in 4	90% or 9 in 10
Ripogenus Dam 38 (1931–68)	32°	May 7	May 14	May 21	May 28	Jun 4	128	Sept 12	Sept 19	Sept 26	Oct 3	Oct 10
	28°	Apr 25	May 2	May 9	May 16	May 23	154	Sept 26	Oct 3	Oct 10	Oct 17	Oct 24
	24°	Apr 14	Apr 21	Apr 28	May 5	May 12	180	Oct 11	Oct 18	Oct 25	Nov 1	Nov 8
	20°	Apr 4	Apr 11	Apr 18	Apr 25	May 2	202	Oct 23	Oct 30	Nov 6	Nov 13	Nov 20
	16°	Mar 25	Apr 1	Apr 8	Apr 15	Apr 22	226	Nov 6	Nov 13	Nov 20	Nov 27	Dec 4
	0°	Mar 4	Mar 11	Mar 18	Mar 25	Apr 1	268	Nov 28	Dec 4	Dec 11	Dec 18	Dec 24
Rockland 32 (1937–68)	32°	Apr 26	May 3	May 10	May 17	May 24	143	Sept 16	Sept 23	Sept 30	Oct 7	Oct 14
	28°	Apr 12	Apr 19	Apr 26	May 3	May 10	170	Sept 29	Oct 6	Oct 13	Oct 20	Oct 27
	24°	Mar 29	Apr 5	Apr 12	Apr 19	Apr 26	200	Oct 15	Oct 22	Oct 29	Nov 5	Nov 12
	20°	Mar 17	Mar 24	Mar 31	Apr 7	Apr 14	228	Oct 31	Nov 7	Nov 14	Nov 21	Nov 28
	16°	Mar 7	Mar 14	Mar 21	Mar 28	Apr 4	250	Nov 12	Nov 19	Nov 26	Dec 3	Dec 10
	0°	Jan 30	Feb 9	Feb 20	Mar 3	Mar 13	309	Dec 7	Dec 16	Dec 26	Jan 5	Jan 14
Rumford Power Plant 38 (1931–68)	32°	May 2	May 9	May 16	May 23	May 30	134	Sept 13	Sept 20	Sept 27	Oct 4	Oct 11
	28°	Apr 15	Apr 22	Apr 29	May 6	May 13	163	Sept 25	Oct 2	Oct 9	Oct 16	Oct 23
	24°	Mar 31	Apr 7	Apr 14	Apr 21	Apr 28	195	Oct 12	Oct 19	Oct 26	Nov 2	Nov 9
	20°	Mar 19	Mar 26	Apr 2	Apr 9	Apr 16	221	Oct 26	Nov 2	Nov 9	Nov 16	Nov 23
	16°	Mar 14	Mar 18	Mar 25	Apr 1	Apr 8	243	Nov 9	Nov 16	Nov 23	Nov 30	Dec 7
	0°	Feb 14	Feb 22	Mar 3	Mar 12	Mar 20	293	Dec 6	Dec 13	Dec 21	Dec 29	Jan 5
Rumford 26 (1943–68)	32°	May 9	May 16	May 23	May 30	Jun 6	122	Sept 8	Sept 15	Sept 22	Sept 29	Oct 6
	28°	Apr 25	May 2	May 9	May 16	May 23	148	Sept 20	Sept 27	Oct 4	Oct 11	Oct 18
	24°	Apr 9	Apr 16	Apr 23	Apr 30	May 7	179	Oct 5	Oct 12	Oct 19	Oct 26	Nov 2
	20°	Mar 25	Apr 1	Apr 8	Apr 15	Apr 22	206	Oct 17	Oct 24	Oct 31	Nov 7	Nov 14
	16°	Mar 16	Mar 23	Mar 30	Apr 6	Apr 13	230	Nov 1	Nov 8	Nov 15	Nov 22	Nov 29
	0°	Feb 17	Feb 25	Mar 6	Mar 15	Mar 23	285	Dec 1	Dec 8	Dec 16	Dec 24	Dec 31
Sanford 16 (1953–68)	32°	May 8	May 15	May 22	May 29	Jun 5	121	Sept 6	Sept 13	Sept 20	Sept 27	Oct 4
	28°	Apr 27	May 4	May 11	May 18	May 25	140	Sept 14	Sept 21	Sept 28	Oct 5	Oct 12
	24°	Apr 11	Apr 18	Apr 25	May 2	May 9	173	Oct 1	Oct 8	Oct 15	Oct 22	Oct 29
	20°	Mar 27	Apr 3	Apr 10	Apr 17	Apr 24	204	Oct 17	Oct 24	Oct 31	Nov 7	Nov 14
	16°	Mar 12	Mar 19	Mar 26	Apr 2	Apr 9	238	Nov 5	Nov 12	Nov 19	Nov 26	Dec 3
	0°	Feb 7	Feb 17	Feb 28	Mar 11	Mar 21	297	Dec 3	Dec 12	Dec 22	Jan 1	Jan 10
Squa Pan Dam 13 (1956–68)	32°	Jun 4	Jun 11	Jun 18	Jun 25	July 2	75	Aug 18	Aug 25	Sept 1	Sept 8	Sept 15
	28°	May 13	May 20	May 27	Jun 3	Jun 10	112	Sept 2	Sept 9	Sept 16	Sept 23	Sept 30
	24°	Apr 27	May 4	May 11	May 18	May 25	147	Sept 21	Sept 28	Oct 5	Oct 12	Oct 19
	20°	Apr 17	Apr 24	May 1	May 8	May 15	171	Oct 5	Oct 12	Oct 19	Oct 26	Nov 2
	16°	Apr 5	Apr 12	Apr 19	Apr 26	May 3	196	Oct 18	Oct 25	Nov 1	Nov 8	Nov 15
	0°	Mar 13	Mar 20	Mar 27	Apr 3	Apr 10	246	Nov 15	Nov 21	Nov 28	Dec 5	Dec 11
Unity 15 (1949–63)	32°	Apr 21	May 1	May 8	May 15	May 22	145	Sept 16	Sept 23	Sept 30	Oct 7	Oct 14
	28°	Apr 7	Apr 14	Apr 21	Apr 28	May 5	178	Oct 2	Oct 9	Oct 16	Oct 23	Oct 30
	24°	Mar 25	Apr 2	Apr 9	Apr 16	Apr 23	209	Oct 21	Oct 28	Nov 4	Nov 11	Nov 18
	20°	Mar 14	Mar 21	Mar 28	Apr 4	Apr 11	233	Nov 2	Nov 9	Nov 16	Nov 23	Nov 30
	16°	Mar 8	Mar 15	Mar 22	Mar 29	Apr 5	249	Nov 12	Nov 19	Nov 26	Dec 3	Dec 10
	0°	Feb 8	Feb 16	Feb 25	Mar 6	Mar 14	301	Dec 8	Dec 15	Dec 23	Dec 31	Jan 7
Van Buren 22 (1913–34)	32°	May 25	Jun 1	Jun 8	Jun 15	Jun 22	92	Aug 25	Sept 1	Sept 8	Sept 15	Sept 22
	28°	May 14	May 21	May 28	Jun 4	Jun 11	122	Sept 13	Sept 20	Sept 27	Oct 4	Oct 11
	24°	Apr 28	May 5	May 12	May 19	May 26	148	Sept 23	Sept 30	Oct 7	Oct 14	Oct 21
	20°	Apr 9	Apr 16	Apr 23	Apr 30	May 7	185	Oct 11	Oct 18	Oct 25	Nov 1	Nov 8
	16°	Mar 30	Apr 6	Apr 13	Apr 20	Apr 27	206	Oct 22	Oct 29	Nov 5	Nov 12	Nov 19
	0°	Mar 9	Mar 16	Mar 23	Mar 30	Apr 6	252	Nov 17	Nov 23	Nov 30	Dec 7	Dec 13
Waterville 38 (1931–68)	32°	Apr 28	May 5	May 12	May 19	May 26	139	Sept 14	Sept 21	Sept 28	Oct 5	Oct 12
	28°	Apr 13	Apr 20	Apr 27	May 4	May 11	169	Sept 29	Oct 6	Oct 13	Oct 20	Oct 27
	24°	Mar 28	Apr 4	Apr 11	Apr 18	Apr 25	198	Oct 12	Oct 19	Oct 26	Nov 2	Nov 9
	20°	Mar 15	Mar 22	Mar 29	Apr 5	Apr 12	225	Oct 26	Nov 2	Nov 9	Nov 16	Nov 23
	16°	Mar 8	Mar 15	Mar 22	Mar 29	Apr 5	246	Nov 9	Nov 16	Nov 23	Nov 30	Dec 7
	0°	Feb 12	Feb 20	Mar 1	Mar 10	Mar 18	293	Dec 4	Dec 11	Dec 19	Dec 27	Jan 3
W. Buxton 16 (1953–68)	32°	May 7	May 14	May 21	May 28	Jun 4	123	Sept 7	Sept 14	Sept 21	Sept 28	Oct 5
	28°	Apr 23	Apr 30	May 7	May 14	May 21	150	Sept 20	Sept 27	Oct 4	Oct 11	Oct 18
	24°	Apr 6	Apr 13	Apr 20	Apr 27	May 4	182	Oct 5	Oct 12	Oct 19	Oct 26	Nov 2
	20°	Mar 27	Apr 3	Apr 10	Apr 17	Apr 24	201	Oct 14	Oct 21	Oct 28	Nov 4	Nov 11
	16°	Mar 15	Mar 22	Mar 29	Apr 5	Apr 12	233	Nov 3	Nov 10	Nov 17	Nov 24	Dec 1
	0°	Feb 16	Feb 24	Mar 5	Mar 14	Mar 22	286	Dec 1	Dec 8	Dec 16	Dec 24	Dec 31
Woodland 38 (1931–68)	32°	May 3	May 10	May 17	May 24	May 31	131	Sept 11	Sept 18	Sept 25	Oct 2	Oct 9
	28°	Apr 22	Apr 29	May 6	May 13	May 20	155	Sept 24	Oct 1	Oct 8	Oct 15	Oct 22
	24°	Apr 7	Apr 14	Apr 21	Apr 28	May 5	183	Oct 7	Oct 14	Oct 21	Oct 28	Nov 4
	20°	Mar 24	Mar 31	Apr 7	Apr 14	Apr 21	213	Oct 23	Oct 30	Nov 6	Nov 13	Nov 20
	16°	Mar 12	Mar 19	Mar 26	Apr 2	Apr 9	237	Nov 4	Nov 11	Nov 18	Nov 25	Dec 2
	0°	Feb 19	Feb 27	Mar 8	Mar 17	Mar 25	283	Dec 1	Dec 8	Dec 16	Dec 24	Dec 31

Prepared by the Climatologist for Maine. Means have been adjusted to take into account years of non-occurrence. A freeze is a numerical substitute for the former term "killing frost" and is the occurrence of a minimum temperature at or below the threshold temperature of 32°, 28°, etc. Freeze data tabulations in greater detail are available and can be reproduced at cost.

*NORMALS BY CLIMATOLOGICAL DIVISIONS

Taken from "Climatography of the United States No. 81-4, Decennial Census of U. S. Climate"

TEMPERATURE (°F)

STATIONS (By Divisions)	JAN	FEB	MAR	APR	MAY	JUNE	JULY	AUG	SEPT	OCT	NOV	DEC	ANN
NORTHERN													
BRASSUA DAM	10.5	12.5	22.8	•	49.9	59.0	64.5	62.6	53.8	43.0	30.2	15.5	38.4
CARIBOU WSO	•	•	•	•	•	•	•	•	•	•	•	•	•
FORT FAIRFIELD	•	•	•	•	•	•	•	•	•	•	•	•	•
GREENVILLE	13.3	15.1	24.5	37.4	50.4	59.8	64.9	63.0	54.8	44.2	32.3	17.9	39.8
HOULTON	14.1	16.3	26.5	39.5	52.4	61.4	67.4	65.0	56.2	45.3	33.4	18.9	41.4
JACKMAN	•	•	•	•	•	•	•	•	•	•	•	•	•
MIDDLE DAM	15.8	17.6	27.9	41.1	53.4	62.5	67.9	66.0	57.6	46.2	34.3	20.0	42.5
MILLINOCKET	•	•	•	•	•	•	•	•	•	•	•	•	•
MOOSEHEAD	•	•	•	•	•	•	•	•	•	•	•	•	•
PRESQUE ISLE	12.9	14.6	24.7	38.4	51.5	60.4	66.0	64.0	55.6	44.6	32.1	17.2	40.2
RIPOGENUS DAM	12.8	13.6	23.5	37.8	50.5	60.8	66.6	64.8	56.5	45.6	33.0	17.8	40.3
SQUA PAN DAM	•	•	•	•	•	•	•	•	•	•	•	•	•
UPPER DAM	•	•	•	•	•	•	•	•	•	•	•	•	•
DIVISION	13.3	14.9	25.0	38.4	51.4	60.8	66.3	64.3	55.8	44.9	32.6	17.9	40.5
SOUTHERN INTERIOR													
FARMINGTON	18.2	20.2	29.7	42.4	54.9	63.7	68.8	66.8	58.5	47.7	35.7	21.8	44.0
GARDINER	20.4	21.5	30.6	42.8	54.1	63.1	68.7	67.3	59.0	48.9	37.9	24.7	44.9
LEWISTON	20.7	22.2	31.1	42.8	54.6	64.0	70.0	68.6	60.2	49.7	37.9	25.1	45.6
MADISON	17.9	19.4	29.0	41.5	53.5	62.7	68.1	66.5	58.4	47.6	35.8	21.9	43.5
OLD TOWN FAA AIRPORT	19.2	21.0	30.5	42.5	54.2	62.5	68.1	66.5	58.2	47.9	37.0	23.4	44.3
RUMFORD POWER PLANT	18.9	20.4	29.3	41.6	53.8	63.1	68.3	66.3	58.1	47.6	35.9	22.8	43.8
WATERVILLE PUMP STA	19.6	21.3	31.2	43.6	55.4	64.3	69.8	68.3	60.0	49.2	37.7	23.7	45.3
WOODLAND	17.7	18.8	28.6	40.8	52.2	61.4	68.0	66.5	57.7	47.0	36.0	22.1	43.1
DIVISION	19.0	20.6	30.0	42.3	54.1	63.0	68.6	67.0	58.7	48.1	36.6	23.0	44.3
COASTAL													
BAR HARBOR	23.8	24.5	32.3	42.6	53.0	60.9	66.9	66.0	58.9	49.4	39.8	27.2	45.4
EASTPORT	22.9	23.5	30.8	40.2	48.7	55.8	61.5	61.9	56.7	48.5	39.0	26.8	43.0
ELLSWORTH	•	•	•	•	•	•	•	•	•	•	•	•	•
MACHIAS	•	•	•	•	•	•	•	•	•	•	•	•	•
PORTLAND WSO	21.8	22.8	31.4	42.5	53.0	62.1	68.1	66.8	58.7	48.6	38.1	25.8	45.0
DIVISION	23.1	24.0	31.9	42.1	52.1	60.3	66.3	65.6	58.6	49.1	39.1	26.7	44.9

PRECIPITATION (In.)

STATIONS (By Divisions)	JAN	FEB	MAR	APR	MAY	JUNE	JULY	AUG	SEPT	OCT	NOV	DEC	ANN
NORTHERN													
BRASSUA DAM	2.72	2.41	2.92	3.20	3.09	3.77	3.56	3.29	3.31	3.47	3.45	2.98	38.17
CARIBOU WSO	2.11	2.02	2.38	2.63	3.03	4.07	4.04	3.67	3.53	3.36	3.04	2.43	36.31
FORT FAIRFIELD	2.68	2.66	2.64	2.93	3.54	4.09	3.33	3.33	3.52	3.61	3.54	3.15	39.20
GREENVILLE	3.17	2.82	3.45	3.57	3.55	4.03	4.14	3.54	3.81	3.88	3.95	3.43	42.34
HOULTON	2.72	2.52	2.69	2.84	2.77	3.48	3.26	2.83	3.29	3.79	3.61	3.19	36.99
JACKMAN	2.44	2.23	2.65	2.79	3.08	3.76	3.59	3.35	3.33	3.09	3.03	2.72	36.06
MIDDLE DAM	2.59	2.54	3.03	3.07	2.54	4.29	4.20	3.43	3.86	3.86	3.51	2.74	40.04
MILLINOCKET	3.24	2.84	3.22	3.44	3.12	3.89	3.47	3.73	3.67	3.72	4.11	3.50	41.95
MOOSEHEAD	2.57	2.27	2.80	3.09	2.96	3.78	4.05	3.52	3.31	3.52	3.47	2.85	38.19
PRESQUE ISLE	2.19	2.07	2.34	2.52	2.85	3.79	3.85	3.31	3.32	3.41	2.99	2.41	35.05
RIPOGENUS DAM	2.62	2.47	2.81	3.46	3.36	4.28	4.20	3.84	3.77	3.95	3.75	2.89	41.40
SQUA PAN DAM	2.52	2.33	2.57	2.87	2.89	4.06	3.71	3.38	3.46	3.42	3.34	2.68	37.23
UPPER DAM	2.56	2.22	2.76	2.85	3.18	3.95	3.68	3.17	3.44	2.98	3.36	2.68	36.83
DIVISION	2.82	2.48	2.88	3.16	3.14	3.95	3.81	3.59	3.57	3.69	3.64	3.06	39.79
SOUTHERN INTERIOR													
FARMINGTON	3.59	2.98	3.85	3.76	3.72	3.96	3.88	3.06	3.85	3.86	4.50	3.62	44.63
GARDINER	3.01	3.29	3.75	3.96	3.31	3.31	3.39	2.90	3.77	3.77	4.59	3.82	43.95
LEWISTON	4.01	3.37	4.13	3.78	3.33	3.23	3.39	2.76	3.65	3.51	4.46	3.96	43.58
MADISON	3.29	2.83	3.28	3.30	3.23	3.33	3.18	2.63	3.45	3.20	3.92	3.28	38.92
OLD TOWN FAA AIRPORT	3.37	2.92	3.24	3.49	3.18	3.45	3.08	2.72	3.67	3.88	4.15	3.49	40.64
RUMFORD POWER PLANT	3.02	2.50	3.44	3.24	3.56	3.42	3.70	2.87	3.64	3.45	4.02	3.28	40.14
WATERVILLE PUMP STA	3.01	2.57	2.95	3.30	3.25	3.06	3.22	2.95	3.83	3.49	4.09	3.19	38.91
WOODLAND	3.70	3.21	3.48	3.53	3.26	3.46	3.29	3.02	3.85	4.25	4.83	3.97	43.85
DIVISION	3.46	2.93	3.45	3.50	3.37	3.43	3.38	2.92	3.75	3.70	4.26	3.49	41.64
COASTAL													
BAR HARBOR	4.46	3.79	4.39	3.91	3.83	3.34	3.10	3.12	4.22	4.38	5.25	4.38	48.17
EASTPORT	4.11	3.44	3.68	3.50	3.09	3.55	3.07	2.86	3.53	3.83	4.48	3.74	42.67
ELLSWORTH	3.98	3.26	3.76	3.55	3.07	3.07	3.10	2.56	2.96	3.82	3.62	3.93	43.28
MACHIAS	4.84	4.21	4.22	4.08	3.55	3.75	3.37	3.09	4.44	4.11	5.27	4.48	49.41
PORTLAND WSO	4.37	3.80	4.34	3.73	3.41	3.18	2.86	2.42	3.52	3.20	4.17	3.85	42.85
DIVISION	4.22	3.60	3.96	3.65	3.42	3.24	3.07	2.80	3.77	3.68	4.70	3.96	44.07

* Normals for the period 1931-1960. Divisional normals may not be the arithmetical average of individual stations published, since additional data for shorter period stations are used to obtain better areal representation.

CONFIDENCE - LIMITS

In absence of trend or record changes, the chances are 9 out of 10 that the true mean will lie in the interval formed by adding and subtracting the values in the following table from the means for any station in the State. Because of the wider variation in mean precipitation, the corresponding monthly means and annual mean must be substituted for "p" in the precipitation table below to obtain mean precipitation confidence limits.

1.4	1.3	1.4	.9	.7	.6	.6	.7	.6	.9	1.0	1.3	.4
$27\sqrt{p}$	$26\sqrt{p}$	$25\sqrt{p}$	$28\sqrt{p}$	$29\sqrt{p}$	$30\sqrt{p}$	$24\sqrt{p}$	$27\sqrt{p}$	$29\sqrt{p}$	$35\sqrt{p}$	$30\sqrt{p}$	$23\sqrt{p}$	$27\sqrt{p}$

COMPARATIVE DATA

Data in the following table are the mean temperature and average precipitation for Farmington, Maine, for the period 1906-1930 and are included in this publication for comparative purposes.

16.2	17.3	28.6	41.5	53.7	62.6	68.3	65.1	57.9	47.4	34.3	21.5	42.9
2.95	2.81	3.48	3.47	3.70	3.57	3.31	4.10	3.69	3.65	3.40		41.82

CLIMATE, NORMALS, AND EXTREMES

Station: CARIBOU, MAINE — MUNICIPAL AIRPORT
Standard time used: EASTERN Latitude: 46° 52' N Longitude: 68° 01' W Elevation (ground): 624 feet

Temperature

Month	Normal Daily maximum	Normal Daily minimum	Normal Monthly	Extremes Record highest	Year	Extremes Record lowest	Year
(a)	(b)	(b)	(b)	32		32	
J	19.8	1.1	10.5	51	1950	-32	1951
F	22.3	2.6	12.5	48	1970	-41	1955
M	31.8	13.8	22.8	73	1962	-20	1967+
A	45.0	27.4	31.4	80	1944	-2	1964
M	60.7	39.1	49.9	91	1960	19	1950
J	68.6	48.6	59.0	96	1944	30	1958
J	75.1	53.9	64.5	95	1955+	36	1969
A	72.6	51.8	62.6	95	1944	34	1965+
S	64.1	43.5	53.8	91	1945	23	1950
O	52.0	34.0	43.0	79	1968+	14	1940
N	36.5	23.5	30.2	68	1956	-3	1956
D	23.5	7.5	15.5	58	1962	-25	1971
YR	47.8	28.9	38.4	96	JUN 1944	-41	FEB 1955

Normal heating degree days (Base 65°) / Precipitation / Snow, Ice pellets

Month	Heating degree days	Precip Normal total	Precip Max monthly	Year	Precip Min monthly	Year	Precip Max in 24 hrs	Year	Snow Mean total	Snow Max monthly	Year	Snow Max in 24 hrs	Year
(a)	(b)	(b)	32		32		32		32	32		32	
J	1690	2.11	3.59	1945	0.12	1945	1.15	1945	22.4	39.5	1961	15.6	1945
F	1470	2.02	5.13	1955	0.56	1964	1.35	1952	23.0	41.0	1960+	23.0	1955
M	1308	2.38	4.13	1953	0.66	1965	1.65	1940	19.7	47.1	1940	17.4	1940
A	858	2.63	4.50	1964	0.54	1942	1.61	1953	6.0	24.4	1961	10.9	1961
M	468	2.91	6.11	1948	0.64	1959	2.25	1967	0.9	10.9	1967	5.8	1967
J	183	3.07	7.11	1940	1.25	1965	2.37	1957	0.0				
J	78	4.04	6.83	1957+	1.97	1942	2.92	1957+	0.0	T	1963+	T	1963+
A	115	3.67	8.45	1958	0.93	1957	4.14	1958	0.0	T			
S	336	3.53	8.14	1968	0.86	1958	4.07	1968	T	3.4	1963	3.6	
O	682	3.36	6.35	1970	0.63	1955	4.07	1970	2.5	12.1	1963	9.4	1954
N	1044	3.04	7.74	1951	0.45	1963	1.77	1969	12.0	32.2	1949	14.7	1951
D	1535	2.43	5.01	1963	0.74	1950	2.10	1950	21.4	49.4	1954	14.3	1954
YR	9767	36.31	8.45	AUG 1958	0.12	JAN 1944	6.23	SEP 1954	109.3	49.4	DEC 1954	18.2	FEB 1952

Relative humidity / Wind / Sky cover / Days / Solar (annual summary)

	01	07	13	19	Mean wind speed	Prevailing dir.	Mean sky cover	Clear	Partly cloudy	Cloudy	Precip .01"+	Snow 1.0"+	Thunderstorms	Heavy fog	Max 90°+	Max 32°&below	Min 32°&below	Min 0°&below	Avg solar radiation
YR	82	80	62	72	11.2	WSW	7.1	57	101	207	159	31	20	27	2	98	190	42	317

Ø Data accumulated through 1962.

% Through 1964. The station did not operate 24 hours daily. Fog and thunderstorm data may be incomplete.

Station: PORTLAND, MAINE — PORTLAND INTERNATIONAL JETPORT
Standard time used: EASTERN Latitude: 43° 39' N Longitude: 70° 19' W Elevation (ground): 43 feet

Temperature

Month	Normal Daily maximum	Normal Daily minimum	Normal Monthly	Extremes Record highest	Year	Extremes Record lowest	Year
(a)	(b)	(b)	(b)	31		31	
J	31.8	11.7	21.8	64	1950	-26	1971
F	33.5	12.1	22.8	64	1957	-39	1943
M	40.5	22.4	31.4	86	1946	-21	1950
A	52.5	32.4	42.5	85	1957	8	1954
M	63.7	42.1	53.0	92	1944	23	1956
J	73.1	51.1	62.1	97	1941	33	1944
J	79.5	56.7	68.1	98	1949+	40	1965+
A	77.4	55.9	66.8	100	1948	33	1965
S	70.0	47.2	58.7	95	1945	23	1941
O	59.8	37.4	48.6	88	1963	18	1966+
N	47.6	28.6	38.1	73	1950	-3	1963
D	36.3	16.3	26.3	62	1962	-21	1963
YR	55.6	34.4	45.0	100	AUG 1948	-39	FEB 1943

Normal heating degree days (Base 65°) / Precipitation / Snow, Ice pellets

Month	Heating degree days	Precip Normal total	Precip Max monthly	Year	Precip Min monthly	Year	Precip Max in 24 hrs	Year	Snow Mean total	Snow Max monthly	Year	Snow Max in 24 hrs	Year
(a)	(b)	(b)	31		31		31		31	31		31	
J	1339	4.37	9.41	1958	0.76	1970+	2.04	1966	18.4	38.2	1966	14.2	1966
F	1182	3.80	6.97	1971	1.26	1968	3.21	1965	21.0	61.2	1969	21.5	1964
M	1042	4.27	9.81	1953	0.80	1953	3.47	1956	14.1	46.6	1956	8.7	1967
A	675	3.73	6.48	1961	0.71	1941	2.43	1954	3.0	15.7	1945	8.1	1945
M	372	3.41	6.23	1967	0.49	1948	2.33	1952	0.0	0.0		0.0	
J	111	3.18	5.87	1969	0.70	1935	5.58	1967	0.0	0.0		0.0	
J	12	2.86	5.87	1951	0.61	1965	2.23	1947	0.0	0.0		0.0	
A	53	2.42	8.30	1946	0.30	1964+	4.18	1946	0.0	0.0		0.0	
S	195	3.20	9.81	1954	0.26	1948	7.49	1954	0.3	3.8	1959	3.8	1959
O	508	3.12	9.81	1962	2.10	1963	7.71	1962	1.1	15.3	1969	8.5	1969
N	807	4.17	9.81	1951	0.98	1969	3.82	1969	3.1	54.8	1970	22.8	1970
D	1215	3.85	12.27	1969					15.2	61.2	1969	22.8	1969
YR	7511	42.85	12.27	DEC 1969	0.26	OCT 1947	7.71	OCT 1962	75.4	61.2	FEB 1969	22.8	DEC 1970

Relative humidity / Wind / Sky cover / Days / Solar (annual summary)

	01	07	13	19	Mean wind speed	Prevailing dir.	Mean sky cover	Clear	Partly cloudy	Cloudy	Precip .01"+	Snow 1.0"+	Thunderstorms	Heavy fog	Max 90°+	Max 32°&below	Min 32°&below	Min 0°&below	Avg solar radiation
YR	83	80	60	74	8.8	N	6.1	109	101	155	125	18	18	53	5	46	161	15	342

Ø For period November 1940 through the current year.
Means and extremes above are from existing and comparable exposures. Annual extremes have been exceeded at other sites in the locality as follows:
Highest temperature 103 in July 1911; maximum monthly precipitation 12.29 in January 1935; minimum monthly precipitation 0.09 in October 1924; maximum snowfall in 24 hours 23.3 in January 1935.

REFERENCE NOTES APPLYING TO ALL "NORMALS, MEANS, AND EXTREMES" TABLES

(a) Length of record, years, based on January data. Other months may be for more or fewer years if there have been breaks in the record.

(b) Climatological standard normals (1931-1960).

* Less than one half.

\+ Also on earlier dates, months, or years.

T Trace, an amount too small to measure.

Below zero temperatures are preceded by a minus sign. The prevailing direction for wind in the Normals, Means, and Extremes table is from records through 1963.

‡ 70° at Alaskan stations.

Unless otherwise indicated, dimensional units used in this bulletin are: temperature in degrees F.; precipitation, including snowfall, in inches; wind movement in miles per hour; and relative humidity in percent. Heating degree day totals are the sums of negative departures of average daily temperatures from 65° F. Cooling degree day totals are the sums of positive departures of average daily temperatures from 65° F. Sleet was included in snowfall totals beginning with July 1948. The term "ice pellets" includes solid grains of ice (sleet) and particles consisting of snow pellets encased in a thin layer of ice. Heavy fog reduces visibility to 1/4 mile or less.

Sky cover is expressed in a range of 0 for no clouds or obscuring phenomena to 10 for complete sky cover. The number of clear days is based on average cloudiness 0-3, partly cloudy days 4-7, and cloudy days 8-10 tenths.

Solar radiation data are the averages of direct and diffuse radiation on a horizontal surface. The langley denotes one gram calorie per square centimeter.

& Figures instead of letters in a direction column indicate direction in tens of degrees from true North; i.e., 09-East, 18-South, 27-West, 36-North, and 00-Calm. -Resultant wind is the vector sum of wind directions and speeds divided by the number of observations. If figures appear in the direction column under "Fastest mile" the corresponding speeds are fastest observed 1-minute values.

To 8 compass points only.

MEAN SEASONAL SNOWFALL, INCHES

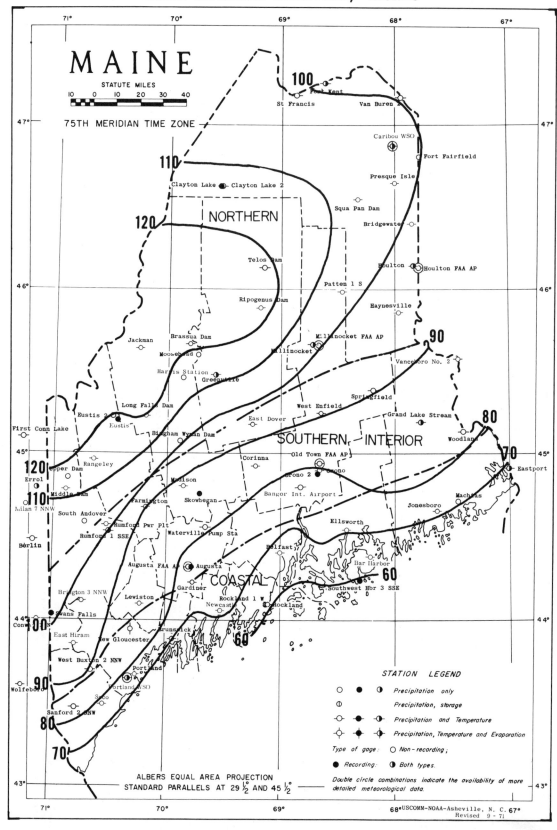

Data are based on the period 1931-68. Isolines are drawn through points of approximately equal value. Caution should be used in interpolating on these maps, particularly in mountainous areas.

MEAN MAXIMUM TEMPERATURE (°F.), JANUARY

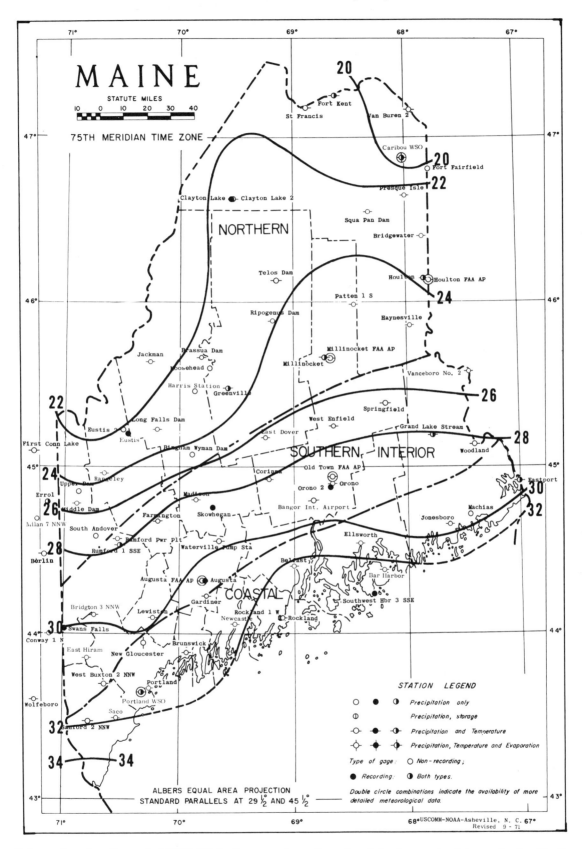

Data are based on the period 1931-52. Isolines are drawn through points of approximately equal value. Caution should be used in interpolating on these maps. particularly in mountainous areas.

MEAN MINIMUM TEMPERATURE (°F.), JANUARY

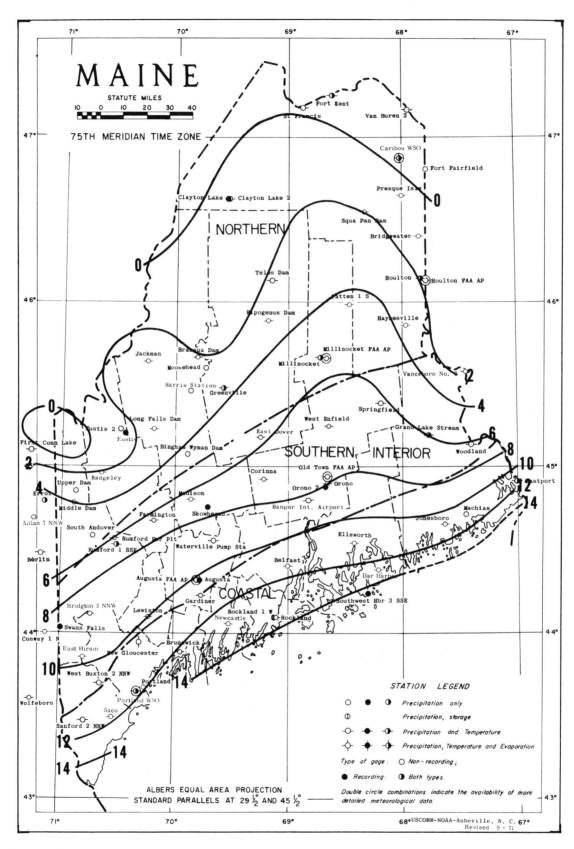

Data are based on the period 1931-52. Isolines are drawn through points of approximately equal value. Caution should be used in interpolating on these maps, particularly in mountainous areas.

MEAN MAXIMUM TEMPERATURE (°F.), JULY

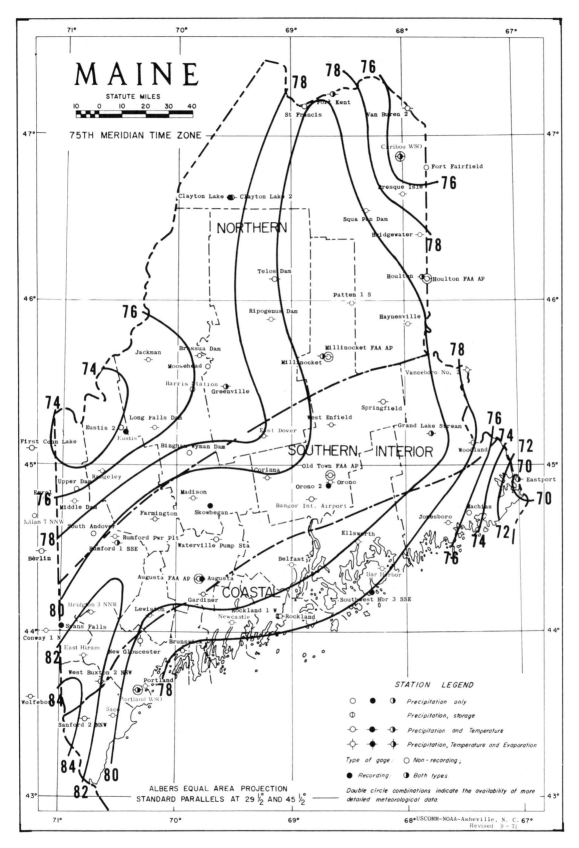

MAINE

STATUTE MILES

10 0 10 20 30 40

75TH MERIDIAN TIME ZONE

NORTHERN

SOUTHERN INTERIOR

COASTAL

ALBERS EQUAL AREA PROJECTION
STANDARD PARALLELS AT 29 ½° AND 45 ½°

STATION LEGEND

○ ● ◑ *Precipitation only*

◑ *Precipitation, storage*

◌○ ◌● ◌◑ *Precipitation and Temperature*

◇○ ◇● ◇◑ *Precipitation, Temperature and Evaporation*

Type of gage: ○ *Non-recording;*

Recording: ● *Both types*

Double circle combinations indicate the availability of more detailed meteorological data.

USCOMM-NOAA-Asheville, N. C.
Revised 9 - 71

Data are based on the period 1931-52. Isolines are drawn through points of approximately equal value. Caution should be used in interpolating on these maps, particularly in mountainous areas.

MEAN MINIMUM TEMPERATURE (°F.), JULY

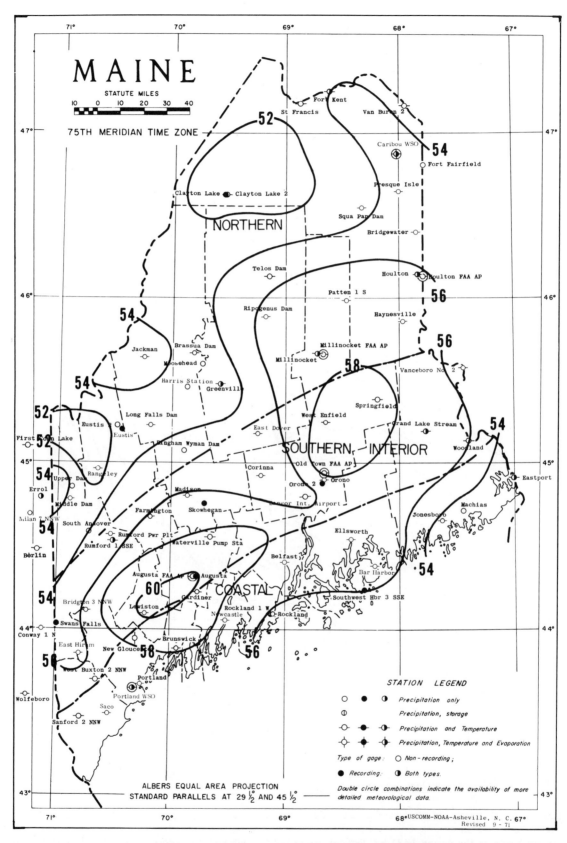

Data are based on the period 1931-52. Isolines are drawn through points of approximately equal value. Caution should be used in interpolating on these maps, particularly in mountainous areas.

MEAN ANNUAL PRECIPITATION, INCHES

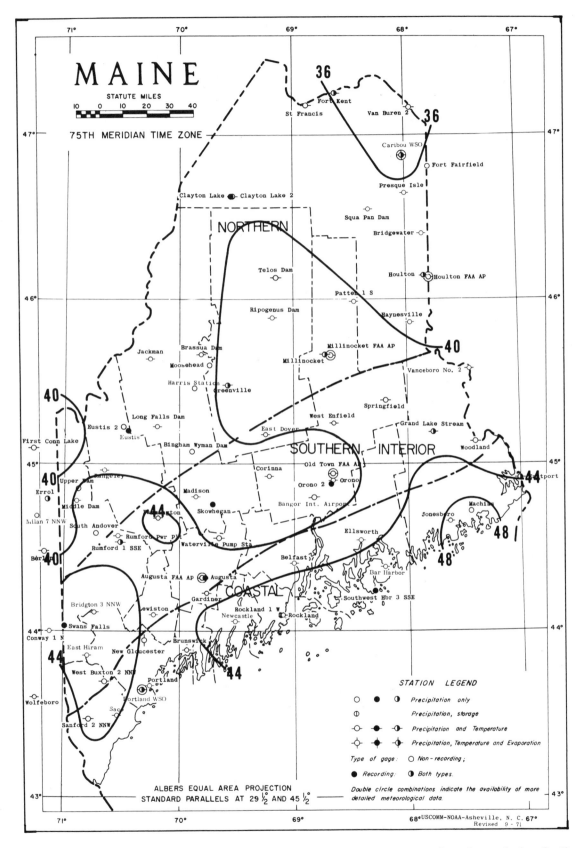

MAINE

STATUTE MILES

75TH MERIDIAN TIME ZONE

NORTHERN

SOUTHERN   INTERIOR

COASTAL

St Francis
Fort Kent
Van Buren 2
Caribou WSO
Fort Fairfield
Presque Isle
Clayton Lake   Clayton Lake 2
Squa Pan Dam
Bridgewater
Telos Dam
Houlton   Houlton FAA AP
Patten 1 S
Ripogenus Dam
Haynesville
Brassua Dam
Millinocket FAA AP
Jackman
Millinocket
Moosehead
Vanceboro No. 2
Harris Station
Greenville
Springfield
Long Falls Dam
West Enfield
Grand Lake Stream
Eustis 2
East Dover
First Conn Lake
Eustis
Bingham   Wyman Dam
Woodland
Upper Dam
Corinna
Old Town FAA AP
Rangeley
Errol
Madison
Orono 2   Orono
Eastport
Middle Dam
Skowhegan
Machias
Milan 7 NNW
Farmington
Bangor Int. Airport
Jonesboro
South Andover
Rumford Pwr Pl
Rumford 1 SSE
Waterville Pump Sta
Ellsworth
Berlin
Belfast
Bar Harbor
Augusta FAA AP   Augusta
Southwest Hbr 3 SSE
Bridgton 3 NNW
Gardiner
Lewiston
Rockland 1 W
Newcastle
Rockland
Swans Falls
Conway 1 N
Brunswick
East Hiram
New Gloucester
West Buxton 2 NNE
Portland
Wolfeboro
Portland WSO
Saco
Sanford 2 NNW

STATION LEGEND

○ ● ◑ *Precipitation only*

⊕ *Precipitation, storage*

-○- -●- -◑- *Precipitation and Temperature*

-◇- -◆- -◈- *Precipitation, Temperature and Evaporation*

Type of gage: ○ *Non-recording;*

● *Recording:* ◑ *Both types.*

Double circle combinations indicate the availability of more detailed meteorological data.

ALBERS EQUAL AREA PROJECTION
STANDARD PARALLELS AT 29½ AND 45½

USCOMM–NOAA–Asheville, N. C.
Revised 9 - 71

Data are based on the period 1931-55. Isolines are drawn through points of approximately equal value. Caution should be used in interpolating on these maps, particularly in mountainous areas.

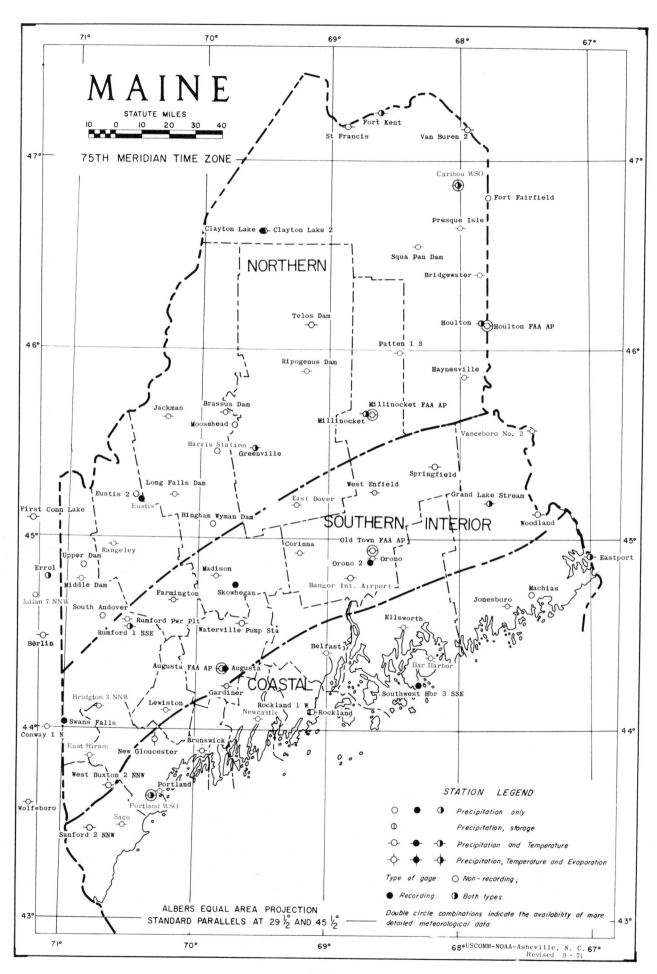

MAINE

STATUTE MILES

10   0   10   20   30   40

75TH MERIDIAN TIME ZONE

Fort Kent

St Francis

Van Buren 2

Caribou WSO

Fort Fairfield

Presque Isle

Clayton Lake   Clayton Lake 2

NORTHERN

Squa Pan Dam

Bridgewater

Telos Dam

Houlton   Houlton FAA AP

Patten 1 S

Ripogenus Dam

Haynesville

Jackman

Brassua Dam

Moosehead

Millinocket FAA AP

Millinocket

Vanceboro No. 2

Harris Station

Greenville

Springfield

West Enfield

Grand Lake Stream

Long Falls Dam

Eustis 2

Woodland

Eustis

First Conn Lake

Bingham   Wyman Dam

East Dover

SOUTHERN   INTERIOR

45°

Corinna

Old Town FAA AP

Eastport

Rangeley

Orono 2   Orono

Upper Dam

Madison

Errol

Machias

Middle Dam

Bangor Int. Airport

Jonesboro

Milan 7 NNW

Farmington

Skowhegan

South Andover

Ellsworth

Rumford Pwr Plt

Belfast

Berlin

Rumford 1 SSE

Waterville Pump Sta

Bar Harbor

Augusta FAA AP   Augusta

COASTAL

Southwest Hbr 3 SSE

Gardiner

Bridgton 3 NNW

Lewiston

Rockland 1 W

Newcastle

Rockland

Swans Falls

Conway 1 N

Brunswick

East Hiram

New Gloucester

West Buxton 2 NNW

Portland

Wolfeboro

Portland WSO

Saco

Sanford 2 NNW

STATION LEGEND

○ ● ◑   *Precipitation only*

⊕   *Precipitation, storage*

◌○ ◑◑ ◑   *Precipitation and Temperature*

◈ ◆ ◑   *Precipitation, Temperature and Evaporation*

Type of gage: ○ *Non-recording;*

● *Recording;*   ◑ *Both types.*

Double circle combinations indicate the availability of more detailed meteorological data.

ALBERS EQUAL AREA PROJECTION
STANDARD PARALLELS AT 29 ½ AND 45 ½

USCOMM-NOAA-Asheville, N. C.
Revised 9 - 71

THE CLIMATE OF
MARYLAND

by
W.J. Moyer

May 1968

GEOGRAPHY

The State of Maryland is on the east coast of the United States and lies in an east-west position between longitudes 75° and 79° W., spanning a distance of 240 miles. The latitude varies from about 38° to nearly 40° N., with a latitudinal width of approximately 125 miles in eastern portions which gradually narrows to about 1 1/2 miles in the Appalachian Mountain region near Hancock and increases again to 35 miles at the extreme western boundary.

The total area of the State is 12,303 square miles, of which 9,887 square miles are land, 2,310 square miles are in the Chesapeake Bay and its tidal river waters, and 106 square miles are in Chincoteague Bay.

The Chesapeake Bay, elongated in a northerly direction, extends for about two-thirds of its length deep into Maryland. It virtually separates the State into two provinces except for a narrow neck of land about 10 miles wide in Cecil County, which bridges the gap between Chesapeake Bay and the State of Pennsylvania.

That portion of the State of Maryland east of Chesapeake Bay is commonly referred to as the Eastern Shore. The five southernmost counties between the Potomac River and Chesapeake Bay are commonly referred to as Southern Maryland. To the north and northwest of Southern Maryland, an area made up of six counties and located on the Piedmont, is an area commonly referred to as Northern-Central Maryland. The remainder of the State including roughly the Appalachian Mountain area or the three western counties is termed Western Maryland.

PHYSIOGRAPHY

Although Maryland ranks as one of the smaller States with respect to size, it encompasses an extremely wide range of physiographic features which contribute to a comparatively wide range of climatic conditions. It extends across three well-defined physiographic belts which parallel the Atlantic coast in varying widths from New England to the southeastern United States. These physiographic provinces are the Coastal Plain, Piedmont province, and Appalachian province.

The land rises more or less gradually from the Atlantic Ocean across the Coastal Plain (which virtually includes the Eastern Shore and Southern Maryland) and then more rapidly across the Piedmont Plateau (northern-central Maryland) and the ridges of the Appalachian Mountains and finally reaches its highest point at 3,340 feet mean sea level on Backbone Mountain in the Allegheny Plateau of Garrett County.

CLIMATIC CONTROLS

There are a number of factors which control the climates of Maryland. The most important factors include (1) the distribution of land and water masses, (2) mountain barriers, (3) topographic features, (4) semipermanent pressure centers, (5) prevailing winds at the surface and at upper levels, (6) storm tracks, including tropical and extra-tropical cyclones, (7) latitude, (8) altitude, and (9) ocean currents.

Distribution of Land and Water Masses: Since the general flow of the atmosphere in temperate latitudes is from west to east, the expansive North American Continent immediately to the west predisposes the Maryland area to a continental type of climate. This type of climate in middle latitudes is marked by well-defined seasons. Winter is the dormant season for plant growth based on low temperatures rather than drought. In spring and fall the changeableness of the weather is a striking feature. It is occasioned by a rapid succession of warm and cold fronts associated with cyclones and anticyclones which generally move from a westerly direction. Summers are warm to hot. The higher atmospheric humidity along the Atlantic coastal area causes the summer heat to be more oppressive and the winter cold more penetrating than for drier climates of the interior of the continent.

Mountain Barriers and Topographic Features: At times in winter the Appalachian Mountains afford a degree of protection from the icy blasts of cold arctic air, particularly when a high pressure area attended by a cold wave approaches from the west. The modifying influences of the mountain barrier attending the passage of a storm area from the Ohio Valley is sometimes quite marked. The warming of the air as it descends the eastern slopes of the mountains may at times exceed 10° F.

In rare cases, however, the mountains may tend to cause lower temperatures in the Maryland area. With a High over New England and a Low over the Ohio Valley, cold winds may travel southwestward and are held east of the mountains by their elevation. Consequently, cold northeast winds are forced over Maryland and farther south. Any flow of this cold air up the eastern slopes of the mountain barrier would tend to lower air temperatures even more in Western Maryland mountain regions.

In the Piedmont the undulating surface configurations and slopes makes cold air drainage an important consideration for crop growth. The cold air layer next to the ground becomes denser and flows from the ridges and higher elevations into the valleys and lower elevations. In the spring and fall it is not uncommon to find lower areas clearly frosted, while higher elevations are not affected. Sometimes, on clear, calm nights, the bottom portion of a tree or bush may be nipped while the upper portion is unaffected.

The Allegheny Mountains contribute to the higher precipitation and heavy snowfall on the Allegheny Plateau. The formation of precipitation in the form of rain or snow is increased in storms or air masses which ascend the mountains from the Ohio Valley. At times the ascent of air masses up the slopes of the mountain barrier is the "trigger action" required to induce precipitation which falls on the Allegheny Plateau. Descending air on the leeward slopes is warmed with the effect of dissipating the clouds and forming of a "rain shadow" to the east of the mountains.

Prevailing Winds at the Surface and at Upper Levels: The prevailing winds at the Surface are determined by the frequency and intensity of anticyclones and cyclones which persist or move over the area. The preponderance of anticyclonic circulation over the northern portion of the continent in winter and the migration to the southeast of anticyclones brings a high percentage of cold northwesterly winds to the Maryland area. Consequently, from October through June the prevailing winds are from the northwesterly quadrant.

This pattern changes in summer as the semipermanent Atlantic High moves northward and westward and dominates the circulation of air over the eastern United States. At this time a flow of warm, moist air spreads over the area with winds from the southwesterly quadrant most of the time. In summer the northern portion of the continent is dominated by low pressure, and the mean storm tracks are displaced far to the north of Maryland.

Mean wind speeds at the surface vary from 9 to 10 m.p.h. in summer and fall to 10 to 12 m.p.h. in winter and early spring. The highest mean speeds are associated with the frequent passages of well-developed cyclones and anticyclones, which bring strong winds and changeable weather of early spring.

Maryland lies just south of the mean position of the upper westerlies in winter and well to the south of the axis of the zonal westerlies in summer. The movement of cyclones and anticyclones over the Maryland area as in other regions is influenced to a large extent by the speed and direction of the upper level winds, which flow around the hemisphere in a wavelike pattern.

In the Maryland area any arrangement of cyclones and anticyclones which would cause a drift of air from the continent to the ocean tends to bring fair weather, while any arrangement which would cause a drift of air from the ocean to the continent would result in increasing cloudiness and a tendency for rain. Therefore, for Maryland, winds from northeast to south tend to be rain producing winds and southwest to north winds are usually associated with dry, fair weather.

A well-developed high pressure system

over New England or the St. Lawrence Valley and a well-developed low pressure system over Georgia, Tennessee, or the Ohio Valley is the most favorable situation for rain in Maryland, while the reverse would certainly produce clear, dry weather.

Storm Tracks: Nearly all migrating cyclones and anticyclones crossing the United States travel from west to east. By far the greater number of cyclones travel in a northeastward direction in a path about 300 to 500 miles north of Maryland, but their influence extends southward to the Atlantic coast and does affect Maryland. Storms which originate in the Gulf of Mexico, the southeastern United States, or adjacent Atlantic coastal regions, frequently move northeastward or northward along the Atlantic coast and bring violent, destructive weather to the Maryland region. As these storms approach the Maryland area from the south, strong easterly to northeasterly winds bring widespread rains and cause high tides on the Atlantic coast and the west side of the Chesapeake Bay. This type of storm is commonly termed a "northeaster". Tropical cyclones or hurricanes which develop in the West Indies, the Caribbean, or the Gulf of Mexico sometimes move into, but rarely pass entirely over the State. These systems cause cloudy weather and heavy rains.

Latitude: The mean temperature difference of 4° to 5° F. between northern and southern latitudes of eastern Maryland in winter and 1° to 2° F. in summer is largely, but not entirely, due to the variation in solar radiational heating. The increase in the annual variation of solar radiation at higher latitudes causes a greater yearly range in mean temperature, 42° F. in the northern portions of the State as compared to 38° F. in the southern portions.

Altitude: Elevations that range from sea level in eastern portions to over 3,000 feet on the Allegheny Plateau have a significant effect on temperature conditions in Maryland. In general, the topography has the effect of reducing the temperature about 1° F. per 300 feet, particularly in summer. In winter, the mean temperature decrease with altitude averages slightly less than 1° F. per 300 feet. In the winter season the effect of elevation is sometimes a critical factor in determining whether the precipitation will fall in the form of rain or snow. Even in Baltimore the elevation difference of 300 to 500 feet sometimes contributes to precipitation in the form of rain in lower portions of the city and heavy snowfall in the higher districts, due to a slight decrease in temperature with elevation.

Ocean Influence: Being in an east coast situation with prevailing westerly winds, the Gulf Stream and the Atlantic Ocean are only moderately effective in influencing the temperatures of Maryland. Nevertheless, the relatively frequent easterly winds associated with cyclonic storms to the southeast bring about movement of air off the mild waters and, consequently, tend to raise the normal winter temperatures and to lower the summer temperatures. Mean winter temperatures of the Coastal Plain and Piedmont Plateau sections of Maryland are approximately 5° F. higher than for regions of the continental interior at the same latitude.

Within the State the effect of the Gulf Stream and the Atlantic Ocean is to lower mean temperatures along the Atlantic coastal area by about 1° to 2° F. in summer as compared to temperatures at the same latitude in the center of the Delmar Peninsula. More specifically, average daily maximum temperatures are 3° to 4° F. lower along the Atlantic coast and average daily minimum temperatures are 1° to 2° F. higher than points in the central portion of the Delmar Peninsula at the same latitude. This depression of the mean temperature in the coastal area in summer is due largely to a sea breeze, which develops as a result of pronounced surface heating of the interior portions of the peninsula. On some hot, summer days the maximum temperatures on the coast may be as much as 10° to 15° F. lower than at points inland. It is this climatic feature that makes Ocean City and adjacent points a popular resort and vacation area in summer.

In winter there is virtually no difference between mean temperatures at coastal and inland points at the same latitude on the Delmar Peninsula.

TEMPERATURE

The mean annual temperature ranges from 48° F. in the Garrett County area to 58° F. in the lower Chesapeake Bay area. The maps of January and July mean monthly maximum and minimum temperatures illustrate the general increase of temperature from west to east across the State.

The winter climate on the Piedmont and Coastal plain sections of Maryland is intermediate between the cold of the Northeast and the mild weather of the South. Extremely cold air masses from the interior of the continent are moderated somewhat by passage over the Appalachian Mountains and in some instances by a short trajectory over the nearby ocean and bays. Weather on the Allegheny Plateau is frequently 10° to 15° F. colder than it is in eastern portions of the State and, at times, extremely low temperatures occur in winter.

The average frost penetration ranges from about 5 inches or less in extreme southern portions of Maryland to more than 18 inches on the Allegheny Plateau. In extreme cold winters maximum frost penetration may be double the average depth.

Summer is characterized by considerable warm weather including at least several hot, humid periods. However, nights are usually quite comfortable. The average length of the freeze-free season based on a minimum temperature higher than 32° F. ranges from more than 225 days in extreme southern portions to fewer than 130 days on the Allegheny Plateau in Garrett County.

The lowest temperature ever recorded in Maryland, -40° F., occurred at Oakland in Garrett County on January 13, 1912, in one of the coldest months ever known in the New England and Middle Atlantic States. This is also the lowest temperature ever recorded south of the Mason Dixon Line.

The highest temperature ever recorded, 109° F., occurred at Boettcherville, Allegany County, on July 3, 1898, Keedysville and Cumberland on August 6, 1918, Cumberland on August 7, 1918, Cumberland and Frederick on July 10, 1936. Extremely high temperatures in Maryland are generally associated with dry weather and sometimes with the occurrence of droughts which reach serious proportions in some parts of the State.

On the average, temperatures of 90° F. or higher occur on 15 to 20 days per year along the Atlantic Coast and from 15 to 25 days along the shores of Chesapeake Bay. Elsewhere, the range is from a maximum of 30 to 40 days in the central portions to a minimum of 2 to 10 days on the Allegheny Plateau. There is a tendency for 35 or more days with 90° F. or higher to occur in the lower elevations around Cumberland due to the heating effect from air masses flowing downslope from the Allegheny Plateau.

The average number of days per year with minimum temperature of 32° F. or lower ranges from a maximum of 160 on the Allegheny Plateau to a minimum of near 60 along the shores of the southern Chesapeake Bay area.

PRECIPITATION

Based on the period, 1931-1960, the average annual precipitation ranges from as much as 49 inches at places in the Allegheny Plateau and southern Eastern Shore area, at extreme ends of the State, to as little as 36 inches in the Cumberland area located in the "rain shadow" to the east of the Allegheny Plateau. Elsewhere over the State, the annual precipitation generally ranges between 40 and 46 inches. Distribution is quite uniform throughout the year, averaging between 2 and 4 inches each month except for a late spring and summer maximum of 4 to 5 1/2 inches.

Although the heaviest precipitation occurs in the summer, this is the season when severe droughts are most frequent. Summer precipitation is less dependable and more variable than in winter. Although rainfall amounts are generally sufficient to grow good crops, the unequal distribution of summer showers and occasional dry periods at critical stages in crop development make irrigation necessary for maximum crop yields in some years.

The seasonal increase in use of water by plants and evaporation (evapotranspiration) during the summer together with the occurrence of a dry period results in a rapid loss of soil moisture and contributes to the development of drought conditions.

The greatest rainfall on record for Maryland for a 24-hour period is 14.75 inches which fell on July 26-27, 1897, at Jewell in the extreme southern portion of Anne Arundel County.

SNOWFALL

Average annual snowfall over Maryland ranges from a minimum of 8 to 10 inches along the coastal areas of the Southern Eastern Shore division to a maximum well over 70 in the Garrett County area. Actually, there is a variation in the annual average from about 30 inches to over 100 inches in this county. During the 1960-1961 winter season, Bittinger reported a total snowfall of 149.4 inches.

Snow flurries fall as early as September on the Allegheny Plateau, and in October in extreme eastern portions of the State. The last snowfall in eastern portions usually occurs in April and on the Allegheny Plateau in May. Even in the warmest winters snow falls in Maryland; however, averages for a climatological division may be less than 1 inch for the season.

Late season snowfalls in March and April are sometimes quite heavy. On April 3, 1915, a late snowstorm dropped snowfall amounts up to 15 inches on the Delmar Peninsula at Sudlersville and Dover. At Salisbury a total of 10 inches fell in this storm.

The most devastating snowstorm to hit the Maryland area occurred on March 19-20, 1958. This storm took a heavy toll of powerlines and overhead wires and caused heavy damage to trees and shrubs, due to the heavy, wet condition of the snow.

FLOODS

All of the State lies in the Atlantic drainage except for a portion of Garrett County in the western end of Maryland which drains into the Ohio Basin. The largest river in the State, the Potomac, forms the southern boundary through most of its length. The far eastern area is drained by many small streams and tidal estuaries into Chesapeake Bay and the Atlantic Ocean.

Minor, or local flood damage, can be expected every year in streams above the tidewater areas. Floods do occur in all months of the year, but the greatest frequency is in late winter and spring. Snow-melt at times is a factor. Intense convectional storms in summer occasionally cause local flash floods. Storms of tropical origin passing through the area in late summer and fall produce high water and occasionally damaging floods, mostly in tidewater areas. These are due to the heavy rains or strong easterly winds accompanying the storm, or a combination of both. Flooding from wind-driven tides at times extends upstream in the Potomac to the District of Columbia area. High water also results from persistent northeast winds along the coast caused by extra-tropical storms.

Major floods in the Potomac River are relatively infrequent, occurring on an average of about once in 5 to 10 years. Notable flood years include 1870, 1877, 1889, 1902, 1924, 1929, 1932, 1936, 1937, 1942 and 1955. High water in tidal reaches in recent years associated with hurricanes occurred in 1933 and 1954.

THUNDERSTORMS

Thunderstorms occur at a given station on an average of 30 days per year in extreme eastern portions of Maryland and 40 days per year in western portions. They occur in all months of the year, but during the 4-month cold season from November through February an average of less than one storm per month is observed. An average of one thunderstorm per year occurs in each of the months, March and October. May, June, July, and August make up the thunderstorm season and include from 75 to 80 percent of the thunderstorms which occur annually. July is the peak of the season with about 25 percent of the annual total number of thunderstorms.

As few as 10 and as many as 50 thunderstorms have been observed in a given locality during the year. At Baltimore as few as 1 and as many as 16 days have been recorded with thunderstorms in July.

HAIL

Hail at a given station occurs on an average of 1 day per year in extreme eastern portions and about 2 days per year in extreme western portions. The total number of days on which hail is observed at one or more stations in Maryland averages about 18 to 20 per year. Hail has been observed in all months of the year; however, occurrences in the 7-month period from September through March are infrequent. The number of days with hail at one or more stations increases from an average of 1 in April to about 5 in July, the peak of the hail season, and then decreases to an average of 3 in August.

Although spring thunderstorms are much fewer in number than summer thunderstorms, they have a much greater tendency to occur with hail. Virtually all hailstorms occur between 2 p.m. and 9 p.m. Severe, devastating hailstorms occur somewhere in the State about once every 5 years on the average. Hail risk to crops in Maryland is rather low compared to other parts of the country, except in Allegany County where hail occasionally damages apple crops.

TORNADOES

Tornadoes occur infrequently in Maryland, and of the ones that do occur most are small and result in nominal losses. Eighty-four tornadoes were reported in Maryland during the 52-year period, 1916-1967, an average between 1 and 2 tornadoes per year. Thirty-one deaths and damages estimated at about 2.5 million dollars resulted from these tornadoes. About 20 percent of the tornadoes occur on the Eastern Shore, 25 percent in southern Maryland, 40 percent in north-central Maryland, and 15 percent in western Maryland. Approximately 70 percent of the tornadoes occur between 2 p.m. and 9 p.m. with a preponderance from 3 p.m. to 6 p.m.

Most tornadoes in Maryland tend to travel in the usual southwest to northeast direction, but few have been reported to travel southeastward or in a southerly direction. Usually paths are not more than a few miles in length; however, 10 to 15 percent of these storms maintain paths 20 miles or more in length.

RELATIVE HUMIDITY

Average relative humidity is lowest in the winter and early spring from February through April and highest in the late summer and early fall from August through October. At Baltimore, the relative humidity averages about 60 percent in February, March, and April and about 75 percent in August, September, and October.

CLIMATE AND HUMAN ACTIVITIES

The climate of Maryland is a dependable natural resource which is favorable for a wide diversification of agricultural, industrial, and other human activities.

Precipitation, the most important climatic resource, contributes to a water supply which is more than adequate for the present and forseeable future needs. This is a vital factor since the population increase, growth of industry, and expansion of farm irrigation systems are all added drains on this indispensable natural commodity. The main problem is the distribution of water in significant quantities at the time and place of demand. In the Coastal Plain the ground water is sufficient for virtually all needs; however, in the Piedmont and Appalachian Mountain areas the main water supplies are in the surface water resources (streams and lakes) and must be stored in reservoirs to maintain sustained yields during infrequent dry periods.

Another favorable climatic factor is the temperature of the Atlantic coast and Chesapeake Bay which contributes to a large number of summer resorts and recreational activities together with the multitude of pleasure craft that operate in these areas. Rest and recreation are both favored by the summer afternoon temperatures which are generally cooler on the shore and over the water than they are inland.

BIBLIOGRAPHY

Ashbaugh, Byron L. and Brancato, G. N., Maryland's Weather. Educational Series No. 38, issued in cooperation with the United States Weather Bureau. State of Maryland Department of Research and Education, Solomons, Maryland, June 1954.

Flora, Snowden D., Hailstorms of the United States. University of Oklahoma Press, Norman, Oklahoma, 1956.

Flora, Snowden D., Tornadoes of the United States. Revised edition, University of Oklahoma Press, Norman, Oklahoma, 1954.

Hoyt, William G. and Langbein, Walter B., Floods. Princeton University Press, Princeton, New Jersey, 1955.

Maryland Weather Service, Physiography and Meteorology of the State. Volume I, 1899 and Volume II, 1907, Johns Hopkins University Press, Baltimore, Maryland.

U. S. Department of Commerce, Climatic Guide for Baltimore. Climatography of the United States, Number 40-18. U. S. Weather Bureau, Washington, D. C., May 1956.

——————————, Climatic Maps of the United States. Environmental Science Services Administration, Environmental Data Service, Washington, D. C.

——————————, Climatic Summary of the United States. Bulletin W, Sections 91 and 92; 1930 edition. Climatography of the United States, Series 10, U. S. Weather Bureau, Washington, D. C.

——————————, Climatic Summary of the United States. Bulletin W Supplement, 1931-1952, Maryland. Climatography of the United States, Series 11-15, U. S. Weather Bureau, Washington, D. C.

——————————, Climatic Summary of the United States. Bulletin W Supplement, 1951-1960, Maryland. Climatography of the United States, Series 86-15, U. S. Weather Bureau, Washington, D. C.

——————————, Climatological Data by Sections, Maryland. (Monthly and Annual). Environmental Science Services Administration, Environmental Data Service, Washington, D. C.

——————————, Climatological Data, National Summary. (Monthly and Annual). Environmental Science Services Administration, Environmental Data Service, Washington, D. C.

——————————, Hourly Precipitation Data, Maryland. (Monthly and Annual). Environmental Science Services Administration, Environmental Data Service, Washington, D. C.

——————————, Local Climatological Data, Baltimore and Frederick, Maryland. (Monthly and Annual). Environmental Science Services Administration, Environmental Data Service, Washington, D. C. (Frederick publication discontinued 1963).

——————————, Rainfall Frequency Atlas of the United States, U. S. Weather Bureau Technical Paper No. 40, Washington, D. C., 1961.

——————————, Storm Data. (Monthly). Environmental Science Services Administration, Environmental Data Service, Washington, D. C.

——————————, Summary of Hourly Observations, Baltimore, Maryland. Climatography of the United States, Series 3D-18. Environmental Science Services Administration, Environmental Data Service, Washington, D. C.

U. S. Geological Survey and U. S. Corps of Engineers, Topographic Charts for Maryland and Delaware.

Vokes, Harold E., Geography and Geology of Maryland. Bulletin 19, Department of Geology, Mines and Water Resources, Board of Natural Resources, State of Maryland, Baltimore, 1957, 243 pp.

STATION	Freeze threshold temperature	Mean date of last Spring occurrence	Mean date of first Fall occurrence	Mean No. of days between dates	Years of record Spring	No. of occurrences in Spring	Years of record Fall	No. of occurrences in Fall
ABERDEEN PHILLIPS FLD	32	04-15	10-26	194	27	27	27	27
	28	03-26	11-10	229	27	27	27	27
	24	03-12	11-25	258	27	24	27	22
	20	02-24	12-13	292	27	22	26	14
	16	02-07	12-24	320	27	17	26	7
ANNAPOLIS USN ACADEMY	32	03-30	11-19	234	30	30	30	27
	28	03-13	12-05	267	30	27	30	21
	24	02-20	12-15	298	30	22	30	15
	20	02-05	12-21	319	30	18	30	13
	16	01-20	12-28	342	30	12	30	5
BALTIMORE WB AIRPORT	32	04-11	10-28	200	10	10	10	10
	28	03-26	11-09	228	10	10	10	10
	24	03-19	11-23	249	10	10	10	10
	20	02-25	11-28	276	10	10	10	10
	16	02-11	12-12	304	10	9	10	9
BALTIMORE WB CITY	32	03-26	11-19	238	30	30	30	30
	28	03-16	11-27	256	30	30	30	30
	24	03-05	12-07	277	30	30	30	29
	20	02-22	12-14	295	30	29	30	25
	16	02-07	12-23	319	30	24	30	16
BALTIMORE HAMILTON	32	04-19	11-02	197	10	10	10	10
	28	03-25	11-09	229	10	10	9	9
	24	03-17	11-29	257	10	10	9	9
	20	03-03	12-04	276	10	10	9	9
	16	02-13	12-14	304	10	10	9	8
BELTSVILLE	32	04-27	10-12	168	10	10	10	10
	28	04-14	10-23	192	10	10	10	10
	24	03-31	11-06	220	10	10	10	10
	20	03-17	11-18	246	10	10	10	10
	16	02-20	11-28	281	10	10	10	10
BELTSVILLE PLANT STA 1	32	04-24	10-15	174	10	10	10	10
	28	04-12	10-28	199	10	10	10	10
	24	04-06	11-10	218	10	10	10	10
	20	03-16	11-16	245	10	10	10	10
	16	02-27	11-30	276	10	10	10	10
BELTSVILLE PLANT STA 2	32	04-27	10-10	166	10	10	10	10
	28	04-16	10-25	192	10	10	10	10
	24	04-08	11-06	212	10	10	10	10
	20	03-23	11-14	236	10	10	10	10
	16	03-02	11-28	271	10	10	10	10
BELTSVILLE PLANT STA 4	32	04-21	10-23	185	10	10	10	10
	28	04-01	11-03	216	10	10	10	10
	24	03-22	11-18	241	10	10	10	10
	20	03-08	11-29	266	10	10	10	10
	16	02-14	12-08	297	10	10	10	10
BELTSVILLE PLANT STA 5	32	04-27	10-05	161	10	10	10	10
	28	04-20	10-22	185	10	10	10	10
	24	04-08	11-06	212	10	10	10	10
	20	03-26	11-13	232	10	10	10	10
	16	03-02	11-28	271	10	10	10	10
BELTSVILLE PLANT STA 6	32	04-18	10-23	188	10	10	10	10
	28	04-03	11-02	213	10	10	10	10
	24	03-22	11-13	236	10	10	10	10
	20	03-08	11-19	256	10	10	10	10
	16	02-14	12-08	297	10	10	10	10
BENSON POLICE BARRACKS	32	04-14	10-15	184	10	10	10	10
	28	04-07	11-03	210	10	9	10	10
	24	03-23	11-11	233	9	9	10	10
	20	03-11	11-24	258	9	9	10	10
	16	02-23	12-10	290	9	9	10	10
BENTLEY SPRINGS 1 WNW	32	04-28	10-14	169	29	29	29	29
	28	04-15	10-28	196	29	29	29	29
	24	03-29	11-12	228	28	28	29	29
	20	03-20	11-22	247	28	28	29	29
	16	03-09	12-03	269	28	28	29	28
BLACKWATER REFUGE	32	04-11	10-27	199	10	10	10	10
	28	03-28	11-13	230	10	10	10	10
	24	03-09	11-25	261	10	10	10	10
	20	02-17	12-08	294	10	10	10	8
	16	02-07	12-17	313	10	9	10	6
CAMBRIDGE 4 W	32	04-14	10-28	197	28	28	27	27
	28	03-24	11-10	231	28	28	27	26
	24	03-10	12-04	269	28	26	27	20
	20	02-09	12-17	311	28	19	26	14
	16	02-01	12-21	323	28	16	26	12
CHARLOTTE HALL 2 ESE	32	04-15	10-24	192	22	22	22	22
	28	03-31	11-07	221	20	20	22	22
	24	03-23	11-23	252	20	20	22	22
	20	03-04	12-01	272	21	21	21	21
	16	02-14	12-14	303	21	19	20	16
CHELTENHAM 1 NW	32	04-22	10-19	180	26	26	26	26
	28	04-09	11-02	207	26	26	25	25
	24	03-15	11-18	248	26	23	25	24
	20	02-24	12-06	285	26	21	25	16
	16	02-09	12-20	314	26	17	25	10
CHESTERTOWN	32	04-10	11-02	206	24	24	25	25
	28	03-25	11-15	235	24	24	25	24
	24	03-09	12-02	268	24	21	25	21
	20	02-16	12-19	306	24	19	25	12
	16	02-02	12-24	325	24	15	25	8
CHEWSVILLE BRIDGEPORT	32	04-28	10-09	164	30	30	30	30
	28	04-16	10-20	187	30	30	30	30
	24	04-01	11-05	218	30	30	30	29
	20	03-10	11-26	261	30	27	30	26
	16	02-21	12-08	290	30	24	30	19
CLEAR SPRING 1 ENE	32	04-23	10-16	176	24	24	16	16
	28	04-06	11-01	209	24	24	16	16
	24	03-21	11-11	235	24	23	16	15
	20	03-03	11-29	271	24	20	16	14
	16	02-14	12-11	300	24	17	16	12
COLEMAN 3 WNW	32	04-04	11-07	217	29	29	30	30
	28	03-23	11-26	248	28	28	28	25
	24	03-08	12-10	277	28	24	29	18
	20	02-18	12-20	305	27	21	29	14
	16	01-30	12-25	329	26	13	29	8
COLLEGE PARK	32	04-26	10-17	174	28	28	29	29
	28	04-14	10-29	198	28	28	29	29
	24	03-22	11-15	238	28	27	29	27
	20	03-02	11-29	272	29	25	28	21
	16	02-15	12-16	304	29	21	28	14
CONOWINGO DAM	32	04-16	10-20	187	25	25	25	25
	28	04-02	11-08	220	25	25	25	25
	24	03-21	11-17	241	25	25	25	25
	20	03-09	12-02	268	25	25	25	25
	16	02-27	12-14	290	25	25	25	23
CONOWINGO POLICE BRKS	32	04-15	10-14	182	10	10	10	10
	28	04-04	10-30	209	10	10	10	10
	24	03-21	11-14	238	10	10	9	9
	20	03-14	12-01	262	10	10	10	10
	16	02-22	12-09	290	10	10	10	10
CRISFIELD HAMMOCK PT	32	03-27	11-19	237	28	28	30	29
	28	03-06	12-05	274	27	24	30	21
	24	02-14	12-17	306	27	20	30	14
	20	01-25	12-27	336	27	13	30	6
	16	01-10	⊕	⊕	27	7	30	3
CUMBERLAND	32	04-24	10-11	170	30	30	30	30
	28	04-12	10-28	199	30	30	30	30
	24	03-21	11-10	234	30	30	28	30
	20	03-04	12-01	272	30	25	30	23
	16	02-18	12-14	299	30	23	30	16

FREEZE DATA

STATION	Freeze threshold temperature	Mean date of last Spring occurrence	Mean date of first Fall occurrence	Mean No. of days between dates	Years of record Spring	No. of occurrences in Spring	Years of record Fall	No. of occurrences in Fall	STATION	Freeze threshold temperature	Mean date of last Spring occurrence	Mean date of first Fall occurrence	Mean No. of days between dates	Years of record Spring	No. of occurrences in Spring	Years of record Fall	No. of occurrences in Fall
DALECARLIA RESVR D C	32	04-13	10-28	198	10	10	10	10	GLENN DALE BELL STA	32	04-29	10-11	165	30	30	30	30
	28	04-01	11-06	219	10	10	10	10		28	04-21	10-22	184	30	30	30	30
	24	03-15	11-17	247	10	10	10	10		24	04-04	11-04	214	30	30	30	30
	20	02-24	11-29	278	10	10	10	10		20	03-23	11-16	238	30	30	30	30
	16	02-09	12-11	305	10	9	9	8		16	03-07	11-28	266	30	30	30	30
DENTON 1 WNW	32	04-17	10-23	189	28	28	29	29	GREENBELT	32	04-20	10-13	176	10	10	10	10
	28	03-30	11-07	222	27	27	29	29		28	04-12	11-02	204	10	10	10	10
	24	03-17	11-20	248	27	27	29	29		24	03-24	11-13	234	10	10	10	10
	20	03-05	12-05	275	25	25	27	26		20	03-12	11-23	256	10	10	10	10
	16	02-20	12-14	297	25	25	27	24		16	02-17	12-04	290	10	10	10	10
DUNDALK	32	04-02	11-03	215	10	10	10	10	HAGERSTOWN	32	04-20	10-09	172	10	10	10	10
	28	03-21	11-20	244	10	10	10	10		28	04-04	10-25	204	10	10	10	10
	24	03-07	11-28	266	10	10	10	10		24	03-23	11-15	237	10	10	10	10
	20	02-14	12-10	299	10	10	10	9		20	03-10	11-25	260	10	10	10	10
	16	02-04	12-21	320	10	8	10	6		16	02-17	12-05	291	10	10	10	10
EASTON POLICE BARRACKS	32	04-14	10-29	198	30	30	30	30	HANCOCK FRUIT LAB	32	05-06	10-03	150	10	10	10	10
	28	03-29	11-08	224	30	30	30	30		28	04-21	10-09	171	10	10	10	10
	24	03-11	11-29	263	30	27	30	25		24	04-08	10-31	206	10	10	10	10
	20	02-21	12-15	297	30	24	30	17		20	03-30	11-08	223	10	10	10	10
	16	01-25	12-22	331	30	14	30	11		16	03-12	11-22	255	10	10	10	10
ELKTON	32	04-20	10-17	180	30	30	30	30	HUNTINGTOWN	32	04-17	10-27	193	16	16	16	16
	28	04-08	10-30	205	30	30	30	30		28	04-02	11-04	216	16	16	16	16
	24	03-23	11-14	236	30	30	30	30		24	03-20	11-26	251	16	16	16	16
	20	03-12	11-25	258	30	30	30	30		20	03-05	12-05	275	16	16	16	16
	16	02-27	12-09	285	30	30	30	28		16	02-21	12-14	296	16	15	16	12
EMMITSBURG	32	04-19	10-27	191	25	25	25	25	KEEDYSVILLE	32	04-28	10-09	164	29	29	29	29
	28	04-03	11-13	224	25	25	25	23		28	04-19	10-21	185	29	29	29	29
	24	03-13	11-28	260	25	22	25	20		24	04-01	11-05	218	30	30	29	29
	20	03-04	12-16	287	25	21	25	11		20	03-19	11-21	247	30	30	29	29
	16	02-13	12-21	311	24	16	25	10		16	03-06	12-04	273	30	30	29	29
FALLSTON	32	04-20	10-27	190	23	23	22	22	LA PLATA 1 W	32	04-21	10-21	183	29	29	29	29
	28	04-01	11-09	213	23	23	22	21		28	04-07	11-02	209	28	28	29	29
	24	03-18	11-21	248	23	21	22	19		24	03-20	11-15	240	28	28	29	29
	20	03-03	12-14	286	23	19	22	9		20	03-06	11-26	265	29	29	28	28
	16	02-13	12-20	310	23	15	22	7		16	02-20	12-07	290	29	27	28	26
FORT GEORGE G MEADE	32	04-29	10-11	165	10	10	10	10	LAUREL 3 W	32	04-14	10-27	196	26	26	26	26
	28	04-17	10-22	188	10	10	10	10		28	04-04	11-08	218	25	25	26	26
	24	03-30	11-07	222	10	10	10	10		24	03-18	11-22	249	25	25	26	26
	20	03-19	11-15	241	10	10	10	10		20	03-07	11-30	268	25	25	26	26
	16	03-02	12-02	275	10	10	10	10		16	02-22	12-12	293	24	22	25	22
FREDERICK POLICE BRKS	32	04-17	10-15	181	10	10	10	10	LEONARDTOWN 4 SSW	32	04-13	10-31	201	15	15	14	14
	28	04-03	11-06	217	10	10	10	10		28	03-30	11-11	226	15	15	14	14
	24	03-19	11-12	238	10	10	10	10		24	03-20	11-27	252	14	14	14	14
	20	03-11	11-28	262	10	10	9	9		20	03-02	12-10	283	14	14	14	14
	16	02-10	12-08	301	10	9	9	9		16	02-06	12-19	316	14	12	14	10
FREDERICK WFMD	32	04-25	10-14	172	30	30	30	30	MILLINGTON	32	04-18	10-18	183	30	30	30	30
	28	04-15	10-29	197	30	30	30	29		28	04-04	11-03	213	30	30	30	30
	24	03-29	11-12	228	30	30	30	29		24	03-22	11-16	239	30	30	30	30
	20	03-12	11-27	260	30	29	30	25		20	03-12	11-28	261	30	30	30	30
	16	02-22	12-11	292	30	26	30	23		16	02-23	12-10	290	30	30	30	29
FRIENDSVILLE	32	05-18	10-02	137	18	18	17	17	OAKLAND 1 SE	32	05-21	09-23	125	30	30	30	30
	28	04-29	10-14	168	18	18	17	17		28	05-10	10-05	148	30	30	30	30
	24	04-20	10-23	186	18	18	17	17		24	04-25	10-13	171	30	30	30	30
	20	03-31	11-08	222	16	16	16	16		20	04-13	10-25	195	30	30	30	30
	16	03-23	11-24	246	16	16	16	16		16	03-26	11-10	229	30	30	30	30
FROSTBURG	32	05-02	10-06	157	30	30	30	30	OWINGS FERRY LANDING	32	04-14	10-26	195	30	30	30	30
	28	04-21	10-17	179	30	30	30	30		28	03-27	11-10	228	30	30	30	30
	24	04-08	11-05	211	30	30	30	30		24	03-18	11-23	250	30	30	30	30
	20	03-27	11-17	235	30	30	30	30		20	03-01	12-02	276	30	30	30	29
	16	03-14	11-29	260	30	30	30	30		16	02-18	12-13	298	30	27	30	25
GERMANTOWN	32	04-24	10-21	180	21	21	21	21	OXFORD	32	03-25	11-16	236	17	17	16	16
	28	04-09	10-31	205	20	20	21	21		28	03-15	12-01	261	17	17	16	16
	24	03-26	11-14	233	19	19	18	18		24	03-03	12-10	282	17	17	16	16
	20	03-11	11-26	260	18	18	17	17		20	02-15	12-22	310	17	16	16	11
	16	03-02	12-07	280	18	18	17	16		16	01-25	12-29	338	17	11	16	5

STATION	Freeze threshold temperature	Mean date of last Spring occurrence	Mean date of first Fall occurrence	Mean No. of days between dates	Years of record Spring	No. of occurrences in Spring	Years of record Fall	No. of occurrences in Fall
PICARDY	32	04-29	10-08	162	28	28	28	28
	28	04-18	10-18	183	28	28	28	28
	24	03-30	11-08	223	28	28	28	28
	20	03-17	11-22	250	28	28	28	28
	16	03-07	12-01	269	28	28	28	27
PIKESVILLE POLICE BRKS	32	04-16	11-01	199	10	10	10	10
	28	03-24	11-10	231	10	10	10	10
	24	03-18	11-17	244	10	10	10	10
	20	03-10	12-02	267	10	10	10	10
	16	02-16	12-11	298	10	10	10	10
POCOMOKE CITY 1 S	32	04-18	10-23	188	10	10	10	10
	28	04-03	11-05	216	10	10	10	10
	24	03-17	11-20	248	10	10	10	10
	20	02-14	11-28	287	10	10	9	10
	16	02-03	12-11	311	10	8	10	9
PRESTON 1 S	32	04-14	10-22	191	10	10	10	10
	28	03-30	11-03	218	10	10	10	10
	24	03-14	11-25	256	10	10	10	10
	20	02-22	12-08	289	10	10	10	10
	16	02-05	12-23	321	10	8	10	6
PRINCESS ANNE	32	04-23	10-17	177	26	26	26	26
	28	04-07	10-30	206	26	26	26	26
	24	03-24	11-14	235	26	26	26	26
	20	03-08	11-28	265	26	26	26	26
	16	02-14	12-09	298	26	26	25	23
RANDALLSTOWN POL BRKS	32	04-17	10-22	188	10	10	10	10
	28	04-04	11-05	215	10	10	10	10
	24	03-19	11-15	241	10	10	10	10
	20	03-11	11-25	259	10	10	10	10
	16	02-17	12-04	290	10	10	10	10
ROCK HALL	32	04-09	10-23	197	10	10	10	10
	28	03-29	11-07	223	10	10	9	9
	24	03-15	11-21	251	10	10	9	9
	20	02-16	12-07	294	10	10	9	8
	16	02-11	12-15	307	10	10	9	7
ROCKVILLE	32	04-13	10-26	196	10	10	10	10
	28	03-25	11-10	230	10	10	10	10
	24	03-16	11-24	253	10	10	10	10
	20	02-26	12-04	281	10	10	10	10
	16	02-09	12-16	310	10	9	10	8
ROYAL OAK	32	04-01	11-02	215	10	10	10	10
	28	03-22	11-15	238	10	10	10	10
	24	03-07	12-01	269	10	10	10	10
	20	02-12	12-14	305	10	9	10	8
	16	02-04	12-21	320	10	8	10	5
SALISBURY	32	04-17	10-26	192	29	29	29	29
	28	03-29	11-11	227	29	29	29	29
	24	03-15	11-23	253	28	28	29	29
	20	02-27	12-06	282	28	28	29	27
	16	02-12	12-16	307	28	26	28	24
SALISBURY POLICE BRKS	32	04-16	10-18	185	10	10	10	10
	28	04-06	10-31	208	10	10	10	10
	24	03-24	11-16	237	10	10	10	10
	20	03-12	11-27	260	10	10	10	10
	16	02-17	12-11	297	10	10	10	10
SALISBURY FAA AIRPORT	32	04-15	10-26	194	10	10	10	10
	28	03-30	11-07	222	10	10	10	10
	24	03-18	11-20	247	10	10	10	10
	20	02-27	12-05	281	10	10	10	10
	16	02-10	12-14	307	10	9	10	9
SINES DEEP CREEK	32	05-12	10-03	144	10	10	10	10
	28	04-29	10-09	163	10	10	10	10
	24	04-12	10-20	191	10	10	10	10
	20	03-31	11-02	216	10	10	10	10
	16	03-22	11-14	237	10	10	10	10
SNOW HILL	32	04-23	10-24	184	28	28	28	28
	28	04-08	11-07	213	28	28	28	28
	24	03-20	11-19	244	26	26	28	28
	20	03-02	11-30	273	26	25	28	28
	16	02-16	12-11	298	26	25	27	24
SOLOMONS	32	03-30	11-23	238	29	29	26	26
	28	03-15	12-01	261	28	28	27	27
	24	03-02	12-10	283	28	27	27	24
	20	02-16	12-20	307	28	25	27	17
	16	01-25	12-26	335	28	17	27	11
TAKOMA PARK MISS AVE	32	04-14	10-31	200	30	30	30	30
	28	03-28	11-11	228	30	30	30	30
	24	03-15	11-23	253	30	30	29	29
	20	03-03	12-04	276	30	30	29	29
	16	02-16	12-14	301	30	28	29	25
TOWSON	32	04-17	10-14	180	10	10	10	10
	28	04-07	11-03	210	10	10	10	10
	24	03-22	11-09	232	10	10	10	10
	20	03-15	11-25	255	10	10	10	10
	16	02-18	12-08	293	10	10	10	10
UNIONVILLE	32	05-05	10-07	155	10	10	10	10
	28	04-20	10-16	179	10	10	10	10
	24	04-08	11-04	210	10	10	10	10
	20	03-31	11-10	224	10	10	10	10
	16	03-07	11-25	263	10	10	10	10
U. S. SOLDIERS HOME DC	32	04-07	11-04	211	10	10	10	10
	28	03-24	11-13	234	10	10	10	10
	24	03-11	11-28	262	10	10	10	10
	20	02-20	12-05	288	10	10	10	10
	16	02-05	12-16	314	10	8	10	8
VIENNA	32	04-08	10-21	196	10	10	10	10
	28	03-23	11-09	231	10	10	10	10
	24	03-13	11-27	259	10	10	10	10
	20	02-11	12-08	300	10	9	10	10
	16	02-07	12-23	319	10	9	10	6
WALDORF POLICE BRKS	32	04-18	10-17	182	10	10	10	10
	28	04-09	10-25	199	10	10	10	10
	24	03-26	11-10	229	10	10	10	10
	20	03-11	11-19	253	10	10	10	10
	16	02-22	11-30	281	10	10	10	10
WASHINGTON DC WB CITY	32	03-29	11-09	225	10	10	10	10
	28	03-18	11-25	252	10	10	10	10
	24	03-02	12-02	275	10	10	10	10
	20	02-10	12-14	307	10	9	10	9
	16	01-26	12-22	330	10	7	10	5
WATERLOO POLICE BRKS	32	04-19	10-21	185	10	10	10	10
	28	04-04	10-31	210	10	10	10	10
	24	03-26	11-09	228	10	10	10	10
	20	03-11	11-24	258	10	10	10	10
	16	02-19	12-06	290	10	10	10	10
WESTERNPORT	32	04-27	10-10	166	30	30	30	30
	28	04-16	10-25	192	30	30	30	30
	24	03-27	11-10	228	30	30	30	30
	20	03-15	11-26	256	30	30	30	30
	16	03-03	12-06	278	30	29	30	28
WESTMINSTER 2 SSE	32	04-22	10-16	177	30	30	30	30
	28	04-12	10-31	202	30	30	30	30
	24	03-22	11-14	237	29	29	30	30
	20	03-16	11-27	256	29	29	29	29
	16	03-02	12-06	279	29	29	29	29
WOODSTOCK	32	04-28	10-10	165	30	30	30	30
	28	04-17	10-27	193	30	30	30	30
	24	03-30	11-06	221	30	30	30	30
	20	03-18	11-19	246	30	30	30	30
	16	03-02	12-04	277	30	30	30	30

FREEZE DATA

STATION	Freeze threshold temperature	Mean date of last Spring occurrence	Mean date of first Fall occurrence	Mean No. of days between dates	Years of record Spring	No. of occurrences in Spring	Years of record Fall	No. of occurrences in Fall	STATION	Freeze threshold temperature	Mean date of last Spring occurrence	Mean date of first Fall occurrence	Mean No. of days between dates	Years of record Spring	No. of occurrences in Spring	Years of record Fall	No. of occurrences in Fall
DELAWARE									DELAWARE								
BRIDGEVILLE 1 NW	32	04-22	10-20	181	30	30	30	30	MILLSBORO	32	04-23	10-21	181	23	23	21	21
	28	04-06	11-03	211	30	30	30	30		28	04-09	11-06	211	23	23	21	21
	24	03-21	11-16	240	30	30	30	30		24	03-22	11-19	242	23	23	21	21
	20	03-10	12-01	266	30	30	30	30		20	03-11	11-29	263	23	23	22	21
	16	02-21	12-12	294	30	28	30	28		16	02-25	12-09	287	23	23	22	21
DELAWARE CITY REEDY PT	32	04-15	10-26	194	22	22	23	23	NEWARK PUMPING STA	32	04-25	10-16	174	10	10	10	10
	28	03-25	11-12	232	22	22	23	23		28	04-12	10-25	196	10	10	10	10
	24	03-17	11-25	253	22	22	23	23		24	03-27	11-16	234	10	10	10	10
	20	03-04	12-05	276	23	23	23	22		20	03-13	11-26	258	10	10	10	10
	16	02-20	12-16	299	24	22	23	20		16	03-02	12-13	286	10	10	10	10
DOVER	32	04-12	10-29	200	30	30	30	30	NEWARK UNIVERSITY FARM	32	04-20	10-16	179	10	10	10	10
	28	03-26	11-14	233	30	30	30	30		28	04-05	11-05	214	10	10	10	10
	24	03-16	11-28	257	30	30	30	30		24	03-23	11-12	234	10	10	10	10
	20	02-27	12-08	284	30	30	30	29		20	03-11	11-25	259	10	10	10	10
	16	02-16	12-16	303	30	28	30	26		16	02-20	12-07	290	10	10	10	10
GEORGETOWN 5 SW	32	04-21	10-15	177	10	10	10	10	WILMGTON NCASTLE WB AP	32	04-09	11-01	206	10	10	10	10
	28	04-12	11-03	205	10	10	10	10		28	03-26	11-08	227	10	10	10	10
	24	03-24	11-14	235	10	10	10	10		24	03-15	11-17	247	10	10	10	10
	20	03-14	11-28	259	10	10	10	10		20	03-03	12-08	280	10	10	10	10
	16	02-22	12-12	293	10	10	10	9		16	02-08	12-13	308	10	9	10	10
LEWES 1 SW	32	04-14	10-28	197	10	10	10	10	WILMGTON PORTER RESVR	32	04-17	10-28	194	30	30	30	30
	28	03-25	11-09	229	10	10	10	10		28	03-30	11-11	226	30	30	30	30
	24	03-18	11-24	251	10	10	10	10		24	03-18	11-25	252	30	30	30	30
	20	03-01	12-06	280	10	10	10	9		20	03-06	12-03	272	30	30	30	30
	16	02-08	12-16	311	10	9	10	8		16	02-26	12-13	290	30	30	30	27
MILFORD 2 WSW	32	04-17	10-23	189	25	25	26	26									
	28	04-01	11-05	218	25	25	26	26									
	24	03-19	11-18	244	25	25	26	26									
	20	03-08	12-02	269	25	25	26	25									
	16	02-20	12-12	295	25	24	26	24									

Data in the above table are based on the period 1931-1960, or that portion of this period for which data are available.

⊕ When the frequency of occurrence in either spring or fall is one year in ten, or less, mean dates are not given.

Means have been adjusted to take into account years of non-occurrence.

A freeze is a numerical substitute for the former term "killing frost" and is the occurrence of a minimum temperature at or below the threshold temperature of 32°, 28°, etc.

Freeze data tabulations in greater detail are available and can be reproduced at cost.

*NORMALS BY CLIMATOLOGICAL DIVISIONS

Taken from "Climatography of the United States No. 81-4, Decennial Census of U. S. Climate"

TEMPERATURE (°F)   PRECIPITATION (In.)

STATIONS (By Divisions)	Jan.	Feb.	Mar.	Apr.	May	June	July	Aug.	Sept.	Oct.	Nov.	Dec.	Annual	Jan.	Feb.	Mar.	Apr.	May	June	July	Aug.	Sept.	Oct.	Nov.	Dec.	Annual
MARYLAND																										
SOUTHERN EASTERN SHORE																										
CRISFIELD SOMERS COVE	39.3	39.9	46.5	56.9	66.9	75.3	79.4	78.0	72.6	62.2	51.2	41.2	59.1	3.56	3.15	4.01	3.66	3.69	3.31	5.05	5.05	3.83	3.37	3.24	2.92	44.84
SALISBURY	.	.	.	.	.	.	.	.	.	.	.	.	.	3.66	3.21	3.49	3.34	3.62	3.49	4.39	6.01	4.44	3.50	3.21	3.13	46.13
SNOW HILL	38.1	38.7	44.7	54.2	64.1	72.6	76.7	75.2	69.2	58.8	48.4	39.3	56.7	3.94	3.43	4.62	3.61	3.61	3.80	5.12	5.67	4.50	3.91	3.55	3.41	49.17
DIVISION	38.3	38.8	45.0	54.8	64.6	73.1	77.3	75.9	70.0	59.6	49.1	39.6	57.2	3.63	3.19	4.14	3.41	3.52	3.45	4.81	5.43	4.23	3.53	3.31	3.12	45.77
CENTRAL EASTERN SHORE																										
EASTON POLICE BARRACKS	36.6	37.2	44.3	54.6	64.7	73.2	77.3	75.6	69.2	58.3	47.9	37.9	56.4	3.47	2.95	3.95	3.48	3.88	3.36	4.74	5.03	3.95	3.18	3.53	3.13	44.65
DIVISION	36.9	37.4	44.3	54.8	64.8	73.2	77.2	75.7	69.5	58.8	48.3	38.3	56.6	3.54	2.91	3.90	3.40	3.95	3.52	4.85	5.18	4.08	3.23	3.54	3.13	45.23
LOWER SOUTHERN																										
OWINGS FERRY LANDING	37.0	37.6	44.6	55.2	64.8	73.0	76.9	75.1	68.9	58.6	48.2	38.3	56.5	3.48	2.62	3.49	3.48	4.10	3.26	4.27	5.31	4.17	3.29	3.23	3.05	43.75
SOLOMONS	38.5	39.0	45.3	55.5	66.0	74.6	78.8	77.8	72.1	61.2	50.4	40.3	58.3	3.55	2.78	3.61	3.50	3.76	3.45	5.57	5.00	3.59	3.11	3.33	2.97	44.22
DIVISION	37.3	38.0	44.8	55.2	64.9	73.1	77.1	75.5	69.4	58.8	48.3	38.5	56.7	3.57	2.77	3.66	3.50	3.93	3.50	5.01	5.27	3.95	3.37	3.23	3.10	44.86
UPPER SOUTHERN																										
ANNAPOLIS	36.1	36.4	43.1	53.8	64.1	73.2	77.5	76.0	69.9	59.2	48.1	38.1	56.3	3.14	2.57	3.62	3.31	3.83	3.51	4.14	4.50	3.46	2.63	2.78	2.85	40.34
BALTIMORE WB AIRPORT R	34.8	35.7	43.1	54.2	64.4	72.5	76.8	75.0	68.1	57.0	45.5	35.8	55.2	3.43	2.89	3.82	3.60	3.98	3.29	4.22	5.19	3.33	3.18	3.13	2.99	43.05
COLLEGE PARK	36.0	36.8	44.0	54.6	64.6	72.7	76.8	75.1	68.6	57.6	46.6	37.0	55.0	3.25	2.70	3.73	3.39	4.21	3.81	4.14	5.15	3.80	3.16	3.09	3.04	43.47
GLENN DALE BELL STA	35.1	35.9	43.2	53.7	63.8	72.0	75.9	74.3	67.8	56.8	45.9	36.0	55.0	3.43	2.83	3.84	3.54	4.23	4.00	5.02	5.07	3.86	3.36	3.22	2.97	45.37
DIVISION	35.6	36.3	43.4	54.0	64.2	72.5	76.7	75.1	68.5	57.7	46.7	36.9	55.6	3.32	2.72	3.70	3.45	4.07	3.72	4.33	5.01	3.76	3.14	3.06	2.97	43.25
NORTHERN EASTERN SHORE																										
COLEMAN 3 WNW	35.3	36.1	41.2	54.4	63.5	73.1	77.5	76.0	70.0	59.3	48.0	37.3	56.0	3.61	2.93	3.86	3.43	4.17	3.64	4.29	4.97	3.71	3.08	3.41	3.18	44.28
MILLINGTON	35.4	36.0	43.4	54.4	64.9	72.2	76.4	74.7	68.6	57.8	47.1	36.8	55.5	3.65	2.93	3.86	3.28	3.93	3.16	4.24	4.89	3.62	2.98	3.50	3.12	43.16
ROCK HALL	35.4	36.0	43.2	53.9	64.2	72.7	77.1	75.2	69.0	58.2	47.3	37.0	55.8	3.43	2.76	3.76	3.37	3.96	3.40	4.53	4.69	3.71	2.97	3.13	3.11	42.82
DIVISION	35.5	36.1	42.8	54.0	64.2	72.8	77.2	75.5	69.3	58.6	47.6	37.3	55.9	3.60	2.91	3.89	3.43	4.11	3.40	4.48	4.75	3.58	3.00	3.34	3.14	43.63
NORTHERN CENTRAL																										
BALTIMORE WB CITY R	37.3	37.8	44.7	55.7	66.1	74.7	79.1	77.3	70.6	60.0	48.8	39.0	57.6	3.43	2.98	3.94	3.71	4.15	3.87	4.39	4.60	3.63	3.25	3.10	3.16	44.21
ELKTON	33.8	34.6	42.1	52.8	63.3	71.8	76.2	74.5	67.8	56.5	45.3	34.2	54.1	3.46	2.99	4.19	3.60	4.25	3.96	4.35	5.02	3.56	3.23	3.55	3.19	45.35
FREDERICK WFMD	32.9	33.7	41.1	52.2	62.6	71.3	75.6	73.9	66.8	55.6	44.2	34.2	53.7	2.98	2.55	3.53	3.67	3.91	3.60	3.92	4.30	3.42	3.13	2.93	3.04	40.83
TOWSON	.	.	.	.	.	.	.	.	.	.	.	.	.	3.53	2.98	4.13	3.82	4.18	3.67	4.17	4.50	3.76	3.49	3.41	3.43	45.07
WESTMINSTER 2 SSE	33.1	33.9	41.3	52.3	62.5	70.7	74.9	73.2	66.5	55.7	44.7	34.4	53.6	3.36	2.82	4.05	3.61	3.95	4.13	4.34	4.79	3.71	3.46	3.29	3.31	44.82
WOODSTOCK	33.8	34.4	41.8	52.7	63.0	71.0	75.1	73.4	66.5	55.6	44.5	34.7	53.9	3.22	2.77	3.62	3.53	4.06	3.80	4.20	4.21	3.97	3.27	3.09	3.02	42.76
DIVISION	33.7	34.5	41.8	52.8	63.1	71.2	75.4	73.8	67.1	56.3	45.1	35.0	54.2	3.30	2.75	3.86	3.58	4.18	3.87	4.29	4.60	3.71	3.32	3.23	3.19	43.88
APPALACHIAN MOUNTAIN																										
CHEWSVILLE BRIDGEPORT	32.4	33.4	40.6	51.5	61.9	70.3	74.4	72.5	65.7	54.8	43.6	33.7	52.9	2.50	2.01	3.08	3.05	4.01	3.55	3.60	4.00	3.03	3.02	2.62	2.61	37.08
FROSTBURG	30.8	31.4	38.1	49.8	60.0	67.8	71.3	69.9	63.5	53.4	41.7	31.9	50.8	3.16	2.50	3.95	3.57	4.33	4.35	3.71	4.08	3.19	3.33	2.61	2.86	41.64
HANCOCK FRUIT LAB	.	.	.	.	.	.	.	.	.	.	.	.	.	2.73	2.10	3.44	3.08	3.80	3.52	3.41	2.90	3.28	2.55	2.50	3.19	37.19
WESTERNPORT	33.9	35.0	41.9	53.0	62.7	70.3	74.0	72.8	66.5	56.0	44.5	34.9	53.8	3.04	2.53	3.77	3.37	4.17	4.10	3.59	4.18	3.12	2.96	2.28	2.63	39.74
DIVISION	32.8	33.8	40.9	52.1	62.2	70.3	74.2	72.5	65.7	55.1	43.6	33.8	53.1	2.80	2.24	3.51	3.25	3.98	3.83	3.66	3.89	3.08	3.07	2.56	2.68	38.55
ALLEGHENY PLATEAU																										
OAKLAND 1 SE	29.2	29.8	36.1	47.2	56.9	64.9	68.0	67.0	60.8	50.6	39.3	30.2	48.3	4.38	3.46	4.55	4.14	4.71	4.68	4.84	4.50	3.28	3.12	3.14	3.71	48.51
SINES DEEP CREEK 2	28.2	28.5	34.8	46.2	56.3	64.2	67.7	66.1	59.6	49.3	38.1	29.3	47.4	4.34	3.65	4.53	4.15	4.67	4.82	4.86	4.49	3.41	3.10	3.07	3.76	48.85
DIVISION	28.3	28.7	35.1	46.6	56.6	64.6	67.9	66.6	60.2	50.0	38.6	29.4	47.7	4.05	3.29	4.31	4.00	4.54	4.55	4.57	4.37	3.25	3.12	2.98	3.49	46.52
DELAWARE																										
NORTHERN																										
WILMGTON NCASTLE WB AP	33.4	33.8	41.3	52.1	62.7	71.4	76.0	74.3	67.6	56.6	45.4	35.1	54.1	3.40	2.95	4.02	3.33	3.53	4.07	4.25	5.59	3.95	2.91	3.53	3.03	44.56
WILMGTON PORTER RESVR	.	.	.	.	.	.	.	.	.	.	.	.	.	3.47	2.79	4.01	3.74	4.10	3.91	4.47	5.32	3.43	3.10	3.67	3.13	45.14
DIVISION M	33.9	34.3	41.7	52.3	62.9	71.5	76.0	74.3	68.0	57.2	46.0	35.6	54.5	3.48	2.85	3.95	3.56	4.00	3.83	4.37	5.24	3.48	3.14	3.48	3.10	44.48
SOUTHERN																										
BRIDGEVILLE 1 NW	36.5	37.1	43.6	53.6	63.7	72.2	76.2	74.5	68.4	58.0	47.5	37.8	55.8	3.54	2.89	3.98	3.43	3.78	3.60	5.06	5.55	4.02	3.19	3.21	3.03	45.28
DOVER	36.4	36.8	43.7	54.1	64.4	72.9	77.1	75.4	69.4	58.7	47.9	37.9	56.2	3.70	3.03	4.12	3.42	4.15	3.46	4.67	5.73	3.81	3.27	3.67	3.11	46.14
DIVISION	36.8	37.2	43.8	53.7	63.9	72.4	76.6	74.9	68.9	58.4	47.9	38.1	56.1	3.76	3.08	4.19	3.50	3.89	3.61	4.84	5.62	4.09	3.40	3.57	3.20	46.75
WASHINGTON, D. C. NATIONAL AIRPORT	36.9	37.8	44.8	55.7	65.8	74.2	78.2	76.5	69.7	59.0	47.7	38.1	57.0	3.03	2.47	3.21	3.15	4.14	3.21	4.15	4.90	3.83	3.07	2.84	2.78	40.78

* Normals for the period 1931 - 1960. Divisional normals may not be the arithmetical average of individual stations published, since additional data for shorter period stations are used to obtain better areal representation.

TEMPERATURE (°F)   PRECIPITATION (In.)

Jan.	Feb.	Mar.	Apr.	May	June	July	Aug.	Sept.	Oct.	Nov.	Dec.	Annual	Jan.	Feb.	Mar.	Apr.	May	June	July	Aug.	Sept.	Oct.	Nov.	Dec.	Annual

CONFIDENCE LIMITS

In the absence of trend or record changes, the chances are 9 out of 10 that the true mean will lie in the interval formed by adding and subtracting the values in the following table from the means for any station in the State. Because of the wider variation in mean precipitation, the corresponding monthly means and annual mean must be substituted for "p" in the precipitation table below to obtain mean precipitation confidence limits.

Jan.	Feb.	Mar.	Apr.	May	June	July	Aug.	Sept.	Oct.	Nov.	Dec.	Annual	Jan.	Feb.	Mar.	Apr.	May	June	July	Aug.	Sept.	Oct.	Nov.	Dec.	Annual
1.5	1.3	1.5	.9	.8	.7	.5	.6	.8	.9	.8	1.0	.4	$.25\sqrt{p}$	$.18\sqrt{p}$	$.27\sqrt{p}$	$.29\sqrt{p}$	$.34\sqrt{p}$	$.25\sqrt{p}$	$.30\sqrt{p}$	$.44\sqrt{p}$	$.43\sqrt{p}$	$.40\sqrt{p}$	$.35\sqrt{p}$	$.23\sqrt{p}$	$.32\sqrt{p}$

COMPARATIVE DATA

Data in the following table are the mean temperature and average precipitation for Woodstock College, Maryland, for the periods 1881-1905 and 1906-1930 and are included in this publication for comparative purposes:

	Jan.	Feb.	Mar.	Apr.	May	June	July	Aug.	Sept.	Oct.	Nov.	Dec.	Annual	Jan.	Feb.	Mar.	Apr.	May	June	July	Aug.	Sept.	Oct.	Nov.	Dec.	Annual
1881-1905	31.1	32.3	40.4	51.7	63.3	71.2	75.7	72.3	66.1	54.3	43.6	34.7	53.1	3.38	3.72	3.88	2.72	3.85	3.47	4.17	3.61	3.65	3.12	2.66	3.07	41.30
1906-1930	33.4	34.3	43.8	53.6	63.0	70.8	75.4	73.2	67.6	56.3	44.8	35.1	54.3	3.35	2.51	3.11	3.63	3.39	4.10	4.00	4.54	3.22	2.68	2.50	2.99	40.02

NORMALS, MEANS, AND EXTREMES

BALTIMORE, MARYLAND — FRIENDSHIP INTERNATIONAL AIRPORT

LATITUDE 39° 11' N
LONGITUDE 76° 40' W
ELEVATION (ground) 148 Feet

Month	Normal Daily max	Normal Daily min	Normal Monthly	Record highest	Year	Record lowest	Year	Normal degree days	Precip. Normal total	Max monthly	Year	Min monthly	Year	Max in 24 hrs	Year	Snow Mean total	Max monthly	Year	Max 24 hrs	Year	RH 1AM	7AM	1PM	7PM	Wind mean hourly	Prevailing dir.	Fastest mile speed	Dir.	Year	Mean sky cover	Pct sunshine
(a)	(b)	(b)	(b)	17		17		(b)	(b)	17		17		17		17	17				14	14	14	14	17	13	17	17		17	17
J	44.2	25.3	34.8	75	1963	-7	1963	936	3.43	5.27	1964	0.29	1955	1.67	1953	6.3	21.4	1966	12.1	1966	70	73	64	64	10.4	WNW	63	NE	1958	6.1	52
F	45.5	25.8	35.7	76	1967	-7	1967	820	2.89	4.95	1960	1.21	1954	1.75	1952	7.0	20.1	1967	15.5	1958	73	75	57	64	10.9	NW	68	W	1956	6.2	54
M	53.5	32.5	43.1	85	1963	6	1960	679	3.82	6.80	1953	1.21	1958	2.68	1966	7.6	21.6	1966	13.0	1962	75	57	56	57	11.5	WNW	80	SE	1952	6.3	56
A	65.9	42.6	54.2	94	1960	20	1965	327	3.60	8.15	1952	1.72	1965	2.80	1952	T	0.4	1964+	T	1964+	73	50	51	56	11.5	WNW	70	S	1954	6.3	54
M	75.9	52.6	64.4	98	1962	32	1966	90	3.98	7.10	1960	0.43	1964	3.64	1960	T	0.1	1963	T	1963	77	52	50	58	9.2	SW	65	SW	1961+	6.1	62
J	83.5	61.4	72.5	100	1959	42	1966	0	3.29	9.16	1963	0.15	1954	3.35	1963	0.0	0.0		0.0		82	53	52	62	8.7	SW	57	SW	1952	5.5	65
J	87.2	66.4	76.8	102	1966+	52	1963+	0	4.22	8.18	1960	0.30	1955	5.86	1952	0.0	0.0		0.0		82	53	53	64	8.7	SW	57	NW	1962	5.5	67
A	85.0	65.0	75.0	102	1953	44	1952	0	5.19	18.35	1955	0.77	1951	7.82	1955	0.0	0.0		0.0		84	54	55	62	8.1	S	54	SW	1955	5.2	65
S	78.4	57.6	68.1	99	1962+	35	1963	48	3.33	8.50	1966	0.21	1967	4.33	1960	0.0	0.0		0.0		84	56	55	70	9.1	S	56	SE	1954	5.2	62
O	68.4	45.6	57.0	92	1954+	25	1966+	264	3.18	5.66	1955	0.68	1963	3.43	1965	T	T	1954	T	1954	80	58	55	73	9.1	WNW	73	E	1952	4.7	62
N	56.5	34.4	45.5	83	1961	13	1955	585	3.13	7.68	1952	0.21	1965	1.65	1960	1.3	8.4	1967	8.4	1967	74	58	53	64	9.9	NW	58	S	1952	5.8	53
D	45.7	25.9	35.8	74	1951	-5	1960	905	2.99	5.31	1967	0.20	1960	—	—	5.4	20.4	1966	14.1	1960	73	76	58	67	9.9	NW	57	S	1953	6.2	49
YR	65.8	44.6	55.2	102	JUL 1966+	-7	JAN 1963	4654	43.05	18.35	AUG 1955	0.15	OCT 1963	7.82	AUG 1955	25.9	21.6	MAR 1960	15.5	FEB 1958	76	78	54	64	9.9	WNW	80	SW	JUN 1952+	5.8	58

Mean number of days — Baltimore:

Month	Clear	Partly cloudy	Cloudy	Precip .01+	Snow 1.0+	Thunderstorms	Heavy fog	Temp max 90+	Temp max 32-	Temp min 32-	Temp min 0-
J	9	8	14	10	2	*	4	0	5	25	*
F	8	7	13	9	2	*	4	0	4	22	*
M	8	9	14	12	2	1	3	0	1	17	0
A	7	9	14	11	*	2	2	*	0	6	0
M	9	11	11	11	0	4	1	2	0	*	0
J	9	13	8	10	0	6	1	7	0	0	0
J	9	13	9	9	0	5	1	11	0	0	0
A	9	14	8	8	0	5	2	8	0	0	0
S	12	11	7	6	0	2	3	3	0	*	0
O	14	8	9	6	*	1	3	*	0	2	0
N	9	11	10	8	2	*	3	0	*	13	0
D	9	10	12	9	2	*	4	0	5	23	1
YR	111	111	143	112	9	27	31	31	15	103	1

Means and extremes in the above table are from comparable exposures for periods through the current year except as noted below. Annual extremes have been exceeded at other locations as follows: Highest temperature 107 in July 1936; maximum monthly snowfall 33.9 in February 1899; maximum snowfall in 24 hours 24.5 in January 1922.

WILMINGTON, DELAWARE — GREATER WILMINGTON AIRPORT

LATITUDE 39° 40' N
LONGITUDE 75° 36' W
ELEVATION (ground) 78 Feet

Month	Normal Daily max	Normal Daily min	Normal Monthly	Record highest	Year	Record lowest	Year	Normal degree days	Precip Normal total	Max monthly	Year	Min monthly	Year	Max in 24 hrs	Year	Snow Mean total	Max monthly	Year	Max 24 hrs	Year	RH 1AM	7AM	1PM	7PM	Wind mean hourly	Prevailing dir.	Fastest mile speed	Dir.	Year	Mean sky cover	Pct sunshine
(a)	(b)	(b)	(b)	20		20		(b)	20	20		20		20		20	20				20	20	20	20	19	16	19	19		20	20
J	41.3	25.5	33.4	75	1957	-4	1957	980	3.40	5.55	1949	0.59	1955	1.60	1962	6.1	17.2	1966	11.2	1966	75	76	62	70	9.7	WNW	46	NW	1957	6.6	
F	42.4	25.2	33.8	74	1961	-4	1961	874	2.95	4.90	1966	1.80	1966	2.29	1966	6.5	18.7	1967	9.9	1958	76	76	60	69	9.9	NW	46	NW	1956	6.4	
M	50.5	32.0	41.3	86	1960	17	1963+	735	4.02	5.72	1953	0.81	1966	2.75	1958	4.5	20.3	1958	15.6	1958	74	74	56	65	11.0	NW	43	SE	1952	6.6	
A	62.9	41.6	52.1	89	1960	22	1965	387	3.53	5.57	1964	0.22	1963	2.56	1961	T	1.1	1963	1.1	1963	75	74	53	62	10.4	WNW	45	SW	1957	6.6	
M	73.4	52.0	62.7	95	1966	32	1966	112	4.07	7.35	1948	0.44	1948	2.01	1964	T	T	1963	T	1963	76	76	53	65	9.2	SW	40	SW	1957	6.3	
J	81.8	61.0	71.4	99	1952	42	1955	6	—	6.34	1955	—	1955	2.67	1955	0.0	0.0		0.0		79	79	52	66	8.2	SW	42	SW	1960	5.8	
J	86.2	65.5	76.0	102	1966+	50	1952	0	4.25	7.51	1958	0.16	1955	6.24	1955	0.0	0.0		0.0		85	81	53	67	7.6	NW	48	NW	1960+	6.0	
A	84.2	64.3	74.3	101	1955+	46	1965	0	5.59	12.09	1966	1.42	1966	4.00	1955	0.0	0.0		0.0		87	84	55	72	7.2	NW	40	W	1955	5.9	
S	77.9	57.3	67.6	99	1953	38	1952	51	2.95	9.53	1960	1.16	1967+	5.62	1960	0.0	0.0		0.0		86	86	55	73	7.9	NW	54	NW	1954	5.5	
O	67.1	45.9	56.6	91	1951	26	1952	270	2.91	5.17	1966	0.21	1965	3.88	1962	T	0.3	1962	0.3	1962	80	81	56	70	8.9	NW	58	NW	1954	5.2	
N	55.9	35.7	45.9	85	1950	14	1951	588	3.53	7.32	1951	0.94	1951	3.83	1953	1.5	11.9	1953	11.9	1953	79	81	56	70	8.9	WNW	46	NW	1950	6.1	
D	43.5	26.7	35.1	72	1966	3	1962	927	3.03	5.79	1955	0.19	1948	1.99	1948	4.8	12.4	1966	12.4	1966	83	78	60	71	9.0	WNW	44	NW	1962	6.3	
YR	63.8	44.4	54.1	102	JUL 1966+	-4	FEB 1961+	4930	44.56	12.09	AUG 1966	0.16	JUL 1955	6.24	JUL 1955	23.5	21.5	DEC 1966	15.6	MAR 1958	80	79	56	68	8.9	NW	58	OCT 1954		6.1	

Mean number of days — Wilmington:

Month	Clear	Partly cloudy	Cloudy	Precip .01+	Snow 1.0+	Thunderstorms	Heavy fog	Temp max 90+	Temp max 32-	Temp min 32-	Temp min 0-
J	7	8	16	11	2	*	4	0	6	25	*
F	7	7	14	10	2	1	4	0	4	22	*
M	8	9	14	12	1	1	3	0	1	17	0
A	7	9	14	11	*	3	2	0	0	3	0
M	6	11	14	12	0	5	2	1	0	*	0
J	6	12	12	10	0	6	3	5	0	0	0
J	7	12	12	9	0	6	3	8	0	0	0
A	9	12	11	9	0	5	4	4	0	0	0
S	10	11	9	8	0	2	4	2	0	*	0
O	12	10	9	8	*	1	4	*	0	2	0
N	8	8	13	9	1	1	4	0	*	12	0
D	8	8	15	9	1	*	4	0	5	23	*
YR	96	109	160	117	6	30	41	20	16	104	

Means and extremes in the above table are from the existing or comparable location(s). Annual extremes have been exceeded at other locations as follows: Highest temperature 107 in August 1918; lowest temperature -15 in February 1934; maximum monthly precipitation 14.91 in August 1911; minimum monthly precipitation 0.06 in October 1924; maximum precipitation in 24 hours 6.53 in August 1945; maximum monthly snowfall 27.0 in January 1935; maximum snowfall in 24 hours 22.0 in December 1909.

NORMALS, MEANS, AND EXTREMES

LATITUDE 38° 51' N
LONGITUDE 77° 03' W
ELEVATION (ground) 14 Feet

WASHINGTON, D. C.
WASHINGTON NATIONAL AIRPORT

Month	Temp Normal Daily max	Daily min	Monthly	Extremes Record highest	Year	Record lowest	Year	Normal degree days	Precip Normal total	Max monthly	Year	Min monthly	Year	Max in 24 hrs	Year	Snow Mean total	Max monthly	Year	Max 24 hrs	Year
J	44.3	29.5	36.9	71	1967	3	1963	871	3.03	5.08	1949	0.31	1955	1.73	1948	5.4	21.3	1966	13.8	1966
F	46.1	29.4	37.8	71	1967	4	1961	762	2.47	5.71	1961	0.85	1954	1.63	1961	5.5	19.0	1967	14.4	1958
M	53.8	35.8	44.8	86	1963	16	1962	626	3.21	7.43	1953	0.64	1945	3.43	1960	2.7	17.1	1960	7.9	1960
A	65.8	45.6	55.7	89	1964+	30	1965+	288	3.15	5.97	1952	0.26	1942	1.77	1948	T	T	1964	0.4	1957
M	75.5	56.0	65.8	96	1963	36	1966	74	4.14	10.69	1953	1.06	1963	4.32	1953	0.0	0.0		0.1	1963
J	83.4	64.9	74.2	100	1964	47	1966	0	3.21	6.87	1963	1.24	1954	3.67	1947	0.0	0.0		0.0	
J	87.0	69.3	78.2	101	1966	56	1966	0	4.15	11.06	1945	0.93	1966	2.97	1952	0.0	0.0		0.0	
A	85.0	67.9	76.5	99	1962	51	1962	0	4.90	14.31	1955	0.55	1962	6.39	1955	0.0	0.0		0.0	
S	78.6	60.7	69.7	95	1961	39	1963	33	3.83	6.87	1966	0.20	1967	4.15	1966	0.0	0.0		0.0	
O	68.5	49.6	59.0	91	1967	30	1965+	217	3.07	8.18	1942	T	1963	4.98	1967	T	T		T	
N	56.5	38.9	47.7	81	1961	21	1967	519	2.84	6.70	1963	0.37	1965	2.60	1963	0.9	6.9	1957+	6.9	1967
D	45.6	30.5	38.1	72	1966+	10	1963+	834	2.78	5.93	1951	0.22	1955	1.85	1951	4.2	16.2	1962	11.4	1957
YR	65.8	48.2	57.0	101 JUL. 1966		3 JAN. 1963		4224	40.78	14.31	AUG. 1955	T	OCT. 1963	6.39	AUG. 1955	18.7	21.3	JAN. 1966	14.4	FEB. 1958

Relative humidity (Standard time used: EASTERN), Wind, Sunshine, Sky cover, Mean number of days, Solar radiation:

Month	RH 1 AM	7 AM	1 PM	7 PM	Wind mean hourly	Prevailing dir	Fastest mile speed	Dir	Year	Pct poss sunshine	Mean sky cover	Clear	Partly cloudy	Cloudy	Precip .01"+	Snow 1.0"+	Thunderstorms	Heavy fog	Max 90+	Max 32-	Min 32-	Min 0-	Avg daily solar (langleys)
J	66	69	54	59	10.3	NW	56	NW	1957	50	6.5	8	7	16	10	2	*	2	0	5	24	0	166
F	67	70	56	59	10.5	S	57	SW	1961+	51	6.4	7	7	14	9	2	1	1	0	3	20	0	236
M	65	70	49	54	11.1	NW	60	E	1951	56	6.5	8	9	14	12	1	1	1	0	*	9	0	350
A	66	69	48	51	10.7	NW	56	W	1952	57	6.5	6	10	14	11	*	3	1	0	0	*	0	405
M	72	72	49	56	9.4	S	48	NW	1951	57	6.3	6	11	13	11	0	5	*	3	0	0	0	514
J	76	75	51	59	8.9	S	57	W	1954	66	5.7	9	11	10	10	0	5	*	9	0	0	0	565
J	75	76	51	60	8.4	S	54	E	1951	63	5.9	8	12	11	10	0	6	*	13	0	0	0	502
A	77	79	53	61	8.3	S	49	NE	1955	63	5.8	10	9	12	9	0	5	1	9	0	0	0	465
S	78	79	55	65	8.5	S	56	SE	1952	63	5.4	13	8	9	8	0	2	2	4	0	0	0	371
O	76	79	49	62	8.7	S	78	SE	1954	62	5.0	13	7	11	7	*	1	2	*	0	1	0	302
N	69	72	50	58	9.3	SSW	60	SE	1952	53	5.8	9	8	13	8	*	*	3	0	*	5	0	196
D	67	72	57	62	9.3	NW	62	SW	1957	49	6.3	9	6	16	9	1	*		0	2	20	0	128
YR	71	73	52	59	9.5	S	78	SE	OCT. 1954	57	6.0	105	105	155	112	5	28	14	38	11	81	0	350

Means and extremes in the above table are from existing or comparable location(s). Annual extremes have been exceeded at other locations as follows:
Highest temperature 106 in July 1930+; lowest temperature -15 in February 1899; maximum monthly precipitation 17.45 in September 1934; maximum precipitation in 24 hours 7.31 in August 1928; maximum monthly snowfall 35.2 in February 1899; maximum snowfall in 24 hours 25.0 in January 1922.

(a) Length of record, years.
(b) Climatological standard normals (1931-1960).
* Less than one half.
+ Also on earlier dates, months or years.
T Trace, an amount too small to measure.
(b) Below-zero temperatures are preceded by a minus sign.
The prevailing direction for wind in the Normals, Means, and Extremes table is from records through 1963.
To 8 compass points only.

REFERENCE NOTES APPLYING TO ALL "NORMALS, MEANS, AND EXTREMES" TABLES

Unless otherwise indicated, dimensional units used in this bulletin are: temperature in degrees F.; precipitation, including snowfall, in inches; wind movement in miles per hour; and relative humidity in percent. Degree day totals are the sums of the negative departures of average daily temperatures from 65°F. Sleet was included in snowfall totals beginning with July 1948. Heavy fog reduces visibility to 1/4 mile or less.

Sky cover is expressed in a range of 0 for no clouds or obscuring phenomena to 10 for complete sky cover. The number of clear days is based on average cloudiness 0-3; partly cloudy days 4-7; and cloudy days 8-10 tenths.

Figures instead of letters in a direction column indicate direction in tens of degrees from true North; i.e., 09-East, 18-South, 27-West, 36-North, and 00-Calm. Resultant wind is the vector sum of wind directions and speeds divided by the number of observations. If figures appear in the direction column under "Fastest mile" the corresponding speeds are fastest observed 1-minute values.

Solar radiation data are the averages of direct and diffuse radiation on a horizontal surface. The langley denotes one gram calorie per square centimeter. Averages in the lower table for some months may be for more than the listed number of years. Solar radiation data in this bulletin are from several locations in the vicinity of Washington. Instruments have been at the Observational Test and Development Center, Sterling, Virginia, since October 1960, elevations (m.s.l.) 276 ft. to 7-13-64 and 281 ft. thereafter.

Mean Maximum Temperature (°F.), January

MARYLAND AND DELAWARE

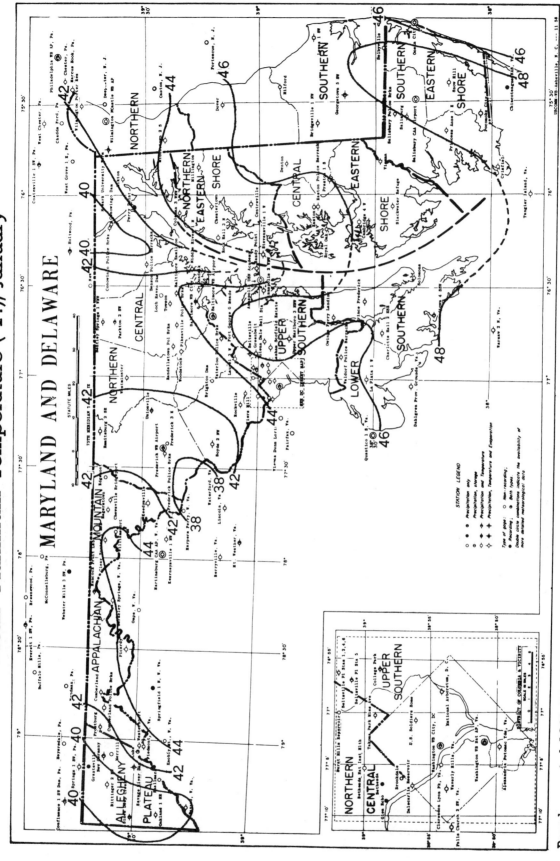

Based on period 1931-52

Isolines are drawn through points of approximately equal value. Caution should be used in interpolating on these maps, particularly in mountainous areas.

Mean Minimum Temperature (°F.), January

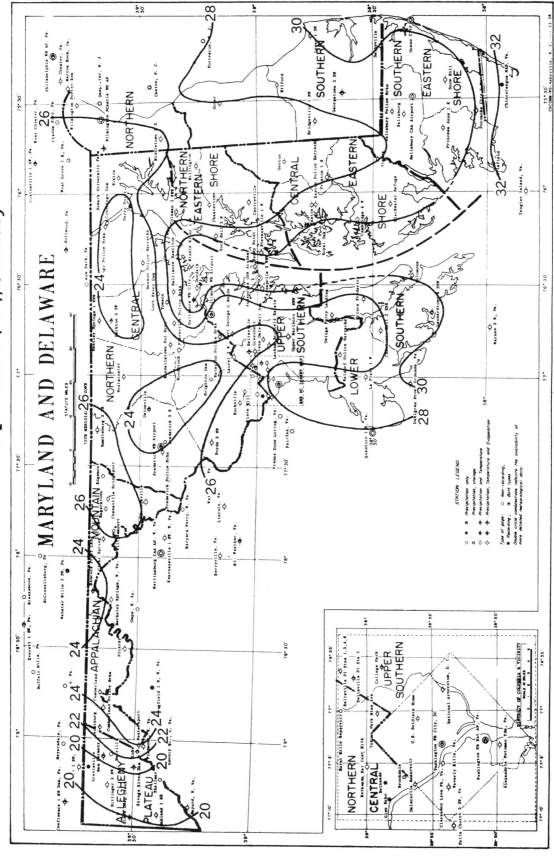

Based on period 1931-52

Isolines are drawn through points of approximately equal value. Caution should be used in interpolating on these maps, particularly in mountainous areas.

Mean Maximum Temperature (°F.), July

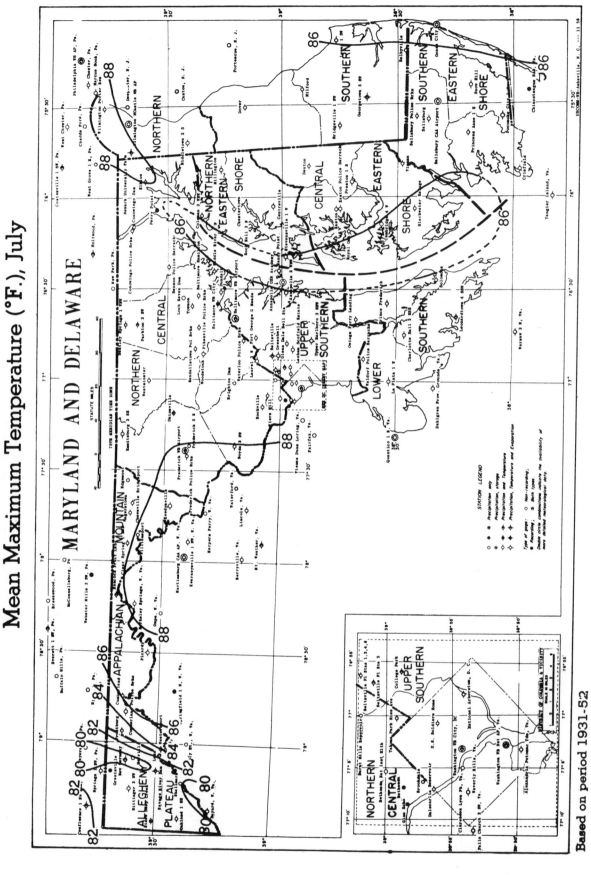

MARYLAND AND DELAWARE

Based on period 1931-52

Isolines are drawn through points of approximately equal value. Caution should be used in interpolating on these maps, particularly in mountainous areas.

STATION LEGEND

Mean Minimum Temperature (°F.), July

MARYLAND AND DELAWARE

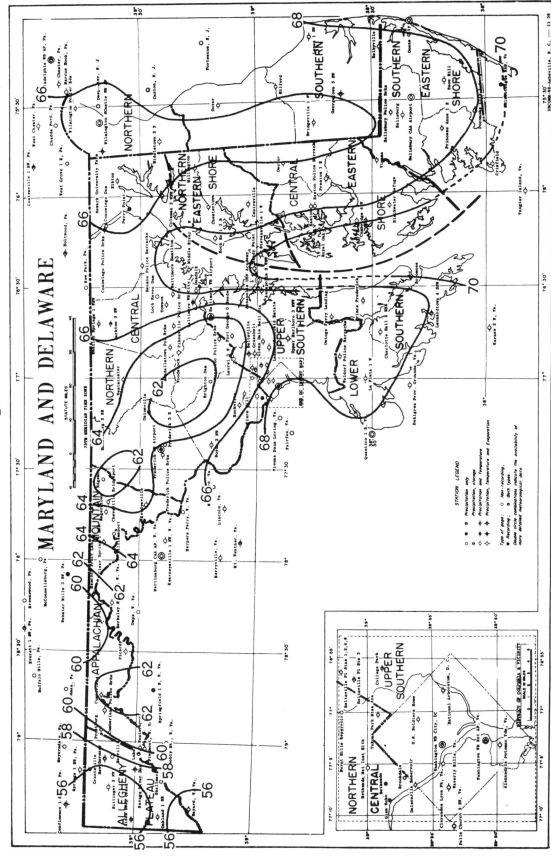

Based on period 1931-52

Isolines are drawn through points of approximately equal value. Caution should be used in interpolating on these maps, particularly in mountainous areas.

Mean Annual Precipitation, Inches

MARYLAND AND DELAWARE

Based on period 1931-55

Isolines are drawn through points of approximately equal value. Caution should be used in interpolating on these maps, particularly in mountainous areas.

STATION LEGEND

THE CLIMATE OF
MASSACHUSETTS

by
Robert E. Lautzenheiser

June 1969

PHYSICAL DESCRIPTION: -- Massachusetts occupies 8,266 square miles, nearly one-eighth of New England's total area. It is the most populous of the New England states. Most of the State lies just above the 42d parallel of latitude. Its north-south width is, roughly, 50 miles, except 100 miles in the eastern, Atlantic coast, portion. The east-west extension is barely 150 miles, excepting "the Cape". This is the familiar name of the long arm of land which reaches around the southern and eastern shores of Cape Cod Bay. Including the Cape, the State is nearly 200 miles in length.

The land surface is mountainous along the western border and generally hilly elsewhere. However, the Cape and some other sections of the coastal area consist of flat land with numerous marshes and some small lakes and ponds. In the west Mt. Greylock rises 3,491 feet above sea level, the highest peak in Massachusetts. The elevation is mostly over 1,000 feet west of the Connecticut River Valley. A number of peaks reach above 2,000 feet. Most of central Massachusetts lies between 500 and 1,000 feet. Eastern Massachusetts and the Connecticut River Valley are mostly less than 500 feet.

GENERAL CLIMATIC FEATURES: -- Climatic characteristics of Massachusetts include: (1) Changeableness in the weather, (2) large ranges of temperature, both daily and annual, (3) great differences between the same seasons in different years, (4) equable distribution of precipitation, and (5) considerable diversity from place to place. The regional New England climatic influences are modified in Massachusetts by varying distances from the ocean, elevations, and types of terrain. These modifying factors divide the State into three climatological divisions.

Massachusetts lies in the "prevailing westerlies", the belt of generally eastward air movement which encircles the globe in middle latitudes. Embedded in this circulation are extensive masses of air originating in higher or lower latitudes and interacting to produce storm systems. Relative to most other sections of the country, a large number of such storms pass over or near Massachusetts. The majority of air masses affecting this State belong to three types: (1) Cold, dry air pouring down from subarctic North America, (2) warm, moist air streaming up on a long overland journey from the Gulf of Mexico and subtropical waters eastward, and (3) cool, damp air moving in from the North Atlantic. Because the atmospheric flow is usually offshore, Massachusetts is more influenced by the first two types than it is by the third. In other words, the adjacent ocean constitutes an important modifying factor, particularly on the immediate coast, but does not dominate the climate.

The procession of contrasting air masses and the relatively frequent passage of storm centers bring about a roughly twice-weekly alternation from fair to cloudy or stormy conditions, attended by often abrupt changes in temperature, moisture, sunshine, wind direction, and speed. There is no regular or persistent rhythm to this sequence, and it is interrupted by periods during which the weather patterns continue the same for several days, infrequently for several weeks. Massachusetts weather, however, is cited for variety rather than monotony. Changeability is also one of its features on a longer time-scale. That is, the same month or season will exhibit varying characteristics over the years, sometimes in close alternation, sometimes arranged in similar groups for successive years. A "normal" month, season, or year is indeed the exception rather than the rule.

The basic climate, as outlined above, obviously does not result from the predominance of any single controlling weather regime, but is rather the integrated effect of a variety of weather patterns. Hence, "weather averages" in Massachusetts usually are not sufficient for important planning purposes without further climatological analysis.

CLIMATOLOGICAL DIVISIONS: -- The Western Division contains about one-fourth of the State, including the Berkshires. A strip roughly 10 to 20 miles wide along the coast comprises the Coastal Division. The remainder is known conveniently as the Central Division. It covers more than 50 percent of the State.

TEMPERATURE: -- The average annual temperature ranges from about 46° F. in the Western Division to 49° F. in the Central, and to around 50° F. in the Coastal Division. Averages vary within these divisions -- elevation, slope, and other environmental aspects, including urbanization, each has an effect. The highest temperature of record is 106° F. observed at several locations; the lowest -34° F. January 18, 1957, at Birch Hill Dam.

Summer temperatures are delightfully comfortable for the most part, and summer averages are nearly uniform over the State. Long-period averages for July range from 67° to 70° F. in the Western Division and in the islands off the coast, and from 70° to 74° F. elsewhere. Hot days with maxima of 90° or higher generally average from 5 to 15 per year, varying not only from place to place but from year to year. They range, in frequency of occurrence, from only a few in cool summers to 25 or more in an occasional hot summer. The Cape and offshore islands are exceptions, averaging less than one

day with a reading of 90° F. or higher per year. Average temperature data may mask marked differences in day to night changes. Islands and the Cape may have a daily range of but 10° to 15° F., but inland areas may vary 20° to 30° F. The diurnal range can reach 40° to 50° F. during cool, dry weather in valleys and marshes. Frost may be a threat even in warmer months in these susceptible areas.

Average temperatures vary from place to place more in winter than in summer. They range in the low 20's for January in the Western Division, in the middle and upper 20's in the Central, and near 30° in the Coastal Division. The diurnal temperature range in winter, though less than in summer, is still greater inland than along the coast. Days with subzero readings are rare on offshore islands. They average only a few per year near the coast, but increase in number of occurrences farther inland to from 5 to 15 annually.

The growing season for vegetation subject to injury from freezing temperatures averages from 120 to 140 days in the Western Division, 140 to 160 days in the Central, and 160 to 200 days in the Coastal Division. Exceptions include bogs and other easily frosted areas which have a shorter season. The average date of the last freezing temperature in spring ranges from the latter part of May at frost-favored places inland to as early as mid-April at Nantucket. Some stations have reported their last freezing temperature well into June in an unusually cold spring. The freeze-free season usually ends in October.

PRECIPITATION: -- Massachusetts is fortunate in having its precipitation rather evenly distributed through the year. In this respect, the State is located in one of the relatively few areas of the world that does not have its "rainy" and "dry" seasons. Storm systems are the principal year-round moisture producers. But in the summer, when this activity ebbs, bands or patches of thunderstorms or showers tend to make up the difference. Though brief and often of small extent, the thunderstorms produce the heaviest local rainfall, and sometimes cause minor washouts of roads and soils. Variations in monthly totals are extreme, ranging from no measurable precipitation to more than 25 inches. Such large fluctuations are rare, however, as most monthly totals fall in the range of 50 to 200 percent of normal. As prolonged droughts are infrequent, irrigation water is available during the fairly common shorter dry spells of summer. Total precipitation averages from 40 to 50 inches per year at stations having long-period records. Local influences cause considerable variation in the totals from station to station. Division averages vary but little. The Coastal Division (the driest) receives annually only about 2 inches of precipitation less than the Western Division (the wettest). Storms of a coastal nature make the Coastal Division the wettest in the winter season. Much of the winter precipitation is in the form of rain or wet snow there. Occasionally freezing rain occurs to coat exposed surfaces with troublesome ice. Inland sections get the heavier rain in the warm season due, principally, to the higher frequency and greater intensity of convective showers and thunderstorms. The mountainous character of much of the Western Division is an additional cause for the heaviest annual totals being recorded in that part of the State. As an illustration of variation in a short distance, consider the normal annual totals for Boston and Blue Hill, less than 12 miles apart. Boston, near sea-level and on the coast, has 42.77 inches. Blue Hill, at about 600 feet higher elevation and not on the immediate coast, has 47.50 inches.

Measurable amounts of precipitation fall on an average of 1 day in 3. Frequency is higher in upland districts. For example, at Pittsfield the average is 4 days out of 10. As much as 6 inches of rain in 24 hours is a rare occurrence in Massachusetts. Some stations have never received that much in a single day. However, Westfield recorded 18.15 inches in 24 hours during hurricane Diane.

SNOWFALL: -- Average annual amounts of snowfall increase rapidly from the coast westward. About 25 to 30 inches fall over Cape Cod, but up to 60 to 80 inches are recorded in the western part of the State. Topography has a marked influence on snowfall, causing much variation even in short distances. As an example, consider Pittsfield located on the western side of the Berkshires, and Greenfield, less than 40 miles distant but on the eastern side. A storm from the west gives Pittsfield much more snow than Greenfield, but a "northeaster" will give Greenfield the heavier amount.

The average number of days with 1 inch or more of snowfall varies from about 8 to 15 in the Coastal Division to mostly 20 to 30 in the Western Division. Most winters will have at least one snowstorm of 5 inches or more. These storms tax snow removal facilities and delay transportation. The heaviest amount recorded in 24 hours was 28.2 inches at Blue Hill Observatory in Milton on February 24-25, 1969. This storm brought drifts 5 to 10 feet in eastern Massachusetts. On February 16, 1958, Boston measured 19.4 inches of snowfall for a new record. Variations, both seasonal and from place to place, cover a wide range. Less than 4 inches for an entire season has been recorded at some Cape stations, while totals of over 100 inches have been received at most stations in the Central and Western Divisions.

The average number of days with snow on the ground also increases from shore to interior and with rise in elevation. There is little lasting snow cover in the coastal lowlands. In the Western Division the cover usually extends well into spring. Maximum snow depths usually occur in the middle part of February. Water stored in the snow over the watersheds makes an important contribution to the water supply. Melting is usually too gradual to threaten serious flooding.

FLOODS: -- The Connecticut River, the largest river system in New England, drains most of the western half of the State. Second in size in Massachusetts is the Merrimack River which occupies the northeast portion. The rest of the rivers are relatively small, most of them with headwaters in the State and flowing southward through Connecticut and Rhode Island, or directly to the coast in the east and southeast. Streams rise quickly in the highlands, and are relatively slow rising in the flat coastal areas in the east.

Flooding occurs most often in spring, caused by combination of rain and melting snow. The Connecticut River shows a regular annual rise as the result of the melting of high elevation snow in northern and central New England, but extensive flooding does not occur unless the rise is accompanied by heavy rains. High flows and major floods occur from rainfall alone but less frequently. Some of the severest floods caused by heavy rains have been those associated with hurricanes or storms of tropical origin in late summer or fall, normally the low water season.

First recorded major flood in the Connecticut River was in 1683. Severe floods in the more recent years have occurred in Massachusetts in November 1927, March 1936, July and September 1938, December 1948, August and October 1955, and March 1968.

OTHER CLIMATIC FEATURES: -- The percentage of possible sunshine averages from 50 to 60 in most sections. Higher elevations are cloudier, reducing the Berkshire average to between 45 and 50 percent. The average annual number of clear days is between 90 and 120 for most of the State, with less in the Berkshires. Heavy fog is frequent and sometimes persistent south of Cape Cod. Nantucket Island has heavy fog on nearly 1 day out of 4. Fog frequency diminishes along the Massachusetts coast north of the Cape. Duration of fog also diminishes inland. But the shorter duration heavy ground fogs of early morning occur frequently at susceptible places inland. These, plus the fewer occurrences of other heavy fog, produce a frequency that also approaches this 1 day out of 4 in many localities. The number of days with fog varies from as low as about 15 up to nearly 100 per year over the State.

WINDS AND STORMS: -- The prevailing wind, on a yearly basis, comes from a westerly direction. It is more northwesterly in winter and southwesterly in summer. Topography has a strong influence on the prevailing direction. Points in the Connecticut Valley, for example, may have prevailing north or south winds, paralleling the direction of the Valley. Along the coast in spring and summer the sea breeze is important. These onshore winds, blowing from the cool ocean, may come inland for 10 miles or so. They tend to retard the spring growth, but they are pleasantly cooling in summer. Boston is famed for its sea breeze along with its beans and cod. The easterly winds are most frequent in May. Coastal storms or "northeasters" are one of the State's most serious weather hazards. They generate very strong winds and heavy rain or snow. They can produce abnormally high, wind-driven tides. These cause heavy damage to coastal installations. In winter, these storms produce the heaviest snow. Occasionally in summer or fall a storm of tropical origin affects Massachusetts. Often these will be similar to the northeasters described above. But the few which retain full hurricane force, as in 1938, 1944, 1954, and 1960, cause widespread and enormous damage. Maximum loss of life and property is usually concentrated along the shore, though hurricane winds and rains may also seriously damage and flood inland areas. Storms of tropical origin seriously affect Massachusetts about once in 2 years, on the average. Two such storms in the same year may be expected once in 8 or 10 years. As many as four such storms in a single year are unlikely to occur in a lifetime.

Tornadoes are not common phenomena, yet, on a per unit area basis, Massachusetts ranks fairly high among the states. One or more may occur in Massachusetts each year. The disastrous "Worcester Tornado" of June 9, 1953, serves as a reminder that the State is not immune to extremely violent developments of the most dangerous type of local storm. Fortunately, most tornadoes in Massachusetts have been very small or took little toll of life or property. As most of the State is forested and much of it not thickly settled, undoubtedly some tornadoes have occurred without being recorded. Four out of five tornadoes occur between May 15 and September 15, and about 75 percent between 2 p.m. and 7 p.m. The peak month is July and the peak of occurrence is 2 to 5 p.m. The chance of a tornado striking any given spot is extremely small.

Thunderstorms and hailstorms have a similar frequency maximum from midspring to early fall. Thunderstorms occur on about 20 to 30 days a year, and the most severe are attended by hail. Hailstorms can severely injure or even ruin field crops, break glass, dent automobiles, and damage other vulnerable exposed objects. Fortunately, the size of an area struck by a hailstorm is usually small. Ice storms (glaze)

of winter produce perilous conditions for transportation. These dangerous storms, however, are usually of brief duration. But a few widespread and prolonged icestorms have occurred, which, besides affecting all forms of travel and transport, have broken trees and limbs, utility lines and poles. In designing structures such as steel towers, one should include consideration of the possible ice load and its resultant magnification of the wind stress, because the ice increases the area exposed to the wind.

CLIMATE AND ECONOMY: -- Activities in Massachusetts are profoundly influenced by its climate. The usually adequate precipitation favors tree growth. Two-thirds of the State is forest-covered. Forests are a major scenic attraction, especially the fall foliage. They are effective in preventing erosion and floods. Forests also support important lumbering and wood-processing industries.

Climate is a significant factor in the State's agriculture. It favors the production of high value specialized crops. Indeed, Massachusetts ranks very high in the Nation in cash receipts per acre from farm marketing. Cranberries, which require exacting climatic and soil conditions, plus an abundance of water for protection from frost, thrive in the southeastern coastal area: a large portion of the Nation's cranberry supply is grown there. Tobacco flourishes in the Connecticut Valley, where the warm summer weather benefits the crop. Dairying and poultry raising find favorable climate and convenient markets. Apples are the most prolific of the tree fruits, and the development and production of quality produce is an important commercial pursuit. Maple syrup is produced in the Berkshire area. In central and eastern portions of the State numerous commercial truck gardens are cultivated.

Massachusetts is fortunate in that it usually enjoys ample precipitation, runoff, and ground water supplies. These are indispensable to the State's industrial production which includes electrical goods, chemicals, textiles, lumber, paper, food, plastics, rubber, and leather, as well as a great diversity of smaller industries.

The climate is particularly important to a major industry, the tourist and vacation trade. This amounts to one-third to one-half billion dollars annually. Much of this is concentrated in the summer when pleasant temperatures prevail at seaside and mountain resorts. Fall leaf coloration draws visitors from afar. Winter sports are rapidly developing, particularly in the Berkshires.

In summary, the climate of Massachusetts contributes greatly to its prominence as an industrial, agricultural, and vacation State. It is a rich, natural asset, invigorating to healthy persons and favorable to further economic development.

GENERAL REFERENCES

1. National Planning Association: The Economic State of New England (1954).

2. U. S. Dept. of Agriculture: Atlas of American Agriculture (1936).

3. ---: Climate and Man (Yearbook of Agriculture for 1941), Part 5, Climatic Data, with special reference to agriculture in the United States.*

4. ---: Soil (Yearbook of Agriculture for 1957).

5. U. S. Geological Survey: Hydrology of Massachusetts. (prepared in cooperation with Mass. Dept. of Pub. Works), Water-Supply Paper No. 1105 (1949).*

SPECIALIZED REFERENCES

1. Brooks, C. F.: "New England Snowfall", Monthly Weather Review, Vol. 45 (1917).

2. ---: "The Rainfall of New England. General Statement", Journ. N. Eng. Water Works Assoc., Vol. 44 (1930).

3. Brown, Rodger A.:"Twisters in New England", unpublished manuscript, Antioch College, 1957.*

4. Church, P. E. : "A Geographical Study of New England Temperatures", Geogr. Review, Vol. 26 (1936).

5. Eustis, R. S. : "Winds over New England in relation to topography", Bull. Amer. Met. Soc., Vol. 23 (1942).

6. Galway, Joseph G. : "A Statistical Study of New England Snowfall", unpublished manuscript of U. S. Weather Bureau (1954).*

7. Goodnough, X. H. : "Rainfall in New England". Journ. N. Eng. Water Works Assoc., Vols. 29 (1915), 35 (1921) and 40 (1926).*

8. Lautzenheiser, R. E.: "Snowfall, Snowfall Frequencies,& Snow cover Data for New England", Proceedings of the Eastern Snow Conference (1968).

9. Perley, S. : Historic Storms of New England (1891).*

10. Stone, R. G. : "Distribution of snow depths over New York and New England", Trans. Amer. Geophy. Union (1940).

11. --- : "The average length of the season with snow cover of various depths in New England", Trans. Amer. Geophy, Union (1944).

12. Upton, W. : "Characteristics of the New England Climate", Annals Harvard Astron. Obser. (1890).

13. Weber, J. H. : "The Rainfall of New England. Historical Statement. Annual Rainfall. Seasonal Rainfall. Mean Monthly Rainfall of Southern New England. Maximum and Minimum Rainfall of Southern New England", Journ. N. Eng. Water Works Assn., Vol. 44 (1930).

14. White, G. V.: "Rainfall in New England", Journ. N. Eng. Water Works Assn., Vols. 56 (1942) and 57 (1943).*

15. Weather Bureau Technical Paper No. 16 - Maximum 24-Hour Precipitation in the United States. Washington, D. C. 1952

16. Weather Bureau Technical Paper No. 25 - Rainfall Intensity-Duration-Frequency Curves For selected stations in the United States, Alaska, Hawaiian Islands, and Puerto Rico.

17. Weather Bureau Technical Paper No. 40 - Rainfall Frequency Atlas for the United States.

18. Weather Bureau Technical Paper No. 49 - 2-to-10-day precipitation for return periods of 2 to 100 years in the contiguous United States.

* References marked with an asterisk are useful sources of data; the others are principally studies of the important climatic elements.

BIBLIOGRAPHY

(A) Climatic Summary of the United States (Bulletin W) 1930 edition, Section 86, U. S. Weather Bureau

(B) Climatic Summary of the United States, New England - Supplement for 1931 through 1952 - (Bulletin W Supplement). U. S. Weather Bureau

(C) Climatic Summary of the United States, New England - Supplement for 1951 through 1960 (Bulletin W Supplement), Climatography of the United States, Series 86-23. U. S. Weather Bureau

(D) Climatological Data - New England, U. S. Weather Bureau

(E) Climatological Data National Summary. U.S. Weather Bureau

(F) Hourly Precipitation Data - New England, U. S. Weather Bureau

(G) Local Climatological Data, U. S. Weather Bureau, for Blue Hill Observatory, Boston Nantucket, Pittsfield, Worcester, Massachusetts

FREEZE DATA

STATION	Freeze threshold temperature	Mean date of last Spring occurrence	Mean date of first Fall occurrence	Mean No. of days between dates	Years of record Spring	No. of occurrences in Spring	Years of record Fall	No. of occurrences in Fall
ADAMS	32	05-19	09-23	127	30	30	30	30
	28	05-01	10-04	156	30	30	30	30
	24	04-16	10-19	186	30	30	30	30
	20	04-01	11-06	219	30	30	30	30
	16	03-22	11-20	243	30	30	30	30
AMHERST	32	05-05	10-05	153	30	30	30	30
	28	04-18	10-15	180	30	30	30	30
	24	04-01	10-29	211	30	30	30	30
	20	03-21	11-18	242	30	30	30	30
	16	03-14	11-30	261	30	30	30	30
BIRCH HILL DAM	32	05-21	09-17	119	12	12	13	13
	28	05-08	09-27	142	12	12	13	13
	24	04-20	10-09	172	12	12	13	13
	20	04-07	10-21	197	12	12	13	13
	16	03-29	11-06	222	12	12	13	13
BLUE HILL WB	32	04-22	10-21	182	23	23	23	23
	28	04-08	11-03	209	23	23	23	23
	24	03-29	11-17	233	23	23	23	23
	20	03-19	11-28	254	23	23	23	23
	16	03-13	12-04	266	23	23	23	23
BOSTON WB AIRPORT	32	04-05	11-08	217	10	10	10	10
	28	03-23	11-17	239	10	10	10	10
	24	03-17	11-29	257	10	10	10	10
	20	03-12	12-05	268	10	10	10	10
	16	02-26	12-19	296	10	10	10	9
BROCKTON	32	04-29	10-14	168	29	29	30	30
	28	04-09	10-30	204	29	29	30	30
	24	03-27	11-09	227	29	29	30	30
	20	03-19	11-22	248	30	30	30	30
	16	03-11	12-02	266	30	30	30	30
CHESTNUT HILL	32	04-19	10-23	187	30	30	30	30
	28	04-02	11-07	219	29	29	30	30
	24	03-21	11-20	244	29	29	30	30
	20	03-15	12-03	263	29	29	30	30
	16	03-09	12-09	275	29	29	30	30
CLINTON	32	04-28	10-10	165	30	30	29	29
	28	04-15	10-29	197	30	30	30	30
	24	04-03	11-08	219	30	30	30	30
	20	03-24	11-25	246	30	30	30	30
	16	03-16	12-03	262	30	30	30	30
EAST WAREHAM	32	05-01	10-09	161	30	30	30	30
	28	04-19	10-22	186	30	30	29	29
	24	04-05	11-07	216	30	30	30	30
	20	03-24	11-19	240	30	30	30	30
	16	03-10	12-01	266	30	30	30	30
EDGARTOWN	32	04-28	10-19	174	15	15	15	15
	28	04-12	11-05	207	15	15	15	15
	24	03-26	11-22	241	15	15	15	15
	20	03-18	11-28	255	15	15	15	15
	16	03-03	12-12	284	15	14	14	14
FALL RIVER	32	04-21	10-23	185	30	30	30	30
	28	04-04	11-09	219	30	30	30	30
	24	03-22	11-19	242	30	30	30	30
	20	03-16	11-27	256	30	30	30	30
	16	03-07	12-07	275	30	30	30	30
FITCHBURG 2 S	32	05-02	10-06	157	30	30	30	30
	28	04-17	10-17	183	30	30	30	30
	24	04-02	10-31	212	30	30	30	30
	20	03-24	11-18	239	30	30	30	30
	16	03-16	11-28	257	30	30	30	30
FRAMINGHAM	32	04-27	10-09	165	30	30	30	30
	28	04-11	10-24	196	30	30	30	30
	24	03-28	11-09	226	30	30	30	30
	20	03-18	11-22	249	30	30	30	30
	16	03-13	12-03	265	30	30	30	30
HAVERHILL	32	04-21	10-19	181	30	30	29	29
	28	04-07	10-30	206	30	30	29	29
	24	03-24	11-17	238	30	30	29	29
	20	03-17	11-28	256	30	30	29	29
	16	03-10	12-04	269	30	30	29	29
HOOSAC TUNNEL	32	05-15	10-02	140	30	30	30	30
	28	04-26	10-11	168	30	30	30	30
	24	04-10	10-30	203	30	30	30	30
	20	03-29	11-13	229	30	30	30	30
	16	03-23	11-26	248	30	30	30	30
HYANNIS 2 NNE	32	04-26	10-23	180	29	29	29	29
	28	04-12	11-03	205	29	29	29	29
	24	03-27	11-19	237	29	29	29	29
	20	03-17	12-02	260	29	29	29	29
	16	03-10	12-09	274	28	28	29	29
KNIGHTVILLE DAM	32	05-20	09-26	129	12	12	13	13
	28	05-04	10-05	154	12	12	13	13
	24	04-14	10-15	184	12	12	13	13
	20	04-01	10-31	213	12	12	13	13
	16	03-25	11-21	241	12	12	13	13
LAWRENCE	32	04-25	10-15	173	30	30	30	30
	28	04-09	10-29	203	30	30	30	30
	24	03-29	11-16	232	30	30	30	30
	20	03-20	11-25	250	30	30	30	30
	16	03-11	12-03	267	30	30	30	30
LOWELL	32	04-23	10-16	176	30	30	30	30
	28	04-09	10-29	203	30	30	30	30
	24	03-26	11-16	235	30	30	30	30
	20	03-19	11-26	252	30	30	30	30
	16	03-12	12-04	267	30	30	30	30
MIDDLETON	32	04-20	10-14	177	25	25	26	26
	28	04-03	11-01	212	25	25	26	26
	24	03-23	11-19	241	25	25	26	26
	20	03-18	12-02	259	25	25	26	26
	16	03-08	12-10	277	25	25	26	25
NANTUCKET WB AIRPORT	32	04-15	11-10	209	14	14	15	15
	28	03-31	11-24	238	14	14	15	15
	24	03-18	12-05	262	14	14	15	15
	20	03-09	12-15	281	14	14	15	14
	16	02-25	12-22	300	14	14	15	11
NEW BEDFORD	32	04-05	11-14	223	30	30	30	30
	28	03-23	11-26	248	30	30	30	30
	24	03-14	11-29	260	30	30	30	30
	20	03-07	12-07	275	30	30	30	30
	16	03-01	12-17	291	30	30	30	25
PITTSFIELD WB AIRPORT	32	05-12	09-27	138	30	30	30	30
	28	04-26	10-08	165	30	30	30	30
	24	04-13	10-22	192	30	30	30	30
	20	04-01	11-10	223	30	30	30	30
	16	03-24	11-23	244	30	30	30	30
PLYMOUTH	32	04-25	10-17	175	23	23	23	23
	28	04-10	11-04	208	23	23	23	23
	24	03-30	11-21	236	23	23	23	23
	20	03-19	11-30	256	23	23	23	23
	16	03-10	12-09	274	23	23	23	23
PROVINCETOWN 3 N	32	04-12	11-07	209	23	23	24	24
	28	03-31	11-21	235	23	23	23	23
	24	03-18	12-04	261	23	23	21	21
	20	03-09	12-08	274	23	23	21	20
	16	02-26	12-16	293	21	21	20	16
ROCKPORT 1 ESE	32	04-16	10-26	193	23	23	24	24
	28	04-04	11-16	226	23	23	24	24
	24	03-23	11-27	249	21	21	24	24
	20	03-17	12-03	261	21	21	24	24
	16	03-08	12-09	276	21	21	24	24

FREEZE DATA

STATION	Freeze threshold temperature	Mean date of last Spring occurrence	Mean date of first Fall occurrence	Mean No. of days between dates	Years of record Spring	No. of occurrences in Spring	Years of record Fall	No. of occurrences in Fall
SANDWICH	32	04-16	11-02	200	16	16	15	15
	28	04-04	11-18	228	15	15	15	15
	24	03-25	11-29	249	15	15	16	16
	20	03-13	12-07	269	15	15	16	16
	16	03-03	12-18	290	15	15	16	15
SHELBURNE FALLS	32	05-12	10-01	142	28	28	28	28
	28	04-25	10-16	174	28	28	28	28
	24	04-10	10-31	204	28	28	28	28
	20	03-28	11-13	230	28	28	28	28
	16	03-18	11-22	249	28	28	28	28
SPRINGFIELD GEN ELEC	32	04-21	10-15	177	30	30	30	30
	28	04-06	10-26	203	30	30	30	30
	24	03-26	11-14	233	30	30	30	30
	20	03-17	11-25	253	30	30	30	30
	16	03-10	12-06	271	30	30	30	30
STOCKBRIDGE	32	05-18	09-25	130	30	30	30	30
	28	05-03	10-05	155	30	30	30	30
	24	04-17	10-18	184	30	30	30	30
	20	04-04	11-02	212	30	30	30	30
	16	03-23	11-20	242	30	30	30	30
TAUNTON	32	05-13	09-29	139	30	30	30	30
	28	04-26	10-11	168	30	30	30	30
	24	04-13	10-24	194	30	30	30	30
	20	03-28	11-06	223	30	30	30	30
	16	03-16	11-25	254	30	30	30	30
TULLY DAM	32	05-25	09-23	121	11	11	12	12
	28	05-07	10-01	147	11	11	12	12
	24	04-15	10-13	181	11	11	12	12
	20	04-03	10-25	205	11	11	12	12
	16	03-26	11-16	235	11	11	12	12
TURNERS FALLS	32	05-11	10-02	144	30	30	30	30
	28	04-23	10-15	175	30	30	30	30
	24	04-08	10-29	204	30	30	30	30
	20	03-26	11-11	230	30	30	30	30
	16	03-18	11-26	253	30	30	30	30
WALPOLE	32	05-15	09-29	137	12	12	13	13
	28	04-21	10-09	171	12	12	13	13
	24	04-08	10-26	201	12	12	13	13
	20	03-29	11-18	234	12	12	13	13
	16	03-19	11-29	255	12	12	13	12
WEST CUMMINGTON	32	06-03	09-13	102	16	16	17	17
	28	05-18	09-27	132	15	15	17	17
	24	04-28	10-05	160	16	16	17	17
	20	04-12	10-21	192	16	16	16	16
	16	03-31	11-10	224	16	16	16	16
WESTON	32	05-12	09-30	141	30	30	30	30
	28	04-24	10-07	166	30	30	30	30
	24	04-10	10-24	197	30	30	30	30
	20	03-29	11-08	224	30	30	30	30
	16	03-17	11-27	255	30	30	30	30
WESTOVER FIELD	32	05-04	10-04	153	19	19	19	19
	28	04-14	10-17	186	19	19	19	19
	24	03-31	11-03	217	19	19	19	19
	20	03-23	11-16	238	19	19	19	19
	16	03-15	12-01	261	19	19	19	19
WORCESTER WB AIRPORT	32	04-25	10-15	173	13	13	13	13
	28	04-14	10-30	199	13	13	13	13
	24	03-31	11-11	225	13	13	13	13
	20	03-23	11-22	244	13	13	13	13
	16	03-17	11-30	258	13	13	13	13
WORCESTER	32	05-07	10-02	148	30	30	30	30
	28	04-23	10-12	172	30	30	30	30
	24	04-07	10-27	203	30	30	30	30
	20	03-27	11-12	230	30	30	30	30
	16	03-18	11-27	254	30	30	30	30

Data in the above table are based on the period 1931-1960, or that portion of this period for which data are available.

⊕ When the frequency of occurrence in either spring or fall is one year in ten, or less, mean dates are not given.

Means have been adjusted to take into account years of non-occurrence.

A freeze is a numerical substitute for the former term "killing frost" and is the occurrence of a minimum temperature at or below the threshold temperature of 32°, 28°, etc.

Freeze data tabulations in greater detail are available and can be reproduced at cost.

*NORMALS BY CLIMATOLOGICAL DIVISIONS

Taken from "Climatography of the United States No. 81-4, Decennial Census of U. S. Climate"

STATIONS (By Divisions)	TEMPERATURE (°F) JAN	FEB	MAR	APR	MAY	JUNE	JULY	AUG	SEPT	OCT	NOV	DEC	ANN	PRECIPITATION (In.) JAN	FEB	MAR	APR	MAY	JUNE	JULY	AUG	SEPT	OCT	NOV	DEC	ANN
WESTERN																										
ADAMS	22.8	23.2	31.8	44.4	55.5	64.5	69.4	67.1	59.4	49.1	38.0	26.2	46.0	3.38	2.54	3.47	4.23	4.17	4.12	4.62	3.43	4.51	3.22	4.26	3.45	45.40
CHESTERFIELD	.	.	.	.	.	.	.	.	.	.	.	.	.	3.58	3.13	4.22	3.84	4.42	4.47	4.11	4.04	4.80	4.07	4.37	3.70	48.75
HEATH	.	.	.	.	.	.	.	.	.	.	.	.	.	4.00	3.30	4.29	4.17	4.70	4.15	4.32	4.01	4.71	3.98	4.40	3.96	49.99
HOOSAC TUNNEL	22.0	23.1	31.5	43.7	55.3	64.2	68.7	66.6	59.4	49.2	37.6	25.2	45.5	3.87	3.13	4.00	4.20	4.36	3.99	4.24	3.49	4.41	3.37	4.23	3.82	47.11
PERU	.	.	.	.	.	.	.	.	.	.	.	.	.	3.74	3.28	4.50	3.80	4.34	4.41	4.59	4.03	4.87	3.86	4.26	3.86	49.54
PITTSFIELD WB AIRPORT	21.8	22.3	30.7	43.0	54.6	63.1	67.8	66.2	58.1	48.1	37.3	25.3	44.9	2.97	2.51	3.22	3.87	3.87	4.28	4.89	3.90	4.50	3.25	3.91	3.25	44.42
PLAINFIELD	.	.	.	.	.	.	.	.	.	.	.	.	.	3.51	3.01	4.07	3.89	4.80	4.17	4.58	4.01	4.79	4.13	4.31	3.59	48.86
SOUTH EGREMONT	.	.	.	.	.	.	.	.	.	.	.	.	.	3.27	2.53	3.33	3.56	3.54	3.87	4.37	4.05	4.08	3.19	3.38	3.45	42.62
STOCKBRIDGE	23.3	23.8	32.2	44.5	55.3	63.2	67.7	65.6	58.5	49.0	38.6	26.7	45.7	3.27	2.55	3.44	3.75	3.86	4.12	4.60	4.15	4.62	3.11	3.86	3.20	44.53
WEST OTIS	.	.	.	.	.	.	.	.	.	.	.	.	.	3.37	2.61	3.78	3.53	3.73	4.29	4.39	4.51	4.36	3.62	4.04	3.49	45.72
DIVISION	22.7	23.3	31.7	44.2	55.6	64.2	68.8	66.6	59.1	49.1	38.0	26.0	45.8	3.39	2.69	3.56	3.90	4.00	3.96	4.41	3.73	4.43	3.25	4.06	3.41	44.79
CENTRAL																										
AMHERST	25.1	26.1	35.0	46.7	58.0	66.8	71.7	69.6	62.0	51.8	40.4	28.3	48.5	3.40	2.65	3.76	3.70	3.78	4.05	3.83	3.86	4.33	3.05	3.80	3.35	43.56
ASHLAND	.	.	.	.	.	.	.	.	.	.	.	.	.	4.13	3.45	4.35	3.85	3.30	3.55	3.55	3.93	4.02	3.07	4.24	3.86	45.30
BLUE HILL WB	27.0	27.4	34.8	45.7	56.7	65.1	70.9	69.4	62.4	52.7	42.0	30.0	48.7	4.49	3.73	4.54	4.00	3.48	3.75	3.27	4.05	3.95	3.75	4.53	3.96	47.50
BOYLSTON	.	.	.	.	.	.	.	.	.	.	.	.	.	4.26	3.46	4.48	3.94	3.77	3.88	3.61	3.98	3.98	3.38	4.38	3.76	46.88
CHESTNUT HILL	29.8	30.6	38.1	48.8	59.6	68.1	73.5	71.8	64.8	54.8	44.6	32.8	51.4	4.01	3.21	4.16	3.77	3.51	3.61	3.28	3.81	3.87	3.36	4.17	3.64	44.40
CLINTON	26.2	26.6	34.5	45.7	57.1	65.7	70.9	69.2	62.0	52.2	41.7	29.9	48.5	4.33	3.47	4.43	3.82	3.65	3.74	3.64	4.10	3.88	3.36	4.35	3.80	46.57
FITCHBURG 2 S	25.9	26.7	35.2	46.7	58.4	67.3	72.3	70.6	62.6	52.4	41.2	29.2	49.0	4.10	3.21	4.37	4.06	3.82	3.94	3.46	3.25	4.11	3.43	4.28	3.71	45.74
FRAMINGHAM	27.5	28.4	36.8	48.0	59.2	68.0	73.2	71.3	63.7	53.2	42.7	30.6	50.2	4.08	3.39	4.45	3.84	3.27	3.61	3.39	3.91	4.06	3.05	4.24	3.88	45.17
FRANKLIN	.	.	.	.	.	.	.	.	.	.	.	.	.	3.86	3.19	4.08	3.80	3.31	3.52	3.45	4.12	3.97	3.20	4.27	3.72	44.49
GARDNER	.	.	.	.	.	.	.	.	.	.	.	.	.	3.54	2.82	3.83	3.57	3.64	4.00	3.75	3.46	4.38	3.41	3.93	3.27	43.60
GROTON	.	.	.	.	.	.	.	.	.	.	.	.	.	3.79	2.97	3.93	3.67	3.54	3.58	3.46	3.24	3.84	3.17	4.00	3.44	42.63
HARDWICK	.	.	.	.	.	.	.	.	.	.	.	.	.	3.39	2.69	3.72	3.91	3.73	4.07	4.30	4.13	4.34	3.35	3.91	3.25	44.79
HAVERHILL	28.0	29.1	37.1	48.4	59.8	68.4	73.8	71.8	64.3	54.0	43.2	31.2	50.8	3.71	2.93	3.82	3.38	3.21	3.08	3.58	2.93	3.72	2.99	3.85	3.31	40.51
HOLYOKE	.	.	.	.	.	.	.	.	.	.	.	.	.	3.56	2.95	3.94	3.98	3.68	4.27	3.70	3.89	4.18	3.18	3.88	3.55	44.76
HUBBARDSTON	.	.	.	.	.	.	.	.	.	.	.	.	.	3.14	2.50	3.60	3.47	3.68	4.16	4.06	3.47	4.45	3.39	3.95	2.97	42.84
LAWRENCE	26.0	27.0	35.3	46.6	57.8	66.8	72.3	70.4	62.8	52.6	41.8	29.5	49.1	3.58	2.89	3.85	3.68	3.14	3.13	3.52	3.20	3.81	3.03	3.90	3.23	40.96
LOWELL	26.7	27.9	36.1	47.5	59.1	68.1	73.6	71.6	63.8	53.2	42.0	30.0	50.0	4.02	3.16	4.22	3.69	3.31	3.36	3.41	3.52	3.71	3.16	4.18	3.60	43.34
MANSFIELD	.	.	.	.	.	.	.	.	.	.	.	.	.	4.09	3.31	4.30	3.98	3.34	3.48	3.46	4.12	3.91	3.40	4.51	3.64	45.54
MILFORD	.	.	.	.	.	.	.	.	.	.	.	.	.	3.82	2.98	3.97	3.84	3.30	3.82	3.67	4.19	4.26	3.18	4.23	3.59	44.85
NORTHBRIDGE 2	.	.	.	.	.	.	.	.	.	.	.	.	.	3.99	3.34	4.47	3.99	3.38	3.56	3.65	4.31	4.04	3.34	4.28	3.89	46.24
PETERSHAM 3 N	.	.	.	.	.	.	.	.	.	.	.	.	.	3.64	3.02	3.95	3.88	3.56	4.27	3.75	3.61	4.23	3.26	4.23	3.51	44.91
SHELBURNE FALLS	22.9	23.4	32.4	44.5	55.8	65.1	69.9	67.9	60.2	49.7	38.4	26.1	46.4	3.80	3.22	4.13	4.14	4.09	3.84	4.01	3.53	4.20	3.67	4.48	3.90	47.01
SOUTHBRIDGE 3 SW	.	.	.	.	.	.	.	.	.	.	.	.	.	4.07	3.40	4.59	4.21	3.72	4.03	4.14	4.66	4.33	3.62	4.53	3.86	49.16
SPRINGFIELD GEN ELEC	28.5	29.7	38.0	49.8	61.0	69.3	74.0	72.2	64.8	55.0	43.5	31.4	51.4	3.64	2.69	3.76	4.00	3.88	4.17	3.99	4.39	4.13	3.16	3.80	3.50	45.11
STERLING	.	.	.	.	.	.	.	.	.	.	.	.	.	4.14	3.21	4.38	4.02	3.81	3.83	3.57	3.66	4.08	3.59	4.46	3.87	46.62
TURNERS FALLS	24.7	26.0	34.9	47.1	58.9	67.7	72.3	70.3	62.6	52.0	40.3	28.0	48.7	3.58	2.77	3.68	3.72	4.03	3.70	3.98	3.60	4.02	3.07	3.99	3.40	43.54
WESTFIELD	.	.	.	.	.	.	.	.	.	.	.	.	.	3.60	3.01	4.05	4.08	3.78	4.16	3.88	4.34	4.30	3.50	4.20	3.59	46.49
WESTON	27.7	28.8	36.5	47.4	58.2	66.5	72.0	70.4	62.9	53.0	42.4	30.5	49.7	3.60	2.96	3.89	3.75	3.39	3.62	3.24	4.15	3.82	3.25	4.08	3.37	43.02
WORCESTER WB AIRPORT	24.0	24.9	32.8	44.6	55.2	64.6	69.8	68.3	60.8	50.5	39.2	27.2	46.8	3.71	2.92	4.11	3.93	3.79	3.84	3.63	4.24	3.92	3.47	4.26	3.59	45.41
WORCESTER	25.9	26.8	34.6	45.8	57.1	65.6	70.6	68.8	61.5	51.6	40.7	28.8	48.2	3.81	2.98	4.22	3.96	3.84	3.78	3.58	4.08	3.89	3.54	4.16	3.53	45.37
DIVISION	26.3	27.3	35.5	46.9	58.2	66.8	72.0	70.1	62.6	52.4	41.5	29.4	49.1	3.86	3.10	4.09	3.84	3.58	3.71	3.60	3.79	3.95	3.23	4.14	3.60	44.49
COASTAL																										
BOSTON WB AIRPORT	29.9	30.3	37.7	47.9	58.8	67.8	73.7	71.7	65.3	55.0	44.9	33.3	51.4	3.94	3.32	4.22	3.77	3.34	3.48	2.88	3.66	3.46	3.14	3.93	3.63	42.77
BROCKTON	.	.	.	.	.	.	.	.	.	.	.	.	.	3.56	3.11	3.89	3.68	3.01	3.11	2.95	3.72	3.69	3.09	3.85	3.30	40.96
EAST WAREHAM	29.0	29.1	36.1	45.7	55.9	64.7	71.0	69.8	62.8	52.9	43.2	31.9	49.3	4.30	3.59	4.80	4.28	3.45	3.26	2.88	4.29	3.84	3.44	4.60	4.20	46.85
FALL RIVER	29.7	30.2	37.3	47.3	57.7	66.6	72.6	71.2	64.3	54.4	43.9	32.7	50.7	4.12	3.41	4.50	3.97	3.41	3.20	3.01	4.47	3.52	3.23	4.36	4.08	45.28
HATCHVILLE	.	.	.	.	.	.	.	.	.	.	.	.	.	4.20	3.61	4.59	4.19	3.35	3.16	2.69	4.64	3.88	3.54	4.39	3.91	46.15
IPSWICH	.	.	.	.	.	.	.	.	.	.	.	.	.	4.41	3.46	4.32	3.79	3.45	3.38	3.27	3.24	3.86	3.34	4.35	3.82	44.69
MIDDLEBORO	.	.	.	.	.	.	.	.	.	.	.	.	.	4.12	3.47	4.41	4.09	3.42	3.09	3.17	4.04	3.74	3.35	4.46	3.88	45.24
MIDDLETON	.	.	.	.	.	.	.	.	.	.	.	.	.	3.92	3.00	3.93	3.42	3.24	3.05	2.96	3.24	3.64	3.34	3.91	3.63	41.28
NANTUCKET WB AIRPORT	33.0	31.4	36.1	44.3	52.6	61.3	68.0	68.1	62.9	54.3	45.9	36.1	49.5	4.22	3.76	4.54	3.76	2.88	2.92	2.71	3.68	3.51	3.70	4.05	3.93	43.66
NEW BEDFORD	31.7	31.7	38.2	47.3	57.4	65.9	72.0	71.2	65.0	55.9	45.9	34.8	51.4	3.84	3.15	4.02	3.73	3.11	2.87	2.24	4.09	3.33	3.05	3.98	3.64	41.05
NEWBURYPORT	.	.	.	.	.	.	.	.	.	.	.	.	.	3.82	2.92	3.92	3.71	3.44	3.15	3.67	3.02	3.89	3.20	4.23	3.71	42.68
PEMBROKE	.	.	.	.	.	.	.	.	.	.	.	.	.	3.85	3.10	3.98	4.05	3.16	3.29	3.07	4.33	3.80	3.45	4.25	3.64	43.97
PLYMOUTH	30.9	31.2	37.9	47.5	57.7	66.4	72.2	70.7	64.0	54.7	45.0	33.8	51.0	4.22	3.46	4.53	4.13	3.43	3.34	2.92	4.22	3.90	3.58	4.50	4.05	46.28
SPOT POND	.	.	.	.	.	.	.	.	.	.	.	.	.	4.72	3.73	4.84	4.32	3.73	3.60	3.19	3.98	4.10	3.68	4.77	4.14	48.80
TAUNTON	28.6	29.4	36.7	46.7	57.3	65.9	71.5	69.8	62.7	52.5	42.6	31.4	49.6	3.67	2.99	3.77	3.89	3.55	3.21	3.44	4.36	3.76	3.50	4.32	3.52	43.98
DIVISION	30.0	30.1	36.8	46.2	56.3	65.0	71.0	69.9	63.2	53.7	44.1	33.0	49.9	4.04	3.37	4.19	3.86	3.23	3.17	2.85	3.85	3.64	3.33	4.11	3.73	43.37

* Normals for the period 1931-1960. Divisional normals may not be the arithmetical averages of individual stations published, since additional data for shorter period stations are used to obtain better areal representation.

TEMPERATURE Jan.	Feb.	Mar.	Apr.	May	June	July	Aug.	Sept.	Oct.	Nov.	Dec.	Annual	PRECIPITATION Jan.	Feb.	Mar.	Apr.	May	June	July	Aug.	Sept.	Oct.	Nov.	Dec.	Annual

CONFIDENCE - LIMITS

In the absence of trend or record changes, the chances are 9 out of 10 that the true mean will lie in the interval formed by adding and subtracting the values in the following table from the means for any station in the State. Because of the wider variation in mean precipitation, the corresponding monthly means and annual mean must be substituted for "p" in the precipitation table below to obtain mean precipitation confidence limits.

1.5	1.3	1.4	.9	.7	.6	.5	.7	.7	.8	.8	1.1	.3	$.20\sqrt{p}$	$.20\sqrt{p}$	$.25\sqrt{p}$	$.27\sqrt{p}$	$.30\sqrt{p}$	$.37\sqrt{p}$	$.31\sqrt{p}$	$.34\sqrt{p}$	$.42\sqrt{p}$	$.37\sqrt{p}$	$.34\sqrt{p}$	$.25\sqrt{p}$	$.31\sqrt{p}$

COMPARATIVE DATA

Data in the following table are the mean temperature and average precipitation for Blue Hill, Massachusetts, for the period 1906-1930 and are included in this publication for comparative purposes.

25.9	25.2	34.4	44.5	55.3	64.2	70.0	68.0	61.9	51.7	40.1	29.1	47.5	3.45	3.87	3.79	4.06	3.77	3.62	4.35	4.41	3.47	3.34	3.72	4.20	46.05

NORMALS, MEANS, AND EXTREMES

BOSTON, MASSACHUSETTS
LOGAN INTERNATIONAL AIRPORT

LATITUDE 42° 22' N
LONGITUDE 71° 02' W
ELEVATION (ground) 15 Feet

Month	Temperature Normal Daily maximum	Normal Daily minimum	Normal Monthly	Extremes Record highest	Year	Record lowest	Year	Normal degree days	Precipitation Normal total	Max monthly	Year	Max 24 hrs	Year	Min monthly	Year	Snow Mean total	Max monthly	Year	Max 24 hrs	Year
(a)	(b)	(b)	(b)	4		4		(b)	(b)	17		17		17		33	33		33	
J	36.8	23.0	29.9	62	1967	-4	1968+	1088	3.94	9.54	1958	2.07	1943	0.92	1955	12.9	32.5	1948	12.8	1943
F	37.4	23.0	30.3	58	1967	-4	1967	972	3.32	5.87	1958	2.65	1958	1.15	1958	11.3	28.7	1962	19.4	1958
M	44.5	30.7	37.6	66	1966	6	1967	846	4.22	11.00	1953	4.13	1968	1.48	1962	8.2	31.2	1956	17.7	1960
A	56.7	40.1	47.9	82	1965	22	1964	513	3.34	7.82	1958	2.09	1964	1.24	1958	0.7	3.1	1967	3.1	1963
M	67.5	50.0	58.8	93	1964	37	1967	208	3.34	13.38	1954	5.74	1954	0.53	1964	T	T	1967+	T	1967+
J	76.3	59.2	67.8	94	1965	46	1965	36	3.48	8.63	1959	2.46	1960	0.48	1959	0.0	0.0		0.0	
J	80.9	65.4	73.7	98	1952+	54	1965	0	2.88	8.12	1959	2.42	1959	0.52	1952	0.0	0.0		0.0	
A	80.0	63.3	71.7	93	1968+	47	1965	9	3.66	17.09	1955	8.40	1955	1.25	1966	0.0	0.0		0.0	
S	73.4	57.1	65.3	92	1953	38	1966+	60	3.46	8.31	1954	5.64	1954	0.35	1957	T	T	1963	T	1963
O	64.7	47.2	55.0	85	1963	28	1966+	316	3.14	8.68	1962	4.26	1962	0.96	1963	T	1.3	1925	8.0	1940
N	51.9	37.8	44.9	69	1950	21	1967	603	3.43	7.74	1944	3.33	1963	1.72	1952	1.3	10.0	1898	13.0	1960
D	40.1	26.5	33.3	70	1966	-3	1966	983	3.63	6.58	1957	2.27	1967	1.03	1955	7.3	26.8	1947	—	—
YR	59.0	43.6	51.4	98 JUL. 1968		-4 JAN. 1968+		5634	42.77	17.09 AUG. 1955		8.40 AUG. 1955		0.35 SEP. 1957		41.7	32.5 JAN. 1948		19.4 FEB. 1958	

Relative humidity / Wind / Mean number of days / Average daily solar radiation columns follow.

Month	Rel. humidity 1 AM	7 AM	1 PM	7 PM	Wind Mean hourly speed	Prevailing direction	Fastest mile Speed	Direction	Year	Pct. of possible sunshine	Mean sky cover sunrise to sunset	Clear	Partly cloudy	Cloudy	Precip .01 in or more	Snow 1.0 in or more	Thunderstorms	Heavy fog	Max 90 and above	Max 32 and below	Min 32 and below	Min 0 and below	Avg solar radiation langleys
	4				11	15	11	11		33	33	33	33	33	17	33	33	33	4	4	4	4	21
J	67	68	63	63	14.8	NW	52	SW	1960	53	6.3	9	7	15	12	3	*	2	0	1	25	1	139
F	64	67	57	60	14.7	WNW	54	E	1960	57	6.1	8	7	13	11	3	*	2	0	1	25	*	204
M	67	68	59	63	14.5	NW	48	ENE	1958	57	6.2	8	8	15	12	2	1	1	0	0	18	0	302
A	68	68	61	61	13.8	SW	52	NW	1963	56	6.5	7	8	15	12	*	2	4	0	0	2	0	375
M	70	69	55	55	12.9	SW	50	NE	1967	58	6.5	6	10	15	11	0	2	2	*	0	0	0	455
J	72	68	60	60	11.9	SW	44	NNW	1958	63	6.2	7	12	11	11	0	2	2	3	0	0	0	503
J	75	73	54	54	11.3	SW	45	N	1964	65	6.1	7	12	12	10	0	3	1	3	0	0	0	491
A	78	75	58	58	11.2	SW	65	NNW	1958	65	5.7	9	11	11	10	0	3	2	2	0	0	0	420
S	79	79	60	60	11.4	SW	57	S	1960	61	5.4	11	11	9	9	0	2	2	1	0	0	0	344
O	73	75	59	59	11.5	SW	45	NW	1963	61	5.5	11	8	12	9	0	2	2	*	0	1	1	244
N	73	74	62	62	12.5	SW	58	NW	1958	54	6.4	8	8	14	11	1	1	1	0	1	8	1	143
D	67	70	60	60	14.2	WNW	49	NW	1962	54	6.1	8	8	14	10	2	*	1	0	6	20	1	120
YR	71	72	58	58	13.1	SW	65 NNW AUG. 1958			60	6.1	100	106	159	128	11	19	23	8	28	99	2	312

(a) For period April 1964 through current year.

Ø Means and extremes in the above table are from existing or comparable location(s). Annual extremes have been exceeded at other locations as follows: Highest temperature 104 in July 1911; lowest temperature -18 in February 1934; maximum monthly precipitation trace in March 1915; maximum monthly snowfall 35.7 in January 1904; fastest mile wind 87 S in September 1938.

Data for the year 1958 are fastest observed 1-minute speeds with directions to 16 compass points; otherwise data are fastest mile with directions to 8 compass points.

$ From instruments at City Office locations. Equipment decommissioned 11-13-68.

BLUE HILL OBSERVATORY
MILTON, MASSACHUSETTS

LATITUDE 42° 13' N
LONGITUDE 71° 07' W
ELEVATION (ground) 629 Feet

Month	Temperature Normal Daily maximum	Normal Daily minimum	Normal Monthly	Extremes Record highest	Year	Record lowest	Year	Normal degree days	Precipitation Normal total	Max monthly	Year	Max 24 hrs	Year	Min monthly	Year	Snow Mean total	Max monthly	Year	Max 24 hrs	Year
(a)	(b)	(b)	(b)	83		83		(b)	(b)	83		83		83		83	83		80	
J	34.4	19.5	27.0	68	1950	-16	1957	1178	4.49	10.97	1958	3.10	1889	0.89	1955	15.9	56.3	1948	20.0	1898
F	35.4	19.3	27.4	67	1957	-21	1934	1053	3.73	8.29	1886	4.85	1886	1.04	1895	16.3	45.2	1967	22.2	1958
M	43.8	34.8	34.8	89	1938	-5	1934	936	4.16	10.96	1968	6.62	1968	0.06	1915	11.8	45.0	1956	27.2	1956
A	54.9	36.5	45.7	93	1938+	1	1923	579	4.00	8.71	1929	2.69	1954	0.92	1892	3.2	21.5	1894	13.0	1887
M	66.5	46.5	56.7	93	1919	27	1944	267	3.75	9.16	1929	3.85	1922	0.50	1929	0.1	0.1	1917	T	1917
J	74.7	55.4	65.1	99	1919	36	1945	69	3.75	10.78	1922	3.92	1922	0.53	1912	0.0	0.0		0.0	
J	80.2	61.6	70.9	99	1952+	46	1890	0	3.27	11.67	1938	4.45	1938	0.13	1952	0.0	0.0		0.0	
A	78.1	60.0	69.1	101	1949	43	1965	22	4.05	18.78	1955	9.93	1955	1.05	1955	0.0	0.0		0.0	
S	71.1	53.0	62.1	99	1953	28	1914	108	3.95	11.04	1933	5.86	1961	0.45	1914	T	T	1934	4.2	1934
O	61.2	42.5	52.7	88	1963	20	1936+	381	3.75	10.84	1962	6.02	1895	0.22	1924	0.2	4.2	1925	4.2	1925
N	49.8	34.2	42.0	81	1950	6	1932	690	4.53	9.29	1945	5.06	1927	0.62	1917	2.7	23.0	1898	16.0	1898
D	37.4	22.5	30.0	68	1966	-19	1933	1085	3.96	9.01	1936	3.03	1888	0.92	1945	10.5	45.2	1945	21.0	1921
YR	57.3	40.0	48.7	101 FEB. 1949		-21 1934		6368	47.50	18.78 AUG. 1955		9.93 AUG. 1955		0.06 MAR. 1915		60.7	56.3 JAN. 1948		27.2 MAR. 1960

Month	Rel. humidity 1 AM	7 AM	1 PM	7 PM	Wind Mean hourly speed	Prevailing direction	Fastest mile Speed	Direction	Year	Pct. of possible sunshine	Precip .01 in or more	Snow 1.0 in or more	Heavy fog	Max 90 and above	Max 32 and below	Min 32 and below	Min 0 and below	Avg solar radiation langleys
	18	30	29	30	56	56	8	8		83	83	79		83	83	83	83	34
J	73	77	62	69	17.4	W	61	WNW	1898	45	13	4		0	14	29	2	157
F	72	76	56	66	17.3	NW	67	NE	1961	49	13	4		0	13	26	1	232
M	74	76	56	65	16.6	NW	64	WNW	1966	48	13	3		0	5	24	*	319
A	76	72	51	62	16.0	NW	65	WNW	1963	52	12	1		*	0	5	0	395
M	79	78	58	72	13.8	S	60	NE	1967	55	12	0		0	*	*	0	477
J	84	84	58	78			44	NE	1967					1	0	0	0	515
J	85	81	57	74	13.0	SW	43	SSW	1963	57	11	0		2	0	0	0	502
A	87	83	59	78	12.9	SW	66	WSW	1968	59	10	0		1	0	0	0	449
S	86	84	59	78	13.3	SW	92	SSE	1960	58	9	*		*	0	*	0	358
O	81	81	60	72	15.2	NW	50	SW	1967	56	10	1		0	0	14	0	267
N	78	80	61	69	16.8	W	67	S	1965	47	11	1		0	0	11	1	162
D	74	76	61	69	16.4	W	68	S	1963	46	11	3		0	11	26	0	136
YR	79	78	58	71	15.4	NW	92 SSE SEP. 1960			52	135	16		5	44	132	4	331

Ø For period May 1960 through the current year. Maximum wind speeds in earlier records for 5-minute periods reached an extreme of 121 m.p.h. in September 1938 with peak gusts up to 186 m.p.h.

$ Through 1964.

NORMALS, MEANS, AND EXTREMES

NANTUCKET, MASSACHUSETTS — MEMORIAL AIRPORT

LATITUDE 41° 15' N
LONGITUDE 70° 04' W
ELEVATION (ground) 43 Feet

Temperature

Month	Normal Daily maximum (b)	Normal Daily minimum (b)	Normal Monthly (b)	Record highest	Year	Record lowest ø	Year	Normal degree days (b)
J	39.2	26.8	33.0	63	1967	2	1968	992
F	38.1	24.6	31.4	54	1967	5	1967	941
M	42.5	29.6	36.1	60	1968	10	1967	896
A	50.6	38.0	44.3	69	1968	23	1967	621
M	59.9	45.3	52.6	77	1967	32	1967	384
J	67.8	54.8	61.3	86	1966	40	1967	129
J	74.3	61.7	68.0	90	1968	50	1968	12
A	74.4	61.7	68.1	86	1968	46	1968	22
S	69.3	55.4	62.3	83	1966	36	1966	93
O	60.8	47.8	54.3	77	1967	22	1965	332
N	52.3	39.5	45.9	68	1967	20	1967	573
D	42.5	29.6	36.1	58	1966	4	1968	896
YR	56.0	43.0	49.5	90 JUL. 1968		2 JAN. 1968		5891

Precipitation

Month	Normal total (b)	Max monthly	Year	Min monthly	Year	Max in 24 hrs	Year
J	4.22	8.24	1955	1.21	1968+	2.82	1953
F	3.76	8.07	1952	1.77	1966	2.32	1953
M	4.54	8.88	1959	0.97	1962	2.92	1968
A	3.76	8.41	1953	1.51	1950	4.48	1953
M	2.88	10.38	1967	0.59	1962	6.53	1967
J	2.92	5.01	1949	0.01	1949	3.02	1967
J	2.71	7.45	1958	0.15	1968+	2.05	1961
A	3.68	12.92	1946	0.28	1947	3.67	1946
S	3.51	7.45	1958	0.42	1958	4.42	1950
O	3.70	7.45	1962	0.37	1946	3.21	1948
N	4.05	7.83	1953	1.06	1966	4.95	1966
D	3.93	6.40	1964	1.31	1955	2.48	1952
YR	43.66	12.92 AUG. 1946		0.01 JUN. 1949		6.53 MAY 1967	

Snow, Sleet

Month	Mean total	Max monthly	Year	Max in 24 hrs	Year
J	9.1	38.9	1965	17.8	1964
F	10.8	36.4	1952	20.1	1952
M	7.3	40.2	1960	16.1	1960
A	0.9	9.5	1955	8.0	1955
M	0.0	0.0		0.0	
J	0.0	0.0		0.0	
J	0.0	0.0		0.0	
A	0.0	0.0		0.0	
S	0.0	T	1965+	T	1965+
O	0.3	2.7	1967	2.5	1967
N	6.9	24.7	1963	15.5	1963
YR	35.3	40.2 MAR. 1960		20.1 FEB. 1952	

Relative humidity (Standard time used: EASTERN)

Month	1 AM	7 AM	1 PM	7 PM
J	79	80	67	75
F	75	78	65	75
M	84	82	69	82
A	87	80	66	83
M	90	82	71	87
J	97	90	78	92
J	94	88	76	92
A	97	91	77	94
S	93	89	72	90
O	86	85	68	85
N	81	81	72	78
D	76	79	69	76
YR	86	84	71	84

Wind

Month	Mean hourly speed	Prevailing direction	Fastest mile Speed	Direction	Year
J	14.7	NW	59	NW	1958
F	15.2	NW	66	WNW	1964
M	15.2	N	73	N	1956
A	14.6	NW	63	N	1956
M	13.0	SW	49	NE	1967+
J	12.0	SW	39	NE	1947
J	11.0	SW	42	S	1960
A	10.8	SW	72	SE	1954
S	11.8	SW	73	SE	1954
O	13.4	SW	69	E	1947
N	14.1	WNW	70	W	1956
D			57		
YR	13.2	SW	73 MAR. 1956+	N	

Means and number of days

Month	Mean sky cover	Pct. of possible sunshine	Clear	Partly cloudy	Cloudy	Precip .01 in or more	Snow/Sleet 1.0 in or more	Thunderstorms	Heavy fog	Max 90° & above	Max 32° & below	Min 32° & below	Min 0° & below
J	6.9	42	6	7	18	13	2	*	5	0	5	25	0
F	6.8	49	6	8	16	12	2	*	5	0	8	26	0
M	6.5	56	7	8	15	12	2	1	6	0	1	20	0
A	6.7	56	7	8	15	12	*	2	8	0	0	10	0
M	6.7	59	7	8	16	11	0	3	6	0	0	1	0
J	6.2	62	7	8	15	9	0	3	12	0	0	0	0
J	6.7	61	6	9	16	8	0	4	15	*	0	0	0
A	6.4	60	9	8	15	9	0	3	13	0	0	0	0
S	6.1	63	10	8	13	8	0	1	7	0	0	0	0
O	5.8	58	8	7	13	9	*	1	5	0	0	1	0
N	6.4	42	7	7	17	12	2	*	4	0	0	8	0
D	7.0	42	7	8	17	12	2		4	0	4	20	0
YR	6.6	55	87	92	186	125	9	20	98	*	18	112	0

ø For period September 1965 through current year.

Means and extremes in the above table are from existing or comparable location(s). Annual extremes have been exceeded at other locations as follows: Highest temperature 95 in August 1948; lowest temperature -6 in February 1918; maximum precipitation in 24 hours 5.73 in August 1889; fastest mile wind 91 E in March 1914.

PITTSFIELD, MASSACHUSETTS — MUNICIPAL AIRPORT 1967

LATITUDE 42° 26' N
LONGITUDE 73° 18' W
ELEVATION (ground) 1170 Feet

Temperature

Month	Normal Daily maximum (b)	Normal Daily minimum (b)	Normal Monthly (b)	Record highest	Year	Record lowest	Year	Normal degree days (b)
J	30.5	13.1	21.8	61	1950	-22	1957	1339
F	31.6	13.0	22.6	61	1943	-25	1943	1196
M	39.6	21.8	30.7	81	1945	-10	1950	1063
A	53.4	32.6	43.0	86	1962+	10	1964	660
M	66.2	42.9	54.6	93	1953	24	1966	326
J	74.6	51.5	63.1	95	1966	33	1964+	105
J	79.2	56.4	67.8	95	1966	40	1966	25
A	77.6	54.7	66.2	95	1955	32	1965+	59
S	69.4	46.8	58.1	95	1953	18	1963	219
O	59.0	37.3	48.1	85	1963	18	1966+	524
N	46.0	28.5	37.3	76	1956+	-9	1938	831
D	33.8	16.8	25.3	62	1962	-20	1942	1231
YR	55.1	34.6	44.9	95 JUL. 1966+		-25 FEB. 1943		7578

Precipitation

Month	Normal total (b)	Max monthly	Year	Min monthly	Year	Max in 24 hrs	Year
J	2.97	4.77	1953	1.22	1955	1.83	1960
F	2.51	5.04	1951	1.02	1957	1.49	1960
M	3.22	6.83	1953	1.00	1962	1.38	1962
A	3.87	6.66	1945	1.81	1966	1.59	1948
M	3.87	7.23	1966	1.75	1962	1.75	1953
J	4.28	10.32	1962	1.78	1962	3.07	1963
J	4.89	8.34	1945	1.22	1944	2.79	1944
A	3.90	8.20	1955	0.60	1953	3.41	1944
S	3.25	7.95	1960	0.95	1948	3.11	1953
O	3.25	7.04	1955	0.42	1963	2.80	1955
N	3.91	6.36	1950	1.35	1946	2.80	1950
D	3.25	9.34	1948	0.50	1948	4.25	1948
YR	44.42	10.32 JUN. 1962		0.42 OCT. 1963		4.25 DEC. 1948	

Snow, Sleet

Month	Mean total	Max monthly	Year	Max in 24 hrs	Year
J	19.0	36.2	1964	14.0	1964
F	20.7	35.0	1967	14.5	1967
M	13.7	40.2	1956	13.4	1956
A	5.0	15.8	1955	5.9	1955
M	0.0	0.0	1963	3.0	1963
J	0.0	0.0		0.0	
J	0.0	0.0		0.0	
A	0.0	0.0		0.0	
S	0.0	0.0		0.0	
O	0.2	2.0	1952	0.5	1952
N	5.1	18.5	1956	4.2	1961
D	14.1	27.7	1966	17.0	1966
YR	78.0	40.2 MAR. 1956		17.0 DEC. 1966	

Means and number of days

Month	Mean sky cover	Clear	Partly cloudy	Cloudy	Precip .01 in or more	Snow/Sleet 1.0 in or more	Thunderstorms	Heavy fog	Max 90° & above	Max 32° & below	Min 32° & below	Min 0° & below
J	7.5	3	9	19	15	6	0	2	0	29	29	29
F	7.5	3	9	17	14	6	*	1	0	19	30	6
M	6.4	4	10	17	14	4	1	3	0	16	26	4
A	7.6	4	9	17	14	1	4	1	0	1	16	0
M	6.9	4	11	14	13	0	4	3	*	0	3	0
J	5.8	4	14	12	12	0	7	4	*	0	0	0
J	5.6	4	15	12	10	0	7	7	1	0	0	0
A	5.7	5	13	13	10	0	5	8	1	0	0	0
S	5.3	6	11	13	9	0	3	5	*	0	0	0
O	7.2	4	8	18	13	1	1	2	0	3	10	0
N	7.1	4	8	18	14	2	*	0	0	15	20	3
D						5		3	0		28	
YR	6.5	57	127	181	152	24	28	45	2	64	163	14

$ DATA ACCUMULATED PRIOR TO 1961.

NORMALS, MEANS, AND EXTREMES

LATITUDE 42° 16' N
LONGITUDE 71° 52' W
ELEVATION (ground) 986 Feet

WORCESTER, MASSACHUSETTS
WORCESTER MUNICIPAL AIRPORT

Temperature

Month	Normal Daily maximum	Normal Daily minimum	Normal Monthly	Extreme Record highest	Year	Extreme Record lowest	Year
(Years of record)	(b)	(b)	(b)	13	13	13	13
J	31.2	16.8	24.0	59	1957	-19	1957+
F	32.8	16.9	24.9	63	1967+	-12	1967+
M	40.5	25.0	32.8	72	1968+	-4	1967
A	53.6	35.6	44.6	88	1962	14	1965
M	66.1	44.2	55.2	92	1962	28	1956
J	74.2	54.9	64.6	93	1956	40	1964+
J	79.0	60.5	69.8	94	1964	46	1963
A	77.4	59.1	68.3	90	1968+	38	1965
S	69.9	51.7	60.8	89	1957	30	1957
O	59.9	41.0	50.5	85	1963+	24	1966+
N	47.1	31.3	39.2	70	1961+	10	1958+
D	34.2	20.1	27.2	66	1966	-13	1962
YR	55.5	38.1	46.8	94	JUL. 1964	-19	JAN. 1957

Precipitation

Month	Normal degree days	Normal total	Max monthly	Year	Min monthly	Year	Max in 24 hrs.	Year
(Years of record)	(b)	(b)	13	13	13	13	13	13
J	1271	3.71	8.11	1958	1.60	1957	2.32	1964
F	1123	2.92	5.44	1960	1.26	1968	2.04	1965
M	998	4.11	7.67	1968	1.62	1957	3.89	1968
A	612	3.93	6.61	1966	1.62	1966	2.35	1959
M	304	3.79	7.01	1959	0.86	1959	3.03	1967
J	78	3.84	7.78	1968	1.77	1968	1.79	1956
J	6	3.63	8.11	1959	1.00	1957	2.64	1959
A	34	4.24	4.90	1961	1.08	1956	2.04	1959
S	147	3.92	6.92	1958	0.83	1957	4.79	1960
O	450	3.47	8.56	1962	1.46	1963	3.77	1959
N	774	4.26	8.20	1965	2.90	1963	2.36	1966
D	1172	3.59	6.92	1957	1.86	1957	2.35	1958
YR	6969	45.41	8.56 OCT.	1962	0.83 SEP.	1957	4.79 SEP.	1960

Snow, Sleet

Month	Mean total	Max monthly	Year	Max in 24 hrs.	Year
(Years of record)	13	13	13	13	13
J	17.8	44.0	1966	18.7	1961
F	19.6	45.2	1962	24.0	1962
M	19.5	36.5	1960	16.6	1960
A	4.4	11.3	1967+	10.2	1967+
M	T	T	1967+	T	
J	0.0	0.0		0.0	
J	0.0	0.0		0.0	
A	0.0	0.0		0.0	
S	0.0	0.0		0.0	
O	0.9	4.7	1962	4.7	1962
N	3.1	15.3	1968	8.3	1961
D	13.6	22.2	1967	15.6	1961
YR	78.9	45.2 FEB.	1962	24.0 FEB.	1962

Relative humidity (Standard time used: EASTERN)

Month	1 AM	7 AM	1 PM	7 PM
(Years of record)	13	13	13	13
J	71	73	60	67
F	69	72	57	63
M	69	70	55	61
A	69	67	50	56
M	71	66	47	56
J	77	71	53	65
J	78	73	55	67
A	80	77	56	70
S	81	80	58	72
O	77	75	54	68
N	76	77	60	69
D	75	74	63	71
YR	75	73	56	66

Wind

Month	Mean hourly speed	Prevailing direction	Fastest mile Speed	Fastest mile Direction	Fastest mile Year
(Years of record)	13	8	13	13	
J	13.1	WSW	60	25	1959
F	12.7	WNW	76	32	1956
M	12.0	W	76	29	1956
A	11.7	W	54	05	1956
M	10.9	SW	48	27	1956
J	9.5	SW	39	25	1958
J	8.9	SW	43	32	1957
A	8.6	SW	34	27	1956
S	9.4	SW	35	05	1961
O	10.2	WSW	43	20	1958
N	11.0	SW	54	20	1956
D	11.5	WSW	51	23	1957
YR	10.8	W	76	29	MAR. 1956+

Mean number of days

Month	Mean sky cover sunrise to sunset	Pct. of possible sunshine	Sunrise to sunset Clear	Partly cloudy	Cloudy	Precipitation .01 inch or more	Snow, Sleet 1.0 inch or more	Thunderstorms	Heavy fog	Max. 90° and above	Max. 32° and below	Min. 32° and below	Min. 0° and below
(Years of record)	13		13	13	13	13	13	13	13	13	13	13	13
J	6.0		9	8	14	11	4	0	6	0	17	30	3
F	6.1		8	6	14	11	4	*	5	0	14	28	2
M	6.4		8	7	16	11	4	1	7	0	6	26	*
A	6.4		7	9	14	11	1	1	8	0	*	12	0
M	6.4		6	11	14	12	0	3	7	*	0	1	0
J	6.2		6	11	13	11	0	4	9	1	0	0	0
J	6.2		8	10	13	11	0	5	7	1	0	0	0
A	6.0		8	10	13	10	0	3	7	*	0	0	0
S	5.6		10	9	11	9	0	1	8	*	0	0	0
O	5.5		11	9	11	9	*	1	8	0	0	5	0
N	6.6		7	9	14	12	2	*	8	0	2	17	0
D	6.1		7	9	15	12	4	0	6	0	15	28	1
YR	6.2		95	108	162	129	19	20	85	2	54	147	7

Means and extremes in the above table are from the existing or comparable location(s). Annual extremes have been exceeded at other locations as follows: Highest temperature 102 in July 1911; lowest temperature -24 in February 1943; maximum monthly precipitation 18.68 in August 1955; minimum monthly precipitation 0.04 in March 1915; maximum precipitation in 24 hours 8.67 in August 1955.

REFERENCE NOTES APPLYING TO ALL "NORMALS, MEANS, AND EXTREMES" TABLES

(a) Length of record, years.
(b) Climatological standard normals (1931-1960).
* Less than one half.
+ Also on earlier dates, months or years.
T Trace, an amount too small to measure.
Below-zero temperatures are preceded by a minus sign.

The prevailing direction for wind in the Normals, Means, and Extremes table is from records through 1963.

To 8 compass points only.

Unless otherwise indicated, dimensional units used in this bulletin are: temperature in degrees F.; precipitation, including snowfall, in inches; wind movement in miles per hour; and relative humidity in percent. Degree day totals are the sums of the negative departures of average daily temperatures from 65°F. Sleet was included in snowfall totals beginning with July 1948. Heavy fog reduces visibility to 1/4 mile or less.

Sky cover is expressed in a range of 0 for no clouds or obscuring phenomena to 10 for complete sky cover. The number of clear days is based on average cloudiness 0-3; partly cloudy days 4-7; and cloudy days 8-10 tenths.

Solar radiation data are the averages of direct and diffuse radiation on a horizontal surface. The langley denotes one gram calorie per square centimeter. Averages in the lower table for some months may be for more than the listed number of years.

& Figures instead of letters in a direction column indicate direction in tens of degrees from true North; i.e., 09 -East, 18 -South, 27 -West, 36 -North, and 00 -Calm. Resultant wind is the vector sum of wind directions and speeds divided by the number of observations. If figures appear in the direction column under "Fastest mile" the corresponding speeds are fastest observed 1-minute values.

Mean Maximum Temperature (°F.), January

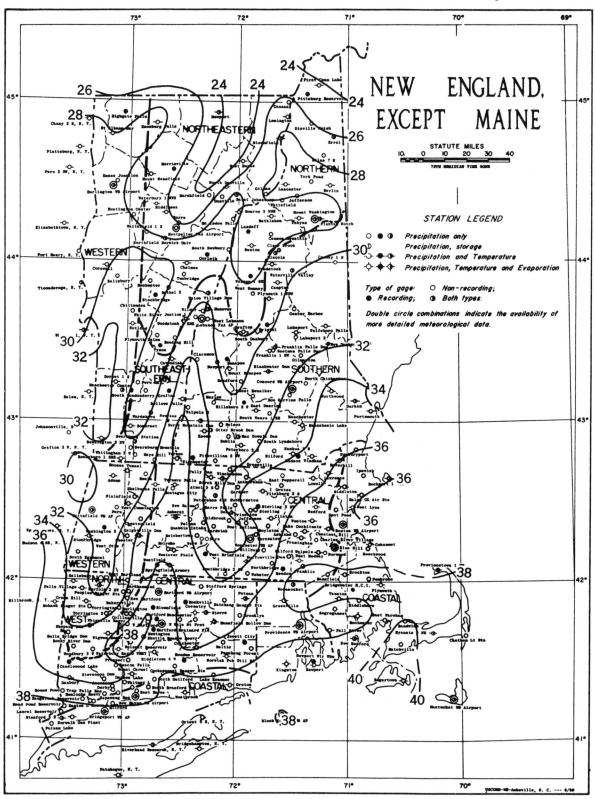

Based on period 1931-52

Isolines are drawn through points of approximately equal value. Caution should be used in interpolating on these maps, particularly in mountainous areas.

Mean Minimum Temperature (°F.), January

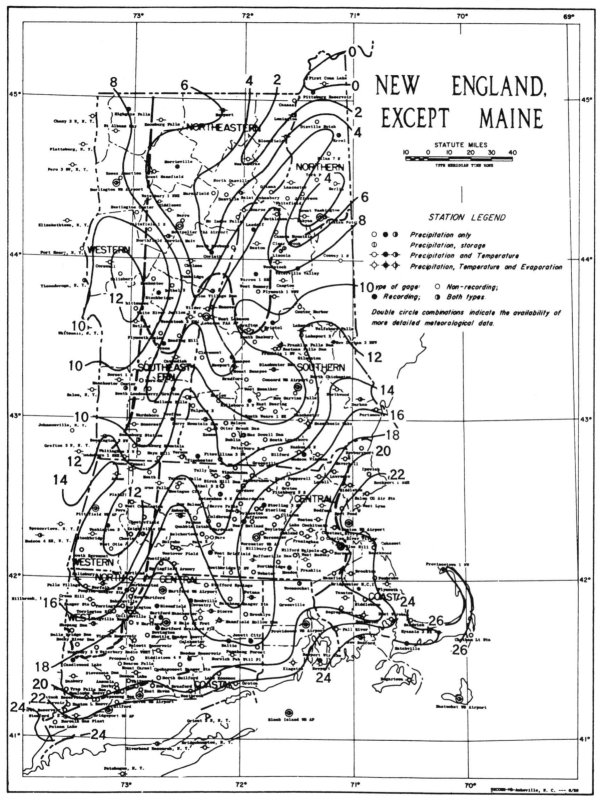

NEW ENGLAND, EXCEPT MAINE

STATUTE MILES
10 0 10 20 30 40
75TH MERIDIAN TIME ZONE

STATION LEGEND

○ ● ◐ Precipitation only
◑ Precipitation, storage
○-●-◐ Precipitation and Temperature
◇-◆-◈ Precipitation, Temperature and Evaporation

10 Type of gage: ○ Non-recording;
● Recording; ◐ Both types.

Double circle combinations indicate the availability of
more detailed meteorological data.

Based on period 1931-52

Isolines are drawn through points of approximately equal value. Caution should be used
in interpolating on these maps, particularly in mountainous areas.

Mean Maximum Temperature (°F.), July

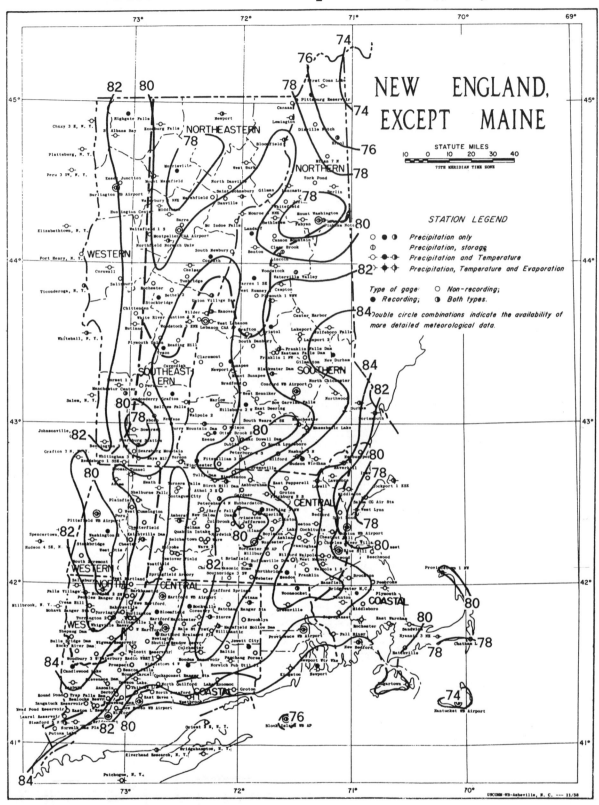

Based on period 1931-52

Isolines are drawn through points of approximately equal value. Caution should be used in interpolating on these maps, particularly in mountainous areas.

Mean Minimum Temperature (°F.), July

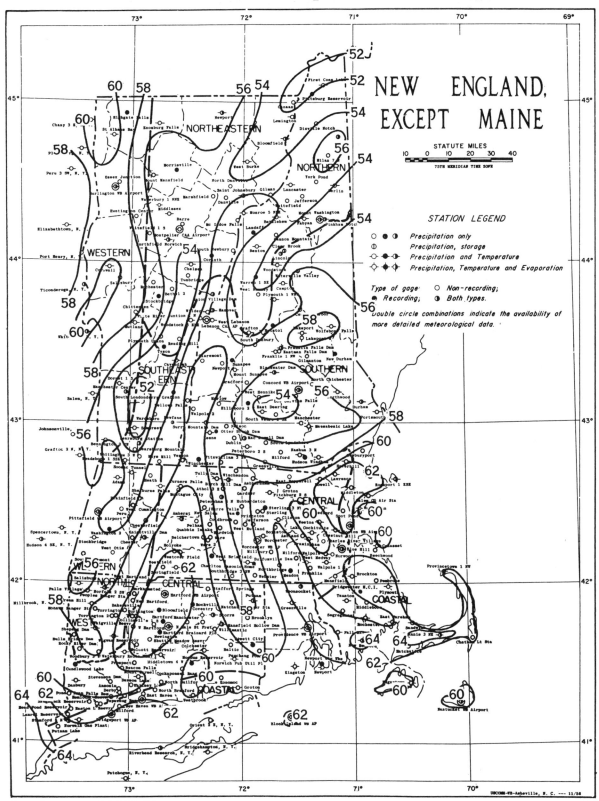

NEW ENGLAND, EXCEPT MAINE

STATUTE MILES

75TH MERIDIAN TIME ZONE

STATION LEGEND

○ ◐ ● Precipitation only
◑ Precipitation, storage
◔ ◑ ● Precipitation and Temperature
◇ ◈ ◆ Precipitation, Temperature and Evaporation

Type of gage: ○ Non-recording;
● Recording; ◑ Both types.

Double circle combinations indicate the availability of more detailed meteorological data.

USCOMM-WB-Asheville, N. C. --- 11/58

Based on period 1931-52

Isolines are drawn through points of approximately equal value. Caution should be used in interpolating on these maps, particularly in mountainous areas.

Mean Annual Precipitation, Inches

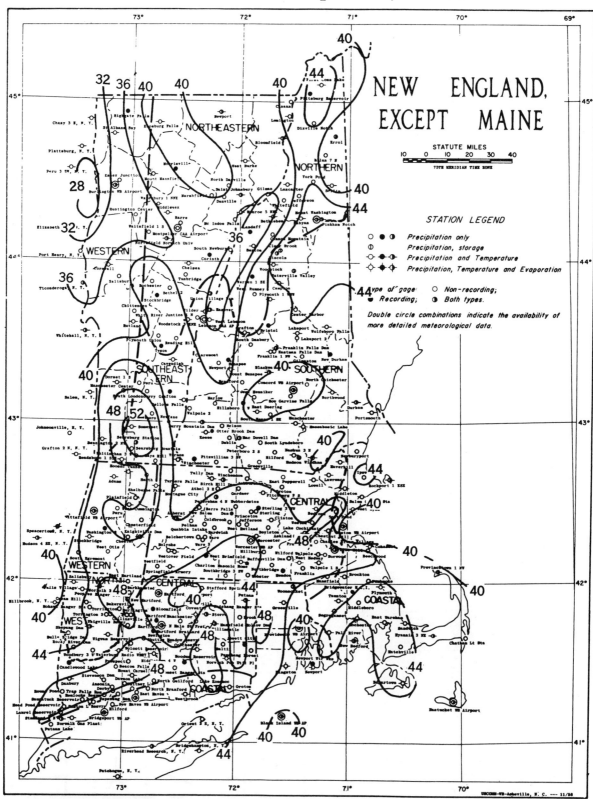

NEW ENGLAND, EXCEPT MAINE

STATION LEGEND

Based on period 1931-55

Isolines are drawn through points of approximately equal value. Caution should be used in interpolating on these maps, particularly in mountainous areas.

NEW ENGLAND, EXCEPT MAINE

STATUTE MILES

10   0   10   20   30   40

75 TH MERIDIAN TIME ZONE

STATION LEGEND

Soil Temperatures

Precipitation only

Precipitation, storage

Precipitation and Temperature

Precipitation, Temperatue and Evaporation

Type of gage   Non-recording,

Recording   Both types

Double circle combinations indicate the availability of more detailed meteorological data.

ALBERS EQUAL AREA PROJECTION
STANDARD PARALLELS AT 29 1/2 AND 45 1/2

USCOMM-ESSA-Asheville, N. C.
Revised 12 - 68

THE CLIMATE OF
MICHIGAN

by
Norton D. Strommen

August 1967

Michigan is located in the heart of the Great Lakes region and is composed of two large peninsulas. Many smaller peninsulas jut from these two peninsulas into the world's largest bodies of fresh water to give most of Michigan a quasi-marine type climate in spite of its midcontinent location.

The Upper Peninsula is long and narrow, lying primarily between 45° and 47° north latitude. It averages only 75 miles in width and extends from Northern Wisconsin eastward over 300 miles into Northern Lake Huron. Lake Superior lies to the north while the northern portion of Lake Michigan forms the boundary to the southeast. Isle Royale, separated from the mainland, is located in Lake Superior about 50 miles northwest of the tip of the Keweenaw Peninsula. Isle Royale, about 10 miles wide and 25 miles long is a popular National Park during the summer months.

The Lower Peninsula, shaped like a mitten and occupying about 70 percent of Michigan's total land area, extends northward nearly 300 miles from the Indiana and Ohio border or about 42° north latitude to the eastern end of the Upper Peninsula. The recently completed Mackinaw Bridge which spans the Straits of Mackinac now joins the two large peninsulas together where Lake Michigan flows into Lake Huron. Lake Michigan extends the entire length of the Lower Peninsula on the west while Lakes Huron, St. Clair and Erie form the eastern boundary. The total coastline for the state exceeds 3100 miles.

In addition, Michigan has over 11,000 smaller lakes with a total surface area of over 1,000 square miles. These lakes are scattered throughout 81 of the 83 counties while more than 36,000 miles of streams wind their way across the state.

While latitude, by determining the amount of solar insolation, is the major climatic control, the Great Lakes and variations in elevation play an important role in the amelioration of Michigan's climate. Because of its mid-latitude location, prevailing winds are from a westerly direction. During the summer months winds are predominantly from the southwest when the semipermanent Bermuda High Pressure Center is located over the southeastern United States. During the winter months the prevailing winds are west to northwest, but change quite frequently for short periods as migrating cyclones and anticyclones move through the area. One exception occurs in the eastern portion of the Upper Peninsula where easterly winds prevail during the late fall and early winter months. This is the result of early winter anticyclones moving eastward across Canada and major storm tracks beginning to push southward.

TOPOGRAPHY.---The eastern half of the Upper Peninsula varies from level to gently rolling hills with elevation generally between 600 and 1000 feet above sea level. The gently rolling hills located along the Lake Superior shoreline contain the famous Pictured Rocks and Tahquamenon Fall areas. The western table-lands rise to elevations generally between 1400 and 1600 feet with Porcupine Mountain, the state's highest point, 2023 feet, located in Ontonagan County overlooking Lake Superior. The rugged hills extend northeastward from Ontonagan County through the center of the Keweenaw Peninsula and play an important role in the larger precipitation amounts received in this area. The Lower Peninsula features range from quite level terrain in the Southeast to gently rolling hills in the southwest with elevations generally between 800 and 1000 feet. A series of sand dunes along the Lake Michigan shoreline rise to heights of nearly 400 feet above the lake level. These are the result of the prevailing westerly winds which blow across the lake. Tablelands cover the northern part of the Lower Peninsula and reach a maximum elevation of 1700 feet in Osecola County near Cadillac. In the northwestern section of the Lower Peninsula a number of finger-like peninsulas extend into Grand Traverse Bay and Lake Michigan.

LAKE INFLUENCE.---The lake effect imparts many interesting departures to Michigan's climate which one would not ordinarily expect to find at a midcontinental location. Because of the lake waters' slow response to temperature changes and the dominating westerly winds, the arrival of both summer and winter are retarded. In the spring, the cooler temperatures slow the development of vegetation until the danger of frost is past. In the fall, the warmer lake waters temper the first outbreaks of cold air allowing additional time for crops to mature or reach a stage which is free from damage by frost. This lake effect is best seen by comparing stations at similiar latitudes in Wisconsin and Michigan. In July we find Madison, Wisconsin's mean temperature is 71.2°, while Lansing has a mean of 71.1° and Muskegon's mean is 69.9° F. In January, this trend is reversed with Madison's mean temperature 17.5°, while Lansing has a mean of 24.3° and Muskegon's mean is 26.0° F.

With the first cold air outbreaks in the fall, Michigan experiences a considerable increase in cloudiness. When cold air passes over the warmer lake water, a shallow layer of unstable, moisture-ladden air developes in the lower levels of the atmosphere. This air, when forced to rise, produces the increased cloudiness and frequent

snow flurry activity observed in the fall and early winter months. A comparison of percent of possible sunshine in December shows Lansing receiving 27 percent while Madison, Wisconsin receives 42 percent. This difference decreases slowly as the lake waters cool, but does not completely disappear until the latter part of February.

Precipitation frequencies also show large variation from one side of the lake to the other. In January, Milwaukee, Wisconsin experiences measurable precipitation about 20 percent of the time or, on an average, once every five days. While Muskegon, with 45 percent, can expect, on an average, measureable precipitation every other day. In June this trend is reversed with Milwaukee's frequency of measurable precipitation almost 25 percent and Muskegon's frequency down to less than 15 percent. This difference in precipitation frequencies decreases inland from the lakes.

On warm summer days when prevailing winds are generally light, the lake's shore area frequently develops a localized wind pattern which may extend inland for only a few miles. This is frequently referred to as the 'lake breeze'. It developes when the much warmer air over the land masses begins to rise, allowing the cooler air over the lakes to move inland. At night this pattern may be reversed creating what is known as a 'land breeze'. A wind of this type may also be observed, but on a much smaller scale along the shores of the larger inland lakes.

GROWING SEASON.---The length of Michigan's growing season or freeze-free period does not decrease in the normal manner from south to north. Instead, isolines for the length of the growing season follow closely the contours of the lake shores. The shortest average growing season, about 60 days, occurs in the interior section of the Western Upper Peninsula. The growing season increases to between 140 and 160 days, as one goes towards the lake shores. A similar pattern exists in the Lower Peninsula where the growing season in the northern table-lands averages only 70 days, but increases rapidly to 140 days near the Lakes. Michigan's maximum average growing season, 170 days, is found in the southwest and southeastern corners of the state.

PRECIPITATION. --- Michigan averages about 31 inches of precipitation per year. About 55-60 percent of the annual total is recorded during the normal growing season. Summer precipitation falls primarily in the form of showers or thunderstorms, while a more steady type of precipitation of lighter intensity dominates the winter months. The annual number of thunderstorms observed decreases from about 40 in the south to around 25 in the Upper Peninsula area with nearly 50 percent of these recorded during the summer months, June through August. The maximum five-minute rainfall total of 0.86 of an inch was recorded at Detroit on August 17, 1926. The greatest monthly total, 16.24 inches, occurred at Battle Creek in June of 1883. Battle Creek also failed to record any precipitation in February 1877. A number of other stations in southern and central Michigan have recorded no rainfall for a month, but these records were established during the late summer or early fall months. The frequency of floods is quite low in Michigan with the greatest likelihood occurring in late winter or early spring when sudden warming and rain may be combined with snowmelt. Mild meteorological drought conditions are not uncommon in Michigan, but meteorological droughts reaching severe conditions are infrequent and generally of short duration. The normally even distribution of precipitation and higher humidities observed in Michigan are helpful in reducing the high demands for moisture, as experienced in other areas of the Upper Midwest.

SNOWFALL.---Michigan receives some of the heaviest snowfall totals east of the Rockies except for isolated points in the New England States. The maximum average annual snowfall amounts of over 170 inches, are located along the escarpment which rises abruptly to an elevation of over 1400 feet above Lake Superior, at the western end of the Upper Peninsula. Another area with amounts exceeding 120 inches is centered in the western section of the tableland region of the Lower Peninsula. The prevailing westerlies, passing over the Great Lakes, become moisture ladden in the lower levels and when forced upward by the land masses, drop much of their excessive moisture in the form of snow squalls in these areas. The record seasonal snowfall total of 276.5 inches occurred at Calumet during the winter of 1949-1950. A single storm from January 15-20, 1950 accounted for 46.1 inches of this total. The 24-hour snowfall record of 27 inches was established on October 23, 1929, at Ishpeming and equaled at Dunbar on March 29, 1947.

TEMPERATURES.---The coldest temperature recorded in Michigan was -51° at Vanderbilt on February 9, 1934. The highest, 112°, occurred July 13, 1936 at Mio. Temperatures below -40° have been recorded in most interior sections of the State, but seldom have readings of -20° been observed in the immediate vicinity of the Great Lakes. This modification in temperature extremes by the lakes enables Michigan to produce successfully a variety of crops more ideally suited to the climate of the southern United States.

STORMS AND TORNADOES.---Damaging or dangerous storms do not occur as frequently in Michigan as in the states to the south and west. Recorded tornado occurrences in Michigan have averaged four per year for the period 1916-1965 with an average of about nine per year during the last decade. The increase in the last decade is attributed primarily to better reporting services and networks. About 90 percent of these tornadoes occurred in the southern one-half of the Lower Peninsula. In recent times, a most destructive series of tornadoes were observed on Palm Sunday, April 11, 1965, causing over $51,000,000 in damages to southern Michigan. Damaging wind storms and blizzards are not as frequent but do cause considerable damage from time to time. Hail is most frequently observed in the spring months, but the total damage caused by hail is small. A higher frequency of hail is noted in the fall months over the northwestern section of the Lower Peninsula. This is attributed mainly to the strong lake influence in this region.

WATER SUPPLY AND AGRICULTURE.--- Michigan is particularly fortunate in its supply of both surface and ground water. Surface water supplies are constantly replenished by a mean annual rainfall averaging about 31 inches, well-distributed throughout the year and with little annual variation. Because of moderately high humidities, evaporation rates are quite low so that less water is lost in this manner. Heavy industrial demands are made upon the water supply of Michigan, but few industries have had to go any great distance to find adequate supplies. Aside from the availability of the lake water, industry can normally meet its water requirement needs from depths running from 25 to 400 feet, with the majority of wells from 50 to 200 feet deep. There is, of course, also an abundance of water to meet the needs of cities and individuals. Because of its climate, soils and marketing conditions, a large variation in agricultural practices and many varieties of crops are found in Michigan. Primary crops in Michigan with respect to agriculture are hay and pasture, corn, wheat, field beans, oats, truck crops, potatoes, sour cherries, apples, sugar beets and peaches. Hay and tillable pasture occupy about 28 percent of the total tillable land and are grown generally throughout the State, as are oats. Corn production is concentrated in the southern half of the Lower Peninsula, and field bean growing is concentrated in the Saginaw Valley and "Thumb" areas. Tuscola, Sanilac, Saginaw, Huron and Bay are the leading sugar beet counties in the State.

The fruit belt of Michigan is located in the southwestern and western border areas of the Lower Peninsula, along the shores of Lake Michigan where, because of the prevailing westerly winds, the tempering influence of the lake water is strongest. Peaches are concentrated in five southwest and western counties, while apple orchards extend north to the Grand Traverse area. This latter area is noted for its production of sour cherries. Cherry growing is best adapted to areas having relatively short, cool growing seasons free from extremely low winter temperatures and free from frosts at blossom time. A number of fingerlike peninsulas extend into Grand Traverse Bay with further amelioration of the climate, and cherry growing is heavily concentrated on these peninsulas.

RECREATION.---Because of the Great Lakes' influence on Michigan's climate the changing topographic features, and thousands of miles of lake shore, we find a robust year-around tourist business in the State. The cooler temperatures and abundant natural beauty provide an excellent summertime vacation attraction. The topography and tempered wintertime temperatures in the heavy snowfall areas are most ideal for skiing and tobogganing. It provides the best conditions in the Midwest for the many winter sports enthusiasts from the heavily populated areas to the south.

BIBLIOGRAPHY

(1) Climatic Summary of the United States (Bulletin W) 1930 edition, Sections 63, 64 and 65. U. S. Weather Bureau

(2) Climatic Summary of the United States, Michigan - Supplement for 1931 through 1952 (Bulletin W Supplement). U. S. Weather Bureau

(3) Climatic Summary of the United States - Supplement for 1951 through 1960 - Michigan. Climatography of the U. S. No. 86-16. U. S. Weather Bureau

(4) Climatological Data - Michigan U. S. Weather Bureau

(5) Climatological Data National Summary U. S. Weather Bureau

(6) Hourly Precipitation Data - Michigan U. S. Weather Bureau

(7) Local Climatological Data, U. S. Weather Bureau, for Alpena, Detroit, Escanaba (through 1961), Flint, Grand Rapids, Lansing, Marquette, Muskegon and Sault Ste. Marie, Michigan

REFERENCES

A. Hill, Elton B. and Russell G. Mawby, Types of Farming in Michigan. Special Bulletin 206 (Revised 1954), Agricultural Experiment Station, Michigan State College.

B. Partridge, Newton L., "Two Cold January Nights Against a Background of Thirty Winters", Quarterly Bulletin, Agricultural Experiment Station, Michigan State College. Vol. 24, May 1942, pp. 337-342.

C. Michigan Department of Economic Development, Michigan Water Resources for Industry.

D. McNamee, Robert L., The Surface Waters of Michigan, Engineering Research Bulletin No. 16, June 1930, Department of Engineering Research of Michigan.

E. Day, Maurice W., "Snow Damage to Conifer Plantations", Quarterly Bulletin, Agricultural Experiment Station, Michigan State College. Vol. 23, November 1940, No. 2, pp. 97-98.

F. Partridge, Newton L., "Extent of Injury to Fruit Trees during the Winter of 1941-42 and What Can Be Done About It", Quarterly Bulletin, Agricultural Experiment Station, Michigan State College. Vol. 25, No. 3, February 1943, pp. 255-263.

G. The Water Resources Commission, State of Michigan, Water State Wide, Combined Second and Third Annual Reports, 1950-1952, pp. 15-17.

H. Wisler, C. O., G. J. Stramd and L. B. Laird, "Water Resources of the Detroit Area".

I. Stramd, G. J., C. O. Wisler and L. B. Laird, Water Resources of the Grand Rapids Area.

J. Water Resources Commission, Water Resources of the Clinton River Basin.

K. Baten, W. D. and A. H. Eichmeier, A Summary of Weather Conditions at East Lansing, Michigan Prior to 1950, Special Bulletin, Agricultural Experiment Station, Michigan State College.

L. Maximum Recorded United States Point Rainfall for 5 Minutes to 24 Hours at 296 First-Order Stations-U. S. Department of Commerce, Weather Bureau, Technical Paper No. 2, Washington, D. C., Revised 1963.

M. Mean Number of Thunderstorm Days in the United States. Prepared by the Climatological Services Division, U. S. Weather Bureau Technical Paper No. 19, Washington, D. C., December 1952.

N. Tornado Occurrences in the United States, U. S. Department of Commerce, Weather Bureau, Technical Paper No. 20, Washington, D. C., Revised 1960.

O. Michigan Snow Depths. Michigan Weather Service, Cooperating with the Weather Bureau, United States Department of Commerce, A. H. Eichmeier, State Climatologist, June 1964.

P. Michigan Freeze Bulletin, Research Report 26-Farm Science, Michigan State University, Agricultural Experiment Station, East Lansing, May 1965.

Q. Climatic Frequency of Precipitation at Central Region Stations, Central Region, Technical Memorandum #8, Kansas City, Missouri, November 1966, Environmental Science Services Administration, Weather Bureau.

FREEZE DATA

STATION	Freeze threshold temperature	Mean date of last Spring occurrence	Mean date of first Fall occurrence	Mean No. of days between dates	Years of record Spring	No. of occurrences in Spring	Years of record Fall	No. of occurrences in Fall
ADRIAN 2 NNE	32	05-03	10-07	157	30	30	30	30
	28	04-19	10-23	187	30	30	30	30
	24	04-05	11-05	214	30	30	30	30
	20	03-24	11-22	243	30	30	30	30
	16	03-16	11-30	259	30	30	30	30
ALLEGAN	32	05-03	10-08	158	30	30	30	30
	28	04-21	10-27	189	30	30	30	30
	24	04-07	11-12	219	30	30	30	30
	20	03-26	11-21	240	30	30	30	30
	16	03-15	12-03	263	30	30	30	29
ALMA	32	05-09	10-04	148	30	30	30	30
	28	04-27	10-22	178	30	30	30	30
	24	04-13	11-08	209	29	29	30	30
	20	03-30	11-20	235	29	29	30	30
	16	03-20	11-27	252	29	29	30	30
ALPENA WB AIRPORT	32	05-29	09-16	110	30	30	29	29
	28	05-16	10-01	138	29	29	28	28
	24	04-29	10-21	175	29	29	28	28
	20	04-14	11-08	208	29	29	28	28
	16	04-02	11-23	235	29	29	27	27
ALPENA SEWAGE PLANT	32	05-06	10-10	157	30	30	30	30
	28	04-22	11-01	193	30	30	30	30
	24	04-09	11-14	219	30	30	30	30
	20	04-01	11-24	237	30	30	30	30
	16	03-24	11-30	251	30	30	30	30
ANN ARBOR UNIV OF MICH	32	04-29	10-18	172	30	30	29	29
	28	04-14	11-03	203	30	30	29	29
	24	04-02	11-14	226	30	30	29	29
	20	03-23	11-25	247	30	30	30	29
	16	03-15	12-01	261	30	30	29	29
ATLANTA 3 WSW	32	05-31	09-12	104	28	28	30	30
	28	05-18	09-28	133	27	27	30	30
	24	05-07	10-17	163	27	27	29	29
	20	04-21	10-30	192	27	27	28	28
	16	04-06	11-18	226	28	28	29	29
BAD AXE	32	05-15	10-04	142	29	29	30	30
	28	05-01	10-16	168	28	28	30	30
	24	04-17	11-05	202	28	28	29	29
	20	04-02	11-19	231	28	28	29	29
	16	03-22	12-01	254	28	28	28	27
BALDWIN STATE FOREST	32	06-01	09-14	105	23	23	21	21
	28	05-17	09-29	135	22	22	21	21
	24	05-08	10-11	156	22	22	21	21
	20	04-22	11-02	194	22	22	21	21
	16	04-13	11-13	214	22	22	20	20
BATTLE CREEK WBCK	32	05-03	10-08	158	30	30	30	30
	28	04-20	10-26	189	30	30	30	30
	24	04-08	11-10	216	30	30	30	30
	20	03-26	11-19	238	30	30	30	30
	16	03-17	11-29	257	30	30	30	30
BAY CITY SEWAGE PLANT	32	05-01	10-14	166	30	30	30	30
	28	04-17	11-03	200	30	30	29	29
	24	04-04	11-14	224	30	30	30	30
	20	03-26	11-25	244	30	30	30	30
	16	03-19	12-01	257	30	30	30	30
BEECHWOOD 7 WNW	32	06-01	09-13	104	10	10	10	10
	28	05-25	09-22	120	9	9	10	10
	24	05-08	10-12	157	10	10	10	10
	20	05-02	10-28	179	10	10	10	10
	16	04-15	11-07	206	10	10	10	10
BENTON HARBOR AIRPORT	32	05-07	10-06	152	20	20	20	20
	28	04-22	10-23	184	20	20	20	20
	24	04-10	11-09	213	20	20	20	20
	20	03-26	11-24	243	19	19	20	20
	16	03-13	12-01	263	19	19	20	20
BERGLAND DAM	32	06-01	09-11	102	10	10	10	10
	28	05-19	09-27	131	11	11	10	10
	24	05-08	10-05	150	11	11	10	10
	20	04-29	11-04	189	11	11	10	10
	16	04-15	11-08	207	11	11	10	10
BIG RAPIDS WATERWORKS	32	05-17	09-25	131	29	29	30	30
	28	05-02	10-10	161	29	29	30	30
	24	04-20	10-29	192	30	30	30	30
	20	04-04	11-14	224	30	30	30	30
	16	03-26	11-24	243	30	30	30	30
BLOOMINGDALE	32	05-07	10-09	155	30	30	30	30
	28	04-23	10-26	186	30	30	30	30
	24	04-10	11-08	212	30	30	30	30
	20	03-28	11-20	237	30	30	30	30
	16	03-17	12-03	261	30	30	30	29
BOYNE FALLS ST NURSERY	32	06-06	09-09	95	10	10	10	10
	28	05-24	09-25	124	10	10	10	10
	24	05-16	10-09	146	10	10	10	10
	20	04-23	10-30	190	10	10	10	10
	16	04-09	11-17	222	10	10	10	10
CADILLAC	32	05-27	09-11	107	30	30	30	30
	28	05-15	10-02	140	30	30	30	30
	24	05-03	10-13	163	29	29	30	30
	20	04-23	11-04	195	30	30	30	30
	16	04-13	11-15	216	30	30	30	30
CARO STATE HOSPITAL	32	05-21	09-25	127	28	28	30	30
	28	05-09	10-08	152	28	28	30	30
	24	04-23	10-26	186	28	28	30	30
	20	04-08	11-10	216	28	28	30	30
	16	03-27	11-24	242	28	28	30	30
CHARLOTTE	32	05-10	10-03	146	28	28	30	30
	28	04-30	10-19	172	28	28	30	30
	24	04-13	11-03	204	26	26	28	28
	20	04-03	11-13	224	26	26	29	29
	16	03-20	11-24	249	28	28	29	29
CHATHAM EXP FARM	32	06-04	09-14	102	30	30	30	30
	28	05-22	09-30	131	30	30	30	30
	24	05-10	10-19	162	30	30	30	30
	20	04-24	11-05	195	30	30	30	30
	16	04-09	11-16	221	30	30	30	30
CHEBOYGAN RR LIGHT STA	32	05-18	10-04	139	29	29	30	30
	28	05-02	10-22	173	29	29	30	30
	24	04-17	11-09	206	29	29	28	28
	20	04-07	11-21	228	29	29	29	29
	16	03-29	11-28	244	29	29	29	29
COLDWATER STATE SCHOOL	32	05-06	10-05	152	30	30	30	30
	28	04-23	10-19	179	30	30	30	30
	24	04-07	11-05	212	30	30	30	30
	20	03-28	11-17	234	30	30	30	30
	16	03-18	11-29	256	30	30	30	30
DEER PARK ST FOREST	32	06-11	09-01	82	23	23	22	22
	28	05-25	09-30	128	23	23	23	23
	24	05-11	10-20	162	23	23	23	23
	20	04-25	11-03	192	24	24	23	23
	16	04-14	11-11	211	24	24	23	23
DETROIT FAA AP CITY	32	04-24	10-22	181	30	30	30	30
	28	04-09	11-06	211	30	30	30	30
	24	03-28	11-22	239	30	30	30	30
	20	03-19	11-28	254	30	30	30	30
	16	03-10	12-06	271	30	30	30	28
DETROIT WBAP WILLOW RN	32	04-21	10-20	182	10	10	10	10
	28	04-09	11-04	209	10	10	10	10
	24	03-29	11-14	230	10	10	10	10
	20	03-19	11-23	249	10	10	10	10
	16	03-13	11-29	261	10	10	10	10

FREEZE DATA

STATION	Freeze threshold temperature	Mean date of last Spring occurrence	Mean date of first Fall occurrence	Mean No. of days between dates	Years of record Spring	No. of occurrences in Spring	Years of record Fall	No. of occurrences in Fall
DOWAGIAC 2 E	32	05-05	10-05	153	14	14	14	14
	28	04-21	10-28	190	14	14	13	13
	24	04-03	11-07	218	14	14	14	14
	20	03-26	11-18	237	14	14	14	14
	16	03-17	11-28	256	13	13	14	14
DUNBAR FOREST EXP STA	32	05-26	09-25	122	19	19	19	19
	28	05-12	10-14	155	19	19	19	19
	24	04-24	11-01	191	19	19	19	19
	20	04-15	11-13	212	19	19	19	19
	16	04-01	11-20	233	19	19	19	19
EAST JORDAN	32	05-28	09-27	122	29	29	27	27
	28	05-15	10-17	155	29	29	28	28
	24	04-29	11-05	190	28	28	27	27
	20	04-13	11-20	221	29	29	29	29
	16	03-31	11-26	240	29	29	30	30
EAST LANSING EXP FARM	32	05-09	10-04	148	10	10	10	10
	28	04-25	10-20	178	10	10	10	10
	24	04-09	11-03	208	10	10	10	10
	20	03-28	11-17	234	10	10	9	9
	16	03-18	11-26	253	10	10	9	9
EAST TAWAS U S FOREST	32	05-23	09-26	126	30	30	29	29
	28	05-06	10-07	154	30	30	30	30
	24	04-24	10-22	181	28	28	30	30
	20	04-12	11-11	213	29	29	26	26
	16	03-28	11-24	241	28	28	28	28
EAU CLAIRE 4 NE	32	05-01	10-19	171	30	30	29	29
	28	04-16	11-03	201	30	30	30	30
	24	04-03	11-11	222	30	30	30	30
	20	03-21	11-22	246	30	30	30	30
	16	03-15	12-01	261	30	30	30	28
ELBERTA 4 SE	32	05-21	10-07	139	24	24	28	28
	28	05-05	10-22	170	23	23	28	28
	24	04-18	11-11	207	23	23	27	27
	20	04-04	11-20	230	23	23	27	27
	16	03-24	12-02	253	22	22	27	27
ESCANABA	32	05-10	10-07	150	30	30	30	30
	28	04-25	10-22	182	30	30	30	30
	24	04-13	11-05	206	30	30	30	30
	20	04-02	11-18	230	30	30	30	30
	16	03-28	11-23	240	30	30	30	30
EVART	32	05-22	09-14	115	10	10	10	10
	28	05-16	09-25	132	10	10	10	10
	24	05-01	10-09	161	10	10	10	10
	20	04-18	11-01	197	10	10	9	9
	16	04-09	11-14	219	10	10	9	9
FAYETTE	32	05-18	10-06	141	27	27	29	29
	28	05-06	10-23	170	27	27	30	30
	24	04-23	11-10	201	27	27	27	27
	20	04-08	11-19	225	27	27	28	28
	16	03-28	11-27	244	27	27	28	28
FIFE LAKE 4 SW	32	06-06	09-06	92	29	29	29	29
	28	05-22	09-30	131	29	29	29	29
	24	05-12	10-13	154	30	30	29	29
	20	05-03	11-02	183	30	30	30	30
	16	04-19	11-11	206	30	30	30	30
FLINT WB AIRPORT	32	05-09	10-07	151	30	30	30	30
	28	04-23	10-24	184	30	30	30	30
	24	04-07	11-09	216	30	30	30	30
	20	03-24	11-21	242	30	30	30	30
	16	03-18	11-30	257	30	30	30	30
GERMFASK WILDLIFE REF	32	05-24	09-20	119	22	22	22	22
	28	05-10	10-10	153	22	22	22	22
	24	04-23	10-25	185	22	22	21	21
	20	04-13	11-11	212	22	22	22	22
	16	04-03	11-21	232	21	21	22	22
GLADWIN	32	05-19	09-24	128	29	29	30	30
	28	05-05	10-09	157	29	29	30	30
	24	04-23	10-22	182	29	29	30	30
	20	04-04	11-10	220	29	29	30	30
	16	03-27	11-23	241	29	29	30	30
GRAND HAVEN FIRE DEPT	32	05-02	10-13	164	30	30	30	30
	28	04-19	11-05	200	30	30	29	29
	24	04-02	11-16	228	30	30	29	29
	20	03-24	11-26	247	30	30	30	30
	16	03-15	12-06	266	30	30	30	29
GRAND MARAIS 1 SSE	32	05-30	09-23	116	30	30	30	30
	28	05-14	10-13	152	30	30	29	29
	24	04-30	10-28	181	29	29	30	30
	20	04-19	11-12	207	29	29	30	30
	16	04-08	11-23	229	29	29	30	30
GRAND RAPIDS WB APT	32	05-03	10-08	158	10	10	10	10
	28	04-17	10-23	189	10	10	10	10
	24	04-09	11-07	212	10	10	10	10
	20	03-27	11-19	237	10	10	10	10
	16	03-22	11-26	249	10	10	10	10
GRAND RAPIDS CITY	32	04-23	10-30	190	25	25	25	25
	28	04-11	11-09	212	25	25	25	25
	24	03-29	11-18	234	25	25	25	25
	20	03-19	11-30	256	25	25	25	25
	16	03-11	12-10	274	25	25	25	23
GRAYLING MILITARY RES	32	05-30	09-16	109	30	30	30	30
	28	05-16	10-01	138	30	30	29	29
	24	05-02	10-16	167	30	30	30	30
	20	04-22	11-03	195	30	30	30	30
	16	04-09	11-15	220	30	30	30	30
GREENVILLE 2 NNE	32	05-06	10-05	152	29	29	27	27
	28	04-23	10-20	180	29	29	29	29
	24	04-12	11-05	207	28	28	30	30
	20	03-31	11-15	229	30	30	30	30
	16	03-20	11-28	253	30	30	29	29
GROSSE POINTE FARMS	32	04-26	10-24	181	14	14	11	11
	28	04-12	11-04	206	13	13	10	10
	24	03-31	11-23	237	13	13	11	11
	20	03-20	11-29	254	13	13	11	11
	16	03-14	12-05	266	13	13	11	10
GULL LAKE BIOL. STA.	32	05-06	10-05	152	30	30	30	30
	28	04-26	10-23	180	30	30	30	30
	24	04-14	11-07	207	30	30	30	30
	20	03-31	11-16	230	29	29	30	30
	16	03-18	11-28	255	30	30	30	30
HALE LOUD DAM	32	05-22	09-26	127	30	30	30	30
	28	05-12	10-07	148	30	30	30	30
	24	04-24	10-23	182	29	29	30	30
	20	04-16	11-11	209	29	29	30	30
	16	04-02	11-21	233	29	29	30	30
HARBOR BEACH 3 NW	32	05-08	10-15	160	29	29	27	27
	28	04-14	10-29	188	29	29	28	28
	24	04-10	11-14	218	29	29	27	27
	20	03-28	11-23	240	28	28	27	27
	16	03-18	11-30	257	28	28	27	27
HARRISVILLE	32	05-19	10-02	136	29	29	30	30
	28	05-06	10-19	166	29	29	30	30
	24	04-21	11-03	196	29	29	30	30
	20	04-07	11-16	223	28	28	30	30
	16	03-31	11-25	239	28	28	29	29
HART	32	05-21	09-28	130	30	30	30	30
	28	05-08	10-18	162	30	30	30	30
	24	04-22	11-07	199	30	30	30	30
	20	04-06	11-18	226	30	30	30	30
	16	03-26	11-28	247	30	30	30	30

FREEZE DATA

STATION	Freeze threshold temperature	Mean date of last Spring occurrence	Mean date of first Fall occurrence	Mean No. of days between dates	Years of record Spring	No. of occurrences in Spring	Years of record Fall	No. of occurrences in Fall
HASTINGS FISHERIES	32	05-12	10-01	142	30	30	29	29
	28	04-28	10-15	170	30	30	29	29
	24	04-17	11-03	200	30	30	29	29
	20	04-01	11-15	228	30	30	29	29
	16	03-20	11-23	248	30	30	28	28
HIGGINS LAKE	32	05-25	09-20	118	30	30	30	30
	28	05-19	10-02	136	30	30	30	30
	24	05-10	10-17	160	30	30	30	30
	20	04-26	10-30	187	30	30	30	30
	16	04-12	11-13	215	29	29	29	29
HILLSDALE	32	05-10	09-29	142	30	30	30	30
	28	04-29	10-12	166	30	30	30	30
	24	04-14	11-01	201	30	30	30	30
	20	03-30	11-14	229	30	30	30	30
	16	03-19	11-25	251	29	29	30	30
HOLLAND WJBL	32	05-05	10-09	157	30	30	30	30
	28	04-24	10-31	190	30	30	30	30
	24	04-03	11-13	224	30	30	30	30
	20	03-24	11-23	244	30	30	30	30
	16	03-14	12-03	264	30	30	30	29
HOUGHTON FAA AIRPORT	32	05-23	09-28	128	30	30	30	30
	28	05-08	10-14	159	30	30	30	30
	24	04-24	10-31	190	30	30	30	30
	20	04-15	11-10	209	30	30	30	30
	16	04-05	11-16	225	30	30	30	30
HOUGHTON	32	05-14	10-09	148	17	17	17	17
	28	04-29	10-27	181	17	17	16	16
	24	04-15	11-08	207	17	17	16	16
	20	04-05	11-15	224	17	17	16	16
	16	04-02	11-22	234	17	17	16	16
HOUGHTON LAKE 3 NW	32	06-02	09-07	97	30	30	30	30
	28	05-16	09-28	135	30	30	29	29
	24	05-04	10-16	165	30	30	29	29
	20	04-22	10-31	192	30	30	29	29
	16	04-10	11-12	216	30	30	30	30
IONIA 5 NW	32	05-09	09-26	140	20	20	20	20
	28	04-27	10-09	165	20	20	20	20
	24	04-19	11-03	198	19	19	21	21
	20	03-30	11-14	229	18	18	21	21
	16	03-21	11-24	248	17	17	20	20
IRON MOUNTAIN WTR WKS	32	05-30	09-16	109	30	30	30	30
	28	05-09	09-25	130	30	30	30	30
	24	05-06	10-06	153	30	30	30	30
	20	04-22	10-27	188	30	30	30	30
	16	04-07	11-10	217	30	30	30	30
IRONWOOD	32	05-22	09-23	124	28	28	29	29
	28	05-09	10-06	150	29	29	29	29
	24	04-29	10-19	173	30	30	29	29
	20	04-17	11-02	199	30	30	29	29
	16	04-06	11-10	218	30	30	29	29
ISHPEMING	32	05-27	09-22	118	30	30	30	30
	28	05-17	10-06	142	30	30	30	30
	24	05-05	10-22	170	30	30	30	30
	20	04-20	11-05	199	30	30	30	30
	16	04-11	11-11	214	30	30	30	30
JACKSON FAA AIRPORT	32	05-07	10-04	150	30	30	30	30
	28	04-22	10-20	181	30	30	30	30
	24	04-09	11-07	212	30	30	30	30
	20	03-27	11-17	235	30	30	30	30
	16	03-17	11-28	256	30	30	30	30
KALAMAZOO STATE HOSP	32	05-04	10-07	156	30	30	30	30
	28	04-22	10-27	188	30	30	30	30
	24	04-09	11-11	216	30	30	30	30
	20	03-28	11-19	236	30	30	30	30
	16	03-17	11-30	258	30	30	30	29
KENTON U S FOREST	32	06-15	08-13	59	20	20	20	20
	28	06-01	09-09	100	20	20	20	20
	24	05-23	09-26	126	20	20	19	19
	20	05-11	10-10	152	19	19	19	19
	16	04-28	10-26	181	20	20	18	18
LAKE CITY EXP FARM	32	05-26	09-15	112	30	30	30	30
	28	05-15	09-27	135	30	30	30	30
	24	05-04	10-16	165	30	30	30	30
	20	04-17	11-01	198	30	30	29	29
	16	04-06	11-13	221	30	30	29	29
LANSING WB AIRPORT	32	05-04	10-07	156	26	26	25	25
	28	04-22	10-25	186	25	25	25	25
	24	04-05	11-07	216	25	25	25	25
	20	03-26	11-23	242	25	25	25	25
	16	03-18	11-29	256	25	25	25	25
LAPEER	32	05-14	10-02	141	30	30	30	30
	28	05-03	10-16	166	30	30	30	30
	24	04-16	11-02	200	29	29	30	30
	20	04-02	11-15	227	28	28	30	30
	16	03-21	11-27	251	28	28	29	29
LUDINGTON 4 SE	32	05-11	10-18	160	29	29	29	29
	28	04-22	11-04	196	29	29	30	30
	24	04-08	11-17	223	29	29	29	29
	20	03-29	11-25	241	29	29	30	30
	16	03-18	12-02	259	28	28	30	30
LUPTON 1 SW	32	06-10	09-01	83	10	10	10	10
	28	05-20	09-19	122	10	10	10	10
	24	05-12	10-01	142	10	10	10	10
	20	04-27	10-15	171	9	9	10	10
	16	04-13	11-07	208	9	9	9	9
MACKINAW CITY NO 2	32	05-15	10-14	152	30	30	29	29
	28	04-29	10-29	183	30	30	29	29
	24	04-15	11-15	214	30	30	29	29
	20	04-05	11-25	234	30	30	30	30
	16	03-31	12-01	245	30	30	30	30
MANISTEE	32	05-11	10-10	152	28	28	30	30
	28	04-25	11-02	191	27	27	30	30
	24	04-11	11-18	221	27	27	30	30
	20	03-31	11-24	238	27	27	30	30
	16	03-20	12-03	258	25	25	30	30
MANISTIQUE WATERWORKS	32	05-24	09-24	123	23	23	24	24
	28	05-13	10-09	149	23	23	24	24
	24	04-26	11-01	189	22	22	24	24
	20	04-11	11-11	214	22	22	24	24
	16	04-04	11-22	232	22	22	23	23
MARQUETTE WB CITY	32	05-13	10-19	159	30	30	30	30
	28	04-26	11-02	190	30	30	30	30
	24	04-14	11-13	213	30	30	30	30
	20	04-03	11-19	230	30	30	30	30
	16	03-27	11-24	242	30	30	30	30
MIDLAND DOW CHEMICAL	32	05-08	10-03	148	30	30	30	30
	28	04-25	10-24	182	30	30	30	30
	24	04-09	11-07	212	30	30	30	30
	20	03-29	11-20	236	30	30	30	30
	16	03-18	11-30	257	30	30	30	30
MILFORD GM PROVING GRN	32	05-09	10-08	152	30	30	30	30
	28	04-25	10-25	183	30	30	30	30
	24	04-09	11-07	212	30	30	30	30
	20	03-31	11-19	233	30	30	30	30
	16	03-19	11-28	254	30	30	30	30
MIO HYDRO PLANT	32	05-31	09-12	104	30	30	30	30
	28	05-21	09-27	129	30	30	30	30
	24	05-10	10-07	150	30	30	30	30
	20	04-28	10-22	177	30	30	30	30
	16	04-11	11-10	213	29	29	30	30

FREEZE DATA

STATION	Freeze threshold temperature	Mean date of last Spring occurrence	Mean date of first Fall occurrence	Mean No. of days between dates	Years of record Spring	No. of occurrences in Spring	Years of record Fall	No. of occurrences in Fall
MONROE SEWAGE PLANT	32	04-27	10-20	176	30	30	30	30
	28	04-12	11-02	204	30	30	30	30
	24	03-30	11-15	230	30	30	30	30
	20	03-23	11-26	248	30	30	30	30
	16	03-11	12-01	265	30	30	30	30
MONTAGUE	32	05-19	09-29	133	10	10	10	10
	28	05-06	10-14	161	10	10	10	10
	24	04-17	10-28	194	10	10	10	10
	20	04-10	11-12	216	10	10	10	10
	16	03-28	11-24	241	10	10	9	9
MOTT ISLAND ISLE ROYAL	32	05-23	10-11	141	14	14	17	17
	28	05-17	10-24	160	11	11	17	17
	24	04-27	11-03	190	6	6	13	13
	20	04-23	11-09	200	3	3	12	12
	16	03-29	11-20	236	3	3	10	10
MOUNT CLEMENS AF BASE	32	04-25	10-20	178	29	29	29	29
	28	04-04	11-04	205	29	29	29	29
	24	03-31	11-18	232	29	29	29	29
	20	03-23	11-25	247	29	29	29	29
	16	03-16	12-04	263	29	29	27	27
MT PLEASANT UNIVERSITY	32	05-15	10-02	140	29	29	30	30
	28	05-02	10-17	168	30	30	30	30
	24	04-16	10-31	198	30	30	30	30
	20	04-02	11-15	227	30	30	30	30
	16	03-23	11-26	248	30	30	30	30
MUNISING	32	06-08	09-15	99	30	30	30	30
	28	05-24	10-05	134	30	30	29	29
	24	05-12	10-27	168	30	30	30	30
	20	04-22	11-09	201	30	30	30	30
	16	04-08	11-16	222	30	30	30	30
MUSKEGON WB AIRPORT	32	05-06	10-14	161	30	30	30	30
	28	04-22	10-28	189	30	30	30	30
	24	04-09	11-14	219	30	30	30	30
	20	03-26	11-25	244	30	30	30	30
	16	03-17	12-06	264	30	30	30	29
NEWAYGO HARDY DAM	32	05-26	09-24	121	30	30	30	30
	28	05-14	10-06	145	30	30	30	30
	24	05-02	10-26	177	30	30	30	30
	20	04-19	11-06	201	30	30	30	30
	16	04-07	11-16	223	30	30	30	30
NEWBERRY STATE HOSP	32	05-25	09-27	125	30	30	28	28
	28	05-12	10-11	152	30	30	28	28
	24	04-27	10-28	184	30	30	28	28
	20	04-15	11-07	206	30	30	28	28
	16	04-02	11-21	233	30	30	28	28
ONAWAY STATE PARK	32	05-26	09-23	120	28	28	29	29
	28	05-12	10-08	149	27	27	29	29
	24	04-24	10-26	185	28	28	29	29
	20	04-12	11-11	213	28	28	29	29
	16	04-01	11-20	233	28	28	29	29
ONAWAY 12 S	32	06-08	09-01	85	23	23	24	24
	28	05-23	09-24	124	25	25	24	24
	24	05-11	10-09	151	25	25	24	24
	20	04-30	10-24	177	25	25	24	24
	16	04-11	11-07	210	25	25	24	24
ONTONAGON 2 SSE	32	05-28	09-28	123	12	12	12	12
	28	05-11	10-12	152	12	12	12	12
	24	05-01	11-01	184	12	12	12	12
	20	04-11	11-15	218	12	12	12	12
	16	03-31	11-22	236	12	12	12	12
OWOSSO SEWAGE PLANT	32	05-11	09-30	142	30	30	30	30
	28	04-30	10-18	171	30	30	30	30
	24	04-14	11-05	205	29	29	30	30
	20	03-30	11-17	232	29	29	30	30
	16	03-20	11-28	253	29	29	30	30
PAW PAW 2 E	32	05-11	10-04	146	30	30	30	30
	28	04-27	10-16	172	30	30	30	30
	24	04-16	11-07	205	30	30	30	30
	20	04-02	11-14	226	30	30	30	30
	16	03-20	11-26	251	30	30	28	27
PELLSTON FAA AIRPORT	32	06-03	09-03	92	19	19	19	19
	28	05-20	09-25	128	19	19	19	19
	24	05-10	10-06	149	19	19	19	19
	20	04-25	10-31	189	19	19	19	19
	16	04-09	11-12	217	19	19	19	19
PONTIAC STATE HOSPITAL	32	05-07	10-12	158	30	30	30	30
	28	04-21	10-27	189	30	30	30	30
	24	04-09	11-08	213	30	30	30	30
	20	03-26	11-22	241	30	30	30	30
	16	03-19	11-29	255	30	30	30	30
PORT HURON SEWAGE PL	32	05-04	10-15	164	28	28	29	29
	28	04-18	10-30	195	26	26	29	29
	24	04-05	11-14	223	27	27	28	28
	20	03-21	11-25	249	28	28	28	28
	16	03-16	11-30	259	28	28	27	27
REXTON	32	06-08	09-02	86	25	25	24	24
	28	05-25	09-21	119	25	25	25	25
	24	05-15	10-04	142	24	24	25	25
	20	05-05	10-20	168	25	25	24	24
	16	04-18	11-07	203	25	25	24	24
ROSCOMMON FOR EXP STA	32	06-11	08-16	66	15	15	16	16
	28	05-27	09-16	112	15	15	16	16
	24	05-17	10-02	138	15	15	16	16
	20	05-05	10-15	163	15	15	16	16
	16	04-25	10-31	189	15	15	15	15
SAGINAW FAA AIRPORT	32	05-05	10-08	156	30	30	30	30
	28	04-23	10-25	185	30	30	30	30
	24	04-11	11-09	212	30	30	30	30
	20	03-24	11-18	239	30	30	30	30
	16	03-18	11-29	256	30	30	30	30
ST IGNACE	32	05-19	10-07	141	16	16	16	16
	28	05-04	10-20	169	16	16	16	16
	24	04-18	11-07	203	16	16	16	16
	20	04-10	11-17	221	16	16	16	16
	16	03-28	11-28	245	16	16	16	16
SAINT JOHNS	32	05-13	09-28	138	21	21	22	22
	28	05-03	10-16	166	19	19	22	22
	24	04-22	11-04	196	19	19	21	21
	20	03-31	11-13	227	19	19	22	22
	16	03-23	11-27	249	19	19	22	22
SANDUSKY	32	05-11	10-11	153	19	19	21	21
	28	04-25	10-25	183	19	19	20	20
	24	04-07	11-07	214	19	19	19	19
	20	03-29	11-21	237	19	19	20	20
	16	03-19	11-29	255	19	19	20	20
SAULT STE MARIE WB AP	32	05-19	09-28	132	30	30	30	30
	28	05-04	10-18	167	30	30	30	30
	24	04-19	11-02	197	30	30	30	30
	20	04-11	11-14	217	30	30	30	30
	16	04-01	11-22	235	30	30	30	30
SOUTH HAVEN EXP FARM	32	05-02	10-16	167	30	30	30	30
	28	04-18	11-03	199	30	30	30	30
	24	04-02	11-18	230	30	30	30	30
	20	03-20	11-27	252	30	30	30	30
	16	03-14	12-07	268	30	30	30	28
STAMBAUGH 1 S	32	05-27	09-16	112	25	25	25	25
	28	05-15	09-25	133	25	25	25	25
	24	05-01	10-12	164	25	25	25	25
	20	04-18	10-28	193	25	25	25	25
	16	04-05	11-10	219	25	25	25	25

FREEZE DATA

STATION	Freeze threshold temperature	Mean date of last Spring occurrence	Mean date of first Fall occurrence	Mean No. of days between dates	Years of record Spring	No. of occurrences in Spring	Years of record Fall	No. of occurrences in Fall	STATION	Freeze threshold temperature	Mean date of last Spring occurrence	Mean date of first Fall occurrence	Mean No. of days between dates	Years of record Spring	No. of occurrences in Spring	Years of record Fall	No. of occurrences in Fall
STANDISH 2 S	32	05-16	09-24	131	21	21	21	21	WATERSMEET	32	06-10	08-14	65	22	22	22	22
	28	05-05	10-12	160	19	19	21	21		28	05-30	09-09	102	22	22	22	22
	24	04-20	10-24	187	19	19	20	20		24	05-17	09-26	132	22	22	22	22
	20	04-08	11-09	215	20	20	18	18		20	05-04	10-10	159	22	22	22	22
	16	03-23	11-20	242	19	19	18	18		16	04-20	10-31	194	22	22	22	22
STEPHENSON 5 W	32	05-28	09-19	114	10	10	10	10	WAYNE	32	05-05	10-08	156	26	26	24	24
	28	05-17	09-29	135	10	10	10	10		28	04-24	10-23	182	26	26	25	25
	24	05-03	10-23	173	10	10	10	10		24	04-07	11-05	212	26	26	25	25
	20	04-14	11-01	201	10	10	10	10		20	03-23	11-18	240	26	26	25	25
	16	04-03	11-18	229	10	10	10	10		16	03-15	11-30	260	26	26	25	25
THREE RIVERS	32	05-06	10-06	153	30	30	30	30	WEST BRANCH 2 N	32	05-29	09-15	109	30	30	29	29
	28	04-24	10-20	179	30	30	30	30		28	05-18	09-30	135	30	30	30	30
	24	04-12	11-05	207	30	30	30	30		24	05-07	10-14	160	30	30	30	30
	20	03-28	11-16	233	29	29	30	30		20	04-25	10-30	188	30	30	30	30
	16	03-17	11-30	258	29	29	30	29		16	04-14	11-10	210	30	30	30	30
TRAVERSE CITY FAA AP	32	05-21	10-05	137	30	30	30	30	WHITEFISH POINT	32	05-20	10-13	146	21	21	20	20
	28	05-06	10-22	169	30	30	30	30		28	05-02	10-23	174	21	21	20	20
	24	04-20	11-10	204	30	30	30	30		24	04-21	11-11	204	20	20	20	20
	20	04-05	11-22	231	30	30	30	30		20	04-09	11-19	224	18	18	20	20
	16	03-24	11-27	248	30	30	30	30		16	04-04	11-30	240	18	18	'19	19
VANDERBILT TROUT STA	32	06-14	08-10	57	29	29	29	29	WILLIS 5 SSW	32	05-08	10-03	148	30	30	30	30
	28	05-30	09-11	104	29	29	29	29		28	04-25	10-15	173	30	30	30	30
	24	05-21	10-02	134	29	29	29	29		24	04-10	10-28	201	30	30	30	30
	20	05-13	10-16	156	29	29	29	29		20	03-27	11-14	232	30	30	30	30
	16	04-24	10-31	190	29	29	29	29		16	03-16	11-26	255	30	30	30	30

Data in the above table are based on the period 1931-1960, or that portion of this period for which data are available.

⊕   When the frequency of occurrence in either spring or fall is one year in ten, or less, mean dates are not given.

Means have been adjusted to take into account years of non-occurrence.

A freeze is a numerical substitute for the former term "killing frost" and is the occurrence of a minimum temperature at or below the threshold temperature of 32°, 28°, etc.

Freeze data tabulations in greater detail are available and can be reproduced at cost.

*NORMALS BY CLIMATOLOGICAL DIVISIONS

Taken from "Climatography of the United States No. 81-4, Decennial Census of U. S. Climate"

STATIONS (By Divisions)	TEMPERATURE (°F) Jan.	Feb.	Mar.	Apr.	May	June	July	Aug.	Sept.	Oct.	Nov.	Dec.	Annual	PRECIPITATION (In.) Jan.	Feb.	Mar.	Apr.	May	June	July	Aug.	Sept.	Oct.	Nov.	Dec.	Annual
WEST UPPER																										
IRON MOUNTAIN WATERWKS	15.7	17.0	25.7	40.8	53.4	63.1	67.7	65.4	56.9	46.9	31.9	20.0	42.0	1.33	1.19	1.67	2.51	3.06	3.77	3.50	3.99	3.13	2.11	2.33	1.37	29.96
IRONWOOD	14.1	15.6	24.8	40.6	53.5	62.9	68.2	66.2	57.5	46.9	30.1	18.3	41.6	1.88	1.78	1.99	2.59	3.62	4.89	4.05	4.21	3.42	2.48	3.14	2.01	36.06
ISHPEMING	15.5	16.3	24.0	38.3	51.0	61.4	66.1	64.5	56.3	45.9	30.6	19.6	40.8	1.47	1.28	1.72	2.49	3.11	3.79	4.04	3.59	3.45	2.63	2.74	1.41	31.72
MARQUETTE CO	19.5	19.7	26.7	39.3	49.9	60.1	66.7	65.8	57.7	48.0	33.8	24.1	42.6	1.89	1.65	1.91	2.70	2.96	3.46	3.20	3.03	3.28	2.33	3.28	1.92	31.61
STAMBAUGH	14.6	16.0	24.7	39.7	52.4	61.8	66.5	64.4	55.7	45.8	29.8	18.5	40.8	1.24	1.21	1.53	2.27	3.23	4.18	3.58	3.65	3.30	2.12	2.21	1.23	29.75
DIVISION	15.4	16.2	24.4	39.0	51.2	60.7	66.1	64.6	56.3	46.2	31.0	19.8	40.9	1.88	1.54	1.75	2.34	3.17	4.00	3.43	3.70	3.43	2.41	2.75	1.79	32.19
EAST UPPER																										
CHATHAM EXPERIMENT FARM	17.6	18.0	24.6	38.3	50.1	60.2	65.7	64.4	56.4	46.5	32.6	21.9	41.4	2.08	1.62	1.74	2.30	3.12	3.54	3.43	3.21	4.23	2.89	3.30	2.01	33.47
ESCANABA CO	18.4	18.7	26.2	39.1	50.3	61.1	67.2	65.6	57.4	47.6	34.2	23.3	42.4	1.45	1.21	1.66	2.23	2.97	3.16	3.53	3.11	3.14	2.11	2.35	1.40	28.32
GRAND MARAIS	19.5	18.7	24.2	36.9	47.8	57.2	62.7	63.1	55.9	46.6	33.8	24.0	40.9	2.19	1.74	1.73	1.97	2.68	3.29	2.67	2.58	3.33	2.34	3.02	2.20	29.74
MUNISING	19.1	18.9	25.2	38.3	49.1	59.2	65.0	64.5	56.9	46.9	33.5	23.1	41.6	2.51	1.94	1.95	2.25	2.94	3.45	3.19	3.18	3.64	2.91	3.52	2.60	34.08
NEWBERRY STATE HOSPITAL	17.7	17.7	24.9	38.7	50.4	60.2	65.7	64.5	56.0	45.9	32.7	22.2	41.4	•	•	•	•	•	•	•	•	•	•	•	•	•
SAULT STE MARIE AP	15.8	15.7	23.8	38.0	49.6	59.0	64.6	64.0	55.8	46.3	33.3	20.9	40.6	2.07	1.50	1.81	2.16	2.77	3.30	2.48	2.89	3.81	2.82	3.33	2.28	31.22
DIVISION	18.3	18.1	24.9	38.1	49.3	59.2	65.0	64.4	56.5	46.7	33.7	22.9	41.4	2.13	1.70	1.87	2.24	2.94	3.36	2.99	3.01	3.68	2.67	3.10	2.18	31.87
NORTHWEST LOWER																										
CADILLAC 3 WSW	19.4	18.6	26.1	40.7	52.7	63.1	67.4	65.8	57.5	46.9	33.9	23.1	42.9	1.83	1.59	2.01	2.68	2.97	3.15	2.78	3.01	3.53	2.88	2.89	1.85	31.17
EAST JORDAN	22.1	20.9	28.2	42.5	54.0	64.2	69.0	67.5	59.6	49.6	36.9	26.3	45.1	•	•	•	•	•	•	•	•	•	•	•	•	•
FIFE LAKE 2 S	20.2	19.4	26.8	41.5	53.3	63.5	67.6	66.0	58.2	47.9	34.9	24.0	43.6	1.90	1.52	1.71	2.41	2.90	3.23	3.18	3.08	3.19	3.02	2.86	1.93	30.93
LAKE CITY EXP FARM	19.7	19.2	26.6	41.7	53.5	63.5	67.8	66.2	58.0	47.6	34.4	23.3	43.5	1.34	1.21	1.48	2.39	2.97	3.19	3.16	2.85	3.20	2.74	2.44	1.36	28.33
MANISTEE	24.9	24.7	31.5	43.8	54.3	64.4	69.7	69.0	61.6	51.2	38.8	28.7	46.9	•	•	•	•	•	•	•	•	•	•	•	•	•
TRAVERSE CITY FAA AP	23.0	22.2	29.0	42.1	53.0	64.2	59.9	68.7	60.5	50.0	37.1	26.9	45.6	1.82	1.40	1.61	2.26	2.89	2.82	2.79	2.88	3.48	2.80	2.86	1.75	29.36
WELLSTON TIPPEY DAM	•	•	•	•	•	•	•	•	•	•	•	•	•	2.06	1.55	1.83	2.63	3.08	3.18	2.69	3.31	3.70	3.01	3.01	1.92	31.97
DIVISION	21.7	20.9	28.0	41.7	52.8	63.7	68.2	67.0	59.2	48.9	36.3	25.8	44.5	1.87	1.48	1.69	2.40	2.92	3.06	2.73	3.01	3.57	2.87	2.91	1.89	30.39
NORTHEAST LOWER																										
ALPENA AP	19.7	18.6	25.7	39.1	50.6	60.6	65.9	64.5	56.0	46.3	34.6	24.1	42.1	1.95	1.61	2.02	2.63	2.91	2.67	2.86	3.03	3.61	2.42	2.37	1.84	29.92
ALPENA SEWAGE PLANT	21.0	20.0	27.6	39.8	50.9	61.4	67.1	65.9	58.3	47.9	36.3	25.6	43.5	1.56	1.40	1.88	2.23	2.61	2.59	2.52	2.54	3.15	2.03	2.15	1.70	26.36
ATLANTA 3 ENE	19.7	19.2	26.9	41.2	53.0	63.1	67.5	65.6	57.7	47.8	35.0	23.7	43.4	•	•	•	•	•	•	•	•	•	•	•	•	•
CHEBOYGAN RR LIGHT STA	20.7	20.1	27.4	40.8	52.1	62.4	68.5	67.1	59.2	49.0	36.3	25.4	44.1	•	•	•	•	•	•	•	•	•	•	•	•	•
GRAYLING MILITARY RES	19.2	19.1	26.4	41.3	53.8	63.8	68.0	66.3	58.2	48.1	34.4	23.1	43.5	1.42	1.21	1.50	2.57	3.51	3.69	3.82	3.37	3.83	3.00	2.87	1.56	32.35
HALE LOUD DAM	20.0	19.4	27.0	41.0	52.3	62.7	67.3	65.9	57.5	48.7	35.4	24.4	43.5	1.67	1.41	1.99	2.57	2.87	2.68	2.67	3.12	3.03	2.46	2.32	1.64	28.43
HARRISVILLE	22.6	22.4	28.8	41.1	51.1	61.7	67.3	66.7	59.6	49.5	37.0	26.4	44.5	•	•	•	•	•	•	•	•	•	•	•	•	•
HIGGINS LAKE	19.5	19.1	26.6	40.7	52.8	63.3	67.9	66.3	58.1	47.7	34.3	23.3	43.3	2.03	1.61	2.00	2.58	2.92	3.42	3.35	3.06	3.51	2.96	3.08	1.99	32.51
HOUGHTON LAKE 3 NW	20.7	20.4	28.1	42.6	54.8	64.3	68.1	66.5	58.6	48.6	35.3	24.5	44.4	1.49	1.29	1.69	2.31	2.93	3.12	2.88	2.67	2.94	2.99	2.36	1.67	28.34
MACKINAW CITY NO 2	21.0	19.3	25.8	38.5	49.5	59.7	66.9	66.4	58.7	49.0	36.6	26.2	43.1	1.59	1.38	1.49	2.12	2.85	2.85	2.37	2.77	3.80	2.58	2.51	1.57	27.88
MIO HYDRO PLANT	19.0	18.3	26.5	40.6	52.4	62.8	66.9	65.2	56.9	46.9	34.6	23.1	42.8	1.30	1.24	1.51	2.15	2.59	2.75	2.90	2.97	3.01	2.11	1.97	1.29	25.79
ONAWAY BLACK L FOREST	20.0	20.0	27.5	41.2	53.3	63.2	68.3	66.8	58.4	48.4	35.4	24.4	43.9	•	•	•	•	•	•	•	•	•	•	•	•	•
VANDERBILT TROUT STA	18.7	17.9	25.4	40.2	52.3	60.5	66.4	64.8	57.0	46.9	34.1	22.9	42.3	1.75	1.34	1.67	2.29	2.70	2.78	2.59	2.89	3.94	2.67	2.67	1.82	29.11
WEST BRANCH 2 N	19.2	19.5	27.0	41.2	53.4	63.2	67.2	65.5	57.0	47.0	33.8	22.9	43.1	1.46	1.40	1.82	2.35	3.16	3.09	2.95	3.05	3.01	2.80	2.60	1.71	29.40
DIVISION	20.2	19.7	27.0	40.9	52.5	62.6	67.4	66.0	58.0	48.0	35.3	24.3	43.5	1.65	1.39	1.72	2.37	2.91	2.95	2.82	2.93	3.46	2.58	2.48	1.67	28.93
WEST CENTRAL LOWER																										
HART	24.4	24.2	31.5	44.6	55.3	65.9	70.5	69.1	61.1	51.0	38.3	28.1	47.0	2.49	1.95	1.82	2.60	3.03	3.19	2.77	3.02	3.16	2.62	3.10	2.05	31.80
LUDINGTON 4 SE	24.9	24.6	31.6	43.6	53.6	63.7	69.1	68.4	61.1	51.0	38.4	28.4	46.5	2.30	1.87	1.81	2.46	2.69	3.04	2.23	2.76	3.04	2.61	2.85	2.09	29.75
MUSKEGON AP	26.0	25.7	32.9	45.2	55.8	66.7	71.3	70.3	63.0	52.5	39.6	29.9	48.2	2.10	1.80	2.05	2.58	2.97	2.70	2.41	2.91	3.06	2.55	2.94	2.00	30.07
NEWAYGO HARDY DAM	22.5	22.2	30.1	43.7	55.6	65.9	70.1	68.6	60.4	49.5	36.7	26.2	46.0	2.14	1.90	2.19	2.77	3.27	3.12	2.76	3.20	3.08	2.68	2.84	2.06	32.01
SCOTTVILLE 1 NE	•	•	•	•	•	•	•	•	•	•	•	•	•	2.29	1.77	1.88	2.44	2.86	3.03	2.19	2.87	2.88	2.69	2.76	1.96	29.62
DIVISION	23.6	23.3	30.9	43.9	55.0	65.2	69.9	68.6	60.7	50.3	37.6	27.3	46.4	2.23	1.87	1.97	2.60	3.02	3.06	2.65	2.96	3.20	2.75	3.00	2.06	31.37
CENTRAL LOWER																										
ALMA	23.9	24.0	32.4	46.0	57.4	67.9	72.3	70.4	62.5	51.7	38.4	27.4	47.9	1.45	1.37	1.80	2.40	3.04	3.08	2.51	3.34	3.15	2.42	2.21	1.59	28.36
BIG RAPIDS WATERWORKS	22.9	23.0	30.7	44.2	55.7	65.6	69.8	68.2	60.0	49.7	37.0	26.5	46.1	2.00	1.78	2.11	2.71	3.02	3.56	2.67	2.78	3.34	2.74	2.84	1.93	31.48
GLADWIN	21.1	21.1	29.3	43.3	54.7	64.8	69.1	67.5	59.2	48.8	35.9	25.1	45.0	1.73	1.48	1.89	2.60	3.08	3.36	2.82	3.10	3.23	2.82	2.49	1.92	30.52
GREENVILLE	24.3	24.6	32.8	46.2	57.7	67.9	72.3	70.5	62.2	51.6	38.2	27.6	48.0	•	•	•	•	•	•	•	•	•	•	•	•	•
MIDLAND DOW CHEMICAL	24.8	24.8	32.8	46.2	57.9	68.1	72.4	70.7	62.9	52.4	39.3	28.3	48.4	1.74	1.75	2.03	2.50	3.01	3.09	2.36	2.86	3.13	2.69	2.39	1.94	29.49
MOUNT PLEASANT COLLEGE	23.0	23.3	31.4	45.3	57.0	67.1	71.6	69.9	62.0	51.1	37.5	26.6	47.2	1.65	1.47	1.81	2.41	2.98	3.08	2.66	2.97	3.24	2.65	2.33	1.70	28.95
DIVISION	23.2	23.3	31.3	45.0	56.6	66.7	71.1	69.4	61.2	50.7	37.5	26.7	46.9	1.73	1.57	1.92	2.56	3.07	3.24	2.61	3.08	3.21	2.63	2.51	1.81	29.94
EAST CENTRAL LOWER																										
BAD AXE RADIO STATION	23.1	22.8	30.6	43.6	54.9	65.2	70.2	69.1	61.3	50.8	37.9	27.0	46.4	•	•	•	•	•	•	•	•	•	•	•	•	•
BAY CITY SEWAGE PLANT	25.0	25.1	32.9	46.3	58.1	68.6	73.1	71.3	63.2	52.6	39.2	28.3	48.6	1.44	1.44	1.75	2.22	2.93	3.23	2.19	2.68	2.84	2.39	2.12	1.50	26.73
CARO STATE HOSPITAL	23.9	23.8	32.4	45.5	56.7	67.1	71.4	69.6	62.0	51.6	38.5	27.6	47.5	1.50	1.37	1.79	2.21	2.87	3.11	2.91	2.93	2.90	2.58	2.25	1.67	28.09
GRAND RAPIDS WB AP	24.4	24.5	32.4	45.7	57.0	67.4	71.9	70.5	62.2	51.2	38.2	28.0	47.8	1.91	1.75	2.28	2.94	3.46	3.31	2.73	2.70	2.98	2.61	2.49	2.03	31.19
SAGINAW FAA AIRPORT	23.4	23.3	31.8	44.9	56.3	66.9	71.6	69.7	61.5	50.7	37.8	26.9	47.1	1.63	1.74	1.87	2.38	2.89	2.94	2.47	2.88	3.01	2.52	2.29	1.78	28.40
DIVISION	23.7	23.5	31.6	44.7	55.8	66.4	71.1	69.6	61.8	51.2	38.4	27.3	47.1	1.67	1.63	1.81	2.39	2.98	3.07	2.67	2.69	2.86	2.51	2.26	1.79	28.33
SOUTHWEST LOWER																										
ALLEGAN SEWAGE PLANT	26.5	26.7	34.5	47.4	58.5	68.5	72.6	71.2	63.6	52.9	39.8	29.7	49.3	2.32	1.93	2.38	3.20	3.57	3.53	2.52	3.45	3.06	2.73	2.66	2.20	33.55
BLOOMINGDALE	26.5	27.1	34.9	47.6	58.5	68.2	72.5	71.0	63.5	52.9	39.7	29.5	49.3	2.57	2.22	2.65	3.27	3.79	3.86	2.65	3.27	3.21	2.98	2.98	2.56	36.01
EAU CLAIRE 4 NE	26.4	27.3	35.5	48.2	59.1	69.5	74.2	72.8	65.2	54.3	39.9	29.3	50.1	2.13	1.77	2.32	3.31	4.07	4.18	3.08	3.20	3.14	3.29	2.75	2.25	35.49
GRAND HAVEN FIRE DEPT	26.6	26.5	33.5	45.2	55.7	65.9	71.0	69.9	62.9	52.5	40.0	30.0	48.3	2.22	1.84	2.12	2.68	3.34	3.24	2.67	2.92	3.07	2.58	2.89	2.18	31.75
GRAND RAPIDS AP	24.0	24.3	32.7	45.7	56.8	67.5	72.3	70.4	62.2	51.1	37.3	27.0	47.6	1.91	1.75	2.28	2.94	3.46	3.31	2.73	2.70	2.98	2.61	2.49	2.03	31.19
GULL LAKE EXP FARM	24.9	25.6	33.8	46.7	58.0	68.3	72.6	71.2	63.3	52.3	38.6	27.9	48.6	2.03	1.78	2.30	3.16	3.73	4.23	2.87	3.66	3.09	3.11	2.56	1.80	34.32
HOLLAND	27.2	27.2	34.8	47.0	58.0	67.9	72.6	71.3	64.0	53.6	40.6	30.3	49.5	2.16	1.90	2.37	2.93	3.42	3.12	2.95	3.01	3.35	2.97	2.96	2.34	33.48
KALAMAZOO STATE HOSP	•	•	•	•	•	•	•	•	•	•	•	•	•	2.26	2.01	2.53	3.08	3.76	3.73	3.31	3.08	3.02	3.01	2.60	2.09	34.48
KENT CITY 2 SW	•	•	•	•	•	•	•	•	•	•	•	•	•	1.93	1.90	2.08	2.80	3.42	3.28	2.61	3.24	3.15	2.66	2.91	2.03	32.01
PAW PAW 2 E	25.9	26.7	34.7	47.5	58.5	68.6	72.9	71.4	63.9	52.9	39.5	28.9	49.3	2.31	1.99	2.79	3.39	4.17	4.15	3.31	3.03	3.18	3.22	2.72	2.24	36.50
SOUTH HAVEN EXP FARM	27.7	28.1	35.0	46.3	56.2	66.7	71.4	70.5	64.3	54.2	41.5	31.0	49.4	1.98	1.76	2.12	2.98	3.52	3.43	2.60	3.00	3.04	3.11	2.65	2.06	32.25
DIVISION	26.5	27.0	34.7	47.1	57.9	68.1	72.7	71.3	64.0	53.3	40.0	29.6	49.4	2.21	1.91	2.40	3.12	3.76	3.74	2.89	3.18	3.14	3.02	2.72	2.21	34.30
SOUTH CENTRAL LOWER																										
ALBION	•	•	•	•	•	•	•	•	•	•	•	•	•	2.18	1.93	2.59	3.39	4.16	4.21	2.97	3.16	3.17	2.93	2.48	1.96	35.13
BATTLE CREEK	25.0	25.7	33.8	46.9	58.4	68.6	72.8	71.4	63.5	52.3	38.5	28.1	48.8	2.04	1.78	2.37	3.00	3.80	4.03	2.87	3.10	2.84	2.96	2.40	1.90	33.09
CHARLOTTE	24.9	25.3	33.6	46.6	57.9	67.9	72.2	70.5	62.8	51.9	38.3	27.4	48.3	1.81	1.66	2.09	2.95	3.93	4.09	2.56	3.15	3.03	2.83	2.33	1.81	32.24
COLDWATER STATE SCHOOL	26.0	26.7	34.9	47.6	58.5	68.7	72.7	71.1	63.9	53.0	39.3	28.8	49.3	1.94	1.84	2.48	3.37	3.68	4.22	3.43	3.04	3.25	2.57	2.48	1.82	34.12
HASTINGS FISHERIES	25.5	26.1	34.2	47.4	58.9	68.8	72.9	71.3	63.8	52.8	39.2	28.5	49.1	1.68	1.66	2.10	2.87	3.76	3.97	2.84	3.47	3.21	2.88	2.25	1.72	32.41
HILLSDALE	25.3	26.2	34.3	47.3	58.4	68.1	72.1	70.4	63.0	52.3	38.6	28.1	48.7	2.33	2.16	2.84	3.57	3.76	4.16	3.18	2.95	3.00	2.89	2.82	2.14	35.80
JACKSON FAA AIRPORT	25.2	26.0	34.0	47.7	58.2	68.2	72.4	70.9	63.2	52.3	38.6	28.2	48.7	1.79	1.75	2.20	3.11	3.63	4.05	2.74	2.69	2.62	2.54	2.26	1.77	31.15
LANSING AP-CAPITAL	24.3	24.2	32.4	45.7	57.1	67.4	71.7	70.2	62.0	51.3	37.9	27.5	47.6	1.96	1.95	2.40	2.87	3.73	3.34	2.58	3.05	2.60	2.50	2.21	1.99	31.18
OWOSSO SEWAGE PLANT	25.0	25.3	33.4	46.4	57.6	67.7	72.0	70.4	63.1	52.4	39.1	28.2	48.4	1.68	1.66	2.00	2.70	3.46	3.44	2.56	2.88	2.77	2.44	2.08	1.70	29.37
THREE RIVERS	26.4	27.4	35.5	48.2	59.2	69.3	73.3	71.6	64.4	53.2	39.7	29.1	49.8	1.96	1.79	2.52	3.33	4.04	4.41	3.30	3.32	3.37	2.85	2.57	2.11	35.57
DIVISION	25.2	25.7	33.8	46.9	58.0	68.2	72.4	70.7	63.2	52.3	38.8	28.2	48.6	1.84	1.75	2.24	3.01	3.72	3.89	2.88	3.15	2.96	2.67	2.36	1.84	32.31

*NORMALS BY CLIMATOLOGICAL DIVISIONS

Taken from "Climatography of the United States No. 81-4, Decennial Census of U. S. Climate"

TEMPERATURE (°F)   PRECIPITATION (In.)

STATIONS (By Divisions)	Jan.	Feb.	Mar.	Apr.	May	June	July	Aug.	Sept.	Oct.	Nov.	Dec.	Annual	Jan.	Feb.	Mar.	Apr.	May	June	July	Aug.	Sept.	Oct.	Nov.	Dec.	Annual
SOUTHEAST LOWER																										
ADRIAN	26.3	27.3	35.3	47.7	59.4	69.9	74.0	72.3	64.2	52.9	39.2	29.0	49.8	1.99	1.82	2.46	3.25	3.59	3.51	3.09	2.64	2.77	2.68	2.24	1.89	31.93
DETROIT AP-CITY	26.9	27.2	34.8	47.6	59.0	69.7	74.4	72.8	65.1	53.8	40.4	29.9	50.1	2.05	2.08	2.42	3.00	3.53	2.83	2.82	2.86	2.44	2.63	2.21	2.08	30.95
DETROIT AP-WAYNE	26.5	27.1	34.9	47.2	58.4	68.4	72.9	71.9	64.3	54.0	40.4	29.9	49.7	1.93	1.95	2.41	3.05	3.54	3.31	2.69	2.84	2.32	2.57	2.27	1.92	30.80
DETROIT AP-WILLOW RUN	26.6	27.4	35.3	47.7	58.7	69.6	74.2	72.6	64.8	53.9	40.0	29.4	50.0	1.94	1.95	2.40	3.05	3.48	3.28	2.68	2.84	2.29	2.53	2.25	1.98	30.67
FLINT AP-BISHOP	24.5	24.8	32.7	45.3	55.4	65.6	71.2	69.6	62.0	51.4	38.2	27.6	47.5	1.70	1.76	2.14	2.66	3.39	3.12	2.91	3.17	2.94	2.37	2.18	1.80	30.14
MILFORD GM PROVING GRND	23.7	23.9	32.2	45.0	56.5	66.7	71.3	69.9	62.0	51.3	37.4	26.7	47.2	1.98	2.09	2.53	3.31	3.80	3.33	2.69	3.16	2.90	2.74	2.44	2.11	33.08
MONROE SEWAGE PLANT	27.2	27.9	35.5	47.6	59.1	69.6	73.9	72.3	65.2	54.0	40.4	29.9	50.2	2.01	1.87	2.32	3.00	3.37	3.52	2.93	2.97	2.68	2.44	1.91	1.80	30.82
MOUNT CLEMENS AF BASE	25.7	25.8	33.5	45.8	57.1	67.6	72.8	71.2	63.4	52.6	39.5	28.8	48.7	1.73	1.96	2.16	2.62	3.09	2.74	2.22	2.56	2.25	2.16	1.95	1.95	27.39
PONTIAC STATE HOSPITAL	24.7	25.2	33.3	46.3	57.8	67.9	72.4	71.0	63.5	52.6	38.8	27.8	48.4	1.83	2.01	2.19	2.88	3.51	3.37	2.82	2.83	2.65	2.71	2.23	1.96	30.99
WILLIS 5 SSW	25.5	25.3	34.4	46.6	57.9	67.8	72.0	70.4	63.0	52.1	39.0	28.2	48.6	1.93	1.77	2.31	3.11	3.39	3.36	2.63	3.06	2.58	2.50	2.18	1.88	30.70
DIVISION	25.6	26.1	34.1	46.6	57.9	68.2	72.7	71.1	63.7	52.7	39.3	28.6	48.9	1.87	1.87	2.26	2.96	3.45	3.26	2.77	2.92	2.66	2.59	2.19	1.92	30.72

TEMPERATURE   PRECIPITATION

Jan.	Feb.	Mar.	Apr.	May	June	July	Aug.	Sept.	Oct.	Nov.	Dec.	Annual	Jan.	Feb.	Mar.	Apr.	May	June	July	Aug.	Sept.	Oct.	Nov.	Dec.	Annual

CONFIDENCE LIMITS

In the absence of trend or record changes, the chances are 9 out of 10 that the true mean will lie in the interval formed by adding and subtracting the values in the following table from the means for any station in the State. Because of the wider variation in mean precipitation, the corresponding monthly means and annual mean must be substituted for "p" in the precipitation table below to obtain mean precipitation confidence limits.

| 1.4 | 1.4 | 1.5 | 1.2 | .9 | .8 | .8 | .9 | .9 | 1.1 | 1.1 | 1.2 | .5 | $.21\sqrt{p}$ | $.19\sqrt{p}$ | $.22\sqrt{p}$ | $.24\sqrt{p}$ | $.26\sqrt{p}$ | $.26\sqrt{p}$ | $.29\sqrt{p}$ | $.30\sqrt{p}$ | $.27\sqrt{p}$ | $.27\sqrt{p}$ | $.24\sqrt{p}$ | $.18\sqrt{p}$ | $.25\sqrt{p}$ |

COMPARATIVE DATA

Data in the following table are the mean temperature and average precipitation for Chatham Experimental Farm, Michigan, for the period 1906 - 1930 and are included in this publication for comparative purposes.

| 14.0 | 14.0 | 23.5 | 36.3 | 47.8 | 58.5 | 63.7 | 61.1 | 55.1 | 44.3 | 32.5 | 20.2 | 39.2 | 2.04 | 1.59 | 1.75 | 2.13 | 2.61 | 3.62 | 3.21 | 2.93 | 3.78 | 3.33 | 2.88 | 2.01 | 31.88 |

NORMALS, MEANS, AND EXTREMES

ALPENA, MICHIGAN — PHELPS COLLINS FIELD

LATITUDE 45° 04' N
LONGITUDE 83° 34' W
ELEVATION (ground) 689 Feet

Month	Normal Daily max (b)	Normal Daily min (b)	Normal Monthly (b)	Extreme Record highest	Year	Extreme Record lowest	Year	Normal degree days (b)	Precip. Normal total (b)	Precip. Max monthly	Year	Precip. Min monthly	Year	Precip. Max in 24 hrs	Year	Snow Mean total	Snow Max monthly	Year	Snow Max in 24 hrs	Year	RH 1AM	RH 7AM	RH 1PM	RH 7PM	Wind Mean hourly speed	Prevailing dir.	Fastest mile Speed	Dir.	Year	Pct. sunshine	Mean sky cover	Clear	Partly cloudy	Cloudy	Precip .01+	Snow 1.0+	Thunderstorms	Heavy fog	Max 90+	Max 32 & below	Min 32 & below	Min 0 & below
(a)				8		8	(b)	(b)	(b)	8		8				7	8				1	7	7	7	6		8	8		7	7	7	7	7	7	7	7	7	7	7	7	7
J	28.9	10.4	19.7	51	1961	-28	1963	1404	1.95	2.79	1965	0.16	1965	0.87	1965	19.0	32.5	1965	8.7	1965	73	77	67	72	7.7	SW	33	NW	1965	43	7.6	4	8	19	13	5	0	0	0	22	31	10
F	28.8	8.3	18.6	52	1966+	-28		1299	1.61	2.79	1962	0.36	1964	0.99	1962	14.6	30.1	1962	11.0	1962	73	75	65	70	7.5	W	33	SW	1965+	48	7.3	6	5	17	12	5	0	0	0	18	28	9
M	36.1	15.3	25.7	68	1962	-27		1218	2.02	3.35	1963	0.86	1962	1.10	1961	12.9	18.1	1964	10.1	1964	71	80	55	61	8.7	NW	42	SW	1962	51	7.0	6	8	17	12	4	1	2	0	11	21	5
A	50.5	27.3	39.1	82	1962	0	1965	777	2.63	3.43	1966	1.43	1962	1.20	1965	2.9	6.8	1965	4.6	1965	75	78	50	58	8.7	SE	42	NW	1963	52	7.0	6	12	12	11	1	1	3	2	1	8	0
M	64.5	36.7	50.6	94	1962	20	1965	446	2.63	5.46	1963	1.05	1966	2.21	1963	0.3	1.5	1961	1.3	1961	69	78	50	57	8.4	SW	36	NW	1966	64	6.5	6	12	13	10	1	3	3	3		1	
J	74.6	46.6	60.6	97		28	1961	156	2.91	3.78	1963	1.12	1963	1.85	1960	0.0					80	78	57	60	6.7	SW	39	N	1961	63	5.5	6	11	13	10	0	6	3	2			
J	80.4	51.4	65.9	98	1966	34	1965	68	2.86	4.30	1964	0.91	1966	2.34	1960	0.0		1965			81	82	49	58	6.4	W	27	NE	1964	73	5.3	9	14	11	8	2	4	2				
A	78.5	50.5	64.5	95	1960	31	1960	105	3.03	4.08	1963	1.76	1966	1.75	1964	0.0		1962			84	84	53	61	6.3	SW	32	SW	1964+	61	5.4	7	13	11	9	2	4	4				
S	69.4	42.6	56.0	93	1960	26	1965+	273	3.61	5.99	1961	1.92	1963	2.54	1961	0.0		1963			82	89	58	67	6.8	SW	30	S	1964+	56	6.4	7	10	14	12	0	4	2			3	
O	58.9	33.6	46.3	86	1963	16	1964	580	2.42	2.15	1960	0.95	1960	1.18	1966	0.6	1.1	1962	1.1	1962	72	83	65	71	7.5	SW	37	E	1966	42	6.4	7	9	15	11	0	2	2			9	
N	43.5	25.7	34.6	75	1950	-11	1963	912	2.37	7.45	1966	0.78	1962	1.93	1966	7.5	30.5	1966	16.1	1966	82	82	64	71	7.4	SW	31	NE	1965	32	8.2	3	6	21	14	4	2	0			19	4
D	31.9	16.3	24.1	62	1966+	-11		1268	1.84	2.38	1966	0.96	1966	1.03	1966	16.4	25.5	1963	8.4	1960	77	80	77	77	7.4	SW				34	8.3	2	7	22	16	6	1	2				
YR	53.9	30.4	42.1	98	JUL. 1966	-28	JAN. 1963	8506	29.92	7.45	NOV. 1966	0.16	JAN. 1961	2.54	SEP. 1961	75.2	32.5	JAN. 1965	16.1	NOV. 1966	77	81	59	68	7.3	SW	43	W	APR. 1963	56	6.8	73	107	185	145	22	38	25	7	73	188	28

Means and extremes in the above table are from the existing location. Annual extremes have been exceeded at other locations as follows:
Highest temperature 104 in July 1936; maximum monthly precipitation 13.18 in October 1877; minimum monthly precipitation 0.10 in May 1891; maximum monthly snowfall 51.9 in January 1887; maximum snowfall in 24 hours 16.8 in January 1898.

DETROIT, MICHIGAN — CITY AIRPORT

LATITUDE 42° 24' N
LONGITUDE 83° 00' W
ELEVATION (ground) 619 Feet

Month	Normal Daily max (b)	Normal Daily min (b)	Normal Monthly (b)	Extreme Record highest	Year	Extreme Record lowest	Year	Normal degree days (b)	Precip. Normal total (b)	Precip. Max monthly	Year	Precip. Min monthly	Year	Precip. Max in 24 hrs	Year	Snow Mean total	Snow Max monthly	Year	Snow Max in 24 hrs	Year	RH 1AM	RH 7AM	RH 1PM	RH 7PM	Wind Mean hourly speed	Prevailing dir.	Fastest mile Speed	Dir.	Year	Pct. sunshine	Mean sky cover	Clear	Partly cloudy	Cloudy	Precip .01+	Snow 1.0+	Thunderstorms	Heavy fog	Max 90+	Max 32 & below	Min 32 & below	Min 0 & below
(a)				33		33	(b)	(b)	(b)	32		32		32		32	32		32		29	33	29	33	33	14	32	32		32	32	32	32	32	32	32	33	33	33	33	33	33
J	33.0	20.7	26.9	67	1950	-13	1963	1181	2.05	4.38	1950	0.23	1961	1.63	1960	7.9	21.1	1939	8.9	1957	76	80	70	75	11.5	SW	57	SW	1949	32	7.8	4	6	21	13	4	*	1	0	16	26	1
F	34.8	20.4	27.2	68	1944	-16	1934	1058	2.08	4.95	1938	0.38	1947	2.43	1965	8.0	15.8	1965	10.0	1965	77	80	67	71	11.1	NW	49	SW	1953+	47	7.0	4	6	18	12	3	*	1	0	13	23	1
M	42.3	27.3	34.8	82	1945	-1	1954	936	2.42	4.40	1938	0.47	1946	1.85	1954	5.5	15.5	1954	10.8	1954	77	80	59	67	11.5	NW	68	W	1943	49	6.9	5	8	18	12	2	1	*	5	*	8	1
A	56.4	38.8	47.6	87	1942+	14	1954	522	3.00	6.89	1947	0.74	1948	2.94	1947	1.2	6.8	1943	4.2	1942	72	75	53	59	11.1	S	56	SW	1949	52	6.4	5	10	15	12	1	2	3	1	*	3	0
M	68.1	49.4	59.0	93	1962+	30	1948	220	3.53	7.51	1943	0.58	1943	2.53	1948	T					71	74	49	53	10.2	S	61	NW	1942	59	6.4	7	11	13	11	1	5	6	*	0	0	0
J	79.1	60.3	69.7	104	1934	38	1945	42	2.83	6.58	1960	1.01	1959	2.65	1954	0.0					74	75	53	57	9.0	S	56	NW	1964+	65	6.0	8	13	9	9	0	6	9	1	4	0	0
J	83.9	64.8	74.4	105	1934	47	1965	0	2.82	7.03	1937	0.81	1936	2.80	1957	0.0					75	76	50	55	8.3	SW	77	NW	1960	70	5.3	10	12	9	9	0	6	7	1	6	0	0
A	81.9	63.6	72.8	101	1947	43	1940	87	2.86	7.51	1940	1.07	1956	3.65	1956	0.0					80	83	54	58	8.1	SW	57	W	1945	65	5.4	9	11	11	9	0	6	7	1	3	0	0
S	74.2	56.6	65.4	100	1953+	32	1942	360	2.44	5.90	1936	0.58	1954	2.56	1961	0.0					80	83	56	64	8.9	SW	66	S	1962	56	5.6	9	11	10	9	0	6	5	1	1	0	0
O	62.8	45.7	53.8	92	1963	24	1965	738	2.63	7.80	1954	0.50	1924	3.72	1954	T					77	80	57	64	9.5	S	56	W	1942	56	5.6	7	10	14	11	0	7	3	1	0	0	0
N	47.1	34.7	40.4	81	1950	-5	1958	1088	2.21	4.14	1948	0.57	1939	2.18	1951	2.6	9.2	1950	5.6	1951	77	80	64	71	11.4	SW	66	NW	1943	37	7.7	4	9	17	11	1	1	1	1	0	2	*
D	35.7	24.1	29.9	64	1966+	-5	1960	1088	2.08	4.60	1957	0.43	1943	2.45	1943	6.6	24.0	1951	6.8	1951	80	80	70	75	11.2	SW				35	7.7	4	6	21	13	3	*	1	2	0	12	0
YR	58.2	42.0	50.1	105	JUL. 1934	-16	FEB. 1934	6232	30.95	8.05	MAY 1943	0.23	JAN. 1961	3.72	OCT. 1954	31.8	24.0	DEC. 1951	10.0	DEC. 1951	76	78	58	65	10.2	S	77	NW	JUL. 1960	54	6.5	80	108	177	133	11	48	52	15	48	125	2

Means and extremes in the above table are from the existing location. Annual extremes have been exceeded at other locations as follows:
Lowest temperature -24 in December 1872; maximum monthly precipitation 8.76 in July 1878; minimum monthly precipitation 0.04 in February 1887; maximum precipitation in 24 hours 4.75 in July 1925; maximum snowfall 38.4 in February 1908; maximum monthly snowfall in 24 hours 24.5 in April 1886; fastest mile of wind 95 from Northwest in June 1890.

NORMALS, MEANS, AND EXTREMES

DETROIT, MICHIGAN — WILLOW RUN AIRPORT

LATITUDE 42° 14' N
LONGITUDE 83° 32' W
ELEVATION (ground) 711 Feet

Temperature, Degree Days, Precipitation

Month	Normal Daily maximum	Normal Daily minimum	Normal Monthly	Extremes Record highest	Year	Record lowest	Year	Normal degree days	Precip Normal total	Max monthly	Year	Min monthly	Year	Max in 24 hrs	Year
(a)	(b)	(b)	(b)					(b)	(b)						
J	33.4	19.8	26.6	59	1965	-13	1963	1190	1.94	4.17	1950	0.25	1961	1.74	1960
F	34.7	20.0	27.4	59	1966	-6	1963	1053	1.95	4.31	1954	0.44	1958	2.08	1954
M	43.6	27.0	35.3	79	1962	3	1962	921	2.40	4.39	1954	0.30	1958	2.35	1954
A	57.8	37.5	47.7	87	1965+	18	1965+	519	3.05	6.67	1947	1.30	1962	2.50	1947
M	70.0	47.3	58.7	96	1962	23	1966	229	3.28	6.95	1947	1.18	1958	3.13	1956
J	81.1	58.1	69.6	94	1962	33	1966	45		5.46	1960	1.55	1952		
J	86.0	62.4	74.2	98	1966	42	1965	0	2.68	4.96	1966	1.62	1954	3.17	1951
A	84.0	61.1	72.6	99	1964	40	1965	0	2.84	8.71	1947	1.15	1947	4.78	1947
S	75.9	53.6	64.8	91	1964	30	1961	90	2.53	6.53	1961	0.69	1961	2.98	1954
O	64.4	43.1	53.9	91	1963	20	1965	357	2.25	6.13	1956	0.39	1956	2.02	1951
N	48.3	31.7	40.0	73	1965	15	1964	750	2.25	4.19	1948	0.74	1964	2.02	1964
D	35.9	22.9	29.4	65	1966	-6	1963	1104	1.98	4.81	1965	0.44	1958	2.85	1965
YR	59.6	40.4	50.0	99	AUG. 1964	-13	JAN. 1963	6258	30.67	8.71	AUG. 1947	0.25	JAN. 1961	4.78	AUG. 1947

Snow/Sleet, Relative Humidity, Wind, Sky Cover

Month	Snow Mean total	Snow Max monthly	Year	Snow Max 24 hrs	Year	RH 1:00 A.M.	RH 7:00 A.M.	RH 1:00 P.M.	RH 7:00 P.M.	Wind Mean hourly speed	Prevailing direction	Mean sky cover
(a)	19	20		20		5	5	5	5	20	17	20
J	7.3	15.5	1957	7.0	1957	77	78	72	72	11.9	SW	7.6
F	7.4	19.0	1962+	8.7	1965	78	78	70	69	11.8	SW	7.4
M	5.7	15.0	1954	8.1	1956	78	79	67	66	12.2	SW	7.3
A	1.1	5.5	1961	3.7	1961	74	79	56	52	12.2	SW	6.4
M	T	0.5	1960	0.5	1960	75	81	53	50	10.8	SW	5.9
J	0.0	0.0		0.0		78	81	53	53	9.3	SW	5.5
J	0.0	0.0		0.0		78	83	51	51	8.7	WNW	5.5
A	0.0	0.0		0.0		83	88	49	49	8.8	SW	5.5
S	0.0	0.2	1954	0.2	1954	85	88	56	56	9.2	SW	5.7
O	T	15.1	1966	8.5	1966	78	82	65	65	10.0	SW	6.6
N	4.1	15.1	1951	8.5	1951	78	82	71	71	11.9	SW	
D	7.3	25.1	1951	8.7	1951	81	82	73	78	11.6	SW	
YR	33.1	25.1	DEC. 1951	8.7	FEB. 1965	79	82	63	58	10.7	SW	6.6

Mean number of days

Month	Clear	Partly cloudy	Cloudy	Precip .01 in or more	Snow/Sleet 1.0 in or more	Thunderstorms	Heavy fog	Max 90° and above	Max 32° and below	Min 32° and below	Min 0° and below
(a)	20	20	20	19	19	20	20	6	5	5	5
J	5	6	20	11	3	*	2	0	19	30	5
F	5	6	17	10	3	*	2	0	14	27	4
M	5	7	18	12	2	2	2	0	5	23	2
A	5	7	18	12	1	4	1	*	0	11	0
M	7	10	14	12	0	5	1	1	0	1	0
J	9	12	9	9	0	6	2	5	0	0	0
J	9	13	9	8	0	5	2	7	0	0	0
A	9	12	10	8	0	6	3	3	0	0	0
S	11	9	11	9	0	4	2	1	0	0	0
O	10	8	12	8	0	2	2	*	0	0	0
N	4	7	19	7	1	1	3	0	1	16	1
D	4	7	20	10	1	*	4	0	15	27	1
YR	80	105	180	125	12	33	28	17	53	142	7

Means and extremes in the above table are from the existing or comparable locations. Annual extremes have been exceeded at other locations as follows:
Highest temperature 100 in July 1955 and earlier.

DETROIT, MICHIGAN — METROPOLITAN AIRPORT

LATITUDE 42° 14' N
LONGITUDE 83° 20' W
ELEVATION (ground) 633 Feet

Temperature, Degree Days, Precipitation

Month	Normal Daily maximum	Normal Daily minimum	Normal Monthly	Extremes Record highest	Year	Record lowest	Year	Normal degree days	Precip Normal total	Max monthly	Year	Min monthly	Year	Max in 24 hrs	Year
(a)	(b)	(b)	(b)	8		8		(b)	(b)	8		8		8	
J	33.5	19.1	26.5	62	1965	-13	1963	1194	1.93	3.63	1960	0.27	1961	1.40	1960
F	35.0	19.2	27.1	58	1966	-6	1963	1061	1.95	2.54	1965	0.67	1963	1.23	1965
M	43.6	26.1	34.9	77	1963	17	1964	933	2.41	3.59	1965	0.92	1965	1.13	1965
A	57.7	36.6	47.2	85	1962+	25	1964	534	3.54	4.09	1963	1.80	1960	1.97	1965
M	69.7	47.1	58.4	92	1965+	36	1965	239	3.31	4.09	1960	1.15	1965	1.72	1965
J	79.9	56.8	68.4	94	1965	41	1965	57		6.60	1966	2.12	1959	2.62	1960
J	84.7	61.1	72.9	98	1966	40	1964	0	2.69	5.24	1964	1.11	1964	3.21	1966
A	83.8	60.0	71.9	97	1964	40	1963	0	2.84	7.70	1962	1.44	1964	3.21	1966
S	76.1	52.5	64.3	94	1960	30	1962	96	2.32	5.83	1959	0.43	1962	2.07	1961
O	64.8	43.1	54.0	91	1963	18	1959	353	2.57	4.14	1961	0.35	1960	2.11	1959
N	48.4	31.8	40.1	72	1964	9	1964	738	2.27	6.00	1966	0.80	1964	1.28	1961
D	36.5	23.2	29.9	66	1966	-9	1960	1088	1.92	6.00	1965	0.46	1965	3.71	1965
YR	59.6	39.7	49.7	98	JUL. 1966	-13	JAN. 1963	6293	30.80	7.70	AUG. 1964	0.27	JAN. 1961	3.71	DEC. 1965

Snow/Sleet, Relative Humidity, Wind, Sky Cover, Sunshine

Month	Snow Mean total	Snow Max monthly	Year	Snow Max 24 hrs	Year	RH 1:00 A.M.	RH 7:00 A.M.	RH 1:00 P.M.	RH 7:00 P.M.	Wind Mean hourly speed	Prevailing direction	Pct of possible sunshine	Mean sky cover
(a)	8	8		8		8	8	8	8	8	5	7	8
J	7.7	13.4	1959	5.3	1965	76	77	68	72	11.2	WSW	44	7.3
F	9.4	17.6	1962	10.3	1965	77	79	66	71	10.9	WSW	34	7.5
M	6.6	16.1	1965	5.5	1965	77	80	63	62	11.1	WSW	50	7.4
A	1.6	7.4	1965	4.2	1961	77	81	57	57	11.1	WSW	45	7.2
M	T	1.6	1963+	1.6	1963+	81	80	57	57	9.7	WSW	65	5.4
J	0.0	0.0		0.0		81	82	51	57	8.5	SW	66	5.3
J	0.0	0.0		0.0		82	84	54	57	8.3	SW	67	5.8
A	0.0	0.0		0.0		86	88	56	68	8.1	SW	61	6.3
S	0.0	T	1965+	T	1965+	84	83	68	54	8.5	SW	61	5.7
O	T	11.8	1966	5.2	1966+	80	83	65	65	10.0	WSW	45	7.5
N	3.4	11.8	1966	5.2	1966	79	80	73	73	10.4	SW	22	7.5
D	8.3	17.3	1962	5.7	1966	80	80	75	75	10.4	SW	24	7.6
YR	37.0	17.4	FEB. 1962	10.3	FEB. 1965	80	82	60	66	9.8	SW	52	6.6

Mean number of days

Month	Clear	Partly cloudy	Cloudy	Precip .01 in or more	Snow/Sleet 1.0 in or more	Thunderstorms	Heavy fog	Max 90° and above	Max 32° and below	Min 32° and below	Min 0° and below
(a)	8	8	8	8	8	8	8	8	8	8	8
J	5	7	19	12	3	*	3	0	17	30	8
F	4	7	17	11	2	*	3	0	13	27	4
M	4	8	18	14	1	2	2	0	6	24	2
A	4	11	18	14	0	4	1	0	0	11	0
M	9	11	10	10	0	4	1	1	0	0	0
J	9	14	11	10	0	6	2	4	0	0	0
J	10	12	11	11	0	5	2	5	0	0	0
A	9	10	12	10	0	7	3	3	0	0	0
S	10	8	12	8	0	5	2	3	0	0	0
O	9	8	18	9	0	2	3	*	0	6	0
N	4	6	20	12	1	1	3	0	1	18	0
D	5	6	20	12	2	0	3	0	14	27	2
YR	77	110	178	130	13	34	26	13	50	143	7

∅ Data from Detroit City Airport through April 1966.

NORMALS, MEANS, AND EXTREMES

ESCANABA, MICHIGAN — POST OFFICE BUILDING, 1961

LATITUDE 45° 45' N
LONGITUDE 87° 03' W
ELEVATION (ground) 594 Feet

Temperature and Degree Days

Month	Normal Daily max	Normal Daily min	Normal Monthly	Record highest	Year	Record lowest	Year	Normal degree days
J	25.0	10.0	17.5	53	1942	-23	1915	1473
F	26.2	9.0	17.6	53	1958+	-31	1934	1327
M	34.4	18.4	26.2	79	1946	-20	1917	1203
A	45.8	30.6	38.2	82	1957	1	1923	804
M	57.8	41.8	49.8	91	1925+	22	1911	471
J	69.0	52.3	60.7	95	1931	32	1927	166
J	75.3	58.4	66.9	100	1916	41	1912	62
A	73.2	56.6	64.9	100	1955	35	1915	95
S	65.2	49.5	57.4	96	1953	25	1926	247
O	54.2	39.9	47.1	86	1922	10	1925	555
N	39.6	27.9	33.9	69	1953+	-9	1950	933
D	28.8	16.0	22.4	57	1961	-20	1927	1321
YR	49.5	34.2	41.9	100 AUG	1955+	-31 FEB	1934	8657

Precipitation and Snow, Sleet

Month	Precip Normal total	Max monthly	Year	Max 24 hrs	Year	Min monthly	Year	Snow Mean total	Snow Max monthly	Year	Snow Max 24 hrs	Year
J	1.53	2.98	1933	1.25	1961	.23	1961	13.5	31.6	1929	14.5	1915
F	1.37	3.88	1937	1.30	1958	.18	1958	10.7	26.6	1953	11.0	1938
M	1.78	3.83	1950	2.24	1937	.39	1946	10.7	32.4	1916	11.0	1916
A	2.10	5.23	1954	2.10	1932	.76	1925+	3.5	21.2	1923	11.4	1910
M	2.60	5.23	1949	4.83	1953	1.06	1910	.6	4.0	1954	3.0	1954
J	2.80	7.91	1915			.38	1939	.0	.0		.0	
J	3.22	9.93	1951	3.50	1951	.69	1930	.0	.0		.0	
A	2.89	7.46	1921	3.87	1937	1.03	1948	.0	.3	1942	.3	1942
S	3.12	7.58	1937	5.05	1937	.07	1948	T	.3	1933	.3	1933
O	2.04	4.70	1919	1.98	1930	.07	1952+	.2	2.2	1942	2.0	1911
N	2.20	5.69	1945	2.00	1945	.42	1930	5.2	18.1	1911	14.0	1911
D	1.43	3.17	1927	1.27	1911	.18	1943	10.5	27.4	1927	10.5	1932+
YR	27.08	9.93 JULY	1951	5.05 SEPT	1937	.07 OCT	1952+	55.7	32.4 MAR	1916	14.5 JAN	1915

Relative Humidity, Wind, Sunshine, Sky Cover

Month	RH 1AM	RH 7AM	RH 1PM	RH 7PM	Wind mean hourly	Prevailing dir	Fastest speed	Fastest dir	Year	Pct sunshine	Mean sky cover
J	76	78	70	74	10.2		52	NE	1960	41	6.7
F	76	79	68	73	10.5		56	NE	1960	47	6.2
M	77	81	63	69	10.9		68	NW	1948	55	5.9
A	77	79	63	67	11.1		47	NW	1949	56	5.8
M	79	79	65	69	10.8		57	NW	1957	58	5.6
J	82	80	65	69	9.6		50	SE	1946	61	5.4
J	84	85	68	76	9.0		49	NW	1949	66	4.9
A	87	87	67	75	9.2		43	NW	1945	61	4.9
S	84	84	67	75	10.0		47	NE	1939	53	5.6
O	81	81	67	72	10.7		52	NE	1950	45	6.1
N	77	81	75	75	11.2		56	NW	1950	34	7.3
D	79	81	77		10.6		47	NW	1939	35	7.1
YR	79	81	67	72	10.3		68 N MAR		1948	53	6.0

Mean Number of Days

Month	Clear	Partly cloudy	Cloudy	Precip .01+	Snow/Sleet 1.0+	Thunderstorms	Heavy fog	Max 90+	Max 32-	Min 32-	Min 0-
J	7	7	17	11	4	*	1	0	24	31	8
F	7	7	13	10	4	*	1	0	21	28	7
M	9	9	13	10	3	1	2	0	10	28	2
A	8	8	11	11	1	2	2	0	1	17	*
M	10	10	11	11	*	4	2	*	0	3	0
J	9	12	9	11	0	6	1	*	0	*	0
J	11	12	8	11	0	7	1	*	0	0	0
A	11	12	9	10	0	6	2	*	0	0	0
S	8	10	11	12	*	4	2	*	0	1	0
O	6	10	14	11	2	2	2	0	6	6	*
N	5	6	19	11	4	1	1	0	19	20	3
D	6	7	18	11	4	*	1	0	19	29	3
YR	102	108	155	129	18	33	19	*	81	163	20

(a) Length of record, years. (b) Normal values are based on the period 1921-1950, and are means adjusted to represent observations taken at the present standard location.

EFFECTIVE WITH JANUARY 1962 DATA, NEW CLIMATOLOGICAL STANDARD NORMALS BASED ON THE PERIOD 1931-1960 WILL BE USED. THESE NEW NORMALS ARE AVAILABLE IN "DECENNIAL CENSUS OF U.S. CLIMATE -- MONTHLY NORMALS OF TEMPERATURE, PRECIPITATION AND DEGREE DAYS -- MICHIGAN", CLIMATOGRAPHY OF THE UNITED STATES NO. 81 -- 16.

Means and extremes in the above table are from the existing location (or last comparable location). Annual extremes have been exceeded at prior locations as follows: Lowest temperature -32 in February 1875; maximum monthly precipitation 12.06 in August 1875; maximum monthly snowfall 40.4 in January 1886; maximum snowfall in 24 hours 18.5 in April 1909.

FLINT, MICHIGAN — BISHOP AIRPORT

LATITUDE 42° 58' N
LONGITUDE 83° 44' W
ELEVATION (ground) 770 Feet

Temperature and Degree Days

Month	Normal Daily max	Normal Daily min	Normal Monthly	Record highest	Year	Record lowest	Year	Normal degree days
J	31.9	17.1	24.5	60	1966	-11	1966	1256
F	32.8	16.7	24.8	54	1966	-7	1966	1126
M	41.3	24.1	32.7	74	1966	-7	1966	1001
A	56.8	34.7	45.8	84	1964	13	1964	591
M	68.0	44.7	56.4	88	1965	22	1966	286
J	78.5	54.8	66.6	94	1964	33	1966	75
J	83.5	58.9	71.2	94	1965	40	1965	9
A	81.6	57.6	69.6	97	1965+	39	1964	24
S	73.4	50.4	62.0	94	1964	31	1964	132
O	62.1	40.6	51.4	84	1963	19	1964	428
N	46.3	30.6	38.5	67	1964	-11	1966	804
D	34.6	20.6	27.6	62	1965	-11	1963	1153
YR	57.5	37.5	47.5	97 AUG	1965+	-11 JAN	1966+	6885

Precipitation and Snow, Sleet

Month	Precip Normal total	Max monthly	Year	Max 24 hrs	Year	Min monthly	Year	Snow Mean total	Snow Max monthly	Year	Snow Max 24 hrs	Year
J	1.70	3.56	1947	1.34	1959	0.07	1961	10.0	26.0	1959	7.8	1959
F	1.76	5.28	1954	2.85	1965			9.8	19.7	1965	11.3	1965
M	2.14	4.33	1948+	2.33	1948	0.62	1958	7.0	16.1	1948	4.7	1948
A	2.66	5.28	1947	2.05	1947	0.92	1958	2.1	6.1	1947	4.7	1952
M	3.39	7.35	1945	2.23	1958	0.90	1943	0.0	0.0		0.0	1961
J	3.12	7.92	1957	3.55	1957	0.83	1944	0.0	0.0		0.0	
J	2.91	8.68	1956	3.72	1957	0.54	1955	0.0	0.0		0.0	
A	3.17	8.68	1950	2.73	1942	0.33	1944	0.0	0.0		0.0	
S	2.37	4.59	1954	6.04	1950	0.66	1952	0.0	0.8	1962	0.8	1962
O	2.18	4.21	1954	1.94	1951	0.44	1949	T	16.2	1951	13.4	1951
N	1.80	3.75	1949	2.11	1951			1.6	24.9	1951	7.5	1951
D				1.54	1946			7.8				
YR	30.14	8.68 AUG	1956+	6.04 SEP	1950	0.07 JAN	1961	40.5	26.0 JAN	1959	13.4 NOV	1951

Relative Humidity, Wind, Sky Cover

Month	RH 1AM	RH 7AM	RH 1PM	RH 7PM	Wind mean hourly	Prevailing dir	Mean sky cover
J	72	73	67	69	11.9	SW	7.6
F	73	74	63	66	11.6	WNW	7.4
M	73	75	62	65	12.0	WNW	7.0
A	71	75	58	60	12.0	WNW	7.2
M	75	76	51	54	9.0	WSW	6.6
J	79	81	53	53	8.1	SW	5.7
J	82	84	51	51	9.2	SW	5.6
A	86	88	61	61	10.5	SW	5.9
S	86	89	56	56	10.7	SW	6.0
O	79	80	66	67	11.7	SSW	8.0
N	78	78	72	72	11.7	SW	7.8
D	76		72	75		SW	
YR	76	78	60	64	10.5	SW	6.8

Mean Number of Days

Month	Clear	Partly cloudy	Cloudy	Precip .01+	Snow/Sleet 1.0+	Thunderstorms	Heavy fog	Max 90+	Max 32-	Min 32-	Min 0-
J	4	6	21	12	3	*	2	0	17	30	4
F	4	7	18	12	3	*	1	0	12	26	4
M	5	8	19	13	2	1	1	0	6	25	1
A	5	10	17	11	1	3	1	0	*	11	0
M	6	12	11	11	0	5	1	0	0	1	0
J	8	13	10	9	0	6	2	2	0	0	0
J	8	13	9	9	0	6	2	4	0	0	0
A	8	13	9	9	0	5	2	3	0	0	0
S	8	9	12	9	0	3	2	0	0	1	0
O	3	7	13	12	1	1	2	0	1	10	0
N	2	7	21	12	2	*	2	0	13	16	1
D				13			2	0		23	
YR	68	109	188	129	13	33	19	10	51	145	7

Temperature and degree day normals revised December 1966.
∅ For period October 1963 through the current year.
Means and extremes in the above table are from the existing location. Annual extremes have been exceeded at other locations as follows: Highest temperature 108 in June 1936; lowest temperature -28 in February 1916; minimum monthly precipitation 0.04 in August 1899. New degree day normals will be used beginning July 1967.

NORMALS, MEANS, AND EXTREMES

GRAND RAPIDS, MICHIGAN
KENT COUNTY AIRPORT

LATITUDE 42° 53' N
LONGITUDE 85° 31' W
ELEVATION (ground) 784 Feet

Month	Temperature Normal Daily maximum	Daily minimum	Monthly	Extremes Record highest	Year	Record lowest	Year	Normal degree days	Precipitation Normal total	Maximum monthly	Year	Minimum monthly	Year	Maximum in 24 hrs.	Year	Snow,Sleet Mean total	Maximum monthly	Year	Maximum in 24 hrs.	Year
J	31.3	16.7	24.0	59	1966	-16	1966	1271	1.91	3.99	1965	1.22	1936	1.28	1964	22.9	25.9	1966	12.1	1964
F	32.2	16.4	24.3	54	1966	-13	1966	1139	1.75	2.28	1966	0.73	1964	1.01	1965	14.4	16.7	1965	7.5	1966
M	41.5	23.9	32.7	73	1965	-1	1964	1001	2.28	3.24	1964	0.65	1965	1.51	1966	19.0	36.0	1965	6.9	1965
A	56.6	34.8	45.7	81	1966	10	1964	579	2.94	5.28	1965	2.49	1965	1.36	1964	3.1	5.4	1965	3.7	1965
M	68.9	44.9	56.8	89	1965	22	1966	323	3.46	3.96	1964	1.53	1965	1.97	1964	0.0	0.0	1966	0.0	1966
J	79.4	55.6	67.5	94	1965	37	1964	71	3.31	4.12	1966	2.36	1966	2.65	1964	0.0	0.0		0.0	
J	84.5	60.0	72.3	97	1966	47	1965	8	2.70	5.36	1966	1.95	1964	1.61	1964	0.0	0.0		0.0	
A	82.6	58.2	70.4	100	1964	41	1964	27	2.70	5.29	1964	1.59	1966	1.59	1966	0.0	0.0		0.0	
S	74.3	50.0	62.2	89	1966+	32	1965	135	2.98	6.02	1965	0.92	1964	1.45	1965	T	T	1965	T	1965
O	62.7	39.5	51.1	82	1964	20	1966	435	2.61	6.33	1966	0.60	1964	1.45	1966	0.6	16.6	1964	8.3	1964
N	46.2	28.4	37.3	73	1964	10	1964	831	2.49	7.81	1966	2.13	1966	2.66	1966	10.5	16.6	1964	8.3	1964
D	34.3	19.7	27.0	64	1966	-6	1963	1178	2.03	4.23	1965	1.54	1963	1.62	1966	13.5	21.6	1963	6.6	1965
YR	57.9	37.4	47.6	100	AUG. 1964	-16	JAN. 1966	6998	31.19	7.81	NOV. 1966	0.60	OCT. 1964	2.66	NOV. 1966	83.4	36.0	MAR. 1965	12.1	JAN. 1964

Means and extremes in the above table are from the existing location (or last comparable location). Annual extremes have been exceeded at other locations as follows:
Highest temperature 108 in July 1936; lowest temperature -24 in February 1899; maximum monthly precipitation 13.22 in June 1892; minimum monthly precipitation 0.02 in February 1877; maximum precipitation in 24 hours 4.58 in June 1905; maximum monthly snowfall 54.0 at Weather Bureau Office (Post Office Building) in December 1951; maximum snowfall in 24 hours 14.0 in January 1914; fastest mile of wind 80 from Southwest in November 1940.

HOUGHTON LAKE, MICHIGAN
ROSCOMMON COUNTY AIRPORT

LATITUDE 44° 22' N
LONGITUDE 84° 41' W
ELEVATION (ground) 1149 Feet

Means and extremes in the above table are from existing or comparable locations. Annual extremes have been exceeded at other locations as follows:
Highest temperature 107 in July 1936 and earlier; lowest temperature -48 in February 1918; maximum monthly precipitation 8.38 in October 1951; minimum monthly precipitation 0.18 in December 1960; maximum precipitation for one observational day 5.18 in July 1957; maximum monthly snowfall 43.5 in March 1923; maximum snowfall for one observational day 14.5 in March 1942.

- 206 -

NORMALS, MEANS, AND EXTREMES

LANSING, MICHIGAN — CAPITAL AIRPORT

LATITUDE 42° 47' N
LONGITUDE 84° 36' W
ELEVATION (ground) 841 Feet

Month	Normal Daily maximum	Normal Daily minimum	Normal Monthly	Extremes Record highest	Year	Extremes Record lowest	Year	Normal degree days	Precip. Normal total	Max monthly	Year	Min monthly	Year	Max in 24 hrs	Year	Snow,Sleet Mean total	Max monthly	Year	Max in 24 hrs	Year	RH 1AM	RH 7AM	RH 1PM	RH 7PM	Wind mean hourly	Prevail. dir.	Fastest mile speed	Dir	Year	% poss. sunshine	Mean sky cover
(a)	(b)	(b)	(b)					(b)	(b)																						
J	31.1	17.4	24.3	59	1966	-16	1966	1262	1.96	3.61	1949	0.45	1961	1.59	1949	9.4	19.2	1962	7.0	1954	84	86	77	82	12.5	SW	49	W	1961	34	7.8
F	31.9	16.5	24.2	54	1966	-19	1966	1142	1.95	4.21	1954	0.39	1963	2.40	1954	9.6	16.9	1962	9.0	1965	84	85	70	75	12.0	W	59	W	1962	42	7.4
M	40.7	24.1	32.4	74	1965	-2	1965	1011	2.40	3.26	1954	0.92	1962	1.59	1954	6.9	18.6	1965	6.2	1954	84	84	67	67	12.3	W	59	W	1961	47	7.4
A	55.7	35.5	45.7	85	1964	19	1964	579	2.87	4.53	1964	1.38	1962	1.18	1964	3.4	11.5	1952	7.9	1952	79	81	55	62	12.5	W	49	W	1961	54	6.1
M	68.1	46.3	57.4	96	1964	30	1966	273	3.73	4.98	1952	1.00	1961	2.26	1963	T	0.3	1954	0.3	1954	79	84	50	57	9.6	S	45	W	1962+	66	5.6
J	78.4	56.3	67.4		1966	39	1965	69	3.34	6.09	1963	1.15	1959	4.35	1963	0.0	0.0		0.0		80	77	48	52	8.8	S	63	SE	1963	71	5.2
J	83.4	59.9	71.7	99	1966	37	1965		3.61	5.08	1959	0.50	1965	2.16	1959	0.0	0.0		0.0		80	87	53	57	8.8	S	56	NE	1959	72	5.1
A	81.6	58.0	70.2	100	1964	36	1964		3.05	3.05	1964	1.60	1964	2.26	1965	0.0	0.0		0.0		87	88	56	62	9.3	S	57	W	1962	65	5.5
S	73.6	51.0	62.0	98	1964	27	1966	22	2.58	7.63	1965	1.22	1966	2.30	1951	0.0	T	1962	T	1962	90	92	61	72	9.3	S	47	W	1966	61	5.5
O	61.4	41.1	51.3	89	1963	15	1964	138	2.60	4.99	1959	0.35	1952	1.98	1951	T	1.6	1962	1.0	1962	83	87	70	75	10.2	S	54	W	1960	52	7.0
N	45.3	30.5	37.9	73	1964	-1	1966	431	2.50	4.60	1959	0.51	1962	1.43	1966+	5.9	16.8	1966+	11.0	1951	86	87	79	80	11.7	SW	56	SW	1961	30	7.7
D	34.0	21.0	27.5	65	1966	-17	1966	813	2.21	4.70	1949	0.37	1960	1.56	1949	9.9	27.8	1951	11.0	1951	86	87	80	85	11.9	S	54	SE	1963	30	7.7
YR	57.1	38.2	47.6	100	AUG. 1964	-19	FEB. 1966	6909	31.18	7.63	SEP. 1965	0.35	OCT. 1952	4.35	JUN. 1963	45.2	27.8	DEC. 1951	11.0	NOV. 1951	83	84	62	68	10.9	S	63	SE 1963		54	6.5

Mean number of days — Sunrise to sunset (Clear / Partly cloudy / Cloudy); Precipitation .01 inch or more; Snow,Sleet 1.0 inch or more; Thunderstorms; Heavy fog; Temperatures Max. 90° and above / 32° and below, Min. 32° and below / 0° and below; Average daily solar radiation (langleys) [X].

Month	Clear	Partly cloudy	Cloudy	Precip .01+	Snow 1.0+	Thunder	Heavy fog	Max 90+	Max 32-	Min 32-	Min 0-	Solar radiation
J	4	6	21	15	5	*	3	0	17	30	3	132
F	4	6	18	12	4	*	2	0	13	27	3	216
M	5	7	19	13	3	1	2	0	6	26	1	305
A	5	8	17	12	1	3	1	*	*	14	0	366
M	8	10	13	11	*	3	1	1	0	4	0	504
J	9	11	10	10	0	7	2	4	0	*	0	564
J	10	12	9	9	0	5	3	6	0	0	0	541
A	11	11	9	10	0	6	3	4	0	0	0	462
S	10	9	11	10	0	4	3	1	0	0	0	373
O	10	10	11	7	0	2	2	0	0	2	0	258
N	4	6	20	11	3	*	2	0	3	18	1	137
D	4	6	20	15	5	*	2	0	14	25	2	109
YR	84	103	178	137	23	33	23	17	54	157	11	331

Ø For period September 1963 through the current year.

Means and extremes in the above table are from existing or comparable locations. Annual extremes have been exceeded at other locations as follows:
Highest temperature 102 in August 1918 and earlier; lowest temperature -33 in February 1875; maximum monthly precipitation 11.35 in June 1883; minimum monthly precipitation 0.00 in August 1894 and earlier; maximum precipitation in 24 hours 5.47 in June 1905; maximum monthly snowfall 33.6 in January 1943; maximum snowfall in 24 hours 18.1 in November 1921.

X Data are from Michigan State University. Pyrheliometer 899 ft. above sea level to June 1959 and 878 ft. thereafter.

MARQUETTE, MICHIGAN — U. S. POST OFFICE

LATITUDE 46° 34' N
LONGITUDE 87° 24' W
ELEVATION (ground) 677 Feet

Month	Normal Daily maximum	Normal Daily minimum	Normal Monthly	Record highest	Year	Record lowest	Year	Normal degree days	Precip. Normal total	Max monthly	Year	Min monthly	Year	Max in 24 hrs	Year	Snow,Sleet Mean total	Max monthly	Year	Max in 24 hrs	Year	RH 1AM	RH 7AM	RH 1PM	RH 7PM	Mean hourly wind	Prevail. dir.	Fastest mile speed	Dir	Year	% poss. sun	Mean sky cover
(a)	(b)	(b)	(b)	30		30		(b)	(b)		29		29		29	29		29		29	11	27	27	27	27	27	29	29	29	29	29
J	25.7	13.2	19.5	57	1944	-21	1963	1411	1.89	3.93	1956	0.51	1956	1.39	1938	19.1	43.7	1950	11.5	1950	74	75	68	72	8.9	W	52	SW	1938	33	8.1
F	26.6	12.8	19.7	56	1961	-13	1966	1268	1.65	3.29	1947	0.31	1947	1.18	1951	18.9	31.5	1960	16.0	1960	73	73	67	72	8.8	NW	42	SW	1959+	41	7.8
M	33.4	20.0	26.7	76	1946	-13	1962+	1187	1.91	4.46	1966	0.30	1966	1.56	1939	16.6	32.8	1966	16.0	1966	70	73	61	65	8.4	N	57	SW	1938	50	7.3
A	46.8	29.3	39.3	86	1938	4	1954	771	2.70	5.73	1947	0.64	1949	2.43	1960	8.5	28.4	1950	11.7	1945+	73	73	63	65	8.2	NW	50	SW	1964	55	6.7
M	58.8	40.9	49.9	98	1964	17	1954	468	2.96	7.70	1960	0.56	1948	2.22	1960	T	3.2	1960	7.8	1945+	76	76	65	66	7.4	NW	59	SW	1943	61	6.3
J	69.6	50.6	60.1	101	1963	34	1945	177	3.46	8.86	1939	1.37	1957	4.65	1939	T	T	1945+	0.0		78	76	64	66	—	—	36	SE	1958	67	5.7
J	75.7	57.6	66.7	99	1963+	43	1960	59	3.20	10.20	1949	0.75	1957	3.94	1949	0.0	0.0	1949	0.0		77	79	63	66	7.1	W	52	SE	1960	67	5.7
A	65.8	57.3	65.8	102	1947	43	1961	81	3.03	5.40	1961	0.98	1951	2.51	1960	0.0	0.0	1960	0.0		80	81	65	71	7.4	W	45	SW	1943	52	6.0
S	65.7	49.7	57.7	98	1939	30	1942	240	3.28	5.73	1942	1.21	1948	1.90	1959	0.2	0.1	1942	6.0	1959	78	81	66	69	8.1	SW	47	NW	1951+	47	6.6
O	55.2	40.6	48.0	86	1950	21	1942	527	2.33	7.13	1959	0.74	1959	4.06	1959	2.1	9.1	1959	6.0	1959	79	80	71	75	8.7	SW	50	SW	1958	26	8.4
N	39.3	33.8	33.8	73	1950	-1	1959	936	3.28	5.08	1942	0.79	1948	2.29	1943	15.9	36.6	1959	14.2	1943	74	76	70	75	9.0	SW	45	NW	1941	27	8.2
D	29.3	18.9	24.1	59	1962	-10	1962	1268	1.92	3.58	1942	0.69	1942	1.23	1940	21.5	45.6	1950	16.1	1960	76	77	74	74	9.1	W	45	SW	1941	27	8.2
YR	50.0	35.2	42.6	102	AUG. 1947	-21	JAN. 1963	8393	31.61	10.20	JUL. 1949	0.21	OCT. 1956	4.65	JUN. 1939	103.8	45.6	FEB. 1960	16.1	FEB. 1947	75	76	65	70	8.3	W	59	SW	JUN. 1958	50	7.0

Mean number of days — Sunrise to sunset (Clear / Partly cloudy / Cloudy); Precipitation .01 inch or more; Snow,Sleet 1.0 inch or more; Thunderstorms; Heavy fog [%]; Temperatures Max. 90° and above / 32° and below, Min. 32° and below / 0° and below.

Month	Clear	Partly cloudy	Cloudy	Precip .01+	Snow 1.0+	Thunder	Heavy fog	Max 90+	Max 32-	Min 32-	Min 0-
J	3	6	22	17	13	*	1	0	24	31	5
F	4	6	19	13	11	0	1	0	28	28	3
M	5	7	19	12	11	*	2	0	14	28	1
A	6	9	15	11	6	1	3	0	2	17	0
M	6	10	15	11	1	2	6	*	*	6	0
J	7	10	13	12	0	4	5	1	0	0	0
J	8	13	10	10	0	5	5	2	0	0	0
A	6	10	15	11	0	6	6	1	0	0	0
S	5	8	17	11	*	4	6	0	0	4	*
O	4	5	23	14	1	2	5	0	7	19	0
N	2	3	25	16	9	*	3	0	19	29	1
D	3	5	23	17	16	0	1	0	19	30	1
YR	62	100	203	156	31	28	44	5	87	160	11

% Through 1964. The station did not operate 24 hours daily. Fog and thunderstorm data therefore may be incomplete.

Means and extremes in the above table are from the existing or comparable location(s). Annual extremes have been exceeded at other locations as follows:
Highest temperature 108 in July 1901; lowest temperature -27 in February 1888 and earlier date(s); maximum monthly precipitation 12.73 in September 1881; minimum monthly precipitation 0.12 in February 1877; maximum precipitation in 24 hours 5.14 in June 1878; maximum monthly snowfall 54.3 in February 1890; maximum snowfall in 24 hours 17.0 in February 1890; fastest mile of wind 91 from South in May 1934.

NORMALS, MEANS, AND EXTREMES

MUSKEGON, MICHIGAN
MUSKEGON COUNTY AIRPORT

LATITUDE 43° 10' N
LONGITUDE 86° 14' W
ELEVATION (ground) 627 Feet

Means and extremes are from the existing location. Annual extremes have been exceeded at other locations as follows:
Lowest temperature -30 in February 1899; minimum monthly precipitation 0.04 in November 1904; maximum precipitation in 24 hours 5.08 in June 1921.

SAULT STE. MARIE, MICHIGAN
MUNICIPAL AIRPORT

LATITUDE 46° 28' N
LONGITUDE 84° 22' W
ELEVATION (ground) 721 Feet

Means and extremes in the above table are from the existing location. Annual extremes have been exceeded at other locations as follows:
Lowest temperature -37 in February 1934 and earlier date(s); maximum monthly precipitation 9.35 in September 1916; minimum monthly precipitation 0.09 in March 1915; maximum precipitation in 24 hours 5.64 in September 1916; maximum monthly snowfall 52.5 in January 1935.

REFERENCE NOTES APPLYING TO ALL "NORMALS, MEANS, AND EXTREMES" TABLES

(a) Length of record, years.
(b) Climatological standard normals (1931-1960).
° Less than one half.
+ Also on earlier dates, months or years.
* T Trace, an amount too small to measure.
Below-zero temperatures are preceded by a minus sign.
The prevailing direction for wind in the Normals, Means, and Extremes table is from records through 1963.
To 8 compass points only.

Unless otherwise indicated, dimensional units used in this bulletin are: temperature in degrees F.; precipitation, including snowfall, in inches; wind movement in miles per hour; and relative humidity in percent. Degree day totals are the sums of the negative departures of average daily temperatures from 65° F. Sleet was included in snowfall totals beginning with July 1948. Heavy fog reduces visibility to 1/4 mile or less.

Sky cover is expressed in a range of 0 for no clouds or obscuring phenomena to 10 for complete sky cover. The number of clear days is based on average cloudiness 0-3; partly cloudy days 4-7; and cloudy days 8-10 tenths.

Solar radiation data are the averages of direct and diffuse radiation on a horizontal surface. The langley denotes one gram calorie per square centimeter. Averages in the lower table for some months may be for more than the listed number of years.

& Figures instead of letters in a direction column indicate direction in tens of degrees from true North; i.e., 09 - East, 18 - South, 27 - West, 36 - North, and 00 - Calm. Resultant wind is the vector sum of wind directions and speeds divided by the number of observations. If figures appear in the direction column under 'Fastest mile' the corresponding speeds are fastest observed 1-minute values.

Mean Maximum Temperature (°F.), January

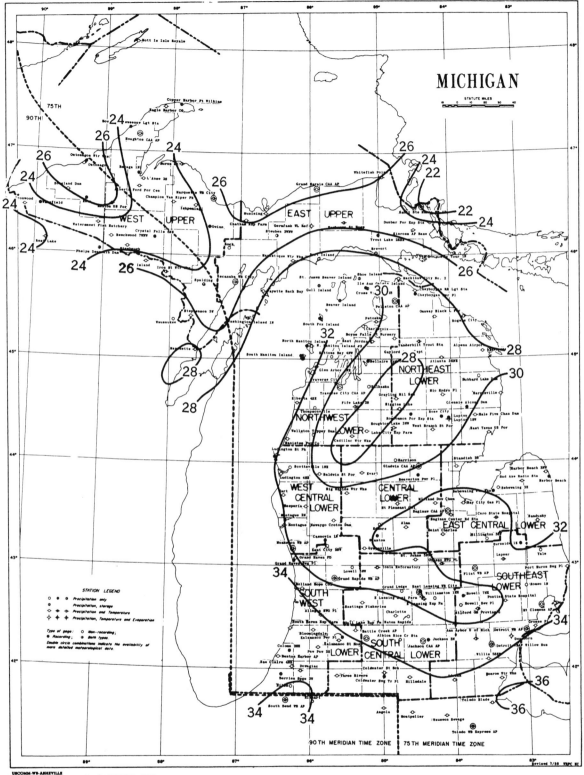

MICHIGAN

Based on period 1931-52

Isolines are drawn through points of approximately equal value. Caution should be used in interpolating on these maps, particularly in mountainous areas.

Mean Minimum Temperature (°F.), January

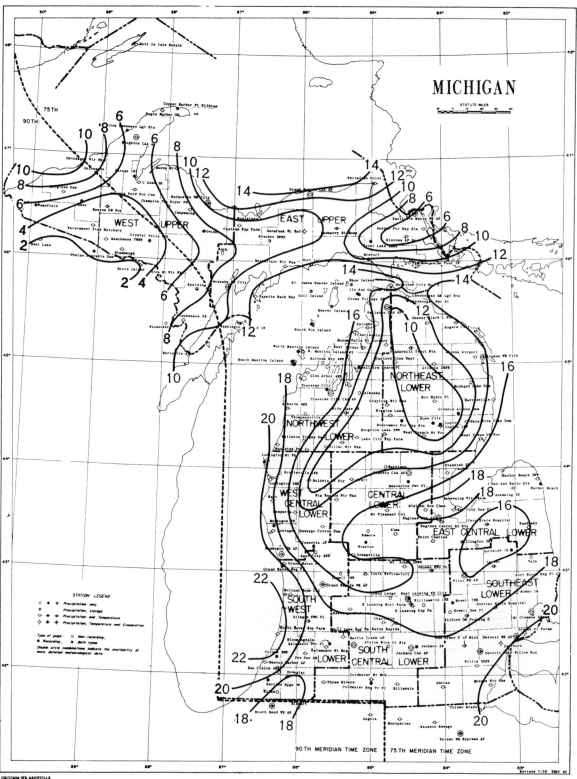

Based on period 1931-52

Isolines are drawn through points of approximately equal value. Caution should be used in interpolating on these maps, particularly in mountainous areas.

Mean Maximum Temperature (°F.), July

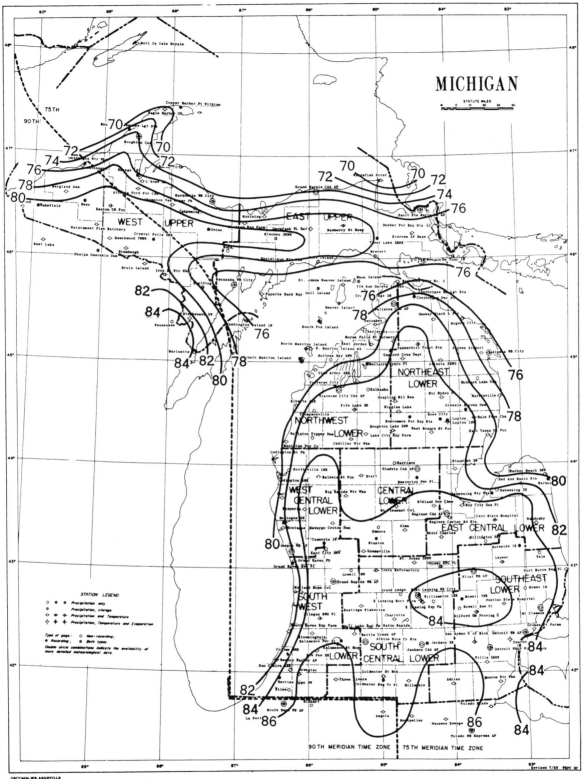

Based on period 1931-52

Isolines are drawn through points of approximately equal value. Caution should be used in interpolating on these maps, particularly in mountainous areas.

Mean Minimum Temperature (°F.), July

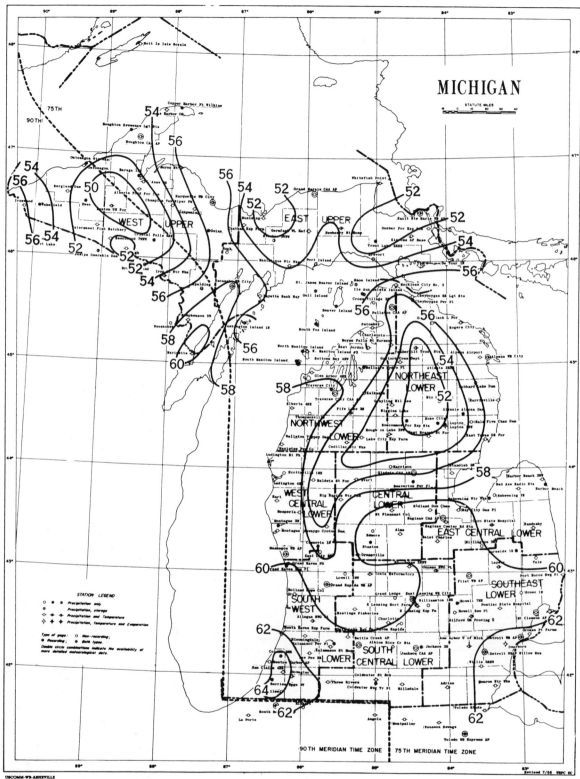

MICHIGAN

Based on period 1931-52

Isolines are drawn through points of approximately equal value. Caution should be used in interpolating on these maps, particularly in mountainous areas.

Mean Annual Precipitation, Inches

MICHIGAN

UBCOMM-WB-ASHEVILLE

Based on period 1931-55

Isolines are drawn through points of approximately equal value.   Caution should be used
in interpolating on these maps, particularly in mountainous areas.

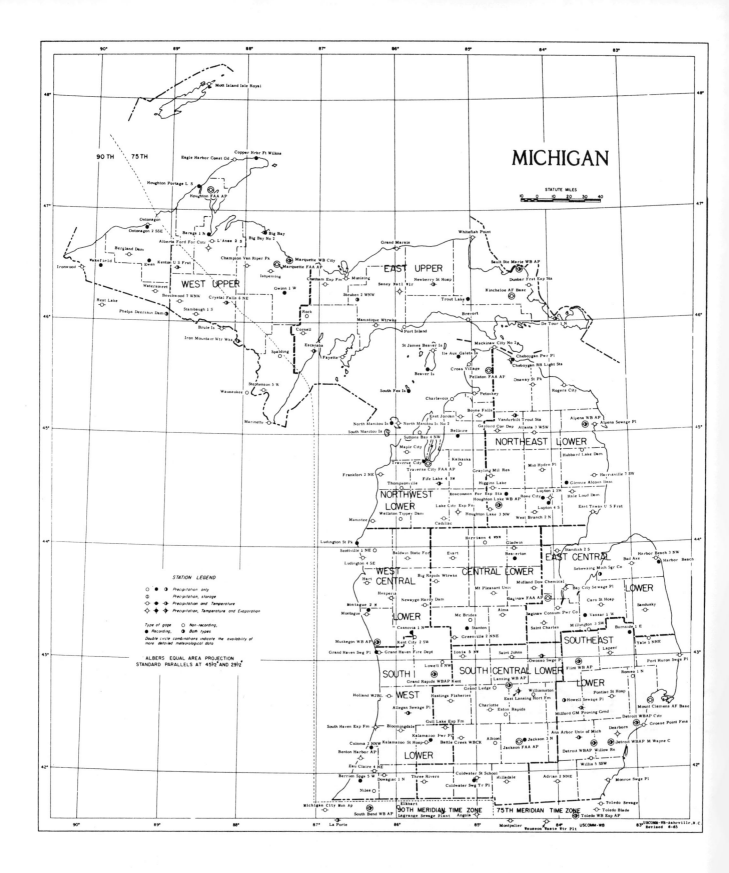

MICHIGAN

STATION LEGEND

Precipitation only
Precipitation, storage
Precipitation and Temperature
Precipitation, Temperature and Evaporation

Type of gage ○ Non-recording,
● Recording, ◐ Both types
Double circle combinations indicate the availability of
more detailed meteorological data

ALBERS EQUAL AREA PROJECTION
STANDARD PARALLELS AT 45½° AND 29½°

THE CLIMATE OF
MISSISSIPPI

by
Ralph Sanders

December 1959

The State of Mississippi extends on the west from the Mississippi River to about longitude 88°W., between latitude 31° and 35°N.; and from the Pearl River on the west to about 88 1/2°W. longitude below latitude 31°N. The southern boundary of this area, a sort of "panhandle", is Mississippi Sound, which is an arm of the Gulf of Mexico. Land areas near the coast line, in contrast to those of Louisiana, are sharply defined, with the land rising to elevations of 10 to 20 feet behind the beaches. The coast is cut by numerous bays. A string of islands parallels the coast a few miles offshore. The coastal strip has developed into a popular summer resort. The waters of Mississippi Sound provide a natural air conditioning to ameliorate the summer heat. Thus Biloxi has an average of only 55 days with temperature 90°F. or higher, while only 40 miles inland Wiggins averages 105 such days.

A triangular area, comprising nearly one-third of the State, with its apex in Rankin County and its base on the coast is composed of rolling hills at from 200 to 500 feet above sea level. The soil in this area lacks natural fertility, except for alluvial deposits along its streams. For this reason it is not now important in the State's cultivated agriculture, but the generous rainfall and long growing season are climatically favorable for timber production. Although many years ago this area was denuded of its virgin timber, new

growth is rapidly taking over, giving impetus to a growing wood pulp and paper industry. The primary industries on the coast are fishing and ship building, with year-round outdoor work possible.

The "Delta" region in the northwest extends from the Yazoo-Tallahatchie River system westward to the Mississippi River. Here the land is level, very fertile, and devoted to intensive cotton cultivation on extensive plantations. The hot and drier summers are well suited to cotton culture. In the northeast lies the fertile upland prairie, where cotton cultivation again predominates for the same reason.

Between the Delta and the upland prairie the land is broken by a series of ridges and valleys which are oriented in a general southwest-northeast direction. These extend from the Tennessee border to the lower Mississippi River. From Vicksburg to Natchez these ridges stop abruptly at the river forming high bluffs along its left bank. The valleys form natural paths for the northeastward passage of tornadoes and for southward drainage of cold winter air. On clear, cold nights temperatures in these valleys are lower, sometimes as much as 20°F., than on the nearby hilltops. Flood control dams in the Yazoo-Tallahatchie basin have in recent years formed large lakes in the north, and these have added to the fishing and recreational resources of the State.

Mississippi is abundantly supplied with water

both above and underground, annual average precipitation ranging from 50 to 65 inches. At many places overflowing artesian wells can be obtained at depths ranging from 400 to 1,000 feet.

All of the State is in the Gulf of Mexico drainage. Main rivers which flow directly into the Gulf include the Tombigbee in the northeast portion (except for a limited area where some minor tributaries flow northward into the Tennessee River in Tennessee) and the Pascagoula and Pearl which forms the southwestern boundary. The Mississippi River which forms most of the western boundary has for its principal tributaries in the State, the Yazoo, and the Big Black Rivers.

The flood season in Mississippi is from November through June (the period of greatest rainfall), with March and April being the months of greatest frequency. The season of high flows in the main Mississippi River is during the first 6 months of the year. In other streams flooding sometimes occurs during the summer from persistent thundershower rains, or during the late summer and early fall from heavy rains associated with tropical storms originating in the Gulf of Mexico and passing through the State.

Local overflows occur on many streams three or four times a year. Severe general flooding occurs about once in 3 years along the larger streams. The Mississippi River floods about once in 2 years from upstream runoff. The only important contribution to the Mississippi within the State is from the Yazoo. A system of levees prevents major damage from Mississippi River floods.

Some years in which major flooding occurred in Mississippi are as follows: Tombigbee, 1892, 1944, 1948, 1951, 1955; Pascagoula, 1900, 1916, 1920; Pearl, 1874, 1900, 1902, 1935, 1938, 1946, 1947, 1950; Yazoo, 1882, 1927, 1932, 1933, 1946; Mississippi, 1913, 1916, 1922, 1927, 1929, 1937, 1950.

In its broader aspects the climate of Mississippi is determined by the huge land mass to the north, its subtropical latitude, and the Gulf of Mexico to the south, but modifications are introduced by the varied topography.

The prevailing southerly winds provide a moist, semitropical climate, with conditions often favorable for afternoon thundershowers. When the pressure distribution is altered so as to bring westerly or northerly winds, periods of hotter and drier weather interrupt the prevailing moist condition. The high humidity, combined with hot days and nights in the interior from May to September, produces discomfort at times. The principal relief is by thunderstorms, sometimes accompanied by locally violent and destructive winds.

In the colder season the State is alternately subjected to warm tropical air and cold continental air, in periods of varying length. However, cold spells seldom last over 3 or 4 days. The ground rarely freezes, and then mostly only in the extreme north and only a few inches deep. Although slowly warmed by its southward journey, the cold air occasionally brings large and rather sudden drops in temperature. In winter the Atlantic High is also sometimes located far enough west to serve as a barrier to cold air approaching the State. Most frequently this produces a pattern of warm, clear weather over the southern part of the State with cold, rainy weather to the north of the "front", but occasionally the entire State will be under the balmy influence of this subtropical anticyclone.

Mississippi is south of the average track of winter cyclones, but occasionally one moves over the State. In some winters a succession of such cyclones will develop in the Gulf of Mexico or in Texas and move over or near the State. The winter of 1957-58 was a classic example of a continuing series of Gulf cyclones. The State is occasionally in the path of tropical storms or hurricanes.

Mean annual precipitation ranges from about 50 inches in the northwest to 65 inches in the southeast.

During the freeze-free season rainfall ranges from 23 to 25 inches in the Delta districts to 36 to 38 inches in the southeast.

This distribution discourages the growth of crops with critical water requirements, such as corn in much of the "Delta" and the northeast prairie but is beneficial in particular for cotton. Conversion from crops to cattle in large areas in the north is due, at least in part, to insufficient or poorly distributed rainfall in that area. Irrigation is being increasingly practiced because the generous rainfall does not always come in the time of greatest need.

During the winter the precipitation maximum is centered over the northern and western counties (16 to 18 inches) with the minimum (13 inches) on the coast. In summer the maximum shifts to the coastal counties (19 to 21 inches) and the minimum to the Delta counties (9 to 11 inches). The spring and fall patterns are very similar to the summer pattern. The fall months are the driest of the year, precipitation ranging from about 8 to 13 inches. This favors harvesting of crops. Fall is the most agreeable season of the year, with cool nights and mild, clear, sunny days persisting for several days, and even weeks, at a time.

The greatest rate of rainfall recorded in Mississippi is 0.83 inch in 5 minutes at Vicksburg or a rate of 9.96 inches per hour. Of course this rate is never sustained for any length of time, and the greatest of record in 1 hour is 3.66 inches at Meridian. From the beginning of record through 1950, the heaviest 24-hour rainfall was 12.35 inches at Merrill during a hurricane on July 5, 1916. Generally, however, the most intense rains are associated with thunderstorms; tropical storms usually cause heavier rains over longer periods of time.

While snowfall is not of much economic importance, it is not such a rare event in Mississippi as is generally believed. During the 60 years from 1898 through 1957, measurable snow or sleet has fallen on some part of the State in all but 3 years, or 95 percent of the years. North of 34°N. latitude measurable snow has fallen in 65 to 85 percent of the years; between 34° and 33°N., it has fallen in 50 to 65 percent of the years; between 33° and 32°N. in 30 to 55 percent of the years; between 32° and 31°N. in 10 to 30 percent of the years; and below 31°N. to the coast, measurable snow has fallen in about 10 percent of the years. During these 60 years snow or sleet has fallen in January in 37 years, in February in 31 years, in March in 15 years, in November in 5 years, and in December in 29 years.

The normal annual temperature ranges from 62°F. in the northern border counties to 68°F. in the coastal counties. The lowest January normal is 43°F. in the north-central area, ranging upward to 54°F. in the coastal district. The highest July normal is 84°F. in the upper Delta, ranging downward to 80° to 81° in parts of central and north-central districts. The lowest temperature ever observed was 16°F. below zero at Batesville in February 1951 and at French Camp in February 1899. Temperatures of 90°F. or higher occur on an average of 55 days per year on the immediate Gulf coast under the ameliorating effect of the relatively cooler Gulf waters. There is a rapid increase in number of days 90°F. or higher inland from the coast, reaching a maximum of 105 such days in Stone County. Temperatures of 32°F. (freezing) or lower occur on an average of 11 days a year on the immediate Gulf coast, increasing to a maximum of 60 days in Panola County. Temperatures exceed 100°F. at one or more stations each summer. They drop to zero or lower in Mississippi on an average of once in 5 years and to 32°F. or lower on the Gulf coast almost every winter.

Thunderstorms occur on an average of 50 to 60 days a year in the northern districts and 70 to 80 days a year near the coast. Thundershowers occur more frequently in July than any other month, with the least in December. Those in late fall, winter, and early spring are more apt to be attended by high winds than in summer. However, in the interior in summer after a spell of unusually high temperatures, thunderstorms may develop with local violence. During late fall, winter, and early spring, thunderstorms may occur at any time of day, as they are usually associated with passing weather systems. During the warm season about 65 percent of the thundershowers occur between noon and 6 p.m. and 85 percent between noon and midnight in the interior. Near the coast 30 percent occur between 6 a.m. and noon, 60 percent between noon and 6 p.m., or about 90 percent between 6 a.m. and 6 p.m.

A hazard to life and property in Mississippi is the tropical cyclone which occurs from June to November. While these storms generally move into the State on the coast, they have on occasion entered as far north as Meridian and Greenville after crossing part of Alabama or Louisiana. These latter storms are usually weakened considerably by passage over land. Loss of life and property due to high winds are mostly confined to the coastal areas with interior losses generally from rain damage to crops and from floods. However, the hurricane of September 26-27, 1906, which moved inland between Pascagoula and Mobile, caused great damage as far inland as Brookhaven and Waynesboro; about 10 percent of the virgin timber in the area was destroyed. The hurricane of July 5-7, 1916, pursued one of the most unusual courses ever observed. It moved inland near Pascagoula late on July 5 on a northwest course as it decreased in intensity. It passed over Jackson to Cleveland, where it made a slow right turn during the night of July 6-7; thence it moved southeastward passing over Macon to near Selma, Ala., where a left turn then carried it over Birmingham and Huntsville on the 7th and 8th; another left turn took it west of Nashville, Tenn., into the Ohio Valley on the 10th. Attending heavy rains for 3 days "caused enormous losses of staple crops and resulted in great floods in the rivers of eastern Mississippi, Alabama, and Georgia."

With such a short coast line only 16 tropical storms or hurricanes have crossed it during the 84 years from 1875 through 1958; 7 of these were hurricanes. However, Mississippi has been affected by high winds, high tides, or heavy rains of 68 tropical storms or hurricanes during this period. Hurricanes which move inland over southeast Louisiana may be as damaging on the Mississippi coast as those which cross the coast line. This is especially true of those moving from the southeast because of the usually more severe winds in the northeast quadrant and because of the high seas which move across Mississippi Sound and pile up on the shore. Those which move westward offshore often cause tide and wind damage on the coast. Those which move northeastward across or south of the Louisiana Delta and move inland between Mobile and Panama City are usually less damaging because

winds are offshore and tides are subnormal. Hurricanes which move inland on the Alabama coast may affect Mississippi only slightly because of less intense and offshore winds in their western portions.

Of the 68 tropical cyclones which affected Mississippi, 18 (or 26 percent) were of hurricane intensity with winds of 74 m.p.h. or higher at some point. Twice in 85 years 3 tropical storms have occurred in the same year—1901 and 1932; in both years only one of these was of hurricane intensity within the State. Six tropical cyclones have occurred in June, 4 in July, 13 in August, 33 in September, 11 in October, and 1 in November. One-half of all hurricanes affecting Mississippi occur in September and twice as many tropical cyclones occur in August and September as during June, July, October, and November combined. Using criteria of central pressure 29 inches or lower and time period 1900-1956, the Hydrometeorological Section has prepared an analysis which shows a frequency of 40 hurricanes per 100 years on the middle Gulf coast.

MISSISSIPPI HURRICANES, TROPICAL STORMS AND TROPICAL DEPRESSIONS 1875-1958*

Month	Hurricanes	Tropical Storms	Tropical Depressions	Total
June	1	3	2	6
July	1	1	2	4
August	4	5	4	13
September	10	17	6	33
October	2	4	5	11
November	0	1	0	1
Total	18	31	19	68

*Those Tropical Cyclones which affected the State without crossing the coastline have been classified according to the maximum intensity of the storm at the time it affected the State. In some cases, winds observed within the State did not reach the hurricane or tropical Storm intensity observed near the center of the storm over the Gulf or in other States.

A tabulation of all tornadoes reported in Mississippi from 1916 through 1958 shows 307 on 189 days. Of these 307 tornadoes, 102 resulted in one or more deaths and another 67 resulted in injury to one or more persons. Thus 169 of the reported 307 caused injury and/or death.

The largest number of reported tornadoes occurred in March (79), while April (55), February (41, and May (32) rank high. These 4 months account for 67 percent of all tornadoes. Tornadoes occur in all months, but October (3), August (4) and September (5) have had the least. In the other months tornadoes were reported as follows: January 13, June 14, July 8, November 30, and December 23. Thus, there is a secondary maximum of occurrence in November and December.

The heaviest casualties occurred in April with 404 killed and at least 1,032 injured while March deaths were 228 and injuries at least 984. Thus, April with over three-fifths as many tornadoes as March had almost twice the number of casualties.

During the 43 years of record, 869 persons have been killed and over 3,000 injured in tornadoes in Mississippi. During the months from July through October only 5 of the 20 tornadoes, or 1 out of 4 resulted in any casualties.

Tornadoes in Mississippi occur at any hour of the day or night, but are least likely between 4 a.m. and 7 a.m. and most likely between 11 a.m.

and 9 p.m. Nearly one-half of all tornadoes in Mississippi occur between 2 p.m. and 9 p.m., with the peak occurring between 6 and 7 p.m. This is about 3 hours later than the time of maximum occurrence in Louisiana.

Tornadoes may occur at any place in Mississippi, but are least likely in the tier of counties within the "panhandle" below 31°N. latitude; but this is the area where hurricanes are most likely to occur. In a study made by Conner and Kraft at New Orleans it is shown that for north and central Mississippi deaths from tornadoes are 108 per 10,000 square miles per million population. This is twice the comparable rate in Louisiana and is as great as in any state, even those with a much higher frequency of occurrence.

The worst tornado of record in Mississippi was the one which struck Tupelo on April 5, 1936, leaving 216 dead and 700 injured. With only 4 other tornadoes that year, deaths were 224 and injured 723, making this the worst year of record. The largest number of tornadoes in any year since 1915 was 26 in 1957 (6 dead, 110 injured) and next highest was 25 in 1933 (58 dead, 264 injured). Tornadoes occurred on more days in 1957 (15) than in any other year. Five tornadoes in 1942 cost 76 lives with 557 injured, but a single tornado in 1908 (before systematic records began) which moved from Lamar to Wayne Counties killed 100 persons and injured another 649, although it missed the City of Hattiesburg by only a few miles. Some other notable tornadoes in Mississippi are listed below:

March 2, 1906, Meridian; 23 dead;

April 20, 1920, Oktibbeha Co., Miss. to Franklin Co., Ala.; 87 dead;

April 6, 1935, Wilkinson to Amite Co.; 11 dead, 75 injured;

March 16, 1942, central to northeast Mississippi (several tornadoes); 75 dead, 525 injured;

Dec. 5, 1953, Vicksburg; 38 dead, 270 injured.

While much detail has been given in this summary to hurricanes and tornadoes, there is a bright face on the coin and these good features have been mentioned throughout this report. It is well to reiterate that Mississippi has a climate characterized by absence of severe cold in winter and extreme heat in summer; that the ground rarely freezes and outdoor activities are generally favored year-round; cold spells are usually of short duration and the growing season is long; rainfall is plentiful, but so is sunshine; and dry spells most frequently come at harvest time when needed most. While tornadoes and hurricanes cost heavily in lives and property, these affect only a small part of the State at any time and protective measures can be taken against them.

REFERENCES

(1) U. S. Weather Bureau, Technical Paper No. 2, Maximum recorded United States Rainfall; April 1947

(2) U. S. Weather Bureau, Technical Paper No. 15, Maximum Station Precipitation, Part XVll, Mississippi; April 1956

(3) U. S. Weather Bureau, Technical Paper No. 16, Maximum 24-Hour Precipitation in the United States; January 1952

(4) U. S. Weather Bureau, Technical Paper No. 19, Mean Number of Thunderstorm Days in the United States, December 1952

(5) U. S. Weather Bureau, Technical Paper No. 20, Tornado Occurrences in the United States; September 1952

(6) U. S. Weather Bureau, Technical Paper No. 25, Rainfall Intensity-Duration-Frequency Curves; December 1955

(7) U. S. Weather Bureau, Technical Paper No. 30, Tornado Deaths in the United States; March 1957

(8) U. S. Weather Bureau, Technical Paper No. 31, Monthly Normal Temperatures, Precipitation and Degree Days; November 1956

(9) U. S. Weather Bureau and Corps of Engineers, U. S. Army, Hydrometeorological Report No. 5, Thunderstorm Rainfall; 1947

(10) U. S. Weather Bureau, Hydrometeorological Report No. 33, Seasonal Variation of the Probable Maximum Precipitation East of 105th Meridian for Areas from 10 to 1,000 Square Miles and Durations 6, 12, 24, and 48 Hours; April 1956

(11) U. S. Weather Bureau, unpublished Report Hur. 2-4, August 30, 1957

(12) U. S. Weather Bureau, unpublished study by Conner and Kraft, Death-Dealing Tornadoes

(13) U. S. Weather Bureau, Freeze Risk Data, unpublished

(14) U. S. Weather Bureau, unpublished tabulation days 90° or above; 32° or below

BIBLIOGRAPHY

(A) Climatic Summary of the United States (Bulletin W) 1930 edition, Sections 78 and 79. U. S. Weather Bureau

(B) Climatic Summary of the United States, Mississippi - Supplement for 1931 through 1952 (Bulletin W Supplement). U. S. Weather Bureau

(C) Climatological Data - Mississippi. U. S. Weather Bureau

(D) Climatological Data National Summary. U. S. Weather Bureau

(E) Hourly Precipitation Data - Mississippi. U. S. Weather Bureau

(F) Local Climatological Data, U. S. Weather Bureau for Jackson, Meridian and Vicksburg, Mississippi.

FREEZE DATA

STATION	Freeze threshold temperature	Mean date of last Spring occurrence	Mean date of first Fall occurrence	Mean No. of days between dates	Years of record Spring	No. of occurrences in Spring	Years of record Fall	No. of occurrences in Fall
ABERDEEN	32	03-22	11-05	228	30	30	30	30
	28	03-06	11-17	256	30	30	30	30
	24	02-16	12-07	294	30	28	30	25
	20	02-02	12-17	318	30	23	30	18
	16	01-16	12-24	342	30	15	30	12
BATESVILLE	32	03-30	10-27	211	30	30	30	30
	28	03-13	11-11	244	30	30	30	30
	24	02-28	11-26	272	30	30	30	29
	20	02-09	12-14	307	30	25	30	20
	16	01-22	12-22	334	30	21	30	12
BAY ST LOUIS	32	02-24	12-06	285	30	29	30	23
	28	01-29	12-20	325	30	24	30	16
	24	01-13	12-28	349	30	13	30	5
	20	01-07	@	@	30	7	30	2
	16	@	@	@	30	2	30	0
BELZONI	32	03-15	11-09	239	20	20	21	21
	28	02-24	11-25	275	20	20	21	20
	24	02-11	12-14	306	20	18	21	14
	20	01-21	12-20	333	20	12	21	10
	16	01-14	12-27	347	20	8	21	5
BILOXI CITY	32	02-21	12-12	294	30	28	30	20
	28	01-29	12-20	325	30	21	30	15
	24	01-13	12-27	348	30	14	30	7
	20	01-05	@	@	30	8	30	2
	16	@	@	@	30	4	30	0
BOONEVILLE	32	03-26	11-08	226	30	30	29	29
	28	03-13	11-17	249	30	30	29	28
	24	02-23	12-04	284	30	29	29	27
	20	02-12	12-15	307	30	28	29	20
	16	01-25	12-24	333	30	22	29	12
BROOKHAVEN	32	03-19	11-15	241	30	30	29	28
	28	02-26	12-02	279	30	29	29	26
	24	02-09	12-17	311	30	27	29	18
	20	01-19	12-22	338	30	18	29	13
	16	01-07	@	@	30	10	29	4
CANTON	32	03-20	11-03	229	28	28	26	26
	28	02-28	11-23	268	29	29	25	23
	24	02-09	12-11	305	29	24	26	19
	20	01-22	12-19	331	29	18	26	13
	16	01-13	12-26	348	29	12	26	6
CLARKSDALE	32	03-18	11-09	236	29	29	30	30
	28	02-28	11-28	273	29	29	30	29
	24	02-13	12-14	304	29	27	30	22
	20	01-31	12-19	322	30	23	30	17
	16	01-15	12-26	345	30	16	30	9
COLUMBIA	32	03-15	11-09	239	30	30	29	29
	28	02-26	11-28	275	30	29	29	27
	24	02-09	12-23	307	30	27	29	20
	20	01-18	12-23	339	30	17	29	10
	16	01-04	@	@	30	7	30	2
COLUMBUS	32	03-24	11-09	229	30	30	30	30
	28	03-08	11-21	258	30	30	30	29
	24	02-18	12-08	293	30	28	30	24
	20	02-04	12-18	317	30	24	30	17
	16	01-17	12-25	342	30	15	30	10
CORINTH	32	03-28	11-05	223	29	29	30	30
	28	03-12	11-20	253	29	29	30	30
	24	02-24	12-02	281	29	29	30	29
	20	02-13	12-18	308	29	27	30	18
	16	01-28	12-24	330	29	23	30	12
CRYSTAL SPRINGS	32	03-16	11-19	249	30	30	30	29
	28	02-28	12-06	281	30	30	30	27
	24	02-12	12-18	309	30	26	30	20
	20	01-19	12-22	336	30	17	30	13
	16	01-09	12-28	353	30	12	30	7
DUCK HILL	32	03-28	10-29	214	30	30	30	30
	28	03-15	11-07	238	30	30	30	30
	24	02-23	11-27	277	30	30	30	28
	20	02-08	12-16	311	30	26	30	18
	16	01-20	12-21	335	30	18	30	13
EDINBURG	32	03-29	10-31	216	30	30	29	29
	28	03-09	11-13	249	29	29	27	27
	24	02-23	11-30	280	29	28	29	24
	20	02-02	12-13	313	29	23	28	21
	16	01-18	12-22	338	29	16	28	11
EUPORA	32	03-31	11-03	217	24	24	24	24
	28	03-16	11-11	240	24	24	24	24
	24	02-26	11-26	274	24	24	24	23
	20	02-11	12-11	304	23	21	24	17
	16	01-23	12-18	330	23	14	24	13
FOREST	32	03-22	11-10	233	30	30	30	30
	28	03-02	11-21	264	30	30	30	28
	24	02-10	12-06	299	30	25	30	24
	20	01-30	12-17	321	30	22	30	18
	16	01-18	12-23	340	30	16	30	10
GREENVILLE	32	03-17	11-13	241	30	30	30	30
	28	03-01	11-26	270	30	30	30	28
	24	02-12	12-12	304	30	26	30	23
	20	01-28	12-19	325	30	21	30	17
	16	01-12	12-23	345	30	13	30	12
GREENWOOD CAA AP	32	03-17	11-03	231	30	30	30	30
	28	02-26	11-22	269	30	30	30	28
	24	02-14	12-11	300	30	27	30	23
	20	01-25	12-18	327	30	22	30	16
	16	01-14	12-26	346	30	15	30	9
HATTIESBURG	32	03-14	11-10	242	30	30	30	30
	28	02-23	11-30	280	30	29	30	25
	24	01-31	12-15	318	30	24	30	22
	20	01-16	12-21	339	30	16	30	12
	16	01-06	@	@	30	8	30	3
HERNANDO	32	03-26	11-10	229	28	28	28	28
	28	03-11	11-20	254	27	27	27	27
	24	02-24	12-06	284	27	26	27	25
	20	02-10	12-20	312	27	24	27	15
	16	01-28	12-22	328	27	22	27	11
HOLLY SPRINGS 2 N	32	03-29	11-04	219	30	30	30	30
	28	03-17	11-15	243	30	30	30	30
	24	02-27	12-01	277	30	30	30	29
	20	02-13	12-14	304	30	27	30	21
	16	01-29	12-21	327	30	23	30	15
JACKSON WB AP	32	03-10	11-13	248	30	30	30	20
	28	02-20	11-25	278	30	25	25	25
	24	02-13	12-04	294	23	23	18	18
	20	01-31	12-10	313	18	18	15	15
	16	01-24	12-17	328	10	10	7	7
KOSCIUSKO	32	03-27	11-06	225	30	30	30	30
	28	03-10	11-20	255	30	30	30	28
	24	02-23	12-05	285	30	29	29	24
	20	02-05	12-17	314	30	24	29	19
	16	01-16	12-22	341	30	14	29	12
LAUREL	32	03-16	11-13	242	30	30	30	30
	28	02-25	12-03	281	30	28	30	25
	24	02-07	12-15	311	30	25	30	20
	20	01-19	12-21	336	30	17	30	14
	16	01-06	12-27	355	30	9	30	6
LEAKESVILLE	32	03-17	11-09	237	28	28	29	29
	28	02-26	11-30	277	28	23	29	25
	24	02-02	12-13	314	28	23	29	20
	20	01-20	12-21	336	26	17	29	11
	16	01-07	12-29	356	28	9	29	5
LOUISVILLE	32	03-24	11-10	231	30	30	30	30
	28	03-07	11-27	265	30	30	30	29
	24	02-19	12-08	292	30	29	29	24
	20	02-06	12-18	315	30	24	29	18
	16	01-19	12-25	340	30	16	29	11
MACON 2 NE	32	03-22	11-08	231	30	30	30	30
	28	03-04	11-21	262	30	30	30	30
	24	02-15	12-09	297	30	28	30	23
	20	01-30	12-18	322	30	22	30	15
	16	01-17	12-25	342	30	15	30	10

FREEZE DATA

STATION	Freeze threshold temperature	Mean date of last Spring occurrence	Mean date of first Fall occurrence	Mean No. of days between dates	Years of record Spring	No. of occurrences in Spring	Years of record Fall	No. of occurrences in Fall
MAGNOLIA	32	03-18	11-12	239	30	30	30	30
	28	02-23	12-05	284	30	28	30	24
	24	02-03	12-16	316	30	25	30	19
	20	01-15	12-24	343	30	16	30	10
	16	01-07	⊕	⊕	30	10	30	3
MERIDIAN WB AP	32	03-13	11-14	246	30	30	30	30
	28	02-25	12-02	281	30	30	30	26
	24	02-06	12-16	313	30	24	30	20
	20	01-22	12-22	334	30	19	30	13
	16	01-11	12-28	351	30	14	30	5
MONTICELLO	32	03-21	11-04	228	30	30	30	30
	28	02-27	11-23	269	30	29	30	28
	24	02-12	12-10	301	30	28	30	22
	20	01-20	12-18	332	30	17	30	15
	16	01-06	12-28	356	30	9	30	6
MOORHEAD	32	03-19	11-03	229	28	28	30	30
	28	03-02	11-20	263	28	28	30	28
	24	02-15	12-06	294	28	26	30	25
	20	01-30	12-17	321	28	22	30	19
	16	01-14	12-25	345	28	13	30	12
NATCHEZ	32	03-12	11-17	249	29	29	30	29
	28	02-21	12-04	286	29	28	30	26
	24	02-01	12-16	318	29	21	30	20
	20	01-16	12-23	341	29	14	30	13
	16	01-06	12-29	357	29	8	30	5
PEARLINGTON	32	03-04	11-28	269	28	28	25	22
	28	02-12	12-15	307	28	25	25	16
	24	01-22	12-22	335	28	17	25	10
	20	01-12	12-28	350	28	11	25	5
	16	01-05	⊕	⊕	28	6	25	0
PONTOTOC	32	03-27	11-05	223	30	30	30	30
	28	03-13	11-18	250	30	30	30	29
	24	02-24	12-03	282	30	29	30	28
	20	02-11	12-15	307	30	28	30	20
	16	01-27	12-22	329	30	22	30	15
POPLARVILLE EXP STA	32	03-06	12-02	272	30	30	30	26
	28	02-16	12-14	301	30	28	30	21
	24	01-22	12-21	332	30	18	30	13
	20	01-10	12-27	351	30	11	30	6
	16	01-05	⊕	⊕	30	8	30	2
PORT GIBSON	32	03-26	11-03	223	30	30	30	30
	28	03-10	11-18	253	30	30	30	29
	24	02-21	12-04	286	30	28	30	24
	20	01-29	12-17	322	30	22	30	19
	16	01-16	12-26	343	30	16	30	9
ROLLING FORK	32	03-18	11-02	228	16	16	17	17
	28	03-03	11-28	269	15	15	16	15
	24	02-14	12-08	297	15	13	16	14
	20	01-19	12-20	335	15	11	16	9
	16	01-08	12-25	352	15	6	16	7
ROSEDALE	32	03-23	11-09	231	18	18	18	18
	28	03-07	11-24	262	18	18	18	17
	24	02-16	12-08	296	18	16	18	16
	20	02-01	12-18	320	18	14	18	12
	16	01-20	12-23	337	18	13	18	8

STATION	Freeze threshold temperature	Mean date of last Spring occurrence	Mean date of first Fall occurrence	Mean No. of days between dates	Years of record Spring	No. of occurrences in Spring	Years of record Fall	No. of occurrences in Fall
SCOTT	32	03-18	11-07	234	30	30	30	30
	28	03-03	11-24	267	30	30	30	28
	24	02-14	12-09	298	30	27	30	24
	20	01-29	12-18	323	30	23	30	17
	16	01-15	12-25	345	30	14	30	10
SHUBUTA	32	03-21	11-05	230	27	27	27	26
	28	03-05	11-22	262	27	27	26	24
	24	02-12	12-04	295	28	26	26	21
	20	01-20	12-15	329	27	17	27	17
	16	01-06	12-26	355	28	7	26	7
STATE COLLEGE	32	03-21	11-09	233	30	30	30	30
	28	03-04	11-26	266	30	29	30	29
	24	02-17	12-11	297	30	28	30	25
	20	02-06	12-20	316	30	25	30	18
	16	01-16	12-25	342	30	15	30	11
STONEVILLE EXP STA	32	03-20	11-04	229	20	20	20	20
	28	03-07	11-16	255	20	20	20	19
	24	02-13	12-07	297	20	18	20	18
	20	01-26	12-19	327	20	13	20	9
	16	01-16	12-26	344	20	10	20	5
TUNICA	32	03-22	11-05	227	17	17	15	15
	28	03-09	11-21	257	17	17	15	14
	24	02-18	12-06	291	17	16	14	12
	20	02-05	12-20	318	17	14	14	9
	16	01-18	12-25	342	17	9	14	6
TUPELO	32	03-26	11-11	230	30	30	30	30
	28	03-09	11-18	254	30	29	30	30
	24	02-19	12-06	291	30	28	30	25
	20	02-08	12-18	313	30	26	30	17
	16	01-19	12-24	339	30	18	30	12
TYLERTOWN	32	03-07	11-16	254	10	10	10	10
	28	02-15	12-05	293	10	9	10	9
	24	01-18	12-18	335	9	6	10	5
	20	01-06	12-22	349	9	3	10	4
	16	01-04	⊕	⊕	9	2	10	1
UNIVERSITY	32	03-28	11-07	224	30	30	30	30
	28	03-10	11-22	258	30	30	30	29
	24	02-24	12-06	285	29	28	30	28
	20	02-10	12-17	309	29	25	30	20
	16	01-30	12-21	325	29	21	30	14
UTICA	32	03-21	11-08	233	30	30	29	29
	28	02-28	11-27	272	30	30	29	26
	24	02-12	12-12	303	30	27	29	22
	20	01-26	12-19	327	30	21	29	15
	16	01-10	12-25	349	29	12	29	10
WATER VALLEY	32	03-27	11-05	223	30	30	30	30
	28	03-15	11-16	246	30	30	30	29
	24	02-24	12-01	279	30	30	30	27
	20	02-09	12-16	310	30	27	30	20
	16	01-24	12-21	331	30	20	30	14
WAYNESBORO 1E	32	03-24	11-09	230	29	29	29	28
	28	03-02	11-22	265	29	28	29	27
	24	02-12	12-08	300	29	26	29	21
	20	01-19	12-19	334	28	16	29	14
	16	01-08	12-27	353	28	11	28	7
WIGGINS FRUITLAND PK	32	03-07	11-20	258	26	26	26	25
	28	02-22	12-07	288	26	25	25	18
	24	01-31	12-14	317	25	20	25	14
	20	01-17	12-22	339	24	14	25	9
	16	01-06	⊕	⊕	24	7	24	1
WOODVILLE	32	03-13	11-21	253	27	27	28	27
	28	02-21	12-09	291	27	26	26	21
	24	02-05	12-17	316	26	21	26	16
	20	01-10	12-24	348	26	11	26	10
	16	01-07	12-29	356	26	8	26	5
YAZOO CITY	32	03-21	11-04	229	30	30	30	30
	28	03-01	11-23	267	30	30	30	27
	24	02-16	12-06	293	30	29	30	24
	20	01-27	12-17	323	30	21	30	16
	16	01-12	12-25	347	30	13	30	12

Data in the above table are based on the period 1921-1950, or that portion of this period for which data are available.

⊕ When the frequency of occurrence in either spring or fall is one year in ten, or less, mean dates are not given.

Means have been adjusted to take into account years of non-occurrence.

A freeze is a numerical substitute for the former term "killing frost" and is the occurrence of a minimum temperature at or below the threshold temperature of 32°, 28°, etc.

Freeze data tabulations in greater detail are available and can be reproduced at cost.

*MEAN TEMPERATURE AND PRECIPITATION

STATION	JANUARY		FEBRUARY		MARCH		APRIL		MAY		JUNE		JULY		AUGUST		SEPTEMBER		OCTOBER		NOVEMBER		DECEMBER		ANNUAL	
	Temperature	Precipitation	Temperature	Precipitation	Temperature	Precipitation	Temperature	Precipitation	Temperature	Precipitation	Temperature	Precipitation	Temperature	Precipitation	Temperature	Precipitation	Temperature	Precipitation	Temperature	Precipitation	Temperature	Precipitaion	Temperature	Precipitation	Temperature	Precipitation
UPPER DELTA																										
CLARKSDALE	45.8	6.23	48.4	5.02	56.0	5.80	65.1	4.68	73.4	4.25	81.4	3.76	83.6	3.64	83.1	2.28	77.1	2.34	66.8	2.33	53.8	4.45	46.5	5.14	65.1	49.92
CLEVELAND		6.04		4.73		6.68		4.91		4.02		3.71		4.42		2.56		2.60		2.58		4.40		5.37		52.02
SCOTT	44.6	6.16	47.2	4.76	54.6	6.21	63.4	4.88	71.6	4.22	79.5	3.26	81.6	3.72	81.2	2.45	75.1	2.43	65.0	2.44	52.4	4.36	45.6	5.19	63.5	50.08
SWAN LAKE 1 S		5.96		4.80		6.13		4.64		3.78		3.96		4.14		2.76		2.43		2.21		4.49		5.31		50.61
DIVISION	45.1	6.33	47.7	4.89	55.1	6.07	64.0	4.74	71.9	4.08	79.7	3.34	82.0	3.75	81.8	2.38	75.6	2.53	65.4	2.44	52.8	4.24	46.0	5.10	63.9	49.89
NORTH CENTRAL																										
BATESVILLE	44.3	6.33	46.8	4.79	53.9	6.25	62.9	5.04	70.7	4.06	78.8	3.28	81.3	4.12	80.8	3.45	74.6	2.64	63.8	2.56	51.7	4.49	44.9	5.55	62.9	52.56
GRENADA		5.78		5.06		6.81		4.27		3.70		3.49		3.94		3.44		2.55		2.25		4.63		5.25		51.17
HOLLY SPRINGS 2 N	43.3	6.72	45.3	5.45	52.9	6.17	61.9	4.70	69.9	4.20	78.2	3.65	80.9	4.69	80.4	3.20	74.1	3.63	63.9	2.88	51.3	4.73	44.1	5.09	62.2	55.11
UNIVERSITY	44.8	6.16	47.1	5.18	54.6	6.09	63.4	4.77	70.9	3.90	78.9	3.71	81.2	4.21	81.2	2.98	75.3	3.15	65.0	2.54	52.4	4.72	45.7	5.32	63.4	52.73
WATER VALLEY	45.3	5.89	47.6	5.14	54.6	6.69	62.8	5.11	70.3	3.96	78.5	3.67	81.2	4.97	80.8	3.13	75.0	2.89	64.7	2.53	52.9	4.49	45.9	5.46	63.3	53.93
DIVISION	44.2	6.27	46.4	5.14	53.8	6.15	62.6	4.90	70.4	4.06	78.6	3.56	81.2	4.25	80.8	3.14	74.7	2.96	64.4	2.65	52.0	4.53	45.0	5.30	62.8	52.91
NORTHEAST																										
BOONEVILLE	43.2	6.17	44.8	5.78	52.3	6.33	61.2	4.76	69.5	4.14	77.6	3.48	80.2	4.24	79.9	3.39	73.9	3.01	63.6	2.84	50.9	4.65	43.6	5.63	61.7	54.42
CORINTH	44.0	6.34	46.0	5.34	53.5	5.71	62.8	4.68	71.1	4.09	79.4	3.62	82.0	4.04	81.5	3.43	75.2	3.05	64.5	2.74	51.7	4.61	44.7	4.89	63.0	52.54
FULTON 3 W		5.55		5.92		6.78		4.60		4.20		3.31		4.81		3.34		2.83		2.32		4.18		5.07		52.91
PONTOTOC	45.1	5.83	47.2	5.14	54.6	6.53	62.9	4.73	70.8	3.71	78.7	3.83	80.9	4.95	80.8	3.27	75.1	3.11	65.1	2.22	52.4	4.47	45.5	5.15	63.3	52.94
TUPELO	44.5	6.00	46.6	5.59	53.8	7.00	62.7	4.14	71.3	4.17	79.5	3.72	81.8	4.49	81.5	2.73	75.3	2.83	64.5	2.65	52.0	4.17	45.3	5.39	63.2	52.88
DIVISION	44.2	6.06	46.2	5.45	53.5	6.39	62.4	4.59	70.7	4.03	78.8	3.67	81.2	4.46	80.9	3.20	74.8	2.99	64.3	2.61	51.8	4.49	44.8	5.23	62.8	53.17
LOWER DELTA																										
BELZONI	47.8	5.78	50.4	5.01	57.5	5.54	65.2	4.88	73.0	3.57	80.4	3.67	82.5	3.56	82.2	2.64	76.3	2.05	66.3	2.31	54.4	4.58	48.4	5.45	65.4	49.04
GREENVILLE	47.0	6.77	49.3	5.06	56.4	5.90	64.6	5.21	72.5	3.79	80.2	2.99	82.4	4.29	82.3	2.53	76.6	2.87	66.7	2.27	54.4	4.47	47.7	5.54	65.0	51.69
GREENWOOD FAA AIRPORT	47.0	5.86	49.5	4.95	56.6	5.87	64.8	4.69	72.7	3.90	80.4	3.90	82.5	4.30	82.2	2.95	76.2	2.63	65.6	2.31	53.5	4.58	47.4	5.65	64.9	51.59
MOORHEAD	46.6	5.59	49.1	4.66	56.2	5.94	64.6	5.00	72.5	3.43	80.3	3.74	82.4	4.24	81.3	2.81	75.7	2.46	64.9	2.21	53.1	4.40	47.2	5.12	64.5	49.60
STONEVILLE EXP STA	46.7	5.77	48.9	4.84	55.9	5.71	64.0	4.74	71.6	3.96	79.5	3.61	81.6	3.96	81.3	2.51	75.6	2.41	65.3	2.47	53.6	4.13	47.4	5.56	64.3	49.67
YAZOO CITY	48.3	5.29	50.6	5.10	57.6	5.43	65.3	4.72	72.9	3.81	80.1	3.77	82.3	4.46	82.2	2.96	76.9	2.10	66.7	2.53	54.8	4.48	48.9	5.18	65.6	49.83
DIVISION	47.3	5.85	49.7	4.94	56.7	5.73	64.7	4.89	72.5	3.77	80.1	3.61	82.2	4.16	81.9	2.72	76.2	2.42	65.9	2.39	54.0	4.41	47.8	5.44	64.9	50.33
CENTRAL																										
CANTON	48.9	5.53	50.9	4.89	57.3	5.93	64.5	4.47	72.1	4.03	79.4	3.47	81.5	5.05	81.3	3.00	76.1	1.93	66.1	1.70	54.8	3.93	49.5	5.54	65.2	49.47
EDINBURG	48.3	5.66	50.1	5.59	56.4	6.22	63.6	4.85	70.8	4.35	78.3	3.62	80.4	5.78	80.0	3.41	75.0	2.70	64.8	1.85	53.6	3.88	48.2	5.67	64.1	53.58
EUPORA	46.8	5.80	48.7	5.05	55.3	5.98	63.0	4.10	70.6	3.51	78.5	3.94	80.7	5.12	80.4	3.19	74.9	2.50	64.9	2.36	53.0	3.86	46.7	5.01	63.6	50.42
KOSCIUSKO	46.9	5.78	48.7	5.47	55.8	5.77	63.4	4.66	71.2	4.11	78.7	3.70	80.8	5.59	80.7	3.00	75.3	2.91	65.4	2.14	53.4	4.22	47.3	5.26	64.0	52.61
DIVISION	47.8	5.74	49.7	5.20	56.2	6.04	63.7	4.51	71.3	3.95	78.8	3.61	80.9	5.38	80.7	3.12	75.4	2.59	65.3	2.07	53.8	4.12	48.0	5.31	64.3	51.64
EAST CENTRAL																										
ABERDEEN	46.8	5.66	48.7	5.68	55.8	6.22	63.8	4.40	71.7	3.60	79.7	3.82	82.0	5.01	81.8	3.12	75.7	2.72	65.0	2.33	53.1	3.84	46.8	5.00	64.2	51.40
COLUMBUS 4 NNE	47.6	5.91	49.4	5.42	56.3	5.69	64.1	4.11	71.9	3.28	79.9	3.38	81.8	5.17	81.8	2.97	76.3	3.36	65.7	2.39	53.7	3.66	47.6	4.86	64.7	50.20
LOUISVILLE	47.3	5.92	49.2	5.92	56.0	6.16	63.6	4.37	71.5	4.11	79.2	3.30	81.2	6.04	81.2	3.32	75.3	3.07	65.4	1.98	53.5	4.12	47.3	5.14	64.2	53.45
MACON 2 NE	47.9	5.76	49.8	5.88	56.9	6.20	64.7	4.83	72.7	3.88	80.4	3.01	82.3	5.63	82.2	3.37	76.8	3.04	66.7	2.24	54.5	4.11	47.8	5.30	65.2	53.25
STATE COLLEGE	47.5	5.46	49.5	5.07	56.2	5.70	64.2	3.78	71.9	3.46	79.6	3.13	81.5	5.49	81.1	3.28	76.0	2.59	65.9	2.25	53.9	3.59	47.6	4.88	64.6	48.68
DIVISION	47.3	5.58	49.1	5.55	56.1	6.03	63.8	4.25	71.7	3.73	79.5	3.30	81.5	5.42	81.3	3.21	75.8	2.93	65.4	2.21	53.5	3.87	47.2	5.03	64.4	51.11
SOUTHWEST																										
BROOKHAVEN	51.2	5.67	53.1	5.27	59.0	6.33	65.7	5.65	73.2	5.12	80.0	4.25	81.6	5.63	81.7	3.56	77.1	3.43	68.1	2.37	56.7	4.00	51.3	5.91	66.6	57.19
JACKSON WB AIRPORT	48.3	5.09	51.0	5.09	56.8	6.23	64.9	4.82	72.3	4.09	79.9	3.72	82.1	4.61	81.4	3.28	76.7	2.10	66.4	2.18	55.2	4.05	49.0	5.60	65.4	50.86
NATCHEZ	51.5	5.75	51.8	5.00	59.7	6.41	66.4	5.29	73.3	6.54	80.2	3.80	81.9	4.47	82.0	3.59	77.4	2.67	68.1	2.21	57.0	4.32	52.2	5.77	66.8	55.82
PORT GIBSON	49.5	5.93	52.0	5.58	58.2	6.00	64.8	5.96	72.0	5.17	78.9	4.07	81.0	5.24	80.8	3.84	75.6	2.75	65.8	2.18	54.7	4.52	49.6	5.56	65.2	56.80
UTICA	50.6	5.70	52.7	5.15	58.6	5.88	65.5	5.72	72.8	4.85	80.0	3.51	81.9	4.79	81.9	3.14	77.2	2.69	67.7	1.91	56.3	4.05	51.1	5.38	66.4	52.77
VICKSBURG WB CITY	49.2	5.44	52.1	5.14	58.1	6.03	65.8	4.84	72.9	4.22	79.5	3.31	81.5	3.96	81.5	2.71	77.0	1.77	67.7	2.22	56.8	4.61	50.5	5.38	66.1	49.63
DIVISION	50.4	5.77	52.7	5.28	58.6	5.99	65.5	5.31	73.2	5.09	79.7	3.87	81.6	5.30	81.5	3.48	77.2	2.77	67.3	2.16	56.1	4.28	50.9	5.70	66.2	55.00
SOUTH CENTRAL																										
COLLINS		5.17		5.01		5.88		5.29		4.78		4.16		6.10		3.75		3.23		2.32		3.47		5.55		54.71
COLUMBIA	52.8	5.35	54.9	5.18	60.5	6.54	67.2	5.68	74.4	4.69	81.2	4.60	82.4	6.58	82.2	4.36	77.5	4.14	68.0	2.63	57.5	3.98	52.9	6.26	67.6	59.99
MONTICELLO 2 S	51.6	5.82	53.6	5.20	59.2	6.41	65.8	5.68	72.9	4.99	79.8	3.76	81.6	5.76	81.5	3.74	76.8	3.19	67.3	2.35	56.4	3.69	51.6	6.01	66.5	56.60
DIVISION	52.1	5.46	54.1	5.13	59.6	6.37	66.4	5.60	73.5	4.72	80.2	4.38	81.7	6.24	81.5	4.15	77.0	3.58	67.6	2.47	56.8	3.75	52.1	5.94	66.9	57.79
SOUTHEAST																										
HATTIESBURG	52.2	4.99	54.1	5.10	59.7	6.83	66.3	5.54	73.7	5.04	80.3	4.16	81.6	6.85	81.4	5.16	76.7	3.56	67.0	2.29	56.6	4.20	51.8	6.20	66.8	59.94
HICKORY		5.70		5.22		6.55		5.70		4.05		3.40		6.50		3.73		3.30		2.09		4.05		6.01		56.30
LAUREL	50.9	5.29	52.5	5.47	58.8	7.10	65.6	5.83	73.2	4.37	80.1	4.17	81.6	7.06	81.7	4.25	76.9	3.36	67.6	2.17	56.2	3.88	50.5	5.63	66.3	58.58
LEAKESVILLE	52.0	4.90	53.8	4.52	59.3	6.66	65.9	5.82	73.1	4.64	79.7	4.26	81.4	7.27	81.2	5.21	76.9	4.06	67.2	2.32	56.4	4.57	51.7	5.88	66.6	60.11
MERIDIAN WB AIRPORT	47.5	5.09	50.7	5.38	56.5	6.58	64.3	5.46	71.5	3.99	79.1	4.33	80.9	5.96	80.5	3.72	75.9	2.61	64.7	2.38	54.1	4.03	48.2	5.54	64.5	55.07
SHUBUTA		5.18		5.43		7.13		5.46		4.57		4.04		7.57		3.35		3.49		1.93		4.00		6.10		58.25
WAYNESBORO 1 E	50.7	4.71	52.7	5.14	58.4	7.27	65.2	5.62	72.4	4.04	79.3	3.96	80.7	6.62	80.6	4.67	76.0	3.57	66.4	1.88	55.4	4.14	50.4	6.20	65.7	57.82
WAYNESBORO 3 WNW		4.72		5.16		7.26		5.62		4.03		4.07		6.65		4.70		3.53		1.90		4.14		6.21		57.99
DIVISION	50.9	5.07	52.7	5.12	58.6	6.74	65.4	5.56	72.8	4.44	79.7	4.21	81.2	6.62	81.1	4.33	76.4	3.42	66.8	2.11	55.8	4.05	50.5	5.83	66.0	57.70
COASTAL																										
BAY SAINT LOUIS	53.8	4.41	55.9	4.58	60.9	5.88	68.0	5.83	75.3	4.65	81.0	4.62	82.1	8.09	82.1	5.68	78.7	5.11	70.0	2.40	59.2	3.72	54.1	5.53	68.4	60.50
BILOXI CITY	54.2	3.82	55.5	3.94	60.5	6.09	67.3	4.57	74.7	4.76	80.7	5.00	81.8	7.65	82.0	5.35	78.3	5.47	70.1	2.27	59.6	3.56	54.6	5.11	68.3	57.59
MERRILL		4.73		4.49		6.62		5.97		4.83		4.68		8.46		5.70		4.81		2.34		4.49		5.86		62.98
PEARLINGTON 2 NNE		4.33		5.07		6.25		5.48		4.37		5.39		8.82		6.60		5.10		2.44		4.07		5.56		63.48
POPLARVILLE EXP STA	54.1	4.80	56.0	5.06	60.8	6.77	67.1	5.45	74.3	4.84	80.4	4.92	81.7	7.19	81.6	5.70	77.8	4.60	69.7	2.23	59.4	4.51	54.5	6.06	68.1	62.13
DIVISION	53.9	4.35	55.7	4.59	60.7	6.20	67.3	5.40	74.4	4.66	80.4	4.97	81.6	7.83	81.6	5.83	78.0	5.05	69.6	2.36	59.1	3.95	54.2	5.52	68.0	60.71

* Averages for period 1931-1955, except for stations marked WB which are "normals" based on period 1921-1950. Divisional means may not be the arithmetical average of individual stations published, since additional data from shorter period stations are used to obtain better areal representation.

CONFIDENCE LIMITS

In the absence of trend or record changes, the chances are 9 out of 10 that the true mean will lie in the interval formed by adding and subtracting the values in the following table from the means for any station in the State:

1.7	1.05	1.5	.86	1.5	.75	.8	1.08	.7	.74	.5	.91	.5	1.01	.5	.53	.9	.55	.9	.62	1.1	1.27	1.3	.70	.3	3.03

COMPARATIVE DATA

Data in the following table are the mean temperature and average precipitation for Mississippi State University, Mississippi, for the period 1906-1930 and are included in this publication for comparative purposes:

46.5	4.82	48.9	4.34	56.2	6.34	63.5	5.39	70.9	5.08	78.8	4.51	81.4	4.20	80.9	3.81	76.6	2.97	65.0	2.70	54.1	5.76	46.8	5.21	64.1	55.13

NORMALS, MEANS, AND EXTREMES

LATITUDE 32° 20' N
LONGITUDE 90° 13' W
ELEVATION (ground) 305 feet

Jackson, Mississippi — Hawkins Field

Month	Normal Daily max	Normal Daily min	Normal Monthly	Extreme Record highest	Year	Extreme Record lowest	Year	Normal degree days	Precip Normal total	Precip Max monthly	Year	Precip Min monthly	Year	Precip Max 24 hr	Year	Snow Mean total	Snow Max monthly	Year	Snow Max 24 hr	Year	RH 12 mid	RH 6 a.m.	RH 12 noon	RH 6 p.m.	Wind Mean hourly	Prevailing dir	Fastest speed	Fastest dir	Year	Mean sky cover	Pct sunshine
(b)				18		18		(b)	(b)	18		18		18		18	18		18		12	12	12	12	10	9	10	10		12	10
J	58.6	37.9	48.3	85	1949	-5	1940	535	5.09	12.01	1950	1.31	1943	4.35	1950	1.3	10.8	1940	10.8	1940	82	86	65	71	9.7	SSE	56	NW	1952	7.3	39
F	61.5	40.4	51.0	82	1957+	-5	1951	405	5.09	9.28	1948	1.15	1947	4.02	1946	0.2	1.8	1951	1.8	1951	80	85	59	64	9.5	SSE	45	S	1955	6.4	48
M	68.1	45.5	56.8	88	1946	17	1943	299	6.23	11.26	1949	2.14	1955	3.79	1949	T	T	1957+	T	1957+	78	83	54	58	9.5	SSE	68	S	1952	6.4	54
A	76.0	53.7	64.9	93	1943	30	1940	81	4.82	10.19	1942	0.08	1942	4.85	1953	0.0	0.0		0.0		80	87	53	58	9.0	SSE	49	NW	1953	5.8	64
M	83.7	60.8	72.3	99	1951	42	1954	0	4.09	9.27	1944	0.16	1951	4.63	1951	0.0	0.0		0.0		82	87	53	59	7.0	SSE	56	NW	1949	5.6	67
J	91.4	68.3	79.9	103	1953+	48	1956	0	3.72			0.52	1944	4.78	1944	0.0	0.0		0.0		84	88	53	60	6.1	SSE	52	SE	1957	4.9	71
J	93.6	70.5	82.1	104	1954+	57	1947	0	4.61	13.13	1955	0.98	1947	3.46	1955	0.0	0.0		0.0		86	90	55	65	5.8	SE	54	N	1957	5.8	73
A	93.1	69.7	81.4	106	1943	54	1956	0	3.28	11.39	1942	0.23	1946	4.48	1942	0.0	0.0		0.0		84	90	51	59	5.8	NW	51	NW	1948	4.5	73
S	88.7	64.7	76.6	103	1954	41	1942	0	2.10	8.29	1956	0.04	1956	2.85	1956	0.0	0.0		0.0		80	88	51	61	6.3	SE	39	NE	1954	4.1	69
O	79.7	53.1	66.4	96	1954	28	1957	69	2.18	5.35	1941	0.04	1952	1.91	1945	0.0	0.0		0.0		80	88	47	60	6.4	SE	59	S	1954	4.0	69
N	67.2	43.1	55.2	86	1955+	18	1950	310	4.05	10.24	1948	1.01	1945	4.01	1948	T	T	1957+	T	1957+	78	84	52	62	5.9	SSE	59	NW	1951	4.8	59
D	59.4	38.6	49.0	84	1951	16	1955+	503	5.60	10.24	1949	2.53	1945	7.50	1942	T	T	1954+	T	1954+	79	84	58	67	8.8	SSE	68	S	1951	6.0	48
Year	76.8	53.9	65.4	106	Aug. 1943+	-5	Jan. 1940	2202	50.86	15.76	Nov. 1948	T	Oct. 1952	7.50	Dec. 1942	1.5	10.8	Jan. 1940	10.6	Jan. 1940	81	87	54	62	7.6	SSE	68	S	Mar. 1952	5.6	61

Mean number of days (Jackson):

Month	Clear	Partly cloudy	Cloudy	Precip .01+	Snow/Sleet 1.0+	Thunderstorms	Heavy fog	Max 90+	Max 32-	Min 32-	Min 0-
J	7	8	16	11	*	2	2	0	1	11	*
F	7	8	13	10	0	3	1	0	*	6	*
M	8	8	15	10	0	5	1	0	*	3	0
A	8	10	12	9	0	6	1	*	0	*	0
M	10	14	7	9	0	7	*	8	0	0	0
J	7	15	9	11	0	9	*	25	0	0	0
J	7	13	9	11	0	12	*	26	0	0	0
A	13	12	6	7	0	8	*	25	0	0	0
S	16	7	7	5	0	4	1	13	0	0	0
O	15	8	8	6	0	2	2	3	0	1	0
N	8	7	10	8	0	2	2	0	0	5	0
D	9	7	15	10	*	3	3	0	*	10	0
Year	117	118	130	107	*	62	15	99	1	36	*

LATITUDE 32° 20' N
LONGITUDE 88° 45' W
ELEVATION (ground) 294 feet

Meridian, Mississippi — Key Field

Month	Normal Daily max	Normal Daily min	Normal Monthly	Extreme Record highest	Year	Extreme Record lowest	Year	Normal degree days	Precip Normal total	Precip Max monthly	Year	Precip Min monthly	Year	Precip Max 24 hr	Year	Snow Mean total	Snow Max monthly	Year	Snow Max 24 hr	Year	RH 12 mid	RH 6 a.m.	RH 12 noon	RH 6 p.m.	Wind Mean hourly	Prevailing dir	Fastest speed	Fastest dir	Year	Mean sky cover	Pct sunshine
(b)				68		68		(b)	(b)	68		68		68		68	68		68		18	66	44	58	60	56	55			27	
J	58.5	36.5	47.5	83	1950	-7	1940	561	5.09	18.77	1937	0.72	1927	4.41	1895	0.6	14.7	1909	7.6	1909	86	86	61	69	6.4	N	42		1918	6.2	
F	62.1	39.3	50.7	85	1899	-6	1899	413	5.38	13.30	1903	1.63	1895	8.04	1936	0.3	17.4	1909	4.5	1915	84	84	65	65	6.8	N	47		1926	6.0	
M	68.1	44.6	56.5	93	1929	15	1943	309	6.58	13.70	1910	0.86	1906	5.77	1929	0.1	4.0	1915	4.0	1915	84	84	58	58	6.4	S	44		1927	5.5	
A	76.7	51.8	64.3	91	1943	28	1891	85	5.46	16.44	1938	0.79	1915	9.50	1900	T	T	1910	T	1910	84	84	51	57	6.1	SW	44		1923	5.1	
M	83.8	59.2	71.5	98	1911	39	1909	9	3.99	12.15	1917	0.27	1951	4.89	1909	0.0	0.0		0.0		90	84	51	60	5.4	SW	40		1933	5.0	
J	90.7	67.4	79.1	102	1954	46	1894	0	4.33	20.06	1900	0.21	1917	3.89	1928	0.0	0.0		0.0		92	84	52	63	4.8	SW	40			4.9	
J	92.0	69.8	80.9	105	1930	55	1947	0	5.96	14.24	1916	0.56	1929	6.35	1916	0.0	0.0		0.0		91	87	56	70	4.6	SW	36		1916	4.8	
A	91.9	69.1	80.5	105	1943+	49	1891	0	3.72	12.10	1919	0.79	1919	4.95	1906	0.0	0.0		0.0		90	89	54	69	4.3	SW	36		1928	4.6	
S	87.4	64.4	75.9	104	1925	39	1899	90	2.61	11.69	1906	0.09	1903	6.89	1915	0.0	0.0		0.0		88	88	53	69	4.7	NE	32		1918+	3.9	
O	78.3	51.1	64.7	104	1952	24	1952	69	2.38	8.76	1932	T	1952	5.46	1932	0.0	0.0		0.0		88	88	52	69	4.7	NE	31		1919	4.7	
N	66.7	41.4	54.1	87	1935	16	1950	338	4.03	13.93	1948	0.87	1896	5.00	1929	0.4	0.4	1939	0.4	1939	86	86	52	69	5.0	N	31		1918	6.0	
D	59.5	36.8	48.2	81	1931	5	1894	528	4.54	11.34	1919	0.17	1896	7.15	1919	0.2	6.9	1924	6.9	1924	86	86	61	73	6.3	N	38		1919	4.7	
Year	76.3	52.6	64.5	105	Aug. 1943+	-7	Jan. 1940	2333	55.07	20.06	June 1900	T		9.50	April 1900	1.2	17.4	Feb. 1909	7.6	Jan. 1909	88	86	55	66	5.5	SW	47		Mar. 1926	5.2	

Mean number of days (Meridian):

Month	Clear	Partly cloudy	Cloudy	Precip .01+	Snow/Sleet 1.0+	Thunderstorms	Heavy fog	Max 90+	Max 32-	Min 32-	Min 0-
J	10	8	13	10	*	2	1	0	1	11	*
F	9	9	11	10	*	3	1	0	*	7	*
M	11	9	11	10	0	4	1	0	*	6	0
A	11	11	9	8	0	5	1	0	*	*	0
M	12	12	8	9	0	5	*	4	0	0	0
J	10	13	7	9	0	7	*	15	0	0	0
J	8	14	8	12	0	12	0	18	0	0	0
A	14	11	6	10	0	10	1	18	0	0	0
S	17	7	7	4	0	4	2	10	0	0	0
O	17	7	7	5	0	2	2	10	0	1	0
N	10	7	9	7	0	2	2	0	0	5	0
D	10	7	14	10	0	2	2	0	*	10	0
Year	136	118	111	109	*	63	9	66	1	37	*

Means and extremes in the above table are from the existing or comparable location(s). Annual extremes have been exceeded at prior locations as follows: Highest temperature 107 in July 1930 and earlier date(s); minimum monthly precipitation 0.00 in October 1924 and earlier date(s); maximum monthly snowfall 11.7 in January 1904; maximum snowfall in 24 hours 11.7 in January 1904.

ELEVATION (ground) 234 Feet

Month	Temp Normal Daily max	Daily min	Monthly	Extremes Record highest	Year	Record lowest	Year	Normal degree days	Precip Normal total	Max monthly	Year	Min monthly	Year	Max in 24 hr	Year	Snow Mean total	Max monthly	Year	Max 24 hr	Year	RH 12mid	RH 6am	RH 12N	RH 6pm	Wind mean hourly speed	Prevailing dir	Fastest mile speed	Dir	Year	Mean sky cover	Pct poss. sunshine	Clear	Partly cloudy	Cloudy	Precip .01+	Snow 1.0+	Thunder-storms	Heavy fog	90° & above	Max 32° & below	Min 32° & below	Min 0° & below
(yrs)	(b)	(b)	(b)	84		84		(b)	(b)	86		86		78		73	73		68		10	63	33	63	78	75	46	46		68	64	85	85	85	85	73	67	68	84	84	84	84
J	57.6	40.8	49.2	82	1890	-3	1886	507	5.44	13.83	1882	0.37	1914	7.03	1901	1.0	10.1	1919	10.0	1919	84	82	65	67	8.5	N	35	NW	1942+	5.8	44	8	9	14	11	*	2	2	0	1	7	0
F	60.9	43.3	52.1	84	1918	-1	1899	374	5.14	11.45	1903	0.44	1899	6.92	1927	0.4	6.7	1899	6.0	1923	84	80	63	65	8.7	N	47	NW	1927+	5.9	48	8	7	13	10	*	3	1	0	*	4	*
M	67.4	48.7	58.1	92	1929	17	1943	273	6.03	14.52	1875	0.53	1910	9.97	1951	T	2.0	1924	2.0	1924	83	79	59	59	9.0	S	40	NW	1933+	5.7	57	10	9	12	10	*	5	1	*	*	1	0
A	75.1	56.5	65.8	92	1887	31	1881	71	4.84	22.24	1874	0.20	1930	8.75	1953	0.0	0.0		0.0		87	82	57	60	8.4	SE	38	NW	1957+	5.3	64	11	9	10	9	0	6	*	*	0	0	0
M	82.1	63.8	72.9	97	1911	43	1911	0	4.22	13.23	1872	0.16	1874	6.23	1935	0.0	0.0		0.0		92	84	59	64	7.2	SE	49	NW	1957+	5.1	69	11	12	8	9	0	6	*	2	0	0	0
J	88.4	70.6	79.5	101	1881	52	1903	0	3.31	11.33	1900	0.03	1930	5.96	1928	0.0	0.0		0.0		93	85	59	67	6.6	SW	38	SW	1957+	4.8	73	11	13	7	8	0	9	*	13	0	0	0
J	90.0	72.9	81.5	102	1930	59	1924	0	3.96	10.88	1907	0.22	1924	7.99	1907	0.0	0.0		0.0		93	87	61	70	6.4	SW	46	N	1939	5.5	69	9	14	8	9	0	11	*	18	0	0	0
A	90.3	72.6	81.5	101	1943	54	1891	0	2.71	11.01	1887	0.06	1874	5.31	1948	0.0	0.0		0.0		93	88	59	71	6.2	SW	41	SE	1948+	4.9	72	11	14	6	9	0	9	*	18	0	0	0
S	86.3	67.7	77.0	104	1925	41	1942	0	1.77	10.51	1880	0.04	1936	5.24	1880	0.0	0.0		0.0		91	87	57	70	6.8	N	47	S	1939	4.2	72	13	10	7	7	0	4	1	9	0	0	0
O	77.6	57.8	67.7	94	1884	31	1884	51	2.22	17.26	1918	T	1918	6.44	1881	0.0	0.0		0.0		92	84	55	66	6.9	N	31	SE	1923	3.5	71	13	8	6	6	0	2	1	1	0	*	0
N	66.2	47.3	56.8	86	1896	20	1950	268	4.61	16.25	1948	T	1924	5.32	1948	T	0.2	1938	0.2	1938	89	83	55	64	7.8	SE	47	SW	1940	4.5	60	13	9	9	10	*	2	1	0	*	2	0
D	58.7	42.2	50.5	82	1951+	10	1917	456	5.38	13.77	1911	0.99	1899	5.82	1884	0.3	8.9	1929	8.0	1929	86	82	64	68	8.1	SE	49	SW	1953	5.6	45	12	7	14	10	*	2	2	0	*	5	0
Year	75.1	57.0	66.1	104	Sept. 1925	-3	Feb. 1899	2000	49.63	22.24	Apr. 1874	T	Nov. 1924+	9.97	Mar. 1951	1.7	10.1	Jan. 1919	10.0	Jan. 1919	89	84	59	66	7.6	SE	49	SW	Dec. 1953+	5.1	63	130	120	115	108	*	61	9	61	1	19	*

(a) Length of record, years.
(b) Normal values are based on the period 1921-1950, and are means adjusted to represent observations taken at the present standard location.
* Less than one-half.

° No record.
A Airport data.
C City Office data.
+ Also on earlier dates, months, or years.
T Trace, an amount too small to measure.

REFERENCE NOTES APPLYING TO ALL "NORMALS, MEANS, AND EXTREMES" TABLES.

Sky cover is expressed in a range of 0 for no clouds or obscuring phenomena to 10 for complete sky cover. The number of clear days is based on average cloudiness 0-3 tenths; partly cloudy days on 4-7 tenths; and cloudy days on 8-10 tenths. Monthly degree day totals are the sum of the negative departures of average daily temperatures from 65°F. Sleet was included in snowfall totals beginning with July 1948. Heavy fog also includes data referred to at various times in the past as "Dense" or "Thick". The upper visibility limit for heavy fog is 1/4 mile. Data in these tables are based on records through 1957.

Mean Maximum Temperature (°F.), January

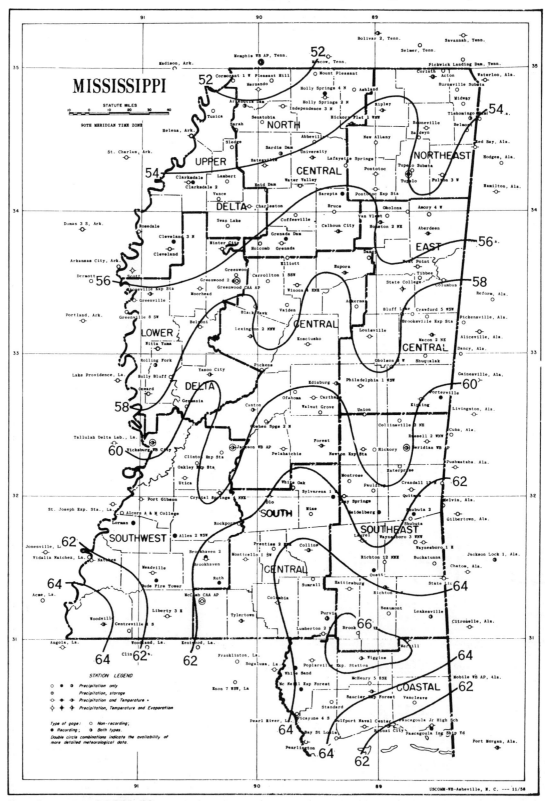

Based on period 1931-52

Isolines are drawn through points of approximately equal value. Caution should be used in interpolating on these maps.

Mean Minimum Temperature (°F.), January

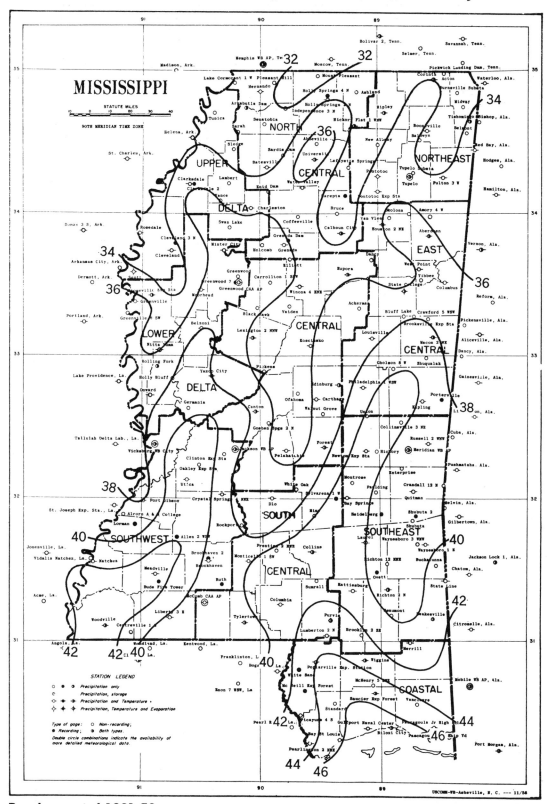

Based on period 1931-52

Isolines are drawn through points of approximately equal value.   Caution should be used
in interpolating on these maps.

Mean Maximum Temperature (°F.), July

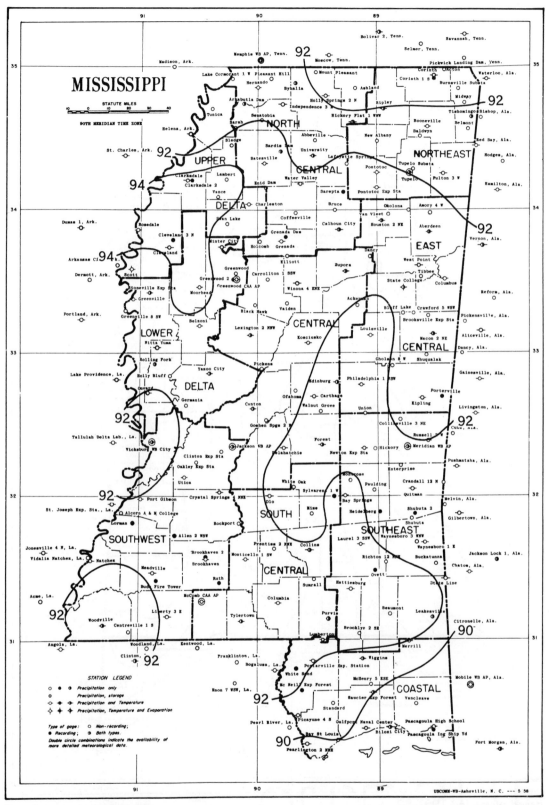

Based on period 1931-52

Isolines are drawn through points of approximately equal value. Caution should be used in interpolating on these maps.

Mean Minimum Temperature (°F.), July

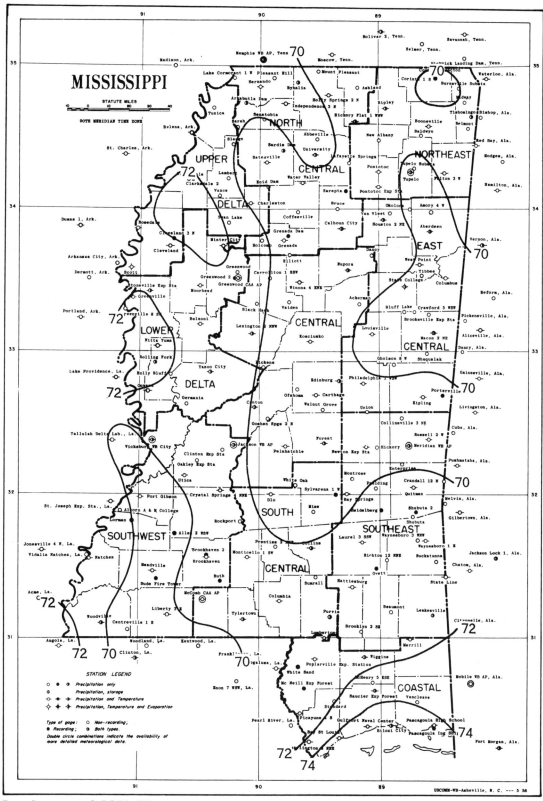

Based on period 1931-52

Isolines are drawn through points of approximately equal value. Caution should be used in interpolating on these maps,

Mean Annual Precipitation, Inches

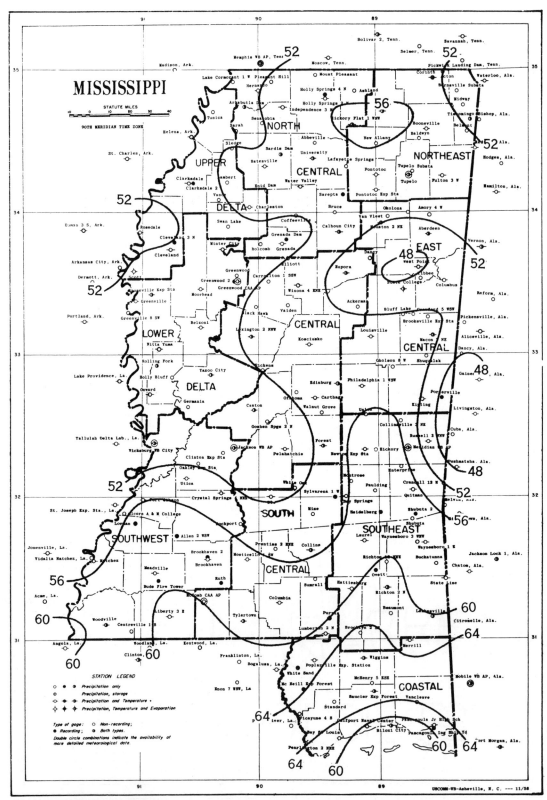

Based on period 1931-55

Isolines are drawn through points of approximately equal value. Caution should be used in interpolating on these maps.

THE CLIMATE OF
NEW HAMPSHIRE

by
Robert E. Lautzenheiser

November 1959

PHYSICAL DESCRIPTION: -- New Hampshire occupies 9,304 square miles, nearly one-seventh of New England's total area. From below the 43d parallel of latitude it extends nearly 200 miles northward to beyond the 45th parallel. At its southern border, New Hampshire extends westward from the Atlantic coastline for nearly 100 miles. It narrows to less than 20 miles in width at its northern tip. The eastern border lies near 71°W. longitude. Its western border is the Connecticut River, except in the extreme north.

The terrain is hilly to mountainous. Elevations of less than 500 feet above sea level are found only in the coastal area of the southeast, the Merrimac River Valley, and the central and southern portions of the Connecticut River Valley. Elsewhere the general elevation is from 500 to 1,500 feet, excepting up to near 2,500 feet in the extreme north. Numerous hills and mountains extend to heights of 2,000 to 4,000 feet above sea level over most of the State except in the southeast. There most hills are not more than 500 to 1,000 feet in height. Many White Mountain peaks rise above 4,000 feet. The elevation of eight peaks in the Presidential Range exceed 1 mile. Mt. Washington reaches 6,288 feet above sea level. This is the highest mountain in the northeastern United States. The extreme climate on top of Mt. Washington makes this location valuable for cold weather research and testing. (A complete weather observa-

tory is maintained there). However, these extreme conditions are not representative of the more temperate climate of the State in general. Therefore, Mt. Washington data are omitted from extremes and average weather statistics for New Hampshire. Some flatland is found near the coast and in the river valleys, with about 10 percent of the total State area being classified as farmland. The glacier of the great Ice Age accounts for much of the topography, including many of the numerous lakes. About 1,300 lakes and ponds add to the State's attractions. The largest is Lake Winnepesaukee, which covers an area of 71 square miles in the central part of the State. Inland waters cover about 280 square miles. The Atlantic coast is 18 miles in length and has several fine beaches.

The two principal rivers in the State are the Connecticut and the Merrimack Rivers, both of which flow in a southerly direction. The larger of the two, the Connecticut, rises in extreme northern New Hampshire and forms the border between it and Vermont. Other rivers include the Androsscoggin and Saco Rivers which rise in the east slopes of the White Mountains and flow eastward into Maine.

Approximately 85 percent of New Hampshire is forested. Considerable area, especially in the north, is sparsely settled. The mountains, hills, lakes, streams, and forests combine to make New Hampshire a state of scenic beauty.

GENERAL CLIMATIC FEATURES: -- Characteristics of

New Hampshire climate are: (1) Changeableness of the weather, (2) large range of temperature, both daily and annual, (3) great differences between the same seasons in different years, (4) equable distribution of precipitation, and (5) considerable diversity from place to place. The regional climatic influences are modified in New Hampshire by varying distances from the ocean, elevations, and types of terrain. The State has been divided into two climatological divisions (Northern and Southern), which take into account the main features of these modifying factors. To take all local factors into consideration would require an impractical number of such areal divisions.

New Hampshire lies in the "prevailing westerlies", the belt of generally eastward air movement which encircles the globe in middle latitudes. Embedded in this circulation are extensive masses of air originating in higher or lower latitudes and interacting to produce low-pressure storm systems. Relative to most other sections of the country, a large number of such storms pass over or near New Hampshire. The majority of air masses affecting this State belong to three types: (1) Cold, dry air pouring down from subarctic North America, (2) warm, moist air streaming up on a long overland journey from the Gulf of Mexico and eastward, and (3) cool, damp air moving in from the North Atlantic. Because the atmospheric flow is usually offshore, New Hampshire is more influenced by the first two types than it is by the third. In other words, the adjacent ocean constitutes an important modifying factor, particularly on the immediate coast, but does not dominate the climate.

The procession of contrasting air masses and the relatively frequent passage of storms bring about approximately twice-weekly alternation from fair to cloudy or stormy conditions, often attended by abrupt changes in temperature, moisture, sunshine, wind direction and speed. There is no regular or persistent rhythm to this sequence, and it is interrupted by periods during which the weather patterns continue the same for several days, infrequently for several weeks. New Hampshire weather, however, is cited for variety rather than monotony. Changeability is also one of its features on a longer time-scale. That is, the same month or season will exhibit varying characteristics over the years, sometimes in close alternation, sometimes arranged in similar groups for successive years. A "normal" month, season, or year is indeed the exception rather than the rule.

The basic climate, as outlined above, obviously does not result from the predominance of any single controlling weather regime, but is rather the integrated effect of a variety of weather patterns. Hence, "weather averages" in New Hampshire usually are not sufficient for important planning purposes without further climatological analysis.

The Northern Division contains approximately one-third of the State, including the northern and west-central areas. Its southern border is roughly parallel to the coast, except where it bends northward near the Connecticut River. This Division represents that area least affected by the ocean influences and most affected by higher elevations as well as by its more northerly latitude. The Southern Division comprises the remaining area. Its lower elevation and latitude tend to cause higher temperatures, though this is modified seasonally by ocean influences. A strip near the coast could be a third division, but its small size hardly merits delineation.

TEMPERATURE: -- The annual temperature averages near 41°F. in the Northern Division and near 46°F. in the Southern. Within the Northern Division it ranges from about 38°F. in the extreme north to about 44°F. in the extreme south. Averages vary within the divisions also from causes other than latitude. Elevation, slope, and other environmental effects, including urbanization, each has an effect. As an extreme example of the effect of altitude is Mt. Washington, whose summit has an annual average of 27°F., compared to averages of 40° to 42°F. at other stations in the general area. The highest temperature of record is 106°F. observed July 4, 1911 at Nashua; the lowest, -46°F., January 28, 1925, at Pittsburg.

Summer temperatures are delightfully comfortable for the most part. They are reasonably uniform over the State, excepting topographical extremes. Hot days with maxima of 90°F. or higher average from only a few per year in the extreme north to 5 to 15 per year over most of the rest of the State. The frequency varies from place to place and from year to year. They range, in frequency of occurrence, from only a few in cool summers to as many as 30 to 40 in the Southern Division in the warmest summers. The diurnal range may reach 40°F. or more during cool, dry weather in valleys and lowlands. Freezing temperatures may be a threat even in the warmer months in a few of the more susceptible areas.

Average temperatures vary from place to place more in the winter than in summer. Days with subzero readings are relatively few along the immediate coast but are common inland. They average from 25 to 50 in number per year in most of the Northern Division and from 10 to 25 in the Southern Division.

The growing season for vegetation subject to injury from freezing temperature averages from 90 to 120 days in the Northern Division. In the Southern Division the average is 120 to 140 days except up to 160 to 180 days in the extreme southeast, a coastal effect. Local topography causes exceptions to the above averages. Swampy areas, particularly, may have a shorter season. The average date of the last freezing temperature in spring ranges from early in June at the colder locations to late in April at a few southern stations. For most of the State the growing season begins in May, and usually ends in the latter part of September.

PRECIPITATION: -- New Hampshire is fortunate in having its precipitation rather evenly distributed through the year. Low pressure, or frontal, storm systems are the principal year-round moisture producers. This activity ebbs somewhat in summer, but thunderstorms are of increased activity at this time, tending to make up the difference. Though brief and often of small extent, the thunderstorms produce the heaviest local rainfall intensities, and sometimes cause minor washouts of roads and soils. Rains of 1 to 2 inches in 1 hour can be expected at least once in a 10-year period.

Variations in monthly precipitation totals are extreme, ranging from no measurable amount to 10 inches or more. A large majority of monthly totals falls in the range of from 50 to 200 percent of normal. As prolonged droughts are infrequent, irrigation water is available during the fairly common shorter dry spells of summer. Similarly, widespread floods are infrequent. However, hurricane Edna, in September 1954, brought the second occurrence that year of heavy flooding and washing rains in southern New Hampshire. Other floods of note were in 1785, 1826, 1852, 1870, 1895, 1896, 1927, the outstandingly disastrous flood of March 1936, and the flood of September 1938. Floods occur most often in the spring when they are caused by a combination of rain and melting snow. At other times of the year high flows and major flooding from rainfall alone occur less frequently. The mean annual runoff in the streams ranges from 14 inches in the north-central Connecticut River Valley to 50 inches in the White Mountains area.

Total annual precipitation averages near 44 inches in the Northern Division and 41 inches in the Southern. The distribution is quite uniform over the Southern Division, ranging from about 37 to 46 inches. The mountainous character of much of central and northern New Hampshire, and the generally higher elevations there, account for the greater annual totals and variability from place to place. As an extreme example, Bethlehem, elevation 1,470 feet, has an annual average of only about one-half that of Mt. Washington (70 inches), where the gage elevation is 6,262 feet above sea level. These stations are only 20 miles apart.

Considerable rain or wet snow falls along the coast in winter, while farther inland snow is more generally the rule. Occasionally freezing rain occurs, coating exposed surfaces with troublesome ice. This problem is less frequent in northern New Hampshire. Most areas can expect at least one occurrence of glaze in the season.

Measurable amounts of precipitation fall on an average of 1 day in 3. Frequency is higher at higher elevations and in extreme northern New Hampshire, up to 140 to 150 days per year. At the Mt. Washington station measurable amounts occur on more than one-half the days. As much as 6 inches of rain in 24 hours is rare in New Hampshire. Most stations have never recorded that much in a single day. However, Warren, N. H., recorded 6.31 inches in 6 hours, and the 24-hour maximum is 8.00 inches at New Durham.

SNOWFALL: -- Average annual amounts of snowfall in the Southern Division increase from around 50 inches near the coast to 60 to 80 inches inland. Totals vary greatly in the Northern Division. Along the Connecticut River in the southern portion, totals average near 60 inches but increase to over 100 inches at the higher elevations of the northern and western portions. The summit of Mt. Washington receives nearly 185 inches. As an example of great variation in a short distance, Bethlehem, only about 20 miles to the west, receives only about 70 inches per year.

The number of days with 1 inch or more of snowfall varies from near 20 per season over much of the Southern Division up to 30 to 40 in the Northern Division and even to 50 or more at the highest elevations. Most winters will have several snowstorms of 5 inches or more. Storms of this magnitude temporarily disrupt transportation.

On November 22-23, 1943, a single storm dropped 56 inches of snow at Randolph, N. H., and over 50 inches at other nearby stations. This was the heaviest snowstorm of record in New Hampshire. However, snowstorms of as much as 20 inches or more are unusual in any part of the State. Heaviest 24-hour falls of record at many stations do not exceed 20 inches.

Snowfall is highly variable from year to year or for the same month in different years, as well as from place to place. Totals for the least snowy seasons range from one-fourth to one-half of the greatest seasonal amounts. In 24 years of record at Mt. Washington, for example, seasonal snowfall totals ranged from a maximum of 317 inches to a minimum of 135 inches. At Concord, in 64 seasons, the totals ranged from 29 to 103 inches. Month to month variations are much greater. Concord's maximum monthly total is 59.0 inches in February, 1893, but only 1.4 inches for the same month in 1941.

Snow cover is continuous through the whole winter season as a rule. Most frequent exceptions are found along the immediate coast and sometimes in extreme southern New Hampshire. Snow cover reaches its maximum depth, on the average, during the latter half of February in the Southern Division. In the Northern Division, the greatest depth comes in early March, excepting the higher elevations where the date is deferred to the middle of March. Some stations have a tendency for a secondary maximum in January or even at the end of December. Water stored in the snow makes an important contribution to a continuous water supply. The spring melting is usually too gradual to produce serious flooding.

OTHER CLIMATIC FEATURES: -- Sunshine averages over 50 percent of the possible amount in the Southern Division. The percentage is near 50 in the lower elevations of the Northern Division. Higher elevations and peaks are cloudier, especially in winter, reducing the percentage to less than 50 percent generally. Mt. Washington reports an average of only 33 percent.

Heavy fog occurrence varies remarkably with location and topography, but not enough data are available to describe this in detail. Persistent fogs are sometimes experienced along the coast and on the higher elevations inland. Duration of fogs diminishes inland over flat and valley locations. But the shorter duration heavy ground fogs of early morning occur frequently at susceptible places in these areas. The number of days with fog probably varies from about 20 to 90 per year over the State, except that it is much higher on the highest mountain peaks.

WINDS AND STORMS: -- The prevailing wind, on a yearly basis, comes from a westerly direction. It is predominantly from the northwest in winter and from the southwest in summer. Topography has a strong influence on prevailing direction. Points in major river valleys, for example, may have prevailing directions paralleling the valley. Along the coast in spring and summer the sea breeze is important. These onshore winds, from the cool ocean, may come inland for 10 miles or so; infrequently, they may reach as far as 30 miles into the interior. They tend to retard spring growth, but are pleasingly cool in summer.

Coastal storms or "northeasters" can be a serious weather hazard in southeastern New Hampshire, decreasing in importance northward. They generate very strong winds and heavy rain or snow. They can produce abnormally high wind-driven tides. These can heavily damage coastal installations and beaches. Some of the heaviest snowstorms result from these storms. Occasionally in summer or fall storms of tropical origin affect New Hampshire. These may often be similar (except for snow) to the northeasters described above. Only a very few retain near or full hurricane force and cause widespread damage. Damage is usually confined to the effects of high tides and heavy rainfall. Hurricane Carol traversed the whole State from south to north on August 31, 1954, with near hurricane force winds in the southern portion. Storms of tropical origin affect or threaten New Hampshire about once in 2 to 3 years, on the average. Two such storms in the same year would not be expected more than once in 10 years.

Tornadoes are not common phenomena, yet, on a per unit area basis, New Hampshire may rank fairly high among the states. Many years may have one or more. The Sunapee tornado of 1821 has been described as perhaps the worst ever experienced in New England. Fortunately, most tornadoes are very small, affecting a very localized area. Due to the extent of forested or sparsely settled areas, a large percentage of tornadoes in New Hampshire are neither seen, recorded, nor do appreciable damage. They may occur even in the northern portion of the State. One such notable occurrence was at Berlin in 1929. About 80 percent of tornadoes occur between May 15 and September 15. About 78 percent strike between 2 and 7 p.m. The peak months are June and July and the peak hour of occurrence 4 to 5 p.m. The chance of a tornado striking any given spot is extremely small.

Thunder and hailstorms have a similar frequency maximum from midspring to early fall. Thunderstorms occur on 15 to 30 days per year. The most severe are attended by hail. Hailstorms can severely injure or even ruin field crops, break glass, dent automobiles, and damage other vulnerable exposed objects. However, this danger is minimized because the size of an area struck by hail is usually small. Glaze (ice) storms of winter can produce perilous conditions for travel. These are usually of brief duration. A few widespread and prolonged ice storms have occurred. Besides affecting travel and transport, they break trees and limbs, utility lines and poles. In designing structures such as steel towers, possible ice load should be considered. An ice load also magnifies the wind stress by increasing the area exposed to the wind.

CLIMATE AND ECONOMY: -- Activities in New Hampshire are profoundly influenced by climate. Tree growth is especially favored. Covering more than four-fifths of the area, forests constitute a major scenic attraction. The spectacular coloration of foliage in the autumn is of special interest, drawing countless visitors. Forests also provide material for forest product industries. These include lumbering, paper-making, wood products manufacturing, and related industries. The ample supply of rainfall provides not only for the growth of trees but also the huge amounts of water required in the making of paper and other manufactures.

Favored industries include textiles and leather goods. A great diversity of other interests take advantage of the abundant water supply. Approximately two-thirds of the State's electrical power requirements are met by water power through a well developed hydroelectric system.

Climate is a significant factor in the State's agriculture. It favors the production of high value specialized crops. New Hampshire, therefore, ranks well in the Nation in cash receipts per acre from farm marketing. The principal farm specialties are poultry raising, dairying, tree fruits, and truck gardening. Broiler production is a large factor in the poultry industry. Apples are the most prolific of the tree fruits, with quality production an important commercial pursuit. Top quality maple syrup is produced in commercial quantity. Considerable acreage is devoted to pasture, hay, oats, and, in the southern portion, corn. Potatoes are also grown. The best soils are found largely in the river valleys, such as those of the Connecticut and Merrimac Rivers.

Climate is particularly important to a major industry, the tourist and vacation trade. Abundant game and teeming lakes and streams draw sportsmen from far and near. Much of the vacation trade is in the summer and fall when pleasant temperatures prevail at coastal and lake resorts. Skiing, with related winter sports, is developing into a very important winter attraction, made possible by the abundant snowfall.

In summary, the climate of New Hampshire contributes greatly to its industrial, agricultural, and vacation activities. The climate is a rich, natural asset, invigorating to persons in average health, and is favorable to further economic development.

SELECTED REFERENCES

General:

1. National Planning Association: The Economic State of New England (1954).

2. U. S. Dept. of Agriculture: Atlas of American Agriculture (1936)

3. --- : Climate and Man (Yearbook of Agriculture for 1941), Part 5, Climatic data, with special reference to agriculture in the United States.*

4. --- : Soil (Yearbook of Agriculture for 1957).

5. --- : Climatological Data New England (issued monthly and annually, Jan. 1888 ---; pub. under various other titles previous to Jan. 1921).*

Specialized:

1. Brooks, C. F.: "New England Snowfall", Monthly Weather Review, Vol. 45 (1917).

2. --- : "The Rainfall of New England. General Statement", Journ. N. Eng. Water Works Assoc., Vol. 44 (1930).

3. Brown, Rodger A.: "Twisters in New England", unpublished manuscript, Antioch College, 1957.*

4. Church, P. E.: "A Geographical Study of New England Temperatures", Geogr. Review, Vol. 26 (1936).

5. Eustis, R. S.: "Winds over New England in relation to topography", Bull. Amer. Met. Soc., Vol. 23 (1942).

6. Galway, Joseph G.: "A Statistical Study of New England Snowfall", unpublished manuscript of U. S. Weather Bureau (1954).*

7. Goodnough, X. H.: "Rainfall in New England", Journ. N. Eng. Water Works Assoc., Vols. 29 (1915), 35 (1921) and 40 (1926).*

8. Palmer, Robert S.: "Agricultural Drought in New England". Technical Bulletin 97, Agricultural Experiment Station, U. of New Hampshire, Durham, N. H. (1958).

9. Perley, S.: Historic Storms of New England (1891).*

10. Stone, R. G.: "Distribution of snow depths over New York and New England", Trans. Amer. Geophy. Union (1940).

11. --- : "The average length of the season with snow cover of various depths in New England", Trans. Amer. Geophy. Union (1944).

12. Upton, W.: "Characteristics of the New England Climate", Annals Harvard Astron. Obser. (1890).

13. U. S. Weather Bureau: Tabulations of frequencies of various climatic elements for various selected stations. Available on microfilm at library of Weather Bureau State Climatologist, 1900 Post Office Bldg., Boston 9, Mass.

14. Weber, J. H.: "The Rainfall of New England. Historical Statement. Annual Rainfall. Seasonal Rainfall. Mean Monthly Rainfall of Southern New England. Maximum and Minimum Rainfall of Southern New England." Journ. N. Eng. Water Works Assn., Vol. 44 (1930).

15. White, C. V.: "Rainfall in New England", Journ. N. Eng. Water Works Assn., Vols. 56 (1942) and 57 (1943).*

16. Weather Bureau Technical Paper No. 15 - Maximum Station Precipitation for 1, 2, 3, 6, 12, and 24 Hours.

17. Weather Bureau Technical Paper No. 16 - Maximum 24-Hour Precipitation in the United States. Washington, D. C. 1952.

18. Weather Bureau Technical Paper No. 25 - Rainfall Intensity-Duration-Frequency Curves. For selected Stations in the United States, Alaska, Hawaiian Islands, and Puerto Rico.

*References marked with an asterisk are useful sources of data; the others are principally studies of the important climatic elements.

BIBLIOGRAPHY

(A) Climatic Summary of the United States (Bulletin W) 1930 edition, Section 84 (New Hampshire and Vermont). U. S. Weather Bureau

(B) Climatic Summary of the United States, New England - Supplement for 1931 through 1952 (Bulletin W Supplement). U. S. Weather Bureau

(C) Climatological Data - New Hampshire. U. S. Weather Bureau

(D) Climatological Data National Summary. U. S. Weather Bureau

(E) Hourly Precipitation Data - New England. U. S. Weather Bureau

(F) Local Climatological Data, U. S. Weather Bureau, for Concord and Mt. Washington, New Hampshire.

FREEZE DATA

STATION	Freeze threshold temperature	Mean date of last Spring occurrence	Mean date of first Fall occurrence	Mean No. of days between dates	Years of record Spring	No. of occurrences in Spring	Years of record Fall	No. of occurrences in Fall
BERLIN	32	05-29	09 15	109	30	30	30	30
	28	05-11	09 28	140	30	30	30	30
	24	04-25	10 13	171	30	30	30	30
	20	04-15	10 27	195	30	30	30	30
	16	04-05	11 09	217	30	30	30	30
BETHLEHEM	32	05-19	09 25	129	29	29	30	30
	28	05-06	10 06	153	29	29	30	30
	24	04-22	10 20	181	29	29	30	30
	20	04-15	11 01	201	29	29	30	30
	16	04-03	11 10	220	29	29	30	30
CONCORD	32	05-11	09 30	142	30	30	30	30
	28	04-27	10 12	168	30	30	30	30
	24	04-13	10 22	191	30	30	30	30
	20	03-31	11 05	219	30	30	30	30
	16	03-23	11 20	242	30	30	30	30
FIRST CONN LAKE (P.O.Pittsburg)	32	06-09	09 09	93	30	30	30	30
	28	05-23	09 22	123	30	30	30	30
	24	05-06	10 06	153	30	30	30	30
	20	04-25	10 21	179	30	30	30	30
	16	04-19	10 30	195	30	30	30	30
FRANKLIN 1 NW	32	05-21	09 25	127	30	30	30	30
	28	05-13	10 05	146	30	30	30	30
	24	04-27	10 16	172	30	30	30	30
	20	04-15	10 27	195	30	30	30	30
	16	03-31	11 12	227	30	30	30	30
GLENCLIFF	32	05-17	09 27	133	30	30	30	30
	28	05-01	10 09	161	30	30	30	30
	24	04-23	10 27	187	30	30	30	30
	20	04-12	11 06	208	30	30	30	30
	16	04-01	11 13	226	30	30	30	30
KEENE	32	05-25	09 19	117	30	30	30	30
	28	05-12	10 02	143	30	30	30	30
	24	04-28	10 13	168	30	30	30	30
	20	04-15	10 26	194	30	30	30	30
	16	03-31	11 11	224	30	30	30	30
MANCHESTER	32	04-29	10 13	167	22	22	22	22
	28	04-16	10 28	195	22	22	22	22
	24	04-02	11 10	222	22	22	22	22
	20	03-20	11 23	248	22	22	22	22
	16	03-15	12 02	262	22	22	22	22
NASHUA 3 N	32	05-22	09 23	124	25	25	26	26
	28	05-08	10 06	151	25	25	26	26
	24	04-21	10 20	182	25	25	26	26
	20	04-05	10 31	209	25	25	26	26
	16	03-22	11 15	238	25	25	26	26
PINKHAM NOTCH	32	05-23	09 22	121	21	21	21	21
	28	05-09	09 30	144	21	21	21	21
	24	04-28	10 19	174	21	21	21	21
	20	04-20	10 28	192	21	21	21	21
	16	04-05	11 08	217	21	21	21	21
PLYMOUTH	32	05-26	09 17	114	30	30	30	30
	28	05-13	09 30	140	30	30	30	30
	24	04-28	10 14	170	30	30	30	30
	20	04-16	10 25	192	30	30	30	30
	16	03-29	11 08	224	30	30	30	30
WOLFEBORO FALLS	32	05-19	09 28	132	15	15	16	16
	28	05-06	10 07	154	15	15	16	16
	24	04-27	10 21	177	15	15	16	16
	20	04-14	11 06	206	15	15	16	16
	16	04-02	11 18	231	15	15	16	16

Data in the above table are based on the period 1921-1950, or that portion of this period for which data are available.

Means have been adjusted to take into account years of non-occurrence.

A freeze is a numerical substitute for the former term "killing frost" and is the occurrence of a minimum temperature at or below the threshold temperature of 32°, 28°, etc.

Freeze data tabulations in greater detail are available and can be reproduced at cost.

*MEAN TEMPERATURE AND PRECIPITATION

STATION	JAN Temp	JAN Precip	FEB Temp	FEB Precip	MAR Temp	MAR Precip	APR Temp	APR Precip	MAY Temp	MAY Precip	JUN Temp	JUN Precip	JUL Temp	JUL Precip	AUG Temp	AUG Precip	SEP Temp	SEP Precip	OCT Temp	OCT Precip	NOV Temp	NOV Precip	DEC Temp	DEC Precip	ANN Temp	ANN Precip
NEW HAMPSHIRE																										
NORTHERN																										
BERLIN	16.0	2.77	17.1	2.35	26.8	3.31	40.2	3.12	52.4	3.21	61.7	4.07	66.6	3.60	64.2	3.19	56.1	3.96	46.0	3.01	33.9	3.56	20.1	2.98	41.8	39.13
BETHLEHEM	17.2	1.73	18.5	1.49	27.9	2.27	41.0	2.67	54.1	3.42	62.9	4.23	67.6	4.08	65.7	3.56	57.6	4.04	47.1	3.13	34.1	3.10	20.4	2.25	42.8	35.97
ERROL		2.86		2.46		3.06		3.21		3.50		4.12		3.80		3.26		3.72		2.98		3.59		2.93		39.49
FIRST CONN LAKE	11.4	3.12	11.9	2.82	21.4	3.20	35.5	3.45	48.8	3.99	58.5	4.79	63.4	4.97	61.2	3.71	53.4	4.37	43.1	3.77	30.7	3.90	16.0	3.42	37.9	45.51
MILAN 7 N		2.80		2.24		2.94		3.09		3.22		4.13		3.55		3.14		3.40		2.84		3.20		2.75		37.30
MONROE 5 NNE	15.0	2.29	15.8	1.99	26.7	2.37	40.6	3.00	53.5	3.31	63.1	3.73	67.9	3.54	65.7	3.30	57.2	3.57	46.5	3.02	34.0	3.09	19.5	2.44	42.1	35.65
MOUNT WASHINGTON	5.4	5.10	5.5	4.75	12.0	5.55	22.0	5.87	35.3	5.32	44.4	6.26	49.3	6.36	47.7	6.15	41.1	6.56	31.7	5.90	19.9	6.58	8.9	5.85	27.0	70.25
PINKHAM NOTCH	16.8	4.43	17.5	3.98	26.1	5.82	37.7	4.93	50.2	5.02	59.0	5.08	63.7	4.45	61.6	4.69	54.3	5.03	45.0	4.95	32.6	5.80	19.9	4.79	40.4	58.97
YORK POND		3.16		2.74		3.80		3.56		3.90		4.66		3.95		3.46		4.39		3.34		3.80		3.26		44.02
DIVISION	16.0	2.88	17.0	2.54	26.3	3.43	39.2	3.50	51.8	3.94	61.0	4.45	65.9	4.25	63.7	3.80	55.9	4.21	45.8	3.57	33.3	4.01	19.8	3.28	41.3	43.86
SOUTHERN																										
CONCORD WB AIRPORT	20.1	2.91	21.2	2.30	31.8	3.04	43.0	3.08	54.8	3.04	64.1	3.62	69.0	3.57	66.5	3.10	58.8	3.39	48.0	2.80	36.7	3.57	24.0	2.81	44.8	37.23
DURHAM	24.4	3.53	25.4	2.97	34.3	4.16	45.0	3.76	56.0	3.40	65.1	3.39	70.5	3.37	68.7	3.36	61.1	3.52	50.7	3.04	39.5	4.18	27.5	3.59	47.4	47.27
FITZWILLIAM 3 SW		3.14		2.54		3.65		3.69		3.94		4.36		4.23		3.97		3.99		3.44		3.86		3.11		43.92
FRANKLIN 1 NW	20.8	3.20	21.9	2.69	31.4	3.68	43.6	3.49	55.7	4.16	65.2	3.76	70.3	3.46	68.0	3.15	60.1	3.79	49.0	2.86	37.1	4.00	24.3	3.45	45.6	41.69
HANOVER	19.2	2.75	20.7	2.34	30.7	2.78	43.3	3.19	55.3	3.37	64.6	3.44	69.4	4.30	67.2	3.13	59.2	3.23	48.4	2.78	36.1	3.36	22.7	2.70	44.7	37.37
KEENE	23.0	3.27	24.2	2.52	33.3	3.33	44.9	3.58	56.3	3.72	65.0	3.94	69.8	3.75	67.7	3.32	60.2	3.77	49.9	2.71	38.3	3.63	26.0	3.03	46.6	40.57
LINCOLN		3.60		2.89		3.63		3.69		3.77		4.07		4.28		4.03		4.59		3.64		4.14		3.76		46.09
MANCHESTER	24.1	3.70	25.0	2.79	33.5	4.00	44.6	3.77	56.2	3.83	65.2	3.73	70.4	3.48	67.8	3.58	60.0	3.74	50.0	2.92	39.2	4.18	27.7	3.54	47.0	43.20
NASHUA 3 N	23.9	3.44	24.9	2.71	33.9	3.78	45.1	3.57	56.4	3.63	65.4	3.89	70.5	3.33	68.2	3.52	60.4	3.75	50.3	3.29	39.0	3.91	27.2	3.31	47.1	42.13
NEWPORT		2.94		2.44		3.26		3.55		3.62		3.70		3.68		3.24		3.61		2.64		3.54		2.86		39.08
WEST LEBANON		2.62		2.20		2.68		3.00		3.22		3.35		3.98		3.02		3.22		2.66		3.17		2.51		35.63
WOLFEBORO FALLS		3.68		3.10		4.01		3.56		3.75		3.48		3.42		2.96		3.57		2.79		4.24		3.85		42.41
DIVISION	21.7	3.30	22.7	2.66	31.9	3.57	43.7	3.55	55.4	3.67	64.6	3.72	69.6	3.70	67.3	3.33	59.5	3.76	49.2	2.91	37.6	3.85	25.1	3.31	45.7	41.33

* Averages for period 1931-1955, except for stations marked WB which are "normals" based on period 1921-1950. Divisional means may not be the arithmetical average of individual stations published, since additional data from shorter period stations are used to obtain better areal representation.

CONFIDENCE LIMITS

In the absence of trend or record changes, the chances are 9 out of 10 that the true mean will lie in the interval formed by adding and subtracting the values in the following table from the means for any station in the State:

1.7	.38	1.4	.24	1.5	.60	1.1	.50	.9	.48	.7	.69	.6	.59	.8	.50	.8	.79	.9	.57	1.0	.68	1.3	.49	.4	2.27

COMPARATIVE DATA

Data in the following table are the mean temperature and average precipitation for Franklin, New Hampshire for the period 1906-1930 and are included in this publication for comparative purposes:

20.0	2.82	20.0	2.87	30.9	3.23	43.0	3.28	54.6	3.16	63.7	3.65	69.4	3.87	66.4	3.60	59.2	3.57	48.5	2.90	36.2	3.42	24.3	2.97	44.7	39.34

NORMALS, MEANS, AND EXTREMES

CONCORD, NEW HAMPSHIRE MUNICIPAL AIRPORT

LATITUDE 43° 12' N
LONGITUDE 71° 30' W
ELEVATION (ground) 339 Feet

Temperature, Degree Days, Precipitation, Snow

Month	Daily maximum (b)	Daily minimum (b)	Monthly (b)	Record highest	Year	Record lowest	Year	Normal degree days (b)	Precip. normal total (b)	Precip. max monthly	Year	Precip. min monthly	Year	Precip. max in 24 hrs	Year	Snow mean total	Snow max monthly	Year	Snow max 24 hrs	Year
J	31.6	8.6	20.1	72	1876	-35	1878	1392	2.91	5.87	1886	1.00	1871	2.10	1888	17.5	46.7	1935	19.0	1944
F	32.5	9.8	21.2	68	1880	-37	1943	1226	2.30	5.90	1896	0.40	1877	2.06	1937	17.0	59.0	1893	15.0	1929
M	42.9	20.7	31.8	85	1946	-16	1943	1029	3.04	9.80	1936	T	1915	2.59	1936	11.7	38.3	1936	12.9	1939
A	55.5	30.4	43.0	92	1941	7	1874	660	3.08	7.44	1904	0.42	1941	2.97	1923	4.4	35.0	1874	18.3	1933
M	68.8	40.7	54.8	98	1911	21	1947	316	3.04	8.26	1954	0.32	1899	3.05	1943	0.0	0.0	—	5.0	1944
J	77.9	50.2	64.1	101	1919	32	1916	82	3.62	10.10	1944	0.10	1913	4.47	1944	0.0	0.0	—	0.0	—
J	82.6	55.3	69.0	102	1911	38	1898	11	3.57	10.29	1915	0.91	1910	5.10	1887	0.0	0.0	—	0.0	—
A	80.2	52.8	66.5	100	1955	33	1942	57	3.10	9.00	1882	0.35	1882	5.32	1954	0.0	0.0	—	0.0	—
S	72.4	45.2	58.8	98	1953	20	1941	192	3.39	10.97	1888	0.21	1914	5.97	1932	0.0	0.0	—	0.0	—
O	61.7	34.3	48.0	92	1879	16	1947	527	2.80	7.76	1890	0.05	1924	4.04	1890	0.5	3.0	1884	1.0	1925
N	47.5	25.8	36.7	80	1950	-17	1875	849	3.57	7.59	1937	0.50	1939	4.94	1938	4.3	25.0	1873	13.3	1938
D	34.3	13.6	24.0	65	1932	-24	1875	1271	2.81	7.64	1936	0.58	1943	2.79	1952	11.9	43.0	1876	14.6	1946
Year	57.3	32.3	44.8	102	July 1911	-37	Feb. 1943	7612	37.23	10.97	Sept. 1888	T	Mar. 1915	5.97	Sept. 1932	68.2	59.0	Feb. 1893	19.0	Jan. 1944

Relative Humidity, Wind, Sky, Mean Number of Days

Month	RH 1:00 a.m.	RH 7:00 a.m.	RH 1:00 p.m.	RH 7:00 p.m.	Wind mean hourly	Prevailing dir.	Fastest mile speed	Fastest mile dir.	Year	Pct. possible sunshine	Mean sky cover	Clear	Partly cloudy	Cloudy	Precip. .01"+	Snow,sleet 1.0"+	Thunderstorms	Heavy fog	Max 90°+	Max 32°-	Min 32°-	Min 0°-
J	79	78	60	69	6.2	NW	43	NW	1945	48	5.6	11	8	12	11	4	0	2	0	17	30	6
F	74	74	56	66	6.6	NW	42	N	1950	55	5.0	11	7	10	9	4	0	1	0	14	27	5
M	74	75	54	63	6.9	NW	71	NW	1950	55	5.4	13	9	10	10	4	1	2	0	4	25	1
A	70	72	47	61	6.9	NW	52	NW	1945	53	5.3	12	10	9	11	1	1	2	0	0	14	0
M	75	76	47	62	6.4	NW	48	NW	1957	52	5.0	12	11	10	10	*	3	3	1	0	1	0
J	78	77	49	66	6.3	NW	38	SW	1954	56	4.9	12	11	7	10	0	5	5	2	0	0	0
J	80	87	51	69	5.3	NW	37	NW	1944	58	4.8	13	12	7	10	0	6	6	3	0	0	0
A	82	84	52	72	4.6	NW	56	E	1954	58	4.8	14	11	7	10	0	5	7	2	0	0	0
S	86	86	55	75	4.8	NW	61	SE	1944	51	5.3	13	9	9	9	0	2	5	1	0	1	0
O	82	84	59	71	5.3	NW	39	NW	1950	53	6.0	12	8	11	9	1	1	3	0	1	8	0
N	84	86	59	71	6.0	NW	72	NE	1950	47	5.9	9	7	13	10	2	0	3	0	2	20	0
D	77	79	59	69	6.1	NW	51	NE	1944	43	5.9	10	7	14	10	4	0	2	0	12	28	3
Year	83	79	53	69	5.8	NW	72	NE, Nov. 1950		52	5.2	139	106	120	120	20	24	53	9	50	154	15

MOUNT WASHINGTON OBSERVATORY, GORHAM, NEW HAMPSHIRE

LATITUDE 44° 16' N
LONGITUDE 71° 18' W
ELEVATION (ground) 6262 feet

Temperature, Degree Days, Precipitation, Snow

Month	Daily maximum (b)	Daily minimum (b)	Monthly (b)	Record highest	Year	Record lowest	Year	Mean degree days	Precip. normal total (b)	Precip. max monthly	Year	Precip. min monthly	Year	Precip. max in 24 hrs	Year	Snow mean total	Snow max monthly	Year	Snow max 24 hrs	Year
J	14.1	-3.3	5.4	42	1950	-46	1934	1803	5.10	9.89	1936	1.80	1951	2.10	1936	26.3	48.5	1935	9.5	1944
F	13.9	-4.4	4.8	42	1937	-46	1943	1645	4.75	10.45	1937	2.64	1953	4.25	1951	27.5	57.7	1953	32.0	1953
M	19.6	1.3	—	48	1946	-38	1950	1621	5.55	15.57	1936	2.15	1946	3.80	1937	30.3	73.0	1953	19.5	1953
A	28.7	15.3	22.0	60	1941	-20	1954	1262	5.32	10.03	1934	2.23	1941	3.77	1934	21.9	43.6	1953	10.5	1951
M	41.6	29.0	35.3	63	1937	-1	1956	947	5.32	10.95	1941	1.78	1934	2.69	1945	10.2	19.5	1956	3.8	1946
J	50.6	38.2	44.4	71	1953	8	1945	602	6.26	10.70	1944	3.30	1937	3.80	1946	0.9	7.0	1933	3.8	—
J	54.1	44.5	49.3	71	1953	24	1940	490	6.36	10.70	1957	2.69	1955	3.14	1947	0.1	1.1	1957	1.1	1957
A	53.1	41.7	47.1	71	1953	24	1953	535	6.15	13.14	1955	2.77	1955	5.20	1955	0.1	1.1	1934	T	—
S	47.1	35.1	41.7	66	1953	11	1942	725	5.90	14.07	1938	2.74	1938	4.55	1948	1.5	7.8	1949	5.9	1949
O	37.9	25.4	31.5	59	1956	-5	1939	1009	6.58	11.58	1937	0.75	1947	4.60	1954	9.0	31.5	1934	14.6	1943
N	26.4	13.0	19.9	51	1951	-19	1936	1317	6.58	10.95	1945	2.31	1939	3.44	1950	23.7	51.8	1943	26.5	1943
D	16.8	1.0	—	45	1934	-46	1933	1707	5.85	13.35	1957	1.49	1957	3.38	1952	33.0	83.7	1954	32.0	1954
Year	33.8	20.1	27.0	71	July 1953	-46	Feb. 1943	13663	70.20	15.57	Mar. 1936	0.75	Oct. 1947	5.20	Aug. 1955	184.5	83.7	Dec. 1954	32.0	Dec. 1954

Relative Humidity, Wind, Sky, Mean Number of Days

Month	RH 1:00 a.m.	RH 7:00 a.m.	RH 1:00 p.m.	RH 7:00 p.m.	Wind mean hourly	Prevailing dir.	Fastest mile speed	Fastest mile dir.	Year	Pct. possible sunshine	Mean sky cover	Clear	Partly cloudy	Cloudy	Precip. .01"+	Snow,sleet 1.0"+	Thunderstorms	Heavy fog	Max 90°+	Max 32°-	Min 32°-	Min 0°-
J	81	81	80	80	47.3	W	170	W	1943	33	7.5	6	5	20	18	9	0	24	0	29	31	17
F	83	84	84	86	46.4	W	144	SE	1937	33	7.5	6	5	18	17	9	0	25	0	27	28	16
M	86	86	86	86	44.5	W	180	W	1942	37	7.6	7	6	18	18	10	1	25	0	26	31	11
A	89	89	86	87	38.0	W	231	SE	1934	34	7.6	5	7	20	18	7	2	24	0	18	29	0
M	88	85	86	88	28.2	W	164	SE	1945	34	8.0	4	7	20	16	3	4	26	0	6	19	0
J	90	90	83	88	25.5	W	136	W	1949	30	8.0	3	8	20	16	1	5	25	0	1	6	0
J	92	93	91	91	24.6	W	110	—	1933	30	8.0	2	8	21	15	1	5	25	0	0	1	0
A	93	92	88	92	25.6	W	142	ENE	1954	30	8.0	5	9	19	15	1	1	25	0	0	0	0
S	90	90	88	88	29.3	W	157	SE	1938	36	7.2	7	7	19	14	3	1	26	0	3	23	1
O	87	87	86	86	34.2	W	161	SE	1943	36	7.7	5	6	21	19	10	1	25	0	10	28	6
N	86	86	86	86	39.8	W	160	W	1950	34	7.8	4	6	21	17	10	0	26	0	23	28	4
D	87	86	84	84	46.9	W	175	W	1942	29	7.8	4	5	21	18	9	0	26	0	28	31	14
Year	88	86	85	87	36.4	W	231	SE, Apr. 1934		33	7.6	52	77	236	203	57	16	303	0	168	242	64

REFERENCE NOTES APPLYING TO ALL "NORMALS, MEANS, AND EXTREMES" TABLES.

(a) Length of record, years.
(b) Normal values are based on the period 1921-1950, and are means adjusted to represent observations taken at the present standard location.
* Less than one-half.

- No record.
+ Airport data.
‡ City Office data.
§ Also on earlier dates, months, or years.
T Trace, an amount too small to measure.

Sky cover is expressed in a range of 0 for no clouds or obscuring phenomena to 10 for complete sky cover. The number of clear days is based on average cloudiness 0-3 tenths; partly cloudy days on 4-7 tenths; and cloudy days on 8-10 tenths. Monthly degree day totals are the sum of the negative departures of average daily temperatures from 65°F. Sleet was included in snowfall totals beginning with July 1948. Heavy fog also includes data referred to at various times in the past as "Dense" or "Thick". The upper visibility limit for heavy fog is 1/4 mile. Data in these tables are based on records through 1957.

Mean Annual Precipitation, Inches

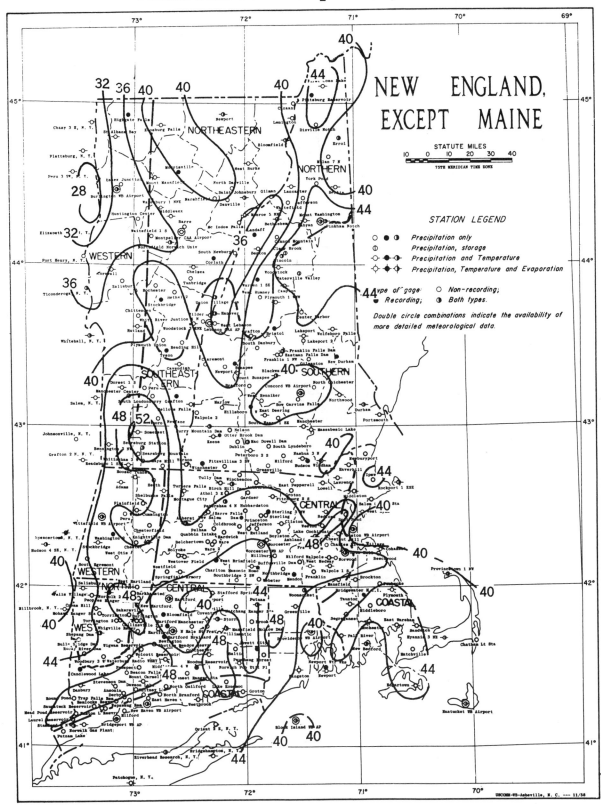

NEW ENGLAND, EXCEPT MAINE

STATUTE MILES
10 0 10 20 30 40
75TH MERIDIAN TIME ZONE

STATION LEGEND

○ ● ◐ *Precipitation only*
① *Precipitation, storage*
⦵ ◑ ⦶ *Precipitation and Temperature*
⧂ ◆ ⬦ *Precipitation, Temperature and Evaporation*

Type of gage: ○ *Non-recording;*
● *Recording;* ◐ *Both types.*

Double circle combinations indicate the availability of more detailed meteorological data.

Based on period 1931-55

Isolines are drawn through points of approximately equal value. Caution should be used in interpolating on these maps, particularly in mountainous areas.

Mean Maximum Temperature (°F.), January

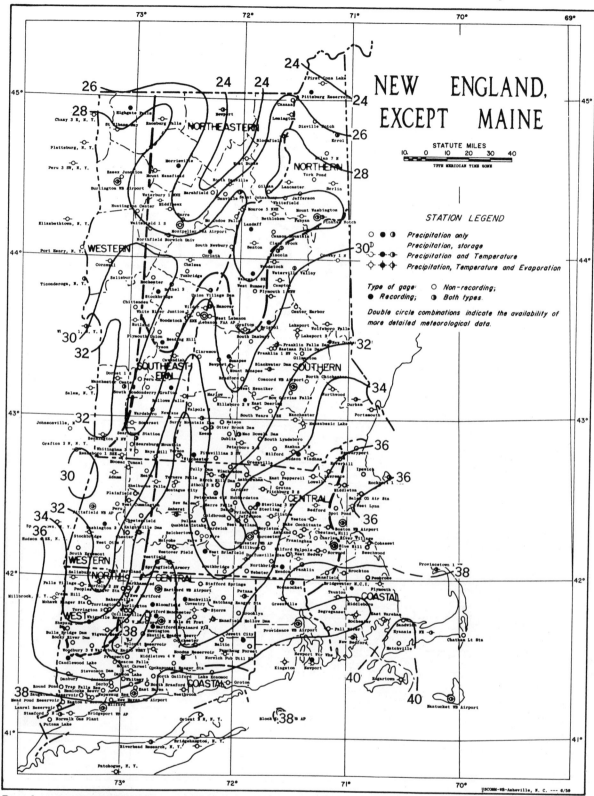

NEW ENGLAND, EXCEPT MAINE

STATUTE MILES

STATION LEGEND

Based on period 1931-52

Isolines are drawn through points of approximately equal value. Caution should be used in interpolating on these maps, particularly in mountainous areas.

Mean Minimum Temperature (°F.), January

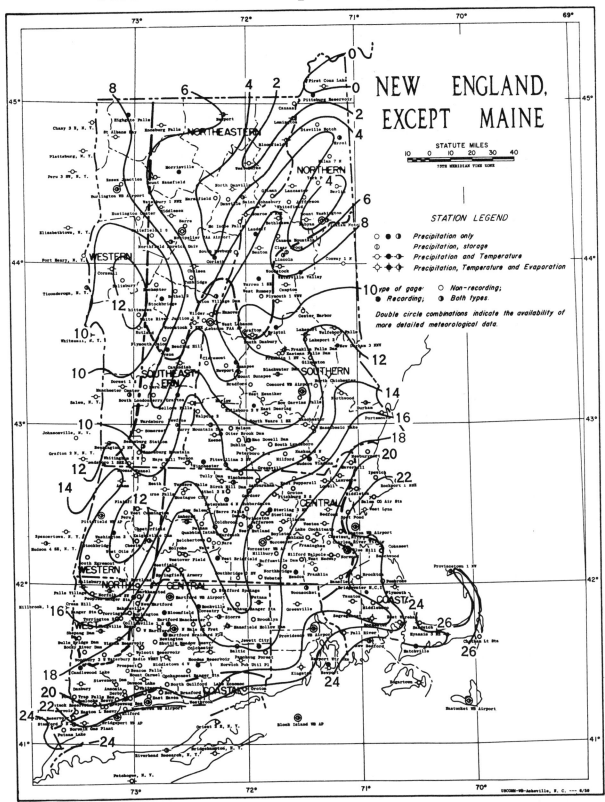

Based on period 1931-52

Isolines are drawn through points of approximately equal value. Caution should be used in interpolating on these maps, particularly in mountainous areas.

Mean Maximum Temperature (°F.), July

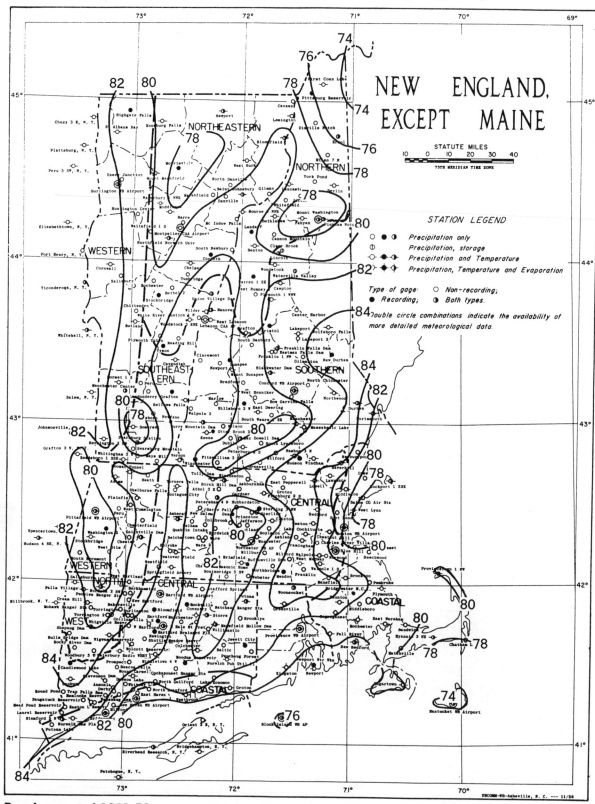

Based on period 1931-52

Isolines are drawn through points of approximately equal value. Caution should be used in interpolating on these maps, particularly in mountainous areas.

Mean Minimum Temperature (°F.), July

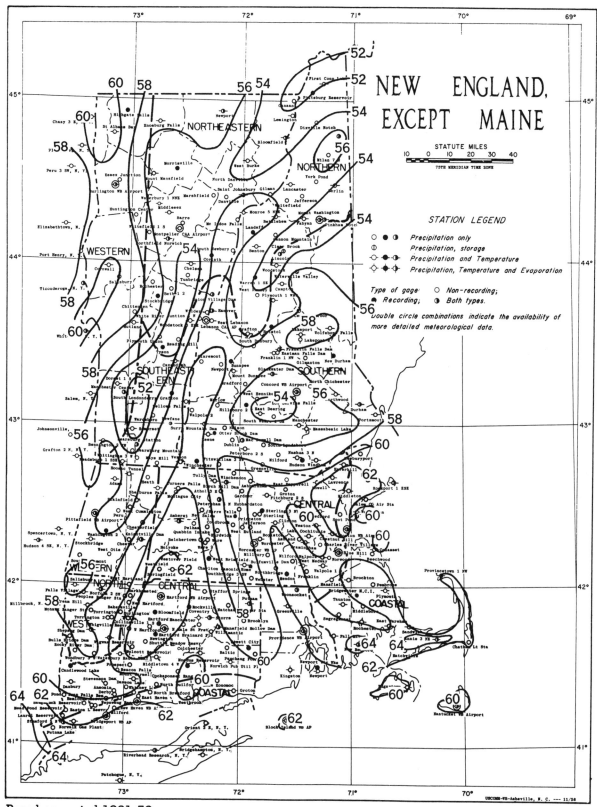

Based on period 1931-52

Isolines are drawn through points of approximately equal value. Caution should be used in interpolating on these maps, particularly in mountainous areas.

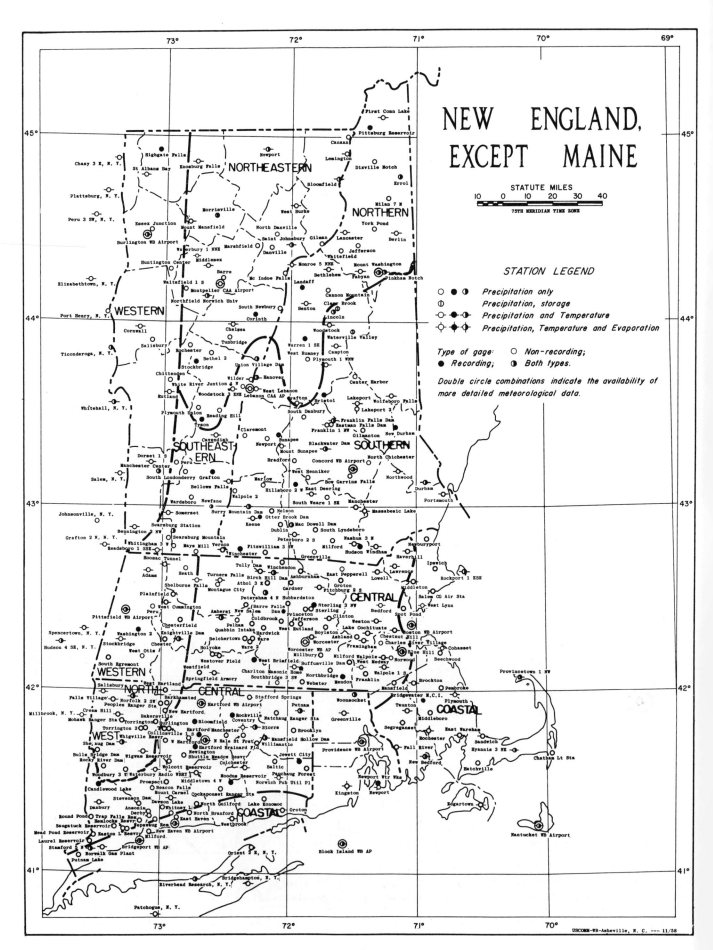

NEW ENGLAND, EXCEPT MAINE

STATUTE MILES

10   0   10   20   30   40

75TH MERIDIAN TIME ZONE

STATION LEGEND

○ ● ◑	*Precipitation only*
◐	*Precipitation, storage*
○- ●- ◑-	*Precipitation and Temperature*
◇ ◆ ◈	*Precipitation, Temperature and Evaporation*

Type of gage: ○ *Non-recording;*
Recording;   ◐ *Both types.*

Double circle combinations indicate the availability of more detailed meteorological data.

NORTHEASTERN

NORTHERN

WESTERN

SOUTHEASTERN

SOUTHERN

CENTRAL

NORTH CENTRAL

WESTERN

COASTAL

COASTAL

First Conn Lake
Pittsburg Reservoir
Canaan
Newport
Lemington
Dixville Notch
Errol
Highgate Falls
Chazy 3 E, N.Y.
St Albans Bay
Enosburg Falls
Bloomfield
Milan 7 N
Plattsburg, N.Y.
Morrisville
West Burke
York Pond
Peru 3 SW, N.Y.
Essex Junction
Mount Mansfield
North Danville
Saint Johnsbury
Gilman
Lancaster
Berlin
Burlington WB Airport
Waterbury 1 NNE
Marshfield
Danville
Whitefield
Jefferson
Huntington Center
Middlesex
Barre
Monroe 5 NNE
Bethlehem
Mount Washington
Fabyan
Pinkham Notch
Elizabethtown, N.Y.
Waitsfield 1 S
Mc Indoe Falls
Landaff
Montpelier CAA Airport
Northfield Norwich Univ
South Newbury
Cannon Mountain
Clear Brook
Port Henry, N.Y.
Cornwall
Corinth
Benton
Lincoln
Chelsea
Woodstock
Waterville Valley
Salisbury
Rochester
Tunbridge
Warren 1 SE
Ticonderoga, N.Y.
Bethel 2
West Rumney
Campton
Stockbridge
Union Village Dam
Plymouth 1 WNW
Chittenden
Wilder
Hanover
Center Harbor
White River Junction 4 N
West Lebanon
Rutland
Woodstock
NNE Lebanon CAA AP
Grafton
Bristol
Lakeport
Wolfeboro Falls
Whitehall, N.Y.
South Danbury
Lakeport 2
Plymouth Union
Reading Hill
Franklin Falls Dam
Tyson
Eastman Falls Dam
New Durham
Claremont
Franklin 1 NW
Gilmanton
Cavendish
Newport
Sunapee
Blackwater Dam
Dorset 1 S
Peru
Mount Sunapee
Concord WB Airport
North Chichester
Manchester Center
Bradford
West Henniker
Northwood
Salem, N.Y.
South Londonderry
Grafton
Marlow
Bow Garvins Falls
Bellows Falls
Hillsboro 2 W
East Deering
Durham
Wardsboro
Newfane
Walpole 2
South Weare 1 SE
Manchester
Portsmouth
Johnsonville, N.Y.
Somerset
Surry Mountain Dam
Nelson
Massabesic Lake
Searsburg Station
Keene
Otter Brook Dam
Bennington 3 NW
Dublin
Mac Dowell Dam
South Lyndeboro
Grafton 2 N, N.Y.
Whitingham 3 W
Searsburg Mountain
Peterboro 2 S
Nashua 3 N
Newburyport
Readsboro 1 SSE
Mays Mill
Vernon
Fitzwilliam 3 SW
Milford
Hudson
Windham
Haverhill
Hoosac Tunnel
Winchester
Greenville
Ipswich
Adams
Heath
Tully Dam
Winchendon
Lawrence
Rockport 1 ESE
Shelburne Falls
Turners Falls
Birch Hill Dam
Ashburnham
East Pepperell
Lowell
Middleton
Plainfield
Montague City
Athol 3 E
Gardner
Groton
Fitchburg 2 S
Salem CG Air Sta
West Cummington
Petersham 4 N
Hubbardston
Sterling 3 NW
Bedford
West Lynn
Pittsfield WB Airport
Peru
Amherst
New Salem
Barre Falls Dam
Sterling
Spot Pond
Spencertown, N.Y.
Chesterfield
Pelham
Princeton
Jefferson
Clinton
Boston WB Airport
Washington 2
Knightville Dam
Quabbin Intake
Coldbrook
West Rutland
Lake Cochituate
Chestnut Hill
Hudson 4 SE, N.Y.
Stockbridge
Chester
Belchertown
Hardwick
Ware
Boylston
Weston
Ashland
Charles River Village
West Otis
Holyoke
Ware 2
Worcester
Framingham
Milford
Blue Hill
Cohasset
South Egremont
Westover Field
Worcester WB AP
Walpole
Norwood
Beechwood
Westfield
Millbury
West Medway
Provincetown 1 NW
West Briarfield
Buffumville Dam
WESTERN
Springfield Armory
Charlton Masonic Home
Northbridge
Walpole 1 S
Brockton
Salisbury
West Hartford
Southbridge 3 SW
Webster
Mendon
Franklin
Pembroke
Barkhamsted
Stafford Springs
Woonsocket
Mansfield
Bridgewater M.C.I.
Plymouth
Falls Village
Norfolk 2 SW
Hartford WB Airport
Putnam
Taunton
Middleboro
Peoples Ranger Sta
New Hartford
Natchaug Ranger Sta
Greenville
Segreganset
Cream Hill
Rockville
Coventry
East Wareham
Millbrook, N.Y.
Bakersville
Burlington
Bloomfield
Storrs
Brooklyn
Rochester
Mohawk Ranger Sta
Torrington
Manchester
Mansfield Hollow Dam
Sandwich
Torrington 2
Collinsville
N Hale St Prst
Williamantic
Fall River
Hyannis 3 NE
Shepaug Dam
W Hartford
Hartford Brainard Fld
Jewett City
New Bedford
Chatham Lt Sta
Whigville Resvr
Newington
Providence WB Airport
Bulls Bridge Dam
Wigvam Reservoir
Shuttle Meadow Resvr
Rocky River Dam
Woodbury 3 E Waterbury Radio WBRY
Colchester
Baltic
Woodbury 3 E
Wolcott Reservoir
Moodus Reservoir
Newport Wtr Wks
Prospect
Middletown 4 W
Norwich Pub Util Pl
Candlewood Lake
Beacon Falls
Mount Carmel
Cockaponset Ranger Sta
Kingston
Newport
Stevenson Dam
Dawson Lake
Lake Konomoc
Danbury
Ansonia
Whitney L
North Guilford
Groton
Edgartown
Round Pond
Trap Falls Res.
Derby
North Branford
Saugatuck Reservoir
Hemlocks Resvr
Pepaucag Res
East Haven
Westbrook
Mead Pond Reservoir
Easton L Resvr
New Haven WB Airport
Laurel Reservoir
Milford
Stamford 5 N
Bridgeport WB AP
Nantucket WB Airport
Norwalk Gas Plant
Orient 3 E, N.Y.
Block Island WB AP
Putnam Lake
Bridgehampton, N.Y.
Riverhead Research, N.Y.
Patchogue, N.Y.

THE CLIMATE OF
NEW JERSEY

by
Donald V. Dunlap

March 1967

New Jersey, though one of the smaller states, has a varied topography. In the northwestern part a section comprising about one-fifth of the area of the State is known as the Highlands and Kittatinny Valley. This region is traversed by several low mountain ridges extending northeasterly across the State with valleys and rolling hills between. The highest of these ranges is the Kittatinny, which rises from the banks of the Delaware River at the famous Delaware Water Gap. To the eastward the region is studded with numerous lakes, some of the largest of which are Lakes Hopatcong, Mohawk, and Greenwood. Elevations up to 1,800 feet above sea level are found in the Kittatinny Mountains near the New York State line.

South and east of the Highlands is a region of about equal area known as the Red Sandstone Plain, or the Piedmont of New Jersey. It is generally hilly in its northwestern part, becoming rolling and then flat toward the south and southeast. At its northeastern corner are the Palisades, cliffs which rise abruptly from the Hudson River to heights of 200 to 500 feet. The seacoast section extends from Sandy Hook to Cape May, or about 125 miles. This area is characterized by long stretches of sandy beaches,

now occupied largely by summer resorts. Tidewater marshes become numerous toward the south.

In the southern interior a region known as the Pines is covered with scrubby forests of pine and some oak. The land is low and some of it is swampy. Here are found the large cranberry bogs of New Jersey. In fact, most of the State that lies south of a line connecting Jersey City and Trenton is low and flat with few elevations higher than 100 feet above mean sea level, these being mainly in Monmouth County.

About 30 percent of the area of New Jersey drains into the Delaware River and Delaware Bay, which form the western boundary. Nearly half of Sussex County, in the northwest, drains northward through the Wallkill River into the Hudson River of New York. The remainder of the State drains directly into the Atlantic Ocean through the Passaic, Hackensack, and Raritan Rivers in the north, and a number of small rivers and streams in the south.

Over the southern interior the soil changes from sandy near the coast to clay and marl in the western part. However, there is no steady transition, the change being effected mostly by alternating stretches of the different soils and combinations of them. In the most productive sections in the southwestern part, light to medium sandy loams predominate. Immense quantities of garden truck for commercial canning, especially tomatoes, are grown in Cumberland, Salem, Gloucester, Camden, and western Burlington Counties.

The extreme length of the State is 166 miles and its greatest width only about 65. The difference in climate is quite marked between the southern tip at Cape May and the northern extremity in the Kittatinny Mountains. The former locality is almost surrounded by water and is fairly well removed from the influence of the frequent storms that cross the Great Lakes region and move out the St. Lawrence Valley. The northern extremity is well within the zone of influence of these storms and, in addition, lies at elevations varying from 800 to 1,800 feet. The influence of these high elevations on the temperature is considerable. The differences between these two localities are particularly marked in the winter, Cape May having a normal January temperature about the same as that of southwestern Virginia, while that of Layton, in the extreme northwest, is similar to that of the northern area of Ohio. Since the prevailing winds are mostly offshore, the ocean influence does not have full effect.

Temperature differences between the northern and southern parts of the State are greatest in winter and least in summer. Nearly every station has registered readings of 100° F. or higher at some time, and all of them have records of zero or below. The highest temperature of record is 110° F. observed July 10, 1936 at Runyon; the lowest, -34° F., January 5, 1904, at River Vale.

In the northern highland area, the average date of last freeze (32° F.) in spring is about May 2, and that of the first in Fall, October 12. On the seacoast corresponding dates are April 6 and November 9, while in the central and southern interior the dates are April 23 and October 19. Freeze-free days in the northern highlands average 163, with 217 along the seacoast and 179 in the central and southern interior.

Northern New Jersey is near enough to the paths of the storms which cross the Great Lakes region and pass down the St. Lawrence Valley to receive part of its precipitation from that source. However, the heaviest general rains are produced by coastal storms of tropical origin. The centers of these storms usually pass some distance offshore, with heaviest rainfall and strongest wind near the coast. On several occasions tropical storms have moved inland along the south Atlantic coast, and then moved northward either through or to the west of New Jersey. Noteworthy storms of this type in recent years include hurricanes Able in 1952, Hazel in 1954, and Connie and Diane in 1955.

The damage by high tides to coastal installations during the passage of a tropical storm is often severe, whether the storm passes offshore or inland.

The average annual precipitation ranges from about 40 inches along the southeast coast to 51 inches in north-central parts of the State. In other sections the annual averages are mostly between 43 and 47 inches. Rainfall is well distributed during the warm months. Heavy 24-hour falls of 7 or 8 inches are occasionally recorded.

Brief periods of drought during the growing season are not uncommon, but prolonged droughts are relatively rare, occurring on the average once in 15 years.

Flooding in New Jersey is usually caused by heavy general rains, at times associated with storms of tropical origin. Local flooding results from ice gorging.

Important flooding occurred along the Raritan River in 1934, 1935, 1938, 1940, and 1948. The Passaic River flooded seriously in 1903, 1936, and 1952, and the Delaware River in 1903, 1936, and 1955.

The season during which measurable quantities of snow are likely to fall extends from about October 15 to April 20 in the Highlands, and from about November 15 to March 15 in the vicinity of Cape May. Average seasonal amounts range from about 13 inches at Cape May to nearly 50 inches in the Highlands. Snowfalls of 10 or more inches in a single storm are occasional occurrences.

The number of days a month with measurable precipitation averages 8 for each of the fall months, September, October, and November, and 9 to 12 for the other months of the year; the average yearly number is 120. Midday relative humidity averages 68 percent along the seacoast and 57 percent or less at inland locations.

Normally, sunshine varies from slightly over one-half of the possible amount in the northern counties to about 60 percent in the south. The prevailing wind is from the northwest from October to April, inclusive, and from the southwest for the other months of the year.

Thirty-four tornadoes were reported in New Jersey for the 50-year period ending with 1965. Two deaths and damage estimated at 2.5 million dollars resulted from these tornadoes. Damage from hail in the 22-year period ending with 1965 totaled about $500,000, with half of that amount resulting from a single storm which passed through Hunterdon and Warren Counties on June 10, 1956.

Damage from windstorms other than tornadoes amounted to about 1.5 million during the period 1938-1965. The most destructive of these storms was that of November 25, 1950, with the September 1944 hurricane second in the amount of damage, followed by hurricane Hazel in October 1954 in third place.

The most damaging storm on record was the wind storm accompanied by a tidal surge March 6-8, 1962, with the loss of 21 lives and damage which was estimated at 80 million dollars. Tidal damage covered the entire coastline of New Jersey, including Delaware and Raritan Bays.

The leading farm products are eggs, milk, vegetables, and poultry. The farm economy of the State is dependent upon an adequate water supply which must be met by irrigation in many areas. During periods of drought, as in the summers of 1962-1966, there is insufficient water for irrigation usage. The storage of even a small percentage of the State's runoff water would meet the requirements of industrial and agricultural users. Precipitation is both plentiful and reliable, thus guaranteeing an adequate water supply for industrial uses.

The resort industry along the seacoast serves the New York City and Philadelphia populace, as well as New Jerseyites. The mean daily maximum temperature for the summer months of June, July, and August at Atlantic City is 77.7° F., giving evidence of the seabreeze effect along the immediate coast of New Jersey. Numerous lakes in the Highlands also provide summer resort facilities, with a moderate climate during the summer months.

The invigorating climate of the north and central portions of the State, with marked changes in weather, generally neither extreme nor severe, provides an excellent setting for industrial and commercial interests, as evidenced by the concentration of population in the northeastern counties.

REFERENCES

(1) Biel, E. R. and A. V. Havens, "Science and Land", 1948-49 Annual Report, Rutgers University, N. J. Agricultural Experiment Station, Meteorology Section.

(2) Blood, Richard D. W., "A Study of Certain Basic and Applied Wind Problems of the Climate of New Jersey", M. S. Thesis, Rutgers Univ., Sept. 1953.

(3) Cantlon, J. E., "Vegetation and Micro-climates on North and South Slopes of Cushetunk Mountain, New Jersey", Ph.D. Thesis, Rutgers Univ., 1950.

(4) Havens, A. V., "A Preliminary Dynamic Climatology for New Jersey and Its Application to Human Activities", M. S. Thesis, Rutgers Univ., 1948.

(5) Janifer, Clarence S., Jr., "A Study of Certain Thermal Features of the Climate of New Jersey", M. S. Thesis, Rutgers Univ., May 1952.

(6) Marlatt, Wm. E., "A Comparison of Evapotranspiration Computed from Climatic Data with Field and Lysimeter Measurements"; M. S. Thesis, Rutgers, May 1958.

(7) McGuire, James K., and Wayne C. Palmer, "The 1957 Drought in the Eastern United States", Monthly Weather Review, Vol. 85, No. 9, Sept. 1957.

(8) Rutgers University Press, "The Economy of New Jersey", a survey published June 20, 1958, including a chapter on "The Climate of New Jersey" by Dr. E. R. Biel.

(9) Weather Bureau Technical Paper No. 15 - Maximum Station Precipitation for 1,2,3, 6,12, and 24 Hours.

(10) Weather Bureau Technical Paper No. 16 - Maximum 24-Hour Precipitation in the United States. Washington, D. C. 1952.

(11) Weather Bureau Technical Paper No. 25 - Rainfall Intensity - Duration - Frequency Curves. For selected stations in the United States, Alaska, Hawaiian Islands and Puerto Rico.

(12) Weather Bureau Technical Paper No. 29 - Rainfall Intensity - Frequency Regime. Washington, D. C.

BIBLIOGRAPHY

(A) Climatic Summary of the United States (Bulletin W)-1930 edition, Section 90. U. S. Weather Bureau

(B) Climatic Summary of the United States, New Jersey - Supplement for 1931 through 1952 (Bulletin W Supplement). U. S. Weather Bureau

(C) Climatic Summary of the United States, New Jersey - Supplement for 1951 through 1960 (Bulletin W Supplement). U. S. Weather Bureau

(D) Climatological Data - New Jersey. U. S. Weather Bureau

(E) Climatological Data National Summary. U. S. Weather Bureau

(F) Hourly Precipitation Data - New Jersey. U. S. Weather Bureau

(G) Local Climatological Data, U. S. Weather Bureau, for Atlantic City, Newark and Trenton, New Jersey.

FREEZE DATA

STATION	Freeze threshold temperature	Mean date of last Spring occurrence	Mean date of first Fall occurrence	Mean No. of days between dates	Years of record Spring	No. of occurrences in Spring	Years of record Fall	No. of occurrences in Fall
ATLANTIC CITY	32	03-31	11-11	225	10	10	10	10
	28	03-18	11-29	256	10	10	10	10
	24	03-04	12-04	275	10	10	10	10
	20	02-15	12-14	302	10	10	10	10
	16	02-09	12-18	312	10	9	10	9
BELLEPLAIN ST FOREST	32	04-27	10-14	170	29	29	29	29
	28	04-13	10-24	194	29	29	29	29
	24	03-31	11-04	218	29	29	28	28
	20	03-19	11-20	246	29	29	28	28
	16	03-08	12-03	270	29	29	28	28
BELVIDERE	32	04-26	10-13	170	30	30	30	30
	28	04-12	10-27	198	30	30	30	30
	24	03-23	11-14	236	30	30	30	30
	20	03-14	11-24	255	30	30	30	30
	16	03-07	12-05	273	30	30	30	30
BOONTON 1 SE	32	04-28	10-09	164	30	30	30	30
	28	04-13	10-22	192	30	30	30	30
	24	03-26	11-07	226	30	30	30	30
	20	03-15	11-23	253	30	30	30	30
	16	03-07	12-06	274	30	30	29	29
BURLINGTON	32	04-12	10-30	201	30	30	30	30
	28	03-27	11-12	230	29	29	30	30
	24	03-17	11-27	255	30	30	30	30
	20	03-08	12-07	274	30	30	30	30
	16	02-27	12-16	292	30	30	30	26
CANOE BROOK	32	05-04	10-10	159	10	10	10	10
	28	04-17	10-19	185	10	10	10	10
	24	04-06	11-06	214	10	10	10	10
	20	03-21	11-11	235	10	10	10	10
	16	03-15	11-25	255	10	10	10	10
CAPE MAY 3 W	32	03-31	11-17	231	21	21	20	20
	28	03-19	11-30	256	21	21	19	19
	24	03-05	12-09	279	20	20	19	19
	20	02-21	12-16	298	21	21	19	16
	16	02-05	12-24	322	21	17	18	10
CHARLOTTEBURG	32	05-17	09-25	131	30	30	30	30
	28	04-30	10-09	162	30	30	30	30
	24	04-15	10-25	193	30	30	30	30
	20	03-29	11-06	222	30	30	30	30
	16	03-19	11-23	249	30	30	30	30
CLAYTON	32	04-12	10-29	200	10	10	10	10
	28	03-25	11-07	227	10	10	10	10
	24	03-20	11-14	239	10	10	10	10
	20	03-04	11-28	269	10	10	10	10
	16	02-11	12-15	307	10	10	10	9
ELIZABETH	32	04-23	10-19	179	30	30	30	30
	28	04-04	11-02	212	30	30	30	30
	24	03-22	11-17	240	30	30	30	30
	20	03-11	11-29	263	30	30	30	30
	16	03-01	12-09	283	30	30	30	29
FLEMINGTON 1 NE	32	04-29	10-13	167	30	30	30	30
	28	04-16	10-23	190	30	30	30	30
	24	03-30	11-09	224	30	30	30	30
	20	03-15	11-21	251	30	30	30	30
	16	03-09	12-06	272	30	30	30	29
FREEHOLD	32	04-23	10-18	178	29	29	30	30
	28	04-07	11-02	209	29	29	30	30
	24	03-24	11-17	238	29	29	30	30
	20	03-16	11-29	258	29	29	30	30
	16	03-05	12-07	277	29	29	30	29
HAMMONTON 2 NNE	32	04-21	10-16	178	28	28	29	29
	28	04-06	10-31	208	28	28	29	29
	24	03-21	11-17	241	27	27	28	28
	20	03-10	11-28	263	27	27	28	28
	16	03-02	12-07	280	26	26	28	27
HIGHTSTOWN 1 N	32	04-22	10-12	173	10	10	10	10
	28	04-06	11-03	211	10	10	10	10
	24	03-20	11-16	241	10	10	10	10
	20	03-08	12-06	273	10	10	10	10
	16	02-24	12-13	292	10	10	10	10
INDIAN MILLS 2 W	32	05-01	10-08	160	29	29	30	30
	28	04-19	10-21	185	29	29	30	30
	24	04-02	10-31	212	30	30	30	30
	20	03-20	11-17	242	30	30	30	30
	16	03-09	12-01	267	30	30	30	30
JERSEY CITY	32	04-06	11-07	215	29	29	29	29
	28	03-24	11-21	242	28	28	28	28
	24	03-16	11-28	257	28	28	28	28
	20	03-08	12-06	273	28	28	28	28
	16	02-26	12-14	291	28	28	28	25
LAMBERTVILLE	32	04-23	10-15	175	29	29	29	29
	28	04-10	11-02	206	29	29	28	28
	24	03-23	11-15	237	28	28	27	27
	20	03-13	11-24	256	28	28	27	27
	16	03-02	12-08	281	29	29	26	25
LAURELTON 1 E	32	04-28	10-14	169	29	29	29	29
	28	04-10	10-28	201	28	28	28	28
	24	03-26	11-11	230	27	27	28	28
	20	03-15	11-27	257	27	27	28	28
	16	03-03	12-05	277	27	26	29	29
LAYTON 3 NW	32	05-21	09-27	129	30	30	30	30
	28	05-09	10-09	153	30	30	30	30
	24	04-26	10-17	174	30	30	30	30
	20	04-12	10-31	202	30	30	30	30
	16	03-24	11-13	234	30	30	30	30
LITTLE FALLS	32	04-23	10-14	174	30	30	30	30
	28	04-06	10-27	204	30	30	30	30
	24	03-25	11-11	231	30	30	30	30
	20	03-12	11-28	261	30	30	30	30
	16	03-04	12-09	280	30	30	30	30
LONG BRANCH 2 N	32	04-13	10-29	199	30	30	30	30
	28	03-27	11-14	232	30	30	30	30
	24	03-18	11-26	253	30	30	30	30
	20	03-06	12-06	275	30	30	30	30
	16	02-26	12-13	290	30	30	30	29
LONG VALLEY	32	05-08	10-01	146	28	28	29	29
	28	04-23	10-14	174	27	27	28	28
	24	04-04	10-29	208	27	27	28	28
	20	03-21	11-11	235	26	26	28	28
	16	03-14	11-28	259	26	26	27	27
MOORESTOWN	32	04-16	10-19	186	30	30	30	30
	28	03-30	11-04	219	30	30	30	30
	24	03-17	11-19	247	30	30	29	29
	20	03-08	12-05	272	30	30	29	29
	16	02-26	12-14	291	30	30	29	26
MORRIS PLAINS 1 W	32	05-09	10-09	153	15	15	16	16
	28	04-14	10-21	190	15	15	16	16
	24	03-31	11-08	222	15	15	16	16
	20	03-20	11-18	243	15	15	16	16
	16	03-09	12-06	272	15	15	16	16
NEWARK WB AIRPORT	32	04-03	11-08	219	10	10	10	10
	28	03-23	11-21	243	10	10	10	10
	24	03-18	12-01	258	10	10	10	10
	20	03-03	12-12	284	10	10	10	10
	16	02-15	12-16	304	10	10	10	10
NEW BRUNSWICK EXP STA	32	04-18	10-19	184	30	30	30	30
	28	04-04	11-03	213	30	30	30	30
	24	03-21	11-18	242	30	30	30	30
	20	03-12	12-02	265	30	30	30	30
	16	03-05	12-11	281	30	30	30	29

FREEZE DATA

STATION	Freeze threshold temperature	Mean date of last Spring occurrence	Mean date of first Fall occurrence	Mean No. of days between dates	Years of record Spring	No. of occurrences in Spring	Years of record Fall	No. of occurrences in Fall	STATION	Freeze threshold temperature	Mean date of last Spring occurrence	Mean date of first Fall occurrence	Mean No. of days between dates	Years of record Spring	No. of occurrences in Spring	Years of record Fall	No. of occurrences in Fall
NEWTON	32	05-04	10-05	154	30	30	29	29	SANDY HOOK LB STA	32	03-29	11-21	237	28	28	27	27
	28	04-20	10-20	183	30	30	29	29		28	03-22	11-29	252	29	29	27	26
	24	04-06	11-03	211	30	30	30	30		24	03-12	12-09	272	29	29	27	26
	20	03-23	11-13	235	30	30	30	30		20	03-04	12-13	284	29	29	27	23
	16	03-15	11-30	260	30	30	30	30		16	02-17	12-21	307	29	27	27	19
PATERSON	32	04-13	10-27	197	29	29	28	28	SHILOH	32	04-15	10-25	193	29	29	27	27
	28	03-30	11-12	227	28	28	28	28		28	03-30	11-09	224	28	28	28	28
	24	03-20	11-22	247	27	27	28	28		24	03-19	11-25	251	29	29	28	28
	20	03-12	12-02	265	27	27	28	28		20	03-06	12-04	273	28	28	28	28
	16	03-05	12-10	280	27	27	27	26		16	02-26	12-14	291	27	27	28	25
PEMBERTON 3 E	32	04-24	10-15	174	30	30	30	30	SOMERVILLE	32	04-27	10-13	169	28	28	28	28
	28	04-08	10-29	204	30	30	30	30		28	04-15	10-28	196	28	28	29	29
	24	03-23	11-15	237	30	30	30	30		24	03-30	11-07	222	28	28	29	29
	20	03-13	11-26	258	30	30	30	30		20	03-16	11-22	251	28	28	30	30
	16	03-04	12-09	280	30	30	30	29		16	03-10	12-05	270	29	29	30	30
PHILLIPSBURG	32	04-24	10-16	175	30	30	30	30	SUSSEX 1 SE	32	05-06	10-05	152	28	28	30	30
	28	04-09	10-31	205	30	30	30	30		28	04-23	10-17	177	28	28	30	30
	24	03-21	11-15	239	30	30	30	30		24	04-01	10-30	212	28	28	30	30
	20	03-15	11-28	258	30	30	29	29		20	03-21	11-15	239	28	28	29	29
	16	03-05	12-08	278	30	30	29	28		16	03-12	11-29	262	29	29	29	29
PLAINFIELD	32	04-21	10-17	179	30	30	30	30	TRENTON WB CITY	32	04-04	11-08	218	30	30	30	30
	28	04-04	11-03	213	30	30	30	30		28	03-26	11-21	240	30	30	30	30
	24	03-23	11-15	237	30	30	30	30		24	03-14	11-30	261	30	30	30	30
	20	03-14	11-30	261	30	30	30	30		20	03-06	12-09	278	30	30	30	30
	16	03-03	12-10	282	30	30	30	30		16	02-23	12-15	295	30	30	30	26
PLEASANTVILLE 1 N	32	04-27	10-15	171	28	28	27	27	TUCKERTON	32	04-17	10-20	186	23	23	26	26
	28	04-14	10-29	198	28	28	27	27		28	03-31	11-06	220	23	23	25	25
	24	03-31	11-10	224	28	28	27	27		24	03-18	11-24	251	23	23	25	25
	20	03-17	11-25	253	27	27	27	27		20	03-11	11-30	264	23	23	26	26
	16	03-06	12-04	273	27	27	27	27		16	03-03	12-12	284	22	22	24	21
RIDGEFIELD	32	04-16	10-21	188	30	30	29	29									
	28	03-31	11-07	221	30	30	29	29									
	24	03-20	11-24	249	30	30	29	29									
	20	03-13	12-02	264	30	30	29	29									
	16	03-03	12-12	284	30	30	29	28									

Data in the above table are based on the period 1931-1960, or that portion of this period for which data are available.

⊕ When the frequency of occurrence in either spring or fall is one year in ten, or less, mean dates are not given.

Means have been adjusted to take into account years of non-occurrence.

A freeze is a numerical substitute for the former term "killing frost" and is the occurrence of a minimum temperature at or below the threshold temperature of 32°, 28°, etc.

Freeze data tabulations in greater detail are available and can be reproduced at cost.

*NORMALS BY CLIMATOLOGICAL DIVISIONS

Taken from "Climatography of the United States No. 81-4, Decennial Census of U. S. Climate"

	TEMPERATURE (°F)													PRECIPITATION (In.)												
STATIONS (By Divisions)	Jan.	Feb.	Mar.	Apr.	May	June	July	Aug.	Sept.	Oct.	Nov.	Dec.	Annual	Jan.	Feb.	Mar.	Apr.	May	June	July	Aug.	Sept.	Oct.	Nov.	Dec.	Annual
NORTHERN																										
BELVIDERE	29.9	31.3	39.3	51.1	62.2	70.6	75.1	72.9	66.0	54.9	42.8	31.9	52.3	3.63	3.12	4.08	4.14	4.03	4.44	5.30	4.99	3.87	3.42	3.86	3.66	48.54
BOONTON 1 SE	28.9	29.5	37.3	49.0	59.1	67.9	72.8	70.8	63.5	53.3	42.5	31.5	50.5	3.24	2.73	3.86	3.86	4.16	3.89	4.24	4.93	4.35	3.40	4.02	3.42	46.10
CANOE BROOK	29.0	29.6	37.6	48.7	58.9	68.0	73.0	71.1	64.2	53.7	43.0	31.2	50.7	3.66	2.96	4.30	4.02	4.15	4.03	4.55	5.23	4.32	3.41	4.10	3.65	48.38
CHARLOTTEBURG	28.6	29.1	36.6	48.0	58.4	66.5	70.9	69.2	62.4	52.3	42.0	30.8	49.6	3.67	3.17	4.47	4.21	4.15	4.24	4.73	4.96	4.41	3.84	4.49	4.12	50.46
CHATHAM	*	*	*	*	*	*	*	*	*	*	*	*	*	3.72	3.17	4.42	4.12	4.17	4.01	4.58	5.25	4.39	3.37	4.19	3.96	49.35
ELIZABETH	32.8	33.3	40.9	51.9	62.2	70.8	75.8	74.1	67.2	56.7	45.7	34.7	53.8	3.97	3.28	4.49	3.95	4.06	3.76	4.51	4.97	4.34	3.53	3.88	3.60	48.34
FLEMINGTON 1 NE	31.5	32.3	40.1	51.2	61.8	70.6	75.4	73.4	66.6	55.8	44.3	33.3	53.0	3.32	2.78	3.99	3.80	4.01	3.80	4.52	5.02	3.59	3.32	3.74	3.42	45.31
JERSEY CITY	32.5	32.7	39.8	50.5	61.2	70.1	75.3	73.8	66.9	56.7	45.7	34.8	53.3	3.37	2.90	4.03	3.59	3.75	3.45	4.17	4.62	3.81	3.18	3.42	3.29	43.58
LAMBERTVILLE	31.7	32.5	39.8	51.1	61.6	70.1	74.8	73.0	66.2	55.5	44.4	33.7	52.9	3.34	2.72	4.06	3.84	3.84	4.08	4.51	5.14	3.29	3.00	3.71	3.38	44.11
LAYTON 3 NW	26.9	27.8	35.9	47.7	58.3	67.2	71.6	69.7	62.3	51.7	40.2	28.8	49.0	2.87	2.49	3.47	3.91	3.94	4.09	4.90	4.42	4.08	3.40	3.64	2.92	44.13
LITTLE FALLS	31.1	31.7	39.3	50.7	61.2	69.7	74.8	72.8	65.8	55.4	44.6	33.6	52.6	3.65	3.07	4.30	4.26	4.33	4.25	4.54	5.09	4.55	3.70	4.28	3.90	49.92
LONG VALLEY														3.54	3.02	4.28	3.97	4.17	4.13	5.20	5.06	3.92	3.57	4.25	3.79	48.90
NEWARK WB AIRPORT	33.3	33.7	41.5	52.3	62.5	72.3	77.3	75.4	68.3	57.6	45.9	35.3	54.6	3.33	2.80	4.09	3.51	3.65	3.44	3.67	4.43	3.76	3.11	3.37	3.22	42.38
NEW MILFORD														3.44	2.81	3.99	3.84	3.85	3.76	4.38	4.45	3.75	3.19	3.66	3.46	44.58
NEWTON	27.5	28.1	36.4	48.3	59.0	67.4	72.2	70.1	63.0	52.5	41.2	29.7	49.6	3.04	2.57	3.45	3.83	3.95	4.30	4.87	4.54	3.99	3.42	3.63	3.20	44.79
PATERSON	31.6	32.1	39.5	51.1	61.9	70.7	75.7	73.6	66.4	56.1	45.1	34.1	53.2	3.60	3.15	4.45	4.07	4.30	4.21	4.69	4.87	4.21	3.47	4.04	3.83	48.83
PHILLIPSBURG	30.5	31.4	39.2	50.7	61.4	70.0	74.3	72.4	65.5	54.5	43.1	32.7	52.2	3.68	3.06	4.09	4.07	4.15	4.26	5.29	5.04	4.00	3.42	3.67	3.64	48.37
PLAINFIELD	31.9	32.5	39.9	50.9	61.2	69.9	74.9	73.1	66.2	55.9	44.8	34.0	52.9	3.61	3.02	4.31	3.87	4.14	4.33	4.61	5.19	4.33	3.46	3.83	3.59	48.29
SOMERVILLE	31.3	32.0	39.5	50.6	61.1	69.8	74.6	72.8	66.0	55.3	43.9	33.1	52.5	3.27	2.75	3.97	3.65	3.81	3.88	4.74	5.01	3.77	3.35	3.66	3.33	45.19
SUSSEX 1 SE	27.7	28.9	37.0	49.1	59.9	68.5	72.9	70.8	63.5	53.2	42.0	30.2	50.3	3.37	2.72	3.40	3.95	3.77	4.09	5.07	5.04	3.92	3.28	3.81	3.36	45.78
WOODCLIFF LAKE	*	*	*	*	*	*	*	*	*	*	*	*	*	3.57	2.89	4.13	3.98	3.95	3.52	4.44	4.71	3.93	3.28	3.87	3.59	45.86
DIVISION	30.2	30.9	38.6	50.0	60.5	69.2	74.0	72.1	65.0	54.5	43.5	32.4	51.7	3.47	2.92	4.10	3.91	4.07	4.01	4.65	4.91	4.06	3.42	3.88	3.56	46.96
SOUTHERN																										
ATLANTIC CITY WB AP	34.8	34.7	41.1	51.0	61.3	70.0	75.1	73.7	67.2	57.2	46.7	36.6	54.1	3.56	3.13	3.91	3.41	3.51	2.83	3.72	4.90	3.31	3.20	3.66	3.22	42.36
BELLEPLAIN ST FOREST														3.47	3.18	4.42	3.47	3.83	3.39	4.84	5.75	3.86	3.69	4.03	3.65	47.58
BURLINGTON	34.2	34.9	42.2	53.3	64.2	72.8	77.2	75.5	68.0	58.0	46.6	36.0	55.3	3.33	2.90	4.34	3.76	3.65	4.24	4.84	4.84	3.65	3.35	3.66	3.06	44.16
FREEHOLD	32.7	33.0	40.0	50.8	61.6	70.2	74.9	73.1	66.5	56.1	45.5	34.7	53.3	3.58	3.14	4.24	3.58	4.03	3.68	4.24	6.76	3.78	3.59	3.96	3.45	46.03
HAMMONTON 2 NNE	34.2	34.5	41.9	52.3	62.7	71.6	76.1	74.2	67.7	57.2	46.7	37.1	54.7	3.57	3.25	4.26	3.60	3.92	3.95	4.56	5.64	3.77	3.69	3.74	3.73	47.68
HIGHTSTOWN 1 N	32.4	32.9	40.2	50.9	61.4	69.9	74.8	72.9	66.2	55.9	45.2	34.3	53.1	3.22	2.73	3.82	3.38	3.72	3.83	4.46	4.52	3.99	3.32	3.39	3.03	43.41
INDIAN MILLS 2 W	33.3	33.8	40.7	51.4	62.0	70.2	74.7	72.9	66.3	55.9	45.2	34.8	53.4	3.62	3.11	4.28	3.42	3.88	3.90	4.27	5.46	3.60	3.41	3.71	3.23	45.89
LAURELTON 1 E	33.2	33.3	40.2	50.5	61.1	69.6	74.4	72.4	65.8	55.5	45.2	35.0	53.0	3.76	3.36	4.34	3.76	3.87	3.40	4.48	4.85	3.75	4.00	4.16	3.32	47.05
MOORESTOWN	33.0	33.4	40.7	51.4	61.7	70.2	75.1	73.1	66.4	55.9	45.0	34.8	53.4	3.12	2.73	3.81	3.44	4.07	3.56	4.17	4.75	3.75	3.06	3.61	3.17	43.98
NEW BRUNSWICK EXP STA	32.3	32.8	40.2	51.1	61.6	70.1	75.0	73.2	66.5	56.1	45.3	34.2	53.2	3.34	2.77	3.75	3.48	3.75	3.63	4.53	4.70	4.06	3.16	3.64	3.17	43.98
PEMBERTON 3 E	33.9	34.3	41.6	52.3	62.7	71.0	75.5	73.8	67.4	56.8	46.3	35.6	54.3	3.30	2.88	3.85	3.45	3.76	3.70	4.65	5.05	5.77	3.27	3.51	3.08	44.27
SHILOH	34.3	34.9	42.2	52.7	63.0	71.7	76.4	74.6	68.0	57.3	46.3	36.1	54.8	3.48	2.89	3.76	3.11	3.96	3.41	4.29	4.95	3.94	3.16	3.73	3.06	43.74
TRENTON WB CITY	33.1	33.4	40.7	51.7	62.3	71.0	76.0	73.9	67.1	56.8	45.8	35.2	53.9	3.10	2.59	3.84	3.21	3.62	3.60	4.18	4.77	3.50	2.84	3.12	2.87	41.28
DIVISION	33.4	33.7	40.9	51.4	62.1	70.7	75.4	73.6	67.0	56.5	45.7	35.3	53.8	3.45	2.99	4.08	3.45	3.78	3.62	4.34	5.06	3.79	3.35	3.74	3.26	44.91
COASTAL																										
ATLANTIC CITY	36.0	35.7	41.1	49.9	59.5	69.0	74.2	73.7	68.4	58.7	48.3	38.3	54.4	3.75	3.38	4.01	3.38	3.16	3.04	3.86	5.14	3.40	3.39	3.79	3.48	43.78
LONG BRANCH 2 N	33.0	33.0	39.8	49.6	59.8	69.1	74.4	72.8	66.2	56.2	45.7	35.3	52.9	3.78	3.54	4.37	3.63	3.49	3.41	4.25	5.25	3.92	3.71	3.81	3.63	46.79
DIVISION	34.5	34.3	40.5	50.0	60.0	69.4	74.6	73.6	67.8	58.0	47.4	37.0	53.9	3.52	3.12	3.92	3.32	3.26	3.07	3.88	4.94	3.67	3.30	3.59	3.34	42.93

* Normals for the period 1931-1960. Divisional normals may not be the arithmetical average of individual stations published, since additional data for shorter period stations are used to obtain better areal representation.

	TEMPERATURE													PRECIPITATION												
	Jan.	Feb.	Mar.	Apr.	May	June	July	Aug.	Sept.	Oct.	Nov.	Dec.	Annual	Jan.	Feb.	Mar.	Apr.	May	June	July	Aug.	Sept.	Oct.	Nov.	Dec.	Annual

CONFIDENCE LIMITS

In the absence of trend or record changes, the chances are 9 out of 10 that the true mean will lie in the interval formed by adding and subtracting the values in the following table from the means for any station in the State. Because of the wider variation in mean precipitation, the corresponding monthly means and annual mean must be substituted for "p" in the precipitation table below to obtain mean precipitation confidence limits.

| 1.4 | 1.2 | 1.3 | .8 | .7 | .6 | .5 | .6 | .7 | .8 | .8 | .9 | .4 | $.25\sqrt{p}$ | $.20\sqrt{p}$ | $.25\sqrt{p}$ | $.25\sqrt{p}$ | $.29\sqrt{p}$ | $.30\sqrt{p}$ | $.37\sqrt{p}$ | $.37\sqrt{p}$ | $.47\sqrt{p}$ | $.37\sqrt{p}$ | $.34\sqrt{p}$ | $.25\sqrt{p}$ | $.32\sqrt{p}$ |

COMPARATIVE DATA

Data in the following table are the mean temperature and average precipitation for New Brunswick, New Jersey, for the period 1906-1930 and are included in this publication for comparative purposes:

| 30.7 | 30.7 | 40.2 | 50.0 | 60.2 | 68.8 | 73.7 | 71.6 | 66.1 | 55.1 | 43.7 | 32.9 | 52.0 | 3.33 | 3.30 | 3.27 | 3.70 | 3.83 | 3.77 | 5.16 | 5.17 | 3.20 | 3.54 | 2.63 | 3.53 | 44.43 |

NORMALS, MEANS, AND EXTREMES

ATLANTIC CITY, NEW JERSEY — NATIONAL AVIATION FACILITIES EXPERIMENTAL CENTER

LATITUDE 39° 27' N
LONGITUDE 74° 35' W
ELEVATION (ground) 64 Feet

Month	Normal Daily max	Normal Daily min	Normal Monthly	Extreme Record highest	Year	Extreme Record lowest	Year	Normal degree days	Precip Normal total	Precip Max monthly $	Year	Precip Min monthly $	Year	Precip Max in 24 hrs	Year	Snow Mean total	Snow Max monthly	Year	Snow Max 24 hrs	Year	RH 1:00 A.M.	RH 7:00 A.M.	RH 1:00 P.M.	RH 7:00 P.M.	Wind Mean hourly	Prevailing direction	Fastest Speed	Fastest Dir	Fastest Year	Mean sky cover	Pct. sunshine	Clear	Partly cloudy	Cloudy	Precip .01"+	Snow 1.0"+	Thunderstorms	Heavy fog	Max 90°+	Max 32°-	Min 32°-	Min 0°-
(b)	(b)	(b)	(b)	2		2		(b)	(b)	23		23		23		22	22		22		2	2	2	2	8	5					9	8	8	8	23	22	8	8	2	2	23	2
J	42.9	26.6	34.8	65	1966+	8	1965	936	3.56	7.71	1948	0.26	1955	2.86	1944	5.5	15.9	1961	14.4	1964	73	73	58	69	11.2	WNW				5.9	58	8	8	13	11	2	*	5	0	9	28	2
F	43.3	26.1	34.7	65	1965	-8	1966	848	3.13	5.98	1958	1.46	1958	2.59	1962	3.9	14.5	1958	10.0	1947	74	80	58	70	11.2	WNW				5.9	52	9	8	13	10	2	1	4	0	7	24	1
M	49.7	32.4	41.1	67	1965	13	1965	741	3.91	6.80	1953	0.62	1945	3.37	1952	3.3	13.4	1960	9.5	1958	74	78	51	64	12.2	WNW				5.9	56	9	9	14	11	1	2	4	0	2	21	0
A	60.3	41.7	51.0	85	1965	24	1965	420	3.41	7.95	1952	1.24	1945	4.15	1952	0.4	3.2	1965	3.2	1965	76	76	51	66	12.6	WNW				5.9	50	7	10	13	11	*	3	3	0	0	8	0
M	71.0	51.5	61.3	93	1965	25	1965	133	2.66	5.51	1948	0.40	1957	2.91	1952	0.0	0.0		0.0		77	77	53	66	10.7	S				6.2	58	6	13	12	10	0	4	3	2	0	1	0
J	79.2	60.7	70.0	98	1966+	39	1965	15	2.83	5.73	1951	0.10	1957	2.91	1959	0.0	0.0		0.0		81	85	53	69	9.9	S				6.1	62	8	10	12	9	0	6	3	4	0	0	0
J	83.8	66.3	75.1	104	1966	46	1965	0	3.72	13.09	1959	1.30	1957	6.46	1959	0.0	0.0		0.0		87	82	51	70	9.5	S				6.1	65	8	13	10	9	0	7	4	7	0	0	0
A	82.2	65.1	73.7	95	1966	40	1966	0	4.90	11.02	1948	0.34	1943	6.40	1966	0.0	0.0		0.0		88	86	55	74	9.7	S				6.2	62	8	14	9	9	0	6	5	4	0	0	0
S	76.0	58.4	67.2	92	1966	35	1965	39	3.31	7.50	1943	0.46	1953	3.98	1954	0.0	0.0		0.0		87	87	57	78	9.7	ENE				5.6	60	8	9	13	7	0	4	5	2	0	0	0
O	66.5	47.8	57.2	85	1965	26	1964	251	3.20	8.60	1944	0.15	1963	2.93	1958	0.0	0.0		0.0		83	86	51	71	9.7	W				5.4	65	11	7	10	7	0	2	4	0	0	1	0
N	55.5	37.9	46.7	76	1965	11	1964	549	3.66	7.60	1946	0.72	1946	2.75	1951	T	5.2	1953	3.2	1953	80	83	57	75	10.5	WNW				5.8	58	7	10	14	9	*	1	3	0	3	15	0
D	45.1	28.1	36.6	72	1966	6	1965	880	3.22	6.57	1948	0.62	1955	2.75	1951	2.8	8.6	1960	7.5	1960	83	80	57	75	11.2	WNW				6.0	48	10	7	14	9	1	*	5	0	3	23	0
YR	63.0	45.2	54.1	104	JUL. 1966	-8	JAN. 1965	4812	42.36	13.09	JUL. 1959	0.10	JUN. 1954	6.46	JUL. 1959	16.1	15.9	JAN. 1961	14.4	JAN. 1964	81	80	54	70	11.0	S				5.9	58	105	104	156	112	4	24	44	22	19	120	3

Ø For period November 1964 through the current year.

Means and extremes in the above table are from existing or comparable location(s). Annual extremes have been exceeded at other locations as follows: Lowest temperature -9 in February 1934; maximum monthly precipitation 14.87 in August 1882; minimum monthly precipitation .01 in September 1941; maximum precipitation in 24 hours 9.21 in October 1903; maximum snowfall 27.9 in February 1899; maximum snowfall in 24 hours 18.0 in February 1902. % Based on U.S. Naval Air Station and Weather Bureau Airport Station records. $ Beginning with August 1943.

The prevailing direction for wind in the Normals, Means, and Extremes table is from records through 1963.

NEWARK, NEW JERSEY — NEWARK AIRPORT

LATITUDE 40° 42' N
LONGITUDE 74° 10' W
ELEVATION (ground) 7 Feet

Month	Normal Daily max	Normal Daily min	Normal Monthly	Extreme Record highest	Year	Extreme Record lowest	Year	Normal degree days	Precip Normal total	Precip Max monthly	Year	Precip Min monthly	Year	Precip Max in 24 hrs	Year	Snow Mean total	Snow Max monthly	Year	Snow Max 24 hrs	Year	RH 1:00 A.M.	RH 7:00 A.M.	RH 1:00 P.M.	RH 7:00 P.M.	Wind Mean hourly	Prevailing direction	Fastest Speed	Fastest Dir	Fastest Year	Mean sky cover	Pct. sunshine	Clear	Partly cloudy	Cloudy	Precip .01"+	Snow 1.0"+	Thunderstorms	Heavy fog	Max 90°+	Max 32°-	Min 32°-	Min 0°-
(a)	23		23	23		23		(b)	(b)	13		13		13		25	25		25		23	23	23	23	23	22				21		25	25	25	25	25	25	25	23	23	23	23
J	39.5	25.0	32.3	74	1950	0	1957+	1014	3.33	5.57	1953	0.81	1955	1.78	1962	7.6	22.2	1961	13.7	1961	69	72	58	64	11.4	NE				6.4		8	7	16	11	2	*	3	0	7	25	23
F	40.7	24.7	32.7	76	1949	-7	1943	904	2.80	4.47	1956	1.89	1959	2.45	1961	7.5	23.3	1961	20.0	1961	68	70	56	60	11.5	NW				6.3		8	7	13	10	2	1	2	0	5	22	23
M	48.8	32.1	40.5	89	1945	7	1943	760	4.09	6.29	1954	1.12	1966	2.01	1966	5.3	17.6	1956	17.6	1956	68	65	49	60	11.2	NW				6.1		8	8	14	12	1	1	2	*	1	14	14
A	60.9	41.7	51.3	91	1960+	23	1954	411	3.51	6.41	1958	0.90	1963	2.36	1958	0.6	4.1	1957	4.1	1957	69	65	49	49	11.0	WNW				6.5		8	11	14	12	1	4	1	1	0	1	2
M	72.1	51.9	62.0	98	1962	33	1945	127	3.65	4.86	1964	0.52	1964	2.36	1964	T	0.0		0.0		75	71	51	52	9.4	SW				6.4		8	13	11	10	0	5	2	2	0	0	1
J	81.3	61.2	71.3	102	1952+	43	1945	9	3.44	3.89	1962	0.49	1966	1.52	1959	0.0	0.0		0.0		78	73	52	62	8.9	SW				5.9		7	13	11	9	0	6	1	6	0	0	0
J	86.1	66.5	76.3	105	1949	52	1945+	0	3.67	7.95	1955	0.89	1966	3.15	1961	0.0	0.0		0.0		78	74	51	63	8.7	SW				6.0		7	13	11	9	0	5	1	5	0	0	0
A	83.8	64.9	74.4	103	1948	51	1948	0	4.43	11.84	1955	0.50	1955	4.17	1955	0.0	0.0		0.0		80	78	53	66	9.0	SW				5.6		10	12	9	8	0	5	1	3	0	0	0
S	77.0	57.6	67.3	105	1953+	35	1947	39	3.76	7.86	1966	1.30	1964	4.71	1966	0.0	0.0		0.0		80	80	53	67	9.4	SW				5.2		12	11	8	8	0	3	1	2	0	0	0
O	66.2	47.0	56.6	92	1953	30	1948	276	3.11	6.70	1955	0.21	1963	2.65	1966	T	0.3	1952	0.3	1952	78	76	55	67	10.1	SW				6.0		12	8	13	10	0	1	2	0	0	*	8
N	53.5	37.3	45.4	85	1950	15	1955	588	3.37	5.68	1963	1.48	1965	2.09	1963	0.4	2.9	1945	2.9	1945	75	73	58	65	10.6	SW				6.0		9	8	13	11	2	*	2	*	0	6	22
D	42.0	27.4	34.7	72	1946	-1	1946	939	3.22	5.74	1957	0.27	1955	1.83	1960	8.3	29.1	1947	26.0	1947	71	73	58	65	10.2	SW				6.2		9	8	14	10	2	*	2	0	6	22	*
YR	62.7	44.8	53.7	105	SEP. 1953+	-7	FEB. 1943	5067	42.38	11.84	AUG. 1955	0.21	OCT. 1963	4.71	SEP. 1966	29.7	29.1	DEC. 1947	26.0	DEC. 1947	74	73	53	63	10.2	SW				6.1		99	114	152	122	8	25	22	25	18	94	*

(b) Climatological standard normals (1931-1960) Revised December 1966.

Means and extremes in the above table are from existing or comparable location(s). Annual extremes have been exceeded at other locations as follows: Lowest temperature -14 in February 1934; maximum monthly precipitation 22.48 in August 1843; minimum monthly precipitation 0.07 in June 1949.

NORMALS, MEANS, AND EXTREMES

LATITUDE 40° 13' N
LONGITUDE 74° 46' W
ELEVATION (ground) 56 Feet

Month	Temperature Normal Daily max	Daily min	Monthly	Extremes Record highest	Year	Record lowest	Year	Normal degree days	Precip. Normal total	Precip. Max monthly	Year	Precip. Min monthly	Year	Precip. Max 24 hrs	Year	Snow Mean total	Snow Max monthly	Year	Snow Max 24 hrs	Year	Mean sky cover	Pct. poss. sunshine	Days Clear	Partly cloudy	Cloudy	Precip .01"+	Snow 1.0"+	Thunderstorms	Heavy fog	Wind mean hourly	Prevailing dir	Fastest speed	Dir.	Year	Max 90°+	Max 32°-	Min 32°-	Min 0°-
(record yrs)	34 (b)	34 (b)	34 (b)	34	34	34	34	34 (b)	34 (b)	34		34		34		34	34		34		23	34	34	34	34	34	34	32		32	32	34	34	34	34	34	34	34
J	40.0	26.2	33.1	72	1950	-3	1936	989	3.10	6.00	1936	0.52	1955	2.03	1936	5.8	16.1	1961	10.1	1961	6.3	51	8	8	15	12	2	*	2	9.8	NW	48	NW	1958	0	8	24	*
F	40.9	25.9	33.4	73	1949	-14	1934	885	2.59	5.56	1939	1.28	1943	2.45	1966	6.7	23.1	1934	13.0	1934	6.1	55	8	8	12	10	2	*	1	10.2	NW	49	W	1960+	0	5	21	*
M	48.8	32.5	40.7	86	1945	8	1943	753	3.84	7.53	1953	1.17	1966	2.55	1953	4.4	21.5	1958	21.5	1958	6.0	56	8	9	14	12	1	1	1	10.7	NW	43	NW	1958	0	1	14	0
A	61.3	42.0	51.7	91	1941	24	1954	399	3.21	5.93	1952	0.83	1963	2.46	1952	0.4	4.2	1956	4.2	1956	5.9	58	9	9	12	12	*	2	1	10.4	S	43	S	1957	*	0	4	0
M	72.3	52.3	62.3	96	1962	34	1947	121	3.62	8.03	1948	0.25	1964	2.68	1966	T	T	1963	0.0	1963	6.3	62	7	11	13	12	0	4	1	9.0	S	37	NW	1953+	1	0	*	0
J	80.7	61.3	71.0	100	1952+	43	1938	12	3.60	9.00	1938	0.06	1949	4.79	1938	0.0	0.0		0.0		5.9	65	7	12	11	11	0	6	1	8.4	S	36	NW	1955	5	0	0	0
J	85.2	66.7	76.0	106	1936	53	1963+	0	4.18	10.19	1944	0.37	1944	4.85	1964	0.0	0.0		0.0		5.8	67	8	12	11	10	0	7	1	7.8	S	46	SW	1945+	7	0	0	0
A	82.8	65.0	73.9	100	1955	48	1940	0	4.77	14.10	1955	0.47	1964	4.76	1955	0.0	0.0		0.0		5.8	64	8	12	11	10	0	6	2	7.6	S	41	N	1947	4	0	0	0
S	76.2	57.7	67.0	100	1953	36	1947	57	3.50	10.49	1934	0.19	1941	4.01	1960	0.0	0.0		0.0		5.6	62	10	9	11	8	0	3	2	7.5	S	56	NW	1960	1	0	0	0
O	65.7	47.0	56.3	94	1941	27	1940+	264	2.84	6.77	1941	0.05	1963	3.46	1966	0.1	1.6	1962	1.6	1962	5.0	62	12	9	10	8	0	1	2	8.3	S	60	E	1951	*	0	*	0
N	53.6	38.0	45.8	83	1950	15	1938	576	3.16	6.97	1936	0.75	1936	2.37	1963	1.0	13.0	1938	7.7	1938	6.0	54	9	9	12	9	*	1	2	9.2	NW	64	E	1950	0	*	7	*
D	42.2	28.2	35.2	72	1966	-2	1942+	924	2.87	6.08	1948	0.19	1955	2.67	1948	4.9	21.5	1960	16.6	1960	6.1	50	8	9	14	10	2	*	2	9.3	NW	48	NW	1962	0	6	21	*
YR	62.5	45.3	53.9	106 JUL. 1936		-14 FEB. 1934		4980	41.28	14.10 AUG. 1955		0.05 OCT. 1963		4.85 JUL. 1964		23.3	23.1 FEB. 1934		16.6 DEC. 1960		5.9	60	101	116	148	121	7	33		9.0	S	64	E	NOV. 1950	18	20	90	*

Means and extremes in the above table are from the existing or comparable location(s). Annual extremes have been exceeded at other locations as follows:
Maximum monthly precipitation 15.22 in July 1880; maximum monthly snowfall 34.0 in February 1899; fastest mile wind 73 in July 1914.

REFERENCE NOTES APPLYING TO ALL "NORMALS, MEANS, AND EXTREMES" TABLES.

(a) Length of record, years.
(b) Climatological standard normals (1931-1960).
* Less than one half.
+ Also on earlier dates, months or years.
T Trace, an amount too small to measure.
Below-zero temperatures are preceded by a minus sign.
The prevailing direction for wind in the Normals, Means, and Extremes table is from records through 1963.
To 8 compass points only.

Unless otherwise indicated, dimensional units used in this bulletin are: temperature in degrees F.; precipitation, including snowfall, in inches; wind movement in miles per hour; and relative humidity in percent. Degree day totals are the sum of the negative departures of average daily temperatures from 65° F. Sleet was included in snowfall totals beginning with July 1948. Heavy fog reduces visibility to 1/4 mile or less.

Sky cover is expressed in a range of 0 for no clouds or obscuring phenomena to 10 for complete sky cover. The number of clear days is based on average cloudiness 0-3; partly cloudy days 4-7; and cloudy days 8-10 tenths.

& Figures instead of letters in a direction column indicate direction in tens of degrees from true North; i.e., 09 - East, 18 - South, 27 - West, 36 - North, and 00 - Calm. Resultant wind is the vector sum of wind directions and speeds divided by the number of observations. If figures appear in the direction column under "Fastest mile" the corresponding speeds are fastest observed 1-minute values.

¢ Temperature extremes and relative humidity means in the Normals, Means, and Extremes table are for comparable locations through 1964. Summaries for the present location of temperature sensors will be published when more data are accumulated.

Mean Annual Precipitation, Inches

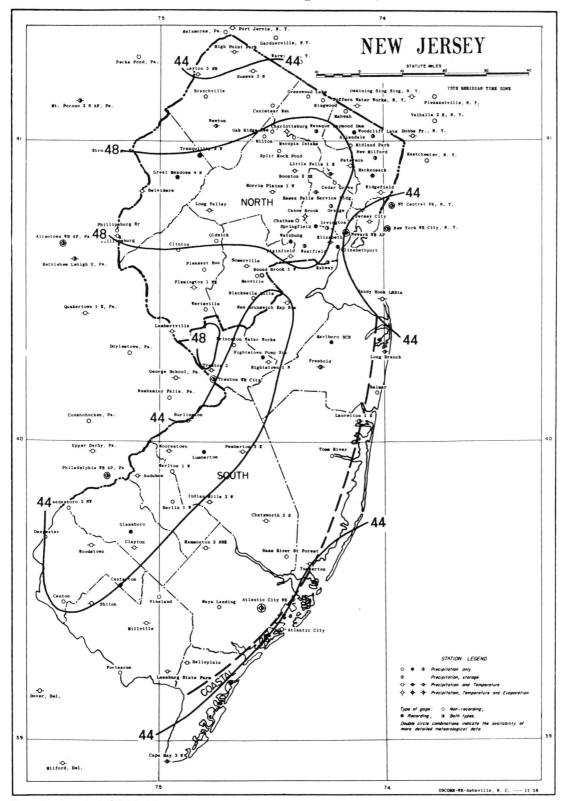

Based on period 1931-55

Isolines are drawn through points of approximately equal value. Caution should be used in interpolating on these maps.

Mean Maximum Temperature (°F.), January

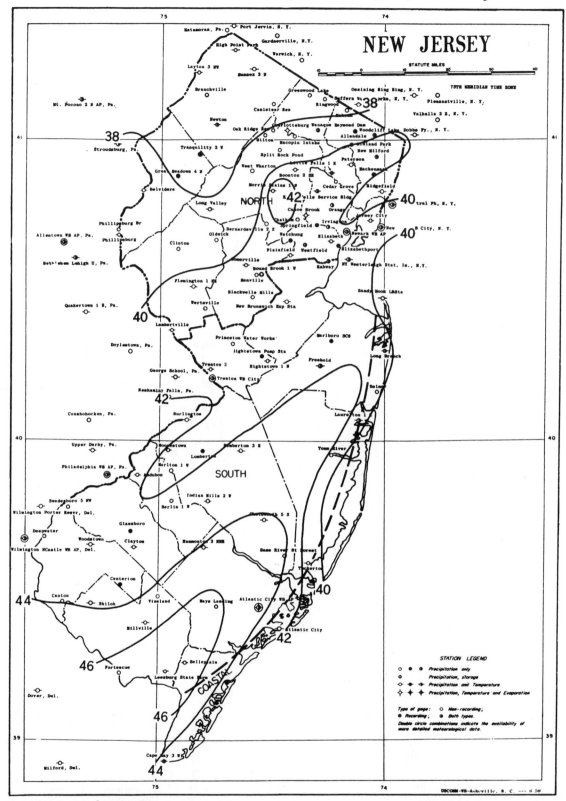

Based on period 1931-52

Isolines are drawn through points of approximately equal value. Caution should be used in interpolating on these maps,

Mean Minimum Temperature (°F.), January

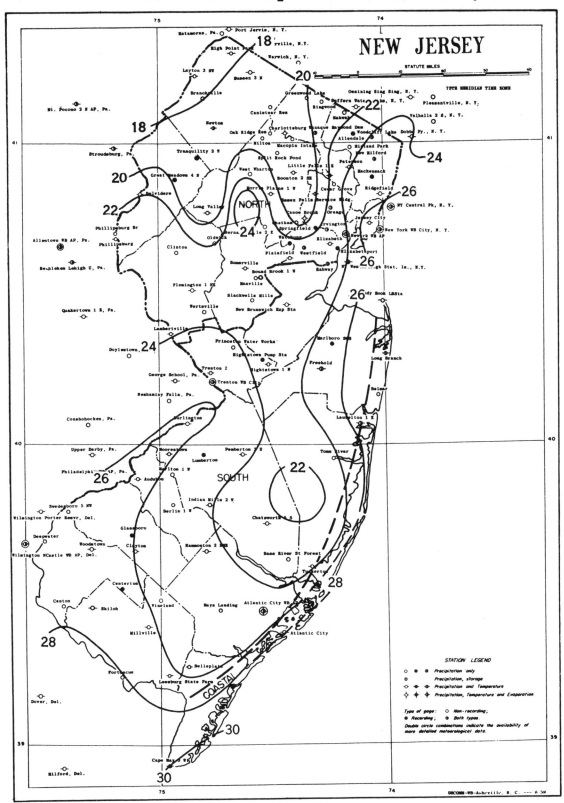

Based on period 1931-52

Isolines are drawn through points of approximately equal value. Caution should be used in interpolating on these maps.

Mean Maximum Temperature (°F.), July

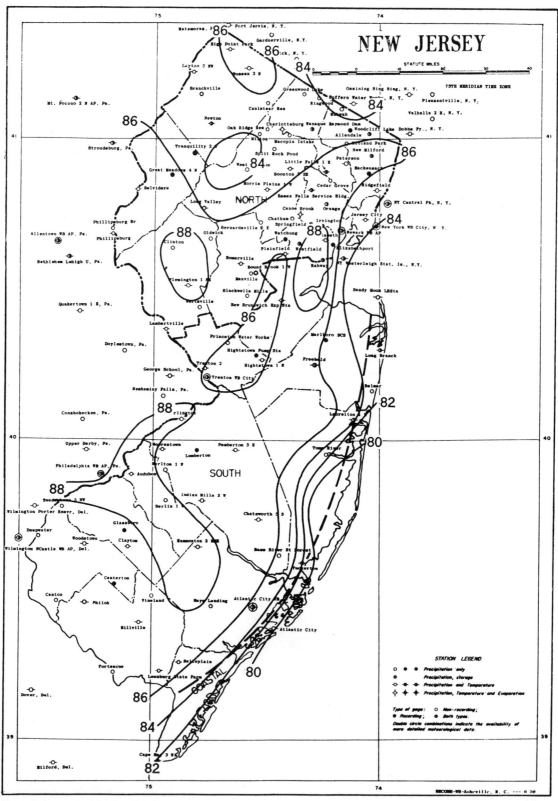

Based on period 1931-52

Isolines are drawn through points of approximately equal value. Caution should be used in interpolating on these maps.

Mean Minimum Temperature (°F.), July

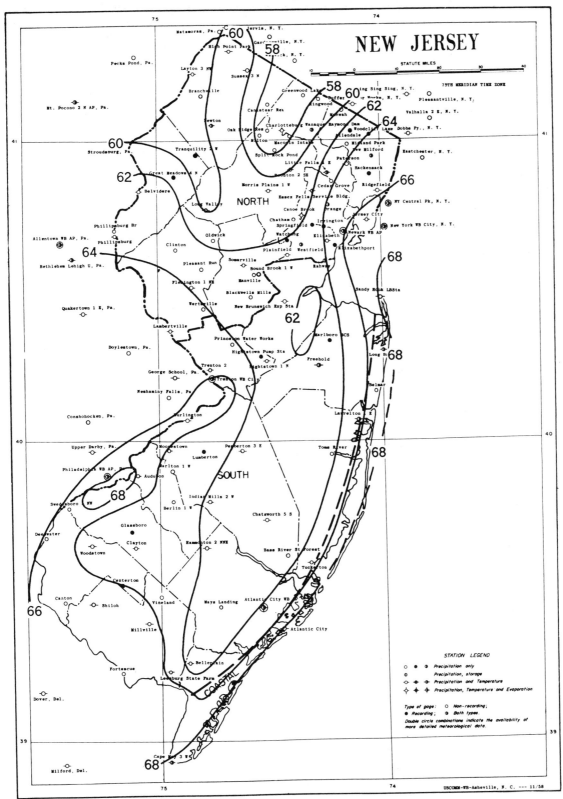

Based on period 1931-52

Isolines are drawn through points of approximately equal value. Caution should be used in interpolating on these maps.

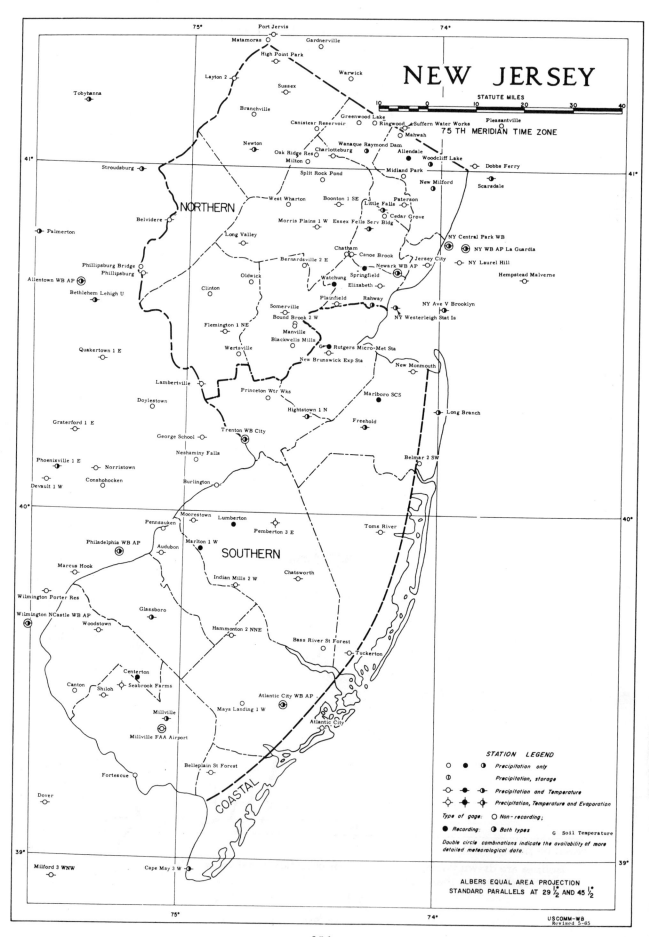

NEW JERSEY

STATUTE MILES

75 TH MERIDIAN TIME ZONE

NORTHERN

SOUTHERN

COASTAL

STATION LEGEND

○ ● ◐ Precipitation only
◉ Precipitation, storage
-○- -●- -◐- Precipitation and Temperature
-◇- -◆- -✦- Precipitation, Temperature and Evaporation

Type of gage: ○ Non-recording;
● Recording: ◐ Both types   G Soil Temperature

Double circle combinations indicate the availability of more detailed meteorological data.

ALBERS EQUAL AREA PROJECTION
STANDARD PARALLELS AT 29½ AND 45½

USCOMM-WB
Revised 5-65

Port Jervis
Matamoras
Gardnerville
High Point Park
Warwick
Layton 2
Sussex
Tobyhanna
Branchville
Greenwood Lake
Canistear Reservoir
Ringwood
Suffern Water Works
Pleasantville
Mahwah
Newton
Wanaque Raymond Dam
Allendale
Oak Ridge Res
Charlotteburg
Woodcliff Lake
Stroudsburg
Milton
Split Rock Pond
Midland Park
Dobbs Ferry
West Wharton
Boonton 1 SE
New Milford
Scarsdale
Belvidere
Paterson
Little Falls
Cedar Grove
Morris Plains 1 W
Essex Fells Serv Bldg
Palmerton
Long Valley
NY Central Park WB
Chatham
NY WB AP La Guardia
Phillipsburg Bridge
Bernardsville 2 E
Canoe Brook
Jersey City
NY Laurel Hill
Phillipsburg
Oldwick
Watchung
Springfield
Allentown WB AP
Clinton
Elizabeth
Hempstead Malverne
Bethlehem Lehigh U
Plainfield
Rahway
Somerville
Bound Brook 2 W
NY Ave V Brooklyn
Flemington 1 NE
Manville
NY Westerleigh Stat Is
Quakertown 1 E
Blackwells Mills
Wertsville
Rutgers Micro-Met Sta
New Brunswick Exp Sta
New Monmouth
Lambertville
Princeton Wtr Wks
Marlboro SCS
Doylestown
Hightstown 1 N
Long Branch
Freehold
Graterford 1 E
Trenton WB City
George School
Neshaminy Falls
Belmar 2 SW
Phoenixville 1 E
Norristown
Burlington
Conshohocken
Devault 1 W
Moorestown
Lumberton
Pennsauken
Toms River
Philadelphia WB AP
Pemberton 3 E
Audubon
Marlton 1 W
Marcus Hook
Indian Mills 2 W
Chatsworth
Wilmington Porter Res
Glassboro
Wilmington NCastle WB AP
Woodstown
Hammonton 2 NNE
Bass River St Forest
Tuckerton
Centerton
Canton
Seabrook Farms
Shiloh
Atlantic City WB AP
Millville
Mays Landing 1 W
Millville FAA Airport
Atlantic City
Belleplain St Forest
Fortescue
Dover
Milford 3 WNW
Cape May 3 W

THE CLIMATE OF
NEW YORK

by
Dr. A. Boyd Pack

June 1972

PHYSICAL DESCRIPTION. New York State contains 49,576 square miles, inclusive of 1,637 square miles of inland water, but exclusive of the boundary-water areas of Long Island Sound, New York Harbor, Lake Ontario, and Lake Erie. The major portion of the State lies generally between latitudes 42° and 45° N. and between longitudes 73° 30' and 79° 45' W. However, in the extreme southeast, a triangular portion extends southward to about latitude 40° 30' N., while Long Island lies eastward to about longitude 72° W.

The principal highland regions of the State are the Adirondacks in the northeast and the Appalachian Plateau (Southern Plateau) in the south. The latter Plateau is subdivided by the deep channel of Seneca Lake, which extends from the Lake Plain of Lake Ontario southward to the Chemung River Valley, into the Western and Eastern Plateaus. The former extends from the eastern Finger Lakes across the hills of southwestern New York to the narrow Lake Plain bordering Lake Erie; the latter extends from the eastern Finger Lakes to the Hudson River Valley and includes the Catskill Mountains.

A minor highland region occurs in southeastern New York where the Hudson River has cut a Valley between the Palisades on the west, near the New Jersey border, and the Taconic Mountains on the east, along the Connecticut and Massachusetts border. Just west of the Adirondacks and the upper Black River Valley in

Lewis County is another minor highland known as Tug Hill.

Much of the eastern border of the State consists of a long, narrow lowland region which is occupied by Lake Champlain, Lake George, and the middle and lower portions of the Hudson Valley. Another lowland region, the Great Lakes Plain, on the northern and western boundaries of the State adjoins the St. Lawrence River, Lake Ontario, and Lake Erie. This latter region is widest south of the eastern end of Lake Ontario, but does narrow to a width of less than 5 miles in the western portion of the State. A third lowland region, which contains Lake Oneida and a deep valley cut by the Mohawk River, connects the Hudson Valley and the Great Lakes Plain. Long Island, which is a part of the Atlantic Coastal Plain, comprises the fourth lowland region of the State.

Approximately 40 percent of New York State has an elevation of more than 1,000 feet above sea level. In northwestern Essex County, confined to an area of 500 or 600 square miles, are a number of peaks with an elevation of between 4,000 to 5,000 feet. The highest point, Mount Marcy, reaches a height of 5,344 feet above sea level. Nearby Mount MacIntyre ranges to a height of 5,112 feet. With the exception of the Blue Ridge of North Carolina and the White Mountains of New Hampshire, these are the loftiest mountains in eastern North America.

The Appalachian Plateau merges variously

into the Great Lakes Plain of western New York with gradual- to steep-sloping terrain. This Plateau is penetrated by the valleys of the Finger Lakes which, resembling the appearance of outstretched fingers on the hand, extend southward from the Great Lakes Plain. The major Finger Lakes going from west to east, are Canandaigua, Keuka, Seneca, Cayuga, and Skaneateles. Other prominent lakes in the State include Lake George in the central part of the eastern boundary, Lake Oneida in central New York between Syracuse and Rome, and Chautauqua Lake in the extreme southwest. Sacandaga and Pepacton Reservoirs are sizeable manmade bodies of water in the eastern portion of the State. Innumerable smaller lakes and ponds dot the landscape, with more than 1,500 in the Adirondack region alone.

Rivers of New York State may be divided into those that are tributary to the Great Lakes and St. Lawrence River and those that flow in a general southward direction. The first group includes rivers such as the Genesee, Oswego, Black, Oswegatchie, Grass, Raquette, Saranac, and Ausable. The Chemung, Susquehanna, Delaware, and Hudson River systems which are part of the Atlantic slope drainage and the Allegheny River which is part of the Ohio Basin drainage comprise the second group.

GENERAL CLIMATIC FEATURES. The climate of New York State is broadly representative of the humid continental type which prevails in the Northeastern United States, but its diversity is not usually encountered within an area of comparable size. The geographical position of the State and the usual course of air masses, governed by the large-scale patterns of atmospheric circulation, provide general climatic controls. Differences in latitude, character of the topography, and proximity to large bodies of water have pronounced effects on the climate.

The planetary atmospheric circulation brings a great variety of air masses to New York State. Masses of cold, dry air frequently arrive from the northern interior of the continent. Prevailing winds from the south and southwest transport warm, humid air which has been conditioned by the Gulf of Mexico and adjacent subtropical waters. These two air masses provide the dominant continental characteristics of the climate. The third great air mass flows inland from the North Atlantic Ocean and produces cool, cloudy, and damp weather conditions. This maritime influence is important to New York's climatic regime, especially in the southeastern portion of the State, but it is secondary to that of the more prevalent air mass flow from the continent.

Nearly all storm and frontal systems moving eastward across the continent pass through or in close proximity to New York State. Storm systems often move northward along the Atlantic coast and have an important influence on the weather and climate of Long Island and the lower Hudson Valley. Frequently, areas deep in the interior of the State feel the effects of such coastal storms.

Lengthy periods of either abnormally cold or warm weather result from the movement of great high pressure (anticyclonic) systems into and through the Eastern United States. Cold winter temperatures prevail over New York whenever Arctic air masses, under high barometric pressure, flow southward from central Canada or from Hudson Bay. High pressure systems often move just off the Atlantic coast, become more or less stagnant for several days, and then a persistent air flow from the southwest or south affects the State. This circulation brings the very warm, often humid weather of the summer season and the mild, more pleasant temperatures during the fall, winter, and spring seasons.

TEMPERATURE. Many atmospheric and physiographic controls on the climate result in a considerable variation of temperature conditions over New York State. The average annual mean temperature ranges from about 40° in the Adirondacks to near 55° in the New York City area. In January, the average mean temperature is approximately 16° in the Adirondacks and St. Lawrence Valley, but increases to about 26° along Lake Erie and in the lower Hudson Valley and to 31° on Long Island. The highest temperature of record in New York State is 107°, observed at Lewiston, Elmira, Poughkeepsie, and New York City. The record coldest temperature is -52° at Stillwater Reservoir (northern Herkimer County). Some 30 communities have recorded temperatures of -40° or colder, most of them occurring in the northern one-half of the State and the remainder in the Western Plateau Division and in localities just south of the Mohawk Valley.

The winters are long and cold in the Plateau Divisions of the State. In the majority of winter seasons, a temperature of -25° or lower can be expected in the northern highlands (Northern Plateau) and -15° or colder in the southwestern and east-central highlands (Southern Plateau). The Adirondack region records from 35 to 45 days with below zero temperatures in normal to severe winters, with a somewhat fewer number of such days occurring near Lake Champlain and the St. Lawrence River. In the Southern Plateau and in the upper Hudson Valley Division, below zero minimums are observed on about 15 days in most winters and on more than 25 days in notably cold seasons.

Winter temperatures are moderated considerably in the Great Lakes Plain of western New York. The moderating influence of Lakes Erie and Ontario is comparable to that produced by the Atlantic Ocean in the southern portion of the Hudson Valley. In both regions, the coldest temperature in most winters will range between 0° and -10°. Long Island and New York City experience below zero minimums in 2 or 3 winters out of 10, with the low temperature generally near -5°.

The summer climate is cool in the Adirondacks, Catskills, and higher elevations of the Southern Plateau. The New York City area and lower portions of the Hudson Valley have rather warm summers by comparison, with some periods of high, uncomfortable humidity. The remainder of New York State enjoys pleasantly warm summers, marred by only occasional, brief intervals of sultry conditions. Summer daytime temperatures usually range from the upper 70s to mid-80s over much of the State, producing an atmospheric environment favorable to many athletic, recreational, and other outdoor activities.

Temperatures of 90° or higher occur from late May to mid-September in all but the normally cooler portions of the State. The New York City area and most of the Hudson Valley record an average of from 18 to 25 days with such temperatures during the warm season, but in the Northern and Southern Plateaus the normal quota does not exceed 2 or 3 days. While temperatures of 100° are rare, many long-term weather stations, especially in the southern one-half of the State, have recorded maximums in the 100° to 105° range on one or more occasions. Minimum, or nighttime, temperatures drop to the 40s and upper 30s with some frequency during the summer season in the interior portions of the Plateau Divisions. It is not uncommon for temperatures to approach the freezing level in the Adirondacks and Southern Plateau during June and the latter one-half of August, but rarely in July.

The moderating effect of Lakes Erie and Ontario on temperatures assumes practical importance during the spring and fall seasons. The lake waters warm slowly in the spring, the effect of which is to reduce the warming of the atmosphere over adjacent land areas. Plant growth is thereby retarded, allowing a great variety of freeze-sensitive crops, especially tree and vine fruits, to reach critical early stages of development when the risk of freeze injury is minimized or greatly reduced. In the fall season, the lake waters cool more slowly than the land areas and thus serve as a heat source. The cooling of the atmosphere at night is moderated or reduced, the occurrence of freezing temperatures is delayed, and the growing season is lengthened for freeze-sensitive crops and vegetation.

The average length of the freeze-free season in New York State varies from 100 to 120 days in the Adirondacks, Catskills, and higher elevations of the Western Plateau Division to 180 to 200 days on Long Island. The important fruit and truck crop areas in the Great Lakes Plain enjoy a frost-free growing season of from 150 to 180 days in duration. A freeze-free season of similar length also prevails in the Hudson Valley from Albany southward to Westchester and Orange Counties, another zone of valuable crop production. The Southern Plateau, St. Lawrence Valley, and Lake Champlain regions have an average duration of 120 to 150 days between the last spring and first fall freezes.

PRECIPITATION. Moisture for precipitation in New York State is transported primarily from the Gulf of Mexico and Atlantic Ocean through circulation patterns and storm systems of the atmosphere. Distribution of precipitation within the State is greatly influenced by topography and proximity to the Great Lakes or Atlantic Ocean. Average annual amounts in excess of 50 inches occur in the western Adirondacks, Tug Hill area, and the Catskills, while slightly less than that amount is noted in the higher elevations of the Western Plateau southeast of Lake Erie. Areas of least rainfall, with average accumulations of about 30 inches, occur near Lake Ontario in the extreme western counties, in the lower half of the Genesee River Valley, and in the vicinity of Lake Champlain.

New York State has a fairly uniform distribution of precipitation during the year. There are no distinctly dry or wet seasons which are regularly repeated on an annual basis. Minimum precipitation occurs in the winter season, with an average monthly accumulation ranging from about 3.5 inches on Long Island to 2.2 inches in the Finger Lakes and Lake Champlain regions. Maximum amounts are noted in the summer season throughout the State except along the Great Lakes where slight peaks of similar magnitude occur in both the spring and fall seasons. Average monthly amounts in the summer vary from 3.0 inches in the lowlands south of Lake Ontario (Great Lakes Division) to 4.0 inches in the Eastern Plateau, Hudson Valley, and Coastal Divisions. New York's precipitation tends to be distributed most uniformly over the year in counties along the coast and the Great Lakes.

Variations in precipitation amounts from month to month or for the same month in different years can be wide for any individual area. Usually such variations range from near 1 inch to about 6 inches; in extreme cases, the variation is from less than 1 inch to 10 inches or more. Almost any calendar month has the potential of having the lightest, or heaviest, monthly accumulation of precipitation within a calendar year at a given location. The greatest monthly precipitation of record in New York State was a total of 25.27 inches at West Shokan (Ulster County) in October 1955. On the other hand, wide areas of the State measured less than 0.3 inch of rain in October 1963. Within relatively short distances, precipitation in the same month may be strikingly different. An extreme example occurred in August 1971 with a total of 16.7 inches falling at New York City's Borough of Richmond (Staten Island), but only 2.9 inches at Riverhead, about

90 miles away in eastern Long Island.

The amount and distribution of precipitation are normally sufficient for the maintenance of the State's water resources for municipal and industrial supplies, transportation, and recreation. Rainfall is usually adequate during the growing season for economic crops, lawns, gardens, shrubs, forests, and woodlands. Severe droughts are rare, but deficiencies of precipitation may occur from time to time which cause at least temporary concern over declining water supplies and moisture stress in crops and other vegetation. In some years, a pronounced shortage of precipitation during the spring or fall months results in a considerable fire hazard in the State's woodlands.

SNOWFALL. The climate of New York State is marked by abundant snowfall. With the exception of the Coastal Division, the State receives an average seasonal amount of 40 inches or more. The average snowfall is greater than 70 inches over some 60 percent of New York's area. The moderating influence of the Atlantic Ocean reduces the snow accumulation to 25 to 35 inches in the New York City area and on Long Island. About one-third of the winter season precipitation in the Coastal Division occurs from storms which also yield at least 1 inch of snow. The great bulk of the winter precipitation in upstate New York comes as snow.

Topography, elevation, and proximity to large bodies of water result in a great variation of snowfall in the State's interior, even within relatively short distances. Maximum seasonal snowfall, averaging more than 175 inches, occurs on the western and southwestern slopes of the Adirondacks and Tug Hill. A secondary maximum of 150 to 180 inches prevails in the southwestern highlands, some 10 to 30 miles inland from Lake Erie. Three separate areas of the Eastern Plateau record heavy snow accumulations, averaging from 100 to 120 inches: (1) the uplands of southeastern Onondaga County and adjoining counties; (2) the Cherry Valley section of northern Otsego and southern Herkimer Counties; and (3) the Catskill highlands in Ulster, Delaware, and Sullivan Counties. Minimum seasonal snowfall of 40 to 50 inches occurs upstate in (1) Niagara County, near the south shore of Lake Ontario, (2) the Chemung and mid-Genesee River Valleys of western New York, and (3) near the Hudson River in Orange, Rockland, and Westchester Counties upstream to the southern portion of Albany County.

In northern New York, the Adirondack region has an average seasonal snowfall in excess of 90 inches, but amounts decrease to 60 to 70 inches in the lowlands of the St. Lawrence Valley and to about 60 inches in the vicinity of Lake Champlain.

Snow produced in the lee of Lakes Erie and Ontario is a prominent and very important aspect of New York's climate. As cold air crosses the unfrozen lake waters, it is warmed in the lower layers, picks up moisture, and reaches the land in an unstable condition. Precipitation in the form of snow is released as the airstream moves inland and over the gradually sloping higher terrain. Heavy snow squalls frequently occur, generating from 1 to 2 feet of snow and occasionally 4 feet or more. Snowfall produced by this "lake-effect" usually extends into the Mohawk Valley and often inland as far as the southern Finger Lakes and nearby southern tier of counties. Counties to the lee of Lake Erie are subject to heavy lake-effect snows in November and December, but as the lake surface gradually freezes by midwinter, these snows become less frequent. Areas near Lake Ontario, especially those to the southeast and east, are exposed to severe snow squalls well into February because the Lake generally retains considerable open water throughout the winter months.

In the heavy snowbelts near Lake Erie and Ontario as well as in the plateau regions of eastern and northern New York, monthly snowfall amounts in excess of 24 inches are experienced in most winters; accumulations of more than 50 inches within 2 consecutive months are not uncommon. Monthly accumulations of between 3 to 10 inches usually occur in New York City and Long Island during the winter season, but occasionally the amounts may exceed 20 inches as a result of recurring coastal storms (northeasters).

A durable snow cover generally begins to develop in the Adirondacks and northern lowlands by late November and remains on the ground until various times in April, depending upon late winter snowfall and early spring temperatures. The Southern Plateau, Great Lakes Plain in southern portions of western upstate New York, and the Hudson Valley experience a continuous snow cover from about mid-December to mid-March, with maximum depths usually occurring in February. Bare ground may occur briefly in the lower elevations of these regions during some winters. From late December or early January through February, the Atlantic coastal region of the State experiences alternating periods of measurable snow cover and bare ground.

FLOODS. Although major floods are relatively infrequent, appreciable damage usually occurs every year in one or more localities of New York State. Floods that arise from a variety of causes have been recorded in all seasons. The greatest potential and frequency for floods occur in the early spring when substantial rains combine with rapid snowmelting to produce a heavy runoff. Since the turn of the century, several historic floods from this cause have occurred in the major river basins of southern and eastern New York. In northern New York, the normally colder early spring temperatures are conducive to a slower rate of snowmelt. In combination with other factors, major spring floods have been less frequent along streams draining into the St. Lawrence River. Ice jams sometimes contribute to serious flooding in very localized areas.

Damaging floods are caused at other times of the year by prolonged periods of heavy rainfall. Examples in recent years were those in southwestern New York in September 1967, in the lower Hudson Valley in May 1968, and in the Catskills in July 1969. In combination with heavy showers and thundershowers, the rugged terrain of the Adirondacks and Southern Plateau is conducive to occasional severe flash floods on smaller streams. The metropolitan New York City area and other heavily urbanized areas of the State are becoming increasingly subject to severe flooding of highways, streets, and low-lying ground. Replacement of the natural soil cover with cement, asphalt, and other impervious materials encourages such floods from rains of not more than moderately heavy intensity, that formerly were easily absorbed.

The shores of Long Island, especially those facing the Atlantic Ocean, are subject to tidal flooding during storm surges. Winds generated by hurricanes and great coastal storms may drive tidal waters well inland, causing extensive property damage and beach erosion. The great storm of November 1950, hurricane Carol in August 1954, and the historic Atlantic storm of March 1962 are some examples of severe, but infrequent, occurrences of this type of flooding.

WINDS AND STORMS. The prevailing wind is generally from the west in New York State. A southwest component becomes evident in winds during the warmer months while a northwest component is characteristic of the colder one-half of the year. Occasionally, well-developed storm systems moving across the continent or along the Atlantic coast are accompanied by very strong winds which cause considerable property damage over wide areas of the State. A unique effect of strong cyclonic winds from the southwest is the rise of water to abnormally high levels at the northeastern end of Lake Erie.

Thunderstorms occur on an average of about 30 days in a year throughout the State. Destructive winds and lightning strikes in local areas are common with the more vigorous warmseason thunderstorms. Locally, hail occurs with more severe thunderstorms, but extensive, crippling losses to property and crops are rare.

Tornadoes are not common. About three or four of these storms strike limited, localized areas of New York State in most years. The paths of destruction, mostly in rural, semirural, or wooded areas, are usually short and narrow. Tornadoes occur generally between late May and late August.

Storms of freezing rain occur on one or more occasions during the winter season and often affect a wide area of the State in any one incident. While such storms are usually limited to a thin but dangerous coating of ice on highways, sidewalks, and exposed surfaces, crippling destruction of utility lines, transmission towers, and trees over an extensive portion of the State may result on rare occasions. Such a destructive ice storm affected east-central and southeastern New York in December 1964.

Hurricanes and tropical storms periodically cause serious and heavy losses in the vicinity of Long Island and southeastern upstate New York. Only one such storm in recent years (October 1954) has brought serious damage to the interior portion of the State.

The greatest storm hazard in terms of area and number of people affected is heavy snow. Coastal northeaster storms occur with some frequency in most winters. Snow yields of from 12 to 24 inches or more from such storms have fallen over the southeastern one-quarter of the State, including Long Island, and will often extend into western and northern interior New York. Snow squalls along the Great Lakes have been previously cited. These may persist over a period of 1 week or more, bringing snow amounts in excess of 40 inches to local areas that lie to the eastern lee of Lakes Erie and Ontario. During heavy snow squalls, surface visibility is reduced to zero. Blizzard conditions of heavy snow, high winds, and rapidly falling temperature occur occasionally, but are much less characteristic of New York's climate than in the plains of Midwestern United States.

OTHER CLIMATIC ELEMENTS. The climate of the State features much cloudy weather during the months of November, December, and January in upstate New York, especially those regions that adjoin the Great Lakes and Finger Lakes and include the southern tier of counties. From June through September, however, about 60 to 70 percent of the possible sunshine hours is received. In the Atlantic coastal region, the sunshine hours increases from 50 percent of possible in the winter to about 65 percent of possible in the summer.

The Atlantic Coastal Plain and lower Hudson Valley experience conditions of high temperature and high humidity with some frequency and duration during the summer. By comparison, such conditions occur less frequently in the broad interior of New York State where they are usually shortened by the arrival of cooler, drier air masses from the northwest.

The occurrence of heavy dense fog is variable over the State. The valleys and ridges of the Southern Plateau are most subject to periods of fog, with occurrences averaging about 50 days in a year. In the Great Lakes Plain and northern valleys, the frequency decreases to only 10 to 20 days annually. In those portions of the State with greater maritime influence on the climate, the frequency of dense fog in a year ranges from about 35 days on the south shore of Long Island to 25 days in the Hudson Valley.

CLIMATE AND THE ECONOMY. New York State's diversified economy, involving agriculture, industry, commerce, and recreation, is greatly influenced by the climate. Human activities, whether in labor pursuits or recreation, are stimulated by an invigorating winter climate and a generally comfortable atmospheric environment during summer.

The general climate as well as regional variations in climate throughout New York State support diversified agriculture. Dairying is the largest, most widespread enterprise. Precipitation and temperature conditions favor the growth of alfalfa and grasses for hay and of corn for silage throughout rural New York, except where limitations are imposed by soils and topography. Corn for grain is produced on some 850,000 acres, mostly in the Great Lakes Plain, Southern Plateau, and Hudson Valley; climatic conditions couple with technology to realize an average statewide yield of 70 to 80 bushels per acre. The amount and distribution of rainfall, warm (rather than hot) daytime temperatures, and frequent cool nights in western and central New York are important environmental factors that aid in the growing of 450,000 acres of small grains. Dry beans, snap beans, and sugar beets are additional valuable crops which thrive well in New York's climate.

A nationally important production area of apples and other tree fruits is found along Lake Ontario, largely the result of favorable climatic conditions induced by the nearby Lake. The climate over the Great Lakes Plain is also benevolent for a wide variety of vegetable crops. New York is a leading producer of grapes, with suitable weather conditions for viticulture existing in the western Great Lakes counties and on the sloping terrain along the Finger Lakes where good air drainage and moderating influence of lake waters produce a suitable temperature regime. The lower Hudson Valley has a climate which also supports important acreage of tree fruits and truck crops.

The warmer climate of eastern Long Island permits a significant production of potatoes for the early season market. Late-season potato varieties are grown in the cooler climate of the Southern Plateau and of northeastern New York. The uplands northwest of the Catskill Mountains have a cool climate very suitable for cauliflower production.

The sugar maple tree (Acer saccharum Marsh.) finds a climate optimum for growth in New York State. Thus, the production of syrup and other maple products constitutes a valuable segment of the agricultural and forestry economy.

Ample precipitation, dependable runoff, and adequate ground water supplies contribute to vast water resources in the Empire State. These water resources have supported the growth of many large metropolitan areas, the establishment of diverse industries, and the development of waterways and impoundments for transportation, power, recreation, and municipal supplies.

Though rigorous and sometimes severe, New York's winter climate is an asset to the economy. Abundant snowfall has made possible the development of skiing and snowmobiling into very important activities for winter sports and recreation. The climate at other times of the year is a prominent factor in attracting tourists and vacationers to the State.

In summary, the climate contributes greatly to the agricultural, industrial, commercial, and recreational economy. It has been an unquestionable asset to the historical development of New York State and to its economic expansion of recent decades. Undoubtedly, the climate will continue its important role in the remainder of this century and beyond.

BIBLIOGRAPHY

Dethier, Bernard E., "Precipitation in New York State," Cornell University Agricultural Experiment Station, <u>Bulletin</u> 1009, New York State College of Agriculture, Ithaca, N.Y., July 1966, 78 pp.

Dethier, B. E., and McGuire, J. K., "The Climate of the Northeast, Probability of Selected Weekly Precipitation Amounts in the Northeast Region of the U. S.," Cornell University Agricultural Experiment Station, <u>Agronomy Mimeo</u> 61-4, Ithaca, N. Y., Nov. 1961, 302 pp.

Dethier, B. E., and Pack, A. B., "The Climate of Northern New York," Division of Meteorology, Department of Agronomy, Cornell University, <u>Agronomy Mimeo</u> 67-14, Ithaca, N. Y., Dec. 1967, 17 pp.

Dethier, B. E., and Vittum, M. T., "Growing Degree Days for New York State," Cornell University Agricultural Experiment Station, <u>Bulletin</u> 1017, Ithaca, N. Y., Nov. 1967, 50 pp.

Dickerson, W. H., "The Climate of the Northeast. Heating Degree Days," West Virginia Agricultural Experiment Station, <u>Bulletin</u> 483T, Morgantown, W. Va., June 1963, 26 pp.

Dunlap, D. C., "The Climate of the Northeast, Probabilities of Extreme Snowfalls and Snow Depths," New Jersey Agricultural Experiment Station, <u>Bulletin</u> 821, Rutgers University, The State University of New Jersey, New Brunswick, N. J., 1970, 15 pp.

Fieldhouse, Donald J., and Palmer, Wayne C., "The Climate of the Northeast. Meteorological and Agricultural Drought," University of Delaware Agricultural Experiment Station, <u>Bulletin</u> 353, Newark, Del., Feb. 1965, 71 pp.

Frederick, Ralph H., Johnson, Ernest C., and MacDonald, H. A., "Spring and Fall Freezing Temperatures in New York State," New York State College of Agriculture, <u>Cornell Miscellaneous Bulletin</u> 33, Ithaca, N. Y., Aug. 1959, 15 pp.

Havens, A. V., and McGuire, J. K., "The Climate of the Northeast. Spring and Fall Low-Temperature Probabilities," New Jersey Agricultural Experiment Station, <u>Bulletin</u> 801, Rutgers, The State University, New Brunswick, N. J., June 1961, 32 pp.

Johnson, Ernest C., "Climate of New York, Climates of the States," <u>Climatography of the United States</u> No. 60-30, U. S. Weather Bureau, Washington, D. C., Feb. 1960, 20 pp.

Miller, J. F., and Frederick, R. H., "The Precipitation Regime of Long Island, New York," U. S. Department of the Interior, <u>U. S. Geological Survey Professional Paper</u> 627-A, Washington, D. C., 1969, A1-A21 pp.

Mordoff, Richard Alan, "The Climate of New York State," Cornell University Extension, <u>Bulletin</u> 764, Ithaca, N. Y., Dec. 1949, 72 pp.

NOAA, Environmental Data Service, "Albany, Binghamton, Buffalo, New York City (Central Park, J. F. Kennedy International Airport, and La Guardia Airport), Rochester, and Syracuse, N. Y.," <u>Local Climatological Data</u>, Asheville, N. C., monthly plus annual summary.

_____, <u>Climatological Data--National Summary</u>, Asheville, N. C., monthly plus annual summary.

_____, <u>Climatological Data--New York</u>, Asheville, N. C., monthly plus annual summary.

_____, <u>Hourly Precipitation Data--New York</u>, Asheville, N. C., monthly plus annual summary.

Pack, A. Boyd, "Average Seasonal Snowfall in New York State (map)." <u>What's Cropping Up in Agronomy</u>, Vol. XIV, No. 1, New York State College of Agriculture, Cornell University, Ithaca, N. Y., Jan. 1970, 3 pp.

_____, "The Water Content of Snowstorms in New York State: Variations Among Different Physiographic Regions," in <u>Proceedings of The Twenty-Sixth Annual Eastern Snow Conference</u>, Feb. 6-7, 1969, Portland, Maine, Vol. 14, pp. 46-54.

Pack, A. Boyd and Dethier, B. E., "The Climate of Western New York," Division of Meteorology, Department of Agronomy, Cornell University, <u>Agronomy Mimeo</u> 69-12. Ithaca, N. Y., 1969, 19 pp.

U.S. Weather Bureau, <u>Climatic Summary of the United States</u> (Bulletin W), Sections 80, 81, 82, and 83, Washington, D.C., 1930, 113 pp.

_____, "Climatic Summary of the United States--Supplement for 1931 through 1952, New York" (Bulletin W Supplement), <u>Climatography of the United States</u> No. 11-26, Washington, D.C., 1953, 66 pp.

_____, "Climatic Summary of the United States--Supplement for 1951 through 1960, New York" (Bulletin W Supplement), <u>Climatography of the United States</u> No. 86-26, Washington, D.C., 1964, 111 pp.

_____, "Decadal Census of Weather Stations, New York," <u>Key to Meteorological Records Documentation</u> No. 6.11, Washington, D.C., 1963, 8 pp.

_____, "Maximum Station Precipitation for 1, 2, 3, 6, 12, and 24 Hours, Part X: New York," <u>Technical Paper</u> No. 15, Washington, D.C., Dec. 1954, 113 pp.

_____, "Monthly Averages for State Climate Divisions, 1931-1960, New York," <u>Climatography of the United States</u> No. 85-26, Washington, D.C., 1963, 4 pp.

_____, "Monthly Normals of Temperature, Precipitation, and Heating Degree Days, New York," <u>Climatography of the United States</u> No. 81-26, Washington, D.C., 1962, 2 pp.

_____, "Rainfall Frequency Atlas of the United States for Durations from 30 Minutes to 24 Hours and Return Periods from 1 to 100 Years," <u>Technical Paper</u> No. 40, Washington, D.C., May 1961, 115 pp.

_____, "Summary of Hourly Observations, Albany, Binghamton, Buffalo, New York City (J.F. Kennedy International Airport and La Guardia Airport), Rochester, and Syracuse, New York, 1951-1960," <u>Climatography of the United States</u> No. 82-30, Washington, D.C., 1962 and 1963, 105 pp.

FREEZE DATA

STATION	Freeze threshold temperature	Mean date of last Spring occurrence	Mean date of first Fall occurrence	Mean No. of days between dates	Years of record Spring	No. of occurrences in Spring	Years of record Fall	No. of occurrences in Fall	STATION	Freeze threshold temperature	Mean date of last Spring occurrence	Mean date of first Fall occurrence	Mean No. of days between dates	Years of record Spring	No. of occurrences in Spring	Years of record Fall	No. of occurrences in Fall
ADDISON	32	05-21	09-30	131	23	23	20	20	CANTON 4 SE	32	05-09	09-26	140	29	29	29	29
	28	05-06	10-10	158	23	23	20	20		28	04-28	10-10	166	29	29	29	29
	24	04-22	10-24	185	22	22	20	20		24	04-15	10-24	192	29	29	29	29
	20	04-08	11-05	211	22	22	20	20		20	04-03	11-08	219	29	29	29	29
	16	03-25	11-22	242	22	22	20	20		16	03-26	11-19	238	29	29	29	29
ALBANY WSO AP	32	04-27	10-13	169	29	29	29	29	CARMEL 1 SW	32	05-03	10-10	160	29	29	28	28
	28	04-14	10-26	195	29	29	29	29		28	04-22	10-22	183	29	29	28	28
	24	03-31	11-09	224	29	29	29	29		24	04-09	11-04	209	29	29	28	28
	20	03-19	11-22	248	29	29	29	29		20	03-24	11-18	240	29	29	28	28
	16	03-13	11-29	261	29	29	29	29		16	03-16	11-30	259	29	29	28	28
ALBION 3 ENE	32	05-01	10-15	166	12	12	12	12	CHASM FALLS	32	05-25	09-14	112	17	17	17	17
	28	04-22	11-03	196	12	12	12	12		28	05-10	10-02	145	17	17	17	17
	24	04-02	11-18	230	12	12	12	12		24	04-30	10-11	164	17	17	17	17
	20	03-25	11-28	248	12	12	12	12		20	04-16	10-27	193	17	17	17	17
	16	03-20	12-08	263	12	12	12	12		16	03-31	11-14	228	17	17	17	17
ALEXANDRIA BAY	32	05-02	10-10	160	14	14	14	14	CHAZY	32	05-15	09-26	134	26	26	24	24
	28	04-18	10-27	192	14	14	14	14		28	04-30	10-08	161	26	26	23	23
	24	04-08	11-11	217	14	14	14	14		24	04-17	10-27	192	26	26	22	22
	20	03-31	11-20	234	14	14	14	14		20	04-06	11-08	216	26	26	19	19
	16	03-22	11-28	251	14	14	14	14		16	03-24	11-22	243	26	26	19	19
ALFRED	32	05-27	09-19	115	29	29	29	29	COOPERSTOWN	32	05-22	09-22	123	24	24	22	22
	28	05-14	10-05	143	29	29	29	29		28	05-12	10-05	145	24	24	22	22
	24	04-27	10-14	171	29	29	29	29		24	04-26	10-18	176	23	23	22	22
	20	04-15	11-01	201	29	29	28	28		20	04-11	11-03	207	23	23	21	21
	16	04-03	11-16	227	29	29	27	27		16	04-01	11-16	229	22	22	20	20
ALLEGANY STATE PARK	32	06-07	09-05	90	26	26	26	26	CORTLAND	32	05-13	09-30	140	29	29	29	29
	28	05-23	09-29	130	26	26	26	26		28	04-25	10-19	177	29	29	29	29
	24	05-08	10-15	160	26	26	26	26		24	04-12	11-07	209	29	29	29	29
	20	04-20	11-03	197	26	26	26	26		20	03-29	11-21	237	29	29	29	29
	16	04-07	11-16	223	26	26	26	26		16	03-20	11-28	253	29	29	29	29
ANGELICA	32	06-01	09-12	103	30	30	27	27	DANNEMORA	32	05-13	10-02	142	29	29	29	29
	28	05-18	09-29	134	30	30	26	26		28	05-01	10-12	164	29	29	28	28
	24	05-02	10-12	163	30	30	26	26		24	04-23	10-26	185	29	29	28	28
	20	04-20	10-26	188	30	30	26	26		20	04-11	11-07	210	29	29	27	27
	16	04-04	11-10	220	30	30	26	26		16	04-04	11-12	223	29	29	27	27
AUBURN WTR WKS	32	05-02	10-21	172	28	28	27	27	DANSVILLE	32	05-10	10-07	150	27	27	27	27
	28	04-18	11-08	204	28	28	26	26		28	04-24	10-27	187	27	27	27	27
	24	04-05	11-19	228	28	28	26	26		24	04-13	11-11	212	26	26	27	27
	20	03-24	11-28	249	27	27	26	26		20	03-29	11-25	242	25	25	27	27
	16	03-16	12-03	262	26	26	26	25		16	03-16	11-30	260	25	25	27	26
BAINBRIDGE	32	05-15	09-26	134	14	14	14	14	DELHI 2 SW	32	05-29	09-23	117	26	26	24	24
	28	05-07	10-11	157	14	14	14	14		28	05-15	10-04	143	26	26	24	24
	24	04-21	10-23	185	14	14	13	13		24	04-30	10-16	169	26	26	24	24
	20	04-03	11-06	217	14	14	13	13		20	04-17	10-30	196	26	26	24	24
	16	03-27	11-23	241	14	14	13	13		16	03-29	11-14	229	26	26	24	24
BEDFORD HILLS	32	04-29	10-13	167	24	24	23	23	ELMIRA	32	05-06	10-09	156	30	30	30	30
	28	04-15	10-26	194	24	24	23	23		28	04-19	10-25	188	30	30	30	30
	24	03-30	11-08	223	23	23	21	21		24	04-08	11-06	212	30	30	30	30
	20	03-19	11-23	249	22	22	21	21		20	03-22	11-24	247	30	30	30	30
	16	03-12	12-04	267	22	22	21	21		16	03-13	12-04	266	30	30	30	29
BINGHAMTON	32	05-04	10-06	154	29	29	28	28	FARMINGDALE 2 NE	32	04-30	10-19	173	11	11	11	11
	28	04-19	10-20	184	29	29	28	28		28	04-16	10-27	194	11	11	11	11
	24	04-04	11-07	217	29	29	28	28		24	03-25	11-14	234	9	9	11	11
	20	03-23	11-21	243	29	29	28	28		20	03-15	12-01	261	9	9	10	10
	16	03-14	12-02	263	29	29	28	28		16	03-07	12-08	276	9	9	9	9
BRIDGEHAMPTON	32	04-20	11-02	196	20	20	21	21	FRANKLINVILLE	32	05-31	09-13	105	14	14	13	13
	28	04-05	11-17	225	20	20	21	21		28	05-18	09-29	134	12	12	13	13
	24	03-19	11-29	256	19	19	21	21		24	05-06	10-14	161	11	11	13	13
	20	03-10	12-05	270	19	19	21	21		20	04-17	10-31	197	11	11	12	12
	16	03-06	12-12	282	19	19	21	19		16	04-05	11-11	219	11	11	12	12
BROCKPORT 2 NW	32	05-07	10-20	166	16	16	14	14	FREDONIA	32	05-07	10-24	171	30	30	30	30
	28	04-21	11-03	196	16	16	14	14		28	04-20	11-04	198	30	30	30	30
	24	04-07	11-11	218	16	16	14	14		24	04-05	11-21	230	30	30	30	30
	20	03-24	11-26	247	16	16	14	14		20	03-24	12-01	253	30	30	30	30
	16	03-11	11-30	264	16	16	13	13		16	03-14	12-07	268	30	30	30	29
BUFFALO WSO AP	32	04-30	10-25	179	29	29	29	29									
	28	04-20	11-09	202	29	29	29	29									
	24	04-03	11-20	231	29	29	29	29									
	20	03-23	11-30	251	29	29	29	28									
	16	03-15	12-07	267	29	29	29	28									

FREEZE DATA

STATION	Freeze threshold temperature	Mean date of last Spring occurrence	Mean date of first Fall occurrence	Mean No. of days between dates	Years of record Spring	No. of occurrences in Spring	Years of record Fall	No. of occurrences in Fall
GENEVA	32	05-06	10-11	158	30	30	29	29
	28	04-20	10-27	190	30	30	29	29
	24	04-08	11-12	218	30	30	29	29
	20	03-26	11-25	244	30	30	29	29
	16	03-13	12-03	264	30	30	29	29
GLENHAM	32	04-29	10-07	161	19	19	19	19
	28	04-18	10-24	189	19	19	19	19
	24	03-30	11-05	220	19	19	19	19
	20	03-22	11-21	244	19	19	19	19
	16	03-14	12-01	261	19	19	19	19
GLOVERSVILLE	32	05-14	09-30	139	23	23	23	23
	28	04-30	10-10	163	23	23	23	23
	24	04-18	10-26	192	23	23	23	23
	20	04-08	11-06	213	23	23	23	23
	16	03-27	11-19	237	23	23	22	22
GOUVERNEUR	32	05-19	09-18	122	13	13	13	13
	28	05-05	09-26	144	13	13	13	13
	24	04-23	10-10	170	13	13	12	12
	20	04-09	10-26	200	13	13	12	12
	16	03-31	11-13	227	13	13	12	12
HEMLOCK	32	05-08	10-12	158	30	30	30	30
	28	04-25	10-26	185	30	30	30	30
	24	04-11	11-08	210	30	30	30	30
	20	03-31	11-22	236	30	30	30	30
	16	03-20	12-01	255	30	30	30	29
INDIAN LAKE 2 SW	32	06-11	08-27	78	27	27	27	27
	28	05-27	09-20	117	28	28	27	27
	24	05-14	10-04	143	28	28	27	27
	20	04-30	10-16	169	28	28	26	26
	16	04-15	10-26	194	28	28	26	26
ITHACA CORNELL UNIV	32	05-12	10-04	145	29	29	29	29
	28	04-23	10-21	181	29	29	29	29
	24	04-12	11-05	207	29	29	29	29
	20	03-29	11-21	237	29	29	29	29
	16	03-18	11-29	255	29	29	29	29
JAMESTOWN	32	05-16	10-07	144	26	26	24	24
	28	04-25	10-23	181	25	25	22	22
	24	04-10	11-11	215	25	25	22	22
	20	03-31	11-23	237	24	24	20	20
	16	03-17	12-06	264	24	24	19	18
LAKE PLACID CLUB	32	06-04	09-11	99	29	29	29	29
	28	05-18	09-22	127	28	28	29	29
	24	05-03	10-09	159	28	28	29	29
	20	04-22	10-21	182	28	28	29	29
	16	04-12	11-02	204	28	28	28	28
LAWRENCEVILLE	32	05-14	09-22	130	18	18	18	18
	28	05-09	10-09	153	16	16	18	18
	24	04-26	10-17	174	16	16	18	18
	20	04-13	11-01	202	16	16	17	17
	16	03-31	11-14	228	16	16	17	17
LEWISTON 1 N	32	05-07	10-11	157	15	15	16	16
	28	04-14	10-29	197	15	15	16	16
	24	04-05	11-17	226	15	15	16	16
	20	03-20	11-25	250	15	14	16	16
	16	03-10	12-07	271	15	14	16	16
LOCKPORT 2 NE	32	05-09	10-13	157	27	27	27	27
	28	04-23	10-31	191	27	27	27	27
	24	04-10	11-14	218	27	27	27	27
	20	03-28	11-23	240	26	26	26	26
	16	03-16	12-03	262	25	24	26	26
LOWVILLE	32	05-18	09-21	126	29	29	30	30
	28	05-05	10-04	152	29	29	30	30
	24	04-22	10-20	180	29	29	30	30
	20	04-08	11-03	209	29	29	30	30
	16	03-29	11-13	229	29	29	30	30
MINEOLA	32	04-07	11-16	223	13	13	12	12
	28	03-26	11-29	249	13	13	12	12
	24	03-15	12-06	267	13	13	12	12
	20	03-09	12-14	280	13	13	12	11
	16	03-05	12-17	288	13	13	12	11
MOHONK LAKE	32	04-29	10-22	177	27	27	27	27
	28	04-16	11-04	201	27	27	27	27
	24	04-05	11-16	225	27	27	27	27
	20	03-25	11-23	243	27	27	26	26
	16	03-17	12-01	259	27	27	26	26
MORRISVILLE	32	05-21	09-20	122	26	26	25	25
	28	05-10	10-03	146	25	25	24	24
	24	04-26	10-19	176	25	25	24	24
	20	04-12	10-30	201	25	25	23	23
	16	03-30	11-14	229	25	25	23	23
NEWCOMB 4 WNW	32	05-27	09-16	112	10	10	11	11
	28	05-12	09-27	138	10	10	10	10
	24	04-29	10-10	164	10	10	10	10
	20	04-17	10-29	195	10	10	10	10
	16	04-04	11-09	219	9	9	10	10
NY CENTRAL PK WSO	32	04-07	11-12	219	29	29	29	29
	28	03-24	11-24	244	29	29	29	29
	24	03-15	12-02	262	29	29	29	29
	20	03-10	12-09	273	29	29	29	27
	16	03-03	12-16	289	29	29	29	23
NORWICH 1 NE	32	05-19	09-24	127	28	28	30	30
	28	05-07	10-06	152	28	28	30	30
	24	04-24	10-22	182	27	27	30	30
	20	04-10	10-30	203	27	27	30	30
	16	03-26	11-18	238	27	27	30	30
OGDENSBURG 3 NE	32	05-08	10-08	153	27	27	27	27
	28	04-25	10-24	182	27	27	27	27
	24	04-13	11-02	203	26	26	27	27
	20	03-31	11-15	229	24	24	27	27
	16	03-25	11-26	247	24	24	26	26
ONEONTA	32	05-19	09-28	133	27	27	27	27
	28	05-01	10-11	163	27	27	27	27
	24	04-20	10-25	188	27	27	27	27
	20	04-01	11-09	221	27	27	27	27
	16	03-25	11-21	241	27	27	27	27
OSWEGO EAST	32	04-24	10-24	184	29	29	29	29
	28	04-10	11-09	214	28	28	29	29
	24	03-28	11-20	237	28	28	29	29
	20	03-19	11-28	254	28	28	29	29
	16	03-13	12-04	266	28	28	29	29
PATCHOGUE 2 N	32	04-28	10-12	167	12	12	13	13
	28	04-11	10-27	199	12	12	13	13
	24	03-31	11-15	228	12	12	13	13
	20	03-22	11-24	247	12	12	13	13
	16	03-10	12-08	273	12	12	13	13
PENN YAN 2 SW	32	05-13	10-05	146	26	26	22	22
	28	04-25	10-18	176	25	25	21	21
	24	04-12	11-03	205	24	24	21	21
	20	03-30	11-15	231	23	23	21	21
	16	03-19	11-28	254	23	23	21	21
PORT JERVIS	32	05-10	10-04	146	29	29	29	29
	28	04-24	10-17	176	29	29	29	29
	24	04-09	10-28	202	29	29	28	28
	20	03-26	11-10	229	29	29	28	28
	16	03-16	11-27	256	29	29	28	28

FREEZE DATA

STATION	Freeze threshold temperature	Mean date of last Spring occurrence	Mean date of first Fall occurrence	Mean No. of days between dates	Years of record Spring	No. of occurrences in Spring	Years of record Fall	No. of occurrences in Fall
RIVERHEAD RESEARCH	32	04-15	11-08	208	13	13	13	13
	28	04-02	11-22	233	13	13	13	13
	24	03-16	12-05	265	13	13	13	13
	20	03-10	12-10	275	13	13	13	13
	16	03-02	12-18	292	13	13	13	11
ROCHESTER WSO AP	32	04-28	10-21	176	29	29	27	27
	28	04-16	11-05	203	29	29	27	27
	24	03-31	11-19	233	29	29	27	27
	20	03-20	11-28	253	29	29	27	27
	16	03-12	12-05	267	29	29	27	26
ROXBURY	32	05-28	09-18	113	27	27	25	25
	28	05-12	09-30	141	26	26	25	25
	24	04-28	10-10	165	24	24	24	24
	20	04-20	10-26	189	24	24	24	24
	16	04-04	11-04	214	24	24	24	24
SALISBURY	32	05-29	09-18	113	30	30	30	30
	28	05-12	09-29	140	30	30	30	30
	24	04-25	10-13	171	30	30	30	30
	20	04-15	10-31	199	30	30	30	30
	16	04-05	11-08	217	30	30	30	30
SCARSDALE	32	04-21	10-20	181	26	26	27	27
	28	04-10	10-30	204	26	26	26	26
	24	03-26	11-20	240	26	26	23	23
	20	03-15	11-30	260	26	26	23	23
	16	03-10	12-07	272	26	26	23	22
SETAUKET	32	04-10	11-09	214	30	30	28	28
	28	03-26	11-23	242	30	30	28	28
	24	03-15	12-05	265	30	30	28	27
	20	03-09	12-09	275	30	30	28	27
	16	02-28	12-19	294	30	29	28	21
SODUS CENTER	32	05-06	10-09	155	16	16	11	11
	28	04-22	10-27	187	16	16	9	9
	24	04-12	11-05	207	16	16	9	9
	20	03-28	11-22	239	15	15	9	9
	16	03-21	12-03	257	15	15	9	8
SOUTH WALES EMERY PK	32	05-17	10-01	137	19	19	20	20
	28	05-05	10-14	163	19	19	20	20
	24	04-16	11-02	200	19	19	20	20
	20	04-03	11-15	226	19	19	20	20
	16	03-25	11-24	244	19	19	20	20
SPIER FALLS	32	05-10	10-07	150	29	29	27	27
	28	04-21	10-20	182	29	29	27	27
	24	04-12	11-04	206	29	29	27	27
	20	03-29	11-16	232	29	29	27	27
	16	03-20	11-26	252	29	29	27	27
STILLWATER RES	32	05-28	09-19	114	24	24	23	23
	28	05-15	09-28	136	23	23	23	23
	24	05-05	10-11	160	23	23	23	23
	20	04-22	10-27	187	23	23	23	23
	16	04-12	11-07	210	23	23	23	23
SYRACUSE WSO AP	32	04-30	10-15	168	29	29	29	29
	28	04-15	10-29	197	29	29	29	29
	24	04-04	11-14	224	29	29	29	29
	20	03-21	11-23	247	29	29	29	29
	16	03-14	12-03	265	29	29	29	28
TUPPER LAKE SUNMOUNT	32	05-28	09-16	111	26	26	24	24
	28	05-14	09-28	137	26	26	24	24
	24	04-26	10-10	166	25	25	24	24
	20	04-17	10-28	194	25	25	23	23
	16	04-06	11-10	217	25	25	22	22
UTICA FAA AP	32	05-14	10-01	140	20	20	18	18
	28	04-28	10-17	172	20	20	18	18
	24	04-12	11-01	203	20	20	18	18
	20	03-28	11-10	227	20	20	18	18
	16	03-20	11-25	251	20	20	18	18
UTICA 3 W	32	05-14	10-02	141	23	23	21	21
	28	04-30	10-19	173	23	23	21	21
	24	04-12	11-01	203	23	23	21	21
	20	03-28	11-13	230	23	23	21	21
	16	03-20	11-27	252	23	23	21	21
WANAKENA R SCHOOL	32	06-02	09-13	103	29	29	26	26
	28	05-14	09-28	137	29	29	26	26
	24	05-02	10-07	158	28	28	26	26
	20	04-22	10-22	183	28	28	26	26
	16	04-11	11-05	208	28	28	26	26
WATERTOWN	32	05-07	10-04	151	30	30	30	30
	28	04-26	10-18	176	30	30	30	30
	24	04-13	10-28	198	30	30	30	30
	20	03-31	11-13	228	30	30	30	30
	16	03-23	11-24	246	30	30	30	30
WEST POINT	32	04-18	10-26	191	25	25	26	26
	28	04-03	11-08	219	25	25	26	26
	24	03-21	11-25	249	24	24	26	26
	20	03-14	12-04	266	24	24	26	25
	16	03-09	12-08	274	24	24	26	25

Data in the above table are based on the period 1921-1950, or that portion of this period for which data are available.

Means have been adjusted to take into account years of non-occurrence.

A freeze is a numerical substitute for the former term "killing frost" and is the occurrence of a minimum temperature at or below the threshold temperature of 32°, 28°, etc.

Freeze data tabulations in greater detail are available and can be reproduced at cost.

*NORMALS BY CLIMATOLOGICAL DIVISIONS

Taken from "Climatography of the United States No. 81-4, Decennial Census of U. S. Climate"

STATIONS (By Divisions)	TEMPERATURE (°F)													PRECIPITATION (In.)												
	JAN	FEB	MAR	APR	MAY	JUNE	JULY	AUG	SEPT	OCT	NOV	DEC	ANN	JAN	FEB	MAR	APR	MAY	JUNE	JULY	AUG	SEPT	OCT	NOV	DEC	ANN
WESTERN PLATEAU																										
ALFRED	23.5	23.4	30.9	43.5	54.7	63.6	67.4	65.7	59.2	49.0	37.1	25.9	45.3	2.28	2.08	3.28	3.05	3.78	3.69	3.46	3.29	2.99	2.85	2.83	2.33	35.91
ALLEGANY STATE PARK	25.6	25.3	32.6	44.8	55.4	63.8	67.5	66.0	59.8	49.9	38.3	27.6	46.4	2.84	2.75	3.46	3.49	4.27	4.27	4.35	3.62	4.05	3.44	3.77	3.07	43.38
ANGELICA	24.5	24.2	32.0	44.5	55.3	64.5	68.6	66.7	59.9	49.5	37.9	26.7	46.2	2.11	1.87	2.84	2.72	3.37	3.46	3.38	2.89	2.88	2.52	2.59	2.09	32.72
CORNING	·	·	·	·	·	·	·	·	·	·	·	·	·	1.84	1.69	2.74	2.98	3.92	3.18	3.71	3.85	2.95	2.73	2.41	2.19	34.19
ELMIRA	27.1	27.0	34.7	47.2	58.4	67.6	72.0	69.7	62.4	51.7	40.5	29.7	49.0	1.88	1.92	2.98	3.00	3.94	3.37	3.58	3.97	2.84	2.70	2.49	2.22	34.89
HASKINVILLE	·	·	·	·	·	·	·	·	·	·	·	·	·	1.87	1.92	2.81	2.85	3.64	3.46	3.82	3.28	2.81	2.71	2.48	2.06	33.71
OLEAN	·	·	·	·	·	·	·	·	·	·	·	·	·	2.40	2.22	3.05	3.39	3.85	3.77	3.92	3.00	3.30	2.87	2.88	2.46	37.11
DIVISION	24.7	24.5	32.1	44.7	55.7	64.7	68.7	66.9	60.2	49.8	38.2	27.1	46.4	2.33	2.26	3.13	3.14	3.82	3.68	3.69	3.44	3.20	2.94	2.91	2.51	37.05
EASTERN PLATEAU																										
BAINBRIDGE	·	·	·	·	·	·	·	·	·	·	·	·	·	2.77	2.57	3.20	3.34	3.77	3.87	4.56	3.71	3.44	3.26	3.08	3.08	40.65
BINGHAMTON WSO	23.8	23.8	31.3	43.5	55.1	63.5	68.4	66.5	59.5	49.8	38.0	26.8	45.8	2.50	2.18	2.89	2.94	3.49	3.85	3.71	3.57	2.95	3.09	2.51	2.56	36.24
COOPERSTOWN	22.7	23.0	31.0	44.1	55.3	64.4	68.7	66.6	59.7	49.4	38.0	25.7	45.7	2.96	2.74	3.22	3.33	3.93	3.77	4.27	3.98	3.70	3.49	3.32	3.07	41.78
CORTLAND	23.7	23.5	31.3	44.3	55.8	65.2	70.1	68.1	60.6	50.3	38.6	26.7	46.5	2.81	2.83	3.54	3.21	3.78	3.70	4.16	3.74	3.33	3.36	2.99	3.21	40.66
DELHI 2 SW	23.7	23.9	31.6	44.2	55.4	64.2	68.4	66.6	59.8	49.5	38.1	26.2	46.0	2.75	2.66	3.09	3.32	3.89	3.92	4.62	4.21	3.90	3.45	3.56	2.86	42.23
FREEHOLD 2 NW	26.0	27.2	35.8	48.5	59.7	68.9	73.4	71.1	63.3	52.5	41.2	29.3	49.7	2.46	2.33	3.09	3.10	3.58	3.29	3.63	3.13	3.92	3.65	3.38	3.03	38.59
MORRISVILLE	20.8	20.7	29.0	42.3	53.5	62.9	67.4	65.4	58.1	47.7	36.6	24.0	44.0	2.53	2.55	2.98	3.07	3.65	3.57	3.74	3.81	3.44	3.62	2.96	2.80	38.72
NORWICH 1 NE	22.3	22.3	30.8	44.0	54.7	64.0	68.4	66.8	59.2	48.6	37.5	25.1	45.3	2.81	2.57	3.39	3.41	3.71	3.89	4.25	3.56	3.52	3.45	3.00	3.00	40.79
ROXBURY	23.4	23.9	31.8	44.1	55.2	63.8	68.3	66.5	59.2	49.0	38.0	26.2	45.8	3.01	2.62	3.44	3.58	3.98	3.87	4.04	3.90	3.75	3.63	3.51	2.94	42.27
SHERBURNE 2 S	·	·	·	·	·	·	·	·	·	·	·	·	·	2.39	2.16	2.95	3.20	3.45	3.59	4.13	3.50	3.67	3.26	2.94	2.83	38.07
DIVISION	23.7	24.0	32.1	44.8	56.0	65.0	69.5	67.6	60.3	50.0	38.5	26.5	46.5	2.81	2.61	3.24	3.38	3.78	3.70	4.17	3.89	3.60	3.40	3.24	2.97	40.79
NORTHERN PLATEAU																										
BIG MOOSE 3 E	·	·	·	·	·	·	·	·	·	·	·	·	·	4.02	3.65	3.74	3.79	4.00	3.88	4.59	3.76	4.73	4.39	4.48	4.48	49.51
HIGHMARKET	·	·	·	·	·	·	·	·	·	·	·	·	·	4.12	3.77	4.09	4.04	4.44	3.62	4.53	4.07	4.98	5.11	4.70	4.55	52.02
HOFFMEISTER 3 W	·	·	·	·	·	·	·	·	·	·	·	·	·	4.12	3.52	4.42	4.28	4.33	4.56	5.04	4.07	4.93	4.62	4.66	4.40	52.95
HOPE	·	·	·	·	·	·	·	·	·	·	·	·	·	3.81	3.32	3.78	3.80	3.67	3.74	4.46	3.53	4.07	3.68	4.08	3.95	45.89
INDIAN LAKE 2 SW	16.5	16.9	25.4	38.3	50.8	59.9	64.1	62.2	55.0	44.5	32.8	19.6	40.5	3.16	2.78	3.29	3.11	3.35	3.72	4.32	3.18	4.28	3.63	3.44	4.32	41.94
LAKE PLACID CLUB	14.9	15.6	24.8	38.6	51.3	60.7	64.9	62.7	55.2	44.8	32.1	18.4	40.3	3.13	2.97	3.24	2.70	3.25	3.67	3.96	3.41	3.74	2.92	2.98	3.19	39.16
LOWVILLE	18.3	19.2	28.3	42.7	54.9	64.3	68.7	66.8	59.2	48.5	35.9	22.3	44.1	2.86	2.51	3.01	3.05	3.25	2.80	3.27	3.08	3.25	3.41	3.54	3.36	37.39
LYONS FALLS	·	·	·	·	·	·	·	·	·	·	·	·	·	3.53	2.92	3.30	3.03	3.72	3.12	4.11	3.27	4.18	4.09	4.12	3.88	43.54
SOUTH EDWARDS 1 E	·	·	·	·	·	·	·	·	·	·	·	·	·	3.03	2.88	3.05	3.35	3.89	3.47	4.10	3.79	4.26	3.94	3.67	3.69	43.12
STILLWATER RESERVOIR	14.4	14.5	23.9	38.5	51.7	61.1	65.5	63.9	56.3	45.4	32.8	18.2	40.5	3.81	3.63	4.08	3.97	4.06	3.91	4.82	3.89	4.56	4.45	4.32	4.41	49.91
TUPPER LAKE SUNMOUNT	16.3	17.0	25.7	39.3	52.1	61.3	65.4	63.4	55.8	45.4	33.1	19.6	41.2	2.44	2.29	2.60	2.64	3.36	3.38	3.97	3.32	3.65	3.12	2.79	2.67	36.74
WANAKENA RANGER SCHOOL	16.8	17.4	26.2	40.3	52.9	61.6	65.7	63.9	56.6	46.1	33.9	20.2	41.8	3.06	2.83	3.17	3.24	3.61	3.53	4.18	3.48	4.02	3.81	3.50	3.50	41.93
DIVISION	16.4	17.0	25.8	39.6	52.3	61.4	65.7	63.7	56.3	45.8	33.5	19.9	41.5	3.26	2.97	3.35	3.24	3.58	3.58	4.21	3.59	4.06	3.71	3.64	3.63	42.82
COASTAL																										
BRIDGEHAMPTON	32.0	31.9	37.6	46.6	56.1	65.3	71.3	70.7	64.4	55.1	45.3	34.8	50.9	4.20	3.59	4.61	3.62	3.44	2.88	2.92	4.42	3.67	3.55	4.66	4.10	45.66
NEW YORK CNTRL PK WSO	33.2	33.4	40.5	51.4	62.4	71.4	76.8	75.1	68.5	58.3	47.0	35.9	54.5	3.31	2.84	4.01	3.43	3.67	3.31	3.70	4.44	3.87	3.14	3.39	3.26	42.37
NY JOHN F KENNEDY INAP	31.8	31.6	38.7	49.0	60.2	70.1	75.9	74.5	67.8	57.6	46.2	34.9	53.2	3.23	2.93	4.15	3.48	3.67	3.35	4.04	4.97	4.16	3.21	3.51	3.23	43.93
N Y LA GUARDIA WSO	33.6	33.6	40.8	51.2	62.1	71.5	76.8	75.4	68.8	58.6	47.4	36.4	54.7	3.31	3.09	4.23	3.57	3.58	3.38	3.71	5.08	3.92	3.37	3.59	3.39	44.22
SCARSDALE	30.5	31.2	38.5	49.7	60.5	69.3	74.3	72.7	65.6	55.3	44.2	33.1	52.1	3.36	2.78	4.39	4.10	4.21	3.79	4.51	4.90	4.40	3.81	4.10	3.73	48.08
SETAUKET	33.0	32.8	39.2	49.5	59.8	68.4	73.8	72.5	66.3	57.1	46.8	35.8	52.9	3.87	3.19	4.26	3.70	3.55	3.40	3.55	4.10	3.91	3.36	4.12	3.64	44.65
DIVISION	32.2	32.4	39.2	49.4	59.9	69.1	74.5	73.1	66.5	56.5	45.9	35.0	52.8	3.64	3.14	4.37	3.75	3.71	3.39	3.78	4.68	3.91	3.46	3.97	3.67	45.47
HUDSON VALLEY																										
ALBANY WSO	22.7	23.7	33.0	46.2	57.9	67.3	72.1	70.0	61.6	50.8	39.1	26.5	47.6	2.47	2.20	2.72	2.77	3.47	3.25	3.49	3.07	3.58	2.77	2.70	2.59	35.08
ALBANY	25.7	26.7	35.7	48.4	59.9	69.0	74.0	71.7	63.7	53.1	41.7	29.4	49.9	2.51	2.26	2.86	2.90	3.62	3.74	4.29	3.30	4.00	2.84	2.90	2.73	37.95
BEDFORD HILLS	29.6	30.5	38.2	49.8	60.8	69.3	74.3	72.4	65.1	54.8	43.1	32.1	51.7	3.32	2.85	4.00	4.02	4.07	3.99	5.00	4.57	4.11	3.70	3.92	3.99	47.54
CARMEL 1 SW	26.4	27.0	35.1	47.2	57.8	66.5	71.7	70.0	62.8	52.8	41.4	29.4	49.0	3.24	2.75	3.76	3.70	4.35	3.90	4.71	4.61	4.24	3.63	4.12	3.59	46.70
CONKLINGVILLE DAM	·	·	·	·	·	·	·	·	·	·	·	·	·	3.45	3.13	3.60	3.47	3.28	3.55	3.97	3.46	3.56	3.33	3.73	3.63	42.16
MECHANICVILLE 2 S	·	·	·	·	·	·	·	·	·	·	·	·	·	2.51	1.97	2.55	3.01	3.47	3.82	3.97	3.17	3.94	3.06	2.93	2.68	37.08
MOHONK LAKE	26.0	26.8	34.5	46.7	58.1	66.1	70.8	68.9	61.9	52.0	40.6	28.8	48.4	3.48	3.08	3.91	4.44	4.33	3.98	4.71	4.20	4.26	3.88	4.03	3.66	47.96
POUGHKEEPSIE	27.3	28.5	37.3	49.5	60.7	69.6	74.6	72.4	64.6	54.0	42.6	30.5	51.0	2.89	2.45	3.09	3.65	3.57	3.48	4.13	3.89	3.67	3.04	3.35	3.00	40.21
SCHENECTADY	23.6	24.6	33.7	46.9	59.0	68.3	73.2	70.8	62.4	51.4	39.8	27.6	48.4	2.57	2.29	2.87	2.84	3.28	3.50	3.46	3.26	3.55	2.91	2.68	2.62	35.83
SMITHS BASIN	·	·	·	·	·	·	·	·	·	·	·	·	·	2.64	2.14	2.71	3.25	3.19	3.95	4.52	3.03	3.64	2.87	3.29	2.60	37.83
SPIER FALLS	22.1	23.2	32.7	46.0	58.1	67.4	71.8	69.9	62.1	51.2	39.2	26.2	47.5	3.09	2.64	3.14	3.33	3.26	3.42	3.88	3.17	3.23	2.95	3.38	3.19	38.68
WARWICK	·	·	·	·	·	·	·	·	·	·	·	·	·	3.00	2.53	3.68	3.73	3.79	4.04	4.36	4.50	3.94	3.30	4.03	3.39	44.29
WEST POINT	28.2	29.3	37.5	49.6	60.6	69.6	74.9	73.0	65.5	54.9	43.2	31.3	51.5	3.34	2.96	4.16	4.08	4.20	3.85	4.40	4.08	4.18	3.56	4.10	3.80	46.71
DIVISION	25.2	26.2	34.7	47.3	58.7	67.5	72.4	70.3	62.6	52.0	40.6	28.5	48.8	3.06	2.62	3.42	3.62	3.82	3.79	4.30	3.83	3.92	3.29	3.61	3.23	42.51

*NORMALS BY CLIMATOLOGICAL DIVISIONS

Taken from "Climatography of the United States No. 81-4,   Decennial Census of U. S. Climate"

TEMPERATURE (°F)   PRECIPITATION (In.)

STATIONS (By Divisions)	JAN	FEB	MAR	APR	MAY	JUNE	JULY	AUG	SEPT	OCT	NOV	DEC	ANN	JAN	FEB	MAR	APR	MAY	JUNE	JULY	AUG	SEPT	OCT	NOV	DEC	ANN
MOHAWK VALLEY																										
CANAJOHARIE	•	•	•	•	•	•	•	•	•	•	•	•	•	2.31	1.98	2.45	2.82	3.50	3.49	3.84	3.51	3.53	2.86	2.61	2.44	35.34
DELTA	•	•	•	•	•	•	•	•	•	•	•	•	•	3.72	3.28	3.42	3.68	3.76	3.63	3.90	3.61	3.74	3.94	3.83	3.82	44.33
DOLGEVILLE	•	•	•	•	•	•	•	•	•	•	•	•	•	3.25	2.78	3.13	3.46	3.92	3.94	4.74	3.69	4.39	3.61	3.36	3.12	43.39
FRANKFORT LOCK 19	•	•	•	•	•	•	•	•	•	•	•	•	•	2.77	2.37	2.61	3.27	3.50	3.63	4.17	3.50	4.20	3.59	3.42	2.70	39.73
GLOVERSVILLE	21.4	22.6	31.2	44.7	56.8	66.1	70.5	68.3	60.3	49.5	37.7	25.0	46.2	3.41	2.89	3.35	3.61	3.83	3.90	4.27	3.66	4.07	3.64	3.44	3.17	43.24
HINCKLEY	•	•	•	•	•	•	•	•	•	•	•	•	•	4.26	3.46	3.99	4.06	4.14	3.84	4.36	3.99	4.46	4.34	4.32	4.16	49.38
LITTLE FALLS CITY RES	21.0	21.7	30.3	44.1	56.2	65.3	70.1	68.2	60.7	49.8	37.5	24.4	45.8	2.85	2.30	2.82	3.37	3.64	3.92	4.51	3.93	4.36	3.56	3.22	2.83	41.31
LITTLE FALLS MILL ST	•	•	•	•	•	•	•	•	•	•	•	•	•	3.32	2.89	3.19	3.58	3.70	3.79	4.42	3.77	4.27	3.59	3.48	3.24	43.24
NEW LONDON LOCK 22	•	•	•	•	•	•	•	•	•	•	•	•	•	3.13	2.98	3.03	3.44	3.65	3.42	3.96	3.54	3.57	3.90	3.64	3.29	41.55
SALISBURY	18.4	18.9	27.8	41.4	53.2	62.4	66.9	64.9	57.3	46.8	35.1	21.8	42.9	3.71	3.18	3.33	3.94	4.14	4.10	4.94	4.08	4.61	4.23	4.01	3.53	47.80
TRENTON FALLS	•	•	•	•	•	•	•	•	•	•	•	•	•	4.00	3.35	3.70	4.06	4.11	4.20	4.73	3.92	4.47	4.38	4.32	4.17	49.41
TRIBES HILL	•	•	•	•	•	•	•	•	•	•	•	•	•	3.21	2.53	3.03	3.16	3.65	3.23	3.82	3.20	3.77	3.20	3.00	2.90	38.70
DIVISION	20.7	21.5	30.2	43.8	55.8	65.2	69.8	67.8	60.2	49.3	37.5	24.4	45.5	3.32	2.76	3.14	3.55	3.74	3.92	4.46	3.88	4.13	3.73	3.54	3.23	43.40
CHAMPLAIN VALLEY																										
DANNEMORA	17.9	19.1	27.9	42.0	55.1	64.5	69.1	67.3	59.7	48.5	35.4	21.6	44.0	2.23	2.18	2.28	2.75	3.26	3.45	3.35	3.09	3.28	2.88	2.49	2.48	33.72
WHITEHALL	•	•	•	•	•	•	•	•	•	•	•	•	•	2.89	2.55	2.86	3.07	3.09	3.42	3.90	3.20	3.43	2.84	3.27	2.92	37.44
DIVISION	18.4	19.6	29.1	43.3	55.9	65.3	69.9	67.9	59.9	48.7	36.5	22.7	44.8	2.26	2.11	2.34	2.72	3.05	3.27	3.39	3.00	3.25	2.72	2.60	2.44	33.15
ST. LAWRENCE VALLEY																										
CANTON 4 SE	17.3	18.6	28.7	43.3	55.6	65.1	69.7	67.7	59.6	48.7	36.4	21.9	44.4	2.41	2.32	2.50	3.01	3.35	3.11	3.59	3.18	3.46	3.03	2.90	2.79	35.65
LAWRENCEVILLE	17.2	18.5	28.4	43.0	55.8	65.2	69.9	67.9	60.0	48.7	36.3	21.5	44.4	1.70	1.82	1.96	2.71	3.39	3.13	3.49	3.23	3.50	3.12	2.36	2.10	32.51
OGDENSBURG 3 NE	17.9	19.7	29.2	44.0	55.9	65.6	70.7	69.0	60.9	50.2	37.7	23.1	45.3	2.09	1.97	2.26	2.54	2.92	2.60	2.93	2.86	2.96	2.55	2.58	2.44	30.70
DIVISION	17.5	18.9	28.7	43.1	55.3	64.9	69.6	67.6	59.7	48.8	36.6	22.2	44.4	2.32	2.26	2.49	2.93	3.32	2.96	3.49	3.06	3.48	3.11	2.86	2.74	35.02
GREAT LAKES																										
BATAVIA	•	•	•	•	•	•	•	•	•	•	•	•	•	2.25	2.24	2.68	2.93	3.08	2.62	2.94	2.96	2.73	2.59	2.59	2.30	31.91
BUFFALO WSO	24.5	24.1	31.5	43.5	54.8	64.8	69.8	68.4	61.4	50.8	39.1	27.7	46.7	2.84	2.72	3.24	3.01	2.95	2.54	2.57	3.05	3.13	3.00	3.60	3.00	35.65
CLYDE LOCK 26	•	•	•	•	•	•	•	•	•	•	•	•	•	2.33	2.59	3.02	3.18	3.71	3.06	3.40	3.00	3.20	3.61	3.07	2.69	36.86
FREDONIA	28.0	27.5	34.3	46.5	57.4	67.9	72.2	71.0	64.7	53.9	42.1	31.4	49.7	2.66	2.13	3.02	3.15	3.12	3.16	3.00	3.27	3.80	3.47	3.33	2.64	36.75
LOCKPORT 2 NE	25.5	25.3	32.6	44.9	55.9	66.2	71.2	69.6	62.5	51.8	40.0	28.9	47.9	2.44	2.38	2.54	2.69	3.19	2.37	2.61	3.02	2.93	2.73	2.67	2.40	31.97
MACEDON	•	•	•	•	•	•	•	•	•	•	•	•	•	2.03	2.40	2.73	2.88	3.25	2.75	2.98	2.84	2.73	3.25	2.49	2.33	32.66
NEWARK	•	•	•	•	•	•	•	•	•	•	•	•	•	1.99	2.18	2.62	2.71	3.35	2.79	3.13	2.72	2.70	3.39	2.60	2.23	32.41
OSWEGO EAST	25.1	25.4	32.6	44.0	54.6	64.5	70.5	69.4	62.3	52.2	40.8	29.0	47.5	2.70	2.62	2.80	2.72	2.97	2.28	2.74	2.51	2.78	3.26	3.01	2.72	33.56
ROCHESTER WSO	25.2	24.9	32.3	45.1	56.7	66.9	71.6	69.7	62.4	51.8	40.1	28.7	48.0	2.40	2.53	3.01	2.67	2.77	2.56	2.84	2.72	2.53	2.58	2.51	2.38	31.50
SOUTH WALES EMERY PARK	24.2	23.9	31.5	44.1	54.9	64.7	69.3	67.5	60.5	50.2	38.4	27.3	46.4	3.33	2.93	3.61	3.50	3.47	3.27	3.29	2.86	3.55	3.50	3.91	3.53	40.75
WATERTOWN	20.4	21.4	30.8	44.5	56.2	66.1	70.5	69.4	61.6	50.6	38.7	24.8	46.3	3.03	2.57	2.67	3.11	3.39	2.72	3.33	3.21	3.71	3.68	3.78	3.65	38.85
DIVISION	25.5	25.4	32.8	45.2	56.3	66.4	71.1	69.6	62.6	52.0	40.3	28.9	48.0	2.67	2.49	2.93	2.97	3.16	2.76	2.94	2.88	3.15	2.99	3.06	2.83	34.83
CENTRAL LAKES																										
AUBURN WATER WORKS	26.0	25.3	32.2	44.4	55.7	66.4	72.1	70.8	63.7	52.9	42.4	30.8	48.6	2.27	2.38	2.98	2.69	2.93	2.86	3.18	2.45	2.50	2.91	2.69	2.58	32.42
BALDWINSVILLE	•	•	•	•	•	•	•	•	•	•	•	•	•	2.94	3.16	3.46	3.04	4.36	3.22	3.19	3.16	3.21	3.52	3.25	3.09	38.60
CAYUGA LOCK 1	•	•	•	•	•	•	•	•	•	•	•	•	•	2.11	2.29	2.80	2.98	3.31	2.93	2.74	2.94	3.26	2.78	2.43		34.06
HEMLOCK	25.4	24.9	32.2	44.9	56.7	66.8	71.2	69.3	62.0	51.6	40.2	29.0	47.9	1.84	1.88	2.93	2.69	3.25	3.03	2.87	2.86	2.58	2.73	2.39	1.99	31.04
MAYS POINT LOCK 25	•	•	•	•	•	•	•	•	•	•	•	•	•	2.22	2.39	2.83	2.74	3.28	2.98	3.54	2.66	2.91	3.34	2.74	2.54	34.17
PENN YAN 2 SW	26.3	26.2	33.6	46.4	57.7	67.4	72.2	70.6	63.4	52.5	40.8	29.4	48.9	2.06	1.91	2.80	2.81	3.32	2.99	3.72	2.85	2.31	2.79	2.27	1.99	31.82
SKANEATELES	•	•	•	•	•	•	•	•	•	•	•	•	•	2.43	2.61	3.22	3.10	3.53	3.33	3.64	2.90	3.19	3.44	3.02	2.77	37.18
SYRACUSE WSO	24.0	24.3	32.6	46.0	57.7	67.3	72.2	70.2	62.4	51.8	40.2	27.8	48.0	3.15	3.13	3.60	3.08	3.27	2.96	3.09	3.25	2.84	3.18	2.90	3.15	37.60
WATERLOO	•	•	•	•	•	•	•	•	•	•	•	•	•	2.20	2.27	2.82	2.89	3.27	2.87	3.49	2.65	2.70	3.11	2.51	2.21	32.99
DIVISION	25.7	25.5	33.0	45.7	56.9	66.8	71.4	69.6	62.5	51.9	40.5	29.0	48.2	2.08	2.12	2.83	2.82	3.16	3.08	3.30	3.01	2.69	2.84	2.51	2.22	32.66

* Normals for the period 1931-1960. Divisional normals may not be the arithmetical average of individual stations published, since additional data for shorter period stations are used to obtain better areal representation.

CONFIDENCE - LIMITS

In absence of trend or record changes, the chances are 9 out of 10 that the true mean will lie in the interval formed by adding and subtracting the values in the following table from the means for any station in the State. Because of the wider variation in mean precipitation, the corresponding monthly means and annual mean must be substituted for "p" in the precipitation table below to obtain mean precipitation confidence limits.

| 1.6 | 1.4 | 1.5 | 1.1 | .9 | .7 | .6 | .8 | .9 | 1.0 | 1.0 | 1.2 | .5 | $.21\sqrt{p}$ | $.17\sqrt{p}$ | $.21\sqrt{p}$ | $.21\sqrt{p}$ | $.25\sqrt{p}$ | $.27\sqrt{p}$ | $.26\sqrt{p}$ | $.27\sqrt{p}$ | $.31\sqrt{p}$ | $.31\sqrt{p}$ | $.26\sqrt{p}$ | $.24\sqrt{p}$ | $.25\sqrt{p}$ |

COMPARATIVE DATA

Data in the following table are the mean temperature and average precipitation for Geneva Experiment Station, New York, for the period 1906-1930 and are included in this publication for comparative purposes.

| 25.2 | 24.6 | 34.4 | 45.8 | 56.7 | 66.3 | 71.6 | 69.5 | 63.7 | 51.7 | 40.1 | 28.9 | 48.2 | 2.01 | 1.65 | 2.24 | 3.01 | 3.27 | 3.56 | 3.27 | 3.12 | 2.80 | 2.62 | 2.40 | 2.02 | 31.97 |

NORMALS, MEANS, AND EXTREMES

Station: ALBANY, NEW YORK — ALBANY COUNTY AIRPORT
Standard time used: EASTERN Latitude: 42° 45' N Longitude: 73° 48' W Elevation (ground): 275 feet

Station: BINGHAMTON, NEW YORK — BROOME COUNTY AIRPORT
Standard time used: EASTERN Latitude: 42° 13' N Longitude: 75° 59' W Elevation (ground): 1590 feet

Ø For period February 1965 through current year.
Means and extremes above are from existing and comparable exposures. Annual extremes have been exceeded at other sites in the locality as follows:
Highest temperature 104 in July 1911; maximum monthly precipitation 13.48 in October 1869; minimum monthly precipitation 0.08 in January 1860; maximum precipitation in 24 hours 4.75 in October 1903; maximum snowfall in 24 hours 30.4 in March 1888.

NORMALS, MEANS, AND EXTREMES

Station: BUFFALO, NEW YORK — GREATER BUFFALO INTL AIRPORT
Standard time used: EASTERN Latitude: 42° 56' N Longitude: 78° 44' W Elevation (ground): 705 feet

Month	Normal Daily max	Normal Daily min	Normal Monthly	Normal heating degree days (Base 65°)	Precip Normal total	Snow, Ice pellets Mean total
J	30.8	18.2	24.5	1256	2.84	22.3
F	31.0	17.2	24.1	1145	2.72	17.8
M	38.6	24.0	31.5	1039	3.24	12.1
A	52.9	34.0	43.5	645	3.01	2.8
M	65.5	44.1	54.8	329	2.95	0.2
J	75.1	54.5	64.8	78	2.54	0.0
J	80.1	59.4	69.8	19	2.57	0.0
A	78.6	58.1	68.4	37	3.05	0.0
S	71.5	51.2	61.4	141	3.13	T
O	60.1	41.7	50.8	440	3.00	0.2
N	46.5	31.1	39.1	777	3.60	12.0
D	34.3	21.1	27.7	1156	3.00	20.2
YR	55.4	37.9	46.7	7062	35.65	88.0

Means and extremes above are from existing and comparable exposures. Annual extremes have been exceeded at other sites in the locality as follows: Highest temperature 99 in August 1948; lowest temperature -21 in February 1934; maximum monthly precipitation 10.63 in August 1885; minimum monthly precipitation .05 in August 1876; maximum precipitation in 24 hours 4.28 in August 1893.

Station: NEW YORK, NEW YORK — CENTRAL PARK OBSERVATORY
Standard time used: EASTERN Latitude: 40° 47' N Longitude: 73° 58' W Elevation (ground): 132 feet

Month	Normal Daily max	Normal Daily min	Normal Monthly	Normal heating degree days (Base 65°)	Precip Normal total	Snow, Ice pellets Mean total
J	39.5	26.9	33.2	986	3.31	7.7
F	40.3	26.4	33.4	885	2.84	8.6
M	47.8	33.2	40.5	760	4.01	5.1
A	60.5	43.1	51.4	408	3.31	1.0
M	71.4	53.4	62.4	118	3.31	T
J	80.2	62.5	71.2	0	3.31	0.0
J	85.3	68.2	76.8	0	3.70	0.0
A	83.8	66.8	75.1	0	4.44	0.0
S	76.7	60.1	68.1	33	3.87	0.0
O	66.3	50.3	58.3	233	3.14	T
N	54.5	41.4	47.0	540	3.39	1.1
D	42.1	29.7	35.9	902	3.26	6.1
YR	62.2	46.7	54.5	4871	42.37	29.8

Means and extremes above are from existing and comparable exposures. Annual extremes have been exceeded at other sites in the locality as follows: Maximum monthly snowfall 37.9 inches in February 1894; fastest mile of wind 113 from the SE in October 1954.

NORMALS, MEANS, AND EXTREMES

Station: YORK, NEW YORK   JOHN F. KENNEDY INTL AIRPORT   Standard time used: EASTERN   Latitude: 40° 39' N   Longitude: 73° 47' W   Elevation (ground): 13 feet

Month	Normal Daily maximum	Normal Daily minimum	Normal Monthly	Record highest	Year	Record lowest	Year	Normal heating degree days (Base 65°)	Precip Normal total	Precip Max monthly	Year	Precip Min monthly	Year	Precip Max in 24 hrs	Year	Snow Mean total	Snow Max monthly	Year	Snow Max in 24 hrs	Year
J	38.0	25.6	31.8	64	1967	11	1968	1029	3.23	5.77	1949	0.21	1964	1.60	1958	7.4	17.4	1965	13.0	1964
F	38.5	24.7	31.6	62	1965	-2	1963	935	2.93	5.48	1960	1.73	1968	2.87	1958	7.8	21.1	1968	13.0	1969
M	45.5	31.4	38.7	77	1967	10	1967	815	4.15	7.93	1953	1.35	1966	2.27	1966	5.0	21.1	1960	8.1	1960
A	57.5	40.4	49.0	84	1963	26	1968	480	3.67	6.60	1964	1.12	1963	2.12	1971	0.6	T	1971	T	1967
M	68.0	50.9	59.9	95	1962	34	1966	167	3.35	6.14	1951	0.38	1968	2.88	1968	T	T		T	
J	79.3	60.9	70.1	99	1964	45	1964	12	3.35	4.27	1949	T	1949	1.81	1968	0.0	0.0		0.0	
J	84.9	67.1	75.9	104	1966	55	1969	0	4.04	8.48	1954	0.46	1954	3.21	1969	0.0	0.0		0.0	
A	82.9	66.0	74.5	99	1968	46	1955	0	4.16	17.41	1955	0.87	1964	6.59	1955	0.0	0.0		0.0	
S	76.2	59.3	67.8	94	1961	40	1963	36	4.16	9.60	1951	0.70	1941	5.83	1960	0.0	0.0		0.0	
O	66.0	49.3	57.8	73	1971	25	1969	248	3.51	8.49	1958	0.09	1963	3.21	1963	T	T	1962	T	1967
N	53.2	39.1	46.2	68	1962	20	1962	564	3.23	7.89	1967	1.30	1949	2.93	1963	0.2	2.1	1958	2.1	1960
D	41.3	28.4	34.9	68	1964	5	1962	933	3.23	6.16	1969	1.46	1971	2.05	1968	5.5	16.4	1960	8.2	1960
YR	61.1	45.3	53.2	104 JUL. 1966		-2 FEB. 1963		5219	43.93	17.41 AUG. 1955		T JUN. 1949		6.59 AUG. 1955		27.1	25.3 FEB. 1961		19.9 FEB. 1969	

Relative humidity (Local time) Hour 01 / 07 / 13 / 19 — YR: 74 / 73 / 57 / 67. Mean sky cover sunrise to sunset YR: 6.0. Wind mean speed YR: 12.0. Fastest mile speed 52, direction 26, JAN. 1966.

Days sunrise to sunset — Clear YR 96, Partly cloudy YR 122, Cloudy YR 147. Precipitation .01 inch or more YR 117. Thunderstorms YR 23. Heavy fog YR 31. Temperatures: Max 90° and above YR 11; Max 32° and below YR 21; Min 32° and below YR 91.

Ø For the period June 1961 through the current year.
$ Greatest calendar day through March 1969.
# Greatest calendar day August 1966 through March 1969.

Station: NEW YORK, NEW YORK   LA GUARDIA AIRPORT   Standard time used: EASTERN   Latitude: 40° 46' N   Longitude: 73° 54' W   Elevation (ground): 11 feet

Month	Normal Daily maximum	Normal Daily minimum	Normal Monthly	Record highest	Year	Record lowest	Year	Normal heating degree days (Base 65°)	Precip Normal total	Precip Max monthly	Year	Precip Min monthly	Year	Precip Max in 24 hrs	Year	Snow Mean total	Snow Max monthly	Year	Snow Max in 24 hrs	Year
J	39.6	27.5	33.6	68	1967	10	1968	973	3.31	5.77	1949	0.76	1970	2.12	1941	6.8	18.3	1948	9.0	1964
F	40.4	26.8	33.6	68	1971	-2	1963	879	3.09	5.76	1960	1.37	1968	2.90	1941	8.4	18.5	1947	17.4	1941
M	47.8	33.8	40.8	77	1963	10	1967	750	4.23	8.73	1953	1.87	1966	3.25	1953	5.4	18.9	1956	15.3	1958
A	59.3	43.5	51.2	84	1964	25	1969	414	3.51	7.36	1961	1.21	1963	2.52	1958	0.8	6.4	1961	6.4	1961
M	70.0	53.4	62.1	95	1962	38	1966	124	3.58	7.42	1948	1.43	1964	3.02	1964	0.0	0.0		0.0	
J	80.7	63.0	72.5	97	1964	47	1967	6	3.38	6.16	1968	0.03	1949	3.26	1946	0.0	0.0		0.0	
J	84.8	68.8	76.8	107	1966	57	1969	0	3.71	9.27	1960	0.69	1954	3.82	1971	0.0	0.0		0.0	
A	82.9	67.8	75.4	94	1968	53	1965	0	5.08	16.05	1955	0.24	1964	7.11	1955	0.0	0.0		0.0	
S	61.3	61.3	68.8	94	1965	41	1963	27	5.02	8.04	1944	0.62	1941	4.52	1959	0.0	0.0		0.0	
O	65.9	51.3	58.6	85	1967	30	1969	223	3.37	7.92	1943	0.06	1963	3.36	1943	T	1.2	1962	1.2	1953
N	53.0	41.0	47.4	76	1971	22	1962	528	3.59	5.82	1955	1.03	1955	3.68	1963	0.7	7.0	1953	6.7	1947
D	42.4	30.3	36.4	66	1970	3	1962	887	3.39		1967	0.31	1955	3.44	1941	6.7	26.1	1947	22.8	1947
YR	62.0	47.3	54.7	107 JUL. 1966		-2 FEB. 1963		4811	44.22	16.05 AUG. 1955		0.03 JUN. 1949		7.11 AUG. 1955		28.6	26.1 DEC. 1947		22.8 DEC. 1947

Relative humidity (Local time) Hour 01 / 07 / 13 / 19 — YR: 67 / 69 / 53 / 59. Mean sky cover sunrise to sunset YR: 6.1. Wind mean speed YR: 12.4. Fastest mile speed 70, direction 26, SEP. 1960.

Days sunrise to sunset — Clear YR 95, Partly cloudy YR 117, Cloudy YR 153. Precipitation .01 inch or more YR 119. Thunderstorms YR 24. Heavy fog YR 14. Temperatures: Max 90° and above YR 14; Max 32° and below YR 22; Min 32° and below YR 78.

Ø For period May 1962 through the current year.
Means and extremes above are from existing and comparable exposures. Annual extremes have been exceeded at other sites in the locality as follows:
Lowest temperature -7 in February 1943.

ROCHESTER, NEW YORK

Station: ROCHESTER—MONROE COUNTY AP Standard time used: EASTERN Latitude: 43° 07' N Longitude: 77° 40' W Elevation (ground): 547 feet

Ø For period July 1963 through the current year.
Means and extremes above are from existing and comparable exposures. Annual extremes have been exceeded at other sites in the locality as follows:
Highest temperature 102 in July 1936; lowest temperature -22 in February 1934; minimum monthly precipitation 0.08 in October 1924; maximum precipitation in 24 hours 4.19 in August 1893; maximum snowfall in 24 hours 29.8 in March 1900.

SYRACUSE, NEW YORK

Station: HANCOCK AIRPORT Standard time used: EASTERN Latitude: 43° 07' N Longitude: 76° 07' W Elevation (ground): 410 feet

Ø For period July 1963 through the current year.
Means and extremes above are from existing and comparable exposures. Annual extremes have been exceeded at other sites in the locality as follows:
Highest temperature 102 in July 1936; minimum monthly precipitation 15.92 in May 1920; minimum monthly precipitation 0.19 in May 1920; maximum precipitation in 24 hours 4.79 in June 1922; maximum snowfall in 24 hours 27.2 in January 1925; fastest mile wind 69 SW in December 1921.

REFERENCE NOTES APPLYING TO ALL "NORMALS, MEANS, AND EXTREMES" TABLES

Unless otherwise indicated, dimensional units used in this bulletin are: temperature in degrees F; precipitation, including snowfall, in inches; wind movement in miles per hour; and relative humidity in percent. Heating degree day totals are the sums of negative departures of average daily temperatures from 65° F. Cooling degree day totals are the sums of positive departures of average daily temperatures from 65° F. Sleet was included in snowfall totals beginning with July 1948. The term "ice pellets" includes solid grains of ice (sleet) and particles consisting of snow pellets encased in a thin layer of ice. Heavy fog reduces visibility to 1/4 mile or less.

Sky cover is expressed in a range of 0 for no clouds or obscuring phenomena to 10 for complete sky cover. The number of clear days is based on average cloudiness 0-3, partly cloudy days 4-7, and cloudy days 8-10 tenths.

Solar radiation data are the average of direct and diffuse radiation on a horizontal surface. The langley denotes one gram calorie per square centimeter.

(a) Length of record, years, based on January data, unless otherwise indicated. Other months may be for more or fewer years if there have been breaks in the record.
(b) Climatological standard normals (1931-1960).
— Less than one half.
+ Also on earlier dates, months, or years.
T Trace, an amount too small to measure.
Below zero temperatures are preceded by a minus sign. The prevailing direction for wind in the Normals, Means, and Extremes table is from records through 1963.
✦ >70° at Alaskan stations.

& Figures instead of letters in a direction column indicate direction in tens of degrees from true North; i.e., 09 - East, 18 - South, 27 - West, 36 - North, and 00 - Calm. Resultant wind is the vector sum of wind directions and speeds divided by the number of observations. If figures appear in the direction column under "Fastest mile" the corresponding speeds are fastest observed 1-minute values.

To 8 compass points only.

MEAN SEASONAL SNOWFALL, INCHES

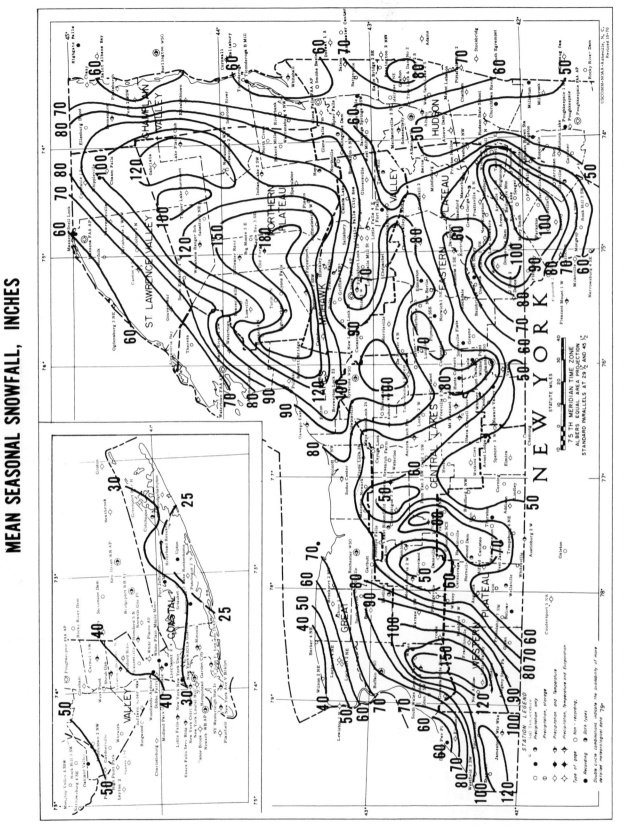

Data are based on the period 1931-68. Isolines are drawn through points of approximately equal value. Caution should be used in interpolating on these maps, particularly in mountainous areas.

MEAN ANNUAL PRECIPITATION, INCHES

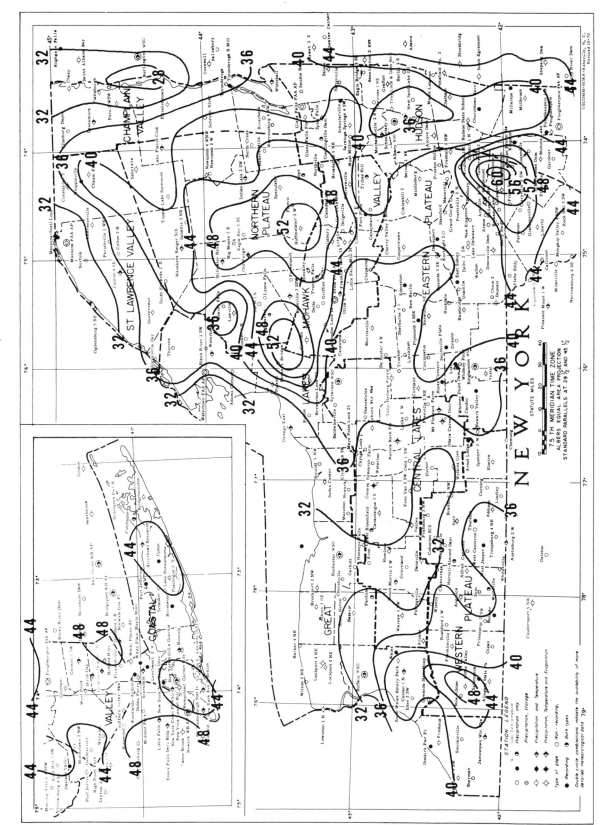

Data are based on the period 1931-55. Isolines are drawn through points of approximately equal value. Caution should be used in interpolating on these maps, particularly in mountainous areas.

MEAN MAXIMUM TEMPERATURE (°F.), JANUARY

ST. LAWRENCE VALLEY
CHAMPLAIN VALLEY
NORTHERN PLATEAU
MOHAWK
VALLEY
HUDSON
LAKES
GREAT
CENTRAL LAKES
EASTERN PLATEAU
WESTERN PLATEAU
COASTAL
VALLEY
NEW YORK
STATION LEGEND
Soil Temperature
Precipitation only
Precipitation, storage
Precipitation and Temperature
Precipitation, Temperature and Evaporation
Type of gage
Non - recording,
Recording
Both types
Double circle combinations indicate the availability of more detailed meteorological data
STATUTE MILES
10 0 10 20 30 40
75 TH MERIDIAN TIME ZONE
ALBERS EQUAL AREA PROJECTION
STANDARD PARALLELS AT 29½ AND 45½
USCOMM-NOAA-Asheville, N. C.
Revised 10-70

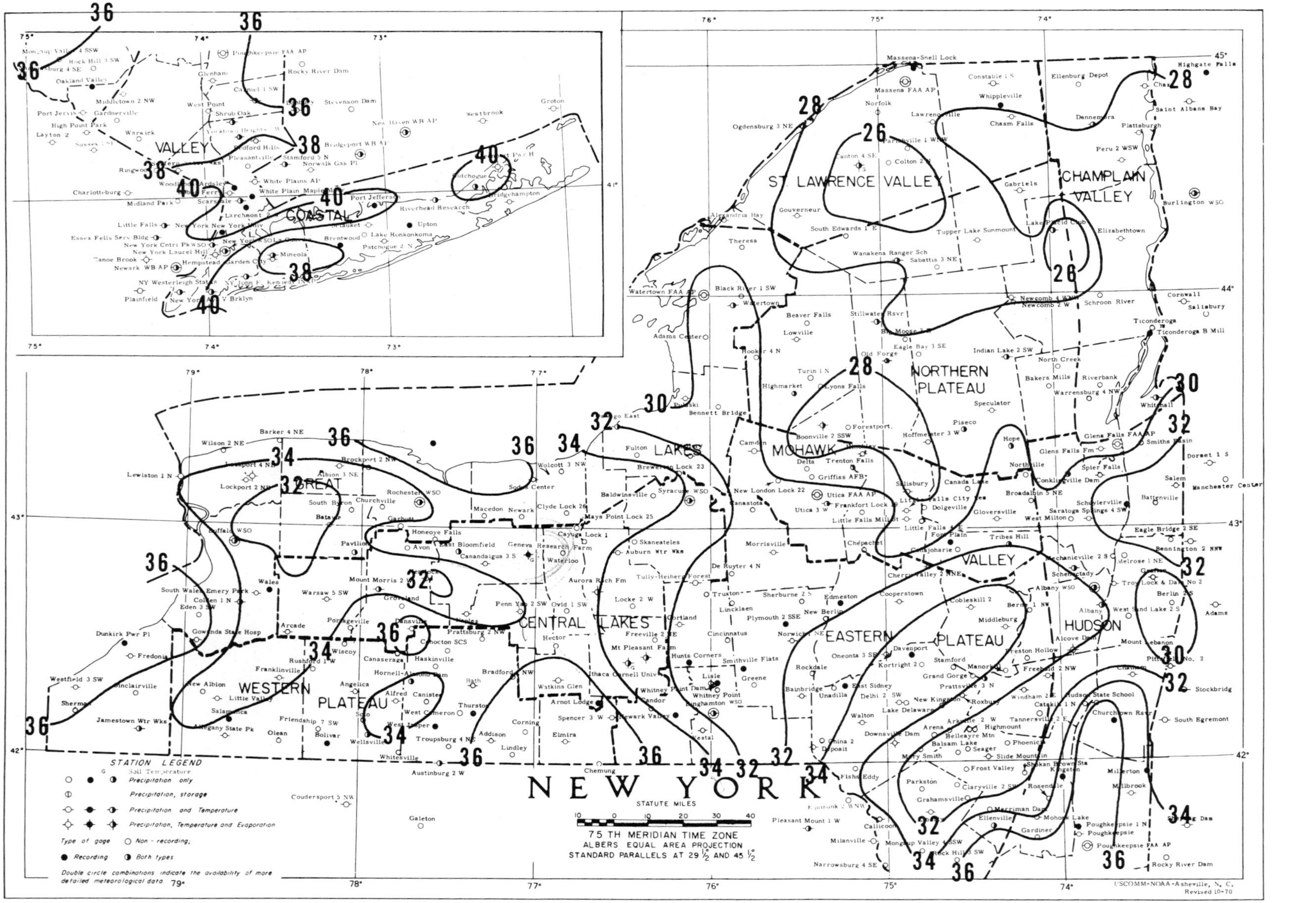

Data are based on the period 1931-52. Isolines are drawn through points of approximately equal value. Caution should be used in interpolating on these maps, particularly in mountainous areas.

MEAN MINIMUM TEMPERATURE (°F.), JANUARY

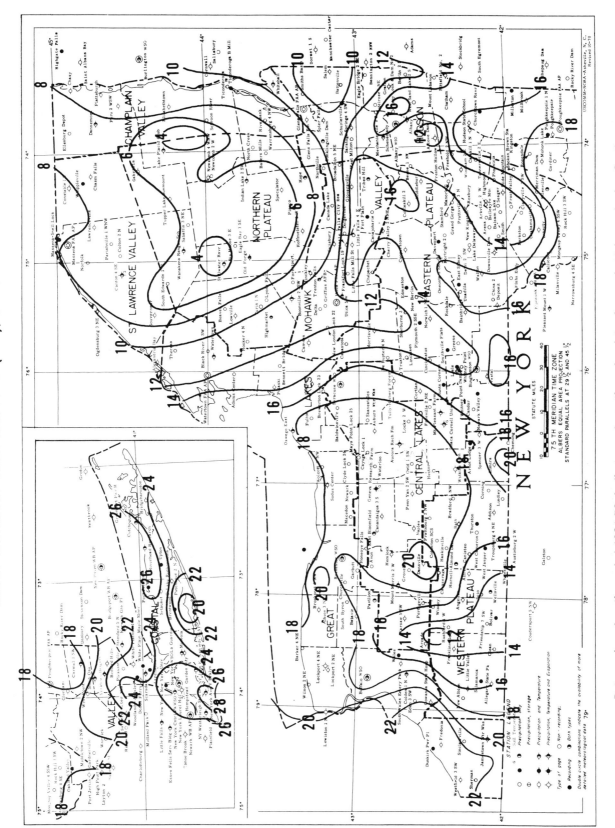

Data are based on the period 1931-52. Isolines are drawn through points of approximately equal value. Caution should be used in interpolating on these maps, particularly in mountainous areas.

MEAN MAXIMUM TEMPERATURE (°F.), JULY

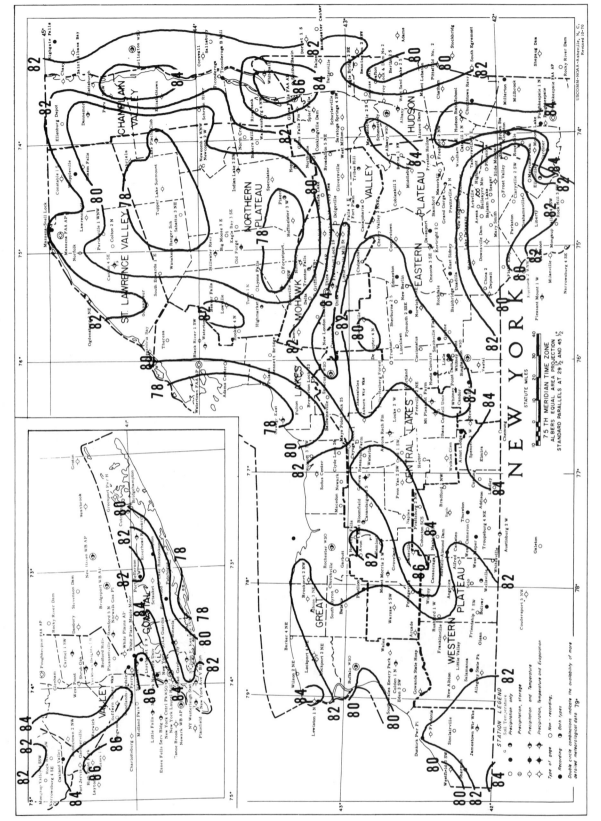

Data are based on the period 1931-52. Isolines are drawn through points of approximately equal value. Caution should be used in interpolating on these maps, particularly in mountainous areas.

MEAN MINIMUM TEMPERATURE (°F.), JULY

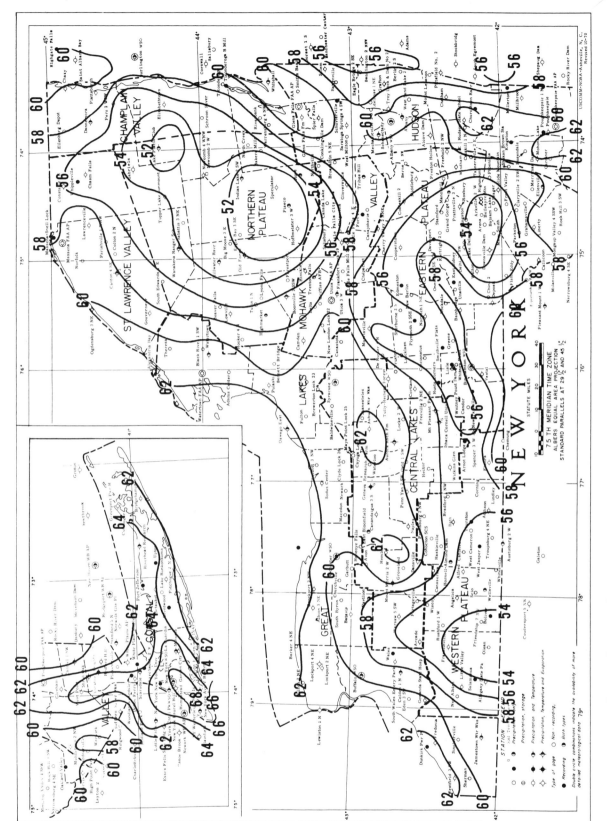

Data are based on the period 1931-52. Isolines are drawn through points of approximately equal value. Caution should be used in interpolating on these maps, particularly in mountainous areas.

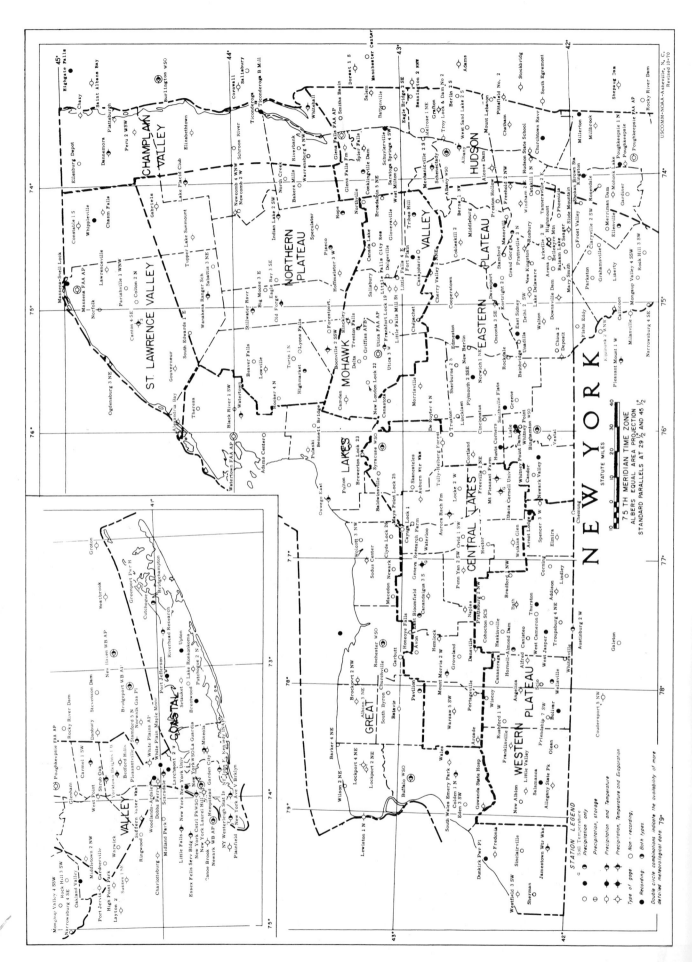

NEW YORK

75TH MERIDIAN TIME ZONE
ALBERS EQUAL AREA PROJECTION
STANDARD PARALLELS AT 29½ AND 45½

STATUTE MILES

STATION LEGEND

Type of gage ○ Non-recording
                ● Recording

Precipitation only

Precipitation, storage

Precipitation and Temperature

Precipitation, Temperature and Evaporation

T Soil Temperature

● Both types

Double circle combinations indicate the availability of more
detailed meteorological data

CHAMPLAIN VALLEY

ST. LAWRENCE VALLEY

NORTHERN PLATEAU

MOHAWK VALLEY

EASTERN PLATEAU

HUDSON VALLEY

LAKES

CENTRAL LAKES

GREAT LAKES

WESTERN PLATEAU

COASTAL

VALLEY

USCOMM-NOAA-Asheville, N.C.
Revised 10-70

THE CLIMATE OF
NORTH CAROLINA

by
Albert V. Hardy

June 1970

North Carolina lies between 33 1/2° and 37° north latitude and between 75° and 84 1/2° west longitude. The span of longitude is greater than that of any other state east of the Mississippi River. The greatest length from east to west is 503 miles. The greatest breadth from north to south is 187 miles. The total area is 52,712 square miles; 49,142 square miles of land and 3,570 square miles of water.

The range of altitude is also the greatest of any eastern state. North Carolina rises from sea level along the Atlantic coast to 6,684 feet at the summit of Mount Mitchell, the highest peak in the eastern United States. Mount Mitchell is in the heart of the Blue Ridge Range. This Range, along with the Great Smokies, lies partly in North Carolina and partly in Tennessee and forms the highest part of the Appalachian Mountains.

The three principal physiographic Divisions of the eastern United States are particularly well developed in North Carolina. Beginning in the east, they are: The Coastal Plain, the Piedmont, and the Mountains. For assembly and study of climate and crop statistics, the Mountain Division is subdivided into a northern and a southern sector, while the Coastal Plain and Piedmont Divisions are each subdivided into northern, central, and southern sectors. Thus a total of eight climatological subdivisions are recognized.

The land and water areas of the Coastal Plain Division comprise nearly half the area of the State. The tidewater portion is generally flat and swampy, while the interior is gently sloping and, for the most part, naturally well drained. Throughout the Coastal Plain the soils consist of soft sediment, with little or no underlying hard rock near the surface. The average slope is from about 200 feet above mean sea level at the "fall line", or western boundary, to generally 30 feet or less over the tidewater area.

The Piedmont Division rises gently from about 200 feet at the fall line to near 1,500 feet at the base of the Mountains; its area is about one-third of the State. The land is mostly gently rolling, with a great deal of hard rock near the surface. There are several ranges of steeper hills. The principal of these are the Uwharrie Range in and around Randolph County, and the Kings Mountain Range in Cleveland and Gaston Counties.

The Mountain Division is the smallest of the three, little more than one-fifth of the State's area. In elevation it ranges downward from Mount Mitchell's peak to about 1,000 feet above mean sea level in the lowest valleys. There are more than 40 peaks higher than 6,000 feet and about 80 others over 5,000 feet high. The surface of the Mountain Division is rocky, the soils being mainly of weathered and eroded rocky materials.

North Carolina has the most varied climate of any eastern state. This is due mainly to its wide range in elevation and distance from the ocean. Lesser influences are latitude, inland bodies of water, soil surface, and plant cover. In all seasons of the year the average temperature varies more than 20° from the lower coast to the highest mountain elevations. The average annual temperature at Southport on the lower coast is nearly as high as that of interior northern Florida, while the average on the summit of Mount Mitchell is lower than that of Buffalo, New York.

Altitude also has an important effect on rainfall. The rainiest part of the eastern United States, with an annual average of more than 80 inches, is in southwestern North Carolina where moist southerly winds are forced upward in passing over the mountain barrier. Less than 50 miles north, in the valley of the French Broad River, is the driest place south of Virginia and east of the Mississippi. Sheltered by mountain ranges on all sides, this point has an average annual precipitation of only 37 inches. East of the Mountains, average annual rainfall ranges mostly 40 to 55 inches.

In winter the greater part of North Carolina is partially protected by the mountain ranges from the frequent outbreaks of cold which move southeastward across the Central States. Such outbreaks often spread southward all the way to the Gulf of Mexico without attaining strength and depth to cross the Appalachian Range. When cold waves do break across they are usually modified by the crossing and the descent on the eastern slopes. The temperature drops to around 10° over central North Carolina once or twice during an average winter. Near the coast a comparable figure is some 10 degrees higher, and in the upper mountains 10 degrees lower. Temperatures as low as 0° are rare outside the mountains, but have occurred at one time or another throughout the western part of the State. The lowest temperature of record is -29°, recorded January 30, 1966, at Mount Mitchell 2 SSW.

Winter temperatures in the eastern Coastal Plain are modified by the proximity of the Atlantic Ocean. This effect raises the average winter temperature and reduces the average day-to-night range. The Gulf Stream, contrary to popular opinion, has little direct effect on North Carolina temperatures, even on the immediate coast. The Stream lies some 50 miles offshore at its nearest point. The southern reaches of the cold Labrador Current pass between the Gulf Stream and the North Carolina coast. This offsets any warming effect the Stream might otherwise have on coastal temperatures. The meeting of the two opposing currents does provide a breeding ground for rough weather. Not infrequently low pressure storms having their origin there develop major proportions, causing rain on the North Carolina coast and over states to the north.

In spring the storm systems that bring cold weather southward reach North Carolina less forcefully than in winter, and temperatures begin to modify. Day-to-day variations in temperature are less pronounced, and warm weather is more likely to occur in conjunction with fair weather.

The rise in average temperatures is greater in May than in any other month. Artificial heating is generally discontinued at some time during May.

Occasional mild invasions of air from the north continue to occur during the summer, but their effect on temperatures is slight and of short duration. Ordinarily, such outbreaks serve principally to clear the air of excess humidity. The increase in sunshine which follows usually brings temperatures back up quickly. When the drying of the air is sufficient to keep cloudiness at a minimum for several days, temperatures may occasionally reach 100° or a little higher in interior sections at elevations below 1,500 feet. The highest of record is 109°. Ordinarily, however, summer cloudiness develops to limit the sun's heating while temperatures are still in the 90° range. An entire summer, or even several consecutive summers, may pass without a high of 100° being recorded in the State. The average daily maximum reading in midsummer is below 90° for most localities.

Differences in temperatures over the various parts of the State are no less pronounced in summer than in winter. The warmest days, however, are found in the interior rather than on the coast in summer. The average daily maximum at midsummer is near 92° at Fayetteville and Goldsboro. On the extreme south coast it is only 86° at that season. The average mid-July maximum is only 67° atop Mount Mitchell, while over widely populated areas in the Mountain Division the figure is about 75°. Lowest morning temperatures average about 20° lower than afternoon highs except along the immediate coast, where the daily range is most often between 10° and 15°.

Autumn is the season of most rapidly changing temperature, the daily downward trend being greater than the corresponding rise in spring. The dropoff is most rapid in October and continues almost as fast in November. Average daily temperatures by the end of that month are generally within 5° of the annual low.

There are no distinct wet and dry seasons in North Carolina. There is some seasonal variation in average precipitation. Summer rainfall is normally the greatest, and July the wettest month. Since the rain at this time of year comes mostly with thunderstorms and convective showers, it is also more variable than at other seasons. Daily showers are not uncommon, nor are periods of one or two weeks without rain. Autumn is the driest season, and October the driest month. Precipitation in winter and spring occurs mostly with migratory low pressure storms. It appears with greater regularity and more even distribution than summer showers.

Winter precipitation usually occurs with southerly through easterly winds, and is seldom associated with very cold weather. Snow and sleet occur on an average of once or twice a year near the coast, and not much more often over the southeastern half of the State. Such occurrences are nearly always connected with northeasterly winds, generated when high pressure over the interior or the northeastern United States causes a flow of cold air down parallel to the coastline, while offshore low pressure brings in cool, moist air from the North Atlantic. Over the Mountains and western Piedmont frozen precipitation sometimes occurs with interior low pressure storms. In the extreme west it can happen with a cold front passage from northwest. Average winter snowfall ranges from about 1 inch per year on the Outer Banks and the lower coast, to about 9 inches in the northern Piedmont and southern Mountains. Some of the higher mountain peaks and upper slopes receive an average of nearly 50 inches a year.

The greatest North Carolina 24-hour snowfall of record was 31 inches at Nashville in March 1927. The record fall for an entire season was 148 inches at Mount Mitchell in 1930-31. The greatest 24-hour rainfall of record, 22.22 inches, fell at Altapass in July 1916. The greatest annual total of 129.60 inches was recorded at Rosman in 1964.

Relative humidity may vary greatly from day to day and even from hour to hour, especially in winter. Observed percentages range from 100 down to 10 or lower. The average relative humidity, however, does not vary greatly from season to season, there being a slight tendency for highest averages in winter and lowest in spring. The lowest relative humidities are found over the southern Piedmont, where the year around average is about 65 percent. The highest are along the immediate coast, where the average may be as high as 75 to 80 percent. The lowest amount of actual moisture is found in the higher mountain areas, but the lower temperatures there bring the relative humidity up as high or higher than other interior areas.

Sunshine is abundant, the average annual percentage of possible ranging from 60 to 65 percent at most recording points. Observations of sky condition taken at airport locations in recent years indicate an average of about 112 days per year clear, 105 partly cloudy, and 148 cloudy. Measurable rain falls on about 120 days. Prevailing winds blow from southwest 10 months of the year, and from northeast during September and October. The average wind speed for interior locations is about 8 m.p.h., for coastal points about 12. The highest wind of record (fastest mile) is 110 m.p.h. at Hatteras in September 1944.

North Carolina rivers fall into two groups: Those that flow into the Atlantic Ocean and those that drain westward into the Mississippi River system. The two are separated by a ridge averaging 2,200 feet above mean sea level. A second chain of mountains ranging up to 6,000 feet marks the western boundary of the State.

Most of the State, including the Coastal Plain, the Piedmont, and the eastern and southern slopes, drain into the Atlantic Ocean. The principal rivers involved are the Roanoke, Tar, Neuse, Cape Fear, Yadkin, and Catawba. The Roanoke rises in the Allegheny Mountains west of Roanoke, Virginia, and flows southeasterly through Virginia and North Carolina a distance of 400 miles, about 150 miles of which is in North Carolina. Flood peaks which used to beset the lower Roanoke are now much subdued by John H. Kerr Dam just upstream from the North Carolina-Virginia State border, and other dams both upstream and downstream. The Tar, Neuse, and Cape Fear Rivers rise in the North Carolina Piedmont and flow southeastwardly to the Atlantic. The Yadkin and Catawba Rivers rise in the Blue Ridge Mountains of western North Carolina, but reach the Atlantic via the Pee Dee and Santee Rivers of South Carolina.

The main stream draining the extreme western part of North Carolina is the French Broad River, which rises in the mountains southwest of Asheville. It flows first northward, then westward through the Great Smoky Mountains into Tennessee, there to join the Holston River near Knoxville to form the Tennessee River, which in turn feeds into the Ohio. Other parts of the southwestern mountain area drain into the Little Tennessee. The northern mountains are drained by streams flowing into the Ohio River system. All eventually reach the Mississippi.

Intense rainstorms occur in the precipitous mountain terrain, especially in the southern portion. Streams here rise quickly to flood, and almost as quickly subside when rain ends. Because of the unusually abundant rainfall and infrequent freezing of the rivers, North Carolina streams when properly dammed furnish a reliable flow of water for hydroelectric development. Extensive use has been made of this resource in the area drained by tributaries of the Tennessee River as developed by the Tennessee Valley Authority. Waters draining eastward have been harnessed less extensively. A survey taken several years ago showed North Carolina as the fourth state in the nation in installed capacity of hydroelectric generators. In addition to public power projects, there were estimated to be nearly 100 private hydroelectric generating plants.

Floods occur frequently, affecting some part of North Carolina each year. Loss of life is rare, and the economic loss not generally large, but the cost of floods is increasing as river lowgrounds are developed. Floods may occur at any season, but are most frequent in early spring, summer, and early fall. Rains associated with West Indian hurricanes are the main cause of summer and fall floods. In mid-August 1940, severe floods occurred as a result of hurricane rains. Later in the same month intense rains of local origin caused severe flooding in western North Carolina. Major floods also occurred in September 1945, October 1929, August 1928, and July 1916.

The greatest economic loss entailed in North Carolina because of stormy weather is that due to summer thunderstorms. These usually affect only limited areas, but hail and wind occurring with some of them account for an average yearly loss of about $5 million. Three to five people are killed in the State by lightning during the average year. Farm livestock, especially cattle, are killed in larger numbers, and there is a considerable loss of property due to fires set by lightning. In any given locality, 40 to 50 days with thunderstorms may be expected in a year.

North Carolina is outside the principal tornado area of the United States. A total of 192 tornadoes were reported during the 53-year period 1916-1968, an average of less than 4 per year. In recent years the number reported has increased to almost twice that figure, probably due to rising population and more effective reporting. A large percentage of North Carolina tornadoes affect only a small area.

Tropical hurricanes come close enough to influence North Carolina weather about twice in an average year. Only about once in 10 years, on the average, does this type storm strike the State with sufficient force to do much damage to inland property. Coastal properties occasionally suffer damage from associated high tides and seas.

The wide variety of climate in North Carolina leads to a wide variety of vegetation. The average annual freeze-free period or "growing season" ranges from near 130 days in the highest mountain areas to around 290 days on the central section of the Outer Banks. At Hatteras entire seasons may pass without either frost or freezing temperature occurring. Tropical fruits can be grown there with a little care. Both orange and grapefruit trees have reached bearing age on the island.

Corn and hay are the most widely planted farm crops, growing in every North Carolina county. Most of these crops are used as feed for livestock, but an increasing percentage of corn is being grown for market. Small grains are almost as widely planted as corn, but the acreage is smaller. The most widely planted of all "money crops" is tobacco. Three-fourths of the North Carolina counties grow some tobacco, and in about one-half the counties it is the chief cash crop. The gross value of tobacco produced in the State is nearly twice that of all other field crops combined, and is the greatest of any single crop in any state except cotton in Texas. The widespread growing of tobacco in North Carolina is partly the result of suitable soils and partly of climate. While some types of tobacco are grown from Florida to Canada, the particular quality of tobacco which abounds in North Carolina can be produced over a rather limited area.

Other important field crops, listed in order of the total value of production in North Carolina in 1968, are: Corn, peanuts, soybeans, cotton, and hay. Sweet potatoes, Irish potatoes, and various fruits and vegetables are also grown commercially in some areas.

Livestock production plays an increasingly important part in the farm economy of North Carolina. The mild climate lends itself to the economical production of poultry and eggs, which account for more than half the total value of livestock products. Cattle and dairy products run second in value. Hogs account for nearly one-fifth of the total. Altogether, livestock and livestock products total nearly one-third of the entire value of farm output.

Favorable climatic factors are causing many manufacturers to establish plants in North Carolina. Manufacturing interests in the State employ more people than agriculture, and the gross value of output is greater. Textiles are the most important of the manufactured products. Tobacco is second, followed by food, furniture, and lumber in that order.

The importance of North Carolina as a vacation area is increasing. The mountains, because of their mild summer temperatures, provide a welcome escape from the heat of lower lands of both this and other states. Midsummer afternoon temperatures average below 80° at elevations of 3,000 feet or higher. Nights are crisp and cool, but seldom too cold for light camping. Mountain streams and forests furnish plenty of fishing and hunting, and there are lakes with boating facilities. Hiking is coming into increased favor, with numerous mountain trails to follow. There are peaks which provide challenge to amateur mountain climbers. Ski slopes are operating with increasing success at several higher mountain locations in midwinter.

Summer temperatures on the North Carolina beaches are considerably modified from the heat of the interior. The entire coastline has developed into a summer vacation area. Fishing, ranging from deep sea to inland lake and stream, is one of the most popular of sports. Midsummer ocean water temperatures range near 80° along the North Carolina coast, and practically the entire shoreline is suitable for ocean bathing. For those who prefer to swim in protected inland waters, hundreds of miles of beaches are found bordering the "sounds", or waters lying within the offshore islands.

The North Carolina beach country, especially the southern portion, is becoming increasingly popular for year-round vacationing. Both air and water temperatures are generally mild. The average ocean water temperature at Southport in January is higher than that along the northern coast of Maine in July.

The best known winter vacation area, however, is the "sandhills" section of the southern Piedmont. Here, due to the sandy character of the soils, the average winter daytime temperature is higher and the corresponding relative humidity lower than in other similarly exposed sections of the State. Winter rains are readily absorbed by the sandy soils without "mucking", and snow is a rarity, so that horse racing, golf and other outdoor sports are widely enjoyed. Pinehurst and Southern Pines, in the center of this area, have become popular places for retirement of those seeking a moderate climate.

Another favored area for winter living is the

"thermal belt" section of the southern mountains. Thermal belts are small areas along the mountain slopes that tend to have higher minimum temperatures than other adjacent or nearby areas less favorably situated. This is because cold air on quiet nights tends to drain into lower valleys, while higher elevations above the thermals belts are cold because of their elevation. In this particular area the situation is augmented by the existence of many long, southward-facing slopes which absorb much winter sunshine, at the same time being protected to a degree from cold air outbreaks by higher mountains to the north and west. As a result "thermal belts" part way up the mountain sides may support vulnerable vegetation long after frost has killed all green both above and below. Often in the dead of winter the contrast of a belt of green flanked by brown both upslope and in the valley is most striking. The term "thermal belt" is sometimes mistakenly applied to the entire area where this phenomenon is most common.

REFERENCES

(1) Carney, C. B., Climate of North Carolina, Climates of the States, Climatography of the United States No. 60-31. ESSA, Environmental Data Service, 1960.

(2) Carney, C. B., and A. V. Hardy. Weather and Climate in North Carolina. N. C. Agricultural Experiment Station Bull. No. 396. Reprint, Raleigh, 1964.

(3) Goerch and Ehringhaus. North Carolina Almanac and State Industrial Guide. Almanac Publishing Company, Raleigh, 1953.

(4) Hardy, A. V. Low Temperature Probabilities in North Carolina. N. C. Agricultural Experiment Station Bull. No. 423. N. C. State Univ., Raleigh, 1964.

(5) Hardy, A. V., C. B. Carney, and H. V. Marshall, Jr. Climate of North Carolina Research Stations. N. C. Agr. Exp. Sta. Bull. No. 433. N.C.S.U., Raleigh, 1967.

(6) North Carolina Department of Agriculture. North Carolina Agricultural Statistics; Annuals, 1953 and 1968, and Special Livestock Issue, 1968.

(7) North Carolina Department of Conservation and Development. Topography, Geology, and Mineral Resources of North Carolina; Educational Series No. 2, 1952.

(8) _____ Geology and Mineral Resources of North Carolina; Educational Series No. 3, 1953.

(9) North Carolina State Planning Board. Report on Water Resources. Raleigh, October 1937.

(10) Weather Bureau Technical Paper No. 15 - Maximum Station Precipitation for 1, 2, 3, 6, 12, and 24 Hours. ESSA, Weather Bureau.

(11) Weather Bureau Technical Paper No. 16 - Maximum 24-Hour Precipitation in the U. S. ESSA, Weather Bureau.

(12) Weather Bureau Technical Paper No. 25 - Rainfall Intensity - Duration - Frequency Curves. ESSA, Weather Bureau.

(13) Weather Bureau Technical Paper No. 29 - Rainfall Intensity - Frequency Regime. ESSA, Weather Bureau.

BIBLIOGRAPHY

(A) Climatic Summary of the United States (Bulletin W) 1930 Edition, Sections 95, 96 and 97. ESSA, Weather Bureau.

(B) Climatic Summary of the United States, North Carolina - Supplement for 1931 through 1952. (Bulletin W Supplement). ESSA, Weather Bureau.

(C) Climatic Summary of the United States, North Carolina - Supplement for 1951 through 1960. (Bulletin W Supplement). ESSA, Weather Bureau.

(D) Climatological Data, North Carolina. ESSA, Environmental Data Service.

(E) Climatological Data, National Summary. ESSA, Environmental Data Service.

(F) Hourly Precipitation Data, North Carolina. ESSA, Environmental Data Service.

(G) Local Climatological Data for Asheville, Charlotte, Greensboro, Hatteras, Raleigh, Wilmington, and Winston-Salem, North Carolina. ESSA, Environmental Data Service.

FREEZE DATA

STATION	Freeze threshold temperature	Mean date of last Spring occurrence	Mean date of first Fall occurrence	Mean No. of days between dates	Years of record Spring	No. of occurrences in Spring	Years of record Fall	No. of occurrences in Fall
ALBEMARLE	32	04-14	10-23	192	30	30	30	30
	28	03-31	11-05	210	30	30	30	30
	24	03-13	11-16	249	30	30	30	30
	20	02-27	12-01	277	30	30	30	27
	16	02-06	12-16	313	30	24	30	21
ANDREWS 2 E	32	04-29	10-15	169	21	21	25	25
	28	04-14	10-24	193	21	21	25	25
	24	03-30	11-05	219	21	21	25	25
	20	03-10	11-15	250	21	21	25	25
	16	02-27	11-30	276	20	20	25	23
ASHEBORO 2 W	32	04-07	11-02	209	25	25	25	25
	28	03-23	11-16	238	25	25	25	24
	24	03-05	11-29	270	25	25	24	23
	20	02-18	12-08	293	25	23	24	20
	16	02-03	12-20	320	25	20	24	15
ASHEVILLE WB CITY	32	04-12	10-24	195	30	30	30	30
	28	03-29	11-04	220	30	30	30	30
	24	03-14	11-21	252	30	30	30	30
	20	02-27	12-04	280	30	30	30	26
	16	02-10	12-16	309	30	26	30	17
BANNER ELK	32	05-14	10-03	142	29	29	30	30
	28	04-30	10-10	163	29	29	29	29
	24	04-11	10-19	191	29	29	27	27
	20	03-28	11-04	221	29	29	27	27
	16	03-17	11-17	245	29	29	27	26
BREVARD	32	04-25	10-15	173	12	12	13	13
	28	04-14	10-21	190	12	12	13	13
	24	04-01	10-31	214	12	12	12	12
	20	03-04	11-20	262	11	11	12	12
	16	02-11	12-05	297	10	8	10	9
CAROLEEN	32	04-10	10-30	202	30	30	29	29
	28	03-24	11-12	233	30	30	29	29
	24	03-09	11-24	260	30	30	29	28
	20	02-24	12-07	286	30	30	29	24
	16	01-29	12-15	320	28	20	27	19
CHAPEL HILL 2 W	32	03-31	11-04	218	30	30	30	30
	28	03-17	11-18	246	30	30	30	30
	24	03-04	11-29	269	30	30	30	30
	20	02-16	12-12	299	30	28	30	23
	16	01-30	12-23	328	30	22	30	13
CHARLOTTE WB CITY	32	03-21	11-15	239	30	30	30	30
	28	03-10	11-29	263	30	30	30	30
	24	02-21	12-11	293	30	28	30	24
	20	02-07	12-21	317	30	26	30	17
	16	01-17	12-26	342	30	15	30	10
CLINTON	32	03-25	11-02	223	14	14	14	14
	28	03-14	11-16	248	14	14	14	14
	24	02-27	12-02	278	14	14	14	13
	20	02-09	12-17	311	14	12	14	9
	16	01-14	12-26	346	14	6	14	4
CONCORD	32	04-04	10-31	211	18	18	18	18
	28	03-20	11-15	240	18	18	18	18
	24	03-05	11-25	266	18	18	18	18
	20	02-15	12-07	295	17	15	18	16
	16	01-31	12-19	322	17	13	18	10
CULLOWHEE	32	04-27	10-16	172	30	30	30	30
	28	04-15	10-23	191	30	30	30	30
	24	03-30	10-31	215	30	30	29	29
	20	03-13	11-14	246	30	30	29	29
	16	03-01	11-24	269	30	29	29	28
DURHAM	32	04-13	10-24	195	29	29	28	28
	28	03-25	11-08	228	29	29	28	28
	24	03-06	11-22	261	28	28	28	28
	20	02-22	12-04	285	28	28	28	25
	16	02-13	12-15	306	28	26	27	19
EDENTON	32	03-29	11-11	227	30	30	30	30
	28	03-12	11-19	252	30	30	30	30
	24	02-28	12-03	278	30	30	29	28
	20	02-15	12-17	305	30	28	29	20
	16	01-22	12-27	339	29	15	29	8
ELIZABETH CITY	32	03-29	11-10	226	29	29	30	30
	28	03-12	11-20	253	29	29	30	30
	24	02-26	12-08	285	29	28	30	26
	20	02-17	12-18	304	28	26	29	19
	16	01-25	12-27	336	28	18	29	9
FAYETTEVILLE	32	03-28	11-06	222	30	30	29	29
	28	03-13	11-16	248	30	30	29	29
	24	03-01	11-30	274	30	30	29	28
	20	02-07	12-17	313	30	27	29	17
	16	01-18	12-25	341	30	16	29	11
GOLDSBORO	32	03-29	11-07	224	30	30	30	30
	28	03-17	11-15	243	30	30	29	29
	24	02-27	11-30	276	30	29	28	27
	20	02-13	12-13	303	30	28	27	19
	16	01-27	12-27	334	30	20	26	8
GREENVILLE NO. 2	32	03-28	11-05	222	30	30	30	30
	28	03-14	11-17	248	30	30	30	30
	24	03-01	12-03	277	30	30	30	30
	20	02-13	12-12	302	30	29	30	24
	16	01-26	12-26	334	30	21	30	11
HATTERAS WB CITY	32	02-25	12-18	296	30	29	30	20
	28	02-04	12-26	324	30	25	30	9
	24	01-14	*	*	29	11	30	4
	20	01-08	*	*	29	8	30	2
	16	*	*	*	29	1	30	0
HENDERSON 3 SW	32	04-14	10-24	193	29	29	29	29
	28	04-03	11-05	216	29	29	28	28
	24	03-14	11-22	253	29	29	28	28
	20	02-28	12-05	280	29	28	28	26
	16	02-12	12-13	304	29	25	28	21
HENDERSONVILLE	32	04-24	10-19	177	30	30	30	30
	28	04-07	10-29	205	30	30	30	30
	24	03-23	11-09	231	30	30	30	30
	20	03-06	11-25	264	30	30	30	29
	16	02-21	12-05	287	30	29	29	27
HICKORY	32	04-04	10-31	210	30	30	30	30
	28	03-22	11-13	236	30	30	30	30
	24	03-09	11-30	266	30	30	30	29
	20	02-19	12-08	292	30	29	30	25
	16	02-05	12-19	317	30	26	30	17
HIGHLANDS	32	04-22	10-21	182	29	29	29	29
	28	04-12	10-30	202	29	29	29	29
	24	04-04	11-14	225	28	28	28	28
	20	03-16	11-26	255	28	28	27	24
	16	03-04	12-08	278	27	27	27	20
HOT SPRINGS 2	32	04-16	10-23	190	30	30	28	28
	28	04-05	10-31	209	30	30	28	28
	24	03-17	11-14	241	30	30	27	27
	20	03-05	11-27	267	30	29	27	25
	16	02-15	12-12	300	28	25	26	19
KINSTON	32	03-25	11-08	228	26	26	25	25
	28	03-12	11-19	252	25	25	24	23
	24	02-24	11-30	279	25	24	24	23
	20	02-11	12-13	305	25	23	24	13
	16	01-17	12-25	342	25	12	24	10
LENOIR	32	04-20	10-18	181	29	29	30	30
	28	04-07	10-31	208	29	29	30	30
	24	03-19	11-15	240	29	29	30	30
	20	03-04	11-29	271	29	29	30	29
	16	02-15	12-07	296	29	26	30	26
LEXINGTON	32	04-07	11-03	211	12	12	13	13
	28	03-28	11-20	237	12	12	13	13
	24	02-28	12-03	278	12	12	13	13
	20	02-14	12-14	303	12	10	13	10
	16	02-01	12-23	325	12	10	13	6
LOUISBURG	32	04-12	10-23	195	29	29	30	30
	28	03-31	11-06	221	29	29	30	30
	24	03-11	11-18	251	29	29	30	30
	20	02-25	12-02	280	29	28	30	29
	16	02-10	12-13	306	29	26	30	23

FREEZE DATA

STATION	Freeze threshold temperature	Mean date of last Spring occurrence	Mean date of first Fall occurrence	Mean No. of days between dates	Years of record Spring	No. of occurrences in Spring	Years of record Fall	No. of occurrences in Fall
LUMBERTON	32	03-26	11-06	225	30	30	30	30
	28	03-13	11-18	249	30	30	30	30
	24	02-22	12-03	284	30	28	30	27
	20	02-07	12-16	312	30	24	30	19
	16	01-13	12-27	347	30	12	30	9
MANTEO	32	03-13	11-26	258	22	22	24	23
	28	02-28	12-07	282	22	22	24	21
	24	02-10	12-22	315	22	18	24	12
	20	01-20	12-25	340	21	11	24	9
	16	01-13	*	*	21	8	24	2
MARION	32	04-16	10-18	186	24	24	29	29
	28	03-28	11-05	221	23	23	29	29
	24	03-14	11-16	247	20	20	28	28
	20	03-05	11-28	268	19	19	27	25
	16	02-14	12-08	296	17	16	26	21
MARSHALL 2 NE	32	04-24	10-15	174	30	30	30	30
	28	04-11	10-25	198	30	30	29	29
	24	03-26	11-05	224	30	30	29	29
	20	03-06	11-21	260	30	29	29	27
	16	02-21	12-04	286	30	28	29	25
MONCURE	32	04-21	10-21	183	30	30	30	30
	28	04-06	11-02	210	30	30	30	30
	24	03-17	11-14	242	30	30	30	30
	20	03-04	11-24	265	30	30	30	30
	16	02-17	12-09	295	30	28	30	25
MONROE 4 SE	32	04-10	10-27	201	30	30	30	30
	28	03-27	11-05	223	30	30	30	30
	24	03-12	11-19	252	30	30	30	30
	20	02-22	11-30	282	30	29	30	24
	16	01-28	12-17	324	30	22	30	17
MORGANTON	32	04-14	10-23	192	29	29	28	28
	28	03-31	11-01	215	29	29	27	27
	24	03-16	11-17	246	29	29	27	27
	20	02-28	12-01	276	28	28	26	25
	16	02-11	12-13	304	26	23	26	20
MT AIRY	32	04-23	10-18	178	29	29	28	28
	28	04-07	10-28	204	29	29	28	28
	24	03-20	11-10	234	29	29	28	28
	20	03-08	11-26	263	29	29	28	26
	16	02-21	12-06	289	29	28	28	25
NASHVILLE	32	04-06	10-28	205	29	29	29	29
	28	03-21	11-10	234	29	29	28	28
	24	03-07	11-21	259	29	29	28	28
	20	02-22	12-08	290	29	28	28	25
	16	02-06	12-19	317	29	25	28	17
NEW BERN	32	03-20	11-16	241	30	30	29	29
	28	03-06	11-26	265	30	30	29	29
	24	02-18	12-11	297	30	29	29	23
	20	01-30	12-22	326	30	22	29	15
	16	01-13	12-28	350	30	12	29	7
NEW HOLLAND	32	03-23	11-13	235	29	29	30	30
	28	03-08	11-23	260	29	29	30	30
	24	02-21	12-11	292	28	28	29	25
	20	01-27	12-24	330	28	21	29	12
	16	01-15	12-27	346	27	11	29	7
OXFORD 2 SW	32	04-09	10-28	202	30	30	30	30
	28	03-23	11-15	237	30	30	30	30
	24	03-08	11-25	262	30	30	30	30
	20	02-23	12-08	288	30	29	30	25
	16	02-07	12-19	315	30	27	30	18
PARKER 1 E	32	05-02	10-09	160	30	30	29	29
	28	04-23	10-18	179	30	30	29	29
	24	04-08	10-28	203	30	30	29	29
	20	03-27	11-13	231	30	30	28	28
	16	03-10	11-26	261	30	30	28	25
PINEHURST	32	04-03	11-01	212	30	30	30	30
	28	03-22	11-15	239	30	30	30	30
	24	03-08	11-30	267	30	30	30	29
	20	02-17	12-09	294	30	28	30	24
	16	02-01	12-20	323	30	22	30	16
RALEIGH WB CITY	32	03-24	11-16	237	30	30	30	30
	28	03-09	11-27	262	30	30	30	29
	24	02-27	12-08	284	30	30	30	25
	20	02-10	12-20	313	30	28	30	18
	16	01-23	12-25	336	30	18	30	11
REIDSVILLE	32	04-06	11-03	211	30	30	30	30
	28	03-21	11-16	240	30	30	30	30
	24	03-08	11-27	264	30	30	30	30
	20	02-24	12-07	287	30	29	30	24
	16	02-07	12-19	315	30	26	30	19
ROCK HOUSE	32	04-12	10-27	198	30	30	29	29
	28	04-05	11-04	213	30	30	29	29
	24	03-19	11-23	249	29	29	29	27
	20	03-11	12-04	268	29	29	29	24
	16	02-21	12-14	297	29	28	29	17
SALISBURY	32	04-07	10-28	205	30	30	30	30
	28	03-27	11-08	226	30	30	30	30
	24	03-12	11-19	252	30	30	30	29
	20	02-27	12-02	278	30	29	30	27
	16	02-10	12-17	310	30	27	30	20
SHELBY	32	04-05	11-04	213	14	14	15	15
	28	03-22	11-15	238	14	14	15	15
	24	03-03	11-27	269	14	14	15	15
	20	02-10	12-10	303	14	12	15	12
	16	01-23	12-21	332	14	9	15	7
SLOAN 3 S	32	04-08	11-03	209	28	28	26	26
	28	03-22	11-13	235	28	28	25	25
	24	03-06	11-24	264	27	27	25	25
	20	02-11	12-10	302	26	22	23	19
	16	01-20	12-21	336	25	11	23	9
SMITHFIELD	32	04-08	10-25	200	28	28	30	30
	28	03-23	11-10	232	28	28	30	30
	24	03-10	11-19	254	28	28	30	30
	20	02-23	12-08	287	28	27	30	26
	16	02-05	12-19	317	28	23	30	17
SOUTHPORT	32	03-15	11-23	254	30	30	30	30
	28	02-25	12-03	281	30	30	30	27
	24	02-09	12-18	312	30	28	30	17
	20	01-16	12-25	343	30	15	30	11
	16	01-08	*	*	30	9	30	4
STATESVILLE 2 W	32	04-07	11-05	212	25	25	26	26
	28	03-21	11-19	243	25	25	26	26
	24	03-09	11-30	266	25	25	25	24
	20	02-18	12-09	294	25	24	25	21
	16	02-11	12-20	312	24	20	25	15
TARBORO	32	03-30	11-01	216	30	30	30	30
	28	03-17	11-16	244	30	30	30	30
	24	03-01	11-30	274	30	30	30	30
	20	02-17	12-13	299	30	29	30	24
	16	01-26	12-23	331	30	18	30	12
TRYON	32	04-10	10-30	203	30	30	29	29
	28	03-26	11-08	228	28	28	28	28
	24	03-09	11-25	261	26	26	25	25
	20	02-20	12-05	288	26	24	24	22
	16	01-30	12-15	319	26	19	22	14
WADESBORO	32	03-31	11-11	225	11	11	11	11
	28	03-15	11-26	256	11	11	8	8
	24	02-24	12-03	282	8	8	8	7
	20	02-09	12-10	304	7	6	8	6
	16	01-29	12-22	327	7	5	7	4
WATERVILLE	32	04-02	10-31	212	20	20	20	20
	28	03-20	11-14	239	20	20	20	20
	24	03-10	11-29	265	20	20	20	18
	20	02-24	12-10	289	20	19	20	15
	16	02-10	12-17	311	20	17	20	11
WAYNESVILLE 1 E	32	05-01	10-11	163	25	25	25	25
	28	04-15	10-21	188	24	24	24	24
	24	04-01	11-02	215	23	23	22	22
	20	03-17	11-11	238	22	22	21	21
	16	03-05	11-29	270	22	22	21	20

FREEZE DATA

STATION	Freeze threshold temperature	Mean date of last Spring occurrence	Mean date of first Fall occurrence	Mean No. of days between dates	Years of record Spring	No. of occurrences in Spring	Years of record Fall	No. of occurrences in Fall	STATION	Freeze threshold temperature	Mean date of last Spring occurrence	Mean date of first Fall occurrence	Mean No. of days between dates	Years of record Spring	No. of occurrences in Spring	Years of record Fall	No. of occurrences in Fall
WELDON	32	04-05	11-02	210	30	30	30	30	WILMINGTON WB CITY	32	03-08	11-24	262	30	30	30	30
	28	03-22	11-13	237	30	30	30	30		28	02-22	12-08	289	30	29	30	26
	24	03-06	11-25	264	30	30	30	30		24	02-03	12-22	322	30	22	30	15
	20	02-21	12-08	291	30	29	30	26		20	01-16	12-27	345	30	14	30	8
	16	02-04	12-24	323	30	23	30	13		16	01-07	⊕	⊕	30	8	30	4
WILLARD 1 N	32	04-10	11-05	209	28	28	24	24	WINSTON SALEM WB AP	32	04-09	10-29	203	30	30	30	30
	28	03-21	11-13	237	26	26	21	21		28	03-26	11-09	229	30	30	30	30
	24	03-06	11-22	262	26	26	21	21		24	03-09	11-21	257	30	30	30	30
	20	02-24	12-14	294	24	24	19	15		20	02-26	12-06	283	30	29	30	25
	16	01-31	12-21	324	20	16	19	10		16	02-09	12-15	310	30	27	30	21

Data in the above table are based on the period 1921-1950, or that portion of this period for which data are available.

⊕ When the frequency of occurrence in either spring or fall is one year in ten, or less, mean dates are not given.

Means have been adjusted to take into account years of non-occurrence.

A freeze is a numerical substitute for the former term "killing frost" and is the occurrence of a minimum temperature at or below the threshold temperature of 32°, 28°, etc.

Freeze data tabulations in greater detail are available and can be reproduced at cost.

*NORMALS BY CLIMATOLOGICAL DIVISIONS

Taken from "Climatography of the United States No. 81-4, Decennial Census of U. S. Climate"

TEMPERATURE (°F)   PRECIPITATION (In.)

STATIONS (By Divisions)	TEMPERATURE (°F)													PRECIPITATION (In.)												
	JAN	FEB	MAR	APR	MAY	JUNE	JULY	AUG	SEPT	OCT	NOV	DEC	ANN	JAN	FEB	MAR	APR	MAY	JUNE	JULY	AUG	SEPT	OCT	NOV	DEC	ANN
SOUTHERN MOUNTAINS																										
ANDREWS 2 E	39.3	40.6	46.3	55.5	63.7	71.0	73.8	72.8	67.6	57.2	46.0	39.5	56.1	6.45	6.67	6.47	5.27	4.24	5.00	6.32	4.91	3.34	3.14	4.11	6.10	62.02
ASHEVILLE WB AP	37.6	38.9	44.6	54.6	63.0	70.1	72.4	71.5	65.9	55.8	44.6	37.9	54.7	4.17	4.03	4.82	4.02	3.66	3.52	5.85	4.94	3.62	3.13	2.77	3.62	48.15
ASHEVILLE	39.7	40.6	46.2	56.0	64.4	71.8	74.4	73.5	67.8	57.6	46.5	40.0	56.5	3.17	3.04	3.74	3.19	2.87	3.52	4.31	3.63	2.78	2.49	2.22	2.92	37.88
BREVARD														5.75	5.55	6.18	5.03	4.83	4.94	6.51	6.05	4.46	4.31	4.25	5.63	63.49
CANTON 1 SW														3.46	3.53	4.00	3.30	3.00	3.13	4.70	4.40	2.83	2.61	2.35	3.00	40.31
CAROLEEN	42.8	44.6	50.7	60.6	69.2	77.0	79.3	78.2	72.5	61.6	50.0	42.3	60.7	4.34	4.25	5.12	4.04	3.55	3.68	4.98	5.02	3.95	3.25	3.18	4.39	49.75
CULLOWHEE	40.7	42.0	47.1	55.9	63.8	70.9	73.9	73.2	67.9	57.7	46.4	40.3	56.7	4.59	4.66	4.94	3.79	3.60	3.55	5.07	4.31	2.91	2.76	2.88	4.40	47.46
ENKA														3.19	3.17	3.91	3.20	3.21	3.42	4.89	4.09	2.81	2.70	2.49	2.91	39.99
FLETCHER 3 W	39.8	40.8	46.3	55.6	63.6	70.8	73.4	72.5	66.7	56.6	45.8	39.7	56.0	4.17	4.03	4.82	4.02	3.66	3.52	5.85	4.94	3.62	3.13	2.77	3.62	48.15
HENDERSONVILLE 1 NE	39.9	41.2	46.9	56.0	63.9	71.2	73.7	72.8	67.1	56.9	46.1	39.6	56.3	4.76	4.58	5.48	4.54	4.15	4.82	6.16	5.52	4.04	3.75	3.60	4.72	56.12
HIGHLANDS 2 S	39.3	39.9	45.0	54.2	61.9	68.3	70.1	69.4	65.1	56.2	46.1	39.9	54.6	7.17	6.76	7.66	6.28	5.40	6.55	9.32	7.51	4.93	5.06	5.16	7.33	79.13
HOT SPRINGS 2	40.3	41.7	47.2	57.2	65.7	73.2	76.0	75.3	70.2	59.6	47.5	40.6	57.9	3.43	3.69	4.23	3.69	3.37	4.39	5.41	4.43	2.96	2.48	2.64	2.95	43.67
MARSHALL	37.7	38.7	44.3	54.1	62.9	70.4	73.4	72.4	66.8	56.1	44.6	37.7	54.9	3.22	3.29	3.84	3.18	2.98	3.64	4.72	3.56	2.52	2.19	2.29	2.79	38.22
MORGANTON	42.3	43.5	49.1	58.7	66.8	74.7	77.1	76.0	70.2	59.9	49.2	42.0	59.1	4.27	3.95	4.89	3.90	3.64	3.69	5.12	5.42	4.46	3.55	3.03	4.11	50.03
MURPHY														5.65	5.85	5.93	4.95	3.72	4.45	5.61	4.52	3.01	2.83	3.65	5.07	55.24
SWANNANOA 2 E														3.54	3.43	4.08	3.34	2.96	3.54	4.34	4.56	3.51	3.19	2.65	3.24	42.38
TAPOCO														5.38	5.71	5.76	4.74	4.15	4.67	5.70	5.03	3.04	2.94	3.64	4.94	55.70
TRYON	44.3	45.7	51.4	60.1	67.9	74.9	77.1	76.0	70.7	61.2	51.2	44.0	60.4	5.29	5.34	6.35	4.83	4.33	4.41	6.21	6.48	4.52	4.35	3.90	5.48	61.49
WAYNESVILLE 1 E	38.9	39.9	45.2	54.2	61.8	68.6	71.3	70.5	64.7	55.1	44.7	38.8	54.5	4.65	4.32	4.95	3.58	3.51	3.41	5.01	4.02	2.74	2.86	2.88	3.99	45.92
DIVISION	39.5	40.5	46.0	55.2	63.3	70.5	73.0	72.2	66.9	57.0	46.1	39.5	55.8	4.83	4.71	5.49	4.45	4.00	4.47	5.94	5.45	3.84	3.60	3.54	4.71	55.03
NORTHERN MOUNTAINS																										
BANNER ELK	34.3	34.9	40.0	49.1	57.0	64.0	66.7	65.8	60.8	51.5	40.9	34.8	50.0	3.99	4.12	4.63	4.33	4.13	4.52	6.11	5.34	3.89	3.27	3.44	3.82	51.59
BOONE	35.1	35.8	40.8	50.2	58.8	66.2	68.9	68.2	62.4	52.9	42.3	36.1	51.5	4.06	4.04	5.10	4.65	4.51	4.39	6.34	5.41	4.38	3.83	4.02	4.07	54.80
ELKIN														3.58	3.37	4.36	3.78	3.95	4.29	5.07	5.61	3.86	3.28	3.02	3.57	47.74
JEFFERSON														3.48	3.44	4.11	3.83	4.09	3.82	5.50	5.82	4.00	3.23	3.22	3.52	48.06
LENOIR	41.7	42.9	48.6	58.1	66.2	73.4	76.1	75.2	69.9	59.6	48.7	41.5	58.5	3.90	4.03	4.63	4.14	3.93	4.33	5.04	5.65	3.85	3.63	2.85	3.77	49.75
MOUNT AIRY	39.6	40.9	47.2	57.1	65.7	73.3	76.2	75.0	69.3	58.9	47.4	39.6	57.5	3.66	3.44	4.33	3.83	4.18	4.15	5.39	4.80	3.56	3.28	3.00	3.47	47.06
NORTH WILKESBORO														3.92	3.75	4.66	3.78	3.96	4.08	5.80	5.55	3.78	3.72	3.06	3.72	49.78
DIVISION	36.8	37.8	43.2	52.7	61.0	68.2	70.9	69.9	64.5	54.9	44.1	37.2	53.4	3.89	3.99	4.79	4.24	4.26	4.40	5.85	5.60	4.11	3.52	3.37	3.82	51.84
NORTHERN PIEDMONT																										
CHAPEL HILL 2 W	43.3	44.7	51.0	60.7	69.0	76.5	79.0	77.8	72.7	62.0	51.6	43.4	61.0	3.67	3.66	3.82	3.78	3.23	3.48	5.71	5.01	3.91	2.79	3.18	3.55	45.79
DURHAM	42.2	43.4	49.9	59.6	68.2	75.8	78.4	77.2	71.4	60.7	50.0	41.9	59.9	3.30	3.36	3.57	3.17	3.74		5.35	4.73	3.44	2.68	2.90	3.04	42.65
GRAHAM 2 ENE														3.50	3.51	3.74	3.67	3.50	3.86	5.08	5.01	3.94	3.02	2.88	3.24	44.95
GREENSBORO WB AIRPORT	39.7	41.0	47.4	57.4	66.0	74.8	77.3	76.2	70.1	59.3	47.9	39.9	58.2	3.40	3.30	3.69	3.43	3.29	3.47	4.79	4.61	3.66	2.71	2.68	3.13	42.16
HENDERSON 2 SW	41.7	43.0	49.6	59.6	67.9	75.7	78.4	77.1	71.3	60.6	50.4	41.9	59.8	3.51	3.22	3.99	3.76	3.80	4.03	5.95	4.94	3.45	2.68	3.10	3.15	45.58
HIGH POINT														3.60	3.57	3.92	3.80	3.57	3.68	5.52	5.06	4.03	2.59	2.88	3.47	45.69
LAKE MICHIE														3.44	3.48	3.82	3.61	3.69	3.54	5.52	4.82	3.79	2.86	2.88	2.93	44.38
LOUISBURG	41.7	43.1	49.6	59.6	68.4	76.2	79.1	77.7	72.0	60.9	50.1	41.7	60.0	3.54	3.69	3.83	3.58	3.61	3.93	5.85	5.35	4.05	2.63	3.25	3.24	46.59
MANGUMS STORE 4 WSW														3.38	3.14	3.80	3.68	3.18	3.86	5.09	4.71	3.30	2.59	3.09	3.16	42.98
OXFORD 2 SW	41.9	43.1	49.3	59.4	67.8	75.4	78.1	76.8	71.3	60.6	50.5	42.1	59.7	3.45	3.21	3.70	3.77	3.93	4.45	5.44	4.77	3.56	2.74	3.07	3.17	45.36
REIDSVILLE 2 NW	41.9	43.1	45.9	59.6	68.1	75.8	78.4	77.3	71.5	61.2	50.6	42.3	59.9	3.37	3.26	3.80	3.74	3.79	3.51	4.34	4.24	4.09	3.21	2.94	3.08	43.37
ROUGEMONT														3.41	3.21	3.86	3.92	3.72	4.21	5.72	4.91	3.68	3.01	3.16	3.09	45.90
ROXBORO														3.48	3.26	3.57	3.71	3.78	4.00	4.97	4.56	4.12	2.78	2.95	3.08	44.26
DIVISION	41.6	42.9	49.3	59.3	67.9	75.6	78.2	76.9	71.3	60.6	50.0	41.7	59.6	3.51	3.44	3.83	3.67	3.58	3.81	5.27	4.73	3.82	2.85	2.96	3.25	44.72
CENTRAL PIEDMONT																										
ASHEBORO 2 W	43.3	44.9	51.2	60.9	68.7	76.0	78.4	77.2	71.7	61.6	51.4	43.5	60.7	3.67	3.49	4.07	3.64	3.69	3.81	5.54	4.85	3.65	2.97	2.69	3.31	45.38
HICKORY	41.3	42.4	48.6	58.6	67.3	75.1	77.4	76.3	70.9	60.6	49.0	41.1	59.1	4.11	4.05	4.73	3.78	3.62	3.88	4.99	5.65	3.81	3.40	3.08	4.05	49.15
MONCURE 3 SE	41.5	43.0	49.5	59.5	68.9	78.5	77.4	71.7	60.5	49.6	41.3		59.7	3.56	3.68	3.96	3.79	3.90	3.62	6.84	5.51	4.29	2.89	3.16	3.45	48.65
NEUSE 2 NE														3.53	3.68	3.68	3.77	3.53	4.07	5.49	5.58	4.14	2.89	3.20	3.45	47.01
RALEIGH DURHAM WB AP	41.6	43.0	49.5	59.3	67.6	75.1	77.9	77.0	71.2	60.5	50.0	41.9	59.5	3.22	3.23	3.35	3.52	3.52	3.70	5.49	5.20	3.85	2.71	2.77	3.02	43.58
RALEIGH 3 W														3.42	3.39	3.85	3.63	3.86	4.07	5.56	4.98	4.36	2.85	3.07	3.24	46.28
RALEIGH STATE COLLEGE	43.3	44.3	50.7	60.6	69.2	76.9	79.4	78.0	72.5	62.3	51.8	43.6	61.1	3.33	3.49	3.72	3.78	3.80	3.94	5.90	5.35	4.57	2.77	3.01	3.20	46.86
RANDLEMAN														3.62	3.72	3.83	3.65	3.68	3.66	5.78	5.12	4.09	2.90	2.86	3.36	46.27
SALISBURY	42.2	43.8	50.1	60.3	69.2	76.9	79.2	78.0	72.0	61.0	49.8	41.8	60.4	3.91	3.65	4.39	3.76	3.79	3.77	5.59	4.89	3.91	3.12	2.96	3.61	47.35
STATESVILLE 2 NNE	42.3	44.0	49.7	59.7	68.2	75.9	78.0	76.8	71.7	61.6	50.5	42.2	60.1	4.06	3.86	4.69	3.73	3.52	3.64	5.07	4.63	3.73	3.30	3.04	3.81	47.08
DIVISION	42.4	43.9	50.1	60.0	68.3	76.0	78.5	77.3	71.7	61.3	50.4	42.4	60.2	3.73	3.66	4.17	3.71	3.71	3.79	5.60	5.07	4.00	3.01	2.99	3.52	46.96

*NORMALS BY CLIMATOLOGICAL DIVISIONS

Taken from "Climatography of the United States No. 81-4, Decennial Census of U. S. Climate"

TEMPERATURE (°F)   PRECIPITATION (In.)

STATIONS (By Divisions)	JAN	FEB	MAR	APR	MAY	JUNE	JULY	AUG	SEPT	OCT	NOV	DEC	ANN	JAN	FEB	MAR	APR	MAY	JUNE	JULY	AUG	SEPT	OCT	NOV	DEC	ANN
SOUTHERN PIEDMONT																										
ALBEMARLE 4 N	43.5	45.0	50.9	60.4	69.1	76.8	78.9	77.7	72.1	61.5	50.9	42.9	60.8	3.65	3.59	4.13	3.87	3.20	3.82	6.22	4.61	4.43	2.77	2.80	3.46	46.55
CHARLOTTE WB AP	42.7	44.2	50.0	60.3	69.0	77.1	79.2	78.7	72.9	62.5	50.4	42.7	60.8	3.53	3.55	4.39	3.49	3.11	3.61	4.88	4.22	3.49	2.96	2.53	3.62	43.38
GASTONIA	44.2	45.9	52.0	62.0	70.5	78.0	79.8	78.8	73.2	62.4	51.7	44.0	61.9	3.91	4.09	4.66	3.95	3.64	3.21	4.93	4.57	4.10	3.51	2.87	3.94	47.38
MONROE 4 SE	44.8	46.3	52.1	61.1	69.7	77.4	79.5	78.3	72.9	62.4	52.1	44.4	61.8	3.53	3.57	3.85	3.74	3.08	3.67	5.31	5.04	3.81	2.64	2.68	3.41	44.33
MOUNT GILEAD 4 W	•	•	•	•	•	•	•	•	•	•	•	•	•	3.20	3.16	3.61	3.59	2.87	3.78	5.72	4.58	4.00	2.66	2.76	3.19	43.12
PINEHURST SRN-PINES	44.1	45.6	51.8	61.7	70.0	77.2	79.2	78.2	72.7	62.4	51.9	43.9	61.6	3.55	3.79	4.18	4.01	3.66	4.30	6.90	5.74	4.04	3.09	3.07	3.55	49.88
SHELBY 2 NNE	•	•	•	•	•	•	•	•	•	•	•	•	•	4.04	4.11	4.60	3.69	3.52	3.64	5.09	4.68	3.46	3.31	3.02	4.15	47.31
SOUTHERN PINES	44.9	46.3	52.2	61.5	69.8	77.4	78.9	77.7	72.9	62.7	52.6	45.1	61.8	3.47	3.71	4.27	4.08	3.80	4.78	7.11	5.50	4.35	3.30	2.92	3.38	50.67
DIVISION	43.8	45.3	51.4	61.0	69.6	77.2	79.3	78.2	72.8	62.3	51.6	43.7	61.4	3.60	3.66	4.28	3.79	3.42	3.86	5.71	5.00	4.03	2.97	2.79	3.53	46.64
SOUTHERN COASTAL PLAIN																										
ELIZABETHTOWN LOCK 2	•	•	•	•	•	•	•	•	•	•	•	•	•	2.78	3.25	3.61	3.23	2.98	4.63	6.14	6.02	4.58	3.03	2.75	3.03	46.03
FAYETTEVILLE	45.3	46.0	52.4	62.2	70.4	78.0	80.3	79.1	74.2	63.4	53.0	44.6	62.4	3.02	3.39	3.89	3.80	3.64	4.16	6.06	6.00	4.39	2.25	2.83	3.01	46.44
LUMBERTON 6 NW	45.9	47.3	53.7	62.8	70.6	77.4	79.5	78.5	73.6	63.2	52.8	45.4	62.6	2.86	3.50	3.87	3.78	3.25	4.56	6.21	4.91	4.23	2.52	2.69	3.10	45.48
RED SPRINGS	•	•	•	•	•	•	•	•	•	•	•	•	•	2.90	3.39	3.69	3.65	3.37	4.54	5.47	5.37	4.07	2.35	2.83	3.29	44.92
SLOAN 3 S	46.0	46.8	52.9	61.6	68.9	75.9	78.9	77.7	73.1	62.9	53.3	45.7	62.0	3.12	3.52	3.82	2.95	3.61	5.17	7.22	5.71	5.04	2.57	2.83	3.21	48.77
SOUTHPORT 5 N	48.7	49.1	54.2	62.7	70.8	77.6	80.2	79.7	75.9	66.5	57.0	49.5	64.3	3.00	3.65	3.94	2.57	3.17	3.87	6.56	5.71	7.08	3.27	3.17	3.50	49.49
WILLARD 1 N	47.3	48.5	54.1	62.5	70.3	77.2	79.6	78.6	74.1	64.0	54.3	46.9	63.1	2.86	3.54	3.67	2.76	3.66	4.73	7.56	5.98	5.28	2.78	2.88	3.27	48.97
WILMINGTON WB AP	47.9	48.7	54.2	62.5	70.5	77.7	80.0	79.4	75.2	65.4	55.4	48.2	63.8	2.85	3.42	4.03	2.86	3.52	4.26	7.68	6.86	6.29	3.01	3.09	3.42	51.29
DIVISION	46.8	47.8	53.5	62.4	70.3	77.3	79.7	78.7	74.1	64.0	54.2	46.6	63.0	3.06	3.51	3.90	3.23	3.56	4.50	6.94	5.94	5.23	2.77	2.96	3.32	48.92
CENTRAL COASTAL PLAIN																										
GOLDSBORO 1 SSW	44.7	46.0	52.5	62.3	70.6	78.0	80.5	79.1	74.0	63.3	53.0	44.6	62.4	3.32	3.37	3.85	3.79	3.85	4.78	7.50	5.62	4.35	2.87	3.10	3.21	49.61
GREENVILLE	•	•	•	•	•	•	•	•	•	•	•	•	•	3.34	3.40	4.08	3.47	3.38	4.08	6.79	5.58	4.84	2.77	3.03	3.26	47.54
KINSTON 5 SE	45.3	46.4	52.9	62.1	70.2	77.4	80.0	79.4	74.2	63.8	53.5	45.2	62.5	3.03	3.46	3.50	3.18	3.64	4.62	7.09	5.83	4.79	2.58	2.95	3.17	47.84
NEW BERN 3 NW	46.6	47.7	53.5	62.8	70.8	77.8	80.2	79.3	74.7	64.7	54.7	46.8	63.3	3.27	3.85	3.96	3.10	3.91	4.54	8.17	6.98	6.68	3.13	3.65	4.13	55.41
NEW HOLLAND	47.0	47.2	52.8	61.0	68.9	75.9	78.6	78.2	73.9	64.6	55.1	47.2	62.6	3.52	3.80	3.65	3.07	3.77	4.74	7.18	6.71	5.86	3.35	3.71	4.01	53.37
SMITHFIELD	44.1	45.4	51.8	61.3	69.4	77.0	79.4	78.3	72.9	62.3	52.0	43.9	61.5	3.42	3.57	3.86	3.97	3.56	4.57	6.04	6.08	4.19	2.81	3.03	3.24	48.34
DIVISION	45.5	46.5	52.5	61.7	70.0	77.3	79.8	79.0	74.2	64.0	54.0	45.8	62.5	3.39	3.64	3.79	3.35	3.69	4.46	7.04	6.16	5.33	3.12	3.38	3.55	50.90
NORTHERN COASTAL PLAIN																										
CAPE HATTERAS WB	46.6	46.5	51.0	59.3	68.0	75.2	78.0	77.6	74.1	65.4	56.2	48.2	62.2	3.90	3.93	4.16	2.99	3.98	4.14	6.15	6.42	5.89	4.24	4.09	4.58	54.47
EDENTON	43.5	44.7	50.9	60.5	69.2	76.9	79.5	78.0	72.8	62.1	52.0	43.8	61.2	3.63	3.40	3.87	3.46	3.39	4.13	6.84	6.28	4.75	2.81	3.31	3.01	48.88
ELIZABETH CITY	44.6	45.2	51.2	60.4	68.9	76.4	79.5	78.2	73.3	63.1	53.3	45.0	61.6	3.61	3.76	3.84	3.42	3.29	3.82	7.11	6.41	5.08	3.18	3.50	3.25	50.27
ENFIELD	•	•	•	•	•	•	•	•	•	•	•	•	•	3.36	3.38	3.69	3.30	3.54	3.74	5.61	5.20	3.50	2.62	2.64	2.89	43.47
HATTERAS	47.6	47.4	52.0	60.1	68.5	75.8	78.9	78.8	75.4	66.6	57.5	49.2	63.2	3.92	4.01	4.21	2.99	4.07	4.06	6.28	6.58	6.02	4.47	4.31	4.87	55.79
MANTEO	45.3	45.6	51.0	59.5	68.4	75.9	79.2	78.3	74.0	64.2	54.8	46.7	61.9	3.25	3.49	3.31	2.59	2.95	3.27	5.84	5.52	4.66	3.08	3.22	3.13	44.31
NASHVILLE	42.8	44.2	50.8	60.4	69.0	76.6	79.2	77.8	72.4	61.8	51.2	42.9	60.8	3.46	3.88	3.86	3.69	3.61	3.87	6.86	6.02	4.31	2.71	2.99	3.17	47.01
ROCKY MOUNT POWER PL	•	•	•	•	•	•	•	•	•	•	•	•	•	3.25	3.29	3.69	3.53	3.59	4.10	6.55	5.80	4.45	2.82	3.10	3.00	47.17
ROCKY MOUNT 8 ESE	•	•	•	•	•	•	•	•	•	•	•	•	•	3.44	3.78	3.84	3.36	3.49	4.39	6.12	5.68	4.18	2.75	3.22	3.04	47.17
SCOTLAND NECK	•	•	•	•	•	•	•	•	•	•	•	•	•	3.43	3.67	3.80	3.54	3.60	4.76	5.98	5.44	4.22	3.04	3.02	3.14	47.64
TARBORO 1 S	43.2	44.3	51.3	61.2	69.9	77.2	80.0	78.8	73.5	62.4	51.7	43.3	61.4	3.47	3.49	3.93	3.40	3.57	4.49	6.08	6.17	4.28	2.87	3.16	3.20	48.11
WELDON	42.4	43.4	50.0	60.0	69.0	77.0	79.9	78.3	72.6	61.5	51.1	42.5	60.6	3.29	3.37	3.43	3.36	3.71	3.38	5.66	5.26	3.97	2.51	2.82	2.89	43.65
WILLIAMSTON 1 ENE	•	•	•	•	•	•	•	•	•	•	•	•	•	3.25	3.45	3.80	3.54	3.55	4.28	6.74	6.04	4.41	2.69	3.10	2.98	47.83
DIVISION	44.0	44.9	50.9	60.2	68.8	76.3	79.2	78.1	73.2	62.8	52.8	44.5	61.3	3.50	3.57	3.78	3.30	3.55	4.00	6.28	6.02	4.69	3.00	3.29	3.32	48.30

* Normals for the period 1931-1960. Divisional normals may not be the arithmetical averages of individual stations published, since additional data for shorter period stations are used to obtain better areal representation.

TEMPERATURE   PRECIPITATION

JAN	FEB	MAR	APR	MAY	JUNE	JULY	AUG	SEPT	OCT	NOV	DEC	ANN	JAN	FEB	MAR	APR	MAY	JUNE	JULY	AUG	SEPT	OCT	NOV	DEC	ANN

CONFIDENCE-LIMITS

In the absence of trend or record changes, the chances are 9 out of 10 that the true mean will lie in the interval formed by adding and subtracting the values in the following table from the means for any station in the State. Because of the wider variation in mean precipitation, the corresponding monthly means and annual mean must be substituted for "p" in the precipitation table below to obtain mean precipitation confidence limits.

| 1.5 | 1.4 | 1.4 | .6 | .6 | .5 | .5 | .5 | .9 | .7 | .8 | 1.3 | .3 | $.25\sqrt{p}$ | $.32\sqrt{p}$ | $.27\sqrt{p}$ | $.25\sqrt{p}$ | $.31\sqrt{p}$ | $.30\sqrt{p}$ | $.39\sqrt{p}$ | $.36\sqrt{p}$ | $.50\sqrt{p}$ | $.40\sqrt{p}$ | $.37\sqrt{p}$ | $.29\sqrt{p}$ | $.35\sqrt{p}$ |

COMPARATIVE DATA

Data in the following table are the mean temperature and average precipitation for Salisbury, North Carolina, for the period 1906-1930 and are included in this publication for comparative purposes.

| 41.9 | 43.7 | 51.1 | 59.6 | 67.6 | 75.2 | 78.3 | 77.2 | 72.3 | 61.1 | 50.2 | 42.6 | 60.1 | 4.20 | 4.07 | 4.65 | 3.86 | 4.16 | 4.76 | 5.28 | 5.18 | 3.83 | 3.13 | 2.45 | 4.04 | 49.61 |

NORMALS, MEANS, AND EXTREMES

Station: ASHEVILLE, NORTH CAROLINA ASHEVILLE AIRPORT Standard time used: **EASTERN** Latitude: **35° 26' N** Longitude: **82° 32' W** Elevation (ground): **2140 feet**

Month	Daily max	Daily min	Monthly	Rec. high	Year	Rec. low	Year	Norm. heat. DD (Base 65°)	Precip norm. total	Precip max mo.	Year	Precip min mo.	Year	Precip max 24h	Year	Snow mean total	Snow max mo.	Year	Snow max 24h	Year	RH 01	RH 07	RH 13	RH 19	Mean speed	Prev. dir	Fast. speed	Fast. dir	Year	Pct sun	Sky cover	Clear	Partly	Cloudy	Precip ≥.01	Snow ≥1.0	Thunder	Heavy fog	90+	32− max	32− min	0−	Solar
(a)	(b)	(b)	(b)	5		5		(b)	(b)	5		5		5		5	5		5		5	5	5	5	5		5	5	5	5	5	5	5	5	5	5	5	5	5	5	5	5	5
J	47.5	27.7	37.6	71	1967	-7	1966	849	4.17	3.37	1966	2.02	1967	1.40	1969	6.4	17.6	1966	7.6	1966	82	86	60	70	9.4		38	34	1969	55	6.1	9	7	15	10	2	*	3	0	4	25		*
F	49.5	27.8	38.9	68	1965	-2	1967	731	4.03	6.56	1966	0.62	1968	3.17	1966	9.2	25.5	1969	11.7	1969	77	82	56	65	10.4		49	35	1969	58	5.7	10	5	13	10	2	0	2	0	2	22		*
M	56.1	33.0	44.6	81	1968+	14	1965	633	4.82	6.65	1968	2.59	1966	5.13	1968	3.7	13.0	1969	10.9	1969	79	84	49	58	9.6		46	35	1969	58	5.7	10	5	13	10	2	0	2	0	*	18		*
A	67.1	42.1	54.6	88	1967	24	1966	312	4.02	5.47	1966	1.11	1966	2.17	1966	T	T	1966	T	1966	83	89	54	61	9.6		40	32	1967	67	6.0	10	9	12	10	0	2	4	0	*	8		*
M	75.6	50.3	63.0	91	1969	30	1966	111	3.66	6.79	1967	2.92	1968	2.23	1967	0.0	0.0		0.0		92	93	56	68	7.0		40	32	1967	67	6.0	9	9	13	10	0	3	4	0	*	1		0
J	82.1	58.1	70.1	96	1969	35	1964	9	3.52	5.06	1968	2.46	1968	2.43	1967	0.0	0.0		0.0		96	97	62	72	5.8		28	34	1968	64	6.2	6	10	14	12	0	9	9	2	0	0		0
J	83.5	61.3	72.4	94	1969	46	1967	0	5.85	7.53	1969	3.24	1966	4.02	1969	0.0	0.0		0.0		97	97	66	76	5.7		30	35	1966	61	6.9	6	11	16	15	0	13	12	4	0	0		0
A	82.5	60.5	71.5	94	1968	43	1968	0	4.94	11.28	1967	3.31	1968	4.12	1967	0.0	0.0		0.0		97	99	63	80	5.7		30	33	1965	60	6.2	7	11	13	15	0	13	12	2	0	0		0
S	77.5	54.2	65.9	85	1969	43	1968	75	3.62	4.69	1965	2.53	1967	2.65	1966	0.0	0.0		0.0		97	99	65	83	5.5		23	36	1967+	60	6.2	7	11	13	14	0	12	14	1	0	0		0
O	68.3	43.2	55.8	83	1969	24	1965	294	3.13	5.37	1966	2.63	1969	2.95	1968	0.0	0.0		0.0		98	99	65	83	5.5		23	36	1967+	57	6.2	7	10	13	9	0	3	13	0	*	7		0
N	56.3	32.8	44.6	77	1968	13	1967	612	2.77	3.32	1967	1.30	1965	1.38	1969	2.2	9.6	1968	5.7	1968	92	96	58	77	7.1		30	34	1967	63	6.3	13	8	10	9	0	1	7	0	*	7		*
D	48.4	27.4	37.9	68	1965	14	1968+	840	3.62	6.13	1967	0.16	1965	1.36	1969	2.7	10.9	1969	5.6	1969	83	86	58	71	8.5		33	35	1965	57	5.6	10	8	13	8	1	1	3	0	1	16		*
YR	66.2	43.2	54.7	96 JUN.	1969	-7 JAN.	1966	4466	48.15	11.28 AUG.	1967	0.16 DEC.	1965	5.13 MAR.	1968	24.2	25.5 FEB.	1969	11.7 FEB.	1969	89	91	59	71	7.7		49	35 FEB.	1969	62	5.9	104	103	158	127	6	51	79	6	8	117		*

Ø Corrected after 1968 issue.

Means and extremes above are from existing and comparable exposures. Annual extremes have been exceeded at other sites in the locality as follows: Highest temperature 99 in July 1936; maximum monthly precipitation 13.75 in August 1940; minimum monthly precipitation T in October 1963; maximum precipitation in 24 hours 7.92 in October 1918; maximum monthly snowfall 28.9 in March 1960; maximum snowfall in 24 hours 15.8 in March 1942.

Station: CAPE HATTERAS, NORTH CAROLINA WEATHER BUREAU BUILDING Standard time used: **EASTERN** Latitude: **35° 16' N** Longitude: **75° 33' W** Elevation (ground): **7 feet**

Month	Daily max	Daily min	Monthly	Rec. high	Year	Rec. low	Year	Norm. heat. DD (Base 65°)	Precip norm. total	Precip max mo.	Year	Precip min mo.	Year	Precip max 24h	Year	Snow mean total	Snow max mo.	Year	Snow max 24h	Year	RH 01	RH 07	RH 13	RH 19	Mean speed	Prev. dir	Fast. speed	Fast. dir	Year	Pct sun	Sky cover	Clear	Partly	Cloudy	Precip ≥.01	Snow ≥1.0	Thunder	Heavy fog	90+	32− max	32− min	0−	Solar
(a)	(b)	(b)	(b)	12		12		(b)	(b)	12		12		12		12	12		12		12	12	12	12	12	6	10	10		12	12	12	12	12	12	12	12	12	12	12	12	12	18
J	53.2	39.9	46.6	72	1966	17	1966	580	3.90	9.07	1966	1.95	1965	3.31	1960	0.3	3.5	1962	3.5	1962	78	81	68	77	12.6	NNE	42	NNE	1961	55	6.0	9	8	14	11	*	1	3	0	*	11	0	232
F	53.5	39.5	46.5	73	1962	14	1958	518	3.93	7.48	1966	1.65	1962	2.77	1968	0.7	4.0	1958	4.0	1958	78	79	66	76	13.2	NNE	58	SSW	1966	54	6.3	9	8	14	11	*	1	3	0	*	9	0	304
M	58.0	44.0	51.0	77	1968	19	1967	440	4.16	7.82	1959	0.98	1967	2.86	1962	0.7	8.5	1960	7.0	1960	79	79	61	76	12.6	SW	79	NNW	1962	54	6.3	9	8	15	11	*	1	3	0	*	4	0	420
A	66.3	52.2	59.3	85	1967	26	1964	177	2.99	7.51	1966	1.20	1967+	5.60	1963	0.0	0.0		0.0		81	78	61	77	12.4	SW	61	SSW	1961	66	5.7	10	7	14	10	*	2	2	0	0	*	0	540
M	74.8	61.1	68.0	88	1962	39	1968	25	3.98	7.51	1966	0.61	1962	3.28	1958	0.0	0.0		0.0		86	82	66	79	12.4	SW	61	SSW	1961	65	5.7	9	9	12	9	0	4	2	0	0	0	0	594
J	81.7	68.6	75.2	94	1959	44	1966	0	4.14	10.80	1962	1.04	1967	6.63	1962	0.0	0.0		0.0		88	83	70	82	11.0	SSW	43	NNE	1966	66	6.3	7	10	13	9	0	5	1	0	0	0	0	607
J	83.8	72.2	78.0	95	1969	55	1963	0	6.15	9.99	1958	0.45	1958	5.53	1967	0.0	0.0		0.0		89	85	70	83	10.4	SW	58	S	1960	66	6.1	7	10	13	9	0	6	1	1	0	0	0	591
A	83.5	71.6	77.6	94	1968	57	1964+	0	6.42	11.68	1962	1.80	1968	8.11	1962	0.0	0.0		0.0		89	86	70	83	11.5	SW	58	S	1964	69	6.7	6	9	16	13	0	9	*	1	0	0	0	522
S	79.8	68.3	74.1	90	1959	48	1967	0	5.89	8.75	1968	2.90	1965	4.77	1962	0.0	0.0		0.0		88	86	70	83	11.0	NE	45	S	1964	69	6.7	8	9	13	11	0	9	*	1	0	0	0	471
O	71.3	59.4	65.4	86	1959	33	1964+	78	4.24	9.58	1968	1.41	1969	4.08	1964	0.0	0.0		0.0		84	83	67	79	11.0	NE	72	S	1960	67	5.6	11	9	12	11	0	4	*	0	0	0	0	353
N	62.8	49.5	56.2	80	1959	22	1967	273	4.09	14.63	1962	1.23	1966	4.02	1962	T	T	1959	T	1959	82	81	65	79	11.5	NNE	60	NNW	1968	65	6.3	12	8	10	9	*	2	1	0	0	2	0	273
D	54.8	41.5	48.2	75	1967	19	1962+	521	4.58	8.63	1962	2.24	1965	2.97	1962	T	T	1968+	T	1968+	77	80	66	77	11.7	NNE	54	SW	1968	61	5.6	10	8	13	8	*	1	1	0	*	7	0	210
YR	68.6	55.7	62.2	95 JUL.	1969	14 FEB.	1958	2612	54.47	14.63 NOV.	1962	0.45 JUL.	1958	8.11 AUG.	1962	1.7	8.5 MAR.	1960	7.0 MAR.	1960	83	81	66	79	11.6	SW	72	S SEP.	1960	63	5.9	106	100	159	120	1	45	16	3	1	33	0	426

Means and extremes above are from existing and comparable exposures. Annual extremes have been exceeded at other sites in the locality as follows: Highest temperature 97 in June 1952; lowest temperature 8 in December 1880; maximum monthly precipitation 20.95 in June 1949; minimum monthly precipitation T in November 1890; maximum precipitation in 24 hours 14.73 in June 1949; maximum monthly snowfall 12.0 in December 1917; maximum snowfall in 24 hours 12.0 in December 1917; fastest mile wind 110 W in September 1944.

NORMALS, MEANS, AND EXTREMES

Station: CHARLOTTE, NORTH CAROLINA DOUGLAS MUNICIPAL AIRPORT Standard time used: EASTERN Latitude: 35 13' N Longitude: 80 56' W Elevation (ground): 736 feet

Month	Temperature — Normal Daily maximum	Daily minimum	Monthly	Extremes Record highest	Year	Record lowest	Year	Normal heating degree days (Base 65°)	Precipitation Normal total	Maximum monthly	Year	Minimum monthly	Year	Maximum in 24 hrs.	Year	Snow, Ice pellets Mean total	Maximum monthly	Year	Maximum in 24 hrs	Year	Relative humidity Hour 01	Hour 07	Hour 13	Hour 19	Wind Mean speed	Prevailing direction	Fastest mile Speed	Direction	Year	Pct. of possible sunshine	Mean sky cover sunrise to sunset	Mean number of days Sunrise to sunset Clear	Partly cloudy	Cloudy	Precipitation .01 inch or more	Snow, Ice pellets 1.0 inch or more	Thunderstorms	Heavy fog	Temperatures Max. 90 and above	32 and below	Min. 32 and below	0 and below	Average daily solar radiation langleys
(a)	(b)	(b)	(b)	9		9		(b)	(b)	30		30		30		30	30		30		9	9	9	9	20	14	20	20		19	20	21	21	21	30	30	30	30	9	9	9	9	
J	51.4	34.0	42.7	74	1967	4	1966	691	3.53	7.44	1962	1.24	1956	3.57	1962	2.3	11.7	1962	10.2	1965	72	79	55	61	7.9	SW	56	NE	1960	57	6.2	9	9	13	10	1	1	3	0	2	21	0	
F	53.7	34.7	44.2	78	1962	7	1967+	582	3.55	6.86	1944	0.87	1968	2.92	1955	1.6	13.2	1969	12.0	1969	67	76	52	55	8.4	NE	54	SW	1958	59	6.0	9	6	13	11	*	1	3	0	*	17	0	
M	60.0	40.0	50.0	86	1967	18	1965	481	4.39	8.69	1944	2.11	1949	3.64	1952	1.1	19.3	1960	8.0	1960	68	79	48	50	8.8	SW	47	W	1956	64	5.9	10	8	13	11	*	2	2	0	0	9	0	
A	71.0	49.6	60.3	91	1963+	28	1961	156	3.49	7.64	1958	0.97	1942	3.20	1962	T	T	1954	0.0	1954	70	81	49	51	9.0	S	53	NW	1958	69	5.7	9	10	11	9	0	3	1	1	0	1	0	
M	79.4	58.6	69.0	95	1964+	32	1963	22	3.11	5.29	1957+	0.11	1941	2.67	1955	0.0	0.0		0.0		77	84	52	58	7.5	SW	48	NW	1958	69	5.9	9	10	12	9	0	6	1	3	0	*	0	
J	87.6	66.6	77.1	99	1964	46	1966	0	3.61	8.26	1961	0.67	1954	3.77	1949	0.0	0.0		0.0		81	87	57	63	6.9	SW	57	NW	1957	71	6.0	7	11	12	9	0	8	1	7	0	0	0	
J	88.8	69.5	79.2	99	1966	53	1961	0	4.88	9.12	1941	1.26	1948	3.00	1949	0.0	0.0		0.0		84	89	58	67	6.5	SW	59	NW	1962	69	6.3	7	11	13	12	0	10	1	12	0	0	0	
A	87.8	69.5	78.7	100	1963	53	1965	0	4.22	9.98	1948	0.88	1968	4.40	1967	0.0	0.0		0.0		84	90	58	67	6.5	S	54	NW	1954	71	5.8	8	12	11	10	0	8	2	10	0	0	0	
S	81.9	63.8	72.9	94	1966+	39	1967	6	3.49	10.89	1945	0.02	1954	4.74	1959	0.0	0.0		0.0		83	90	55	66	6.9	NE	47	NNE	1956	68	5.6	10	9	11	7	0	3	2	3	0	0	0	
O	72.6	52.2	62.5	87	1962	24	1962	124	2.96	7.66	1947	T	1953	4.84	1955	0.0	0.0		0.0		79	88	50	64	7.1	NW	47	NW	1960	71	4.4	15	6	10	7	0	1	2	0	0	8	0	
N	60.7	40.1	50.4	85	1961	20	1969	438	2.53	8.17	1948	0.60	1960	2.79	1962	0.1	2.5	1968	2.5	1968	75	84	51	61	7.3	SSW	47	NW	1957	63	5.2	12	7	11	7	*	1	2	0	0	8	0	
D	51.3	34.1	42.7	74	1966	2	1962	691	3.62	7.41	1945	0.43	1965	2.59	1958	0.3	4.4	1945	2.9	1958	74	79	55	62	7.2	SW	57	NE	1954	60	5.8	10	7	14	10	*	*	4	0	0	19	0	
YR	70.5	51.1	60.8	100	AUG. 1963	2	DEC. 1962	3191	43.38	10.89	SEP. 1945	T	OCT. 1953	4.84	OCT. 1955	5.4	19.3	MAR. 1960	12.0	FEB. 1969	76	84	53	60	7.5	SW	59	NW	JUL. 1962	66	5.7	115	104	146	110	2	42	24	36	3	76	0	

Means and extremes above are from existing and comparable exposures. Annual extremes have been exceeded at other sites in the locality as follows: Highest temperature 104 in September 1954; lowest temperature -5 in February 1899; maximum monthly precipitation 16.55 in July 1916; maximum precipitation in 24 hours 6.59 in July 1944; maximum snowfall in 24 hours 14.0 in February 1902.

Station: GREENSBORO, HIGH POINT, WINSTON SALEM AP, N. C. Standard time used: EASTERN Latitude: 36° 05' N Longitude: 79° 57' W Elevation (ground): 897 feet

Month	Temperature — Normal Daily maximum	Daily minimum	Monthly	Extremes Ø Record highest	Year	Record lowest	Year	Normal heating degree days (Base 65°)	Precipitation Normal total	Maximum monthly	Year	Minimum monthly	Year	Maximum in 24 hrs.	Year	Snow, Ice pellets Mean total	Maximum monthly	Year	Maximum in 24 hrs	Year	Relative humidity Hour 01	Hour 07	Hour 13	Hour 19	Wind Mean speed	Prevailing direction	Fastest mile Speed	Direction	Year	Pct. of possible sunshine	Mean sky cover sunrise to sunset	Mean number of days Sunrise to sunset Clear	Partly cloudy	Cloudy	Precipitation .01 inch or more	Snow, Ice pellets 1.0 inch or more	Thunderstorms	Heavy fog	Temperatures Max. 90 and above	32 and below	Min. 32 and below	0 and below	Average daily solar radiation langleys
(a)	(b)	(b)	(b)	6		6		(b)	(b)	41		41		41		41	41		41		6	6	6	6	41	15	21	21		41	41	41	41	41	41	41	41	41	6	6	6	6	16
J	49.6	29.8	39.7	76	1967	3	1966	784	3.40	8.24	1937	1.04	1956	3.06	1936	3.4	22.9	1966	14.0	1940	75	80	56	64	8.3	SW	40	NW	1943+	51	6.2	9	7	15	10	1	*	5	0	3	22	0	210
F	51.8	30.2	41.0	75	1965	6	1967	672	3.30	7.04	1929	0.84	1941	3.00	1934	2.0	9.8	1936	8.6	1960+	71	76	50	58	8.8	SW	51	W	1956	56	6.0	9	6	13	11	*	1	3	0	*	13	0	275
M	58.9	35.8	47.4	89	1968	15	1967	552	3.69	7.21	1952	1.21	1967	3.07	1932	1.8	21.3	1960	11.1	1960	69	77	46	50	9.4	SW	54	SW	1955+	60	5.8	9	9	12	10	0	1	1	0	1	0	0	375
A	69.3	45.4	57.4	91	1967	27	1964	234	3.43	6.19	1936	0.55	1942	2.70	1944	T	T	1961+	T	1961+	82	84	54	62	7.8	SW	61	NW	1967	66	5.7	9	11	11	10	0	7	2	2	0	1	0	460
M	78.8	54.9	66.9	94	1964	35	1966	47	3.29	6.04	1930	0.37	1956	3.04	1956	0.0	0.0		0.0		87	87	57	66	7.0	SW	56	W	1940	66	5.8	7	13	11	11	0	9	1	8	0	0	0	528
J	86.1	63.4	74.8	98	1964	44	1966	0	3.47	7.99	1965	0.32	1933	4.20	1969	0.0	0.0		0.0		90	91	62	73	6.6	SW	63	N	1932	63	5.7	6	13	12	13	0	11	2	11	0	0	0	526
J	87.7	66.9	77.3	100	1966	50	1963	0	4.79	9.81	1959	0.98	1953	4.43	1944	0.0	0.0		0.0		90	91	62	73	6.6	SW	45	N	1952	64	5.9	6	12	13	12	0	11	2	11	0	0	0	476
A	86.4	65.9	76.2	98	1968	48	1965	0	4.61	12.53	1939	1.19	1962	4.47	1949	0.0	0.0		0.0		90	91	56	69	6.9	NE	40	N	1934	64	5.9	9	10	11	8	0	3	3	1	0	0	0	409
S	80.9	59.3	70.1	93	1966	38	1963	33	3.66	13.26	1947	0.13	1939	7.49	1947	0.0	0.0		0.0		83	87	53	70	7.3	NE	43	N	1960	68	4.5	14	8	9	7	0	1	2	0	0	0	332
O	71.5	47.0	59.3	86	1968+	25	1968	192	2.71	9.60	1959	0.26	1963	6.24	1954	0.0	0.0		0.0		76	81	49	63	7.7	SW	40	SW	1955	60	5.3	12	6	12	8	*	*	4	0	0	10	0	241
N	59.8	36.0	47.9	79	1968	18	1967	513	2.68	6.72	1948	0.37	1931	3.32	1962	0.2	5.9	1968	5.0	1968	78	81	56	67	7.7	SW	45	SSW	1954	54	5.9	10	7	14	9	*	*	4	0	*	19	0	193
D	50.2	29.6	39.9	73	1966	8	1963	778	3.13	6.34	1967	0.33	1955	3.60	1958	1.4	14.3	1930	14.3	1930	78	81	56	67	7.7	SW																
YR	69.3	47.0	58.2	100	JUL. 1966	3	JAN. 1966	3805	42.16	13.26	SEP. 1947	0.13	SEP. 1939	7.49	SEP. 1947	8.8	22.9	JAN. 1966	14.3	DEC. 1930	80	84	54	64	7.8	SW	63	N	JUL. 1932	62	5.7	112	110	143	117	2	47	31	30	3	89	0	381

Ø For period July 1963 through the current year.
Means and extremes above are from existing and comparable exposures. Annual extremes have been exceeded at other sites in the locality as follows: Highest temperature 102 in July 1954+; lowest temperature -7 in January 1940.

Station (top table)

Standard time used: EASTERN Latitude: 35° 52' N Longitude: 78° 47' W Elevation (ground): 434 feet

Ø For period November 1964 through the current year.
Means and extremes above are from existing and comparable exposures. Annual extremes have been exceeded at other sites in the locality as follows:
Highest temperature 105° in July 1952; lowest temperature -2 in February 1899; maximum monthly precipitation 13.63 in August 1908; minimum monthly precipitation 0.06 in November 1931+; maximum precipitation in 24 hours 6.66 in September 1929; maximum monthly snowfall 20.0 in January 1893; maximum snowfall in 24 hours 17.8 in March 1927.

Station: WILMINGTON, NORTH CAROLINA   NEW HANOVER COUNTY AIRPORT

Standard time used: EASTERN Latitude: 34° 16' N Longitude: 77° 55' W Elevation (ground): 28 feet

Ø For period July 1963 through the current year.
Means and extremes above are from existing and comparable exposures. Annual extremes have been exceeded at other sites in the locality as follows:
Highest temperature 104 in June 1952; lowest temperature 5 in February 1899; maximum monthly precipitation 21.12 in July 1886; minimum monthly precipitation .02 in October 1943; maximum precipitation in 24 hours 9.52 in September 1938; maximum monthly snowfall 12.1 in February 1896; maximum snowfall in 24 hours 11.1 in February 1896.

REFERENCE NOTES APPLYING TO ALL "NORMALS, MEANS, AND EXTREMES" TABLES

(a) Length of record, years, based on January data.
Other months may be for more or fewer years if there have been breaks in the record.

(b) Climatological standard normals (1931-1960).
* Less than one half.
+ Also on earlier dates, months, or years.
T Trace, an amount too small to measure.
Below zero temperatures are preceded by a minus sign.
The prevailing direction for wind in the Normals, Means, and Extremes table is from records through 1963.
≥ 70° at Alaskan stations.

Unless otherwise indicated, dimensional units used in this bulletin are: temperature in degrees F.; precipitation, including snowfall, in inches; wind movement in miles per hour; and relative humidity in percent. Heating degree day totals are the sums of negative departures of average daily temperatures from 65° F. Cooling degree day totals are the sums of positive departures of average daily temperatures from 65° F. Sleet was included in snowfall totals beginning with July 1948. The term "Ice pellets" includes traces of ice (sleet) and particles consisting of snow pellets encased in a thin layer of ice. Heavy fog reduces visibility to 1/4 mile or less.

Sky cover is expressed in a range of 0 for no clouds or obscuring phenomena to 10 for complete sky cover. The number of clear days is based on average cloudiness 0-3, partly cloudy days 4-7, and cloudy days 8-10 tenths.

Solar radiation data are the averages of direct and diffuse radiation on a horizontal surface. The langley denotes one gram calorie per square centimeter.

& Figures instead of letters in a direction column indicate direction in tens of degrees from true North; i.e., 09 - East, 18 - South, 27 - West, 36 - North, and 00 - Calm. Resultant wind is the vector sum of wind directions and speeds divided by the number of observations. If figures appear in the direction column under "Fastest mile" the corresponding speeds are fastest observed 1-minute values.

To 8 compass points only.

Mean Maximum Temperature (°F.), January

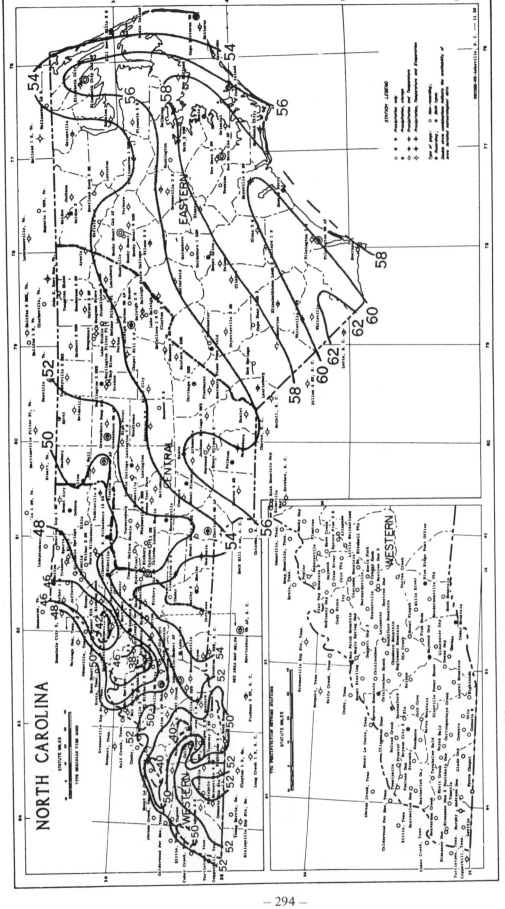

Based on period 1931-52

Isolines are drawn through points of approximately equal value. Caution should be used in interpolating on these maps, particularly in mountainous areas.

Mean Minimum Temperature (°F.), January

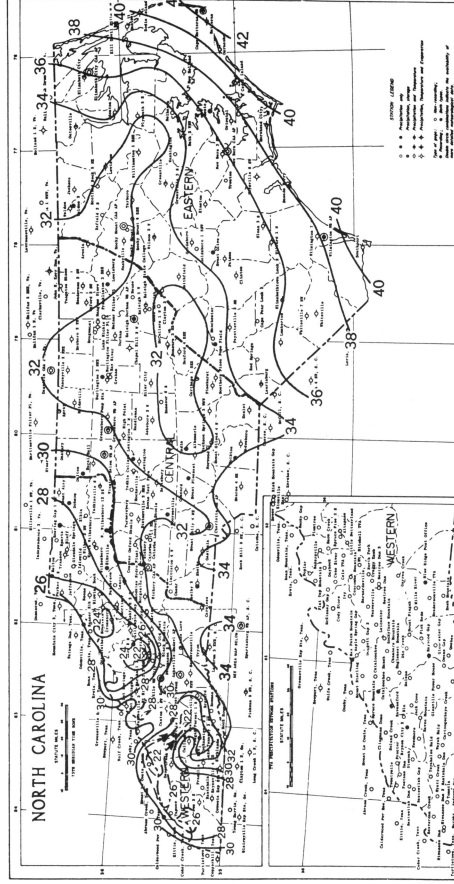

Based on period 1931-52

Isolines are drawn through points of approximately equal value. Caution should be used in interpolating on these maps, particularly in mountainous areas.

Mean Maximum Temperature (°F.), July

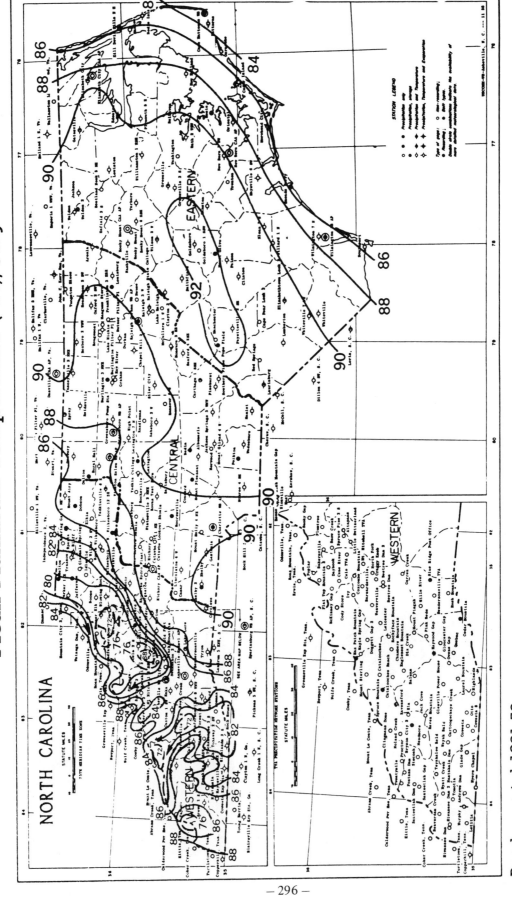

Based on period 1931-52

Isolines are drawn through points of approximately equal value. Caution should be used in interpolating on these maps, particularly in mountainous areas.

Mean Minimum Temperature (°F.), July

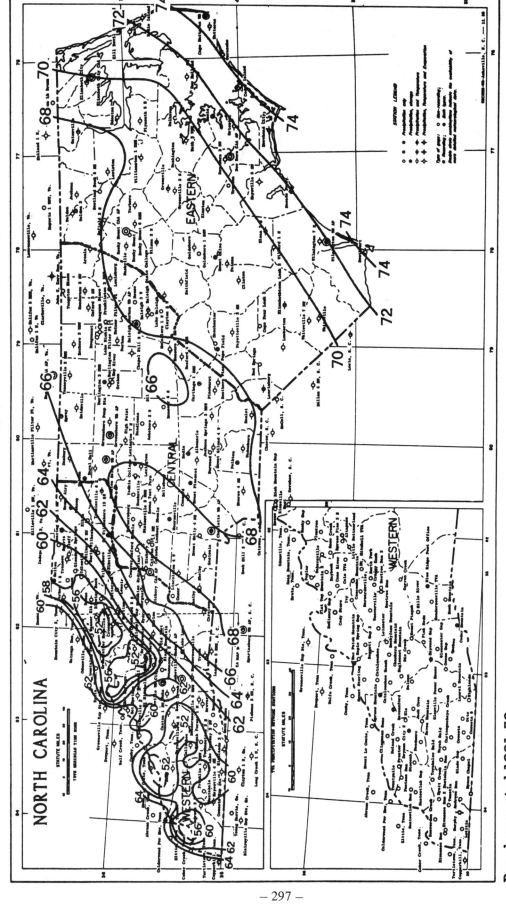

Based on period 1931-52

Isolines are drawn through points of approximately equal value.   Caution should be used
in interpolating on these maps, particularly in mountainous areas.

Mean Annual Precipitation, Inches

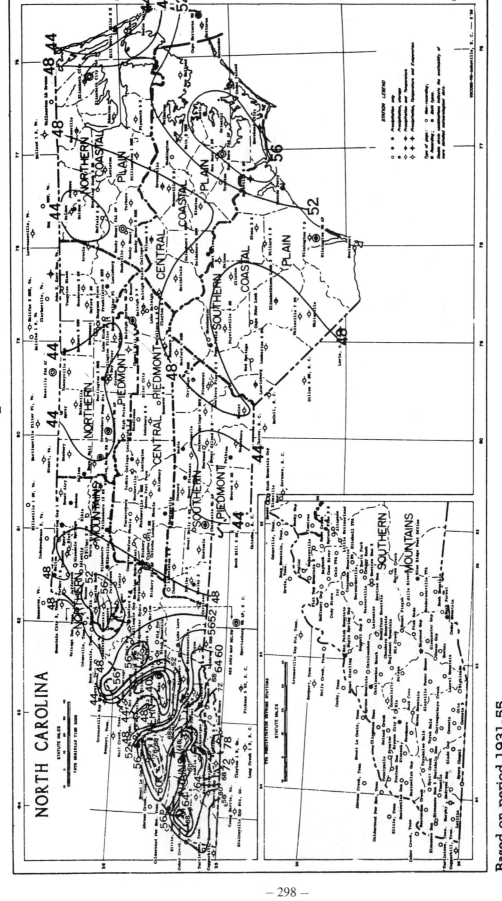

Based on period 1931-55

Isolines are drawn through points of approximately equal value. Caution should be used in interpolating on these maps, particularly in mountainous areas.

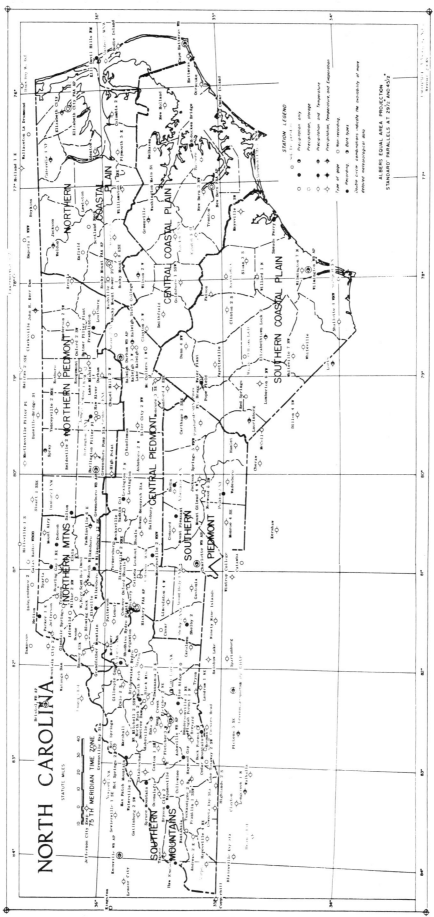

THE CLIMATE OF
OHIO.

by
L.T. Pierce

December 1959

The climate of Ohio is remarkably varied. Less than one-half of its area is occupied by typical plains, while most of eastern and much of southern Ohio is decidedly hilly. Topography ranges in elevation from 430 feet above sea level at the junction of the Great Miami and Ohio Rivers up to 1,550 feet on a summit near Bellefontaine. In addition to this high point there are innumerable other hills which rise above 1,400 feet (mean sea level). These are located mainly along the dividing line between the Ohio River and Lake Erie drainage basins. Large areas in the State have elevations above 1,000 feet. An extensive area in northwestern Ohio is occupied by a flat lake plain--once the bottom of glacial Lake Maumee which was much larger than the present Lake Erie. The greater part of eastern Ohio is within the Allegheny Plateau, an unglaciated area consisting of picturesque hills many of which rise above 1,300 feet and between which there are many winding rivers and streams.

The Ohio River, which forms the southern and southeastern boundaries of Ohio, and its tributaries drain the greater portion of the State. A number of streams drain northward into Lake Erie. Although this area comprises nearly a third of the State, the divide between the two drainages is only 20 to 40 miles from the lake shore for a distance of more than 100 miles until it dips south of the arrowhead-shaped Maumee Basin. The largest streams in this region are the Maumee, Sandusky, and Cuyahoga Rivers. Principal tributaries flowing southward into the Ohio River include the Muskingum in the east, the Scioto in the central section, and the Great Miami in the west. A small portion in the west-central region drains westward into the Wabash River basin of Indiana.

The availability of ample water supply, together with Ohio's favorable geographical location and its large mineral resources, have favored the development of large industries. In fact, Ohio is one of the major industrial states of the Union. Due in large part to the accessibility of ample supplies of coal and water, this is one of the principal steel producing states. It ranks first in such diversified lines as machine tools, rubber, business machines, clay products, and several others. While Cleveland, Cincinnati, Toledo, Akron, and Dayton are the largest industrial centers, many other cities have become sites for important and diversified manufacturing plants.

Ohio's agriculture is characterized by its diversity. Much of the western half lies in the great Corn Belt which includes the bed of the Old Lake Maumee in the northwest. Central and northeastern sections have rolling terrain and are consequently devoted largely to dairing and general farming. The hilly southeastern and extreme southern section, however, is principally a general farming and woodland area, although considerable tobacco and truck crops are grown along the Ohio River. The largest sources of farm income are dairy, hogs, and poultry, followed by wheat, soybeans, and beef cattle. Tobacco is grown extensively in southern counties, sugar beets in the northwest, and grapes principally along the shore of Lake Erie. Truck crops are grown mostly in the west and the north, their distribution being controlled primarily by climatic and topographic considerations.

Located west of the Appalachian Mountains, Ohio has a climate essentially continental in nature, characterized by moderate extremes of heat and cold, and wetness and dryness. Summers are moderately warm and humid, with occasional days when temperatures exceed 100°F.; winters are reasonably cold, with an average of about 2 days of subzero weather; and autumns are predominately cool, dry, and invigorating. Spring is the wettest season and vegetation grows luxuriantly. Annual precipitation is slightly in excess of the national average and is well distributed, though with peaks in early spring and summer. In spite of the relatively small range in latitude and the compact shape of Ohio, rainfall varies considerably in amount and seasonal distribution. This is accounted for not only by the presence of Lake Erie on the north, but also by its topography and proximity to rain producing storm paths. Annual precipitation averages about 38 inches, being most generous in spring (about 4 inches in April) and least in the fall (about 2.5 inches in October). Greatest amounts are measured in the southwest where Wilmington has an average of 44.36 inches; and the lake shore is driest, Gilbralter Island having a normal of only 29.06 inches. The southern half of the State is visited more frequently by productive rainstorms which, together with the general roughness of terrain, accounts for the larger total precipitation. The lifting of moist air masses over the hills tends to increase the yield of rainfall, especially in winter and spring. There is a marked tendency during the cold season for northeastern counties to receive snowfall amounts substantially in excess of those measured elsewhere. Northerly winds have a long fetch across Lake Huron and the widest part of Lake Erie, thus picking up moisture and heat from the lakes. This moisture is then forced to condense as the air is lifted abruptly over the divide a short distance from the lake. Average snowfall ranges from 60 inches in parts of Lake and adjoining counties down to 16 inches or less along the Ohio River.

The normal annual temperature for the State ranges from 49.6°F. at Hiram in Portage County up to 56.9°F. at Portsmouth on the Ohio River. Variations over the State are due mainly to differences in latitude and topography, but the immediate Lake shore area experiences a moderating effect due to its proximity to a large body of water. This latter effect accounts in large part for the concentration of grape culture along the lake shore. Widest temperature ranges are found generally among the eastern hills. The lowest ever recorded in the State was -39°F. at Milligan in Perry County on February 10, 1899, while the record high of 113°F. has been observed at scattered stations, among them Wilmington in the southwest. In an average year, 90-degree heat may be expected about 20 times in summer with 100°F. or more once or twice. Readings of zero or lower are generally to be expected on 2 to 4 days each winter, and these are just as likely to occur in the south as the north. However, 1 winter out of 6 or 8 will pass without experiencing zero readings anywhere in the State.

The growing season, as defined by the period 32°F. or higher, ranges widely because of latitude and proximity to Lake Erie. The longest is about 200 days on the lake shore and the shortest is in the northeastern valleys within the Ohio River drainage. Dates of the average last freezing temperature in spring range from April 15 to May 18 and the mean first freeze date in fall varies from September 30 to November 6, the latter being on the western lake shore.

Damaging windstorms are mostly associated with heavy thunderstorms or line squalls. Three or four tornadoes may be expected to strike in Ohio each year. Most tornadoes, however, are of limited effect having paths that are short and narrow.

Most floods in Ohio are caused by unusual precipitation. The storms causing floods may bring rainfall of unusual intensity or of unusual duration and extent. Some floods may be caused by a series of ordinary storms which follow one another in rapid succession. Others may result from rain falling at relatively high temperatures on snow-covered areas. At times, though infrequent, flood conditions are caused or aggravated by ice gorges, especially in the tributary streams. Severe thunderstorms frequently cause local flash flooding. General flooding occurs most frequently during January to March and rarely occurs during August to October.

The frequency of floods in Ohio is about 1 in 3 years. Notable flood years in Ohio have been 1832, 1882, 1883, 1884, 1898, 1907, 1913, 1927, 1929, 1937, 1943, and 1959.

A system of flood control reservoirs in the Muskingum and Miami River basins controls the flow of those rivers. The lower 93 miles of the Muskingum is canalized by a system of 11 locks and dams. There is a series of 30 navigational locks in the Ohio River from East Liverpool to North Bend, Ohio.

In general Ohio's climate is a good one in which to live, and one favorable for agricultural and industrial development. While this is not a "vacation land" in the usual sense, people are coming more and more to enjoy recreational facilities afforded by the numerous parks, flood control reservoirs, and the shore of Lake Erie.

REFERENCES

(1) "Ohio, An Empire Within An Empire", published by the Ohio Development and Publicity Commission.

(2) "Climatological History of Ohio" - 1924 W. H. Alexander: Bulletin No. 26 Engineering Experiment Station Ohio State University.

(3) Weather Bureau Technical Paper No. 16 - Maximum 24-Hour Precipitation in the United States. Washington, D. C. 1952.

(4) Weather Bureau Technical Paper No. 29 - Rainfall Intensity-Duration-Frequency Regime. Washington, D. C.

(5) "Climatic Features of Ohio", G. W. Mindling (manuscript).

(6) Weather Bureau Technical Paper No. 15 - Maximum Station Precipitation for 1, 2, 3, 6, 12, and 24 Hours.

(7) Weather Bureau Technical Paper No. 25 - Rainfall Intensity-Duration-Frequency Curves. For selected stations in the United States, Alaska, Hawaiian Islands, and Puerto Rico.

BIBLIOGRAPHY

(A) Climatic Summary of the United States (Bulletin W) 1930 edition, Section 68, 69, 70 and 71. U. S. Weather Bureau

(B) Climatic Summary of the United States Ohio - Supplement for 1931 through 1952 (Bulletin W Supplement). U. S. Weather Bureau

(C) Climatological Data - Ohio. U. S. Weather Bureau

(D) Climatological Data National Summary. U. S. Weather Bureau

(E) Hourly Precipitation Data - Ohio. U. S. Weather Bureau

(F) Local Climatological Data, U. S. Weather Bureau, for Akron-Canton, Cincinnati, Cleveland, Columbus, Dayton, Sandusky, Toledo and Youngstown, Ohio.

STATION	Freeze threshold temperature	Mean date of last Spring occurrence	Mean date of first Fall occurrence	Mean No. of days between dates	Years of record Spring	No. of occurrences in Spring	Years of record Fall	No. of occurrences in Fall
AKRON CANTON WB AP	32	04-29	10-20	173	30	30	30	30
	28	04-14	11-05	205	30	30	30	30
	24	04-01	11-17	231	30	30	30	30
	20	03-20	11-28	253	30	30	30	30
	16	03-11	12-05	269	30	30	30	28
ASHLAND	32	05-07	10-11	157	30	30	30	30
	28	04-19	10-24	187	30	30	30	30
	24	04-05	11-04	213	30	30	30	30
	20	03-26	11-21	240	30	30	30	30
	16	03-15	12-02	262	30	30	30	29
ATHENS 1 E	32	04-28	10-10	165	20	20	19	19
	28	04-15	10-24	193	19	19	19	19
	24	03-25	11-07	227	18	18	19	19
	20	03-16	11-20	249	18	18	19	19
	16	03-09	12-02	267	18	18	19	18
BATAVIA 4 N	32	04-28	10-13	168	30	30	30	30
	28	04-14	10-26	195	30	30	30	30
	24	03-27	11-07	225	30	30	30	30
	20	03-15	11-18	249	30	30	30	30
	16	03-04	12-03	274	30	30	30	28
BELLEFONTAINE SEWAGE	32	05-04	10-15	164	28	28	29	29
	28	04-22	10-23	184	28	28	28	28
	24	04-06	11-06	214	28	28	28	28
	20	03-24	11-17	238	28	28	28	28
	16	03-12	11-26	259	28	28	28	27
BOWLING GREEN SWG PL	32	05-06	10-08	155	30	30	30	30
	28	04-24	10-26	185	30	30	30	30
	24	04-08	11-05	211	30	30	30	30
	20	03-25	11-19	239	30	30	30	30
	16	03-12	11-30	263	30	30	30	29
BUCYRUS	32	05-03	10-09	159	30	30	30	30
	28	04-22	10-22	184	30	30	30	30
	24	04-06	11-05	213	30	30	30	30
	20	03-25	11-17	237	30	30	30	30
	16	03-11	11-29	263	30	30	30	29
CADIZ	32	05-01	10-14	166	28	28	28	28
	28	04-18	10-26	191	28	28	28	28
	24	04-03	11-10	220	28	28	27	27
	20	03-24	11-19	240	28	28	27	27
	16	03-14	11-29	260	27	27	27	27
CALDWELL 4 W	32	04-29	10-15	169	16	16	17	17
	28	04-16	10-28	195	16	16	17	17
	24	04-04	11-12	222	16	16	17	17
	20	03-26	11-23	242	16	16	17	17
	16	03-10	11-30	265	16	16	17	17
CAMBRIDGE ST HOSP	32	05-04	10-10	159	30	30	30	30
	28	04-18	10-23	188	30	30	30	30
	24	04-03	11-07	219	30	30	30	30
	20	03-16	11-19	249	30	30	30	30
	16	03-08	12-03	271	30	30	30	28
CANFIELD 1 S	32	05-15	10-04	142	29	29	30	30
	28	04-28	10-16	170	30	30	30	30
	24	04-15	10-30	198	30	30	30	30
	20	04-02	11-20	232	30	30	30	30
	16	03-18	11-30	256	29	29	30	30
CANTON	32	05-05	10-14	162	28	28	27	27
	28	04-18	10-28	193	28	28	27	27
	24	04-05	11-16	225	28	28	27	27
	20	03-19	11-22	247	28	28	27	27
	16	03-10	12-06	271	28	28	27	25
CATAWBA ISLAND 1 SW	32	04-18	10-30	195	30	30	30	30
	28	04-06	11-12	220	30	30	30	30
	24	03-28	11-23	240	29	29	30	30
	20	03-17	12-03	261	29	29	30	29
	16	03-07	12-09	277	29	29	30	28
CHILLICOTHE	32	04-25	10-15	173	28	28	27	27
	28	04-10	10-27	200	28	28	27	27
	24	03-25	11-09	230	27	27	27	27
	20	03-12	11-20	253	27	27	27	27
	16	03-05	12-06	276	27	27	26	25
CHIPPEWA LAKE	32	05-14	10-07	147	30	30	30	30
	28	04-29	10-22	175	30	30	30	30
	24	04-14	11-07	206	30	30	30	30
	20	03-31	11-21	235	30	30	30	30
	16	03-16	11-29	258	30	30	30	29
CINCINNATI ABBE OBS	32	04-15	10-25	192	30	30	30	30
	28	03-29	11-07	223	30	30	30	30
	24	03-19	11-21	247	30	30	30	30
	20	03-08	12-02	269	30	30	30	29
	16	02-28	12-11	286	30	30	30	26
CIRCLEVILLE	32	04-28	10-12	166	27	27	27	27
	28	04-18	10-24	188	26	26	27	27
	24	03-29	11-08	224	25	25	27	27
	20	03-12	11-19	252	25	25	27	27
	16	03-03	11-30	273	24	24	26	24
CLEVELAND WB AP	32	04-21	11-02	195	30	30	30	30
	28	04-07	11-13	220	30	30	30	30
	24	03-25	11-24	245	30	30	30	30
	20	03-15	12-04	263	30	30	30	28
	16	03-05	12-11	281	30	30	30	27
COLUMBUS WB CITY	32	04-17	10-30	196	30	30	30	30
	28	04-01	11-12	225	30	30	30	30
	24	03-22	11-22	245	30	30	30	30
	20	03-10	12-06	271	30	30	30	28
	16	03-05	12-10	280	30	30	30	27
COSHOCTON	32	05-04	10-11	160	30	30	30	30
	28	04-18	10-25	190	30	30	30	30
	24	04-01	11-09	222	30	30	30	30
	20	03-19	11-20	246	30	30	30	30
	16	03-10	12-03	268	30	30	30	29
DELAWARE	32	05-01	10-09	161	30	30	29	29
	28	04-19	10-23	187	30	30	29	29
	24	04-04	11-04	214	30	30	29	29
	20	03-21	11-16	239	30	30	29	29
	16	03-10	11-30	265	30	30	29	28
DENNISON	32	05-12	10-03	144	30	30	30	30
	28	04-29	10-19	172	30	30	30	30
	24	04-14	11-02	203	29	29	30	30
	20	03-27	11-11	229	29	29	30	30
	16	03-13	11-26	258	29	29	30	30
FINDLAY SWG PLANT	32	05-03	10-11	160	30	30	30	30
	28	04-22	10-27	188	30	30	30	30
	24	04-06	11-07	215	30	30	30	30
	20	03-22	11-19	242	30	30	30	30
	16	03-12	12-01	264	30	30	30	29
GERMANTOWN 3 NE	32	04-27	10-13	168	30	30	30	30
	28	04-14	10-27	196	30	30	30	30
	24	03-25	11-09	229	30	30	30	30
	20	03-14	11-22	252	30	30	30	30
	16	03-06	12-05	274	29	29	30	28
GREENVILLE SWG PLT	32	04-30	10-14	167	29	29	30	30
	28	04-17	10-28	194	29	29	30	30
	24	03-30	11-09	224	30	30	30	30
	20	03-19	11-18	244	30	30	30	30
	16	03-07	12-04	272	30	30	30	27
HAMILTON WATER WORKS	32	04-30	10-14	167	29	29	29	29
	28	04-14	10-26	195	29	29	29	29
	24	03-25	11-11	231	29	29	29	29
	20	03-12	11-21	255	29	29	29	29
	16	03-04	12-08	280	29	29	29	26
HILLSBORO	32	04-29	10-14	168	30	30	30	30
	28	04-16	10-28	195	30	30	30	30
	24	04-03	11-08	219	30	30	30	30
	20	03-17	11-19	247	30	30	30	30
	16	03-07	12-01	269	30	30	30	29
HIRAM	32	05-06	10-14	161	29	29	30	30
	28	04-22	10-31	192	29	29	29	29
	24	04-11	11-13	216	29	29	29	29
	20	03-29	11-20	236	29	29	29	29
	16	03-17	12-01	259	29	29	29	28

FREEZE DATA

STATION	Freeze threshold temperature	Mean date of last Spring occurrence	Mean date of first Fall occurrence	Mean No. of days between dates	Years of record Spring	No. of occurrences in Spring	Years of record Fall	No. of occurrences in Fall
IRONTON	32	04-23	10-16	176	30	30	30	30
	28	04-11	10-30	202	30	30	30	30
	24	03-26	11-11	231	30	30	30	30
	20	03-10	11-24	258	30	30	30	30
	16	03-02	12-07	280	30	30	30	27
JACKSON	32	04-30	10-10	162	29	29	29	29
	28	04-17	10-21	187	29	29	29	29
	24	04-01	11-03	216	29	29	29	29
	20	03-19	11-12	238	29	29	20	28
	16	03-09	11-25	261	29	29	28	27
JEFFERSON	32	05-03	10-16	166	11	11	11	11
	28	04-21	10-30	192	11	11	11	11
	24	04-07	11-14	222	11	11	11	11
	20	03-28	11-28	244	12	12	11	11
	16	03-14	12-04	265	12	12	11	10
KENTON 2 W	32	05-07	10-08	155	30	30	30	30
	28	04-23	10-22	182	30	30	30	30
	24	04-05	11-03	212	30	30	30	30
	20	03-24	11-16	237	30	30	30	30
	16	03-11	11-30	264	30	30	30	29
LANCASTER	32	05-07	10-05	152	30	30	30	30
	28	04-24	10-20	179	30	30	30	30
	24	04-09	10-30	205	30	30	30	30
	20	03-21	11-15	239	30	30	30	30
	16	03-11	11-22	256	30	30	30	30
LIMA SWG PLT	32	05-03	10-11	161	29	29	29	29
	28	04-19	10-26	190	29	29	29	29
	24	04-01	11-06	219	29	29	29	29
	20	03-20	11-20	245	29	29	29	29
	16	03-09	12-04	269	29	29	29	27
LONDON 4 W	32	05-02	10-14	165	30	30	30	30
	28	04-19	10-25	189	30	30	30	30
	24	04-02	11-10	222	29	29	30	30
	20	03-22	11-17	239	29	29	30	30
	16	03-10	11-29	263	29	29	30	30
MANSFIELD 6 W	32	05-09	10-06	149	30	30	29	29
	28	04-27	10-18	175	30	30	29	29
	24	04-13	11-03	205	30	30	29	29
	20	03-30	11-14	229	30	30	29	29
	16	03-16	11-27	256	30	30	29	28
MARIETTA WATER WORKS	32	04-27	10-14	171	28	28	29	29
	28	04-11	10-30	202	28	28	29	29
	24	03-27	11-12	230	28	28	29	29
	20	03-10	11-27	261	28	28	29	29
	16	03-06	12-07	276	28	28	29	27
MARION WATER WKS	32	04-28	10-19	174	30	30	30	30
	28	04-13	10-31	201	30	30	30	30
	24	03-30	11-13	228	30	30	30	30
	20	03-20	11-24	250	30	30	30	29
	16	03-09	12-08	274	30	30	30	27
MARYSVILLE	32	05-05	10-10	157	30	30	30	30
	28	04-24	10-22	181	30	30	30	30
	24	04-04	11-03	213	30	30	30	30
	20	03-22	11-15	238	30	30	30	30
	16	03-10	11-27	262	30	30	30	29
MC CONNELSVILLE LK 7	32	05-03	10-13	163	29	29	28	28
	28	04-18	10-26	191	29	29	28	28
	24	04-03	11-08	219	29	29	28	28
	20	03-18	11-18	245	29	29	28	28
	16	03-08	12-01	268	29	29	28	28
MILLERSBURG	32	05-12	10-04	145	28	28	30	30
	28	04-29	10-16	170	29	29	30	30
	24	04-13	10-30	200	29	29	30	30
	20	03-24	11-15	235	29	29	30	30
	16	03-12	11-28	262	29	29	30	29
MILLPORT 2 NW	32	05-18	09-30	135	29	29	29	29
	28	05-02	10-13	164	29	29	30	30
	24	04-22	10-27	187	29	29	30	30
	20	04-06	11-10	218	30	30	30	30
	16	03-18	11-23	250	30	30	30	30

STATION	Freeze threshold temperature	Mean date of last Spring occurrence	Mean date of first Fall occurrence	Mean No. of days between dates	Years of record Spring	No. of occurrences in Spring	Years of record Fall	No. of occurrences in Fall
MONTPELIER	32	05-04	10-11	159	19	19	19	19
	28	04-17	10-23	189	20	20	19	19
	24	04-04	11-10	220	20	20	19	19
	20	03-27	11-24	242	20	20	19	19
	16	03-15	11-28	258	18	18	19	19
MT HEALTHY EXP FARM	32	04-27	10-18	175	30	30	30	30
	28	04-10	11-02	206	30	30	30	30
	24	03-25	11-15	235	30	30	30	30
	20	03-13	11-23	254	30	30	29	29
	16	03-04	12-05	276	30	30	29	26
NAPOLEON	32	05-03	10-12	162	29	29	29	29
	28	04-18	10-25	190	29	29	29	29
	24	04-03	11-07	218	29	29	29	29
	20	03-20	11-19	245	29	29	29	29
	16	03-10	12-02	267	29	29	29	28
NEWARK WATR WKS	32	05-06	10-01	149	16	16	16	16
	28	04-27	10-12	168	16	16	16	16
	24	04-14	10-29	197	16	16	16	16
	20	03-26	11-12	230	16	16	16	16
	16	03-14	11-25	256	16	16	16	16
NEW BREMEN	32	04-30	10-15	169	19	19	20	20
	28	04-12	10-30	200	19	19	20	20
	24	03-29	11-09	225	19	19	20	20
	20	03-20	11-19	244	19	19	20	20
	16	03-06	12-02	271	19	19	20	17
NORWALK	32	05-10	10-10	153	29	29	29	29
	28	04-26	10-21	178	29	29	29	29
	24	04-10	11-03	208	30	30	29	29
	20	03-25	11-18	238	30	30	29	28
	16	03-14	12-02	263	30	30	30	29
OBERLIN	32	05-07	10-12	158	30	30	30	30
	28	04-24	10-28	187	30	30	30	30
	24	04-06	11-09	217	30	30	30	30
	20	03-22	11-23	246	30	30	30	30
	16	03-14	12-04	265	29	29	30	28
OTTAWA	32	05-02	10-14	165	28	28	29	29
	28	04-18	10-28	193	28	28	29	29
	24	04-03	11-09	220	28	28	29	29
	20	03-20	11-21	247	28	28	29	29
	16	03-12	12-01	264	28	28	29	28
PAULDING	32	05-07	10-07	153	28	28	30	30
	28	04-26	10-22	179	29	29	30	30
	24	04-10	11-04	209	28	28	30	30
	20	03-26	11-14	233	28	28	30	30
	16	03-14	11-26	258	28	28	30	29
PEEBLES 1 S	32	05-04	10-07	156	30	30	30	30
	28	04-23	10-18	178	30	30	30	30
	24	04-06	10-30	207	30	30	30	30
	20	03-21	11-12	236	30	30	30	30
	16	03-08	11-23	261	30	30	29	29
PLYMOUTH	32	05-07	10-05	152	18	18	18	18
	28	04-25	10-22	180	18	18	17	17
	24	04-04	11-04	214	18	18	18	18
	20	03-28	11-19	236	18	18	18	18
	16	03-16	12-01	260	18	18	18	18
PORTSMOUTH	32	04-18	10-21	186	30	30	30	30
	28	03-29	11-04	221	30	30	30	30
	24	03-13	11-18	249	30	30	30	30
	20	03-06	12-02	271	30	30	30	29
	16	02-25	12-12	290	30	30	30	25
PUT IN BAY STONE LAB	32	04-16	11-06	205	30	30	30	30
	28	04-03	11-20	231	30	30	30	30
	24	03-26	11-28	247	30	30	30	30
	20	03-13	12-07	269	30	30	30	28
	16	03-08	12-15	282	30	30	30	26
SANDUSKY WB CITY	32	04-17	10-30	197	30	30	30	30
	28	04-05	11-10	219	30	30	30	30
	24	03-24	11-22	244	30	30	30	30
	20	03-13	12-02	265	30	30	30	29
	16	03-04	12-10	281	30	30	30	28

FREEZE DATA

STATION	Freeze threshold temperature	Mean date of last Spring occurrence	Mean date of first Fall occurrence	Mean No. of days between dates	Years of record Spring	No. of occurrences in Spring	Years of record Fall	No. of occurrences in Fall
SIDNEY (2)	32	05-01	10-12	164	27	27	28	28
	28	04-16	10-27	194	27	27	28	28
	24	03-29	11-07	223	27	27	28	28
	20	03-17	11-19	247	27	27	28	28
	16	03-08	11-30	267	26	26	27	26
TIFFIN	32	05-02	10-12	163	30	30	29	29
	28	04-19	10-26	190	30	30	29	29
	24	04-01	11-08	222	30	30	29	29
	20	03-23	11-23	245	30	30	29	29
	16	03-12	12-04	267	30	30	29	28
TOLEDO WB AP	32	04-24	10-25	184	30	30	30	30
	28	04-09	11-07	213	30	30	30	30
	24	03-29	11-19	235	30	30	30	30
	20	03-20	11-29	255	30	30	30	29
	16	03-08	12-07	274	30	30	30	28
UPPER SANDUSKY	32	05-04	10-07	156	30	30	30	30
	28	04-22	10-24	185	30	30	30	30
	24	04-05	11-08	217	30	30	30	30
	20	03-23	11-16	238	30	30	30	30
	16	03-11	12-01	264	30	30	30	29
URBANA JR COLLEGE	32	05-02	10-12	163	30	30	30	30
	28	04-19	10-26	190	30	30	30	30
	24	04-01	11-07	220	30	30	29	29
	20	03-18	11-20	247	30	30	29	29
	16	03-08	12-02	269	30	30	29	27
VAN WERT	32	05-06	10-09	156	26	26	27	27
	28	04-26	10-24	181	26	26	27	27
	24	04-11	11-05	208	26	26	27	27
	20	03-23	11-15	236	26	26	27	27
	16	03-14	11-28	259	26	26	27	27
VICKERY 2 NW	32	05-04	10-14	163	30	30	30	30
	28	04-21	10-28	190	30	30	30	30
	24	04-03	11-10	220	30	30	30	30
	20	03-18	11-26	253	29	29	30	30
	16	03-10	12-05	270	29	29	30	29
WARREN	32	05-12	10-07	148	29	29	30	30
	28	04-23	10-25	184	29	29	30	30
	24	04-09	11-10	215	29	29	30	30
	20	03-26	11-26	245	30	30	30	30
	16	03-13	12-06	268	30	30	30	27
WASHINGTON C HOUSE	32	04-28	10-12	167	25	25	24	24
	28	04-20	10-25	188	25	25	24	24
	24	03-31	11-08	223	25	25	25	25
	20	03-17	11-15	242	25	25	25	25
	16	03-07	11-28	266	25	25	25	24
WAUSEON SEWAGE	32	05-06	10-13	160	30	30	30	30
	28	04-20	10-26	189	30	30	30	30
	24	04-07	11-08	214	30	30	30	30
	20	03-23	11-20	242	30	30	30	30
	16	03-12	11-30	263	30	30	30	29
WAVERLY	32	04-29	10-09	162	27	27	28	28
	28	04-18	10-18	183	27	27	28	28
	24	03-31	11-03	217	27	27	28	28
	20	03-17	11-16	244	27	27	28	28
	16	03-08	11-28	265	27	27	28	27
WILMINGTON	32	05-02	10-09	160	30	30	30	30
	28	04-24	10-21	181	30	30	30	30
	24	04-02	11-06	218	30	30	30	30
	20	03-20	11-15	240	30	30	30	30
	16	03-09	11-27	264	30	30	30	29
WOOSTER EXP STA	32	05-11	10-05	147	29	29	29	29
	28	04-25	10-19	177	29	29	29	29
	24	04-10	11-01	205	29	29	29	29
	20	03-24	11-16	236	29	29	29	29
	16	03-12	12-01	263	29	29	29	28

Data in the above table are based on the period 1921-1950, or that portion of this period for which data are available.

Means have been adjusted to take into account years of non-occurrence.

A freeze is a numerical substitute for the former term "killing frost" and is the occurrence of a minimum temperature at or below the threshold temperature of 32°, 28°, etc.

Freeze data tabulations in greater detail are available and can be reproduced at cost.

*MEAN TEMPERATURE AND PRECIPITATION

STATION	JANUARY		FEBRUARY		MARCH		APRIL		MAY		JUNE		JULY		AUGUST		SEPTEMBER		OCTOBER		NOVEMBER		DECEMBER		ANNUAL	
	Temperature	Precipitation	Temperature	Precipitation	Temperature	Precipitation	Temperature	Precipitation	Temperature	Precipitation	Temperature	Precipitation	Temperature	Precipitation	Temperature	Precipitation	Temperature	Precipitation	Temperature	Precipitation	Temperature	Precipitaion	Temperature	Precipitation	Temperature	Precipitation
NORTHWEST																										
BOWLING GREEN SEWAGE P	28.9	2.07	29.5	1.69	37.7	2.76	48.9	3.11	60.4	3.62	70.8	3.70	74.8	3.11	72.7	2.50	65.9	2.72	54.9	2.48	41.1	2.04	30.5	1.86	51.3	31.66
FINDLAY SEWAGE PLANT	28.5	2.38	29.1	2.03	37.2	3.35	48.2	3.22	59.7	3.71	70.2	4.18	74.2	3.57	72.3	2.86	65.1	2.57	53.7	2.74	40.4	2.31	30.1	2.15	50.7	35.07
LIMA SEWAGE PLANT	29.5	2.57	30.2	1.96	38.5	3.40	49.4	3.32	60.5	3.45	70.8	4.15	74.5	3.29	72.5	3.41	65.6	3.00	54.5	2.88	41.3	2.65	31.1	2.20	51.5	36.28
MONTPELIER	27.5	2.28	28.2	2.03	36.7	3.00	48.6	3.34	59.9	3.62	70.4	3.89	74.4	3.04	72.4	3.10	65.0	2.91	53.9	2.59	40.0	2.39	29.3	2.01	50.5	34.20
NAPOLEON	28.3	2.52	29.0	2.24	37.4	3.10	48.8	3.37	60.4	3.79	70.8	4.20	74.8	3.29	72.6	2.98	65.3	3.15	54.2	2.62	40.6	2.29	29.9	2.28	51.0	35.83
PAULDING	28.2	2.42	29.1	2.04	37.7	3.24	49.0	3.38	59.9	3.64	70.5	4.40	74.5	3.41	72.6	2.46	65.6	2.92	54.1	2.75	40.3	2.37	29.5	2.13	50.9	35.16
TOLEDO WB EXPRESS AP	26.4	2.25	27.3	1.86	35.8	2.86	46.5	3.25	58.2	2.95	68.6	3.55	73.1	2.65	71.0	2.69	64.2	3.02	52.7	2.32	39.8	2.15	28.9	2.29	49.4	31.84
VAN WERT	29.1	2.57	30.1	1.91	38.6	3.49	49.8	3.45	60.9	4.05	71.2	4.19	75.0	3.47	72.8	2.60	66.0	3.07	54.9	2.94	41.0	2.48	30.6	2.12	51.7	36.34
WAUSEON SEWAGE	27.3	2.28	27.8	1.96	36.2	2.86	47.8	3.27	59.3	3.65	70.0	3.81	73.8	2.93	71.7	2.62	64.4	2.98	53.4	2.58	39.6	2.28	29.1	2.07	50.0	33.29
DIVISION	28.4	2.35	29.0	1.95	37.3	3.06	48.6	3.20	60.0	3.60	70.6	3.99	74.5	3.14	72.5	3.72	65.4	2.83	54.2	2.65	40.5	2.27	30.0	2.08	50.9	33.85
NORTH CENTRAL																										
BUCYRUS SEWAGE PLANT	29.1	2.85	29.6	2.25	38.2	3.42	48.9	3.00	60.1	3.21	70.3	4.37	73.7	3.03	71.4	3.29	64.4	2.68	53.4	2.43	40.4	2.37	30.5	2.22	50.8	35.12
CATAWBA ISLAND 1 SW		2.16		1.77		2.76		3.11		3.26		3.47		2.83		2.76		2.40		2.25		2.22		2.07		31.06
FREMONT		2.50		1.99		2.91		3.32		3.54		4.10		3.48		2.88		2.82		2.43		2.26		2.05		34.28
NORWALK	29.2	2.41	29.5	1.87	37.6	2.94	48.1	3.24	59.6	3.67	69.8	4.05	73.3	3.61	71.2	3.43	64.8	3.22	54.1	2.30	41.2	2.32	30.9	2.00	50.8	35.06
OBERLIN	29.7	2.48	29.7	2.03	37.6	2.98	48.4	3.24	59.6	3.41	69.8	3.73	73.5	3.05	71.6	3.23	65.2	3.05	54.4	2.40	41.4	2.26	31.3	2.17	51.0	34.03
SANDUSKY WB CITY	28.8	2.29	29.4	1.92	37.5	2.89	47.9	2.96	59.3	3.32	69.9	3.73	74.6	3.45	72.8	2.81	66.5	3.26	55.0	2.10	42.2	2.27	31.5	2.16	51.3	33.16
TIFFIN	29.5	2.54	30.0	2.07	38.2	3.29	49.2	3.18	60.4	3.67	70.7	4.04	74.4	3.24	72.2	3.34	65.6	2.91	54.6	2.44	41.3	2.42	31.0	2.10	51.4	35.24
UPPER SANDUSKY	29.6	2.74	30.3	2.20	38.8	3.42	49.6	3.07	60.9	3.40	71.0	4.25	74.5	3.22	72.6	3.22	65.8	2.80	54.6	2.38	41.2	2.43	31.1	2.21	51.7	35.34
DIVISION	29.3	2.46	29.6	1.98	37.7	3.02	48.4	3.14	59.8	3.45	70.3	3.96	74.3	3.14	72.3	3.12	65.7	2.80	54.7	2.35	41.4	2.25	31.1	2.06	51.2	33.73
NORTHEAST																										
AKRON CANTON WB AP	27.4	2.74	28.1	2.13	36.5	3.16	47.1	3.20	58.5	3.75	68.4	3.84	72.4	4.20	70.6	3.26	64.6	3.63	53.0	2.39	40.4	2.38	30.1	2.58	49.7	37.26
CHIPPEWA LAKE		2.88		2.36		3.49		3.49		3.71		4.14		3.80		3.03		2.89		2.61		2.60		2.41		37.41
CLEVELAND WB AIRPORT	28.5	2.38	28.6	2.12	36.8	2.89	47.3	2.73	59.1	2.73	69.4	3.05	73.7	3.04	71.9	2.64	65.5	3.13	54.4	2.42	41.7	2.66	30.9	2.29	50.6	32.0[illegible]
HIRAM	28.1	2.88	27.7	2.19	35.8	3.40	47.1	3.52	58.2	3.81	67.8	3.83	71.7	3.31	70.0	3.24	63.8	2.89	53.2	2.86	40.1	2.78	29.9	2.49	49.5	37.2[illegible]
MINERAL RIDGE WATER WK		2.58		1.87		2.93		3.21		3.44		3.27		3.55		3.15		2.54		2.61		2.27		2.05		33.4[illegible]
WARREN	29.9	2.87	29.6	2.19	37.6	3.28	48.5	3.50	59.6	3.64	69.2	3.55	73.1	3.38	71.2	3.27	64.6	2.83	53.8	2.72	41.6	2.69	31.6	2.34	50.9	36.2[illegible]
YOUNGSTOWN WB AIRPORT	27.5	3.32	28.2	2.72	36.7	3.45	47.2	3.64	58.3	4.09	68.0	3.71	72.3	4.16	70.4	3.46	64.7	3.58	53.9	2.77	40.6	3.49	30.0	2.94	49.8	41.3[illegible]
DIVISION	29.1	2.75	28.9	2.22	36.8	3.27	47.7	3.36	58.9	3.61	68.8	3.57	72.8	3.45	71.1	3.19	64.6	2.95	53.9	2.80	41.1	2.68	31.0	2.40	50.4	36.2[illegible]
WEST CENTRAL																										
BELLEFONTAINE SEWAGE		2.98		1.94		3.29		3.24		3.31		4.38		3.37		3.48		2.62		2.60		2.31		2.30		35.8[illegible]
GREENVILLE SEWAGE PL	29.7	3.02	30.3	2.28	38.8	3.69	49.5	3.43	60.4	4.00	70.2	4.23	73.9	3.46	71.9	2.86	64.9	3.32	53.9	2.58	40.7	2.72	30.8	2.38	51.3	37.9[illegible]
KENTON 2 W	28.7	2.49	29.5	1.80	37.6	3.13	48.4	3.00	59.5	3.22	69.6	4.05	73.5	3.23	71.4	2.85	64.9	2.98	53.7	2.42	40.6	2.08	30.2	2.01	50.6	33.2[illegible]
LAKEVIEW 3 NE		2.59		2.00		3.10		3.13		3.31		4.64		3.34		3.30		2.95		2.44		2.23		2.11		35.1[illegible]
NEW CARLISLE		3.25		2.56		3.52		3.32		3.80		4.27		3.26		3.54		2.76		2.48		2.83		2.48		38.07
PIQUA		3.06		2.36		3.61		3.45		3.67		4.30		3.52		2.82		3.05		2.53		2.54		2.41		37.37
PLEASANT HILL 1 NW		3.00		2.18		3.42		3.36		3.52		3.77		3.28		3.11		2.78		2.39		2.48		2.34		35.63
TIPP CITY		3.19		2.34		3.34		3.30		3.77		4.13		3.33		3.13		3.05		2.41		2.62		2.37		36.98
URBANA GRIMES FIELD	30.3	3.25	31.3	2.36	39.6	3.53	50.1	3.46	60.8	3.64	70.6	4.14	74.3	3.52	72.0	3.58	65.2	2.97	54.2	2.53	41.3	2.55	31.4	2.38	51.8	37.91
VERSAILLES		2.94		2.37		3.57		3.68		3.85		4.49		3.15		3.22		2.99		2.48		2.68		2.40		37.82
DIVISION	29.9	2.96	30.8	2.21	39.1	3.40	49.8	3.32	60.8	3.59	70.8	4.19	74.5	3.42	72.5	3.14	65.7	3.05	54.5	2.54	41.2	2.46	31.1	2.28	51.7	36.56
CENTRAL																										
CIRCLEVILLE		3.46		2.55		4.02		3.83		3.61		4.11		4.52		3.19		2.78		1.93		2.77		2.51		39.28
COLUMBUS VALLEY CROSS	31.3	3.16	32.1	2.24	40.4	3.36	50.6	3.63	61.3	3.54	70.6	3.92	74.0	3.89	72.1	3.10	65.7	2.55	54.6	1.99	41.6	2.50	32.1	2.39	52.2	36.27
COLUMBUS WB AIRPORT	29.7	2.94	31.2	2.27	39.8	3.43	50.2	3.44	60.8	3.97	70.7	4.33	74.4	3.85	72.4	3.21	66.5	2.91	54.5	2.18	41.9	2.86	31.7	2.49	52.0	37.98
COLUMBUS WB CITY	31.1	2.81	32.6	2.15	41.1	3.22	51.4	3.19	62.2	3.23	72.0	3.66	75.8	3.53	73.8	3.01	67.8	2.58	56.1	2.00	43.2	2.54	33.3	2.44	53.4	34.36
DELAWARE	30.7	3.07	31.3	2.31	39.9	3.48	50.2	3.41	61.0	3.44	71.0	4.36	74.4	3.36	72.5	3.09	65.7	2.81	54.6	2.16	41.4	2.43	31.5	2.46	52.0	36.38
LANCASTER 2 NW	32.6	3.35	33.3	2.68	41.3	4.07	51.4	3.72	61.7	3.81	71.3	4.19	74.4	4.22	72.7	3.73	66.1	3.13	55.2	2.17	42.7	2.66	33.4	2.81	53.0	40.54
LA RUE		2.78		2.08		3.45		3.12		3.35		4.19		3.05		3.16		2.29		2.39		2.12		2.19		34.17
LONDON WATER WORKS	30.1	3.36	31.0	2.53	39.5	3.68	49.9	3.86	60.7	4.07	70.8	4.00	74.3	4.08	72.6	3.05	66.1	2.81	55.0	2.31	41.0	2.76	31.1	2.69	51.8	39.20
MARION WATER WORKS	30.5	2.89	31.2	2.18	39.7	3.40	50.3	3.21	61.3	3.31	71.1	4.42	74.7	3.10	72.7	3.46	66.2	2.65	55.4	2.54	42.1	2.41	31.8	2.25	52.3	35.82
MARYSVILLE	29.7	3.12	30.5	2.27	39.0	3.47	49.6	3.43	60.6	3.51	70.3	4.27	73.9	3.28	72.0	3.59	65.2	2.82	54.3	2.37	40.7	2.37	30.7	2.53	51.4	37.03
PROSPECT 3 N		2.82		2.14		3.22		3.22		3.29		4.18		2.89		2.93		2.57		2.29		2.21		2.11		33.87
WASHINGTON COURT HOUSE	31.5	3.38	32.3	2.44	40.5	3.99	50.9	3.84	61.8	3.81	71.5	3.97	74.9	3.72	73.1	3.00	66.6	2.71	55.3	2.00	41.8	2.70	32.1	2.48	52.7	38.04
DIVISION	31.4	3.20	32.1	2.36	40.4	3.59	50.8	3.60	61.7	3.65	71.4	4.16	74.8	3.77	73.0	3.25	66.3	2.74	55.3	2.18	42.0	2.55	32.2	2.48	52.6	37.53
CENTRAL HILLS																										
ASHLAND 2 ENE		2.60		1.96		3.30		3.06		3.47		3.53		3.29		3.54		2.75		2.15		2.23		2.05		33.93
ASHLAND 3 NW	29.3	2.79	29.6	2.08	37.9	3.50	48.7	3.21	60.0	3.50	69.8	3.74	73.5	3.57	71.8	3.67	65.1	2.76	54.4	2.27	40.7	2.40	30.6	2.23	51.0	35.72
COSHOCTON SEWAGE PL	31.6	3.49	32.0	2.44	40.4	3.79	51.2	3.65	62.2	3.97	71.9	4.37	75.0	4.13	73.1	4.00	66.5	2.86	55.5	2.33	42.4	2.88	32.6	2.54	52.9	40.45
WOOSTER EXP STATION	28.9	2.81	29.1	2.04	37.3	3.29	47.8	3.11	58.7	3.75	68.7	3.92	72.4	3.76	70.7	3.51	63.8	2.70	53.0	2.32	40.1	2.19	30.3	2.21	50.1	35.61
WOOSTER 2 SE	29.4	2.76	29.6	1.99	37.6	3.25	48.4	3.23	59.6	3.72	69.6	3.99	73.3	3.76	71.5	3.51	64.[illegible]	2.67	53.9	2.33	40.7	2.23	30.7	2.25	50.8	35.69
DIVISION	29.7	3.01	29.9	2.15	38.0	3.42	48.6	3.26	59.7	3.61	69.5	4.17	73.1	3.77	71.3	3.54	64.7	2.83	53.8	2.30	40.6	2.45	30.9	2.36	50.8	36.87
NORTHEAST HILLS																										
CANFIELD 1 S		2.50		1.81		3.13		3.05		3.65		3.62		3.41		3.00		2.50		2.61		2.30		2.12		33.70
DENNISON	31.9	3.22	32.2	2.36	40.2	3.72	50.7	3.61	61.0	3.92	70.3	4.27	73.3	4.04	71.5	3.75	65.0	2.89	54.2	2.55	42.0	2.76	32.6	2.60	52.1	39.69
ELLSWORTH		2.53		1.84		3.28		3.58		3.92		3.80		3.41		3.05		2.88		2.82		2.49		2.14		35.74
MILLPORT 2 NW	29.0	3.10	29.1	2.15	37.2	3.57	47.7	3.29	58.2	3.58	67.6	3.71	71.1	3.98	69.7	3.33	62.9	3.13	52.0	2.58	39.9	2.40	30.4	2.30	49.6	37.12
NEWCOMERSTOWN		3.34		2.29		3.64		3.50		3.82		4.26		3.91		4.09		2.69		2.36		2.64		2.30		38.84
DIVISION	30.3	3.04	30.4	2.20	38.5	3.58	49.1	3.40	60.0	3.68	69.3	4.01	72.8	4.01	71.1	3.43	64.5	2.93	53.6	2.56	41.1	2.52	31.4	2.43	51.0	37.79
SOUTHWEST																										
CHILO DAM 34		3.98		2.97		4.58		3.93		3.43		3.96		3.61		3.41		2.56		2.07		2.95		2.86		40.31
CINCINNATI ABBE OBS	33.1	3.44	34.8	2.55	43.3	4.07	53.6	3.64	63.6	3.54	72.8	4.05	76.6	3.70	74.8	3.38	68.9	2.88	57.4	2.19	44.6	3.06	35.0	2.84	54.9	39.34
CINCINNATI WB CITY	34.6	3.44	36.0	2.55	44.5	4.07	54.9	3.64	64.9	3.54	74.4	4.05	78.1	3.70	76.2	3.38	70.1	2.88	59.1	2.19	46.1	3.06	36.6	2.84	56.3	39.34
DAYTON WB AIRPORT	29.7	2.96	31.4	2.11	39.8	3.24	50.5	3.07	61.0	3.54	70.9	3.90	75.0	3.29	72.9	2.86	66.9	2.80	55.1	2.30	41.9	2.61	31.7	2.47	52.2	35.15
EATON		3.61		2.64		3.81		3.51		3.61		4.29		3.37		3.15		3.18		2.53		2.87		2.52		39.09
FERNBANK DAM 37		3.73		2.78		4.27		3.60		3.45		3.97		3.53		2.92		2.86		2.26		2.93		2.59		38.92
FRANKLIN		3.65		2.66		3.78		3.58		3.88		4.18		3.44		3.17		3.26		2.38		2.92		2.68		39.58
HAMILTON		3.70		2.68		4.04		3.64		3.65		4.06		3.77		2.85		3.54		2.33		2.90		2.69		39.85
HAMILTON WTR WKS SOUTH		3.70		2.68		4.05		3.62		3.60		3.99		3.68		2.80		3.63		2.34		2.81		2.69		39.59
HILLSBORO	33.0	4.11	33.7	3.16	42.0	4.77	52.1	4.08	62.1	3.92	70.9	4.38	74.3	4.04	72.7	4.03	66.6	3.23	56.0	2.35	43.1	2.96	33.7	3.10	53.4	44.13
KINGS MILLS		3.73		2.79		4.09		3.95		3.98		4.29		3.84		3.34		3.33		2.18		3.12		2.77		41.41
MIAMISBURG		3.78		2.68		3.95		3.77		4.01		4.47		3.25		3.10		3.44		2.64		3.12		2.79		41.00
MIDDLETOWN		3.78		2.85		4.00		3.59		3.98		4.14		3.76		3.02		3.25		2.39		2.92		2.78		40.46
WEST MANCHESTER 3 SW		3.46		2.55		3.63		3.42		3.53		4.26		3.21		3.24		2.98		2.61		2.84		2.66		38.39
WILMINGTON		4.26		3.33		4.71		4.29		3.88		4.17		4.25		3.47		3.30		2.29		3.16		3.25		44.36
XENIA 4 SSW	31.7	3.15	32.7	2.42	41.1	3.44	51.2	3.55	61.0	3.71	71.5	4.43	74.6	3.66	72.5	3.77	66.4	2.90	55.3	2.24	42.5	2.83	32.4	2.40	52.7	39.50
DIVISION	33.0	3.66	34.0	2.76	42.2	4.10	52.6	3.69	62.9	3.71	72.5	4.13	76.0	3.63	74.2	3.21	67.7	3.03	56.6	2.35	43.4	2.92	33.9	2.73	54.1	39.92

*MEAN TEMPERATURE AND PRECIPITATION

STATION	JAN Temp	JAN Precip	FEB Temp	FEB Precip	MAR Temp	MAR Precip	APR Temp	APR Precip	MAY Temp	MAY Precip	JUN Temp	JUN Precip	JUL Temp	JUL Precip	AUG Temp	AUG Precip	SEP Temp	SEP Precip	OCT Temp	OCT Precip	NOV Temp	NOV Precip	DEC Temp	DEC Precip	ANN Temp	ANN Precip
SOUTH CENTRAL																										
CHILLICOTHE	33.9	3.54	34.5	2.71	42.7	3.94	53.4	3.71	63.7	3.73	72.9	4.01	76.2	4.08	74.3	3.57	67.8	2.94	56.6	1.82	43.8	2.56	34.6	2.67	54.5	39.28
GALLIPOLIS 5 V		3.85		2.98		4.27		3.59		3.90		3.98		4.15		3.66		2.45		2.12		2.62		2.90		40.97
IRONTON		4.08		2.85		4.57		3.76		3.97		4.57		4.51		3.74		2.86		2.09		2.76		2.92		42.68
JACKSON 2 NW	34.5	3.93	35.1	2.88	43.3	4.39	53.6	3.77	63.1	3.91	71.5	4.07	74.7	3.44	73.1	3.54	66.7	2.42	55.4	2.23	43.5	2.73	34.6	3.05	54.1	41.36
NORTH KENOVA DAM 28		4.02		2.91		4.28		3.48		3.82		4.23		4.47		3.29		2.66		2.03		2.67		2.82		40.68
PORTSMOUTH	36.8	4.22	37.7	3.08	45.6	4.63	55.9	3.67	65.9	4.10	74.6	3.94	77.8	4.73	76.2	3.62	69.9	2.54	58.6	2.11	46.1	2.54	37.3	2.88	56.9	42.16
FORTSMOUTH US GRANT BR		4.14		3.00		4.57		3.65		4.07		3.83		4.88		3.68		2.61		2.16		2.92		2.92		42.33
RACINE DAM 23		3.89		2.98		4.31		3.58		3.76		4.27		4.52		3.88		2.87		2.09		2.71		2.88		41.74
WAVERLY		3.78		2.79		4.13		3.54		3.87		3.89		4.06		3.54		2.52		1.99		2.54		2.79		39.54
DIVISION	35.3	3.91	36.1	2.87	44.2	4.36	54.4	3.69	64.1	3.93	73.0	4.10	76.1	4.23	74.6	3.67	68.2	2.95	57.0	2.09	44.5	2.67	35.6	2.90	55.3	41.27
SOUTHEAST																										
ATHENS		3.42		2.65		3.93		3.49		4.01		4.23		4.44		4.04		2.87		1.92		2.63		2.71		40.34
CAMBRIDGE STATE HOSP		3.16		2.36		3.75		3.46		3.62		4.10		4.15		3.84		2.97		2.36		2.75		2.51		39.03
CLARINGTON LOCK 14		3.41		2.66		3.69		3.59		4.23		4.76		4.36		4.22		2.67		2.49		2.61		2.85		41.54
MARIETTA LOCK 1		3.42		2.74		3.94		3.38		3.79		4.19		4.56		4.42		2.74		2.08		2.57		2.76		40.59
MC CONNELSVILLE LOCK 7	34.1	3.35	34.3	2.62	42.1	3.96	52.2	3.67	62.5	4.16	71.4	4.62	74.7	4.12	73.4	3.43	67.1	2.98	56.4	2.27	43.3	2.44	34.1	2.59	53.8	40.14
PHILO	33.4	3.16	33.9	2.48	42.0	3.89	52.3	3.64	62.7	4.07	71.9	4.41	75.0	4.38	73.3	3.78	66.9	2.91	56.1	2.27	43.2	2.62	33.8	2.71	53.7	40.32
PHILO 3 SW	33.1	2.72	33.6	2.07	41.8	3.32	52.5	3.01		3.34	62.8	3.63	71.8	3.71	75.1	3.71	67.3	2.74	56.8	2.10	43.7	2.31	34.0	2.30	53.8	35.67
ZANESVILLE LOCK 10		2.96		2.21		3.46		3.34		3.66		4.19		4.06		3.50		2.92		2.12		2.55		2.30		37.28
DIVISION	33.2	3.22	33.5	2.46	41.6	3.75	52.0	3.47	62.3	3.91	71.3	4.28	74.5	4.17	72.9	3.77	66.5	2.85	55.7	2.20	42.9	2.55	33.6	2.62	53.3	39.26

* Averages for period 1931-1955, except for stations marked WB which are "normals" based on period 1921-1950. Divisional means may not be the arithmetical average of individual stations published, since additional data from shorter period stations are used to obtain better areal representation.

CONFIDENCE LIMITS

In the absence of trend or record changes, the chances are 9 out of 10 that the true mean will lie in the interval formed by adding and subtracting the values in the following table from the means for any station in the State. Because of the wider variation in mean precipitation, the corresponding monthly means and annual mean must be substituted for "p" in the precipitation table below to obtain mean precipitation confidence limits.

1.8	$.38\sqrt{p}$	1.6	$.25\sqrt{p}$	1.8	$.32\sqrt{p}$	1.1	$.26\sqrt{p}$	1.1	$.34\sqrt{p}$	1.0	$.34\sqrt{p}$	.8	$.35\sqrt{p}$	.9	$.31\sqrt{p}$	1.0	$.30\sqrt{p}$	1.1	$.32\sqrt{p}$	1.1	$.30\sqrt{p}$	1.4	$.24\sqrt{p}$	.4	$.32\sqrt{p}$

COMPARATIVE DATA

Data in the following table are the mean temperature and average precipitation for Wooster Experiment Station, Ohio for the period 1906-1930 and are included in this publication for comparative purposes:

27.5	3.33	28.4	2.32	37.9	3.61	48.8	3.29	58.4	3.65	67.2	3.86	71.6	3.86	70.1	3.71	64.3	3.24	52.3	2.75	40.7	2.57	30.1	2.81	49.8	39.00

NORMALS, MEANS, AND EXTREMES

AKRON, OHIO — AKRON CANTON AIRPORT

Month	Normal Daily max	Normal Daily min	Normal Monthly	Extremes Record highest	Year	Extremes Record lowest	Year	Normal degree days	Precip Normal total	Precip Max monthly	Year	Precip Min monthly	Year	Precip Max 24 hrs	Year	Snow Mean total	Snow Max monthly	Year	Snow Max 24 hrs	Year	Rel hum 1 a.m.	7 a.m.	1 p.m.	7 p.m.	Wind mean hourly	Prevailing dir	Mean sky cover	Days clear	Partly cloudy	Cloudy
J	34.8	20.0	27.4	73	1906	-14	1897	1166	2.74	8.70	1950	0.96	1950	1.89	1952	8.6	22.0	1904	11.0	1914	82	83	74	79	11.8	SW	7.7	3	6	22
F	35.9	20.3	28.1	71	1932	-20	1899	1033	2.13	5.79	1893	0.64	1893	2.35	1956	7.9	25.3	1906	8.0	1910	83	83	70	76	11.8	SW	7.3	4	6	18
M	45.1	27.9	36.5	82	1910	-6	1913	834	3.16	10.46	1913	0.49	1913	4.75	1913	7.3	17.1	1906	9.0	1897	82	82	64	72	12.1	SW	6.9	5	8	18
A	57.1	37.0	47.1	89	1948	13	1923	534	3.16	7.58	1929	0.83	1926	5.13	1929	2.3	11.3	1901	7.4	1901	80	81	60	68	11.0	SW	6.7	6	8	16
M	69.3	47.5	58.5	94	1895	32	1903	235	3.75	9.60	1924	0.47	1956	2.55	1916	0.1	3.3	1908	3.3	1901	78	81	59	65	9.3	SW	6.2	7	11	13
J	79.0	57.7	68.4	100	1952	36	1945	50	3.84	11.12	1923	1.54	1924	3.75	1908	0.0	0.0		0.0		85	81	56	66	8.3	SW	5.9	6	13	11
J	83.1	61.6	72.4	102	1936	41	1904	0	4.20	10.85	1894	1.14	1943	5.96	1942	0.0	0.0		0.0		85	82	53	64	7.4	SW	5.2	9	14	8
A	81.3	59.8	70.6	104	1918	39	1915	17	3.26	7.98	1894	0.28	1947	3.62	1947	0.0	0.0		0.0		86	88	53	67	7.0	S	5.4	9	13	9
S	71.3	53.9	64.6	99	1953	29	1942	83	3.63	10.28	1928	0.22	1926	4.65	1952	T	T	1942	T	1942	88	88	54	71	8.7	S	5.4	10	13	9
O	62.8	43.2	53.0	89	1927	20	1952	378	2.39	8.42	1924	0.22	1954	2.77	1913	0.3	6.8	1950	3.9	1952	81	84	56	72	8.8	SW	5.6	10	7	12
N	47.9	32.9	40.4	82	1950	2	1929	738	2.38	5.13	1939	0.41	1921	2.00	1913	3.5	22.3	1950	15.6	1951	81	84	65	74	11.3	SW	7.1	7	9	18
D	36.8	23.3	30.1	68	1941	-11	1951	1082	2.58	6.14	1955	0.31	1927	2.07	1951	7.7	19.6	1951	8.5	1951	82	84	72	79	10.9	SW	7.7	7	7	21
Year	59.0	40.4	49.7	104 Aug. 1918		-20 Feb. 1899		6203	37.26	11.98 Sept. 1926		0.22 Oct. 1924		5.96 July 1943		36.1	26.6 Apr. 1901		15.6 1901		83	83	61	71	9.8	SW	6.4	77	113	175

CINCINNATI, OHIO — ABBE OBSERVATORY

Month	Normal Daily max	Normal Daily min	Normal Monthly	Extremes Record highest	Year	Extremes Record lowest	Year	Normal degree days	Precip Normal total	Precip Max monthly	Year	Precip Min monthly	Year	Precip Max 24 hrs	Year	Snow Mean total	Snow Max monthly	Year	Snow Max 24 hrs	Year	Rel hum 1 a.m.	7 a.m.	1 p.m.	7 p.m.	Wind mean hourly	Prevailing dir	Mean sky cover	Pct sunshine	Days clear	Partly cloudy	Cloudy
J	40.9	25.2	33.1	77	1950	-17	1936	989	3.44	13.68	1937	0.97	1931	2.71	1937	5.1	20.2	1918	9.6	1951	80	82	69	74	8.5	SW	5.2	41	6	7	18
F	43.0	26.6	34.8	76	1932	-9	1951	846	2.55	6.24	1955	0.48	1947	2.41	1940	3.8	11.6	1918	7.2	1945	77	80	56	68	8.6	SW	4.9	45	7	8	14
M	52.7	33.8	43.3	88	1929	6	1925	682	4.07	10.94	1913	1.06	1945	3.76	1937	2.7	13.0	1937	7.0	1937	75	80	56	60	9.3	SW	6.3	53	6	9	15
A	63.9	43.3	53.6	90	1925	18	1923	347	3.64	8.62	1947	0.84	1915	3.43	1920	T	5.2	1920	4.7	1916	74	78	52	58	8.6	SW	6.0	55	8	11	11
M	74.2	52.9	63.6	95	1927	32	1957+	132	3.54	8.81	1928	0.76	1928	4.77	1945	T	5.2	1952+	T	1952+	80	83	52	60	7.0	SW	5.8	63	9	12	10
J	83.2	62.4	72.8	102	1944	40	1929	13	4.05	9.07	1944	0.59	1928	3.46	1934	0.0	0.0		0.0		83	83	55	63	6.6	W	5.8	69	8	12	10
J	87.4	65.9	76.6	109	1934	50	1924	0	3.70	10.02	1940	0.33	1926	4.07	1940	0.0	0.0		0.0		84	84	52	60	5.5	SW	5.2	76	11	13	7
A	85.4	64.2	74.8	103	1936	43	1915	0	3.38	6.54	1951	0.54	1954	3.37	1932	0.0	0.0		0.0		82	82	49	59	5.4	SW	4.9	74	11	12	8
S	79.7	58.0	68.9	101	1954+	32	1942	56	2.88	5.86	1955	0.71	1955	2.70	1917	0.0	4.7	1925	2.5	1925	86	86	51	63	5.7	SW	4.5	70	13	10	8
O	68.2	46.5	57.4	92	1951	20	1925	263	2.19	6.46	1919	0.21	1924	2.15	1950+	0.1	0.1	1950+	0.1		82	82	51	64	6.2	SW	4.8	68	12	13	10
N	53.0	36.2	44.6	83	1950	-13	1929	612	3.06	6.94	1948	0.31	1917	2.99	1936	1.4	8.9	1936	8.9	1936	76	81	59	66	8.2	SW	6.4	46	8	13	11
D	42.4	27.5	35.0	71	1956+	-17	1917	930	2.84	6.94	1923	0.56	1955	3.34	1917	4.2	16.3	1917	11.0	1917	79	81	66	72	8.4	SW	6.1	39	7	7	17
Year	64.5	45.2	54.9	109 July 1934		-17 Jan. 1936		4870	39.34	13.68 Jan. 1937		0.17 Oct. 1924		4.77 May 1933		17.3	20.2 Jan. 1918		11.0 Dec. 1917		79	83	58	64	7.4	SW	5.6	57	109	113	143

NORMALS, MEANS, AND EXTREMES

Cleveland, Ohio — Cleveland Hopkins Airport

Month	Norm Daily Max	Norm Daily Min	Norm Monthly	Rec High	Yr	Rec Low	Yr	Norm Deg Days	Precip Norm	Precip Max Mo	Yr	Precip Min Mo	Yr	Precip Max 24h	Yr	Snow Mean	Snow Max Mo	Yr	Snow Max 24h	Yr	RH 1a	RH 7a	RH 1p	RH 7p	Wind Mean Hr	Prev Dir	Fast Spd	Dir	Yr	Sky Cover	% Sun	Clear	Pt Cldy	Cloudy	Days Precip	Days Snow	Tstorm	Fog	Max 90+	Max 32−	Min 32−	Min 0−
(b)	17			17		17		(b)	(b)	18				16		16	16		16		16	16	16	16	16	8	16	16	16	16	27	16	16	16	16	16	16	16	16	16	16	16
J	36.0	20.9	28.5	73	1950	−9	1942	1132	2.38	7.01	1950	0.79	1946	1.74	1952	10.6	18.7	1945	9.3	1945	80	82	73	79	12.5	S	59	SW	1950	8.1	27	3	5	23	16	4	*	2	0	16	27	2
F	36.4	20.8	28.6	69	1957	−8	1951+	1019	2.12	4.64	1950	0.75	1947	1.77	1944	9.7	20.7	1950	7.5	1944	82	81	69	72	12.5	S	65	W	1957	7.7	34	3	6	19	15	4	*	2	0	12	24	1
M	45.2	28.6	36.8	83	1945	−5	1948	814	2.89	6.07	1954	1.50	1957	2.76	1948	10.9	26.3	1954	14.9	1954	77	80	62	61	13.0	WNW	74	W	1948	7.1	45	6	7	18	16	4	2	1	0	9	21	1
A	57.1	37.5	47.3	88	1942	19	1954	531	2.73	5.90	1950	1.18	1946	1.92	1950	3.1	T	1957+	7.6	1957	77	78	56	57	13.0	S	65	W	1948	6.9	50	5	8	17	14	1	2	1	*	5	8	*
M	69.9	48.2	59.1	92	1944	29	1944	223	2.73	6.04	1947	1.04	1954	3.73	1950	0.0	0.0		0.0		79	76	55	66	12.0	S	68	SW	1957	6.9	64	5	11	15	14	0	4	1	*	0	0	0
J	80.4	58.3	69.4	101	1947	38	1945	46	3.05	6.07	1947	1.38	1946	2.79	1946	0.0	0.0		0.0		79	76	54	61	10.5	S	56	S	1953+	6.0	69	6	13	11	11	0	6	1	5	0	0	0
J	84.7	62.6	73.7	103	1941	46	1946+	0	3.04	5.37	1943	1.23	1952	2.73	1950	0.0	0.0		0.0		81	79	52	58	9.5	S	65	S	1956	5.3	69	10	12	9	10	0	7	1	7	0	0	0
A	82.7	61.1	71.9	101	1948	44	1952	10	2.64	5.19	1947	1.61	1945	3.07	1947	0.0	0.0		0.0		83	82	53	63	8.9	S	61	W	1956	5.1	65	10	13	9	9	0	6	1	6	0	0	0
S	75.8	55.1	65.5	100	1953	32	1942	75	3.13	6.37	1945	1.57	1953	1.85	1943	0.0	0.0		0.0		83	82	53	63	9.5	S	45	W	1953+	5.3	63	10	12	10	9	0	3	1	2	0	*	0
O	64.3	44.5	54.4	90	1946	25	1946	340	2.42	8.50	1954	0.61	1952	3.44	1954	0.6	6.4	1954	6.1	1954	81	82	57	70	10.2	S	43	W	1946	5.7	55	7	8	12	10	*	1	1	*	0	2	0
N	49.2	34.2	41.7	82	1950	−7	1950	699	2.66	6.44	1950	1.14	1956	1.94	1950	6.5	22.3	1950	15.0	1950	78	80	64	74	12.9	S	59	S	1948	8.0	33	3	6	20	15	2	1	1	0	2	13	*
D	37.5	24.3	30.9	69	1941	−9	1951+	1057	2.29	5.60	1951	0.96	1955	1.26	1955	11.1	28.9	1944	8.7	1944	78	80	70	77	12.9	S	49	W	1949	8.0	28	3	6	22	16	5	1	1	0	10	25	1
Year	59.9	41.3	50.6	103	July 1941	−9	Dec. 1951+	6006	32.08	9.50	Oct. 1954	0.61	Oct. 1952	3.73	May 1955	52.5	28.9	Dec. 1944	15.0	Nov. 1950	79	80	60	69	11.1	S	74	W	Mar. 1948	6.6	49	75	106	184	156	38	14	14	20	38	119	4

Means and extremes in the above table are from the existing or comparable location(s). Annual extremes have been exceeded at prior locations as follows: Lowest temperature −17 in January 1873; maximum monthly precipitation 9.77 in January 1902; minimum monthly precipitation 0.17 in August 1881; maximum precipitation in 24 hours 4.97 in September 1901; maximum monthly snowfall 30.5 in February 1908; maximum snowfall in 24 hours 17.4 in November 1913; fastest mile of wind 78 from Southwest in May 1940.

Columbus, Ohio — Port Columbus Airport

Month	Norm Daily Max	Norm Daily Min	Norm Monthly	Rec High	Yr	Rec Low	Yr	Norm Deg Days	Precip Norm	Precip Max Mo	Yr	Precip Min Mo	Yr	Precip Max 24h	Yr	Snow Mean	Snow Max Mo	Yr	Snow Max 24h	Yr	RH 1a	RH 7a	RH 1p	RH 7p	Wind Mean Hr	Prev Dir	Fast Spd	Dir	Yr	Sky Cover	% Sun	Clear	Pt Cldy	Cloudy	Days Precip	Days Snow	Tstorm	Fog	Max 90+	Max 32−	Min 32−	Min 0−
(b)	79			79		79		(b)	(b)	79				79		73	73		73		18	18	18	18	52	8	55			68	63	79	79	79	79	73	79	72	79	79	79	79
J	37.8	21.6	29.7	74	1950	−20	1884	1094	2.94	10.71	1937	0.50	1940	2.92	1952	6.7	25.4	1918	11.9	1910	81	82	71	76	9.8	NW	63	NW	1938	7.1	63	6	8	18	14	4	1	1	0	10	25	1
F	39.6	22.8	31.2	73	1957	−20	1899	948	2.27	7.65	1893	0.43	1907	2.27	1893	5.2	29.2	1910	9.0	1914	82	81	66	73	9.9	NW	58	SW	1946	6.8	44	6	8	14	12	3	1	1	0	8	22	1
M	49.5	30.0	39.8	85	1945	15	1943	781	3.43	8.09	1913	0.28	1910	3.33	1895	3.1	25.2	1906	9.6	1906	78	81	58	66	10.3	NW	68	W	1942	6.7	49	6	9	14	14	2	2	1	*	3	17	*
A	61.4	39.9	50.2	89	1915	18	1881	444	3.44	7.08	1893	0.33	1889	3.28	1906	0.8	16.9	1886	6.1	1886	77	78	55	63	9.9	SSW	76	W	1920	6.5	54	7	9	13	13	*	4	*	*	0	3	0
M	72.5	49.1	60.8	96	1895	28	1881	180	3.97	9.59	1882	0.33	1939	2.50	1892	T	0.0	1923	0.3	1923	79	79	55	62	8.3	S	62	NW	1920	5.7	68	7	12	9	13	*	6	*	*	0	0	0
J	82.2	58.2	70.2	102	1944	39	1913	31	4.33	8.52	1902	0.74	1950	3.54	1902	0.0	0.0		0.0		82	81	55	62	7.3	SSW				5.3		8	12	9	12	0	8	*	5	0	0	0
J	86.2	62.6	74.4	106	1936	44	1940	0	3.85	9.77	1896	0.49	1940	3.87	1947	0.0	0.0		0.0		83	84	52	60	6.6	NNW	84	NW	1916	4.9	71	12	13	7	11	0	8	*	8	0	0	0
A	84.0	60.8	72.4	103	1918	42	1887	8	3.21	7.16	1924	0.33	1924	3.71	1915	0.0	0.0		0.0		84	86	52	65	6.4	SE	78	W	1918	4.7	66	12	10	9	9	0	6	1	5	0	0	0
S	78.3	54.5	66.5	100	1939	31	1942	69	2.91	7.13	1890	0.42	1908	3.91	1938	0.0	0.0		0.0		86	86	53	65	7.0	NNW	60	SW	1939	5.5	66	10	12	8	9	0	4	1	2	0	*	0
O	66.0	43.0	54.5	91	1939	17	1952	337	2.18	8.64	1881	0.10	1924	3.18	1910	0.1	3.0	1925	1.5	1925	83	82	57	69	7.9	SSW	61	N	1952	6.5	44	10	10	11	10	*	1	1	*	0	2	0
N	50.8	33.0	41.9	80	1950	−5	1880	693	2.86	7.54	1897	0.18	1917	2.81	1881	1.7	14.5	1950	8.1	1950	81	80	64	76	9.4	S	58	W	1920	7.2	35	7	15	13	11	1	1	1	0	2	13	*
D	39.4	23.9	31.7	70	1956	−14	1951	1032	2.49	6.12	1923	0.46	1955	2.05	1921	4.3	14.5	1890	8.1	1957	83	82	69	76	9.3	SSW	58	W		7.2		5	18	18	13	3	1	1	0	8	23	1
Year	62.3	41.6	52.0	106	July 1936	−20	Feb. 1899+	5615	37.88	10.71	Jan. 1937	0.10	Oct. 1924	3.91	Sept. 1938	21.9	29.2	Feb. 1910	11.9	Jan. 1910	82	82	58	67	8.5	S	84	NW	July 1916	5.9	55	101	118	146	140	6	43	6	21	31	107	3

NORMALS, MEANS, AND EXTREMES

DAYTON, OHIO — MUNICIPAL AIRPORT

LATITUDE 39° 54' N
LONGITUDE 84° 12' W
ELEVATION (ground) 1002 Feet

Temperature and Normal degree days

Month	Normal Daily max	Normal Daily min	Normal Monthly	Record highest	Year	Record lowest	Year	Normal degree days
(a)	(b)	(b)	(b)	43	43	43	43	(b)
J	37.3	22.1	29.7	71	1950	-16	1936	1094
F	39.2	23.5	31.4	72	1932	-16	1951	941
M	48.8	30.8	39.8	84	1929	-4	1943	781
A	60.5	40.4	50.5	89	1915	18	1923	435
M	71.2	50.5	61.0	93	1941+	27	1947	179
J	80.8	60.9	70.9	99	1944	40	1929	39
J	85.3	64.7	75.0	106	1936	49	1940	0
A	83.0	62.8	72.9	103	1918	42	1915	5
S	77.1	56.6	66.9	101	1954	31	1942+	74
O	65.1	45.0	55.1	89	1950	21	1942+	324
N	49.9	33.9	41.9	79	1950	-5	1930	683
D	38.8	24.5	31.7	70	1936	-15	1951	1032
Year	61.4	43.0	52.2	106 July 1936		-16 Feb. 1951+		5597

Precipitation

Month	Normal total	Max monthly	Year	Max in 24 hrs	Year	Min monthly	Year
J	2.96	12.41	1952	3.41	1929	0.62	1937
F	2.11	4.57	1951	2.73	1914	0.14	1950
M	3.24	8.89	1941	4.40	1956	0.77	1941
A	3.07	6.69	1913	3.41	1920	0.83	1947
M	3.54	7.59	1922	2.84	1923	0.94	1938
J	3.90	—	1935	3.23	1932	0.52	1924
J	3.29	7.21	1955	3.29	1955	0.33	1955
A	2.86	6.64	1917	3.08	1917	0.38	1912
S	2.80	6.44	1925	4.56	1925	0.47	1925
O	2.30	7.03	1944	2.00	1929	0.10	1919
N	2.10	6.50	1917	3.18	1938	0.34	1927
D	2.47	6.84	1955	2.94	1932	0.36	1923
Year	35.15	12.41 Jan. 1952		4.56 Sept. 1925		0.10 Oct. 1944	

Snow, Sleet

Month	Mean total	Max monthly	Year	Max in 24 hrs	Year
J	6.2	24.4	1918	6.6	1929
F	4.0	15.2	1914	9.0	1912
M	3.1	8.9	1956	5.7	1955
A	T	4.5	1920	3.4	1920
M	0.0	0.5	1923	0.5	1923
J	0.0	0.0		0.0	
J	0.0	0.0		0.0	
A	0.0	0.0		0.0	
S	0.0	0.0		0.0	
O	0.1	2.8	1925	1.5	1925
N	1.8	12.7	1950	10.0	1950
D	4.5	14.1	1951	7.0	1951
Year	20.1	24.4 Jan. 1918		10.0 Nov. 1950	

Relative humidity, Wind, Sunshine and Sky cover

Month	RH 1am	RH 7am	RH 1pm	RH 7pm	Wind mean hourly	Prevailing dir	Fastest speed	Fastest dir	Fastest year	Pct. possible sunshine	Mean sky cover
J	81	81	73	76	11.4	S	60	NW	1926+	38	7.2
F	80	80	69	73	11.5	NNW	72	W	1916	43	6.8
M	77	79	60	67	12.1	NW	75	W	1916	50	6.6
A	76	76	57	62	12.5	SW	80	W	1917	61	6.4
M	80	76	55	61	10.4	SW	80	SW	1917	61	5.8
J	80	78	55	60	8.7	SSW	60	W	1950	68	5.5
J	78	78	52	56	7.7	SW	74	NW	1926	72	5.1
A	79	82	52	61	7.2	N	70	SW	1919	68	5.0
S	78	83	53	65	7.4	S	65	SW	1950	59	4.8
O	77	83	53	63	8.9	S	56	SW	1954	59	5.4
N	78	81	71	76	11.2	SSW	44	W	1919	36	7.1
D	81	82	71	76	11.0	SW	70	W	1920	36	—
Year	79	80	59	67	9.9	SW	78	NW	June 1950	57	6.0

Mean number of days

Month	Clear	Partly cloudy	Cloudy	Precip .01"+	Snow/Sleet 1.0"+	Thunderstorms	Heavy fog	Max 90°+	Max 32°-	Min 32°-	Min 0°-
J	6	7	18	13	2	1	2	0	11	25	2
F	6	7	15	11	2	1	2	0	7	22	2
M	7	8	16	13	1	3	2	0	2	17	*
A	8	10	13	13	*	4	1	0	*	5	0
M	8	11	12	12	0	7	1	1	0	*	0
J	8	13	12	11	0	9	*	4	0	0	0
J	11	13	7	10	0	9	1	8	0	0	0
A	11	13	7	10	0	7	1	5	0	0	0
S	12	10	8	9	0	4	2	2	0	*	0
O	12	8	11	8	*	2	1	0	*	2	0
N	8	7	15	11	1	1	1	0	2	13	*
D	6	7	18	12	2	*	2	0	8	22	1
Year	101	114	150	132	6	48	14	20	31	106	4

Means and extremes in the above table are from the existing or comparable location(s). Annual extremes have been exceeded at prior locations as follows: Highest temperature 108 in July 1901; lowest temperature -28 in February 1899.

SANDUSKY, OHIO — POST OFFICE BUILDING

LATITUDE 41° 27' N
LONGITUDE 82° 43' W
ELEVATION (ground) 603 feet

Temperature and Normal degree days

Month	Normal Daily max	Normal Daily min	Normal Monthly	Record highest	Year	Record lowest	Year	Normal degree days
(a)	(b)	(b)	(b)	80	80	80	80	(b)
J	35.7	21.8	28.8	73	1950	-16	1879	1122
F	36.4	22.4	29.4	72	1944	-15	1899	997
M	45.2	29.8	37.5	81	1910	-3	1885	853
A	56.3	39.4	47.9	90	1942+	12	1923	513
M	68.3	50.7	59.5	93	1941+	32	1923+	217
J	79.0	60.7	69.9	104	1934	40	1894	41
J	83.7	65.5	74.6	105	1936	50	1918	0
A	81.7	63.9	72.8	105	1918	45	1946	0
S	75.4	57.5	66.5	93	1953+	34	1956+	66
O	63.5	46.5	55.0	82	1950	22	1925	327
N	46.5	35.2	42.1	70	1899	0	1880	684
D	37.6	25.3	31.5	—	1886	-13	1880	1039
Year	59.3	43.2	51.3	105 July 1936+		-16 Jan. 1879		5859

Precipitation

Month	Normal total	Max monthly	Year	Max in 24 hrs	Year	Min monthly	Year
J	2.29	6.58	1902	1.71	1937	0.60	1937
F	1.92	3.53	1920	2.98	1887	0.27	1887
M	2.89	8.89	1910	2.96	1913	0.28	1910
A	2.96	6.24	1915	2.21	1929	0.35	1915
M	3.32	9.04	1934	3.83	1938	0.64	1934
J	3.73	12.51	1919	5.95	1937	0.91	1919
J	3.45	9.71	1916	3.87	1943	0.26	1916
A	2.81	8.02	1882	4.20	1906	0.23	1894
S	3.26	6.72	1928	4.28	1950	0.73	1928
O	2.10	6.22	1897	2.76	1920	0.09	1897
N	2.27	6.43	1904	2.26	1927	0.09	1904
D	2.16	6.27	1934	1.74	1881	0.63	1927
Year	33.16	12.51 June 1937		5.95 June 1937		0.09 Nov. 1904	

Snow, Sleet

Month	Mean total	Max monthly	Year	Max in 24 hrs	Year
J	8.2	29.8	1893	11.8	1910
F	6.5	22.1	1893	10.1	1900
M	4.5	16.1	1916	8.7	1955
A	1.2	12.0	1957	9.0	1886
M	0.0	0.0		0.0	
J	0.0	0.0		0.0	
J	0.0	0.0		0.0	
A	0.0	0.0		0.0	
S	0.0	0.0		0.0	
O	T	1.6	1917	1.5	1917
N	1.9	12.3	1950	12.9	1950
D	6.1	20.2	1951	9.0	1886
Year	28.2	29.8 Jan. 1893		12.9 Nov. 1950	

Relative humidity, Wind, Sunshine and Sky cover

Month	RH 7am	RH 1:30pm	RH 7pm	Wind mean hourly	Prevailing dir	Fastest speed	Fastest dir	Fastest year	Pct. possible sunshine	Mean sky cover
J	81	68	78	10.8	SW	56	SW	1952	35	7.4
F	81	70	74	10.9	SW	64	NE	1881	41	7.0
M	79	64	74	11.3	SW	64	NW	1918	49	6.7
A	79	61	68	10.6	SW	55	SW	1919	52	6.3
M	73	59	67	8.9	SW	48	SW	1924	59	5.8
J	73	58	68	8.1	SW	77	NW		69	5.3
J	72	54	67	7.6	SW	69	N	1879	71	4.7
A	77	57	68	7.5	SW	63	NE	1885	68	4.8
S	77	59	70	8.3	SW	52	NW	1897	63	5.5
O	77	66	73	9.3	SW	54	SW	1885	55	5.7
N	81	71	75	10.9	SW	68	NW	1919	37	7.1
D	81	—	—	10.6	NW	56	NW	1895	29	7.7
Year	76	62	71	9.6	SW	77	NW	June 1924	53	6.1

Mean number of days

Month	Clear	Partly cloudy	Cloudy	Precip .01"+	Snow/Sleet 1.0"+	Thunderstorms	Heavy fog	Max 90°+	Max 32°-	Min 32°-	Min 0°-
J	5	8	18	14	4	0	0	0	13	26	2
F	5	8	15	12	3	0	1	0	11	24	1
M	6	9	16	13	2	2	1	0	5	21	0
A	6	11	13	12	0	3	0	0	0	5	0
M	9	13	8	11	0	5	0	0	0	0	0
J	13	13	—	10	0	7	0	3	0	0	0
J	14	14	6	9	0	7	0	5	0	0	0
A	13	13	5	10	0	5	1	2	0	0	0
S	11	11	8	12	0	3	1	0	0	0	0
O	10	8	12	13	1	1	1	0	1	1	0
N	5	8	17	13	1	0	1	0	2	11	0
D	3	5	20	13	3	1	1	0	10	23	0
Year	91	121	153	138	13	34	6	13	41	111	3

TOLEDO, OHIO — EXPRESS AIRPORT

LATITUDE 41° 36' N
LONGITUDE 83° 48' W
ELEVATION (ground) 676 Feet

Month	Temperature Normal Daily max (b)	Daily min (b)	Monthly (b)	Extremes Record highest	Year	Record lowest	Year	Normal degree days (b)	Precip Normal total (b)	Max monthly	Year	Min monthly	Year	Max in 24 hrs	Year
				87	87			(b)	(b)	87		87		87	
J	34.7	18.0	26.4	71	1950+	-16	1897	1197	2.25	6.63	1913	0.51	1945	1.56	1948
F	36.0	18.6	27.3	71	1944	-16	1885	1058	1.86	6.84	1887	0.08	1877	2.28	1883
M	45.6	26.0	35.8	83	1910	-10	1948	905	2.86	7.99	1913	0.05	1910	2.69	1913
A	58.0	35.0	46.5	89	1942+	12	1875	555	3.25	7.13	1929	0.55	1876	2.93	1929
M	70.4	46.5	58.2	95	1911	28	1944	245	2.95	8.04	1943	0.45	1881	3.57	1913
J	80.7	56.2	68.6	101	1934	38	1945+	60	3.55	7.86	1911	0.12	1952	3.44	1944
J	85.3	60.9	73.1	105	1936	44	1945	0	2.65	6.65	1896	0.35	1916	2.47	1929
A	83.0	58.9	71.0	103	1918	41	1946	12	2.69	7.64	1894	0.60	1894	4.58	1920
S	75.8	52.5	64.2	100	1953+	29	1942	102	3.02	8.07	1926	0.37	1946	5.98	1918
O	63.9	41.5	52.7	92	1951	21	1952+	387	2.32	8.49	1881	0.32	1892	3.10	1881
N	48.9	30.7	39.8	80	1950	-2	1947+	756	2.15	5.58	1927	0.04	1904	2.68	1871
D	36.9	20.9	28.9	70	1889	-15	1872	1119	2.39	5.74	1923	0.33	1951	2.04	1885
Year	59.9	38.8	49.4	105	July 1936	-16	Jan 1897+	6394	31.84	8.49	Oct 1881	0.04	Nov 1904	5.98	Sept 1918

Month	Snow,Sleet Mean total 73	Max monthly 73	Year	Max 24hr 73	Year	RelHum 1a 69	7a 69	1p 27	7p 69	Wind mean hrly 77	Prev dir 9	Fastest speed 46	Dir	Year	Pct sun 58	Sky cover 87
J	8.0	26.2	1918	9.8	1885	80	80	72	76	12.6	WSW	66	SW	1949	33	7.0
F	6.9	25.1	1900	19.0	1900	80	80	68	73	13.0	SW	69	SW	1919	40	6.5
M	5.1	17.1	1888	8.4	1888	78	78	62	70	13.0	SW	87	SW	1948	47	6.6
A	1.3	12.0	1957	9.8	1957	79	79	56	64	12.8	SW	72	SW	1956	50	5.7
M	0.1	4.0	1923	4.0	1923	79	79	54	63	11.0	ENE	58	SW	1917	55	5.5
J	0.0	0.0		0.0		76	76	54	63	9.8	SW	56	W	1930	65	4.8
J	0.0	0.0		0.0		83	76	52	59	9.1	SW	76	SW	1922	71	4.1
A	0.0	0.0		0.0		85	80	54	63	8.8	SW	76	W	1924	71	4.4
S	0.0	0.0		0.0		84	82	55	67	9.7	SW	62	S	1919	62	4.4
O	T	1.2	1925	1.1	1925	81	81	56	62	10.5	SW	57	SW	1916	55	5.1
N	2.0	12.0	1932	11.5	1932	81	81	66	73	12.4	SW	76	SW	1919	36	6.6
D	7.0	25.5	1909	10.8	1909	82	82	72	76	12.3	SW	69	SW	1920	30	7.1
Year	30.4	26.2	Jan 1918	19.0	Feb 1900	82	79	60	68	11.2	SW	87	SW	Mar 1948	53	5.6

Mean number of days (Toledo):

Month	Clear 87	Partly cloudy 87	Cloudy 87	Precip .01"+ 87	Snow 1.0"+ 67	Tstorms 87	Heavy fog 86	Temp Max 90°+ 84	Max 32°- 84	Min 32°- 84	Min 0°- 84
J	6	8	17	13	3	*	1	0	14	27	2
F	6	8	14	11	2	*	1	0	12	25	2
M	8	9	14	12	2	2	1	0	5	22	*
A	9	11	12	12	1	3	*	*	1	7	0
M	10	11	10	11	0	5	1	0	0	0	0
J	11	12	7	11	0	7	1	3	0	0	0
J	14	12	5	9	0	7	1	5	0	0	0
A	14	10	6	9	0	5	1	4	0	0	0
S	13	8	7	9	0	3	1	1	0	0	0
O	12	8	11	9	*	2	1	*	0	2	0
N	7	11	15	11	1	1	1	0	2	14	0
D	5	8	18	12	1	*	1	0	10	24	1
Year	115	114	136	131	10	35	7	13	43	121	5

YOUNGSTOWN, OHIO — MUNICIPAL AIRPORT

LATITUDE 41° 16' N
LONGITUDE 80° 40' W
ELEVATION (ground) 1178 Feet

Month	Temp Normal Daily max (b)	Daily min (b)	Monthly (b)	Record highest 14	Year	Record lowest 14	Year	Normal degree days (b)	Normal total (b)	Max monthly 14	Year	Min monthly 14	Year	Max 24hr 14	Year
J	34.9	20.0	27.5	71	1950	-12	1957	1163	3.32	7.64	1950	0.99	1946	2.16	1952
F	36.2	20.2	28.2	65	1957+	-11	1951	1030	2.72	5.26	1950	1.28	1953	1.84	1950
M	45.6	27.3	36.7	86	1945	-4	1948	877	3.45	5.68	1951	2.13	1957	2.47	1954
A	57.7	36.5	47.2	86	1948	11	1944	534	3.64	6.43	1957	1.30	1946	1.75	1957
M	69.7	47.2	58.5	88	1944	28	1950+	241	4.09	9.87	1946	1.38	1954	2.85	1946
J	79.0	56.3	68.0	99	1952	35	1952	53	3.71	6.97	1957	1.35	1957	2.96	1957
J	83.4	60.4	72.3	100	1954	44	1953	0	4.16	7.07	1956	1.57	1957	2.70	1955
A	81.4	59.3	70.4	99	1953	43	1953+	19	3.58	7.86	1945	0.51	1947	2.98	1947
S	75.5	53.9	64.7	100	1954	29	1957	83	2.77	8.17	1954	1.43	1951	1.31	1951
O	63.1	44.8	53.9	87	1957	21	1952	355	3.49	5.39	1950	1.54	1954	2.67	1954
N	48.0	33.2	40.6	79	1950	-9	1955+	732	3.49	5.39	1951	0.95	1951	1.48	1944
D	36.8	23.2	30.0	66	1956		1951	1085	2.94	4.36					
Year	59.3	40.3	49.8	100	July 1954	-12	Jan 1957	6172	41.33	9.87	May 1948	0.43	Oct 1953	4.31	Oct 1954

Month	Snow Mean total 14	Max monthly 14	Year	Max 24hr 14	Year	RelHum 1a 10	7a 10	1p 10	7p 10	Wind mean hrly 9	Prev dir 9	Sky cover 11
J	11.2	30.1	1948	17.5	1948	83	84	74	80	11.7	SW	8.2
F	8.7	18.9	1950	6.4	1947	82	82	71	78	11.2	W	7.4
M	10.0	19.9	1951	9.4	1951	80	81	63	71	12.1	SW	7.3
A	2.5	4.7	1957	4.1	1951	80	81	57	63	11.3	SW	6.9
M	T	0.1	1956+	0.1	1956+	84	84	52	63	9.6	WSW	6.5
J	0.0	0.0		0.0		84	84	53	66	9.6	SW	5.6
J	0.0	0.0		0.0		82	82	53	65	7.7	SW	5.3
A	0.0	0.0		0.0		85	85	53	62	7.3	W	5.1
S	T	T	1956+	T	1956+	85	87	56	68	8.4	SW	5.5
O	0.4	1.8	1954	1.8	1954	82	86	56	71	9.2	SW	6.5
N	7.2	20.7	1950	20.7	1950	86	86	74	72	11.7	SW	7.9
D	11.6	21.7	1944	14.8	1944	84	84	79	79	11.9	SW	
Year	51.6	30.6	Nov 1950	20.7	Nov 1950	82	83	61	71	10.1	SW	6.7

Mean number of days (Youngstown):

Month	Clear 14	Partly cloudy 14	Cloudy 14	Precip .01"+ 14	Snow 1.0"+ 14	Tstorms 14	Heavy fog 14	Max 90°+ 14	Max 32°- 14	Min 32°- 14	Min 0°- 14
J	3	5	23	19	3	*	3	0	14	27	2
F	3	5	19	16	3	*	3	0	14	26	1
M	5	8	18	17	3	2	3	0	6	23	*
A	5	8	17	14	1	4	2	*	1	9	0
M	7	9	16	13	0	6	3	0	0	1	0
J	8	10	11	11	0	7	3	2	0	0	0
J	10	13	8	10	0	7	3	4	0	0	0
A	11	10	10	10	0	6	3	3	0	*	0
S		10	10	9	0	3	3	2	0	4	0
O	7	6	20	15	*	1	2	0	0	4	0
N	3	7	21	18	1	*	3	0	3	17	1
D					4	*	3	0	11	26	1
Year	74	105	186	166	16	38	30	11	42	135	4

REFERENCE NOTES APPLYING TO ALL "NORMALS, MEANS, AND EXTREMES" TABLES.

(a) Length of record, years.
(b) Normal values are based on the period 1921-1950, and are means adjusted to represent observations taken at the present standard location.
* Less than one-half.

- No record.
† Airport data.
‡ City Office data.
+ Also on earlier dates, months, or years.
T Trace, an amount too small to measure.

Sky cover is expressed in a range of 0 for no clouds or obscuring phenomena to 10 for complete sky cover. The number of clear days is based on average cloudiness 0-3 tenths; partly cloudy days on 4-7 tenths; and cloudy days on 8-10 tenths. Monthly degree day totals are the sum of the negative departures of average daily temperatures from 65°F. Sleet was included in snowfall totals beginning with July 1948. Snowfall also includes data referred to at various times in the past as "Dense" or "Thick". The upper limit for heavy fog is 1/4 mile. Data in these tables are based on records through 1957.

Mean Annual Precipitation, Inches

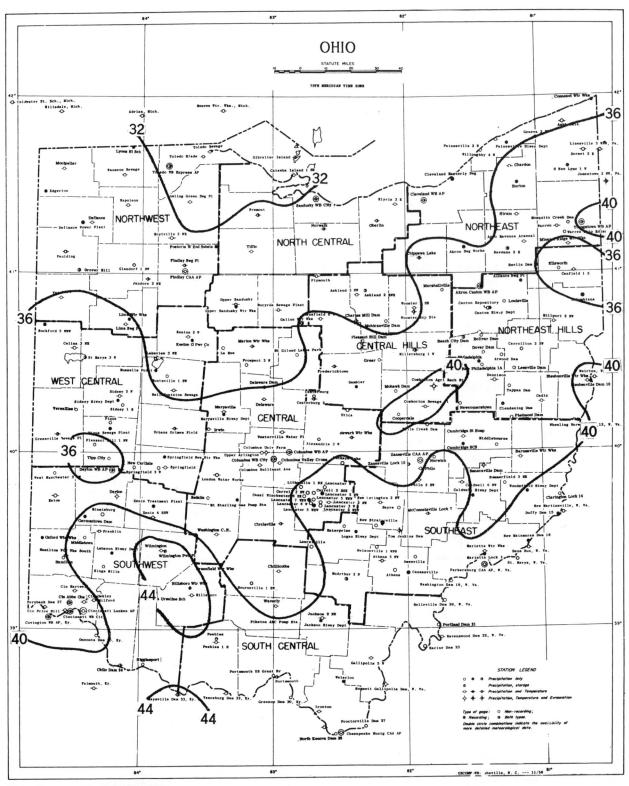

Based on period 1931-55

Isolines are drawn through points of approximately equal value. Caution should be used in interpolating on these maps, particularly in mountainous areas.

Mean Maximum Temperature (°F.), January

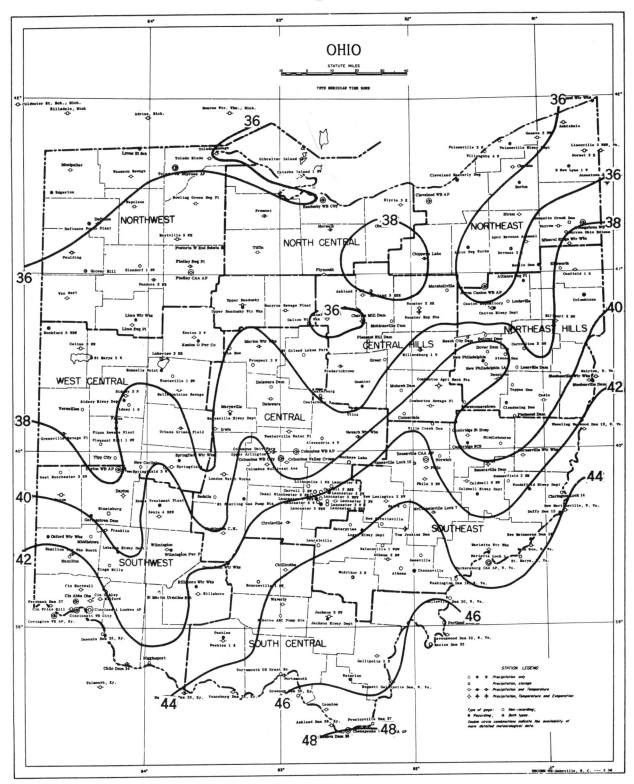

Based on period 1931-52

Isolines are drawn through points of approximately equal value. Caution should be used
in interpolating on these maps, particularly in mountainous areas.

Mean Minimum Temperature (°F.), January

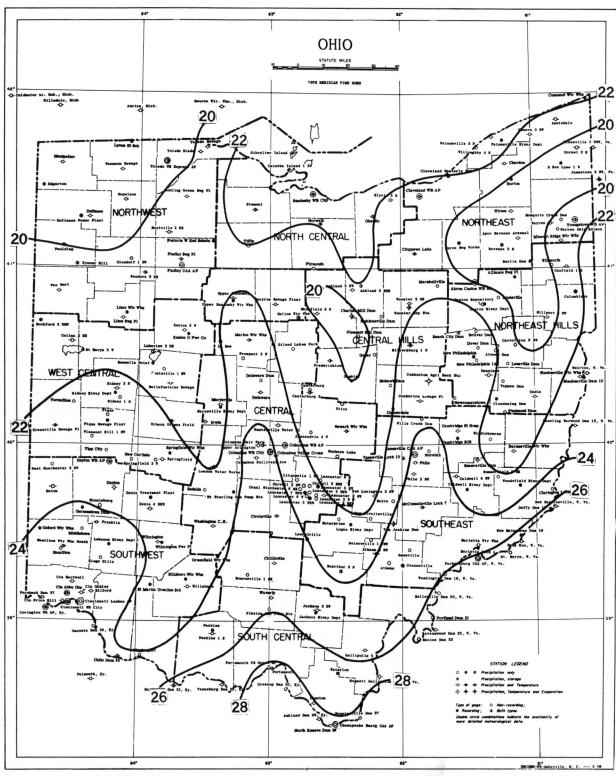

Based on period 1931-52

Isolines are drawn through points of approximately equal value. Caution should be used in interpolating on these maps, particularly in mountainous areas.

Mean Maximum Temperature (°F.), July

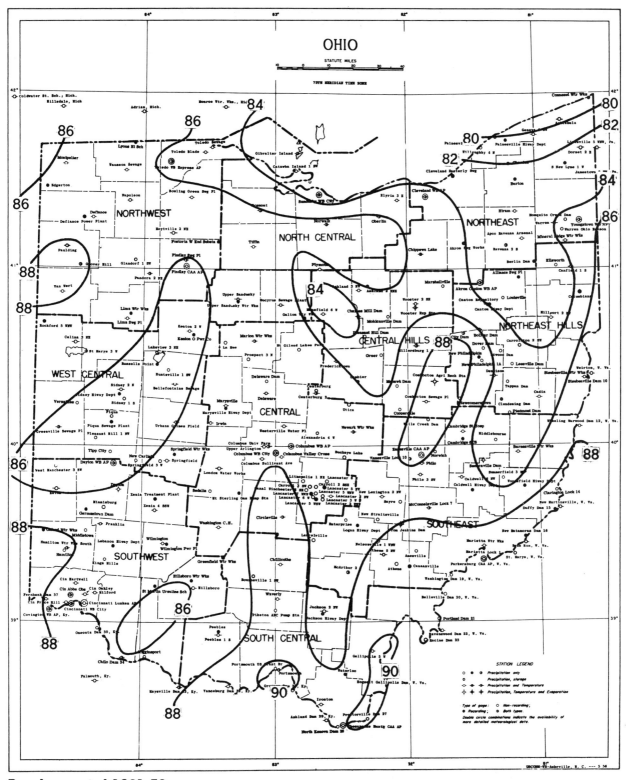

Based on period 1931-52

Isolines are drawn through points of approximately equal value. Caution should be used in interpolating on these maps, particularly in mountainous areas.

Mean Minimum Temperature (°F.), July

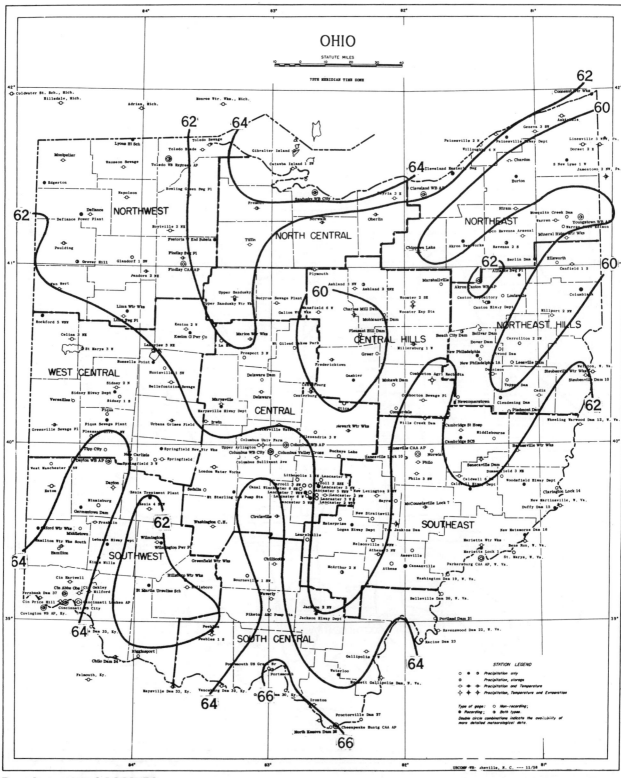

Based on period 1931-52

Isolines are drawn through points of approximately equal value. Caution should be used in interpolating on these maps, particularly in mountainous areas.

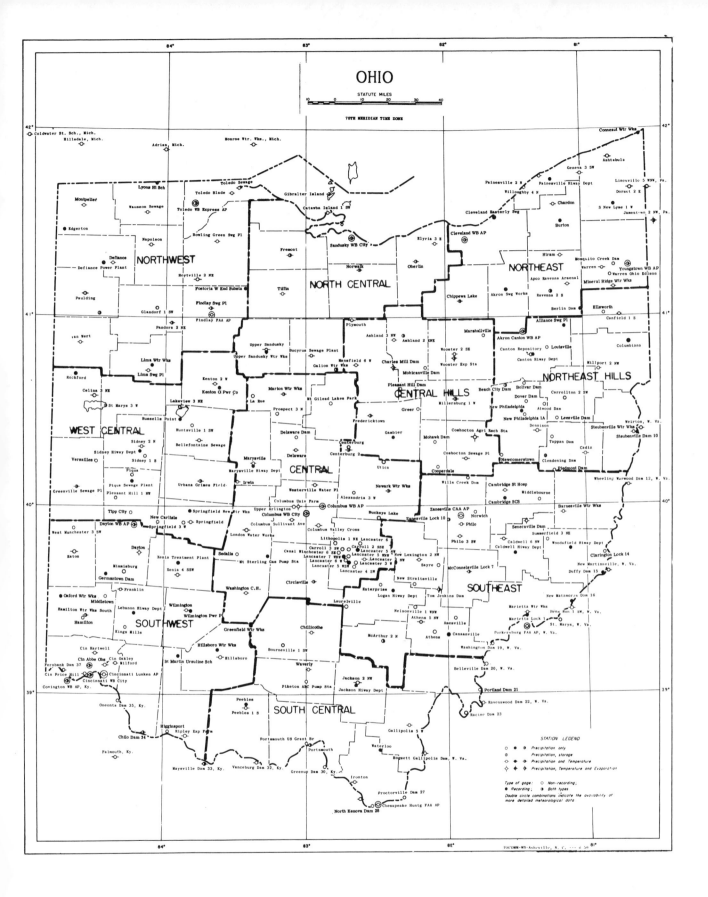

OHIO

STATUTE MILES

75TH MERIDIAN TIME ZONE

NORTHWEST

NORTH CENTRAL

NORTHEAST

NORTHEAST HILLS

CENTRAL HILLS

WEST CENTRAL

CENTRAL

SOUTHEAST

SOUTHWEST

SOUTH CENTRAL

STATION LEGEND

○ ⊙ Precipitation only
⊙ Precipitation, storage
○ ◐ ● Precipitation and Temperature
◇ ◈ ◆ Precipitation, Temperature and Evaporation

Type of gage: ○ Non-recording;
● Recording; ⊙ Both types
Double circle combinations indicate the availability of
more detailed meteorological data

ESCOMM-WB-Asheville, N. C.

– 317 –

THE CLIMATE OF
PENNSYLVANIA

by
Paul W. Dailey, Jr.

March 1971

The erratic course of the Delaware River is the only natural boundary of Pennsylvania. All others are arbitrary boundaries that do not conform to physical features. Notable contrasts in topography, climate, and soils exist. Within this 45,126-square-mile area lies a great variety of physical land forms of which the most notable is the Appalachian Mountain system composed of two ranges, the Blue Ridge and the Allegheny. These mountains divide the Commonwealth into three major topographical sections. In addition, two plain areas of relatively small size also exist, one in the southeast and the other in the northwest.

In the extreme southeast is the Coastal Plain situated along the Delaware River and covering an area 50 miles long and 10 miles wide. The land is low, flat, and poorly drained, but has been improved for industrial and commercial use because of its proximity to ocean transportation via the Delaware River. Philadelphia lies almost in the center of this area.

Bordering the Coastal Plain and extending 60 to 80 miles northwest to the Blue Ridge is the Piedmont Plateau, with elevations ranging from 100 to 500 feet and including rolling or undulating uplands, low hills, fertile valleys, and well-drained soils. These features, combined with the prevailing climate, have aided this area in becoming the leading agricultural section of the State. Good pastures, productive land, and short distances to markets have resulted in dairy farm-

ing becoming one of the leading agricultural activities. Another activity is the growing of fruit, primarily apples and peaches. Gentle hillside slopes provide an excellent place for fruit trees, as cold air drainage helps to prevent unseasonable freezing temperatures on these slightly elevated lands. The area has many orchards, with Adams County leading all others within the region in the production of apples. The climate and soils in the Lancaster County area are especially well suited for the growing of cigar leaf tobacco, as is pointed up by the fact that Pennsylvania is the leading producer of cigar leaf of any type in the Nation.

Just northwest of the Piedmont and between the Blue Ridge and Allegheny Mountains is the Ridge and Valley Region, in which forested ridges alternate with fertile and extensively farmed valleys. Vegetables, grown primarily for canning, are the leading crop. This has led to a well-developed canning industry, which is concentrated in the middle Susquehanna Valley. The Ridge and Valley Province is 80 to 100 miles wide and characterized by parallel ridges and valleys oriented northeast-southwest. The mountain ridges vary from 1,300 to 1,600 feet above sea level, with local relief 600 to 700 feet.

North and west of the Ridge and Valley Region and extending to the New York and Ohio borders is the area known as the Allegheny Plateau. This is the largest natural division of the State and occupies more than half the area. It is crossed by many deep narrow valleys and drained by the Delaware, Susquehanna, Allegheny, and Monongahela River systems. Elevations are generally 1,000 to 2,000 feet above sea level; however, some mountain peaks extend to 3,000 feet. The area is heavily wooded and among the most rugged in the State. Numerous lakes and swamps characterize this once glaciated area, creating a very picturesque landscape; this is particularly outstanding in the more northerly counties. The combination of lakes and forests at elevations high enough to keep summer temperatures comfortable and its location close to heavily populated cities have made the Pocono Mountain area the leading tourist and recreational center in Pennsylvania.

Bordering Lake Erie is a narrow 40-mile strip of flat, rich land 3 to 4 miles wide called the Lake Erie Plain. Fine alluvial soils and favorable climate permit intensive vegetable and fruit cultivation, which is typical of the much larger area surrounding Lake Erie.

Eastern and central Pennsylvania drains into the Atlantic Ocean, while the western portion of the State lies in the Ohio River Basin, except the Lake Erie Plain in the northwest, which is drained by a number of small streams into Lake Erie. The Delaware River, which forms the eastern boundary, drains the eastern portion and flows into Delaware Bay. The Susquehanna River drains the central portion and flows into Chesapeake Bay. In the western portion, the Allegheny and the Monongahela Rivers have their confluence at Pittsburgh and form the Ohio River.

Floods may occur during any month of the year in Pennsylvania, although they occur with greater frequency in the spring months of March and April. They may result from heavy rains during any season. Generally, the most widespread flooding occurs during the winter and spring when associated with heavy rains, or heavy rains combined with snowmelt. Serious local flooding sometimes results from ice jams during the spring thaw. Heavy local thunderstorm rains cause severe flash flooding in many areas. Storms of tropical origin sometimes deposit flood-producing rains, especially in the eastern portion of the State.

Floods may be expected at least once in most years. For instance, flood stage at Pittsburgh is exceeded on the average of 1.3 times per year, based on the long-term record. However, floods of notable severity and magnitude for the State occur about once in 8 years.

Some years in which major flooding occurred along principal rivers are as follows: Schuylkill, 1902, 1935, 1942, 1955, 1969; Delaware, 1903, 1936, 1955, 1967; Susquehanna, 1865, 1889, 1894, 1902, 1904, 1936, 1964; Allegheny, 1865, 1889, 1892, 1905, 1907, 1910, 1913, 1936, 1942, 1947, 1964; Monongahela, 1888, 1907, 1918, 1936; Ohio, 1907, 1936, 1942, 1954.

Pennsylvania is generally considered to have a humid continental type of climate, but the varied physiographic features have a marked effect on the weather and climate of the various sections within the State. The prevailing westerly winds carry most of the weather disturbances that affect Pennsylvania from the interior of the continent, so that the Atlantic Ocean has only limited influence upon the climate of the State. Coastal storms do, at times, affect the day-to-day weather, especially in eastern sections. It is here that storms of tropical origin have the greatest effect within the State, causing floods in some instances.

Throughout the State temperatures generally remain between 0° and 100° and average from near 47° annually in the north-central mountains to 57° annually in the extreme southeast. The highest temperature of record in Pennsylvania of 111° was observed at Phoenixville July 9 and 10, 1936, while the record low of -42° occurred at Smethport January 5, 1904.

Summers are generally warm, averaging about 68° along Lake Erie to 74° in southeastern counties. High temperatures, 90° or above, occur on the average of 10 to 20 days per year in most sections; but occasionally southeastern localities may experience a season with as many as 30 days, while the extreme northwest averages as few as 4 days annually. Only rarely does a summer pass without excessive temperatures being reported somewhere in the State. However, there are places such as immediately adjacent to Lake Erie and at some higher elevations where readings of 100° have never been recorded. Daily temperatures during the warm season usually have a range of about 20° over much of the State, while the daily range in winter is several degrees less. During the coldest months temperatures average near the freezing point with daily minimum readings sometimes near 0° or below. Freezing temperatures occur on the average of 100 or more days annually with the greatest number of occurrences in mountainous regions. Records show that freezing temperatures have occurred somewhere in the State during all months of the year and below 0° readings from November to April, inclusive.

Precipitation is fairly evenly distributed throughout the year. Annual amounts generally range between 34 to 52 inches, while the majority of places receive 38 to 46 inches. Greatest amounts usually occur in spring and summer months, while February is the driest month, having about 2 inches less than the wettest months. Precipitation tends to be somewhat greater in eastern sections due primarily to coastal storms which occasionally frequent the area. During the warm season these storms bring heavy rain, while in winter heavy snow or a mixture of rain and snow may be produced. Thunderstorms, which average between 30 to 35 per year, are concentrated in the warm months and are responsible for most of the summertime rainfall, which averages from 11 inches in the northwest to 13 inches in the east. Occasionally dry spells may develop and persist for several months during which time monthly precipitation may total less than one-quarter inch. These periods almost never affect all sections of the State at the same time, nor are they confined to any particular season of the year. Winter precipitation is usually 3 to 4 inches less than summer rainfall and is produced most frequently from northeastward-moving storms. When temperatures are low enough these storms sometimes cause heavy snow which may accumulate to 20 inches or more. Annual snowfall ranges between wide limits from year to year and place to place. Some years are quite lean as snowfall may total less than 10 inches while other years may produce upwards to 100 inches mostly in northern and mountainous areas. Annual snowfall averages from about 20 inches in the extreme southeast to 90 inches in parts of McKean County. Measurable snow generally occurs between November 20 and March 15 although snow has been observed as early as the beginning of October and as late as May, especially in northern counties. Greatest monthly amounts usually fall in December and January, however, greatest amounts from individual storms generally occur in March as the moisture supply increases with the annual march of temperature.

As mentioned earlier, hurricanes or low pressure systems with a tropical origin seldom affect the State. Damages, as a result of hurricane winds, are rare and usually confined to extreme eastern portions. However, nature's most violent storm, the tornado, does occur in Pennsylvania. At least one tornado has been noted in almost all counties since the advent of severe storms records in 1854. On the average, 5 or 6 tornadoes are observed annually in Pennsylvania, and the State ranks 27th nationally. June is the month of highest frequency, followed closely by July and August. Principal areas of tornado concentration are in the extreme northwest, the Southwest Plateau, and the Southeastern Piedmont. The frequency in the latter area is the highest in the State per square mile, similar to what is observed in portions of Midwestern United States. Many of the tornadoes in Pennsylvania have caused relatively minor damages. However, several have claimed lives and dealt severe local economic setbacks. The most destructive activity occurred June 23, 1944, when 3 tornadoes raked the southwestern portion of the Commonwealth, killing 45 persons, injuring another 362, and causing over $2 million in property damage.

The topographic features of Pennsylvania divide the State into four rather distinct climatic areas:

(1)   The Southeastern Coastal Plain and Piedmont Plateau,

(2)   The Ridge and Valley Province,

(3)   The Allegheny Plateau, and

(4)   The Lake Erie Plain.

In the Southeastern Coastal Plain and Piedmont Plateau summers are long and at times uncomfortably hot. Daily temperatures reach 90° or above on the average of 25 days during the summer season; however, readings of 100° or above are comparatively rare. From about July 1 to the middle of September this area occasionally experiences uncomfortably warm periods, 4 to 5 days to a week in length, during which light wind movement and high relative humidity make conditions oppressive. In general, the winters are comparatively mild, with an average of less than 100 days with minimum temperatures below the freezing point. Temperatures 0° or lower occur at Philadelphia, on an average, 1 winter in 4, and at Harrisburg 1 in 3. The freeze-free season averages 170 to 200 days.

Average annual precipitation in the area ranges from about 30 inches in the lower Susquehanna Valley to about 46 in Chester County. Under the influence of an occasional severe coastal storm, a normal month's rainfall, or more, may occur within a period of 48 hours. The average seasonal snowfall is about 30 inches, and fields are ordinarily snow covered about one-third of the time during the winter season.

The Ridge and Valley Province is not rugged enough for a true mountain type of climate, but it does have many of the characteristics of such a climate. The mountain-and-valley influence on the air movements causes somewhat greater temperature extremes than are experienced in the southeastern part of the State where the modifying coastal and Chesapeake Bay influence hold them relatively constant, and the daily range of temperature increases somewhat under the valley influences.

The effects of nocturnal radiation in the valleys and the tendency for cool airmasses to flow down them at night result in a shortening of the growing season by causing freezes later in spring and earlier in fall than would otherwise occur. The growing (freeze-free) season in this section is longest in the middle Susquehanna Valley, where it averages about 165 days, and shortest in Schuylkill and Carbon Counties, averaging less than 130 days.

The annual precipitation in this area has a mean value of 3 or 4 inches more than in the southeastern part of the State, but its geographic distribution is less uniform. The mountain ridges are high enough to have some deflecting influence on general storm winds, while summer showers and thunderstorms are often shunted up the valleys.

Seasonal snowfall of the Ridge and Valley Province varies considerably within short distances. It is greatest in Somerset County, averaging 88 inches in the vicinity of Somerset, and least in Huntingdon, Mifflin, and Juniata Counties, averaging about 37 inches.

The Allegheny Plateau is fairly typical of a continental type of climate, with changeable temperatures and more frequent precipitation than other parts of the State. In the more northerly sections the influence of latitude, together with higher elevation and radiation conditions, serve to make this the coldest area in the State. Occasionally, winter minimum temperatures are severe. The daily temperature range is fairly large, averaging about 20° in midwinter and 26° in midsummer. In the southern counties the daily temperature range is a few degrees higher and the same may be said of the normal annual range. Because of the rugged topography the freeze-free season is variable, ranging between 130 days in the north to 175 days in the south.

Annual precipitation has a mean of about 41 inches, ranging from less than 35 inches in the northern parts of Tioga and Bradford Counties to more than 45 inches in parts of Crawford, Warren, and Wayne Counties. The seasonal snowfall averages 54 inches in northern areas, while southern sections receive several inches less. Fields are normally snow covered three-fourths of the time during the winter season. With rapidly flowing streams in the Ohio Drainage system (except the Monongahela), it is fortunate that this part of the State is not subject to torrential rains such as sometimes occur along the Atlantic slope. Although average annual precipitation is about equal to that for the State as a whole, it usually occurs in smaller amounts at more frequent intervals; 24-hour rains exceeding 2.5 inches are comparatively rare.

Although the Lake Erie Plain is of relatively small size, it has a unique and agriculturally advantageous climate typical of the coastal areas surrounding much of the Great Lakes. Both in spring and autumn the lake water exerts a retarding influence on the temperature regime and the freeze-free season is extended about 45 days. In the autumn this prevents early freezing temperatures, which is a critical factor in the growing of fruit and vegetables.

Annual precipitation totals about 34.5 inches, which is fairly evenly distributed throughout the year. Snowfall exceeds 54 inches per year, with heavy snows sometimes experienced late in April.

BIBLIOGRAPHY

Benfer, Neil A., "An Economic Geography of the Canning Industry in the Middle Susquehanna Valley Lowland," Master's thesis, Pennsylvania State University, University Park, Pa., 1951.

Bingham, Christopher, "Probabilities of Weekly Averages of the Daily Temperature Maximum, Minimum, and Range," The Climate of the Northeast, Bulletin 659, The Connecticut Agricultural Experiment Station, New Haven, Conn., Sept. 1963, 28 pp.

Chester County Planning Commission, "Chester County Natural Environment and Planning," West Chester, Pa., July 1963.

Chestnutwood, Charles M., "The Geographical Bases of Pennsylvania's Tourist Industry," Master's thesis, Pennsylvania State University, University Park, Pa., 1954.

Cornell University Agricultural Station, "Precipitation Amounts in the Northeast Region of the U.S.," The Climate of the Northeast, Agronomy Mimeo 61-4, Ithaca, N. Y., Nov. 1961, 302 pp.

Dailey, Paul W., Jr., "Tornadoes in Pennsylvania," Information Report No. 63, Institute for Research on Land and Water Resources, Pennsylvania State University, University Park, Pa., June 1970.

Dethier, B. E. and Vittum, M. T., "Growing Degree Days," The Climate of the Northeast, Bulletin 801, New York State Agricultural Experiment Station, Geneva, N. Y., Aug. 1963.

Dickerson, W. H., "Heating Degree Days," The Climate of the Northeast, Bulletin 483T, West Virginia University Agricultural Experiment Station, Morgantown, W. Va., June 1963.

Fuller, Theodore E., "A Geographical Analysis of the Agricultural Regions of Adams County, Pennsylvania," Master's thesis, Pennsylvania State University, University Park, Pa., 1956.

Havens, A. V. and McGuire, J. K., "Spring and Fall Low-Temperature Probabilities," The Climate of the Northeast, Bulletin 801, New Jersey Agricultural Experiment Station, Rutgers University, New Brunswick, N. J., June 1961, 32 pp.

Heppell, Roger C., "Agricultural Geography of the Cigar Tobacco Industry of the Lancaster, Pennsylvania Region," Master's thesis, Pennsylvania State University, University Park, Pa., 1953.

Kauffman, Nelson M., "Climates of the States--Pennsylvania," Climatography of the United States No. 60-36, U.S. Weather Bureau, U.S. Government Printing Office, Washington, D. C., 1960, 19 pp.

_____, and Butler, R. G., "In Pennsylvania--Late Spring and Early Fall Freezes," Pennsylvania Crop Reporting Service, Harrisburg, Pa., Aug. 1961, 11 pp.

Murphy, Raymond E. and Marion, Pennsylvania--A Regional Geography, The Pennsylvania Book Service, Harrisburg, Pa., 1937.

NOAA, Environmental Data Service, Climatological Data--National Summary, Silver Spring, Md., monthly plus annual summary.

_____, Environmental Data Service, "Climatological Data--Pennsylvania," Silver Spring, Md., monthly plus annual summary.

_____, Environmental Data Service, Hourly Precipitation Data, Silver Spring, Md., monthly plus annual summary.

_____, "Local Climatological Data" for Allentown, Erie, Harrisburg, Philadelphia, Pittsburgh, Reading, Scranton, Shippingport, and Williamsport, Pennsylvania, Silver Spring, Md., monthly plus annual summary.

_____, "Substation Climatological Summary," Climatography of the United States, No. 20-36, in cooperation with the Pennsylvania Department of Commerce for Altoona, Bethlehem, Brookville, Chambersburg, Emporium, Franklin, Freeland, George School, Johnstown, Lancaster, Lawrenceville, Lebanon, Madera, Montrose, Mount Pocono, New Castle, Selinsgrove-Sunbury, Slippery Rock, Somerset, State College, Towanda, Uniontown, Warren, and York, Penna.;Silver Spring, Md., irregular.

BIBLIOGRAPHY

Pennsylvania Crop Reporting Service, "Pennsylvania Crop and Livestock Annual Summary," Harrisburg, Pa., annual, 1959-

Pennsylvania Writers Project, Pennsylvania--A Guide to the Keystone State, Oxford University Press, New York, N. Y., 1940.

Sharpe, William E.; Lee, Richard; and Jones, E. Bruce, "Summary of Temperature Means of State College, Pennsylvania, for 1931-1967," Information Report No. 59, Institute for Research on Land and Water Resources, Pennsylvania State University, University Park, Pa., October 1968.

Tukey, L. D.; Kauffman, N. M.; and Weiser, E. V., Jr., "Regional Weather Summary of Pennsylvania," Progress Report 254, Climatic Series 1, Southeastern Area, Pennsylvania State University Agricultural Experiment Station, University Park, Pa., Jan. 1965, 39 pp.

_____, "Regional Weather Summary of Pennsylvania," Progress Report 260, Climatic Series 2, Lower Susquehanna Area, Pennsylvania State University Agricultural Experiment Station, University Park, Pa., July 1965, 40 pp.

_____, "Regional Weather Summary of Pennsylvania," Progress Report 266, Climatic Series 3, Northeastern Area, Pennsylvania State University Agricultural Experiment Station, University Park, Pa., June 1966, 32 pp.

U.S. Department of Agriculture, Soil Survey, Soil Conservation Service, county summaries for Adams (May 1967), Carbon (Nov. 1962), Chester and Delaware (May 1963), Clinton (Aug. 1966), Columbia (Mar. 1967), Erie (Dec. 1960), Fulton (Nov. 1969), Indiana (Jan. 1968), Jefferson (Aug. 1964), Lancaster (Oct. 1959), Lehigh (Nov. 1963), Montgomery (Apr. 1967), Pike (June 1969), Potter (July 1958), Westmoreland (Nov. 1968), and York (May 1963), Pennsylvania. Washington, D. C.

_____, Yearbook of Agriculture, 1941: Climate and Man, U.S. Government Printing Office, Washington, D. C., 1941, 1248 pp.

U.S. Weather Bureau, "Maximum Station Precipitation for 1, 2, 3, 6, 12, and 24 hours, Part XVI, Pennsylvania," Technical Paper No. 15, Washington, D. C., 1956, 146 pp.

U.S. Weather Bureau, "Maximum 24-Hour Precipitation in the United States," its Technical Paper No. 16, Washington, D. C., 1952, 284 pp.

_____, Climatic Summary of the United States (Bulletin W), Sections 87, 88, and 89, Washington, D. C., 1930, 69 pp.

_____, Climatic Summary of the United States, Pennsylvania, (Bulletin W Supplement), 1931-1952, Washington, D. C., 1953, 77 pp.

_____, Climatic Summary of the United States, Pennsylvania, (Bulletin W Supplement), 1951-1960, Washington, D. C., 1961, 95 pp.

_____, "Rainfall Intensity-Duration-Frequency Curves for Selected Stations in the United States, Alaska, Hawaiian Islands, and Puerto Rico," its Technical Paper No. 25, Washington, D. C., 1955, 53 pp.

_____, "Rainfall Intensity-Frequency Regime, Part 3--the Middle Atlantic Region," its Technical Paper No. 29, Washington, D. C., 1958, 38 pp.

FREEZE DATA

STATION	Freeze threshold temperature	Mean date of last Spring occurrence	Mean date of first Fall occurrence	Mean No. of days between dates	Years of record Spring	No. of occurrences in Spring	Years of record Fall	No. of occurrences in Fall
ALTOONA HORSESHOE CV	32	05-06	10-08	155	30	30	30	30
	28	04-21	10-21	183	30	30	30	30
	24	04-12	11-05	207	30	30	30	30
	20	03-31	11-18	232	30	30	30	30
	16	03-16	12-01	259	30	30	29	29
BETHLEHEM LEHI UNIV	32	04-14	10-23	192	28	28	29	29
	28	04-01	11-07	221	28	28	28	28
	24	03-21	11-23	247	28	28	28	28
	20	03-10	12-03	268	28	28	28	28
	16	03-06	12-11	280	28	28	28	25
BROOKVILLE AP	32	05-27	09-20	116	27	27	30	30
	28	05-14	10-03	143	27	27	30	30
	24	04-29	10-18	172	27	27	29	29
	20	04-15	11-05	203	27	27	29	29
	16	03-29	11-17	234	28	28	29	29
BUTLER	32	05-21	10-03	135	23	23	24	24
	28	04-29	10-15	170	23	23	24	24
	24	04-15	11-01	200	23	23	24	24
	20	04-08	11-12	218	23	23	24	24
	16	03-17	11-28	256	23	23	24	24
CARLISLE	32	04-27	10-11	167	30	30	30	30
	28	04-17	10-24	190	30	30	30	30
	24	03-30	11-09	224	30	30	30	30
	20	03-16	11-26	255	30	30	30	30
	16	03-08	12-06	273	30	30	30	29
CHAMBERSBURG	32	05-01	10-12	164	30	30	30	30
	28	04-15	10-26	195	30	30	30	30
	24	03-28	11-14	231	30	30	30	30
	20	03-16	11-28	257	30	30	30	30
	16	03-08	12-09	276	30	30	30	26
CLAYSVILLE 3 W	32	05-15	09-30	138	30	30	29	29
	28	04-30	10-13	166	30	30	29	29
	24	04-17	10-28	194	30	30	29	29
	20	04-03	11-14	225	30	30	29	29
	16	03-20	11-25	250	30	30	29	29
COATESVILLE 1 SW	32	04-29	10-16	170	30	30	30	30
	28	04-12	10-28	200	30	30	30	30
	24	03-28	11-07	224	30	30	30	30
	20	03-18	11-24	250	30	30	30	30
	16	03-09	12-07	273	30	30	30	29
CORRY	32	05-24	09-23	123	28	28	28	28
	28	05-09	10-11	155	28	28	28	28
	24	04-23	10-27	188	28	28	28	28
	20	04-10	11-13	217	28	28	28	28
	16	03-29	11-27	244	28	28	28	27
COUDERSPORT	32	05-29	09-12	107	18	18	18	18
	28	05-17	09-30	136	17	17	17	17
	24	05-02	10-10	161	17	17	17	17
	20	04-15	10-22	190	17	17	17	17
	16	04-01	11-11	224	16	16	18	18
DERRY	32	05-08	10-08	153	30	30	30	30
	28	04-21	10-20	183	30	30	29	29
	24	04-12	11-05	207	29	29	30	30
	20	03-27	11-20	238	29	29	30	30
	16	03-16	12-02	262	29	29	30	30
EBENSBURG	32	05-15	10-01	139	29	29	29	29
	28	05-01	10-16	168	29	29	28	28
	24	04-19	10-25	189	30	30	29	29
	20	04-04	11-10	220	30	30	30	30
	16	03-26	11-24	243	30	30	30	30
EMPORIUM 1 E	32	05-19	09-29	133	29	29	30	30
	28	05-08	10-16	160	29	29	29	29
	24	04-24	10-24	183	29	29	29	29
	20	04-12	11-08	209	29	29	29	29
	16	03-26	11-19	238	29	29	29	29
ERIE WSO	32	04-20	11-07	200	30	30	30	30
	28	04-06	11-14	223	30	30	30	30
	24	03-28	11-30	247	30	30	30	30
	20	03-16	12-07	267	30	30	28	28
	16	03-10	12-10	276	30	30	27	27
FRANKLIN	32	05-14	10-05	144	30	30	30	30
	28	04-27	10-22	178	30	30	30	30
	24	04-16	11-06	204	30	30	30	30
	20	04-06	11-18	226	30	30	30	30
	16	03-18	11-30	257	30	30	30	30
FREELAND	32	05-07	10-09	155	30	30	30	30
	28	04-25	10-23	181	30	30	30	30
	24	04-15	11-03	201	30	30	30	30
	20	04-06	11-16	224	30	30	30	30
	16	03-26	11-25	244	30	30	30	30
GEORGE SCHOOL	32	04-25	10-16	174	30	30	30	30
	28	04-12	10-31	202	30	30	30	30
	24	03-25	11-11	231	30	30	30	30
	20	03-15	11-27	258	30	30	30	30
	16	03-05	12-07	278	30	30	30	30
GETTYSBURG	32	04-23	10-19	179	30	30	30	30
	28	04-08	11-03	209	30	30	30	30
	24	03-21	11-18	241	30	30	30	30
	20	03-13	12-01	263	30	30	30	30
	16	03-04	12-11	282	30	30	30	28
GORDON	32	05-17	09-27	133	30	30	30	30
	28	05-04	10-09	158	30	30	30	30
	24	04-20	10-25	187	30	30	30	30
	20	04-04	11-06	217	30	30	30	30
	16	03-20	11-20	245	30	30	30	30
GREENSBURG 2 S	32	05-07	10-06	152	30	30	30	30
	28	04-22	10-19	180	30	30	30	30
	24	04-10	11-05	208	30	30	30	30
	20	03-29	11-21	237	30	30	30	30
	16	03-14	11-29	260	30	30	30	30
GREENVILLE	32	05-22	09-30	131	29	29	30	30
	28	05-03	10-12	162	30	30	30	30
	24	04-21	10-31	193	29	29	30	30
	20	04-07	11-13	220	29	29	30	30
	16	03-23	12-01	253	29	29	28	26
HANOVER	32	04-20	10-21	184	30	30	30	30
	28	04-07	11-04	212	30	30	30	30
	24	03-22	11-21	244	30	30	30	30
	20	03-13	12-03	265	30	30	29	29
	16	03-06	12-11	281	29	29	29	26
HAWLEY	32	05-19	09-26	130	30	30	31	31
	28	05-05	10-08	155	30	30	31	31
	24	04-21	10-21	183	30	30	30	30
	20	04-09	11-01	206	30	30	30	30
	16	03-25	11-18	238	30	30	30	30
HAWLEY 1 S DAM	32	05-18	09-18	123	14	14	15	15
	28	05-11	10-01	143	11	11	15	15
	24	04-21	10-15	177	6	6	12	12
	20	04-07	10-21	196	5	5	10	10
	16	03-30	11-06	221	5	5	7	7
HOLTWOOD	32	04-05	11-08	217	30	30	30	30
	28	03-23	11-22	243	30	30	30	30
	24	03-11	12-03	267	30	30	29	29
	20	03-06	12-10	280	30	30	29	29
	16	02-24	12-19	299	30	30	30	22
HUNTINGDON	32	05-14	10-05	145	30	30	30	30
	28	04-28	10-18	172	30	30	30	30
	24	04-15	10-28	197	30	30	30	30
	20	03-29	11-13	229	30	30	30	30
	16	03-14	11-25	255	30	30	30	30
INDIANA	32	05-14	10-07	145	25	25	24	24
	28	04-30	10-11	164	25	25	24	24
	24	04-17	10-30	196	25	25	25	25
	20	04-06	11-13	222	24	24	24	24
	16	03-20	11-28	253	24	24	22	22
IRWIN	32	05-11	10-07	149	28	28	29	29
	28	05-01	10-16	168	28	28	29	29
	24	04-17	10-30	195	28	28	27	27
	20	03-29	11-15	231	27	27	27	27
	16	03-14	11-29	261	27	27	26	26

FREEZE DATA

STATION	Freeze threshold temperature	Mean date of last Spring occurrence	Mean date of first Fall occurrence	Mean No. of days between dates	Years of record Spring	No. of occurrences in Spring	Years of record Fall	No. of occurrences in Fall
JOHNSTOWN 1	32	04-28	10-14	169	29	29	30	30
	28	04-20	10-27	190	29	29	30	30
	24	03-30	11-13	228	28	28	30	30
	20	03-21	11-26	250	28	28	30	30
	16	03-10	12-06	271	28	28	30	30
KANE 1 NNE	32	06-02	09-13	104	18	18	19	19
	28	05-16	10-01	138	18	18	19	19
	24	05-01	10-11	163	18	18	19	19
	20	04-22	10-19	180	18	18	19	19
	16	04-11	11-09	212	18	18	19	19
LANCASTER 2 NE PUMP	32	05-04	10-09	158	30	30	30	30
	28	04-24	10-22	180	29	29	30	30
	24	04-07	11-05	212	29	29	30	30
	20	03-21	11-19	242	29	29	30	30
	16	03-08	12-07	274	30	30	30	29
LAWRENCEVILLE 1 S	32	05-22	09-28	129	30	30	30	30
	28	05-05	10-11	159	29	29	30	30
	24	04-20	10-22	186	29	29	30	30
	20	04-06	11-05	214	30	30	30	30
	16	03-20	11-23	247	30	30	30	30
LEBANON	32	04-22	10-17	179	30	30	30	30
	28	04-08	10-30	205	30	30	30	30
	24	03-22	11-16	239	30	30	30	30
	20	03-12	12-02	265	30	30	30	29
	16	03-06	12-10	279	30	30	30	28
LEWISTOWN	32	04-25	10-16	174	11	11	12	12
	28	04-12	10-21	192	11	11	12	12
	24	04-04	11-09	220	9	9	12	12
	20	03-22	11-29	252	9	9	12	12
	16	03-12	12-11	274	9	9	12	11
LOCK HAVEN	32	05-04	10-12	162	25	25	27	27
	28	04-18	10-24	189	25	25	27	27
	24	04-03	11-08	219	25	25	27	27
	20	03-20	11-21	246	25	25	27	27
	16	03-12	12-02	265	25	25	27	27
MAUCH CHUNK 1 SW	32	05-09	10-01	145	28	28	28	28
	28	04-25	10-16	174	29	29	29	29
	24	04-07	10-28	204	29	29	29	29
	20	03-27	11-12	230	28	28	29	29
	16	03-15	11-28	258	28	28	29	29
MEADVULLE 1 S	32	05-18	10-04	139	22	22	23	23
	28	04-30	10-21	174	22	22	23	23
	24	04-17	11-07	204	22	22	23	23
	20	04-04	11-20	230	22	22	23	23
	16	03-23	12-02	254	22	22	23	22
MERCER AP	32	05-19	10-08	142	11	11	11	11
	28	05-05	10-17	165	11	11	11	11
	24	04-22	11-08	200	11	11	11	11
	20	04-03	11-18	230	11	11	11	11
	16	03-23	11-28	250	11	11	11	11
MIDDLETOWN OLMSTED F	32	04-24	10-24	184	10	10	10	10
	28	04-01	11-08	221	10	10	10	10
	24	03-22	11-22	245	10	10	10	10
	20	03-12	12-01	264	10	10	10	10
	16	03-05	12-14	284	10	10	10	10
MIDLAND DAM LOWER 7	32	05-02	10-19	170	13	13	13	13
	28	04-13	11-05	205	13	13	13	13
	24	03-27	11-21	239	13	13	13	13
	20	03-16	11-28	256	13	13	13	13
	16	03-10	12-10	275	13	13	13	12
MONTROSE 3 E	32	05-15	10-02	139	28	28	27	27
	28	04-30	10-17	170	28	28	26	26
	24	04-18	10-30	195	28	28	27	27
	20	04-09	11-11	216	28	28	28	28
	16	03-29	11-22	238	27	27	28	28
MT POCONO	32	05-17	09-30	136	24	24	27	27
	28	05-05	10-14	162	23	23	27	27
	24	04-15	10-28	196	23	23	24	24
	20	04-04	11-06	216	24	24	24	24
	16	03-24	11-15	237	23	23	23	23

STATION	Freeze threshold temperature	Mean date of last Spring occurrence	Mean date of first Fall occurrence	Mean No. of days between dates	Years of record Spring	No. of occurrences in Spring	Years of record Fall	No. of occurrences in Fall
NEW CASTLE 1 N	32	05-15	10-05	143	30	30	30	30
	28	04-29	10-21	175	30	30	30	30
	24	04-15	11-05	204	30	30	30	30
	20	03-28	11-21	238	30	30	30	30
	16	03-14	12-03	264	30	30	30	29
PALMERTON	32	05-04	10-09	159	30	30	30	30
	28	04-17	10-19	186	30	30	30	30
	24	04-01	11-07	221	30	30	30	30
	20	03-18	11-21	248	30	30	30	30
	16	03-09	12-02	268	30	30	30	29
PHILADELPHIA SHAWMNT	32	04-18	10-26	192	30	30	30	30
	28	03-31	11-09	223	30	30	30	30
	24	03-18	11-23	250	30	30	30	30
	20	03-09	12-06	272	30	30	30	29
	16	03-01	12-12	287	30	30	30	28
PHILADELPHIA CITY	32	03-30	11-17	232	30	30	30	30
	28	03-19	11-29	255	30	30	30	30
	24	03-11	12-06	270	30	30	30	29
	20	03-04	12-13	284	30	30	30	26
	16	02-17	12-20	306	30	27	30	20
PHOENIXVILLE 1 E	32	05-03	10-11	161	30	30	30	30
	28	04-16	10-23	190	30	30	28	28
	24	03-30	11-06	221	30	30	28	28
	20	03-17	11-24	252	30	30	28	28
	16	03-08	12-04	271	30	30	27	26
PITTSBURGH ALLEG CO AP	32	04-20	10-23	187	15	15	16	16
	28	04-07	11-10	217	15	15	16	16
	24	03-28	11-21	238	15	15	16	16
	20	03-21	11-29	254	15	15	16	16
	16	03-14	12-06	267	15	15	16	16
PITTSBURGH WSO	32	04-16	11-03	200	30	30	30	30
	28	03-31	11-16	230	30	30	30	30
	24	03-21	11-26	250	30	30	30	30
	20	03-13	12-07	270	30	30	30	27
	16	03-03	12-14	286	30	30	30	26
PORT CLINTON 1 S	32	05-06	10-10	158	28	28	29	29
	28	04-19	10-22	186	28	28	28	28
	24	04-04	11-05	217	29	29	28	28
	20	03-17	11-19	247	29	29	27	27
	16	03-08	12-02	269	29	29	27	27
QUAKERTOWN	32	05-08	10-04	149	30	30	30	30
	28	04-21	10-18	180	30	30	30	30
	24	04-04	10-27	207	30	30	30	30
	20	03-18	11-16	243	30	30	29	29
	16	03-13	12-02	264	30	30	29	28
READING 3 N	32	04-13	10-29	198	30	30	30	30
	28	03-27	11-17	235	30	30	30	30
	24	03-16	11-30	260	30	30	30	30
	20	03-09	12-08	274	30	30	29	28
	16	03-04	12-15	286	30	30	27	25
RETREAT 1 SW	32	05-07	10-09	155	16	16	18	18
	28	04-22	10-20	180	16	16	18	18
	24	04-04	11-03	214	16	16	18	18
	20	03-28	11-15	231	17	17	18	18
	16	03-17	11-30	258	17	17	18	18
RIDGWAY	32	05-28	09-24	119	29	29	28	28
	28	05-14	10-06	145	28	28	28	28
	24	04-30	10-19	172	28	28	28	28
	20	04-17	10-30	196	28	28	28	28
	16	04-03	11-18	229	26	26	28	28
SCRANTON	32	04-24	10-14	174	30	30	30	30
	28	04-11	10-31	203	30	30	30	30
	24	03-26	11-19	237	30	30	30	30
	20	03-17	11-29	257	30	30	30	30
	16	03-10	12-06	271	30	30	30	29
SELINSGROVE FAA AP	32	05-04	10-09	158	29	29	30	30
	28	04-20	10-22	185	29	29	30	30
	24	04-03	11-07	218	28	28	30	30
	20	03-21	11-19	244	28	28	29	29
	16	03-12	12-02	264	28	28	29	29

FREEZE DATA

STATION	Freeze threshold temperature	Mean date of last Spring occurrence	Mean date of first Fall occurrence	Mean No. of days between dates	Years of record Spring	No. of occurrences in Spring	Years of record Fall	No. of occurrences in Fall	STATION	Freeze threshold temperature	Mean date of last Spring occurrence	Mean date of first Fall occurrence	Mean No. of days between dates	Years of record Spring	No. of occurrences in Spring	Years of record Fall	No. of occurrences in Fall
SOMERSET WATER WORKS	32	05-24	09-21	120	29	29	28	28	WARREN	32	05-14	10-06	146	30	30	30	30
	28	05-14	10-04	143	28	28	28	28		28	04-30	10-22	175	30	30	30	30
	24	04-30	10-15	168	28	28	28	28		24	04-15	11-09	208	30	30	30	30
	20	04-15	10-27	195	28	28	28	28		20	04-04	11-19	229	30	30	30	30
	16	03-27	11-12	230	28	28	28	28		16	03-21	11-28	252	30	30	30	30
SPRINGS 1 SW	32	05-25	09-29	126	25	25	27	27	WELLSBORO 3 S	32	05-27	09-22	118	30	30	29	29
	28	05-06	10-09	156	25	25	27	27		28	05-12	10-05	146	30	30	30	30
	24	04-23	10-22	183	26	26	27	27		24	04-24	10-16	176	30	30	30	30
	20	04-13	11-04	206	26	26	27	27		20	04-14	10-27	196	30	30	30	30
	16	03-29	11-18	234	26	26	27	27		16	03-27	11-14	232	30	30	29	29
STATE COLLEGE	32	04-29	10-12	166	30	30	30	30	WEST CHESTER	32	04-18	10-25	189	30	30	29	29
	28	04-17	10-21	187	30	30	30	30		28	04-04	11-05	215	30	30	29	29
	24	04-05	11-10	219	30	30	30	30		24	03-21	11-23	247	30	30	29	29
	20	03-23	11-19	241	30	30	30	30		20	03-12	12-02	265	30	30	29	29
	16	03-12	12-05	268	30	30	30	30		16	03-05	12-12	282	30	30	29	26
TOWANDA	32	05-12	10-02	143	30	30	30	30	WILLIAMSPORT WSO	32	05-03	10-13	164	30	30	30	30
	28	04-29	10-15	169	30	30	30	30		28	04-17	10-25	191	30	30	30	30
	24	04-14	10-25	194	30	30	30	30		24	03-30	11-09	224	30	30	30	30
	20	03-30	11-13	228	30	30	30	30		20	03-19	11-24	250	30	30	30	30
	16	03-17	11-27	254	30	30	30	30		16	03-10	12-05	271	30	30	30	29
UNIONTOWN	32	04-30	10-15	167	30	30	30	30	YORK 3 SW PUMP STA	32	05-02	10-09	160	30	30	30	30
	28	04-18	10-27	193	30	30	30	30		28	04-20	10-23	186	30	30	30	30
	24	04-04	11-12	222	30	30	30	30		24	04-02	11-08	220	30	30	30	30
	20	03-19	11-26	252	30	30	30	30		20	03-17	11-21	249	30	30	30	30
	16	03-11	12-04	268	30	30	30	29		16	03-08	12-07	274	30	30	30	30

Data in the above table are based on the period 1921-1950, or that portion of this period for which data are available.

Means have been adjusted to take into account years of non-occurrence.

A freeze is a numerical substitute for the former term "killing frost" and is the occurrence of a minimum temperature at or below the threshold temperature of 32°, 28°, etc.

Freeze data tabulations in greater detail are available and can be reproduced at cost.

*NORMALS BY CLIMATOLOGICAL DIVISIONS

Taken from "Climatography of the United States No. 81-4, Decennial Census of U. S. Climate"

TEMPERATURE (°F)   **PRECIPITATION (In.)**

STATIONS (By Divisions)	JAN	FEB	MAR	APR	MAY	JUNE	JULY	AUG	SEPT	OCT	NOV	DEC	ANN	JAN	FEB	MAR	APR	MAY	JUNE	JULY	AUG	SEPT	OCT	NOV	DEC	ANN
POCONO MOUNTAINS																										
FREELAND	25.4	25.8	33.4	45.8	57.7	63.9	70.3	68.4	61.6	51.1	38.8	27.7	47.5	3.09	2.72	3.82	4.10	4.84	4.20	5.38	4.27	3.95	3.68	4.15	3.77	47.97
GOULDSBORO	•	•	•	•	•	•	•	•	•	•	•	•	•	3.10	2.89	3.84	4.09	4.60	3.89	4.86	4.41	3.81	3.83	4.10	3.48	46.90
HAWLEY	•	•	•	•	•	•	•	•	•	•	•	•	•	2.79	2.62	3.42	3.58	4.03	3.47	4.35	4.28	3.70	3.38	3.33	3.02	41.97
HOLLISTERVILLE	•	•	•	•	•	•	•	•	•	•	•	•	•	2.93	2.85	3.66	3.97	4.08	3.93	4.81	4.00	3.50	3.49	3.56	3.20	43.98
LAKEVILLE 2 NNE	•	•	•	•	•	•	•	•	•	•	•	•	•	2.72	2.76	3.59	3.81	4.12	3.84	4.91	4.35	3.73	3.69	3.56	3.12	44.20
MATAMORAS	•	•	•	•	•	•	•	•	•	•	•	•	•	3.02	2.65	3.45	3.92	3.96	4.29	4.24	3.99	3.42	3.65	3.29		43.79
PAUPACK 2 WNW	•	•	•	•	•	•	•	•	•	•	•	•	•	2.89	2.77	3.61	3.83	4.07	3.67	4.62	4.38	3.64	3.60	3.60	3.20	43.88
PLEASANT MOUNT 1 W	•	•	•	•	•	•	•	•	•	•	•	•	•	3.34	2.95	3.83	4.09	4.32	4.06	4.63	4.12	4.18	3.82	3.96	3.50	46.80
SCRANTON	28.4	28.6	36.5	48.5	59.6	68.3	70.5	70.7	63.4	53.0	41.8	30.8	50.0	2.24	2.12	2.88	3.43	3.62	3.73	4.81	3.86	3.03	2.98	3.01	2.44	38.15
STROUDSBURG	28.0	28.8	36.6	48.7	59.3	68.0	72.7	70.5	63.0	52.5	41.4	30.6	50.0	3.36	2.86	4.12	4.15	4.01	4.21	4.70	4.57	4.29	3.65	4.15	3.89	47.96
TOBYHANNA	•	•	•	•	•	•	•	•	•	•	•	•	•	3.60	3.09	4.46	4.34	4.52	4.33	5.23	5.25	4.40	4.32	4.30	3.97	51.81
WILKES-BARRE 4 NE	•	•	•	•	•	•	•	•	•	•	•	•	•	2.31	2.27	2.89	3.38	3.73	3.66	4.64	4.03	3.36	3.28	3.07	2.75	39.37
W-BARRE-SCRANTON WSO	27.7	28.3	36.2	48.4	59.6	68.2	72.4	70.0	62.5	51.0	39.6	29.4	49.4	2.29	1.99	2.82	3.27	3.95	3.91	4.79	3.58	2.97	3.50	2.94	2.47	38.48
DIVISION	25.9	26.2	34.0	46.3	57.3	65.6	70.0	68.1	60.9	50.8	39.4	28.3	47.7	2.93	2.61	3.67	3.90	4.16	3.97	4.81	4.31	3.81	3.61	3.70	3.26	44.74
EAST CENTRAL MOUNTAIN																										
ALLENTOWN WSO	29.0	29.2	37.6	49.3	60.4	69.4	74.1	72.0	64.7	53.8	41.9	31.3	51.1	3.17	2.64	3.79	3.76	4.08	4.07	4.82	4.47	3.75	2.97	3.33	3.27	44.12
BETHLEHEM LEHIGH UNIV	31.1	32.1	39.8	51.5	62.2	71.1	75.5	73.5	66.8	56.2	44.5	33.5	53.2	3.11	2.59	3.60	3.63	3.85	3.72	4.66	4.44	3.56	2.99	3.36	3.25	42.76
PALMERTON	29.2	29.6	37.4	49.0	59.7	68.4	73.1	70.9	63.5	52.9	41.6	31.3	50.6	2.96	2.49	3.56	3.49	3.88	3.88	5.10	4.75	3.99	3.24	3.49	3.24	44.07
PORT CLINTON	29.6	30.0	37.5	49.0	59.7	68.2	72.7	70.6	63.4	52.8	41.8	31.4	50.6	3.29	2.81	4.20	4.11	4.51	3.66	4.74	4.45	4.32	3.57	3.75	3.58	46.99
TAMAQUA 4 N DAM	•	•	•	•	•	•	•	•	•	•	•	•	•	3.32	2.89	4.17	4.26	4.74	4.12	5.03	4.94	4.19	3.96	4.18	3.71	49.51
DIVISION	29.2	29.7	37.3	49.1	59.9	68.5	73.0	70.9	63.9	53.3	41.9	31.2	50.7	3.16	2.67	3.87	3.85	4.32	3.88	5.01	4.54	3.85	3.34	3.67	3.44	45.60
SOUTHEASTERN PIEDMONT																										
COATESVILLE 1 SW	31.4	31.6	39.2	50.7	61.3	70.4	74.9	72.8	65.7	54.7	43.3	32.9	52.4	3.30	2.78	4.11	3.49	4.17	4.42	4.26	5.06	3.49	3.19	3.55	3.39	45.21
CONSHOHOCKEN	•	•	•	•	•	•	•	•	•	•	•	•	•	3.25	2.86	3.99	3.52	3.97	3.77	4.45	4.62	3.68	3.12	3.61	3.20	44.04
DOYLESTOWN	•	•	•	•	•	•	•	•	•	•	•	•	•	3.09	2.54	3.60	3.36	4.10	3.81	4.69	4.70	3.57	3.12	3.48	3.15	43.21
EPHRATA	32.1	32.9	40.7	52.0	62.8	71.4	75.8	73.7	66.6	55.6	44.3	33.8	53.5	2.98	2.52	3.65	3.41	3.83	4.20	4.74	4.71	3.76	3.20	3.30	3.07	43.37
GEORGE SCHOOL	31.8	32.4	39.8	51.0	61.5	70.1	74.8	72.9	66.2	55.6	44.4	33.5	52.8	3.34	2.70	4.07	3.64	3.95	3.75	4.94	4.74	3.66	3.18	3.76	3.35	45.08
GRATERFORD	•	•	•	•	•	•	•	•	•	•	•	•	•	3.25	2.71	3.99	3.49	4.26	3.84	4.65	4.43	3.47	3.18	3.55	3.28	44.30
HOLTWOOD	32.8	33.2	40.7	52.1	63.4	72.5	77.5	75.7	68.8	57.4	45.3	34.7	54.5	2.63	2.31	3.34	3.09	3.25	3.50	3.79	4.21	3.21	2.66	2.94	2.73	37.66
LANCASTER 2 NE PMP STA	31.6	32.5	40.6	51.6	62.2	70.3	74.4	72.4	65.4	54.0	43.1	32.8	52.6	2.89	2.48	3.81	3.60	3.90	3.90	4.89	4.94	3.51	3.22	3.16	2.99	43.29
LEBANON 3 W	30.3	30.9	38.7	50.4	61.5	70.0	74.5	72.9	65.5	54.5	42.6	31.9	52.1	3.10	2.59	3.75	3.65	4.23	3.93	4.39	4.27	3.89	3.55	3.36	3.28	43.99
MARCUS HOOK	35.8	36.5	43.9	54.6	65.6	74.6	79.0	77.1	70.2	59.8	47.9	38.0	56.9	3.36	2.82	3.98	3.64	3.73	3.54	3.88	5.06	3.22	2.87	3.51	3.12	42.73
NESHAMINY FALLS	•	•	•	•	•	•	•	•	•	•	•	•	•	3.26	2.83	4.27	3.40	4.71	4.98	3.56	5.33	3.46	3.33	3.66	3.17	45.34
PHILADELPHIA WSO	32.3	33.2	41.0	52.0	62.6	71.0	75.6	73.6	66.7	55.7	44.3	33.9	53.5	3.32	2.80	3.80	3.40	3.74	4.05	4.16	4.63	3.46	2.78	3.40	2.94	42.48
PHOENIXVILLE 1 E	32.9	33.5	41.5	52.8	63.2	71.7	76.3	74.1	67.3	56.3	45.1	34.4	54.1	3.44	2.82	4.22	3.50	4.17	3.82	4.47	4.98	3.55	3.21	3.79	3.40	45.37
QUAKERTOWN	30.0	30.6	38.6	50.0	60.4	68.7	73.4	70.6	63.8	53.2	42.5	31.9	51.1	3.38	2.71	4.00	3.89	4.16	4.19	4.63	4.85	3.74	3.47	3.71	3.45	46.18
READING 3 N	32.7	33.4	41.3	52.6	63.3	72.2	76.9	74.8	67.5	57.2	45.1	34.7	54.3	3.07	2.64	3.78	3.42	3.79	3.72	4.26	4.05	3.32	2.84	3.40	3.14	41.43
WEST CHESTER 1 W	33.0	33.5	41.0	52.1	62.5	71.0	75.5	73.5	66.9	56.4	45.2	34.7	53.8	3.43	2.90	4.18	3.43	4.17	3.85	4.73	5.13	3.80	3.16	3.88	3.37	46.03
WEST GROVE 1 SE	•	•	•	•	•	•	•	•	•	•	•	•	•	3.39	2.75	4.24	3.56	4.21	3.69	4.59	5.27	3.46	3.00	3.52	3.35	45.03
YORK HAVEN	•	•	•	•	•	•	•	•	•	•	•	•	•	2.72	2.36	3.47	3.47	4.06	3.68	3.84	3.94	3.04	3.24	3.16	2.91	39.89
DIVISION	32.4	33.0	40.6	51.8	62.6	71.1	75.7	73.7	66.8	56.1	44.7	34.2	53.6	3.21	2.67	3.88	3.51	3.99	3.90	4.40	4.66	3.52	3.16	3.48	3.15	43.53
LOWER SUSQUEHANNA																										
ARENDTSVILLE	30.7	31.3	39.0	50.8	61.5	70.2	74.6	72.8	65.5	54.4	42.5	32.1	52.1	2.96	2.49	3.91	3.74	4.29	3.94	3.85	4.34	3.51	3.57	3.30	3.15	43.05
BLOSERVILLE 1 N	•	•	•	•	•	•	•	•	•	•	•	•	•	2.87	2.26	3.68	3.87	4.48	4.08	4.27	4.85	3.46	3.52	3.52	2.98	43.84
CARLISLE	31.6	32.7	40.9	52.6	63.3	71.7	75.8	73.7	66.5	55.4	43.5	33.0	53.4	3.07	2.52	3.78	3.73	4.13	3.87	3.96	4.05	3.11	3.31	3.26	3.13	41.92
CHAMBERSBURG 1 ESE	31.6	32.6	40.1	51.4	62.1	70.8	74.9	73.0	65.9	54.7	43.0	33.0	52.8	3.02	2.32	3.77	3.46	4.10	4.08	3.95	4.08	3.30	3.15	3.12	2.98	41.33
GETTYSBURG	32.7	33.6	41.2	52.6	63.0	71.5	75.7	73.8	66.9	56.1	44.6	34.2	53.8	2.92	2.48	3.84	3.51	4.07	3.52	4.15	4.22	3.32	3.31	3.26	2.95	41.55
HANOVER	32.8	33.4	40.8	52.3	63.0	71.5	75.8	73.8	66.9	56.1	44.8	34.3	53.8	3.01	2.61	3.82	3.52	4.03	3.76	4.09	4.14	3.08	3.27	3.07	3.10	41.50
HARRISBURG WSO	31.3	32.6	40.3	51.8	62.7	71.3	76.2	74.1	66.9	55.7	43.4	33.0	53.3	2.76	2.31	3.43	3.02	3.90	3.42	3.51	3.65	2.82	2.97	2.95	2.91	37.65
MERCERSBURG	•	•	•	•	•	•	•	•	•	•	•	•	•	2.73	2.35	3.63	3.35	4.27	4.00	3.57	3.56	3.11	3.33	2.98	2.87	39.75
NEW PARK	•	•	•	•	•	•	•	•	•	•	•	•	•	3.15	2.60	3.94	3.77	4.39	4.10	4.35	5.22	3.64	3.46	3.46	3.27	45.35
SPRING GROVE	•	•	•	•	•	•	•	•	•	•	•	•	•	2.97	2.55	3.76	3.53	4.09	3.71	3.88	4.56	3.24	3.30	3.08	3.08	41.75
YORK 3 SSW PUMP STA	32.8	33.7	41.5	52.6	63.3	71.5	75.7	73.9	67.1	55.8	44.4	34.0	53.9	2.97	2.38	3.62	3.52	4.34	3.81	3.84	4.75	3.35	3.38	3.17	2.87	42.00
DIVISION	31.8	32.7	40.3	51.8	62.5	71.0	75.3	73.4	66.3	55.3	43.6	33.3	53.1	2.92	2.39	3.72	3.50	4.17	3.78	3.86	4.16	3.18	3.26	3.18	3.03	41.15
MIDDLE SUSQUEHANNA																										
BEAR GAP	•	•	•	•	•	•	•	•	•	•	•	•	•	2.80	2.47	3.73	3.55	4.41	4.01	4.83	4.01	3.50	3.51	3.71	3.22	43.75
NEWPORT	•	•	•	•	•	•	•	•	•	•	•	•	•	2.90	2.37	3.55	3.56	4.07	3.71	4.19	4.11	3.03	3.44	3.05	2.95	41.35
SHAMOKIN	•	•	•	•	•	•	•	•	•	•	•	•	•	2.82	2.63	3.73	3.49	4.49	3.85	4.58	3.91	3.43	3.32	3.54	3.27	43.06
SUNBURY	•	•	•	•	•	•	•	•	•	•	•	•	•	3.04	2.58	3.62	3.65	4.60	3.45	3.85	3.86	3.27	3.23	3.36	3.22	41.73
WILLIAMSPORT WSO	28.8	29.2	37.4	49.4	60.4	69.2	73.6	71.6	64.1	53.1	41.1	30.4	50.7	2.67	2.51	3.73	3.55	4.08	3.23	4.18	3.62	3.27	3.32	3.45	3.04	40.65
DIVISION	29.4	30.2	38.0	49.8	60.7	69.2	73.4	71.6	64.5	53.5	41.8	31.1	51.1	2.70	2.28	3.60	3.54	4.28	3.48	3.87	3.88	3.19	3.23	3.31	2.90	40.26

Taken from "Climatography of the United States No. 81-4, Decennial Census of U. S. Climate"

TEMPERATURE (°F) PRECIPITATION (In.)

STATION	JAN	FEB	MAR	APR	MAY	JUNE	JULY	AUG	SEPT	OCT	NOV	DEC	ANN	JAN	FEB	MAR	APR	MAY	JUNE	JULY	AUG	SEPT	OCT	NOV	DEC	ANN
UPPER SUSQUEHANNA																										
LAWRENCEVILLE	26.1	26.2	33.9	46.4	57.6	66.7	70.6	68.6	61.7	51.0	39.0	28.2	48.0	1.77	1.67	2.79	2.68	3.91	3.30	3.97	3.95	3.03	2.79	2.36	1.97	34.19
TOWANDA 1 ESE	27.4	27.6	35.4	47.5	58.3	67.0	71.3	69.5	62.5	51.8	40.6	29.5	49.0	1.86	1.93	2.86	3.09	4.02	3.18	3.84	3.33	3.49	2.95	2.62	2.17	35.34
DIVISION	25.1	25.2	32.9	45.3	56.5	65.3	69.7	67.8	60.7	50.2	38.6	27.4	47.1	2.22	2.15	3.07	3.28	4.10	3.38	3.98	3.93	3.43	3.11	2.91	2.51	38.07
CENTRAL MOUNTAINS																										
CLEARFIELD	•	•	•	•	•	•	•	•	•	•	•	•	•	3.19	2.62	4.00	3.97	4.39	4.13	4.30	3.78	3.04	3.04	3.11	3.09	42.66
LOCK HAVEN	29.4	30.1	38.0	50.8	61.4	69.7	73.5	71.7	64.6	53.5	41.3	31.0	51.3	2.57	2.21	3.67	3.39	4.31	3.48	4.33	3.89	3.14	3.14	3.12	2.77	40.02
RENOVO 1 W	•	•	•	•	•	•	•	•	•	•	•	•	•	2.46	2.19	3.46	3.38	4.31	3.72	4.20	3.18	2.67	3.03	3.02	2.66	38.28
RIDGWAY	25.9	25.0	32.5	45.1	56.1	64.8	68.2	66.8	59.9	49.2	37.8	27.3	46.6	2.94	2.39	3.44	3.75	4.17	3.99	4.44	3.50	3.25	3.00	3.06	2.87	40.80
STATE COLLEGE	28.7	29.3	36.6	48.7	59.9	68.2	72.1	70.2	62.9	52.6	41.0	30.5	50.1	2.67	2.19	3.70	3.51	4.39	3.57	3.78	3.53	2.62	2.93	3.05	2.76	38.70
DIVISION	26.8	27.0	34.5	46.8	57.8	66.1	70.0	68.3	61.3	50.8	39.1	28.5	48.1	2.82	2.38	3.71	3.60	4.41	3.80	4.26	3.77	3.03	3.06	3.09	2.86	40.79
SOUTH CENTRAL MOUNTAI																										
ALTOONA HORSESHOE CRVE	28.5	29.1	36.4	48.6	59.1	67.3	70.9	69.4	62.8	52.7	40.6	30.2	49.6	3.23	2.47	4.27	4.20	4.47	4.57	4.54	3.72	3.06	3.21	3.06	3.03	43.83
BUFFALO MILLS	•	•	•	•	•	•	•	•	•	•	•	•	•	2.69	2.03	3.53	3.28	3.94	3.65	3.68	3.62	2.77	2.97	2.53	2.40	37.09
EBENSBURG	26.9	27.2	34.2	46.1	56.8	65.1	68.8	67.3	61.1	50.8	38.6	28.6	47.6	3.49	2.90	4.06	4.01	4.61	4.71	4.50	4.09	3.09	2.97	3.11	3.46	45.00
HUNTINGDON	30.0	30.6	37.8	49.6	60.4	68.8	72.8	71.1	64.1	53.2	41.7	31.4	51.0	2.69	2.11	3.72	3.54	4.14	3.76	4.14	3.96	3.00	2.83	3.02	2.73	39.64
HYNDMAN	•	•	•	•	•	•	•	•	•	•	•	•	•	2.55	1.97	3.42	3.10	3.59	3.55	3.41	3.83	2.57	2.65	2.41	2.42	35.47
JOHNSTOWN	30.8	31.4	38.6	50.2	60.9	69.6	73.1	71.5	62.7	54.3	42.5	32.4	51.5	3.64	3.07	4.24	4.25	4.52	4.45	4.70	4.10	2.99	2.74	2.77	3.30	44.77
DIVISION	29.5	30.1	37.4	49.1	59.7	68.0	71.8	70.2	63.3	52.8	41.0	31.0	50.3	2.98	2.41	3.85	3.68	4.22	4.11	4.20	3.84	2.88	2.92	2.84	2.92	40.85
SOUTHWEST PLATEAU																										
ACMETONIA LOCK 3	•	•	•	•	•	•	•	•	•	•	•	•	•	3.20	2.66	3.70	3.76	3.98	4.20	4.12	3.85	2.76	2.79	2.69	2.79	40.50
BEAVER FALLS	•	•	•	•	•	•	•	•	•	•	•	•	•	2.67	2.18	3.10	3.21	3.72	3.74	3.65	3.45	2.67	2.42	2.31	2.24	35.36
BUTLER	30.3	30.5	37.9	49.7	59.9	68.1	72.6	70.6	64.1	53.3	41.6	31.6	50.9	2.96	2.57	3.45	3.76	3.98	4.06	3.95	3.46	3.08	2.95	2.77	2.82	39.81
CHARLEROI LOCK 4	•	•	•	•	•	•	•	•	•	•	•	•	•	3.00	2.44	3.68	3.69	3.90	3.84	3.90	3.78	2.89	2.47	2.58	2.62	38.79
CLAYSVILLE 3 W	31.1	31.6	38.8	49.9	60.1	68.7	71.9	70.4	64.3	53.1	41.4	32.0	51.1	3.07	2.51	3.66	3.49	4.08	4.28	4.29	3.73	3.20	2.53	2.61	2.79	40.24
CONFLUENCE 1 NW	•	•	•	•	•	•	•	•	•	•	•	•	•	3.62	2.87	4.06	3.68	4.41	4.68	4.65	4.40	3.05	2.89	2.75	3.17	44.23
CONNELLSVILLE	•	•	•	•	•	•	•	•	•	•	•	•	•	2.96	2.35	3.37	3.59	4.46	4.44	4.84	4.30	3.21	2.95	2.45	2.45	41.37
CORAOPOLIS NEVILLE IS	•	•	•	•	•	•	•	•	•	•	•	•	•	2.66	2.13	3.23	3.33	3.81	3.77	3.88	3.38	2.69	2.50	2.37	2.31	36.06
CREEKSIDE	•	•	•	•	•	•	•	•	•	•	•	•	•	3.24	2.71	3.67	3.98	4.12	4.59	4.81	3.92	3.20	3.03	2.98	3.05	43.30
DONORA	34.3	34.9	41.8	53.3	63.4	72.3	75.7	74.2	67.9	57.0	45.2	35.8	54.7	2.75	2.30	3.57	3.51	3.82	3.92	4.04	3.91	2.76	2.52	2.35	2.42	37.87
GREENSBORO LOCK 7	•	•	•	•	•	•	•	•	•	•	•	•	•	3.10	2.52	3.64	3.55	4.01	4.26	4.64	4.20	2.96	2.58	2.52	2.57	40.55
MC KEESPORT	•	•	•	•	•	•	•	•	•	•	•	•	•	2.85	2.34	3.42	3.30	3.59	4.08	3.67	3.64	2.76	2.38	2.39	2.48	36.90
NATRONA LOCK 4	•	•	•	•	•	•	•	•	•	•	•	•	•	3.11	2.60	3.71	3.72	3.98	4.29	4.26	3.96	2.83	2.86	2.64	2.67	40.63
NEW CASTLE 1 N	30.0	30.0	37.6	49.4	59.8	69.1	72.6	71.1	64.7	53.7	42.0	31.8	51.0	2.86	2.38	3.38	3.61	4.00	4.05	4.27	3.57	2.88	2.73	2.66	2.56	38.95
PITTSBURGH WSO 2	28.9	29.2	36.8	49.0	59.8	68.4	72.1	70.8	64.2	53.1	40.8	30.7	50.3	2.97	2.19	3.32	3.08	3.91	3.78	3.88	3.31	2.54	2.52	2.24	2.40	36.14
PITTSBURGH CITY WSO	32.1	32.9	40.2	52.1	62.7	71.4	74.9	73.0	66.2	54.7	42.8	33.4	53.0	2.82	2.31	3.52	3.37	3.75	3.95	3.60	3.50	2.67	2.50	2.34	2.54	36.87
SCHENLEY LOCK 5	•	•	•	•	•	•	•	•	•	•	•	•	•	3.19	2.78	4.10	3.80	4.10	4.18	3.89	3.73	2.98	2.77	2.71	2.85	41.08
UNIONTOWN	34.0	34.6	41.4	52.7	62.5	70.7	73.9	72.4	66.5	55.8	44.3	35.1	53.7	3.27	2.58	3.76	3.70	4.52	4.60	4.30	4.00	3.00	2.90	2.71	2.72	42.06
VANDERGRIFT	•	•	•	•	•	•	•	•	•	•	•	•	•	2.68	2.38	3.46	3.55	3.87	4.34	4.09	4.04	3.02	3.04	2.53	2.54	39.54
WAYNESBURG 1 E	32.6	33.1	40.2	51.2	61.0	69.4	72.7	71.3	64.9	54.0	42.6	33.3	52.2	3.08	2.47	3.65	3.59	4.24	4.21	4.32	3.91	3.08	2.66	2.46	2.61	40.28
DIVISION	30.7	31.1	38.2	49.8	60.0	68.6	72.1	70.6	64.2	53.5	41.8	32.1	51.1	3.12	2.58	3.74	3.75	4.22	4.38	4.33	3.95	3.02	2.84	2.68	2.80	41.41
NORTHWEST PLATEAU																										
CORRY	26.0	25.8	33.3	45.9	56.8	66.1	69.9	66.4	61.7	50.9	38.9	28.5	47.5	3.39	2.93	3.80	4.06	4.03	4.36	3.89	3.43	3.58	3.47	4.00	3.52	44.46
ERIE WSO	27.3	26.4	33.6	45.5	56.4	66.7	71.1	69.8	63.3	52.6	41.2	30.7	48.7	2.67	2.32	2.88	3.56	3.54	3.05	3.67	2.98	3.56	3.30	3.36	2.61	37.50
FARRELL SHAPON	30.0	30.3	38.5	50.7	61.7	71.3	74.8	72.9	66.1	54.8	42.4	32.2	52.1	2.83	2.29	2.95	3.39	3.84	3.71	3.58	3.89	2.71	2.45	2.51	2.41	36.56
FRANKLIN	27.9	27.2	34.8	46.4	58.0	66.9	70.8	69.2	62.6	51.8	40.2	30.0	48.8	3.06	2.43	3.24	3.82	4.35	3.87	4.33	3.11	3.13	3.03	3.07	2.57	40.01
GALETON	•	•	•	•	•	•	•	•	•	•	•	•	•	2.59	2.15	3.47	3.29	4.06	3.51	3.90	3.74	3.16	3.12	3.01	2.61	38.61
GREENVILLE	28.4	28.7	36.4	48.2	59.0	68.4	72.1	70.2	63.7	52.7	40.8	30.7	49.9	3.08	2.52	3.29	3.85	4.06	4.13	4.11	3.79	2.96	3.17	3.00	2.58	40.54
JAMESTOWN 2 NW	•	•	•	•	•	•	•	•	•	•	•	•	•	2.76	2.26	2.97	3.63	4.01	3.82	3.92	3.54	3.21	2.97	2.77	2.36	38.22
MEADVILLE 1 S	26.8	26.2	34.2	46.4	57.2	66.7	70.4	68.7	62.1	51.4	39.5	29.1	48.2	2.91	2.52	3.27	3.59	4.01	4.31	4.25	3.46	3.01	3.31	3.09	2.71	40.44
WARREN	27.6	27.3	34.4	46.4	57.6	66.7	70.5	69.0	62.6	52.0	40.1	29.8	48.7	2.86	2.56	3.51	3.61	4.27	4.67	4.29	3.50	3.43	3.41	3.59	3.03	42.73
DIVISION	27.0	26.7	34.3	46.4	57.3	66.5	70.3	68.6	62.1	51.5	39.5	29.1	48.3	3.04	2.58	3.38	3.75	4.15	4.09	4.14	3.46	3.22	3.15	3.24	2.84	41.04

* Normals for the period 1931-1960. Divisional normals may not be the arithmetical average of individual stations published, since additional data for shorter period stations are used to obtain better areal representation.

*NORMALS BY CLIMATOLOGICAL DIVISIONS

Taken from "Climatography of the United States No. 81-4, Decennial Census of U. S. Climate"

TEMPERATURE (°F)													PRECIPITATION (In.)												
JAN	FEB	MAR	APR	MAY	JUNE	JULY	AUG	SEPT	OCT	NOV	DEC	ANN	JAN	FEB	MAR	APR	MAY	JUNE	JULY	AUG	SEPT	OCT	NOV	DEC	ANN

CONFIDENCE - LIMITS

In the absence of trend or record changes, the chances are 9 out of 10 that the true mean will lie in the interval formed by adding and subtracting the values in the following table from the means for any station in the State. Because of the wider variation in mean precipitation, the corresponding monthly means and annual mean must be substituted for "p" in the precipitation table below to obtain mean precipitation confidence limits.

| 1.6 | 1.4 | 1.5 | 1.0 | .9 | .7 | .5 | .7 | .9 | .9 | .8 | 1.1 | .4 | $.22\sqrt{p}$ | $.19\sqrt{p}$ | $.22\sqrt{p}$ | $.24\sqrt{p}$ | $.28\sqrt{p}$ | $.27\sqrt{p}$ | $.26\sqrt{p}$ | $.28\sqrt{p}$ | $.35\sqrt{p}$ | $.38\sqrt{p}$ | $.29\sqrt{p}$ | $.25\sqrt{p}$ | $.29\sqrt{p}$ |

COMPARATIVE DATA

Data in the following table are the mean temperature and average precipitation for Huntingdon, Pennsylvania, for the period 1906-1930 and are included in this publication for comparative purposes.

| 28.7 | 30.0 | 39.4 | 49.8 | 59.9 | 68.1 | 72.3 | 70.3 | 64.9 | 53.2 | 41.8 | 31.2 | 50.8 | 3.30 | 2.77 | 3.14 | 3.73 | 3.62 | 4.16 | 3.88 | 3.67 | 3.39 | 2.87 | 2.19 | 2.83 | 39.55 |

NORMALS, MEANS, AND EXTREMES

ALLENTOWN, PENNSYLVANIA — ALLENTOWN-BETHLEHEM-EASTON AP

Standard time used: EASTERN
Latitude: 40° 39' N Longitude: 75° 26' W Elevation (ground): 387 feet Year: 1970

Means and extremes above are from existing and comparable exposures. Annual extremes have been exceeded at other sites in the locality as follows:
Maximum monthly snowfall 43.2 in January 1925.

ERIE, PENNSYLVANIA — ERIE INTERNATIONAL AIRPORT

Standard time used: EASTERN
Latitude: 42° 05' N Longitude: 80° 11' W Elevation (ground): 731 feet Year: 1970

Means and extremes above are from existing and comparable exposures. Annual extremes have been exceeded at other sites in the locality as follows:
Highest temperature 99 in September 1953; lowest temperature -16 in February 1875; maximum monthly precipitation 13.27 in July 1947; minimum monthly precipitation 0.02 in October 1924; maximum precipitation in 24 hours 10.42 in July 1947; maximum snowfall in 24 hours 26.5 in December 1944.

NORMALS, MEANS, AND EXTREMES

Station: HARRISBURG, PENNSYLVANIA — HARRISBURG STATE AIRPORT — Standard time used: EASTERN — Latitude: 40° 13' N — Longitude: 76° 51' W — Elevation (ground): 338 feet — Year: 1970

	Temperature								Normal heating degree days (Base 65°)	Precipitation									Snow, Ice pellets									Relative humidity				Wind &					Pct. of possible sunshine	Mean sky cover sunrise to sunset	Mean number of days											Average daily solar radiation - langleys
	Normal			Extremes							Maximum monthly		Minimum monthly		Maximum in 24 hrs.		Mean total	Maximum monthly			Maximum in 24 hrs.		Hour 01	Hour 07	Hour 13	Hour 19	Mean speed	Prevailing direction	Fastest mile					Sunrise to sunset			Precipitation .01 inch or more	Snow, Ice pellets 1.0 inch or more	Thunderstorms	Heavy fog	Temperatures Max.		Temperatures Min.			
Month	Daily maximum	Daily minimum	Monthly	Record highest	Year	Record lowest	Year			Normal total														(Local time)						Speed	Direction	Year			Clear	Partly cloudy	Cloudy					90° and above	32° and below	32° and below	0° and below		
(a)	(b)	(b)	(b)	32		32		(b)	(b)	(b)		32		32		32		32	32		32		32	32	32	32	32	32	14	32		32		32	20	32	32	32	32	32	32	32	32	32	32	32
J	38.9	23.6	31.3	73	1950	-5	1968	1045	2.76	4.78	1964	0.70	1955	1.94	1944	9.5	34.0	1961	21.0	1945	69	71	58	64	8.4	WNW	47	W	1956	48	6.6	7	7	17	11	2	*	3	0	9	26	1
F	41.4	23.7	32.6	75	1954	-2	1967	907	2.31	4.44	1966	0.53	1968	1.93	1965	9.5	30.2	1964	19.6	1964	67	70	55	61	9.3	WNW	60		1939	54	6.7	7	7	14	10	2	*	2	0	5	23	*
M	50.0	30.5	40.3	86	1945	8	1943	766	3.43	5.47	1944	1.20	1949	2.11	1952	7.1	22.6	1960	10.7	1960	67	71	52	57	9.7	WNW	68		1955	57	6.7	7	8	16	12	2	1	2	0	1	18	0
A	63.0	40.5	51.8	92	1957+	21	1969+	396	3.02	5.97	1952	0.57	1966	2.18	1940	0.3	3.1	1959	3.1	1959	68	70	49	56	9.3	WNW	56	SW	1952	57	6.9	5	9	16	12	*	2	1	*	0	3	0
M	74.4	51.0	62.7	97	1942+	31	1966	124	3.90	8.06	1953	0.51	1964	3.11	1944	T	T	1966+	T	1966+	73	73	50	57	7.8	W	46		1940	60	6.5	6	10	15	12	0	6	1	1	0	*	0
J	82.7	59.9	71.3	100	1966+	41	1961	12	3.42	5.80	1970	0.07	1966	2.81	1946	0.0	0.0		0.0		78	76	52	60	6.8	W	48		1963	65	6.1	6	12	12	11	0	7	1	6	0	0	0
J	87.0	65.3	76.2	107	1966	49	1945	0	3.51	9.72	1969	0.78	1955	5.36	1969	0.0	0.0		0.0		78	78	51	60	6.3	W	47	NW	1954	68	6.1	7	11	13	10	0	7	1	9	0	0	0
A	84.7	63.5	74.1	101	1944	46	1944	0	3.65	9.07	1955	0.93	1957	3.35	1955	0.0	0.0		0.0		81	82	53	65	6.1	W	45		1939	67	5.8	8	11	12	10	0	6	1	7	0	0	0
S	77.6	56.2	66.9	102	1953	30	1943	63	2.82	6.12	1966	0.72	1941	4.36	1952	0.0	0.0		0.0		82	84	54	67	6.2	WNW	40	W	1967	63	5.5	10	8	12	8	0	3	2	2	0	*	0
O	66.7	44.6	55.7	97	1941	23	1969	298	2.97	6.82	1943	0.04	1963	2.61	1943	T	1.2	1940	1.2	1940	79	82	53	66	6.7	W	50	E	1954	58	5.6	11	7	13	8	*	1	3	*	0	0	0
N	52.2	34.5	43.4	84	1950	13	1955	648	2.95	5.93	1963	0.53	1930	2.93	1943	1.9	15.4	1953	15.4	1953	74	77	56	65	7.9	WNW	58	SE	1950	44	6.2	8	8	16	10	1	*	2	0	1	12	0
D	40.7	25.3	33.0	71	1946	-8	1960	992	2.91	6.46	1969	0.23	1955	2.01	1950	8.3	28.3	1969	12.9	1969	71	73	59	63	8.2	WNW	61	NW	1953	45	6.9	6	8	17	10	0	0	2	0	6	24	*
YR	63.3	43.2	53.3	107	JUL. 1966	-8	DEC. 1960	5251	37.65	9.72	JUL. 1969	0.04	OCT. 1963	5.36	JUL. 1969	36.6	34.0	JAN. 1961	21.0	JAN. 1945	74	75	54	62	7.7	WNW	68	W	MAR. 1955	59	6.4	86	106	173	124	10	33	21	26	22	108	1

To 8 compass points only.

Means and extremes above are from existing and comparable exposures. Annual extremes have been exceeded at other sites in the locality as follows:
Lowest temperature -14 in January 1912; maximum monthly precipitation 10.67 in August 1933; minimum monthly precipitation 0.02 in October 1924;
maximum precipitation in 24 hours 7.46 on May 31-June 1, 1889.

Station: PHILADELPHIA, PENNSYLVANIA — INTERNATIONAL AIRPORT — Standard time used: EASTERN — Latitude: 39° 53' N — Longitude: 75° 15' W — Elevation (ground): 5 feet — Year: 1970

	Temperature								Normal heating degree days (Base 65°)	Precipitation									Snow, Ice pellets									Relative humidity				Wind &					Pct. of possible sunshine	Mean sky cover sunrise to sunset	Mean number of days											Average daily solar radiation - langleys
	Normal			Extremes							Maximum monthly		Minimum monthly		Maximum in 24 hrs.		Mean total	Maximum monthly			Maximum in 24 hrs.		Hour 01	Hour 07	Hour 13	Hour 19	Mean speed	Prevailing direction	Fastest mile					Sunrise to sunset			Precipitation .01 inch or more	Snow, Ice pellets 1.0 inch or more	Thunderstorms	Heavy fog	Temperatures Max.		Temperatures Min.			
Month	Daily maximum	Daily minimum	Monthly	Record highest	Year	Record lowest	Year			Normal total														(Local time)						Speed	Direction	Year			Clear	Partly cloudy	Cloudy					90° and above	32° and below	32° and below	0° and below		
(a)	(b)	(b)	(b)	11		11		(b)	(b)	(b)		28		28		24		28	28			28		11	11	11	11	30	23	30	30				28	30	30	30	30	30	30	30	11	11	11	11
J	40.3	24.3	32.3	69	1967	-5	1963	1014	3.32	6.06	1949	0.45	1955	2.27	1968	5.6	19.7	1961	13.2	1961	70	72	59	66	10.3	WNW	61	NE	1958	50	6.6	8	7	16	11	2	*	3	0	9	27	*
F	41.8	24.6	33.2	67	1961	-4	1961	890	2.80	4.64	1958+	1.37	1954	1.96	1966	6.3	18.4	1967	13.0	1958	68	71	57	61	11.1	NW	59	NW	1956	53	6.4	7	7	14	9	2	1	2	0	5	24	*
M	50.3	31.6	41.0	80	1960	9	1960	744	3.80	6.27	1953	0.68	1966	2.39	1968	4.1	13.4	1958	10.0	1958	69	72	54	59	11.4	N	56	N	1966	57.5	6.3	8	8	15	11	1	1	2	0	1	17	0
A	62.6	41.4	52.0	92	1960	24	1969	390	3.40	6.58	1947	1.13	1963	3.36	1970	0.1	3.0	1965	3.0	1965	69	70	49	55	11.0	SW	59	SW	1958	56	6.5	8	8	15	11	*	2	2	0	1	17	0
M	81.6	60.4	71.0	100	1964	44	1967+	12	4.05	7.40	1962	0.11	1949	3.62	1962	T	T	1963	T	1963	75	75	52	57	9.7	WSW	56	SW	1957	58	6.5	6	11	14	11	0	4	2	1	0	*	0
J	85.9	65.2	75.6	104	1966	51	1966	0	4.16	8.33	1969	0.64	1967	4.26	1969	0.0	0.0		0.0		80	77	53	60	8.8	WSW	73	NW	1958	64	6.2	7	11	12	10	0	5	1	4	0	0	0
J	83.7	63.5	73.6	97	1968	45	1965	0	4.63	9.70	1955	0.49	1964	4.81	1955	0.0	0.0		0.0		83	80	55	64	8.1	WSW	47	W	1959+	62	6.2	7	11	13	9	0	6	1	6	0	0	0
A	77.2	56.2	66.7	95	1970	35	1969	0	3.46	8.78	1960	0.44	1968	5.45	1960	0.0	0.0		0.0		82	81	55	64	7.8	SW	67	E	1955	62	5.9	8	11	12	10	0	5	1	4	0	0	0
S	66.5	44.9	55.7	88	1967	25	1969+	60	2.78	5.21	1943	0.09	1963	3.78	1966	T	T	1970+	T	1970+	82	83	55	70	8.3	SW	49	NE	1960	61	5.6	10	9	11	8	0	2	2	0	0	0	0
O	54.0	34.5	44.3	80	1961	17	1969	291	3.40	6.67	1963	1.05	1965	3.46	1950	0.8	8.8	1953	8.7	1953	79	81	52	68	8.8	WSW	66	SW	1958	60	5.5	11	8	12	7	*	*	3	0	0	11	0
N	42.3	25.5	33.9	71	1966	3	1962	621	2.94	7.23	1963	0.25	1955	1.77	1951	4.7	18.8	1966	14.6	1966	72	74	60	66	9.6	WSW	60	SW	1958	52	6.3	7	10	13	9	*	1	2	0	6	24	0
YR	63.3	43.7	53.5	104	JUL. 1966	-5	JAN. 1963	5101	42.48	9.70	AUG. 1955	0.09	OCT. 1963	5.45	SEP. 1960	21.6	19.7	JAN. 1961	14.6	DEC. 1960	75	76	55	63	9.6	WSW	73	NW	JUN. 1958	58	6.2	94	110	161	115	6	26	27	17	21	108	*

To 8 compass points only.

Means and extremes above are from existing and comparable exposures. Annual extremes have been exceeded at other sites in the locality as follows:
Highest temperature 106 in August 1918; lowest temperature -11 in February 1934; maximum monthly precipitation 12.10 in August 1911; maximum precipitation
in 24 hours 5.89 in August 1898; maximum monthly snowfall 31.5 in February 1899; maximum snowfall in 24 hours 21.0 in December 1909; fastest mile of wind
88 from North in July 1931.

NORMALS, MEANS, AND EXTREMES

Station: PITTSBURGH, PENNSYLVANIA — GREATER PITTSBURGH AIRPORT

Standard time used: EASTERN Latitude: 40° 30' N Longitude: 80° 13' W Elevation (ground): 1137 feet Year: 1970

Month	Normal Daily maximum	Normal Daily minimum	Normal Monthly	Record highest	Year	Record lowest	Year	Normal heating degree days (Base 65°)	Normal total precipitation
	(b)	(b)	(b)	11	11	11		(b)	(b)
J	36.5	21.2	28.9	66	1967	-18	1963	1119	2.97
F	37.6	20.7	29.2	66	1961	-5	1963	1002	2.19
M	46.1	27.4	36.8	80	1966	-1	1960	874	3.32
A	60.4	37.3	49.0	87	1970	15	1964	480	3.08
M	71.4	48.1	59.8	91	1962	26	1970+	195	3.91
J	79.6	56.9	68.4	96	1966	36	1966	39	4.61
J	83.3	60.5	72.1	98	1963	42	1963	0	3.88
A	81.5	59.6	70.8	95	1965	40	1965	9	3.31
S	75.5	52.8	64.2	94	1962+	31	1959	105	2.54
O	63.7	42.4	53.1	87	1963	16	1965	375	2.52
N	47.5	32.0	40.8	82	1961	6	1959	726	2.24
D	38.1	23.2	30.7	68	1966	-1	1962	1063	2.40
YR	60.3	40.3	50.3	98 JUL. 1966		-18 JAN. 1963		5987	36.14

Means and extremes above are from existing and comparable exposures. Annual extremes have been exceeded at other sites in the locality as follows:
AIRPORT - Highest temperature 102 in July 1936; maximum monthly precipitation 10.25 in June 1951; maximum monthly snowfall 32.3 in November 1950; and maximum snowfall in 24 hours 17.5 in November 1950.
CITY OFFICE - Highest temperature 103 in July 1936; lowest temperature -20 in February 1899; minimum monthly precipitation 0.06 in October 1874; maximum precipitation in 24 hours 4.08 in September 1876; and maximum monthly snowfall 36.3 in December 1890.

Station: PITTSBURGH, PENNSYLVANIA — FEDERAL BUILDING

Standard time used: EASTERN Latitude: 40° 27' N Longitude: 80° 00' W Elevation (ground): 747 feet Year: 1970

Month	Normal Daily maximum	Normal Daily minimum	Normal Monthly	Record highest	Year	Record lowest	Year	Normal heating degree days (Base 65°)	Normal total precipitation
	(b)	(b)	(b)	36	36	36		(b)	(b)
J	39.5	24.7	32.1	77	1950	-13	1963	1020	2.82
F	41.0	24.7	32.9	74	1937	-5	1958	899	2.31
M	49.0	31.7	40.4	83	1950	18	1950	777	3.52
A	62.5	41.7	52.1	90	1970	18	1950	394	3.37
M	73.4	52.2	62.7	92	1962+	29	1966	131	3.75
J	81.6	61.1	71.4	99	1952	39	1966	14	3.95
J	85.0	64.8	74.9	100	1936	45	1965	0	3.60
A	84.0	64.1	74.0	100	1948	50	1965	6	3.34
S	76.6	57.5	67.0	91	1953	37	1957	74	2.67
O	64.7	45.1	54.9	91	1950	24	1954	330	2.50
N	51.1	34.5	42.8	83	1961	5	1939	666	2.34
D	40.5	26.2	33.4	71	1941	-7	1955	980	2.54
YR	62.3	43.7	53.0	103 JUL. 1936		-13 JAN. 1963		5291	36.87

Means and extremes above are from existing and comparable exposures. Annual extremes have been exceeded at other sites in the locality as follows:
Lowest temperature -20 in February 1899; maximum monthly precipitation 9.51 in July 1887; minimum monthly precipitation 0.06 in October 1874; maximum precipitation in 24 hours 4.08 in September 1876; maximum monthly snowfall 36.3 in December 1890.

NORMALS, MEANS, AND EXTREMES

Station: AVOCA, PENNSYLVANIA WILKES-BARRE SCRANTON AIRPORT Standard time used: EASTERN Latitude: 41° 20' N Longitude: 75° 44' W Elevation (ground): 930 feet Year: 1970

Month	Temperature: Normal: Daily maximum	Normal: Daily minimum	Normal: Monthly	Extremes: Record highest	Year	Record lowest	Year	Normal heating degree days (Base 65°)	Precipitation: Normal total	Maximum monthly	Year	Minimum monthly	Year	Maximum in 24 hrs.	Year	Snow, Ice pellets: Mean total	Maximum monthly	Year	Maximum in 24 hrs.	Year	Relative humidity: Hour 01	Hour 07	Hour 13	Hour 19 (Local time)	Wind &: Mean speed	Prevailing direction	Fastest mile: Speed	Direction #	Year	Pct. of possible sunshine	Mean sky cover sunrise to sunset	Mean number of days: Sunrise to sunset: Clear	Partly cloudy	Cloudy	Precipitation .01 inch or more	Snow, Ice pellets 1.0 inch or more	Thunderstorms	Heavy fog	Temperatures: Max.: 90° and above ‡	Max.: 32° and below	Min.: 32° and below	Min.: 0° and below	Average daily solar radiation - langleys
(a)	(b)	(b)	(b)	15		15		(b)	(b)	15		15		15		15	15		15		15	15	15	15	15	8	15	15		15	15	15	15	15	15	15	15	15	15	15	15	15	
J	33.6	21.7	27.7	67	1967	-10	1968+	1156	2.29	3.40	1964	0.52	1970	1.52	1959	10.7	27.9	1964	20.1	1964	72	75	64	67	8.8	SW	43	SE	1964	43	7.3	5	7	19	12	3	*	3	0	16	28	2	
F	35.1	21.4	28.3	60	1961	-11	1961	1028	1.99	3.64	1956	0.30	1968	1.60	1962	12.0	22.0	1964	13.3	1961	72	76	63	65	9.3	SW	60	W	1956	47	7.2	4	8	16	12	3	*	3	0	12	26	2	
M	43.7	28.7	36.2	78	1968+	-4	1967	893	2.82	3.91	1967	1.45	1969	2.20	1964	11.5	29.7	1967	15.5	1960	71	75	58	62	9.2	NW	49	S	1970	49	7.1	5	8	18	13	3	1	3	0	5	24	*	
A	57.4	39.4	48.4	89	1962	15	1964	498	3.27	5.81	1957	1.30	1963	1.59	1970	3.2	12.0	1956	8.4	1956	68	72	51	55	9.6	SW	47	NW	1957+	52	6.8	6	8	16	12	1	3	2	0	0	9	0	
M	69.4	49.7	59.6	93	1962	28	1957	195	3.95	4.64	1968	0.77	1959	1.42	1961	0.1	0.6	1966	0.6	1966	72	74	50	55	8.9	WSW	40	NW	1956	58	6.6	6	11	14	12	0	4	1	*	0	1	0	
J	77.4	58.9	68.2	97	1964	37	1958	33	3.91	5.00	1964	0.27	1966	2.83	1969	0.0	0.0		0.0		79	81	54	60	7.9	SW	43	W	1956	62	6.1	7	12	11	10	0	6	1	3	0	0	0	
J	81.6	63.2	72.4	101	1966	45	1957	0	4.79	6.81	1969	1.23	1968+	2.33	1969	0.0	0.0		0.0		81	83	54	62	7.3	WSW	42	NW	1960	60	6.3	6	12	13	12	0	8	1	3	0	0	0	
A	78.8	61.2	70.0	94	1959	43	1957	19	3.58	5.23	1965	1.38	1957	3.18	1966	0.0	0.0		0.0		83	86	56	65	7.2	SW	50	NE	1956	61	6.0	7	13	11	11	0	6	2	2	0	0	0	
S	71.4	53.5	62.5	95	1959	30	1957	132	2.97	7.78	1960	0.82	1964	3.09	1960	T	T	1956	T	1956	84	88	58	70	7.4	SW	38	SW	1957	58	5.9	8	10	12	9	0	3	3	*	0	*	0	
O	59.7	42.3	51.0	84	1959	21	1969	434	3.50	5.46	1962	0.03	1963	2.61	1962	0.4	4.4	1962	4.4	1962	80	85	56	67	7.9	WSW	36	SE	1956	56	5.9	9	9	13	9	*	1	2	0	0	4	0	
N	40.2	32.9	39.6	76	1956	10	1958	762	2.94	4.62	1963	1.31	1956	2.23	1963	2.4	10.9	1968	8.0	1968	75	79	63	69	8.7	WSW	45	S	1957	37	7.5	4	7	19	12	1	1	2	0	1	15	0	
D	34.8	23.9	29.4	65	1966	-7	1969	1104	2.47	4.04	1957	0.35	1958	1.46	1957	11.4	33.9	1969	12.4	1969	76	78	67	71	8.9	SW	47	SW	1957	36	7.7	4	6	21	13	3	*	2	0	14	27	1	
YR	57.4	41.4	49.4	101	JUL. 1966	-11	FEB. 1961	6254	38.48	7.78	SEP. 1960	0.03	OCT. 1963	3.18	AUG. 1966	51.7	33.9	DEC. 1969	20.1	JAN. 1964	76	79	58	64	8.4	SW	60	W	FEB. 1956	53	6.7	71	111	183	137	14	32	25	8	47	134	5	

To 8 compass points only.

Means and extremes above are from existing and comparable exposures. Annual extremes have been exceeded at other sites in the locality as follows: Highest temperature 103 in July 1936; lowest temperature -19 in February 1934; maximum monthly precipitation 11.76 in August 1955; maximum precipitation in 24 hours 5.09 in September 1924; maximum monthly snowfall 38.0 in March 1916.

Station: WILLIAMSPORT, PENNSYLVANIA WILLIAMSPORT-LYCOMING CO. AP Standard time used: EASTERN Latitude: 41° 15' N Longitude: 76° 55' W Elevation (ground): 524 feet Year: 1970

Month	Temperature: Normal: Daily maximum	Normal: Daily minimum	Normal: Monthly	Extremes: Record highest	Year	Record lowest	Year	Normal heating degree days (Base 65°)	Precipitation: Normal total	Maximum monthly	Year	Minimum monthly	Year	Maximum in 24 hrs.	Year	Snow, Ice pellets: Mean total	Maximum monthly	Year	Maximum in 24 hrs.	Year	Relative humidity: Hour 01	Hour 07	Hour 13	Hour 19 (Local time)	Wind &: Mean speed	Prevailing direction	Fastest mile: Speed	Direction	Year	Pct. of possible sunshine	Mean sky cover sunrise to sunset	Mean number of days: Sunrise to sunset: Clear	Partly cloudy	Cloudy	Precipitation .01 inch or more	Snow, Ice pellets 1.0 inch or more	Thunderstorms	Heavy fog	Temperatures: Max.: 90° and above ‡	Max.: 32° and below	Min.: 32° and below	Min.: 0° and below	Average daily solar radiation - langleys
(a)	(b)	(b)	(b)	26		26		(b)	(b)	26		26		26		26	26		26		20	26	26	26	10	3	17	17			26	26	26	26	26	26	10	10	26	26	26	26	
J	36.1	21.4	28.8	69	1967	-14	1948	1122	2.67	4.95	1964+	0.95	1970	1.49	1964	9.7	33.5	1964	23.1	1964	76	77	62	69	9.1	W	66	27	1951+		7.2	5	8	18	13	3	*	1	0	12	28	2	
F	37.6	20.8	29.2	71	1954	-11	1948	1002	2.51	4.81	1960	0.57	1968	2.05	1966	10.5	25.2	1964	16.5	1964	75	76	59	66	9.9	WNW	60	14	1951		7.0	5	7	16	12	3	*	1	0	7	26	1	
M	46.6	28.1	37.4	83	1945	-1	1967	856	3.73	5.52	1950	1.29	1949	2.52	1964	9.3	29.5	1967	13.9	1967	75	78	55	62	9.2	W	58	11	1954		6.8	6	8	17	13	2	1	2	0	2	22	*	
A	60.2	38.6	49.4	91	1960	18	1969+	468	3.55	6.30	1957	0.99	1946	1.90	1952	1.1	7.9	1961	5.0	1961	75	76	50	57	9.3	W	62	18	1950		6.9	6	8	16	14	*	2	1	*	0	8	0	
M	72.0	48.7	60.4	95	1969+	28	1966	177	4.08	9.45	1946	0.80	1964	4.15	1946	T	T	1970+	T	1970+	81	79	51	58	8.1	W	55	18	1953		6.8	5	11	15	13	0	4	2	1	0	1	0	
J	80.5	57.9	69.2	102	1952	36	1945	24	3.23	5.74	1957	0.66	1966	2.26	1961	0.0	0.0		0.0		88	83	53	61	7.0	W	62	29	1951		6.1	6	13	11	11	0	7	3	4	0	0	0	
J	85.3	61.8	73.6	100	1966+	43	1965+	0	4.18	8.30	1958	0.99	1955	2.53	1958	0.0	0.0		0.0		89	87	53	64	6.8	W	78	20	1951		6.2	6	13	12	12	0	9	3	6	0	0	0	
A	83.1	60.0	71.6	100	1955	38	1965	9	3.62	7.67	1955	0.96	1951	3.46	1950	0.0	0.0		0.0		91	90	56	70	6.5	W	60	29	1949		6.2	6	14	11	11	0	7	5	3	0	0	0	
S	75.4	52.8	64.1	102	1953	28	1947	111	3.27	6.41	1968	0.50	1964	3.07	1968	0.0	0.0		0.0		91	92	57	75	6.6	WNW	59	16	1951		6.3	7	11	12	9	0	3	9	1	0	*	0	
O	64.0	42.1	53.1	91	1951	20	1969+	375	3.32	7.78	1955	0.19	1963	4.38	1955	T	0.3	1962	0.3	1962	87	89	56	73	7.1	W	75	11	1954		6.1	8	10	13	9	0	1	6	*	0	4	0	
N	49.5	32.7	41.1	83	1950	12	1955+	717	3.45	6.49	1948	0.89	1946	3.46	1956	3.0	13.7	1953	12.1	1953	81	83	61	72	8.5	W	77	09	1950		7.3	4	7	19	13	1	1	3	0	1	16	0	
D	37.9	22.9	30.4	65	1966	-15	1950	1073	3.04	5.63	1969	0.84	1955	2.46	1952	10.1	35.5	1969	16.5	1969	78	79	64	72	9.1	W	58	16	1953		7.5	4	8	19	13	3	1	2	0	9	27	1	
YR	60.7	40.7	50.7	102	SEP. 1953+	-15	DEC. 1950	5934	40.65	9.45	MAY 1946	0.19	OCT. 1963	4.38	OCT. 1955	43.7	35.5	DEC. 1969	23.1	JAN. 1964	82	82	56	67	8.1	W	78	20	JUL. 1951		6.7	68	118	179	142	12	34	37	15	30	131	4	

REFERENCE NOTES APPLYING TO ALL "NORMALS, MEANS, AND EXTREMES" TABLES

(a) Length of record, years, based on January data. Other months may be for more or fewer years if there have been breaks in the record.
(b) Climatological standard normals (1931-1960).
* Less than one half.
\+ Also on earlier dates, months, or years.
T Trace, an amount too small to measure.
Below zero temperatures are preceded by a minus sign.
The prevailing direction for wind in the Normals, Means, and Extremes table is from records through 1963.
‡ ≤ 70° at Alaskan stations.

Unless otherwise indicated, dimensional units used in this bulletin are: temperature in degrees F.; precipitation, including snowfall, in inches; wind movement in miles per hour; and relative humidity in percent. Heating degree day totals are the sums of negative departures of average daily temperatures from 65° F. Cooling degree day totals are the sums of positive departures of average daily temperatures from 65° F. Sleet was included in snowfall totals beginning with July 1948. The term "Ice pellets" includes solid grains of ice (sleet) and particles consisting of snow pellets encased in a thin layer of ice. Heavy fog reduces visibility to 1/4 mile or less.

Sky cover is expressed in a range of 0 for no clouds or obscuring phenomena to 10 for complete sky cover. The number of clear days is based on average cloudiness 0-3, partly cloudy days 4-7, and cloudy days 8-10 tenths.

Solar radiation data are the averages of direct and diffuse radiation on a horizontal surface. The langley denotes one gram calorie per square centimeter.

& Figures instead of letters in a direction column indicate direction in tens of degrees from true North; i.e., 09 - East, 18 - South, 27 - West, 36 - North, and 00 - Calm. Resultant wind is the vector sum of wind directions and speeds divided by the number of observations. If figures appear in the direction column under "Fastest mile" the corresponding speeds are fastest observed 1-minute values.

Mean Annual Precipitation, Inches

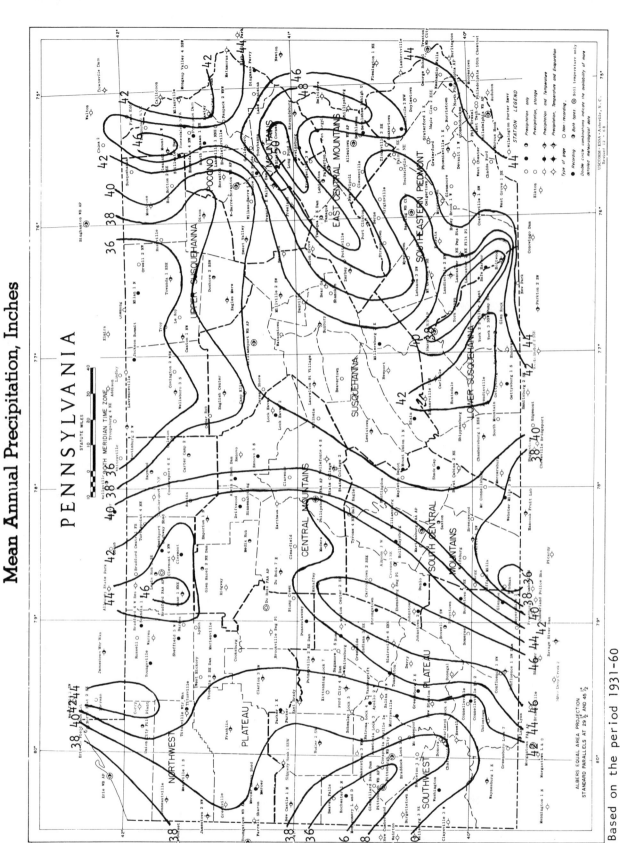

Based on the period 1931-60

Isolines are drawn through points of approximately equal value. Caution should be used in interpolating on these maps, particularly in mountainous areas.

Mean Maximum Temperature (°F), January

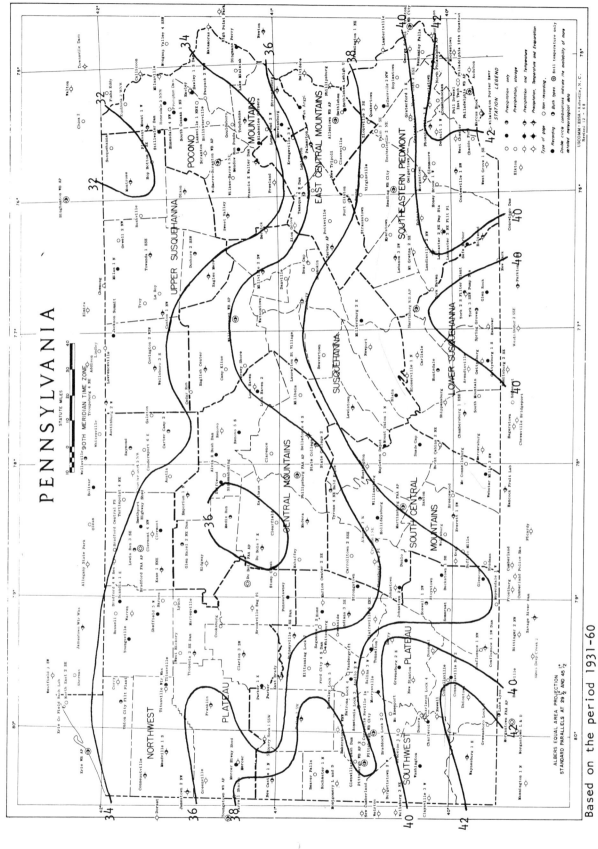

PENNSYLVANIA

Based on the period 1931-60

Isolines are drawn through points of approximately equal value. Caution should be used in interpolating on these maps, particularly in mountainous areas.

Mean Minimum Temperature (°F), January

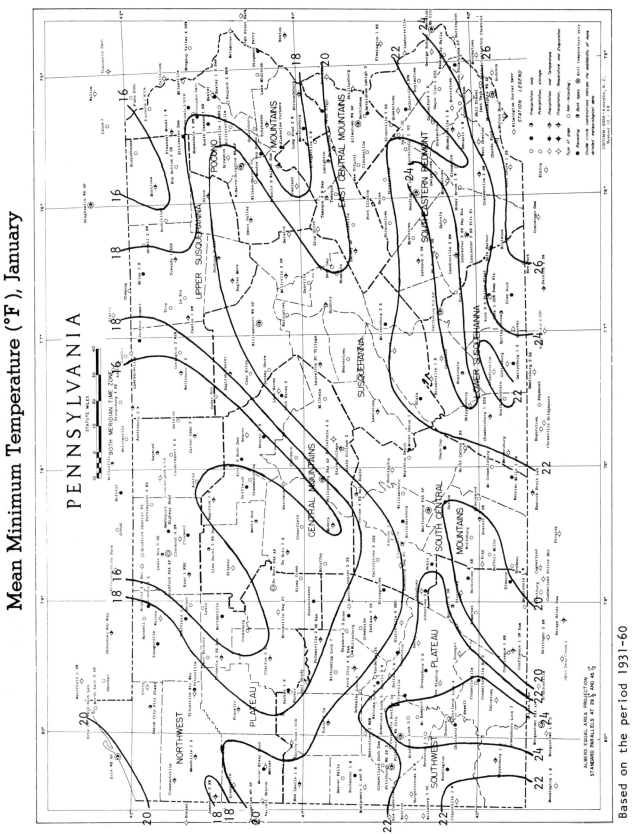

PENNSYLVANIA

Based on the period 1931-60

Isolines are drawn through points of approximately equal value. Caution should be used in interpolating on these maps, particularly in mountainous areas.

— 336 —

Mean Maximum Temperature (°F), July

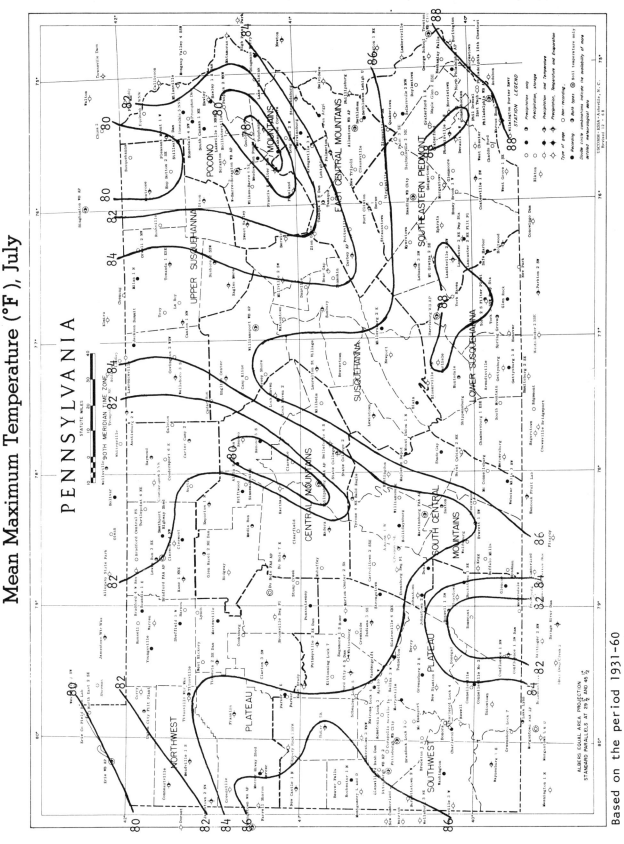

Based on the period 1931-60

Isolines are drawn through points of approximately equal value. Caution should be used in interpolating on these maps, particularly in mountainous areas.

Mean Minimum Temperature (°F), July

PENNSYLVANIA

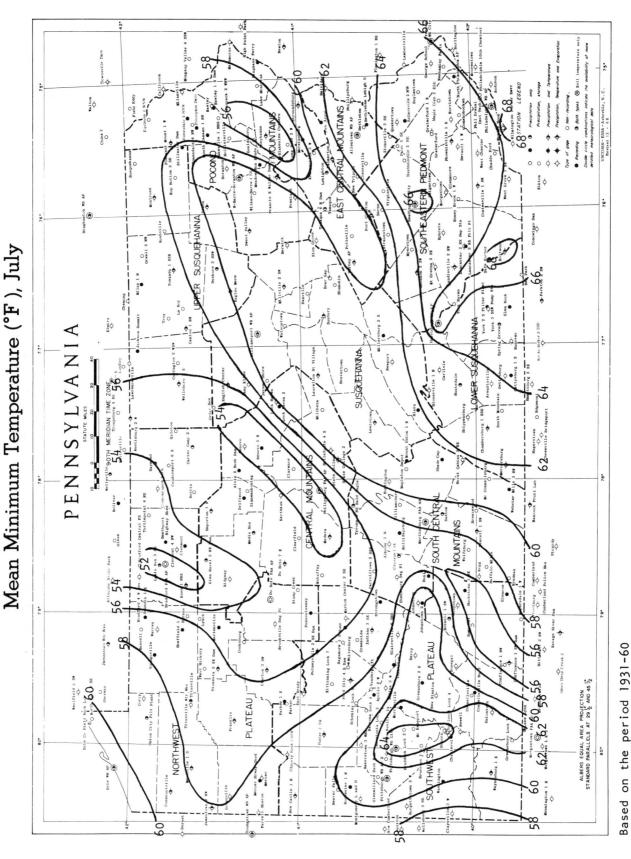

ALBERS EQUAL AREA PROJECTION
STANDARD PARALLELS AT 29½ AND 45½

Based on the period 1931-60

Isolines are drawn through points of approximately equal value. Caution should be used in interpolating on these maps, particularly in mountainous areas.

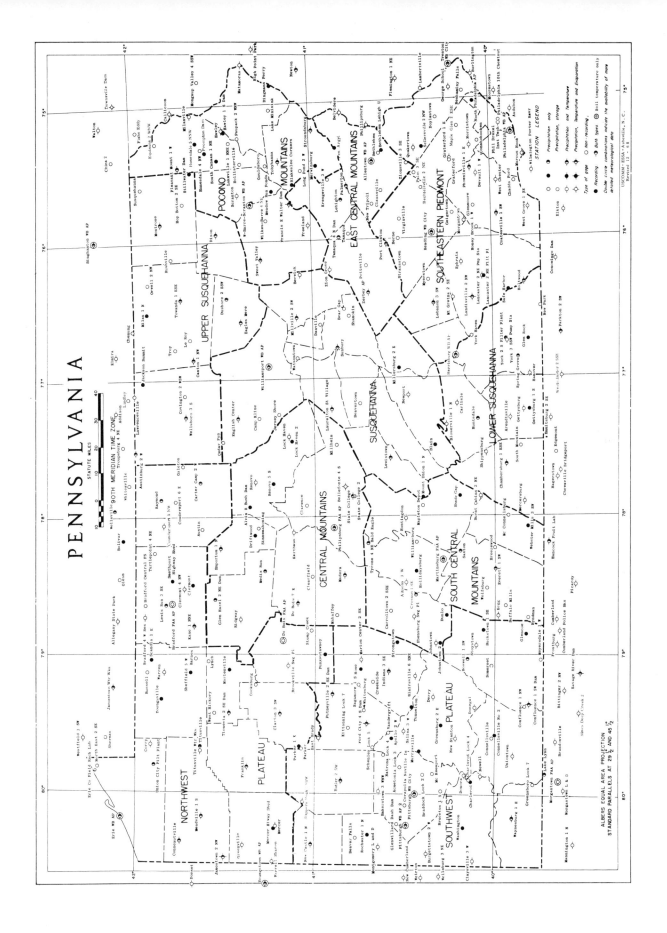

PENNSYLVANIA

STATUTE MILES

ALBERS EQUAL AREA PROJECTION
STANDARD PARALLELS AT 29½ AND 45½

— 339 —

THE CLIMATE OF
RHODE ISLAND

by
A. Boyd Pack

November 1959

PHYSICAL DESCRIPTION: -- Rhode Island, the smallest of the states, shares the southeastern corner of New England with a portion of Massachusetts. The State extends for 50 miles in a north-south direction and has an average width of about 30 miles. The total area, including Block Island some 10 miles offshore, is 1,497 square miles of which Narragansett Bay occupies about 25 percent.

There are three topographical divisions of the State. A narrow coastal plain occurs along the south shore and around Narragansett Bay with an elevation of less than 100 feet. A second division lies to the north and east of the Bay with gently rolling uplands of up to 200 feet elevation. The western two-thirds of Rhode Island consists of predominantly hilly uplands of mostly 200 to 600 feet elevation but rising to a maximum of 800 feet above sea level in the northwest corner of the State.

Narragansett Bay has a very irregular shoreline, indented by numerous small bays or coves and the mouths of the Taunton and Blackstone Rivers. The Bay contains several islands of which the one known as Aquidneck, or Rhode Island, is the largest. The shore line facing Long Island Sound is about 20 miles long and has many fine beaches. No point in the State is more than 25 miles from the ocean.

The Blackstone River in northeastern Rhode Island is the principal river. A number of smaller rivers or brooks originating in the western uplands of the State or in southeastern Massachusetts empty into Narragansett Bay or Long Island Sound.

GENERAL CLIMATIC FEATURES: -- The chief characteristics of Rhode Island's climate may be summarized as follows: (1) Equable distribution of precipitation among the four seasons; (2) large ranges of temperature both daily and annual; (3) great differences in the same season of different years; and (4) considerable diversity of the weather over short periods of time. These characteristics are modified by nearness to the Bay or ocean, elevation, and nature of the terrain.

Rhode Island lies in the "prevailing westerlies", the belt of generally eastward air movement which encircles the globe in middle latitudes. Embedded in this circulation are extensive masses of air originating in higher and lower latitudes and interacting to produce storm systems. A large number of these systems and air-mass fronts pass near or over Rhode Island in a year.

Air masses affecting the State belong to three types: (1) Cold, dry air pouring down from sub-arctic North America; (2) warm, moist air streaming up on a long overland journey from the Gulf of Mexico and adjacent waters; and (3) cool, damp air moving from the North Atlantic. Because the atmospheric flow is usually from continental areas, Rhode Island is more influenced by the first two types than it is by the third. The ocean constitutes an important modifying factor, particularly in southeast sections of the State, but does not dominate the climate as it would if the prevailing circulation was onshore.

The procession of contrasting air masses and the relatively frequent passage of "Lows" bring about a roughly twice-weekly alternation from fair to cloudy or stormy weather, usually attended by abrupt changes in temperature, moisture, sunshine, wind direction, and speed. There is no regular or persistent rhythm to this sequence, and it is sometimes interrupted by periods of several days, or infrequently of a few weeks with the same weather pattern.

Day-to-day variety rather than monotony is the main feature of Rhode Island's weather. Changeability is also one of its features on a longer time scale. That is, the same month or season will exhibit varying characteristics over the years, sometimes in close alternation and sometimes arranged in similar groups for successive years. A "Normal" month, season, or year is the exception rather than the rule.

As just outlined, the basic climate does not result from the predominance of any single controlling weather regime. It is composed of a large variety of weather patterns. Hence, weather averages in Rhode Island are not useful for important planning purposes and should be supplemented by more detailed climatological analysis.

TEMPERATURE: -- The mean annual temperature ranges from 48° to 49°F., except near the south shore, Narragansett Bay, and in the large built-up area around Providence, where it is 50° to 51°F. Southwestern Rhode Island, from 4 to 10 miles inland, exhibits a coolness not suggested by the nearness to the ocean or the general elevation of 50 to 150 feet. Here the annual mean temperature is not more than 48°F., making the section as cool as the cooler areas of the northwest interior.

The average daily minimum temperature in January and February is 19° to 20°F. over about two-thirds of the State, increasing to near 25°F. in immediate coastal sections. The number of days with minimum temperature of zero or below average one or less per year in the Bay and coastal areas. The number increases to about 5 per year in most of the interior. In a particularly cold winter month as many as 6 to 8 days with such temperatures are observed in southwestern Rhode Island, a few miles inland from the coast.

A maximum temperature of 32°F. or lower occurs on an average of 20 to 25 days per year along the shoreline and 30 to 40 days in the remainder of the state.

Summer temperatures are considerably influenced by proximity to the coastal waters and the frequent onshore flow of air during the warmer months. The average July maximum temperature is about 80°F., except in the northwestern interior where it is a few degrees higher. The greatest number of hot days occur in the metropolitan areas and in parts of the northern interior. Here, about 8 to 10 days of temperatures 90°F. or higher may be expected per year with a variation of from 2 to 5 days in cool summers to 20 or more in exceptionally warm summers. Near the immediate coast the occurrence of 90°F. temperatures is limited to 1 day in the average summer if it occurs at all. Temperatures of 100°F. or higher have been recorded in the northern interior in an occasional year. Among eight weather stations scattered over the State and operated for 14 years by the Rhode Island Agricultural Experiment Station, a temperature of 100°F. was noted only once.

The length of the freeze-free season, as limited by the occurrence of temperatures of 32° or lower, averages from 155 to 180 days in most of the State.

Exceptions are in the southwestern interior with an average length of 130 to 145 days and in the immediate Bay area with 200 days or more. Near the southeastern shore of the Bay the first autumn freezing temperature is considerably delayed compared to the rest of the State. From year to year there is a good deal of variation in the length of period free of temperatures 32°F. or lower.

Grasses and hardy crops begin growing about mid-April in the interior and by early April in areas modified by oceanic influence. The growing season for freeze-sensitive crops starts about a month later in both areas. It comes to a conclusion by early October in the interior and by the end of October or early November in the southeastern corner. Climatic differences of temperature in this small State are very striking in the fall season. Autumnal coloration of foliage will be past its peak of brilliance in the northwestern interior before leaves have begun to noticeably turn color in the Newport area of the southeast.

PRECIPITATION: -- The climate of Rhode Island is characterized by the rather even distribution of precipitation throughout the year. Storm centers and their accompanying fronts are the principal year-round producers of precipitation. Storms moving up the Atlantic coast generally yield the heaviest amounts of rain and snow. Bands and patches of thunderstorms or convective showers contribute considerable precipitation in the summer and make up the difference resulting from decreased activity of the storm centers. In comparison with the general storms, these are of brief duration, but they yield the heaviest local rainfall.

Variations in precipitation from month to month are sometimes extreme in Rhode Island. A month having 5 inches or more may be preceded or followed by one with less than 2 inches of precipitation, in any season. Months with less than 1 inch generally over the State are known to occur as well as those with precipitation in excess of 8 inches. Such large fluctuations, however, are not characteristic of the precipitation supply in the State. Consequently, prolonged droughts are infrequent. So are widespread floods.

Annual precipitation averages 42 to 46 inches over most of the State, with a tendency for decreasing amounts from west to east. It varies from about 40 inches in the immediate southeastern Bay area and on Block Island to 48 inches in the western uplands. Total precipitation in the freeze-free season of April through October shows similar differences over the State with an average of 22 to 24 inches near the Bay and 26 to 29 inches in the western interior.

While there are no pronounced wet and dry months as in other climates, the months of May through July are relatively dry in proximity of the Bay. The average total precipitation for each of these months is 2.5 to 3.0 inches. October and February have an average total of slightly more than 3 inches over most of the State. The remaining months each yield from 3.5 to 4.0 inches.

Measurable precipitation falls on an average of 1 day in 3 or on approximately 120 days per year. Periods of 5 days or more of successive daily precipitation occur a few times during most years. On the other hand, extended periods of little or no precipitation are observed nearly every summer or early fall. Such a period may last from 10 to 20 days.

Twenty-four hour periods with rainfall of more than 4 inches have been observed in all parts of Rhode Island on rare occasions. Such heavy rainfalls have occurred most frequently during the summer and early fall months.

SNOWFALL: -- The average annual snowfall in Rhode Island increases from about 20 inches on

Block Island and along the southeast shores of Narragansett Bay to from 40 to 55 inches in the western third of the State. Areas near the western and northern shores of the Bay, including greater Providence, have an average of from 25 to a little more than 30 inches of snow per year. In mild winters the snowfall in the southeast may total only 10 to 15 inches, while a light annual total in the western interior would amount to 25 to 30 inches. The more snowy winters will yield a total of 35 to 40 inches to areas like Block Island and 60 inches or more to the western uplands.

Most of the snow falls in January and February, with each month averaging about 10 inches in the Providence area, for example, and 5 inches at Block Island. However, there are occasional winters when in coastal sections, particularly, heavier monthly amounts will occur in December or March.

In the western and northern portions of the State the first snowfall of 1 inch or more usually occurs in mid or late November. The southeastern Bay area does not observe measurable snow before December in the great majority of years. In about half of the years of record the total winter snowfall has been less than 1.5 inches by January 1 on Block Island. The last measurable snowfall usually occurs by late March in the populous areas of the State, although an April snowstorm is by no means rare.

The average number of days with 1 inch or more of snow on the ground also increases from the shore areas to the western interior. In the latter, a snow cover prevails most of the time from mid or late December to about mid-March. Near the Bay a snow cover does not last more than a few days unless a heavy snowstorm is followed by prolonged cold temperatures.

WINDS AND STORMS: -- The prevailing wind in Rhode Island is northwesterly from December through March and southwesterly in the remaining months. An important feature of the climate is the sea breeze which affects a considerable portion of the State's area. From approximately late spring to midautumn this cool onshore wind blows during the afternoon hours and penetrates from 5 to 10 miles inland. Since much of Rhode Island is within 10 miles of the Sound or Bay, the relatively cool summer maximum temperatures can be accounted for.

Thunderstorms occur on an average of about 20 days per year, although the number in some years may be as many as 30 in the western third of the State. Often these storms are accompanied by high winds and occasionally by destructive hail. They may cause considerable damage to property and crops. Storms of glaze or freezing rain are usually a part of the winter's weather, especially in the Bay area. Extensive disruption of telephone and power lines result, as well as the serious crippling of surface transportation.

Coastal storms or "northeasters", aside from hurricanes, are the most serious weather hazard in Rhode Island. They generate very strong winds and heavy rains, and produce the greatest snowfalls in the winter. Heavy water damage results along the shores of Long Island Sound and Narragansett Bay when these storms occur at the time of high or rising tide.

Hurricanes or storms of tropical origin occasionally affect the State during the summer or fall months as they move on a path well out to sea. However, within the past 20 years 4 hurricanes (1938, 1944, and two in 1954) have passed over Rhode Island. They have caused enormous damage to property and heavy loss of life. In contrast, tornadoes have been recorded only twice in the State within the last 150 years.

While the frequency of floods is not high, a few major ones have occurred in the last 35 years, particularly along the Blackstone River. Localized thunderstorms with heavy and intense rainfall on occasions cause damaging flash floods in the small as well as the larger streams of the State. Major floods have occurred in November 1927, March 1936, July and September 1938, and August 1955.

OTHER CLIMATIC ELEMENTS: -- The percentage of possible sunshine averages 55 to 60 percent, ranging from about 50 percent in the winter months to a little over 60 percent during the summer. The average number of clear and cloudy days per year are about equal with 125 to 130 each. The highest number of clear days per month usually occur in September or October, while the maximum number of cloudy days are noted in December and January.

Heavy fog is observed on an average of about 50 days per year on Block Island and in the southeastern areas of the Bay. This number decreases to 30 or 35 along the western and northern shores of the Bay and to about 25 days in the western interior. Fog occurs on an average of 1 day in 4 during the late spring and early summer months on Block Island. In an especially foggy month as many as 20 days of heavy fog may be observed at this island outpost.

The humidity tends to be lowest in the early spring season and highest in late summer or early fall. While an occasional summer day may be uncomfortable from a combination of high humidity and temperature in the interior and urban areas of Rhode Island, the frequency of such days is much less than in the Southern or Midwestern States.

CLIMATE AND THE STATE ECONOMY: -- The long coastline with numerous beaches and harbors make Rhode Island a strong attraction for bathers, fishermen, and sailing enthusiasts. These recreational activities, favored by the climate, contribute greatly to the economy of this State. Rhode Island is primarily an industrial state with agriculture secondary in importance. The agriculture is of an intensive type and about 10 percent of the land area is devoted to crop and pasture production. An index of crop production shows that crop yields in Rhode Island are above the national average.

The climate plays a significant role in the State's agriculture. A summer mean temperature of close to 70°F. is favorable for the production of alfalfa, mixed grass hays, and pasture. Thus dairying is the leading agricultural enterprise. Poultry raising ranks second and represents about 20 percent of the total farm income.

Potatoes are the major single crop as southern Rhode Island ranks among the important potato producing areas of New England. The cool summer temperatures and adequate precipitation are ideal for this crop. The mean maximum temperature in July, for example, is only slightly warmer than it is in the great potato area of Aroostook County, Maine.

A considerable variety of truck and fruit crops are produced, and together they account for about 10 percent of the total farm income. The truck crops are grown mostly in close proximity to Narragansett Bay. The mild nights, delayed fall freezes, and a reduced freeze hazard in the mid and late spring make this area very satisfactory for these crops. Central Rhode Island contains most of the commercial apple and peach orchards, while various small fruits are produced in scattered areas of the eastern half of the State. Field corn and a small acreage of small grains are grown for livestock feeding.

Forests cover about 75 percent of the total land area. While the income from wood and wood products is relatively small, the forests are a valuable resource in erosion and flood control. They are also a tourist attraction during the autumnal coloration. A wide range of deciduous and coniferous trees are supported by the temperature and precipitation features of the climate. The greenhouse and nursery industry is becoming increasingly important in the agricultural economy of the State. A climate favorable to a wide variety of ornamental trees, shrubs, and flowers has helped to bring

this enterprise to third rank in the agricultural economy of Rhode Island.

The growth and development of industry has been aided by the mild, temperate climate. Comfortable summer temperatures, relatively mild winters, and ample rainfall make the climate tolerable to the many aspects of the industrial economy.

SELECTED REFERENCES

General:

1. National Planning Association: The Economic State of New England. 1954.

2. U. S. Dept. of Agriculture: Atlas of American Agriculture. 1936.

3. --------------------------: Climate and Man. Yearbook of Agriculture for 1941.

Specialized:

1. Brooks, C. F.: New England Snowfall. Monthly Weather Review, Vol. 45, 1917.

2. -------------: The Rainfall of New England., General Statement. Journal New Eng. Water Works Assoc., Vol. 44, 1930.

3. Church, P. E.: A Geographical Study of New England Temperatures. Geographical Review, Vol. 26, 1936.

4. Eustis, R. S.: Winds over New England in Relation to Topography. Bulletin Amer. Met. Soc., Vol. 23, 1942.

5. Goodnough, X. H.: Rainfall in New England, Jour. New Eng. Water Works Assoc., Vols. 29, 1915; 35, 1921; 40, 1926.

6. Harris, B. K., and Odland, T. E.: Rhode Island Weather. Rhode Island Agricultural Experiment Station, Bull. 299 (1948)

7. Perley, S.: Historic Storms of New England. 1891.

8. Stone, R. G.: Distribution of snow depths over New York and New England. Trans. Amer. Geophy. Union. 1940.

9. -------------: The average length of the season with snow cover of various depths in New England. Trans. Amer. Geophy. Union. 1944.

4. --------------------------: Local Climatological Data with Comparative Data. Issued monthly and annually for Bridgeport, New Haven, and Hartford, Conn.

5. Upton, W.: Characteristics of the New England Climate. Annals Harvard Astron. Observatory. 1890.

10. Westveld, Marinus, et al: Natural Forest Vegetation Zones in New England. Jour. of Forestry, Vol. 54, 332-338. 1956.

11. Weber, J. H.: The Rainfall of New England. Historical Statement. Annual Rainfall. Seasonal Rainfall. Mean Monthly Rainfall of Southern New England. Maximum and Minimum Rainfall of Southern New England. Jour. New Eng. Water Works Assoc., Vol. 44, 1930.

12. White, G. V.: Rainfall in New England. Jour. New Eng. Water Works Assoc. Vols. 56, 1942; 57, 1943.

13. Brown, R. A.: Twisters in New England. Unpublished compilation of historical records of tornadoes. 1957.

14. Weather Bureau Technical Paper No. 15 - Maximum Station Precipitation for 1, 2, 3, 6, 12, and 24-Hours.

15. Weather Bureau Technical Paper No. 16 - Maximum 24-Hour Precipitation in the United States. Washington, D. C. 1952.

16. Weather Bureau Technical Paper No. 25 - Rainfall Intensity-Duration-Frequency Curves. For selected stations in the United States. Alaska, Hawaiian Islands, and Puerto Rico.

BIBLIOGRAPHY

(A) Climatic Summary of the United States (Bulletin W) 1930 edition, Massachusetts, Rhode Island and Connecticut. Section 86. U. S. Weather Bureau

(B) Climatic Summary of the United States, New England. Supplement for 1931 through 1952 (Bulletin W Supplement). U. S. Weather Bureau

(C) Climatological Data - Rhode Island. U. S. Weather Bureau

(D) Climatological Data National Summary. U. S. Weather Bureau

(E) Hourly Precipitation Data - New England. U. S. Weather Bureau

(F) Local Climatological Data, U. S. Weather Bureau, for Block Island and Providence, Rhode Island.

FREEZE DATA

STATION	Freeze threshold temperature	Mean date of last Spring occurrence	Mean date of first Fall occurrence	Mean No. of days between dates	Years of record Spring	No. of occurrences in Spring	Years of record Fall	No. of occurrences in Fall
RHODE ISLAND								
BLOCK ISLAND WB CITY	32	04-09	11-16	221	30	30	30	30
	28	03-23	11-27	249	30	30	30	30
	24	03-14	12-03	264	30	30	30	30
	20	03-08	12-09	276	30	30	30	27
	16	02-28	12-15	290	30	30	30	24
KINGSTON	32	05-08	10-05	150	30	30	30	30
	28	04-24	10-16	176	30	30	30	30
	24	04-08	10-28	203	30	30	29	29
	20	03-24	11-16	237	30	30	29	29
	16	03-12	11-26	259	30	30	29	29
PROVIDENCE WB CITY	32	04-13	10-27	197	30	30	30	30
	28	04-01	11-11	224	29	29	30	30
	24	03-18	11-24	251	29	29	30	30
	20	03-12	12-02	265	29	29	30	30
	16	03-05	12-07	276	29	29	30	29
CONNECTICUT								
HARTFORD	32	04-22	10-19	180	30	30	30	30
	28	04-06	11-02	210	30	30	30	30
	24	03-24	11-17	239	30	30	30	30
	20	03-13	11-29	261	30	30	30	30
	16	03-08	12-07	273	30	30	30	29
NEW HAVEN WB AP	32	04-15	10-27	195	30	30	30	30
	28	03-28	11-10	226	30	30	30	30
	24	03-18	11-23	249	30	30	30	30
	20	03-10	12-03	268	30	30	30	29
	16	03-05	12-10	280	30	30	30	27
PUTMAN	32	05-15	09-24	133	15	15	17	17
	28	04-29	10-08	162	15	15	17	17
	24	04-13	10-19	189	15	15	17	17
	20	03-30	11-06	221	15	15	16	16
	16	03-19	11-18	244	15	15	16	16

Data in the above table are based on the period 1921-1950, or that portion of this period for which data are available.

Means have been adjusted to take into account years of non-occurrence.

A freeze is a numerical substitute for the former term "killing frost" and is the occurrence of a minimum temperature at or below the threshold temperature of 32°, 28°, etc.

Freeze data tabulations in greater detail are available and can be reproduced at cost.

*MEAN TEMPERATURE AND PRECIPITATION

STATION	JANUARY Temperature	JANUARY Precipitation	FEBRUARY Temperature	FEBRUARY Precipitation	MARCH Temperature	MARCH Precipitation	APRIL Temperature	APRIL Precipitation	MAY Temperature	MAY Precipitation	JUNE Temperature	JUNE Precipitation	JULY Temperature	JULY Precipitation	AUGUST Temperature	AUGUST Precipitation	SEPTEMBER Temperature	SEPTEMBER Precipitation	OCTOBER Temperature	OCTOBER Precipitation	NOVEMBER Temperature	NOVEMBER Precipitation	DECEMBER Temperature	DECEMBER Precipitation	ANNUAL Temperature	ANNUAL Precipitation
RHODE ISLAND																										
BLOCK ISLAND WB AP	31.9	3.67	30.9	3.25	37.1	3.54	44.9	3.37	54.2	2.96	63.0	2.86	69.1	2.55	69.0	3.46	63.5	2.98	54.5	3.10	45.3	3.53	35.1	3.36	49.9	38.63
KINGSTON	29.5	4.12	29.7	3.27	36.7	4.26	45.6	3.83	55.7	3.49	64.2	3.06	70.1	2.57	68.9	4.66	62.2	3.74	52.5	3.21	42.2	4.74	31.7	3.75	49.1	44.70
PROVIDENCE WB AIRPORT	28.7	3.75	28.6	2.84	36.8	3.58	46.0	3.37	56.8	3.02	65.6	3.17	71.0	3.06	6v.4	3.63	62.7	3.19	52.7	2.83	42.6	3.74	31.6	3.45	49.4	39.63
DIVISION	30.7	3.99	30.5	3.15	37.4	4.16	46.6	3.78	56.7	3.31	65.4	3.02	71.5	2.55	70.3	4.33	63.6	3.50	54.3	3.15	43.9	4.44	33.3	3.61	50.4	42.99
CONNECTICUT																										
NORTHWEST																										
CREAM HILL	25.0	3.55	25.4	2.99	33.7	3.88	45.4	3.76	56.9	4.23	65.4	4.70	70.4	4.59	68.5	4.24	61.2	4.24	51.6	3.34	39.3	4.14	27.5	3.60	47.5	47.26
FALLS VILLAGE	25.5	3.09	26.0	2.38	34.6	3.39	45.7	3.55	57.2	3.97	65.7	4.58	70.3	4.01	68.4	3.92	60.9	4.14	50.5	3.03	39.8	3.74	27.7	2.93	47.7	42.73
DIVISION	24.7	3.56	25.1	2.91	33.6	3.91	45.1	3.78	56.6	4.31	64.9	4.78	69.8	4.20	67.8	4.33	60.4	4.21	50.5	3.37	38.7	4.26	27.3	3.57	47.0	47.19
CENTRAL																										
COLLINSVILLE 1 S		4.05		3.25		4.66		4.14		4.30		4.30		3.87		4.91		3.96		3.37		4.55		3.97		49.33
HARTFORD BRAINARD FLD	27.0	3.74	27.5	3.03	36.9	3.53	47.4	3.55	58.9	3.77	67.7	3.76	72.7	3.93	70.4	3.67	63.1	3.41	52.6	2.70	41.7	3.85	30.1	3.49	49.7	42.43
HARTFORD WB AIRPORT	27.0	3.15	28.1	2.57	37.2	3.81	48.0	3.56	59.7	3.66	68.9	3.62	73.8	3.56	71.4	3.54	63.8	3.44	52.9	2.80	41.3	3.48	29.6	3.29	50.1	40.48
MIDDLETOWN WB	26.8	3.91	27.1	3.01	36.0	3.92	46.3	3.96	57.8	4.14	66.7	3.80	71.6	3.23	69.7	3.54	63.0	3.49	52.3	2.81	41.3	4.40	29.8	3.93	49.0	44.14
MIDDLETOWN 4 W	28.3	4.19	28.7	3.21	36.5	4.71	47.4	4.38	58.7	4.48	67.3	4.18	72.7	3.72	70.6	4.51	63.2	4.40	53.8	3.48	42.4	5.06	31.0	4.20	50.1	50.52
STORRS	26.5	3.62	26.7	2.82	34.5	4.34	45.2	3.84	56.4	4.04	65.1	3.65	70.2	3.56	68.4	5.13	61.2	4.07	51.7	3.35	40.6	4.23	29.1	3.56	48.0	46.21
DIVISION	28.0	3.80	28.4	2.92	36.5	4.33	47.1	3.92	58.1	4.03	66.8	3.82	72.1	3.69	70.0	4.62	62.7	4.04	53.0	3.29	41.7	4.47	30.4	3.79	49.6	46.72
COASTAL																										
BRIDGEPORT WB AP	29.2	3.43	29.0	2.97	36.9	3.60	46.3	3.49	57.2	3.60	66.9	3.47	72.8	3.97	71.7	4.43	65.2	3.55	54.4	2.83	43.5	3.59	32.3	3.08	50.5	42.01
LAKE KONOMOC		4.36		3.44		4.99		4.18		4.19		3.55		3.73		4.83		4.49		3.86		4.97		4.29		50.88
NEW HAVEN WB AIRPORT	29.1	3.89	29.1	3.30	37.1	4.12	46.1	3.89	56.7	3.87	65.8	3.81	71.2	3.66	69.8	4.11	63.6	3.46	53.3	3.00	42.9	3.94	31.9	3.94	49.7	44.99
DIVISION	30.1	3.98	30.4	3.09	37.8	4.71	47.9	3.91	58.4	4.00	67.3	3.61	72.8	3.59	71.1	4.95	64.3	4.08	54.4	3.42	43.4	4.30	32.4	3.85	50.9	47.49

* Averages for period 1931-1955, except for stations marked WB which are "normals" based on period 1921-1950. Divisional means may not be the arithmetical average of individual stations published, since additional data from shorter period stations are used to obtain better areal representation.

CONFIDENCE LIMITS

In the absence of trend or record changes, the chances are 9 out of 10 that the true mean will lie in the interval formed by adding and subtracting the values in the following table from the means for any station in the State:

1.5	.45	1.3	.29	1.4	.50	.9	.52	.8	.54	.7	.68	.5	.63	.7	.78	.7	.64	.9	.61	.9	.80	1.1	.55	.4	2.41

COMPARATIVE DATA

Data in the following table are the mean temperature and average precipitation for Kingston, Rhode Island, for the period 1906 - 1930 and are included in this publication for comparative purposes

27.7	4.79	26.9	4.24	35.1	4.22	44.1	4.47	54.3	3.99	63.2	3.41	68.8	3.44	67.3	4.50	61.6	3.46	52.1	3.80	40.8	4.13	30.6	5.04	47.7	49.49

NORMALS, MEANS, AND EXTREMES

BLOCK ISLAND, RHODE ISLAND — STATE AIRPORT

LATITUDE 41° 10' N LONGITUDE 71° 35' W ELEVATION (ground) 110 feet

Month	(a)	Temperature Normal: Daily max	Daily min	Monthly	Extremes: Record highest	Year	Record lowest	Year	Normal degree days	Precipitation: Normal total	Max monthly	Year	Min monthly	Year	Max in 24 hrs	Year	Snow,Sleet: Mean total	Max monthly	Year	Max in 24 hrs	Year
J	77	38.0	25.0	31.9	59	1885	-4	1895	1026	3.67	8.57	1895	0.73	1927	2.90	1915	4.9	19.0	1935	11.1	1927
F	77	36.8	25.0	30.9	58	1890	-10	1934	955	3.25	9.73	1882	0.83	1882	4.54	1886	6.2	22.2	1901	12.0	1949
M	77	43.0	31.1	37.1	68	1907	6	1896	865	3.54	8.54	1899	0.33	1915	4.29	1899	4.0	24.1	1956	11.0	1956
A	77	51.0	38.7	44.9	78	1915	14	1923	605	3.37	8.37	1909	0.85	1942	3.73	1897	0.8	13.2	1887	T	1938
M	77	60.7	47.6	54.2	84	1941	31	1948	335	2.96	8.37	1948	0.41	1923	3.19	1906	T	T	1941	T	1941
J	77	69.4	56.5	63.0	92	1945	43	1881	96	2.86	12.93	1881	T	1957+	4.93	1881	0.0	0.0			
J	77	75.3	62.9	69.1	92	1899	52	1895	6	2.55	8.57	1895	0.22	1898	6.22	1894	0.0	0.0			
A	77	75.3	62.6	69.0	95	1948	49	1911	21	3.46	9.79	1911	0.57	1953	4.86	1916	0.0	0.0			
S	77	69.6	57.4	63.5	90	1953+	38	1914	88	2.98	8.74	1896	0.08	1941	4.88	1953	0.0	0.0			
O	77	60.5	48.4	54.5	79	1946+	29	1936	330	3.10	9.73	1944	0.22	1946	6.63	1955	T	0.3	1925	0.3	1925
N	77	51.4	39.1	45.3	70	1881	14	1932+	591	3.53	9.73	1944	0.19	1917	4.54	1897	0.4	7.5	1898	6.0	1898
D	77	41.4	28.7	35.1	64	1953	-6	1917	927	3.36	9.42	1936	0.83	1955	2.55	1953	3.6	19.9	1904	8.5	1904
Year	77	56.0	43.7	49.9	95	Aug.1948	-10	Feb.1934	5843	38.63	12.93	June 1881	T	June 1957+	6.63	Oct.1955	19.9	24.1	Mar.1956	12.0	Feb.1949

Month	Relative humidity: 1:00a.m.	7:00a.m.	1:00p.m.	7:00p.m.	Wind: Mean hourly speed	Prevailing direction	Fastest mile speed	Direction	Year	Pct possible sunshine	Mean sky cover sunrise-sunset	Days Clear	Partly cloudy	Cloudy	Precip .01"+	Snow 1.0"+	Thunderstorms	Heavy fog	Max 90°+	Max 32°-	Min 32°-	Min 0°-
(a)	10	69	39	62						48		77	77	77	77	77	77	77	77	77	77	77
J	78	79	74	77						45		10	8	13	12	2	*	3	0	8	22	*
F	79	79	72	77						54		10	8	10	12	2	*	3	0	8	22	*
M	79	80	72	81						47		11	9	11	12	1	1	4	0	3	16	*
A	79	83	73	84						56		10	9	11	11	0	1	4	0	*	2	0
M	85	85	76	88						58		10	11	10	11	0	2	8	0	0	0	0
J	90	87	78	90						60		11	12	8	9	0	3	8	*	0	0	0
J	92	88	77	90						62		12	11	8	8	0	4	8	*	0	0	0
A	88	85	75	89						62		12	11	9	9	0	3	5	0	0	0	0
S	87	85	73	85						60		12	10	8	8	0	1	4	0	0	0	0
O	80	81	70	80						59		12	9	11	10	0	*	2	0	0	*	*
N	79	79	73	77						50		9	11	12	11	1	*	2	0	0	5	*
D	78	78	72	75						44		9	9	13	11	1	*	2	0	4	18	*
Year	83	82	74	83						56		125	115	122	122	7	17	53	*	23	85	*

PROVIDENCE, RHODE ISLAND — T. F. GREEN AIRPORT

LATITUDE 41° 44' N LONGITUDE 71° 26' W ELEVATION (ground) 55 feet

Month	(a)	Temperature Normal: Daily max	Daily min	Monthly	Extremes: Record highest	Year	Record lowest	Year	Normal degree days	Precipitation: Normal total	Max monthly	Year	Min monthly	Year	Max in 24 hrs	Year	Snow,Sleet: Mean total	Max monthly	Year	Max in 24 hrs	Year
J	53	36.8	20.6	28.7	68	1932	-9	1957+	1125	3.75	7.12	1953	0.78	1955	2.55	1955	9.6	31.9	1948	15.0	1948
F	53	36.8	20.4	28.6	69	1930	-17	1934	1014	2.84	5.80	1909	1.18	1905	2.39	1924	9.7	26.2	1907	14.5	1907
M	53	45.0	28.6	36.8	83	1945	1	1950	874	3.58	8.31	1953	1.07	1915	2.56	1956	9.3	31.6	1956	14.7	1956
A	53	54.8	37.2	46.0	91	1938	11	1923	570	3.37	6.70	1953	0.72	1942	2.41	1947	1.0	9.6	1939	8.7	1939
M	53	66.3	47.3	56.8	95	1944	30	1938	258	3.02	7.25	1949	0.57	1938	3.10	1939	T	T	1917+	T	1917+
J	53	75.0	56.1	65.6	101	1952	39	1931	58	3.17	7.21	1938	0.04	1949	3.04	1954+	0.0	0.0			
J	53	80.0	61.9	71.0	101	1954	49	1954	0	3.06	6.92	1938	0.24	1952	3.83	1922	0.0	0.0			
A	53	78.5	60.3	69.4	100	1948	45	1940	26	3.63	12.24	1946	0.78	1916	5.47	1955	0.0	0.0			
S	53	72.0	53.3	62.7	99	1953	33	1914	107	3.19	9.74	1944	0.48	1914	6.17	1932	T	T	1957+	T	1957+
O	53	62.1	43.2	52.7	90	1949	25	1936	381	2.83	7.00	1955	0.15	1924	3.31	1939	T	10.2	1945	6.9	1945
N	53	50.9	34.7	42.8	82	1950	12	1932	672	3.74	8.50	1917	0.31	1917	3.71	1953	1.2	26.7	1945	11.8	1945
D	53	39.4	23.7	31.6	68	1912	-12	1917	1035	3.45	9.44	1936	0.58	1955	2.78	1936	5.6	28.7	1945	11.8	1945
Year	53	58.1	40.6	49.4	102	Aug.1948	-17	Feb.1934	6125	39.63	12.24	Aug.1946	0.04	June 1949	6.17	Sept.1932	33.0	31.9	Jan.1948	15.0	Jan.1948

Month	Relative humidity: 1:00a.m.	7:00a.m.	1:00p.m.	7:00p.m.	Wind: Mean hourly speed	Prevailing direction	Fastest mile speed	Direction	Year	Pct possible sunshine	Mean sky cover sunrise-sunset	Days Clear	Partly cloudy	Cloudy	Precip .01"+	Snow 1.0"+	Thunderstorms	Heavy fog	Max 90°+	Max 32°-	Min 32°-	Min 0°-
(a)	18	18	18	18	44	9	49	49	49	53	53	53	53	53	53	53	53	53	53	53	53	53
J	71	73	60	67	11.8	NW	70	NW	1919	49	5.9	10	7	14	12	3	*	2	0	10	26	1
F	73	73	56	65	12.5	SW	63	SW	1953	55	5.5	11	8	9	12	3	1	1	0	7	24	1
M	72	72	51	61	12.5	NW	69	SW	1929	58	5.5	11	9	11	12	1	1	3	0	2	18	0
A	75	74	55	63	12.1	N	57	N	1929	57	5.7	9	9	12	11	*	1	3	*	*	5	0
M	81	81	57	68	10.8	SW	56	SW	1932	60	5.6	9	11	11	11	0	3	3	*	0	*	0
J	84	76	56	70	10.0	SW				63	5.4	10	12	9	10	0	4	3	1	0	0	0
J	87	79	56	69	9.1	SW	56	SW	1924	63	5.2	10	12	9	9	0	5	3	1	0	0	0
A	89	80	57	73	8.9	SW	49	SW	1920	62	5.0	11	11	9	9	0	4	4	1	0	0	0
S	87	83	56	72	9.7	SW	95	SW	1938	62	4.7	12	12	8	8	0	2	3	*	0	0	0
O	84	85	54	74	10.3	SW	60	W	1925	60	4.9	11	10	10	8	*	1	4	0	0	1	*
N	80	80	57	71	11.4	SW	51	SW	1924	51	5.5	10	11	10	10	1	*	5	0	1	11	0
D	73	74	58	68	11.4	W	85	WNW	1916	50	5.7	10	8	13	11	2	*	2	0	6	23	*
Year	79	77	56	70	10.8	SW	95	SW	Sept.1938	58	5.4	125	110	130	123	10	21	32	9	26	108	2

Mean Maximum Temperature (°F.), January

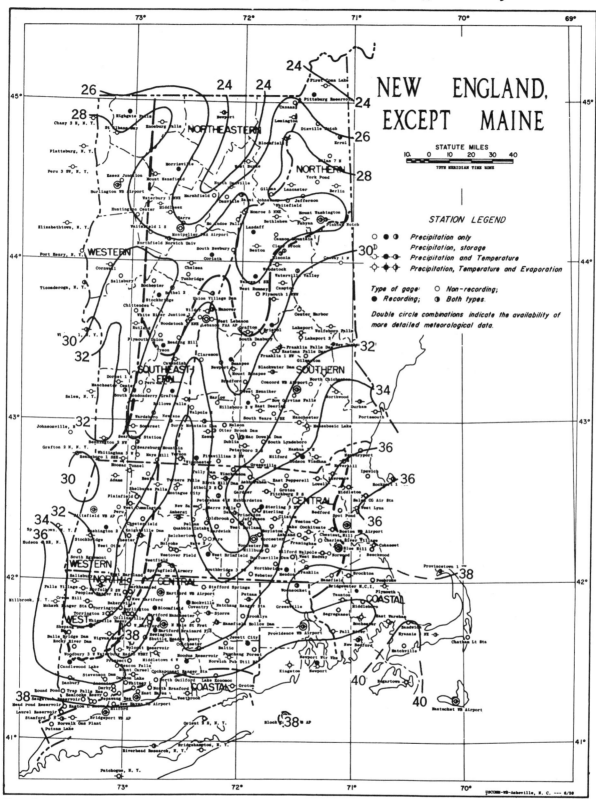

Based on period 1931-52

Isolines are drawn through points of approximately equal value. Caution should be used in interpolating on these maps, particularly in mountainous areas.

Mean Minimum Temperature (°F.), January

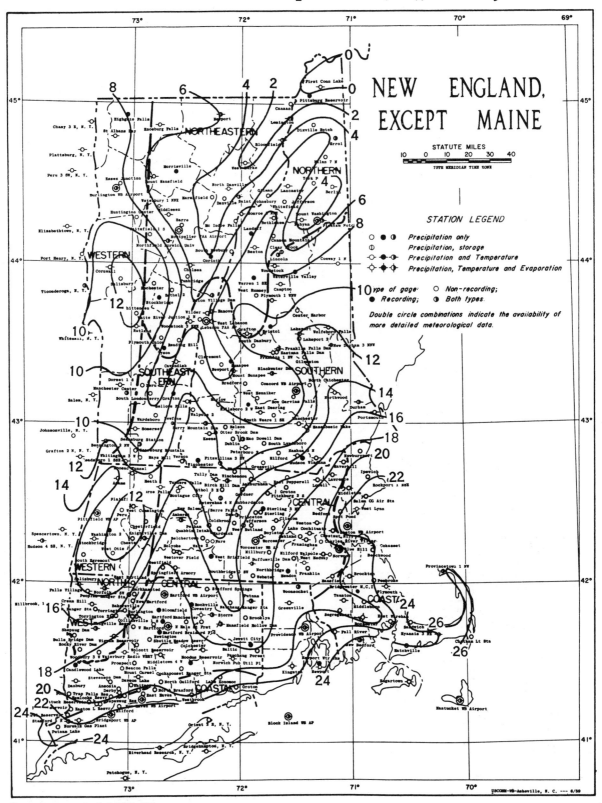

NEW ENGLAND, EXCEPT MAINE

STATUTE MILES

75TH MERIDIAN TIME ZONE

STATION LEGEND

○ ● ◐ Precipitation only
○ ● ◐ Precipitation, storage
○ ● ◐ Precipitation and Temperature
◇ ◆ ◈ Precipitation, Temperature and Evaporation

Type of gage: ○ Non-recording; ● Recording; ◐ Both types.

Double circle combinations indicate the availability of more detailed meteorological data.

Based on period 1931-52

Isolines are drawn through points of approximately equal value. Caution should be used in interpolating on these maps, particularly in mountainous areas.

Mean Maximum Temperature (°F.), July

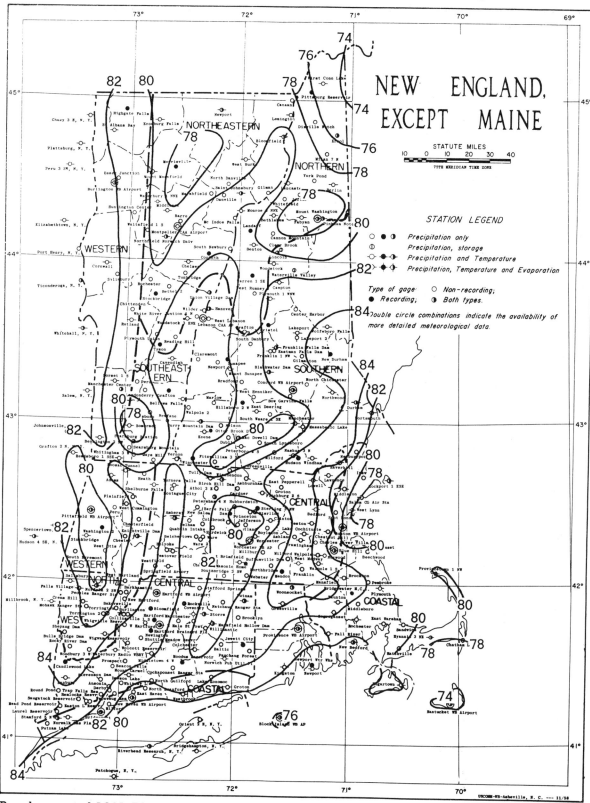

NEW ENGLAND, EXCEPT MAINE

STATUTE MILES
10 0   10 20 30 40
75TH MERIDIAN TIME ZONE

STATION LEGEND

○ ● ◑ Precipitation only
◉ Precipitation, storage
◎ ● ◐ Precipitation and Temperature
◉ ● ◑ Precipitation, Temperature and Evaporation

Type of gage: ○ Non-recording;
● Recording; ◑ Both types.

Double circle combinations indicate the availability of more detailed meteorological data.

Based on period 1931-52

Isolines are drawn through points of approximately equal value. Caution should be used in interpolating on these maps, particularly in mountainous areas.

Mean Minimum Temperature (°F.), July

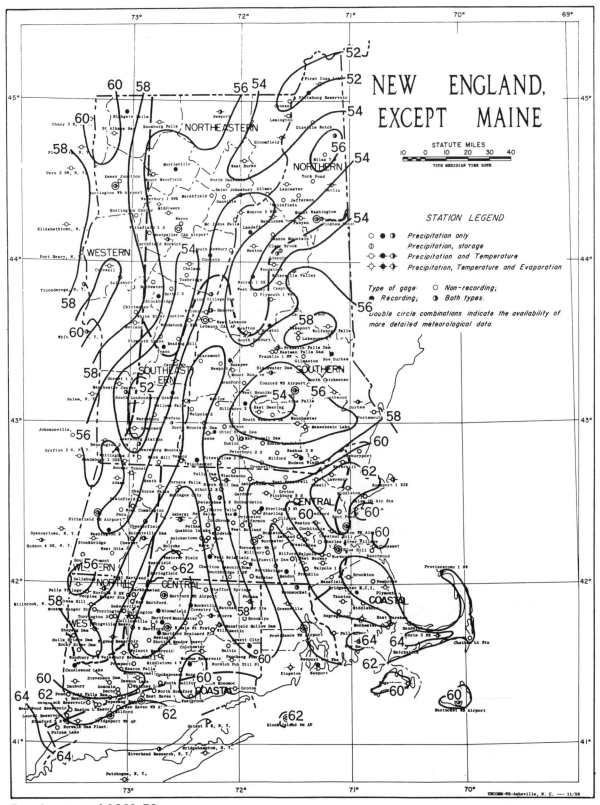

NEW ENGLAND, EXCEPT MAINE

STATUTE MILES
10   0   10   20   30   40
75TH MERIDIAN TIME ZONE

STATION LEGEND

○ ◐ ● *Precipitation only*
⊕ *Precipitation, storage*
◌-◐-● *Precipitation and Temperature*
◇ ◈ ◆ *Precipitation, Temperature and Evaporation*

Type of gage: ○ *Non-recording;*
● *Recording;* ◐ *Both types.*

Double circle combinations indicate the availability of more detailed meteorological data.

Based on period 1931-52

Isolines are drawn through points of approximately equal value.   Caution should be used in interpolating on these maps, particularly in mountainous areas.

USCOMM-WB-Asheville, N. C. --- 11/58

Mean Annual Precipitation, Inches

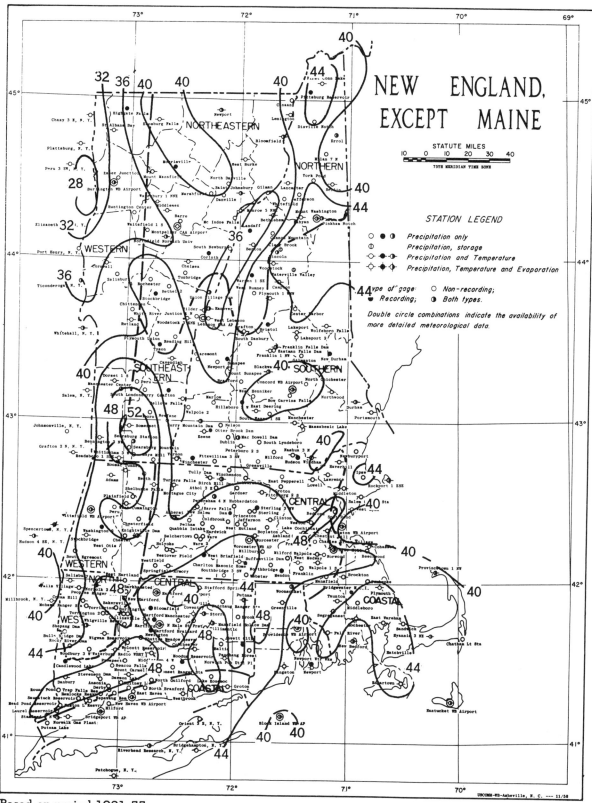

NEW ENGLAND, EXCEPT MAINE

STATUTE MILES
10 0 10 20 30 40
75TH MERIDIAN TIME ZONE

STATION LEGEND

○ ◑ Precipitation only
◐ Precipitation, storage
○—●—◐ Precipitation and Temperature
◇—●—◆ Precipitation, Temperature and Evaporation

Type of gage: ○ Non-recording;
● Recording; ◐ Both types.

Double circle combinations indicate the availability of
more detailed meteorological data.

Based on period 1931-55

Isolines are drawn through points of approximately equal value. Caution should be used
in interpolating on these maps, particularly in mountainous areas.

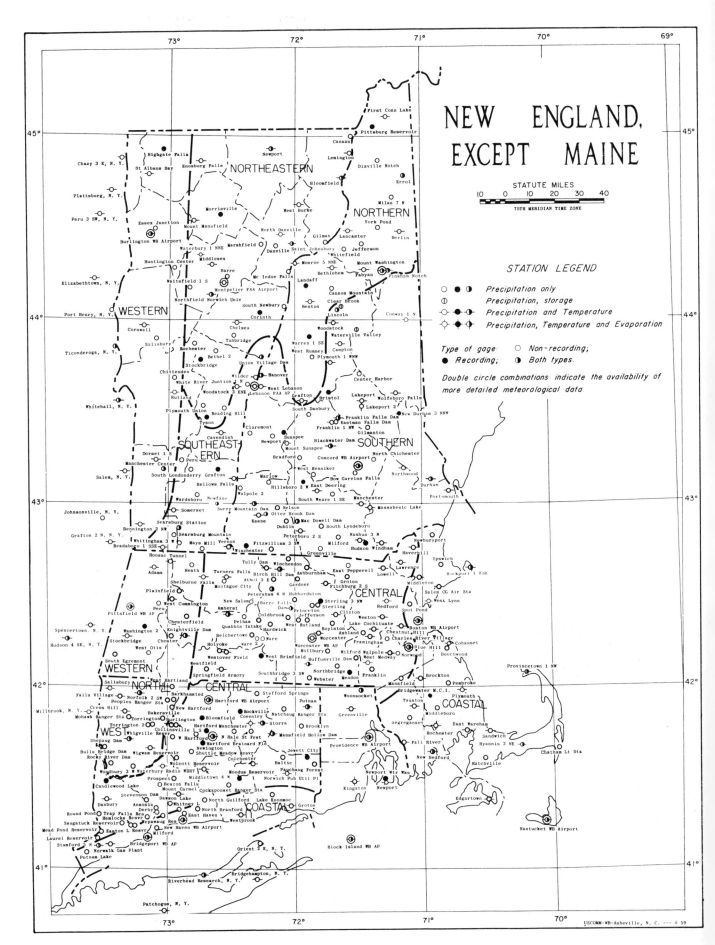

NEW ENGLAND, EXCEPT MAINE

STATUTE MILES
10   0   10   20   30   40
75TH MERIDIAN TIME ZONE

STATION LEGEND

○ ● ◑ ◐   Precipitation only
◍   Precipitation, storage
○ ● ◑ ◐   Precipitation and Temperature
○ ● ◑ ◐   Precipitation, Temperature and Evaporation

Type of gage: ○ Non-recording;
● Recording;   ◑ Both types.

Double circle combinations indicate the availability of
more detailed meteorological data.

USCOMM-WB-Asheville, N. C. --- 6 59

THE CLIMATE OF

SOUTH CAROLINA

by
H. Landers

June 1970

South Carolina is located on the southeastern coast of the United States between the southern part of the Appalachian Mountains and the Atlantic Ocean. Its north-south extent is 220 miles, from 32° to 35.2° north latitude. The mountains in the extreme northwestern part of the State are 240 miles from the coastline. The coastline is 185 miles long and oriented southwest to northeast.

South Carolina shares some common topographic features with several eastern seaboard states. All of these features have a southwest to northeast orientation and extend across the whole State. The Blue Ridge Range of the Appalachian mountains lies in the extreme northwestern part of the State. Elevations range from 1,000 to 2,000 feet with several peaks going over 3,000 feet. Sassafras Mountain, at 3,554 feet elevation, is the highest point in the State. The Mountain Region covers less than 10 percent of the State's area and to its southeast lies the Piedmont Plateau. The Plateau extends nearly to the center of the State with elevations decreasing northwest to southeast from 1,000 to 500 feet. There is a narrow hilly region where the Plateau descends to the Coastal Plain. In South Carolina this "fall line" region is known as the "Sand Hills", elevations range from 500 to 200 feet. The width of the Sand Hills area is about 30 to 40 miles. Between the Sand Hills and the Atlantic Ocean lies the Coastal Plain. The Plain is broad and nearly level with elevations mostly between 50 and 200 feet. About 40 percent of the area of the State lies in the Coastal Plain.

All of the State's rivers drain southeast from the Mountain Region or Piedmont Plateau toward the ocean. There are three major and one minor river-basin systems. The Santee is the largest and drains the entire center portion of the State. The Savannah drains the western part of the State. Both of these systems extend all the way from the mountains to the ocean. The third major system is the Pee Dee, located in the northeastern section. Its tributaries drain parts of the Piedmont area of South Carolina and North Carolina. The Edisto is a lesser river system lying between the Santee and Savannah. It drains a part of the Piedmont Plateau in western South Carolina known as the "ridge" which extends southeastward to northern Aiken county. Several large lakes and reservoirs have been created by damming the major rivers, mostly in the Plateau area. In addition to the large hydroelectric power plants located at these dam sites, nuclear power plants are now being constructed on the Keowee River near the edge of the Mountain Region.

The major coast indentations are Winyah Bay, Charleston Harbor, St. Helena Sound, Port Royal Sound, and Tybee Roads at the mouth of the Savannah River. There are many low sea islands separated from the mainland by shallow straits, sounds, and coastal streams. The Intracoastal Waterway can be found along much of the coastline.

Several major factors combine to give South Carolina a pleasant, mild, and humid climate. It is located at a relatively low latitude (32° to 35° N.) and most of the State is under 1,000 feet in elevation. It has a long coastline along which moves the warm Gulf Stream current. The mountains to the north and west block or delay many cold air masses approaching from those directions. Even the deep cold air masses which cross the mountains rapidly are warmed somewhat as the air is heated by compression when it descends on the southeastern side. This effect can be seen on the maps of minimum temperature in January and to a lesser degree in July, where a fairly large area of relatively higher temperature appears just southeast of the mountains. It is convenient for climatic discussion to divide the State into areas coinciding closely with the topographic features already discussed. Six areas can be defined, each of which is closely associated with the existing temperature and rainfall patterns:

1. The Outer Coastal Plain, elevations 0-50 feet, width 25-30 miles.

2. The Inner Coastal Plain, elevations 51-200 feet, width 40-45 miles.

3. The Sand Hills, elevations 201-500 feet, width 30-40 miles.

4. The Lower Piedmont Plateau, elevations 501-700 feet, width 45-50 miles.

5. The Upper Piedmont Plateau, elevations 701-1,000 feet, width 30 miles.

6. The Mountain Region, elevations greater than 1,000 feet, width 15 miles.

Some major factors affecting temperatures are elevation, latitude, and distance inland from the coast. All three of these factors work together in South Carolina. Lower temperatures can be expected in the Upper Piedmont and Mountain Region, where latitude, elevation and distance inland all have large values. Higher temperatures will result from smaller values of the three factors, as are found along the southern coast. Annual average temperatures are 10° lower in the extreme Upper Piedmont than along the coast between Charleston and Savannah. Except for small-scale and local irregularities, there is a gradual decrease in annual average temperature northwestward from 68° at the coast to 58° at the edge of the mountains. Within the Mountain Region variations in elevation are great over short horizontal distances. Thus, variations in temperature are due almost entirely to elevation differences. All of the record low wintertime temperatures were set in the Mountain Region or extreme Upper Piedmont. The lowest temperature on record was -13° on January 6, 1940, at

Long Creek, which has an elevation of nearly 2,000 feet. But the highest summertime temperatures ever observed are not found along the south coast. The ocean waters have very small daily and annual changes in temperature when compared with the land surface. The air over the coastal water is cooler than the air over the land in summer and warmer than the air over land in winter and this has a controlling effect on the temperatures of locations on and very near the coast. The July average maximum temperature map reveals that the highest temperatures of 92° to 93° are found in the central part of the State with the coast being 4° to 5° cooler. The July maximum and minimum temperature maps viewed together show a daily range along the coast of about 13°, while the range is about 21° in the center of the State. Even in January at the time of maximum temperatures, the air over the land a short distance from the coast is 2° or 3° warmer than at the coastal stations. The daily range along the coast in January is about 16° and about 23° in the center of the State. The highest temperature on record was 111°. This reading occurred three times; at Calhoun Falls on September 8, 1925, Blackville on September 4, 1925, and Camden on June 28, 1954. The record highest July and August temperatures are 110° and 109°, respectively. Clouds and rainfall have a minor effect on temperature. Maximum temperatures in summer are reduced slightly in areas where afternoon cloudiness and rain are persistent. Such an area is found along the Outer Coastal Plain where sea breezes produce clouds and rain nearly every summer day and dissipate at night. Another minor effect is drainage of cold air, mostly in winter, into some of the river and stream valleys causing minimum temperatures to be somewhat lower than they would be otherwise. One example of this takes place in a rather deep section of the Broad River valley from Lockhart to a little north of Columbia. Generally, not enough temperature stations are available for this small-scale effect to be clearly defined on minimum temperature maps.

The growing season for most cultivated crops is limited by the fall and spring freezes. The freeze-free period, the time elapsing between the last temperature of 32° or less in the spring and the first in the fall, is quite important to agriculture. The average length of the freeze-free period varies from about 200 days in the coldest area to about 280 days along the south coast. In the area where most of the major crops are grown, it is from 210 to 235 days, or 7 to 8 months. The average date of the last freezing temperature in spring ranges from March 10 in the south to April 1 in the north. The fall dates range from late October in the north to November 20 in the south. Freezes have occurred as much as four weeks later than the average date in spring and three weeks earlier than the average date in the fall. The minimum temperature is 32° or less on 50 to 70 winter days in the Upper Piedmont and 10 days near the coast. Counties in the Inner Coastal Plain and Sand Hills area have maximum temperatures of 90° or more on 80 summer days. There are 30 such days along the coast and 10 to 30 in the mountains.

Summers are rather hot and air conditioning is desirable at elevations below 500 feet. Fall and spring are mild and winters are rather cool at elevations above 500 feet. Heating of homes and places of business in varying amounts is necessary in all parts of the State.

Rainfall is adequate in all parts of the State. Annual rainfall averages up to 80 inches in the highest part of the Mountain Region and less than 42 inches in parts of the Inner Coastal Plain and the Sand Hills. The general features on the annual rainfall map are related to the topographical features defined earlier. The Mountain Region is wet with amounts of 56 inches or more, the Upper Piedmont is relatively wet with amounts of 48 to 55 inches, the Lower Piedmont is relatively dry with amounts of 43 to 47 inches, the Outer Coastal Plain is relatively wet with amounts of 48 to 53 inches, and the Inner Coastal Plain is relatively dry with amounts of 38 to 47 inches. The Sand Hills area is less clear cut but is in general a relatively wet strip with a small dry area imbedded in it a few miles south of Columbia. The immediate south coast is also on the dry side. In winter (Dec.-Jan.-Feb.), rainfall decreases from 6 or 7 inches per month in the mountains to 3 or 3 1/2 inches along the coast. In summer (June-July-August) there is a maximum of 6 to 8 inches per month along the Outer Coastal Plain, less in the Inner Coastal Plain, a maximum again of 5 to 7 inches in the Sand Hills, only 3 1/2 to 4 1/2 inches per month in the Lower Piedmont, and the Upper Piedmont and Mountain Region receiving 4 to 7 inches per month. In September and early October the northeast coast gets some additional rain from occasional tropical storms, but fall is actually a dry season with amounts dropping to less than 2 inches along the south coast by November. March is a month of heavy rain in all parts of the State ranging from 4 inches in the Coastal Plain to 7 1/2 inches in the Mountain Region. The pattern of a relatively dry Lower Piedmont and a relatively wet Sand Hills begins to appear in April and continues into May at which time the sea breeze maximum in the Outer Coastal Plain begins to appear. This arrangement persists through the summer and begins to break down in early October. May, however, is the driest month in the spring dry period with less than 3 1/2 inches everywhere except the mountains. In summary, the driest period is in October and November when there is little cyclonic storm activity and "Indian Summer" prevails. Rainfall increases gradually and reaches a peak in March when cyclone and cold front activity are at a maximum. There is a general decrease again to a dry period from late April through early June. From the latter part of June through early September is a wet period primarily due to thunderstorm and shower

activity which reaches its peak in July, the wettest summer month. The summer maximum stretches a little into the fall along the coast due to occasional tropical storm activity. The greatest monthly rainfall on record was 31.13 inches at Kingstree in July 1916. The greatest in 24 hours was 13.25 inches in the same month and year at nearby Effingham.

Solid forms of precipitation include snow, sleet, and hail. Hail is not too frequent but does occur with spring thunderstorms from March through early May. These thunderstorms usually accompany squall lines or cold fronts. Snow and sleet may occur separately or together or mixed with rain during the winter months of December through February. Snow may occur from one to three times in winter. Seldom do accumulations remain very long on the ground except in the mountains. Statewide snows of notable amounts can occur when a cyclonic storm moves northeastward along or just off the coast. Two intense storms of this type brought record snowfalls with a belt of greatest amounts running along the Sand Hills and the southern edge of the Lower Piedmont. These two storms, in February 1899 and in February 1914, blanketed the State with depths of 2 inches along the coast and from 8 to 18 inches along the heavy snow belt, with amounts of 4 to 6 inches in the Upper Piedmont. Nearly all of the State's snowfall records were established on Caesars Head Mountain at an elevation of 3,200 feet. The greatest snowfall in 24 hours at Caesars Head was 22 inches, the greatest in one month was 34 inches, and the greatest annual was 48 inches. All of these records were set in 1969 and the greatest month was February. Freezing rain also occurs from one to three times per winter in the northern half of the State. This rain, which freezes on contact with the ground and other objects, can cause hazardous driving conditions, breakage of limbs and tree tops, and breakage of various types of wires and the poles on which they are strung. One of the most severe cases of ice accumulation from freezing rain was in February 1969 in several north central and northeastern counties. Timber losses were tremendous and power and telephone services were seriously disrupted over a large area.

Severe drouths occur about once in 15 years with less severe and less widespread drouths about once in 7 or 8 years. Irrigation is used for most of the truck vegetable crops and tobacco, and by most of the large peach growers. The field crops of corn, cotton, soy beans, and others are largely unirrigated and dry weather takes a heavy toll when it extends over several weeks during the growing season.

The percent of possible sunshine received varies over the State in a way similar to the variation in cloudiness and precipitation. Values in winter range from 50 to 60 percent, in summer from 60 to 70, with the dry periods in spring and fall receiving 70 to 75 percent. The variation in relative humidity with time of day is considerably greater than day to day and month to month variations. Highest values of 80 to 90 percent or more are reached at about sunrise and the lowest values of 45 to 50 percent occur an hour or two after local noon. There is about a 10 percent difference between winter and summer, with summer being the higher of the two seasons. The prevailing surface winds tend to be either from northeast or southwest due to the presence and orientation of the Appalachian Mountains. Winds of all directions occur throughout the State during the year, but the prevailing directions by seasons are; spring--southwest, summer--south and southwest, autumn--northeast, and in winter--northeast and southwest have almost the same frequency. Average surface wind speeds for all months range between 6 and 10 m.p.h. Prevailing winds at levels above the mountain effect are between southwest and northwest in winter and spring, from south to southwest in summer, and from southwest to west in autumn.

Severe weather comes to South Carolina occasionally in the form of violent thunderstorms, tornadoes and hurricanes. Although thunderstorms are common in the summer months, the really violent ones generally accompany the squall lines and active cold fronts of spring. Generally, they bring high winds, hail and considerable lightning and sometimes spawn a tornado. In the 57-year period prior to 1970 there were 252 tornadoes, which is an average of 4 or 5 per year. Many tornadoes were not detected in the early part of that period due to a much smaller population and lack of the organized effort launched in recent years to cut down on loss of life due to these storms. In the 20-year period from 1950 through 1969 the tornado count was 148, an average of 7 or 8 tornadoes per year. Since a tornado is very small and affects a very small area, the probability of any given locality having a tornado in any given year is close to zero. Sixty percent of the tornadoes occur from March through June with April being the peak month with 25 percent. A smaller maximum is found in August and September which accounts for 21 percent of the total. Many are waterspouts or tornadoes that accompany tropical storms and are detected near the coast with some never reaching land. Tornadoes are rather rare from October through February. Only 13 percent of the total are experienced during this five-month period. The worst tornado day in South Carolina's history was April 30, 1924 when two tornadoes struck. The paths of both were unusually long, each over 100 miles, and together they killed 77 persons, injured 778 more, destroyed 465 homes and many other buildings, resulting in many millions of dollars of damage. Tropical storms or hurricanes affect the State about one year out of two. Most of the occurrences are tropical storms which do little damage, frequently bringing rains at a time when they are needed. Most of the hurricanes affect only the Outer Coastal Plain. If they do come far inland they decrease in intensity quite rapidly. The

most devastating hurricane, as far as loss of life is concerned, was the one that struck south of Savannah on August 27, 1893. The fast moving storm had piled up vast amounts of water to the east of its center and the south coast and sea islands were badly inundated. Winds of 120 m.p.h. were measured at Charleston and were probably higher between Charleston and Beaufort. More than 1,000 persons were drowned and damage estimates exceeded $10 million. The greatest amount of property damage, $27 million, was done at Myrtle Beach by a hurricane on October 15, 1954. No lives were lost and the highest windspeed was measured at 100 m.p.h. A hurricane that crossed the State moved inland between Charleston and Savannah on September 29, 1959, and continued on a northerly track. Coastal winds were estimated at 140 m.p.h., damage estimated at $20 million and seven lives were lost. Considerable flooding accompanies hurricanes which come very far inland and high tides occur along the coast to the north and east of the storm centers.

There is minor flooding somewhere in the State every year. It can occur on any of the many streams and rivers. A certain amount of control can be effected on the larger rivers which have dams. There is a major flood about once every 7 or 8 years. They are not generally a threat to human life but are very damaging to crops and livestock, disrupting logging operations and damaging homes, stores and other structures. The most extensive flooding took place in August 1908 when all the major rivers in the State rose from 9 to 22 feet above flood stage. Flooding in July 1916 did not affect the western rivers, the Saluda and Savannah, but broke several of the 1908 records in the eastern two thirds of the State. At the six points where the records were broken, heights above flood stage ranged from 13 to 30 feet.

There have been many earth tremors in South Carolina over the years. The southern part of the Coastal Plain is indicated as earthquake prone on a recently published seismic map. There has been only one major quake recorded in the area in the last 100 years. After having light tremors on the 27th, 28th and 30th, a major shock hit the Charleston area on August 31, 1886, just before 10 o'clock in the evening. The area of greatest intensity of the shock was about 10 miles west-northwest of the center of the city of Charleston. More than 100 buildings, including frame houses, were demolished. Ninety percent of all brick structures suffered some damage and 14,000 chimneys tumbled down. The pulverized mortar from the demolished masonry structures filled the city air with a fine white dust. There were many cracks in the earth and numerous small craters up to 20 feet in diameter which spewed water and sand in the air to heights of 20 feet. There was considerable loud moaning and praying as many thought the day of judgement had surely arrived. Eighty-three persons were killed and many injured. The major shock was felt over an area of 2,800,000 square miles, from Canada to the Gulf of Mexico and from Bermuda to Iowa, Missouri and Arkansas, and it broke windows in Milwaukee.

South Carolina is an agricultural State although there is a steady decrease of farm acreage with time. At the same time the size of the remaining farms is increasing. Industry is increasing steadily and many textile mills and other manufacturing plants dot the countryside mostly in the Plateau section. Important crops are tobacco, cotton, corn, soybeans, peaches, truck crops, and small grains. There are many wooded areas

and lumbering is an important activity. The port of Charleston is very important in shipping the State's products abroad and bringing in needed items from other countries. The State is naturally suited to many kinds of recreational activities with its beautiful beaches on the coast, abundant streams, rivers, lakes and forests in the interior, and mountains in the northwest. Parks and recreation areas have always been ample and are now being increased in number and improved. About the only kinds of recreation not available are those which require ice or snow.

REFERENCES

(1) Kronberg, N. and Purvis, J. C., "Climate of South Carolina", published in Climates of the States, South Carolina, Climatography of the United States, No. 60-38, 1959, ESSA, Environmental Data Service.

(2) Agricultural Weather Research Series. Data for Agricultural Weather Stations and Reports on Agriculture-Weather Relationships. ESSA, Weather Bureau and South Carolina Agricultural Experiment Station at Clemson University, Clemson, South Carolina.

(3) Climatic Research Series. Reports on various aspects of South Carolina Climate. ESSA, Weather Bureau and South Carolina Agricultural Experiment Station at Clemson University, Clemson, South Carolina.

BIBLIOGRAPHY

A. Climatic Summary of the United States (Bulletin W), 1930 Edition, Sections 98 and 99. ESSA, Weather Bureau.

B. Climatic Summary of the United States, South Carolina-Supplement for 1931 through 1952 (Bulletin W Supplement), ESSA, Weather Bureau.

C. Climatic Summary of the United States, South Carolina-Supplement for 1951 through 1960 (Bulletin W Supplement), ESSA, Weather Bureau.

D. Climatic Summaries of Resort Areas, Myrtle Beach and Lake Harwell, South Carolina, Climatography of the United States, No. 21-38-1 and 2. ESSA, Environmental Data Service.

E. Climatological Data-South Carolina, Vols. 38 through 67, ESSA, Environmental Data Service.

F. Climatological Data - National Summary, ESSA, Environmental Data Service.

G. Climatological Summaries for 21 South Carolina Substations. Climatography of the United States, No. 20-33, 1965, ESSA Weather Bureau.

H. Hourly Precipitation Data-South Carolina, ESSA, Environmental Data Service.

I. Local Climatological Data for Charleston, Columbia, Florence, Greenville and Spartanburg, South Carolina. ESSA, Environmental Data Service.

J. Storm Data. ESSA, Environmental Data Service

K. Summary of the Climatological Data by Sections (Bulletin W), Volume II, Sections 87 and 88. ESSA, Weather Bureau.

FREEZE DATA

STATION	Freeze threshold temperature	Mean date of last Spring occurrence	Mean date of first Fall occurrence	Mean No. of days between dates	Years of record Spring	No. of occurrences in Spring	Years of record Fall	No. of occurrences in Fall
AIKEN	32	03-21	11-18	242	30	30	30	30
	28	03-04	12-01	271	30	30	30	27
	24	02-12	12-16	307	30	26	30	17
	20	01-30	12-23	326	30	22	30	13
	16	01-10	12-27	352	30	11	30	6
ANDERSON	32	03-26	11-13	232	30	30	30	30
	28	03-12	11-27	260	30	30	30	29
	24	02-20	12-09	292	30	30	30	24
	20	02-01	12-17	319	30	22	30	18
	16	01-19	12-24	340	30	13	30	10
BEAUFORT 7 SW	32	03-05	11-25	265	25	25	24	23
	28	02-15	12-09	297	25	24	23	18
	24	01-21	12-19	333	25	17	23	10
	20	01-08	@	@	25	8	23	3
	16	01-03	@	@	25	4	23	2
BISHOPVILLE	32	03-27	11-12	230	16	16	15	15
	28	03-09	11-20	256	15	15	15	15
	24	02-20	12-04	287	13	13	15	13
	20	02-02	12-20	321	13	11	15	7
	16	01-18	12-25	342	13	8	15	4
BLACKVILLE	32	03-17	11-18	247	29	29	30	30
	28	03-01	11-28	271	29	29	30	29
	24	02-04	12-12	311	29	22	29	19
	20	01-19	12-21	337	29	16	29	12
	16	01-08	12-28	354	29	9	29	6
CAESARS HEAD	32	04-11	10-30	201	26	26	26	26
	28	03-31	11-06	220	26	26	25	25
	24	03-18	11-20	247	26	26	25	24
	20	03-11	11-28	262	25	25	25	24
	16	02-21	12-09	291	25	25	25	19
CALHOUN FALLS	32	03-29	11-09	225	29	29	28	28
	28	03-14	11-21	252	29	29	27	27
	24	02-25	12-01	278	29	29	27	24
	20	02-08	12-13	308	29	27	27	19
	16	01-25	12-24	333	29	19	27	10
CAMDEN	32	03-26	11-11	231	29	29	30	30
	28	03-14	11-21	252	29	29	30	29
	24	02-28	12-02	276	29	29	30	27
	20	02-03	12-14	314	29	22	30	20
	16	01-17	12-26	343	29	15	29	9
CHARLESTON WB CITY	32	02-19	12-10	294	30	29	30	21
	28	02-01	12-19	321	30	21	30	15
	24	01-10	12-26	350	30	11	30	7
	20	01-04	@	@	30	6	30	4
	16	@	@	@	30	0	30	0
CHERAW	32	04-01	11-03	217	30	30	30	30
	28	03-15	11-16	247	30	30	30	30
	24	03-04	11-29	271	30	30	30	28
	20	02-14	12-12	301	30	28	30	23
	16	01-15	12-23	342	30	14	30	13
CHESTER	32	04-06	11-03	211	19	19	17	17
	28	03-25	11-15	235	19	19	17	17
	24	03-06	11-26	265	19	19	17	17
	20	02-17	12-10	296	19	18	17	13
	16	01-29	12-19	324	17	13	17	8
CLEMSON COLLEGE	32	04-04	10-30	209	30	30	30	30
	28	03-20	11-12	237	29	29	30	30
	24	03-07	11-25	263	29	29	30	30
	20	02-18	12-07	292	29	27	30	25
	16	02-05	12-19	317	29	24	30	17
COLUMBIA WB CITY	32	03-14	11-21	252	30	30	30	30
	28	02-25	12-02	281	30	29	30	27
	24	02-05	12-17	315	30	23	30	18
	20	01-21	12-26	339	30	16	30	9
	16	01-06	@	@	30	10	30	4
CONWAY	32	03-18	11-12	240	30	30	30	30
	28	03-05	11-24	264	30	30	30	30
	24	02-14	12-13	302	30	27	30	24
	20	01-19	12-24	339	30	17	30	12
	16	01-08	12-29	355	30	9	30	5
DARLINGTON	32	03-25	11-10	229	28	28	27	27
	28	03-11	11-19	253	28	23	26	26
	24	02-25	12-02	279	28	28	26	26
	20	02-01	12-16	318	27	21	26	16
	16	01-14	12-25	344	27	12	26	10
EUTAWVILLE	32	03-19	11-08	234	27	27	28	28
	28	03-03	11-21	263	26	25	27	27
	24	02-10	12-06	299	26	22	27	24
	20	01-18	12-21	338	26	15	26	12
	16	01-06	12-27	355	26	7	26	5
FLORENCE 2 N	32	03-17	11-13	241	28	28	28	28
	28	03-02	11-25	267	28	28	28	27
	24	02-16	12-08	295	28	26	28	22
	20	01-21	12-21	334	28	16	28	14
	16	01-11	12-27	350	28	11	28	6
GEORGETOWN	32	03-19	11-15	242	24	24	21	21
	28	02-24	11-25	274	23	22	21	21
	24	02-06	12-10	308	22	17	21	17
	20	01-20	12-25	339	21	12	21	7
	16	01-07	@	@	21	6	21	3
GREENVILLE WB AP	32	03-23	11-17	239	30	30	30	30
	28	03-05	11-29	269	30	30	30	30
	24	02-22	12-11	291	30	30	30	21
	20	02-05	12-20	318	30	23	30	17
	16	01-16	12-27	344	30	14	30	7
GREENWOOD	32	03-22	11-13	236	30	30	30	30
	28	03-09	11-29	265	30	30	30	30
	24	02-19	12-09	294	30	29	30	22
	20	02-01	12-21	323	30	23	30	14
	16	01-15	12-27	346	30	14	30	7
HEATH SPRINGS	32	03-30	11-08	222	30	30	28	28
	28	03-15	11-26	256	30	30	27	27
	24	02-27	12-03	279	30	30	26	22
	20	02-11	12-16	308	29	27	26	18
	16	01-25	12-25	334	28	17	26	9
KERSHAW	32	03-28	11-11	228	30	30	28	28
	28	03-13	11-20	253	30	30	27	27
	24	02-28	12-05	280	30	30	27	23
	20	02-11	12-14	306	30	26	27	20
	16	01-21	12-23	337	28	15	26	11
KINGSTREE	32	03-17	11-10	239	30	30	30	30
	28	03-03	11-21	263	30	30	30	30
	24	02-10	12-04	297	30	24	30	28
	20	01-17	12-20	337	30	16	30	18
	16	01-09	12-27	352	30	11	30	7
LAKE CITY	32	03-21	11-16	240	12	12	13	13
	28	03-04	11-24	265	12	12	13	13
	24	02-08	12-02	298	12	10	13	12
	20	01-20	12-21	335	12	7	13	7
	16	01-11	12-27	351	12	4	13	2
LANDRUM	32	04-01	11-02	216	30	30	29	29
	28	03-18	11-20	247	30	30	29	29
	24	03-03	11-30	272	30	30	29	29
	20	02-17	12-10	296	30	28	29	23
	16	01-30	12-20	324	29	23	29	15
LAURENS	32	04-01	11-06	219	30	30	28	28
	28	03-19	11-19	245	30	30	28	28
	24	03-02	12-05	278	29	29	28	25
	20	02-08	12-12	307	28	24	28	21
	16	01-24	12-22	332	28	16	28	13
LITTLE MOUNTAIN	32	03-29	11-07	223	30	30	30	30
	28	03-15	11-24	254	30	30	30	30
	24	02-26	12-08	284	30	30	30	24
	20	02-05	12-19	317	30	24	30	18
	16	01-22	12-26	338	30	17	30	8
MARION	32	03-26	11-09	228	17	17	17	17
	28	03-18	11-18	245	16	16	16	16
	24	02-25	12-02	280	15	14	16	14
	20	02-01	12-21	324	15	11	16	8
	16	01-11	12-26	350	14	5	16	5

FREEZE DATA

STATION	Freeze threshold temperature	Mean date of last Spring occurrence	Mean date of first Fall occurrence	Mean No. of days between dates	Years of record Spring	No. of occurrences in Spring	Years of record Fall	No. of occurrences in Fall
NEWBERRY	32	04-01	11-05	219	30	30	29	29
	28	03-17	11-19	247	30	29	29	28
	24	02-28	12-03	278	30	30	29	26
	20	02-04	12-14	313	30	24	29	19
	16	01-16	12-25	343	30	13	29	9
ORANGEBURG 2 SE	32	03-18	11-14	241	30	30	30	30
	28	02-28	11-26	271	30	30	30	28
	24	02-04	12-11	310	30	23	30	25
	20	01-16	12-25	343	30	15	30	12
	16	01-06	12-29	357	30	8	30	6
PINOPOLIS DAM	32	03-13	11-20	252	22	22	17	17
	28	02-22	12-01	282	22	21	17	15
	24	01-31	12-16	318	21	15	17	11
	20	01-13	12-24	345	20	8	17	5
	16	01-03	⊕	⊕	20	3	17	1
SALUDA	32	03-29	11-03	219	25	25	29	29
	28	03-18	11-14	242	25	25	29	29
	24	02-27	11-28	274	25	25	29	28
	20	02-04	12-14	312	23	18	29	20
	16	01-18	12-25	341	22	13	29	9
SANTUCK 4 SE	32	04-09	10-30	204	30	30	30	30
	28	03-27	11-07	224	29	29	30	30
	24	03-07	11-20	258	29	29	30	30
	20	02-22	12-04	286	30	28	30	28
	16	01-31	12-15	318	30	21	30	19
SOCIETY HILL	32	03-31	11-07	221	30	30	28	28
	28	03-17	11-16	245	29	29	28	28
	24	03-02	11-28	271	29	28	28	26
	20	02-15	12-14	302	28	26	28	18
	16	01-20	12-25	339	28	17	28	10
SPARTANBURG WB AP	32	03-29	11-11	227	30	30	29	29
	28	03-11	11-24	258	30	30	29	29
	24	02-28	12-04	279	30	30	29	24
	20	02-13	12-16	305	30	27	29	18
	16	01-22	12-24	336	30	16	29	9
TRENTON 1 NNE	32	03-22	11-15	238	27	27	27	27
	28	03-05	11-28	268	26	26	27	26
	24	02-15	12-13	301	25	23	27	16
	20	01-31	12-23	325	25	19	27	11
	16	01-11	12-27	350	25	9	27	5
WALHALLA	32	04-11	10-28	200	30	30	30	30
	28	03-27	11-06	225	30	30	30	30
	24	03-13	11-22	254	30	30	30	30
	20	03-01	12-07	281	30	29	30	25
	16	02-07	12-17	312	30	26	30	16
WALTERBORO	32	03-21	11-18	242	13	13	13	13
	28	03-06	11-26	265	13	13	13	13
	24	02-05	12-10	308	13	10	13	10
	20	01-28	12-23	329	13	8	13	6
	16	01-10	12-28	352	13	5	13	2
WEDGEFIELD	32	03-19	11-17	243	27	27	27	27
	28	03-07	11-24	262	26	26	27	27
	24	02-18	12-12	297	24	24	27	17
	20	01-29	12-25	331	24	17	27	9
	16	01-14	12-28	349	24	12	27	5
WINTHROP COLLEGE	32	04-01	11-07	220	30	30	30	30
	28	03-18	11-22	249	30	30	30	30
	24	03-03	12-01	274	30	30	30	26
	20	02-19	12-11	296	30	28	30	22
	16	01-26	12-23	331	30	18	30	12
YEMASSEE 4 W	32	03-18	11-17	244	30	30	30	30
	28	02-25	11-24	271	30	30	30	30
	24	02-07	12-10	306	30	26	30	21
	20	01-19	12-23	339	30	17	30	10
	16	01-06	12-28	356	30	8	30	5

Data in the above table are based on the period 1921-1950, or that portion of this period for which data are available.

⊕ When the frequency of occurrence in either spring or fall is one year in ten, or less, mean dates are not given.

Means have been adjusted to take into account years of non-occurrence.

A freeze is a numerical substitute for the former term "killing frost" and is the occurrence of a minimum temperature at or below the threshold temperature of 32°, 28°, etc.

Freeze data tabulations in greater detail are available and can be reproduced at cost.

Taken from "Climatography of the United States No. 81-4, Decennial Census of U. S. Climate"

TEMPERATURE (°F) PRECIPITATION (In.)

STATIONS (By Divisions)	JAN	FEB	MAR	APR	MAY	JUNE	JULY	AUG	SEPT	OCT	NOV	DEC	ANN	JAN	FEB	MAR	APR	MAY	JUNE	JULY	AUG	SEPT	OCT	NOV	DEC	ANN
MOUNTAIN																										
CAESARS HEAD	39.0	39.9	45.3	55.0	62.8	69.6	71.5	70.5	65.8	56.6	46.4	39.6	55.2	6.74	6.15	7.22	5.83	5.40	5.42	7.55	7.74	5.53	5.09	5.21	7.19	75.07
CAESARS HEAD 1 NE	39.0	39.9	45.3	55.0	62.8	69.6	71.5	70.5	65.8	56.6	46.4	39.6	55.2	6.74	6.15	7.22	5.83	5.40	5.42	7.55	7.74	5.53	5.09	5.21	7.19	75.07
DIVISION	40.4	41.3	46.6	56.3	63.9	70.8	72.7	72.0	67.0	57.9	47.8	40.7	56.5	6.34	5.91	6.70	5.43	5.03	5.12	7.14	7.18	5.03	4.97	4.68	6.66	70.19
NORTHWEST																										
ANDERSON	45.1	46.6	52.3	61.7	70.0	77.9	79.5	78.8	73.6	63.7	52.8	45.0	62.3	4.60	4.35	5.15	3.87	2.86	3.52	4.69	4.48	3.19	2.96	3.08	4.62	47.37
CLEMSON UNIVERSITY	44.6	46.2	52.0	61.3	69.4	77.1	79.0	78.3	73.3	62.7	51.6	44.5	61.7	4.83	4.98	5.54	4.36	3.67	3.33	5.13	4.95	3.23	3.29	3.17	5.10	51.58
CRESCENT 1 S	.	.	.	.	.	.	.	.	.	.	.	.	.	5.00	5.07	5.30	4.08	3.33	2.73	4.95	4.78	3.97	3.58	3.27	4.91	50.97
GRNVLE-SPTNBRG WBAP	43.7	45.1	51.4	60.9	69.4	77.0	79.0	78.2	72.7	62.4	51.3	43.6	61.2	4.28	4.07	4.75	3.94	3.22	2.89	4.65	4.51	3.88	3.36	2.84	4.03	46.42
LANDRUM 1 NE	43.6	45.0	50.8	60.1	68.1	75.9	78.0	76.5	71.4	61.3	50.6	43.4	60.4	5.04	4.47	5.75	4.63	4.10	4.24	5.94	5.94	4.29	3.71	3.90	4.77	56.78
LAURENS	45.1	47.0	53.0	62.3	70.4	78.3	80.3	79.3	73.8	63.2	52.0	44.7	62.5	4.04	4.05	4.83	3.72	3.52	3.77	4.89	3.92	3.52	2.64	2.88	4.20	45.98
RAINBOW LAKE	43.1	44.6	50.4	59.7	68.1	75.9	78.2	77.4	71.8	61.2	50.1	42.6	60.3	4.43	4.48	5.20	3.93	3.53	3.52	5.00	4.57	4.01	3.43	3.05	4.54	49.69
SANTUCK	44.9	46.3	52.3	61.4	69.8	77.4	79.5	78.6	73.1	62.9	52.0	44.5	61.9	3.95	3.84	4.45	3.57	3.38	3.15	5.15	4.84	3.43	2.84	2.71	3.77	45.08
WALHALLA	43.4	44.6	50.1	59.4	67.8	75.6	77.5	76.6	71.7	61.2	50.2	42.9	60.1	5.38	5.62	6.44	4.87	3.96	4.61	5.57	5.44	3.92	4.15	3.78	5.47	59.21
WEST PELZER	.	.	.	.	.	.	.	.	.	.	.	.	.	4.70	4.54	5.26	3.88	3.04	3.43	4.93	4.72	3.50	3.53	3.02	4.46	49.01
DIVISION	44.1	45.6	51.4	60.8	69.2	77.0	79.0	78.0	72.7	62.4	51.3	43.9	61.3	4.56	4.42	5.22	4.06	3.48	3.53	5.02	4.80	3.62	3.24	3.14	4.52	49.61
NORTH CENTRAL																										
BLAIR	.	.	.	.	.	.	.	.	.	.	.	.	.	3.67	3.64	4.29	3.73	3.30	2.91	5.69	4.40	3.52	2.37	2.61	3.85	43.98
CAMDEN 2 WSW	45.4	46.6	52.7	61.6	70.6	78.4	80.6	79.6	74.4	63.6	52.4	44.8	62.6	3.60	3.83	4.46	4.09	3.51	4.40	5.86	5.04	4.07	2.74	2.99	3.63	48.22
CATAWBA	.	.	.	.	.	.	.	.	.	.	.	.	.	3.77	3.70	3.93	3.74	3.27	3.29	6.04	4.50	3.13	2.76	2.68	3.77	44.58
KERSHAW	45.1	46.4	52.7	61.9	70.9	77.6	79.7	78.6	73.6	63.5	52.7	44.8	62.8	3.43	3.45	3.82	4.00	2.95	3.94	5.88	5.32	4.02	2.61	2.72	3.51	45.74
WINNSBORO	45.7	46.9	53.1	62.6	71.0	78.4	80.4	79.0	73.9	63.7	53.1	45.2	62.8	3.72	3.56	4.55	3.91	3.17	3.47	5.71	5.29	3.65	2.37	2.47	3.38	45.25
WINTHROP COLLEGE	44.7	46.4	52.4	61.9	70.3	77.6	79.5	78.2	73.2	63.2	52.5	44.3	62.0	4.16	3.88	4.40	3.87	3.27	3.11	6.04	4.64	3.51	3.05	2.78	4.01	46.72
DIVISION	45.2	46.6	52.6	61.9	70.4	77.8	79.9	78.7	73.6	63.5	52.6	44.7	62.3	3.66	3.64	4.21	3.89	3.18	3.60	5.62	4.95	3.77	2.63	2.69	3.57	45.41
NORTHEAST																										
CHERAW	44.6	45.8	51.9	61.5	70.6	78.2	80.7	79.5	74.3	63.6	52.4	44.4	62.3	3.36	3.47	4.14	3.84	3.36	4.09	5.62	5.35	4.26	2.68	2.80	3.27	46.24
CONWAY	48.9	50.1	55.9	64.4	72.2	79.0	81.1	80.3	75.6	65.7	55.7	48.4	64.8	2.75	3.70	4.28	3.06	3.31	5.12	7.52	6.59	5.77	2.70	2.41	3.15	50.36
DARLINGTON	46.7	47.7	54.0	63.0	71.3	78.3	80.4	79.3	74.2	63.7	53.4	46.1	63.2	2.80	3.27	3.76	3.81	3.02	4.03	5.78	4.99	4.54	2.52	2.60	3.19	43.91
EFFINGHAM	.	.	.	.	.	.	.	.	.	.	.	.	.	2.57	3.33	3.92	3.39	3.22	4.05	5.36	5.36	4.27	2.35	2.18	3.11	43.11
FLORENCE FAA AP	47.5	48.8	54.7	63.5	71.9	78.8	80.7	79.9	74.9	65.0	54.6	47.2	64.0	2.64	3.17	3.64	3.61	2.89	4.28	6.24	4.57	4.10	2.18	2.31	3.06	42.69
FLORENCE 2 N	.	.	.	.	.	.	.	.	.	.	.	.	.	2.90	3.39	3.89	3.58	3.27	4.33	6.21	4.66	4.45	2.23	2.57	2.97	44.45
KINGSTREE 1 SE	49.1	50.3	56.4	64.8	72.7	79.6	81.5	80.5	75.8	65.7	55.7	48.5	65.1	2.64	3.55	4.11	3.16	3.21	4.34	5.85	6.42	5.14	2.35	2.35	3.07	46.19
PEE DEE	.	.	.	.	.	.	.	.	.	.	.	.	.	2.68	3.33	3.68	3.79	3.13	3.98	5.81	5.99	4.95	2.80	2.37	3.10	45.61
DIVISION	47.4	48.5	54.5	63.3	71.5	78.4	80.5	79.5	74.7	64.5	54.4	46.9	63.7	2.84	3.43	3.92	3.43	3.17	4.47	6.18	5.42	4.87	2.53	2.52	3.14	45.92
WEST CENTRAL																										
AIKEN	48.9	50.5	56.1	64.5	72.3	79.2	80.8	80.1	75.3	65.7	55.6	48.6	64.8	3.24	3.89	4.33	3.82	3.35	3.45	4.34	4.71	3.79	2.33	2.60	3.71	43.56
CALHOUN FALLS	45.2	47.0	52.7	62.1	70.6	78.2	80.3	79.5	74.1	63.6	52.5	45.1	62.6	4.22	4.09	4.84	3.81	3.47	3.70	4.52	3.73	3.50	2.74	2.56	4.32	45.50
CHAPPELLS	.	.	.	.	.	.	.	.	.	.	.	.	.	3.93	3.94	4.73	3.96	3.42	3.20	4.37	4.52	3.51	2.71	2.54	3.79	44.62
EDGEFIELD 1 ENE	.	.	.	.	.	.	.	.	.	.	.	.	.	3.52	3.87	4.49	4.44	3.16	3.25	4.92	4.14	3.89	2.15	2.32	3.58	43.73
GREENWOOD	44.7	46.0	51.9	61.5	70.5	78.2	80.1	79.2	73.4	63.3	52.2	44.5	62.1	4.51	4.24	4.99	3.87	3.49	3.27	4.65	4.43	3.65	2.71	2.63	3.90	46.34
LITTLE MOUNTAIN	47.3	48.8	54.5	63.5	71.8	79.2	81.0	80.0	75.2	65.5	55.0	47.0	64.1	3.70	3.93	4.36	3.96	3.33	3.24	5.19	4.60	3.80	2.63	2.45	3.61	44.80
NEWBERRY	46.0	47.6	53.3	62.5	71.0	78.5	80.6	79.6	74.2	63.6	52.6	45.2	62.9	3.89	3.81	4.47	3.80	3.41	3.57	5.06	3.93	3.60	2.59	2.47	3.88	44.48
SALUDA 2 W	46.9	48.5	54.0	63.0	71.6	78.8	81.1	80.1	74.7	64.0	53.1	46.1	63.5	3.89	4.05	4.87	4.20	3.09	3.67	4.61	4.72	4.14	2.42	2.44	3.84	45.94
DIVISION	46.6	48.2	53.8	62.9	71.3	78.6	80.5	79.6	74.5	64.4	53.7	46.2	63.4	3.83	3.99	4.59	3.92	3.36	3.48	4.80	4.33	3.70	2.55	2.47	3.86	44.88
CENTRAL																										
COLUMBIA WB AIRPORT	46.9	48.4	54.4	63.6	72.2	79.7	81.6	80.5	75.3	64.7	53.7	46.4	64.0	3.02	3.74	4.26	4.01	3.54	3.85	6.09	5.74	4.31	2.38	2.36	3.52	46.82
COLUMBIA UNI OF S C	48.2	49.7	55.4	64.3	72.6	79.6	81.3	80.3	75.3	65.5	55.0	47.7	64.6	.	.	.	.	.	.	.	.	.	.	.	.	.
ORANGEBURG 2	48.4	49.9	55.8	64.4	72.6	79.2	81.2	80.5	75.4	65.2	54.8	47.6	64.6	2.70	3.72	3.84	3.55	3.34	4.29	4.75	5.58	4.33	2.55	2.35	3.21	44.21
RIMINI	.	.	.	.	.	.	.	.	.	.	.	.	.	2.85	3.84	3.80	3.56	3.28	4.81	4.63	4.72	4.37	2.52	2.32	3.51	44.21
SUMTER	48.3	50.1	56.0	64.6	72.3	79.0	81.0	79.9	75.1	65.2	55.1	47.8	64.5	2.71	3.43	3.88	3.68	3.32	4.17	5.56	5.27	4.39	2.40	2.29	3.28	44.38
DIVISION	47.8	49.4	55.2	64.0	72.2	79.1	80.9	80.0	75.1	64.9	54.6	47.3	64.2	2.69	3.48	3.83	3.53	3.25	3.94	5.47	5.30	4.16	2.40	2.27	3.20	43.52

*NORMALS BY CLIMATOLOGICAL DIVISIONS

Taken from "Climatography of the United States No. 81-4, Decennial Census of U. S. Climate"

TEMPERATURE (°F) PRECIPITATION (In.)

STATIONS (By Divisions)	JAN	FEB	MAR	APR	MAY	JUNE	JULY	AUG	SEPT	OCT	NOV	DEC	ANN	JAN	FEB	MAR	APR	MAY	JUNE	JULY	AUG	SEPT	OCT	NOV	DEC	ANN
SOUTHERN																										
BEAUFORT 7 SW	51.4	52.8	57.6	65.6	73.2	79.3	81.0	80.6	76.5	67.4	58.1	51.3	66.2	2.62	2.97	3.79	2.54	3.48	4.61	6.08	5.91	5.59	2.28	1.88	2.66	44.41
BLACKVILLE 3 W	48.9	50.2	56.0	64.5	72.8	79.1	80.9	80.0	75.6	66.0	55.7	48.3	64.8	2.73	3.65	4.23	3.74	3.53	4.29	5.01	4.91	4.06	2.69	2.45	3.33	44.62
CHARLESTON WB AP	49.8	51.5	56.7	64.8	72.9	79.2	80.6	79.7	75.6	66.2	55.9	50.0	65.2	2.54	3.29	3.93	2.88	3.61	4.98	7.71	6.61	5.83	2.84	2.09	2.85	49.16
CHARLESTON WB CITY	51.5	52.3	57.5	65.6	73.3	79.6	81.5	81.0	77.1	68.1	58.6	51.7	66.5	2.40	3.07	3.62	2.54	3.09	4.25	7.78	6.07	6.32	2.74	1.92	2.74	46.54
SUMMERVILLE 2 WNW	49.3	50.5	56.3	64.4	72.3	78.7	80.7	80.1	75.8	66.2	55.8	48.8	64.9	2.44	3.25	3.74	2.82	3.51	5.03	7.15	6.45	5.22	2.30	2.05	2.73	46.69
YEMASSEE 4 W	50.6	52.0	57.4	65.2	72.6	78.9	80.7	80.3	76.0	66.5	56.6	50.1	65.6	2.57	3.49	3.87	3.41	3.78	5.15	6.78	6.49	4.99	2.92	2.25	2.74	48.44
DIVISION	50.5	51.7	57.1	65.1	72.9	79.2	81.0	80.4	76.3	66.8	56.9	50.1	65.7	2.57	3.35	3.89	3.00	3.56	4.67	6.54	6.05	5.23	2.71	2.13	2.81	46.51

* Normals for the period 1931-1960. Divisional normals may not be the arithmetical averages of individual stations published, since additional data for shorter period stations are used to obtain better areal representation.

TEMPERATURE PRECIPITATION

JAN	FEB	MAR	APR	MAY	JUNE	JULY	AUG	SEPT	OCT	NOV	DEC	ANN	JAN	FEB	MAR	APR	MAY	JUNE	JULY	AUG	SEPT	OCT	NOV	DEC	ANN

CONFIDENCE - LIMITS

In the absence of trend or record changes, the chances are 9 out of 10 that the true mean will lie in the interval formed by adding and subtracting the values in the following table from the means for any station in the State. Because of the wider variation in mean precipitation, the corresponding monthly means and annual mean must be substituted for "p" in the precipitation table below to obtain mean precipitation confidence limits.

1.5	1.4	1.4	.5	.5	.5	.5	.5	.8	.7	.9	1.2	.3	.31√p	.36√p	.34√p	.29√p	.35√p	.31√p	.35√p	.40√p	.44√p	.44√p	.35√p	.35√p	.36√p

COMPARATIVE DATA

Data in the following table are the mean temperature and average precipitation for Winthrop College, South Carolina, for the period 1906-1930 and are included in this publication for comparative purposes.

44.1	45.6	53.0	61.5	69.4	76.7	79.3	78.1	73.4	62.6	51.7	44.3	61.6	3.93	4.07	4.05	3.11	3.93	4.66	5.50	4.75	3.60	2.95	2.39	4.05	46.99

NORMALS, MEANS, AND EXTREMES

CHARLESTON, SOUTH CAROLINA — MUNICIPAL AIRPORT

Standard time used: EASTERN Latitude: 32° 54′ N Longitude: 80° 02′ W Elevation (ground): 40 feet

Means and extremes above are from existing and comparable exposures. Annual extremes have been exceeded at other sites in the locality as follows: Highest temperature 104 in June 1944; lowest temperature 7° in February 1899; maximum monthly precipitation 23.75 in July 1964 at City Office; minimum monthly precipitation 0.01 in October 1942; maximum precipitation in 24 hours 10.57 in September 1933; maximum snowfall 3.9 in February 1899; maximum snowfall in 24 hours 3.9 in February 1899.

COLUMBIA, SOUTH CAROLINA — COLUMBIA METROPOLITAN AIRPORT

Standard time used: EASTERN Latitude: 33° 57′ N Longitude: 81° 07′ W Elevation (ground): 213 feet

Means and extremes above are from existing and comparable exposures. Annual extremes have been exceeded at other sites in the locality as follows: Highest temperature 107 in June 1954; lowest temperature -2 in February 1899; maximum monthly snowfall 11.8 in February 1899; maximum snowfall 11.7 in February 1914.

NORMALS, MEANS, AND EXTREMES

Station: GREENVILLE-SPARTANBURG AP   GREER, S. C.   Standard time used: EASTERN   Latitude: 34° 54' N   Longitude: 82° 13' W   Elevation (ground): 957 feet

Month	Temperature Normal Daily maximum (b)	Normal Daily minimum (b)	Normal Monthly (b)	Extremes Record highest	Year	Record lowest	Year	Normal heating degree days (Base 65°) (b)	Precipitation Normal total (b)	Maximum monthly	Year	Minimum monthly	Year	Maximum in 24 hrs.	Year	Snow, Ice pellets Mean total	Maximum monthly	Year	Maximum in 24 hrs.	Year	Relative humidity Hour 01	07	13	19	Wind Mean speed	Prevailing direction	Fastest mile Speed	Direction	Year	Mean sky cover sunrise to sunset	Pct. of possible sunshine	Mean number of days — Sunrise to sunset Clear	Partly cloudy	Cloudy	Precipitation .01 inch or more	Snow, Ice pellets 1.0 inch or more	Thunderstorms	Heavy fog	Temperatures Max 90° and above	32° and below	Temperatures Min 32° and below	0° and below	Average daily solar radiation - langleys
J	52.4	35.0	43.7	76	1966	-6	1966	665	4.28	5.44	1964	2.39	1965	2.61	1969	2.7	9.1	1966	5.7	1965	70	74	52	61	7.4	NE	44	SW	1967	6.0	54	10	6	16	11	1	*	3	1	*	20	0	*
F	54.4	35.8	45.1	75	1965	-8	1967	557	4.07	6.78	1966	1.00	1968	2.33	1966	2.6	6.9	1969	5.3	1969	67	72	49	50	8.5	NE	44	SW	1966	6.0	56	9	7	12	9	1	*	3	0	*	18	0	0
M	61.7	41.0	51.4	85	1967	17	1969	441	4.75	9.66	1963	1.98	1967	4.45	1963	0.5	1.9	1969	1.9	1969	69	73	45	50	8.4	SW	44	NW	1963	6.1	58	11	9	11	11	*	3	2	0	0	8	0	0
A	71.4	50.3	60.9	91	1967	32	1966	145	3.94	11.30	1964	2.36	1963	3.76	1964	0.0	0.0		0.0		71	76	50	54	8.3	SW	40	NW	1964	5.8	60	13	9	11	10	0	3	1	*	0	*	0	0
M	79.4	59.3	69.4	97	1963	38	1966+	23	3.22	4.97	1967	1.09	1965	1.67	1967	0.0	0.0		0.0		79	84	56	56	6.5	SW	35	NW	1969	5.9	60	8	10	13	10	0	6	2	*	0	0	0	0
J	86.1	67.8	77.0	98	1964	46	1966	0	2.89	9.59	1969	3.84	1966	4.21	1969+	0.0	0.0		0.0		83	84	59	64	6.2	NE	52	NE	1966	6.5	58	8	10	12	11	0	7	1	3	0	0	0	0
J	87.0	71.0	79.0	98	1966	58	1966	0	4.05	7.44	1964	2.46	1963	3.89	1964	0.0	0.0		0.0		85	87	59	69	6.1	WSW	34	NE	1966	6.5	59	5	13	13	12	0	12	3	7	0	0	0	0
A	86.1	70.2	78.2	99	1968	52	1967	0	4.51	7.51	1967	1.16	1967	4.49	1967	0.0	0.0		0.0		87	88	58	67	6.0	NE	31	SW	1964	6.4	64	5	11	11	10	0	7	3	1	0	0	0	0
S	81.0	64.4	72.7	91	1965+	36	1967	9	3.88	7.98	1966	0.93	1964	4.22	1966	0.0	0.0		0.0		85	88	57	71	6.3	NE	31	NE	1964	5.1	63	10	9	11	7	0	2	2	1	0	*	0	0
O	72.4	52.3	62.4	86	1963	27	1965	130	2.84	10.24	1963	0.24	1963	2.83	1963	0.0	0.0		0.0		80	78	51	61	7.0	NE	32	NE	1968	4.2	65	16	6	9	7	0	1	2	0	0	1	0	0
N	61.5	41.0	51.3	81	1961	17	1964	411	2.89	5.07	1968	1.93	1966	1.91	1964	0.3	1.9	1968	1.4	1968	73	78	48	61	7.1	SW	32	NE	1968	4.1	68	12	7	11	9	*	1	2	0	*	6	0	0
D	52.4	34.7	43.6	71	1967+	14	1963	663	4.03	7.40	1967	0.37	1965			0.3	2.1	1963	1.4	1963	73	78	54	64	7.3	NE	47	NE	1963	5.7	54	12	5	14	14	*	1	5	1	*	17	0	0
YR	70.5	51.9	61.2	99	AUG. 1968	-6	JAN. 1966	3044	46.42	11.30	APR. 1964	0.24	OCT. 1963	4.49	AUG. 1967	6.4	9.1	JAN. 1966	5.7	JAN. 1965	77	80	52	62	7.2	NE	52	NE	JUL. 1966	5.6	60	124	95	146	112	3	41	31	14	1	71		*

REFERENCE NOTES APPLYING TO ALL "NORMALS, MEANS, AND EXTREMES" TABLES

(a) Length of record, years, based on January data. Other months may be for more or fewer years if there have been breaks in the record.

(b) Climatological standard normals (1931-1960).

* Less than one half.

+ Also on earlier dates, months, or years.

T Trace, an amount too small to measure.

Below zero temperatures are preceded by a minus sign. The prevailing direction for wind in the Normals, Means, and Extremes table is from records through 1963.

§ > 70° at Alaskan stations.

Unless otherwise indicated, dimensional units used in this bulletin are: temperature in degrees F.; precipitation, including snowfall, in inches; wind movement in miles per hour; and relative humidity in percent. Heating degree day totals are the sums of negative departures of average daily temperatures from 65° F. Cooling degree day totals are the sums of positive departures of average daily temperatures from 65° F. Sleet was included in snowfall totals beginning with July 1948. The term "Ice pellets" includes solid grains of ice (sleet) and particles consisting of snow pellets encased in a thin layer of ice. Heavy fog reduces visibility to 1/4 mile or less.

Sky cover is expressed in a range of 0 for no clouds or obscuring phenomena to 10 for complete sky cover. The number of clear days is based on average cloudiness 0-3, partly cloudy days 4-7, and cloudy days 8-10 tenths.

Solar radiation data are the averages of direct and diffuse radiation on a horizontal surface. The langley denotes one gram calorie per square centimeter.

* Figures instead of letters in a direction column indicate direction in tens of degrees from true North; i.e., 09 - East, 18 - South, 27 - West, 36 - North, and 00 - Calm. Resultant wind is the vector sum of wind directions and speeds divided by the number of observations. If figures appear in the direction column under "Fastest mile" the corresponding speeds are fastest observed 1-minute values.

To 8 compass points only.

January Average Maximum Temperature (°F)

SOUTH CAROLINA

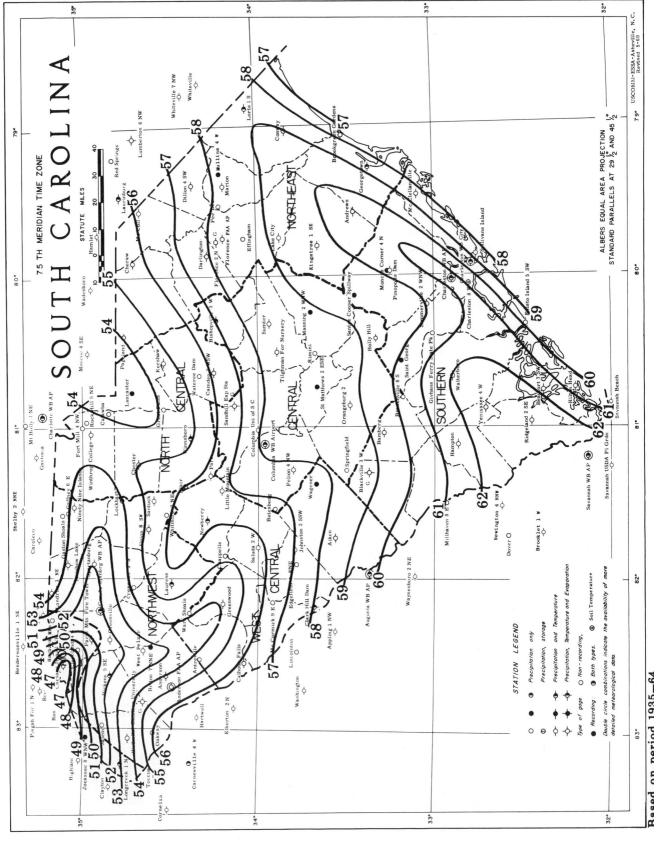

75 TH MERIDIAN TIME ZONE

STATUTE MILES

ALBERS EQUAL AREA PROJECTION
STANDARD PARALLELS AT 29½ AND 45½

USCOMM-ESSA-Asheville, N.C.
Revised 5-69

STATION LEGEND

○ Precipitation only
⊖ Precipitation, storage
⊙ Precipitation and Temperature
⊕ Precipitation, Temperature and Evaporation

Type of page
○ Non-recording

● Recording   ⊛ Both types.
⊚ Soil Temperature

Double circle combinations indicate the availability of more detailed meteorological data.

Based on period 1935—64
Isolines are drawn through points of approximately equal value. Caution should be used

January Average Minimum Temperature (°F)

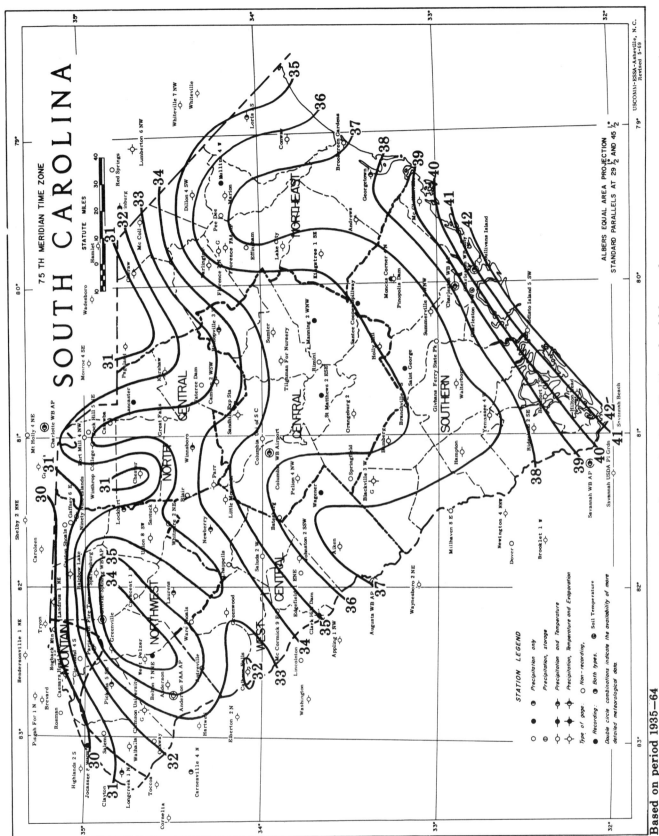

SOUTH CAROLINA

75 TH MERIDIAN TIME ZONE

STATUTE MILES

STATION LEGEND

○ Precipitation only

◐ Precipitation, storage

◒ Precipitation and Temperature

◓ Precipitation, Temperature and Evaporation

Type of gage: ○ Non-recording,

● Recording;   ◉ Both types.

Double circle combinations indicate the availability of more
detailed meteorological data.

⊕ Soil Temperature

ALBERS EQUAL AREA PROJECTION
STANDARD PARALLELS AT 29 1/2 AND 45 1/2

USCOMM—ESSA—Asheville, N.C.
Revised 5-69

Based on period 1935—64
Isolines are drawn through points of approximately equal value. Caution should be used
in interpolating on these maps, particularly in mountainous areas.

July Average Maximum Temperature (°F)

SOUTH CAROLINA

75 TH MERIDIAN TIME ZONE

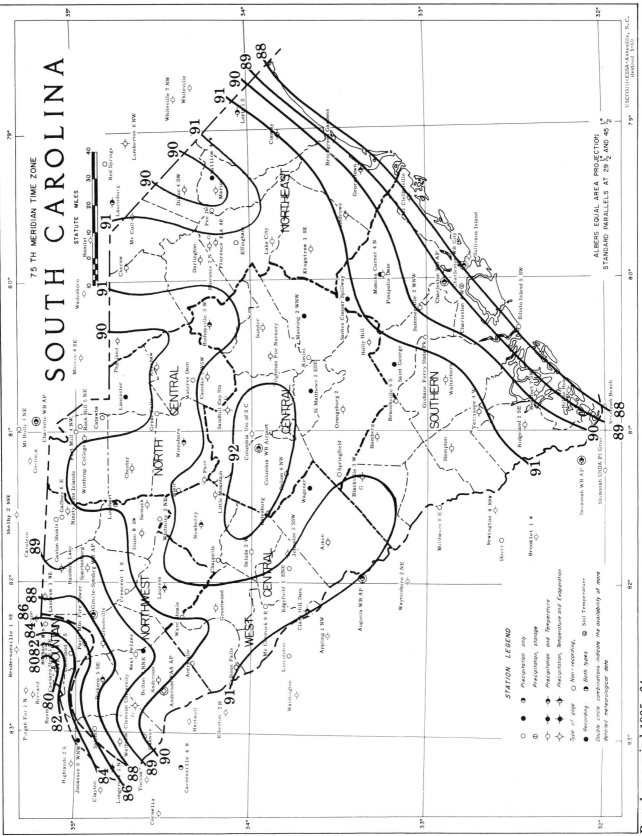

STATUTE MILES

STATION LEGEND

- Precipitation only
- Precipitation, storage
- Precipitation and Temperature
- Precipitation, Temperature and Evaporation
- Type of gage   O Non-recording,
- Recording   ⊚ Soil Temperature
- Both types.

Double circle combinations indicate the availability of more detailed meteorological data

USCOMM-ESSA-Asheville, N.C.
Revised 5-69

ALBERS EQUAL AREA PROJECTION
STANDARD PARALLELS AT 29½ AND 45½

Based on period 1935—64
Isolines are drawn through points of approximately equal value. Caution should be used
in interpolating on these maps. particularly in mountainous areas.

July Average Minimum Temperature (°F)

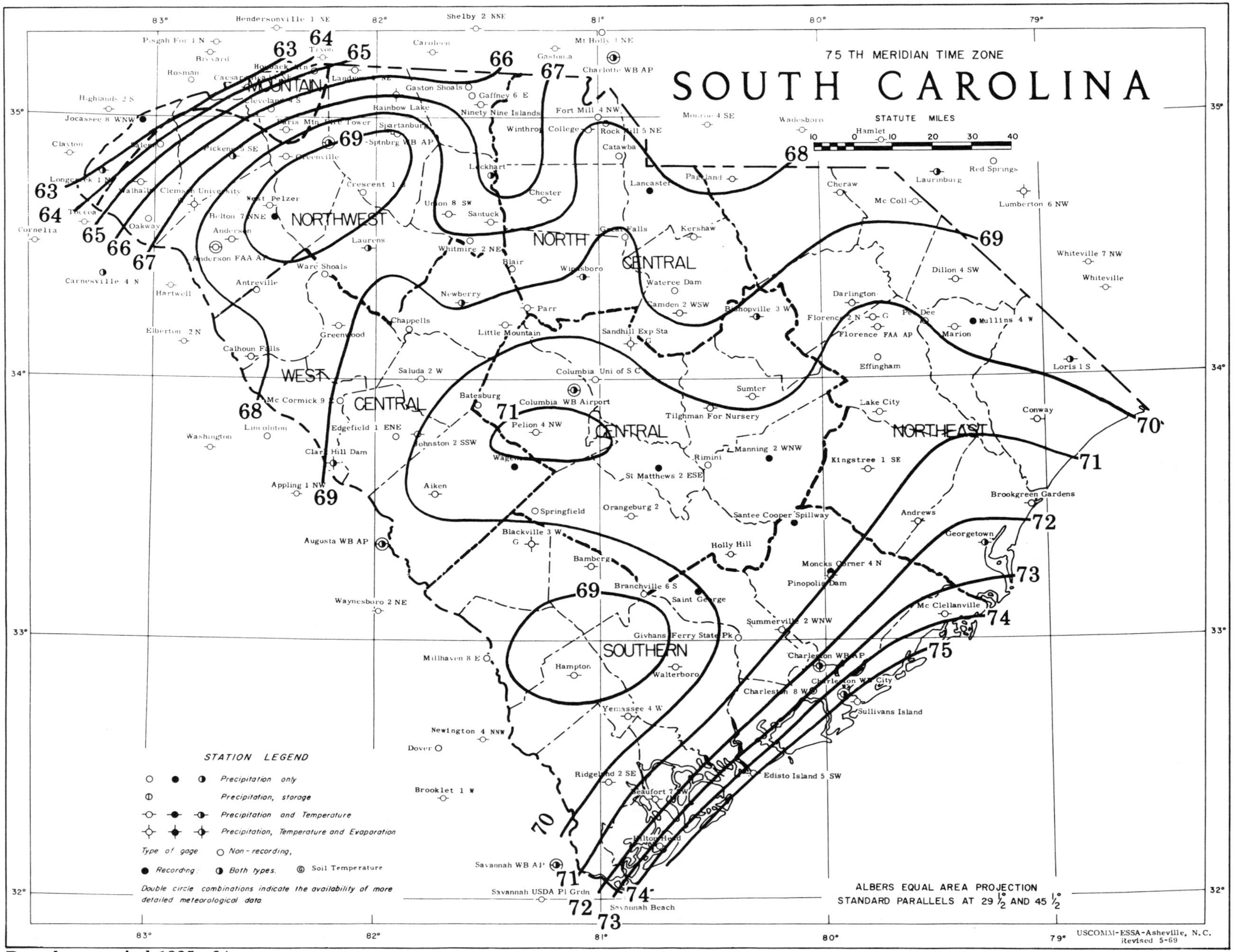

SOUTH CAROLINA

75 TH MERIDIAN TIME ZONE

STATUTE MILES

10 0 10 20 30 40

NORTHWEST

MOUNTAIN

NORTH CENTRAL

WEST CENTRAL

CENTRAL

NORTHEAST

SOUTHERN

USCOMM-ESSA-Asheville, N.C.
Revised 5-69

ALBERS EQUAL AREA PROJECTION
STANDARD PARALLELS AT 29 1/2 AND 45 1/2

STATION LEGEND

○ Precipitation only

◐ Precipitation, storage

⊕ Precipitation and Temperature

✛ Precipitation, Temperature and Evaporation

Type of gage   ● Non-recording

○ Recording   ◉ Soil Temperature

● Double circle combinations indicate the availability of more
detailed meteorological data

Based on period 1935–64

Isolines are drawn through points of approximately equal value. Caution should be used
in interpolating on these maps, particularly in mountainous areas.

– 367 –

Annual Average Rainfall (Inches)

SOUTH CAROLINA

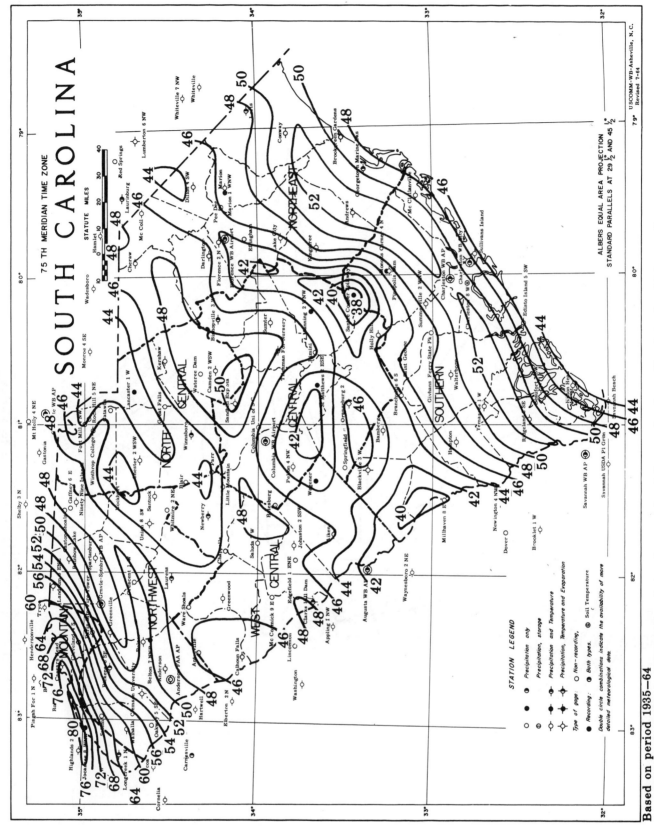

Based on period 1935–64
Isolines are drawn through points of approximately equal value. Caution should be used in interpolating on these maps, particularly in mountainous areas.

ALBERS EQUAL AREA PROJECTION
STANDARD PARALLELS AT 29 ½ AND 45 ½

U.S.COMM-WB-Asheville, N. C.
Revised 7-64

STATION LEGEND

Precipitation only
Precipitation, storage
Precipitation and Temperature
Precipitation, Temperature and Evaporation
Type of gage: Non-recording,
Recording; Both types.
Double circle combinations indicate the availability of more detailed meteorological data.
Soil Temperature

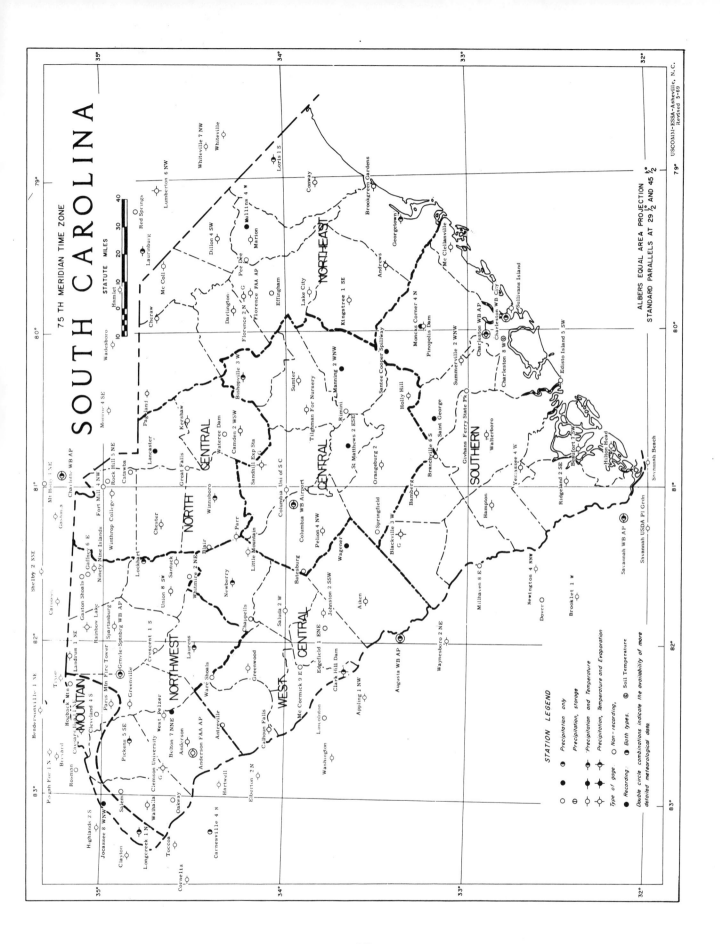

SOUTH CAROLINA

75 TH MERIDIAN TIME ZONE

STATUTE MILES

ALBERS EQUAL AREA PROJECTION
STANDARD PARALLELS AT 29 ½ AND 45 ½

USCOMM-ESSA-Asheville, N.C.
Revised 5-59

STATION LEGEND

○ Precipitation only

⊖ Precipitation, storage

Precipitation and Temperature

Precipitation, Temperature and Evaporation

Type of gage   ○ Non-recording,

Recording: ⊚ Soil Temperature

Double circle combinations indicate the availability of more
detailed meteorological data.

— 369 —

THE CLIMATE OF
TENNESSEE

by
Robert R. Dickson

February 1960

The topography of Tennessee is quite varied, stretching from the lowlands of the Mississippi Valley to the mountain peaks in the east. The westernmost part of the State, between the bluffs overlooking the Mississippi River and the western valley of the Tennessee River, is a region of gently rolling plains sloping gradually from 200 to 250 feet in the west to about 600 feet above sea level in the hills overlooking the Tennessee River. The hilly Highland Rim, in a wide circle touching the Tennessee River Valley on the west and the Cumberland Plateau on the east, together with the enclosed Central Basin makes up the whole of Middle Tennessee. The Highland Rim ranges from about 600 feet in elevation along the Tennessee River to 1,000 feet in the east and rises 300 to 400 feet above the Central Basin which is a rolling plain of about 600 feet average elevation, but with a crescent of hills reaching to over 1,000 feet above sea level south of Nashville. The Cumberland Plateau, with an average elevation of 2,000 feet above sea level, extends roughly northeast-southwest across the State in a belt 30 to 50 miles wide, being bounded on the west by the Highland Rim and overlooking the Great Valley of East Tennessee on the east. The Great Valley, paralleling the Plateau to the west and the Great Smoky Mountains to the east, is a funnel shaped valley varying in width from about 30 miles in the south to about 90 miles in the north. Within the valley which slopes from 1,500 feet in the north to 700 feet above sea level in the south are a series of northeast-southwest ridges. Along the Tennessee-North Carolina border lie the Great Smoky Mountains, the most rugged and elevated portion of Tennessee, with numerous peaks from 4,000 to 6,000 feet above sea level.

Tennessee, except for a small area east of Chattanooga, lies entirely within the drainage of the Mississippi River system. The extreme western section of the State is drained through several relatively small rivers directly into the Mississippi River. Otherwise drainage is into either the Cumberland or Tennessee Rivers, both of which flow northward near the end of their courses to join the Ohio River along the Kentucky - Illinois border. The Cumberland River which drains north-central portions of Tennessee rises in the Cumberland Mountains in Kentucky, flows southwestward, then south into Tennessee reaching the Nashville area before turning northward to re-enter Kentucky. The Tennessee River is formed by the juncture of the Holston and French Broad Rivers at Knoxville in the east. It flows southwesterly along the Great Valley of East Tennessee into Alabama, re-enters Tennessee at the Alabama-Mississippi line, and then flows northward across the State into Kentucky. Besides the headwater streams, other important tributaries include the Clinch, Little Tennessee, Hiwassee, Elk and Duck Rivers.

Most aspects of the State's climate are related to the widely varying topography within its borders. The decrease of temperature with elevation is quite apparent, amounting to, on the average, 3°F. per 1000 feet increase in elevation. Thus higher portions of the State, such as the Cumberland Plateau and the mountains of the east, have lower average temperatures than the Great Valley of East Tennessee, which they flank, and other lower parts of the State. In the Great Valley temperature increases from north to south, reaching a value at the low end comparable to that of Middle and West Tennessee where elevation variations are generally minor consideration. Across the State, the average annual temperature varies from over 62°F. in the extreme southwest to near 45°F. atop the highest peaks of the east. Temperatures at elevated locations (e.g. Bristol and Crossville) can be compared with those at lower elevations (e.g. Nashville and Union City) in the accompanying tables. It is of interest to note that average January temperature atop a 6,000-foot peak in the Great Smokies is equivalent to that in central Ohio, while average July temperature is duplicated along the southern edge of the Hudson Bay in Canada. While most of the State can be described as having a warm, humid summer and a mild winter, this must be qualified to include variations with elevation. Thus with increasing elevation, summers become cooler and more pleasant while winters become colder and more blustery.

This dependence of temperature upon elevation is of considerable importance to a variety of interests. Temperature, together with precipitation, plays an important role in determining what plant and animal life are adaptable to the area. In the Great Smoky Mountains, for example, the variations in elevation from 1,000 to 6,000 feet above sea level with attendant variations in temperature contribute to the remarkable variety of plant life found. The relative coolness of the mountain area has also contributed to the popularity of that area during the warmer part of the year.

Length of growing season (freeze-free period) is linked to topography in a way similar to temperature, varying from 237 days at low-lying Memphis to near 130 days on the highest mountains in the east. Most of the State is included in the range 180 to 220 days. Shorter growing seasons than this are confined to the mountains forming the State's eastern border and to the northern part of the Cumberland Plateau. Longer growing seasons are found in counties bordering the Mississippi River, parts of the Central Basin of the Middle Tennessee, and the southern end of the Great Valley of East Tennessee.

Since the principal source of moist air for this area is the Gulf region, there exists a gradual decrease of average precipitation from south to north. This effect is largely obscured, however, by the overruling influence of topography. Air forced to ascend cools and condenses out a portion of its moisture charge; thus average precipitation is generally greater at higher elevations. This is apparent in all parts of the State. In West Tennessee average annual precipitation ranges from 46 to 54 inches, increasing from Mississippi bottomlands to the slight hills farther east. In Middle Tennessee the variation is from a minimum of 45 inches in the Central Basin to 50 to 55 inches in the surrounding hilly Highland Rim. Over the elevated Cumberland Plateau average annual precipitation is generally from 50 to 55 inches. In contrast, average annual precipitation in the Great Valley of East Tennessee increases from near 40 inches in northern portions to over 50 inches in

the south. The northern minimum, lowest for the entire state, results from the shielding influence of the Great Smoky Mountains to the southeast and the Cumberland Plateau to the northwest. The mountainous eastern border of the State is its wettest part, having average annual precipitation ranging up to about 80 inches on the higher, well-exposed peaks of the Smokies.

Over most of the State greatest precipitation occurs during the winter and early spring due to the more frequent passage of large scale storms over and near the State during those months. A secondary maximum of precipitation occurs in midsummer in response to shower and thundershower activity. This is especially pronounced in the mountains of the east where July rainfall exceeds the precipitation of any other month. Lightest precipitation, observed in the fall, is brought about by the maximum occurrence of slow-moving, rain suppressing high pressure areas during that season. Although all parts of Tennessee are generally well supplied with precipitation, there occurs on the average one or more prolonged dry spells each year during summer and fall. Current studies illustrate the beneficial effects of supplemental irrigation of crops, despite the usually bountiful annual precipitation.

The most important flood season is during the winter and early spring (December through March) when the frequent migratory storms bring general rains of high intensity. During this period both widespread flooding and local flash floods can occur. During summer, heavy thunderstorm rains frequently result in local flash flooding. In the fall, while flood producing rains are rare, decadent hurricanes on occasion cause serious floods in the east. The numerous dams constructed along the Tennessee and Cumberland Rivers are major features in the control of flood waters in the State.

Some of the more notable flood years in Tennessee are 1793, 1867, 1886, 1901, 1902, 1917, 1926, 1927, 1929, 1937 and 1955.

The dams of the Tennessee and Cumberland River Systems and the lakes so formed, in addition to vastly reducing flood damage have facilitated water transportation, provided abundant low cost hydroelectric power and created extensive recreation areas. Fishing, boating, swimming, and camping along the "Great Lakes of the South", together with utilization of the several state and national parks, has made the tourist industry one of major proportions in the State.

Average annual snowfall varies from 4 to 6 inches in southern and western parts of the State and in most of the Great Valley of East Tennessee to more than 10 inches over the northern Cumberland Plateau and the mountains of the east. Over most of the State, due to relatively mild winter temperatures, a snow cover rarely persists for more than a few days.

Water resources of Tennessee have been a major factor in the recent rapid industrial growth. The bountiful and good quality water supply has influenced the location of industry, especially chemical processing plants. Three major waterways, the Mississippi, Cumberland, and Tennessee Rivers, are suitable for commercial traffic. Finally, the availability of low cost hydroelectric power from the multipurpose dams of the Cumberland and Tennessee Rivers and tributaries has been a stimulus to industry of all types. The principal types of manufacturing industries in Tennessee, in order of value added, are chemical and allied products, food and kindred products, textile mill products, primary metals, fabricated metal products, and lumber and lumber products.

Although surpassed in monetary value by industrial activity, agriculture remains a vital feature of Tennessee's economic life. The wide range of climates in Tennessee, from river bottom to mountaintop, coupled with a wide range of soils, has resulted in a large number of crops which prosper in the State. Cotton, in addition to being the foremost money crop, also serves as raw material for the State's extensive textile industry. Primary crops (and major areas in which grown) in order of value of production, are cotton (west), tobacco (north-central and northeast), corn (west and south-central), and hay (middle and east). Livestock and livestock products constitute nearly one-half of the entire farm cash income. Major activity concerns hogs, cattle, calves, dairy products, poultry, and eggs.

Forests represent an additional important segment of Tennessee's natural resources related to the climate of the State. Timberland, containing principally hardwood types, covers approximately one-half the total area of Tennessee. This has led to a highly diversified woodworking industry in which Memphis has become a hardwood flooring

center of the United States. The temperate climate of the State is very favorable for logging operations, allowing full scale activity during 9 months of the year and to a lesser extent during the winter months. The abundant wood supply is utilized by a number of chemical processing industries in the State.

Severe storms are relatively infrequent in Tennessee, being east of the center of tornado activity, south of most blizzard conditions, and too far inland to be often affected by hurricanes. On the average four or five tornadoes are observed in the State each year, with greatest frequency in March when one or two usually occur. Tornado occurrence is not evenly distributed throughout the State, being largely confined to areas west of the Cumberland Plateau. Annual expectancy is less than in bordering locations to the south and west. Damage from tropical storms is rare, occurring only about once every 18 years and blizzard conditions during a recent 40-year period were non-existent. Hailstorms at a given locality are observed two or three times a year and damaging glazestorms occur in the State every 5 or 6 years. Thunderstorms are frequent and severe thunderstorms with damaging winds are experienced at scattered locations throughout the State each year during the warm season.

REFERENCES

(1) Federal Disaster Insurance -- Staff Study for the Committee on Banking and Currency, United States Senate, "Storms", pages 123 - 137, November, 1955.

(2) Klein, W. H., "Principal Tracks and Mean Frequencies of Cyclones & Anticyclones in the Northern Hemisphere", U. S. Weather Bureau Research Paper No. 40, Washington, D. C., 1957.

(3) Tennessee Deaprtment of Agriculture, Agricultural Trends in Tennessee, November 1958.

(4) Tennessee State Planning Commission, Vol. I., Industrial Trends in Tennessee, December 1949; Vol. II Forests, Agriculture and Minerals, December 1948; Vol. III Water Supplies, Fuels, Electric Power and Transportation Facilities, December 1948.

(5) Tharp, Max M. & C. W. Crickman, "Supplemental Irrigation in Humid Regions", Water, 1955 Yearbook of Agriculture, pp 252 - 258, Washington, D. C.

(6) U. S. Weather Bureau, "Tornado Occurrences in the United States", Technical Paper No. 20, Washington, D. C. September 1952.

(7) U. S. Weather Bureau, "Mean Temperature and Precipitation, Tennessee", L. S. 5733, Washington, D. C., August, 1957.

(8) Van Horn, A. G., W. M. Whitaker, R. H. Lush and John R. Carreker, "Irrigation of Pastures for Dairy Cows", University of Tennessee Agricultural Experiment Station Bulletin No. 248, Knoxville, Tennessee, June 1956.

(9) Wessenauer, A. M. "Success of the TVA multiple-purpose river development", Civil Engineering, July 1956.

(10) Weather Bureau Technical Paper No. 15 - Maximum Station Precipitation for 1, 2, 3, 6, 12, and 24 hours. Washington, D.C. September 1956.

(11) Weather Bureau Technical Paper No. 16 - Maximum 24-Hour Precipitation in the United States - Washington, D. C. January, 1952.

(12) Weather Bureau Technical Paper No. 25 - Rainfall Intensity-Duration-Frequency Curves. For selected stations in the United States, Alaska, Hawaiian Islands, and Puerto Rico. Washington, D. C. December 1955.

(13) Weather Bureau Technical Paper No. 29 - Rainfall Intensity-Frequency Regime, Part I - The Ohio Valley. Washington, D. C. June 1957.

(14) Hoyt, William G. and Langbern, Walter B., Floods, Princeton University Press, Princeton, N. J. 1947.

BIBLIOGRAPHY

(A) Climatic Summary of the United States (Bulletin W) 1930 edition, Sections 76 and 77. U. S. Weather Bureau

(B) Climatic Summary of the United States, Tennessee - Supplement for 1931 through 1952 (Bulletin W Supplement). U. S. Weather Bureau

(C) Climatological Data - Tennessee. U. S. Weather Bureau

(D) Climatological Data National Summary. U. S. Weather Bureau

(E) Hourly Precipitation Data - Tennessee. U. S. Weather Bureau

(F) Local Climatological Data, U. S. Weather Bureau for Bristol, Chattanooga, Knoxville, Memphis, Nashville, and Oak Ridge, Tennessee.

FREEZE DATA

STATION	Freeze threshold temperature	Mean date of last Spring occurrence	Mean date of first Fall occurrence	Mean No. of days between dates	Years of record Spring	No. of occurrences in Spring	Years of record Fall	No. of occurrences in Fall
ASHWOOD	32	04-08	10-24	199	30	30	30	30
	28	03-29	11-07	223	30	30	30	30
	24	03-13	11-20	252	30	30	30	30
	20	02-28	12-01	276	29	29	30	27
	16	02-13	12-10	300	29	27	27	23
BOLIVAR	32	04-03	10-23	203	26	26	25	25
	28	03-21	11-04	229	26	26	25	25
	24	03-04	11-17	258	25	25	25	24
	20	02-15	12-02	290	25	23	24	24
	16	02-08	12-14	309	23	21	24	16
BRISTOL WB AP	32	04-16	10-23	190	13	13	13	13
	28	04-02	11-02	213	13	13	13	13
	24	03-24	11-15	236	13	13	13	13
	20	03-05	11-24	264	13	13	13	13
	16	02-19	12-12	296	13	12	13	11
BROWNSVILLE	32	03-28	10-31	217	30	30	30	30
	28	03-16	11-14	243	30	30	30	30
	24	03-01	11-30	274	30	30	30	29
	20	02-16	12-11	298	30	28	29	25
	16	01-31	12-18	321	28	24	25	16
CARTHAGE	32	04-07	10-28	205	30	30	30	30
	28	03-22	11-11	233	30	30	30	30
	24	03-10	11-23	257	30	30	30	29
	20	02-24	12-04	283	30	29	29	26
	16	02-15	12-14	302	29	27	26	20
CHATTANOOGA WB CITY	32	03-26	11-10	229	30	30	30	29
	28	03-14	11-20	252	30	30	29	28
	24	02-23	12-03	283	30	28	28	25
	20	02-17	12-11	297	28	27	24	20
	16	01-26	12-18	326	26	18	20	13
CLARKSVILLE	32	04-04	10-29	208	30	30	30	30
	28	03-22	11-09	232	30	30	30	30
	24	03-06	11-22	261	30	30	30	29
	20	02-24	12-03	282	30	29	29	26
	16	02-12	12-14	304	29	26	26	23
COLDWATER 1 E	32	04-11	10-21	192	29	29	30	30
	28	03-30	10-31	215	29	29	29	29
	24	03-11	11-15	249	29	29	29	28
	20	02-26	11-26	273	29	28	27	25
	16	02-13	12-07	296	28	27	25	21
COPPERHILL	32	04-11	10-23	196	30	30	30	30
	28	03-28	11-04	221	30	30	30	30
	24	03-14	11-19	250	30	30	30	29
	20	02-27	12-02	278	30	29	29	28
	16	02-11	12-12	304	29	27	28	20
COVINGTON	32	03-27	11-03	222	30	30	30	30
	28	03-14	11-12	244	30	30	30	30
	24	03-01	11-27	271	30	30	30	29
	20	02-14	12-10	300	30	27	29	24
	16	01-31	12-18	321	27	23	24	16
CROSSVILLE	32	04-21	10-14	176	29	29	29	29
	28	04-13	10-25	195	29	29	29	29
	24	04-01	11-04	217	29	29	30	30
	20	03-18	11-18	245	29	29	30	29
	16	03-06	11-29	269	29	29	29	28
DALE HOLLOW DAM (Celina)	32	04-13	10-21	191	16	16	16	16
	28	04-05	11-02	211	16	16	16	16
	24	03-16	11-13	242	16	16	16	16
	20	03-04	11-27	268	14	14	16	16
	16	02-27	12-08	284	14	14	15	13
DECATUR	32	04-14	10-24	192	30	30	30	30
	28,	04-03	11-04	216	30	30	30	30
	24	03-19	11-16	242	30	30	30	30
	20	03-03	11-30	272	30	30	30	27
	16	02-19	12-08	291	30	28	27	23
DICKSON	32	04-10	10-23	197	30	30	30	30
	28	03-28	11-07	224	30	30	30	30
	24	03-13	11-18	251	30	30	30	29
	20	02-28	11-29	275	30	29	29	29
	16	02-19	12-13	296	29	28	29	24
DOVER 1 NW	32	04-15	10-19	187	30	30	30	30
	28	03-31	11-03	217	30	30	30	30
	24	03-15	11-12	242	30	30	30	30
	20	03-03	11-26	268	30	30	30	28
	16	02-18	12-09	294	30	28	28	25
FRANKLIN	32	04-12	10-21	192	30	30	30	30
	28	03-28	11-02	219	30	30	30	30
	24	03-13	11-14	246	30	30	30	30
	20	02-27	11-29	275	30	30	30	28
	16	02-14	12-11	301	30	27	28	22
GATLINBURG 2 SW	32	04-29	10-16	171	24	24	25	25
	28	04-13	10-27	197	24	24	25	25
	24	03-27	11-09	227	24	24	25	25
	20	03-11	11-23	257	24	24	25	25
	16	03-01	12-02	276	24	24	24	23
GREENEVILLE EXP STA	32	04-20	10-17	180	18	18	18	18
	28	04-12	10-28	198	18	18	18	18
	24	03-27	11-08	226	18	18	18	18
	20	03-10	11-22	257	18	18	18	18
	16	02-27	12-05	281	18	18	18	18
JACKSON CAA AP	32	04-06	10-23	200	30	30	30	30
	28	03-25	11-05	226	30	30	30	30
	24	03-05	11-22	262	30	30	30	29
	20	02-19	12-07	291	30	28	29	25
	16	02-04	12-17	316	28	24	25	16
KENTON	32	04-01	10-24	206	25	25	24	24
	28	03-19	11-05	231	25	25	25	25
	24	03-07	11-21	259	25	25	25	24
	20	02-23	12-03	283	25	25	24	21
	16	02-06	12-12	309	25	22	21	18
KNOXVILLE WB CO	32	03-31	11-06	220	30	30	30	30
	28	03-18	11-20	247	30	30	30	29
	24	03-03	12-03	275	30	30	29	28
	20	02-20	12-14	297	30	29	28	21
	16	01-30	12-19	323	29	21	21	15
LEWISBURG EXP STA	32	04-14	10-20	189	30	30	30	30
	28	03-31	10-30	214	30	30	30	30
	24	03-15	11-15	245	30	30	30	30
	20	03-04	11-25	266	30	30	30	28
	16	02-16	12-10	296	30	28	28	25
LOUDON	32	04-10	10-24	197	29	29	29	29
	28	03-31	11-07	222	29	29	29	29
	24	03-15	11-19	249	29	29	29	28
	20	03-02	11-27	270	29	29	27	27
	16	02-13	12-12	303	29	26	27	20
LYNNVILLE 4 SW	32	04-13	10-20	190	30	30	30	30
	28	03-29	11-01	217	29	29	30	30
	24	03-18	11-14	240	29	29	29	28
	20	03-04	11-25	267	29	29	28	28
	16	02-15	12-10	298	29	28	28	22
MC MINNVILLE	32	04-08	10-28	203	30	30	30	30
	28	03-28	11-09	226	29	29	30	30
	24	03-06	11-22	257	29	29	30	29
	20	02-28	12-03	279	29	29	29	27
	16	02-13	12-14	304	29	27	27	19
MEMPHIS WB CO	32	03-20	11-12	237	30	30	30	30
	28	03-05	11-29	269	30	30	30	29
	24	02-16	12-09	296	30	28	29	27
	20	02-09	12-17	311	28	26	27	19
	16	01-28	12-19	325	26	22	19	12
MILAN	32	04-04	10-26	206	30	30	30	30
	28	03-25	11-05	226	30	30	30	30
	24	03-06	11-21	260	30	30	30	29
	20	02-19	12-03	287	30	29	29	26
	16	02-07	12-14	310	29	26	26	19
MOSCOW	32	04-02	10-24	205	30	30	30	30
	28	03-23	11-07	230	30	30	30	30
	24	03-04	11-19	260	30	30	30	30
	20	02-15	12-09	297	30	27	30	24
	16	02-03	12-16	317	27	24	24	17

FREEZE DATA

STATION	Freeze threshold temperature	Mean date of last Spring occurrence	Mean date of first Fall occurrence	Mean No. of days between dates	Years of record Spring	No. of occurrences in Spring	Years of record Fall	No. of occurrences in Fall
MURFREESBORO	32	04-05	10-25	203	30	30	30	30
	28	03-25	11-05	225	30	30	30	30
	24	03-09	11-18	255	30	30	30	29
	20	02-23	12-02	282	30	28	30	27
	16	02-11	12-16	309	29	26	30	22
NASHVILLE WB CO	32	03-28	11-07	224	30	30	30	30
	28	03-18	11-17	245	30	30	30	29
	24	03-04	11-28	269	30	30	29	28
	20	02-18	12-11	296	30	29	28	23
	16	02-05	12-17	315	29	26	23	18
NEWBERN	32	03-27	10-31	217	30	30	30	30
	28	03-20	11-13	238	30	30	30	30
	24	03-07	11-26	265	29	29	29	28
	20	02-20	12-07	290	29	27	29	27
	16	02-07	12-19	315	29	26	29	17
NEWPORT	32	04-11	10-24	197	30	30	30	30
	28	03-30	11-05	220	30	30	30	30
	24	03-14	11-18	249	30	30	30	30
	20	02-28	12-02	277	30	29	30	28
	16	02-19	12-12	297	29	27	28	20
NORRIS	32	04-16	10-27	194	16	16	16	16
	28	04-04	11-02	211	16	16	16	16
	24	03-28	11-21	238	16	16	16	16
	20	03-10	11-29	264	16	16	16	16
	16	02-18	12-10	295	16	14	16	13
PALMETTO	32	04-10	10-24	197	30	30	29	29
	28	03-27	11-06	224	30	30	29	29
	24	03-11	11-21	255	30	30	29	28
	20	02-20	11-30	283	30	28	28	26
	16	02-14	12-15	304	28	27	26	19
PARIS	32	04-12	10-31	202	17	17	16	16
	28	03-29	11-06	222	16	16	16	16
	24	03-11	11-18	252	15	15	16	16
	20	02-27	11-30	276	15	15	16	16
	16	02-09	12-14	309	15	14	16	11
ROGERSVILLE	32	04-17	10-24	190	28	28	30	30
	28	03-30	11-03	218	28	28	30	30
	24	03-14	11-18	250	28	28	30	30
	20	03-02	12-01	274	28	28	30	27
	16	02-20	12-08	290	28	27	27	23
RUGBY	32	05-03	10-08	158	30	30	29	29
	28	04-16	10-20	187	30	30	29	29
	24	04-02	10-31	212	30	30	29	29
	20	03-19	11-11	237	30	30	29	29
	16	03-08	11-25	262	29	29	29	27
SAMBURG WL REF (Tiptonville)	32	03-29	11-02	218	22	22	23	23
	28	03-19	11-11	237	22	22	23	23
	24	03-03	11-23	266	22	22	23	22
	20	02-18	12-04	288	22	22	22	21
	16	02-10	12-19	312	22	20	21	14
SAVANNAH	32	04-04	10-28	207	29	29	30	30
	28	03-22	11-12	235	29	29	30	30
	24	03-08	11-27	264	29	28	29	28
	20	02-21	12-02	284	28	27	28	25
	16	02-09	12-15	309	27	25	25	18
SPRINGFIELD EXP STA (Cedar Hill)	32	04-07	10-30	206	30	30	30	30
	28	03-27	11-09	227	30	30	30	30
	24	03-13	11-22	254	30	30	30	29
	20	03-02	12-03	276	30	30	30	27
	16	02-13	12-12	302	30	27	29	24
TULLAHOMA	32	04-10	10-23	196	30	30	30	30
	28	03-25	11-06	226	30	29	30	30
	24	03-13	11-16	248	30	30	30	29
	20	02-26	12-03	280	30	30	30	26
	16	02-12	12-14	305	29	27	30	21
UNION CITY	32	03-31	10-28	211	30	30	30	30
	28	03-22	11-09	232	30	30	30	30
	24	03-05	11-22	262	30	30	30	30
	20	02-21	12-07	289	30	30	30	26
	16	02-09	12-15	309	30	27	26	19
WAYNESBORO	32	04-18	10-15	180	28	28	28	28
	28	04-05	10-29	208	28	28	28	28
	24	03-24	11-07	227	28	28	28	28
	20	03-08	11-22	260	28	28	28	27
	16	02-23	12-07	287	28	27	28	22

Data in the above table are based on the period 1921-1950, or that portion of this period for which data are available.

Means have been adjusted to take into account years of non-occurrence.

A freeze is a numerical substitute for the former term "killing frost" and is the occurrence of a minimum temperature at or below the threshold temperature of 32°, 28°, etc.

Freeze data tabulations in greater detail are available and can be reproduced at cost.

*MEAN TEMPERATURE AND PRECIPITATION

STATION	JANUARY		FEBRUARY		MARCH		APRIL		MAY		JUNE		JULY		AUGUST		SEPTEMBER		OCTOBER		NOVEMBER		DECEMBER		ANNUAL	
	Temperature	Precipitation	Temperature	Precipitation	Temperature	Precipitation	Temperature	Precipitation	Temperature	Precipitation	Temperature	Precipitation	Temperature	Precipitation	Temperature	Precipitation	Temperature	Precipitation	Temperature	Precipitation	Temperature	Precipitation	Temperature	Precipitation	Temperature	Precipitation
EASTERN																										
BLUFF CITY		4.13		3.65		4.42		3.30		3.72		3.66		5.84		4.02		2.88		2.15		2.69		3.53		43.99
BRISTOL WB AIRPORT	38.6	3.50	40.1	3.41	46.7	3.77	56.0	3.25	64.2	3.44	72.3	3.67	74.8	5.14	73.6	3.89	68.5	2.79	57.9	2.39	45.8	2.56	38.7	3.43	56.4	41.24
CHARLESTON		5.70		5.13		5.84		4.25		3.77		3.95		4.63		4.13		2.55		2.72		3.78		5.26		51.71
CHATTANOOGA WB AP	41.6	5.23	44.0	5.11	50.7	6.05	59.7	4.53	67.7	4.16	75.8	4.21	78.3	5.34	77.3	3.70	72.5	2.69	60.8	3.24	49.1	4.03	42.1	5.31	60.0	53.60
CLINTON		5.68		5.44		5.43		3.87		4.17		3.59		5.28		3.75		3.34		2.63		3.99		5.22		52.39
COPPERHILL	40.4	5.88	41.2	5.38	47.7	6.35	56.8	4.26	65.6	3.66	73.4	4.22	76.1	5.77	75.1	4.99	69.3	2.89	58.0	2.97	46.3	3.66	40.1	5.34	57.5	55.37
DANDRIDGE		4.40		4.33		4.87		3.36		3.45		3.18		4.75		3.82		2.59		2.19		2.84		3.72		43.50
EMBREEVILLE		3.65		3.72		4.62		3.46		4.04		3.55		6.36		4.00		3.04		2.48		2.70		3.25		44.87
ERWIN		3.67		3.76		4.86		3.48		4.11		3.90		6.55		4.23		3.07		2.51		2.68		3.27		46.09
KINGSPORT 3 SE		4.03		3.80		4.56		3.31		3.40		3.36		5.05		3.98		2.64		2.19		2.69		3.33		42.34
KINGSTON		5.35		5.28		5.39		3.91		3.64		3.53		4.46		4.10		2.55		2.40		3.64		4.85		49.10
KNOXVILLE WB AP	40.5	4.54	42.5	4.73	49.4	4.83	59.0	3.64	67.4	3.58	75.8	3.47	78.4	4.72	77.0	3.43	72.1	2.53	60.3	2.63	48.4	3.15	41.0	4.26	59.3	45.51
LOUDON 1 E	41.3	5.57	42.5	5.25	47.7	5.54	58.9	3.84	67.7	3.78	76.2	3.42	79.0	5.35	78.0	3.57	72.4	2.74	61.0	2.61	48.0	3.56	40.9	4.80	59.5	50.03
NEWPORT	40.9	4.29	41.7	4.14	48.5	4.82	58.1	3.40	67.1	3.56	75.4	3.73	78.1	5.25	77.0	4.02	71.3	2.45	60.0	2.42	47.2	2.65	40.2	3.46	58.8	44.19
OAK RIDGE	39.6	5.11	41.4	5.21	48.3	5.64	57.7	4.29	66.0	3.78	74.2	4.18	76.7	4.83	75.4	4.45	71.1	2.97	59.3	2.77	47.5	3.94	40.1	5.21	58.1	52.38
OAK RIDGE WB	39.0	6.06	41.0	5.79	48.1	5.69	57.6	4.08	66.0	4.18	74.2	3.91	76.7	5.34	75.3	4.06	70.8	2.55	58.8	2.42	47.0	3.98	39.6	5.66	57.8	53.72
PARKSVILLE		5.46		5.16		5.72		4.27		3.78		3.90		5.73		4.38		2.87		2.74		3.53		4.89		52.43
ROGERSVILLE 1 NE	40.9	4.47	42.2	4.24	49.1	4.65	58.2	3.28	66.5	3.50	74.3	3.10	77.0	4.78	76.3	3.48	71.3	2.60	60.5	2.17	47.8	3.01	40.3	3.63	58.7	42.91
TELLICO PLAINS		5.88		5.16		6.11		4.03		4.34		4.24		5.24		4.73		2.81		2.77		3.84		4.72		53.87
DIVISION	40.9	4.94	42.0	4.67	48.7	5.25	57.9	3.75	66.5	3.74	74.5	3.75	77.2	5.52	76.2	4.13	70.7	2.74	59.6	2.58	47.3	3.19	40.4	4.31	58.5	48.57
CUMBERLAND PLATEAU																										
CROSSVILLE EXP STA	37.8	5.97	38.6	5.61	45.7	5.63	54.9	4.23	63.6	3.90	70.8	3.95	73.8	4.53	73.0	4.55	67.5	3.18	57.2	2.70	45.3	3.82	37.9	5.46	55.5	53.53
MC MINNVILLE	42.5	5.74	43.9	5.65	50.7	5.62	59.4	3.77	67.5	3.83	75.5	4.24	78.1	4.36	77.3	4.15	71.9	3.22	61.2	2.49	49.1	3.50	42.5	5.04	60.0	51.61
ROCK ISLAND 2 NW		5.79		5.59		5.78		3.83		3.91		3.69		4.62		4.03		3.31		2.58		3.70		5.16		51.99
SEWANEE		6.12		5.80		6.29		4.87		4.10		4.20		5.69		4.13		3.13		2.92		4.35		5.39		56.99
TULLAHOMA	41.5	6.19	43.0	5.85	49.9	6.06	58.8	4.39	66.9	3.62	74.9	3.79	77.6	5.12	76.8	3.54	71.1	3.12	60.3	2.56	48.4	3.85	41.8	5.39	59.3	53.48
DIVISION	39.9	6.03	41.1	5.65	48.1	5.87	57.1	4.22	65.5	3.87	73.3	4.20	76.1	5.01	75.2	4.14	69.8	3.15	59.3	2.62	47.1	3.89	40.1	5.38	57.7	54.03
MIDDLE																										
ASHWOOD	41.5	5.92	43.0	5.34	50.1	5.92	58.9	4.35	67.2	4.04	75.8	3.40	78.4	4.39	77.4	3.54	71.4	3.23	60.8	2.51	48.9	3.78	42.0	4.74	59.6	51.16
CARTHAGE	41.9	6.09	43.4	5.06	50.7	5.58	59.8	3.93	68.4	3.96	76.7	4.54	79.6	4.36	78.8	3.89	72.9	3.29	62.0	2.38	49.4	4.04	42.2	4.68	60.5	51.90
CLARKSVILLE	40.5	6.13	42.7	4.14	50.4	5.41	59.9	4.11	68.3	3.61	77.2	3.33	80.3	3.67	79.4	3.34	72.7	3.13	61.7	2.55	49.1	3.72	41.3	4.32	60.3	47.46
COLDWATER 1 E	42.9	6.27	44.5	5.87	51.3	6.43	59.9	4.23	67.9	4.26	76.6	3.15	79.3	4.53	78.5	3.66	72.5	3.25	61.6	2.82	49.4	3.94	43.2	5.33	60.6	53.74
DICKSON	40.8	6.14	42.7	4.60	50.5	6.03	59.9	4.39	67.8	4.26	76.4	3.71	79.1	4.56	78.2	3.47	71.8	2.97	61.2	2.66	48.9	4.02	41.5	4.73	59.9	51.54
DOVER 1 NW	39.9	5.80	42.0	3.92	49.7	5.47	59.2	3.85	66.9	3.71	75.5	3.36	78.6	4.01	77.7	3.38	71.0	3.37	60.3	2.84	48.2	3.79	40.5	4.25	59.1	47.75
FAYETTEVILLE 1 NE		5.70		5.51		5.74		4.04		3.99		3.13		5.12		3.79		3.10		2.61		3.98		5.18		51.89
FRANKLIN SEWAGE PLANT	40.6	5.82	42.4	4.96	49.6	5.53	58.6	4.04	67.2	3.76	76.0	3.27	78.6	3.85	77.9	3.49	71.4	2.67	60.5	2.20	48.2	3.43	40.9	4.27	59.3	47.29
LEWISBURG EXP STA	40.7	5.86	42.2	5.98	49.5	5.92	58.6	4.37	67.2	3.73	75.9	3.60	78.8	4.65	78.1	3.05	71.8	2.94	60.6	2.54	48.0	3.98	41.2	4.64	59.4	51.26
MURFREESBORO	41.7	5.67	43.5	5.29	50.7	5.46	59.7	3.96	68.3	3.86	76.8	2.94	79.5	4.25	78.6	3.74	72.6	3.15	61.5	2.31	49.1	3.97	42.2	4.41	60.4	49.01
NASHVILLE WB AP	39.9	4.93	42.3	4.16	49.8	5.28	59.7	3.69	68.2	3.78	76.9	3.19	80.0	3.96	78.7	3.31	73.2	2.74	61.8	2.52	49.3	3.41	41.6	4.06	60.1	45.03
NASHVILLE		5.63		4.42		5.38		3.83		3.67		3.09		4.17		2.87		2.78		2.18		3.26		4.12		45.40
PALMETTO	41.6	5.63	43.1	5.68	50.5	5.47	59.0	4.00	67.7	3.99	76.2	3.56	78.7	4.98	78.0	3.22	72.3	2.80	61.5	2.64	48.9	3.84	42.2	4.58	60.0	50.39
PERRYVILLE		7.08		5.05		5.81		4.41		4.04		3.67		3.97		3.47		2.90		2.45		4.27		4.81		51.93
SAVANNAH	42.8	6.83	44.7	5.39	52.2	5.91	61.2	4.45	69.1	4.03	77.6	3.86	80.6	3.88	80.0	3.91	73.5	3.12	62.5	2.64	50.4	4.28	43.1	5.04	61.5	53.34
WAYNESBORO	40.2	6.27	41.6	5.74	49.5	6.06	58.4	4.93	66.0	4.07	74.7	4.04	77.5	4.60	76.9	3.91	70.5	3.08	59.5	2.70	47.4	4.54	40.6	5.12	58.6	55.06
DIVISION	41.0	6.16	42.7	5.07	50.1	5.70	59.3	4.19	67.6	3.92	76.3	3.65	79.1	4.22	78.3	3.44	72.0	3.04	61.1	2.56	48.7	3.89	41.5	4.64	59.8	50.48
WESTERN																										
BOLIVAR 2	41.8	6.92	44.0	5.10	51.6	5.67	60.8	4.64	68.5	3.70	77.0	3.88	80.0	4.40	79.2	3.59	72.3	3.70	61.6	2.76	49.4	4.54	42.5	4.72	60.7	53.62
BROWNSVILLE	42.0	6.64	44.3	4.78	52.1	5.74	61.4	4.47	69.6	4.19	78.1	3.78	80.7	4.12	79.7	2.93	72.9	3.62	63.0	2.59	50.7	4.44	43.1	4.76	61.5	52.06
COVINGTON	41.9	6.39	44.2	4.55	51.6	5.94	60.9	4.51	69.5	4.57	78.3	3.70	81.0	3.78	80.0	2.79	73.2	3.37	62.9	2.98	50.5	4.31	42.9	4.79	61.4	51.68
DRESDEN		5.86		4.36		5.59		4.22		4.31		3.77		3.57		2.97		3.22		2.83		3.87		4.59		49.16
JACKSON EXP STA	41.1	6.55	43.5	4.74	50.9	5.56	60.0	4.60	68.2	4.21	76.9	4.09	79.6	4.63	78.8	3.65	69.1	3.58	61.6	2.60	49.2	4.22	42.1	4.53	60.1	52.06
MC KENZIE		6.14		4.29		5.50		3.74		3.89		3.35		3.53		2.49		3.38		2.85		3.91		4.39		47.46
MEMPHIS WB AIRPORT	41.6	5.55	44.5	4.59	52.0	5.59	61.8	4.80	70.1	3.92	78.3	3.33	81.2	3.23	80.3	2.94	74.3	2.55	63.6	3.27	50.6	4.56	43.3	5.09	61.8	49.42
MEMPHIS WB CITY	41.9	5.38	44.5	4.22	52.4	5.20	62.2	4.70	70.5	3.65	78.7	3.25	81.3	3.10	80.3	2.53	74.8	2.50	65.0	3.18	52.1	4.31	44.4	4.79	62.4	46.91
MILAN	41.0	6.60	43.5	4.50	51.0	6.15	60.3	4.63	68.5	4.23	77.3	4.30	80.0	4.58	79.2	3.45	72.5	3.61	61.9	3.03	49.5	4.18	40.3	4.69	60.4	53.95
MOSCOW	42.6	6.18	44.8	5.44	52.4	5.76	61.2	4.53	69.2	4.11	77.3	4.21	80.3	4.01	79.6	3.51	72.9	3.08	62.4	2.78	50.4	4.38	43.3	5.13	61.4	53.12
NEWBERN	40.2	5.27	42.6	4.18	50.5	5.62	60.4	4.00	69.3	3.84	77.9	3.93	80.6	3.89	79.8	3.10	73.2	3.20	62.7	3.05	49.5	3.78	41.5	4.44	60.7	48.30
SAMBURG W L REFUGE	39.1	5.25	41.4	4.27	49.8	5.38	59.9	4.18	69.0	4.38	77.9	3.89	80.8	3.83	79.8	2.80	72.4	3.34	61.6	3.19	48.4	3.85	40.3	4.36	60.0	48.72
SELMER		6.37		5.67		5.45		4.50		3.80		4.17		3.78		3.44		3.16		2.70		4.25		4.87		52.16
UNION CITY	38.2	5.15	40.2	4.23	48.3	5.49	58.6	4.43	68.0	4.14	77.3	3.78	80.3	3.93	79.2	2.87	71.7	3.47	60.8	3.17	47.6	3.97	39.5	4.38	59.1	49.01
DIVISION	40.8	6.08	43.1	4.55	50.9	5.61	60.4	4.32	68.9	4.07	77.7	3.84	80.4	4.01	79.6	3.08	72.7	3.31	62.1	2.89	49.4	4.09	41.7	4.58	60.6	50.43

* Averages for period 1931-1955, except for stations marked WB which are "normals" based on period 1921-1950. Divisional means may not be the arithmetical average of individual stations published, since additional data from shorter period stations are used to obtain better areal representation.

CONFIDENCE LIMITS

In the absence of trend or record changes, the chances are 9 out of 10 that the true mean will lie in the interval formed by adding and subtracting the values in the following table from the means for any station in the State. Because of the wider variation in mean precipitation, the corresponding monthly means and annual mean must be substituted for "p" in the precipitation table below to obtain mean precipitation confidence limits.

1.8	$.48\sqrt{p}$	1.6	$.38\sqrt{p}$	1.7	$.36\sqrt{p}$	.9	$.35\sqrt{p}$	.9	$.37\sqrt{p}$	.9	$.40\sqrt{p}$	.7	$.42\sqrt{p}$	.8	$.38\sqrt{p}$	1.2	$.37\sqrt{p}$	1.1	$.44\sqrt{p}$	1.1	$.37\sqrt{p}$	1.5	$.39\sqrt{p}$	.3	$.39\sqrt{p}$

COMPARATIVE DATA

Data in the following table are the mean temperature and average precipitation for Lewisburg Experiment Station, Tennessee for the period 1906 - 1930 and are included in this publication for comparative purposes :

40.7	4.94	42.3	4.37	50.5	5.81	59.2	4.73	66.8	4.77	75.2	4.37	78.4	4.11	77.6	4.61	72.9	3.09	60.3	3.18	49.3	3.63	41.2	5.20	59.5	52.81

NORMALS, MEANS, AND EXTREMES

BRISTOL, TENNESSEE — TRI-CITY AIRPORT

LATITUDE 36° 29' N
LONGITUDE 82° 24' W
ELEVATION (ground) 1519 feet

Month	Normal Daily max	Normal Daily min	Normal Monthly	Record highest	Year	Record lowest	Year	Normal degree days	Precip. Normal total	Max. monthly	Year	Min. monthly	Year	Max. in 24 hrs	Year	Snow Mean total	Max. monthly	Year	Max. in 24 hrs	Year	RH 1a	RH 7a	RH 1p	RH 7p	Wind mean hourly	Prev. dir.	Clear	Partly cloudy	Cloudy
(yrs)	(b)	(b)	(b)	20		20		(b)	(b)	20		20		20		20	20		14		5				5	5	20		20
J	48.5	28.6	38.6	79	1950	-10	1940	818	3.50	9.18	1957	1.06	1940	2.34	1950	3.3	16.9	1948	9.7	1955	82	85	65	72	6.9	WSW	5	6	9
F	50.4	29.8	40.1	79	1943	-2	1955	697	3.41	7.29	1956	1.12	1943	1.76	1943	3.1	15.8	1947	6.8	1947	78	83	57	64	6.7	W	5	8	15
M	58.0	35.3	46.7	85	1954+	3	1943	576	3.77	9.56	1957	1.33	1955	2.21	1955	1.8	15.2	1947	12.7	1954	73	81	53	64	7.8	WSW	7	8	16
A	68.0	44.0	56.0	93	1942	20	1942	274	3.25	9.71	1950	0.74	1949	2.32	1956	0.2	0.2	1953+	0.2	1953+	72	78	50	56	7.8	WSW	9	9	14
M	76.7	51.7	64.2	94	1941	30	1940	95	3.44	9.71	1950	0.65	1950	1.99	1939	0.0	0.0		0.0		76	84	54	62	6.0	W	6	12	13
J	84.2	60.4	72.3	99	1944	41	1946	0	3.67	6.98	1942	1.30	1954	3.10	1944	0.0	0.0		0.0		84	86	54	65	4.5	W	5	13	12
J	85.5	64.0	74.8	102	1952	45	1947	0	5.14	9.73	1949	0.79	1949	2.90	1946	0.0	0.0		0.0		85	88	55	67	4.1	NE	6	13	12
A	84.7	62.4	73.6	98	1954	46	1945	0	3.89	7.43	1942	0.82	1938	2.67	1954	0.0	0.0		0.0		85	91	54	67	3.9	NE	7	14	10
S	80.5	56.4	68.5	104	1954	33	1942	58	2.79	5.05	1957	0.86	1939	3.21	1944	0.0	0.0		T	1957+	90	90	53	65	4.6	NE	11	9	10
O	71.1	44.7	57.9	95	1941	21	1952	239	2.56	3.72	1938	0.32	1938	1.54	1949	T	T	1952	T	1952	84	90	50	64	4.7	E	12	7	13
N	57.5	34.1	45.8	85	1950	5	1950	576	3.43	5.90	1953	1.07	1953	2.55	1957	1.6	18.1	1957+	16.2	1957+	79	84	55	65	6.5	E	10	7	13
D	48.9	28.4	38.7	78	1951	2	1957	815	3.43	5.72	1947	1.49	1942	2.29	1942	1.8	9.2	1944	8.0	1944	83	83	55	65	6.1	E	7	10	17
Year	67.8	45.0	56.4	104	July 1952	-10	Jan. 1940	4148	41.24	9.73	July 1949	0.32	Oct. 1938	3.21	Sept. 1944	10.8	18.1	Nov. 1952	16.2	Nov. 1952	82	85	55	65	5.7	W	88	117	160

Mean number of days (Bristol): Precipitation .01 in or more — 14, 12, 14, 10, 12, 12, 10, 8, 7, 10, 12, 12 | 132; Snow/Sleet 1.0 in or more — 1, 1, *, 0, 0, 0, 0, 0, 0, 0, *, 1 | 3; Thunderstorms — *, 1, 2, 3, 7, 12, 10, 8, 5, 1, 1, * | 50; Heavy fog — 3, 3, 1, 2, 3, 4, 4, 7, 4, 4, 4, 3 | 40; Max 90° and above — 0, 0, 0, *, 1, 6, 8, 7, 3, *, 0, 0 | 25; Max 32° and below — 3, 2, 1, *, 0, 0, 0, 0, 0, 0, 1, 2 | 8; Min 32° and below — 21, 16, 13, 4, *, 0, 0, 0, 0, 3, 15, 21 | 93; Min 0° and below — *, *, 0, 0, 0, 0, 0, 0, 0, 0, 0, 0 | *.

Mean sky cover (sunrise to sunset): 7.7, 6.9, 6.7, 6.3, 6.0, 5.9, 5.8, 5.6, 5.2, 5.4, 4.9, 6.6 | 6.1

CHATTANOOGA, TENNESSEE — LOVELL FIELD

LATITUDE 35° 02' N
LONGITUDE 85° 12' W
ELEVATION (ground) 670 Feet

Month	Normal Daily max	Normal Daily min	Normal Monthly	Record highest	Year	Record lowest	Year	Normal degree days	Precip. Normal total	Max. monthly	Year	Min. monthly	Year	Max. in 24 hrs	Year	Snow Mean total	Max. monthly	Year	Max. in 24 hrs	Year	RH 1a	RH 7a	RH 1p	RH 7p	Wind mean hourly	Clear	Partly cloudy	Cloudy
(yrs)	(b)	(b)	(b)	79		79		(b)	(b)	79		79		79		79	79		79		17	79	42	79	79	79	79	79
J	51.6	31.6	41.6	78	1949	-7	1886	725	5.23	14.74	1882	1.33	1907	4.44	1949	2.0	15.8	1893	9.8	1892	81	80	62	68	7.6	9	8	15
F	54.6	33.3	44.0	79	1930	-10	1899	588	5.11	12.30	1939	0.62	1941	3.93	1948	1.6	17.3	1895	9.9	1912	80	77	58	62	8.2	9	8	12
M	62.2	39.1	50.7	89	1929	10	1886	467	6.05	14.05	1899	0.93	1886	7.61	1886	0.6	11.0	1927	10.8	1927	77	53	58	58	8.4	10	10	11
A	71.8	47.5	59.7	93	1942	25	1881	179	4.16	15.29	1929	0.54	1942	6.60	1911	0.1	4.1	1910	4.1	1910	80	75	51	59	8.1	9	13	10
M	79.9	55.5	67.7	99	1941	36	1889	45	4.21	12.00	1892	0.29	1903	3.70	1941	T	T	1944	T	1944	86	78	51	59	6.4	10	13	8
J	87.6	64.0	75.8	104	1952	41	1889	0	4.21	11.40	1931	0.39	1949	4.85	1944	0.0	0.0		0.0		89	80	55	64	5.4	8	15	7
J	89.6	67.0	78.3	106	1952	54	1916	0	5.34	13.49	1916	0.20	1933	5.03	1957	0.0	0.0		0.0		83	56	66		5.5	7	16	8
A	88.7	65.9	77.3	105	1947	57	1920	0	3.70	12.38	1920	0.45	1957	3.70	1929	0.0	0.0		0.0		90	85	55	65	5.1	8	11	7
S	84.6	60.3	72.5	104	1925	37	1942	24	3.24	12.19	1957	0.04	1919	4.35	1919	0.0	0.0		0.0		85	83	53	65	5.2	12	11	7
O	73.9	47.7	60.8	94	1954	22	1925	169	2.69	11.91	1925	0.08	1938	4.16	1932	T	T	1954+	T	1954+	83	79	54	65	5.7	15	8	8
N	60.8	37.4	49.1	83	1915	3	1950	477	4.03	13.59	1890	0.16	1890	4.58	1948	0.2	6.5	1906	6.5	1906	82	79	54	62	6.1	12	8	10
D	52.2	31.9	42.1	78	1951	1	1880	710	5.31	11.69	1942	0.44	1889	5.25	1942	1.1	14.8	1886	12.0	1886	83	83	62	69	7.1	9	8	14
Year	71.5	48.4	60.0	106	July 1952	-10	Feb. 1899	3384	53.60	15.29	Apr. 1911	0.04	Sept. 1919	7.61	Mar. 1886	5.6	17.3	Feb. 1895	12.0	Dec. 1886	83	80	55	64	6.7	114	130	121

Mean number of days (Chattanooga): Precipitation .01 in or more — 12, 13, 12, 10, 10, 12, 13, 8, 6, 9, 12 | 131; Snow/Sleet 1.0 in or more — 2, 1, 1, *, 0, 0, 0, 0, *, *, 1 | 5; Thunderstorms — 2, 2, 4, 5, 8, 11, 12, 10, 4, 1, 1, 1 | 59; Heavy fog — 3, 1, 1, 1, 1, 1, 2, 5, 4, 4 | 27; Max 90° and above — 0, 0, *, 2, 9, 13, 15, *, 0, 0 | 40; Max 32° and below — 1, 1, *, 0, 0, 0, 0, 0, *, 1 | 3; Min 32° and below — 14, 11, 6, 1, *, 0, 0, 0, *, 6, 13 | 51; Min 0° and below — *, *, 0, 0, 0, 0, 0, 0, 0, 0 | *.

Mean sky cover (sunrise to sunset): 6.4, 5.6, 5.8, 5.6, 5.4, 5.6, 5.5, 4.8, 4.2, 4.9, 6.2 | 5.5
Pct. of possible sunshine: 43, 45, 50, 58, 55, 64, 61, 65, 65, 53, 41 | 56

Fastest mile (Chattanooga) — speed/direction/year by month: 46 SW 1898; 63 SW 1952; 82 W 1947; 57 W 1910; 63 NW 1953; 67 NW 1953; 48 NNW 1952; 62 SW 1946; 57 SW 1957; 35 SW 1952; 43 SW 1915; 46 SW 1886 | 82 W Mar. 1947

NORMALS, MEANS, AND EXTREMES

KNOXVILLE, TENNESSEE — MCGHEE TYSON AIRPORT
LATITUDE 35° 49' N
LONGITUDE 83° 59' W
ELEVATION (ground) 950 Feet

Month	Temp Normal Daily max (b)	Daily min (b)	Monthly (b)	Extremes Record highest	Year	Record lowest	Year	Normal degree days (b)	Precip Normal total (b)	Max monthly	Year	Min monthly	Year	Max 24 hrs	Year
J	49.8	31.1	40.5	77	1950	-16	1884	760	4.54	16.98	1907	1.29	1882	3.90	1879
F	52.5	32.4	42.5	79	1927	-10	1905	630	4.73	12.52	1898	0.56	1873	4.45	1875
M	60.5	38.3	49.4	88	1929	5	1897	500	4.83	13.35	1910	0.72	1886	5.56	1886
A	70.8	47.2	59.0	93	1925	23	1923	196	3.64	11.32	1874	0.84	1942	5.68	1874
M	79.0	55.8	67.4	96	1941	34	1944	50	3.58	8.81	1941	0.71	1938	3.12	1938
J	87.0	64.5	75.8	102	1944	42	1903	0	3.47	11.83	1944	0.20	1929	3.44	1929
J	89.1	67.6	78.4	104	1930	52	1885	0	4.72	13.18	1901	0.69	1917	6.20	1917
A	87.6	66.1	77.0	102	1948	49	1946	0	3.43	11.33	1954	0.77	1920	3.09	1942
S	83.5	60.6	72.1	103	1954	35	1888	33	2.53	8.61	1879	0.07	1925	5.68	1879
O	72.5	48.1	60.3	94	1884	24	1910	179	2.63	9.51	1925	0.07	1904	3.52	1925
N	59.1	37.6	48.4	84	1948	-5	1950	498	3.15	10.36	1948	0.17	1890	4.06	1948
D	50.3	31.7	41.0	77	1951	-5	1880	744	4.26	12.34	1896	0.95	1901	3.76	1896
Year	70.2	48.4	59.3	104	July 1930	-16	Jan 1884	3590	45.51	17.32	Oct 1904	0.07	Apr 1874	6.20	July 1917

Month	Snow Mean total	Max monthly	Year	Max 24 hrs	Year	RH 1:00 a.m. EST	7:00 a.m.	1:00 p.m.	7:00 p.m.	Wind mean hrly	Prev dir	Fastest speed	Dir	Year
J	2.6	13.7	1918	9.5	1918	79	83	65	69	7.3	NE	56	SW	1955
F	2.5	25.7	1895	8.2	1895	75	81	60	64	7.7	NE	54	SW	1945
M	1.4	13.0	1942	12.1	1942	73	78	54	58	8.1	SW	61	SW	1945
A	0.1	4.4	1901	4.4	1901	73	75	49	54	8.0	SW	71	SW	1944
M	T	T	1945+	T	1945+	82	78	51	58	6.7	SW	59	SW	1944
J	0.0	0.0		0.0		81	81	53	63	5.9	SW	65	SW	1944
J	0.0	0.0		0.0		84	83	55	66	5.6	SW	57	NW	1951
A	0.0	0.0		0.0		85	87	56	68	5.6	SW	49	W	1943
S	0.0	0.0		0.0		82	86	52	65	5.8	NE	56	SW	1943
O	T	T	1937+	T	1937+	78	83	51	64	5.8	NE	42	W	1947
N	0.6	18.2	1952	18.2	1952	78	83	56	61	6.5	NE	41	W	1947
D	1.7	25.4	1886	15.1	1886	80	84	65	69	6.9	NE	52	SW	1954+
Year	8.9	25.7	Feb 1895	18.2	Feb 1895	79	82	56	64	6.7	NE	71	SW	Apr 1944

Month	Pct sunshine	Mean sky cover	Clear	Partly cloudy	Cloudy	Precip .01"+	Snow 1.0"+	Thunderstorms	Heavy fog	Max 90°+	Max 32°below	Min 32°below	Min 0°below
J	42	6.7	7	9	15	13	1	*	2	0	3	17	*
F	48	6.3	8	9	13	12	1	1	1	0	1	13	*
M	53	6.1	9	9	13	12	1	3	1	0	*	8	0
A	60	5.6	10	10	10	11	0	4	1	*	0	1	0
M	64	5.5	9	13	8	11	0	6	1	1	0	*	0
J	65	5.4	9	14	8	12	0	9	1	7	0	0	0
J	60	5.3	9	14	8	11	0	10	1	11	0	0	0
A	60	4.6	13	10	8	8	0	8	2	7	0	0	0
S	63	4.4	13	8	9	7	0	4	3	4	0	*	0
O	63	4.4	11	8	11	8	0	1	3	*	0	1	0
N	54	5.4	8	8	14	11	1	*	2	0	1	8	*
D	41	6.4	8	8	11	12	1	1	2	0	2	16	*
Year	56	5.6	120	121	124	132	3	47	18	32	7	64	*

MEMPHIS, TENNESSEE — MUNICIPAL AIRPORT
LATITUDE 35° 03' N
LONGITUDE 89° 59' W
ELEVATION (ground) 263 Feet

Month	Temp Normal Daily max (b)	Daily min (b)	Monthly (b)	Record highest	Year	Record lowest (c)	Year	Normal degree days (b)	Precip Normal total (b)	Max monthly (d)	Year (d)	Min monthly (d)	Year (d)	Max 24 hrs (d)	Year (d)
J	50.3	32.9	41.6	79	1890	-8	1918+	725	5.55	17.56	1937	0.98	1931	5.75	1930
F	53.2	35.4	44.5	80	1918+	-1	1951	574	4.59	10.07	1939	0.71	1947	4.57	1938
M	62.0	42.0	52.0	87	1918+	12	1943	427	4.59	13.04	1927	0.70	1918	9.30	1919
A	72.0	51.5	61.8	91	1925	27	1928	139	3.92	13.30	1927	0.89	1925	5.26	1949
M	80.4	59.7	70.1	98	1953	38	1881	24	3.33	13.34	1873	0.82	1911	9.67	1953
J	88.4	68.2	78.3	104	1934	50	1944	0	3.33	11.97	1877	0.04	1953	—	—
J	91.1	71.3	81.2	106	1954+	56	1947	0	3.23	7.55	1892	0.01	1943	5.42	1929
A	90.7	69.8	80.3	106	1943+	52	1947	0	2.94	10.60	1915	0.40	1936	4.55	1888
S	85.3	63.2	74.3	105	1954+	36	1949+	17	2.55	10.82	1920	0.01	1897	4.66	1901
O	75.4	51.5	63.6	94	1951	25	1952	126	3.27	10.13	1919	0.05	1908	6.44	1934
N	61.4	39.8	50.6	85	1955	13	1950	432	4.56	14.53	1894	0.04	1935	5.40	1904
D	52.6	33.9	43.3	79	1951	2	1917+	673	5.09	11.97	1944	0.46	1889		
Year	71.9	51.6	61.8	106	July 1952+	-8	Feb 1951	3137	49.42	18.16	June 1877	0.00	Sept 1897	10.48	Nov 1934

Month	Snow Mean total	Max monthly	Year	Max 24 hrs (d)	Year	RH 12:00 a.m. CST	6:00 a.m.	12:00 p.m.	6:00 p.m.	Wind mean hrly	Prev dir	Fastest speed	Dir (d)	Year (d)
J	1.9	15.1	1948	10.1	1948	79	82	67	73	12.3	S	56		1928
F	1.4	10.3	1905	9.8	1886	77	82	63	68	11.6	S	51		1918
M	0.5	18.5	1892	18.0	1892	74	80	57	62	12.2	S	54	S	1952
A	T	T	1920+	T	1920+	74	80	54	57	11.7	S	51		1917
M	0.0	0.0		0.0		81	84	56	59	9.4	S	51		1916
J	0.0	0.0		0.0		82	86	56	59	8.4	S	54		1917
J	0.0	0.0		0.0		82	86	59	63	7.9	S	47		1914
A	0.0	0.0		0.0		80	85	55	58	7.4	NE	51		1914
S	0.0	0.0		0.0		80	85	52	61	8.1	S	56		1924
O	T	T	1910	0.0	1910	76	81	51	66	8.5	S	47		1934
N	0.1	4.9	1929	4.9	1929	78	81	57	66	10.3	W	45	W	1924
D	1.2	11.2	1917	8.5	1917	81	84	66	71	11.0	S	56		1924
Year	5.1	18.5	Mar 1892	18.0	Mar 1892	78	83	57	63	9.9	S	57	S	May 1916

Month	Pct sunshine	Mean sky cover	Clear	Partly cloudy	Cloudy (d)	Precip .01"+ (d)	Snow 1.0"+ (d)	Thunderstorms (d)	Heavy fog (d)	Max 90°+ (c)	Max 32°below (c)	Min 32°below (c)	Min 0°below (c)
J	44	6.5	9	7	15	11	1	2	1	0	3	14	*
F	51	6.0	10	7	12	10	1	2	1	0	2	10	*
M	57	6.0	11	9	12	11	*	4	1	*	*	4	0
A	64	5.2	11	11	10	10	0	5	1	2	0	*	0
M	68	5.3	11	11	9	9	0	6	*	12	0	0	0
J	73	4.2	14	12	7	9	0	8	*	17	0	0	0
J	73	4.1	14	11	6	9	0	8	*	15	0	0	0
A	70	4.1	17	7	6	7	0	7	*	7	0	0	0
S	69	4.1	15	7	7	7	0	3	1	0	0	0	0
O	58	4.5	12	7	11	6	*	2	1	0	*	*	0
N	52	6.2	10	7	14	9	*	2	1	0	2	5	*
D	45		12	7	14	11	1	1	1	0		12	
Year	64	5.6	140	107	118	112	2	50	6	53	7	45	*

NORMALS, MEANS, AND EXTREMES

NASHVILLE, TENNESSEE — BERRY FIELD

LATITUDE 36° 07' N
LONGITUDE 86° 41' W
ELEVATION (ground) 577 feet

Month	Temperature — Normal			Temperature — Extremes				Normal degree days	Precipitation							Snow, Sleet					Relative humidity				Wind						Pct. of possible sunshine	Mean sky cover sunrise to sunset	Sunrise to sunset			Mean number of days							
	Daily max.	Daily min.	Monthly	Rec. high	Year	Rec. low	Year		Normal total	Max. monthly	Year	Min. monthly	Year	Max in 24 hrs	Year	Mean total	Max monthly	Year	Max 24 hr	Year	12:00 mid	6:00 a.m.	12:00 n	9:00 p.m.	Mean hourly	Prevailing dir.	Fastest speed	Dir.	Year			Clear	Partly cloudy	Cloudy	Precip .01+	Snow 1.0+	Tstm	Hvy fog	Max 90+	Max 32−	Min 32−	Min 0−
(yrs)	(b)	(b)	(b)	87		87		(b)	(b)	87		87		87		73	73		73		17	70	70	73	49	14	46	46		61	37	87	87	87	87	14	85	67	87	87	87	87
J	48.9	30.9	39.9	78	1911	−10	1918+	778	4.93	14.75	1937	1.13	1940	4.40	1946	2.6	18.8	1948	8.5	1905	80	84	61	73	9.6	S	56	SW	1949+	42	6.8	7	8	16	12	1	1	1		3	17	*
F	51.6	33.5	42.3	79	1917	−13	1951+	636	4.16	12.37	1880	0.63	1898	5.25	1894	2.4	17.9	1886	16.3	1929	77	80	62	65	9.8	S	57	NW	1953	47	6.5	7	7	13	11	1	2	1		3	13	*
M	60.1	38.9	49.5	89	1907	−3	1899	498	5.28	11.84	1891	0.85	1910	5.28	1955	1.4	21.5	1892	17.0	1892	74	78	56	60	10.1	S	70		1916	54	6.2	9	9	13	12	1	4	1		*	7	*
A	70.5	48.8	59.7	90	1952+	25	1950+	186	3.69	10.31	1874	0.72	1915	5.04	1883	T	T	1910	1.5	1910	74	76	51	55	9.6	S	60			60	5.8	9	10	11	10	*	5	1	1	0	1	0
M	79.1	57.5	68.2	98	1911	36	1947	43	3.78	9.97	1933	0.63	1951	3.57	1944	T	T	1952+	1.5	1952+	79	79	53	53	10.1	S	57			65	5.6	10	12	9	10	0	7	1	1	0	0	0
J	87.7	66.1	76.9	106	1952	42	1894	0	3.19	11.64	1928	0.21	1936	4.39	1928	0.0	0.0		0.0		82	79	52	60	7.6	S	73			69	5.3	12	14	7	9	0	9	1	9	0	0	0
J	90.6	69.3	80.0	107	1952	51	1947	0	3.96	9.43	1878	0.46	1890	5.09	1878	0.0	0.0		0.0		83	81	61	62	6.9	S	59		1878	69	5.2	10	14	7	10	0	10	1	14	0	0	0
A	89.4	68.0	78.7	105	1930	47	1946	0	3.31	9.60	1929	0.51	1923	5.65	1898	0.0	0.0		0.0		83	84	54	62	6.7	S	49		1912	68	5.0	12	12	7	9	0	7	1	12	0	0	0
S	84.6	61.8	73.2	105	1954	36	1949	22	2.74	10.95	1906	0.13	1903	4.93	1895	0.0	0.0		0.0		81	84	58	58	7.1	S-N	47		1920	68	5.4	15	8	7	8	0	4	1	6	0	0	0
O	73.8	49.8	61.8	94	1953	26	1952	154	2.52	8.35	1919	0.03	1924	3.78	1928	T	T	1925	1.2	1950	81	84	51	62	7.6	S	51		1932	65	4.4	15	8	8	7	0	1	1	0	*	1	*
N	59.2	39.3	49.3	85	1935	−1	1950	471	3.41	9.04	1945	0.54	1949	6.05	1900	0.4	1.2	1950	1.2	1950	80	82	58	62	8.0	S	58		1938	55	5.2	11	8	11	8	0	1	1	0	*	8	*
D	50.2	32.6	41.6	76	1951	−13	1950	725	4.06	13.53	1926	0.91	1935	5.10	1926	1.4	10.5	1917	9.0	1916	80	82	65	69	9.2	S	47		1946	42	6.5	8	8	15	11	1	1	2		2	15	*
Year	70.5	49.7	60.1	107	July 1952	−13	Feb. 1951+	3513	45.03	14.75	Jan. 1937	0.03	Oct. 1924	6.05	Nov. 1900	8.2	21.5	March 1892	17.0	March 1892	80	81	56	62	8.5	S	73	NW	June 1953	59	5.6	120	120	125	120	3	52	11	42	7	62	*

OAK RIDGE, TENNESSEE — WEATHER BUREAU OFFICE

LATITUDE 36° 01' N
LONGITUDE 84° 14' W
ELEVATION (ground) 905 feet

Month	Temperature — Normal			Temperature — Extremes				Normal degree days	Precipitation							Snow, Sleet					Wind					Mean number of days							
	Daily max.	Daily min.	Monthly	Rec. high	Year	Rec. low	Year		Normal total	Max. monthly	Year	Min. monthly	Year	Max in 24 hrs	Year	Mean total	Max monthly	Year	Max 24 hr	Year	Mean hourly	Prevailing dir.	Clear	Partly cloudy	Cloudy	Precip .01+	Snow 1.0+	Tstm	Hvy fog	Max 90+	Max 32−	Min 32−	Min 0−
(yrs)	(b)	(b)	(b)	10		10		(b)	(b)	10		10		10		10	10		10		9	9	10	10	10	10	5	9	7	10	10	10	10
J	48.8	29.2	39.0	75	1952	4	1948	806	6.06	13.27	1954	2.44	1955	4.25	1954	2.3	10.7	1948	5.7	1948	5.0	SW	6	6	19	15	1	1	3		1	18	1
F	51.6	30.4	41.0	74	1951+	−1	1951	672	5.79	10.47	1956	1.90	1952	3.05	1948	1.1	4.0	1948	4.0	1948	5.3	W	6	8	15	13	1	2	1		1	12	*
M	59.2	36.0	48.1	83	1948	13	1948	532	5.69	8.09	1950	2.13	1950	2.80	1951	T	1.0	1951	1.0	1951	5.3	SW	8	9	15	13	1	3	1		*	5	0
A	70.2	45.0	57.6	89	1955	34	1950	230	4.08	9.71	1956	1.39	1951	3.74	1956	0.0	0.0		0.0		5.6	SW	10	8	13	10	0	5	2	0	0	0	0
M	78.5	53.4	66.0	93	1955	34	1954	64	4.18	6.83	1953	1.23	1952	2.20	1953	0.0	0.0		0.0		6.1	SW	10	9	11	11	0	9	2	1	0	0	0
J	86.3	62.1	74.2	101	1954	43	1956	0	3.91	4.83	1951	1.18	1952	2.01	1957	0.0	0.0		0.0		3.8	SW	10	12	8	8	0	9	2	11	0	0	0
J	88.3	65.1	76.7	105	1952	51	1952+	0	5.34	8.51	1949	2.72	1957	3.48	1954	0.0	0.0		0.0		3.7	SW	11	10	8	11	0	11	3	14	0	0	0
A	86.9	63.6	75.3	102	1954	51	1956	0	4.06	5.94	1950	0.54	1953	3.20	1955	0.0	0.0		0.0		3.5	E	11	10	8	9	0	9	3	6	0	0	0
S	83.1	58.4	70.8	102	1954+	37	1949	38	2.55	9.10	1957	1.14	1948	2.61	1957	0.0	0.0		0.0		3.7	E	14	7	9	8	0	4	3	*	0	0	0
O	71.7	45.9	58.8	90	1954	21	1952	218	2.42	6.11	1949	0.58	1953	2.13	1949	T	T	1957	T	1957	4.2	SW	12	8	10	7	0	2	2	0	0	1	0
N	58.3	35.7	47.0	81	1948	6	1950	540	3.98	12.22	1948	1.37	1949	2.97	1957	1.0	6.5	1950	6.5	1950	4.4	SW	10	7	13	11	*	2	3	0	0	15	0
D	49.3	29.9	39.6	74	1951	6	1957	787	5.66	10.31	1956	3.83	1950	4.73	1954	0.4	2.1	1954	6.8	1954	4.4	SW	8	7	16	11	1	2	2		1	19	*
Year	69.4	46.2	57.8	105	July 1952	0	Nov. 1950	3885	53.72	13.27	Jan. 1954	0.54	Aug. 1953	4.73	Dec. 1954	5.0	10.7	Jan. 1948	6.5	Nov. 1950	4.4	SW	121	100	144	129	2	57	35	46	2	79	*

NORMALS, MEANS, AND EXTREMES

Month	Temperature Normal Daily max	Temperature Normal Daily min	Temperature Normal Monthly	Extremes Record highest	Year	Extremes Record lowest	Year	Normal degree days	Precip. Normal total	Precip. Max. monthly	Year	Precip. Min. monthly	Year	Precip. Max. in 24 hrs.	Year	Wind Mean hourly speed	Wind Prevailing direction	Days Precip. .01 in. or more	Days Max. 90° and above	Days Max. 32° and below	Days Min. 32° and below	Days Min. 0° and below
(years of record)	13	13	13	13		13			14	14		14		14		10	10	11	13	13	13	13
	(b)	(b)	(b)					(b)	(b)												
J	48.7	30.4	39.6	77	1952	1	1948	787	5.11	12.37	1944	1.11	1944	3.96	1954	5.7	NE	11	0	1	17	0
F	51.2	31.5	41.4	77	1948	3	1951	661	5.21	10.01	1947	1.89	1947	3.23	1948	5.9	NE	14	0	1	13	0
M	59.3	37.3	48.3	87	1945	14	1955	534	5.64	8.91	1957	2.06	1957	2.85	1952	6.8	SW	11	0	*	10	0
A	69.5	45.9	57.7	89	1948+	24	1950	227	4.29	8.54	1952	1.25	1952	2.96	1956	6.8	SW	12	0	0	2	0
M	77.7	54.2	66.0	94	1956	34	1954	64	3.78	7.01	1950	0.90	1951	2.09	1955	4.9	SW	10	2	0	0	0
J	85.7	62.7	74.2	99	1952	41	1946	0	4.18	5.87	1945	1.18	1956	3.08	1945	4.1	SW	9	10	0	0	0
J	87.8	65.6	76.7	103	1952	49	1947	0	4.83	8.13	1950	2.14	1954	2.40	1950	3.8	NE	10	13	0	0	0
A	86.6	64.2	75.4	99	1957+	44	1946	0	4.45	10.31	1950	0.50	1953	2.34	1950	3.2	NE	9	13	0	0	0
S	83.0	59.1	71.1	103	1954+	33	1949	41	2.97	12.84	1944	1.07	1945	7.75	1944	4.2	NE	7	6	0	0	0
O	71.8	46.7	59.3	91	1954	21	1952	206	2.77	6.43	1949	0.60	1953	2.32	1949	4.0	NE	7	*	0	1	0
N	58.4	36.6	47.5	82	1950	4	1950	531	3.94	12.00	1948	1.01	1949	3.20	1948	4.9	SW	10	0	*	12	0
D	49.3	30.9	40.1	76	1951	4	1957	772	5.21	10.28	1954	3.00	1947	4.38	1954	4.9	NE	10	0	1	19	0
Year	69.1	47.1	58.1	103	Sept. 1954+	1	Jan. 1948	3823	52.38	12.84	Sept. 1944	0.50	Aug. 1953	7.75	Sept. 1944	4.9	NE	119	44	3	74	0

(a) Length of record, years.
(b) Normal values are based on the period 1921-1950, and are means adjusted to represent observations taken at the present standard location.

\# No record.
* Airport data.
+ City Office data.
‡ Also on earlier dates, months, or years.
T Trace, an amount too small to measure.
* Less than one-half.

REFERENCE NOTES APPLYING TO ALL "NORMALS, MEANS, AND EXTREMES" TABLES.

Sky cover is expressed in a range of 0 for no clouds or obscuring phenomena to 10 for complete sky cover. The number of clear days is based on average cloudiness 0-3 tenths; partly cloudy days on 4-7 tenths; and cloudy days on 8-10 tenths. Monthly degree day totals are the sum of the negative departures of average daily temperatures from 65°F. Sleet was included in snowfall totals beginning with July 1948. Heavy fog also includes data referred to at various times in the past as "Dense" or "Thick". The upper visibility limit for heavy fog is 1/4 mile. Data in these tables are based on records through 1957.

Mean Annual Precipitation, Inches

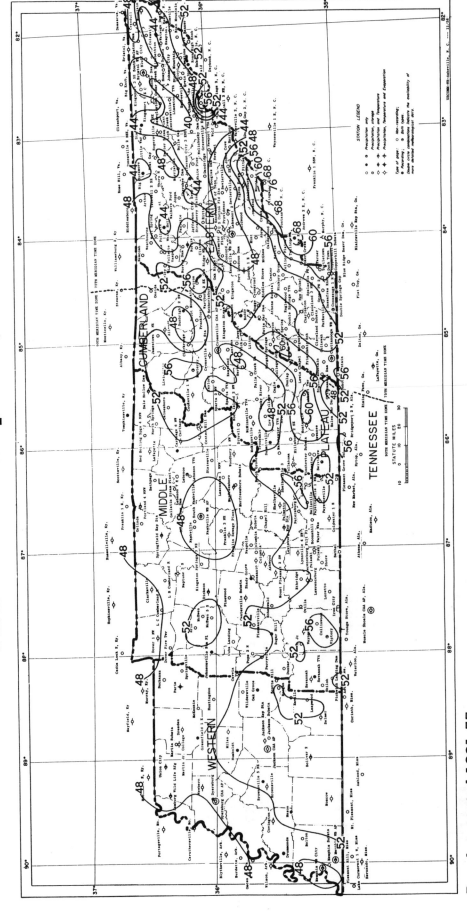

Based on period 1931-55

Isolines are drawn through points of approximately equal value. Caution should be used in interpolating on these maps, particularly in mountainous areas.

Mean Maximum Temperature (°F.), January

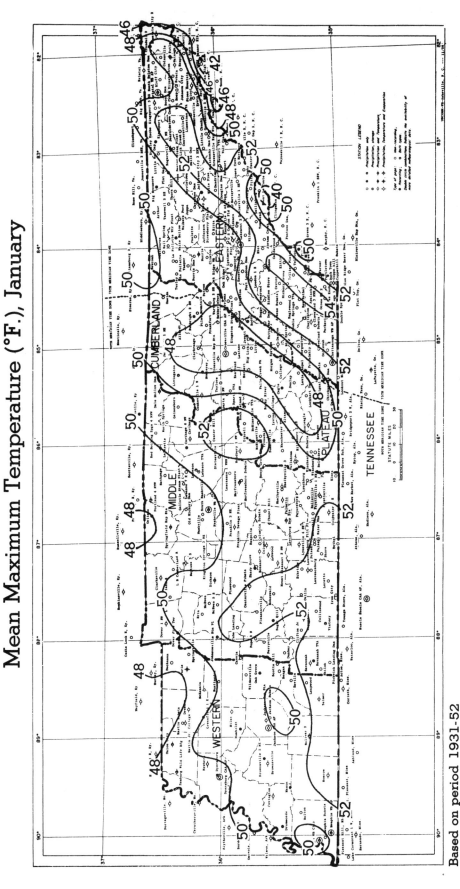

Based on period 1931-52

Isolines are drawn through points of approximately equal value. Caution should be used in interpolating on these maps, particulary in mountainous areas.

Mean Minimum Temperature (°F.), January

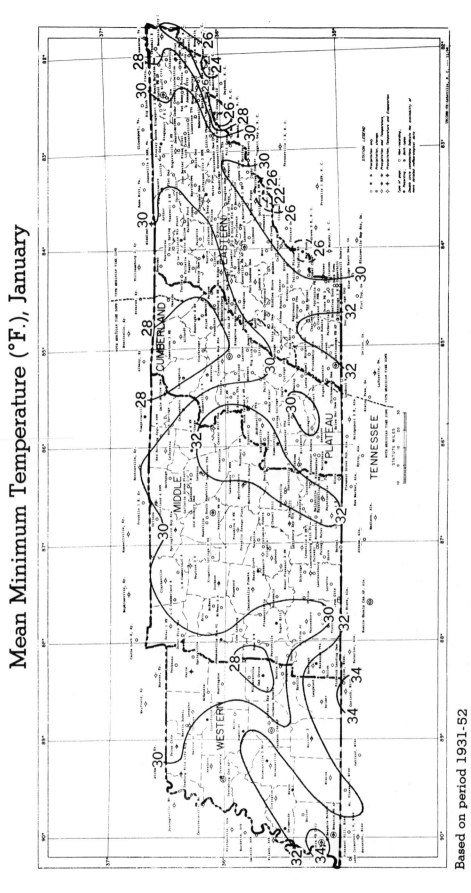

Based on period 1931-52

Isolines are drawn through points of approximately equal value.   Caution should be used
in interpolating on these maps, particulary in mountainous areas.

Mean Maximum Temperature (°F.), July

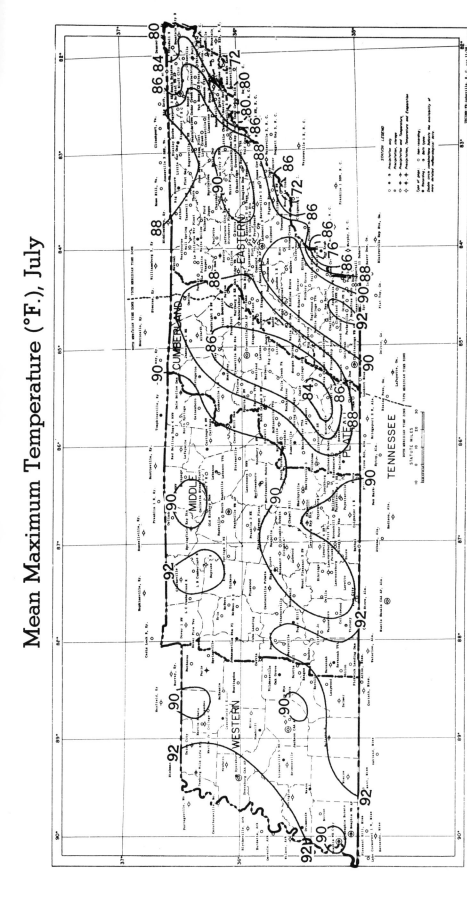

Based on period 1931-52

Isolines are drawn through points of approximately equal value. Caution should be used in interpolating on these maps, particulary in mountainous areas.

Mean Minimum Temperature (°F.), July

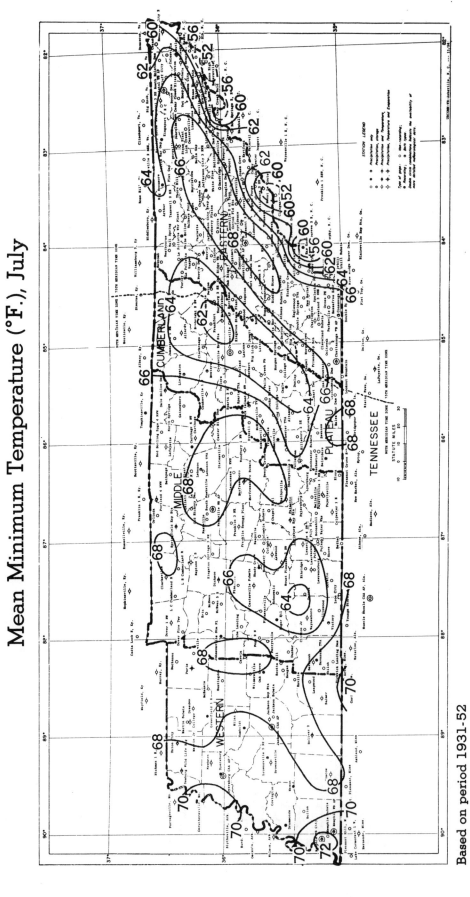

Based on period 1931-52

Isolines are drawn through points of approximately equal value. Caution should be used in interpolating on these maps, particularly in mountainous areas.

THE CLIMATE OF
VERMONT

by
Robert E. Lautzenheiser

December 1959

PHYSICAL DESCRIPTION: -- "The Green Mountain State" occupies 9,609 square miles, fully one-seventh of New England's total area. Though Vermont is the only New England state without a coastline on the Atlantic Ocean, most of its boundary is water. The Connecticut River forms the entire eastern border. Lake Champlain marks over 100 miles of the western boundary. Vermont extends southward from near the 45° parallel of latitude almost 160 miles to about 20 miles south of the 43d parallel. Vermont widens northward from about 40 to 90 miles across.

The terrain is hilly to mountainous. The Green Mountains extend the length of the State. They rise to their highest elevation at Mt. Mansfield, 4,393 feet above sea level. Many peaks in this range rise to over 3,000 feet, as do several others in eastern Vermont. Elevations of less than 500 feet above sea level are mostly confined to the lowlands paralleling Lake Champlain in the west and to the central and southern portions of the Connecticut Valley in the east. Much of the State ranges from 500 to 2,000 feet in elevation. The glacier of the great Ice Age accounts for many topographical features, lakes, and soils. Inland waters cover more than 300 square miles.

Two-thirds of Vermont is forest, contained in National, State, municipal, and private reserves and in farm woodlands. A considerable area, especially in the north, is sparsely settled. The mountains, hills, lakes, streams, and forests combine to make Vermont a state noted for its scenic beauty.

GENERAL CLIMATIC FEATURES: -- Vermont shares with the other New England states in the chief climatic characteristics. These include: (1) Changeableness of the weather, (2) large range of temperature, both daily and annual, (3) great differences between the same seasons in different years, (4) equable distribution of precipitation, and (5) considerable diversity from place to place. The regional climatic influences are modified in Vermont by varying elevations, types of terrain, and distances from the Atlantic Ocean and from Lake Champlain. The State has been divided into three climatological divisions (Western, Northeastern, and Southeastern) which take into account the main features of these modifying factors, in a general way. To take all local factors into consideration would require an impractical number of areal divisions.

Vermont lies in the "prevailing westerlies", the belt of generally eastward air movement which encircles the globe in middle latitudes. Embedded in this circulation are extensive masses of air originating in higher or lower latitudes and interacting to produce low-pressure storm systems. Relative to most other sections of the country, a large number of such storms pass over or near Vermont. The majority of air masses affecting this

State belong to three types: (1) Cold, dry air pouring down from subarctic North America, (2) warm, moist air streaming up on a long overland journey from the Gulf of Mexico and other subtropical waters, and (3) cool, damp air moving in from the North Atlantic. Because the atmospheric flow is usually from a westerly direction, Vermont is more influenced by the first two types than it is by the third. In other words, the Atlantic Ocean sometimes affects Vermont, but does not dominate its climate.

The procession of contrasting air masses and the relatively frequent passage of "Lows" bring about on the average a twice-weekly alternation from fair to cloudy or stormy conditions, attended by often abrupt changes in temperature, moisture, sunshine, wind direction and speed. There is no regular or persistent rhythm to this sequence, and it is interrupted by periods during which the weather patterns continue the same for several days, infrequently for several weeks. Vermont weather, however, is cited for variety rather than monotony. Changeability is also one of its features on a longer time-scale. That is, the same month or season will exhibit varying characteristics over the years, sometimes in close alternation, and sometimes arranged in similar groups for successive years. A "normal" month, season, or year is indeed the exception rather than the rule.

The basic climate, as outlined above, obviously does not result from the predominance of any single controlling weather regime, but is rather the integrated effect of a variety of weather patterns. Hence, "weather averages" in Vermont usually are not sufficient for important planning purposes without further climatological analysis.

The Western Division is a relatively narrow band running the full length of the State west of the Green Mountains. This Division is least affected by Atlantic Ocean influences. Because its northern portion is moderated by Lake Champlain it can be included with southwestern Vermont even though its north-south extension is so long. The Northeastern Division is the largest of the three and includes the northeastern, north-central, and east-central portions of Vermont, excepting a narrow strip in the Connecticut River Valley in the east-central portion. This strip is included as a part of the Southeastern Division because of its lower elevation.

TEMPERATURE: -- The annual mean temperature is near 43°F. in the Northeastern Division, 44°F. in the Southeastern, and 46°F. in the Western. Averages vary also within the divisions. Elevation, slope, and other local environmental aspects, including urbanization, all have an effect. As an extreme example of the effect of altitude, a comparison between the summit station on Mt. Mansfield with Enosburg Falls is interesting. Though these stations are about the same distance from Lake Champlain, the average temperature for the year 1958 on Mt. Mansfield was only slightly above freezing, 32.8°F.; Enosburg Falls, at 3,500 feet lower elevation, was nearly 10° warmer, with 42.0°F.; and Enosburg Falls is about 25 miles north of Mt. Mansfield. The highest temperature of record in the State is 105°F. observed July 4, 1911 at Vernon; the lowest, -50°F., December 30, 1933, at Bloomfield.

Summer temperatures are delightfully comfortable as a rule. They are also reasonably uniform over the State, excepting topographical extremes. Long-period means for July average near 70°F. in the Western Division and near 68°F. in the other Divisions. Average daily minima in July are in the 50's over nearly the entire State. The average daily maxima reach only near 80°F. Hot days with maxima of 90°F. or higher average less than 10 per year at most stations. The frequency varies from place to place and from year to year. In the coolest summers, they range, in frequency of occurrence, from none at many stations to only a few at the warmest stations. In the warmest years many stations still have less than 10, but the frequency ranges up to as high as 30 at the warmer sites. Even after one of these hot days the temperature is likely to fall to 60°F. or lower during the night. The average daily range is 20° to 30° in summer, with the variation averaging a little more in the south than in the north. The diurnal range may reach 40°F. or more during cool, dry weather in valleys and lowlands. Late spring or early fall freeze may be a threat at a few of the more susceptible areas.

Temperatures from place to place vary more in winter than in summer. The Northeastern Division average in January is near 17°F. The Southeastern Division average is near 19°F. and the Western Division, 21°F. The daily temperature range is less in winter than in summer, averaging near 20°F. Days with subzero readings are common at most stations in winter. They number from 10 to 40 per year in the southern portion and from 20 to 50 in the north. The number exceeds 60 at some stations in the coldest winters and may be less than 10 at other stations in the mildest winters.

The growing season for vegetation subject to injury from freezing temperature averages 130 to 150 days in much of the Western Division and along the Connecticut River in the Southeastern Division. Elsewhere, and including the extreme southern portion of the Western Division the season varies from 100 to 130 days. Local topography causes exceptions and some localities have growing seasons as short as 80 to 90 days. The growing season begins in May and ends in September for most of the State.

PRECIPITATION: -- Vermont's precipitation, fortunately, is well distributed through the year. The summer months ordinarily receive adequate amounts for growing crops over the entire State. Winter precipitation is noticeably less than summer rainfall in the northern and western portions of the State. This difference is greater in those areas than in any other part of New England. New England as a whole is noted for the even distribution of its precipitation throughout the year, an effect due to the influence of the Atlantic Ocean. "Wet" or "dry" seasons, climatic characteristics of most parts of the World, are not, normally, conditions with which this section has to contend. This ocean influence is still strongly felt in southeastern Vermont, but it becomes weaker with increasing distance from the ocean. Low-pressure, or frontal, storm systems are the principal year-round moisture producers. When this activity ebbs somewhat in summer, bands or patches of thunderstorms increase in activity, more than making up the difference. Though brief and often of small extent, the thunderstorms produce the heaviest local rainfall intensities. They sometimes cause minor washouts of roads and soils. Rains of 1 to 2 inches in 1 hour can be expected at least once in a 10-year period.

Variations in monthly totals are extreme, ranging from none to over 10 inches. Such large fluctuations are rare. A large majority of monthly totals falls in the range of from 50 to 200 percent of normal. As prolonged droughts are infrequent, irrigation water is available during the fairly common shorter dry spells of summer. Similarly widespread floods are infrequent. However, torrential rains on November 2-3, 1927, caused flood damage estimated at $26 million. Other floods of note occurred in 1801, 1826, 1830, 1886, 1895, 1897, 1909, 1913, 1936, and 1947. Floods occur most often in the spring when they are caused by rainfall and melting snow. Stages of spring over-bank flooding are frequently increased by ice jams. Local flash floods result on occasions from short period summer storms between May and November.

The mean annual runoff in the streams ranges from about 10 inches in portions of the Lake Champlain drainage to 40 inches in southern Vermont. The Connecticut River forms the eastern border and its tributaries drain the major portion of Vermont. In the northwest portion, rivers drain into Lake Champlain or directly to the St. Lawrence. A small area in southwest Vermont drains to the Hudson River.

Total annual precipitation averages nearly 45 inches in the Southeastern Division and nearly 38 inches in the other divisions. Individual means vary considerably from station to station, especially within the Southeastern Division. Bellows Falls, with less than 41 inches per year, and Searsburg Station, with 55 inches, are less than 30 miles apart. The mountainous character of much of the State largely accounts for the variability from place to place.

Occasionally freezing rain occurs, coating exposed surfaces with troublesome ice. Most areas can expect at least one such occurrence in a winter. Frequency of days with measurable precipitation is between 120 and 160 days per year. As much as 6 inches of rain in 24 hours is rare in Vermont. Most stations have never recorded that much in a single day. However, Somerset received 8.77 inches in 24 hours during the flooding rains of November 1927.

SNOWFALL: -- Average annual total snowfall is from 55 to 65 inches in much of the Western Division and also in parts of the Connecticut River Valley. Elsewhere the annual averages vary greatly. They range upward to as much as 100 inches and, at a few stations, 100 to 125 inches. Topographical differences cause large variations in a short distance. As an example, Bennington has only about 55 inches per year, while Somerset, with over 120 inches, is only about 15 miles away but at a much higher elevation.

Snowfall is highly variable from season to season. It also varies for the same month in different years as well as from place to place. Variations in seasonal totals are mostly from about 50 to 150 percent of the long-period average. Totals for the least snowy seasons range from 25 to 50 percent of the greatest seasonal amounts. Month to month variations are much greater. Burlington's maximum monthly total is 34.3 inches in February 1958, but only 1.3 inches fell in that month in 1957.

The average number of days with 1 inch or more of snowfall in a season varies from near 20 to 40. The frequency increases with elevation. Most winters have several snowstorms of 5 inches or more per year. Storms of this magnitude may temporarily disrupt transportation.

One of the heaviest single snowstorms of record was that of March 11-14, 1888, known as the "Great Blizzard". Amounts in the southwestern part of the State ranged from 40 to 50 inches and in the southeastern part, from 30 to 40 inches. Drifts of 15 to 40 feet high were reported. Most of northern Vermont received from 20 to 30 inches in this storm. However, snowfalls of 20 inches or more are unusual in any part of the State. The heaviest 24-hour falls of record at many stations do not exceed 25 inches.

Snow cover is continuous throughout the winter season as a rule. Depth of snow on the ground reaches its maximum for much of the State in the latter part of February. At the highest elevations, however, the date falls in the middle of March. Water stored in the snow is an important contribution to the water supply. Spring melting is usually too gradual to produce serious flooding.

OTHER CLIMATIC FEATURES: -- Sunshine averages near 50 percent of possible on a year-round basis, but varies with topography. Data is not sufficient to describe this in detail. Higher elevations and peaks are much more cloudy, especially in winter, probably reducing the percentage to as low as 40 in local areas. Sunshine is most abundant during the summer season.

Heavy fog occurrence varies remarkably with location and topography but, again, not enough data are available to describe this in detail. Persistent fogs are sometimes experienced on the higher elevations. The duration of fog diminishes over flat and valley locations. But the shorter duration heavy ground fogs of early morning occur frequently at susceptible places in these areas. The number of days with fog probably varies from 10 to 60 per year over the State, except possibly even more on the highest mountain peaks.

WINDS AND STORMS: -- Vermont lies in the region of prevailing westerlies -- wind from the northwest in winter, and from the southwest in the warmer part of the year. But because the rugged topography has a strong influence on the direction of the wind, many areas have prevailing winds paralleling a valley. The major valleys tend to lie in a north-south direction. Thus prevailing winds may be from the north in winter and from the south in the warmer seasons in those areas.

Coastal storms, or "northeasters", are well known to New England. Their influence on Vermont is minimized by its inland location. They remain a factor, however, especially in the Southeastern Division. They generate very strong winds and heavy rain or snow. Some of the heavier snows are produced by these storms.

Storms of tropical origin may occasionally affect Vermont in summer or fall, but only rarely contain destructive winds. The very severe, rapidly moving hurricane of September 1938 is best remembered. Its path crossed the entire State, from near Wilder to Burlington. However, Vermont is far enough inland so that, usually, winds are considerably weakened by the time tropical storms reach the State, and are generally only light to moderate. Rainfall associated with these storms may, however, remain heavy.

Tornadoes are not common phenomena. Yet, on a per unit area basis, Vermont ranks with many other states in frequency of tornado occurrence. One or more of these most violent storms may occur in a year. Historical accounts suggest that the most notable Vermont tornado occurred on June 23, 1782. Entering the State at the southwest corner, it traveled northward and eastward and crossed into New Hampshire near Weathersfield. Fortunately, most tornadoes are very small, affecting a very localized area. Due to the extent of forested or sparsely settled areas, a large percentage that do occur are probably neither seen, recorded, nor do appreciable damage. They may occur even in the northern portion of the State. About 73 percent occur between May 15 and September 15. About 78 percent strike between 2 and 7 p.m. The peak months are June and July and the peak hour is 5 to 6 p.m. The chance of a tornado striking any given spot is extremely small.

Thunder and hailstorms also have a frequency maximum from midspring to early fall. Thunderstorms occur on 20 to 30 days per year. The most severe are attended by hail. Hail can damage or even ruin field crops, break glass, dent automobiles, and damage other vulnerable exposed objects. The size of an area struck by a hailstorm, however, is usually small. Glaze and icestorms of winter can make travel hazardous. These are usually of brief duration. At least one ice storm may be expected each year. A few widespread and prolonged ice storms have occurred. Besides affecting travel and transport, they also break trees or limbs, utility lines and poles. In such structural design as steel towers, possible ice load should be considered. The ice load also magnifies the wind stress by increasing the area exposed to the wind.

CLIMATE AND ECONOMY: -- Activities in Vermont are profoundly influenced by climate. Tree growth is especially favored. Covering two-thirds of the area, forests are a major scenic attraction. The spectacular coloration of foliage in the autumn is of special interest, drawing countless visitors. Lumbering and related wood products are leading industries. The ample supply of rainfall provides not only for timber growth but also the huge amount of water required in making of paper and other manufactures. Favored industries also include the manufacturing of machinery, textiles, and leather, and stone, clay, and glass products. A great diversity of other interests takes advantage of the abundant water supply. A large portion of the State's electrical power comes from a well developed hydroelectric system.

Climate is a significant factor in Vermont agriculture. Principal farm specialties include dairying, poultry raising, tree fruit, and truck gardening. Fresh milk and milk products are the leading farm outputs. These amount to one-third of New England's total dairy production. Apples are the most prolific of the tree fruits, with quality production an important commercial pursuit. Vermont is the leading state in top quality maple syrup and sugar production. Strawberries are an important truck product. A large acreage is devoted to pasture and hay, and to oats and corn.

Climate is particularly important to a major industry, the tourist and vacation trade, amounting to over $100 million annually. Summer camps abound on the shores of many of the State's 400 lakes and ponds. Abundant game and teeming lakes and streams draw sportsmen from far and near. Skiing, with related winter sports, is a very important seasonal attraction, made possible by the abundant snowfall. The winter sports industry has grown rapidly in recent years making Vermont a four-season vacation area.

SELECTED REFERENCES

General:

1. National Planning Association: The Economic State of New England (1954).

2. U. S. Dept. Agriculture: Atlas of American Agriculture (1936)

3. --- : Climate and Man Yearbook of Agriculture for 1941, Part 5, Climatic data, with special reference to agriculture in the United States.*

4. --- : Soil (Yearbook of Agriculture for 1957).

5. --- : Climatological Data, New England (issued monthly and annually, 1888 ---; pub. under various other titles previous to Jan. 1921).*

Specialized:

1. Brooks, C. F.: "New England Snowfall", Monthly Weather Review, Vol. 45 (1917).

2. --- : "The Rainfall of New England - General Statement", Journ. N. Eng. Water Works Assoc., Vol. 44 (1930).

3. Brown, Rodger A.: "Twisters in New England", unpublished manuscript, Antioch College, 1957.*

4. Church, P. E.: "A Geographical Study of New England Temperatures", Geogr. Review, Vol. 26 (1936).

5. Eustis, R. S.: "Winds over New England in relation to topography", Bull. Amer. Met. Soc., Vol. 23 (1942).

6. Galway, Joseph G.: "A Statistical Study of New England Snowfall", unpublished manuscript of U. S. Weather Bureau (1954).*

7. Goodnough, X. H.: "Rainfall in New England", Journ. N. Eng. Water Works Assoc., Vols. 29 (1915), 35 (1921) and 40 (1926).*

8. Palmer, Robert S.: "Agricultural Drought in New England". Technical Bulletin 97, Agricultural Experiment Station, U. of New Hampshire, Durham, N. H. (1958).

9. Perley, S.: Historic Storms of New England (1891).*

* References marked with an asterisk are useful sources of data; the others are principally studies of the important climatic elements.

10. Stone, R. G.: "Distribution of snow depths over New York and New England", Trans. Amer. Geophy. Union (1940).

11. --- : "The average length of the season with snow cover of various depths in New England", Trans. Amer. Geophy. Union (1944).

12. Upton, W.: "Characteristics of the New England Climate", Annals Harvard Astron. Obser. (1890).

13. U. S. Weather Bureau: Tabulations of frequencies of various climatic elements for various selected stations. Available on microfilm at library of Weather Bureau State Climatologist, 1900 Post Office Bldg., Boston 9, Mass.

14. Weber, J. H.: "The Rainfall in New England. Historical Statement. Annual Rainfall. Seasonal Rainfall. Mean Monthly Rainfall of Southern New England. Maximum and Minimum Rainfall of Southern New England". Journ. N. Eng. Water Works Assn., Vol. 44 (1930).

15. White, C. V.: "Rainfall in New England", Journ. N. Eng. Water Works Assn., Vols. 56 (1942) and 57 (1943).*

16. Weather Bureau Technical Paper No. 15 - Maximum Station Precipitation for 1, 2, 3, 6, 12, and 24 Hours.

17. Weather Bureau Technical Paper No. 16 - Maximum 24-Hour Precipitation in the United States. Washington, D. C. 1952.

18. Weather Bureau Technical Paper No. 25 - Rainfall Intensity-Duration-Frequency Curves. For selected stations in the United States, Alaska, Hawaiian Islands, and Puerto Rico.

(A) Climatic Summary of the United States (Bulletim W) 1930 edition, Section 84 (New Hampshire and Vermont). U. S. Weather Bureau

(B) Climatic Summary of the United States, New England - Supplement for 1931 through 1952 (Bulletin W Supplement). U. S. Weather Bureau

(C) Climatological Data - New England, U. S. Weather Bureau

(D) Climatological Data National Summary. U. S. Weather Bureau

(E) Hourly Precipitation Data - New England. U. S. Weather Bureau

(F) Local Climatological Data. U. S. Weather Bureau, for Burlington, Vermont.

FREEZE DATA

STATION	Freeze threshold temperature	Mean date of last Spring occurrence	Mean date of first Fall occurrence	Mean No. of days between dates	Years of record Spring	No. of occurrences in Spring	Years of record Fall	No. of occurrences in Fall	STATION	Freeze threshold temperature	Mean date of last Spring occurrence	Mean date of first Fall occurrence	Mean No. of days between dates	Years of record Spring	No. of occurrences in Spring	Years of record Fall	No. of occurrences in Fall
BELLOWS FALLS	32	05-16	10-02	139	20	20	20	20	DORSET 1 SSW	32	05-29	09-14	108	10	10	10	10
	28	04-27	10-12	168	20	20	20	20		28	05-12	09-24	135	10	10	10	10
	24	04-15	10-28	195	20	20	20	20		24	05-03	10-04	154	10	10	10	10
	20	03-28	11-12	229	20	20	20	20		20	04-19	10-16	180	10	10	10	10
	16	03-22	11-23	246	20	20	20	20		16	04-01	11-03	216	10	10	10	10
BLOOMFIELD	32	06-02	09-16	107	30	30	30	30	ENOSBURG FALLS	32	05-27	09-20	117	25	25	25	25
	28	05-19	09-27	131	30	30	30	30		28	05-12	09-29	140	25	25	25	25
	24	05-06	10-11	158	30	30	30	30		24	04-29	10-11	165	25	25	25	25
	20	04-22	10-20	181	30	30	30	30		20	04-16	10-25	192	25	25	25	25
	16	04-13	11-02	203	30	30	30	30		16	04-06	11-08	216	25	25	25	25
BURLINGTON WB	32	05-08	10-03	148	30	30	30	30	NORTHFIELD NORWICH U	32	05-27	09-17	113	30	30	30	30
	28	04-22	10-18	179	30	30	30	30		28	05-15	09-29	137	30	30	30	30
	24	04-13	11-03	204	30	30	30	30		24	04-28	10-11	165	30	30	30	30
	20	04-03	11-15	227	30	30	30	30		20	04-16	10-25	193	30	30	30	30
	16	03-24	11-25	246	30	30	30	30		16	04-03	11-09	219	30	30	30	30
CAVENDISH	32	05-30	09-13	106	30	30	30	30	RUTLAND	32	05-15	09-24	131	30	30	30	30
	28	05-14	09-29	138	30	30	30	30		28	05-01	10-06	159	30	30	30	30
	24	05-02	10-08	159	29	29	30	30		24	04-19	10-21	185	30	30	30	30
	20	04-18	10-21	186	29	29	30	30		20	04-06	11-04	212	30	30	30	30
	16	04-05	11-04	213	29	29	30	30		16	03-28	11-17	234	30	30	30	30
CHELSEA	32	06-02	09-10	100	30	30	30	30	ST JOHNSBURY	32	05-21	09-23	125	30	30	30	30
	28	05-19	09-25	129	30	30	30	30		28	05-09	10-03	148	30	30	30	30
	24	05-07	10-06	152	30	30	30	30		24	04-24	10-18	177	30	30	30	30
	20	04-21	10-19	181	30	30	30	30		20	04-10	11-01	204	30	30	30	30
	16	04-10	11-02	206	30	30	30	30		16	04-02	11-12	225	30	30	30	30
CORNWALL	32	05-06	10-07	154	30	30	30	30	SOMERSET	32	06-08	08-30	83	30	30	30	30
	28	04-25	10-18	176	30	30	30	30		28	05-24	09-23	122	30	30	30	30
	24	04-13	11-05	206	30	30	30	30		24	05-11	10-06	149	30	30	30	30
	20	04-02	11-12	225	30	30	30	30		20	04-24	10-19	179	30	30	30	30
	16	03-24	11-22	242	30	30	30	30		16	04-17	10-30	196	30	30	30	30
									WOODSTOCK	32	05-28	09-18	113	30	30	30	30
										28	05-14	09-29	138	30	30	30	30
										24	04-29	10-11	165	30	30	30	30
										20	04-15	10-25	193	30	30	30	30
										16	04-04	11-08	218	30	30	30	30

Data in the above table are based on the period 1921-1950, or that portion of this period for which data are available.

Means have been adjusted to take into account years of non-occurrence.

A freeze is a numerical substitute for the former term "killing frost" and is the occurrence of a minimum temperature at or below the threshold temperature of 32°, 28°, etc.

Freeze data tabulations in greater detail are available and can be reproduced at cost.

*MEAN TEMPERATURE AND PRECIPITATION

STATION	JAN Temp	JAN Prec	FEB Temp	FEB Prec	MAR Temp	MAR Prec	APR Temp	APR Prec	MAY Temp	MAY Prec	JUN Temp	JUN Prec	JUL Temp	JUL Prec	AUG Temp	AUG Prec	SEP Temp	SEP Prec	OCT Temp	OCT Prec	NOV Temp	NOV Prec	DEC Temp	DEC Prec	ANN Temp	ANN Prec
VERMONT																										
NORTHEASTERN																										
BLOOMFIELD	16.3	2.40	17.4	2.16	27.3	2.47	40.7	3.18	53.1	3.60	62.4	4.17	67.0	4.23	65.0	3.75	57.3	3.99	46.7	3.29	34.4	3.44	20.1	2.58	42.3	39.26
CHELSEA	16.8	2.48	17.5	2.17	28.0	2.66	40.9	3.14	53.1	3.57	62.4	3.67	67.2	4.09	64.8	3.28	56.9	3.56	46.5	3.02	34.4	3.21	20.8	2.46	42.4	37.31
NEWPORT	15.6	2.35	17.0	2.24	26.6	2.51	40.3	3.00	53.5	3.07	63.1	3.73	67.5	4.21	65.5	3.42	57.5	3.71	46.9	3.07	34.2	3.02	19.7	2.49	42.3	36.82
NORTHFIELD NORWICH UNI	17.8	2.20	19.0	2.06	28.9	2.46	41.4	2.60	53.6	3.14	62.8	3.33	67.5	3.53	65.0	3.20	57.3	3.23	47.3	2.67	35.3	3.04	21.5	2.38	43.1	33.84
ROCHESTER		3.02		2.98		3.61		3.63		3.98		3.81		4.37		3.71		3.92		3.34		3.81		3.26		43.44
SAINT JOHNSBURY	17.5	2.53	19.5	2.17	29.6	2.51	42.8	2.87	55.7	3.30	65.1	3.86	69.6	3.52	67.3	3.35	59.1	3.53	48.5	2.87	35.8	3.14	21.7	2.59	44.4	36.24
WEST BURKE		2.76		2.50		2.76		3.16		3.56		3.84		3.83		3.40		4.00		3.28		3.42		3.08		39.59
DIVISION	16.5	2.41	17.7	2.20	27.8	2.60	41.0	3.06	53.7	3.43	62.9	3.87	67.6	3.95	65.3	3.48	57.4	3.73	47.0	3.13	34.7	3.24	20.4	2.60	42.7	37.70
WESTERN																										
BURLINGTON WB AIRPORT	17.9	1.89	18.1	1.53	29.3	2.19	42.3	2.63	55.4	2.89	65.5	3.57	70.4	3.75	68.1	3.01	59.9	3.14	48.2	2.89	36.4	2.85	22.8	1.88	44.5	32.22
CORNWALL	20.7		21.7		31.6		44.8		57.3		66.3		71.3		69.2		61.2		50.4		24.7				46.4	
RUTLAND	21.5	2.55	22.4	2.11	32.0	2.70	44.6	3.00	56.3	3.71	65.1	4.25	69.5	4.70		3.45	59.6	3.76	49.5	2.96	38.0	3.14	25.3	2.34	45.9	38.67
DIVISION	20.6	2.46	21.5	2.11	31.2	2.74	44.0	3.11	56.2	3.49	65.2	3.78	70.0	4.17	67.7	3.59	60.0	3.65	49.5	3.04	37.6	3.10	24.5	2.55	45.7	37.79
SOUTHEASTERN																										
BELLOWS FALLS		3.22		2.70		3.33		3.78		3.84		3.57		3.80		3.22		3.57		2.85		4.04		3.03		40.95
CAVENDISH	19.0	3.35	20.0	3.10	29.7	3.72	42.6	3.80	55.2	3.85	64.0	3.96	68.5	4.37	65.9	3.29	58.0	3.54	47.2	3.27	35.3	3.85	22.5	3.22	44.0	43.32
MAYS MILL		4.25		3.60		4.78		4.40		4.93		4.16		4.41		4.14		4.44		3.89		4.84		3.90		52.04
READSBORO 1 SSE		4.04		3.24		4.34		4.39		4.62		4.20		4.29		4.05		4.36		3.59		4.27				49.29
SEARSBURG MOUNTAIN		3.97		3.29		4.40		4.76		5.02		4.78		4.68		4.79		5.18		4.42		4.61		4.18		54.08
SEARSBURG STATION		4.58		3.74		4.79		4.87		5.02		4.47		4.61		4.49		4.75		4.14		4.88		4.67		55.01
SOMERSET	17.6	4.56	17.2	3.71	25.4	4.86	37.8	4.69	50.0	4.74	58.6	4.71	62.8	4.29	60.7	4.08	53.7	4.93	44.6	4.11	32.8	4.78	20.2	4.54	40.1	54.00
VERNON		3.38		2.71		3.70		3.82		4.01		3.82		3.78		3.85		4.11		2.92		3.92		3.41		43.43
WHITE RIVER JUNCTION 1		2.86		2.43		2.86		3.27		3.41		3.47		4.06		3.08		3.32		2.80		3.28		2.70		37.54
WHITINGHAM 3 W		4.34		3.46		4.57		4.58		4.77		4.30		4.47		4.15		4.57		3.89		4.57		4.44		52.11
WILDER	19.6	2.59	20.7	2.12	30.4	2.53	43.2	3.12	55.2	3.31	64.1	3.72	68.7	4.18	66.5	3.03	58.7	3.29	48.0	2.75	36.3	3.21	23.1	2.51	44.5	35.96
WOODSTOCK 3 ENE		3.30		2.78		3.49		3.57		3.72		3.34		4.22		3.31		3.49		3.12		3.76		3.19		41.68
DIVISION	19.2	3.60	19.8	3.03	29.3	3.87	41.9	3.90	54.2	3.97	63.2	4.00	67.8	4.14	65.5	3.51	57.7	3.95	47.4	3.38	35.5	4.09	22.6	3.50	43.7	44.94

* Averages for period 1931-1955, except for stations marked WB which are "normals" based on period 1921-1950. Divisional means may not be the arithmetical average of individual stations published, since additional data from shorter period stations are used to obtain better areal representation.

CONFIDENCE LIMITS

In the absence of trend or record changes, the chances are 9 out of 10 that the true mean will lie in the interval formed by adding and subtracting the values in the following table from the means for any station in the State. Because of the wider variation in mean precipitation, the corresponding monthly means and annual mean must be substituted for "p" in the precipitation table below to obtain mean precipitation confidence limits.

1.8	.22√p	1.5	.18√p	1.7	.23√p	1.1	.23√p	.9	.26√p	.7	.30√p	.7	.31√p	.9	.29√p	.8	.33√p	1.0	.32√p	1.1	.29√p	1.3	.28√p	.4	.30√p

COMPARATIVE DATA

Data in the following table are the mean temperature and average precipitation for Northfield, Vermont for the period 1906-1930 and are included in this publication for comparative purposes:

15.9	2.00	15.7	2.37	26.8	3.64	39.4	2.54	51.2	2.76	60.1	3.38	65.5	3.22	62.4	3.12	55.8	3.01	45.5	3.04	33.3	2.88	20.4	2.11	41.0	34.07

NORMALS, MEANS, AND EXTREMES

LATITUDE 44° 28' N
LONGITUDE 73° 09' W
ELEVATION (ground) 331 feet

BURLINGTON, VERMONT
MUNICIPAL AIRPORT

Temperature, Precipitation, Snow/Sleet

Month	Normal Daily maximum (b)	Normal Daily minimum (b)	Normal Monthly (b)	Record highest	Year	Record lowest	Year	Normal degree days (b)	Precip. Normal total (b)	Max. monthly	Year	Min. monthly	Year	Max. in 24 hrs.	Year	Snow Mean total	Max. monthly	Year	Max. in 24 hrs.	Year
J	27.6	8.2	17.9	63	1950	-30	1957	1460	1.89	3.51	1947	0.49	1927	1.75	1934	14.3	33.9	1954	24.2	1934
F	28.2	8.0	18.1	60	1957	-28	1920	1313	1.53	4.18	1909	0.41	1914	1.64	1947	14.8	31.4	1947	12.5	1947
M	38.9	19.6	29.3	84	1946	-24	1938	1107	2.19	4.53	1913	0.22	1915	2.05	1920	12.3	28.0	1956	14.0	1920
A	52.9	31.6	42.3	86	1941	5	1923	681	2.63	5.83	1920	0.70	1941	1.41	1942	4.0	19.1	1924	8.4	1924
M	67.3	43.4	55.4	92	1929	25	1947	307	2.89	5.90	1945	0.36	1914	2.26	1955	0.1	2.8	1908	2.8	1908
J	77.5	53.5	65.5	96	1946	33	1924	72	3.57	9.92	1922	1.09	1949	4.45	1942	0.0	0.0		0.0	
J	82.4	58.4	70.4	100	1911	43	1929	19	3.75	8.08	1932	0.98	1922	3.32	1899	0.0	0.0		0.0	
A	79.8	56.3	68.1	101	1944	38	1909	47	3.01	11.54	1955	0.72	1957	3.59	1955	0.0	0.0		0.0	
S	71.3	48.4	59.9	95	1931	25	1914	172	3.14	8.18	1945	0.68	1927	2.91	1938	0.0	0.0	1957+	T	1957+
O	58.6	37.8	48.2	85	1949+	17	1925	521	2.85	6.75	1918	0.15	1924	4.21	1932	0.3	5.1	1925	4.4	1925
N	44.3	28.4	36.4	75	1948	-3	1938	858	2.85	10.13	1927	0.63	1952	4.49	1927	5.9	23.0	1921	12.9	1921
D	31.1	14.5	22.8	67	1941	-29	1933	1308	1.88	5.29	1920	0.31	1928	2.60	1950	12.6	30.6	1929	14.5	1929
Year	55.0	34.0	44.5	101	Aug. 1944	-30	Jan. 1957	7865	32.22	11.54	Aug. 1955	0.15	Oct. 1924	4.49	Nov. 1927	64.3	33.0	Jan. 1954	24.2	Jan. 1934

Relative Humidity, Wind, Sunshine, Mean Number of Days, Temperature Days

Month	RH 1:00 a.m. EST	RH 7:00 a.m. EST	RH 1:00 p.m. EST	RH 7:00 p.m. EST	Wind Mean hourly speed	Prevailing direction	Fastest mile Speed	Direction	Year	Pct. possible sunshine	Mean sky cover	Clear	Partly cloudy	Cloudy	Precip. .01 in.+	Snow/Sleet 1.0 in.+	Thunder-storms	Heavy fog	Max 90°+	Max 32°-	Min 32°-	Min 0°-
J	79	79	69	77	11.7	S	62	S	1923	34	7.3	5	7	19	13	5	*	1	0	20	29	9
F	79	79	65	74	10.8	S	57	S	1918	44	6.9	5	8	15	12	5	*	1	0	18	27	7
M	78	76	61	70	10.9	S	66	S	1929	48	6.6	6	8	16	13	3	*	1	0	9	27	1
A	78	74	55	65	10.6	S	63	S	1928	47	6.7	6	8	16	13	1	1	1	0	1	15	0
M	81	73	54	65	9.5	S	49	SW	1918	53	6.7	6	10	15	13	0	3	1	*	0	2	0
J	84	76	55	67	8.6	S	50	S	1925	59	6.2	6	12	12	12	0	6	*	1	0	0	0
J	84	76	66	66	8.2	S	42	S	1930	62	5.8	7	13	11	12	0	7	1	2	0	0	0
A	85	80	55	57	8.1	S	53	N	1955	59	5.5	8	10	11	11	0	6	1	1	0	0	0
S	85	83	59	59	9.0	S	56	SE	1938	51	6.0	8	8	13	11	0	3	1	*	0	1	0
O	80	81	59	74	10.2	S	70	SE	1954	41	6.7	7	7	16	12	0	1	1	0	0	7	0
N	80	81	68	76	11.6	S	72	SE	1950	26	7.9	3	7	21	13	2	*	1	0	4	18	*
D	80	80	69	77	11.3	S	57	S	1922	25	7.9	3	7	21	13	4	*	1	0	16	28	4
Year	81	78	60	71	10.0	S	72	SE	Nov. 1950	46	6.7	71	109	186	148	20	27	11	4	68	154	21

Means and extremes in the above table are from the existing or comparable location(s). Annual extremes have been exceeded at prior locations as follows: Maximum monthly snowfall 37.0 in March 1896.

(a) Length of record, years.
(b) Normal values are based on the period 1921-1950, and are means adjusted to represent observations taken at the present standard location.
* Less than one-half.

REFERENCE NOTES APPLYING TO ALL "NORMALS, MEANS, AND EXTREMES" TABLES.

Sky cover is expressed in a range of 0 for no clouds or obscuring phenomena to 10 for complete sky cover. The number of clear days is based on average cloudiness 0-3 tenths; partly cloudy days on 4-7 tenths; and cloudy days on 8-10 tenths. Monthly degree day totals are the sum of the negative departures of average daily temperatures from 65°F. Sleet was included in snowfall totals beginning with July 1948. Heavy fog also includes data referred to at various times in the past as "Dense" or "Thick." The upper visibility limit for heavy fog is 1/4 mile. Data in these tables are based on records through 1957.

- No record.
° Airport data.
† City Office data.
‡ Also on earlier dates, months, or years.
+ Also includes data for present standard location.
T Trace, an amount too small to measure.

Mean Maximum Temperature (°F.), January

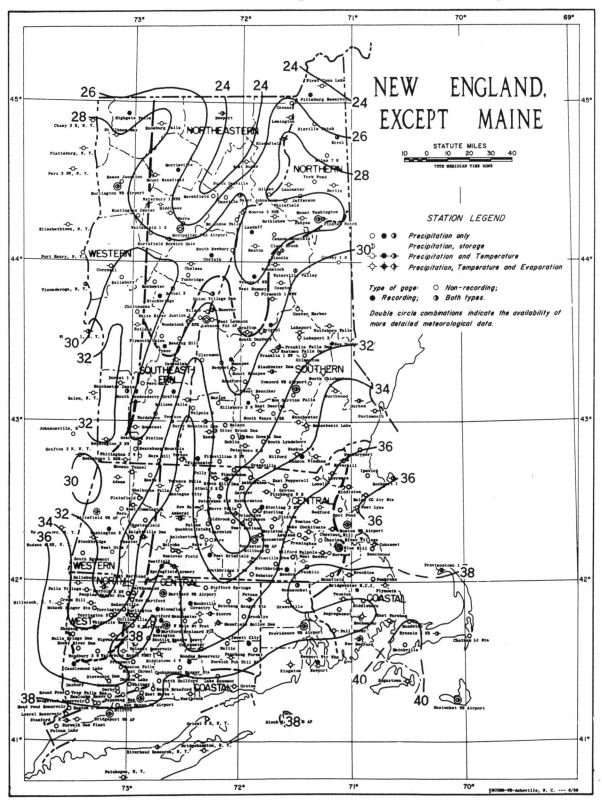

Based on period 1931-52

Isolines are drawn through points of approximately equal value. Caution should be used in interpolating on these maps, particularly in mountainous areas.

Mean Minimum Temperature (°F.), January

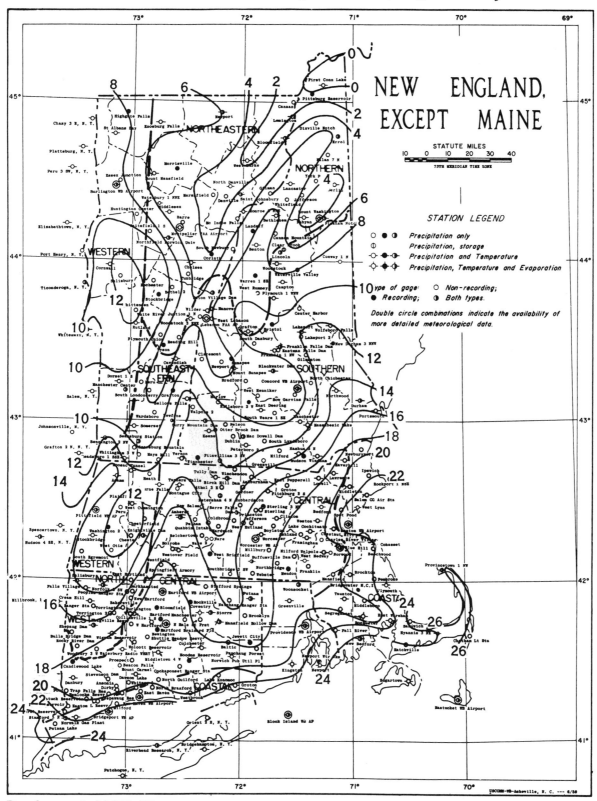

NEW ENGLAND,
EXCEPT MAINE

STATUTE MILES

75TH MERIDIAN TIME ZONE

STATION LEGEND

○ ● ◐ *Precipitation only*
◑ *Precipitation, storage*
◌ ● ◉ *Precipitation and Temperature*
◇ ◆ ◈ *Precipitation, Temperature and Evaporation*

Type of gage: ○ *Non-recording;*
● *Recording;* ◐ *Both types.*

*Double circle combinations indicate the availability of
more detailed meteorological data.*

Based on period 1931-52

Isolines are drawn through points of approximately equal value. Caution should be used
in interpolating on these maps, particularly in mountainous areas.

Mean Maximum Temperature (°F.), July

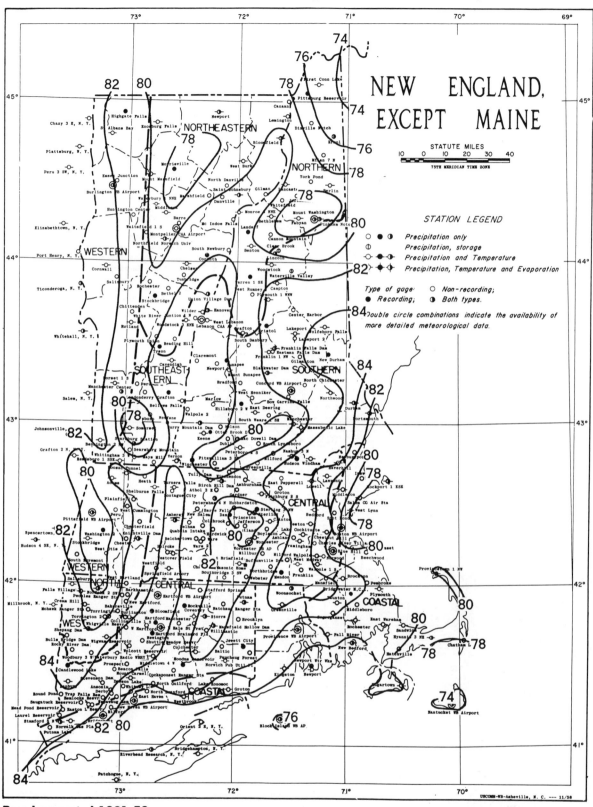

Based on period 1931-52

Isolines are drawn through points of approximately equal value. Caution should be used
in interpolating on these maps, particularly in mountainous areas.

Mean Minimum Temperature (°F.), July

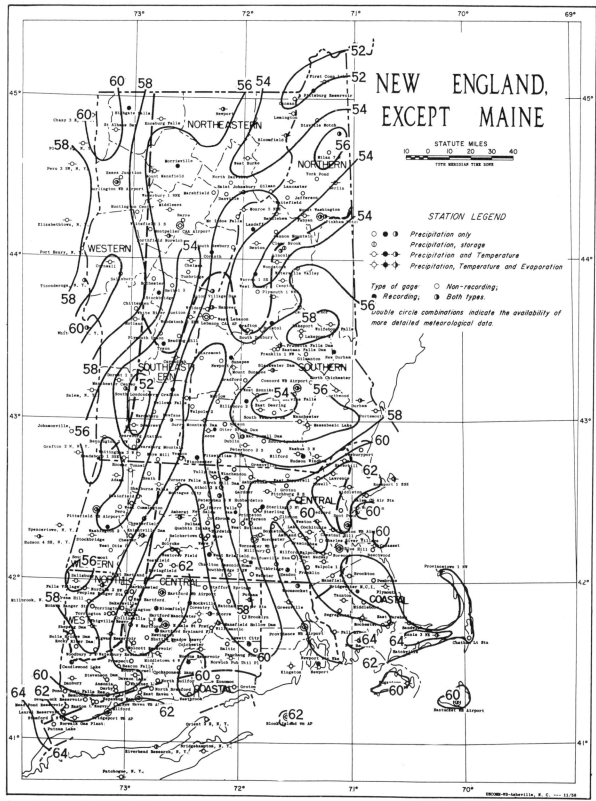

Based on period 1931-52

Isolines are drawn through points of approximately equal value. Caution should be used in interpolating on these maps, particularly in mountainous areas.

Mean Annual Precipitation, Inches

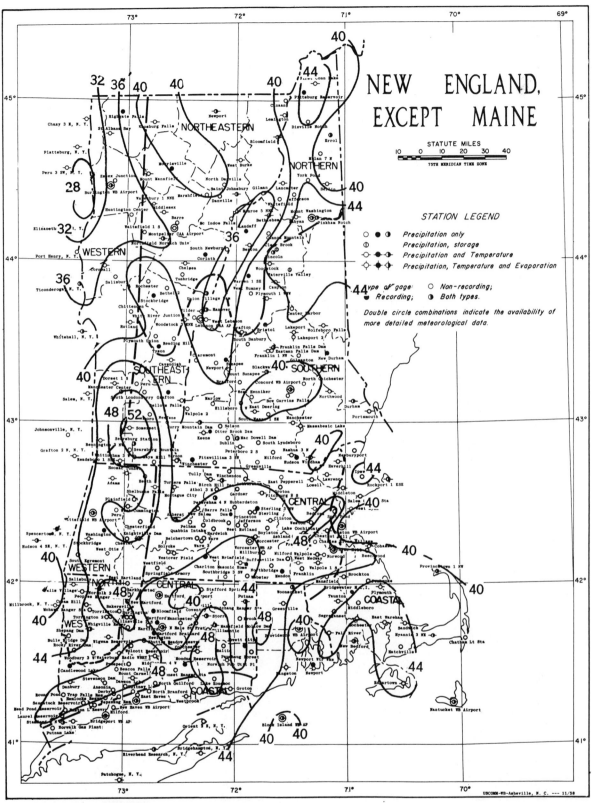

Based on period 1931-55

Isolines are drawn through points of approximately equal value. Caution should be used in interpolating on these maps, particularly in mountainous areas.

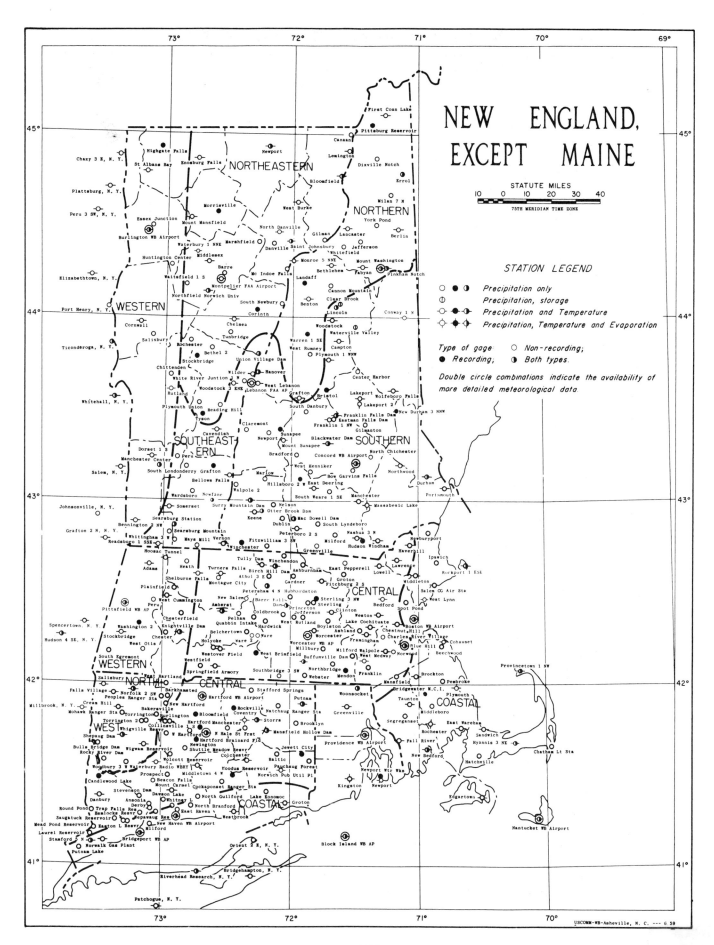

NEW ENGLAND, EXCEPT MAINE

STATUTE MILES

10 0 10 20 30 40

75TH MERIDIAN TIME ZONE

STATION LEGEND

○ ● ◑ *Precipitation only*
◐ *Precipitation, storage*
◌ ◒ ◍ *Precipitation and Temperature*
◇ ◆ ◈ *Precipitation, Temperature and Evaporation*

Type of gage: ○ *Non-recording;*
● *Recording;* ◑ *Both types.*

Double circle combinations indicate the availability of more detailed meteorological data.

NORTHEASTERN
NORTHERN
WESTERN
SOUTHERN
SOUTHEASTERN
CENTRAL
WESTERN
NORTH CENTRAL
COASTAL
WEST
COASTAL

First Conn Lake
Pittsburg Reservoir
Canaan
Lemington
Dixville Notch
Errol
Milan 7 N
York Pond
Berlin
Chazy 3 E, N. Y.
Highgate Falls
St Albans Bay
Enosburg Falls
Newport
Bloomfield
Plattsburg, N. Y.
Peru 3 SW, N. Y.
Morrisville
West Burke
Essex Junction
Mount Mansfield
North Danville
Gilman
Lancaster
Burlington WB Airport
Waterbury 1 NNE
Marshfield
Danville
Saint Johnsbury
Jefferson
Whitefield
Middlesex
Monroe 5 NNE
Mount Washington
Fabyan
Pinkham Notch
Elizabethtown, N. Y.
Huntington Center
Barre
Mc Indoe Falls
Landaff
Bethlehem
Waitsfield 1 S
Montpelier FAA Airport
Northfield Norwich Univ
South Newbury
Benton
Cannon Mountain
Clear Brook
Lincoln
Conway 1 N
Port Henry, N. Y.
Cornwall
Chelsea
Corinth
Woodstock
Waterville Valley
Salisbury
Rochester
Tunbridge
Warren 1 SE
Campton
Ticonderoga, N. Y.
Bethel 2
West Rumney
Plymouth 1 WNW
Stockbridge
Union Village Dam
Chittenden
Wilder
Hanover
Center Harbor
White River Junction 1 N
West Lebanon
Lakeport
Wolfeboro Falls
Whitehall, N. Y.
Rutland
Woodstock 3 ENE
Lebanon FAA AP
Grafton
Bristol
Lakeport 2
Plymouth Union
Reading Hill
South Danbury
New Durham 3 NNW
Tyson
Claremont
Franklin Falls Dam
Eastman Falls Dam
Cavendish
Newport
Sunapee
Franklin 1 NW
Gilmanton
Dorset 1 S
Peru
Mount Sunapee
Blackwater Dam
Southeastern
Manchester Center
Bradford
Concord WB Airport
North Chichester
Salem, N. Y.
South Londonderry
Grafton
West Henniker
Northwood
Bellows Falls
Marlow
Bow Garvins Falls
East Deering
Durham
Wardsboro
Newfane
Hillsboro 2 W
South Weare 1 SE
Manchester
Portsmouth
Walpole 2
Somerset
Surry Mountain Dam
Nelson
Massabesic Lake
Johnsonville, N. Y.
Searsburg Station
Keene
Otter Brook Dam
Mac Dowell Dam
Bennington 2 NW
Searsburg Mountain
Dublin
South Lyndeboro
Grafton 2 N, N. Y.
Whitingham 3 W
Mays Mill
Vernon
Peterboro 2 S
Nashua 3 N
Newburyport
Readsboro 1 SSE
Winchester
Fitzwilliam 3 SW
Milford
Hudson
Windham
Hoosac Tunnel
Greenville
Haverhill
Adams
Heath
Tully Dam
Winchendon
Ipswich
Shelburne Falls
Turners Falls
Birch Hill Dam
Ashburnham
East Pepperell
Lowell
Lawrence
Rockport 1 ESE
Plainfield
Athol 3 E
Groton
Middleton
Montague City
Gardner
Fitchburg 2 S
Salem CG Air Sta
West Cummington
New Salem
Barre Falls Dam
Sterling 3 NW
West Lynn
Peru
Amherst
Petersham 4 N
Hubbardston
Sterling
Bedford
Spot Pond
Pittsfield WB AP
Pelham
Princeton
Jefferson
Clinton
Weston
Lake Cochituate
Boston WB Airport
Spencertown, N. Y.
Chesterfield
Quabbin Intake
Coldbrook
West Rutland
Boylston
Ashland
Chestnut Hill
Washington 2
Knightville Dam
Hardwick
Worcester
Framingham
Charles River Village
Cohasset
Hudson 4 SE, N. Y.
Stockbridge
Chester
Belchertown
Ware 2
Worcester WB AP
Milford
Walpole
Blue Hill
Beechwood
West Otis
Holyoke
Ware
Millbury
West Medway
Norwood
South Egremont
Westover Field
West Brimfield
Buffumville Dam
Northbridge 3
Franklin
Provincetown 1 NW
Westfield
Mendon
WESTERN
Salisbury
Hartland
Springfield Armory
Southbridge 3 SW
Webster
Brockton
Pembroke
Falls Village
Norfolk 2 SW
Barkhamsted
Stafford Springs
Woonsocket
Mansfield
Bridgewater W.C.I.
Taunton
Plymouth
Peoples Ranger Sta
New Hartford
Putnam
Bridgewater
Millbrook, N. Y.
Cream Hill
Hartford WB Airport
Rockville
Coventry
Natchaug Ranger Sta
Greenville
Middleboro
Mohawk Ranger Sta
Bakersville
Bloomfield
Storrs
Segreganset
East Wareham
Torrington 2
Torrington
Burlington
Hartford Manchester
Brooklyn
Rochester
Sandwich
Whigville Resvr
Collinsville
W Hartford
Mansfield Hollow Dam
Fall River
Hyannis 3 NE
Shepaug Dam
N Hale St Frst
New Bedford
Bulls Bridge Dam
Wigwam Reservoir
Hartford Brainard Fld
Jewett City
Hatchville
Chatham Lt Sta
Rocky River Dam
Newington
Shuttle Meadow Resvr
Colchester
Baltic
Woodbury 3 W
Waterbury Radio WBRY
Wolcott Reservoir
Moodus Reservoir
Pachaug Forest
Prospect
Middletown 4 W
Norwich Pub Util Pl
Candlewood Lake
Beacon Falls
Newport Wtr Wks
Edgartown
Stevenson Dam
Mount Carmel
Cockaponset Ranger Sta
Kingston
Newport
Danbury
Dawson Lake
North Guilford
Lake Konomoc
Round Pond
Ansonia
Derby
Whitney L
North Branford
Groton
COASTAL
Trap Falls Res
Hemlocks Resvr
East Haven
Westbrook
Mead Pond Reservoir
Saugatuck Reservoir
Nepaug Res
New Haven WB Airport
Milford
Laurel Reservoir
Easton L Resvr
Stamford 5 N
Bridgeport WB AP
Norwalk Gas Plant
Orient 2 E, N. Y.
Block Island WB AP
Putnam Lake
Bridgehampton, N. Y.
Nantucket WB Airport
Riverhead Research, N. Y.
Patchogue, N. Y.

— 397 —

USCOMM-WB-Asheville, N. C. --- 6-59

THE CLIMATE OF
VIRGINIA

by
Curtis W. Crockett

March 1971

Virginia is located on the east coast of the North American continent between latitudes 36-1/2° and 39-1/2° north. The State is triangular in shape with the longest north-south distance of about 200 miles and the longest east-west distance more than 400 miles. There are 40,815 square miles of area within the State of which 1,200 square miles are inland waters.

The State is composed of 3 natural topographic regions, namely: the Tidewater or coastal plains area, the Piedmont plateau or middle Virginia, and the western mountain region. Natural regions of lesser extent include the "Fall Line", located between Tidewater Virginia and the Piedmont region; the Blue Ridge Mountains that serve as the eastern boundary of the great Shenandoah Valley; the Shenandoah Valley; and the Appalachian plateau, in southwestern Virginia.

Tidewater Virginia extends westward from the Atlantic Coast and west shore of the Chesapeake Bay to the "Fall Line." The "Fall Line" extends from Great Falls in the north, southward through Richmond to Emporia. It is divided into necks or peninsulas by 4 principal rivers and by numerous estuaries that open into the Chesapeake Bay. There are numerous peninsulas, wide estuaries, and many swamp areas. The principal rivers include the Potomac, Rappahannock, York, and the James. Tidewater extends up these rivers to near the "Fall Line." The James and Potomac Rivers are navigable by medium sized ships across Tidewater Virginia.

The Piedmont region is more than 200 miles wide in southern Virginia, but the Virginia section becomes quite narrow in the north. This region from east to west becomes more rolling and hilly with a few isolated mountains and ridges appearing a few miles east of the Blue Ridge. Elevations in general range from about 300 feet above sea level in the east to about 1,000 feet in the west. The James, the largest river crossing this region, divides it into two parts.

West of the Piedmont, the Blue Ridge Mountains traverse the State from southwest to northeast. They range from narrow ridges in the north to a high, wide plateau southwest from Roanoke. Elevations range generally from 1,500 to 3,500 feet. Mt. Rogers, in western Grayson County, towers to 5,719 feet, the highest point in the State.

A great valley west of the Blue Ridge extends from Tennessee through Scott and Washington Counties in the south, northeastward to the northern-most point of the State. It embraces 6 separate valleys of which the largest is the Shenandoah. Elevations range mostly from 1,000 to 2,000 feet. This great valley of Virginia is well drained. The north is drained by the north and south forks of the Shenandoah River, thence into the Potomac; the central portion, by the Cow Pasture and Jackson Rivers flowing southeastward into the James; and the southwestern half of the valley is drained by the Roanoke River, the New River, and three forks of the Holston River. The New River drains northwestward into West Virginia and the Ohio River Basin. The Holston drains southwestward into the Tennessee River.

The Appalachian Plateau in southwestern Virginia is divided into many sharp ridges and deep valleys. Large coal beds underlie the area.

The climate of Virginia is determined by its proximity to the Atlantic Ocean, latitude, and topography. The State is in the zone of prevailing westerly movement of the earth's atmosphere, in or near the mean path of winter storm tracks, and in the mean path of tropical, moist air from the southwest Atlantic and Gulf of Mexico much of the summer and early fall seasons. The mountains provide the usual elevation effects on temperatures, which are distinctly lower in this section, and there are wide variations over short distances as elevations change. Summers in the mountains are comparatively cool, and winters are more severe. In addition, these mountains produce various steering, blocking, and modifying effects on storms and general air movements in their vicinity. Temperature variations within the State due to latitude alone are very small, yearly averages are only 2° to 3° higher in the south than in the north. The longitudinal variations, however, show a sharper contrast, from the mountain extremes in the west toward an ocean influence in the east. The prevalence of winds with a westerly component prevents the extension of ocean influences very far westward from the coast.

Annual temperature averages, by divisions, range from 54° in the Southwestern Mountain Division to nearly 59° in Tidewater Virginia. As might be expected, the highest temperatures of record in the State have been recorded in the Piedmont Plateau, and the lowest in the higher mountain sections of the central and southwest.

The growing season, based on average dates of the last freeze in spring and the first in fall, range from around 140 days in parts of Tazewell County to a little over 250 days in the Norfolk area. Cold air drainage is an important factor in determining the growing season. The first and last freezes of the season usually occur with large surface high-pressure systems where clear skies and light winds are conducive to large radiational losses of heat. The cold air layer next to the ground becomes more dense and flows from the ridges and higher elevations into the valleys and lower elevations. The temperature at nearby locations under such conditions may differ by several degrees.

Virginia lies in the zone of prevailing westerlies where the general motion is from west to east. Southerly and northerly winds are about equally frequent, reflecting the progression of weather systems over the State. The Appalachian mountains, however, act to deflect these winds to some extent with northeasterly and south-

westerly directions occurring frequently. Local winds are also created by such other factors as differential heating, air drainage, local terrain, and proximity to bodies of water. During the cold season a more intense circulation is present with frequent storms and outbreaks of cold polar air. Northerly winds are most common during this season. The storm track is well north of the State during the warm season and southerly wind with light speeds prevail.

Summers in Virginia are usually warm and humid, and several hot and humid periods usually occur each year. Principal sources of moisture are the Gulf of Mexico and the Atlantic Ocean. Relative humidity, the usual measure of moisture, varies inversely with temperature -- high in the morning and low in the afternoon. Average values are not appreciably different over the State, but Tidewater locations have a much higher frequency of humidity and temperature values in the range where human discomfort occurs. For example, consider the frequency of temperature greater than 80° and relative humidity greater than 70 percent. Norfolk has an average of 228 hours each year in this category; Roanoke has only 23.

The annual precipitation based on the period 1931-60 ranges from about 35 to 50 inches. The heaviest amounts occur in the extreme southwest, the southeast, and the south-central areas. Minimum amounts are found in the sheltered valleys west of the Blue Ridge Mountains. Precipitation is well distributed throughout the year without distinct wet and dry periods. Maximum rainfall occurs in the summer months and minimum in the fall months. Precipitation during the cold season is associated with migratory low-pressure storms. The amounts are quite evenly distributed during this season in comparison to the warm season when showers and thundershowers account for most of the rainfall. Excessive rainfall usually occurs in the fall season with the passage of hurricanes. Hurricane rainfall in excess of 8 inches at some location in the State can be expected about once every 6 years.

Snow is common in winter, but normally without damaging consequences. Average seasonal amounts range from less than 10 inches in Tidewater Virginia to around 20 inches west of the Blue Ridge, and up to 30 inches on the mountains. Snow for individual seasons may range from nearly none up to the record amount of 98 inches observed at Mountain Lake in 1913-14. A month with snow of 10 inches or more has occurred about once every 4 years in the Tidewater area, about once every 2 years in Piedmont areas, and almost yearly in mountain locations. Occasionally, a major snowstorm will occur with snow depths up to, but usually much less than, the record of 42 inches which fell at Big Meadows in March 1962. Such a storm with snow depths greater than 10 inches usually causes considerable damage to trees, interrupts electric and telephone service, blocks highways, and generally hinders the normal way of life.

The greater portion of the State lies in the Atlantic drainage. The extreme southwestern portion drains to the Ohio Basin. Floods occur in all months of the year. The greatest frequency occurs in late winter and early spring; snowmelt occasionally is a factor. July is the month of least flooding. A second period of high water shows up in late summer and fall in the Piedmont and Tidewater sections associated mainly with tropical storms. Intense convectional storms in summer occasionally cause local flash floods.

Virginia is also subject to drought periods. Drought may be defined broadly as a prolonged and abnormal moisture deficiency. Some portion of Virginia sustains real damage from drought on the average of 1 year out of 3, but rarely are all crops affected for the entire season. Equitable distribution of ample precipitation has seldom, if ever, occurred over all of Virginia for an entire season.

Almost every year some sections undergo periods of insufficient rainfall during which time crops make little growth or actually sustain damage. Normal precipitation during the months of May through August just about equals the moisture lost through evaporation from the soil and vegetation.

The drought of 1930 is generally considered to have been the most serious in the past century. In more recent years some parts of the State suffered disastrous or near disastrous drought conditions. The variety of Virginia's crops and their widely variable moisture requirements are factors which have sustained Virginia's agriculture during droughts.

Although Virginia is normally favored with abundant rainfall, the frequency of dry spells has prompted the use of supplemental irrigation. Although the percentage of crops under irrigation only totals around 1 percent, the acreage is expected to approximately triple during the decade of the 1970's. The use of irrigation for crops has been found economical mainly for tobacco, truck crops, fruit, and slightly for corn. Virginia's water use by industry, agriculture, electric utilities, and municipalities is enormous, in addition to the extensive use for recreational purposes. Several impoundments have been made on Virginia streams and rivers and many more are in the planning stage. The lakes created by these dams provide a more uniform supply and lead to increased use of water.

There are 11 major river basins or portions thereof in Virginia. Two of the largest, the James and Potomac Rivers, traverse the three main natural topographic divisions of the State. Nine of these rivers originate within the State, the exceptions being the New River whose headwaters lie within a small portion of northwestern North Carolina, and the Potomac which originates in West Virginia.

Of Virginia's 26 million plus acres, approximately 87 percent are classed as agricultural or potentially suited for agriculture, 10 percent are submarginal for agriculture, and the remaining approximate 3 percent consists of inland water areas. To some extent, all significant staple agricultural products of the nation are grown within the State.

Milder winters and longer growing seasons have made possible the extensive truck farming industry in the southeastern counties and on the eastern shore.

Marine life is abundant, and commercial fisheries constitute one of Virginia's important industries. Extensive forests and mineral resources have created diversified manufacturing and allied chemical industries which are scattered throughout Virginia's river valleys, favorably situated to adequate water supply.

Thunderstorms occur on the average of 32 to 50 days each year, the greater number occurring in the mountains of extreme southwestern Virginia, decreasing in number toward the northeastern part of the State. About 85 percent of the annual total occur during the period May to September. Only a small percentage of these can be classed as severe, however. Thunderstorms exact a sizable annual toll of damage, when accompanied by severe lightning, wind, or hail. One of the most destructive single hailstorms was reported near Winchester in Frederick County on June 19, 1944, when losses amounting to more than $500 thousand were sustained mainly to the fruit crops. In general, damage from hail over the years has amounted to more than 5 times the damage from tornadoes.

Tornadoes are local storms of short duration and usually small dimensions, formed of winds rotating at high speeds. Sometimes the tornado is visible as a funnel extending from a thunderstorm cloud. Wind speeds have been estimated up to 300 m.p.h. in tornadoes, and extreme damage occurs wherever the funnel touches the ground. Good sight observations of tornadoes are rare in Virginia as most funnels are hidden by the usually heavy precipitation occurring with the associated thunderstorm. Approximately 4 tornadoes are reported in Virginia each year. These tornadoes occur mainly east of the Blue Ridge, but a few have been observed in the mountains. On May 2, 1929, a series of tornadoes which unquestionably caused the greatest loss in lives and property of any Virginia has ever had, first struck at Rye Cove in Scott County, where 13 fatalities occurred when a rural high school was demolished. At least 4 or 5 other tornadoes occurred later the same day in Rappahannock, Bath, Culpeper, Fauquier, and Loudoun Counties. A total of 22 lives were lost, and property damage approximated $500 thousand.

A hurricane is a tropical storm with winds of at least 74 m.p.h., which blow in a large spiral around a relatively calm center. This center is called the "eye" and is unique to hurricanes. The "eye" is bordered by hurricane force winds and torrential rains. Virginia has been affected by hurricanes since the early settlement days, but most have decreased in intensity before entering the State. Even though a hurricane may not enter the State, it can be much more destructive by passing closely offshore and maintaining its intense circulation. High winds are not the prime cause of destruction. High tides along with waves and currents, and flooding from the torrential rains cause immense damage.

About 80 percent of the hurricanes occur during August, September, and October. An average of about 2 hurricanes (2.3) each year come close enough to affect Virginia, but less than one (0.6) enters the State. A hurricane has entered the State in less than half of the years in the past century. Two have been observed in the State about 1 year in 10.

The three most destructive hurricanes affecting Virginia were Camille in August 1969, and August 22-23, 1933, hurricane, and Hazel in October 1954. The destruction associated with Camille was mostly the results of excessive rainfall (up to 27 inches) which caused flash floods and earth slides on the eastern slopes of the Blue Ridge. Total damage exceeded $100 million, and fatalities numbered 151. The August 1933 hurricane moved from the southeast inland south of Norfolk. Norfolk reported winds of 70 m.p.h. and tides 9.7 feet above mean low water. There were 18 fatalities and monetary damage, adjusted to 1969, amounted to $79 million. Hazel in October 1954 moved almost due north through central Virginia. This hurricane maintained its intense circulation. Richmond reported winds of 68 m.p.h., and Washington National Airport reported winds of 78 m.p.h. Widespread damage occurred, with total damage, adjusted to 1969, of about $25 million, with 13 fatalities.

Middle latitude storms sometimes develop south of Virginia and move northward along the Virginia coast. These storms, although usually weaker than hurricanes, produce similar type of damage. This type of storm, often referred to as a "Northeaster," generally occurs from late fall through the spring months. They often account for considerable damage from high tides, strong east or northeast winds, and heavy rain, mainly in Tidewater Virginia.

BIBLIOGRAPHY

Bailey, M. H., "Monthly Precipitation--Amount Probabilities for Selected Stations in Virginia," ESSA Technical Memorandum WBTM-ER-30, Washington, D. C., 1968, 13 pp.

_____, and Tinga, J. H., "Extreme Temperatures in Virginia," Research Division Report 128, Virginia Polytechnic Institute, Blacksburg, Va., 1968.

Chapman, Dorthy J., "Storms of Tropical Origin That Have Affected the Norfolk Area," Unpublished, Norfolk, Va., 1970.

Environmental Science Services Administration, "Tropical Cyclone Rainfall," ESSA Professional Paper I, Silver Spring, Md., 1967, 67 pp.

Flora, Snowden D., Hailstorms of the U.S., University of Oklahoma Press, Norman, Okla., 1956.

_____, Tornadoes of the United States, University of Oklahoma Press, Norman, Okla., Revised Edition, 1954.

Hoyt, William G., and Langbein, Walter B., Floods, Princeton University Press, Princeton, N. J., 1955.

NOAA, Environmental Data Service, "Climatic Summaries of Resort Areas," Climatography of the United States, No. 21-44, Silver Spring, Md., irregular.

_____, Climatological Data--National Summary, Silver Spring, Md., monthly plus annual summary.

_____, "Climatological Data--Virginia," Silver Spring, Md., Monthly plus annual summary.

_____, "Climatological Substation Summaries," Climatography of the United States, No. 20-44, Silver Spring, Md., irregular.

_____, "Hourly Precipitation Data--Virginia," Silver Spring, Md., monthly plus annual summary.

_____, "Local Climatological Data," Washington National Airport, Lynchburg, Dulles Airport, Norfolk, Roanoke, and Richmond; Silver Spring, Md., monthly plus annual summary.

_____, "Monthly Normals of Temperature, Precipitation, and Heating Degree Days, Virginia," Climatography of the United States, No. 81-38, Silver Spring, Md.

NOAA, Environmental Data Service, Storm Data, Silver Spring, Md., monthly plus annual summary.

NOAA, National Ocean Survey, "Tides in Hampton Roads," Unpublished, Norfolk, Va., 1970.

Rice, K. A., "Climate of Virginia, Climates of the States, Virginia," Climatography of the United States, No. 60-44, U.S. Weather Bureau, Washington, D. C., 1959.

Tannehill, Ivan R., Drought, Its Causes and Effects, Princeton University Press, Princeton, N. J., 1947.

_____, Hurricanes, Princeton University Press, Princeton, N. J., 9th Revised Edition, 1956.

Tinga, J. H., and Bailey, M. H., "Freeze Probabilities in Virginia and Protection Practices," Research Report 119, Virginia Polytechnic Institute, Blacksburg, Va., 1967, 20 pp.

U.S. Department of Agriculture, Soils--Yearbook of Agriculture, 1957, Washington, D. C., 1957, 784 pp.

_____, Water--Yearbook of Agriculture, 1955, Washington, D. C., 1955, 751 pp.

U.S. Weather Bureau, Climatic Summary of the United States (Bulletin W), Sections 91, 93, 94, Washington, D. C., 1930, 64 pp.

_____, Climatic Summary of the United States, Virginia, Supplement for 1931 through 1952 (Bulletin W Supplement), Washington, D. C., 43 pp.

BIBLIOGRAPHY

_____, Climatic Summary of the United States, Virginia, Supplement for 1951 through 1960 (Bulletin W Supplement), Washington, D. C., 67 pp.

_____, The Climatic Handbook for Washington, D. C., its Technical Paper No. 8, Washington, D. C., 1949, 235 pp.

_____, "Summary of Hourly Observations, for Roanoke, Washington National Airport, Norfolk, and Richmond, "Climatography of the United States, No. 82-44, Washington, D. C.

_____, "Tropical Cyclones of the North Atlantic Ocean," its Technical Paper No. 55, Silver Spring, Md., 1965, 148 pp.

Van Bavel, C. H. M., and Lillard, J. H., "Agricultural Drought in Virginia," Technical Bulletin 128, Virginia Polytechnic Institute, Blacksburg, Va., 1957, 38 pp.

Virginia Polytechnic Institute, "A Handbook of Agronomy," its Bulletin No. 97, Revised, Blacksburg, Va., 1966, 167 pp.

Virginia Polytechnic Institute, "Soils of Virginia," its Bulletin No. 203, Revised, Blacksburg, Va., Dec. 1966.

_____, "Soil, Virginia's Basic Natural Resources," its Bulletin No. 253, Blacksburg, Va., Jan. 1962.

FREEZE DATA

PROBABILITY OF SELECTED TEMPERATURES ON OR AFTER GIVEN DATES IN SPRING AND ON OR BEFORE GIVEN DATES IN FALL

STATION	TEMP	PROBABILITY - SPRING 90%	70%	50%	30%	10%	NO. OF DAYS BETWEEN 50% P'S	PROBABILITY - FALL 10%	30%	50%	70%	90%	SPRING Y O	FALL Y O
ASHLAND HANOVER COUNTY	36	APR 20	APR 28	MAY 3	MAY 8	MAY 15	160	SEP 25	OCT 4	OCT 10	OCT 16	OCT 25	22 22	23 23
	32	APR 6	APR 15	APR 21	APR 27	MAY 5	178	OCT 3	OCT 11	OCT 16	OCT 22	OCT 30	22 22	22 22
	28	MAR 22	MAR 30	APR 5	APR 11	APR 19	208	OCT 18	OCT 25	OCT 30	NOV 4	NOV 12	22 22	22 22
	24	MAR 8	MAR 18	MAR 24	MAR 31	APR 10	232	OCT 27	NOV 5	NOV 11	NOV 17	NOV 26	22 22	22 22
	20	FEB 26	MAR 8	MAR 15	MAR 22	APR 2	252	NOV 6	NOV 15	NOV 22	NOV 29	DEC 7	22 22	22 22
	16	JAN 17	FEB 13	FEB 24	MAR 6	MAR 19	284	NOV 19	NOV 29	DEC 5	DEC 13	*	22 20	22 20
BACK BAY WILDLIFE REFUGE	36	MAR 21	MAR 27	MAR 31	APR 4	APR 10	226	OCT 30	NOV 7	NOV 12	NOV 18	NOV 26	17 17	17 17
	32	MAR 4	MAR 12	MAR 17	MAR 23	MAR 31	251	NOV 10	NOV 18	NOV 23	NOV 29	DEC 6	17 17	17 17
	28	FEB 9	FEB 22	MAR 4	MAR 13	MAR 26	277	NOV 20	NOV 29	DEC 6	DEC 13	DEC 26	17 17	17 16
	24	JAN 24	FEB 6	FEB 16	FEB 25	MAR 11	299	NOV 26	DEC 5	DEC 12	DEC 23	*	17 17	17 13
	20	*	JAN 28	FEB 8	FEB 18	MAR 3	—	DEC 9	DEC 18	*	*	*	17 15	17 7
	16	*	*	JAN 11	FEB 2	FEB 18	—	DEC 10	*	*	*	*	17 9	17 3
BALCONY FALLS ROCKBRIDGE COUNTY	36	APR 14	APR 22	APR 28	MAY 4	MAY 12	166	SEP 27	OCT 5	OCT 11	OCT 16	OCT 24	19 19	19 19
	32	APR 7	APR 15	APR 21	APR 26	MAY 4	186	OCT 10	OCT 18	OCT 24	OCT 30	NOV 7	20 20	19 19
	28	MAR 17	MAR 25	MAR 31	APR 6	APR 15	216	OCT 24	OCT 29	NOV 2	NOV 6	NOV 11	20 20	19 19
	24	MAR 3	MAR 14	MAR 21	MAR 28	APR 8	237	OCT 28	NOV 6	NOV 13	NOV 20	NOV 29	20 20	19 19
	20	FEB 17	FEB 28	MAR 8	MAR 16	MAR 27	262	NOV 8	NOV 18	NOV 25	DEC 2	DEC 11	20 20	19 19
	16	JAN 28	FEB 13	FEB 22	MAR 2	MAR 14	289	NOV 23	DEC 2	DEC 8	DEC 14	DEC 22	18 17	19 19
BEDFORD BEDFORD COUNTY	36	APR 14	APR 24	MAY 1	MAY 8	MAY 19	157	SEP 24	OCT 1	OCT 5	OCT 10	OCT 17	30 30	31 31
	32	APR 5	APR 14	APR 20	APR 25	MAY 4	179	SEP 30	OCT 10	OCT 16	OCT 23	NOV 1	30 30	31 31
	28	MAR 20	MAR 30	APR 5	APR 12	APR 21	209	OCT 16	OCT 25	OCT 31	NOV 6	NOV 14	30 30	31 31
	24	FEB 28	MAR 12	MAR 21	MAR 29	APR 11	238	OCT 28	NOV 7	NOV 14	NOV 21	DEC 1	30 30	31 31
	20	FEB 14	FEB 27	MAR 8	MAR 17	MAR 30	264	NOV 15	NOV 22	NOV 27	DEC 2	DEC 9	30 30	31 31
	16	JAN 24	FEB 10	FEB 21	MAR 3	MAR 18	291	NOV 25	DEC 3	DEC 9	DEC 15	DEC 25	31 30	31 30
BERRYVILLE CLARKE COUNTY	36	MAY 5	MAY 12	MAY 17	MAY 22	MAY 29	134	SEP 14	SEP 22	SEP 28	OCT 4	OCT 13	18 18	19 19
	32	APR 15	APR 26	MAY 3	MAY 11	MAY 21	157	SEP 23	OCT 1	OCT 7	OCT 13	OCT 21	18 18	19 19
	28	APR 3	APR 14	APR 22	APR 30	MAY 11	177	OCT 4	OCT 11	OCT 16	OCT 22	OCT 29	18 18	19 19
	24	MAR 24	MAR 31	APR 5	APR 11	APR 18	212	OCT 24	OCT 30	NOV 3	NOV 7	NOV 12	18 18	19 19
	20	MAR 10	MAR 19	MAR 26	APR 2	APR 11	231	OCT 28	NOV 6	NOV 12	NOV 19	NOV 27	18 18	19 19
	16	FEB 22	MAR 4	MAR 11	MAR 18	MAR 28	261	NOV 15	NOV 22	NOV 27	DEC 2	DEC 9	18 18	19 19
BIG MEADOWS PAGE COUNTY	36	MAY 9	MAY 16	MAY 20	MAY 25	JUN 1	127	SEP 9	SEP 18	SEP 24	SEP 30	OCT 9	30 30	31 31
	32	APR 17	APR 30	MAY 9	MAY 17	MAY 30	146	SEP 20	SEP 27	OCT 2	OCT 6	OCT 13	30 30	31 31
	28	APR 10	APR 19	APR 25	MAY 2	MAY 11	176	OCT 3	OCT 11	OCT 18	OCT 24	NOV 2	30 30	31 31
	24	MAR 30	APR 9	APR 15	APR 22	MAY 1	196	OCT 12	OCT 21	OCT 28	NOV 3	NOV 13	30 30	31 31
	20	MAR 19	MAR 29	APR 6	APR 13	APR 23	214	OCT 23	NOV 1	NOV 6	NOV 12	NOV 20	30 30	31 31
	16	MAR 6	MAR 16	MAR 24	MAR 31	APR 11	237	NOV 1	NOV 10	NOV 16	NOV 23	DEC 2	30 30	31 31
BLACKSBURG MONTGOMERY COUNTY	36	APR 20	MAY 2	MAY 10	MAY 18	MAY 30	141	SEP 11	SEP 21	SEP 28	OCT 5	OCT 15	31 31	31 31
	32	APR 14	APR 23	APR 30	MAY 6	MAY 16	161	SEP 22	OCT 1	OCT 8	OCT 15	OCT 24	31 31	31 31
	28	APR 2	APR 12	APR 19	APR 26	MAY 6	183	OCT 3	OCT 13	OCT 19	OCT 26	NOV 4	31 31	31 31
	24	MAR 17	MAR 27	APR 3	APR 10	APR 19	214	OCT 20	OCT 28	NOV 3	NOV 9	NOV 17	31 31	31 31
	20	FEB 22	MAR 9	MAR 19	MAR 29	APR 12	242	OCT 30	NOV 9	NOV 16	NOV 22	DEC 2	31 31	31 31
	16	FEB 7	FEB 22	MAR 3	MAR 12	MAR 25	268	NOV 7	NOV 19	NOV 26	DEC 4	DEC 16	31 30	31 31
BLACKSTONE NOTTOWAY COUNTY	36	APR 10	APR 18	APR 23	APR 29	MAY 7	181	OCT 6	OCT 15	OCT 21	OCT 27	NOV 5	25 25	25 25
	32	MAR 27	APR 3	APR 8	APR 13	APR 20	206	OCT 17	OCT 25	OCT 31	NOV 6	NOV 14	25 25	25 25
	28	MAR 12	MAR 20	MAR 25	MAR 30	APR 6	230	OCT 23	NOV 3	NOV 10	NOV 18	NOV 28	25 25	25 25
	24	FEB 26	MAR 7	MAR 14	MAR 20	MAR 29	255	NOV 10	NOV 19	NOV 24	NOV 29	DEC 6	25 25	25 25
	20	JAN 29	FEB 16	FEB 26	MAR 9	MAR 24	278	NOV 15	NOV 25	DEC 1	DEC 7	DEC 17	25 24	25 25
	16	*	FEB 6	FEB 15	FEB 23	MAR 5	304	DEC 1	DEC 9	DEC 16	DEC 26	*	25 22	25 19
BOHANNAN MATHEWS COUNTY	36	APR 5	APR 15	APR 22	APR 30	MAY 10	184	OCT 8	OCT 17	OCT 23	OCT 29	NOV 6	21 21	21 21
	32	MAR 28	APR 5	APR 11	APR 17	APR 25	207	OCT 25	OCT 31	NOV 4	NOV 8	NOV 14	21 21	21 21
	28	MAR 14	MAR 23	MAR 29	APR 4	APR 13	231	NOV 2	NOV 9	NOV 15	NOV 20	NOV 28	21 21	21 21
	24	MAR 2	MAR 10	MAR 16	MAR 27	MAR 30	263	NOV 10	NOV 18	NOV 24	NOV 30	DEC 8	20 20	21 21
	20	FEB 13	FEB 23	MAR 2	MAR 2	MAR 19	283	NOV 23	DEC 3	DEC 10	DEC 18	*	21 21	22 20
	16	JAN 19	FEB 5	FEB 14	FEB 24	MAR 9	313	DEC 6	DEC 15	DEC 24	*	*	21 20	22 13
BOYKINS SOUTHAMPTON COUNTY	36	APR 11	APR 21	APR 28	MAY 5	MAY 14	170	OCT 3	OCT 11	OCT 15	OCT 20	OCT 27	17 17	16 16
	32	MAR 28	APR 8	APR 16	APR 24	MAY 5	190	OCT 5	OCT 16	OCT 23	OCT 31	NOV 11	17 17	17 17
	28	MAR 9	MAR 18	MAR 24	MAR 30	APR 8	226	OCT 18	OCT 28	NOV 5	NOV 12	NOV 22	17 17	17 16
	24	MAR 2	MAR 11	MAR 17	MAR 23	APR 1	242	OCT 30	NOV 7	NOV 14	NOV 20	DEC 2	16 16	17 16
	20	FEB 5	FEB 16	FEB 23	MAR 2	MAR 12	274	NOV 6	NOV 16	NOV 24	DEC 3	*	16 16	17 15
	16	JAN 8	JAN 28	FEB 8	FEB 18	MAR 4	310	NOV 28	DEC 8	DEC 15	DEC 25	*	16 15	17 13
BRISTOL WASHINGTON COUNTY	36	APR 12	APR 22	APR 30	MAY 7	MAY 18	173	OCT 7	OCT 14	OCT 20	OCT 25	NOV 2	19 19	18 18
	32	APR 9	APR 16	APR 21	APR 26	MAY 3	188	OCT 15	OCT 22	OCT 26	OCT 30	NOV 6	19 19	18 18
	28	MAR 22	MAR 31	APR 6	APR 12	APR 21	208	OCT 22	OCT 27	OCT 31	NOV 3	NOV 8	19 19	18 18
	24	FEB 24	MAR 10	MAR 20	MAR 29	APR 12	231	OCT 24	NOV 1	NOV 6	NOV 11	NOV 17	19 19	18 17
	20	FEB 17	FEB 28	MAR 8	MAR 16	MAR 27	255	OCT 30	NOV 10	NOV 18	NOV 26	DEC 8	19 19	18 18
	16	JAN 21	FEB 10	FEB 21	MAR 4	MAR 20	292	NOV 20	DEC 2	DEC 10	DEC 18	*	19 18	18 17
BUCHANAN BOTETOURT COUNTY	36	APR 19	APR 29	MAY 6	MAY 14	MAY 24	152	SEP 23	SEP 30	OCT 5	OCT 10	OCT 17	31 31	31 31
	32	APR 15	APR 22	APR 27	MAY 2	MAY 10	170	SEP 30	OCT 9	OCT 14	OCT 20	OCT 28	31 31	31 31
	28	MAR 30	APR 8	APR 14	APR 19	APR 28	196	OCT 12	OCT 21	OCT 27	NOV 2	NOV 11	31 31	31 31
	24	MAR 11	MAR 20	MAR 27	APR 2	APR 12	227	OCT 28	NOV 4	NOV 9	NOV 14	NOV 21	31 31	31 31
	20	FEB 20	MAR 3	MAR 11	MAR 19	MAR 31	254	NOV 3	NOV 13	NOV 20	NOV 27	DEC 7	31 31	31 31
	16	FEB 2	FEB 17	FEB 26	MAR 8	MAR 21	279	NOV 15	NOV 25	DEC 2	DEC 9	DEC 19	31 30	31 31
BURKES GARDEN TAZEWELL COUNTY	36	MAY 10	MAY 20	MAY 27	JUN 3	JUN 14	103	AUG 11	AUG 27	SEP 7	SEP 18	OCT 4	31 31	31 31
	32	MAY 1	MAY 9	MAY 15	MAY 21	MAY 29	135	SEP 14	SEP 22	SEP 27	OCT 3	OCT 10	31 31	31 31
	28	APR 15	APR 26	MAY 3	MAY 11	MAY 21	155	SEP 23	SEP 29	OCT 5	OCT 11	OCT 19	30 30	31 31
	24	APR 5	APR 13	APR 19	APR 25	MAY 3	181	OCT 2	OCT 11	OCT 17	OCT 24	NOV 2	30 30	31 31
	20	MAR 20	MAR 30	APR 6	APR 13	APR 23	206	OCT 12	OCT 22	OCT 29	NOV 5	NOV 15	30 30	31 31
	16	FEB 23	MAR 8	MAR 17	MAR 27	APR 9	240	OCT 23	NOV 4	NOV 12	NOV 20	DEC 1	30 30	31 31
CAPE HENRY	36	MAR 14	MAR 23	MAR 29	APR 3	APR 12	232	NOV 2	NOV 10	NOV 16	NOV 22	NOV 30	29 28	29 29
	32	FEB 26	MAR 9	MAR 17	MAR 24	APR 4	259	NOV 18	NOV 26	DEC 1	DEC 7	DEC 15	30 30	28 28
	28	FEB 2	FEB 17	FEB 28	MAR 10	MAR 26	289	NOV 29	DEC 7	DEC 14	DEC 23	*	30 30	28 21
	24	*	FEB 2	FEB 13	FEB 22	MAR 8	—	DEC 5	DEC 14	*	*	*	30 27	28 14
	20	*	*	JAN 31	FEB 9	FEB 20	—	DEC 12	*	*	*	*	30 21	28 9
	16	*	*	*	JAN 15	FEB 9	—	*	*	*	*	*	30 10	28 0

FREEZE DATA

PROBABILITY OF SELECTED TEMPERATURES ON OR AFTER GIVEN DATES IN SPRING AND ON OR BEFORE GIVEN DATES IN FALL

STATION	TEMP	SPRING 90%	SPRING 70%	SPRING 50%	SPRING 30%	SPRING 10%	NO. OF DAYS BETWEEN 50% P'S	FALL 10%	FALL 30%	FALL 50%	FALL 70%	FALL 90%	SPRING Y	SPRING O	FALL Y	FALL O
CATAWBA ROANOKE COUNTY	36	APR 17	APR 27	MAY 4	MAY 11	MAY 20	154	SEP 23	SEP 30	OCT 5	OCT 10	OCT 17	31	31	31	31
	32	APR 6	APR 16	APR 24	MAY 1	MAY 12	178	OCT 3	OCT 12	OCT 19	OCT 25	NOV 4	31	31	31	31
	28	MAR 19	MAR 28	APR 4	APR 11	APR 21	213	OCT 20	OCT 28	NOV 3	NOV 9	NOV 17	31	31	31	31
	24	MAR 6	MAR 16	MAR 23	MAR 30	APR 8	237	OCT 31	NOV 9	NOV 16	NOV 21	NOV 29	31	31	31	31
	20	FEB 16	MAR 1	MAR 10	MAR 19	APR 1	260	NOV 11	NOV 19	NOV 25	DEC 1	DEC 10	31	31	31	31
	16	FEB 4	FEB 20	MAR 1	MAR 11	MAR 24	277	NOV 20	NOV 28	DEC 3	DEC 9	DEC 16	31	30	31	31
CHARLOTTE COURT HOUSE CHARLOTTE COUNTY	36	APR 11	APR 21	APR 28	MAY 4	MAY 14	167	OCT 2	OCT 8	OCT 12	OCT 16	OCT 23	22	22	23	23
	32	APR 1	APR 11	APR 17	APR 24	MAY 4	188	OCT 7	OCT 16	OCT 22	OCT 29	NOV 6	22	22	23	23
	28	MAR 15	MAR 25	APR 1	APR 8	APR 18	213	OCT 17	OCT 25	OCT 31	NOV 5	NOV 13	22	22	23	23
	24	MAR 3	MAR 13	MAR 19	MAR 26	APR 5	240	OCT 29	NOV 8	NOV 14	NOV 21	NOV 30	22	22	23	23
	20	FEB 13	FEB 25	MAR 6	MAR 14	MAR 26	262	NOV 8	NOV 17	NOV 23	NOV 29	DEC 8	22	22	23	23
	16	*	FEB 7	FEB 19	MAR 1	MAR 14	290	NOV 19	NOV 29	DEC 6	DEC 13	DEC 25	22	19	23	22
CHARLOTTESVILLE 2W ALBEMARLE COUNTY	36	APR 4	APR 11	APR 16	APR 21	APR 27	190	OCT 8	OCT 17	OCT 23	OCT 30	NOV 8	31	31	31	31
	32	MAR 27	APR 4	APR 9	APR 14	APR 22	211	OCT 24	OCT 31	NOV 6	NOV 11	NOV 19	31	31	31	31
	28	MAR 6	MAR 17	MAR 24	MAR 31	APR 11	235	OCT 29	NOV 8	NOV 14	NOV 21	DEC 1	31	31	31	31
	24	FEB 20	MAR 4	MAR 12	MAR 21	APR 1	262	NOV 19	NOV 25	NOV 29	DEC 4	DEC 10	31	31	31	31
	20	FEB 9	FEB 21	MAR 1	MAR 9	MAR 21	279	NOV 20	NOV 29	DEC 5	DEC 12	DEC 22	31	31	31	30
	16	JAN 5	FEB 5	FEB 15	FEB 24	MAR 9	306	DEC 2	DEC 11	DEC 18	*	*	31	28	31	20
CHARLOTTESVILLE 1 W ALBEMARLE COUNTY	36	APR 10	APR 19	APR 25	MAY 1	MAY 10	172	SEP 28	OCT 7	OCT 14	OCT 21	OCT 31	19	19	20	20
	32	MAR 31	APR 8	APR 14	APR 19	APR 27	194	OCT 10	OCT 19	OCT 25	OCT 31	NOV 9	19	19	20	20
	28	MAR 18	MAR 26	MAR 31	APR 5	APR 13	218	OCT 23	OCT 30	NOV 4	NOV 8	NOV 15	19	19	20	20
	24	MAR 6	MAR 13	MAR 18	MAR 23	MAR 30	244	NOV 5	NOV 12	NOV 17	NOV 23	NOV 30	19	19	20	20
	20	FEB 6	FEB 21	MAR 3	MAR 13	MAR 27	271	NOV 13	NOV 23	NOV 29	DEC 5	DEC 14	19	19	20	20
	16	JAN 23	FEB 7	FEB 16	FEB 24	MAR 8	299	NOV 27	DEC 6	DEC 12	DEC 19	*	19	18	20	18
CHASE CITY MECKLENBURG COUNTY	36	APR 8	APR 16	APR 23	APR 29	MAY 8	180	OCT 9	OCT 16	OCT 20	OCT 25	NOV 1	18	18	19	19
	32	MAR 27	APR 4	APR 10	APR 15	APR 23	202	OCT 19	OCT 25	OCT 29	NOV 3	NOV 9	18	18	19	19
	28	MAR 6	MAR 17	MAR 24	MAR 31	APR 10	230	OCT 26	NOV 3	NOV 9	NOV 14	NOV 22	18	18	19	19
	24	FEB 21	MAR 4	MAR 13	MAR 21	APR 1	252	NOV 2	NOV 12	NOV 20	NOV 27	DEC 8	18	18	19	19
	20	JAN 28	FEB 13	FEB 25	MAR 8	MAR 25	281	NOV 18	NOV 27	DEC 3	DEC 10	DEC 19	18	18	19	19
	16	*	JAN 28	FEB 10	FEB 21	MAR 9	309	DEC 2	DEC 10	DEC 16	DEC 27	*	18	16	19	14
CHATHAM PITTSYLVANIA COUNTY	36	APR 14	APR 23	APR 30	MAY 6	MAY 15	161	SEP 27	OCT 3	OCT 8	OCT 13	OCT 19	31	31	31	31
	32	APR 4	APR 13	APR 19	APR 25	MAY 3	183	OCT 3	OCT 12	OCT 19	OCT 25	NOV 3	31	31	31	31
	28	MAR 24	MAR 31	APR 5	APR 10	APR 17	209	OCT 19	OCT 26	OCT 31	NOV 5	NOV 12	31	31	31	31
	24	FEB 25	MAR 10	MAR 18	MAR 27	APR 8	241	OCT 30	NOV 8	NOV 14	NOV 21	NOV 30	31	31	31	31
	20	FEB 11	FEB 23	MAR 3	MAR 11	MAR 23	273	NOV 14	NOV 24	DEC 1	DEC 7	DEC 17	31	31	31	31
	16	JAN 16	FEB 6	FEB 17	FEB 27	MAR 14	295	NOV 26	DEC 4	DEC 9	DEC 15	DEC 24	31	29	31	30
CHERITON NORTHAMPTON COUNTY	36	MAR 27	APR 6	APR 14	APR 21	MAY 1	206	OCT 25	NOV 1	NOV 6	NOV 11	NOV 18	31	31	31	31
	32	MAR 16	MAR 25	APR 1	APR 7	APR 17	229	NOV 2	NOV 11	NOV 16	NOV 22	NOV 30	31	31	31	31
	28	FEB 24	MAR 7	MAR 15	MAR 22	APR 2	260	NOV 17	NOV 25	NOV 30	DEC 6	DEC 15	31	31	31	30
	24	FEB 1	FEB 17	FEB 25	MAR 5	MAR 16	290	NOV 25	DEC 5	DEC 12	DEC 21	*	31	29	31	25
	20	*	JAN 28	FEB 8	FEB 18	MAR 5	324	DEC 3	DEC 15	DEC 29	*	*	31	28	31	17
	16	*	*	JAN 27	FEB 9	FEB 23	+0+	DEC 16	*	*	*	*	31	20	31	8
CHILHOWIE SMYTH COUNTY	36	MAY 2	MAY 12	MAY 20	MAY 27	JUN 6	124	SEP 9	SEP 16	SEP 21	SEP 26	OCT 3	19	19	19	19
	32	APR 23	MAY 2	MAY 9	MAY 16	MAY 26	146	SEP 17	SEP 26	OCT 2	OCT 8	OCT 16	19	19	19	19
	28	APR 16	APR 22	APR 27	MAY 2	MAY 9	171	OCT 5	OCT 11	OCT 15	OCT 19	OCT 25	19	19	19	19
	24	MAR 27	APR 5	APR 12	APR 18	APR 27	195	OCT 12	OCT 19	OCT 24	OCT 29	NOV 4	18	18	19	19
	20	MAR 6	MAR 18	MAR 27	APR 5	APR 17	218	OCT 21	OCT 27	OCT 31	NOV 4	NOV 10	18	18	19	19
	16	FEB 17	MAR 3	MAR 13	MAR 23	APR 6	248	OCT 29	NOV 9	NOV 16	NOV 24	DEC 4	18	18	19	19
COLUMBIA GOOCHLAND COUNTY	36	APR 19	APR 29	MAY 6	MAY 13	MAY 23	152	SEP 25	OCT 1	OCT 5	OCT 9	OCT 15	30	30	31	31
	32	APR 12	APR 20	APR 26	MAY 2	MAY 10	172	OCT 1	OCT 9	OCT 15	OCT 20	OCT 28	30	30	31	31
	28	MAR 25	APR 4	APR 11	APR 18	APR 28	198	OCT 12	OCT 20	OCT 26	NOV 1	NOV 9	30	30	31	31
	24	MAR 10	MAR 19	MAR 25	APR 1	APR 10	227	OCT 27	NOV 3	NOV 7	NOV 12	NOV 19	30	30	31	31
	20	FEB 25	MAR 6	MAR 13	MAR 19	MAR 28	254	NOV 9	NOV 17	NOV 22	NOV 28	DEC 5	30	30	31	31
	16	FEB 7	FEB 19	FEB 28	MAR 9	MAR 21	279	NOV 20	NOV 28	DEC 4	DEC 9	DEC 17	30	30	31	31
CULPEPER CULPEPER COUNTY	36	APR 19	APR 27	MAY 3	MAY 8	MAY 16	157	SEP 25	OCT 2	OCT 7	OCT 12	OCT 20	31	31	31	31
	32	APR 5	APR 14	APR 20	APR 26	MAY 4	181	OCT 3	OCT 12	OCT 18	OCT 24	NOV 2	30	30	31	31
	28	MAR 24	MAR 31	APR 6	APR 11	APR 18	207	OCT 14	OCT 23	OCT 30	NOV 5	NOV 14	31	31	31	31
	24	MAR 8	MAR 18	MAR 25	APR 1	APR 11	232	OCT 28	NOV 6	NOV 12	NOV 17	NOV 26	31	31	31	31
	20	FEB 25	MAR 6	MAR 12	MAR 18	MAR 27	254	NOV 3	NOV 14	NOV 21	NOV 28	DEC 8	31	31	31	31
	16	FEB 7	FEB 19	FEB 27	MAR 7	MAR 19	280	NOV 20	NOV 28	DEC 4	DEC 9	DEC 17	31	31	31	31
DALE ENTERPRISE ROCKINGHAM COUNTY	36	APR 23	MAY 3	MAY 10	MAY 17	MAY 28	145	SEP 20	SEP 27	OCT 2	OCT 7	OCT 15	31	31	31	31
	32	APR 12	APR 21	APR 28	MAY 5	MAY 14	166	SEP 27	OCT 5	OCT 11	OCT 16	OCT 24	31	31	31	31
	28	MAR 30	APR 9	APR 17	APR 24	MAY 4	193	OCT 12	OCT 21	OCT 27	NOV 2	NOV 11	31	31	31	31
	24	MAR 14	MAR 25	APR 1	APR 9	APR 20	220	OCT 26	NOV 2	NOV 7	NOV 12	NOV 19	31	31	31	31
	20	MAR 2	MAR 12	MAR 19	MAR 26	APR 5	243	NOV 1	NOV 10	NOV 17	NOV 23	DEC 2	31	31	31	31
	16	FEB 11	FEB 24	MAR 5	MAR 14	MAR 27	270	NOV 12	NOV 23	NOV 30	DEC 8	DEC 19	31	31	31	31
DANVILLE PITTSYLVANIA COUNTY	36	APR 8	APR 18	APR 24	MAY 1	MAY 10	173	OCT 2	OCT 9	OCT 14	OCT 19	OCT 27	31	31	31	31
	32	MAR 24	APR 5	APR 13	APR 21	MAY 3	196	OCT 12	OCT 21	OCT 26	NOV 1	NOV 9	31	31	31	31
	28	MAR 12	MAR 22	MAR 29	APR 5	APR 16	223	OCT 27	NOV 2	NOV 7	NOV 11	NOV 17	31	31	31	31
	24	FEB 21	MAR 6	MAR 14	MAR 23	APR 4	248	NOV 3	NOV 12	NOV 17	NOV 23	DEC 2	31	31	31	31
	20	FEB 5	FEB 18	FEB 28	MAR 9	MAR 22	279	NOV 19	NOV 27	DEC 3	DEC 9	DEC 18	31	31	31	31
	16	*	JAN 28	FEB 9	FEB 20	MAR 6	307	NOV 20	NOV 27	DEC 7	DEC 13	DEC 20	31	27	31	27
DIAMOND SPRINGS	36	MAR 23	APR 2	APR 9	APR 17	APR 27	209	OCT 24	OCT 31	NOV 4	NOV 8	NOV 15	29	29	28	28
	32	MAR 9	MAR 18	MAR 24	MAR 30	APR 8	237	NOV 2	NOV 10	NOV 16	NOV 22	NOV 30	31	31	31	31
	28	FEB 25	MAR 7	MAR 13	MAR 20	MAR 29	262	NOV 20	NOV 26	DEC 1	DEC 6	DEC 11	31	31	31	31
	24	JAN 24	FEB 12	FEB 22	MAR 3	MAR 16	295	NOV 30	DEC 8	DEC 14	DEC 23	*	31	29	31	24
	20	*	JAN 31	FEB 10	FEB 20	MAR 5	323	DEC 8	DEC 16	DEC 30	*	*	31	28	31	16
	16	*	*	JAN 24	FEB 7	FEB 21	+0+	DEC 16	*	*	*	*	31	18	31	7
ELKWOOD CULPEPER COUNTY	36	APR 22	MAY 1	MAY 8	MAY 15	MAY 24	147	SEP 20	SEP 27	OCT 2	OCT 7	OCT 14	22	22	23	23
	32	APR 9	APR 19	APR 26	MAY 3	MAY 13	169	SEP 30	OCT 7	OCT 12	OCT 17	OCT 24	23	23	23	23
	28	APR 2	APR 10	APR 15	APR 20	APR 28	193	OCT 10	OCT 19	OCT 25	OCT 31	NOV 9	23	23	23	23
	24	MAR 15	MAR 24	MAR 30	APR 5	APR 13	218	OCT 22	OCT 30	NOV 5	NOV 11	NOV 19	23	23	23	23
	20	MAR 4	MAR 13	MAR 20	MAR 26	APR 5	241	NOV 2	NOV 10	NOV 16	NOV 23	DEC 1	22	22	23	23
	16	FEB 14	FEB 25	MAR 5	MAR 13	MAR 25	271	NOV 16	NOV 25	DEC 1	DEC 7	DEC 15	22	22	23	23

FREEZE DATA

PROBABILITY OF SELECTED TEMPERATURES ON OR AFTER GIVEN DATES IN SPRING AND ON OR BEFORE GIVEN DATES IN FALL

STATION	TEMP	PROBABILITY - SPRING 90%	70%	50%	30%	10%	NO. OF DAYS BETWEEN 50% P'S	PROBABILITY - FALL 10%	30%	50%	70%	90%	SPRING Y O	FALL Y O
FALLS CHURCH FAIRFAX COUNTY	36	APR 14	APR 26	MAY 4	MAY 13	MAY 25	157	SEP 25	OCT 3	OCT 8	OCT 13	OCT 21	21 21	21 21
	32	APR 8	APR 16	APR 21	APR 27	MAY 4	182	OCT 1	OCT 13	OCT 20	OCT 26	NOV 9	21 21	20 20
	28	MAR 25	APR 1	APR 6	APR 11	APR 18	209	OCT 21	OCT 27	NOV 1	NOV 5	NOV 11	21 21	20 20
	24	MAR 8	MAR 18	MAR 24	MAR 31	APR 9	233	OCT 29	NOV 6	NOV 12	NOV 17	NOV 25	21 21	20 20
	20	FEB 21	MAR 2	MAR 9	MAR 16	MAR 25	262	NOV 12	NOV 20	NOV 26	DEC 2	DEC 10	21 21	20 20
	16	JAN 25	FEB 11	FEB 20	MAR 2	MAR 15	293	NOV 27	DEC 5	DEC 10	DEC 16	DEC 26	21 20	20 19
FARMVILLE CUMBERLAND COUNTY	36	APR 21	MAY 1	MAY 8	MAY 15	MAY 24	147	SEP 19	SEP 27	OCT 2	OCT 7	OCT 15	31 31	31 31
	32	APR 16	APR 24	APR 29	MAY 4	MAY 12	164	SEP 27	OCT 4	OCT 10	OCT 15	OCT 22	31 31	31 31
	28	APR 2	APR 11	APR 17	APR 23	MAY 1	188	OCT 8	OCT 17	OCT 22	OCT 28	NOV 5	31 31	31 31
	24	MAR 17	MAR 26	APR 1	APR 7	APR 15	218	OCT 22	OCT 31	NOV 5	NOV 11	NOV 19	31 31	31 31
	20	MAR 6	MAR 15	MAR 22	MAR 28	APR 7	239	OCT 30	NOV 9	NOV 16	NOV 22	DEC 2	31 31	31 31
	16	FEB 17	FEB 28	MAR 8	MAR 16	MAR 27	263	NOV 9	NOV 19	NOV 26	DEC 3	DEC 13	31 31	31 31
FLOYD FLOYD COUNTY	36	APR 22	MAY 4	MAY 13	MAY 21	JUN 2	129	AUG 23	SEP 8	SEP 19	OCT 1	OCT 17	31 31	31 31
	32	APR 15	APR 26	MAY 4	MAY 12	MAY 23	155	SEP 18	SEP 29	OCT 6	OCT 14	OCT 24	30 30	31 31
	28	APR 1	APR 11	APR 19	APR 26	MAY 6	182	SEP 27	OCT 13	OCT 18	OCT 27	NOV 9	30 30	31 31
	24	MAR 14	MAR 26	APR 3	APR 12	APR 23	213	OCT 15	OCT 26	NOV 2	NOV 9	NOV 20	31 31	31 31
	20	MAR 2	MAR 15	MAR 24	APR 2	APR 16	231	OCT 22	NOV 2	NOV 10	NOV 17	NOV 29	31 31	31 31
	16	FEB 16	MAR 2	MAR 13	MAR 23	APR 6	254	NOV 5	NOV 13	NOV 22	NOV 30	DEC 10	31 31	31 31
FORT LEE PRINCE GEORGE COUNTY	36	APR 16	APR 24	APR 29	MAY 4	MAY 12	165	OCT 2	OCT 7	OCT 11	OCT 15	OCT 21	21 21	20 20
	32	APR 3	APR 12	APR 18	APR 24	MAY 2	191	OCT 14	OCT 21	OCT 26	OCT 31	NOV 7	21 21	20 20
	28	MAR 11	MAR 24	APR 3	APR 12	APR 25	217	OCT 25	NOV 1	NOV 6	NOV 10	NOV 18	21 21	20 20
	24	MAR 4	MAR 13	MAR 20	MAR 26	APR 5	242	NOV 2	NOV 11	NOV 17	NOV 23	DEC 1	20 20	20 20
	20	FEB 6	FEB 20	MAR 1	MAR 10	MAR 24	274	NOV 16	NOV 24	NOV 30	DEC 5	DEC 14	20 20	20 20
	16	JAN 24	FEB 7	FEB 15	FEB 23	MAR 6	304	DEC 1	DEC 10	DEC 16	DEC 23	DEC 7	20 19	20 18
FREDERICKSBURG SPOTSYLVANIA COUNTY	36	APR 20	APR 27	MAY 2	MAY 7	MAY 14	157	SEP 24	OCT 1	OCT 6	OCT 12	OCT 19	30 30	29 29
	32	APR 8	APR 17	APR 22	APR 28	MAY 6	178	OCT 3	OCT 11	OCT 17	OCT 23	OCT 31	30 30	29 29
	28	MAR 23	APR 2	APR 9	APR 16	APR 26	202	OCT 14	OCT 22	OCT 28	NOV 3	NOV 12	30 30	29 29
	24	MAR 9	MAR 19	MAR 26	APR 2	APR 12	230	OCT 28	NOV 5	NOV 10	NOV 16	NOV 24	30 30	30 30
	20	FEB 20	MAR 4	MAR 13	MAR 22	APR 3	258	NOV 13	NOV 21	NOV 26	DEC 1	DEC 9	30 30	30 30
	16	FEB 5	FEB 17	FEB 25	MAR 5	MAR 17	285	NOV 23	DEC 1	DEC 7	DEC 12	DEC 20	30 30	30 30
GALAX CARROLL COUNTY	36	MAY 1	MAY 10	MAY 16	MAY 22	MAY 31	131	SEP 10	SEP 19	SEP 24	SEP 30	OCT 9	15 15	16 16
	32	APR 19	MAY 1	MAY 9	MAY 17	MAY 29	149	SEP 22	SEP 29	OCT 5	OCT 10	OCT 17	15 15	16 16
	28	APR 5	APR 14	APR 21	APR 27	MAY 7	177	OCT 3	OCT 10	OCT 15	OCT 20	OCT 28	15 15	16 16
	24	MAR 27	APR 3	APR 8	APR 13	APR 21	202	OCT 14	OCT 21	OCT 27	NOV 1	NOV 9	15 15	16 16
	20	MAR 4	MAR 16	MAR 25	APR 2	APR 14	229	OCT 30	NOV 5	NOV 9	NOV 14	NOV 22	15 15	16 16
	16	FEB 21	MAR 4	MAR 12	MAR 20	MAR 31	254	NOV 4	NOV 14	NOV 21	NOV 29	DEC 9	15 15	15 15
GLEN LYN GILES COUNTY	36	APR 18	APR 29	MAY 7	MAY 15	MAY 26	150	SEP 20	SEP 28	OCT 4	OCT 10	OCT 18	31 31	31 31
	32	APR 12	APR 20	APR 26	MAY 2	MAY 11	175	OCT 4	OCT 12	OCT 18	OCT 24	NOV 1	31 31	31 31
	28	MAR 30	APR 5	APR 9	APR 14	APR 20	204	OCT 17	OCT 24	OCT 30	NOV 4	NOV 12	31 31	31 31
	24	MAR 6	MAR 17	MAR 24	MAR 31	APR 11	230	OCT 26	NOV 3	NOV 9	NOV 14	NOV 23	31 31	30 30
	20	FEB 17	MAR 3	MAR 13	MAR 23	APR 6	252	NOV 3	NOV 13	NOV 20	NOV 27	DEC 7	31 31	30 30
	16	FEB 10	FEB 23	MAR 5	MAR 14	MAR 27	272	NOV 15	NOV 25	DEC 2	DEC 9	DEC 21	31 31	30 29
HALIFAX HALIFAX COUNTY	36	APR 9	APR 17	APR 23	APR 28	MAY 6	171	OCT 1	OCT 7	OCT 11	OCT 15	OCT 21	23 23	24 24
	32	MAR 28	APR 8	APR 15	APR 23	MAY 3	192	OCT 8	OCT 18	OCT 24	OCT 30	NOV 3	23 23	24 24
	28	MAR 14	MAR 23	MAR 29	APR 5	APR 14	220	OCT 22	OCT 30	NOV 4	NOV 10	NOV 18	23 23	24 24
	24	FEB 26	MAR 9	MAR 17	MAR 25	APR 5	242	OCT 30	NOV 8	NOV 14	NOV 20	NOV 29	23 23	24 24
	20	FEB 7	FEB 20	MAR 1	MAR 10	MAR 24	270	NOV 9	NOV 19	NOV 26	DEC 3	DEC 14	23 23	24 24
	16	JAN 9	FEB 4	FEB 14	FEB 24	MAR 10	298	NOV 26	DEC 3	DEC 9	DEC 16	*	23 21	24 20
HOLLAND NANSEMOND COUNTY	36	APR 16	APR 24	APR 29	MAY 4	MAY 12	168	OCT 3	OCT 9	OCT 14	OCT 19	OCT 26	31 31	31 31
	32	APR 4	APR 13	APR 19	APR 25	MAY 4	189	OCT 12	OCT 20	OCT 25	OCT 31	NOV 8	31 31	31 31
	28	MAR 14	MAR 25	APR 1	APR 8	APR 18	221	OCT 25	NOV 2	NOV 8	NOV 13	NOV 22	31 31	31 31
	24	FEB 19	MAR 4	MAR 13	MAR 22	APR 4	253	NOV 7	NOV 15	NOV 21	NOV 26	DEC 5	31 31	31 31
	20	FEB 2	FEB 18	FEB 26	MAR 7	MAR 20	282	NOV 17	NOV 27	DEC 5	DEC 14	*	31 30	31 28
	16	JAN 16	FEB 5	FEB 15	FEB 25	MAR 11	307	DEC 6	DEC 13	DEC 19	DEC 29	*	31 29	31 23
HOPEWELL PRINCE GEORGE COUNTY	36	APR 8	APR 16	APR 21	APR 26	MAY 4	181	OCT 7	OCT 14	OCT 19	OCT 24	OCT 31	31 31	31 31
	32	MAR 26	APR 3	APR 9	APR 15	APR 23	204	OCT 16	OCT 24	OCT 30	NOV 4	NOV 12	31 31	31 31
	28	MAR 8	MAR 18	MAR 24	MAR 31	APR 10	229	OCT 26	NOV 3	NOV 9	NOV 14	NOV 21	31 31	31 31
	24	FEB 20	MAR 3	MAR 12	MAR 20	APR 1	254	NOV 6	NOV 15	NOV 21	NOV 27	DEC 6	31 31	31 31
	20	FEB 5	FEB 17	FEB 25	MAR 6	MAR 18	285	NOV 21	NOV 30	DEC 7	DEC 14	DEC 24	31 31	31 30
	16	JAN 13	JAN 31	FEB 10	FEB 20	MAR 5	312	DEC 1	DEC 10	DEC 19	*	*	31 29	31 19
HOT SPRINGS BATH COUNTY	36	APR 27	MAY 7	MAY 15	MAY 22	JUN 1	135	SEP 14	SEP 22	SEP 27	OCT 2	OCT 10	31 31	31 31
	32	APR 17	APR 26	MAY 3	MAY 9	MAY 19	157	SEP 25	OCT 2	OCT 7	OCT 13	OCT 20	31 31	31 31
	28	APR 8	APR 17	APR 22	APR 28	MAY 7	181	OCT 3	OCT 13	OCT 20	OCT 27	NOV 6	31 31	31 31
	24	MAR 24	APR 2	APR 8	APR 15	APR 29	210	OCT 21	OCT 29	NOV 4	NOV 10	NOV 19	31 31	31 31
	20	MAR 7	MAR 17	MAR 24	MAR 31	APR 11	236	OCT 29	NOV 8	NOV 15	NOV 21	DEC 1	31 31	31 31
	16	FEB 23	MAR 4	MAR 11	MAR 18	MAR 28	260	NOV 10	NOV 20	NOV 26	DEC 2	DEC 12	31 31	31 31
JOHN H KERR DAM MECKLENBURG COUNTY	36	APR 12	APR 20	APR 25	MAY 1	MAY 9	173	OCT 1	OCT 9	OCT 15	OCT 21	OCT 29	22 22	23 23
	32	MAR 30	APR 10	APR 18	APR 25	MAY 6	188	OCT 7	OCT 16	OCT 22	OCT 29	NOV 7	22 22	23 23
	28	MAR 13	MAR 24	APR 1	APR 9	APR 20	213	OCT 16	OCT 24	OCT 31	NOV 6	NOV 15	22 22	23 23
	24	MAR 4	MAR 11	MAR 17	MAR 22	MAR 30	242	OCT 27	NOV 6	NOV 14	NOV 21	DEC 2	22 22	23 22
	20	FEB 2	FEB 17	FEB 27	MAR 9	MAR 23	274	NOV 20	DEC 1	DEC 8	DEC 17	*	22 18	23 21
	16	*	JAN 29	FEB 13	FEB 24	MAR 11	298	NOV 20	DEC 1	DEC 8	DEC 17	*	22 18	23 21
LANGLEY FIELD	36	MAR 24	APR 2	APR 9	APR 15	APR 25	208	OCT 17	OCT 27	NOV 3	NOV 10	NOV 19	31 31	31 31
	32	MAR 6	MAR 17	MAR 25	APR 2	APR 13	237	NOV 2	NOV 10	NOV 17	NOV 23	DEC 3	31 31	31 30
	28	FEB 16	MAR 1	MAR 10	MAR 20	APR 2	264	NOV 13	NOV 22	NOV 28	DEC 6	DEC 16	31 31	31 30
	24	JAN 27	FEB 11	FEB 20	MAR 1	MAR 14	293	NOV 24	DEC 3	DEC 10	DEC 17	*	31 30	31 27
	20	*	JAN 26	FEB 8	FEB 20	MAR 7	325	DEC 7	DEC 16	DEC 30	*	*	31 26	31 16
	16	*	*	JAN 19	FEB 2	FEB 17		DEC 13	*	*	*	*	31 19	31 9
LAWRENCEVILLE BRUNSWICK COUNTY	36	APR 17	APR 26	MAY 2	MAY 8	MAY 16	161	SEP 30	OCT 6	OCT 10	OCT 13	OCT 19	28 28	29 29
	32	APR 9	APR 18	APR 23	APR 29	MAY 8	181	OCT 7	OCT 15	OCT 21	OCT 26	NOV 3	28 28	29 29
	28	MAR 22	MAR 31	APR 6	APR 12	APR 21	209	OCT 21	OCT 27	NOV 1	NOV 5	NOV 11	28 28	28 28
	24	MAR 3	MAR 14	MAR 21	MAR 28	APR 9	238	NOV 1	NOV 8	NOV 14	NOV 19	NOV 26	29 29	28 28
	20	FEB 16	FEB 28	MAR 8	MAR 17	MAR 29	262	NOV 10	NOV 18	NOV 25	DEC 1	DEC 11	29 29	28 27
	16	JAN 29	FEB 14	FEB 23	MAR 4	MAR 18	286	NOV 17	NOV 28	DEC 6	DEC 14	DEC 31	29 28	28 26

FREEZE DATA

PROBABILITY OF SELECTED TEMPERATURES ON OR AFTER GIVEN DATES IN SPRING AND ON OR BEFORE GIVEN DATES IN FALL

STATION	TEMP	PROBABILITY – SPRING 90%	70%	50%	30%	10%	NO. OF DAYS BETWEEN 50% P'S	PROBABILITY – FALL 10%	30%	50%	70%	90%	SPRING Y O	FALL Y O
LEXINGTON ROCKBRIDGE COUNTY	36	APR 18	APR 29	MAY 6	MAY 13	MAY 24	151	SEP 23	SEP 30	OCT 4	OCT 9	OCT 16	31 31	31 31
	32	APR 11	APR 20	APR 26	MAY 2	MAY 11	168	SEP 27	OCT 5	OCT 11	OCT 17	OCT 25	31 31	31 31
	28	MAR 24	APR 3	APR 10	APR 17	APR 26	200	OCT 12	OCT 21	OCT 27	NOV 2	NOV 10	31 31	31 31
	24	MAR 7	MAR 18	MAR 25	APR 2	APR 13	228	OCT 25	NOV 3	NOV 8	NOV 14	NOV 22	31 31	31 31
	20	FEB 22	MAR 4	MAR 12	MAR 19	MAR 29	253	NOV 5	NOV 14	NOV 20	NOV 26	DEC 4	31 31	31 31
	16	FEB 1	FEB 17	FEB 27	MAR 9	MAR 22	279	NOV 16	NOV 26	DEC 3	DEC 10	DEC 20	31 30	31 31
LINCOLN LOUDOUN COUNTY	36	APR 15	APR 23	APR 29	MAY 5	MAY 13	165	SEP 25	OCT 5	OCT 11	OCT 18	OCT 27	31 31	31 31
	32	APR 3	APR 11	APR 17	APR 23	MAY 2	187	OCT 3	OCT 13	OCT 21	OCT 28	NOV 8	31 31	31 31
	28	MAR 16	MAR 25	APR 1	APR 8	APR 17	218	OCT 23	OCT 31	NOV 5	NOV 10	NOV 17	31 31	31 31
	24	MAR 3	MAR 13	MAR 21	MAR 28	APR 7	244	NOV 6	NOV 14	NOV 20	NOV 25	DEC 3	31 31	31 31
	20	FEB 20	MAR 3	MAR 11	MAR 19	MAR 30	265	NOV 16	NOV 24	DEC 1	DEC 7	DEC 15	31 31	31 31
	16	FEB 5	FEB 18	FEB 27	MAR 8	MAR 22	286	NOV 26	DEC 4	DEC 10	DEC 16	DEC 26	31 31	31 30
LOUISA LOUISA COUNTY	36	APR 19	APR 28	MAY 5	MAY 11	MAY 20	155	SEP 25	OCT 2	OCT 7	OCT 11	OCT 18	31 31	30 30
	32	APR 8	APR 18	APR 24	MAY 1	MAY 10	173	SEP 29	OCT 8	OCT 14	OCT 21	OCT 29	31 31	30 30
	28	MAR 25	APR 3	APR 10	APR 17	APR 26	200	OCT 11	OCT 20	OCT 27	NOV 2	NOV 11	31 31	30 30
	24	MAR 13	MAR 22	MAR 28	APR 3	APR 11	227	OCT 28	NOV 4	NOV 10	NOV 15	NOV 22	31 31	30 30
	20	FEB 26	MAR 8	MAR 15	MAR 22	APR 1	253	NOV 9	NOV 17	NOV 23	NOV 29	DEC 8	31 30	30 30
	16	FEB 6	FEB 20	MAR 1	MAR 10	MAR 22	279	NOV 22	NOV 29	DEC 5	DEC 10	DEC 18	31 30	30 30
LURAY PAGE COUNTY	36	APR 30	MAY 10	MAY 17	MAY 24	JUN 2	136	SEP 13	SEP 23	SEP 30	OCT 7	OCT 17	29 29	29 29
	32	APR 19	APR 30	MAY 8	MAY 16	MAY 27	154	SEP 23	OCT 2	OCT 9	OCT 15	OCT 24	29 29	29 29
	28	APR 5	APR 14	APR 20	APR 26	MAY 5	184	OCT 6	OCT 15	OCT 21	OCT 28	NOV 6	28 28	29 29
	24	MAR 22	MAR 31	APR 7	APR 13	APR 22	212	OCT 24	OCT 31	NOV 5	NOV 10	NOV 17	28 28	29 29
	20	MAR 4	MAR 14	MAR 21	MAR 28	APR 6	239	OCT 28	NOV 8	NOV 15	NOV 22	DEC 2	28 28	29 29
	16	FEB 23	MAR 5	MAR 11	MAR 18	MAR 27	263	NOV 17	NOV 24	NOV 29	DEC 4	DEC 11	28 28	29 29
LYNCHBURG MUNICIPAL AIRPORT	36	APR 9	APR 17	APR 23	APR 29	MAY 8	176	OCT 2	OCT 10	OCT 16	OCT 21	OCT 29	31 31	31 31
	32	MAR 25	APR 4	APR 11	APR 18	APR 28	200	OCT 15	OCT 22	OCT 28	NOV 2	NOV 9	31 31	31 31
	28	MAR 5	MAR 16	MAR 24	MAR 31	APR 11	229	OCT 23	NOV 2	NOV 8	NOV 15	NOV 25	31 31	31 31
	24	FEB 22	MAR 5	MAR 13	MAR 21	APR 1	253	NOV 2	NOV 13	NOV 21	NOV 29	DEC 10	31 31	31 31
	20	FEB 1	FEB 16	FEB 27	MAR 10	MAR 26	279	NOV 19	NOV 27	DEC 3	DEC 8	DEC 17	31 31	31 30
	16	JAN 1	FEB 5	FEB 16	FEB 26	MAR 12	299	NOV 25	DEC 5	DEC 12	DEC 20	*	31 28	31 27
MANASSAS PRINCE WILLIAM COUNTY	36	APR 15	APR 25	MAY 3	MAY 10	MAY 20	156	SEP 22	SEP 30	OCT 6	OCT 12	OCT 20	22 22	23 23
	32	APR 2	APR 12	APR 19	APR 25	MAY 5	183	OCT 1	OCT 13	OCT 19	OCT 25	NOV 3	22 22	23 23
	28	MAR 20	MAR 29	APR 5	APR 12	APR 22	210	OCT 20	OCT 27	NOV 1	NOV 7	NOV 14	22 22	23 23
	24	MAR 9	MAR 17	MAR 22	MAR 28	APR 5	235	OCT 28	NOV 6	NOV 12	NOV 18	NOV 27	22 22	23 23
	20	FEB 21	MAR 4	MAR 11	MAR 19	MAR 30	259	NOV 9	NOV 18	NOV 25	DEC 2	DEC 12	22 22	23 23
	16	FEB 2	FEB 16	FEB 24	MAR 4	MAR 15	286	NOV 22	DEC 1	DEC 7	DEC 14	DEC 23	22 21	23 23
MARTINSVILLE HENRY COUNTY	36	APR 15	APR 27	MAY 6	MAY 15	MAY 27	149	SEP 16	SEP 25	OCT 2	OCT 9	OCT 18	31 31	31 31
	32	APR 8	APR 19	APR 26	MAY 4	MAY 15	173	SEP 28	OCT 9	OCT 16	OCT 23	NOV 3	31 31	31 31
	28	MAR 27	APR 8	APR 16	APR 24	MAY 5	191	OCT 5	OCT 16	OCT 24	OCT 31	NOV 11	31 31	31 31
	24	MAR 5	MAR 19	MAR 29	APR 8	APR 22	225	OCT 22	NOV 1	NOV 9	NOV 16	NOV 26	31 31	31 31
	20	FEB 21	MAR 8	MAR 18	MAR 28	APR 12	246	OCT 31	NOV 11	NOV 19	NOV 27	DEC 9	31 31	31 31
	16	FEB 8	FEB 21	MAR 2	MAR 11	MAR 25	271	NOV 8	NOV 20	NOV 28	DEC 6	DEC 22	31 31	31 29
MONTEREY HIGHLAND COUNTY	36	MAY 2	MAY 14	MAY 23	JUN 1	JUN 13	121	SEP 5	SEP 14	SEP 21	SEP 27	OCT 7	30 30	29 29
	32	APR 22	MAY 4	MAY 12	MAY 20	MAY 31	143	SEP 17	SEP 26	OCT 2	OCT 9	OCT 18	30 30	29 29
	28	APR 6	APR 19	APR 28	MAY 7	MAY 20	167	SEP 26	OCT 5	OCT 12	OCT 18	OCT 28	30 30	29 29
	24	MAR 5	APR 5	APR 15	APR 24	MAY 7	189	OCT 2	OCT 13	OCT 21	OCT 29	NOV 9	29 29	29 29
	20	MAR 12	MAR 23	MAR 31	APR 8	APR 19	219	OCT 18	OCT 28	NOV 5	NOV 12	NOV 23	29 29	29 29
	16	MAR 4	MAR 14	MAR 21	MAR 28	APR 7	241	OCT 30	NOV 10	NOV 17	NOV 24	DEC 4	29 29	29 29
MOUNT WEATHER LOUDOUN COUNTY	36	APR 13	APR 22	APR 28	MAY 5	MAY 14	167	SEP 30	OCT 7	OCT 12	OCT 17	OCT 24	26 26	26 26
	32	APR 6	APR 16	APR 24	MAY 1	MAY 11	178	OCT 2	OCT 12	OCT 19	OCT 26	NOV 5	30 30	30 30
	28	MAR 25	APR 4	APR 11	APR 18	APR 29	203	OCT 14	OCT 24	OCT 31	NOV 7	NOV 16	30 30	31 31
	24	MAR 6	MAR 19	MAR 27	APR 5	APR 18	231	OCT 28	NOV 7	NOV 13	NOV 20	NOV 29	30 30	31 31
	20	FEB 27	MAR 12	MAR 22	MAR 29	APR 10	250	NOV 7	NOV 17	NOV 25	DEC 2	DEC 12	31 31	31 31
	16	FEB 17	MAR 1	MAR 10	MAR 18	MAR 31	269	NOV 22	NOV 29	DEC 4	DEC 9	DEC 16	31 31	31 31
NASSAWADOX NORTHAMPTON COUNTY	36	APR 5	APR 17	APR 23	APR 29	MAY 8	185	OCT 10	OCT 19	OCT 25	OCT 31	NOV 9	15 14	15 15
	32	MAR 24	APR 3	APR 8	APR 13	APR 20	209	OCT 23	OCT 29	NOV 3	NOV 8	NOV 14	15 14	15 15
	28	MAR 8	MAR 17	MAR 23	MAR 30	APR 7	237	NOV 3	NOV 10	NOV 15	NOV 20	NOV 28	15 14	15 15
	24	FEB 17	MAR 3	MAR 10	MAR 17	MAR 26	262	NOV 14	NOV 22	NOV 27	DEC 3	DEC 11	15 14	15 15
	20	JAN 25	FEB 11	FEB 21	MAR 1	MAR 13	293	NOV 26	DEC 5	DEC 11	DEC 18	DEC 30	15 14	15 14
	16	*	JAN 30	FEB 12	FEB 23	MAR 11	+0+	DEC 9	DEC 16	*	*	*	15 13	15 5
NEW CANTON BUCKINGHAM COUNTY	36	APR 25	MAY 3	MAY 9	MAY 15	MAY 23	148	SEP 22	SEP 29	OCT 4	OCT 9	OCT 17	31 31	31 31
	32	APR 16	APR 23	APR 29	MAY 4	MAY 12	168	SEP 29	OCT 8	OCT 14	OCT 20	OCT 29	31 31	31 31
	28	MAR 30	APR 8	APR 15	APR 21	MAY 1	194	OCT 11	OCT 20	OCT 26	OCT 31	NOV 9	31 31	31 31
	24	MAR 15	MAR 25	APR 1	APR 8	APR 19	219	OCT 25	NOV 1	NOV 6	NOV 10	NOV 17	31 31	31 31
	20	MAR 3	MAR 13	MAR 20	MAR 27	APR 6	240	OCT 30	NOV 9	NOV 15	NOV 21	DEC 1	31 31	31 31
	16	FEB 11	FEB 24	MAR 4	MAR 13	MAR 26	270	NOV 14	NOV 23	NOV 29	DEC 6	DEC 15	31 31	31 31
NEWPORT NEWS	36	MAR 15	MAR 26	APR 3	APR 11	APR 22	219	OCT 24	NOV 2	NOV 8	NOV 14	NOV 23	22 22	22 22
	32	MAR 7	MAR 17	MAR 23	MAR 30	APR 8	243	NOV 7	NOV 15	NOV 21	NOV 26	DEC 4	22 22	22 22
	28	FEB 20	MAR 4	MAR 12	MAR 21	APR 2	266	NOV 14	NOV 25	DEC 3	DEC 10	DEC 24	22 22	23 22
	24	JAN 23	FEB 12	FEB 23	MAR 6	MAR 22	292	NOV 25	DEC 4	DEC 12	DEC 20	*	22 21	23 20
	20	*	JAN 25	FEB 7	FEB 17	MAR 3	320	DEC 5	DEC 14	DEC 24	*	*	22 18	23 13
	16	*	*	JAN 15	FEB 6	FEB 25	+0+	DEC 13	*	*	*	*	22 12	23 3
NORFOLK REGIONAL AIRPORT	36	MAR 15	MAR 27	APR 4	APR 12	APR 24	219	OCT 24	NOV 2	NOV 9	NOV 15	NOV 24	31 31	31 31
	32	MAR 5	MAR 15	MAR 22	MAR 29	APR 8	244	NOV 2	NOV 13	NOV 21	NOV 28	DEC 9	31 31	31 31
	28	FEB 16	FEB 27	MAR 7	MAR 15	MAR 26	271	NOV 19	NOV 27	DEC 3	DEC 9	DEC 19	31 31	31 30
	24	JAN 20	FEB 8	FEB 18	FEB 27	MAR 13	298	NOV 27	DEC 6	DEC 13	DEC 23	*	31 29	31 24
	20	*	JAN 18	FEB 3	FEB 14	MAR 1	+0+	DEC 18	*	*	*	*	31 24	31 15
	16	*	*	*	FEB 2	FEB 20	+0+	DEC 17	*	*	*	*	31 15	31 6
PAINTER ACCOMAC COUNTY	36	APR 8	APR 18	APR 25	MAY 2	MAY 13	182	OCT 6	OCT 16	OCT 24	OCT 31	NOV 11	31 31	31 31
	32	MAR 25	APR 4	APR 11	APR 18	APR 29	207	OCT 19	OCT 28	NOV 4	NOV 10	NOV 20	31 31	31 31
	28	MAR 14	MAR 23	MAR 30	APR 6	APR 16	228	OCT 29	NOV 7	NOV 13	NOV 20	NOV 29	31 31	31 31
	24	FEB 21	MAR 4	MAR 11	MAR 19	MAR 30	264	NOV 12	NOV 23	NOV 30	DEC 7	DEC 18	31 31	31 31
	20	FEB 1	FEB 17	FEB 25	MAR 4	MAR 15	291	NOV 27	DEC 6	DEC 13	DEC 22	*	31 29	31 24
	16	*	JAN 26	FEB 7	FEB 17	MAR 2	322	DEC 10	DEC 17	DEC 26	*	*	31 27	31 17

FREEZE DATA

PROBABILITY OF SELECTED TEMPERATURES ON OR AFTER GIVEN DATES IN SPRING AND ON OR BEFORE GIVEN DATES IN FALL

STATION	TEMP	PROBABILITY - SPRING					NO. OF DAYS BETWEEN 50% P'S	PROBABILITY - FALL					SPRING Y O	FALL Y O
		90%	70%	50%	30%	10%		10%	30%	50%	70%	90%		
PARTLOW SPOTSYLVANIA COUNTY	36	APR 30	MAY 9	MAY 16	MAY 22	JUN 1	133	SEP 15	SEP 21	SEP 26	SEP 30	OCT 7	18 18	19 19
	32	APR 21	APR 29	MAY 4	MAY 10	MAY 18	155	SEP 24	OCT 1	OCT 6	OCT 10	OCT 17	18 18	19 19
	28	APR 7	APR 15	APR 21	APR 27	MAY 5	177	OCT 1	OCT 9	OCT 15	OCT 20	OCT 28	17 17	19 19
	24	MAR 24	APR 2	APR 9	APR 16	APR 26	201	OCT 13	OCT 22	OCT 27	NOV 2	NOV 11	18 18	19 19
	20	MAR 6	MAR 17	MAR 25	APR 2	APR 13	226	OCT 22	OCT 31	NOV 6	NOV 12	NOV 20	18 18	19 19
	16	FEB 21	MAR 5	MAR 13	MAR 21	APR 1	252	NOV 2	NOV 13	NOV 20	NOV 29	DEC 8	18 18	19 19
PENNINGTON GAP LEE COUNTY	36	APR 27	MAY 6	MAY 11	MAY 17	MAY 25	143	SEP 19	SEP 26	OCT 1	OCT 5	OCT 12	31 31	29 29
	32	APR 12	APR 22	APR 29	MAY 6	MAY 16	164	SEP 28	OCT 5	OCT 10	OCT 15	OCT 22	31 31	29 29
	28	APR 1	APR 9	APR 15	APR 20	APR 28	194	OCT 14	OCT 21	OCT 26	OCT 31	NOV 8	31 31	29 29
	24	MAR 19	MAR 27	APR 2	APR 8	APR 17	214	OCT 21	OCT 28	NOV 2	NOV 7	NOV 15	31 31	29 29
	20	FEB 22	MAR 6	MAR 15	MAR 23	APR 4	244	OCT 29	NOV 7	NOV 14	NOV 21	NOV 30	31 31	29 29
	16	FEB 12	FEB 25	MAR 6	MAR 15	MAR 29	263	NOV 4	NOV 16	NOV 24	DEC 2	DEC 14	31 31	29 29
PHILPOT DAM HENRY COUNTY	36	APR 14	APR 25	MAY 2	MAY 10	MAY 21	164	OCT 1	OCT 9	OCT 13	OCT 18	OCT 26	17 17	17 17
	32	APR 2	APR 11	APR 17	APR 23	MAY 2	189	OCT 9	OCT 17	OCT 23	OCT 29	NOV 6	17 17	17 17
	28	MAR 17	MAR 25	APR 1	APR 7	APR 16	215	OCT 22	OCT 29	NOV 2	NOV 7	NOV 14	17 17	17 17
	24	MAR 4	MAR 13	MAR 20	MAR 26	APR 4	241	OCT 30	NOV 9	NOV 16	NOV 23	DEC 4	17 17	18 18
	20	FEB 8	FEB 23	MAR 6	MAR 17	APR 2	265	NOV 10	NOV 20	DEC 3	DEC 13	DEC 27	17 17	18 18
	16	JAN 26	FEB 9	FEB 19	MAR 1	MAR 16	295	NOV 28	DEC 6	DEC 11	DEC 17	DEC 27	17 17	18 17
PIEDMONT FIELD STATION ORANGE COUNTY	36	APR 12	APR 23	APR 30	MAY 8	MAY 19	165	SEP 29	OCT 6	OCT 12	OCT 17	OCT 25	24 24	24 24
	32	MAR 31	APR 10	APR 17	APR 24	MAY 4	189	OCT 7	OCT 16	OCT 23	OCT 29	NOV 7	24 24	24 24
	28	MAR 22	APR 1	APR 7	APR 14	APR 23	208	OCT 21	OCT 28	NOV 1	NOV 6	NOV 13	24 24	24 24
	24	MAR 7	MAR 16	MAR 23	MAR 29	APR 8	237	OCT 31	NOV 9	NOV 15	NOV 21	NOV 30	24 24	24 24
	20	FEB 25	MAR 7	MAR 14	MAR 21	APR 1	255	NOV 9	NOV 18	NOV 24	NOV 30	DEC 9	24 24	24 24
	16	FEB 5	FEB 18	FEB 26	MAR 5	MAR 16	285	NOV 24	DEC 2	DEC 8	DEC 14	DEC 23	24 23	24 24
PULASKI PULASKI COUNTY	36	APR 15	APR 28	MAY 8	MAY 17	MAY 30	144	SEP 16	SEP 23	SEP 29	OCT 4	OCT 11	21 21	22 22
	32	APR 9	APR 18	APR 25	MAY 1	MAY 10	167	SEP 24	OCT 3	OCT 9	OCT 15	OCT 23	21 21	22 22
	28	APR 2	APR 11	APR 17	APR 23	MAY 1	185	OCT 2	OCT 12	OCT 19	OCT 26	NOV 5	21 21	22 22
	24	MAR 9	MAR 21	MAR 30	APR 7	APR 19	217	OCT 18	OCT 27	NOV 2	NOV 9	NOV 18	21 21	22 22
	20	FEB 25	MAR 7	MAR 14	MAR 21	APR 1	245	OCT 26	NOV 6	NOV 14	NOV 22	DEC 4	21 21	22 22
	16	FEB 13	FEB 26	MAR 6	MAR 15	MAR 28	267	NOV 10	NOV 20	NOV 28	DEC 5	DEC 15	21 21	22 22
QUANTICO PRINCE WILLIAM COUNTY	36	APR 8	APR 17	APR 23	APR 30	MAY 9	179	OCT 4	OCT 13	OCT 19	OCT 25	NOV 3	31 31	31 31
	32	MAR 24	APR 4	APR 12	APR 20	MAY 1	202	OCT 19	OCT 26	OCT 31	NOV 5	NOV 13	31 31	31 31
	28	MAR 9	MAR 20	MAR 27	APR 3	APR 14	231	NOV 1	NOV 8	NOV 13	NOV 18	NOV 26	31 31	31 31
	24	FEB 23	MAR 5	MAR 12	MAR 19	MAR 29	259	NOV 12	NOV 20	NOV 24	DEC 1	DEC 9	31 31	31 31
	20	FEB 3	FEB 16	FEB 25	MAR 5	MAR 18	284	NOV 21	NOV 30	DEC 6	DEC 12	DEC 21	31 31	31 31
	16	JAN 18	FEB 7	FEB 18	FEB 28	MAR 14	298	NOV 26	DEC 6	DEC 13	DEC 22	*	31 29	31 26
RICHMOND BYRD FIELD	36	APR 4	APR 14	APR 20	APR 26	MAY 6	181	OCT 5	OCT 13	OCT 18	OCT 24	NOV 1	31 31	31 31
	32	MAR 24	APR 2	APR 9	APR 16	APR 26	206	OCT 17	OCT 26	NOV 1	NOV 7	NOV 16	31 31	31 31
	28	MAR 5	MAR 17	MAR 24	APR 1	APR 13	233	OCT 27	NOV 6	NOV 12	NOV 19	NOV 28	31 31	31 31
	24	FEB 22	MAR 4	MAR 11	MAR 18	MAR 28	261	NOV 7	NOV 21	NOV 27	DEC 3	DEC 13	31 31	31 31
	20	FEB 5	FEB 18	FEB 26	MAR 7	MAR 19	283	NOV 17	NOV 29	DEC 6	DEC 12	DEC 24	31 31	31 29
	16	JAN 1	JAN 30	FEB 9	FEB 17	MAR 1	312	DEC 5	DEC 12	DEC 18	DEC 26	*	31 28	31 24
ROANOKE WOODRUM AIRPORT	36	APR 14	APR 23	APR 29	MAY 5	MAY 13	165	SEP 27	OCT 6	OCT 11	OCT 17	OCT 26	31 31	31 31
	32	MAR 28	APR 6	APR 12	APR 19	APR 28	194	OCT 7	OCT 17	OCT 23	OCT 30	NOV 8	31 31	31 31
	28	MAR 13	MAR 22	MAR 28	APR 4	APR 13	223	OCT 24	NOV 1	NOV 6	NOV 12	NOV 20	31 31	31 31
	24	FEB 20	MAR 4	MAR 13	MAR 21	APR 2	255	NOV 10	NOV 18	NOV 23	NOV 28	DEC 6	31 31	31 31
	20	FEB 4	FEB 20	MAR 2	MAR 12	MAR 26	274	NOV 16	NOV 25	DEC 1	DEC 7	DEC 15	31 30	31 31
	16	JAN 23	FEB 9	FEB 19	FEB 27	MAR 12	296	NOV 27	DEC 5	DEC 12	DEC 19	*	31 29	31 26
ROCKY KNOB FLOYD COUNTY	36	APR 21	MAY 1	MAY 9	MAY 16	MAY 27	155	SEP 25	OCT 4	OCT 11	OCT 17	OCT 26	21 21	21 21
	32	APR 6	APR 17	APR 24	MAY 2	MAY 13	182	OCT 8	OCT 17	OCT 23	OCT 29	NOV 7	21 21	20 20
	28	MAR 28	APR 7	APR 14	APR 21	MAY 2	200	OCT 18	OCT 26	OCT 31	NOV 6	NOV 13	20 20	20 20
	24	MAR 16	MAR 26	APR 2	APR 9	APR 19	219	OCT 22	OCT 31	NOV 7	NOV 13	NOV 23	20 20	20 21
	20	FEB 27	MAR 10	MAR 18	MAR 26	APR 5	245	OCT 30	NOV 10	NOV 18	NOV 25	DEC 6	20 20	22 22
	16	FEB 11	FEB 26	MAR 8	MAR 18	APR 2	265	NOV 12	NOV 21	NOV 28	DEC 4	DEC 13	20 20	22 22
ROCKY MOUNT FRANKLIN COUNTY	36	APR 15	APR 26	MAY 3	MAY 11	MAY 22	157	SEP 26	OCT 3	OCT 7	OCT 12	OCT 18	29 29	25 25
	32	APR 3	APR 13	APR 21	APR 29	MAY 10	178	OCT 1	OCT 10	OCT 16	OCT 22	OCT 31	29 29	25 25
	28	MAR 14	MAR 26	APR 4	APR 13	APR 25	212	OCT 15	OCT 25	NOV 2	NOV 9	NOV 20	29 29	25 25
	24	MAR 2	MAR 13	MAR 21	MAR 29	APR 9	239	OCT 30	NOV 8	NOV 15	NOV 21	DEC 1	30 30	25 25
	20	FEB 15	FEB 28	MAR 9	MAR 18	APR 1	263	NOV 15	NOV 22	NOV 27	DEC 2	DEC 9	30 30	25 25
	16	JAN 27	FEB 12	FEB 22	MAR 4	MAR 18	287	NOV 22	NOV 30	DEC 6	DEC 11	DEC 20	30 29	25 24
SALTVILLE SMYTH COUNTY	36	APR 17	APR 26	MAY 2	MAY 8	MAY 17	158	SEP 26	OCT 3	OCT 7	OCT 12	OCT 19	31 31	31 31
	32	APR 3	APR 12	APR 18	APR 24	MAY 2	189	OCT 9	OCT 18	OCT 24	OCT 30	NOV 7	31 31	31 31
	28	MAR 17	MAR 28	APR 4	APR 11	APR 22	210	OCT 18	OCT 25	OCT 31	NOV 5	NOV 12	31 31	31 31
	24	FEB 27	MAR 12	MAR 20	MAR 29	APR 10	238	OCT 29	NOV 7	NOV 13	NOV 19	NOV 28	31 31	31 31
	20	FEB 17	MAR 1	MAR 9	MAR 18	MAR 29	259	NOV 5	NOV 16	NOV 23	NOV 30	DEC 11	31 31	31 31
	16	FEB 2	FEB 18	FEB 27	MAR 9	MAR 22	284	NOV 23	DEC 1	DEC 8	DEC 14	DEC 26	31 30	31 29
STAUNTON AUGUSTA COUNTY	36	APR 13	APR 25	MAY 3	MAY 12	MAY 23	155	SEP 20	SEP 29	OCT 5	OCT 12	OCT 19	31 31	30 30
	32	APR 8	APR 17	APR 24	MAY 1	MAY 10	175	OCT 3	OCT 11	OCT 16	OCT 22	OCT 30	31 31	30 30
	28	MAR 24	APR 4	APR 11	APR 19	APR 30	203	OCT 14	OCT 24	OCT 31	NOV 7	NOV 17	31 31	31 31
	24	MAR 7	MAR 18	MAR 25	APR 2	APR 13	230	OCT 26	NOV 4	NOV 10	NOV 16	NOV 25	31 31	31 31
	20	FEB 26	MAR 9	MAR 16	MAR 24	APR 4	250	NOV 3	NOV 13	NOV 21	NOV 28	DEC 8	31 31	31 31
	16	FEB 9	FEB 22	MAR 2	MAR 10	MAR 21	276	NOV 19	NOV 27	DEC 3	DEC 8	DEC 16	31 30	31 31
STUART PATRICK COUNTY	36	APR 12	APR 22	APR 28	MAY 4	MAY 14	169	SEP 28	OCT 7	OCT 14	OCT 20	OCT 29	29 29	29 29
	32	MAR 20	APR 3	APR 12	APR 21	MAY 5	196	OCT 11	OCT 21	OCT 30	NOV 8	NOV 22	29 29	29 29
	28	MAR 12	MAR 24	APR 2	APR 11	APR 23	218	OCT 21	OCT 30	NOV 6	NOV 13	NOV 22	29 29	27 27
	24	FEB 27	MAR 13	MAR 22	APR 1	APR 15	242	NOV 3	NOV 12	NOV 19	NOV 25	DEC 5	29 29	28 28
	20	FEB 14	FEB 27	MAR 8	MAR 17	MAR 30	267	NOV 15	NOV 24	NOV 30	DEC 5	DEC 14	29 29	28 28
	16	JAN 15	FEB 8	FEB 21	MAR 5	MAR 21	291	NOV 23	DEC 3	DEC 9	DEC 17	*	29 27	28 26
SUFFOLK NANSEMOND COUNTY	36	APR 4	APR 13	APR 19	APR 26	MAY 5	189	OCT 11	OCT 19	OCT 25	OCT 31	NOV 9	22 22	23 23
	32	MAR 24	APR 2	APR 9	APR 15	APR 25	208	OCT 21	OCT 29	NOV 3	NOV 8	NOV 16	23 23	23 23
	28	MAR 10	MAR 17	MAR 22	MAR 27	APR 3	237	OCT 29	NOV 9	NOV 14	NOV 21	NOV 30	23 23	23 23
	24	FEB 18	MAR 1	MAR 9	MAR 17	MAR 29	264	NOV 13	NOV 22	NOV 28	DEC 4	DEC 13	23 23	23 23
	20	FEB 1	FEB 13	FEB 22	MAR 2	MAR 15	297	NOV 28	DEC 8	DEC 16	DEC 27	*	22 22	23 18
	16	*	JAN 26	FEB 6	FEB 15	FEB 28	325	DEC 11	DEC 18	DEC 28	*	*	22 19	23 12

FREEZE DATA

PROBABILITY OF SELECTED TEMPERATURES ON OR AFTER GIVEN DATES IN SPRING AND ON OR BEFORE GIVEN DATES IN FALL

STATION	TEMP	PROBABILITY – SPRING 90%	70%	50%	30%	10%	NO. OF DAYS BETWEEN 50% P'S	PROBABILITY – FALL 10%	30%	50%	70%	90%	SPRING Y	SPRING O	FALL Y	FALL O
SUNNYBANK NORTHUMBERLAND COUNTY	36	MAR 28	APR 9	APR 16	APR 22	APR 30	192	OCT 12	OCT 20	OCT 25	OCT 30	NOV 7	17	16	16	16
	32	MAR 22	MAR 31	APR 5	APR 10	APR 16	208	OCT 17	OCT 25	OCT 30	NOV 5	NOV 12	17	16	16	16
	28	MAR 7	MAR 17	MAR 23	MAR 28	APR 4	240	NOV 2	NOV 11	NOV 18	NOV 24	DEC 4	17	16	16	16
	24	FEB 12	FEB 27	MAR 8	MAR 16	MAR 27	268	NOV 16	NOV 25	DEC 1	DEC 6	DEC 15	17	16	16	16
	20	FEB 1	FEB 14	FEB 22	MAR 1	MAR 11	294	DEC 2	DEC 8	DEC 13	DEC 19	*	17	16	16	14
	16	*	JAN 28	FEB 9	FEB 20	MAR 6	314	DEC 6	DEC 13	DEC 20	*	*	17	15	16	10
TIMBERVILLE ROCKINGHAM COUNTY	36	APR 25	MAY 6	MAY 13	MAY 20	MAY 30	140	SEP 16	SEP 24	SEP 30	OCT 5	OCT 13	22	22	23	23
	32	APR 13	APR 23	APR 30	MAY 6	MAY 16	163	SEP 26	OCT 5	OCT 10	OCT 16	OCT 24	23	23	23	23
	28	APR 4	APR 12	APR 18	APR 23	MAY 1	185	OCT 7	OCT 15	OCT 20	OCT 26	NOV 2	23	23	23	23
	24	MAR 21	MAR 30	APR 4	APR 10	APR 19	214	OCT 24	OCT 31	NOV 4	NOV 9	NOV 16	23	23	23	23
	20	MAR 6	MAR 17	MAR 24	MAR 31	APR 10	234	OCT 27	NOV 6	NOV 13	NOV 20	NOV 29	22	22	23	23
	16	FEB 19	MAR 1	MAR 9	MAR 16	MAR 26	264	NOV 15	NOV 23	NOV 28	DEC 4	DEC 12	22	22	23	23
URBANNA MIDDLESEX COUNTY	36	APR 5	APR 12	APR 17	APR 22	APR 29	190	OCT 9	OCT 18	OCT 24	OCT 30	NOV 8	23	23	24	24
	32	MAR 25	APR 1	APR 5	APR 10	APR 16	213	OCT 23	OCT 30	NOV 4	NOV 9	NOV 16	23	23	24	24
	28	MAR 9	MAR 17	MAR 22	MAR 27	APR 4	237	OCT 30	NOV 8	NOV 14	NOV 20	NOV 29	23	23	24	24
	24	FEB 21	MAR 3	MAR 10	MAR 17	MAR 27	266	NOV 16	NOV 25	DEC 1	DEC 7	DEC 16	23	23	24	24
	20	JAN 23	FEB 13	FEB 22	MAR 2	MAR 13	293	NOV 26	DEC 6	DEC 12	DEC 19	DEC 30	23	23	24	23
	16	JAN 5	JAN 29	FEB 8	FEB 18	MAR 3	318	DEC 6	DEC 15	DEC 23	*	*	23	21	24	15
WALKERTON KING & QUEEN COUNTY	36	APR 18	APR 26	MAY 1	MAY 6	MAY 14	162	SEP 27	OCT 4	OCT 10	OCT 16	OCT 23	31	31	31	31
	32	APR 8	APR 16	APR 21	APR 27	MAY 4	181	OCT 6	OCT 13	OCT 19	OCT 24	NOV 1	31	31	31	31
	28	MAR 21	MAR 30	APR 5	APR 12	APR 21	208	OCT 19	OCT 26	OCT 30	NOV 4	NOV 11	31	31	31	31
	24	MAR 8	MAR 17	MAR 24	MAR 30	APR 9	235	OCT 30	NOV 8	NOV 14	NOV 21	NOV 29	31	31	31	31
	20	FEB 20	MAR 2	MAR 10	MAR 17	MAR 28	262	NOV 14	NOV 21	NOV 27	DEC 2	DEC 10	31	31	31	31
	16	FEB 5	FEB 18	FEB 26	MAR 7	MAR 19	285	NOV 25	DEC 3	DEC 8	DEC 14	DEC 22	31	31	31	30
WARRENTON FAUQUIER COUNTY	36	APR 9	APR 19	APR 26	MAY 3	MAY 13	173	OCT 1	OCT 10	OCT 16	OCT 21	OCT 30	20	20	20	20
	32	APR 2	APR 9	APR 14	APR 19	APR 26	199	OCT 19	OCT 26	OCT 30	NOV 4	NOV 11	20	20	20	20
	28	MAR 22	MAR 30	APR 4	APR 10	APR 17	218	OCT 25	NOV 2	NOV 8	NOV 13	NOV 21	20	20	20	20
	24	MAR 7	MAR 15	MAR 21	MAR 27	APR 5	246	NOV 7	NOV 15	NOV 22	NOV 29	DEC 10	20	20	20	20
	20	FEB 22	MAR 4	MAR 12	MAR 19	MAR 30	263	NOV 18	NOV 25	NOV 30	DEC 4	DEC 11	19	19	20	20
	16	FEB 1	FEB 13	FEB 22	MAR 2	MAR 15	291	NOV 26	DEC 4	DEC 10	DEC 16	DEC 24	19	19	20	20
WARSAW RICHMOND COUNTY	36	APR 12	APR 21	APR 27	MAY 4	MAY 13	169	SEP 27	OCT 6	OCT 13	OCT 19	OCT 29	27	27	26	26
	32	APR 2	APR 10	APR 15	APR 20	APR 27	194	OCT 11	OCT 20	OCT 26	NOV 1	NOV 10	27	27	25	25
	28	MAR 13	MAR 24	MAR 31	APR 7	APR 17	219	OCT 26	NOV 2	NOV 5	NOV 9	NOV 15	27	27	25	25
	24	FEB 24	MAR 8	MAR 16	MAR 23	APR 4	251	NOV 8	NOV 16	NOV 22	NOV 28	DEC 6	27	27	25	25
	20	FEB 7	FEB 21	MAR 2	MAR 11	MAR 24	276	NOV 20	NOV 27	DEC 3	DEC 8	DEC 16	27	27	25	25
	16	JAN 25	FEB 11	FEB 22	MAR 4	MAR 19	296	DEC 2	DEC 10	DEC 15	DEC 22	*	27	26	25	21
WASHINGTON RAPPAHANNOCK COUNTY	36	MAY 3	MAY 10	MAY 15	MAY 19	MAY 26	138	SEP 17	SEP 25	SEP 30	OCT 5	OCT 13	16	16	15	15
	32	APR 17	APR 27	MAY 4	MAY 12	MAY 22	159	SEP 22	OCT 3	OCT 10	OCT 18	OCT 29	16	16	15	15
	28	APR 4	APR 12	APR 17	APR 23	MAY 1	187	SEP 29	OCT 12	OCT 21	OCT 30	NOV 12	16	16	15	15
	24	MAR 19	MAR 28	APR 4	APR 10	APR 19	215	OCT 20	OCT 29	NOV 5	NOV 12	NOV 21	15	15	15	15
	20	MAR 8	MAR 17	MAR 23	MAR 29	APR 7	233	OCT 26	NOV 4	NOV 11	NOV 18	NOV 27	15	15	15	15
	16	FEB 25	MAR 7	MAR 13	MAR 20	MAR 30	253	NOV 2	NOV 14	NOV 21	NOV 30	DEC 16	15	15	15	14
WASHINGTON NATIONAL AIRPORT	36	MAR 31	APR 9	APR 14	APR 20	APR 29	200	OCT 19	OCT 26	OCT 31	NOV 5	NOV 11	31	31	31	31
	32	MAR 15	MAR 24	MAR 30	APR 4	APR 13	223	OCT 25	NOV 2	NOV 8	NOV 13	NOV 21	31	31	31	31
	28	MAR 1	MAR 12	MAR 19	MAR 26	APR 5	250	NOV 13	NOV 19	NOV 24	NOV 29	DEC 6	31	31	31	31
	24	FEB 17	MAR 1	MAR 9	MAR 17	MAR 29	272	NOV 23	DEC 1	DEC 6	DEC 11	DEC 20	31	31	31	30
	20	JAN 30	FEB 16	FEB 25	MAR 6	MAR 18	291	NOV 29	DEC 7	DEC 13	DEC 19	*	31	29	31	28
	16	*	JAN 30	FEB 11	FEB 21	MAR 7	314	DEC 6	DEC 14	DEC 22	*	*	31	27	31	18

FREEZE DATA

PROBABILITY OF SELECTED TEMPERATURES ON OR AFTER GIVEN DATES IN SPRING AND ON OR BEFORE GIVEN DATES IN FALL

STATION	TEMP	PROBABILITY - SPRING 90%	70%	50%	30%	10%	NO. OF DAYS BETWEEN 50% P'S	PROBABILITY - FALL 10%	30%	50%	70%	90%	SPRING Y	O	FALL Y	O
WEST POINT NEW KENT COUNTY	36	APR 11	APR 20	APR 26	MAY 2	MAY 11	170	SEP 30	OCT 8	OCT 13	OCT 18	OCT 25	17	17	17	17
	32	APR 2	APR 9	APR 15	APR 20	APR 28	191	OCT 8	OCT 17	OCT 23	OCT 29	NOV 7	17	17	17	17
	28	MAR 21	MAR 29	APR 2	APR 7	APR 15	216	OCT 21	OCT 29	NOV 4	NOV 10	NOV 18	17	17	17	17
	24	MAR 3	MAR 13	MAR 20	MAR 27	APR 5	243	NOV 3	NOV 12	NOV 18	NOV 23	DEC 2	17	17	17	17
	20	FEB 21	MAR 3	MAR 9	MAR 16	MAR 26	265	NOV 14	NOV 23	NOV 29	DEC 6	DEC 15	16	16	17	17
	16	JAN 31	FEB 12	FEB 20	FEB 28	MAR 12	292	NOV 28	DEC 4	DEC 9	DEC 14	*	16	16	17	15
WILLIAMSBURG YORK COUNTY	36	APR 11	APR 20	APR 26	MAY 2	MAY 10	175	OCT 4	OCT 12	OCT 18	OCT 23	NOV 1	30	30	31	31
	32	APR 1	APR 11	APR 18	APR 26	MAY 6	192	OCT 12	OCT 21	OCT 27	NOV 2	NOV 11	30	30	31	31
	28	MAR 15	MAR 25	APR 1	APR 8	APR 17	224	OCT 28	NOV 5	NOV 11	NOV 16	NOV 24	30	30	31	31
	24	MAR 1	MAR 11	MAR 18	MAR 24	APR 3	249	NOV 8	NOV 16	NOV 22	NOV 28	DEC 6	30	30	31	31
	20	FEB 14	FEB 26	MAR 5	MAR 13	MAR 25	273	NOV 19	NOV 27	DEC 3	DEC 9	DEC 18	30	30	31	31
	16	JAN 19	FEB 7	FEB 16	FEB 25	MAR 10	307	DEC 4	DEC 12	DEC 20	*	*	30	28	31	22
WINCHESTER FREDERICK COUNTY	36	APR 13	APR 23	APR 29	MAY 6	MAY 15	162	SEP 24	OCT 2	OCT 8	OCT 14	OCT 22	31	31	31	31
	32	APR 3	APR 12	APR 18	APR 24	MAY 3	185	OCT 3	OCT 13	OCT 20	OCT 26	NOV 5	31	31	31	31
	28	MAR 22	APR 1	APR 8	APR 15	APR 25	208	OCT 19	OCT 27	NOV 2	NOV 8	NOV 16	31	31	31	31
	24	MAR 8	MAR 17	MAR 24	MAR 31	APR 10	237	NOV 2	NOV 10	NOV 16	NOV 22	NOV 30	31	31	31	31
	20	FEB 23	MAR 6	MAR 13	MAR 20	MAR 31	260	NOV 15	NOV 22	NOV 28	DEC 3	DEC 11	31	31	31	31
	16	FEB 5	FEB 17	FEB 25	MAR 5	MAR 17	286	NOV 24	DEC 2	DEC 8	DEC 14	DEC 24	31	31	31	30
WISE WISE COUNTY	36	APR 29	MAY 12	MAY 18	MAY 25	JUN 3	131	SEP 12	SEP 20	SEP 26	OCT 2	OCT 10	16	15	16	16
	32	APR 12	APR 27	MAY 5	MAY 13	MAY 24	156	SEP 23	OCT 8	OCT 14	OCT 22		16	15	16	16
	28	APR 8	APR 19	APR 25	MAY 1	MAY 9	176	OCT 7	OCT 14	OCT 18	OCT 23	OCT 30	16	15	16	16
	24	MAR 18	MAR 31	APR 6	APR 13	APR 22	201	OCT 12	OCT 19	OCT 24	OCT 29	NOV 5	16	15	16	16
	20	FEB 23	MAR 12	MAR 22	MAR 31	APR 13	230	OCT 25	NOV 2	NOV 7	NOV 12	NOV 19	16	15	16	16
	16	FEB 14	MAR 2	MAR 11	MAR 20	APR 1	254	NOV 5	NOV 14	NOV 20	NOV 25	DEC 4	16	15	16	16
WOODSTOCK SHENANDOAH COUNTY	36	APR 20	APR 30	MAY 6	MAY 13	MAY 22	147	SEP 17	SEP 25	SEP 30	OCT 6	OCT 14	31	31	31	31
	32	APR 8	APR 18	APR 25	MAY 2	MAY 12	169	SEP 27	OCT 5	OCT 11	OCT 17	OCT 25	31	31	31	31
	28	MAR 26	APR 4	APR 11	APR 16	APR 25	199	OCT 8	OCT 19	OCT 26	NOV 2	NOV 13	31	31	31	31
	24	MAR 10	MAR 20	MAR 26	APR 2	APR 12	229	OCT 29	NOV 9	NOV 17	NOV 23	DEC 10	31	31	31	31
	20	FEB 22	MAR 5	MAR 13	MAR 21	APR 1	255	NOV 9	NOV 17	NOV 23	NOV 30	DEC 10	31	30	31	31
	16	FEB 6	FEB 21	MAR 2	MAR 11	MAR 23	276	NOV 19	NOV 27	DEC 3	DEC 9	DEC 17	31	30	31	31
WYTHEVILLE WYTHE COUNTY	36	APR 26	MAY 6	MAY 13	MAY 20	MAY 30	136	SEP 10	SEP 20	SEP 26	OCT 3	OCT 12	31	31	31	31
	32	APR 8	APR 19	APR 27	MAY 5	MAY 17	163	SEP 20	SEP 30	OCT 7	OCT 14	OCT 23	31	31	31	31
	28	MAR 31	APR 10	APR 17	APR 24	MAY 4	186	SEP 30	OCT 12	OCT 20	OCT 28	NOV 9	31	31	31	31
	24	MAR 15	MAR 27	APR 5	APR 14	APR 26	210	OCT 13	OCT 24	NOV 1	NOV 9	NOV 20	31	31	31	31
	20	FEB 24	MAR 9	MAR 18	MAR 28	APR 10	239	OCT 21	NOV 3	NOV 12	NOV 21	DEC 5	31	31	31	31
	16	FEB 8	FEB 24	MAR 6	MAR 15	MAR 29	263	NOV 2	NOV 15	NOV 24	DEC 3	DEC 16	31	30	31	31

Data in the above table are based on the period 1940-70, or that portion of this period for which data are available.

* Preselected probability value greater than the probability of the temperature threshold being reached or the computed date fell before January 1 for Spring or after December 31 for Fall.

Y Number of years with available observations.

O Number of years with observed threshold temperature.

⊢O⊣ Insufficient data

A freeze is a numerical substitute for the former term "killing frost" and is the occurrence of a minimum temperature at or below the threshold occurrence of 32°, 28°, 24°, 20°, and 16°. The 36° temperature has been included in the table since temperatures at ground level may be at freezing or lower. Data are adjusted to take into account the years of non-occurrence.

*NORMALS BY CLIMATOLOGICAL DIVISIONS

Taken from "Climatography of the United States No. 81-4, Decennial Census of U. S. Climate"

STATIONS (By Divisions)	TEMPERATURE (°F)													PRECIPITATION (In.)												
	JAN	FEB	MAR	APR	MAY	JUNE	JULY	AUG	SEPT	OCT	NOV	DEC	ANN	JAN	FEB	MAR	APR	MAY	JUNE	JULY	AUG	SEPT	OCT	NOV	DEC	ANN
TIDEWATER																										
CAPE HENRY WSO	42.6	42.4	47.9	56.9	66.1	74.5	78.1	77.5	73.1	63.2	53.0	44.2	60.0	2.98	3.25	3.36	2.93	3.04	3.54	5.00	5.36	3.95	2.87	2.80	2.76	41.84
COLONIAL BEACH	37.5	38.2	45.0	55.6	65.5	73.8	77.7	76.3	70.1	59.4	48.2	38.9	57.2	3.24	2.43	3.20	3.14	3.13	3.23	4.55	4.49	3.77	2.98	2.44	2.95	39.55
DIAMOND SPRINGS	43.7	44.0	50.0	59.1	67.9	75.7	79.0	77.9	73.2	63.2	53.5	44.6	61.0	3.63	3.45	3.93	3.37	3.66	3.79	6.19	6.58	4.48	3.17	3.33	2.96	48.54
EMPORIA 1 WNW	.	.	.	.	.	.	.	.	.	.	.	.	.	3.16	3.34	3.57	3.43	4.02	4.11	6.18	5.03	4.01	2.46	2.79	2.99	45.09
FREDERICKSBURG NATL PK	37.1	38.0	45.0	55.6	65.3	73.3	77.2	75.5	69.1	58.3	47.3	37.9	56.6	3.20	2.38	3.28	3.10	3.56	3.34	4.89	5.10	3.45	3.35	2.84	2.80	41.29
HOPEWELL	41.5	42.7	49.3	59.7	68.5	76.1	79.2	77.8	72.2	61.3	51.0	41.9	60.1	3.07	2.76	3.16	3.34	3.97	4.23	5.86	5.10	3.73	2.88	2.80	2.78	43.68
LANGLEY AIR FORCE BASE	41.5	42.0	48.3	57.7	67.2	75.4	79.0	77.9	72.7	62.1	51.6	42.6	59.8	3.23	3.02	3.42	3.03	3.48	3.10	5.33	5.13	3.98	2.87	2.79	2.57	41.95
NORFOLK WSO	41.2	41.6	48.0	58.0	67.5	75.6	78.8	77.5	72.6	62.0	51.4	42.5	59.7	3.16	2.36	3.22	3.16	3.36	3.61	5.92	5.97	4.22	2.92	3.05	2.74	44.94
QUANTICO IS	36.5	37.5	44.5	55.5	65.2	73.5	77.3	75.7	72.6	58.3	47.0	37.4	56.8	3.16	2.36	3.22	3.08	3.47	3.26	4.58	4.91	3.29	2.95	2.79	2.80	39.87
WALLACETON LK DRUMMOND	.	.	.	.	.	.	.	.	.	.	.	.	.	3.64	3.65	3.95	3.76	3.98	4.49	6.73	5.92	4.37	3.20	3.45	3.28	50.42
DIVISION	40.3	41.1	47.4	57.2	66.6	74.6	78.1	76.8	71.3	60.8	50.5	41.3	58.8	3.37	3.02	3.53	3.23	3.61	3.57	5.41	5.37	3.88	3.06	2.98	2.84	43.87
EASTERN PIEDMONT																										
CLARKSVILLE	.	.	.	.	.	.	.	.	.	.	.	.	.	3.66	3.32	3.90	3.81	3.61	4.13	4.94	5.31	3.85	2.73	3.26	3.10	45.62
COLUMBIA	37.7	38.8	45.8	56.3	65.9	73.9	77.5	75.9	69.4	58.1	47.2	38.3	57.1	3.24	2.64	3.47	3.49	3.79	3.21	4.63	4.29	3.78	2.94	2.73	2.94	41.15
FARMVILLE 2 N	39.4	40.7	47.0	57.2	65.9	73.8	77.5	75.9	69.6	58.7	48.1	39.3	57.8	3.80	3.25	3.98	3.64	4.02	4.31	4.82	4.69	3.80	2.96	3.09	3.18	45.54
NEW CANTON	38.7	39.7	46.2	56.4	65.1	72.4	76.0	74.4	68.4	57.8	47.7	39.0	56.8	3.60	2.87	3.75	3.68	3.76	3.26	4.57	4.52	3.56	2.88	2.93	3.31	42.69
RICHMOND WSO	38.7	39.9	47.7	58.1	67.0	75.1	78.1	76.0	70.2	58.7	48.5	39.7	58.1	3.46	2.90	3.42	3.15	3.72	3.75	5.61	5.54	3.65	3.00	3.04	2.97	44.21
DIVISION	38.8	39.9	46.6	56.9	66.1	73.9	77.4	75.9	69.7	58.7	48.1	39.2	57.6	3.51	2.97	3.71	3.53	3.69	3.67	5.09	4.78	3.64	2.92	3.00	3.09	43.60
WESTERN PIEDMONT																										
BEDFORD	.	.	.	.	.	.	.	.	.	.	.	.	.	3.46	2.92	4.11	3.48	4.08	4.45	4.48	5.13	3.44	3.00	2.90	3.42	44.87
CHARLOTTESVILLE 2 W	37.2	38.4	45.3	56.6	66.3	73.7	77.3	76.0	69.9	59.5	48.4	38.7	57.3	3.30	2.78	3.90	3.69	3.89	3.98	5.58	4.70	4.24	3.37	3.01	3.43	45.87
CHATHAM 2 NE	39.2	40.4	46.8	57.1	66.2	73.9	76.9	75.5	69.6	58.9	47.8	39.4	57.6	3.44	3.08	3.80	3.56	3.67	3.79	4.38	4.50	4.10	2.88	2.96	3.30	43.46
DANVILLE-BRIDGE ST	41.1	42.2	48.7	59.2	68.0	75.7	78.5	77.4	71.4	60.5	49.4	41.0	59.4	3.51	3.18	3.87	3.65	4.18	3.64	4.55	4.25	4.05	2.87	2.91	3.18	44.10
HALIFAX 1 N	.	.	.	.	.	.	.	.	.	.	.	.	.	3.35	3.13	3.87	3.75	3.84	4.11	4.90	4.35	3.74	2.85	3.03	3.18	44.10
LYNCHBURG WSO	37.6	38.9	45.5	56.2	65.3	73.0	76.3	74.8	68.7	58.2	47.0	38.5	56.7	3.29	2.65	3.61	3.14	3.21	4.06	4.21	4.41	4.36	2.64	2.58	3.14	40.30
RANDOLPH	.	.	.	.	.	.	.	.	.	.	.	.	.	3.63	3.14	3.88	3.91	4.10	3.60	4.64	4.81	4.47	2.87	3.29	3.18	45.52
ROCKY MOUNT	38.7	39.6	45.8	56.0	64.9	72.1	75.7	74.4	67.8	58.2	47.3	39.2	56.6	3.21	2.98	3.84	3.43	4.13	3.83	4.49	4.14	4.06	3.32	2.68	3.10	43.21
DIVISION	39.0	40.0	46.3	56.9	66.0	73.5	76.6	75.3	69.2	58.9	47.9	39.4	57.4	3.42	3.04	3.94	3.68	3.96	4.08	4.82	4.63	4.07	3.11	2.91	3.32	44.98
NORTHERN																										
CULPEPER	36.8	37.9	45.5	56.1	65.5	73.2	76.9	75.1	69.0	58.2	47.0	37.4	56.6	3.03	2.55	3.38	3.64	4.10	3.68	4.80	4.38	3.62	3.14	2.98	2.87	42.17
LINCOLN	35.7	36.7	44.0	55.0	65.3	73.5	77.4	76.0	69.0	58.3	47.0	36.9	56.2	3.03	2.74	3.78	3.53	4.16	3.59	4.26	4.74	3.33	3.47	2.88	3.12	42.63
MOUNT WEATHER	30.4	31.1	38.4	49.3	59.8	68.5	72.5	71.2	64.6	54.0	42.6	32.3	51.2	2.67	2.33	3.26	3.31	4.29	3.56	3.83	4.49	3.77	3.70	2.89	2.63	40.73
RIVERTON	.	.	.	.	.	.	.	.	.	.	.	.	.	2.40	1.99	2.86	2.88	2.87	3.21	3.38	3.76	2.86	2.41	2.37	2.38	35.37
WASHINGTON NAT AP WSO	36.9	37.8	44.8	55.7	65.8	74.2	78.2	76.5	69.7	59.0	47.3	38.1	57.0	3.03	2.47	3.21	3.15	4.14	3.21	4.15	4.90	3.83	3.07	2.84	2.78	40.78
WINCHESTER 1 N	34.9	36.1	43.0	54.2	64.3	72.2	76.1	74.5	68.0	57.3	46.0	36.2	55.2	2.39	2.12	3.16	3.11	4.10	3.70	4.24	4.16	2.97	3.45	2.59	2.48	38.47
WOODSTOCK	35.9	36.9	43.6	54.4	63.8	71.4	75.2	73.5	67.0	56.6	45.9	36.5	55.1	2.26	1.74	2.86	2.83	3.81	3.83	3.80	4.26	2.83	2.90	2.13	2.22	35.47
DIVISION	34.8	35.7	42.7	53.7	63.6	71.5	75.3	73.7	67.2	56.6	45.5	35.8	54.7	2.81	2.35	3.37	3.33	4.13	3.67	4.31	4.56	3.43	3.45	2.83	2.80	41.04
CENTRAL MOUNTAIN																										
BALCONY FALLS	.	.	.	.	.	.	.	.	.	.	.	.	.	3.42	2.98	3.99	3.61	3.71	3.51	4.08	4.62	3.80	3.33	2.99	3.21	43.25
BUCHANAN	38.4	39.9	46.2	56.5	65.3	72.7	76.0	74.9	68.8	58.1	46.8	38.5	56.8	3.25	2.93	3.81	3.35	4.04	4.32	4.41	4.67	3.71	3.42	3.09	3.20	44.20
CATAWBA SANATORIUM	36.4	37.2	43.3	54.1	62.8	69.8	73.0	72.0	66.3	56.5	45.5	37.3	54.5	3.13	2.90	3.61	3.38	4.02	4.56	4.51	3.68	3.17	2.59	2.93		42.50
CLIFTON FORGE	.	.	.	.	.	.	.	.	.	.	.	.	.	2.99	2.85	3.92	3.08	3.98	3.71	3.90	4.37	3.31	2.94	2.74	2.95	40.74
DALE ENTERPRISE	35.0	36.0	42.2	52.8	62.3	69.9	73.4	71.8	66.0	55.7	44.5	35.8	53.8	2.16	1.95	2.87	2.59	4.00	3.79	4.57	4.31	3.14	2.63	2.10	2.23	36.34
HOT SPRINGS	32.6	33.6	40.1	50.9	60.1	67.2	70.3	68.9	62.7	52.6	41.6	33.4	51.2	3.01	2.84	3.93	3.12	3.67	4.17	4.18	4.65	3.22	2.94	2.77	2.88	41.38
LEXINGTON	37.4	38.6	45.1	55.5	64.2	71.6	75.1	73.8	67.7	56.9	46.0	37.7	55.8	3.08	2.75	3.61	3.03	3.58	4.18	4.03		3.39	2.87	2.66	2.91	39.64
NORTH RIVER DAM	.	.	.	.	.	.	.	.	.	.	.	.	.	2.68	2.43	3.34	3.21	3.93	4.21	5.02	4.89	3.69	3.04	2.71	2.61	41.76
ROANOKE	38.9	40.0	46.2	56.6	65.7	73.2	76.4	75.1	68.6	58.3	47.2	39.2	57.1	3.11	2.87	3.64	3.10	4.10	4.06	4.44	4.86	3.68	3.44	2.72	3.10	43.12
ROANOKE WSO	38.1	39.2	45.5	56.4	65.7	73.4	76.5	75.4	69.1	58.2	46.7	38.4	56.9	3.12	2.86	3.53	3.10	3.79	3.80	4.25	4.63	3.26	3.21	2.70	2.98	41.23
STAUNTON SEWAGE PLANT	35.7	36.4	42.6	53.5	63.3	71.1	74.5	72.7	66.8	56.3	45.4	36.6	54.6	2.67	2.11	3.17	2.98	3.74	3.69	4.01	3.82	3.29	2.83	2.41	2.70	37.42
TIMBERVILLE 3 E	.	.	.	.	.	.	.	.	.	.	.	.	.	2.18	2.03	2.96	2.72	3.76	3.52	4.25	4.04	3.40	2.61	2.13	2.17	35.77
DIVISION	36.0	37.0	43.3	53.9	63.1	70.5	73.9	72.5	66.4	56.1	45.0	36.5	54.5	2.90	2.61	3.49	3.07	3.86	3.87	4.35	4.44	3.37	3.02	2.61	2.83	40.43

Taken from "Climatography of the United States No. 81-4, Decennial Census of U. S. Climate"

STATIONS (By Divisions)	TEMPERATURE (°F)													PRECIPITATION (In.)												
	JAN	FEB	MAR	APR	MAY	JUNE	JULY	AUG	SEPT	OCT	NOV	DEC	ANN	JAN	FEB	MAR	APR	MAY	JUNE	JULY	AUG	SEPT	OCT	NOV	DEC	ANN
SOUTHWESTERN MOUNTAIN																										
BURKES GARDEN	33.2	34.3	39.6	49.7	57.8	64.9	67.9	67.0	61.3	51.4	40.6	33.6	50.1	3.90	3.62	4.41	3.48	3.94	4.07	4.81	4.32	3.23	2.39	2.99	3.37	44.53
DAMASCUS	•	•	•	•	•	•	•	•	•	•	•	•	•	3.93	3.91	4.30	3.59	4.23	4.12	5.96	4.43	2.90	2.34	2.81	3.28	45.80
GLEN LYN	•	•	•	•	•	•	•	•	•	•	•	•	•	2.88	2.85	3.53	3.03	3.39	3.61	4.38	3.91	2.56	2.34	2.27	2.48	37.23
MENDOTA	•	•	•	•	•	•	•	•	•	•	•	•	•	4.27	3.90	4.59	3.70	3.91	3.67	5.36	4.38	2.79	2.45	2.93	3.54	45.49
PENNINGTON GAP	36.1	37.6	43.6	53.9	63.1	70.5	73.7	72.8	67.3	56.5	44.0	36.8	54.7	5.28	4.88	5.48	3.92	3.85	4.21	5.60	4.24	2.76	2.44	3.42	4.36	50.44
RADFORD 5 SW	•	•	•	•	•	•	•	•	•	•	•	•	•	2.90	2.62	3.33	2.85	3.45	3.23	4.37	3.37	2.75	2.43	2.20	2.61	36.11
SALTVILLE 1 N	36.6	37.7	43.7	54.7	64.2	71.8	74.7	73.6	67.9	57.0	44.8	36.8	55.3	3.89	3.68	4.27	3.19	4.18	3.80	4.81	4.29	2.92	2.33	2.64	3.24	43.24
WYTHEVILLE 1 S	36.2	37.1	42.9	53.0	61.6	68.9	71.9	70.8	65.0	55.0	44.0	36.3	53.6	2.79	2.71	3.38	2.91	3.61	3.16	4.36	3.88	2.90	2.17	2.24	2.55	36.66
DIVISION	35.6	36.8	42.7	53.2	62.0	69.3	72.3	71.3	65.5	55.2	43.7	36.0	53.6	3.65	3.50	4.15	3.45	3.88	3.99	4.92	4.30	3.11	2.52	2.69	3.27	43.40

* Normals for the period 1931-1960. Divisional normals may not be the arithmetical average of individual stations published, since additional data for shorter period stations are used to obtain better areal representation.

CONFIDENCE - LIMITS

In the absence of trend or record changes, the chances are 9 out of 10 that the true mean will lie in the interval formed by adding and subtracting the values in the following table from the means for any station in the State. Because of the wider variation in mean precipitation, the corresponding monthly means and annual mean must be substituted for "p" in the precipitation table below to obtain mean precipitation confidence limits.

| 1.4 | 1.4 | 1.4 | .7 | .8 | .7 | .6 | .6 | 1.0 | .8 | .8 | 1.2 | .3 | .27$\sqrt{p}$ | .24$\sqrt{p}$ | .25$\sqrt{p}$ | .25$\sqrt{p}$ | .33$\sqrt{p}$ | .29$\sqrt{p}$ | .34$\sqrt{p}$ | .42$\sqrt{p}$ | .38$\sqrt{p}$ | .37$\sqrt{p}$ | .34$\sqrt{p}$ | .26$\sqrt{p}$ | .32$\sqrt{p}$ |

COMPARATIVE DATA

Data in the following table are the mean temperature and average precipitation for Staunton D. and B. Institute, Virginia, for the period 1906-1930 and are included in this publication for comparative purposes.

| 36.3 | 36.9 | 45.4 | 53.8 | 63.5 | 70.4 | 74.3 | 72.7 | 67.9 | 56.1 | 45.5 | 36.9 | 55.0 | 2.83 | 2.15 | 2.93 | 3.01 | 2.79 | 4.21 | 4.48 | 4.20 | 2.79 | 2.89 | 2.19 | 2.61 | 37.08 |

NORMALS, MEANS, AND EXTREMES

Station: LYNCHBURG, VIRGINIA — MUNICIPAL AIRPORT — Standard time used: EASTERN — Latitude: 37° 20' N — Longitude: 79° 12' W — Elevation (ground): 916 feet — Year: 1970

Month	Temperature							Normal heating degree days (Base 65°)	Precipitation								Snow, Ice pellets						Relative humidity				Wind &					Pct. of possible sunshine	Mean sky cover sunrise to sunset	Mean number of days											Average daily solar radiation-langleys
	Normal			Extremes Ø																				Hour 01	Hour 07	Hour 13	Hour 19	Mean speed	Prevailing direction	Fastest mile					Sunrise to sunset			Precipitation .01 inch or more	Snow, Ice pellets 1.0 inch or more	Thunderstorms	% Heavy fog	%	Temperatures				
	Daily maximum	Daily minimum	Monthly	Record highest	Year	Record lowest	Year		Normal total	Maximum monthly	Year	Minimum monthly	Year	Maximum in 24 hrs.	Year	Mean total	Maximum monthly	Year	Maximum in 24 hrs	Year									Speed	Direction	Year			Clear	Partly cloudy	Cloudy						Max. 90 and above	Max. 32 and below	Min. 32 and below	Min. 0 and below	
																								(Local time)						#																	
(a)	(b)	(b)	(b)		7		7	(b)	(b)		26		26		26	26	26		26		7	7	7	20	9	26	26				26	26	26	26	26	26	26	22	22		7	7	7	7	
J	46.2	29.0	37.6	74	1967	-4	1970	849	3.29	4.82	1964	0.76	1956	2.13	1948	5.8	31.8	1966	10.9	1966	70	51	59	8.8	SW	38	S	1959+	51	6.2	9	7	15	11	2	*	4	0	5	24	*
F	48.3	29.5	38.9	71	1965	0	1965	731	2.65	5.23	1961	0.64	1968+	1.90	1965	4.8	14.7	1967	12.8	1960	68	48	54	9.0	SW	50	S	1961	55	6.1	8	7	13	10	2	*	4	0	2	23	*
M	55.7	35.3	45.5	85	1968	7	1965	605	3.61	6.69	1948	0.74	1966	2.43	1949	4.1	24.9	1960	13.4	1969	71	46	52	9.4	SW	43	S	1950	58	6.0	9	9	13	11	1	1	3	0	*	15	0
A	67.5	44.8	56.2	86	1966	24	1970+	267	3.14	4.66	1951	1.15	1963	2.45	1945	0.1	2.8	1957	2.8	1957	75	47	54	9.4	SW	43	S	1961	59	6.1	8	9	13	10	*	3	3	0	0	2	0
M	76.8	53.7	65.3	93	1969	31	1966	78	3.21	6.02	1960	1.36	1957	3.47	1960	0.0	0.0		0.0		79	50	60	8.0	SW	56	N	1958	63	6.0	8	11	12	11	0	6	4	1	0	0	0
J	83.8	62.1	73.0	99	1964	42	1966	0	4.06	8.50	1949	0.67	1964	3.23	1949	0.0	0.0		0.0		81	51	62	7.1	SW	56	SW	1951	66	5.7	8	12	10	10	0	7	2	6	0	0	0
J	86.6	65.9	76.3	99	1966	50	1963	0	4.21	7.92	1950	1.75	1963	2.65	1956	0.0	0.0		0.0		86	57	69	6.8	SW	43	NW	1954+	61	5.9	7	12	12	12	0	9	3	8	0	0	0
A	84.7	64.9	74.8	98	1968	45	1965	0	4.41	11.36	1952	0.93	1963	4.47	1967	0.0	0.0		0.0		88	58	74	6.4	N	46	NE	1952	61	5.8	8	12	11	10	0	8	5	0	0	0	0
S	79.1	58.2	68.7	93	1970+	36	1967	51	3.36	7.48	1957	0.78	1970	3.10	1957	0.0	0.0		0.0		86	53	71	7.2	N	40	NE	1955	63	5.3	11	8	11	8	0	4	2	0	0	0	0
O	69.0	47.4	58.2	85	1970+	21	1969+	223	2.64	5.89	1954	0.79	1963	4.98	1954	T	T	1962+	T	1962+	83	51	68	7.5	N	41	N	1954	62	4.7	14	7	10	7	0	1	3	0	0	4	0
N	57.1	36.9	47.0	81	1968	8	1970	540	2.58	5.10	1963+	0.90	1960	2.65	1951	0.9	11.6	1968	6.7	1968	75	49	60	8.1	SW	43	NW	1949	56	5.5	11	7	12	9	*	1	4	0	*	11	0
D	47.0	29.9	38.5	71	1966	7	1968	822	3.14	6.74	1948	0.33	1965	3.03	1948	3.7	17.9	1965	12.7	1969	74	53	63	8.0	SW	45	SE	1950	53	5.8	11	6	14	9	1	*	4	0	2	21	0
YR	66.8	46.5	56.7	99	JUL. 1966+	-4	JAN. 1970	4166	40.30	11.36	AUG. 1952	0.33	DEC. 1965	4.98	OCT. 1954	19.4	31.8	JAN. 1966	13.4	MAR. 1969	78	51	62	8.0	SW	56	N	MAY 1958+	59	5.8	112	107	146	119	6	41	40	23	9	100	1

Ø For period July 1963 through the current year.

To 8 compass points only.   % Data through 1966. Station operated less than 24 hours per day prior to August 1962. Fog and thunderstorm data may be incomplete.

Means and extremes above are from existing and comparable exposures. Annual extremes have been exceeded at other sites in the locality as follows: Highest temperature 106 in July 1936; lowest temperature -7 in January 1912; maximum monthly precipitation 14.87 in August 1928; minimum monthly precipitation .03 in November 1890; maximum precipitation in 24 hours 7.59 in August 1928; maximum snowfall in 24 hours 16.4 in January 1922; fastest mile wind 62 W in April 1917+.

— 413 —

Station: NORFOLK, VIRGINIA — NORFOLK REGIONAL AIRPORT — Standard time used: EASTERN — Latitude: 36° 54' N — Longitude: 76° 12' W — Elevation (ground): 22 feet — Year: 1970

Month	Temperature							Normal heating degree days (Base 65°)	Precipitation								Snow, Ice pellets						Relative humidity				Wind &					Pct. of possible sunshine	Mean sky cover sunrise to sunset	Mean number of days											Average daily solar radiation-langleys
	Normal			Extremes																				Hour 01	Hour 07	Hour 13	Hour 19	Mean speed	Prevailing direction	Fastest mile					Sunrise to sunset			Precipitation .01 inch or more	Snow, Ice pellets 1.0 inch or more	Thunderstorms	Heavy fog		Temperatures				
	Daily maximum	Daily minimum	Monthly	Record highest	Year	Record lowest	Year		Normal total	Maximum monthly	Year	Minimum monthly	Year	Maximum in 24 hrs	Year	Mean total	Maximum monthly	Year	Maximum in 24 hrs	Year									Speed	Direction	Year			Clear	Partly cloudy	Cloudy						Max. 90 and above	Max. 32 and below	Min. 32 and below	Min. 0 and below	
																								(Local time)						#																	
(a)	(b)	(b)	(b)		22		22	(b)	(b)		22		22		22	22	22		22		22	22	22	22	15	17	17				17	22	22	22	22	22	22	22	22		22	22	22	22	
J	50.2	32.2	41.2	78	1970	10	1966	738	3.33	6.40	1954	1.60	1949	3.80	1967	3.4	14.2	1966	7.3	1966	74	76	60	69	11.7	SW	56	SW	1959	58	6.1	10	6	15	10	1	*	2	0	3	17	0
F	51.0	32.2	41.6	78	1965+	8	1965	655	3.21	5.72	1956	0.86	1950	1.87	1970	1.8	7.5	1963	6.3	1963	73	75	58	67	12.0	NNE	66	SW	1965	58	6.2	8	6	14	10	1	*	3	0	1	14	0
M	57.2	38.7	48.0	85	1968+	20	1950	533	3.45	6.41	1958	1.34	1967	3.16	1958	0.6	7.9	1960	3.9	1960	72	73	53	62	12.4	SW	57	W	1963	63	6.0	9	8	14	11	*	2	2	0	*	7	0
A	68.0	47.9	58.0	97	1960	28	1964	216	3.16	5.80	1953	1.29	1963	2.40	1953	0.1	1.2	1964	1.2	1964	74	74	52	63	11.8	SW	62	N	1956	65	6.0	9	8	13	11	*	2	2	0	*	7	0
M	77.3	57.7	67.5	97	1956	36	1966+	37	3.36	7.77	1961	1.48	1965	2.94	1961	0.0	0.0		0.0		81	77	56	68	10.1	SW	53	SW	1965	67	6.0	8	11	12	10	0	6	2	0	0	0	0
J	84.9	66.3	75.6	101	1964	45	1967	0	3.61	12.83	1950	0.37	1954	6.85	1963	0.0	0.0		0.0		83	79	57	68	9.4	SW	52	W	1970	68	5.7	8	12	10	9	0	6	2	7	0	0	0
J	87.9	69.6	78.8	103	1952	57	1962	0	5.92	12.83	1950	1.69	1961	5.64	1969	0.0	0.0		0.0		85	82	60	72	8.6	SW	63	SW	1956	65	6.0	8	11	12	11	0	8	1	11	0	0	0
A	86.2	68.8	77.5	99	1968	52	1965	0	5.97	11.19	1967	2.04	1970+	11.40	1964	0.0	0.0		0.0		87	85	62	76	8.7	SW	57	NE	1957	65	5.8	8	12	11	11	0	7	2	8	0	0	0
S	80.9	64.3	72.6	98	1954	45	1967	0	4.22	12.26	1964	0.36	1958	6.79	1959	0.0	0.0		0.0		85	84	61	76	9.6	NE	73	W	1960	64	5.7	9	9	12	8	0	3	2	3	0	0	0
O	70.9	53.1	62.0	95	1954	29	1965	136	2.92	8.78	1959	0.93	1967	4.19	1959	0.0	0.0		0.0		83	84	61	76	10.4	NE	77	S	1954	62	5.2	13	6	12	7	0	2	3	*	0	0	0
N	61.0	41.8	51.4	85	1950	20	1950	408	3.05	7.01	1951	0.49	1965	3.35	1952	T	0.6	1950	0.6	1950	77	80	56	70	10.7	SW	52	SE	1962	61	5.4	11	8	11	8	0	*	2	0	0	1	0
D	51.8	33.1	42.5	77	1956+	14	1962	698	2.74	4.84	1967	1.08	1965	2.12	1958	1.5	14.7	1958	11.4	1958	73	76	58	68	10.9	SW	48	NW	1960	58	5.9	10	7	14	8	0	*	2	0	1	15	0
YR	68.9	50.5	59.7	103	JUL. 1952	8	FEB. 1965	3421	44.94	12.83	JUL. 1950	0.36	SEP. 1958	11.40	AUG. 1964	7.4	14.7	DEC. 1958	11.4	DEC. 1958	79	79	58	69	10.5	SW	78	S	OCT. 1954	63	5.8	110	106	149	114	2	37	24	31	5	57	0

To 8 compass points only.

Means and extremes above are from existing and comparable exposures. Annual extremes have been exceeded at other sites in the locality as follows: Highest temperature 105 in August 1918; lowest temperature 2 in February 1895; maximum monthly precipitation 15.61 in August 1942; minimum monthly precipitation 0.04 in October 1874; maximum monthly snowfall 18.6 in December 1892; maximum snowfall in 24 hours 17.7 in December 1892; fastest mile wind 80 W in June 1925.

NORMALS, MEANS, AND EXTREMES

Station: RICHMOND, VIRGINIA   BYRD FIELD   Standard time used: EASTERN   Latitude: 37° 30' N   Longitude: 77° 20' W   Elevation (ground): 164 feet   Year: 1970

Month	Temperature — Normal — Daily maximum	Daily minimum	Monthly	Extremes — Record highest	Year	Record lowest	Year	Normal heating degree days (Base 65°)	Precipitation — Normal total	Maximum monthly	Year	Minimum monthly	Year	Maximum in 24 hrs.	Year	Snow, Ice pellets — Mean total	Maximum monthly	Year	Maximum in 24 hrs.	Year	Relative humidity — Hour 01	Hour 07	Hour 13	Hour 19	Wind — Mean speed	Prevailing direction	Fastest mile — Speed	Direction	Year	Pct. of possible sunshine	Mean sky cover sunrise to sunset	Mean number of days — Clear	Partly cloudy	Cloudy	Precipitation .01 inch or more	Snow, Ice pellets 1.0 inch or more	Thunderstorms	Heavy fog	Temperatures Max. 90° and above	32° and below	Min. 32° and below	0° and below	Average daily solar radiation - langleys
(a)	(b)	(b)	(b)	41		41		(b)	(b)	33		33		33		33	33		33		36	36	36	36	22	15	20	20		20	25	25	25	25	33	33	33	41	41	41	41	41
J	48.3	29.0	38.7	80	1950	-12	1940	815	3.46	5.95	1962	1.08	1951	3.31	1962	5.6	28.5	1940	21.6	1940	77	81	57	69	8.1	S	40	S	1959	52	6.4	9	6	16	10	2	*	3	0	3	21	*
F	50.6	29.2	39.9	83	1932+	-10	1936	703	2.90	5.61	1944	0.98	1968	1.87	1969	3.3	17.1	1967	9.2	1947	74	79	52	63	8.6	NNE	45	SW	1951	54	6.1	9	6	13	9	1	*	2	0	1	19	*
M	59.1	36.3	47.7	93	1938	11	1960+	546	3.42	5.85	1944	0.94	1966	2.04	1942	3.0	19.7	1960	12.1	1962	74	78	48	59	9.0	W	42	SE	1952	59	6.1	8	9	14	11	1	1	2	*	*	10	0
A	70.4	45.8	58.1	96	1960	26	1964+	219	3.15	5.32	1952	0.64	1963	2.07	1952	0.1	2.0	1940	2.0	1940	75	76	46	57	8.9	S	40	NW	1954	61	6.2	7	10	13	10	*	2	2	1	0	2	0
M	79.3	54.6	67.0	100	1941	31	1956	53	3.72	7.73	1946	0.87	1965	2.30	1958	0.0	0.0		0.0		83	78	49	65	7.8	SSW	45	N	1962	65	6.2	7	11	13	10	0	6	2	3	0	*	0
J	86.8	63.4	75.1	104	1952	40	1967	0	3.75	9.24	1938	0.91	1960	4.61	1963	0.0	0.0		0.0		87	81	53	68	7.3	S	52	NW	1952	67	5.9	7	12	11	9	0	7	2	10	0	0	0
J	89.4	66.7	78.1	104	1936+	51	1965+	0	5.61	18.87	1945	0.52	1963	5.73	1969	0.0	0.0		0.0		89	85	56	72	6.8	SSW	56	NW	1955	65	6.1	7	12	12	11	0	9	2	13	0	0	0
A	86.5	65.4	76.0	102	1953	46	1934	0	5.54	14.10	1955	0.52	1943	8.79	1955	0.0	0.0		0.0		90	88	57	76	6.5	S	54	W	1964	64	6.0	7	12	12	10	0	7	3	11	0	0	0
S	81.8	58.6	70.2	103	1954	37	1963	36	3.65	8.49	1945	0.69	1954	3.82	1955	0.0	0.0		0.0		90	89	55	79	6.8	S	45	SE	1952	65	5.6	10	8	12	8	0	3	3	5	0	0	0
O	70.6	46.7	58.7	99	1941	21	1962	214	3.00	8.78	1961	0.30	1963	6.50	1961	T	T	1954	T	1954	87	89	52	77	7.0	NNE	38	SE	1954	60	5.2	12	7	12	7	0	1	4	*	0	2	0
N	59.9	37.1	48.5	86	1950	10	1933+	495	3.04	7.64	1959	0.36	1965	4.07	1956	0.5	7.3	1953	7.3	1953	80	84	50	70	7.5	S	35	S	1957	55	5.7	10	7	13	8	*	1	2	0	*	11	0
D	49.8	29.5	39.7	78	1956	-1	1942	784	2.97	6.88	1957	0.72	1965	3.16	1958	2.2	12.5	1958	7.5	1966	77	81	54	70	7.5	SW	40	SW	1968+	52	6.0	10	6	15	9	1	*	3	0	2	21	*
YR	69.4	46.9	58.1	104	JUN. 1952+	-12	JAN. 1940	3865	44.21	18.87	JUL. 1945	0.30	OCT. 1963	8.79	AUG. 1955	14.7	28.5	JAN. 1940	21.6	JAN. 1940	82	82	53	69	7.6	S	68	SE	OCT. 1954	61	6.0	103	106	156	113	4	37	29	43	6	86	1

♯ To 8 compass points only.

Means and extremes above are from existing and comparable exposures. Annual extremes have been exceeded at other sites in the locality as follows:
Highest temperature 107 in August 1918; minimum monthly precipitation 0.11 in November 1890 and earlier.

Station: ROANOKE, VIRGINIA   WOODRUM AIRPORT   Standard time used: EASTERN   Latitude: 37° 19' N   Longitude: 79° 58' W   Elevation (ground): 1149 feet   Year: 1970

Month	Temperature — Normal — Daily maximum	Daily minimum	Monthly	Extremes Ø — Record highest	Year	Record lowest	Year	Normal heating degree days (Base 65°)	Precipitation — Normal total	Maximum monthly	Year	Minimum monthly	Year	Maximum in 24 hrs.	Year	Snow, Ice pellets — Mean total	Maximum monthly	Year	Maximum in 24 hrs.	Year	Relative humidity — Hour 01	Hour 07	Hour 13	Hour 19	Wind — Mean speed	Prevailing direction	Fastest mile — Speed	Direction	Year	Pct. of possible sunshine	Mean sky cover sunrise to sunset	Mean number of days — Clear	Partly cloudy	Cloudy	Precipitation .01 inch or more	Snow, Ice pellets 1.0 inch or more	Thunderstorms	Heavy fog	Temperatures Max. 90° and above	32° and below	Min. 32° and below	0° and below	Average daily solar radiation - langleys
(a)	(b)	(b)	(b)	6		6		(b)	(b)	23		23		23		23	23		23		6	6	6	6	22	15	9	9		22	23	23	23	23	23	23	23	23	6	6	6	6	6
J	47.0	29.1	38.1	73	1967	-1	1970	834	3.12	5.20	1964	0.60	1956	2.71	1963	7.7	41.2	1966	13.7	1966	65	68	51	56	9.9	WNW	53	SW	1964		6.3	8	7	15	10	2	*	3	0	6	25	*
F	49.1	29.3	39.2	69	1967	1	1970	722	2.86	7.17	1960	0.56	1968	2.54	1954	7.1	27.6	1960	15.7	1960	60	64	46	51	10.4	SE	40	SW	1962		6.1	8	7	13	10	2	1	3	0	3	21	0
M	56.2	34.8	45.5	86	1968	14	1965	614	3.53	5.13	1953	0.43	1966	2.57	1967	4.3	30.3	1960	17.4	1960	60	66	43	47	10.6	WNW	52	SW	1967		6.1	8	9	14	11	1	1	2	0	0	15	0
A	68.2	44.6	56.4	87	1970	24	1969	261	3.10	5.15	1949	0.87	1963	1.79	1952	0.1	0.7	1959	0.7	1959	66	72	48	50	10.1	SE	58	SW	1963		6.0	7	10	13	11	0	3	1	0	0	*	0
M	78.0	53.4	65.7	93	1969	31	1960	65	3.79	8.42	1950	1.27	1951	2.97	1950	0.0	0.0		0.0		75	77	50	60	8.9	SE	46	SW	1962		6.0	7	11	13	12	0	6	2	1	0	*	0
J	85.4	61.3	73.4	97	1968	40	1966	0	3.80	6.67	1949	1.40	1966	3.23	1949	0.0	0.0		0.0		82	81	50	60	6.9	SE	46	SW	1966		5.9	7	13	10	10	0	7	1	5	0	0	0
J	88.0	65.2	76.6	100	1966	50	1965	0	4.25	7.85	1949	1.13	1957	2.72	1966	0.0	0.0		0.0		81	80	54	62	6.7	W	35	SW	1964+		6.0	6	14	11	12	0	9	1	9	0	0	0
A	86.8	64.1	75.4	95	1965	43	1965	0	4.63	9.12	1949	1.16	1965	3.35	1970	0.0	0.0		0.0		85	85	56	67	6.3	SE	37	SW	1968		5.9	8	13	10	11	0	7	2	6	0	0	0
S	80.7	57.5	69.1	94	1970+	37	1967	51	3.26	7.25	1966	0.44	1968	3.45	1959	T	1.0	1953	T	1953	84	86	52	66	6.1	SE	29	SW	1970		5.4	11	8	11	8	0	3	2	2	0	0	0
O	70.0	46.3	58.2	84	1969+	23	1969	229	3.21	8.06	1968	0.27	1963	6.41	1968	T	1.0	1957	1.0	1957	78	81	50	62	6.9	SE	35	SW	1963		4.9	14	7	10	7	*	1	2	0	0	3	0
N	57.3	36.0	46.7	79	1968	11	1970	549	2.70	6.36	1948	0.44	1960	3.00	1962	1.8	13.8	1968	10.0	1968	69	72	49	57	8.8	NW	52	SW	1963		5.7	9	6	15	9	1	*	2	0	0	10	0
D	47.5	29.2	38.4	72	1967	7	1968	825	2.98	7.10	1948	0.18	1965	3.40	1948	5.2	22.6	1966	16.4	1969	66	69	49	57	9.3	NW	40	SW	1970		6.0	9	6	13	8	1	*	3	0	3	21	0
YR	67.8	45.9	56.9	100	JUL. 1966	-1	JAN. 1970	4150	41.23	9.12	AUG. 1949	0.18	DEC. 1965	6.41	OCT. 1968	26.2	41.2	JAN. 1966	17.4	MAR. 1960	73	75	50	58	8.3	SE	58	SW	APR. 1963		5.9	104	116	145	120	7	39	25	24	12	97	*

Ø For period September 1964 through the current year.
Means and extremes above are from existing and comparable exposures. Annual extremes have been exceeded at other sites in the locality as follows:
Highest temperature 105 in July 1936; lowest temperature -12 in December 1917; maximum monthly precipitation 12.91 in August 1940; minimum monthly precipitation .16 in November 1931.

NORMALS, MEANS, AND EXTREMES

Station: WASHINGTON, D.C. **DULLES INTERNATIONAL AIRPORT** **Standard time used:** EASTERN **Latitude:** 38° 57' N **Longitude:** 77° 27' W **Elevation (ground):** 290 feet **Year:** 1970

◎ Normals have not been established for this station.

Means and extremes above are from existing and comparable exposures. Annual extremes have been exceeded at other sites in the locality as follows:
Highest temperature 106 in July 1930+; lowest temperature -15 in February 1899; maximum monthly precipitation 17.45 in September 1934; maximum snowfall 35.2 in February 1899; maximum snowfall 25.0 in January 1922.

Solar radiation data have been recorded at several locations in the vicinity of Washington. Instruments have been at the Observational Test and Development Center, Sterling, Virginia, since October 1960, elevations (m.s.l.) 276 ft. to 7-13-64 and 281 ft. thereafter.

Station: WASHINGTON, D.C. **NATIONAL AIRPORT** **Standard time used:** EASTERN **Latitude:** 38° 51' N **Longitude:** 77° 02' W **Elevation (ground):** 10 feet **Year:** 1970

REFERENCE NOTES APPLYING TO ALL "NORMALS, MEANS, AND EXTREMES" TABLES

(a) Length of record, years, based on January data, other months may be for more or fewer years if there have been breaks in the record.
   ◎ Climatological standard normals (1931-1960).
   (b) Less than one half.
   * Also on earlier dates, months, or years.
   T Trace, an amount too small to measure.

   Below zero temperatures are preceded by a minus sign.
   The prevailing direction for wind in the Normals, Means, and Extremes table is from records through 1963.
   > 70° at Alaskan stations.

& Figures instead of letters in a direction column indicate direction in tens of degrees from true North; i.e., 09 - East, 18 - South, 27 - West, 36 - North, and 00 - Calm. Resultant wind is the vector sum of wind directions and speeds divided by the number of observations. If figures appear in the direction column under 'Fastest mile' the corresponding speeds are fastest observed 1-minute values.

Unless otherwise indicated, dimensional units used in this bulletin are: temperature in degrees F; precipitation, including snowfall, in inches; wind movement in miles per hour; and relative humidity in percent. Heating degree day totals are the sums of negative departures of average daily temperatures from 65° F. Cooling degree day totals are the sums of positive departures of average daily temperatures from 65° F. Sleet was included in snowfall totals beginning with July 1948. The term "Ice pellets" includes solid grains of ice (sleet) and particles consisting of snow pellets encased in a thin layer of ice. Heavy fog reduces visibility to 1/4 mile or less.

Sky cover is expressed in a range of 0 for no clouds or obscuring phenomena to 10 for complete sky cover. The number of clear days is based on average cloudiness 0-3, partly cloudy days 4-7, and cloudy days 8-10 tenths.

Solar radiation data are the averages of direct and diffuse radiation on a horizontal surface. The langley denotes one gram calorie per square centimeter.

— 415 —

Mean Maximum Temperature (°F), January

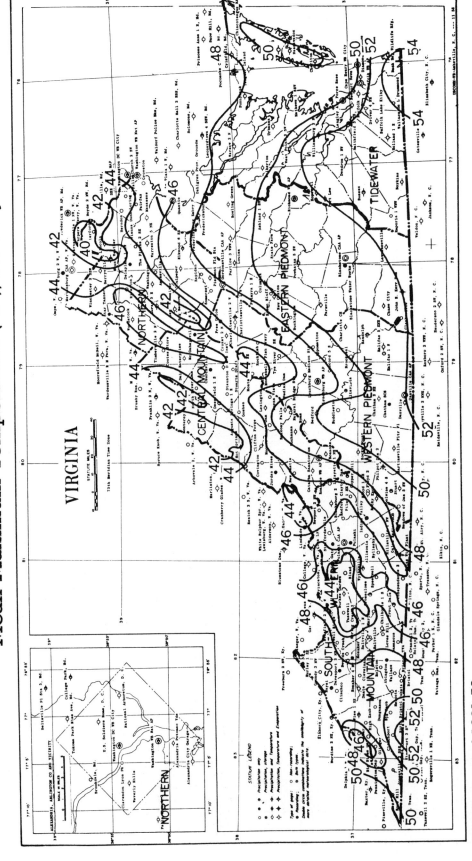

Based on period 1931-52

Isolines are drawn through points of approximately equal value. Caution should be used in interpolating on these maps, particularly in mountainous areas.

Mean Minimum Temperature (°F), January

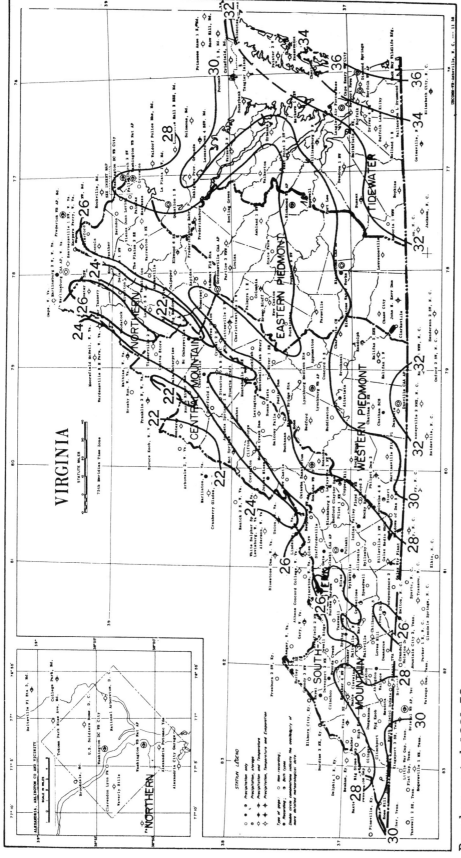

VIRGINIA

Based on period 1931-52

Isolines are drawn through points of approximately equal value. Caution should be used in interpolating on these maps, particularly in mountainous areas.

Mean Maximum Temperature (°F), July

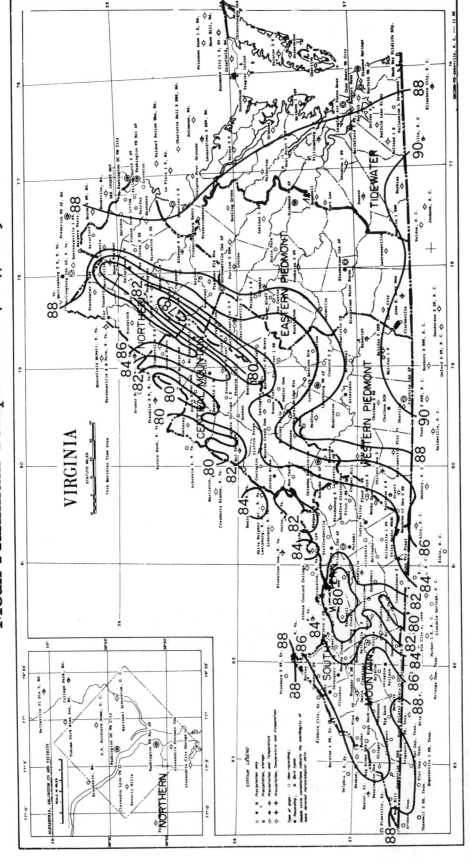

Based on period 1931-52

Isolines are drawn through points of approximately equal value. Caution should be used in interpolating on these maps, particularly in mountainous areas.

Mean Minimum Temperature (°F), July

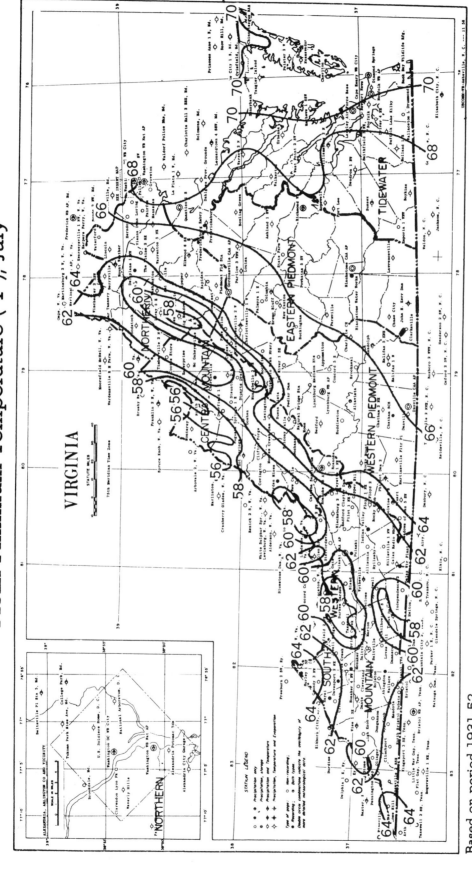

Based on period 1931-52

Isolines are drawn through points of approximately equal value. Caution should be used in interpolating on these maps, particularly in mountainous areas.

Mean Annual Precipitation, Inches

Based on period 1931-55

Isolines are drawn through points of approximately equal value. Caution should be used in interpolating on these maps, particularly in mountainous areas.

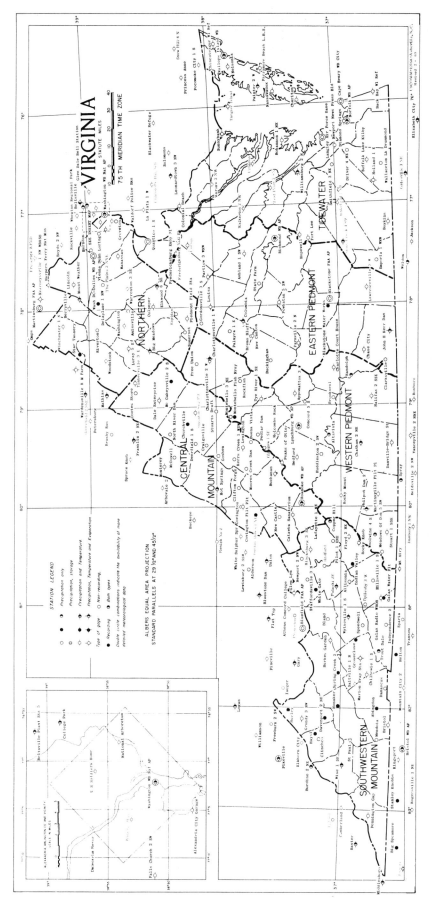

VIRGINIA

STATION LEGEND

ALBERS EQUAL AREA PROJECTION
STANDARD PARALLELS AT 29 1/2° AND 45 1/2°

75 TH MERIDIAN TIME ZONE

NORTHERN

CENTRAL MOUNTAIN

EASTERN PIEDMONT

WESTERN PIEDMONT

TIDEWATER

SOUTHWESTERN MOUNTAIN

THE CLIMATE OF
WEST VIRGINIA

by
Victor T. Horn & James K. McGuire

February 1960

TOPOGRAPHIC FEATURES--The diversity of climatic conditions in West Virginia can be understood best only with some background knowledge of the topography.

West Virginia has an area of over 24,000 square miles, and its main portion is roughly oblong in shape. From southwest to northeast, the oblong is about 200 miles in length; width averages a little over one-half the length. There are two projections: one, the Northeastern Panhandle, which juts eastward between Maryland and Virginia; the other, the Northern Panhandle, is a narrow strip stretching northward along the Ohio River between Ohio and Pennsylvania. The easternmost extremity of the State is about 150 miles from the Atlantic Ocean and the southwestern corner adjacent to Kentucky is nearly 400 miles away from the ocean. As a result, West Virginia lies beyond the immediate climatic effect of the Atlantic, and its climate is much more of the continental than it is of the maritime type. The most important aspect of this type of climate is the marked temperature contrast between summer and winter.

Furthermore, the physical configuration of the State accentuates its interior location. Excluding the Northeastern Panhandle, the State lies in the Allegheny Plateau; but becuase the Appalachian Mountains are the most pronounced feature of the eastern part of the plateau, it is more appropriate to treat the main part of the State in two parts.

The eastern third of the plateau is part of the Appalachian Mountain chain and contains the highest land in the State. Peak elevations in this area range from about 2,500 feet to 4,860 feet (above sea level) at Spruce Knob, the highest point in West Virginia. The central and western thirds of the plateau slope generally westward to the Ohio River which lies at about 550 to 650 feet above sea level. In the north and west, the Allegheny Plateau has been well cut by weather and stream erosion into rounded hills and many fertile and winding valleys. In the south, the plateau has not been eroded as much, and is characterized by flat-topped hills with precipitous slopes. The nature of the terrain and the general topography -- the eastern border of the plateau containing the highest land -- have important climatological effects that will be indicated below.

The foregoing has excluded the Northeastern Panhandle. This is marked by long ridges and valleys, oriented southwest-northeast, intersected by the winding courses of the Potomac River and its tributaries. The main stream of the Potomac with its North Branch forms the northern border of this part of the State. Summit elevations exceed 4,000 feet (above sea level), but the land in general slopes eastward away from the main ridgeline to the west and finally reaches the lowest elevation in the State of 274 feet at Harpers

Ferry. This section lies in the Atlantic Ocean drainage and is drained by the Potomac River. The remainder of the State drains into the Ohio River, whose principal subbasins from north to south are the Monongahela (which flows northward to join the Allegheny River at Pittsburgh, Pa., to form the Ohio River), Little Kanawha, Kanawha, Guyandot, and the Big Sandy. These flow in a general north to west direction from the mountain belt, across the plateau to the main stream which forms most of the State's western border.

CLIMATIC FEATURES--It has been necessary to describe West Virginia's topography in some detail because its physical features considerably modify the effects of the major climatic controls. Briefly, the State's latitudinal position (from about 37° 15' N. Lat. in the south to 40° in the north) places it in the zone of prevailing westerly winds, which are frequently interrupted by northward and southward surges of relatively warm and cold air, respectively. These atmospheric movements are accompanied by the passage of high and low-pressure areas; the latter are the large-dimension storms, known as extratropical cyclones, which are most common in the United States in the colder half-year. West Virginia lies near the average path of the extratropical cyclones that move in a general easterly direction across the United States. In the warmer half-year, the State is affected by the showers and thunderstorms that occur in the broad current of air that tends to sweep northeastward from the Gulf of Mexico.

The State has a moderately severe winter climate, accentuated and prolonged in the mountains, with frequent alternations of fair and stormy weather. Summer is marked by hot and showery weather: the heat is less pronounced in the mountains, but they are more subject to thunderstorms and have fewer clear days the year-round. Little more can be said in the way of general climatic characteristics because there are marked variations in temperature, precipitation, and the other weather elements, due to the rugged topography, occurring not only between the mountains and plateau areas but even between different parts of the same county. For example, appreciable differences exist between the bottoms and upper slopes of the numerous valleys that entrench the Allegheny Plateau.

For climatological purposes, the State has been divided into six divisions. They are: (1) Northeastern, comprising the projection into the Potomac drainage basin; (2) North Central, embracing most of the northern part of the plateau; (3) Northwestern, made up of the adjacent strip along the Ohio and the panhandle extending thence northward; (4) Southwestern, covering the remainder of the Ohio Valley and stretching back over the major portion of the southern plateau; (5) Central, which includes the main mountainous area; and (6) Southern, occupying the small remainder of the plateau and the mountain country at the lower end of the State. The exact position and area of each of these divisions are shown on the maps accompanying this article. They delineate the more important climatic zones, but cannot be taken to represent all the local differences mentioned above.

TEMPERATURE--The maps of January and July mean monthly maximum and minimum temperature illustrate the winter and summer thermal patterns. Despite several considerable differences, the maps share a common feature: there is about as much temperature contrast across the State from east to west as there is twice the distance from north to south. This condition prevails throughout the year, though it varies in magnitude with the seasons and cannot be expected to hold every day. Here the general effect of the topography is clear: locations in

the mountainous belt, regardless of their latitude, tend to have lower temperatures than those in the rest of the State. Average winter minimum temperatures range from the low 20's in the mountains of the Central and Northeastern Divisions, and in the Northern Panhandle, to near 30°F. in the extreme southern and southwestern corners of the State, while average winter maximum readings are in the middle and upper 40's, except in the mountains and in the Northern Panhandle where they are close to 40°F. In summer, maximum temperatures average over 85°F. everywhere except in the mountains, where they are 5° to 10° cooler; average minimum temperatures during this season range from the middle 50's in the mountains to the middle 60's elsewhere.

Spring and autumn mean temperatures average in the 50's, with similar geographical variations. The average date of the last freezing temperature in spring ranges from mid-April in the southwest to mid-May in the mountains; the average first occurrence of 32°F. in the fall similarly varies from late October to late September. A table accompanying this article gives more information for specific places on the occurrence of 32° and other low temperatures.

Despite what has been said about the coolness of the mountains, they can on occasion be as hot as any other part of West Virginia. Temperatures near or over 100°F. have been recorded at all observing stations in the State, up to 112° at Martinsburg in the Northeastern Division. On the other hand, very low temperatures (below -30°) have been observed only in the mountains and in the North Central Division, down to -37°F. at Lewisburg. Of course, these are extremes, and do not represent usual winter conditions. Cold waves, with near or subzero temperatures, come on an average of three times a winter, but as a rule do not last more than 2 or 3 days.

HUMIDITY AND FOG CONDITIONS--Because of the varied topography and associated differences in local climates, it is difficult to generalize about the humidity conditions over the State. Relative humidity averages from the Weather Bureau Office at Parkersburg may be taken as representative of conditions in the Ohio Valley and the western part of the plateau. At this location, nighttime and early morning relative humidity averages about 80 percent, being somewhat less in spring (near 74 percent) and higher in late summer and autumn (about 84 percent). The maximum in late summer and autumn is associated with the occasional occurrence of nocturnal and morning fogs in the river bottoms at this time of year. Midday values are moderate, about 50 to 60 percent for all months, so that there is usually a sharp decrease in the relative humidity from sunrise to noon. Only infrequently will there occur a spell of oppressively hot, muggy weather in the summer, lasting as long as 2 weeks or more, when a steady flow of warm, humid air from the Gulf of Mexico is pumped northward, induced by a more-or-less stationary high pressure center off the Southeastern Coast.

At Charleston, in the southern part of the State, the midday humidities average practically the same as those at Parkersburg. The morning values are about the same in winter and spring, but average higher in summer and fall when there is a higher frequency of fog conditions.

At Elkins, which is representative of the high valleys of the central mountain area, 1 a.m. and 7 a.m. relative humidity averages are quite high (80 to 95 percent), reflecting valley fog conditions in the early morning hours. Since these values are accompanied by moderate air temperatures, the high humidities cause comparatively little discomfort. During the rest of the day, humidities are generally at comfortable levels.

At Petersburg, on the eastern side of the main Appalachian ridge, relative humidities in the morning and midday average somewhat lower than at Elkins for all seasons of the year.

Fog conditions over the State are complicated as to their causes and distribution. The valley fogs, just mentioned, are usually of the radiation type, and occur characteristically when a high-pressure area is centered over or near the State. This situation is most common in late summer and fall. Low cloudiness and fog in the mountains are generally orographic in nature, that is, the result of moist winds moving upslope, so that there is usually a great difference in cloud and fog conditions on opposite sides of a ridge.

PRECIPITATION (INCLUDING SNOWFALL)--The map of mean annual precipitation exhibits some interesting features. It will be noticed that yearly amounts average the greatest in the Central Division -- in excess of 50 (and even 60) inches. West of this belt of heavy precipitation, amounts decrease to about 40 inches along the Ohio River. East of it, there is a much more abrupt decrease to close to 30 inches in the western part of the Northeastern Division, with an increase to about 40 inches in the extreme eastern tip of the State.

This pattern can be directly related to the fact that the rain and snow-producing atmospheric currents generally move across West Virginia on an eastward course. As they approach the mountains, these air currents are subject to orographic lifting, which acts to "trigger" potential precipitation or to intensify the rain or snow that may already be falling. As a result, average annual precipitation increases from the Ohio eastward to the Appalachians. On the other side of the mountains, there is the well-marked "rain shadow" where the air currents descend the leeward slopes and precipitation is correspondingly reduced, to increase only when more favorable topographic influences are encountered farther eastward and where the influence of the ocean and coastal storms is more pronounced.

Mean annual snowfall exhibits the same features, but to a more remarkable degree. The mountain belt receives over 60 inches of snow a year, on the average. Pickens, at an elevation of 2,700 feet (above sea level), located near the middle of the western boundary of the Central Division, had an average annual snowfall of 115 inches for a recent period of 14 years. Amounts over 20 inches have been experienced everywhere else, except in that part of the State west of longitude 81° 30' W. which usually receives about 15 inches. The Northeastern Division averages about 20 to 30 inches yearly; much of this occurs with the coastal storms. These are very heavy producers of snow and occasionally strike this portion, but only infrequently affect the area farther inland.

It is very unusual for a relatively small and compact area the size of West Virginia to exhibit such great differences in snowfall. From Charleston to Pickens there is a sevenfold increase in average annual snowfall over an airline distance of only 75 miles. Furthermore, the heavy snowfall at elevations under 5,000 feet (above sea level) is unusual here in the East, for an area located south of 40° north latitude.

In winter, roads may be blockaded by heavy falls of snow, particularly in the mountain country. The snow, as a general rule, does not remain on the ground for extended periods over most of the State. Except in the higher portions of the plateau and in the mountains themselves, the snow cover does not persist for anything like the duration of the winter. In other words, the snowstorms are usually followed by thawing periods and there is no large-scale melting in the spring of a seasonally accumulated snowpack.

SUNSHINE AND CLOUDINESS--West Virginia lies in a cloudy belt. Percentage of possible sunshine is only about 40 in winter, increasing to somewhat over 60 percent in early autumn. Cloudiness is most pronounced over the mountains. The average annual number of clear days ranges from about 80 in the mountains to about 120 in the western portion. Conversely, cloudy days average fewest (about 140) in the west and increase by 10 to 20 percent in the mountain belt. In addition to cloudiness, the hours of sunshine are reduced by fog, particularly in the river valleys.

WINDS AND STORMS--As stated previously, the prevailing winds blow from westerly directions. There is a tendency outside of the mountain belt for southerly or southwesterly winds during summer and fall. Thunderstorms occur on an average of 40 to 50 days per year, being more frequent in the mountains. June and July are the months of most frequent occurrence. Violent local winds accompanying thunderstorms are experienced every year in some part of the State, but tornadoes are rare. In the 43 years ending with 1958, a total of 13 tornadoes struck the State; almost all the deaths and destruction recorded from such storms during this period were due to one very severe tornado that struck Shinnston and nearby towns on June 23, 1944; all the other tornadoes were comparatively minor. The most outstanding hailstorm reported in the State caused losses of $200,000 to building and crops in the northern part of Preston County on July 18, 1926. The climatological records show that destructive hailstorms occur on an average of about three per year in West Virginia. Hailstorms are most serious in their economic effects on the fruit growing areas of the Northeastern Panhandle and, to a lesser extent, on the burley tobacco growing areas of the southwestern part of the State.

Though hurricanes have damaged the State, principally as a result of heavy rains, it is uncommon for this type of storm to strike West Virginia with full force. The remnants of the hurricanes which have affected the State have been more noted for their accompanying heavy rainfalls than for any high winds produced. In the Northeastern Panhandle, there have been sizeable losses from fruit drop caused by winds accompanying the passage of a hurricane, but such losses were due more to the circumstance that the fruits were at the stage of development when droppage is likely to occur rather than to any unusually high intensity of the wind.

Much more frequent and costly is the damage from intense large-area storms -- that is to say, from exceptionally strong specimens of the ordinary LOWS that affect the State quite frequently during the colder half of the year. The great storm of November 1950 is an example of this sort. Such storms produce high winds and heavy rain or snow; they paralyze commercial and agricultural activities and cause widespread major damage with deaths and injuries. Under proper conditions, they lead to flooding and damage to the river towns.

Warm-season thunderstorms, mostly those of June and July, often yield intense local rainfall and cause flash flooding in the narrow valleys that cut through the plateau and mountain districts. Greatest precipitation amounts recorded in 24 hours or less at officially recognized precipitation-measuring stations have exceeded 5.00 inches in all six climatological divisions and have exceeded 6.00 inches in divisions 1, 2, and 3 (Northeastern,

North Central, and Northwestern); amounts in excess of 10.0 inches (in 24 hours or less) have been accepted for locations in those same three divisions. Perhaps the outstanding example of intense local rainfall due to thunderstorm activity was the occurrence of a deluge of 19.0 inches in 2 hours and 10 minutes at Rockport in Wood County on July 18, 1889. More recently, 31 fatalities and damage exceeding $10 million resulted from flash flooding caused by heavy thunderstorm rainfall (amounts up to 14 inches) on the night of June 24-25, 1950, in parts of Doddridge, Gilmer, Harrison, Lewis, Pleasants, Ritchie, Tyler, and Upshur Counties. The Petersburg-Moorefield area was hard hit by flash floods on the night of June 17-18, 1949, when up to 12 inches of rain fell in 24 hours. The climatological records for the past quarter century show that this kind of severe local flood, caused by heavy thunderstorm rainfall, is likely to occur in some part of the State every year.

In contrast to flash flooding on the smaller streams due principally to heavy local thundershowers in the warm season, flooding in the larger streams is almost exclusively a cold season phenomenon. Of the 58 floods recorded on the Ohio River in the Parkersburg area since 1832, 54 have occurred during the months from December to April, inclusive. The ideal setup for the cold season floods requires the soil to be well saturated from previous rains, a good snow cover, and a more-or-less stationary front lying northeast-southwest across the State. Along this front separating two contrasting air masses, a succession of "waves" may move northeastward, resulting in copious warm rains for a period of at least several days and a rapid melting of the snow cover. Hoyt and Langbein point out that the Ohio River basin is unique in relation to storm tracks across the United States in that it lies directly in the path of many of the large-scale cyclonic storms which, in the cold half-year and under the conditions just outlined, may bring about the interaction of polar and tropical air masses and consequent excessive and prolonged rainfall simultaneously with the melting of any snow cover present. The Potomac Basin is also subject to winter floods, but they are generally of lesser magnitude than those on the Ohio.

The Ohio River exceeds flood stage more frequently than any of its tributary streams, but severe overflow is infrequent. Since the turn of the century, severe and extensive overflow along the Ohio occurred in March 1913 and 1936 and January 1937. Disastrous floods occurred in the Big Sandy and Guyandot River basins in January 1957. Some other notable flood years in tributary basins have been 1901, 1912, 1916, 1917, 1918, 1926, 1929, 1932, 1935, and 1940.

ECONOMIC ASPECTS--There are several ways in which the State's climate may be related to the activities of its citizens. The farm population, 460,000, according to the 1950 census, represents about 23 percent of the total population. There are about 68,000 farms with an average of 107 acres. In 1957, poultry raising accounted for 28 percent of the total cash receipts from farm sales. The two other most important types of agriculture were the raising of cattle, sheep, and hogs (29 percent of the cash receipts) and dairy farming (23 percent). Other activities are fruit growing, the cultivation of field crops, lumbering, and raising greenhouse and nursery products.

All these agricultural activities are dependent, to a greater or lesser degree, on the weather and climate. For example, broiler and turkey production is a major activity in portions of the Northeastern Panhandle where the yearly extremes of heat and cold are not severe. The important commercial fruit-growing business in the Northeastern Panhandle is favored by the combination of relatively cool winters and frost-free conditions on the higher slopes in spring.

There are 10 million acres of forest land in West Virginia, or 65 percent of the total land area, of which approximately 1 million acres are owned by the Federal and State governments. About one-third of the remainder consists of farm woodlots. The rest is held for nonagricultural and industrial purposes. The forests are predominantly hardwood, with coniferous or softwood spruce occupying only about 3 percent of the total wooded area. The moderate climate and abundant rainfall help to account for the rapid growth and healthy development of the hardwood trees.

In recent years, many varied kinds of manufacturing activities have been attracted to West Virginia. In numerous phases of their operations, they rely upon an ample water supply in the State's principal streams which maintain an adequate flow owing to the abundant and generally dependable precipitation. Furthermore, the Ohio River is a major commercial artery, not only for West Virginia, but also for the neighboring States, and its status as such owes much to the rain and snow that fall in the West Virginia headwaters. Worker efficiency is promoted by the climate in that it is characterized by weather changes that stimulate bodily well-being, without being so severe as to strain the physique. Also in recent years, more and more summer vacationists and weekend visitors from nearby States have been attracted by the temperatures that prevail in the West Virginia mountains, especially at night. The post-war years have witnessed a general upsurge in winter sports, and ski-slope developers have taken advantage of favorable snow conditions in the Beckley, Davis, Morgantown, and Terra Alta sections which have some of the few such installations south of the Mason and Dixon line.

All in all, the climate of West Virginia may be summarized as favorable to human activity, with occasional periods in summer and winter that are extreme but rarely prolonged. The State is usually favored by ample precipitation, though by the same token subject to considerable cloudiness; is strongly influenced by its geographical position and topographic features; and is marked by a diversity of local climates the most striking of which is that of the mountain belt.

REFERENCES

(1) Flora, S. D., <u>Tornadoes of the United States</u>, 2nd rev. ed., University of Oklahoma Press, 1954.

(2) <u>Hailstorms of the United States</u>, University of Oklahoma Press, 1956.

(3) Hoyt, W. G. and Langbein, W. B., <u>Floods</u>, Princeton University Press, 1955.

(4) U. S. Department of Agriculture, <u>Climate and Man, The Yearbook of Agriculture for 1941</u>, Government Printing Office, 1942 (esp. pp. 1182-1190 on the Climate of West Virginia).

(5) West Virginia, Dept. of Agriculture, <u>West Virginia Agricultural Statistics</u>, 1956, Charleston, W. Va., 1955.

(6) West Virginia University, Agricultural Extension Service, <u>Cash Receipts from Farm Sales in West Virginia 1956 and 1957</u>, Morgantown, W. Va., 1958.

(7) West Virginia, Geological and Economic Survey, <u>Natural Resources of West Virginia</u>, Morgantown, W. Va., 1952.

(8) West Virginia, Industrial and Publicity Commission, <u>West Virginia</u>, Charleston, W. Va., n.d.

(9) West Virginia, State Planning Board, <u>Water Resources of West Virginia</u>, Morgantown, W. Va., December 1937.

(10) Weather Bureau Technical Paper No. 15 - Maximum Station Precipitation for 1, 2, 3, 6, 12, and 24 Hours.

(11) Weather Bureau Technical Paper No. 16 - Maximum 24-Hour Precipitation in the United States. Washington, D. C. 1952.

(12) Weather Bureau Technical Paper No. 25 - Rainfall Intensity-Duration-Frequency Curves. For selected stations in the United States, Alaska, Hawaiian Islands, and Puerto Rico.

(13) Weather Bureau Technical Paper No. 29 - Rainfall Intensity-Frequency Regime. Washington, D. C.

BIBLIOGRAPHY

(A) Climatic Summary of the United States (Bulletin W) 1930 edition, Sections 72 and 73. U. S. Weather Bureau

(B) Climatic Summary of the United States, West Virginia-Supplement for 1931 through 1952 (Bulletin W Supplement). U. S. Weather Bureau

(C) Climatological Data - West Virginia. U. S. Weather Bureau

(D) Climatological Data National Summary. U. S. Weather Bureau

(E) Hourly Precipitation Data - West Virginia. U. S. Weather Bureau

(F) Local Climatological Data, U. S. Weather Bureau for Charleston, Elkins, Huntington and Parkersburg, West Virginia.

STATION	Freeze threshold temperature	Mean date of last Spring occurrence	Mean date of first Fall occurrence	Mean No. of days between dates	Years of record Spring	No. of occurrences in Spring	Years of record Fall	No. of occurrences in Fall
BAYARD	32	05-30	09-14	107	29	29	30	30
	28	05-18	09-28	133	28	28	30	30
	24	04-29	10-09	163	28	28	30	30
	20	04-17	10-21	188	28	28	30	30
	16	03-29	11-08	223	28	28	30	30
BENSON	32	05-15	10-05	143	26	26	27	27
	28	05-01	10-14	166	26	26	27	27
	24	04-19	10-27	190	26	26	27	27
	20	04-05	11-06	214	26	26	27	27
	16	03-20	11-22	248	26	26	27	27
BENS RUN	32	04-21	10-27	189	30	30	30	30
	28	04-05	11-07	217	30	30	30	30
	24	03-22	11-17	240	29	29	30	30
	20	03-08	11-30	267	29	29	30	30
	16	03-05	12-09	279	29	29	30	27
BLUEFIELD 1	32	04-24	10-15	174	29	29	30	30
	28	04-10	10-28	201	29	29	30	30
	24	03-29	11-11	227	29	29	30	30
	20	03-15	11-24	253	29	29	30	30
	16	03-03	12-05	277	29	29	30	28
BUCKHANNON	32	05-06	10-10	157	30	30	30	30
	28	04-17	10-21	187	30	30	30	30
	24	04-03	11-03	214	30	30	30	30
	20	03-16	11-20	249	30	30	30	30
	16	03-09	12-02	268	30	30	30	29
CAIRO 3 S	32	05-02	10-09	160	26	26	26	26
	28	04-22	10-17	177	26	26	26	26
	24	04-10	10-31	204	25	25	26	26
	20	03-22	11-13	236	25	25	26	26
	16	03-11	11-25	259	24	24	26	26
CHARLESTON 1	32	04-18	10-28	193	28	28	27	27
	28	04-01	11-09	222	28	28	27	27
	24	03-19	11-22	248	27	27	27	26
	20	03-09	12-03	270	27	27	27	25
	16	03-02	12-16	289	27	27	27	21
CLARKSBURG 1	32	05-02	10-12	163	28	28	28	28
	28	04-20	10-24	188	28	28	28	28
	24	04-08	11-03	210	28	28	28	28
	20	03-22	11-15	238	28	28	28	28
	16	03-11	11-28	262	28	28	28	28
ELKINS	32	05-10	10-07	150	30	30	30	30
	28	04-26	10-18	175	30	30	30	30
	24	04-11	10-30	202	30	30	30	30
	20	03-27	11-13	231	30	30	30	30
	16	03-16	11-27	256	29	29	30	30
FAIRMONT	32	04-28	10-15	169	30	30	30	30
	28	04-15	10-29	196	30	30	30	30
	24	03-31	11-09	223	30	30	30	30
	20	03-15	11-24	254	30	30	30	30
	16	03-07	12-05	273	30	30	29	29
FLAT TOP	32	05-09	10-05	149	12	12	13	13
	28	04-26	10-21	178	12	12	13	13
	24	04-09	11-02	206	12	12	13	13
	20	04-01	11-14	227	12	12	13	13
	16	03-24	11-25	246	12	12	13	13
GARY	32	04-26	10-17	174	30	30	30	30
	28	04-16	10-29	196	30	30	30	30
	24	03-30	11-08	223	30	30	30	30
	20	03-13	11-23	255	30	30	30	29
	16	03-02	12-07	280	30	30	30	27
HASTINGS	32	04-30	10-16	169	16	16	16	16
	28	04-16	10-31	198	16	16	16	16
	24	04-04	11-11	221	16	16	16	16
	20	03-16	11-18	247	16	16	16	16
	16	03-12	12-02	265	16	16	16	15
HUNTINGTON 1	32	04-14	10-27	196	29	29	28	28
	28	03-28	11-09	226	29	29	28	28
	24	03-13	11-23	254	29	29	28	28
	20	03-06	12-03	272	29	29	28	26
	16	02-20	12-11	294	29	27	28	23
LOGAN	32	04-19	11-01	195	12	12	13	13
	28	04-04	11-09	219	12	12	13	13
	24	03-17	11-28	256	12	12	13	13
	20	03-08	12-08	275	12	12	13	13
	16	02-24	12-15	294	12	11	13	11
LONDON LOCKS	32	04-18	11-01	197	14	14	13	13
	28	04-05	11-11	219	14	14	12	12
	24	03-20	11-24	249	14	14	12	12
	20	03-12	12-06	268	14	14	12	11
	16	03-04	12-07	277	14	14	12	11
MADISON	32	04-24	10-18	176	17	17	18	18
	28	04-10	10-30	203	17	17	18	18
	24	03-25	11-09	230	17	17	18	18
	20	03-12	11-24	258	17	17	17	17
	16	03-03	12-01	273	17	17	17	17
MANNINGTON 1 N	32	05-10	10-07	150	30	30	30	30
	28	04-26	10-16	173	30	30	30	30
	24	04-14	10-28	197	30	30	30	30
	20	03-27	11-12	230	30	30	30	30
	16	03-17	11-26	253	30	30	30	30
MARTINSBURG CAA AP	32	04-23	10-17	177	29	29	30	30
	28	04-08	11-03	209	29	29	29	29
	24	03-22	11-15	238	29	29	29	29
	20	03-10	11-29	264	29	29	29	29
	16	03-01	12-10	284	29	29	29	25
MOOREFIELD MCNEILL	32	05-14	10-01	140	28	28	28	28
	28	04-30	10-11	164	28	28	27	27
	24	04-17	10-20	186	28	28	26	26
	20	04-02	10-31	212	28	28	26	26
	16	03-17	11-17	245	28	28	26	26
NEW CUMBERLAND DAM 9	32	05-04	10-17	166	30	30	30	30
	28	04-22	10-31	192	30	30	30	30
	24	04-10	11-13	217	30	30	30	30
	20	03-23	11-26	248	30	30	30	30
	16	03-11	12-04	268	30	30	30	28
NEW MARTINSVILLE	32	05-02	10-15	166	29	29	29	29
	28	04-19	10-30	194	29	29	29	29
	24	04-01	11-09	223	29	29	29	29
	20	03-18	11-21	248	28	28	29	29
	16	03-07	11-30	268	28	28	28	28
PARKERSBURG	32	04-16	10-21	189	30	30	30	30
	28	04-02	11-08	220	30	30	30	30
	24	03-17	11-22	250	30	30	30	30
	20	03-08	12-01	268	30	30	30	29
	16	03-01	12-12	286	30	30	30	26
PARSONS	32	05-08	10-11	156	27	27	26	26
	28	04-26	10-21	178	27	27	26	26
	24	04-14	11-02	202	27	27	26	26
	20	03-27	11-13	231	25	25	25	25
	16	03-15	11-24	255	25	25	25	25
PETERSBURG WB CITY	32	04-30	10-05	158	12	12	12	12
	28	04-19	10-19	184	12	12	12	12
	24	04-03	11-03	213	12	12	12	12
	20	03-18	11-16	243	12	12	12	12
	16	03-06	12-03	272	12	12	12	12
PIEDMONT	32	05-01	10-10	162	29	29	30	30
	28	04-15	10-25	192	29	29	30	30
	24	03-28	11-08	226	29	29	30	30
	20	03-13	11-25	257	29	29	30	30
	16	03-05	12-08	277	29	29	30	30
PT PLEASANT	32	04-23	10-22	182	29	29	29	29
	28	04-07	11-05	212	29	29	28	28
	24	03-23	11-15	238	29	29	28	28
	20	03-11	11-24	258	29	29	27	27
	16	03-01	12-07	281	29	28	27	24
RAINELLE	32	05-20	10-01	134	20	20	21	21
	28	04-30	10-11	164	20	20	21	21
	24	04-22	10-19	180	20	20	21	21
	20	04-07	11-01	208	20	20	21	21
	16	03-19	11-14	241	20	20	21	21

FREEZE DATA

STATION	Freeze threshold temperature	Mean date of last Spring occurrence	Mean date of first Fall occurrence	Mean No. of days between dates	Years of record Spring	No. of occurrences in Spring	Years of record Fall	No. of occurrences in Fall	STATION	Freeze threshold temperature	Mean date of last Spring occurrence	Mean date of first Fall occurrence	Mean No. of days between dates	Years of record Spring	No. of occurrences in Spring	Years of record Fall	No. of occurrences in Fall
RAVENSWOOD DAM 22	32	04-28	10-12	167	27	27	28	28	TERRA ALTA 1	32	05-27	09-24	120	24	24	23	23
	28	04-15	10-25	194	27	27	28	28		28	05-15	10-06	144	23	23	22	22
	24	04-01	11-07	220	27	27	28	28		24	04-24	10-21	180	23	23	21	21
	20	03-16	11-18	247	27	27	28	28		20	04-10	11-03	207	22	22	21	21
	16	03-08	12-02	269	25	26	27	26		16	03-27	11-16	235	22	22	20	20
RICHWOOD	32	05-08	10-05	150	19	19	17	17	WARDENSVILLE R M FRM	32	05-12	10-02	143	25	25	27	27
	28	04-26	10-18	175	19	19	16	16		28	04-27	10-10	167	24	24	27	27
	24	04-14	10-28	197	19	19	16	16		24	04-15	10-24	192	24	24	25	25
	20	04-03	11-16	227	19	19	15	15		20	03-27	11-11	229	24	24	24	24
	16	03-19	11-24	250	18	18	15	15		16	03-13	11-23	255	23	23	23	23
ROMNEY	32	05-06	10-05	152	28	28	29	29	WELLSBURG 3 NE	32	05-13	10-04	143	30	30	30	30
	28	04-23	10-16	177	28	28	28	28		28	04-29	10-17	172	30	30	30	30
	24	04-04	10-30	209	28	28	28	28		24	04-19	10-31	196	30	30	30	30
	20	03-17	11-16	244	28	28	28	28		20	04-02	11-13	225	30	30	30	30
	16	03-08	11-30	268	28	28	28	28		16	03-15	11-26	256	30	30	30	30
SPENCER	32	04-30	10-12	165	29	29	29	29	WHEELING WARWD D 12	32	05-05	10-16	164	30	30	30	30
	28	04-14	10-24	193	28	28	29	29		28	04-16	11-04	202	29	29	30	30
	24	04-01	11-06	219	28	28	29	29		24	04-03	11-15	226	29	29	30	30
	20	03-17	11-18	246	28	28	29	29		20	03-20	11-25	250	29	29	30	30
	16	03-07	11-30	268	28	28	29	28		16	03-08	12-07	274	29	29	30	28
SUTTON 3 SE	32	05-02	10-12	163	29	29	30	30	WHITE SULPHUR SPRGS	32	05-11	10-05	147	30	30	29	29
	28	04-18	10-27	192	28	28	29	29		28	04-27	10-14	170	28	28	29	29
	24	04-04	11-04	214	28	28	29	29		24	04-14	10-22	192	27	27	29	29
	20	03-18	11-16	242	28	28	29	29		20	03-30	11-04	219	26	26	29	29
	16	03-11	12-02	266	28	28	29	27		16	03-15	11-19	249	26	26	29	29
									WILLIAMSON	32	04-11	10-29	201	29	29	28	28
										28	03-26	11-10	228	29	29	28	28
										24	03-11	11-27	262	29	29	28	26
										20	03-03	12-06	278	29	29	28	24
										16	02-20	12-13	297	29	28	27	20
									WINFIELD LOCKS	32	04-19	10-24	187	11	11	12	12
										28	04-05	11-12	221	11	11	12	12
										24	03-17	11-24	251	11	11	12	12
										20	03-12	12-04	267	11	11	12	12
										16	03-10	12-10	275	11	11	11	10

Data in the above table are based on the period 1921-1950, or that portion of this period for which data are available.

Means have been adjusted to take into account years of non-occurrence.

A freeze is a numerical substitute for the former term "killing frost" and is the occurrence of a minimum temperature at or below the threshold temperature of 32°, 28°, etc.

Freeze data tabulations in greater detail are available and can be reproduced at cost.

*MEAN TEMPERATURE AND PRECIPITATION

STATION	JANUARY Temperature	JANUARY Precipitation	FEBRUARY Temperature	FEBRUARY Precipitation	MARCH Temperature	MARCH Precipitation	APRIL Temperature	APRIL Precipitation	MAY Temperature	MAY Precipitation	JUNE Temperature	JUNE Precipitation	JULY Temperature	JULY Precipitation	AUGUST Temperature	AUGUST Precipitation	SEPTEMBER Temperature	SEPTEMBER Precipitation	OCTOBER Temperature	OCTOBER Precipitation	NOVEMBER Temperature	NOVEMBER Precipitation	DECEMBER Temperature	DECEMBER Precipitation	ANNUAL Temperature	ANNUAL Precipitation
NORTHWESTERN																										
BELLEVILLE DAM 20		3.60		2.83		4.07		3.63		3.96		4.25		4.24		4.02		2.91		2.14		2.67		2.95		41.27
BENS RUN	35.0	3.65	35.5	2.67	43.1	3.94	53.6	3.59	63.9	4.38	73.3	4.93	76.7	4.25	75.0	4.65	68.6	2.90	57.0	2.35	44.2	2.84	35.2	3.25	55.1	43.40
CAIRO 3 S	35.3	3.91	35.8	2.92	43.4	4.27	53.6	3.67	63.0	4.21	71.5	4.90	74.4	4.78	73.2	4.12	66.9	3.42	56.6	2.58	44.0	3.03	35.0	3.08	54.4	44.89
CRESTON	35.1	4.05	35.4	3.11	42.8	4.43	52.9	3.52	62.8	4.08	71.0	4.91	74.6	4.32	73.3	4.43	67.2	2.96	56.0	2.57	43.5	2.98	35.0	3.30	54.1	44.66
MCMECHEN DAM 13		3.36		2.57		3.88		3.39		3.99		4.12		4.15		3.93		2.90		2.37		2.66		2.66		39.08
NEW CUMBERLAND DAM 9	32.6	3.06	32.5	2.21	40.3	3.60	51.0	3.42	61.7	3.56	70.9	4.10	74.4	3.71	72.8	3.63	66.6	2.98	55.5	2.61	43.1	2.59	33.6	2.44	52.9	37.91
NEW MARTINSVILLE	34.7	3.73	35.0	2.92	42.8	4.05	53.2	3.60	63.8	4.21	72.9	4.54	76.1	4.47	74.6	4.57	68.0	3.10	56.8	2.52	44.1	2.91	34.8	3.40	54.7	44.02
PARKERSBURG WB CITY	34.4	3.17	35.5	2.65	43.6	3.54	53.8	3.08	63.5	3.50	72.4	4.18	75.7	4.16	74.0	4.15	68.4	2.99	56.9	2.12	45.0	2.67	36.1	2.90	54.9	39.11
WASHINGTON DAM 19		3.40		2.65		3.77		3.34		3.62		3.97		3.93		3.98		2.82		1.91		2.45		2.65		38.49
WELLSBURG 3 NE	32.6	3.10	32.6	2.32	40.6	3.91	50.8	3.72	61.0	3.85	70.2	4.25	73.4	4.42	71.9	3.94	65.4	3.38	53.9	2.67	42.0	2.75	33.1	2.52	52.3	40.83
WHEELING WARWOOD DAM 1	32.5	3.10	32.1	2.45	39.7	3.73	50.9	3.40	61.8	3.95	71.3	4.55	75.0	3.71	73.4	3.51	66.8	3.07	55.4	2.30	42.8	2.64	33.3	2.54	52.9	38.95
DIVISION	34.1	3.50	34.3	2.66	41.9	3.93	52.5	3.51	62.7	3.97	71.7	4.54	75.0	4.19	73.5	4.11	67.2	3.07	56.0	2.44	43.6	2.73	34.4	2.90	53.9	41.55
NORTH CENTRAL																										
ABERDEEN		3.97		3.22		4.30		3.96		4.42		4.99		5.30		5.17		3.56		3.03		3.08		3.48		48.48
BENSON	35.1	4.18	35.2	3.40	42.2	4.43	52.5	3.83	62.2	4.27	70.6	5.05	73.8	4.90	72.2	4.35	65.9	3.52	55.3	2.85	43.1	3.12	34.9	3.62	53.6	47.52
BUCKHANNON 2 W	35.1	4.16	34.8	3.46	41.9	4.80	52.1	3.87	61.1	4.51	69.2	5.64	72.0	5.57	70.5	4.93	64.4	3.46	54.3	3.15	43.0	3.25	35.0	3.67	52.8	50.47
CLARKSBURG 1	33.0	3.47	32.7	2.78	40.2	3.58	51.1	3.36	61.6	3.97	70.5	4.40	73.6	4.43	72.0	4.49	65.3	3.28	53.8	2.52	41.5	2.65	33.1	2.89	52.4	41.82
CRAWFORD		3.82		3.10		4.43		3.62		4.41		4.98		5.66		4.88		3.33		2.98		2.99		3.34		47.54
FAIRMONT	34.3	3.59	34.3	2.77	41.9	3.84	53.2	3.52	63.3	4.12	71.8	4.63	75.0	4.39	73.4	4.26	67.1	3.04	55.9	2.61	43.6	2.79	34.6	3.07	54.0	42.63
GLENVILLE	36.8	4.25	37.4	3.49	44.7	4.43	55.0	3.76	64.4	4.37	72.9	5.21	76.0	5.24	74.7	4.77	68.6	3.54	57.4	2.88	45.1	3.29	36.5	3.49	55.8	48.72
GRAFTON 1 NE	34.9	3.89	34.9	2.89	42.1	4.36	51.8	3.79	61.4	4.51	69.6	4.80	72.8	5.01	71.4	4.77	65.6	3.26	55.2	2.99	43.4	3.10	34.9	3.42	53.2	46.79
HORNER		4.06		2.99		4.08		3.65		4.19		5.50		5.11		4.52		3.54		3.04		3.10		3.41		47.19
HOULT LOCK 15		3.78		2.90		3.97		3.60		4.18		4.86		4.29		4.35		3.04		2.61		2.84		3.01		43.43
JANE LEW		3.83		2.94		4.13		3.74		4.33		4.97		5.03		4.74		3.52		2.85		3.08		3.20		46.36
LAKE LYNN		3.28		2.40		3.78		3.51		4.07		4.43		4.08		4.46		3.02		2.65		2.56		2.60		40.84
MANNINGTON 1 N	33.7	4.03	34.0	3.07	41.4	4.45	51.5	4.10	61.3	4.59	69.6	5.24	72.5	4.97	71.0	4.54	64.4	3.29	54.1	2.74	42.3	3.11	33.6	3.43	52.4	47.56
MIDDLEBOURNE 2 ESE		3.41		2.54		3.78		3.62		4.27		4.75		4.73		4.59		3.21		2.51		2.79		2.91		43.11
MORGANTOWN LOCK AND DA		3.58		2.68		4.00		3.64		4.09		4.50		4.08		4.32		3.04		2.67		2.83		2.87		42.30
PHILIPPI		4.02		3.22		4.26		3.66		4.45		4.47		5.14		4.90		3.46		3.18		3.17		3.35		47.28
ROANOKE		3.90		3.11		4.23		3.50		4.23		5.05		4.87		4.72		3.39		3.00		3.08		3.38		46.46
VANDALIA		4.08		3.31		4.37		3.69		4.37		5.08		5.23		4.85		3.46		2.95		3.10		3.47		47.96
WESTON	36.6	4.43	36.5	3.54	43.7	4.76	54.0	3.85	63.8	4.54	72.6	5.45	75.4	5.45	74.0	4.73	67.9	3.70	57.2	3.03	45.0	3.35	36.2	3.78	55.2	50.61
DIVISION	35.1	3.95	35.2	3.11	42.5	4.27	52.8	3.72	62.6	4.30	71.1	4.92	74.1	4.82	72.7	4.58	66.4	3.33	55.8	2.79	43.6	3.04	35.1	3.34	53.9	46.17
SOUTHWESTERN																										
CHARLESTON WB AP	36.4	3.99	38.2	3.50	44.9	4.16	55.0	3.74	63.7	3.78	72.0	3.93	75.4	5.45	73.6	4.55	68.6	2.94	57.4	2.81	45.8	3.17	38.1	2.98	55.8	45.00
CHARLESTON 1		3.80		3.13		4.31		3.29		3.74		3.75		5.39		3.95		2.88		2.37		2.90		2.85		42.36
CLAY 1		4.03		3.25		4.59		3.61		3.92		4.79		5.94		4.67		3.21		2.74		2.96		3.55		47.26
HOGSETT GALLIPOLIS DAM		3.78		2.82		4.43		3.49		3.76		4.06		4.01		3.43		2.56		1.99		2.71		2.82		39.86
HUNTINGTON WB CITY	38.0	3.61	39.0	3.06	47.2	4.08	57.3	3.42	65.7	3.82	74.1	4.34	76.9	4.82	75.4	3.37	70.5	2.88	59.2	2.43	46.7	2.79	39.1	3.17	57.4	41.79
LOGAN		3.86		3.39		4.80		3.39		4.02		4.54		4.84		4.12		2.84		2.31		2.76		3.20		44.07
RAVENSWOOD DAM 22	36.6	3.58	36.9	2.78	44.1	3.82	54.7	3.30	63.9	3.58	72.6	3.90	75.6	4.40	74.1	3.53	68.1	2.56	57.6	2.06	45.0	2.58	36.4	2.85	55.5	38.94
SPENCER	36.1	4.10	36.2	3.40	43.6	4.33	53.9	3.62	63.2	3.92	71.5	4.61	74.7	4.46	73.2	4.17	67.2	2.90	56.2	2.69	44.3	2.85	36.0	3.18	54.7	44.23
WILLIAMSON	38.6	3.67	39.2	3.29	46.1	4.66	56.5	3.42	66.1	3.92	74.6	4.41	77.7	5.19	76.4	4.53	70.2	2.61	58.9	2.12	46.2	2.76	38.7	3.21	57.4	43.79
WINFIELD LOCKS	36.4	4.00	36.6	2.94	44.0	4.17	54.4	3.39	64.4	3.46	73.3	3.95	76.6	4.40	75.0	3.97	69.1	3.02	58.0	2.02	45.6	2.70	36.6	2.77	55.8	40.79
DIVISION	36.9	3.88	37.3	3.07	44.6	4.35	55.1	3.41	64.5	3.83	73.0	4.30	76.2	4.87	74.9	3.90	68.8	2.87	57.7	2.25	45.2	2.76	36.9	3.07	55.9	42.51
CENTRAL																										
ARBOVALE 2		3.38		3.14		4.05		3.09		3.90		4.41		4.68		3.92		2.77		2.42		2.66		2.81		41.23
BAYARD	29.8	4.11	29.4	3.26	36.0	4.52	46.1	3.96	56.0	4.75	64.0	4.89	67.0	4.82	65.3	4.76	58.9	3.29	49.1	3.23	38.2	2.98	29.7	3.35	47.5	47.92
BECKLEY V A HOSPITAL	34.9	3.77	35.1	3.37	41.7	4.58	51.5	3.59	60.2	4.09	67.5	4.67	70.7	5.20	69.4	4.69	63.8	3.08	53.8	2.68	42.4	2.60	34.7	3.25	52.1	45.57
ELKINS AIRPORT	32.2	3.22	32.5	3.05	39.3	3.79	49.1	3.36	58.2	4.25	66.5	5.26	70.0	5.14	68.2	3.83	63.0	3.28	51.7	2.86	40.8	2.87	32.9	3.13	50.4	44.04
FLAT TOP	31.5	3.77	31.7	3.58	37.9	4.67	48.1	3.46	57.4	3.96	64.8	4.55	67.7	4.82	66.7	4.51	61.4	2.63	51.6	2.52	39.2	2.65	31.6	3.32	49.1	44.44
STONY RIVER DAM		3.36		2.95		3.17		3.82		4.59		4.63		4.04		4.59		2.96		3.43		2.84		3.08		43.46
THOMAS		4.45		3.99		5.17		4.37		5.44		5.70		5.78		5.25		3.81		3.52		3.42		4.10		55.00
DIVISION	32.3	4.21	32.2	3.67	38.9	4.94	49.0	3.93	58.3	4.82	66.1	5.11	69.3	5.46	67.8	4.99	62.0	3.31	52.0	3.11	40.3	3.14	32.2	3.66	50.0	50.35
SOUTHERN																										
BLUEFIELD 1	37.1	3.40	37.3	3.32	43.8	4.23	53.5	3.08	62.3	3.91	69.6	3.96	72.2	4.72	71.0	4.27	65.6	2.63	55.8	2.38	44.5	2.59	36.8	2.93	54.1	41.42
GARY	37.1	3.44	37.2	3.25	43.7	4.31	53.8	3.24	63.0	3.96	70.9	4.05	74.0	5.25	73.0	4.08	66.9	2.72	55.9	2.29	44.1	2.26	36.7	3.07	54.7	41.92
UNION	34.2	3.03	34.7	2.67	41.5	3.71	51.6	2.55	60.8	3.45	68.5	3.86	71.7	4.30	70.4	3.46	64.5	2.44	54.0	2.09	41.8	2.24	33.4	2.55	52.3	36.35
WHITE SULPHUR SPRINGS	34.3	3.19	35.0	2.68	42.1	4.11	52.3	2.98	62.1	3.48	69.8	3.70	72.7	4.25	71.2	3.91	64.7	2.65	54.3	2.27	42.1	2.62	33.7	2.69	52.9	38.53
DIVISION	35.8	3.25	36.4	2.96	43.0	4.01	53.2	2.98	62.5	3.71	70.3	3.85	73.2	4.49	72.0	3.93	66.0	2.54	55.5	2.19	43.6	2.41	35.2	2.85	53.9	39.17
NORTHEASTERN																										
HARPERS FERRY		2.69		2.23		3.43		3.44		3.73		3.45		3.88		4.48		3.44		3.65		2.93		2.96		40.31
KEARNEYSVILLE 1 NW	34.2	2.68	35.0	2.09	42.9	3.24	53.0	3.24	63.4	3.77	71.7	3.47	75.7	3.51	73.7	4.28	66.7	3.41	56.0	3.40	44.8	2.84	34.9	2.78	54.3	38.71
MARTINSBURG CAA AP	33.5	2.71	34.0	2.02	41.9	3.38	52.5	3.29	63.5	3.87	72.2	3.17	76.7	3.46	74.4	4.46	66.8	3.18	55.7	3.36	43.9	2.75	34.4	2.82	54.1	38.47
PIEDMONT	32.8	2.98	33.2	2.02	40.9	3.84	51.5	3.22	61.8	3.98	69.7	4.14	73.6	3.42	72.1	4.11	65.3	3.00	55.1	2.82	42.8	2.27	33.1	2.45	52.7	38.25
WARDENSVILLE R M FARM	32.9	2.11	33.3	1.71	40.8	3.01	50.6	2.89	61.1	3.72	69.0	3.61	72.9	3.78	71.1	4.84	64.2	2.77	53.9	2.94	42.7	2.36	32.9	2.17	52.1	35.91
DIVISION	33.8	2.42	34.4	1.86	41.9	3.11	51.9	2.89	62.2	3.69	70.3	3.75	74.2	3.66	72.3	4.19	65.5	2.85	55.0	2.98	43.6	2.42	34.1	2.31	53.3	36.13

* Averages for period 1931-1955, except for stations marked Wb which are "normals" based on period 1921-1950. Divisional means may not be the arithmetical average of individual stations published, since additional data from shorter period stations are used to obtain better areal representation.

CONFIDENCE LIMITS

In the absence of trend or record changes, the chances are 9 out of 10 that the true mean will lie in the interval formed by adding and subtracting the values in the following table from the means for any station in the State.

1.9	.48	1.6	.38	1.9	.44	1.0	.44	.9	.71	.8	.59	.6	.63	.8	.77	1.0	.46	1.0	.67	1.0	.43	1.4	.48	.3	2.32

COMPARATIVE DATA

Data in the following table are the mean temperature and average precipitation for Spencer, West Virginia for the period 1906 - 1930 and are included in this publication for comparative purposes :

33.2	4.22	34.5	3.24	43.5	3.97	52.5	3.45	61.7	3.67	69.0	4.34	72.8	4.31	71.8	4.18	66.1	3.00	54.9	3.35	43.8	2.75	35.4	3.44	53.3	43.92

NORMALS, MEANS, AND EXTREMES

CHARLESTON, WEST VIRGINIA — KANAWHA AIRPORT

LATITUDE 38° 22' N
LONGITUDE 81° 36' W
ELEVATION (ground) 950 feet

Temperature

Month	Normal Daily max	Normal Daily min	Normal Monthly	Extreme Record highest	Year	Extreme Record lowest	Year
(a)	(b)	(b)	(b)	50		49	
J	46.3	26.5	36.4	81	1932	-9	1936+
F	48.7	27.7	38.2	80	1932	-11	1903
M	56.6	33.1	44.9	92	1929	2	1934
A	68.2	41.8	55.0	96	1925	18	1923
M	77.0	50.3	63.7	96	1930+	31	1907
J	84.5	59.5	72.0	98	1931	39	1945
J	87.3	63.5	75.4	105	1931	46	1937
A	85.5	61.7	73.6	108	1918	47	1953
S	81.4	55.7	68.6	108	1932	33	1928+
O	70.6	44.2	57.4	104	1903	18	1952
N	56.6	35.0	45.8	88	1948	6	1950+
D	47.5	28.6	38.1	79	1951+	-17	1917
Year	67.5	44.0	55.8	108	July 1931+	-17	Dec. 1917

Precipitation

Month	Normal degree days	Normal total	Max monthly	Year	Min monthly	Year	Max 24 hr	Year
(a)	(b)	(b)	73		73		58	
J	887	3.99	9.11	1950	1.00	1896	2.45	1945
F	750	3.50	8.10	1887	0.88	1941	2.45	1951
M	632	4.16	8.94	1890	1.28	1910	2.52	1955
A	310	3.74	7.15	1908	1.05	1900	2.72	1948
M	110	3.78	8.61	1919	1.26	1911	3.35	1919
J	8	3.93	11.32	1936	0.42	1894	3.70	1901
J	0	5.45	9.96	1938	0.66	1899	2.75	1945
A	0	4.51	8.88	1890	0.46	1957	2.49	1951
S	60	2.84	7.18	1925	0.00	1908	2.30	1948
O	250	2.81	7.54	1926	0.45	1897	2.58	1915
N	576	3.17	8.81	1921	0.45	1904	2.58	1926
D	834	2.98	7.41	1901	0.78	1925	2.20	1919
Year	4417	45.00	11.32	July 1938	0.00	Oct. 1897	3.70	July 1932

Snow, Sleet

Month	Mean total	Max monthly	Year	Max 24 hr	Year
(a)	53	58		53	
J	4.9	19.5	1948	10.5	1954
F	4.1	19.5	1929	12.7	1929
M	3.0	21.4	1914	12.9	1954
A	0.2	3.0	1918	3.0	1918
M	T	0.0	1923	T	1923
J	0.0	0.0		0.0	
J	0.0	0.0		0.0	
A	0.0	0.0		0.0	
S	T	0.8	1957	0.8	1957
O	0.4	0.8	1950	0.8	1950
N	1.4	25.8	1950	15.1	1950
D	3.2	17.2	1917	10.0	1916
Year	16.8	25.8	Nov. 1950	15.1	Nov. 1950

Relative humidity, Wind, Sky, Days

Month	RH 1:00 a.m.	RH 7:00 a.m.	RH 1:00 p.m.	RH 7:00 p.m.	Wind mean hourly	Prevailing direction	Mean sky cover
(a)	8	8	8	8	8	8	10
J	76	78	64	69	8.8	S	8.3
F	74	78	58	63	8.7	S	7.5
M	70	75	53	56	9.3	WSW	7.4
A	69	74	47	51	8.9	WSW	7.1
M	81	81	50	59	7.7	SW	6.6
J	88	86	55	63	7.1	SW	6.4
J	91	87	52	66	6.3	SW	6.4
A	90	91	53	65	5.5	S	6.2
S	88	90	51	64	5.7	S	6.0
O	82	90	51	60	6.0	S	6.1
N	79	85	54	64	8.3	SW	7.2
D	75	86	60	65	—	SW	7.4
Year	80	84	54	62	7.4	SW	6.9

Mean number of days (Charleston): Clear 63, Partly cloudy 115, Cloudy 187; Precipitation .01 inch or more 152; Snow/Sleet 1.0 inch or more 7; Thunderstorms 45; Heavy fog 109; Max 90° and above 30; Max 32° and below 12; Min 32° and below 97; Min 0° and below *.

ELKINS, WEST VIRGINIA — MUNICIPAL AIRPORT

LATITUDE 38° 53' N
LONGITUDE 79° 51' W
ELEVATION (ground) 1970 feet

Temperature

Month	Normal Daily max	Normal Daily min	Normal Monthly	Extreme Record highest	Year	Extreme Record lowest	Year
(a)	(b)	(b)	(b)	59		59	
J	42.8	21.5	32.2	78	1914	-20	1931
F	43.6	21.4	32.5	77	1932	-21	1934
M	51.5	27.1	39.3	86	1929	-9	1914
A	62.5	35.6	49.1	90	1925	18	1923
M	71.7	43.6	57.7	90	1911	21	1947
J	79.4	53.6	66.5	95	1914	31	1945
J	82.5	57.4	70.0	96	1934	37	1947
A	80.6	55.8	68.2	96	1918	37	1930
S	76.3	49.7	63.0	97	1953	26	1942
O	65.1	38.3	51.7	87	1927	-8	1952
N	52.4	29.3	40.9	80	1946	-16	1917
D	43.4	22.3	32.9	76	1951	-28	1917
Year	62.7	38.0	50.4	99	Aug. 1918	-28	Dec. 1917

Precipitation

Month	Normal degree days	Normal total	Max monthly	Year	Min monthly	Year	Max 24 hr	Year
(a)	(b)	(b)	59		59		59	
J	1017	3.22	8.93	1931	1.36	1931	1.84	1908
F	910	3.05	5.99	1952	0.92	1952	2.75	1929
M	797	3.79	8.32	1910	0.68	1910	3.40	1936
A	477	3.36	7.25	1927	1.37	1900	2.50	1902
M	224	4.25	9.18	1933	1.35	1939	2.69	1954
J	53	5.26	8.35	1939	2.45	1915	4.20	1919
J	9	5.14	11.10	1907	0.61	1912	5.45	1935
A	31	3.83	10.42	1907	0.83	1939	3.40	1921
S	122	3.28	8.43	1954	0.52	1914	3.08	1934
O	412	2.86	6.21	1945	0.26	1924	4.24	1929
N	726	2.87	6.95	1917	0.73	1914	3.15	1900
D	995	3.13	6.95	1901	1.17	1917	2.79	1901
Year	5773	44.04	11.10	July 1907	0.26	Oct. 1924	5.45	July 1935

Snow, Sleet

Month	Mean total	Max monthly	Year	Max 24 hr	Year
(a)	59	59		59	
J	11.0	36.6	1905	14.0	1908
F	10.5	26.8	1947	10.0	1929
M	8.8	29.8	1936	16.0	1936
A	3.0	22.6	1902	16.0	1902
M	T	0.5	1954	0.5	1954
J	0.0	0.0	1919	0.0	1919
J	0.0	0.0		0.0	
A	0.0	0.0		0.0	
S	T	0.8	1925	0.8	1925
O	0.4	5.5	1950	4.6	1917
N	4.5	37.6	1935	18.5	1913
D	9.8	29.8	1944	13.5	1944
Year	47.2	37.6	Nov. 1950	18.8	Nov. 1913

Relative humidity, Wind, Sky, Sunshine

Month	RH 1:00 a.m.	RH 7:00 a.m.	RH 1:00 p.m.	RH 7:00 p.m.	Wind mean hourly	Prevailing direction	Fastest mile speed	Direction	Year	Mean sky cover	Pct possible sunshine
(a)	14	14	14	14	9	9	43	43		27	55
J	83	82	65	77	6.3	WNW	60	NW	1952	7.9	33
F	83	83	65	72	6.6	WNW	54	W	1953	7.5	37
M	80	83	58	67	6.9	WNW	57	W	1945	7.4	42
A	81	81	53	62	6.4	WNW	57	W	1954	7.0	47
M	88	88	56	65	5.2	WNW	56	W	1947	6.7	55
J	94	91	59	67	4.2	WNW	56	W	1948	6.5	56
J	95	94	61	76	3.7	NW	72	NW	1951	6.5	56
A	96	95	59	81	3.6	NW	45	NW	1947	6.1	55
S	95	93	56	80	4.3	SE	38	NW	1949	5.7	55
O	93	85	53	74	5.6	NW	52	NW	1944	5.1	51
N	83	83	59	73	6.0	NW	50	W	1950	7.2	41
D	—	86	66	—	—	NW	—	—	—	7.6	55
Year	88	88	59	73	5.2	WNW	72	NW	July 1951	6.9	48

Mean number of days (Elkins): Clear 71, Partly cloudy 116, Cloudy 178; Precipitation .01 inch or more 171; Snow/Sleet 1.0 inch or more 30; Thunderstorms 46; Heavy fog 67; Max 90° and above 3; Max 32° and below 26; Min 32° and below 132; Min 0° and below 4.

NORMALS, MEANS, AND EXTREMES

LATITUDE 38° 25' N
LONGITUDE 82° 27' W
ELEVATION (ground) 565 feet

Huntington, West Virginia

Month	Temperature Normal — Daily maximum	Daily minimum	Monthly	Extremes — Record highest	Year	Record lowest	Year	Precipitation Normal total (b)	Normal degree days (b)	Max monthly	Year	Min monthly	Year	Max in 24 hrs	Year	Snow, Sleet Mean total	Max monthly	Year	Max in 24 hrs	Year
(a)	(b)	(b)	(b)	17	17	17	17	(b)	(b)	17	17	17	17	17	17	17	17	17	17	17
J	48.0	28.0	38.0	79	1950	-10	1948+	3.61	837	8.93	1950	1.62	1942	2.02	1946	5.4	19.1	1948	8.2	1954
F	49.6	28.3	39.0	78	1950	-1	1951	3.08	728	7.58	1956	0.43	1947	2.04	1945	3.0	9.0	1948	8.9	1948
M	59.3	35.0	47.2	89	1929	-3	1943	4.08	570	7.58	1941	1.98	1941	2.53	1948	2.3	9.0	1944	10.3	1944
A	70.3	44.2	57.3	92	1957+	15	1943	3.42	251	6.09	1950	0.96	1950	2.67	1948	0.2	T	1944	T	1944
M	78.6	52.8	65.7	96	1941	28	1947	3.82	85	5.76	1943	1.04	1941	2.16	1941	0.0	0.0		0.0	1953+
J	86.2	62.0	74.1	100	1941	39	1941	4.34	5	8.88	1946	1.22	1954	2.83	1947	0.0	0.0		0.0	
J	88.6	65.2	76.9	103	1954	47	1945	4.82	0	8.63	1941	0.98	1944	4.23	1954	0.0	0.0		0.0	
A	87.4	63.4	75.4	103	1948	43	1946	3.37	0	4.92	1955	0.89	1953	2.35	1951	0.0	0.0		0.0	
S	83.1	57.9	70.5	104	1953	29	1949	2.88	35	8.90	1949	0.13	1953	2.70	1953	0.0	0.0		0.0	
O	72.2	46.2	59.2	85	1951+	21	1952	2.43	210	3.20	1949	0.79	1950	1.60	1949	1.7	T	1957+	8.5	1957+
N	57.7	35.4	46.7	85	1948	6	1950	2.79	549	5.30	1957	0.79	1951	1.25	1951	2.9	19.6	1950	8.9	1950
D	48.6	29.6	39.1	80	1951	-6	1942	3.17	803	6.15	1942	1.02	1947	2.32	1956	2.9	10.3	1945	6.3	1942
Year	69.1	45.7	57.4	105	July 1954	-10	Jan 1948	41.79	4073	8.93	Jan 1950	0.22	Oct 1953	4.23	July 1954	15.5	19.6	Nov 1950	10.3	Mar 1954

Mean number of days — Huntington

Month	Precipitation .01 or more	Thunderstorms	Heavy fog	Temperature Max 90° and above	Max 32° and below	Min 32° and below	Min 0° and below
(a)	17		17	17	17	17	17
J	15			0	4	22	1
F	12			0	3	19	*
M	14			0	1	15	*
A	13			1	0	4	0
M	13			3	0	*	0
J	11			12	0	0	0
J	11			15	0	0	0
A				13	0	0	0
S	9			6	0	1	0
O	8			0	0	2	0
N	10			0	1	14	*
D	12			0	3	21	*
Year	136			51	12	97	1

Means and extremes in the above table are from the existing or comparable location(s). Annual extremes have been exceeded at prior locations as follows: Highest temperature 108 in July 1930; lowest temperature -24 in February 1899.

LATITUDE 39° 16' N
LONGITUDE 81° 34' W
ELEVATION (ground) 615 feet

Parkersburg, West Virginia

Month	Temperature Normal — Daily maximum	Daily minimum	Monthly	Extremes — Record highest	Year	Record lowest	Year	Precipitation Normal total (b)	Normal degree days (b)	Max monthly	Year	Min monthly	Year	Max in 24 hrs	Year	Snow, Sleet Mean total	Max monthly	Year	Max in 24 hrs	Year
(a)	(b)	(b)	(b)	69	69	69	69	(b)	(b)	69	69	69	69	69	69	69	69	69	69	69
J	42.7	26.1	34.4	78	1950	-16	1936	3.17	949	8.99	1937	0.91	1931	2.97	1913	6.8	26.4	1948	12.5	1936
F	44.2	26.8	35.5	77	1932	-27	1899	2.65	826	7.04	1895	0.99	1895	2.89	1945	6.0	25.7	1894	13.0	1914
M	53.6	33.6	43.6	89	1929	-3	1943	3.54	672	6.95	1890	0.10	1910	2.41	1918	4.2	22.0	1896	11.5	1902
A	64.9	42.6	53.8	93	1925	15	1923	3.08	347	6.75	1948	0.77	1900	3.40	1920	0.7	8.7	1918	8.7	1918
M	74.7	52.4	63.5	96	1914	29	1945	3.50	119	7.50	1933	0.53	1939	3.00	1905	T	0.1	1895	0.1	1895
J	83.0	61.7	72.4	99	1895	47	1947	4.18	13	8.63	1928	1.17	1930	3.58	1952	0.0	0.0		0.0	
J	86.3	65.1	75.7	104	1930	45	1890	4.16	0	11.46	1896	0.89	1901	4.81	1947	0.0	0.0		0.0	
A	84.3	64.0	74.0	106	1918	45	1935	4.15	0	10.42	1935	1.02	1951	3.00	1935	0.0	0.0		0.0	
S	79.2	57.5	68.4	102	1953	32	1942	2.99	56	8.41	1890	0.46	1897	3.40	1897	0.0	0.0		0.0	
O	67.8	45.8	56.9	91	1927	19	1905	2.12	272	6.48	1905	0.07	1897	3.22	1900	0.1	3.7	1925	3.5	1925
N	53.8	36.0	45.0	83	1948	-10	1917	2.67	600	5.59	1927	0.11	1904	2.69	1948	1.8	34.6	1950	16.8	1950
D	44.0	28.1	36.1	75	1951	-27	1899	2.90	896	5.51	1948	0.57	1925			4.5	29.1	1890	18.3	1890
Year	64.9	44.9	54.9	106	Aug 1918	-27	Feb 1899	39.11	4750	11.46	July 1896	0.07	Oct 1897	4.81	July 1947	24.1	34.6	Nov 1950	18.3	Dec 1890

Relative humidity, Wind, Sky cover, and Days — Parkersburg

Month	RH 7:00 a.m. EST	12:00 Noon EST	7:00 p.m. EST	Wind Fastest mile Speed	Direction	Year	Pct of possible sunshine	Mean sky cover sunrise to sunset	Sunrise to sunset Clear	Partly cloudy	Cloudy	Precip .01 or more	Snow/Sleet 1.0 in or more	Thunderstorms	Heavy fog	Max 90°+	Max 32°-	Min 32°-	Min 0°-
(a)	60	31	60	69	69	69	29	68	69	69	69	69	69	69	61	69	69	69	69
J	82	66	74	45	NW	1936	29	7.4	6	6	19	18	9	*	2	0	7	22	1
F	81	63	71	45	NW	1956	36	7.6	6	7	15	13	8	1	1	0	6	21	1
M	79	56	64	47	SE	1932	44	7.3	8	8	15	13	5	1	1	0	2	15	*
A	74	49	58	47	SW	1920	48	6.0	9	10	13	12	1	3	1	*	1	4	0
M	74	51	60	43	NW	1914	56	6.2	10	10	13	12	*	6	1	1	0	*	0
J	78	56	66	49	NW	1934	60	5.4	10	11	9	11	0	7	1	5	0	0	0
J	80	52	67	62	NW	1926	63	5.0	12	11	8	11	0	7	*	8	0	0	0
A	83	57	70	37	N	1955	60	4.8	12	11	8	10	0	4	1	6	0	0	0
S	84	51	71	61	NW	1954	59	5.1	12	10	9	9	0	2	2	3	0	1	0
O	84	52	70	38	SE	1932	54	5.1	11	7	16	11	*	1	5	1	0	2	0
N	81	57	71	66	NW	1954	36	6.6	6	6	19	11	1	*	2	0	1	12	*
D	81	64	76	35	W	1920	29	7.3				13	3	*	1	0	5	21	1
Year	80	56	68	66	NW	Nov 1954	48	6.3	110	103	152	143	6	43	11	23	21	97	2

Reference Notes Applying to all "Normals, Means, and Extremes" Tables

(a) Length of record, years.
(b) Normal values are based on the period 1921-1950, and are means adjusted to represent observations taken at the present standard location.
* Less than one-half.

- No record.
† Airport data.
‡ City Office data.
+ Also on earlier dates, months, or years.
T Trace, an amount too small to measure.

Sky cover is expressed in a range of 0 for no clouds or obscuring phenomena to 10 for complete sky cover. The number of clear days is based on average cloudiness 0-3 tenths; partly cloudy days on 4-7 tenths; and cloudy days on 8-10 tenths. Monthly degree day totals are the sum of the negative departures of average daily temperatures from 65°F. Sleet was included in snowfall totals beginning with July 1948. Sleet also includes data referred to at various times in the past as "Dense" or "Thick". The upper limit for heavy fog is 1/4 mile. Data in these tables are based on records through 1957.

Mean Annual Precipitation, Inches

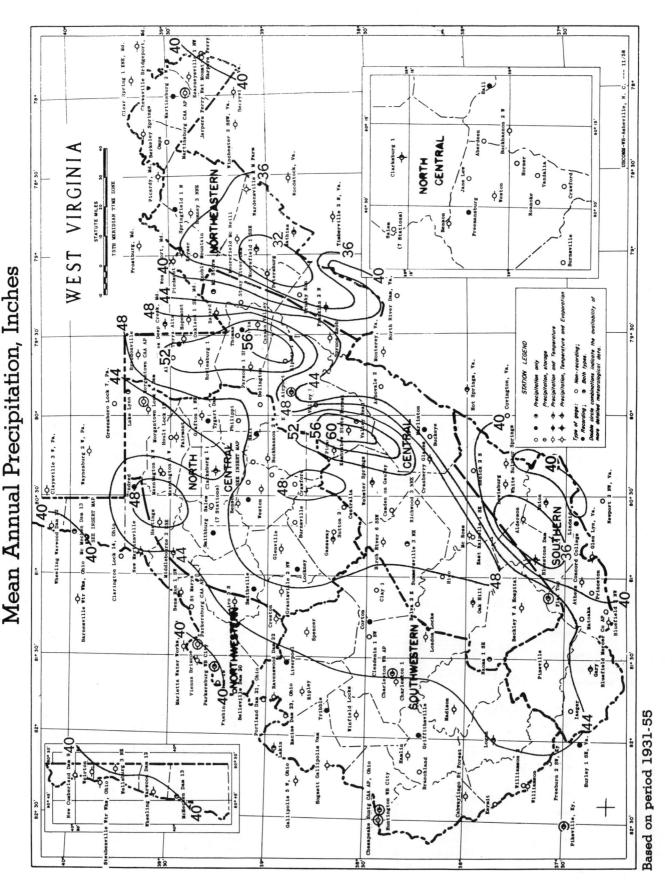

WEST VIRGINIA

STATUTE MILES

75TH MERIDIAN TIME ZONE

NORTH CENTRAL

STATION LEGEND

Based on period 1931-55

Isolines are drawn through points of approximately equal value. Caution should be used in interpolating on these maps, particularly in mountainous areas.

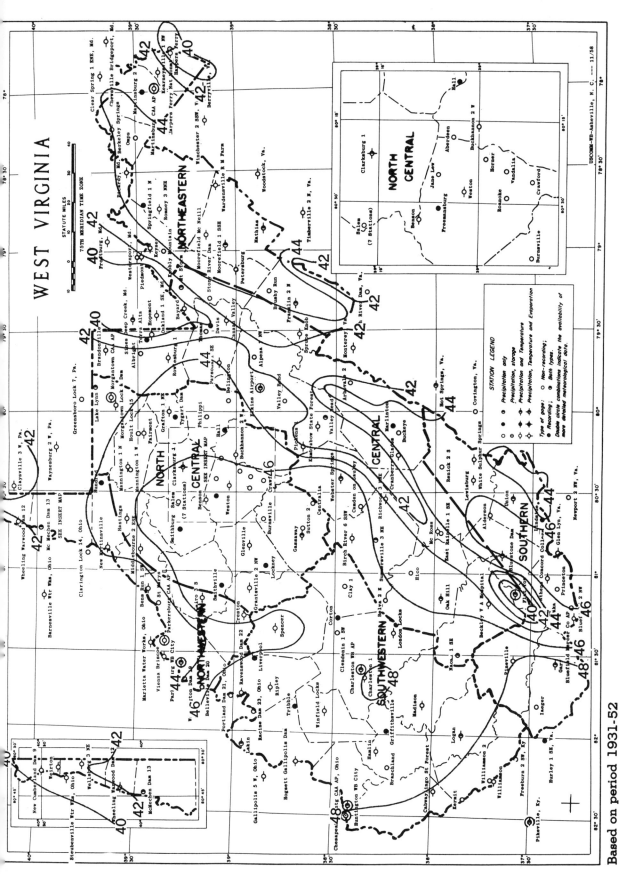

WEST VIRGINIA

75TH MERIDIAN TIME ZONE

STATUTE MILES

Based on period 1931-52

Isolines are drawn through points of approximately equal value. Caution should be used in interpolating on these maps, particularly in mountainous areas.

STATION LEGEND

Precipitation only
Precipitation, storage
Precipitation and Temperature
Precipitation, Temperature and Evaporation

Type of gage: ○ Non-recording;
● Recording;   ⊕ Both types.
Double circle combinations indicate the availability of more detailed meteorological data.

NORTH CENTRAL

NORTHEASTERN

NORTHWESTERN

NORTH CENTRAL

CENTRAL

SOUTHERN

SOUTHWESTERN

USCOMM—WB—Asheville, N. C. — 11/58

Mean Minimum Temperature (°F.), January

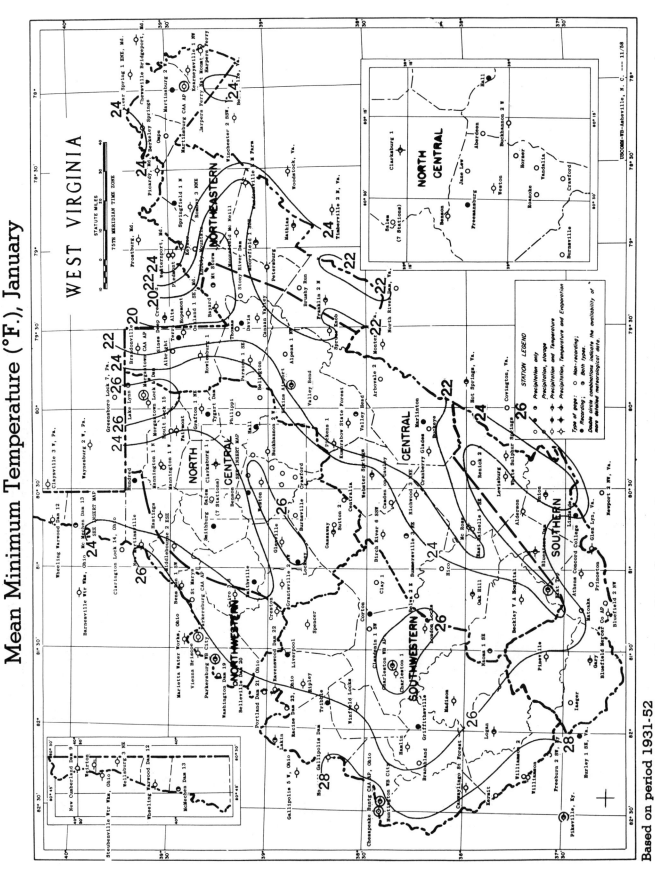

WEST VIRGINIA

Based on period 1931-52

Isolines are drawn through points of approximately equal value. Caution should be used in interpolating on these maps, particularly in mountainous areas.

Mean Maximum Temperature (°F.), July

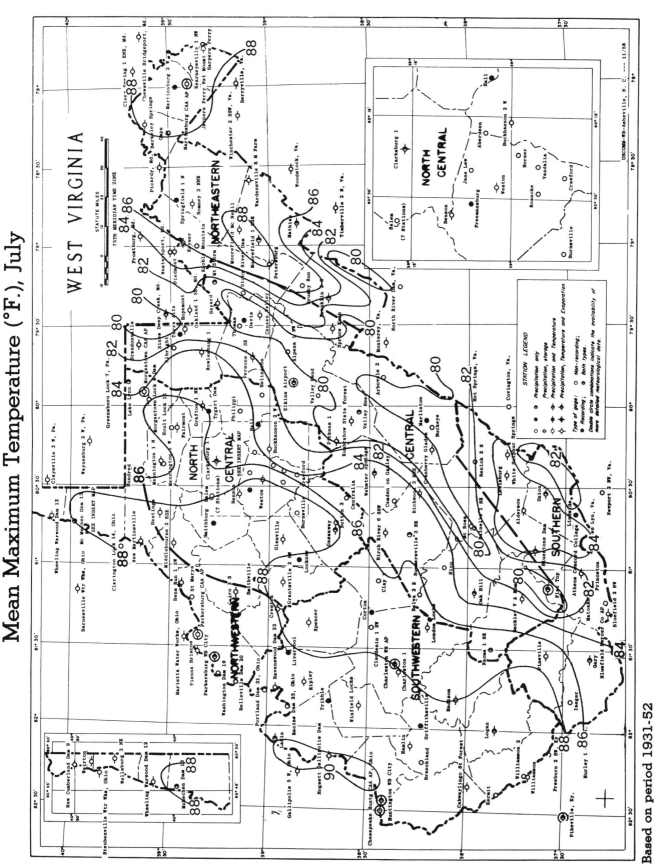

WEST VIRGINIA

Based on period 1931-52

Isolines are drawn through points of approximately equal value. Caution should be used in interpolating on these maps, particularly in mountainous areas.

Mean Minimum Temperature (°F.), July

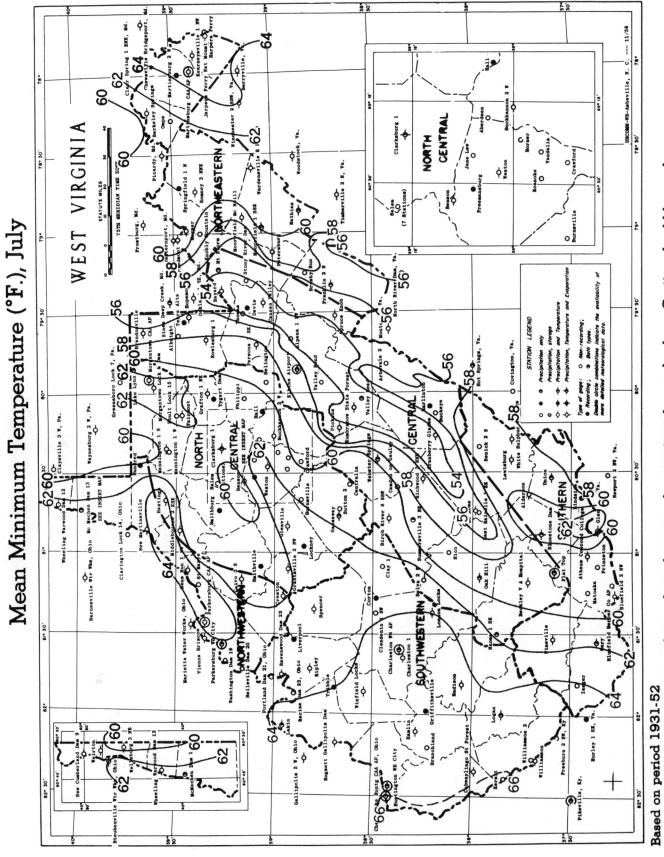

Based on period 1931-52

Isolines are drawn through points of approximately equal value. Caution should be used in interpolating on these maps, particularly in mountainous areas.

— 436 —

THE CLIMATE OF
WISCONSIN

by
Paul J. Waite

February 1960

Wisconsin lies in the upper Midwest between Lake Superior, Upper Michigan, Lake Michigan, and the Mississippi and Saint Croix Rivers. Its greatest length is 320 miles, greatest width 295 miles, and total area 56,066 square miles.

Glaciation has largely determined the topography and soils of the State, excepting the 13,360 square miles of driftless area in southwestern Wisconsin. The various glaciations created a rolling terrain with nearly 9,000 lakes and several areas of marshes and swamps.

Elevations range from about 600 feet above sea level along the Lake Superior and Lake Michigan shores and in the Mississippi flood plain in south-western Wisconsin to nearly 1,950 feet above sea level at Rib and Strawberry Hills.

The Northern Highlands, a plateau extending across northern Wisconsin, is an area of about 15,000 square miles with elevations from 1,000 to 1,800 feet above sea level. This area is the location of many lakes and the origin of most of the major streams in the State. The slope down to the narrow Lake Superior plain is quite steep.

A comparatively flat, crescent shaped lowland lies immediately south of the Northern Highlands embodying nearly one-fourth of Wisconsin.

The eastern ridges and lowlands to the southeast of the central plains is the most densely populated with the highest concentration of industry and most available land in farms.

The western uplands of southwestern Wisconsin west of the ridges and lowlands and south of the central plains contains approximately one-fourth of the State. This area is the roughest section of the State rising 200 to 350 feet above the central plains and 100 to 200 feet above the Eastern Ridges and Lowlands. The Mississippi River bluffs rise 230 to 650 feet.

The Wisconsin climate is typically continental with some modification by Lakes Michigan and Superior. The cold, snowy winters favor a variety of winter sports, and the warm summers appeal to thousands of vacationers each year. About two-thirds of the annual precipitation falls during the growing season (freeze-free period). It is normally adequate for vegetation, although drought is occasionally reported. This climate is most favorable for dairy farming; the primary crops being corn, small grains, hay, and vegetables. The rapid succession of storms moving from west to east and southwest to northeast account for the stimulating climate.

The average annual temperature varies from 39.0°F. at Winter to 49.5°F. at Beloit. The highest temperature ever recorded in Wisconsin was 114°F. at Wisconsin Dells on July 13, 1936, and the lowest temperature on record is -54°F. reported from Danbury on January 24, 1922.

During more than one-half of the winters temperatures fall to -40°F. or lower, and almost **every** winter -30°F. or colder is reported from northern Wisconsin. Summer temperatures above 90°F. or higher average 2 to 4 days in northern counties to about 14 days in southern districts. During marked cool outbreaks in the summer months, the central lowlands occasionally report freezing temperatures. Wisconsin temperatures have averaged about 1°F. warmer since 1931 as compared to the 1891-1930 period.

The freeze-free season averages around 80 days per year in the upper northeast and north-central lowlands to about 180 days in the Milwaukee area. The pronounced moderating effect of Lake Michigan is well illustrated by the fact that the growing season of 140 to 150 days along the east-central coastal area is of the same duration as in the southwestern Wisconsin valleys. The short growing season in the central portion of the State is attributed to a number of factors, among them being an inward cold air drainage and the low heat capacities of the peat and sandy soils. The average date of last spring freeze ranges from early May along the Lake Michigan coastal area and southern counties to early June in the northern-most counties. The first autumn freezes occur in late August and early September in northern, and central lowlands to mid-October along the Lake Michigan coast line. However, July freeze is not unusual in the north and central Wisconsin lowlands.

The long-term mean (1931-55) annual precipitation totals 30 to 34 inches over most of the Western Uplands and Northern Highlands, diminishing to about 28 inches along most of the Wisconsin coastal area bordering Lake Michigan and 28 to 30 inches over most of the Wisconsin Central Plain and Lake Superior Coastal area. The higher average annual precipitation coincides generally with the highest elevations, particularly to the windward slopes of the Western Uplands and Northern Highlands. Thunderstorms average about 30 per year in northern Wisconsin to about 40 per year in southern counties, occurring mostly in the summer. Occasional hail, wind, and lightning damage are reported.

The average seasonal snowfall varies from about 30 inches at Beloit to well over 100 inches in northern Iron County along the steep western slope of the Gogebic Range. The heavy snowfall along the Gogebic Range is a result of the prevailing cold northerly winter winds blowing across the relatively warm Lake Superior. Relatively greater average snowfall is recorded over the Western Uplands and Eastern Ridges than in adjacent low-land areas. The mean dates of the first snowfall of consequence, an inch or more, varies from early November in northern localities to around December 1 in southern Wisconsin counties. Average annual duration of snowcover ranges from 85 days in southern-most Wisconsin to more than 140 days along Lake Superior. The snow cover acts as a protective insulation for grasses, autumn seeded grains, and other wintering vegetation.

The drainage of Wisconsin is into Lake Superior, Lake Michigan, and the Mississippi River. The Mississippi and St. Croix Rivers form most of the western boundary. About one-half of the northwestern portion of the State is drained through the Chippewa River, while the remainder of this region drains directly into the Mississippi or the St. Croix and into Lake Superior. The Wisconsin River has its source at a small lake nearly 1,600 feet above mean sea level on the Upper Michigan boundary and drains most of central Wisconsin.

Most of the Wisconsin River tributaries also spring from the many lakes in the north. Except for the Rock River, a Mississippi River tributary which flows through northern Illinois, eastern Wisconsin drains into Lake Michigan, a large part through Green Bay.

Most of the streams and lakes in Wisconsin are ice-covered from late November to late March. Snow covers the ground in practically all the winter months, except in the extreme southern areas. Flooding is most frequent, and most serious during April, due to the melting of snow associated with spring rains. During this period, flood conditions are often aggravated by ice jams which back up the flood waters. Excessive rains of the thunderstorm type sometimes produce tributary flooding or flash flooding along the smaller streams and creeks. Major flooding occurs on the Mississippi River, on the average, about 3 years in 10.

The most notable floods along the Mississippi occurred in 1880, 1951, and 1952. Important overflow in tributary basins occurred in 1905, 1912, 1916, 1920, 1922, 1934, 1935, 1938, 1941, 1944, 1945, 1946, 1948, 1950, and 1951.

Tornado occurrences have averaged four per year for the period 1916-58 although better observations and public awareness have increased in recent years resulting in more tornadoes being reported. Most of the very destructive Wisconsin tornadoes occur in the northwestern quarter of the State. Wisconsin tornado frequency is highest in June and July, followed in order by April, May, and September.

The fertile soils, gently rolling topography, and climate favor intensified dairy farming in southern and eastern Wisconsin. Most southeastern counties have over 90 percent of the land area in farms as contrasted to the northern third of the State where much less than 50 percent of the land area is in farms.

Farmland constitutes about 43 percent of the total land area of Wisconsin. The primary crops, of hay, oats, and corn utilize 93 percent of the land farmed. Milk is the largest single source of farm income, now surpassing the combined farm income of all other products. Wisconsin farm income for the past several years has ranked among the first 10 states of the Nation. Wisconsin is among the country's leading states in the following commercial vegetables harvested: green peas, sweet corn, cucumbers for pickles, snap beans, beets, cabbage for sauerkraut, carrots, and tomatoes. Marshy areas have been utilized for cranberry growing in central and scattered northern localities, with state production now second in the Nation. Since freeze is a hazard through all summer months, water sources for flooding are a necessity.

About 50 percent of the State is forested and supplies about one-third of the pulpwood used in the paper industry. Since much of the northern areas are of low fertility and with a short growing season, reforestation continues with improved lumber prospect. Naturally, wood and food processing industries have become important as well as machine manufacture.

Wisconsin's stimulating climate has created an energetic population concentrated mostly in the southern and eastern districts of the State, with northern districts sparsely settled. With adequate transportation the recreational facilities of the northern counties are coming into greater usage.

REFERENCES

(1) Finley, Robert W., 1957. Geography of Wisconsin. Regents of the University of Wisconsin.

(2) Martin, Lawrence. 1932. The Physical Geography of Wisconsin. Wisconsin Survey Bulletin No. XXXVI.

(3) Thom, H.C.S. 1957. Probabilities of One-Inch Snowfall Thresholds for the United States. Monthly Weather Review. Vol. 85, No. 8.

(4) U. S. Department of Agriculture. 1941. Climate of the States, Wisconsin. pp. 1199 and 1200.

(5) U. S. Department of Agriculture and Wisconsin Department of Agriculture. 1957. Cranberries of Wisconsin. Bull. No. 70.

(6) U. S. Department of Agriculture and Wisconsin Department of Agriculture. 1954. Wisconsin Agriculture in Mid-Century. Bulletin No. 325.

(7) Weather Bureau Technical Paper No. 16 - Maximum 24-Hour Precipitation in the United States. Washington, D. C. 1952.

(8) Weather Bureau Technical Paper No. 25 - Rainfall Intensity-Duration-Frequency Curves. For selected stations in the United States, Alaska, Hawaiian Islands and Puerto Rico.

BIBLIOGRAPHY

(A) Climatic Summary of the United States (Bulletin W) 1930 edition, Sections 47, 49 and 49. U. S. Weather Bureau

(B) Climatic Summary of the United States, Wisconsin - Supplement for 1931 through 1952 (Bulletin W Supplement). U. S. Weather Bureau

(C) Climatological Data - Wisconsin. U. S. Weather Bureau

(D) Climatological Data National Summary. U. S. Weather Bureau

(F) Hourly Precipitation Data - Wisconsin. U. S. Weather Bureau

(F) Local Climatological Data, U. S. Weather Bureau for Green Bay, La Crosse, Madison and Milwaukee, Wisconsin.

FREEZE DATA

STATION	Freeze threshold temperature	Mean date of last Spring occurrence	Mean date of first Fall occurrence	Mean No. of days between dates	Years of record Spring	No. of occurrences in Spring	Years of record Fall	No. of occurrences in Fall	STATION	Freeze threshold temperature	Mean date of last Spring occurrence	Mean date of first Fall occurrence	Mean No. of days between dates	Years of record Spring	No. of occurrences in Spring	Years of record Fall	No. of occurrences in Fall
AMERY HYDRO PLT	32	05-15	09-19	127	28	28	28	28	DODGEVILLF	32	05-03	10-08	158	10	10	10	10
	28	05-04	10-04	153	28	28	28	28		28	04-23	10-26	186	10	10	10	10
	24	04-23	10-17	177	28	28	28	28		24	04-03	11-05	216	10	10	10	10
	20	04-08	10-27	202	28	28	27	27		20	03-29	11-11	228	10	10	10	10
	16	03-28	11-09	225	28	28	28	28		16	03-19	11-19	245	10	10	9	9
ANTIGO	32	05-17	09-29	135	30	30	30	30	EAU CLAIRE CAA AP	32	05-05	10-04	151	30	30	30	30
	28	05-02	10-10	161	30	30	30	30		28	04-21	10-18	180	30	30	30	30
	24	04-20	10-22	185	30	30	30	30		24	04-08	10-31	205	30	30	30	30
	20	04-10	11-04	209	30	30	30	30		20	03-31	11-07	222	30	30	30	30
	16	03-30	11-13	228	30	30	30	30		16	03-23	11-14	237	30	30	30	30
ASHLAND EXP FARM	32	05-30	09-16	109	30	30	30	30	FOND DU LAC	32	05-11	10-09	151	30	30	30	30
	28	05-19	09-28	131	30	30	30	30		28	04-25	10-18	176	29	29	30	30
	24	05-11	10-11	153	30	30	30	30		24	04-10	10-31	204	29	29	30	30
	20	04-22	10-25	186	30	30	30	30		20	03-31	11-09	223	29	29	30	30
	16	04-10	11-06	211	30	30	30	30		16	03-23	11-20	242	29	29	30	30
BELOIT COLLEGE	32	05-01	10-13	165	30	30	30	30	GRAND RIVER LOCK	32	05-13	09-26	135	30	30	30	30
	28	04-16	10-25	193	30	30	30	30		28	04-30	10-02	155	30	30	30	30
	24	04-01	11-05	218	30	30	30	30		24	04-16	10-17	184	30	30	29	29
	20	03-24	11-16	236	30	30	30	30		20	04-03	10-27	207	30	30	29	29
	16	03-17	11-25	253	30	30	30	30		16	03-22	11-10	233	30	30	29	29
BIG ST GERMAIN DAM	32	06-06	09-05	91	30	30	29	29	GRANTSBURG CAA AP	32	05-24	09-21	121	30	30	30	30
	28	05-25	09-23	121	30	30	29	29		28	05-13	10-01	141	30	30	30	30
	24	05-12	10-08	149	30	30	28	28		24	04-30	10-12	165	30	30	30	30
	20	05-02	10-18	170	30	30	28	28		20	04-19	10-27	190	30	30	30	30
	16	04-21	10-31	194	30	30	28	28		16	04-09	11-04	209	29	29	30	30
BLAIR	32	05-19	09-24	128	25	25	25	25	GREEN BAY WB AP	32	05-06	10-13	161	30	30	30	30
	28	05-07	10-04	149	25	25	25	25		28	04-20	10-26	188	30	30	30	30
	24	04-24	10-14	173	25	25	25	25		24	04-06	11-06	214	30	30	30	30
	20	04-08	10-27	202	25	25	25	25		20	03-28	11-15	232	30	30	30	30
	16	03-27	11-06	224	24	24	25	25		16	03-22	11-23	246	30	30	30	30
BRODHEAD	32	05-09	10-03	147	30	30	30	30	HANCOCK EXP FARM	32	05-17	09-30	135	29	29	30	30
	28	04-26	10-14	171	30	30	30	30		28	05-06	10-07	154	29	29	30	30
	24	04-13	10-29	199	30	30	30	30		24	04-26	10-20	177	28	28	30	30
	20	03-30	11-09	224	30	30	30	30		20	04-15	10-31	199	28	28	30	30
	16	03-22	11-19	242	30	30	30	30		16	04-03	11-06	217	28	28	30	30
BRULE ISL	32	06-10	09-04	87	15	15	14	14	HATFIELD DAM	32	05-23	09-17	116	29	29	29	29
	28	05-20	09-17	120	14	14	14	14		28	05-11	09-26	138	30	30	29	29
	24	05-12	10-02	143	13	13	14	14		24	04-30	10-08	161	30	30	29	29
	20	04-27	10-18	174	13	13	14	14		20	04-19	10-19	183	30	30	29	29
	16	04-14	11-02	202	13	13	14	14		16	04-05	10-30	208	30	30	29	29
BURNETT 2 NW	32	05-12	10-01	142	29	29	28	28	HILLSBORO	32	05-16	09-28	135	30	30	29	29
	28	04-29	10-13	167	29	29	28	28		28	05-03	10-06	156	30	30	29	29
	24	04-14	10-28	197	28	28	28	28		24	04-18	10-18	183	30	30	29	29
	20	03-30	11-06	221	28	28	28	28		20	04-05	10-31	208	30	30	29	29
	16	03-27	11-19	237	28	28	28	28		16	03-27	11-12	230	30	30	29	29
CODDINGTON EXP FARM	32	06-02	08-30	89	29	29	29	29	KEWAUNEE	32	05-06	10-14	161	21	21	22	22
	28	05-18	09-14	119	29	29	29	29		28	04-16	10-24	191	21	21	22	22
	24	05-05	09-28	146	29	29	28	28		24	04-04	11-05	215	21	21	22	22
	20	04-24	10-14	174	29	29	28	28		20	03-29	11-18	235	21	21	20	20
	16	04-11	10-26	198	29	29	28	28		16	03-22	11-26	249	21	21	19	19
CRIVITZ HIGH FALLS	32	05-26	09-21	119	25	25	25	25	LA CROSSE WB CITY	32	05-01	10-08	161	30	30	30	30
	28	05-13	09-29	139	25	25	25	25		28	04-19	10-19	183	30	30	30	30
	24	05-02	10-13	163	25	25	25	25		24	04-05	11-01	210	30	30	30	30
	20	04-20	10-25	188	25	25	24	24		20	03-28	11-11	228	30	30	30	30
	16	04-10	11-04	207	25	25	24	24		16	03-17	11-19	247	30	30	30	30
CUMBERLAND	32	05-15	09-24	133	19	19	19	19	LAKE MILLS	32	05-10	10-08	151	30	30	30	30
	28	05-06	10-08	155	19	19	19	19		28	04-26	10-17	174	30	30	30	30
	24	04-22	10-18	179	19	19	18	18		24	04-11	11-01	205	30	30	30	30
	20	04-10	10-27	200	19	19	18	18		20	04-01	11-11	224	30	30	30	30
	16	04-01	11-06	219	19	19	18	18		16	03-24	11-21	242	30	30	29	29
DANBURY	32	05-31	09-12	105	28	28	29	29	LANCASTER	32	05-07	10-10	156	28	28	29	29
	28	05-18	09-25	130	27	27	29	29		28	04-21	10-22	184	28	28	29	29
	24	05-05	10-09	156	28	28	28	28		24	04-06	11-02	210	28	28	29	29
	20	04-25	10-16	174	27	27	28	28		20	03-29	11-10	226	28	28	29	29
	16	04-13	10-29	199	27	27	27	27		16	03-23	11-16	238	28	28	29	29
DARLINGTON	32	05-12	09-29	140	30	30	29	29	LAONA RS	32	06-03	09-02	92	17	17	19	19
	28	05-03	10-09	159	28	28	28	28		28	05-24	09-27	127	17	17	17	17
	24	04-18	10-20	185	28	28	28	28		24	05-09	10-07	151	19	19	16	16
	20	04-04	11-01	211	28	28	28	28		20	04-23	10-23	183	17	17	16	16
	16	03-24	11-11	233	28	28	28	28		16	04-10	10-30	203	16	16	14	14

FREEZE DATA

STATION	Freeze threshold temperature	Mean date of last Spring occurrence	Mean date of first Fall occurrence	Mean No. of days between dates	Years of record Spring	No. of occurrences in Spring	Years of record Fall	No. of occurrences in Fall
LONG LAKE DAM	32	06-11	08-24	73	30	30	30	30
	28	05-31	09-14	106	30	30	29	29
	24	05-17	09-28	135	30	30	29	29
	20	05-08	10-10	155	30	30	29	29
	16	04-26	10-26	183	30	30	29	29
MADISON WB	32	04-26	10-19	177	30	30	30	30
	28	04-10	10-31	204	30	30	30	30
	24	03-31	11-09	223	30	30	30	30
	20	03-23	11-16	237	30	30	30	30
	16	03-18	11-24	251	30	30	30	30
MANITOWOC	32	04-29	10-19	173	30	30	30	30
	28	04-14	10-31	200	30	30	30	30
	24	04-04	11-10	220	30	30	30	30
	20	03-27	11-19	237	30	30	30	30
	16	03-19	11-26	252	30	30	30	30
MARINETTE	32	05-12	10-03	143	30	30	30	30
	28	04-25	10-16	173	30	30	30	30
	24	04-16	10-30	197	30	30	30	30
	20	04-06	11-10	218	30	30	29	29
	16	03-26	11-17	237	30	30	29	29
MARSHFIELD EXP STA	32	05-17	09-27	133	30	30	30	30
	28	05-03	10-06	155	30	30	30	30
	24	04-20	10-19	182	30	30	30	30
	20	04-08	10-31	205	30	30	30	30
	16	03-29	11-10	227	30	30	30	30
MATHER 3 NW	32	05-18	09-21	126	30	30	30	30
	28	05-07	10-01	147	30	30	30	30
	24	04-23	10-10	170	29	30	30	30
	20	04-09	10-27	201	30	30	29	29
	16	03-30	11-04	219	30	30	28	28
MEDFORD	32	05-19	09-23	126	29	29	27	27
	28	05-10	10-02	145	28	28	27	27
	24	04-27	10-12	168	30	30	27	27
	20	04-15	10-29	197	29	29	27	27
	16	04-01	11-06	219	29	29	27	27
MELLEN 2 NE	32	06-03	09-10	99	22	22	21	21
	28	05-19	09-26	130	22	22	21	21
	24	05-12	10-10	151	21	21	21	21
	20	04-27	10-25	181	20	20	20	20
	16	04-14	11-01	201	20	20	20	20
MENASHA LOCKS	32	05-05	10-08	155	30	30	30	30
	28	04-18	10-25	190	30	30	30	30
	24	04-07	11-07	214	30	30	30	30
	20	03-30	11-14	229	30	30	30	30
	16	03-24	11-22	244	30	30	30	30
MERRILL	32	05-22	09-23	124	29	29	29	29
	28	05-08	10-02	146	28	28	28	28
	24	04-24	10-14	173	29	29	28	28
	20	04-14	10-29	199	28	28	27	27
	16	04-04	11-09	219	28	28	27	27
MILWAUKEE WB CITY	32	04-20	10-25	188	24	24	25	25
	28	04-05	11-02	210	24	24	25	25
	24	03-28	11-11	228	24	24	25	25
	20	03-24	11-21	243	24	24	25	25
	16	03-12	11-28	261	24	24	25	25
MINOCQUA DAM	32	05-28	09-22	117	28	28	28	28
	28	05-15	10-03	140	28	28	28	28
	24	05-04	10-16	166	28	28	28	28
	20	04-23	10-27	187	28	28	28	28
	16	04-14	11-05	205	28	28	27	27
MONDOVI	32	05-17	09-28	134	24	24	23	23
	28	05-03	10-04	154	22	22	23	23
	24	04-20	10-18	181	25	25	21	21
	20	04-12	10-30	202	24	24	21	21
	16	03-30	11-08	223	24	24	20	20
NEILLSVILLE	32	05-14	09-28	138	29	29	27	27
	28	05-04	10-05	154	29	29	28	28
	24	04-20	10-19	182	29	29	28	28
	20	04-05	11-03	212	29	29	27	27
	16	03-29	11-11	227	29	29	27	27
NEW LONDON	32	05-10	09-30	143	29	29	30	30
	28	04-26	10-13	169	29	29	30	30
	24	04-11	10-24	196	28	28	29	29
	20	04-02	11-06	219	28	28	29	29
	16	03-26	11-16	235	29	29	29	29
OCONTO	32	05-17	09-27	133	29	29	30	30
	28	05-02	10-11	162	29	29	29	29
	24	04-18	10-29	194	29	29	29	29
	20	04-03	11-06	217	28	28	29	29
	16	03-29	11-14	229	29	29	28	28
OSHKOSH BUCKSTAFF OB	32	05-09	10-04	148	30	30	30	30
	28	04-26	10-17	174	30	30	30	30
	24	04-12	10-28	199	30	30	30	30
	20	03-30	11-07	222	30	30	30	30
	16	03-22	11-21	244	30	30	30	30
PARK FALLS	32	05-27	09-16	112	30	30	30	30
	28	05-10	10-01	144	30	30	30	30
	24	04-29	10-12	167	30	30	30	30
	20	04-17	10-27	192	30	30	30	30
	16	04-05	11-05	213	30	30	30	30
PINE RIVER	32	05-11	10-03	145	27	27	26	26
	28	05-01	10-11	164	27	27	27	27
	24	04-16	10-26	194	27	27	27	27
	20	04-04	11-05	215	27	27	26	26
	16	03-29	11-17	233	27	27	26	26
PLYMOUTH	32	05-09	10-11	155	30	30	30	30
	28	04-25	10-21	179	30	30	30	30
	24	04-11	11-02	205	29	29	30	30
	20	04-01	11-11	224	27	27	30	30
	16	03-27	11-20	238	26	26	30	30
PORTAGE LOCK	32	04-29	10-11	165	29	29	30	30
	28	04-17	10-23	189	30	30	30	30
	24	04-05	11-04	213	30	30	30	30
	20	03-30	11-12	227	30	30	30	30
	16	03-21	11-19	244	30	30	30	30
PRAIRIE DU CHIEN	32	04-27	10-09	165	30	30	30	30
	28	04-16	10-21	187	30	30	30	30
	24	04-05	11-03	212	30	30	30	30
	20	03-26	11-09	228	30	30	30	30
	16	03-17	11-17	245	30	30	30	30
PRAIRIE DU SAC 2 N	32	04-26	10-16	173	30	30	30	30
	28	04-12	10-27	198	30	30	30	30
	24	03-31	11-08	222	30	30	30	30
	20	03-25	11-15	236	30	30	30	30
	16	03-18	11-24	251	30	30	30	30
PRENTICE 5 W	32	06-05	08-29	85	30	30	30	30
	28	05-22	09-18	119	30	30	29	29
	24	05-04	09-28	144	30	30	29	29
	20	04-24	10-12	171	30	30	29	29
	16	04-11	10-27	199	30	30	29	29
RACINE	32	04-28	10-20	174	30	30	30	30
	28	04-18	10-31	197	30	30	30	30
	24	03-31	11-10	224	30	30	30	30
	20	03-23	11-20	242	30	30	30	30
	16	03-14	11-28	259	30	30	30	30
REST LAKE	32	05-28	09-16	111	24	24	22	22
	28	05-18	09-28	133	23	23	21	21
	24	05-08	10-13	158	23	23	21	21
	20	04-26	10-28	185	24	24	22	22
	16	04-14	11-08	208	24	24	19	19
RICHLAND CENTER	32	05-10	09-28	142	29	29	29	29
	28	05-01	10-09	161	29	29	29	29
	24	04-18	10-24	189	29	29	29	29
	20	04-03	11-03	214	29	29	29	29
	16	03-24	11-15	237	29	29	29	29
RIVER FALLS	32	05-14	09-26	135	30	30	30	30
	28	05-04	10-08	157	30	30	30	30
	24	04-20	10-18	181	30	30	30	30
	20	04-11	10-30	203	30	30	30	30
	16	03-26	11-08	227	30	30	30	30

FREEZE DATA

STATION	Freeze threshold temperature	Mean date of last Spring occurrence	Mean date of first Fall occurrence	Mean No. of days between dates	Years of record Spring	No. of occurrences in Spring	Years of record Fall	No. of occurrences in Fall
SHAWANO	32	05-18	09-26	131	26	26	25	25
	28	05-06	10-07	154	26	26	26	26
	24	04-19	10-21	185	26	26	26	26
	20	04-08	11-02	208	26	26	26	26
	16	03-30	11-16	231	26	26	26	26
SHEBOYGAN	32	04-28	10-19	174	30	30	29	29
	28	04-17	10-31	197	30	30	29	29
	24	04-03	11-09	220	30	30	29	29
	20	03-29	11-19	236	30	30	29	29
	16	03-21	11-27	252	30	30	29	28
SOLON SPRINGS	32	06-01	09-13	104	28	28	26	26
	28	05-19	09-23	127	29	29	27	27
	24	05-09	10-08	152	29	29	26	26
	20	04-27	10-14	170	29	29	26	26
	16	04-15	10-29	197	29	29	26	26
SPARTA	32	05-10	09-29	142	14	14	15	15
	28	04-24	10-08	167	14	14	15	15
	24	04-11	10-23	195	14	14	15	15
	20	03-30	11-06	220	14	14	15	15
	16	03-21	11-15	239	14	14	15	15
SPOONER EXP FARM	32	05-24	09-20	120	29	29	28	28
	28	05-10	09-30	143	29	29	28	28
	24	04-30	10-12	165	30	30	28	28
	20	04-17	10-23	189	30	30	28	28
	16	04-07	11-05	212	30	30	28	28
STANLEY 1 E	32	05-16	09-23	130	28	28	26	26
	28	05-06	10-04	151	28	28	26	26
	24	04-16	10-18	185	28	28	26	26
	20	04-05	11-03	212	28	28	24	24
	16	03-29	11-11	227	27	27	24	24
STEVENS PT	32	05-11	10-01	142	30	30	30	30
	28	04-28	10-12	167	30	30	30	30
	24	04-15	10-26	195	30	30	30	30
	20	04-03	11-07	217	30	30	30	30
	16	03-27	11-16	234	30	30	30	30
STURGEON BAY EXP FRM	32	05-17	10-02	137	27	27	29	29
	28	04-30	10-18	171	27	27	29	29
	24	04-17	11-02	199	27	27	29	29
	20	04-04	11-14	224	27	27	28	28
	16	03-26	11-25	244	27	27	28	28
SUPERIOR BONG AP	32	05-12	10-04	145	30	30	30	30
	28	04-26	10-16	173	30	30	30	30
	24	04-14	10-29	197	30	30	30	30
	20	04-04	11-09	218	30	30	30	30
	16	03-30	11-13	229	30	30	30	30
VIROQUA	32	05-06	10-05	152	24	24	26	26
	28	04-25	10-19	177	25	25	25	25
	24	04-10	10-30	203	26	26	25	25
	20	04-01	11-09	222	27	27	25	25
	16	03-23	11-16	239	27	27	25	25
WASHINGTON ISL	32	05-17	10-17	153	19	19	19	19
	28	05-02	10-31	182	18	18	19	19
	24	04-15	11-09	208	18	18	19	19
	20	04-08	11-18	225	18	18	19	19
	16	03-28	11-28	245	18	18	19	19
WATERTOWN	32	05-04	10-10	159	30	30	30	30
	28	04-21	10-23	186	30	30	30	30
	24	04-02	11-05	217	30	30	30	30
	20	03-27	11-12	230	30	30	30	30
	16	03-18	11-23	249	30	30	30	30
WAUKESHA WTR WRKS	32	05-08	10-08	153	30	30	28	28
	28	04-23	10-20	180	30	30	28	28
	24	04-09	11-02	208	30	30	28	28
	20	03-31	11-10	224	30	30	28	28
	16	03-20	11-21	247	30	30	29	29
WAUPACA	32	05-11	10-02	144	29	29	28	28
	28	04-29	10-13	167	29	29	28	28
	24	04-13	10-26	196	29	29	28	28
	20	04-04	11-08	218	29	29	28	28
	16	03-24	11-20	240	29	29	28	28
WAUSAU	32	05-14	09-30	139	30	30	30	30
	28	05-03	10-10	160	30	30	30	30
	24	04-17	10-24	190	30	30	30	30
	20	04-08	11-09	215	30	30	30	30
	16	03-28	11-14	231	30	30	30	30
WEST BEND	32	05-10	10-08	151	25	25	25	25
	28	04-24	10-20	179	25	25	25	25
	24	04-10	10-31	204	25	25	25	25
	20	03-31	11-10	224	25	25	25	25
	16	03-23	11-20	242	25	25	25	25
WEYERHAUSER	32	05-20	09-22	125	30	30	29	29
	28	05-10	10-01	145	30	30	30	30
	24	04-26	10-15	172	30	30	30	30
	20	04-13	10-28	198	30	30	30	30
	16	04-02	11-04	216	30	30	30	30
WILLIAMS BAY YERKES OBS	32	05-02	10-10	160	30	30	30	30
	28	04-24	10-26	185	29	29	30	30
	24	04-06	11-01	209	29	29	30	30
	20	03-27	11-12	230	29	29	30	30
	16	03-21	11-22	246	29	29	30	30
WINTER PK RES	32	06-03	09-10	99	15	15	15	15
	28	05-24	09-25	124	14	14	15	15
	24	05-16	10-01	139	14	14	15	15
	20	04-28	10-12	168	14	14	15	15
	16	04-12	10-30	200	14	14	15	15
WISCONSIN DELLS	32	05-08	10-03	148	19	19	20	20
	28	04-27	10-15	171	19	19	20	20
	24	04-17	10-25	191	19	19	20	20
	20	04-07	11-04	211	19	19	20	20
	16	03-25	11-15	235	19	19	20	20
WISCONSIN RAPIDS	32	05-19	09-20	124	30	30	30	30
	28	05-08	09-29	144	30	30	30	30
	24	04-26	10-12	169	30	30	29	29
	20	04-15	10-23	191	30	30	29	29
	16	04-01	11-06	219	30	30	29	29

Data in the above table are based on the period 1921-1950, or that portion of this period for which data are available.

Means have been adjusted to take into account years of non-occurrence.

A freeze is a numerical substitute for the former term "killing frost" and is the occurrence of a minimum temperature at or below the threshold temperature of 32°, 28°, etc.

Freeze data tabulations in greater detail are available and can be reproduced at cost.

*MEAN TEMPERATURE AND PRECIPITATION

STATION	JANUARY		FEBRUARY		MARCH		APRIL		MAY		JUNE		JULY		AUGUST		SEPTEMBER		OCTOBER		NOVEMBER		DECEMBER		ANNUAL	
	Temperature	Precipitation	Temperature	Precipitation	Temperature	Precipitation	Temperature	Precipitation	Temperature	Precipitation	Temperature	Precipitation	Temperature	Precipitation	Temperature	Precipitation	Temperature	Precipitation	Temperature	Precipitation	Temperature	Precipitation	Temperature	Precipitation	Temperature	Precipitation
NORTHWEST DIVISION																										
AMERY BLACK BROOK HYDRO	12.3	.81	14.8	.88	26.2	1.46	43.4	2.24	56.1	3.42	65.6	4.99	71.1	3.24	68.6	3.69	59.6	3.00	48.2	1.80	30.9	1.58	17.4	.87	42.9	27.98
ASHLAND EXP FARM	13.7	1.01	15.5	.76	24.8	1.27	38.9	2.26	50.2	3.44	60.7	4.55	69.6	3.85	65.3	3.95	56.8	2.64	46.3	2.18	30.9	1.93	18.7	.91	41.0	28.75
DANBURY 1 SE	10.5	1.07	13.6	.97	25.2	1.65	41.7	2.36	54.3	3.65	64.3	5.13	69.9	3.52	67.0	4.44	57.8	3.18	46.6	2.13	29.7	1.78	15.7	1.04	41.4	30.92
HOLCOMBE		1.04		.86		1.64		2.89		3.76		4.61		3.73		3.68		3.72		2.05		1.64		1.04		30.66
LADYSMITH		.90		.87		1.66		2.53		3.50		4.65		3.63		3.76		3.19		2.18		1.74		1.03		29.64
SOLON SPRINGS	12.0		14.5		25.3		41.3		53.6		63.3		69.0		66.3		57.1		46.8		30.0		16.7		41.3	
SPOONER EXP FARM	12.4	.81	14.9	.70	26.2	1.41	42.7	2.23	55.5	3.28	65.0	4.39	70.5	3.79	67.8	3.91	58.5	3.16	47.5	1.88	30.4	1.63	17.2	.90	42.4	28.09
SUPERIOR 7 SE	12.9		15.5		25.4		39.4		49.6		59.3		67.0		66.2		56.8		46.6		30.7		18.2		40.6	
WEYERHAUSER 1 N	12.9		15.5		26.1		42.5		54.8		63.8		69.1		66.5		56.8		46.9		30.4		17.3		42.0	
WINTER 6 NNW	10.3	.99	11.3	.92	21.9	1.51	39.0	2.40	52.1	3.73	61.6	4.88	66.5	4.06	63.6	4.29	54.7	3.19	43.9	2.04	28.5	1.90	14.7	.93	39.0	30.84
DIVISION	12.4	1.00	15.0	.90	25.4	1.59	41.6	2.52	53.8	3.61	63.4	4.77	69.1	3.76	66.8	4.01	57.7	3.12	47.1	2.06	30.5	1.78	17.4	1.02	41.7	30.14
NORTH CENTRAL DIVISION																										
BIG SAINT GERMAIN DAM	12.4	1.26	14.2	1.04	23.8	1.53	39.0	2.11	52.2	3.56	61.5	4.73	66.1	4.11	63.6	3.91	55.4	3.86	44.9	2.27	28.9	2.17	16.6	1.17	39.9	31.72
FLAMBEAU RESERVOIR		1.10		1.09		1.59		2.48		3.81		5.33		4.31		4.06		3.31		2.22		2.08		1.15		32.53
LONG LAKE DAM	12.6	1.43	13.6	1.19	23.1	1.65	38.6	2.44	51.8	3.38	61.4	4.47	66.0	4.14	63.6	4.02	55.4	3.63	44.7	2.46	29.1	2.36	16.8	1.32	39.7	32.47
MEDFORD	13.5	1.36	15.3	1.20	25.6	1.85	41.9	2.45	54.2	3.96	63.6	5.21	68.4	3.46	66.9	4.15	57.7	3.79	46.6	2.15	30.3	2.17	17.6	1.46	41.8	33.21
MELLEN 2 N	13.5		14.7		24.3		40.0		52.1		62.0		67.7		64.9		56.8		46.3		30.8		18.2		40.9	
MERRILL		1.17		.96		1.44		2.26		3.71		4.64		3.21		3.80		3.66		2.15		2.11		1.06		30.17
PARK FALLS	12.7	1.19	14.3	1.04	24.7	1.61	40.5	2.63	53.4	3.56	62.9	5.68	68.1	4.27	65.4	4.40	56.6	3.33	45.7	2.29	29.2	2.00	16.5	1.19	40.8	33.19
PHELPS DEERSKIN DAM				1.12		1.44		2.38		3.32		4.70		3.85		3.96		3.78		2.43		2.23		1.28		31.88
PRENTICE 5 W	12.9	1.45	14.6	1.27	25.0	2.04	41.3	2.70	53.6	4.01	62.7	5.40	67.3	3.92	64.7	4.21	56.4	3.99	46.2	2.39	30.0	2.10	17.3	1.30	41.0	34.78
RHINELANDER	13.1	1.33	14.6	1.26	24.8	1.64	40.6	2.18	53.5	3.40	63.3	4.81	68.3	3.80	65.6	3.80	57.0	3.50	46.4	2.34	30.3	2.00	17.7	1.20	41.3	31.26
WAUSAU OLD POST OFFICE	16.9	1.43	18.3	1.35	28.8	1.91	44.5	2.66	57.3	3.75	67.1	4.76	72.1	3.55	69.5	4.04	60.9	3.54	49.2	2.38	33.3	2.22	21.0	1.31	44.9	32.90
DIVISION	13.4	1.28	15.2	1.11	25.0	1.68	41.0	2.47	53.6	3.67	63.2	4.95	68.1	3.99	65.8	4.00	57.2	3.59	46.7	2.28	30.5	2.07	17.9	1.20	41.5	32.29
NORTHEAST DIVISION																										
ANTIGO	16.1	1.30	17.4	1.03	27.0	1.51	42.5	2.47	55.2	3.46	64.5	4.58	69.4	3.58	67.0	3.79	58.8	3.60	47.9	2.28	32.0	1.97	19.8	1.08	43.1	30.65
BREAKWATER		1.30				1.72		2.26		3.18		3.82		3.27		3.44		2.99		1.96		2.31		1.22		28.62
BRULE ISLAND		1.40		1.27		1.70		2.16		3.33		4.46		3.58		3.27		3.37		2.08		2.31		1.23		30.16
CRIVITZ HIGH FALLS	16.6	1.30	17.7	1.09	27.2	1.64	42.2	2.60	54.8	2.82	65.0	3.83	70.0	3.05	67.5	3.18	58.9	3.13	48.0	1.96	33.1	2.29	20.3	1.25	43.4	28.14
MARINETTE	20.4	1.59	21.5	1.27	30.0	1.65	43.2	2.37	55.2	2.78	66.0	3.75	71.9	2.71	69.5	3.04	61.5	3.14	50.6	2.17	35.8	2.43	24.4	1.29	45.8	28.19
OCONTO	18.6	1.56	19.8	1.37	29.4	1.74	43.4	2.55	55.1	2.48	65.7	3.41	70.9	2.42	68.7	2.90	60.3	3.07	49.4	2.02	35.0	2.24	23.0	1.29	44.9	27.05
SHAWANO	17.5	1.65	19.2	1.43	29.3	1.77	44.6	2.66	57.0	3.19	66.9	4.19	71.7	2.85	69.0	3.66	60.4	3.13	48.8	2.15	34.1	2.33	21.7	1.58	45.0	30.59
DIVISION	17.0	1.47	18.3	1.25	27.7	1.66	42.6	2.45	54.7	3.07	64.7	4.08	69.6	3.14	67.3	3.29	58.9	3.23	48.2	2.11	33.3	2.25	21.2	1.28	43.6	29.28
WEST CENTRAL DIVISION																										
BLAIR	15.0	1.31	17.3	1.12	29.2	1.91	45.0	2.68	57.4	3.58	67.3	4.54	72.1	3.94	69.5	3.66	60.7	3.70	49.2	2.06	32.8	1.89	19.6	1.05	44.6	31.44
EAU CLAIRE	15.7	1.05	18.4	1.06	29.5	1.90	45.4	2.88	58.8	3.52	68.8	4.61	74.3	3.33	71.6	3.70	62.2	3.43	50.3	2.06	33.3	1.82	20.5	1.06	45.7	30.42
HATFIELD DAM	15.2	.99	17.7	.85	28.3	1.67	44.2	2.73	56.9	3.95	66.3	5.11	71.0	3.36	68.6	3.46	60.0	3.46	48.9	2.26	32.8	1.83	20.2	1.00	44.2	30.67
LA CROSSE WB AP	15.7	1.22	19.3	1.11	31.6	1.86	46.6	2.31	59.0	3.27	68.6	3.87	74.0	3.21	71.4	3.29	62.3	3.82	50.8	1.93	34.3	1.81	20.5	1.22	46.2	28.92
MATHER 3 NW	15.2	1.31	17.2	1.12	28.1	1.91	43.7	2.68	56.2	4.00	65.8	4.67														
RIVER FALLS	13.3	1.00	16.3	.92	27.8	1.80	44.5	2.54	57.2	3.85	66.9	4.69	72.2	3.71	69.7	3.20	60.6	3.30	49.2	1.90	32.0	1.63	18.6	1.17	44.0	29.71
DIVISION	15.2	1.05	18.2	.97	29.0	1.85	45.1	2.70	57.6	3.73	67.3	4.78	72.3	3.62	69.9	3.50	60.9	3.31	49.8	1.98	33.0	1.75	19.9	1.08	44.9	30.32
CENTRAL DIVISION																										
CODDINGTON 1 E	15.2	1.05	16.7	.95	27.3	1.55	42.8	2.87	54.7	3.63	64.1	5.31	68.9	3.26	66.3	3.36	58.1	3.59	47.3	2.41	32.0	2.25	19.1	1.00	42.7	31.23
HANCOCK EXP FARM	16.5	1.06	18.3	.98	28.7	1.51	44.5	2.61	57.0	3.59	67.2	4.64	72.3	3.12	69.5	3.03	60.8	3.61	49.7	2.29	33.2	2.17	20.4	1.06	44.8	29.67
MARSHFIELD EXP FARM	14.8	1.31	16.7	1.10	27.1	1.71	43.1	2.79	55.3	3.69	64.9	4.85	69.8	3.22	67.5	3.90	59.0	3.47	47.9	2.44	31.8	2.02	19.1	1.14	43.1	31.64
MAUSTON		1.15		1.18		1.79		2.72		3.37		4.24		3.52		3.31		3.36		2.10		2.00		1.03		29.77
MONTELLO	19.0	1.21	20.6	1.12	31.1	1.64	45.7	2.81	57.5	3.13	67.3	4.43	71.8	3.10	69.5	3.18	61.3	3.35	50.6	1.95	35.2	2.02	22.6	1.21	46.0	29.15
NEW LONDON	18.2	1.55	19.7	1.32	30.1	1.95	45.1	3.00	57.3	3.25	67.4	4.47	72.3	2.86	69.6	2.95	61.3	3.44	50.2	2.15	34.7	2.29	22.1	1.39	45.7	30.62
STEVENS POINT	16.7	1.54	18.2	1.36	29.0	1.77	44.8	2.76	57.6	3.79	67.7	4.88	72.8	3.03	70.2	3.61	61.5	3.57	49.7	2.17	33.5	2.15	20.7	1.36	45.2	31.99
WAUPACA	18.0	1.21	19.3	1.02	29.7	1.57	45.0	2.79	57.6	3.36	67.4	4.64	72.4	2.88	69.8	3.34	61.7	3.37	50.3	2.13	34.8	2.10	22.1	1.12	45.7	29.53
WISCONSIN RAPIDS	15.4	1.14	17.0	1.07	27.8	1.69	43.4	2.68	56.1	3.69	66.2	4.88	71.2	3.10	68.9	3.39	59.5	3.67	48.2	2.30	32.1	2.17	19.5	1.21	43.8	30.99
DIVISION	16.8	1.23	18.7	1.15	29.0	1.72	44.5	2.84	56.4	3.51	66.4	4.61	71.4	3.19	68.9	3.35	60.4	3.37	49.5	2.19	33.5	2.10	20.8	1.18	44.7	30.44

* Averages for period 1931-1955, except for stations marked WB which are "normals" based on period 1921-1950. Divisional means may not be the arithmetical average of individual stations published, since additional data from shorter period stations are used to obtain better areal representation.

*MEAN TEMPERATURE AND PRECIPITATION

STATION	JANUARY Temperature	JANUARY Precipitation	FEBRUARY Temperature	FEBRUARY Precipitation	MARCH Temperature	MARCH Precipitation	APRIL Temperature	APRIL Precipitation	MAY Temperature	MAY Precipitation	JUNE Temperature	JUNE Precipitation	JULY Temperature	JULY Precipitation	AUGUST Temperature	AUGUST Precipitation	SEPTEMBER Temperature	SEPTEMBER Precipitation	OCTOBER Temperature	OCTOBER Precipitation	NOVEMBER Temperature	NOVEMBER Precipitation	DECEMBER Temperature	DECEMBER Precipitation	ANNUAL Temperature	ANNUAL Precipitation
EAST CENTRAL DIVISION																										
APPLETON	18.5	1.36	19.7	1.32	29.7	1.70	44.0	2.59	56.5	2.86	67.1	4.09	72.5	2.80	70.2	2.82	61.7	3.23	50.3	1.92	34.9	2.13	22.7	1.41	45.7	28.23
FOND DU LAC	20.4	1.43	22.0	1.35	32.0	1.84	46.3	2.45	57.9	2.82	68.0	4.08	73.0	3.25	70.8	3.34	62.6	3.23	51.6	1.99	36.2	2.06	23.7	1.38	47.0	29.22
GREEN BAY WB AP	16.1	1.29	17.3	1.36	28.5	1.76	41.8	2.51	54.4	2.53	64.7	3.57	69.9	2.59	67.8	3.03	60.2	2.87	48.4	1.80	33.5	1.94	20.1	1.26	43.6	26.51
KEWAUNEE	21.2	1.51	21.9	1.40	30.8	1.65	42.1	2.50	51.9	2.42	61.7	3.35	69.2	2.82	68.5	2.78	60.7	2.94	50.4	1.75	36.3	2.26	25.0	1.51	45.0	26.89
MANITOWOC	22.3	1.53	23.2	1.44	31.4	1.90	43.4	2.64	54.1	2.63	64.5	3.82	71.4	2.38	69.9	3.02	61.7	3.20	51.1	2.05	37.1	2.19	25.9	1.45	46.3	28.25
OSHKOSH	19.0	1.42	20.3	1.23	30.2	1.63	44.6	2.59	56.9	2.64	67.5	4.06	72.8	2.78	70.7	3.18	62.3	3.25	50.9	1.85	35.2	2.14	22.7	1.35	46.1	28.12
PLYMOUTH	20.7		21.8		30.7		44.1		55.2		65.4		71.2		69.7		61.5		50.8		36.0		24.1		45.9	
SHEBOYGAN	21.7	1.77	22.6	1.57	31.8	2.01	43.5	2.41	53.7	2.99	64.5	4.01	72.0	2.75	70.8	3.00	63.0	3.11	51.8	2.22	37.1	2.18	25.4	1.74	46.5	29.76
STURGEON BAY EXP FARM	19.1	1.34	19.2	1.33	28.2	1.76	41.2	2.43	52.1	2.46	62.6	3.20	69.0	2.75	67.4	2.86	59.3	3.25	48.7	2.16	35.0	2.36	23.8	1.32	43.8	27.22
DIVISION	20.0	1.46	21.3	1.33	30.6	1.78	43.9	2.58	54.9	2.71	65.3	3.83	71.3	2.90	69.7	3.01	61.5	3.10	50.8	1.98	35.9	2.16	23.9	1.42	45.8	28.26
SOUTHWEST DIVISION																										
DARLINGTON	20.6	1.39	23.5	1.08	33.6	2.07	47.1	2.80	57.9	3.59	67.9	4.94	72.5	3.82	70.0	4.28	62.0	3.63	51.3	2.32	36.1	2.18	23.9	1.42	47.2	33.52
HILLSBORO	18.2	1.23	20.6	1.15	30.8	1.97	45.6	2.85	57.3	3.47	67.2	4.56	72.1	3.67	69.4	3.46	61.1	3.93	50.1	2.24	34.6	2.29	22.0	1.20	45.8	32.02
LANCASTER	19.9	1.32	22.6	1.13	32.7	2.33	47.2	2.73	59.0	3.73	68.7	5.20	73.9	3.86	71.6	3.60	63.4	3.78	52.5	2.32	36.0	2.16	23.6	1.42	47.6	33.58
PRAIRIE DU CHIEN	20.9	1.22	23.5	1.11	34.1	2.19	49.1	2.71	60.9	3.82	70.6	5.11	75.4	3.80	73.1	4.09	64.7	3.76	53.1	2.07	37.1	2.09	24.8	1.32	48.9	33.29
PRAIRIE DU SAC 2 N	19.2	1.18	21.2	1.04	31.1	1.59	45.7	2.41	58.0	3.05	68.2	4.16	73.8	3.45	71.3	3.37	62.5	3.55	51.2	1.91	35.4	1.92	23.1	1.16	46.7	28.79
RICHLAND CENTER	20.1	1.24	22.8	1.17	33.0	2.17	47.3	2.65	58.9	3.48	68.7	5.20	73.4	4.03	71.0	3.73	62.7	4.02	51.5	2.31	36.0	2.13	23.7	1.30	47.4	33.43
VIROQUA	17.2	1.23	19.8	1.13	30.4	1.97	45.6	2.56	57.6	3.69	68.2	4.79	73.4	4.16	71.0	3.32	62.0	3.84	50.7	2.09	33.9	1.90	21.2	1.17	45.9	31.85
DIVISION	19.3	1.22	22.4	1.11	32.2	2.03	47.2	2.81	58.4	3.56	68.4	5.10	73.3	3.94	71.0	3.66	62.5	3.56	51.7	2.14	35.5	1.96	23.0	1.27	47.1	32.36
SOUTH CENTRAL DIV																										
ARLINGTON		1.48		1.20		1.72		2.63		3.35		3.94		3.61		3.59		4.06		2.03		2.05		1.34		31.00
BELOIT	23.3	1.64	25.5	1.29	35.4	2.03	49.0	2.60	60.1	3.46	70.1	4.55	74.9	3.75	72.5	3.80	64.7	3.82	53.9	2.34	38.5	2.33	26.5	1.61	49.5	33.22
BRODHEAD 1 SW	21.4	1.77	23.6	1.36	33.7	2.21	47.6	2.90	59.2	3.42	69.5	4.43	74.5	3.46	72.0	4.25	63.6	3.77	52.3	2.28	36.7	2.36	24.5	1.65	48.2	33.86
LAKE MILLS	21.2	1.66	22.8	1.21	32.4	1.92	46.5	2.67	57.9	3.21	67.8	4.33	73.1	3.53	71.2	2.95	62.9	3.51	52.1	2.25	36.9	2.34	24.5	1.60	47.4	31.18
MADISON WB AP	19.1	1.31	21.9	1.13	32.5	1.83	45.7	2.49	57.5	3.27	67.4	4.02	73.0	3.30	70.7	2.89	62.1	3.99	50.4	2.08	35.3	2.29	23.0	1.40	46.6	30.00
MADISON WB CITY	19.3	1.47	21.9	1.27	32.4	2.03	45.9	2.49	57.7	3.21	67.7	4.02	73.1	3.40	70.9	3.07	62.9	4.11	51.7	2.00	36.2	2.20	23.5	1.44	46.9	30.71
PORTAGE	20.6	1.48	22.7	1.25	32.7	1.95	47.5	2.82	59.6	3.02	69.4	4.21	74.4	3.41	71.8	3.33	63.7	3.90	52.5	1.93	36.9	2.11	24.2	1.36	48.0	30.07
STOUGHTON		1.58		1.24		1.90		2.55		3.21		4.05		3.24		3.32		3.28		2.14		2.10		1.46		30.07
WATERTOWN	22.2	1.71	22.9	1.32	33.3	2.09	46.9	2.73	58.4	3.10	68.5	4.10	73.7	3.40	71.5	3.23	63.1	3.59	52.2	2.26	37.1	2.13	25.2	1.60	48.0	31.26
WISCONSIN DELLS	18.6	1.31	20.8	1.21	30.9	2.02	45.3	2.80	57.6	3.03	67.6	4.24	73.0	3.66	70.6	3.48	61.7	3.87	50.7	2.16	34.8	2.13	22.5	1.24	46.2	31.15
DIVISION	20.8	1.53	22.9	1.23	32.8	1.93	46.9	2.67	58.4	3.23	68.4	4.26	73.5	3.51	71.1	3.45	63.0	3.68	52.1	2.13	36.6	2.14	24.2	1.47	47.6	31.23
SOUTHEAST DIVISION																										
MILWAUKEE WB AP	21.9	1.58	24.2	1.27	33.3	2.19	44.3	2.39	54.3	2.98	64.9	3.22	71.3	2.43	69.9	2.62	62.6	3.33	51.4	1.97	37.3	2.11	25.7	1.48	46.8	27.57
PORT WASHINGTON		1.58		1.35		1.84		2.50		2.73		3.75		2.74		2.80		2.94		2.03		1.95		1.50		27.71
RACINE	24.9	1.99	26.2	1.55	34.4	2.63	45.8	2.58	55.8	3.84	66.9	3.65	73.2	2.96	72.1	3.26	64.6	3.26	53.6	1.93	39.0	2.34	27.6	1.96	48.7	31.95
WAUKESHA	21.3	1.84	22.9	1.43	32.1	2.14	45.2	2.43	56.3	3.53	66.8	3.90	72.3	3.30	70.6	3.08	62.3	2.99	51.3	2.02	36.4	2.29	24.6	1.61	46.8	30.48
WEST BEND	20.9	1.82	22.3	1.43	31.4	2.07	44.9	2.51	56.3	2.91	66.8	4.20	72.2	3.21	70.0	2.94	62.0	3.29	51.1	2.15	36.3	2.15	24.2	1.55	46.5	30.23
WILLIAMS BAY	21.8	1.96	23.4	1.32	32.7	2.42	46.3	2.68	57.7	3.59	68.0	4.08	73.3	3.80	71.5	3.53	63.6	3.36	52.6	2.17	36.9	2.45	24.9	1.75	47.7	33.11
DIVISION	22.1	1.79	24.2	1.32	32.8	2.59	46.0	2.81	56.5	3.36	67.1	4.22	72.6	3.60	71.0	3.14	63.0	2.95	52.5	1.97	37.1	2.17	25.2	1.75	47.5	31.67

* Averages for period 1931-1955, except for stations marked WB which are "normals" based on period 1921-1950. Divisional means may not be the arithmetical average of individual stations published, since additional data from shorter period stations are used to obtain better areal representation.

CONFIDENCE LIMITS

In the absence of trend or record changes, the chances are 9 out of 10 that the true mean will lie in the interval formed by adding and subtracting the values in the following table from the means for any station in the State.

2.0	.21	1.7	.18	1.7	.26	1.3	.47	1.2	.71	1.2	.86	.9	.55	.9	.66	1.1	.57	1.3	.47	1.3	.55	1.7	.22	.6	2.51

COMPARATIVE DATA

Data in the following table are the mean temperature and average precipitation for Hancock, Wisconsin for the period 1906 - 1930 and are included in this publication for comparative purposes:

13.1	1.13	17.1	1.21	29.8	1.63	44.4	3.15	55.9	4.16	65.6	4.53	71.1	3.27	68.6	3.38	61.1	3.95	48.0	2.28	34.0	1.74	19.6	1.19	44.0	31.62

NORMALS, MEANS, AND EXTREMES

GREEN BAY, WISCONSIN — AUSTIN STRAUBEL AIRPORT

LATITUDE 44° 29' N
LONGITUDE 88° 08' W
ELEVATION (ground) 689 Feet

Month	(b)	Temp Daily max	Temp Daily min	Temp Monthly	Normal degree days	Rec high	Yr	Rec low	Yr	Norm precip total	Max monthly precip	Yr	Min monthly precip	Yr	Max 24h precip	Yr	Snow mean total	Max monthly snow	Yr	Max 24h snow	Yr	RH midnight	RH 6AM	RH noon	RH 6PM	Wind mean hourly	Prev dir	Fastest mi speed	Dir	Yr	Mean sky cover	Pct sun	Clear	Partly cloudy	Cloudy	Precip .01+	Snow 1.0+	T-storms	Heavy fog	Max 90+	Max 32-	Min 32-	Min 0-
		(b)	(b)	(b)	(b)	8		8		(b)					8		8			8		8	8	8	8	8		8															
J		24.5	7.7	16.1	1516	44	1953	-31	1951	1.29	2.64	1950	.35	1957	1.05	1950	8.2	15.5	1952	5.2	1952	75	75	68	73	10.6	SW	61	W	1950	6.7	47	7	7	17	10	3	*	2	0	23	31	8
F		26.0	8.6	17.3	1336	50	1954	-24	1950	1.36	3.56	1953	.43	1957	1.51	1953	9.2	18.4	1953	8.2	1950	77	79	69	74	10.1	SW	66	W	1951	6.5	51	7	7	14	9	3	*	4	0	18	31	11
M		36.7	20.2	28.5	1132	56	1945	-26	1950	1.76	2.66	1951	.46	1957	.96	1950	9.2	19.0	1950	9.3	1956	79	79	68	71	11.6	NW	68	NE	1950	6.7	56	7	7	16	11	2	1	2	0	8	28	7
A		51.7	31.9	41.8	696	84	1952	10	1954	2.51	5.52	1953	1.45	1956	1.75	1953	2.5	6.9	1957	3.5	1957	77	78	57	62	11.8	NE	57	SW	1952	6.4	57	7	7	14	11	1	3	1	*	*	16	1
M		65.2	43.6	54.4	347	92	1952	25	1954	2.53	5.28	1957	.89	1951	2.73	1956	.1	1.1	1954	.8	1954	79	83	59	59	9.7	NE	109	SW	1950	6.4	58	7	7	14	12	1	3	3	*	0	3	0
J		75.2	54.1	64.7	107	92	1956#	33	1956#	3.57	4.38	1954	1.90	1953	1.56	1954	.0	.0		.0		85	86	64	64	8.2	SW	70	NE	1957	6.6	64	10	7	11	11	0	6	2	1	0	0	0
J		81.1	58.7	69.9	32	98	1955	44	1952#	2.59	6.50	1950	2.94	1954	2.32	1956	.0	.0		.0		87	89	59	68	7.5	SW	50	SW	1950	5.6	65	10	10	11	10	0	8	1	2	0	0	0
A		78.5	57.1	67.8	58	99	1955	38	1950	3.03	6.50	1951	.76	1955	2.62	1951	.0	.0		T		89	89	59	64	7.5	SW	56	W	1951	5.6	62	9	9	10	10	0	5	3	2	0	0	0
S		69.9	50.4	60.2	183	99	1955#	27	1956	2.87	5.78	1954	.70	1955	1.60	1954	.0	.0		.0		86	85	56	68	9.4	SW	65	SW	1951	5.6	59	10	10	12	10	0	3	3	1	0	0	0
O		57.5	39.3	48.4	515	83	1953	16	1952	1.80	5.00	1954	.39	1953	3.68	1954	T	+.0		T		81	77	56	71	9.9	SW	65	W	1951	5.6	57	10	11	12	8	0	1	3	0	0	2	0
N		40.6	26.4	33.5	945	72	1953	-18	1950	1.94	3.52	1957	.42	1954	1.25	1957	4.2	8.9	1951#	3.8	1951#	79	80	71	77	12.0	SW	73	SW	1950	7.0	40	9	11	10	11	3	1	3	0	7	23	1
D		27.1	13.0	20.1	1392	53	1951	-31	1950	1.26	1.84	1950	.42	1950	.86	1950	7.8	14.1	1950	5.3	1950	79	79	72	77	11.0	SW	52	W	1957#	7.3	37	6	12	19	11	3	*	3	0	18	30	5
Year		52.8	34.3	43.6	8259	99	AUG 1955	-31	JAN 1951	26.51	6.50	JULY 1950	T	OCT 1952	3.68	OCT 1954	40.3	19.0	MAR 1956	9.3	MAR 1956	81	82	62	69	10.2	SW	109	SW	MAY 1950	6.3	55	94	103	168	121	13	37	29	7	77	169	25

Means and extremes in the above table are from the existing or comparable location(s). Annual extremes have been exceeded at prior locations as follows: Highest temperature 104 in July 1936; lowest temperature -36 in January 1888; maximum monthly precipitation 9.70 in May 1918; maximum precipitation in 24 hours 4.41 in June 1914; maximum monthly snowfall 32.1 in March 1923; maximum snowfall in 24 hours 20.0 in January 1893.

LA CROSSE, WISCONSIN — MUNICIPAL AIRPORT

LATITUDE 43° 52' N
LONGITUDE 91° 15' W
ELEVATION (ground) 652 Feet

Month	(b)	Temp Daily max	Temp Daily min	Temp Monthly	Normal degree days	Rec high	Yr	Rec low	Yr	Norm precip total	Max monthly precip	Yr	Min monthly precip	Yr	Max 24h precip	Yr	Snow mean total	Max monthly snow	Yr	Max 24h snow	Yr	RH midnight	RH 6AM	RH noon	RH 6PM	Wind mean hourly	Prev dir	Fastest mi speed	Dir	Yr	Mean sky cover	Pct sun	Clear	Partly cloudy	Cloudy	Precip .01+	Snow 1.0+	T-storms	Heavy fog	Max 90+	Max 32-	Min 32-	Min 0-
		(b)	(b)	(b)	(b)	7		7		(b)					7		7			7		3	7	7	7	7		7															
J		25.0	6.3	15.7	1528	46	1957#	-37	1951	1.22	2.22	1952	.25	1957	.60	1951	8.1	15.1	1952	6.0	1952	78	77	66	72	9.0	NW	35	NNW	1951	6.9		7	7	17	11	8	*	2	0	23	31	10
F		28.6	9.9	19.3	1280	56	1954	-26	1951	1.11	1.79	1951	.19	1951	.87	1951	5.5	11.5	1951	7.1	1952	80	80	65	70	8.8	NW	36	NW	1952	6.7		6	8	15	8	2	1	1	0	14	27	5
M		40.7	22.5	31.6	1035	70	1953	-18	1956#	1.86	3.82	1951	1.03	1957	1.58	1957	6.3	10.2	1951	9.1	1951	77	79	60	64	10.6	NW	37	W	1957#	6.7		6	7	16	12	3	2	1	0	2	16	1
A		56.9	36.3	46.6	552	82	1952	10	1954	2.51	6.79	1954	1.82	1955	3.84	1954	2.3	T	1954	7.3	1954	76	77	54	55	10.8	NW	50	SSW	1951	6.5		8	7	15	12	*	4	1	*	0	*	10	0
M		69.8	48.1	59.0	250	92	1953	28	1954	3.27	5.68	1956	1.75	1956	2.61	1956	T	.0		.0		77	77	53	57	9.0	S	60	E	1952	6.7		7	9	15	12	0	6	1	4	0	0	*	0
J		79.0	58.1	68.6	74	98	1953	41	1951	3.87	7.67	1955	1.66	1955	3.06	1955	.0	.0		.0		82	81	56	57	7.6	S	36	NNW	1951	6.2		7	11	12	11	0	10	4	4	0	0	0	0
J		85.1	62.8	74.0	11	101	1955	49	1951	3.21	9.16	1953	2.47	1956	3.72	1956	.0	.0		.0		84	85	59	61	6.5	S	25	NE	1951	5.9		8	11	12	10	0	7	5	10	0	0	0	0
A		82.1	60.6	71.4	20	103	1955	47	1956	3.82	4.52	1952	.79	1955	1.75	1955	.0	.0		.0		86	88	58	61	6.5	S	36	SSW	1955	6.0		8	11	12	12	0	5	9	3	0	0	0	0
S		72.8	51.7	62.3	152	100	1953	33	1951	3.82	4.47	1954	.42	1952	1.53	1951	T	.2	1952	.2	1952	82	87	54	54	8.6	S	38	WNW	1951	5.5		12	10	8	10	0	7	6	1	0	0	6	0
O		61.3	40.3	50.8	447	86	1953	19	1951	1.93	2.88	1957	.02	1952	1.14	1957	T	.2	1957	.2	1957	75	81	54	60	10.0	S	35	WNW	1953#	5.3		6	6	18	9	0	1	7	0	0	0	7	0
N		42.5	26.1	34.3	921	74	1953	-18	1951	1.81	2.88	1955	.62	1955	.76	1953	6.3	13.0	1952	11.0	1952	72	79	63	69	11.5	S	43	NNW	1957	7.1		6	6	18	9	1	1	2	0	7	22	6
D		28.7	12.5	20.5	1380	60	1951	-18	1951	1.22	1.62	1953	.47	1953			7.0	14.2	1952	7.3	1952	78	80	70	74	9.5	S				7.5		6	6	19	9	2	*	1	0	17	30	5
Year		56.0	36.3	46.2	7650	103	AUG 1955	-37	JAN 1951	28.92	9.16	JULY 1953	.02	OCT 1952	3.84	APR 1954	40.8	23.8	MAR 1951	11.0	NOV 1957	79	81	59	63	9.5	S	60	NNW	JUNE 1954	6.4		90	100	175	112	10	47	22	15	70	154	21

Means and extremes in the above table are from the existing or comparable location(s). Annual extremes have been exceeded at prior locations as follows: Highest temperature 108 in July 1936; lowest temperature -43 in January 1873; maximum monthly precipitation 12.09 in October 1900; minimum monthly precipitation 0.01 in December 1943; maximum precipitation in 24 hours 7.23 in October 1900; maximum monthly snowfall 39.6 in January 1929; maximum snowfall in 24 hours 13.1 in March 1931; fastest mile of wind 69 from Southwest in October 1949.

NORMALS, MEANS, AND EXTREMES

MADISON, WISCONSIN — TRUAX FIELD

LATITUDE 43° 08' N LONGITUDE 89° 20' W ELEVATION (ground) 857 Feet

Temperature / Normal Degree Days

Month	Daily maximum	Daily minimum	Monthly	Record highest	Year	Record lowest	Year	Normal degree days
(years)	(a) 18	(b)		18		18		(b)
J	27.9	10.3	19.1	55	1947#	-37	1951	1423
F	31.0	12.7	21.9	55	1954	-25	1951	1207
M	41.8	23.2	32.5	78	1946#	-11	1941	1008
A	56.5	34.8	45.7	90	1952	10	1953	579
M	69.3	45.6	57.5	97	1953#	27	1947	272
J	79.1	55.7	67.4	97	1953#	33	1945	82
J	85.7	60.2	73.0	102	1940	42	1947	*
A	83.1	58.2	70.7	101	1947#	39	1950#	*
S	75.7	50.6	62.1	99	1953	25	1949	150
O	61.3	39.5	51.3	87	1953	14	1952	459
N	43.7	26.9	35.3	74	1950#	-11	1947	891
D	30.9	15.0	23.0	60	1951	-21	1950	1302
Year	57.0	36.1	46.6	102	JULY 1940	-37	JAN 1951	7417

Precipitation / Snow, Sleet

Month	Normal total	Max monthly	Year	Min monthly	Year	Max in 24 hrs	Year	Snow Mean total	Snow Max monthly	Year	Snow Max 24 hrs	Year
(years)	(b)	18		18		9		9	9		9	
J	1.31	2.43	1950	.41	1957	1.05	1950	8.4	19.7	1951	6.4	1951
F	1.13	2.13	1948	.35	1947	1.55	1953	6.3	12.4	1950	10.3	1950
M	1.83	2.92	1952	.92	1942	1.02	1956	8.1	14.3	1956	7.9	1957
A	2.49	4.86	1947	.96	1946	2.15	1955	1.0	3.2	1949	2.6	1957
M	3.27	3.27	1953	1.02	1953	2.08	1953	.0	.0		.0	
J	4.02	7.36	1951#	2.55	1951#	3.46	1954	.0	.0		.0	
J		10.93	1950	1.38	1946	5.25	1950	.0	.0		.0	
A		6.76	1940	1.15	1942	5.25	1950	.0	.0		.0	
S		5.98	1941	.49	1952	1.63	1957	.2	.8	1952	.8	1952
O		5.38	1951	.65	1956	1.75	1954	.2	8.6	1951	6.8	1951
N		3.56	1948	.51	1941	1.79	1952	3.9	8.6	1951	6.8	1951
D	1.40	2.17	1951	.55	1955	.79	1953	18.4	19.7	1950	6.0	1950
Year	30.00	10.93	JULY 1950	.06	OCT 1952	5.25	JULY 1950	37.6	19.7	JAN 1951	10.3	FEB 1950

Relative humidity / Wind / Sunshine / Sky cover / Sunrise to sunset

Month	RH Midnight CST	RH 6:00 A.M. CST	RH Noon CST	RH 6:00 P.M. CST	Wind Mean hourly speed	Prevailing direction	Fastest mile Speed	Direction	Year	Pct. possible sunshine	Mean sky cover sunrise-sunset	Clear	Partly Cloudy	Cloudy
(years)	18	11	11	11	11	8				11		11	11	11
J	77	75	67	73	11.1	NW	68	E	1947	47	6.9	6	6	18
F	77	78	66	72	11.2	WNW	57	W	1948	52	6.8	7	6	15
M	79	80	53	59	11.8	NW	70	SW	1954	56	6.8	7	7	16
A	77	80	53	59	12.5	NW	73	SW	1947	56	6.5	7	8	14
M	82	80	54	60	10.1	S	59	W	1950	60	6.5	7	10	13
J	82	86	52	58	8.7	S	72	NW	1947	67	6.0	8	11	11
J	84	85	52	59	8.5	S	47	W	1951	73	5.6	10	10	11
A	86	85	51	59	9.6	SW	52	W	1955	70	5.3	10	11	10
S	85	84	51	64	9.9	SW	73	W	1948	65	5.5	10	11	9
O	80	80	52	72	9.5	SW	56	SE	1951	52	5.0	12	6	13
N	79	78	63	72	12.0	WNW	56	SE	1947	41	7.1	6	6	18
D	82	80	70	77	10.9	WNW	65	SW	1949	47	7.3	6	6	18
Year	80	80	58	66	10.7	WNW	77	SW	MAY 1950	60	6.2	96	169	100

Mean number of days

Month	Precip .01 in or more	Snow/Sleet 1.0 in or more	Thunderstorms	Heavy fog	Max 90° and above	Max 32° and below	Min 32° and below	Min 0° and below
(years)	11	12	9	11	18	18	18	18
J	11	3	*	3	0	19	30	9
F	10	3	1	2	0	15	28	5
M	12	3	2	2	0	7	26	1
A	11	1	4	1	*	*	12	0
M	11	*	5	1	*	0	2	0
J	11	0	8	1	1	0	0	0
J	10	0	8	1	4	0	0	0
A	10	0	5	2	3	0	0	0
S	9	0	4	1	1	0	0	0
O	8	1	2	1	0	0	6	*
N	11	3	1	2	0	7	21	1
D	11	3	1	3	0	17	30	4
Year	115	12	39	22	19	64	156	19

Means and extremes in the above table are from the existing or comparable location(s). Annual extremes have been exceeded at prior locations as follows: Highest temperature 107 in July 1936; minimum monthly precipitation T in October 1889 and earlier date(s); maximum precipitation in 24 hours 5.31 in September 1941; maximum monthly snowfall 31.8 in January 1929; maximum snowfall in 24 hours 12.9 in March 1923.

MILWAUKEE, WISCONSIN — GENERAL MITCHELL FIELD

LATITUDE 42° 57' N LONGITUDE 87° 54' W ELEVATION (ground) 672 Feet

Temperature / Normal Degree Days

Month	Daily maximum	Daily minimum	Monthly	Record highest	Year	Record lowest	Year	Normal degree days
(years)	(a) 17	(b)		17		17		(b)
J	29.2	14.5	21.9	62	1944	-24	1951	1336
F	31.8	16.6	24.2	60	1954	-19	1951	1142
M	40.8	25.7	33.3	81	1945	-9	1941	983
A	52.8	34.4	43.8	82	1956#	13	1953	621
M	63.9	44.7	54.3	90	1956#	28	1947	351
J	75.1	54.7	64.9	99	1953#	40	1945	109
J	81.2	61.4	71.3	101	1955#	45	1945	20
A	79.2	60.5	69.9	100	1955#	44	1950	32
S	71.8	53.3	62.6	98	1953	28	1942	134
O	60.3	42.9	51.4	86	1947	21	1948#	428
N	44.7	30.7	37.3	77	1950#	-5	1950#	831
D	32.7	18.7	25.7	63	1946	-12	1951#	1218
Year	55.3	38.2	46.8	101	JULY 1955	-24	JAN 1951	7205

Precipitation / Snow, Sleet

Month	Normal total	Max monthly	Year	Min monthly	Year	Max in 24 hrs	Year	Snow Mean total	Snow Max monthly	Year	Snow Max 24 hrs	Year
(years)	(b)	17		17		17		17	17		17	
J	1.58	2.59	1949	.31	1945	1.13	1947	11.5	28.4	1943	12.3	1947
F	1.27	1.87	1951	.29	1947	1.32	1956	6.3	9.9	1956	5.8	1956
M	2.19	3.67	1952	1.05	1955	1.63	1943	8.1	19.8	1952	7.9	1950
A	2.39	4.91	1951	.81	1942	2.11	1956	.6	5.1	1942	3.4	1954#
M	2.98	5.27	1945	1.72	1949	2.06	1948	.0	.0		.0	
J	3.22	8.28	1943	2.33	1943	3.13	1951	.0	.0		.0	
J		6.69	1952	.95	1946	3.30	1950	.0	.0		.0	
A		4.34	1953	.46	1948	4.05	1953	.0	.0		.0	
S		9.87	1941	.15	1956	2.18	1954	T	T	1942	T	1942
O		4.42	1951	.15	1954	1.39	1954	3.1	9.6	1957#	6.3	1957#
N		3.36	1949	.62	1949	1.93	1942	9.5	26.5	1951	8.2	1951
D	1.48	2.64	1954	.99	1943			39.1	28.4	1943	12.3	1947
Year	27.57	9.87	SEPT 1941	.15	OCT 1956	5.28	SEPT 1941	39.1	28.4	JAN 1943	12.3	JAN 1947

Relative humidity / Wind / Sunshine / Sky cover / Sunrise to sunset

Month	RH Midnight CST	RH 6:00 A.M. CST	RH Noon CST	RH 6:00 P.M. CST	Wind Mean hourly speed	Prevailing direction	Fastest mile Speed	Direction	Year	Pct. possible sunshine	Mean sky cover sunrise-sunset	Clear	Partly Cloudy	Cloudy
(years)	17	17	17	17	17	4				17		17	17	17
J	76	76	70	73	13.4	WNW	62	N	1952	40	7.0	6	6	18
F	78	78	64	71	13.2	WNW	73	W	1957#	42	6.8	7	6	15
M	76	78	64	65	14.1	WNW	73	SW	1954	49	6.8	8	7	16
A	77	80	61	60	14.0	NNE	66	SW	1947	52	6.5	6	8	16
M	77	80	61	65	12.7	NE	72	SW	1950	56	6.5	6	10	15
J	81	81	60	61	10.9	SE	57	S	1953	65	6.1	5	12	13
J	82	83	58	64	10.0	NNE	50	W	1952	70	5.2	10	12	9
A	83	83	58	68	10.0	SSW	54	SW	1949	66	5.1	11	11	9
S	81	83	58	69	11.4	SSW	60	S	1941	63	5.1	10	11	9
O	77	80	57	73	12.1	SSW	60	S	1955	57	4.7	11	8	11
N	77	77	65	74	13.3	WNW	62	SW	1948	41	7.1	6	6	18
D	76	77	74		13.4	WNW	62			38	7.1	6	6	19
Year	78	79	62	69	12.4	WNW	73	SW	MAR 1954	54	6.2	93	102	170

Mean number of days

Month	Precip .01 in or more	Snow/Sleet 1.0 in or more	Thunderstorms	Heavy fog	Max 90° and above	Max 32° and below	Min 32° and below	Min 0° and below
(years)	12	19	17	17	17	17	17	17
J	10	3	*	3	0	17	30	6
F	8	3	1	2	0	13	26	2
M	11	3	1	3	0	7	24	*
A	11	1	4	3	0	*	10	0
M	11	*	4	5	*	0	1	0
J	11	0	5	7	1	0	0	0
J	9	0	7	6	4	0	0	0
A	9	0	6	4	2	0	0	0
S	8	0	4	1	1	0	*	0
O	8	1	2	1	0	0	4	*
N	10	3	1	2	0	4	18	1
D	10	3	*	2	0	13	28	2
Year	119	12	37	26	11	54	139	10

Means and extremes in the above table are from the existing or comparable location(s). Annual extremes have been exceeded at prior locations as follows: Highest temperature 105 in July 1934; lowest temperature -25 in January 1875; maximum monthly precipitation 10.03 in June 1917; minimum monthly precipitation 0.05 in March 1910; maximum precipitation in 24 hours 5.76 in June 1917; maximum monthly snowfall 52.6 in January 1918; maximum snowfall in 24 hours 20.3 in February 1924.

REFERENCE NOTES APPLYING TO ALL "NORMALS, MEANS, AND EXTREMES" TABLES.

(a) Length of record, years.
(b) Normal values are based on the period 1921-1950, and are means adjusted to represent observations taken at the present standard location.
* Less than one-half.
- No record.
† Airport data.
‡ City Office data.
# Also on earlier dates, months, or years.
T Trace, an amount too small to measure.

Sky cover is expressed in a range of 0 for no clouds or obscuring phenomena to 10 for complete sky cover. The number of clear days is based on average cloudiness 0-3 tenths; partly cloudy days on 4-7 tenths; and cloudy days on 8-10 tenths. Monthly degree day totals are the sum of the negative departures of average daily temperatures from 65°F. Sleet was included in snowfall totals beginning with July 1948. Heavy fog also includes data referred to at various times in the past as "Dense" or "Thick". The upper visibility limit for heavy fog is 1/4 mile. Data in these tables are based on records through 1957.

Mean Annual Precipitation, Inches

Based on period 1931-55

Isolines are drawn through points of approximately equal value. Caution should be used in interpolating on these maps.

Mean Maximum Temperature (°F.), January

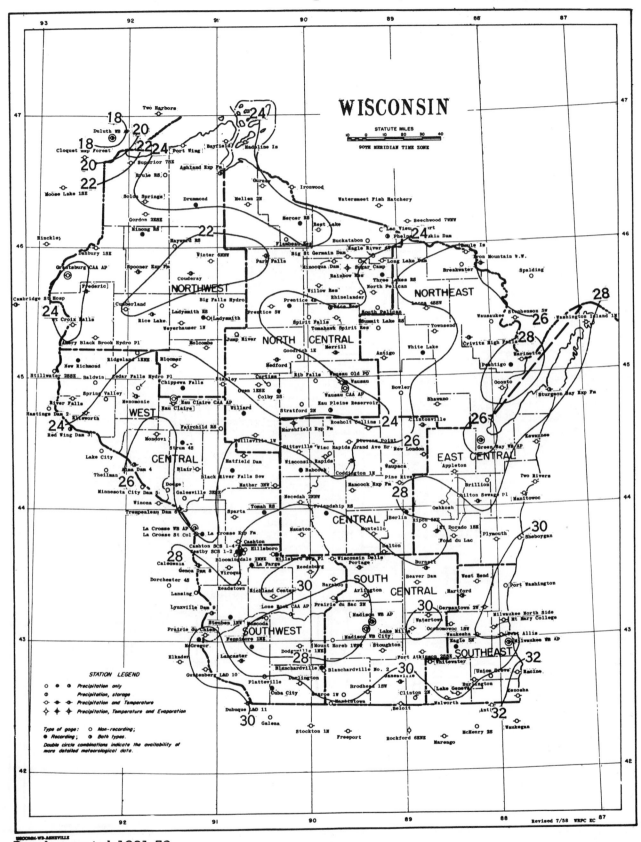

Based on period 1931-52

Isolines are drawn through points of approximately equal value. Caution should be used in interpolating on these maps.

Mean Minimum Temperature (°F.), January

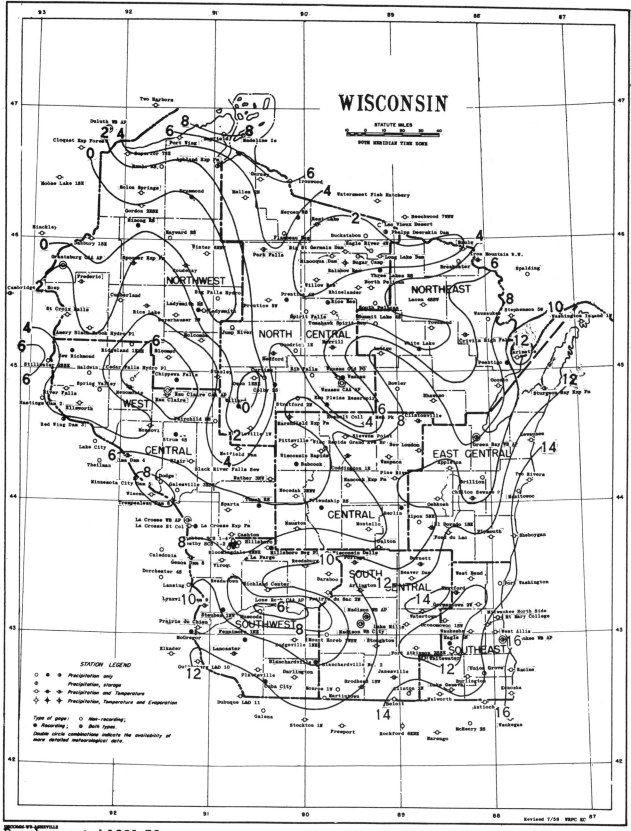

Isolines are drawn through points of approximately equal value. Caution should be used in interpolating on these maps.

Mean Maximum Temperature (°F.), July

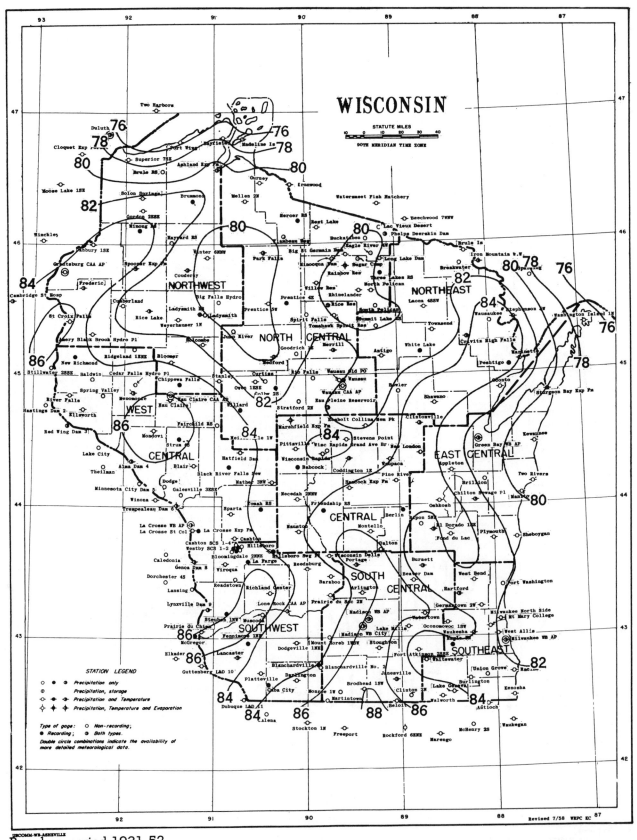

Based on period 1931-52

Isolines are drawn through points of approximately equal value. Caution should be used in interpolating on these maps.

Mean Minimum Temperature (°F.), July

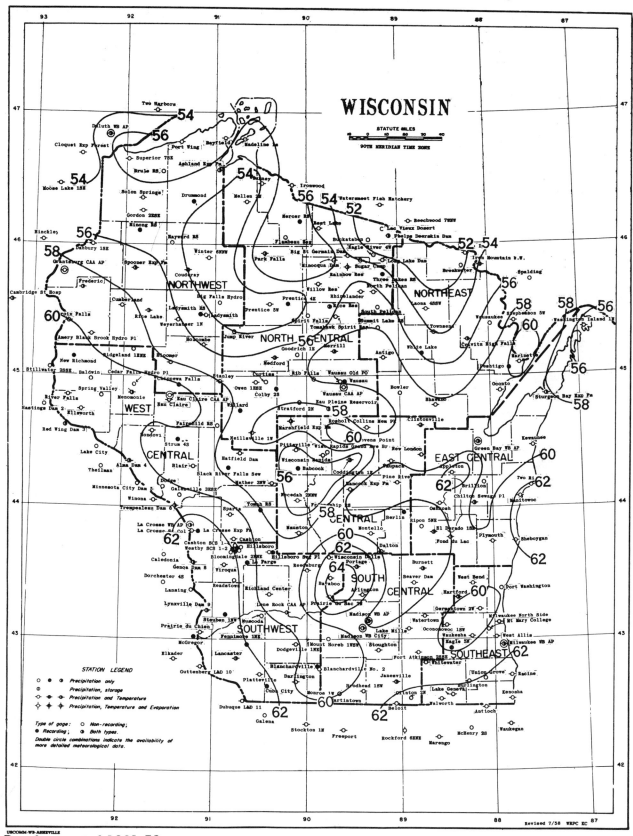

WISCONSIN

STATUTE MILES

90TH MERIDIAN TIME ZONE

Based on period 1931-52

Isolines are drawn through points of approximately equal value. Caution should be used in interpolating on these maps.

STATION LEGEND

Precipitation only

Precipitation, storage

Precipitation and Temperature

Precipitation, Temperature and Evaporation

Type of gage: ○ Non-recording;
● Recording; ◉ Both types.
Double circle combinations indicate the availability of
more detailed meteorological data.

USCOMM-WB-ASHEVILLE

Revised 7/58 WRPC KC 87

WISCONSIN

STATUTE MILES

90TH MERIDIAN TIME ZONE

STATION LEGEND

○ ● ◑	Precipitation only
⦶	Precipitation, storage
○◐ ◒	Precipitation and Temperature
◇ ◆ ◈	Precipitation, Temperature and Evaporation

Type of gage: ○ Non-recording;
● Recording; ◑ Both types.
Double circle combinations indicate the availability of
more detailed meteorological data.

NORTHWEST

NORTH CENTRAL

NORTHEAST

WEST CENTRAL

CENTRAL

EAST CENTRAL

SOUTHWEST

SOUTH CENTRAL

SOUTHEAST

Revised 7/59 WRPC KC

USCOMM-WB-ASHEVILLE

THE CLIMATE OF

PUERTO RICO and the
U.S. VIRGIN ISLANDS

by
Robert J. Calvesbert

June 1970

INTRODUCTION

Puerto Rico and the U. S. Virgin Islands are tropical, hilly islands which lie directly in the path of the easterly trade winds throughout the year.

This single introductory sentence opens the door to understanding a great deal about the climate of these islands. Since all are islands, daily temperature ranges are relatively small, at least close to the coasts due to the tempering effect of the nearby waters. The larger islands are influenced by land and sea breezes. Since they are hilly islands, there are rather sizable variations in rainfall and temperature over relatively short distances. The rugged aspect of the terrain also causes wide local variations in wind speed and direction due to sheltering and channeling effects. Location in the tropics induces warm temperatures and raises some question as to their location with respect to tropical storms and hurricanes. Naturally, we would expect the tropical warmth to be moderated considerably by the trade winds.

PUERTO RICO

Location: Puerto Rico is the easternmost and smallest of the Greater Antilles and is shaped somewhat like a brick with its southeast corner broken off. It lies between 18° 31'N and 17° 55'N latitude and 65° 37'W and 67° 17'W longitude. From its easternmost to its westernmost tip it is about 109 miles, while from its northernmost to its southernmost tip it is about 40 miles. The overall area is a little more than 3,400 square miles which makes it slightly larger than the States of Delaware and Rhode Island combined. In addition to the main island, Puerto Rico also embraces three secondary and a number of smaller islands.

Mona Island lies nearly 40 miles to the west of Puerto Rico proper, covers an area of about 21 square miles and is, for the most part, uninhabited. Vieques Island, the largest of the three secondary islands, lies a short distance to the east of the main island and covers an area of slightly more than 50 square miles. This is

– 453 –

the most densely populated of the three with about 15,000 inhabitants. It is primarily an agricultural community, although efforts are being directed toward development as a tourist resort in coming years. A large portion of Vieques Island at the east and west ends is used as a U. S. military reservation for training exercises. Culebra Island lies to the north of the eastern tip of Vieques Island and covers an area of about 10 square miles. It, too, is partially a military reservation and has a total population of less than 1,000. Vieques and Culebra Islands are the westernmost of the Lesser Antilles which extend from Puerto Rico in a sweeping southeasterly curve to the coast of South America.

Topography: Viewed from the horizon in any direction, Puerto Rico presents a rather rugged profile with peaks to more than 3,000 feet in the east and to more than 4,000 feet in the west. El Toro, rising to 3,535 feet, is the highest in the east and Cerro de Punta reaches an elevation of 4,389 feet in the west. There is a main divide running roughly east-west across the Island. In the eastern third it is closer to the northern coast, and in the western two-thirds it lies closer to the southern coast. The divide is over 3,000 feet above sea level for the most part, with passes where the north-south roads cross the Island at an elevation of about 2,000 feet. With this configuration, there are coastal plains varying from 8 to 12 miles along the northern coast and 2 to 8 miles along the southern coast. There is a gradual rise to the peaks in the north, while in the south the descent to the sea is rather abrupt. In the western end the mountains spread out fanwise filling the width of the Island. In the east the divide curves to the northeast corner where there is a detached group of peaks known as the Luquillos.

Mona Island has a plateau-like surface which descends abruptly to the sea in cliffs 150 to 200 feet high. A small area of about 1,000 acres in the southwestern corner is sandy beach.

Vieques Island is somewhat spindle-shaped, 21 by 6 miles, with its longer axis running east-northeast to west-southwest. It is mostly level, although somewhat hilly in the central sections with elevations rising to 981 feet at Mt. Pirata in the southwest.

Culebra Island with a very irregular shoreline is slightly higher than Mona Island. Mount Resaca at 646 feet in the central part of the Island is the highest point. The entire island is quite hilly, and very little level land is to be found.

U. S. VIRGIN ISLANDS

Location: The U. S. Virgin Islands are composed of three major islands together with a number of smaller islands and cays totaling about 50. The three of primary importance are St. Thomas, where the capital is located; St. Croix, the largest; and St. John, the smallest.

These islands follow Vieques Island and Culebra Island in the path of the Lesser Antilles toward South America. St. Thomas lies some 38 miles east of Puerto Rico and about 1,500 miles southeast of New York. St. John lies a few miles east of St. Thomas and St. Croix is located about 40 miles south of St. Thomas and St. John.

With an area of about 28 square miles, St. Thomas is the second largest of the U. S. Virgin Islands. This island lies between latitudes 18° 23'N and 18° 18'N and longitudes 65° 03'W and 64° 50'W. It is about 5 miles from its northernmost to its southernmost points and a little more than 12 miles from its eastern to western extremities.

The smallest of the three principal islands is St. John, with an area of only about 20 square miles. It is also the least populated. St. John lies between latitudes 18° 23' N and 18° 18'N, and longitudes 64° 48'W and 64° 40'W. This island extends about 5 miles from its northern to southern tips and about 8 miles from its easternmost to westernmost points.

Somewhat apart from the others, the largest of the three islands is St. Croix which has an area of 84 square miles. It lies between latitudes 17° 47'N and 17° 41'N, and longitudes 64° 54'W and 64° 34'W. The Island extends some 19 miles from east to west and 6 miles from north to south.

Topography: St. Thomas has an extremely irregular coastline and is very hilly with practically no flatland. The highest hills are generally found near the center of the Island, with Crown Mountain at 1,550 feet the highest point. The Island is relatively small and many of the peaks rise above 1,000 feet. This results in rather steep slopes over all the island, so that rainfall runoff is quite rapid and there are no permanent streams or rivers.

Like St. Thomas, St. John has an extremely irregular shoreline and a very hilly topography. It has a number of peaks over 1,000 feet, topped by Bordeaux Mountain at 1,297 feet in the eastern portion of the island. Slopes are quite steep over all of the island, and there are very few areas of flatland. There are no permanent rivers or creeks.

St. Croix is the largest of the three U. S. Virgin Islands. The topography is somewhat different from the other two with a broad expanse of low, relatively flatland running along the southern two-thirds of the Island. A range of hills, ranging in elevation from about 500 feet to more than 1,000 feet, topped by Mount Eagle at 1,165 feet, runs along the northern coast. In the eastern end of St. Croix is found another group of slightly lower hills with a maximum elevation of about 860 feet. The relatively small area covered by hills on St. Croix results in rather steep slopes down to the Caribbean in the north and to the level areas to the south.

PUERTO RICO

With nearly 2.8 million inhabitants Puerto Rico has a population density slightly less than that of Rhode Island. With so many people to feed and situated so far from the mainland, it is quite obvious that agriculture is of major importance to the overall economy.

Agricultural products grown include sugarcane (the principal crop), coffee, tobacco, pineapple, bananas, plantains, and many subsistence crops. Chickens are raised for food and eggs, and cattle for milk and meat.

Rainfall is of prime importance to all growing crops and a favorable wet-dry season relationship is vital for sugarcane. In areas where rainfall is deficient, irrigation is necessary.

The Island is earnestly striving to increase the standard of living, provide more jobs and, in general, to make Puerto Rico a more desirable place to live by stimulating the growth of the industrial community of the Island. As a result, industry has grown by leaps and bounds over the past years.

Climate is important to industry with its need for water for cooling and other purposes; its need to exhaust smoke and gasses, sometimes noxious, into the air; its sensitivity to moisture and salt content of the air both in its manufacturing processes and in the deterioration of plant and machinery.

Over the years tourism is playing an increasingly greater role in the overall economic picture with great strides forward being taken each year. The number of tourists has increased markedly year after year, and many cruise ships are making San Juan a port of call during the season.

As in any tourist resort, a favorable climate is one of the greatest assets. Abundant sunshine, comfortable swimming water, refreshing breezes, and a reliable small temperature range are desirable. Careful consideration of anticipated climatic conditions may well be the determining factor in whether a vacation is pleasant or not.

U. S. VIRGIN ISLANDS

Agriculture is not as important in the U. S. Virgin Islands as it is in Puerto Rico. St. Croix is the only one of the U. S. Virgin Islands with any sizable expanse of flatland suitable for farming. Here sugarcane, which was the principal crop, has been abandoned. Subsistence crops are now a minor effort. Some cattle are raised for milk and meat.

In St. Croix, industrial growth has become a significant factor in the island's economy. With the downgrading of agriculture, industrial complexes are being expanded to include the petrochemical industry and refinement of aluminum. Light industrial plants and the manufacture of rum are the other industrial activities in St. Croix and St. Thomas. St. John has no industrial development and remains primarily a National Park.

Tourism is the biggest factor in the Virgin Islands' economy. It has, over the past years, undergone a vast increase in the numbers of cruise ships, especially at St. Thomas and St. Croix. Hotel facilities have been increased on both islands.

One of the principal causes of concern in the U. S. Virgin Islands is the short supply of water. Rainfall, while above 40 inches annually over most of the area, is insufficient. This is due partially to a high evaporation rate and the rapid runoff from the steep slopes on St. Thomas and St. John and, to a certain extent, on St. Croix.

In an effort to utilize available water efficiently, most homes and business establishments catch rainwater on the roofs and pipe it to cisterns. The runway at the airport at St. Thomas is also used as a catchment area. On St. Thomas and St. John it is common to see the entire side of a hill cemented to act as a catchment area. Generally, during the drier portion of the year, it is necessary to carry water by barge from Puerto Rico. Installation of a sea-water distillation unit on St. Thomas and St. Croix has helped alleviate the water shortage but it remains a significant factor in the development of the island's economy.

PRECIPITATION

PUERTO RICO

Rainfall in Puerto Rico varies markedly from place to place over relatively short distances. Measurements of rainfall are made daily at approximately 100 Weather Bureau cooperative climatological stations.

The majority of all Puerto Rico's rainfall is orographic in nature. Moisture laden air from the ocean is carried by the trade winds inland and forced to ascend over the mountains and is cooled, thus causing condensation in the form of rainfall. The majority of these orographic showers are rather brief. Abundant sunshine prevails throughout the year--even during the so called rainy seasons.

There are two rainfall-producing mechanisms in Puerto Rico: easterly waves and cold fronts. During the period from May through November, Puerto Rico experiences easterly waves. These are migratory wave-like disturbances moving from east to west within the basic easterly air current, generally more slowly than this current in which they are imbedded. An easterly wave of slight intensity may produce little more than greater-than-normal cloudiness. However, an intense easterly wave, especially if it is moving slowly, may bring one or several cloudy, rainy days with it which can produce rains sufficient to cause flooding.

Tropical storms and hurricanes occasionally develop in the easterly waves and may cause torrential rains on Puerto Rico. Fortunately visitations by these storms are infrequent, although each year there are generally one or more scares where "every eye" is focused on the sky and "every ear" is glued to the radio for news of the storm.

The other major rainfall-producing situation occurs during the winter months, generally from about November to April. Occasionally during this period, the trailing edge of a cold front which has swept across the continental United States penetrates far enough south to have a definite effect upon Puerto Rico's rainfall. The degree of effect depends upon the intensity and rate of progression of the cold front. A weak or dissipating cold front may cause only cloudier-than-normal skies while a strong, active slowly moving front is capable of bringing heavy and continuing rainfall which may last for several days. Occasionally the front moves over the entire Island, bringing rainfall to most areas. In this case the topographic effects combine with the frontal mechanism. At other times the front may only penetrate a short distance into Puerto Rico, generally over the northwestern corner. In those cases that section of the Island is subjected to heavy rains, while the remainder of the Island enjoys pleasant weather.

The marked difference in rainfall over relatively short distances is illustrated by a glimpse at the annual rainfall map. In the area of El Yunque an annual average for a 10-year period of 183.51 inches is noted. Along the western portion of the southern coast there is an area of less than 40 inches, with the lowest being 34.62 inches at Santa Rita. Thus, over a distance of only about 60 miles there is a difference of nearly 150 inches annually. At the La Mina El Yunque station the total for one year has reached 253.79 inches. In contrast, stations along the south coast have experienced occasional years when less than 15 inches were received.

The geographical distribution of the rainfall over the Island shows four areas of heavy rainfall. The heaviest of these is centered over El Yunque in the Luquillo Mountains in the northeastern section. While a little more than 180 inches, on the average, has been recorded part way down the slope, it may well be that at the peak an estimate of a 200-inch mean annual total would not be unrealistic. The second area of heavy rainfall lies to the southwest of El Yunque in the San Lorenzo area. Here the mean annual rainfall exceeds 120 inches. Farther west, over the highest peaks is another area of heavy rainfall. Guineo Reservoir and Toro Negro Hydroelectric Plant, with mean annual totals of 113.41 and 105.95 inches, respectively, are in the center of this region. Farther west around Maricao, another area of copious rainfall above 100 inches exists. The lowest annual averages are along the southern coast from Aguirre westward.

Rainfall patterns on the outlying islands of Culebra, Vieques, and Mona are not well defined due to a lack of sufficient observations, but existing data indicate about 40 to 50 inches of rain annually.

The effect of topography and winds upon the rainfall is well illustrated in the isohyets (lines of equal rainfall). In general, rainfall along the northern, or windward coast, is greater than along the southern, or leeward coast. As indicated earlier, in the north the slope of the hills to the peaks is a gradual one, while the drop to the ocean on the south is quite precipitous. This is reflected in the isohyets where there is a rather gradual increase in the amount of rainfall on the northern slopes up to the divide. In the

south, where the topographic slope is much greater, rainfall shows a sharp decrease from the ridge to the coast.

The distribution of rainfall over the year does not show an absolute wet season - dry season relationship, but only a relatively dry season and a relatively wet season. The length of the dry season varies somewhat with location of the Island. In the northern portion of Puerto Rico the dry season is generally a little shorter than it is in the southern section of the Island, with a narrow intermediate belt in between. The difference in the length of this relatively dry season is produced by variations in the beginning of the dry season. This is because the onset, in May, of the relatively rainy season is the same over the entire Island. In the north the dry season normally begins in February and ends in April, while in the south the dry season sets in during December.

In most areas there is a transitional period of about 1 month between these seasons. Naturally, there are isolated exceptions to this general rule--the dry season is not <u>always</u> from December to April nor is the wet season <u>always</u> from May to November. Some of the heaviest rainfalls have occurred during the so-called relatively dry season. In practically every instance, however, on the average the driest month is either February or March. For the most part, May may be considered to be the rainiest month in the northern half of the Island, while September or October is the rainiest month in the south.

The number of days with measurable rain follows the isohyetal (rainfall) pattern, with the largest number in areas of greatest rainfall. Once again there is the pattern of an increase from north to south to the area of heaviest rainfall, and then a rather sharp decrease to the south coast where the lowest average rainfall is noted. The average number of rainy days varies from just under 300 in the El Yunque area to less than 100 along the southern coast, and below 50 at the driest locations.

The matter of anticipated rainfall intensity is of the utmost importance to a great many people, including the engineer, farmer, and of course, all those with an interest in flood situations. About once in 100 years each section of Puerto Rico may expect a rainfall of at least 8.50 inches during a 24-hour period. The area of the highest peaks in the southwest may receive as much as 18 inches in a 24-hour period. Twelve-hour rainfall in this area may reach as high as 15 inches, while the 1-hour rainfall may be as much as 6.50 inches.

U. S. VIRGIN ISLANDS

Rainfall in the U. S. Virgin Islands is of the same nature as that in Puerto Rico, falling most frequently in the form of brief showers. The rainfall-producing mechanisms are essentially the same as in Puerto Rico except in the matter of degree.

Orographic lifting of the moisture laden air

over the hilly terrain of these islands is the most frequent cause of rainfall. However, due to the smaller elevations and smaller size of the islands, there is a less marked variation in annual amounts. The higher mean annual totals are between 50 and 60 inches at the higher elevations, and the variation between the greatest and least average value is not as marked as it is in Puerto Rico. Clouds formed by forced ascent of the wind over small and narrow islands, as is the case for St. Thomas and St. Croix, lean to the leeward, so that most of the rain from them falls in the ocean to the lee of the island. Easterly wave passages are important contributors to the rainfall of the Virgin Islands during the months from May through November. Like Puerto Rico, the U. S. Virgin Islands lie in the path of the tropical storms and hurricanes which form over the ocean to the east of the Lesser Antilles. As in Puerto Rico, they are relatively infrequent. While cold frontal passages affect the rainfall regime of the Virgin Islands, the frequency of fronts is less and their intensity is more likely to be diminished and less effective than in Puerto Rico.

Annual rainfall values indicate differences in rainfall from location to location with higher elevations generally receiving greater amounts. On St. Thomas and St. John, on the basis of the limited data available, annual averages of between 40 and 60 inches appear reasonable. On St. Croix there is a more noticeable variation from place to place. This Island has the greatest annual rainfall, in excess of 50 inches in the northwestern corner. There are some indications that stations in a small area along the central portion of the southern coast of St. Croix receive about 40 to 45 inches. A narrow finger of between 25 and 35 inches extends northeast to southwest over the flatlands south of the hills in the western portion of the Island. Annual rainfall averages less than 30 inches in the eastern end of St. Croix, possibly as low as 20 inches.

As in Puerto Rico, there is no sharply defined wet-dry season relationship. Records available for the three islands indicate a relatively wet-relatively dry season distribution similar to that found in the southern portion of Puerto Rico. The relatively dry period extends from about December through June. Occasionally, quite heavy rainfall occurs during the so-called drier months. The driest month on St. Thomas and St. John usually is February or March and the wettest month September or October, as in the southern sections of Puerto Rico. On St. Croix, the month with the heaviest rainfall, on the average, ranges from September through November.

The number of days with measurable rainfall over the Virgin Islands, based on a few known-to-be reliable stations, range from a little less than 200 days annually at the higher rainfall stations to less than 100 days annually at the stations with lowest rainfall.

TEMPERATURE

PUERTO RICO

Mean temperatures in Puerto Rico have a very small range between the warmest and coldest months. The smallest range is generally found in areas near the coast, with only 4.5° at Patillas. In the interior a slightly larger range is observed, with 7.3° at Humacao the largest. The normal range in downtown San Juan is only 5.7° between the warmest month, August - 80.8°, and the coolest months, January and February - 75.1°.

These small annual temperature ranges are due in part to the fact that Puerto Rico is an island surrounded by waters whose temperature changes but little from the warmest to coolest season (82.5° in September to 77.9° in February and March). It is also due to its location, only about 1,100 miles north of the equator, and the resultant small differences in the energy received from the sun from season to season. The difference between the length of the longest day (13 hours 13 minutes) and the shortest days (11 hours 2 minutes) is only a little over 2 hours. Also there is not much change during the year in the height of the sun above the southern horizon. The coolest average temperatures are found in the area of the higher peaks, while the warmer average temperatures are observed along the coastal regions. The lowest mean annual temperature is 67.0° at Guineo Reservoir, while the highest is 81.0° at Guayama on the southern coast. Mean annual maximum temperatures range from 89.2° at Dos Bocas to 74.6° at Guineo Reservoir. There are two areas of high average maximum temperature. One is along the southern or leeward coast and the other is in the Dos Bocas area. Mean annual minimum temperatures range from 74.4° at the Roosevelt Roads Naval Station at the eastern end of the Island to 59.3° at Guineo Reservoir. Once again the higher values are found along the coastal areas.

The daily range of temperature or the difference between the daytime maximum and nighttime minimum, varies with location on the island. In the areas along the northern and eastern coasts, the mean daily range is usually between 10° and 15°, with increasing values inland away from the tempering effect of the oceans. The mean daily range is between 15° and 20° in the southeast and in excess of 20° in the west and southwest. Maximum values of a little over 25° are found in the Caguas Valley and at Utuado, where the highest annual mean daily range, 26.3° is found. There is no significant variation in this pattern from month to month. San Juan has the lowest annual mean daily range, 9.6°.

Afternoon temperatures in the 90's are not unusual in some sections of Puerto Rico. The number of days with temperatures of 90° or higher during a year may reach 200 or more annually along the southern coast and more than 100 at some stations in lower portions of the interior and along the western coast. Along the northern and eastern coasts, however, extremely warm temperatures are rather infrequent due to the tempering effect of the waters of the ocean. At San Juan City the annual average number of days with 90° or more is only 9 and other stations close to the water record less than 100 annually. Due to the effect of altitude there are several stations where the annual number of 90° days is usually less than 1 every 2 years, such as Garzas Dam, Barranquitas, Cidra, and Aibonito. At Guineo Reservoir the maximum temperature has never gone above 85°. There are some locations where the record maximum temperature is 100° or more. The highest temperature ever recorded in Puerto Rico was 103° at San Lorenzo in August 1906. At San Juan City the record high of 96° occurred in March 1958. At this time the Island was experiencing one of the worst droughts in history, and the soil over the island was practically without moisture. An interruption in the normal easterly windflow brought the extremely warm, dry air from the south over San Juan. Practically all of San Juan's higher temperatures occur during periods of southerly windflow, when air is warmed during passage overland.

Over practically the entire Island the lowest mean maximum temperatures occur in January, although there are a few areas where they are found in February. The warmest daytime temperatures are recorded in the northern section of the island and the Cayey-Patillas-Guayama area of the southeast in September, in the southwest in July, and elsewhere in August. At San Juan it is interesting to note that July daily maximum temperatures do not generally reach the extremes that they do during June and August.

Freezing temperatures are unknown in Puerto Rico, even at the higher elevations. In fact, the coolest temperature ever recorded is 40° at Aibonito in March 1911. Most of the hilly areas have experienced temperatures in the 40's, while in the coastal areas and foothill sections the coolest temperatures on record are generally between 50° and 60°. In downtown San Juan, which is literally surrounded by water, the coolest temperature recorded is 62°. Almost without exception the lowest mean minimum temperatures occur during February. Over most of Puerto Rico the highest mean minimum temperatures have been recorded during August with a few scattered areas reporting them in June, July, or even in September.

U. S. VIRGIN ISLANDS

As in Puerto Rico, one of the most striking features of the temperature regime in the U. S. Virgin Islands is the relatively small variation in temperature from the coolest to the warmest months, ranging from about 5° to 7°.

Due to the small size of the islands and the location of all stations within a few miles of the water, the mean daily range is quite small. It varies from 9.1° at Charlotte Amalie to 15.1° at Wintberg. For these same reasons extremes of temperature are not as great as they are in Puerto Rico, and relatively few days have temperatures of 90° or above. Since the extent of land areas is so small, the overland air passage is quite short and there is not sufficient time for extreme heating to take place, regardless of the wind direction. On St. Croix, Annas Hope has had a temperature as high as 99°.

During the warmest months, maximum temperatures average about 87° to 89°, with nighttime temperatures falling to about 74° to 78°, and a little lower at the higher elevations. In the winter, daily maximum temperatures are generally in the low 80's and nighttime minima in the high 60's or low 70's.

The highest mean maximum temperatures are found in August, while the lowest mean maxima fall either in January or February. The lowest mean minimum temperatures are observed in January and February, and the highest mean minimum temperatures are generally in July or August.

WIND

One of the outstanding features of the wind in Puerto Rico and the U. S. Virgin Islands is the steadiness of the trade winds. They blow almost without exception from an easterly direction, i.e., between north-northeast and south-southeast. These islands are under the influence of three wind regimes, with the trade winds primary and the others superimposed. Being surrounded by water, the land and sea breeze effect is important in most coastal areas, but it is not as noticeable at places farther in the interior. The trade winds, modified somewhat by the land and sea breeze, pass inland to a formidable barrier of hills where they are lifted over the top or pushed aside through narrow passes and valleys until their basic characteristics have become quite confused. Night winds are lighter than the daytime winds. About daybreak the wind speed begins to pick up, reaching a maximum late in the morning or early afternoon. The return to the lighter nighttime winds begins later in the afternoon, usually about 4 p.m. The afternoon decrease appears to be more leisurely than the morning increase.

The highest mean maximum wind speeds occur during July, with average peak speeds reaching more than 18 m.p.h. at downtown San Juan and at Ramey Air Force Base in northwestern Puerto Rico. Alexander Hamilton Field, St. Croix, V. I., follows with maximum values slightly above 16 m.p.h. while the other stations have mean speeds several miles per hour slower. At several of the stations a secondary maximum appears in March or April, while the lightest winds are during the autumn in either October or November.

A measure of the degree of variability of daily mean maximum wind speed from strongest to weakest months is illustrated in the following table, followed by a similar table for the daily mean minimum wind speed.

MEAN MAXIMUM WIND SPEED (m.p.h.) AND LOCAL TIME

Station and Period of Record	Strongest Month		Weakest Month	
WBO San Juan, P. R. (1931-42)	18.4	2 PM	13.8	2 PM
WBFO San Juan, P. R. (1957-60)	14.8	2-3 PM	11.5	2-3 PM
Ramey AFB, P. R. (1940-55)	18.6	1 PM	11.7	1-2 PM
Santa Isabel AP, P. R. (1940-45 and 1951-54)	13.9	1 PM	10.0	2 PM
Alex. Hamilton Fld., St. Croix, V. I. (1954-58)	16.1	11 AM	11.0	Noon-1 PM
Truman Field, St. Thomas, V. I. (1953-58)	14.9	Noon	9.9	11 AM

MEAN MINIMUM WIND SPEED (m.p.h.) AND LOCAL TIME

Station and Period of Record	Strongest Month		Weakest Month	
WBO San Juan, P. R. (1931-42)	9.3	5 AM	5.5	8 AM
WBFO San Juan, P. R. (1957-60)	5.3	7 AM	3.5	6 AM
Ramey AFB, P. R. (1940-55)	9.1	3 AM	5.1	3 AM
Santa Isabel AP, P. R. (1940-45 and 1951-54)	3.7	1 AM & 3 AM	2.9	1 AM
Alex. Hamilton Fld., St. Croix, V. I. (1954-58)	7.5	4 AM	4.8	4 AM & 7 AM
Truman Field, St. Thomas, V. I. (1953-58)	7.4	3-4 AM	3.7	4 AM

From the preceding tables it is apparent that even with the modifications of the trade winds by the land and sea breezes and local topography, the overall pattern shows much similarity. The six stations for which these summaries were prepared are variously located, with three exposures on the north coast of Puerto Rico, one on the south coast of Puerto Rico, one on the south coast of St. Croix, V. I., and the other in a rather sheltered location on St. Thomas.

ANNUAL PREVAILING WIND DIRECTION

Station	2 AM	8 AM	2 PM	8 PM	Year
			(Local Time)		
WBFO San Juan, P. R.	SE	ESE	ENE	ENE	ENE
Santa Isabel AP, P. R.	NE	NE	SE	NE	NE
Ramey AFB, P. R.	E	E	ENE	E	E
Roosevelt Roads, P. R.	E	E	E	E	E
Alex. Hamilton Field, St. Croix, V. I.	ENE	ENE	ESE	ENE	ENE
Truman Field, St. Thomas, V. I.	E	E	ESE	E	E

This table shows that even though the most frequent direction through the year may be in the northeastern quadrant, the distribution through the day shows a regular variation which might have been expected from the location of the stations. The WBFO San Juan, P. R., and Ramey AFB, P. R., on the northern coast show evidence, with that at San Juan the more pronounced, of a more northerly component of the wind during the afternoon as a result of the sea breeze effect blowing onshore. At San Juan the southerly component during the early morning hours reflects the strength of the land breeze. At Santa Isabel AP, P. R., and on St. Croix, V. I., the situation is the reverse of that at San Juan and Ramey AFB. Northeasterly winds predominate, except in the afternoon when the sea breeze becomes strong enough to bring about an alteration in the direction of the trade winds to a more southerly direction. The station in St. Thomas lies in a very sheltered position on the lee side of the hills to the south. Here again, the sea breeze effect is illustrated by a slight variation in direction at the 2 p. m. observation. However, it is not as marked as at the two other less sheltered south coast exposures.

At Mayaguez on the western end of Puerto Rico, the effect of the land and sea breezes is quite different than it is in other locations on the island. The sea breeze acts in an almost opposite direction and lessens the strength of the trade winds to such an extent that it frequently becomes dominant and a westerly wind is observed. This is quite infrequent in other sections of the islands.

Another indication of the land and sea breeze effects may be found in the following table of "Mean Annual Percentage Frequencies of Wind Direction" at several stations.

MEAN ANNUAL PERCENTAGE FREQUENCIES OF WIND DIRECTION

	WBFO San Juan, Puerto Rico	Ramey AFB, Puerto Rico	Roosevelt Roads, Puerto Rico	Alex. Hamilton Fld., St. Croix, Virgin Islands	Truman Field St. Thomas, Virgin Islands
N	0.9	1.0	3.7	1.7	0.8
NNE	1.7	1.7	3.5	2.4	1.1
NE	6.4	7.5	15.4	14.8	6.0
ENE	28.5	21.8	22.3	31.8	23.3
E	9.7	30.5	24.7	11.7	33.6
ESE	13.4	13.6	8.6	19.5	13.2
SE	9.8	9.9	4.6	7.5	5.6
SSE	6.0	1.9	1.1	2.2	1.0
S	4.2	2.0	1.7	0.7	1.0
SSW	1.8	0.8	0.9	0.3	0.4
SW	1.5	1.1	1.0	0.2	0.1
WSW	0.8	0.3	0.9	0.1	0.1
W	0.3	0.6	0.5	0.1	0.2
WNW	0.3	0.3	0.3	0.1	0.2
NW	0.5	0.7	1.4	0.2	0.3
NNW	0.7	0.4	2.3	0.4	0.3
CALM	13.5	5.9	7.1	6.3	12.7

At two stations in the preceding table, where the degree of shift in wind direction as a result of the sea breeze is the greatest, WBFO San Juan, P. R., and Alexander Hamilton Field, St. Croix, the frequency distribution is bimodal. That is, there are two peaks, with the greatest from east-northeast and the secondary one from east-southeast, while the frequency of east winds is somewhat less. The wind passes through east twice--during the transitional period from east-northeast to east-southeast in the case of Alexander Hamilton Field, and from east-southeast to east-northeast at San Juan, and vice-versa. At Ramey AFB and Truman Field, where the degree of shift in wind direction due to land and sea breeze is smaller, a distribution with a single mode (peak) from the east is observed. At Roosevelt Roads no directional shift occurs.

At Santa Isabel Airport, on the southern coast of Puerto Rico, the annual frequency distribution indicates that the shift from a northeasterly direction to a southeasterly direction begins to take place about 8 a.m. It reaches the peak about 1 to 2 p.m. when nearly half of the observations have a southeasterly wind. The return to northeasterly then takes place slowly, reaching the peak during the early morning hours shortly after midnight.

At Ramey AFB, in the northwestern corner of Puerto Rico, east is the most frequent wind during the night, with a gradual shift to east-northeast beginning at about 9 a.m. The shift back to east begins about 4 to 5 p.m., reaching a maximum in the early morning hours when a larger number of southeasterly winds are observed.

At Mayaguez, during the rainy season, from May through December, the predominant westerly wind does not appear until 11 a.m., which is 1 hour later than in the dry season (January through April) and it disappears at 5 p.m., 1 hour earlier than in the dry season.

Factors which interrupt the trade wind flow are the same as those discussed above as rainfall-producing mechanisms; the frontal passages and easterly wave passages. As the cold front approaches, a shift to a more southerly direction is noted, and then as the front passes a gradual shift through the southwest and northwest quadrants back to northeast. The easterly wave passage normally does not bring westerly winds but is usually characterized by an east-northeast wind ahead of the wave and a change to east-southeast following its passage.

During recent years hourly observations taken at the WBFO San Juan, Alexander Hamilton Field, St. Croix, V. I., and Truman Field, St. Thomas, V. I., indicate that less than 1 percent of all wind observations were above 24 m.p.h., and less than 5 percent were above 18 m.p.h.

Of course it must be realized that since these islands lie in the path of tropical storms and hurricanes, occasional winds of extreme speed are experienced. At San Juan, during the passage of the hurricane known locally as "San Felipe" in September 1928 the Weather Bureau's ane-mometer blew away after reaching an extreme velocity of 160 m.p.h. This is the highest value recorded in Puerto Rico to date. Winds of 110 m.p.h. are expected to occur once every century on the average.

EVAPORATION

Evaporation in Puerto Rico and the U. S. Virgin Islands, due to warm temperatures and rather constant wind flow, is fairly high as indicated by measurements at several locations in Puerto Rico and one at St. Croix, V. I., over a period of years. The San Juan annual average evaporation is 81.59 inches compared to annual values of 81.40 at Lajas and 79.85 inches at Aguirre. Higher elevation interior stations have somewhat less: Gurabo with 62.50 inches, Corozal 53.13 inches, and Adjuntas 50.93 inches. On St. Croix measurements made at the Annas Hope Agricultural Experiment Station averaged 72.69 inches per year.

It is interesting to note that at the coastal stations the evaporation is more than the average annual rainfall. At all locations the distribution throughout the year is similar, with maximum monthly values in the spring and early summer and the lowest monthly averages during November and December.

MOISTURE IN THE AIR

The relative humidity in the islands is rather high, averaging somewhere near 80 percent over the course of the year.

The relatively small differences in the dewpoint temperature over the area indicate that the air mass overlying the area is quite similar with respect to moisture content throughout the year.

Another interesting aspect of the distribution of the dewpoint is the small diurnal change. At San Juan, for example, the annual average dewpoint at 2 a.m., 8 a.m., 2 p.m., and 8 p.m. is 70°, thus illustrating the fact that the changes throughout the day are small. This also appears to be true at the other locations, with a small increase to the 2 p.m. value which is most marked at Santa Isabel along the southern coast. Here the winds most of the time are from overland, but the sea breeze during the afternoon brings slightly more moisture from the ocean.

The highest relative humidity values are generally found during the night when temperatures are the lowest. As the temperatures begin to rise, the relative humidity begins to fall, reaching its lowest point at about the time of the maximum temperature. High relative humidities are recorded during rainfall and when occasional cold fronts or easterly waves pass over.

The combination of high temperatures and fairly high relative humidity would usually result in physical discomfort. However, in Puerto Rico and the U. S. Virgin Islands there are several factors which greatly influence personal comfort and make the area normally very pleasant. The most noteworthy factor which lowers the sensible temperatures is the consistency

MEAN RELATIVE HUMIDITY (%) AND DEW POINT (°F)

(Local Time)	San Juan, Puerto Rico		Ramey AFB, Puerto Rico		Santa Isabel Airport, Puerto Rico		St. Thomas, Virgin Islands		St. Croix, Virgin Islands	
2 AM	RH	DP	RH	DP	RH	DP	RH	DP	RH	DP
March	82	66	83	67	87	63	79	67	84	67
Sept.	89	73	87	73	92	72	84	74	86	74
Annual Avg.	86	70	86	70	90	68	81	70	84	70
8 AM										
March	80	66	80	66	72	64	73	67	76	68
Sept.	83	73	81	72	79	72	76	74	79	75
Annual Avg.	81	70	82	70	77	70	74	70	78	71
2 PM										
March	63	67	65	69	61	67	64	67	64	68
Sept.	70	73	70	75	71	75	68	74	71	75
Annual Avg.	67	70	69	72	66	72	66	71	67	72
8 PM										
March	74	67	78	70	76	65	74	67	80	68
Sept.	81	73	82	74	86	73	80	74	85	74
Annual Avg.	78	70	81	72	82	70	77	70	82	71

of the trade winds. The greatest beneficial effect normally is felt during the afternoon hours when temperatures are highest. Even at night when temperatures fall to their lowest level, a mild breeze is still experienced. This factor is an important one to consider when a home, hospital, industrial plant, or any building for that matter, is being designed. Proper orientation and proper location of parking areas, and care in landscaping may well spell the difference between comfort and discomfort for all concerned.

Also of importance in the matter of comfort is the normal diurnal march of temperature and relative humidity, which progress in opposite directions, with the lowest relative humidities at the time of maximum temperature. Indeed, the frequency of occasions with high temperature at the same time as high relative humidity is quite small. A study of a 5-year period of record shows that only one-tenth of 1 percent of the hourly observations at San Juan had temperatures higher than 84° with simultaneous relative humidities of 80 percent or higher.

Generally speaking, relative humidity values of 90 percent or above during the nighttime are not infrequent. During the day, values from 60 percent to the low 70's are predominant. Extremely low relative humidity is almost never experienced, with the lowest readings generally in the 50's. During the 5-year period noted above, only about three-tenths of 1 percent of the hourly observations at San Juan had relative humidities below 50 percent.

VISIBILITY

Visibility normally presents no problem in Puerto Rico and the U. S. Virgin Islands, with very few occurrences of observations with critical values. During the 10-year period from 1951 through 1960 at San Juan, less than one-half of 1 percent of the observations had values of less than 3 miles. During 1960 at Alexander Hamilton Field, St. Croix, V. I., and Truman Field, St. Thomas, V. I., only about two-tenths of 1 percent of the hourly observations recorded visibilities of less than 3 miles. At Ramey AFB only three-tenths of 1 percent of all observations had visibility less than 3 miles in a 14-year period.

The comparatively rare occurrences of low visibilities are associated with periods of heavy rainfall. Fog does not enter as a factor. However, industrialization is now beginning to cause a decrease in visibility due to air pollution under certain atmospheric conditions.

FOG

Fog, in the sense that it is known elsewhere with visibility reduced to almost zero, is unknown in Puerto Rico and the Virgin Islands. The only type of fog found in the area is radiation fog, which may form in the early morning hours and is rapidly dissipated. This usually occurs in the interior valleys. No serious problems to either air or surface travel are encountered.

HAIL

PUERTO RICO

Hail is a relatively rare event and evokes much curiosity at each occurrence since many persons on the Island have never experienced it. It is believed to fall somewhere in Puerto Rico at some time during each year. Past, incomplete records indicate that at one time or another hail has fallen in just about all areas, but a greater frequency of occurrence is noted in the Las Marias, Coloso, Lares, and Maricao areas.

The hailstones are generally small, usually about the size of peas and in most cases cause no serious damage. However, there have been a few reports of hailstones up to 1 inch in diameter and of hail damage amounting to thousands of dollars to crops and buildings.

Hail has not fallen at the Weather Bureau Office in San Juan, but on May 6, 1926 a severe local hailstorm accompanied by sharp lightning and heavy thunder occurred about 2 miles away, at Santurce. Rough solid ice hailstones, some as large as hens eggs, fell during the storm which lasted about 15 minutes. Damage to paper-covered roofs resulted and trees were uprooted by the high winds which accompanied the storm.

U. S. VIRGIN ISLANDS

In the U. S. Virgin Islands, hail is even less frequent than in Puerto Rico. In January of 1969 a severe local hailstorm with hailstones up to 1 1/2 inches in diameter occurred. This was the first hailstorm on record in the U. S. Virgin Islands.

DROUGHT

PUERTO RICO

Periods of deficient rainfall are not uncommon in Puerto Rico and almost every year some section of the Island suffers to some extent for varying periods of time. Due to the importance of agriculture to Puerto Rico's economy deficient rainfall, even over a restricted area, can have a serious impact. The most susceptible periods for a serious deficiency of rainfall are late fall, winter, and early spring, when rainfall is normally the lowest.

Rainfall distribution in Puerto Rico is most uneven. During the 1957 drought, the 6-month rainfall, December-May, at Rio Blanco Upper totaled 56.31 inches. In contrast, rainfall for those 6 months at Yauco was only 4.36 inches, an amount totally inadequate to satisfy requirements. On the south coast when the farmers speak of the "drought season" they are referring to the months from December through April. Irrigation for growing crops is a must in this season. On the other hand, the sugar plantations take advantage of the dry weather to mature their cane and carry out harvesting and grinding operations. A true drought, resulting from a prolonged rainfall deficiency, will adversely affect the economy of the island once or twice during every decade. One of the most extreme droughts on recent record occurred during 1967 through May 1968. The period 1964 through May 1965 was also one of extreme drought, particularly in the south coastal section. Prior to this a short but severe drought started in late 1956 and ended with June rains of 1957. In all these cases, agriculture and the dairy industry suffered the worst effects when pastures burned up, irrigation supplies became exhausted, and entire agricultural zones in non-irrigated areas lacked sufficient soil moisture to support crops. Reservoirs remained empty. Near the end of the 1967-68 drought the San Juan Metropolitan area went on strict water rationing until the May rains replenished water storage supplies. Water tables over most of the island were lowered to record levels and many wells were lost through intrusion of salt water. Some of the south coast sections received only about 35 percent of their normal rainfall during the year 1964. Individual stations received as little as 13 inches of rain. The island rainfall average for the 1967 drought year was only 43 inches compared to a long term average of 69 inches. This was the lowest rainfall on record since the beginning of observations in 1899.

U. S. VIRGIN ISLANDS

Drought in the Virgin Islands occurs about as often and is just as damaging as it is in Puerto Rico. None of the three islands has any significant running rivers or streams and only St. Croix has an underground water source in a few sections. Water for irrigation is not available in quantity at any time. Large storage reservoirs do not exist so the Virgin Islands are, to some extent, more at the mercy of "Mother Nature" than is Puerto Rico where there are adequate facilities for water storage.

In the Virgin Islands, as noted above, a variety of rainfall catchment areas, such as housetops, sides of hills, and even airport runways, are used the year round in an effort to catch and store as much of the precious rainwater as possible. During times of drought the water supply reaches dangerously low levels and the hauling of potable water in barges from Puerto Rico becomes a necessity. On St. John there is at present, a small private sea water distillation unit, and additional municipal units are in operation at St. Thomas and St. Croix.

FLOODS

The time of greatest likelihood of flooding naturally is at the time of maximum rainfall expectancy, roughly from May through November or December. The hilly nature of Puerto Rico and the Virgin Islands and the steep slope of the waterways from their basins in the mountainous areas to their outlets into the ocean dictates that many of the flood situations will be of the flash flood type. Thus with a relatively short warning period possible, eternal vigilance for possible flood-producing situations must be maintained.

The orographic effect, when carried to maximum development, is capable of depositing large quantities of water over the area with resulting occasional flooding. This situation existed during the May 1960 flood in the Virgin Islands, when thunderstorms formed over the northern Virgin Islands and remained practically stationary. They deposited up to 14.56 inches of rainfall during a 3-day period, with the greatest amounts falling on the last day.

Hurricanes and tropical storms, when they pass over the area or close by, are a serious flood-producing threat. Two of the three major floods in Puerto Rico occurred during hurricanes passages over the Island. The first occurred on August 8, 1899, as hurricane "San Ciriaco" buffeted Puerto Rico, and the other on September 13-14, 1928 in hurricane "San Felipe". In September 1960, hurricane "Donna" passed to the northeast of Puerto Rico without bringing dangerous winds, but the heavy rainfall which followed deposited torrential amounts, which caused widespread floods and accounted for 107 deaths, 136 injuries, and considerable agricultural and property losses.

Cold frontal action may also be responsible for floods. During late November and early December 1960 such activity was sufficient to produce as much as 17.25 inches of rainfall at Yaurel in the southeastern corner of Puerto Rico. This occurred during a 6-day period, while 11 other stations in Puerto Rico received more than 14 inches. In the Virgin Islands the rainfall during this same 6-day period totaled more than 12 inches at several stations on St. Thomas and St. Croix. Agricultural and property damage resulted together with the loss of 4 lives in Puerto Rico. Damages were slight with no loss of life in the Virgin Islands.

Areas of Puerto Rico most susceptible to flooding include the flatlands in the vicinity of the rivers. The annual average is about one serious local flood together with one or more lesser floods either in Puerto Rico or the U. S. Virgin Islands. During 1960, however, there were three major floods: One occurred on St. Thomas and St. John in May; the next in September 1960, associated with hurricane Donna, affected a large portion of Puerto Rico, principally the northern coast from Manati to the east and in the southeastern sections of the Island; and the third flood, during December 1960, affected both Puerto Rico and the Virgin Islands.

CLOUD COVER

Cloud observations taken at WBFO San Juan and Ramey AFB in Puerto Rico, Alexander Hamilton Field on St. Croix, V. I., and at Truman Field, St. Thomas, V. I., show a high degree of similarity. As with wind speed, the minimum cloudiness occurs during the hours of darkness with increasing amounts after sunrise. Maximum cloudiness, averaged over the year, occurs between 10 a.m. and 5 p.m. at Alexander Hamilton Field, and from 11 a.m. to 3 p.m. at Truman Field. At San Juan the maximum occurs a little later between 2 p.m. and 3 p.m., although there is only a small variation from 10 a.m. to 7 p.m. Ramey AFB has a later maximum, with the highest values from 5 to 7 p.m.

The seasonal variation of cloudiness shows a double maximum at all stations, in May or June and again in September or October, with the June maximum somewhat more pronounced. At all four stations the lowest daily average cloudiness is in March, with the nighttime minima reaching their lowest point at that time.

Daytime cloud cover is greatest at Ramey AFB with a peak value of 8.6-tenths at 5 p.m., in June. The San Juan maximum value is 7.5-tenths at 4 p.m., in June. At Alexander Hamilton Field a double maximum is noted, with a value of 6.8-tenths at 8 a.m., and again at 5 p.m., in June with a slight decrease between those hours. The Truman Field record indicates a maximum of 6.4-tenths at 11 a.m., and at noon during May. During the night the minimum value noted is 1.9-tenths at Truman Field at 4 a.m., during March.

Actual observational data are not available for the interior. However, cloudiness in the interior is somewhat higher during the day than at the coastal stations due to the convective buildups, and somewhat lower during the night.

THUNDERSTORMS

While thunderstorms are not completely unknown during the winter months, the period of greatest likelihood falls between May and November. Due to the small size of the Virgin Islands, thunderstorm activity is less frequent there than in Puerto Rico, where the larger land mass and higher hills are more favorable for thunderstorm development especially in the western Cordilleras.

The average number of thunderstorms per year reported over various locations in Puerto

Rico: Ponce-12, San Juan WFBO-10, Roosevelt Roads-11, Mayaguez-35, and Ramey AFB-22. At St. Thomas thunder is heard on an average of about 13 hours per year; at St. Croix, 45 hours.

TROPICAL CYCLONES

Hurricanes and tropical storms are an important feature of Puerto Rico and U. S. Virgin Islands climate during summer and early autumn. The tropical cyclone season in the North Atlantic region extends from June through November. Because of seasonal shifts in favored locations of tropical cyclone development, the Puerto Rico-Virgin Islands area is outside the main paths of these most severe tropical atmospheric disturbances, except from August through the first half of October. A few "off-season" tropical cyclones have, however, slightly "brushed" the area at infrequent intervals.

Those hurricanes and tropical storms which do severely affect Puerto Rico and the Virgin Islands develop over the waters of the southern North Atlantic to the east of the Lesser Antilles. The movements of the storms are usually towards the west and northwest. They may pass either to the south or to the north of the islands, and occasionally directly over them.

A map is included which shows the paths of the more severe hurricanes which have passed directly over Puerto Rico since 1893. These are:

SOME HURRICANE PATHS ACROSS PUERTO RICO

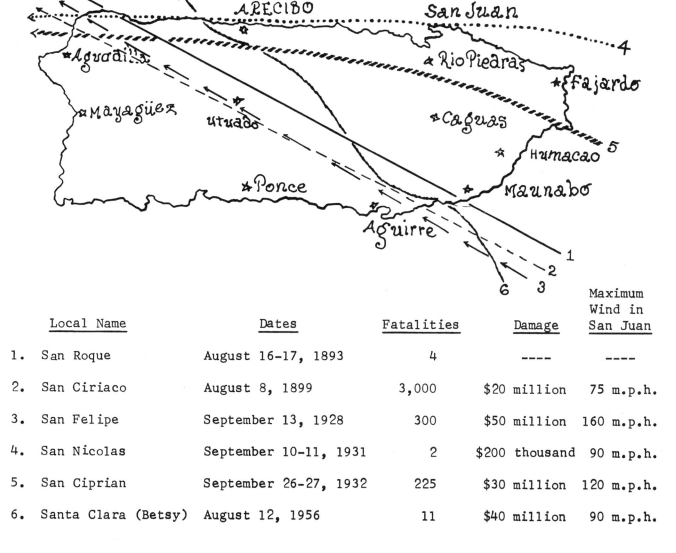

	Local Name	Dates	Fatalities	Damage	Maximum Wind in San Juan
1.	San Roque	August 16-17, 1893	4	----	----
2.	San Ciriaco	August 8, 1899	3,000	$20 million	75 m.p.h.
3.	San Felipe	September 13, 1928	300	$50 million	160 m.p.h.
4.	San Nicolas	September 10-11, 1931	2	$200 thousand	90 m.p.h.
5.	San Ciprian	September 26-27, 1932	225	$30 million	120 m.p.h.
6.	Santa Clara (Betsy)	August 12, 1956	11	$40 million	90 m.p.h.

Additional "near misses" of intense hurricanes or tropical storms, which produce little wind damage, may cause extensive rain (flooding) and/or tide damage.

An outstanding example of the fringe effects of hurricanes on the area occurred during the passage of Donna in September 1960. The hurricane center passed west-northwestward to the north of Puerto Rico and the Virgin Islands on September 5. Winds reached only 62 m.p.h. at St. Thomas and 42 m.p.h. at San Juan, in gusts, and little wind damage was sustained. Torrential rains developed during the evening of the 5th after the center had passed. Rains of up to 10 to 15 inches fell over much of the southeastern half of Puerto Rico during the night and produced the highest flash floods ever known on many streams. Several million dollars of crop and property damage was reported and 107 persons perished-- mostly by drowning.

The extent of destruction and loss of life caused by a tropical storm or hurricane is dependent upon several factors, both meteorological and in the field of public awareness.

The intensity of the storm, exemplified by the strength of the maximum winds (which have reached speeds of near 160 m.p.h. at San Juan), is of major importance, together with the forward speed and size of the hurricane. Longer durations of extreme intensity wind will generally cause more destruction. Slowly moving storms also increase the possibility of excessive rainfall. Fortunately the area covered by extremely high winds in hurricanes in tropical latitudes is usually very small, and very slow forward speeds are rare, so that high winds are usually sustained for only a very few hours at any one place.

Elements affecting the extent of damage and loss of life which may be controlled include: 1) the accuracy of warnings and the time of advance notice, and 2) the public response to issued warnings.

The accuracy of hurricane forecasts has been continually improved and the time of advance notice has increased during recent years, as a result of advanced communication facilities, aircraft reconnaissance observations, satellite pictures, and land-based radar. The structure and behavior of tropical cyclones is also now much better known, and this too has improved accuracy of forecasts and lengthened advance notice.

After warnings have been issued, the utilization of this information in holding damage, injury, and loss of life to a minimum, devolves upon the public and the local authorities. Complete cooperation with the authorities in taking adequate protective and precautionary measures and in keeping informed concerning latest advices are necessary for individual safety and the reduction of destruction.

CONCLUSION

Climate is closely linked with man's everyday life, his degree of comfort and happiness, his ability to work efficiently, and indeed his very survival. The tropical marine climate of Puerto Rico and the Virgin Islands can be one of the most compatible and enjoyable in the world. However, the key to this is man's willingness to plan his activities in consonance with the climate and not in spite of it. Whether it is the choice of an industrial or home site, structural design or orientation, vacation date or trip itinerary - a well founded decision must include complete consideration of the climatic factor. This is true more so in the tropics than perhaps any other areas of the world. Where the weather factor is involved in a tropical environment the balance between human comfort or discomfort, between economic success or failure, or between safe and compatible building design can be a delicate one. Through effective planning and intelligent application of climatic considerations to life in the Caribbean, man can truly say he has found his tropical paradise.

BIBLIOGRAPHY

BOGART, D. B., T. ARNOW, and J. W. Crooks, Recursos Hidrologicos Boletin Num. 1, Problemas de Abasto de Aguas en Puerto Rico y Programa de Investigaciones de Recursos Hidrologicos.

------------------------, Water Resources of Puerto Rico, A Progress Report, 1964.

BRISCOE, C. B., Weather in the Luquillo Mountains of Puerto Rico, Institute of Tropical Forestry, Rio Piedras, Puerto Rico.

CALVESBERT, R. J., 1966, The Climatology of Meteorological Drought in Puerto Rico, Proceedings of Conference on Climatology and Related Fields in the Caribbean, University of the West Indies, Kingston, Jamaica.

COLON, J. A., 1958, A study of surface winds in the vicinity of Mayaguez, Puerto Rico: U. S. Weather Bureau.

CRY, G. W., 1961, North Atlantic Tropical Cyclones 1960: U. S. Weather Bureau, Mariner's Weather Log, pp. 1-7.

DUNN, G. E. & B. I. Miller, (Director and Research Meteorologist, respectively, at the National Hurricane Center, Miami, Fla.) 1960, Atlantic Hurricanes, Louisiana State University Press, Baton Rouge.

BIBLIOGRAPHY

FASSIG, O. L., The Trade Winds of the Eastern Caribbean. (In American Geophysical Union. Transactions, fourteenth annual meeting, 1933).

-------------- The Climate of Porto Rico. (In P. R. Rev. Pub. Health, v. 4, 1928, p. 199).

GRAY, RICHARD W., U. S. Weather Bureau, La Distribucion Anual y Geografica de Lluvia en Puerto Rico, pp. 47, Almanaque Agricola de Puerto Rico 1944.

HARRIS, F. MILES, U. S. Weather Bureau, Almanaque Agricola de Puerto Rico 1947, La Temperature de Puerto Rico, pp. 75.

JONES, Clarence F. and PICO, RAFAEL, Symposium on the Geography of Puerto Rico, University of Puerto Rico Press 1955.

PICO, RAFAEL, The Geographic Regions of Puerto Rico, University of Puerto Rico 1950.

PUIG-THOMAS, JUAN, MANUEL LEDESMA and JOSE L. GARCIA DE QUEVEDO, 1960, Puerto Rico Nuclear Center-Meteorology Report.

QUINONES, MIGUEL A., High Intensity Rainfall and Major Floods in Puerto Rico.

RIEHL, HERBERT, Tropical Meteorology. New York, McGraw-Hill, 1954. p. 392.

SALIVIA, LUIS A., Historia de Los Temporales de Puerto Rico (1508-1949) San Juan, P. R. 1950.

SMEDLEY, DAVID, Climate of Puerto Rico and U. S. Virgin Islands, Climates of the States, Climatography of the United States No. 60-52, 1961. ESSA, Environmental Data Service.

STONE, ROBERT G., The New York Academy of Sciences, Scientific Survey of Porto Rico and the Virgin Islands, Volume XIX - Part 1 - Meteorology of the Virgin Islands.

TANNEHILL, IVAN RAY, Hurricanes, Princeton University Press.

WARD, ROBERT DeC. and CHARLES F. BROOKS, Handbuch der Klimatologie, Band II, Teil I, Westindien, Climatology of the West Indies, 1934.

Agricultural Yearbook Separate No. 1869 - Climate of the States, The West Indies Islands (Including Puerto Rico). ESSA, Weather Bureau

Climatography of the United States No. 30-52, Summary of Hourly Observations, San Juan, P. R., ESSA, Weather Bureau

Department of Tourism and Trade - Government of the Virgin Islands, Annual Report - July 1, 1959--June 30, 1960.

Economic Development Administration, Office of Economic Research, General Economics Division, Selected Statistics on the Visitors and Hotel Industry in Puerto Rico - 1959-60.

Local Climatological Data, Monthly and Annual, San Juan, P. R., Alexander Hamilton Field, St. Croix, V. I. and Santa Isabel Airport, P. R. ESSA, Environmental Data Service.

Miscellaneous Mimeographed Summaries of Climatological Observations for the Puerto Rico-Virgin Islands Area.

Monthly Weather Review. ESSA, Weather Bureau.

Puerto Rico Planning Board, Bureau of Economics and Statistics, Statistical Yearbook, Historical Statistics.

The West Indies and Caribbean Year Book, Thomas Skinner & Co. (Publishers), Limited.

U. S. Air Force, Uniform Summary of Surface Weather Observations-Station #11603 Aguadilla, P. R., Ramey AFB - Period of Record March 1940-March 1955, less May 1946.

U. S. Navy, Summary of Monthly Aerological Records, San Juan, P. R. - Period of Record March 1945-February 1959.

U. S. Navy, Summary of Monthly Aerological Records, Roosevelt Roads, P. R.

Unpublished Series of Temperature and Rainfall Maps. ESSA, Weather Bureau.

Unpublished special summaries for Ramey AFB, Santa Isabel, AP, Puerto Rico, Alexander Hamilton Field, St. Croix, V. I., Harry S. Truman Field, St. Thomas, V. I. and San Juan, Puerto Rico. ESSA, Weather Bureau.

Weather Bureau Technical Paper No. 13 - Mean Monthly and Annual Evaporation from Free Water Surfaces for the United States, Alaska, Hawaii and the West Indies. Washington, D. C. 1950. ESSA, Weather Bureau.

Weather Bureau Technical Paper No. 32, Upper-Air Climatology of the United States. ESSA, Weather Bureau.

Weather Bureau Technical Paper No. 36, North Atlantic Tropical Cyclones. ESSA, Weather Bureau.

BIBLIOGRAPHY

Weather Bureau Technical Paper No. 42, Generalized Estimates of Probable Maximum Precipitation and Rainfall, Frequency Data for Puerto Rico and Virgin Islands for Areas to 400 Square Miles, Durations to 24 Hours and Return Periods from 1 to 100 years. ESSA, Weather Bureau.

Weekly Weather and Crop Bulletin, Puerto Rico. ESSA, Weather Bureau.

Additional Publications:

Climatic Summary of the United States (Bulletin W) 1930 Edition, Section 106. ESSA, Weather Bureau.

Climatic Summary of the United States - Puerto Rico and the Virgin Islands - Supplement for 1931 through 1952 (Bulletin W Supplement). ESSA, Weather Bureau.

Climatic Summary of the United States - Puerto Rico and the United States. Virgin Islands - Supplement for 1951 through 1960 (Bulletin W Supplement). ESSA, Weather Bureau.

Climatological Data - Puerto Rico and the Virgin Islands. ESSA, Environmental Data Service.

Climatological Data National Summary. ESSA, Environmental Data Service.

*NORMALS BY CLIMATOLOGICAL DIVISIONS

Taken from "Climatography of the United States No. 81-4, Decennial Census of U. S. Climate"

TEMPERATURE (°F)   PRECIPITATION (In.)

STATIONS (By Divisions)	JAN	FEB	MAR	APR	MAY	JUNE	JULY	AUG	SEPT	OCT	NOV	DEC	ANN	JAN	FEB	MAR	APR	MAY	JUNE	JULY	AUG	SEPT	OCT	NOV	DEC	ANN
NORTH COASTAL																										
ARECIBO 2 ESE	73.9	73.6	74.6	76.0	78.1	79.4	79.9	80.2	80.2	79.5	77.7	75.6	77.4	4.25	3.27	2.57	3.77	7.40	4.24	4.14	4.96	4.98	5.58	5.58	4.79	55.53
DORADO 4 W	74.5	74.5	75.4	76.8	78.6	79.9	79.8	80.3	80.2	79.4	77.7	75.8	77.7	5.83	3.25	2.48	4.41	6.78	5.26	7.02	6.65	5.69	5.31	6.28	6.18	65.14
GARROCHALES	•	•	•	•	•	•	•	•	•	•	•	•	•	4.93	3.27	2.72	4.17	6.46	4.58	4.71	5.82	5.08	5.29	5.78	5.03	57.84
QUEBRADILLAS	•	•	•	•	•	•	•	•	•	•	•	•	•	4.07	3.29	3.18	4.68	8.04	5.00	3.98	5.47	5.46	5.34	5.72	4.69	58.92
SAN JUAN CITY	75.1	75.1	76.0	77.0	78.8	79.9	80.1	80.8	80.7	80.2	78.6	76.7	78.3	4.13	2.70	2.07	3.89	7.16	5.83	6.02	6.34	6.04	5.24	6.05	4.89	60.36
SAN JUAN WB AIRPORT 2	74.4	74.4	75.3	76.6	78.7	80.0	80.4	80.9	80.5	80.0	78.2	76.2	78.0	4.70	2.90	2.20	3.72	7.12	5.66	6.25	7.13	6.76	5.83	6.49	5.45	64.21
DIVISION	73.9	73.7	74.8	76.2	78.2	79.4	79.7	80.1	79.9	79.2	77.5	75.5	77.7	4.75	3.31	2.72	4.32	7.36	5.66	6.06	6.72	6.08	5.81	5.88	5.28	64.09
SOUTH COASTAL																										
AGUIRRE	77.0	76.8	77.6	78.9	80.4	81.6	82.2	82.5	82.2	81.7	80.4	78.5	80.0	1.07	1.29	.76	2.37	4.94	4.73	4.17	5.28	5.93	5.77	4.36	2.10	42.77
CENTRAL SAN FRANCISCO	•	•	•	•	•	•	•	•	•	•	•	•	•	.74	1.24	.72	1.83	4.20	3.08	2.90	4.55	6.20	5.19	3.40	1.24	35.29
ENSENADA	•	•	•	•	•	•	•	•	•	•	•	•	•	.70	1.28	1.02	1.95	4.13	2.69	2.60	3.72	5.71	4.55	3.88	1.66	33.89
PONCE 4 E	76.1	75.8	76.6	78.2	79.7	80.9	81.5	81.6	81.4	80.7	79.4	77.6	79.1	.86	1.07	.61	2.28	4.49	3.51	2.74	4.69	5.93	5.31	3.70	1.34	36.53
POTOLA	•	•	•	•	•	•	•	•	•	•	•	•	•	.79	1.05	.67	2.07	4.71	3.34	3.00	4.34	6.42	5.70	3.46	1.45	37.00
SABATER	•	•	•	•	•	•	•	•	•	•	•	•	•	.82	1.12	.75	2.22	4.55	4.43	3.70	4.74	5.93	5.99	3.71	1.96	39.92
SANTA ISABEL	•	•	•	•	•	•	•	•	•	•	•	•	•	.86	.95	.50	1.70	4.42	3.92	2.74	4.44	5.91	5.14	3.11	1.38	35.07
SANTA RITA	•	•	•	•	•	•	•	•	•	•	•	•	•	.59	1.40	.93	2.28	4.15	2.77	2.46	4.19	5.72	5.12	3.75	1.75	35.11
YAUCO 1 S	•	•	•	•	•	•	•	•	•	•	•	•	•	.72	1.46	.87	2.02	4.82	3.22	2.75	4.46	6.50	5.68	3.74	1.41	37.65
DIVISION	76.1	75.9	76.7	78.2	79.7	80.9	81.6	81.6	81.4	80.8	79.5	77.7	79.0	.87	1.21	.79	2.11	4.51	3.50	3.02	4.71	6.12	5.51	3.74	1.71	37.76
NORTHERN SLOPES																										
CALERO CAMP	•	•	•	•	•	•	•	•	•	•	•	•	•	3.18	2.68	2.92	4.37	7.68	7.07	4.80	6.14	5.78	6.30	4.45	3.69	59.06
CANOVANAS 2 N	74.0	73.7	74.9	76.6	78.5	79.5	79.6	80.6	79.9	79.2	77.5	75.5	77.5	4.96	3.57	2.82	4.50	8.25	7.08	7.80	7.65	7.23	6.18	6.79	6.01	72.84
FAJARDO	76.0	75.9	77.0	78.4	80.0	81.2	81.7	81.9	81.3	80.5	79.2	77.5	79.2	3.40	2.78	2.32	4.32	7.97	6.12	6.06	6.80	7.75	8.01	6.08	4.40	66.01
ISABELA 4 SW	73.5	73.4	74.5	75.5	77.1	78.3	79.1	79.1	78.9	78.3	77.0	75.1	76.7	3.80	3.34	3.25	4.84	8.44	7.54	5.00	6.68	6.90	6.52	5.42	4.32	66.05
DIVISION	74.3	74.1	75.3	76.8	78.4	79.6	80.0	80.3	79.9	79.3	77.7	75.7	77.5	4.17	3.28	2.80	4.50	8.22	6.56	5.85	6.77	7.11	6.61	5.87	4.85	66.50
SOUTHERN SLOPES																										
CABO ROJO	•	•	•	•	•	•	•	•	•	•	•	•	•	2.33	2.19	3.23	4.92	7.33	4.00	6.47	8.29	8.48	7.20	5.84	3.74	64.02
GUAYABAL RESERVOIR	•	•	•	•	•	•	•	•	•	•	•	•	•	1.36	1.91	1.49	3.76	6.26	4.37	3.98	6.59	9.45	8.76	5.58	2.18	55.69
GUAYAMA	78.5	78.4	79.1	80.3	81.3	82.3	82.7	83.3	82.9	82.5	81.4	79.8	81.0	2.09	1.81	1.27	3.26	6.85	6.80	5.75	6.45	8.69	7.70	5.57	3.50	59.84
HUMACAO	73.0	73.4	75.1	77.2	78.8	80.2	80.2	80.3	79.8	78.8	76.6	74.1	77.3	4.23	3.30	2.87	5.22	10.26	9.53	8.65	9.41	10.87	9.96	8.16	5.59	88.05
JOSEFA	•	•	•	•	•	•	•	•	•	•	•	•	•	1.62	1.57	1.01	2.84	6.02	5.78	4.63	6.18	7.46	6.89	5.89	3.15	53.04
JUANA DIAZ CAMP	76.5	76.3	77.0	78.2	79.5	80.8	81.8	81.5	80.7	80.2	79.1	77.7	79.1	1.22	1.52	.95	2.64	5.00	3.89	3.51	5.56	8.00	6.85	4.50	2.00	45.64
MAYAGUEZ NUCLEAR CTR	74.4	74.8	75.7	76.9	78.3	79.3	79.3	79.6	79.5	78.8	77.2	75.7	77.5	1.91	1.64	3.50	4.68	8.45	8.52	9.60	10.08	11.01	8.50	5.63	2.60	76.12
PATILLAS DAM	•	•	•	•	•	•	•	•	•	•	•	•	•	3.01	2.24	1.69	3.88	8.65	8.57	6.95	7.78	9.53	9.06	6.56	4.36	72.32
SAN GERMAN	74.6	74.6	75.4	76.9	78.6	80.0	79.9	80.0	79.9	79.2	77.6	75.8	77.7	2.52	2.64	3.09	5.25	7.19	3.89	5.91	7.73	9.96	9.21	6.74	3.90	68.03
YABUCOA 1 NE	•	•	•	•	•	•	•	•	•	•	•	•	•	4.38	3.45	2.70	5.13	10.46	8.76	7.94	9.44	10.51	10.47	8.22	6.80	88.26
DIVISION	75.4	75.4	76.3	77.7	79.1	80.3	80.5	80.8	80.4	79.7	78.3	76.6	78.3	2.59	2.27	2.13	4.09	7.79	6.40	6.34	7.79	9.49	8.61	6.33	3.93	67.71
EASTERN INTERIOR																										
CARITE DAM	69.6	69.4	70.4	71.1	72.2	73.3	73.8	74.7	74.6	74.1	73.0	71.2	72.3	4.20	3.40	2.57	5.26	9.66	8.79	8.92	8.65	10.86	9.71	7.18	5.39	84.59
CAYEY 1 E	69.1	69.3	71.0	73.0	74.6	75.5	76.0	76.2	75.8	74.8	73.0	70.9	73.3	3.85	2.66	1.81	3.85	6.66	5.36	6.15	6.35	7.89	7.03	4.99	3.89	59.92
CIDRA 1 E	•	•	•	•	•	•	•	•	•	•	•	•	•	4.17	3.10	2.18	4.61	7.05	6.11	7.25	7.32	8.14	7.07	5.61	4.37	66.98
JAJOME ALTO	•	•	•	•	•	•	•	•	•	•	•	•	•	3.69	3.25	2.49	4.50	8.06	7.54	7.76	7.46	9.79	8.71	6.01	4.90	74.16
JUNCOS 1 NNE	•	•	•	•	•	•	•	•	•	•	•	•	•	2.74	2.51	1.52	3.34	7.41	7.32	6.36	7.85	9.28	7.58	5.43	4.38	65.72
RIO BLANCO UPPER	•	•	•	•	•	•	•	•	•	•	•	•	•	10.67	9.68	6.65	10.50	16.41	14.67	14.00	15.63	17.06	16.05	14.19	13.67	159.18
DIVISION	70.7	70.5	71.8	73.4	75.1	76.1	76.5	76.9	76.7	75.9	74.3	72.2	74.1	4.81	3.94	2.81	5.22	9.57	8.48	8.12	9.11	10.47	9.46	7.63	6.24	85.80
WESTERN INTERIOR																										
COLOSO	74.1	73.9	75.0	76.5	78.4	79.6	80.2	80.3	79.9	79.2	77.8	75.8	77.6	2.38	2.10	2.99	5.95	11.64	11.75	9.77	11.44	10.20	7.98	5.02	2.55	83.77
GARZAS DAM	•	•	•	•	•	•	•	•	•	•	•	•	•	2.90	2.97	3.39	6.15	10.97	7.34	8.18	11.98	12.89	13.63	7.66	3.96	92.02
GUAJATACA DAM	•	•	•	•	•	•	•	•	•	•	•	•	•	3.82	3.22	2.49	6.22	11.21	5.82	7.87	8.99	8.64	4.24			76.84
GUINEO RESERVOIR	•	•	•	•	•	•	•	•	•	•	•	•	•	4.08	4.28	4.37	8.35	16.45	9.08	7.70	12.63	15.76	16.17	8.76	5.00	112.63
LARES 3 SE	71.9	71.7	72.8	73.8	75.3	76.4	76.9	77.4	77.0	76.5	75.3	73.3	74.9	3.22	3.06	4.03	8.28	15.96	8.59	7.54	10.24	12.12	11.80	6.81	3.85	95.88
MARICAO	•	•	•	•	•	•	•	•	•	•	•	•	•	2.67	2.94	4.62	7.56	12.85	8.36	9.51	14.06	14.16	13.70	7.20	3.84	101.91
TORO NEGRO PLANT 2	•	•	•	•	•	•	•	•	•	•	•	•	•	4.34	4.47	3.85	7.47	14.95	7.95	6.46	11.92	15.56	15.65	8.22	4.88	105.72
DIVISION	71.0	70.9	71.6	73.2	75.0	76.4	76.7	77.0	76.6	75.9	74.4	72.3	74.1	3.55	3.37	3.46	6.57	11.65	7.23	6.86	10.12	11.43	10.77	7.02	4.33	86.32
OUTLYING ISLANDS																										
DIVISION	76.7	76.6	77.4	78.8	80.4	81.7	82.3	82.6	82.1	81.5	80.0	77.9	79.8	2.17	1.65	1.60	2.62	4.69	3.83	3.46	4.28	5.27	5.22	4.78	3.36	42.74

*NORMALS BY CLIMATOLOGICAL DIVISIONS

Taken from "Climatography of the United States No. 81-4, Decennial Census of U. S. Climate"

TEMPERATURE (°F)   PRECIPITATION (In.)

STATIONS (By Divisions)	JAN	FEB	MAR	APR	MAY	JUNE	JULY	AUG	SEPT	OCT	NOV	DEC	ANN	JAN	FEB	MAR	APR	MAY	JUNE	JULY	AUG	SEPT	OCT	NOV	DEC	ANN
VIRGIN ISLANDS																										
CHARLOTTE AMALIE 2	77.1	76.8	77.7	78.9	80.3	81.9	82.5	82.9	82.6	81.9	80.5	78.7	80.2	2.87	1.91	1.59	2.53	4.95	3.30	3.47	4.88	6.08	5.51	4.39	3.27	44.75
DIVISION	76.5	76.3	77.2	78.4	79.7	81.3	81.8	82.1	81.4	80.8	79.6	77.7	79.3	2.71	1.78	1.47	2.45	4.54	3.30	3.73	4.49	6.18	5.56	4.84	3.33	44.32

* Normals for the period 1931-1960. Divisional normals may not be the arithmetical averages of individual stations published, since additional data for shorter period stations are used to obtain better areal representation.

ADDITIONAL MEAN DATA FOR VIRGIN ISLANDS NOT INCLUDED IN "NORMALS" TABULATION*

Station		JAN	FEB	MAR	APR	MAY	JUNE	JULY	AUG	SEPT	OCT	NOV	DEC	ANN	JAN	FEB	MAR	APR	MAY	JUNE	JULY	AUG	SEPT	OCT	NOV	DEC	ANN
ALEX HAMILTON AP FAA	MEAN	76.6	76.7	77.5	79.0	80.4	81.8	81.9	82.2	81.5	81.0	79.6	77.9	79.7	2.23	2.19	1.73	2.83	4.31	3.10	3.51	4.58	6.65	5.45	4.65	3.34	44.57
	YEARS	13	13	13	13	13	13	14	14	14	14	14	14		13	13	13	13	13	13	14	14	14	14	14	14	
ANNAS HOPE EXP STA	MEAN	76.0	75.6	76.3	77.5	79.2	80.5	81.0	81.2	80.3	79.7	78.4	76.8	78.5	2.69	1.96	1.75	2.24	3.72	2.95	3.31	4.49	6.39	5.84	5.06	3.28	43.68
	YEARS	30	29	30	31	31	31	31	31	31	30	30	30		41	41	41	41	41	41	41	41	41	41	41	41	
CRUZ BAY	MEAN	76.7	77.0	77.9	79.3	80.2	81.3	81.8	82.3	81.8	80.4	79.5	77.8	79.7	2.38	1.87	1.55	2.15	4.62	3.11	3.80	3.93	5.94	5.09	4.51	2.93	41.88
	YEARS	14	13	13	12	13	14	14	14	13	13	13	14		39	40	40	39	39	39	40	40	40	39	39	39	

*OBTAINED FROM "CLIMATIC SUMMARY OF THE UNITED STATES— SUPPLEMENT FOR 1951 THROUGH 1960."

TEMPERATURE   PRECIPITATION

JAN	FEB	MAR	APR	MAY	JUNE	JULY	AUG	SEPT	OCT	NOV	DEC	ANN	JAN	FEB	MAR	APR	MAY	JUNE	JULY	AUG	SEPT	OCT	NOV	DEC	ANN

CONFIDENCE - LIMITS

In the absence of trend or record changes, the chances are 9 out of 10 that the true mean will lie in the interval formed by adding and subtracting the values in the following table from the means for any station in the State. Because of the wider variation in mean precipitation, the corresponding monthly means and annual mean must be substituted for "p" in the precipitation table below to obtain mean precipitation confidence limits.

| .3 | .3 | .4 | .4 | .4 | .3 | .2 | .2 | .2 | .2 | .2 | .2 | .2 | .44√p | .44√p | .44√p | .44√p | .44√p | .44√p | .44√p | .44√p | .44√p | .44√p | .44√p | .44√p | .44√p |

COMPARATIVE DATA

Data in the following table are the mean temperature and average precipitation for Canovanas 2 N, Puerto Rico, for the period 1906-1930 and are included in this publication for comparative purposes.

| 74.3 | 74.7 | 75.5 | 77.1 | 79.6 | 80.6 | 80.6 | 81.2 | 81.0 | 80.2 | 78.0 | 75.8 | 78.2 | 6.19 | 4.03 | 4.42 | 4.74 | 7.20 | 6.95 | 8.46 | 7.84 | 7.53 | 7.44 | 8.23 | 7.43 | 80.46 |

MEANS AND EXTREMES

Station: AGUADILLA/RAMEY AFB Standard time used: ATLANTIC Latitude: 18° 30' Longitude: 67° 08' Elevation (ground): 247 feet

Month	Temperature MEANS Daily maximum	Daily minimum	Monthly	Extremes Record highest	Year	Record lowest	Year	Normal heating degree days (Base 65°)	Precipitation Normal total	Maximum monthly	Year	Minimum monthly	Year	Maximum in 24 hrs	Year	Snow, Ice pellets Mean total	Maximum monthly	Year	Maximum in 24 hrs	Year	Relative humidity 00-02 Hour	06-08 Hour	12-14 Hour	16-20 Hour (Local time)	Wind Mean speed	Prevailing direction	Fastest mile Speed	Direction	Year	Pct. of possible sunshine	Mean sky cover sunrise to sunset	Mean number of days Sunrise to sunset Clear	Partly cloudy	Cloudy	Precipitation .01 inch or more	Snow, Ice pellets 1.0 inch or more	Thunderstorms	Heavy fog	Temperatures Max. 90° and above	t	32° and below	Min. 32° and below	0° and below	Average daily solar radiation - langleys
(a)	27	27	27	27		27			27	27		27		27		27	27		27		27	27	27	27		18						27	27	21	21	27	27	27	27		
J	80.2	68.6	74.5	87	1963	60	1948	—	2.82	10.44	—	1.12	—	3.48	1959	0.0	0.0		0.0		84	84	68	76	9.8	E	53	ENE	1950			12	0	*	*	0	0	0	0	
F	80.4	68.2	74.4	89	1946	61	1965	—	2.40	6.96	—	.03	—	2.83	1956	0.0	0.0		0.0		83	83	68	76	9.6	E	43	ENE	1949			9	0	*	*	0	0	0	0	
M	81.5	69.0	75.3	90	1952	62	1951+	—	2.64	9.48	—	.23	—	4.43	1963	0.0	0.0		0.0		82	81	66	76	9.8	E	38	ENE	1950			9	0	*	*	*	0	0	0	
A	82.2	70.6	76.5	90	1958+	64	1946	—	4.46	14.14	—	.47	—	5.53	1958	0.0	0.0		0.0		84	81	69	78	9.4	E	39	E	1966			11	0	2	*	*	0	0	0	
M	83.6	72.5	78.2	90	1964+	67	1962+	—	5.40	10.24	—	.70	—	4.40	1945	0.0	0.0		0.0		85	82	72	80	8.0	E	36	E	1963			14	0	8	*	*	0	0	0	
J	84.9	73.4	79.3	93	1963	66	1943	—	6.22	12.34	—	1.67	—	3.61	1956	0.0	0.0		0.0		86	82	72	81	8.7	E	57	ENE	1963			14	0	14	*	*	0	0	0	
J	85.3	74.2	79.8	95	1962	68	1945	—	4.23	8.03	—	.96	—	2.60	1966	0.0	0.0		0.0		86	84	72	80	10.6	E	43	E	1950			14	0	13	*	*	0	0	0	
A	85.8	74.2	80.1	92	1940	68	1940	—	5.16	10.64	—	1.70	—	4.60	1956	0.0	0.0		0.0		87	84	72	81	8.8	E	41	ENE	1950			15	0	12	*	*	0	0	0	
S	86.0	73.8	80.0	92	1963+	66	1942	—	4.74	8.44	—	1.35	—	3.67	1954	0.0	0.0		0.0		87	84	71	81	7.2	E	54	ESE	1949			14	0	14	*	1	0	0	0	
S	85.5	73.2	79.5	92	1963+	64	1958	—	5.43	9.46	—	1.73	—	3.74	1958	0.0	0.0		0.0		87	84	71	82	6.0	E	35	SSE	1965			13	0	12	*	*	0	0	0	
N	83.6	71.9	77.9	90	1963	64	1966	—	3.80	10.62	—	.61	—	4.07	1961	0.0	0.0		0.0		86	84	69	79	7.7	E	40	ENE	1950			12	0	4	*	*	0	0	0	
D	81.6	70.0	75.9	87	1963+	62	1949	—	3.86	10.04	—	.80	—	3.27	1959	0.0	0.0		0.0		85	84	68	78	9.0	E	38	E	1965			13	0	1	*	0	0	0	0	
YR	83.4	71.7	77.7	95	JUL 1962	60	JAN 1948	—	51.16	14.14	APR	.03	FEB	5.53	APR 1958	0.0	0.0		0.0		85	83	70	79	8.7	E	57	ENE	1963			150	0	80	*	1	0	0	0	

THIS TABLE CONSISTS OF DATA EXTRACTED FROM THE USAF PUBLICATION, "REVISED UNIFORM SUMMARY OF
SURFACE WEATHER OBSERVATIONS." YEARS OF RECORD: 1940-1966

JAN. 49—DEC. 68   N 18°15' W 65°38' ZONE: +4 HOURS

NWSED, ROOSEVELT ROADS, P. R.

Month	Temperature Means Daily maximum	Daily minimum	Monthly	Extremes Record highest	Year	Record lowest	Year	Mean degree days	Precipitation Mean	Maximum monthly	Year	Minimum monthly	Year	Maximum in 24 hrs.	Year	Snow, Sleet Mean	Maximum monthly	Year	Maximum in 24 hrs.	Year	Relative humidity Mean 0200 LST	0800 LST	1400 LST	2000 LST	Wind Mean speed (Kts)	Prevailing direction	Peak gust Speed (Kts)	Direction	Year	Mean cloud amount	Mean number of days Precipitation .01 inch or more	Snow, Sleet 1.0 inch or more	Thunder and Thunderstorms	Fog	Temperature Max. 90° and above	32° and below	Min. 32° and below	0° and below	Mean number of 3 hrly. obs. Temperature wb 73° and above	db 93° and above	wb 67° and above	db 80° and above
	13	13	13	13		13		13	13	13		13		13		13	13				13	13	13	13		12				13	13	13	13	13	13	13	13	13	13	13	13	13
J	82.7	71.2	77.2	88	1961	62	1965	0	3.46	10.82	1958	1.69	1965	4.26	1958	0.0	0.0		0.0		79	79	66	77	8.8	ENE	47	ENE	1950	.5	17	0	13	13	*	0	13	13	13	60	0	227	59
F	83.0	71.2	77.3	90	1964	64	1966	0	2.14	7.48	1950+	1.30	1960	1.49	1950+	0.0	0.0		0.0		80	78	65	76	8.5	E	43	ESE	1950+	.5	16	0	1	*	*	0	0	0	0	51	0	205	58
M	84.1	72.1	78.4	89	1965+	64	1949	0	2.33	5.96	1960	0.23	1958	8.25	1949	0.0	0.0		0.0		80	76	63	76	8.4	E	41	ESE	1949	.5	13	0	1	*	0	0	0	0	0	76	0	234	82
A	84.5	73.3	79.2	90	1966	68	1968	0	4.06	9.14	1960	0.87	1967	4.11	1960	0.0	0.0		0.0		81	75	66	78	8.6	E	35	E	1960	.5	15	0	1	0	0	0	0	0	0	123	0	231	92
M	86.9	76.7	82.1	92	1960	66	1967+	0	4.99	9.61	1962	2.30	1961	2.74	1960	0.0	0.0		0.0		82	76	69	79	8.3	E	37	S	1960	.6	19	0	2	*	2	0	0	0	0	205	0	247	126
J																															18	0	3	0	3	0	0	0	0	228	0	239	164
J	87.3	77.3	82.6	92	1962	71	1967+	0	5.13	9.71	1958	2.77	1967	2.97	1958	0.0	0.0		0.0		81	76	69	80	9.0	E	46	ENE	1960+	.6	21	0	5	0	5	0	0	0	0	243	0	248	201
A	87.5	77.5	83.0	93	1968	70	1967	0	5.46	10.04	1960	2.43	1967	2.59	1960	0.0	0.0		0.0		82	77	68	80	8.3	E	44	S	1949	.6	14	0	4	0	5	0	0	0	0	245	0	247	207
S	88.2	76.4	82.5	92	1960+	68	1960	0	6.50	16.27	1960	2.53	1967	7.39	1960	0.0	0.0		0.0		84	79	69	81	6.4	E	76	SW	1960	.6	20	0	7	0	5	0	0	0	0	230	0	239	181
O	87.5	75.2	81.7	92	1963+	66	1965+	0	5.26	10.65	1949	2.01	1957	4.00	1961	0.0	0.0		0.0		84	80	68	80	6.4	E	38	ESE	1958	.6	17	0	5	0	4	0	0	0	0	218	0	247	160
N	86.3	73.7	80.2	93	1957	66	1963	0	5.23	9.82	1957	2.05	1964	5.48	1966	0.0	0.0		0.0		83	81	67	80	6.0	E	41	ESE	1958	.5	17	0	3	0	2	0	0	0	0	169	*	247	115
D	84.1	72.5	78.6	91	1966	63	1964	0	4.89	11.28	1960	1.92	1965	2.76	1965	0.0	0.0		0.0		82	81	68	79	7.3	E	40	NE	1960	.5	19	0	1	0	*	0	0	0	0	121	0	232	80
	85.7	74.4	80.3	93	AUG 1968+	62	FEB 1966+	0	55.72	16.27	SEP 1960	0.23	MAR 1958	8.25	MAR 1949	0.0	0.0		0.0		82	78	67	79	7.7	E	76	SW	SEP 1960	.5	211	0	33	*	23	0	0	0	1969	*	2833	1525

NORMALS, MEANS, AND EXTREMES

ST. CROIX, VIRGIN ISLANDS, U.S.A. — ALEXANDER HAMILTON FIELD, 1953

LATITUDE 17° 42' N LONGITUDE 64° 48' W ELEVATION (ground) 53 feet

Month	Temp Normal Daily max	Daily min	Monthly	Extreme highest	Year	Extreme lowest	Year	Precip Normal total	Max monthly	Year	Min monthly	Year	Max 24 hrs	Year
	(b)	(b)	(b)	6		8		(b)	6		6		6	
J	82.5	69.1	75.8	88	1948	64	1953+	2.97	4.77	1952	1.17	1953	2.88	1952
F	82.7	69.0	75.9	88	1948	62	1951	2.80	8.33	1950	0.93	1953	3.95	1950
M	83.7	69.5	76.6	88	1949	62	1951	2.33	6.63	1949	0.17	1953	3.42	1951
A	84.7	71.5	78.1	89	1951	65	1953+	2.21	5.38	1952	0.22	1953	1.24	1953
M	85.8	73.7	79.8	90	1948	62	1950	4.90	6.21	1950	2.72	1953	2.84	1953
J	87.0	75.0	81.0	91	1953	68	1953	2.93	4.76	1953	1.70	1952	3.47	1952
J	88.2	75.5	81.9	95	1948	69	1949	2.77	6.14	1952	1.51	1952	2.00	1947
A	88.7	75.2	82.0	93	1953	70	1947	4.04	6.67	1947	2.22	1949	2.19	1952
S	87.8	74.1	81.0	92	1953+	68	1949	5.84	15.23	1948	2.61	1947	7.76	1953
O	86.7	73.7	80.2	90	1953+	67	1951	5.34	8.66	1951	2.58	1950	2.35	1948
N	85.1	72.7	78.9	89	1953+	64	1952	4.17	10.35	1953	2.43	1953	4.49	1951
D	83.7	71.0	77.4	87	1953+	63	1949	2.05	4.56	1953	1.08	1953	1.89	1952
Year	85.6	72.5	79.1	95	July 1948	62	March 1951+	42.35	15.23	Sept 1947	0.17	Mar 1951	7.76	Sept 1953

Snow, Sleet, Hail (Mean total / Max monthly / Max 24 hrs): all 0.0

Month	Rel. humidity 2:30 a.m.	8:30 a.m.	2:30 p.m.	8:30 p.m.	Wind mean hourly speed	Prevailing direction	Mean sky cover sunrise–sunset	Clear	Partly cloudy	Cloudy	Precip .01 in or more	Thunderstorms	Heavy fog	Temp 90°+	32°+ below	32° below	0° below
J	81	76	66	80	12.7	ENE	5.6	6	20	6	19	0	0	0	0	0	0
F	84	76	63	80	12.2	ENE	5.6	5	17	6	13	0	0	0	0	0	0
M	83	73	63	79	12.6	ENE	5.7	5	21	5	13	0	0	0	0	0	0
A	85	74	66	81	12.3	ESE	5.9	4	19	7	13	1	0	0	0	0	0
M	87	78	69	82	12.0	ESE	6.2	2	14	11	15	4	0	4	0	0	0
J	84	75	69	82	13.0	ESE	6.9	1	18	11	15	5	0	6	0	0	0
J	86	77	70	84	13.0	ENE	7.0	1	17	13	19	4	0	6	0	0	0
A	87	78	69	85	12.1	ENE	6.1	2	21	8	18	6	0	10	0	0	0
S	88	78	73	86	11.0	ESE	7.0	2	15	13	17	9	0	5	0	0	0
O	90	80	72	87	9.4	ESE	6.5	3	17	11	17	7	0	2	0	0	0
N	86	79	69	83	11.2	ENE	6.0	4	19	7	18	4	0	0	0	0	0
D						ENE	5.8	5	18	8	18	2	0	0	0	0	0
Year	86	77	69	83	11.7	ENE	6.3	39	216	110	196	43	0	27	0	0	0

(a) Length of record, years.
(b) Normal values are based on the period 1921-1950.

SANTA ISABEL, PUERTO RICO — SANTA ISABEL AIRPORT, 1953

LATITUDE 17° 58' N LONGITUDE 66° 24' W ELEVATION (ground) 28 feet

Month	Temp Normal Daily max	Daily min	Monthly	Extreme highest	Year	Extreme lowest	Year	Precip Normal total	Max monthly	Year	Min monthly	Year	Max 24 hrs	Year
	(b)	(b)	(b)	8		8		(b)	8		8		8	
J	83.0	64.9	74.0	89	1947	59	1951	0.74	1.19	1952	0.05	1948	0.91	1952
F	82.0	64.3	73.2	89	1949+	57	1951	0.82	1.30	1950	0.07	1948	0.54	1950
M	82.2	65.1	73.7	89	1948	58	1951	0.48	1.07	1949	0.09	1951	0.44	1949
A	83.0	67.1	75.1	90	1947	62	1953+	1.41	3.73	1952	0.22	1949	1.66	1952
M	84.0	70.4	77.2	94	1952+	60	1950	4.29	5.28	1946	0.31	1947	1.70	1949
J	85.6	71.7	78.7	94	1948	67	1950	3.38	4.02	1952	0.77	1952	1.94	1946
J	86.9	72.0	79.5	96	1948	67	1950	3.41	6.11	1949	0.19	1946	3.97	1949
A	87.1	72.0	79.5	95	1950	68	1953+	4.36	7.02	1952	2.46	1951	3.33	1953
S	87.6	71.8	79.3	93	1953+	67	1951	4.54	12.37	1950	2.17	1948	6.62	1950
O	85.9	70.5	78.2	91	1947	67	1950	4.24	12.10	1948	2.49	1952	4.02	1952
N	84.9	68.6	76.8	91	1948	63	1952	3.96	7.66	1950	2.45	1950	1.98	1952
D	83.8	66.0	74.9	88	1953+	59	1950+	1.02	2.62	1948	0.10	1948	—	—
Year	84.6	68.7	76.7	96	July 1948	57	Feb 1951	32.65	12.37	Sept 1950	0.05	Jan 1948	6.62	Sept 1949

Snow, Sleet, Hail (Mean total / Max monthly / Max 24 hrs): all 0.0

Month	Rel. humidity 2:30 a.m.	8:30 a.m.	2:30 p.m.	8:30 p.m.	Wind mean hourly speed	Prevailing direction	Mean sky cover sunrise–sunset	Clear	Partly cloudy	Cloudy	Precip .01 in or more	Thunderstorms	Heavy fog	Temp 90°+	32°+ below	32° below	0° below
J	83	78	59	76	6.7	ENE	4.3	13	15	3	4	0	0	0	0	0	0
F	84	76	61	75	6.2	NE	4.9	11	14	3	4	0	0	0	0	0	0
M	83	71	61	73	7.1	ESE	4.5	11	14	7	3	0	0	0	0	0	0
A	73	62	57	66	7.1	SE	5.0	11	12	7	7	0	0	0	0	0	0
M	76	63	58	66	7.5	SE	5.3	4	16	11	11	5	0	6	0	0	0
J	75	63	58	68	7.2	SE	6.4	4	16	10	9	6	0	0	0	0	0
J	74	61	58	67	7.4	SE	6.1	7	14	10	9	8	0	10	0	0	0
A	86	71	67	79	6.9	E	6.2	5	16	9	12	12	0	10	0	0	0
S	87	76	70	85	5.8	E	6.5	5	13	12	13	15	0	2	0	0	0
O	90	78	71	86	5.4	NE	6.1	6	16	9	10	18	0	1	0	0	0
N	90	80	66	85	5.1	NE	5.0	9	15	6	8	5	0	0	0	0	0
D	88	80	63	82	5.2	NE	4.6	12	14	5		1	0	0	0	0	0
Year	82	72	62	76	6.4	SE	5.4	105	175	85	99	70	0	35	0	0	0

(a) Length of record, years.
(b) Normal values are based on the period 1921-1950.

NORMALS, MEANS, AND EXTREMES

Station (City Office) — LONGITUDE 66° 06' W · ELEVATION (ground) 47 feet

Month	Normal Daily maximum	Normal Daily minimum	Normal Monthly	Extremes Record highest	Year	Extremes Record lowest	Year
J	79.9	70.3	75.1	90	1958	63	1917+
F	80.0	70.1	75.1	92	1964	62	1911
M	81.0	71.0	76.0	96	1958	65	1917
A	81.8	72.2	77.0	93	1963	65	1923
M	83.6	74.1	78.8	94	1964+	66	1915
J	84.4	75.4	79.9	94	1964	66	1902
J	84.2	76.0	80.1	92	1942	70	1961
A	85.3	76.2	80.8	93	1955	68	1904
S	85.8	75.6	80.7	94	1961+	69	1929+
O	85.5	75.0	80.2	94	1963	68	1903+
N	83.4	73.7	78.6	93	1915	65	1898
D	81.4	72.0	76.7	91	1962+	62	1910+
YR	83.0	73.5	78.3	96	OCT. 1963+	62	DEC. 1917+

Precipitation Normal total (YR 60.36); Snow, Sleet all 0.0.

Mean number of days (record 66 yrs): Precipitation .01 inch or more — J 20, F 14, M 14, A 16, M 17, J 19, J 20, A 18, S 18, O 19, N 20, YR 208. Snow/Sleet 1.0 inch or more 0. Thunderstorms YR 49. Heavy fog 0. Temperatures Max 90 and above YR 10; 32 and below 0; Min 32 and below 0; 0 and below 0.

Maximum precipitation in 24 hrs YR 10.55 DEC. 1910.

Station: SAN JUAN, PUERTO RICO — ISLA VERDE AIRPORT
Standard time used: ATLANTIC · Latitude: 18° 26' N · Longitude: 66° 00' W · Elevation (ground): 13 feet

Month	(a)	Normal Daily maximum	Normal Daily minimum	Normal Monthly	Extremes Record highest	Year	Extremes Record lowest	Year
J	15	81.3	67.4	74.4	90	1962	61	1962
F	15	81.8	67.0	74.4	92	1968+	60	1958
M	15	83.1	67.5	75.3	96	1958	60	1957
A	15	84.0	69.2	76.7	93	1956	65	1968
M	15	85.6	71.5	78.7	95	1968+	66	1962
J	15	87.1	72.9	80.0	94	1967+	69	1967+
J	15	87.1	73.7	80.4	92	1965	70	1958
A	15	87.8	74.0	80.9	93	1967+	70	1956
S	15	87.8	73.7	80.8	92	1960+	67	1960+
O	15	86.9	72.8	79.9	93	1963	67	1960+
N	15	85.0	69.6	77.3	90	1968	63	1964
D	15	82.7	69.6	76.2	90	1965+	63	1964
YR	15	85.0	70.9	78.0	96	SEP. 1958	60	MAR. 1957

Precipitation Normal total (YR 64.21). Maximum precipitation in 24 hrs YR 5.08 JAN. 1969+. Snow, Ice pellets all 0.0.

Relative humidity (Local time) Hour 02, 08, 14, 20 — record 14 yrs. Mean sky cover (sunrise to sunset) around 5.8–6.3.

Prevailing wind ENE. Fastest mile generally NE, speeds 30–40 mph. Mean number of days (record 14 yrs): Precipitation .01 inch or more YR 205; Thunderstorms YR 40; Temperatures Max 90 and above YR 31; others 0. Average daily solar radiation — langleys.

Means and extremes above are from existing and comparable exposures. Annual extremes have been exceeded at other sites in the locality as follows:
Annual extremes have been exceeded at City Office locations as follows (1890 – 1964 record): Highest temperature 96 in October 1963 and earlier;
maximum monthly precipitation 16.88 in May 1936; minimum monthly precipitation 0.05 in February 1941; maximum precipitation in 24 hours 10.55 in
December 1910.

REFERENCE NOTES APPLYING TO ALL "NORMALS, MEANS, AND EXTREMES" TABLES

(a) Length of record, years, based on January data, other months may be for more or fewer years if there have been breaks in the record.

(b) Climatological standard normals (1931-1960).

* Less than one half.

+ Trace, an amount too small to measure.

T Also on earlier dates, months, or years.

Below zero temperatures are preceded by a minus sign.
The prevailing direction for wind is from records through 1963.
The Normals, Means, and Extremes table is from records through 1963.

‡ 70° at Alaskan stations.

Unless otherwise indicated, dimensional units used in this bulletin are: temperature in degrees F.; precipitation, including snowfall, in inches; wind movement in miles per hour; and relative humidity in percent. Degree day totals are the sums of negative departures of average daily temperatures from 65° F. Cooling degree day totals are the sums of positive departures of average daily temperatures from 65° F. Sleet was included in snowfall totals beginning with July 1948. The term "Ice pellets" includes solid grains of ice (sleet) and particles consisting of snow pellets encased in a thin layer of ice. Heavy fog reduces visibility to 1/4 mile or less.

Sky cover is expressed in a range of 0 for no clouds or obscuring phenomena to 10 for complete sky cover. The number of clear days is based on average cloudiness 0-3, partly cloudy days 4-7, and cloudy days 8-10 tenths.

Solar radiation data are the averages of direct and diffuse radiation on a horizontal surface. The langley denotes one gram calorie per square centimeter.

& Figures instead of letters in a direction column indicate direction in tens of degrees from true North; i.e., 09 East, 18 South, 27 West, 36 North, and 00 Calm. Resultant wind is the vector sum of wind directions and speeds divided by the number of observations. Fastest mile is the fastest observed 1-minute value; direction under Fastest mile the corresponding direction.

To 8 compass points only.

% The station does not operate 24 hours daily. Fog and thunderstorm data therefore may be incomplete.

MEAN ANNUAL TEMPERATURE (°F)

PUERTO RICO

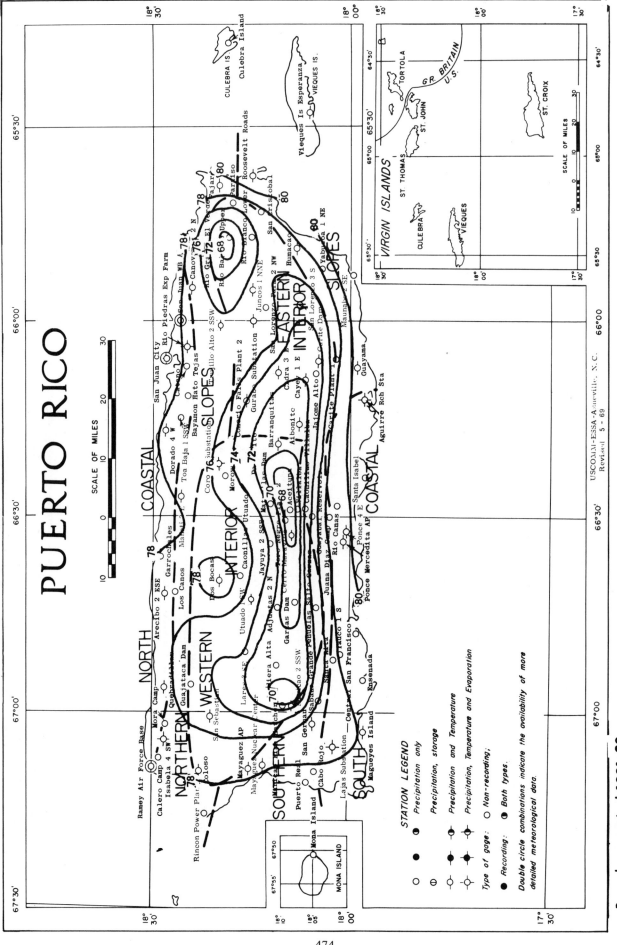

VIRGIN ISLANDS

MONA ISLAND

SCALE OF MILES

STATION LEGEND

Precipitation only

Precipitation, storage

Precipitation and Temperature

Precipitation, Temperature and Evaporation

Type of gage : ○ Non-recording;

Recording : ● Both types.

Double circle combinations indicate the availability of more
detailed meteorological data.

USCOMM-ESSA-Asheville, N.C.
Revised 5 - 69

Based on period 1931-60

Isolines are drawn through points of approximately equal value. Caution should be used
in interpolating on these maps, particularly in mountainous areas.

MEAN ANNUAL PRECIPITATION, INCHES

PUERTO RICO

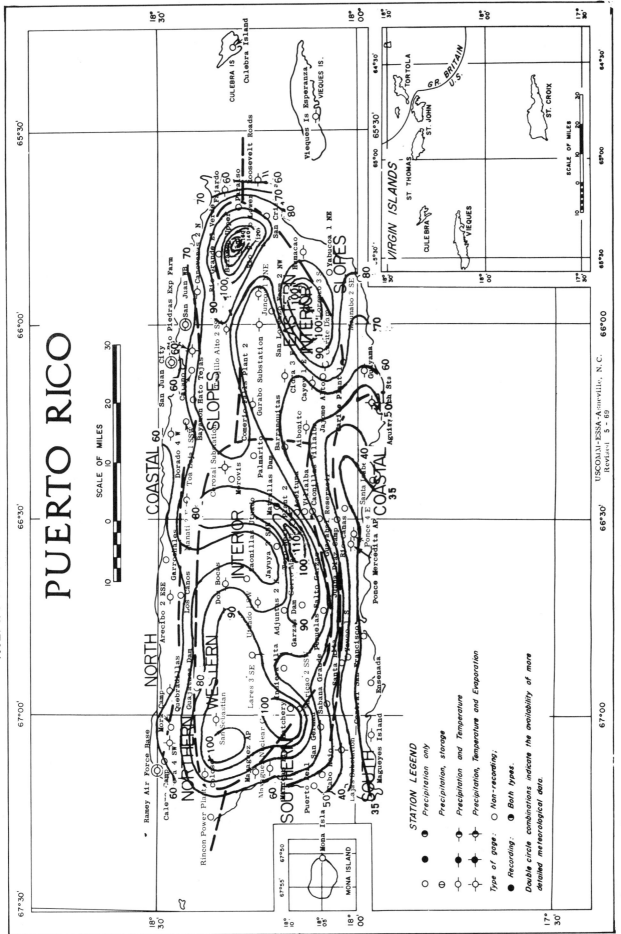

SCALE OF MILES

STATION LEGEND

Precipitation only

Precipitation, storage

Precipitation and Temperature

Precipitation, Temperature and Evaporation

Type of gage: ○ Non-recording;

Recording: ● Both types.

Double circle combinations indicate the availability of more detailed meteorological data.

VIRGIN ISLANDS

SCALE OF MILES

USCOMM-ESSA-Asheville, N.C.
Revised 5 - 69

Based on period 1931–60

Isolines are drawn through points of approximately equal value. Caution should be used in interpolating on these maps, particularly in mountainous areas.

MEAN MAXIMUM TEMPERATURE (°F), JANUARY

PUERTO RICO

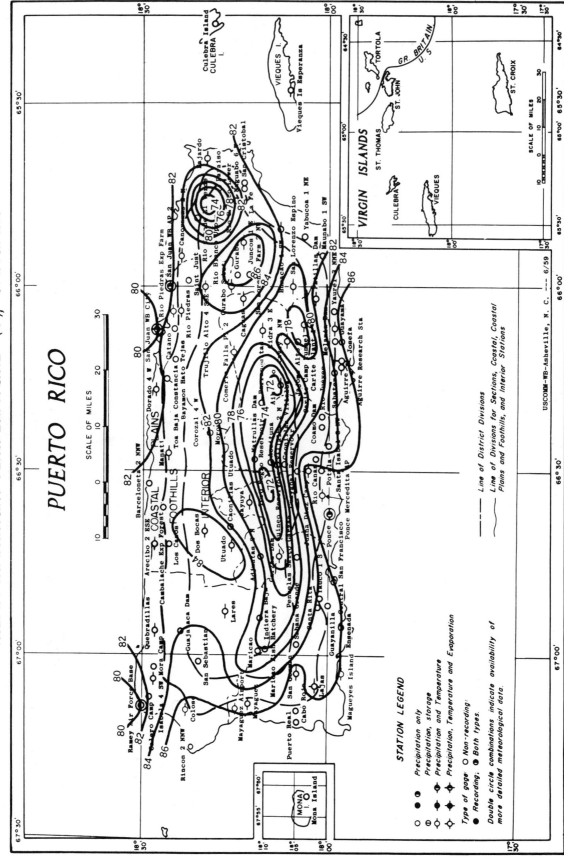

SCALE OF MILES

STATION LEGEND

Precipitation only
Precipitation, storage
Precipitation and Temperature
Precipitation, Temperature and Evaporation

Type of gage: ○ Non-recording:
● Recording; ◐ Both types.

Double circle combinations indicate availability of
more detailed meteorological data.

- - - Line of District Divisions
―― Line of Divisions for Sections, Coastal, Coastal
    Plains and Foothills, and Interior Stations

VIRGIN ISLANDS

SCALE OF MILES

USCOMM-WB-Asheville, N. C. --- 6/59

Based on period 1931-52

Isolines are drawn through points of approximately equal value. Caution should be used in interpolating on these maps, particularly in mountainous areas.

MEAN MINIMUM TEMPERATURE (°F), JANUARY

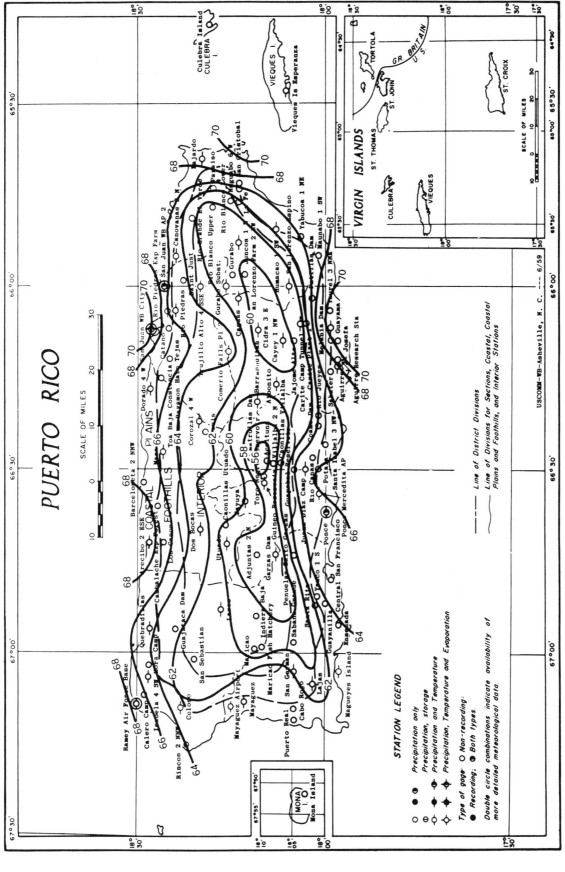

Based on period 1931-52

Isolines are drawn through points of approximately equal value. Caution should be used in interpolating on these maps, particularly in mountainous areas.

MEAN MAXIMUM TEMPERATURE (°F), JULY

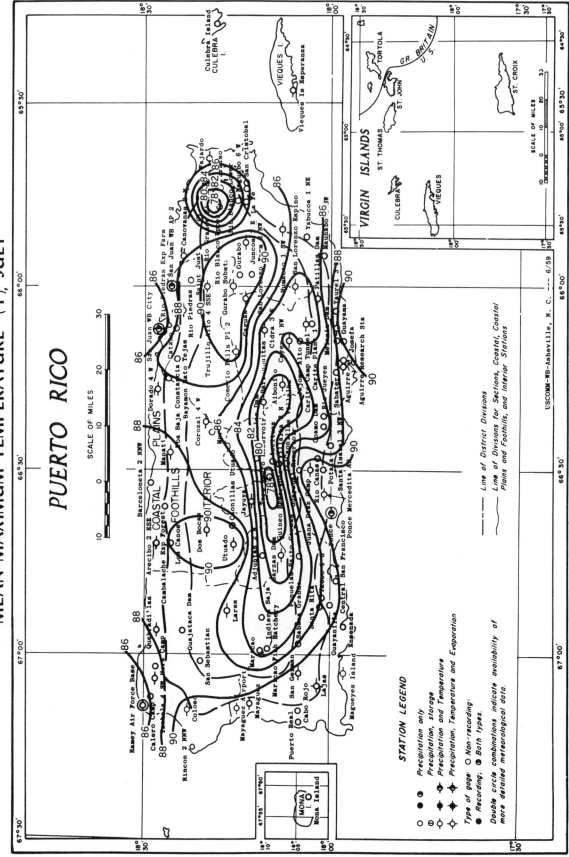

PUERTO RICO

Based on period 1931-52

Isolines are drawn through points of approximately equal value. Caution should be used in interpolating on these maps, particularly in mountainous areas.

USCOMM-WB-Asheville, N. C. -- 6/59

MEAN MINIMUM TEMPERATURE (°F), JULY

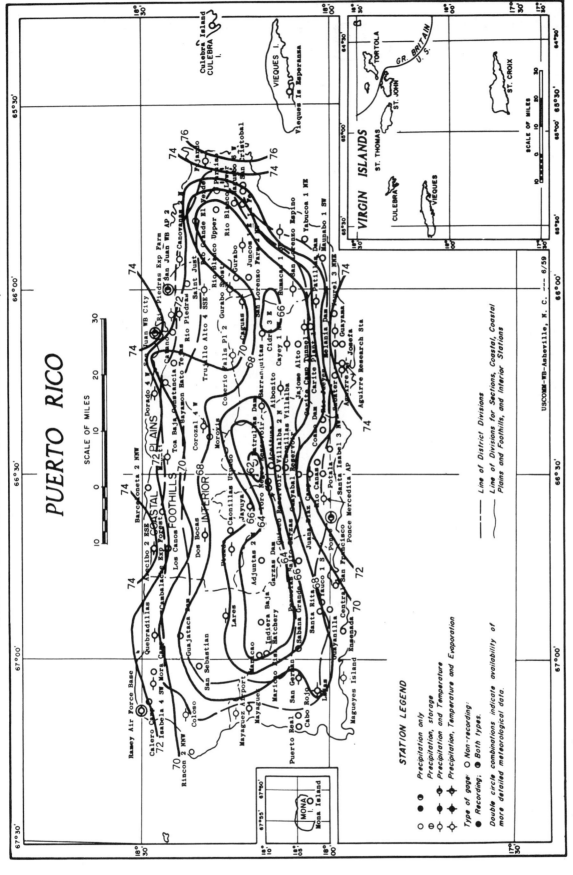

PUERTO RICO

SCALE OF MILES

Based on period 1931-52

Isolines are drawn through points of approximately equal value. Caution should be used in interpolating on these maps, particularly in mountainous areas.

USCOMM-WB-Asheville, N. C. --- 6/59

———— Line of District Divisions

———— Line of Divisions for Sections, Coastal, Coastal
Plains and Foothills, and Interior Stations

STATION LEGEND

Type of gage: ○ Non-recording:
● Recording; ⊕ Both types.

○ ● Precipitation only
⊖ ⊕ Precipitation, storage
⊖̷ ⊕̷ Precipitation and Temperature
◇ ◆ Precipitation, Temperature and Evaporation

Double circle combinations indicate availability of
more detailed meteorological data.

VIRGIN ISLANDS

SCALE OF MILES

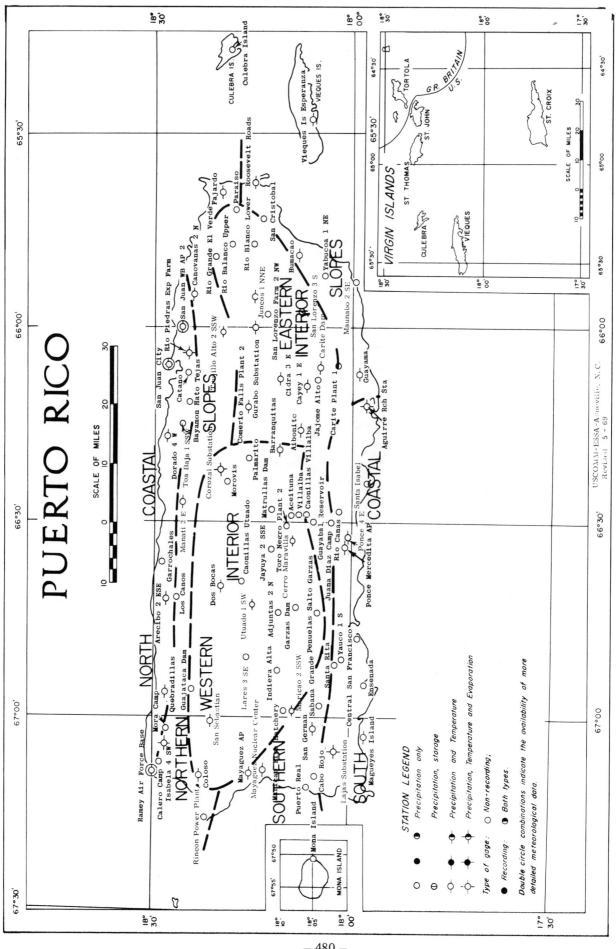

PUERTO RICO

SCALE OF MILES

NORTH COASTAL

WESTERN INTERIOR

NORTHERN

SOUTHERN

SOUTH COASTAL

EASTERN INTERIOR

SLOPES

SLOPES

Ramey Air Force Base
Calero Camp
Mora Camp
Isabela 4 SW
Quebradillas
Arecibo 2 ESE
Garrochales
Los Canos
Manati 2 E
Dorado 4 W
Toa Baja 1 SSW
San Juan City
San Juan WB AP 2
Rio Piedras Exp Farm
Canovanas 2 N
Rio Grande El Verde
Fajardo
Paraiso
Roosevelt Roads
CULEBRA IS.
Culebra Island
Vieques Is Esperanza
VIEQUES IS.

Rincon Power Plant
Coloso
Guajataca Dam
San Sebastian
Lares 3 SE
Utuado 1 SW
Dos Bocas
Caonillas Utuado
Corozal Substation
Morovis
Bayamon
Hato Tejas
Catano
Trujillo Alto 2 SSW
Comerio Falls Plant 2
Gurabo Substation
Juncos 1 NNE
Rio Balanco Upper
Rio Blanco Lower
San Cristobal
Humacao
Yabucoa 1 NE

Mayaguez AP
Mayaguez Nuclear Center
Hatchery
San German
Maricao 2 SSW
Indiera Alta
Adjuntas
Garzas Dam
Cerro Maravilla
Jayuya 2 SSE
Toro Negro Plant 2
Matrullas Dam
Palmarito
Barranquitas
Aceituna
Villalba
Caonillas Villalba
Aibonito
Cidra 3 E
Cayey 1 E
San Lorenzo Farm 2 NW
San Lorenzo 3 S
Carite Dam
Carite Plant 1
Maunabo 2 SE

Puerto Real
Cabo Rojo
Isabela Grande
Sabana Grande
Penuelas
Salto Garzas
Garzas
Juana Diaz Camp
Guayabal Reservoir
Jajome Alto
Guayama

Central San Francisco
Santa Rita
Yauco 1 S
Rio Canas
Ponce Mercedita AP
Ponce 4 E
Santa Isabel

Lajas Substation
Magueyes Island
Ensenada
Aguirre Rch Sta

MONA ISLAND
Mona Island

VIRGIN ISLANDS
ST. THOMAS
ST. JOHN
TORTOLA
GR. BRITAIN
U.S.
CULEBRA
VIEQUES
VIEQUES
ST. CROIX

SCALE OF MILES

USCOMM-ESSA-Asheville, N. C.
Revised 5 - 69

STATION LEGEND

Precipitation only

Precipitation, storage

Precipitation and Temperature

Precipitation, Temperature and Evaporation

Type of gage: ○ Non-recording;

Recording: ● Both types.

Double circle combinations indicate the availability of more detailed meteorological data.

APPENDIX

Number of Times Destruction was Caused
by Tropical Storms, 1901-1955

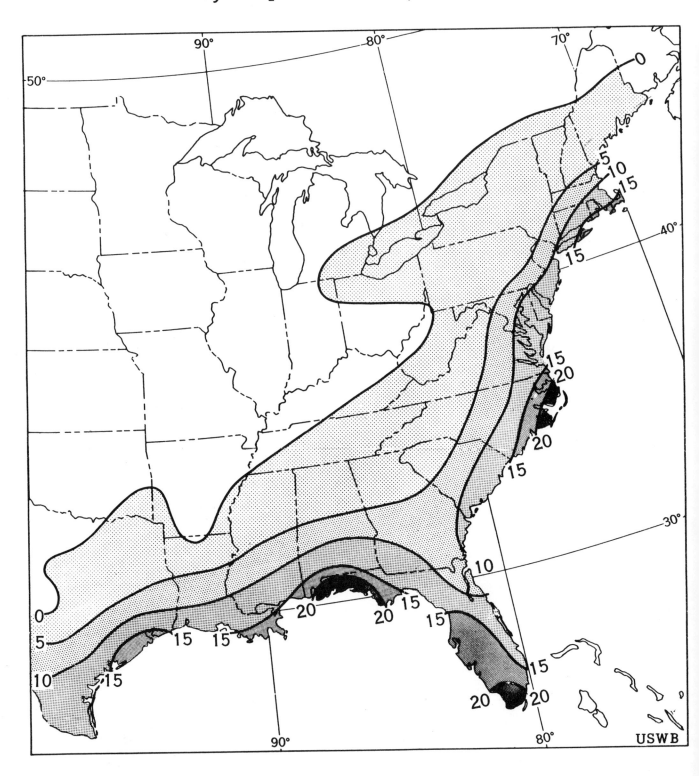

USWB

RISK OF TROPICAL CYCLONES

U.S. Gulf of Mexico Coastline

This histogram and table show the probability (percentage) that a <u>tropical storm</u>, <u>hurricane</u>, or <u>great hurricane</u> will occur in any one year in a 50 mile segment of the coastline.

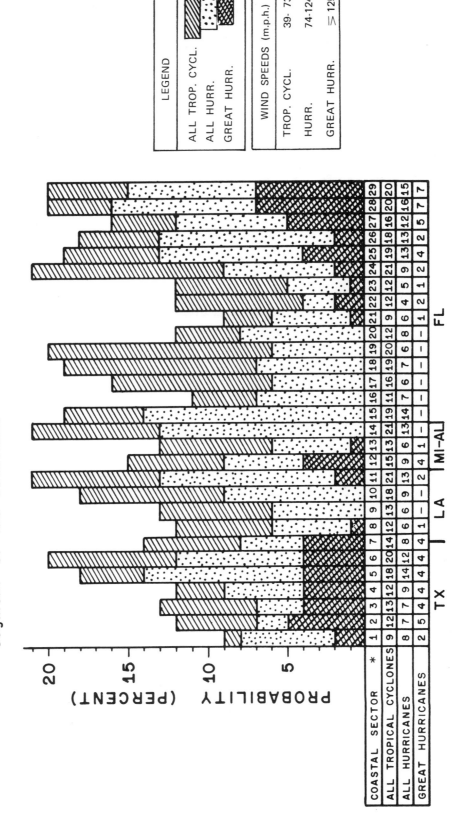

LEGEND

ALL TROP. CYCL.	
ALL HURR.	
GREAT HURR.	

WIND SPEEDS (m.p.h.)

TROP. CYCL.	39- 73
HURR.	74-124
GREAT HURR.	≥ 125

COASTAL SECTOR	1	2	3	4	5	6	7	8	9	10	11	12	13	14	15	16	17	18	19	20	21	22	23	24	25	26	27	28	29
ALL TROPICAL CYCLONES	9	12	13	12	18	20	14	12	13	18	21	15	21	24	21	11	16	19	20	9	12	21	12	21	19	18	16	20	20
ALL HURRICANES	8	7	7	9	14	12	8	6	9	13	9	6	13	14	7	6	7	8	6	8	6	4	5	9	13	13	12	16	20
GREAT HURRICANES	2	5	4	4	4	4	4	1	-	-	2	1	-	1	4	-	-	1	-	1	4	1	2	4	2	5	7		

TX | LA | MI-AL | FL

RISK OF TROPICAL CYCLONES
U.S. Atlantic Coastline

This histogram and table show the probability (percentage) that a tropical storm, hurricane, or great hurricane will occur in any one year in a 50 mile segment of the coastline.

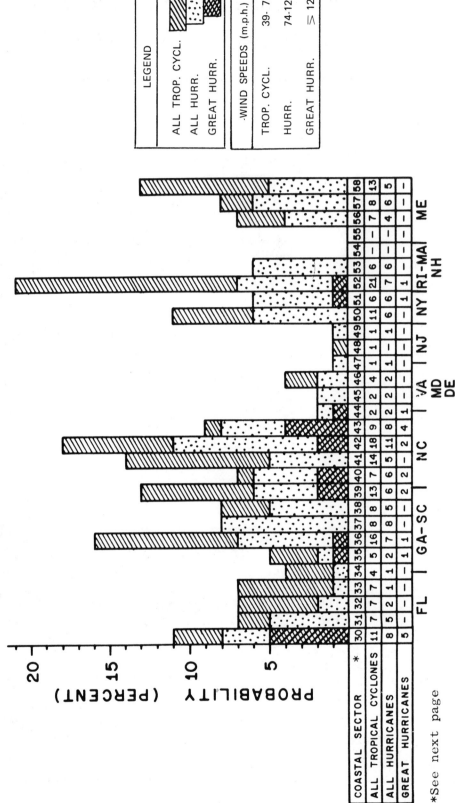

PROBABILITY (PERCENT) — vertical axis marked at 5, 10, 15, 20

COASTAL SECTOR *	30	31	32	33	34	35	36	37	38	39	40	41	42	43	44	45	46	47	48	49	50	51	52	53	54	55	56	57	58
ALL TROPICAL CYCLONES	11	7	7	4	5	16	8	13	7	14	18	9	2	4	1	1	1	1	1	6	11	6	21	6	-	-	7	8	13
ALL HURRICANES	8	5	2	1	2	7	8	5	6	6	5	11	8	2	2	2	1	1	1	6	6	7	6	1	-	1	6	4	5
GREAT HURRICANES	5	-	-	-	1	1	1	-	2	2	2	4	1	-	1	-	1	1	1	-	1	-	-	-	-	-	-	-	-

Coastal sector labels: FL | GA–SC | NC | VA | MD | DE | NJ | NY | RI–MA | NH | ME

*See next page

(Simpson and Lawrence 1971)

LEGEND

ALL TROP. CYCL.	(hatched)
ALL HURR.	(dotted)
GREAT HURR.	(cross-hatched)

WIND SPEEDS (m.p.h.)

TROP. CYCL.	39 – 73
HURR.	74 – 124
GREAT HURR.	≥ 125

EARLIEST AND LATEST TROPICAL CYCLONE OCCURRENCES 1886-1970

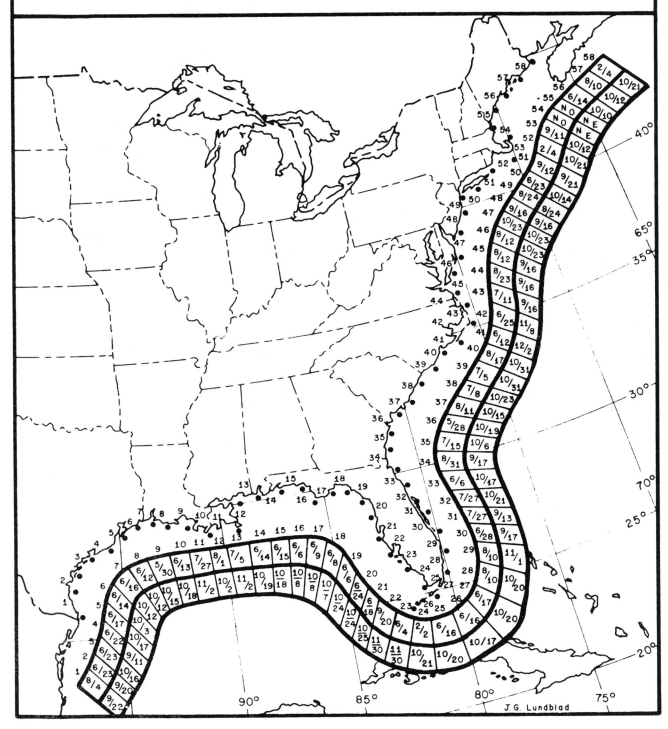

Numerals indicate coastal strips approximately 50 nautical miles in length.
(Simpson and Lawrence 1971)

GENERAL REFERENCE NOTES

Figures and letters following a station name, such as 12 SSW, indicate distance in miles and direction from the Post Office.

The following units are generally used: Temperature in ° F., precipitation and snowfall in inches, wind movement in miles per hour, and relative humidity in percent. Degree day totals are the sums of the negative departures of average daily temperature from 65° F. Below zero temperatures are preceded by a minus sign (-).

Information on the history of changes in locations, elevations, exposures, etc., of substations through 1955 may be found in the publication Substation History for this State; price per copy 35 cents. Similar information for regular National Weather Service Offices is found in the latest issue of Local Climatological Data Annual published for each of these Offices; price 15 cents per copy.

Additional detailed climatological data for regular National Weather Service Offices are found in the issues of the monthly Local Climatological Data published for each Office. Price of each issue of the monthly Local Climatological Data is 10 cents. Additional detailed climatological data for substations and regular National Weather Service Offices, including present locations, elevations, etc., are found in the monthly and annual issues of Climatological Data for this State. Separate copies are priced at 20 cents each.

Long-period detailed data for substations may also be found in the publication Climatological Summary of the United States, 1930 edition (Bulletin W), the Climatological Summary of the United States -- Supplement for 1931 Through 1952 (Bulletin W Supplement), and the Climatological Summary of the United States -- Supplement for 1951 Through 1960 (Bulletin W Supplement). The 1930 edition (Bulletin W) is priced at 10 cents per copy; the Bulletin W Supplements at 25 and 35 cents, respectively.

The Weekly Weather and Crop Bulletin is prepared jointly by the Statistical Reporting Service of the U.S. Department of Agriculture and the Environmental Data Service of NOAA. In it are given the weather of the past week in narrative, tabular, and chart form, together with the weather's effect on crops and farm activities. It also contains special articles on weather and crops. Single issues are priced at 10 cents.

Any of the above publications may be obtained from the Superintendent of Documents, Government Printing Office, Washington, D. C. 20402.

Weekly Weather and Crop Bulletins are published in most States. They are the cooperative effort of the Environmental Data Service of NOAA, the Statistical Reporting Service of the U. S. Department of Agriculture, and other State and Federal agencies. These may be obtained from the Climatologist for this State at the address given below.